ASTRONOMY AND ASTROPHYSICS ABSTRACTS

A Publication of the Astronomisches Rechen-Institut Heidelberg
Member of the Abstracting Board of the International
Council of Scientific Unions

Volume 2
Literature 1969, Part 2

Edited by
S. Böhme · W. Fricke · U. Güntzel-Lingner
F. Henn · D. Krahn · G. Zech

Published for
Astronomisches Rechen-Institut
by
Springer-Verlag Berlin · Heidelberg · New York 1970

Astronomisches Rechen-Institut
Heidelberg

Director: Prof. Dr. W. Fricke

Astronomy and Astrophysics Abstracts
Editor-in-Chief: F. Henn

ISBN 978-3-642-49292-1   ISBN 978-3-642-49290-7 (eBook)
DOI 10.1007/978-3-642-49290-7

Preface

Astronomy and Astrophysics Abstracts, which appears in semi-annual volumes, is devoted to the recording, summarizing and indexing of astronomical publications throughout the world. It aims to present a comprehensive documentation of literature in all fields of astronomy and astrophysics. Every effort will be made to ensure that the average time interval between the date of receipt of the original literature and publication of the abstracts will not exceed eight months. This time interval is near to that achieved by monthly issued abstracting journals, compared to which our system of accumulating abstracts for about six months offers the advantage of greater convenience for the user.

Volume 2 contains literature published in 1969 and received before March 15, 1970; some older literature which was received late and which is not recorded in Volume 1 is also included.

The authors of papers who have sent us abstracts on request have effectively contributed to the success of our service. We should like to express our gratitude to them. We acknowledge with thanks contributions to this volume by Dr. J. Bouška, who surveyed journals and publications in Czech language and supplied us with abstracts in English, by Dr. B. Onderlička, Brno, for providing English abstracts of Russian papers, and by the Commonwealth Scientific and Industrial Research Organization (C.S.I.R.O.), Sydney, for providing titles and abstracts of papers on radio astronomy.

Our warmest thanks also to the ladies of our editorial office, Mrs. Monika Betz, and Mrs. Utta-Barbara Stegemann, who typed the text of this volume on IBM 72 Composers and compiled the pages from abstract slips in a perfect form for offset reproduction, and to Mrs. Eva Rott, for punching all the material for author index and subject index.

Heidelberg, April 1970

Siegfried Böhme
Walter Fricke
Ulrich Güntzel-Lingner
Frieda Henn
Dietlinde Krahn
Gert Zech

Contents

Positional Astronomy. Celestial Mechanics

Space Research

Theoretical Astrophysics

Sun

Earth

Stellar Systems

Introduction

Astronomical bibliographies

Astronomy and Astrophysics Abstracts begins documentation and abstracting as from the year 1969. For information on astronomical literature before this date consultation of one of the following bibliographies is suggested:

(1) J. J. de Lalande, Bibliographie Astronomique, Paris 1803 (this work covers the time from 480 B. C. to the year 1803, VIII + 966 pages).

(2) J. C. Houzeau, A. Lancaster, Bibliographie générale de l'astronomie, Volume I (in two parts), Bruxelles 1882, 1887, Volume II, Bruxelles 1889. The complete title of Volume II is "Bibliographie générale de l'astronomie ou catalogue méthodique des ouvrages, des mémoires et des observations astronomiques, publiés depuis l'origine de l'imprimerie jusqu'en 1880". A new edition of these volumes was prepared by D. W. Dewhirst and published by Holland Press, London 1964. No bibliography was published for the years from 1880 to 1898, although on recommendation and with support of the International Astronomical Union almost complete information was collected for this period; the unpublished material is kept on file by the Observatoire Royal de Belgique, Bruxelles.

(3) Astronomischer Jahresbericht, 1899 gegründet von Walter Wislicenus, herausgegeben vom Astronomischen Rechen-Institut in Heidelberg (formerly in Berlin), Verlag W. de Gruyter, Berlin. For the period from 1899 to 1968 sixty-eight volumes were published, each of which, in general, covers the literature of one year.

(4) Bulletin Signalétique – Section Astronomie, Astrophysique, Physique du Globe. Published by Centre de Documentation du Centre National de la Recherche Scientifique, Paris. This publication is a continuation of "Bibliographie Mensuelle de l'Astronomie" founded in 1933 by the Société Astronomique de France. The publication is continued.

(5) Referativnyj Zhurnal. Founded in 1953 and published by Vsesoyuznyj Institut Nauchnoj i Tekhnicheskoj Informatsii, Akademiya Nauk, Moskva. The publication is continued.

Concept of Astronomy and Astrophysics Abstracts

This abstracting service aims to present a comprehensive documentation of the literature in all fields of astronomy and astrophysics. It appears in semi-annual volumes, two of which cover the literature of a calendar year. The half-yearly period of issue is regarded as an optimal period of time for summarizing papers into subject categories and for the presentation of abstracts as quickly as possible after the publication of the original literature. The time limits at which the documentation begins and ends for a volume are not sharply defined, except in the sense that all literature will be covered which was received by the editors within these limits.

Volume 2 is devoted to the recording, summarizing and indexing of astronomical publications of the year 1969 received from August 1969 to March 15, 1970; it also records a number of papers issued before 1969 but received within the given period of time.

The main characteristics of the concept of Astronomy and Astrophysics Abstracts may be summarized briefly.

(1) Titles of papers are given in the language of their authors whenever possible. If they are not in English but supplied with English translations they will be given in English. Abstracts are presented in English, French or German. Titles of papers in Russian are, as a rule, given in English; occasionally, they are given in German.

(2) Authors' abstracts are used whenever possible. In this volume only very few abstracts have been written by persons other than the authors; in such cases the name of the abstractor is given. As a rule, no popular articles were abstracted, their titles however given, occasionally with the addition "Popular article".

(3) As a rule, each paper has been classified into one of 108 numbered subject categories and allocated a serial number within the category. In this way each item is numbered by six figures, the first three of which indicate the number of the category. Three further figures indicate the serial number within the category, which was allocated in the order of the receipt of the abstract. Reference to an abstract in Volume 1 is indicated by "01" before the number of the category; for example, 01.074.028, denotes Volume 1, category 074, abstract 028. Volume 2 is indicated by "02".

A paper may have been classified into more than one category. Then its abstract has been allocated a number in one of the categories involved, and in the other category (or categories) the paper has been indicated by the title and a reference to the abstract number.

Papers whose authors are not named were treated like those with authors' names, with one exception: reports from correspondents of journals whose names were unknown were not numbered.

(4) There are categories which suggest the presentation of the material in subject groups; these, however, can only be formed immediately before the completion of a volume. For instance, a subject group may be formed by all information received on the same solar eclipse, comet, nova, etc. The unsorted presentation of such material in a subject category would be inconvenient for the user, even if the individual comet, etc. were included in the subject index.

The following subject categories are subdivided into subject groups:

008 Observatories, Institutes. The publications of observatories and astronomical institutes are listed in alphabetical order of the towns of the institutions, each town forming a numbered subject group. For each publication a reference to an abstract number is made.

010 Societies, Associations, Organizations. The publications of each one form a subject group. The groups are presented in alphabetical order.

079 Solar eclipses. All publications related to one solar eclipse form a subject group.

103 Comets, listed objects. All publications related to the same comet form a numbered group.

124 Novae. All publications related to one nova form a subject group.

(5) Border fields of astronomy and astrophysics have been taken into a account by presenting titles of papers occasionally without abstracts. The selection of papers for inclusion has been made according to the degree of relevance to astronomical research.

(6) The text of the publication was typed on IBM 72 Composers in the editorial office, and it was given to the printer in a form ready for offset reproduction. The author index and the subject index were compiled and printed by means of electronic computer (Siemens 2002).

(7) While each volume is scheduled to contain an author index and a subject index, the magnetic tapes containing the

index information will be used to produce separate index volumes (authors and subjects) at intervals of a few years.

Transliteration of the Russian Alphabet

The transliteration of the Russian alphabet in use in Astronomy and Astrophysics Abstracts is presented here.

А	а	a	Р	р	r
Б	б	b	С	с	s
В	в	v	Т	т	t
Г	г	g	У	у	u
Д	д	d	Ф	ф	f
Е	е	e	Х	х	kh
Ё	ё	e	Ц	ц	ts
Ж	ж	zh	Ч	ч	ch
З	з	z	Ш	ш	sh
И	и	i	Щ	щ	shch
Й	й	j	Ъ	ъ	''
К	к	k	Ы	ы	y
Л	л	l	Ь	ь	'
М	м	m	Э	э	eh
Н	н	n	Ю	ю	yu
О	о	o	Я	я	ya
П	п	p			

This transliteration was recommended by the Abstracting Board of the International Council of Scientific Unions in 1969. It is essentially the same as the transliteration proposed by the Academy of Sciences, Moscow, and used by the Referativnyi Zhurnal (See Referativnyi Zhurnal, 51. Astronomiya, 1969 No. 1). It may be noted that the letters can be read and printed by usual data processing machines.

In the literature the names of Russian authors can be found transliterated in different ways. We present the names in the form in which they are given in the literature.

Sources of information

The majority of sources of information for this volume are given in section **001 Periodicals** and in section **008 Observatories, Institutes.** The term "periodical" has been used in its widest sense for publications in a sequence of undetermined duration, even if the intervals of appearance are not regular. Section 001 records 265 periodicals with their full titles and with abbreviations which are in use in Astronomy and Astrophysics Abstracts. It may be noted that the titles of the periodicals are given in their original languages, and that Russian titles have been transliterated applying the transliteration given above. Section 008 records 173 periodicals; these are publication series of observatories and astronomical institutes which have not been included in section 001. The abbreviations of the titles of the periodicals have been given so that in most cases they permit recognition of the full title without recourse to the key in section 001. The steadily growing number of periodicals makes it necessary to use more extensive abbreviations and to abandon the use of very condensed ones.

Other abstracting journals have been consulted in order to examine the degree of completeness of our service. Occasionally, in particular in Physics Abstracts, Referativnyi Zhurnal, and Bulletin Signalétique abstracts of papers were found which had not come to our attention. In such cases Astronomy and Astrophysics Abstracts gives the titles with references to the other abstracting service.

Classification into a scheme of subject categories

The subdivision of astronomy and its border fields into subject categories is facilitated by the fact that the astronomical objects appear to be particularly well suited for the formation of categories. Sun, moon, earth, planets, comets, and meteorites, the various kinds of stars, galaxies, radio sources, quasars, and pulsars etc. suggest natural subdivisions. It may be assumed that such subdivisions can be maintained for long periods of time. Experience shows, however, that progress in research may imply changes in the classification scheme, in particular, in fields where the expansion of knowledge is explosive. Probably one of the best examples which shows clearly the reflection of research progress on astronomical documentation is the Crab nebula, which may be classified as a supernova remnant, an emission nebula, a radio source or pulsar. In this volume subject category 134 is devoted to the Crab nebula. Papers related to the Crab nebula will, however, be found in the subject categories 125 (Supernova remnants), 141 (Radio sources, pulsars), 142 (X ray sources), 143 (Cosmic radiation).

A few explanatory remarks may be in order on some of the subject categories. Section 002 includes short news notes whose titles and authors are given, but the authors of the notes have not been included in the author index. In section 003 books on Astronomy and Astrophysics and its border fields are listed which came to our notice from August to March. In cases where books can be classified into one of the subject categories these books are additionally listed under their categories. References to book reviews are given if the reviews appeared quicKly.

For completeness of documentation, personal notes (section 006) and obituaries (section 007) are listed. In section 012 (Proceedings of Colloquia, Congresses, Meetings, and Symposia) the proceedings etc. are listed with titles and editors. Whenever the volumes were at hand, the papers were classified into their subject categories and, occasionally, supplied with abstracts.

Author index and subject index

The subject category and the serial number forming six figures for each abstract have been used as a means of reference in the author index and the subject index. These references are more precise than page references. They offer considerable advantages in indexing by means of data processing machines, and they are more convenient for the user.

The author index of this volume contains 5783 names. A complete reference comprises six figures, three for the subject category and three for the serial number within the category. In the case of more than one reference to abstracts in one category, the number of the category is given only once and not repeated in the immediately following references. The total number of papers (some do not give names of authors) recorded in this volume is about 5800.

We consider the subject index as only a first approximation to an optimal index covering all fields of astronomy and astrophysics and its border fields. Several iterative steps appear to be necessary until an index has been compiled for one of the subsequent volumes which may then serve as a kind of standard for the near future. The assigning of one or more key words to a paper is undoubtedly a difficult task. Some journals have started giving key words together with the titles of papers. These key words are chosen by the authors themselves and are in many cases identical with our designations of subject categories with no additional specification. In fact, in some cases it may be more useful to refer to a subject category as a whole than to an item number, in particular, if the total number of abstracts in a category is very small, and if more specific key words do not provide a proper description of the paper.

Abbreviations

AAS	American Astronomical Society		Geogr.	Geography, etc.
AAVSO	American Association of Variable Star Observers		Geophys.	Geophysics, etc.
			Ges.	Gesellschaft
Abh.	Abhandlungen		Glav.	Glavnyj (Main)
Abstr.	Abstract		Gos.	Gosudarstvennyj (State)
Abt.	Abteilung		HRD	Herzsprung-Russell diagram
Acad.	Academy, etc.		Hydrogr.	Hydrography, etc.
Accad.	Accademia		IAF	International Astronautical Federation
Adv.	Advances		IAU	International Astronomical Union
AG	Astronomische Gesellschaft		ICSU	International Council of Scientific Unions
AIAA	American Institute of Aeronautics and Astronautics		IEEE	Institute of Electrical and Electronics Engineers
AJB	Astronomischer Jahresbericht		Industr.	Industry, etc.
Akad.	Akademie		Inform.	Information
An.	Anales, etc.		Inst.	Institute, etc.
Ann.	Annals, etc.		Instn.	Institution
Arch.	Archiv, etc.		Ionosph.	Ionosphere, etc.
Ark.	Arkiv		Issled.	Issledovaniya (Research)
ASA	Astronomical Society of Australia		Ist.	Istituto
Asoc.	Asociación		Izv.	Izvestiya (News)
ASP	Astronomical Society of the Pacific		Jb.	Jahrbuch
Ass.	Association		JO	Journal des Observateurs
ASSA	Astronomical Society of Southern Africa		Journ.	Journal
Astrofis.	Astrofisica, etc.		Kl.	Klasse
Astrofiz.	Astrofizika, etc.		Lab.	Laboratory
Astron.	Astronomy, etc.		Mag.	Magazine
Astronaut.	Astronautics, etc.		Mat.	Matematica, etc.
Astrophys.	Astrophysics, etc.		Math.	Mathematics, etc.
ASV	Astronomical Society of Victoria		Mech.	Mechanics, etc.
ASWA	Astronomical Society of Western Australia		Med.	Mededelingen
Atmosph.	Atmosphere, etc.		Medd.	Meddelande, Meddelser
BA	Bulletin Astronomique		Mekhan.	Mekhanika, etc.
BAA	British Astronomical Association		Mém.	Mémoires
BAN	Bulletin of the Astronomical Institutes of the Netherlands		Mem.	Memoirs, Memorandum, etc.
			Meteorol.	Meteorology, etc.
Ber.	Berichte		MIT	Massachusetts Institute of Technology
BIH	Bureau International de l'Heure (Paris)		Mitt.	Mitteilungen
Bol.	Boletin		MN RAS	Monthly Notices of the Royal Astronomical Society
Boll.	Bolletino			
Bull.	Bulletin		MVS Sonneberg	Mitteilungen über Veränderliche Sterne, Sonneberg
Byull.	Byulleten' (Bulletin)			
Circ.	Circular		Nachr.	Nachrichten
Cl.	Classe		Nat.	Naturwissenschaftlich, etc.
Coll.	Collection		Naut.	Nautics, etc.
Commun.	Communication		NBS	National Bureau of Standards
Comun.	Comunicazioni		NRAO	National Radio Astronomy Observatory (Green Bank)
Contr.	Contributions, etc.			
COSPAR	Committee on Space Research		NRL	Naval Research Laboratory (Washington)
C.S.I.R.O.	Commonwealth Scientific Industrial Research Organization		Obs.	Observatory, etc.
			OSA	Optical Society of America
Dep.	Department		Oss.	Osservatorio, Osservazioni, etc.
Diss.	Dissertation		Ped.	Pedagogika, etc. (Pedagogics)
Div.	Division		Phil.	Philosophical
Dokl.	Doklady (Reports)		Phys.	Physics, etc.
ESO	European Southern Observatory		Planet.	Planetary
ESRO	European Space Research Organization		Priklad.	Prikladnoj (Applied)
Fis.	Fisica, etc.		Proc.	Proceedings
Fiz.	Fizika, etc.		Progr.	Progress, etc.
Fys.	Fysica, etc.		Pubbl.	Pubblicazioni
Géod.	Géodésie, etc.		Publ.	Publications
Geod.	Geodesy, etc.		Rap.	Raportoj
Geofis.	Geofisica, etc.		RAS	Royal Astronomical Society
Geofiz.	Geofizika, etc.		RAS Canada	Royal Astronomical Society of Canada
Geofys.	Geofysik, etc.		Rech.	Recherches
Geol.	Geology, etc.		Rend.	Rendiconti

3

Rep.	Report	Techn.	Technics, etc.
Repr.	Reprint	Tekhn.	Tekhnika, etc.
Res.	Research	Teor.	Teoreticheskij
Rev.	Review, etc.	Terr.	Terrestrial, etc.
Ric.	Ricerche	TH	Technische Hochschule
Roy.	Royal, etc.	Theor.	Theoretical
SAF	Société Astronomique de France	Tidssk.	Tidsskrift
SAI	Società Astronomica Italiana	Trans.	Transactions
SAO	Smithsonian Astrophysical Observatory	Trudy	Trudy (Publications)
SAS	Société Astronomique de Suisse	Tsentr.	Tsentral'nyj (Central)
Sci.	Science, etc.	Tsirk.	Tsirkulyar (Circular)
Sect.	Section	TU	Technical University
Ser.	Series, etc.	Uch. Zap.	Uchenye Zapiski (Treatise)
S. I. R.	Service International Rapide des Latitudes	Univ.	University, etc.
Sitz.-Ber.	Sitzungsberichte	URSI	Union Radio Scientifique Internationale
Soc.	Society	Verh.	Verhandlungen
Soobshch.	Soobshcheniya (Communications)	Veröff.	Veröffentlichungen
Sternw.	Sternwarte	Wet.	Wetenschappen
Stud. Cerc.	Studii şi Cercetari	Wiss.	Wissenschaften, etc.
Supl.	Suplemento	Zeitschr.	Zeitschrift
Suppl.	Supplement	ZfA	Zeitschrift für Astrophysik
SuW	Sterne und Weltraum	Zhurn.	Zhurnal (Journal)

Periodicals, Proceedings, Books, Activities

001 Periodicals

Abh. Deutsch. Akad. Wiss. Berlin
Abhandlungen der Deutschen Akademie der Wissenschaften zu Berlin. Klasse Mathematik, Physik und Technik. Publisher: Akademie-Verlag, Berlin.

Acad. Roy. Belgique, Bull. Cl. Sci.
Académie Royale de Belgique, Bulletin de la Classe des Sciences (Koninklijke Academie van België, Mededelingen van de Klasse der Wetenschappen). 5e Série. Palais des Académies, Bruxelles.

Acta Astron.
Acta Astronomica. Publisher: Polska Akademia Nauk, Warszawa - Kraków.

Acta Phys. Austriaca
Acta Physica Austriaca. Publisher: Springer-Verlag, Wien.

Acta Univ. Carolinae Math. Phys.
Acta Universitatis Carolinae, Mathematica et Physica. Administrace: Matematicko-fyzikální fakulta University Karlovy, Praha.

Acta Univ. Lundensis
Acta Universitatis Lundensis. Sectio II: Medica, Mathematica, Scientiae Rerum Naturalium. Editor: Royal Physiographic Society of Lund.

Actas Acad. Nacional Cienc. Lima
Actas de la Academia Nacional de Ciencias Exactas, Fisicas y Naturales de Lima. Lima - Peru.

Adv. Astron. Astrophys.
Advances in Astronomy and Astrophysics. Publisher: Academic Press, New York — London.

AIAA Journ.
AIAA Journal. A Publication of the American Institute of Aeronautics and Astronautics, Easton, Pa.

Ann. d'Astrophys.
Annales d'Astrophysique. Revue internationale bimestrielle publiée par le Centre National de la Recherche Scientifique et éditée par son Service d'Astrophysique, Paris. After Vol. 31 replaced by "Astronomy and Astrophysics".

Ann. Françaises Chronométrie Micromécanique
Annales Françaises de Chronométrie et de Micromécanique, publication annuelle de l'Observatoire de Besançon, du Centre Technique de l'Industrie Horlogère et de la Société Française de Chronométrie et de Micromécanique. Rédaction et administration: Observatoire de Besançon. Publiées avec le concours du Centre National de la Recherche Scientifique et des organismes corporatifs.

Ann. Géophys.
Annales de Géophysique. Revue Internationale trimestrielle, publiée par le Centre National de la Recherche Scientifique, Paris.

Ann. Obs. Astron. Météorol. Toulouse
Annales de l'Observatoire Astronomique et Météorologique de Toulouse. Publisher: Gauthier-Villars, Paris.

Ann. Physics
Annals of Physics. Publisher: Academic Press, New York.

Ann. Physik
Annalen der Physik. 7. Folge. Publisher: Johann Ambrosius Barth, Leipzig.

Ann. Physique
Annales de Physique. Publisher: Masson et Cie., Paris.

Ann. Univ.-Sternw. Wien
Annalen der Universitäts-Sternwarte Wien. In Kommission bei Ferd. Dümmlers Verlag, Bonn.

Annual Rev. Astron. Astrophys.
Annual Review of Astronomy and Astrophysics. Publisher: Annual Reviews Inc., Palo Alto, California.

Anzeiger. Österreich. Akad. Wiss. Math.-Nat. Kl.
Anzeiger. Österreichische Akademie der Wissenschaften. Mathematisch-Naturwissenschaftliche Klasse. Publisher: Springer-Verlag, Wien.

Applied Optics
Applied Optics. Published by the Optical Society of America (in Cooperation with the American Institute of Physics), Washington, D.C.

Arch. Sci. Genève
Archives des Sciences, éditées par la Société de Physique et d'Histoire naturelle de Genève. Publisher: Imprimérie Kundig, Genève.

Ann. Soc. Sci. Bruxelles
Annales de la Société Scientifique de Bruxelles. Série I: Sciences Mathématiques, Astronomiques et Physiques. Published by Institut de Physique, Heverlé-Louvain.

Ark. Astron.
Arkiv för Astronomi. Utgivet av Kungliga Svenska Vetenskapsakademien, Stockholm. Printed by Almqvist & Wiksell, Stockholm.

Ark. Fys.
Arkiv för Fysik. Kungliga Svenska Vetenskapsakademien, Stockholm. Printed by Almqvist & Wiksell, Stockholm.

Artificial Satellites
Artificial Satellites. Publication of Polish Scientific Institutions. Polish Academy of Sciences, National Committee of Geophysics and Geodesy, National Committee for Space Research, Warsaw. Publishing Office: Palac Kultury i Nauki, Warszawa.

Asoc. Argentina Astron. Bol.

Asociacion Argentina de Astronomia. Boletin. Editor: Instituto Argentino de Radioastronomia, Provincia de Buenos Aires, Argentina. Printer: Talleres Gráficos "Renovacion", La Plata, República Argentina.

Astrofizika
Astrofizika. Izdatel'stvo Akademii Nauk Armyanskoj SSR, Erevan. [A translation published as "Astrophysics."]

Astron. Astrophys.
Astronomy and Astrophysics. A European Journal. Published by Springer-Verlag, Berlin - Heidelberg - New York.

Astron. in der Schule
Astronomie in der Schule. Zeitschrift für die Hand des Astronomielehrers. Herausgegeben vom Verlag Volk und Wissen, Berlin. Redaktion: Sternwarte Bautzen.

Astron. Journ.
The Astronomical Journal. Published for the American Astronomical Society by the American Institute of Physics, New York.

Astron. Nachr.
Astronomische Nachrichten. Publisher: Akademie-Verlag, Berlin.

Astron. Soc. Pacific Leaflet
Astronomical Society of the Pacific. Leaflet. Edited by the Astronomical Society of the Pacific, San Francisco, California.

Astron. Tidssk.
Astronomisk Tidsskrift. Edited by Astronomisk Selskab, København; Norsk Astronomisk Selskap, Oslo; Svenska Astronomiska Sällskapet, Stockholm. Printed by John Griegs Boktrykkeri, Bergen.

Astron. Tsirk.
Astronomicheskij Tsirkulyar, izdavaemyj Byuro Astronomicheskikh Soobshchenij Akademii Nauk SSSR. Moskva.

Astron. Vestn.
Astronomicheskij Vestnik. Publishers: Izdatel'stvo "Nauka", Moskva.

Astron. Zhurn. Akad. Nauk SSSR
Astronomicheskij Zhurnal. Akademiya Nauk SSSR. Publishers: Izdatel'stvo "Nauka", Moskva.

Astronaut. Acta
Astronautica Acta. An Archive Journal of the International Academy of Astronautics. Publisher: Pergamon Press, Oxford — New York.

Astronaut. Aeronaut.
Astronautics & Aeronautics. A Publication of the American Institute of Aeronautics and Astronautics. Published monthly by the American Institute of Aeronautics and Astronautics, Easton, Pennsylvania.

Astrophysics
Astrophysics. The Faraday Press cover-to-cover translation of Astrofizika. The Faraday Press, Inc., New York, N. Y.

Astrophys. Journ.
The Astrophysical Journal. Published in collaboration with the American Astronomical Society by the Uni-

versity of Chicago Press, Chicago, Illinois.

Astrophys. Journ. Suppl. Series
The Astrophysical Journal. Supplement Series. Published in collaboration with the American Astronomical Society by the University of Chicago Press, Chicago, Illinois.

Astrophys. Letters
Astrophysical Letters. Published by Gordon and Breach, Science Publisher Ltd., New York - London - Paris.

Astrophys. Norvegica
Astrophysica Norvegica. Edited by The Institute of Theoretical Astrophysics, University of Oslo (Det Norske Videnskaps-Akademi i Oslo). Universitets-forlaget, Oslo.

Astrophys. Space Sci.
Astrophysics and Space Science. An International Journal of Cosmic Physics. Published by D. Reidel Publishing Company, Dordrecht, Holland.

Atti Accad. Nazionale Lincei. Mem.
Atti della Accademia Nazionale dei Lincei. Serie Ottava. Memorie. Classe di Scienze fisiche, matematiche e naturali. Sezione I: Matematica, Meccanica, Astronomia, Geodesia e Geofisica. Published by Accademia Nazionale dei Lincei, Roma.

Atti Accad. Nazionale Lincei Rend.
Atti della Accademia Nazionale dei Lincei. Serie Ottava. Rendiconti. Classe di Scienze fisiche, matematiche e naturali. Published by Accademia Nazionale dei Lincei, Roma.

Australian Journ. Phys.
Australian Journal of Physics. Published by the Commonwealth Scientific and Industrial Research Organization, East Melbourne, Victoria.

Australian Journ. Phys. Astrophys. Suppl.
Australian Journal of Physics, Astrophysical Supplement. Printed by Commenwealth Scientific and Industrial Research Organization, Melbourne, Victoria.

BAV Rundbrief
BAV Rundbrief. Mitteilungsblatt der Berliner Arbeitsgemeinschaft für Veränderliche Sterne. Editor: BAV Berliner Arbeitsgemeinschaft für Veränderliche Sterne eV., Berlin.

Bol. Acad. Cienc. Fis. Mat. Nat.
Boletin de la Academia de Ciencias Fisicas, Matemáticas y Naturales, Republica de Venezuela. Palacio de las Academias, Caracas, Venezuela.

Bol. Liga Latinoamericana Astron.
Boletin de la Liga Latinoamericana de Astronomia.Publicado por la Asociacion Argentina Amigos de la Astronomia, Buenos Aires, Argentina.

Boll. Geod. Sci. Affini
Bolletino di Geodesia e Scienze Affini. Pubblicazione dell' Istituto Geografico Militare, Firenze.

British Astron. Ass. Circ.
British Astronomical Association, Circular. Editorial Office: 97 Hawkswood Drive, Hailsham, Sussex.

Bull. American Astron. Soc.
Bulletin of the American Astronomical Society. Published for the American Astronomical Society by the Ame-

rican Institute of Physics Inc., New York, N. Y.

Bull. Astron. (BA)
Bulletin Astronomique. 3ᵉ Série. Publié par le Centre National de la Recherche Scientifique, Paris. After Vol. 3 (1968) replaced by "Astronomy and Astrophysics".

Bull. Astron. Inst. Czechoslovakia (BAC)
Bulletin of the Astronomical Institutes of Czechoslovakia. Published under the auspices of the Czechoslovak Academy of Sciences by Academia, Praha. Editor: Astronomical Institutes of the Czechoslovak Academy of Sciences, Praha.

Bull. Astron. Inst. Netherlands (BAN)
Bulletin of the Astronomical Institutes of the Netherlands. Publisher: North-Holland Publishing Company, Amsterdam. After Vol. 20 replaced by "Astronomy and Astrophysics".

Bull. Astron. Inst. Netherlands, Suppl. Series
Bulletin of the Astronomical Institutes of the Netherlands. Supplement Series. Published by the Astronomical Institutes. Replaced by "Astronomy and Astrophysics", Supplement Series.

Bull. Géod.
Bulletin Géodésique. Nouvelle Série. Publié par le Bureau Central de l'Association Internationale de Géodésie, Paris.

Bull. Geograph. Survey Inst.
Bulletin of the Geographical Survey Institute. Published by the Geographical Survey Institute, Ministry of Construction, Tokyo, Japan.

Bull. Hor.
Bulletin Horaire du Bureau International de l'Heure. Rédaction: BIH, Observatoire de Paris.

Bull. Mesures Ionosph.
Bulletin de Mesures Ionosphériques. Publié par le Centre National d'Etudes des Télécommunications, Issy-les-Moulineaux.

Bull. Obs. Astron. Beograd
Bulletin de l'Observatoire Astronomique de Beograd. Editor: Observatoire Astronomique de Beograd. Printed by Naučna delo, Beograd.

Bull. Sci. Yougoslavie
Bulletin Scientifique. Conseil des Academies des Sciences et des Arts de la RSF de Yougoslavie. Section A: Sciences Naturelles, Techniques et Médicales. Redaction et Administration: Opaticka ul. 18/II, Zagreb (Yougoslavie).

Bull. Signal.
Bulletin Signalétique. Section 120: Astronomie et Astrophysique, Physique du Globe. Centre de Documentation du Centre National de la Recherche Scientifique, Paris.

Bull. Soc. Roy. Sci. Liège
Bulletin de la Société Royale des Sciences de Liège. L'Université, Liège.

Byull. Abastuman. Astrofiz. Obs.
Abastumanskaya Astrofizicheskaya Observatoriya, Gora Kanobili. Byulleten'. Akademiya Nauk Gruzinskoj SSR. Publishers: Izdatel'stvo "Metsniereba", Tbilisi.

Byull. Stantsij Optichesk. Nablyud. Iskusstv. Sputnikov Zemli
Byulleten' Stantsij Opticheskogo Nablyudeniya Iskusstvennykh Sputnikov Zemli. Published by Astronomicheskij Sovet Akademii Nauk SSSR, Moskva.

Canadian Journ. Phys.
Canadian Journal of Physics. Published by the National Research Council of Canada, Ottawa. Printed in Canada by the University of Toronto Press, Toronto,Ont.

Celestial Mechanics
Celestial Mechanics. An International Journal of Space Dynamics. Publishers: D. Reidel Publishing Company, Dordrecht—Holland.

Ciel et Terre
Ciel et Terre. Bulletin de la Société Belge d'Astronomie, de Métérologie et de Physique du Globe. Publié avec le concours du Ministère de l'Education Nationale, par la Société Belge d'Astronomie, Bruxelles.

Circ. d'Information
Circulaire d'Information. Union Astronomique Internationale. Commission des Etoiles Doubles. Address: Observatoire de Meudon, Meudon, France.

Coelum
Coelum. Periodico bimestrale per la Divulgazione dell' Astronomia. Editor: Osservatorio Astronomico Universitario di Bologna.

Comments Astrophys. Space Phys.
Comments on Astrophysics and Space Physics. A Journal of Critical Discussion of the Current Literature. Publishers: Gordon and Breach, Science Publishers, Inc., New York — London.

Commun. ACM
Communications of the ACM. A Publication of the Association for Computing Machinery. Published monthly by the Association for Computing Machinery, Baltimore, Md.

Comptes Rendus Acad. Sci. Paris
Comptes Rendus hebdomadaires des Séances de l'Académie des Sciences, publié avec le concours du Centre National de la Recherche Scientifique. Imprimérie: Gauthier-Villars, Paris.

Contr. Atmosph. Phys.
Contributions to Atmospheric Physics — Beiträge zur Physik der Atmosphäre. Publisher: Friedrich Vieweg & Sohn, Braunschweig.

COSPAR Inform. Bull.
COSPAR. Information Bulletin. Address: COSPAR Secretariat, Paris.

Deutsche Geod. Kommission Bayer. Akad. Wiss.
Deutsche Geodätische Kommission bei der Bayerischen Akademie der Wissenschaften. Reihe A: Höhere Geodäsie; Reihe B: Angewandte Geodäsie; Reihe C: Dissertationen; Reihe D: Tafelwerke; Reihe E: Geschichte und Entwicklung der Geodäsie. Published by Verlag der Bayerischen Akademie der Wissenschaften, München.

Documentat. Observateurs
Documentation des Observateurs. Rédaction: Station d'Astrophysique de Forcalquier.

Documentat. Observateurs Circ.
Documentation des Observateurs. Circulaire. Rédaction:

Station d'Astrophysique de Forcalquier.

Dokl. Akad. Nauk
Doklady Akademii Nauk SSSR. Seriya Matematika, Fizika. Publishers: Izdatel'stvo "Nauka", Moskva.

Dunsink Obs. Publ.
Dunsik Observatory Publications. The Observatory of the School of Cosmic Physics, Dublin Institute for Advanced Studies, Dublin.

Earth Planet. Sci. Letters
Earth and Planetary Science Letters. A Letter Journal devoted to the Development in Time of the Earth and Planetary System. Publisher: North-Holland Publishing Company, Amsterdam.

El Universo
El Universo. Organo de la Sociedad Astronomica de Mexico, Mexico, D. F.

Endeavour
Eine in vier Sprachen erscheinende Übersicht über Fortschritte der Naturwissenschaft. Published by Imperial Chemical Industries Limited, London.

ESO Bull.
European Southern Observatory, Bulletin. Edited by European Southern Observatory. Office of the Director: Hamburg.

Fortschritte Phys.
Fortschritte der Physik. Publisher: Akademie-Verlag, Berlin.

Gaz. Astron. Mém.
Gazette Astronomique. Mémoires van het Sterrenkundig Genootschap van Antwerpen, (de la Société d'Astronomie d'Anvers), Antwerpen. Printer: «De Voorzorg», A. Van Leuvenhaege, Antwerpen.

Geochim. Cosmochim. Acta
Geochimica et Cosmochimica Acta. Journal of the Geochemical Society. Publishing House: Pergamon Press, Ltd., Oxford.

Geodezja Kartografia
Geodezja i Kartografia. Komitet Geodezji Polskiej Akademii Nauk. Publisher: Państwowe Wydawnictwo Naukowe, Warszawa.

Geomagn. Aeronom.
Geomagnetizm i Aehronomiya. Akademiya Nauk SSSR. Izdatel'stvo "Nauka", Moskva [A translation published as "Geomagnetism and Aeronomy".]

Geophys. Journ.
The Geophysical Journal of the Royal Astronomical Society. Published for the Royal Astronomical Society by Blackwell Scientific Publications, Oxford − Edinburgh.

Gerlands Beiträge Geophys.
Gerlands Beiträge zur Geophysik. Publisher: Akademische Verlagsgesellschaft Geest & Portig K.-G., Leipzig.

Glasnik Mat.
Glasnik Matematički. Published by the Society of Mathematicians and Physicists of the S. R. of Croatia. Publisher: Društvo Matematičara i Fizičara S. R. Hrvats-

ke, Zagreb.

Helvetica Phys. Acta
Helvetica Physica Acta. Publisher: E. Birhäuser, Basel.

Hemel en Dampkring
Maandblad van de Nederlandse Vereniging voor Weer-en Sterrenkunde en van de Vereniging voor Sterrenkunde, Meteorologie, Geophysica en Aanverwante Wetenschappen in Belgie. Publisher: Wolters-Noordhoff N.V., Groningen.

IAU Circ.
International Astronomical Union, Circular. Central Bureau for Astronomical Telegrams, Smithsonian Observatory, Cambridge, Mass.

IBM Journ. Res. Development
IBM Journal of Research and Development. Published bimonthly by International Business Machines Corporation, Armonk, New York.

ICSU Bull.
ICSU Bulletin. International Council of Scientific Unions. Secretariat: 7, Via Cornelio Celso, Rome, Italy.

Icarus
Icarus. International Journal of the Solar System. Publisher: Academic Press, New York − London.

IEEE Spectrum
IEEE Spectrum. Published monthly by the Institute of Electrical and Electronics Engineers, Inc., New York, N.Y.

Inform. Bull. Southern Hemisphere
Information Bulletin of the Southern Hemisphere. Editorial Office: Observatorio Astronómico, La Plata, Argentina.

Inform. Bull. Variable Stars
Commission 27 of the I.A.U. Information Bulletin on Variable Stars. Konkoly Observatory, Budapest.

Infrared Physics
An International Research Journal. Pergamon Press Ltd., Oxford − London − New York.

Irish Astron. Journ.
The Irish Astronomical Journal. A Quarterly Publication under the auspices of the Observatories of Armagh and Dunsink. Subscription address: Managing Editor, Irish Astronomical Journal, Armagh Observatory, Northern Ireland.

Izv. Akad. Nauk Armyan. SSR
Izvestiya Akademii Nauk Armyanskoj SSR. Fizika Erevan.

Izv. Glav. Astron. Obs. Pulkove
Izvestiya Glavnoj Astronomicheskoj Observatorii v Pulkove. Akademiya Nauk SSSR. Izdanie Glavnoj astronomicheskoj observatorii v Pulkove, Leningrad.

Izv. Komissii Fiz. Planet
Izvestiya Komissii po Fizike Planet. Akademiya Nauk SSSR. Astronomicheskij Sovet. Moskva.

Izv. Krymskoj Astrofiz. Obs.
Izvestiya Krymskoj Astrofizicheskoj Observatorii. Akademiya Nauk SSR. Publishers: Izdatel'stvo "Nauka", Moskva.

JETP Letters
JETP Letters. A translation of JETP Pis'ma v Redaktsiyu of the Academy of Sciences in the USSR. Published semi-monthly by the American Institute of Physics, Lancaster, Pennsylvania.

Journ. Ass. Computing Machinery
Journal of the Association for Computing Machinery. Published by the Association for Computing Machinery, Baltimore, Md.

Journ. Astronaut. Sci.
The Journal of the Astronautical Sciences. Published by the American Astronautical Society Inc., Baltimore, Md.

Journ. Astron. Soc. Victoria
The Journal of the Astronomical Society of Victoria. Printed by D. Buscombe Printers, Glen Waverley, Victoria.

Journ. Astron. Soc. Western Australia
The Journal of the Astronomical Society of Western Australia. Edited by the Astronomical Society of Western Australia, Perth, W. A.

Journ. Atmosph. Terr. Phys.
Journal of Atmospheric and Terrestrial Physics. Publishers: Pergamon Press, Oxford - London - New York.

Journ. British Astron. Ass.
Journal of the British Astronomical Association. Subscription address: Office of the Association, Hounslow West, Middlesex.

Journ. British Interplanet. Soc.
Journal of the British Interplanetary Society. Printed by Unwin Brothers, Ltd., London, and published by the British Interplanetary Society.

Journ. Fluid Mechanics
Journal of Fluid Mechanics. Published by Cambridge University Press, London – New York.

Journ. Geophys. Res.
Journal of Geophysical Research. An International Scientific Publication. Published by the American Geophysical Union, Washington, D. C.

Journ. Inst. Navigation
Journal of the Institute of Navigation. Published quarterly by the Institute of Navigation, London.

Journ. Observateurs (JO)
Journal des Observateurs. Publié avec le concours de l'Université d'Aix-Marseille par le Centre National de la Recherche Scientifique, Paris. After Vol. 51 replaced by "Astronomy and Astrophysics".

Journ. Optical Soc. America
Journal of the Optical Society of America. Publisher: American Institute of Physics, New York.

Journ. Phys. A.General Phys.
Journal of Physics A. General Physics. (Proceedings of the Physical Society) Series 2. Published by the Institute of Physics and the Physical Society, London, England, in association with the American Institute of Physics, New York.

Journ. Physique

Journal de Physique. Publication de la Société Française de Physique, Paris.

Journ. Proc. Roy. Soc. New South Wales
Journal and Proceedings of the Royal Society of New South Wales. Published by the Society, Science House, Sydney.

Journ. Quant. Spectrosc. Radiat. Transfer
Journal of Quantitative Spectroscopy & Radiative Transfer. Publisher: Pergamon Press, Oxford – New York.

Journ. Roy. Astron. Soc. Canada
The Journal of the Royal Astronomical Society of Canada, devoted to the Advancement of Astronomy and Allied Sciences. Printed by the University of Toronto Press, Toronto, Ontario.

Kometn. Tsirk. *Kiev*
Kometnyj Tsirkulyar. Gruppa po Issledovaniyu Komet Astrosoveta i Mezhduvedomstvennyj Geofizicheskij Komitet, Akademii Nauk SSSR. Kievskij Universitet im. T. G. Shevchenko.

Komety i Meteory
Komety i Meteory. Akademiya Nauk Tadzhikskoj SSR. Astronomicheskij Sovet Akademii Nauk SSSR. Publishers: Izdatel'stvo "Donish", Dushanbe.

Kosmich. Issled.
Kosmicheskie Issledovaniya. Akademiya Nauk SSSR. Publishers: Izdatel'stvo "Nauka", Moskva.

L'Astronomie
L'Astronomie et Bulletin de la Société Astronomique de France. Revue mensuelle. Rédaction: Société Astronomique de France, Paris.

L'Universo
L'Universo. Rivista dell'Instituto Geografico Militare. Direzione, Redazione e Amministrazione: Istituto Geografico Militare, Firenze.

Magnitnye Polya Solnech. Pyaten
Magnitnye Polya Solnechnykh Pyaten. (Supplements to Solnechnye Dannye. Byulleten' (*Solar Data*)). Publishers: Izdatel'stvo "Nauka", Leningrad.

Math. Rev.
Mathematical Reviews. Published by the American Mathematical Society, Providence, R. I.

Mem. Fac. Sci. Kyoto Univ.
Memoirs of the Faculty of Science, Kyoto University. Series of Physics, Astrophysics, Geophysics, and Chemistry. Printed by Yamashiro Printing Publishing Co. Ltd., Kamigyo, Kyoto.

Mem. Japan Astron. Study Ass.
Memoirs of the Japan Astronomical Study Association. Izumi 59, Yugawara-machi, Kanagawa–ken, Japan.

Mem. Roy. Astron. Soc.
Memoirs of the Royal Astronomical Society. Published for the Royal Astronomical Society by Blackwell Scientific Publications, Oxford – Edinburgh.

Mem. Soc. Astron. Italiana
Memorie della Società Astronomica Italiana. Nuova Serie. Pubblicate sotto gli auspici del Consiglio Nazionale delle Ricerche. Publisher: Tipografia Baccini & Chiappi,

Firenze.

Messtechnik
Messtechnik (Zeitschrift für Instrumentenkunde). Publishers: Verlag Friedrich Vieweg & Sohn GmbH, Braunschweig.

Meteoritics
Meteoritics. The Journal of the Meteoritical Society. Circulation Manager: C. F. Lewis, Center for Meteorite Studies, The Arizona State University, Tempe, Arizona.

Meteoritika
Akademiya Nauk SSSR. Komitet po Meteoritam. Publishers: Izdatel'stvo "Nauka", Moskva.

Mitt. Astron. Ges.
Mitteilungen der Astronomischen Gesellschaft, Hamburg.

Monatsber. Deutsch. Akad. Wiss. Berlin
Monatsberichte der Deutschen Akademie der Wissenschaften zu Berlin. Mitteilungen aus Mathematik, Naturwissenschaft, Medizin und Technik. Publisher: Akademie-Verlag, Berlin.

Monthly Notes Astron. Soc. Southern Africa
Monthly Notes of the Royal Astronomical Society of Southern Africa. Published by the Astronomical Society of Southern Africa, Royal Observatory, Cape Province, South Africa.

Monthly Notices Roy. Astron. Soc.
Monthly Notices of the Royal Astronomical Society. Published for the Royal Astronomical Society by Blackwell Scientific Publications, Oxford – Edinburgh.

MVS Sonneberg
Mitteilungen über Veränderliche Sterne. Edited by Sternwarte Sonneberg (Zentralinstitut für Astrophysik, Bereich Sternphysik) der Deutschen Akademie der Wissenschaften.

Nachr. Akad. Wiss. Göttingen
Nachrichten der Akademie der Wissenschaften in Göttingen. II. Mathematisch-Physikalische Klasse. Vandenhoeck & Ruprecht, Göttingen.

Nachr. Karten-, Vermessungswesen
Nachrichten aus dem Karten- und Vermessungswesen. Editor: Institut für Angewandte Geodäsie (Abt. II des Deutschen Geodätischen Forschungsinstituts). Published by Verlag des Instituts für Angewandte Geodäsie, Frankfurt a. M.

Nature
Nature. A weekly Journal of Science. MacMillan & Co., Ltd., London; St. Martin's Press, Inc., New York.

Naturwissenschaften
Die Naturwissenschaften. Publisher: Springer-Verlag, Berlin – Heidelberg – New York.

Nauchn. Informatsii
Nauchnye Informatsii. Astronomicheskij Sovet Akademii Nauk SSSR, Moskva.

Naučna Misao
Naučna Misao. Drustvo za Unapredivanje i Širenje Nauke, Zagreb. Printer: Novinsko izdavacko poduzeče "Slobodna Dalmacja", Split.

Numerische Math.
Numerische Mathematik. Publisher: Springer-Verlag, Berlin – Heidelberg – New York.

Nuovo Cimento
Il Nuovo Cimento. Rivista Internazionale e Organo della Società Italiana di Fisica, Series A, B. Publisher: Nicola Zanichelli, Editore, Bologna.

Nuovo Cimento Lettere
Lettere al Nuovo Cimento. Rivista internazionale della Società Italiana di Fisica. Serie prima. Editrice Compositori, Bologna.

Nuovo Cimento Rivista
Rivista del Nuovo Cimento a cura della Società Italiana di Fisica. Editrice Compositori, Bologna.

Nuovo Cimento Suppl.
Supplemento al Nuovo Cimento. Nicola Zanichelli, Editore, Bologna.

Observations Artificial Earth Satellites
Observations of Artificial Satellites of the Earth (Nablyudeniya Iskusstvennykh Sputnikov Zemli). Magyar Tudományos Akadémia Csillagvizsgáló Intézete, Budapest.

Observatory
The Observatory. A Review of Astronomy. Publishers: The Editors of "The Observatory", Royal Greenwich Observatory, Herstmonceaux Castle, Hailsham, Sussex, England.

Optik
Optik. Zeitschrift für das gesamte Gebiet der Licht- und Elektronenoptik. Publishers: Wissenschaftliche Verlagsgesellschaft mbH., Stuttgart.

Orion Schaffhausen
Orion. Zeitschrift der Schweizerischen Astronomischen Gesellschaft (SAG). Bulletin de la Société Astronomique de Suisse (SAS). Administration: Generalsekretariat der SAG, Schaffhausen.

Österreich. Zeitschr. Vermessungswesen
Österreichische Zeitschrift für Vermessungswesen. Editor and Publisher: Österreichischer Verein für Vermessungswesen, Wien.

Peremennye Zvezdy, Byull.
Peremennye Zvezdy, Byulleten', izdavaemyj Astronomicheskim Sovetom Akademii Nauk SSSR. Published by Astronomicheskij Sovet Akademii Nauk SSSR, Moskva.

Phil. Mag.
The Philosophical Magazine. A Journal of Theoretical, Experimental and Applied Physics. Eighth Series. Publisher: Taylor & Francis, Ltd., London.

Phil. Trans. Roy. Soc. London
Philosophical Transactions of the Royal Society of London. Series A, Mathematical and Physical Sciences. Published by the Royal Society, London.

Phys. Abstr.
Physics Abstracts. Science Abstracts, Series A. An INSPEC Publication, published by The Institution of Electrical Engineers, London.

Phys. Ber.
Physikalische Berichte. Herausgegeben von der Deutschen Physikalischen Gesellschaft e. V. und von der Deutschen Akademie der Wissenschaften zu Berlin. Friedrich Vieweg & Sohn, Braunschweig.

Phys. Blätter
Physikalische Blätter. Physik-Verlag, Mosbach/Baden.

Phys. Earth Planet. Interiors
Physics of the Earth and Planetary Interiors. A journal devoted to observational and experimental studies of the Earth and Planetary interiors and their theoretical interpretation by the physical sciences. Publisher: North-Holland Publishing Company, Amsterdam, Netherlands.

Phys. Fluids
The Physics of Fluids. Published by the American Institute of Physics, New York.

Phys. Rev.
The Physical Review. A journal of experimental and theoretical physics. Second Series. Published for The American Physical Society by the American Institute of Physics, Lancaster, Pa., and New York, N. Y.

Phys. Rev. Letters
Physical Review Letters. Published weekly by The American Physical Society, New York, N.Y.

Phys. Today
Physics Today. Published by the American Institute of Physics, New York.

Physica
Physica. Publishers: North–Holland Publishing Company, Amsterdam, The Netherland, on request of the Foundation "Physica", Utrecht.

Planet. Space Sci.
Planetary and Space Science. Pergamon Press, Oxford – London – New York.

Pokroky
Pokroky matematiky, fyziky a astronomie. Vydává Jednota čs. matematiků a fyziků. Publisher: Academia, Praha.

Postępy Astron.
Postępy Astronomii. Czasopismo Poświecone Upowszechnianiu Wiedzy Astronomicznej. Polskie Towarzystwo Astronomiczne, Warszawa. Printed in Poland by Państwowe Wydawnictwo Naukowe, Lódź.

Priroda
Priroda. Publishers: Izdatel'stvo "Nauka", Moskva.

Proc. Astron. Soc. Australia
Proceedings of the Astronomical Society of Australia. Published for the Society by Sydney University Press, Sydney.

Proc. Cambridge Phil. Soc.
Proceedings of the Cambridge Philosophical Society (Mathematical and Physical Sciences). Publishers: Cambridge University Press, London.

Proc. IEEE
Proceedings of the IEEE. Published monthly by the Institute of Electrical and Electronics Engineers, Inc. New York.

Proc. Koninkl. Nederl. Akad. Wet.
Koninklijke Nederlandse Akademie van Wetenschappen. Proceedings. Series B, Physical Sciences. Publishers: North-Holland Publishing Company, Amsterdam.

Proc. National Acad. Sci. U. S. A.
Proceedings of the National Academy of Sciences of the Unites States of America. Published monthly by the National Academy of Sciences, Washington, D. C.

Proc. Roy. Soc.
Proceedings of the Royal Society. Series A, Mathematical and Physical Sciences. Published by the Royal Society, London.

Progr. Theor. Phys. Japan
Progress of Theoretical Physics. Published for the Research Institute for Fundamental Physics and the Physical Society of Japan. Publication Office: Progress of Theoretical Physics, Yukawa Hall, Kyoto University, Kyoto, Japan.

Progr. Theor. Phys. Suppl.
Supplement of the Progress of Theoretical Physics. Published for the Research Institute for Fundamental Physics and The Physical Society of Japan. Publication Office: Progress of Theoretical Physics, Yukawa Hall, Kyoto University, Kyoto, Japan.

PTB Mitt.
PTB Mitteilungen. Amts- und Mitteilungsblatt der Physikalisch-Technischen Bundesanstalt, Braunschweig — Berlin.

Publ. Astron. Soc. Japan
Publications of the Astronomical Society of Japan. Published by the Astronomical Society of Japan. Office of the Society: Tokyo Astronomical Observatory, Mitaka, Tokyo. Agent: Maruzen Co. Ltd. (Export Department), Nihonbashi, Tokyo, Japan.

Publ. Astron. Soc. Pacific
Publications of the Astronomical Society of the Pacific. Published in Provo, Utah, by the Astronomical Society of the Pacific, San Francisco, California. Printed by Brigham Young University Press, Provo, Utah.

Publ. Roy. Obs. Edinburgh
The Royal Observatory, Edinburgh. Publication. Her Majesty's Stationery Office, Edinburgh.

Publ. Tartu Astrofiz. Obs.
W. Struve nimelise, Tartu Astrofüüsika Observatooriumi, Publikatsioonid. Eesti NSV Teaduste Akadeemia, Tartu.

Quarterly Journ. Roy. Astron. Soc.
Quarterly Journal of the Royal Astronomical Society. Published for the Royal Astronomical Society by Blackwell Scientific Publications, Oxford.

Referativ. Zhurn. 51. Astron.
Referativnyj Zhurnal. 51. Astronomiya. Vsesoyuznyj Institut Nauchnoj i Tekhnicheskoj Informatsii. Moskva.

Referativ. Zhurn. 52. Geod. i Aehros"emka.
Referativnyj Zhurnal. 52. Geodeziya i Aehros"emka. Vsesoyuznyj Institut Nauchnoj i Tekhnicheskoj Informatsii. Moskva.

Referativ. Zhurn. 62. Issled. kosm. prostranstv.
Referativnyj Zhurnal. 62. Issledovanie Kosmicheskogo

Prostranstva. Vsesoyuznyj Institut Nauchnoj i Tekhnicheskoj Informatsii. Moskva.

Rep. Progr. Phys.
Reports on Progress in Physics. Institute of Physics and
the Physical Society, London.

Rev. Geophys.
Reviews of Geophysics. Published by the American
Geophysical Union Washington, D. C.

Revista Astron.
Revista Astronomica. Organo de la Asociación Argentina
Amigos de la Astronomia, Buenos Aires.

Rev. Modern Phys.
Reviews of Modern Physics. Published for The American
Physical Society by the American Institute of Physics,
Lancaster, Pa., and New York, N. Y.

Rev. Sci. Instruments
Reviews of Scientific Instruments. Published by the
American Institute of Physics, Lancaster, Pa., and New
York, N. Y.

Rezul'taty Nablyud. Sovet. Iskusstv. Sputnikov Zemli
Rezul'taty Nablyudenij Sovetskikh Iskusstvennykh Sputnikov Zemli. Published by Astronomicheskij Sovet Akademii Nauk SSSR, Moskva.

Ric. Sci.
La Ricerca Scientifica. Serie Seconda. Consiglio Nazionale delle Ricerche, Roma.

Říše hvězd
Říše hvězd. Czechoslovak popular astronomical journal.
Publisher: Orbis, Praha.

Roy. Astron. Soc. New Zealand Circ.
Royal Astronomical Society of New Zealand, Variable
Star Section, Circular. Office: Box 33, Lake Tekapo,
New Zealand.

Roy. Astron. Soc. New Zealand Variable Star Sect. Repr.
Royal Astronomical Society of New Zealand. Variable
Star Section. Reprint. Address: P. O. Box 33, Lake Te-
kapo, New Zealand.

Rumanian Sci. Abstr.
Rumanian Scientific Abstracts. Natural Sciences.
Publisher: The Scientific Documentation Centre of the
Academy of the Socialist Republic of Romania, Bucu-
reşti.

Sci. American
Scientific American. Published monthly by Scientific
American, Inc., New York, N. Y.

Sci. Rep. Tôhoku Univ.
The Science Reports of the Tôhoku University. First
Series (Physics, Chemistry, Astronomy). Published by
the Faculty of Science, Tôhoku University, Sendai,
Japan.

Science
Science. American Association for the Advancement
of Science, Washington, D. C.

Science Progrès, La Nature
Science Progrès, La Nature. Revue Mensuelle. Publishers:
Dunod, Editeur, Paris.

Sitz.-Ber. Bayer. Akad. Wiss.
Bayerische Akademie der Wissenschaften. Mathematisch-
Naturwissenschaftliche Klasse. Sitzungsberichte. Publisher: Verlag der Bayerischen Akademie der Wissenschaften, München.

Sitz.-Ber. Deutsch. Akad. Wiss. Berlin
Sitzungsberichte der Deutschen Akademie der Wissen-
schaften zu Berlin. Klasse für Mathematik, Physik und
Technik. Publisher: Akademie-Verlag, Berlin.

Sitz.-Ber. Heidelberger Akad. Wiss.
Sitzungsberichte der Heidelberger Akademie der Wissen-
schaften. Mathematisch-Naturwissenschaftliche Klasse.
Publisher: Springer-Verlag, Heidelberg.

Sitz.-Ber. Österreich. Akad. Wiss.
Sitzungsberichte. Österreichische Akademie der Wissenschaften. Mathematisch-Naturwissenschaftliche Klasse.
Abteilung II: Mathematik, Astronomie, Meteorologie
und Technik. Publisher: Springer-Verlag, Wien.

Sky Telescope
Sky and Telescope. Published by Sky and Telescope
Corporation, Cambridge, Mass.

Smithsonian Contr. Astrophys.
Smithsonian Contributions to Astrophysics. Astrophysical Observatory of the Smithsonian Institution. For sale
by the Superintendent of Documents, U. S. Government
Printing Office, Washington, D. C.

Smithsonian Year
Smithsonian Year. Annual Report of the Smithsonian
Institution, including the financial report of the Exe-
cutive Committee of the Boards of Regents. Published
by the Smithsonian Institution, Washington, D. C.

Solar Physics
Solar Physics. A Journal for Solar Research and the
Study of Solar Terrestrial Physics. Publishers: D. Reidel
Publishing Company, Dordrecht-Holland.

Solnech. Dannye Byull.
Solnechnye Dannye. Byulleten. *(Solar Data).* Publishers:
Izdatel'stvo "Nauka", Leningrad.

Soobshch. Byurakan. Obs.
Soobshcheniya Byurakanskoj Observatorii. Akademiya
Nauk Armyanskoj SSR, Erevan.

Soobshch. Gos. Astron. Inst. Shternberg
Soobshcheniya Gosudarstvennogo Astronomicheskogo
Instituta im P. K. Shternberga. Publishers: Izdatel'stvo
Moskovskogo Universiteta, Moskva.

Southern Stars
Southern Stars. The Journal of the Royal Astronomical
Society of New Zealand (Inc.). Address of the Society:
P.O. Box 3181, Wellington C1, New Zealand.

Soviet Astron. AJ
Soviet Astronomy AJ. A translation of the Astronomical
Journal of the Academy of Sciences of the USSR. Pub-
lished by the American Institute of Physics, Inc., New
York.

Spaceflight
Spaceflight. Published by the British Interplanetary Society, London.

Space Sci. Rev.
Space Science Reviews. Publishers: D. Reidel Publishing Company, Dordrecht-Holland.

Springer Tracts Modern Phys.
Springer Tracts on Modern Physics. (Ergebnisse der exakten Naturwissenschaften). Springer-Verlag, Berlin–Heidelberg–New York.

Sterne
Die Sterne. Zeitschrift für alle Gebiete der Himmelskunde. Johann Ambrosius Barth, Leipzig.

Sternenbote
Sternenbote. Monatsschrift für Österreichs Amateurastronomen. Publisher: Astronomisches Büro, Hermann Mucke, Wien.

Stockholms Obs. Ann.
Stockholms Observatoriums Annaler. Printed by Almqvist & Wiksell, Stockholm.

Strolling Astronomer
The Strolling Astronomer. The Journal of The Association of Lunar and Planetary Observers. Publication Office: The Strolling Astronomer, Box 3AZ, University Park, New Mexico.

Stud. Cerc. Astron.
Studii şi Cercetări de Astronomie. Editura Academiei Republicii Socialiste România. Editorial Office: Observatorul Astronomic, Bucureşti.

Stud. Geophys. Geod.
Studia geophysica et geodaetica. Published for the Geophysical Institute of the Czechoslovak Academy of Sciences by Academia, Praha.

Stud. Univ. Babeş-Bolyai
Studia Universitatis Babeş-Bolyai. Series Mathematica-Physica. Publishers: Intreprinderea Poligrafica, Cluj.

SuW
Sterne und Weltraum. Astronomische Monatsschrift. Verlag Bibliographisches Institut AG, Mannheim.

Tellus
Tellus, a bi-monthly Journal of Geophysics. Svenska Geofysiska Foreningen. Printed in Sweden by Almqvist & Wiksells Boktryckeri AB, Uppsala.

Trans. Astron. Obs. Yale Univ.
Transactions of the Astronomical Observatory of Yale University. Published by the Observatory, New Haven.

Trans. Roy. Soc. Canada
Transactions of the Royal Society of Canada. Published by the Royal Society of Canada, National Research Building, Ottawa.

Trudy Astrofiz. Inst. Alma-Ata
Trudy Astrofizicheskogo Instituta, Alma-Ata. Akademiya Nauk Kazakhskoj SSR. Publishers: Izdatel'stvo "Nauka" Kazakhskoj SSR, Alma Ata.

Trudy Glav. Astron. Obs. Pulkove
Trudy Glavnoj Astronomicheskoj Observatorii v Pulkove. Akademiya Nauk SSSR. Izdanie Glavnoj astronomicheskoj observatorii v Pulkove, Leningrad.

Trudy Inst. Teor. Astron. *Leningrad*

Trudy Instituta Teoreticheskoj Astronomii. Akademiya Nauk SSSR. Publishers: Izdatel'stvo "Nauka", Leningrad.

Trudy Tashkent. Astron. Obs.
Trudy Tashkentskoj Astronomicheskoj Observatorii. Akademiya Nauk Uzbekskoj SSR. Publishers: Izdatel' stvo "FAN" Uzbekskoj SSR, Tashkent.

Tsirk. Astron. Inst. Tashkent
Tsirkulyar Astronomicheskogo Instituta. Akademiya Nauk Uzbekskoj SSR. Izdatel'stvo "FAN" Uzbekskoj SSR, Tashkent.

Tsirk. Astron. Obs. L'vov
Tsirkulyar. Astronomicheskaya Observatoriya. L'vovskij Ordena Lenina Gosudarstvennyj Universitet emeni Ivana Franko. Publisher: Izdatel'stvo L'vovskogo Universiteta, L'vov.

Umschau
Umschau in Wissenschaft und Technik. Umschau-Verlag Frankfurt a. M.

Urania Barcelona
Urania. Revista de Astronomia y Ciencias Afines. Organo de la Sociedad Astronómica de España y América, Barcelona; Unión Nacional de Astronomia y Ciencias Afines, Madrid.

Urania Kraków
Urania. Miesięcznik Polskiego Towarzystwa Miłośników Astronomii, Kraków. Publisher: Krakowska Drukarnia Prasowa, Kraków.

Vasiona
Vasiona. Revue d'Astronomie et d'Astronautique. Bulletin de la Société Astronomique "R. Bosković", Beograd.

VdS Nachrichtenblatt
Nachrichtenblatt der Vereinigung der Sternfreunde e.V. After Vol. 18 No. 3 published in combination with "Sterne und Weltraum". Bibliographisches Institut, Mannheim.

Veröff. Astron. Rechen-Inst. Heidelberg
Veröffentlichungen des Astronomischen Rechen-Instituts Heidelberg. Verlag G. Braun, Karlsruhe.

Veröff. Sternw. Sonneberg
Deutsche Akademie der Wissenschaften zu Berlin. Institut für Sternphysik. Veröffentlichungen der Sternwarte in Sonneberg. Publisher: Akademie-Verlag, Berlin.

Vesmír
Vesmír. Přírodovědecky časopis Čs. akademie věd. Publisher: Academia, Praha.

Vestn. Khar'kov. Univ.
Vestnik Khar'kovskogo Universiteta. Seriya Astronomicheskaya. Publishers: Izdatel'stvo Khar'kovskogo Universiteta, Khar'kov.

Vestn. Kiev. Univ.
Vestnik Kievskogo Universiteta. Seriya Astronomii. Publishers: Izdatel'stvo Kievskogo Universiteta, Kiev.

VJS Naturforsch. Ges. Zürich
Vierteljahresschrift der Naturforschenden Gesellschaft in Zürich. Printer and Publisher: Leeman AG, Zürich.

Weltraumfahrt

Weltraumfahrt. Zeitschrift für Astronautik und Raketen-
technik. Umschau-Verlag, Frankfurt a. M.

Wiss. Zeitschr. Humboldt-Univ. Berlin
Wissenschaftliche Zeitschrift der Humboldt-Universität
zu Berlin. Mathematisch-Naturwissenschaftliche Reihe.
Edited by the Rektor der Humboldt-Universität, Berlin.

Yamamoto Circ.
Yamamoto Circular. Published by the Yamamoto Observatory, Kamitanakami – Kiryutyo, Otu, Siga-ken, Japan.

Zeitschr. Angew. Physik
Zeitschrift für Angewandte Physik. Publisher: Springer-
Verlag, Berlin–Heidelberg–New York.

Zeitschr. Astrophys. (ZfA)
Zeitschrift für Astrophysik. Publisher: Springer-Verlag,
Berlin–Heidelberg–New York. After Vol. 69 (1968)
replaced by "Astronomy and Astrophysics".

Zeitschr. Geophys.
Zeitschrift für Geophysik. Publisher: Physica-Verlag,
Würzburg.

Zeitschr. Naturforschung
Zeitschrift für Naturforschung. Verlag der Zeitschrift für
Naturforschung, Tübingen.

Zeitschr. Physik
Zeitschrift für Physik. Publisher: Springer-Verlag, Berlin–
Heidelberg–New York.

Zemlya i Vselennaya
Zemlya i Vselennaya. Nauchno-Populyarnyj Zhurnal
Akademii Nauk SSSR. Publishers: Izdadel'stvo "Nauka",
Moskva.

Zentralblatt Math. Grenzgebiete
Zentralblatt für Mathematik und ihre Grenzgebiete. Publisher: Springer-Verlag, Berlin - Heidelberg - New York.

Zvaigžņota Debess
Latvijas PSR Zinātņu Akadēmijas Radioastrofizikas
Observatorijas Populārzinatnisks Gadalaiku Izdevums.
Izdevnieciba "Zinātne", Riga.

002 Bibliographical Publications

002.001 Kurzberichte aus der Forschung.
SuW, Vol. 8, 202 - 205, 208 - 209 (1969).
Solare Neutronen *(H. Wöhl);* Ein schrumpfender Neutronenstern? *(H. Hippelein);* Standardlichtquelle im fernen Ultraviolett *(D. Lemke);* Projekt BOMEX, Wettervorhersage *(E. Pitz);* Allmählicher Verlauf von Sternbedeckungen *(J. Classen);* Neue OH-Quellen im Cygnus *(G. Ackermann);* Neue Jupitermonde? *(G. Fugmann);* Ergebnisse von OSO-3-Messungen *(H. Wöhl);* OSO-5 gestartet *(C. Leinert);* Optigami – Eine Hilfe für den Entwurf optischer Systeme *(W. Hofmann);* Mondüberwachung während des Apollo-8-Fluges *(J. Classen);* H$_2$O-Banden im Spektrum des Infrarotsternes NML Cyg? *(G. Ackermann);* Welches Material bedeckt die Marsoberfläche? *(C. Leinert);* Raketenmessungen im fernen Infrarot *(G. Ackermann);* Beobachtungen des galaktischen Zentrums im fernen Infrarot *(W. Hofmann);* 12 Jahre Interkosmos in Osteuropa *(H. Krefft);* Sternphotometrie im Ultravioletten *(D. Lemke);* Sonnenobservatorium San Fernando *(H. Ruhm).*

002.002 News notes.
Sky Telescope, Vol. 38, 166 - 167 (1969).
Standing waves on the moon; Unusual Algol-type variable; Coordinated X-ray and optical observations of a pulsar; Ancient giant craters.

002.003 News notes.
Sky Telescope, Vol. 38, 79 - 80 (1969).
Unusual Cepheid variable; Minor planet Geographos; Historian of the space age; The Vilna meteorite; Enigmatic black cloud.

002.004 News and comments. E. J. Öpik.
Irish Astron. Journ. Vol. 9, 32 - 44 (1969).
Stellar magnitudes, decibels, and Weber-Fechner law; P Cygni; The distribution of OB stars in the Northern Milky Way; Heavily obscured Wolf-Rayet stars in the Cygnus OB2 association; Solid hydrogen accretion on grains; Solid hydrogen and the formation of galaxies and stars; Graphite particles; Interstellar reddening; The ratio of total absorption to colour excess; Distribution of absorption regions in the galaxy; Organic molecules; The interstellar diffuse absorption band at 4430 Å; Reviews of interstellar reddening; A catalogue of dark globules; Three-colour photometry of open galactic star clusters and associations; Polarization of starlight; Ice caps on Venus? ; Technetium.

002.005 Zeitschrift für Astrophysik: Generalregister für die Bände 51 - 69 (1960 - 1968).
Compiled by W. Petri.
Springer-Verlag, Berlin - Heidelberg - New York, 61 pp. (1969).

002.006 Selected bibliography of the literature on the history of astronomy issued in the USSR and in other countries in 1965 - 1968.
N. B. Lavrova, P. G. Kulikovskij.
Istoriko-Astron. Issled. Vyp. (No.) 10, p. 344 - 349 (1969). In Russian.

002.007 Nouvelle de la science.
L'Astronomie, 83e année, 331 - 332, 377 (1969).
Enquête sur les «objets volants non identifiés», le contenu en métaux des amas globulaires, température de la haute atmosphère.

002.008 Nouvelles brèves.
Ciel et Terre, Vol. 85, 330 - 334 (1969).
La mission scientifique d'Apollo-12; La chute du satellite Echo 2; Petites planètes perdues; Nouvelles de la tache rouge de Jupiter.

002.009 Mitteilungen aus Wissenschaft und Literatur.
Sterne, 45. Jahrgang, 122 - 124 (1969). – Deutsche Städte im Weltraum *(J. W. Ekrutt);* Planetoid 1620 Geographos in Erdnähe *(J. W. Ekrutt);* Kommensurabilitäten in den Bahnen der großen Planeten *(F. Schmeidler).*

002.010 News notes.
Sky Telescope, Vol. 38, 222 - 224 (1969).
An enormous meteorite; Radar observations of Martian relief; More astronomical journals; Progress at Palomar; Unusual brightness of Mira; Gamma-ray source; News about comets.

002.011 Mitteilungen aus Wissenschaft und Literatur.
Sterne, 45. Jahrgang, 165 - 169 (1969).
Erste astronomische Ergebnisse von Apollo 11 *(J. Classen);* Zeitschrift "The Moon" gegründet *(D. B. Herrmann);* Zwei neue Pulsare entdeckt *(W. Rehpenning);* Periode eines Pulsars nahm abrupt ab *(W. Rehpenning);* Aus der statistischen Astronomie *(W. Rehpenning);* Abplattung der Sonne und das Merkurperihel *(F. Schmeidler);* Eine stellare Korona um die Galaxis M87 *(T. Schmidt-Kaler).*

002.012 Bulletin of the Astronomical Institute of the Netherlands 1921 – 1969. Author index.
Bull. Astron. Inst. Netherlands, Vol. 20, 337 - 361 (1969).

002.013 Bulletin of the Astronomical Institutes of the Netherlands 1921 – 1969. Subject index.
C. J. van Houten.
Bull. Astron. Inst. Netherlands, Vol. 20, 362 - 394 (1969).

002.014 Bulletin of the Astronomical Institutes of the Netherlands 1921 – 1969. Index of variable stars.
C. J. van Houten.
Bull. Astron. Inst. Netherlands, Vol. 20, 395 - 405 (1969).

002.015 Bulletin of the Astronomical Institutes of the Netherlands 1921 – 1969. Index of minor planets.
C. J. van Houten.
Bull. Astron. Inst. Netherlands, Vol. 20, 406 - 413 (1969).

002.016 Bulletin of the Astronomical Institutes of the Netherlands 1921 – 1969. Index of comets.
C. J. van Houten.
Bull. Astron. Inst. Netherlands, Vol. 20, 414 (1969).

002.017 Notes and observations.
Priroda No. 7, p. 102 - 103 (1969). In Russian. – Fossil meteoritic dust *(I. A. Yudin).*

002.018 Science news.
Priroda, No. 7, p. 106 - 116; No. 8, p. 107 - 116; No. 9, p. 108 - 117 (1969). In Russian. – (7) Das Magnetfeld um die Venus *(A. A. Lukin);* Polnische Raketen in den oberen Schichten der Atmosphäre; Pulsare — magnetische Neutronensterne; Neues über den optischen Pulsar im Crab-Nebel; Erste Photographie eines optischen Pulsars; Die abgeplattete Sonne, Gravitation und Neutrino; Ungewöhnlicher Sprung in der Periode eines Pulsars. (8) Instrumente für den Mond; Flüge zu den Planeten Saturn, Uranus, Neptun und Pluto; Gravitationswellen und Neutronensterne; Das interplanetare Magnetfeld; Ein Zweikanal-Fersehteleskop; Die Parameter von Icarus haben sich geändert. (9) Kosmische Strahlung und geomagnetische Stürme; Interstellare Materie.

002.019 Science news.
Priroda No. 10, p. 106 - 114; No. 11, p. 101 - 111; No. 12, p. 98 - 107 (1969). In Russian. – (10) Wasser im Kosmos; Interstellarer Diamantstaub; Die Oberflächentemperatur der Sonne im Millimeterwellenlängenbereich gemessen; Der Kern unserer Galaxis – eine Infrarot-Quelle; Neues über den inneren Aufbau des Mars. (11) Wo sind die Quellen der Meteorite? Der Mond – eine extraterrestrische Elektrostation; Gravitation deformiert Galaxien. (12) Die Quasare lüften ihr Geheimnis; Mondgestein erzählt ...

002.020 News notes.
Sky Telescope, Vol. 38, 300 - 301 (1969). – Hektor scrutinized; Solar superflare? ; Fine structure of a meteor train; Selection effects on comet discoveries; Some radio observations of the outer planets.

002.021 News notes.
Sky Telescope, Vol. 38, 388 - 389 (1969). – Murchison meteorite; Nitric acid in the upper atmosphere; More about an unusual eclipsing variable; Tektites from Tycho? ; New telescope in Chile.

002.022 Nouvelles brèves.
Ciel et Terre, Vol. 85, 398 - 405 (1969). – Conjunctions triples; L'orbite de la comète Seki-Lines; Nouvelles de Samos 2; Etoiles multiples parmi les étoiles doubles visuelles; Etoiles à grand mouvement propre; Une curiosité astronomique: Trois phases lunaires en février.

002.023 Nouvelles de la science.
L'Astronomie, 83ᵉ année, 414 - 416, 472 - 474 (1969). – Masses enterrées dans la lune ; Nouvelles des comètes; La masse de Pluton; La période des pulsars; A propos du phénomène lumineux observé le 21 décembre 1968; Deux sondes soviétiques traversent l'atmosphère de Vénus; Horloge type esclangon: temps moyen, temps sidéral; Vapeur d'eau dans l'espace interstellaire.

002.024 News from science and other informations.
Zemlya i Vselennaya, No. 4 (1969). In Russian. News about the atmosphere of Venus, p. 11; The flights of "Apollo 9" and "Apollo 10", p. 17 - 18; The flight of the Soviet automatic station "Luna 15", p. 18; Interstellar ammonia, p. 30; Diamonds in interstellar space, p. 31; Channels on the moon? , p. 31; Ice crystals in the atmosphere of Mars, p. 45; "Eternal frost" on Mars, p. 45 - 46; Photographs of the surface of Mars, p. 46 - 47.

002.025 News from science and other informations.
Zemlya i Vselennaya, No. 5 (1969). In Russian. The landing site of "Apollo 11" (*V. V. Mikhajlov*), p. 14 - 15; The first winner of the Yu. A. Gagarin Gold Medal (*I. G. Borisenko*), p. 15; Are the pulsars ejected from the Galaxy? p. 24; Pulsar in Crab nebula (*S. B. Dostovalov*), p. 24 - 25; Craters in the Sahara, p. 41; Photographs of the solar eruption of 8. June 1968 (*S. B. Dostovalov*), p. 49 - 50; The most precise pendulum–clocks of the world, p. 54; The optical Theodolite T-05 with high precision, p. 54; "Molniya 1" in its orbit, p. 59.

002.026 News from science and other informations.
Zemlya i Vselennaya, No. 6 (1969). In Russian. "Intercosmos 1" launched, p. 7; Planets of Barnard's star? p. 20; Gravitational waves detected? p. 37.

002.027 Rassegna delle riviste e notizie brevi.
P. Maffei.
Coelum, Vol. 37, 160 - 164, 227 - 234, 271 - 286 (1969).

002.028 Mitteilungen aus Wissenschaft und Literatur.
Sterne, 45. Jahrgang, 199 - 201 (1969).
Monderschütterungen registriert *(J. Classen);* Erste Untersuchungsergebnisse über das Mondgestein *(J. Classen);* Die Marssonden Mariner 6 und 7 *(J. Classen);* Wasser im interstellaren Raum? *(W. Rehpenning).*

002.029 Kurzberichte aus der Forschung.
SuW, Vol. 8, 235 - 239 (1969). – Astrophysikalischer Satellit TD-1 (*D. Lemke*); Die Dicke der Saturnringe (*T. Schmidt-Kaler*); Der Mond – ein ehemaliger Planet? (*H. Krefft*); Seltene Erden auf der Sonne (*H. Wöhl*); Merkurdurchgänge und Sonnenforschung (*W. Kokott*); Eine Sternbedeckung durch Neptun (*T. Schmidt-Kaler*); Meteoriten bei Pueblito de Allende, Chihuahua, Mexiko (*H. Link*); Bewegungen in solaren Bogen-Filamenten (*H. Wöhl*); OGO-Serie beendet (*C. Leinert*); Interstellare Wolken (*H. Scheffler*); Raumfahrtkongreß in Salzburg (*H. Köhler*).

002.030 Kurzberichte aus der Forschung.
SuW, Vol. 8, 272 - 273 (1969). – Dasar; Kurzlebige chromosphärische Feinstrukturen in Sonnenflecken (*A. Wittmann*); Neue Sammelstelle für Himmelsbeobachtungen (*J. Classen*).

002.031 Kurzberichte aus der Forschung.
SuW, Vol. 8, 286 - 287 (1969). – Erster Gammastrahlen-Stern entdeckt? Ist die Sonne ein magnetischer Rotator? (*D. Labs*); Neue Modelle von Sonnenflecken (*H. Wöhl*); Die Heliumhäufigkeit im solaren Wind (*H. Wöhl*); Planetensysteme sonnennaher Sterne; Pulsierende Röntgenstrahlung vom Crabnebel (*K. Birkle*).

002.032 Hydrographische Bibliographie.
(Ozeanographie, Erdmagnetismus, Nautik).
Separate prints from Deutsche Hydrographische Zeitschr. [edited by Deutsches Hydrographisches Institut, Hamburg], Jahrgang 21 (1968), 127 pp. (1969).

002.033 Astronomischer Jahresbericht. 68. Band: Die Literatur des Jahres 1968.
Herausgegeben vom Astronomischen Rechen-Institut in Heidelberg. Bearbeitet von W. Lohmann, F. Henn, U. Güntzel-Lingner, D. Krahn.
Walter de Gruyter & Co., Berlin. 15 + 762 pp. Price DM 80.00 (1969).

002.034 Astronomy and Astrophysics Abstracts. Vol. 1, Literature 1969, Part 1.
W. Fricke, U. Güntzel-Lingner, F. Henn, D. Krahn, G. Zech (Editors).
Published for Astronomisches Rechen-Institut, Heidelberg by Springer-Verlag, Berlin–Heidelberg–New York. 8 + 435 pp. Price DM 72.–, US$ 19.80 respectively [Subscription price per volume DM 57.60; US$ 15.90] (1969).

002.035 Chronicle.
Urania Kraków, Vol. 40, 205 - 209; 242 - 245; 277 - 282; 332 - 335 (1969). In Polish.
The flight of a space probe to meet the Halley comet; Close approach of minor planet 1620; Geographos to the earth; Unusual bottoms of lunar craters; Is the Earth crater an extinct volcano; An interesting variable; The thickness of Saturn rings; Apollo 10 on the way to moon; Landing sites for the lunar Apollo program; Two new pulsars; Distances to pulsars; An identification of a radiosource with an eclipsing system; Balloon meteor traps; The principal data about the moon; Intralunar waters; The plan of the conquest of Mars.

002.036 Astronomische Dokumentation "Astronomy and Astrophysics Abstracts". W. Fricke.
Mitt. Astron. Ges. No. 27, p. 125 - 126 (1969). – Abstract AG.

003 Books

003.001 **Methods of Geodetic Astronomy for the Intertropical Zone.** L. Cichowicz.
Państwowe Wydawnictwo Naukowe (Polish Academy of Sciences, National Committe of Geophysics and Geodesy), Warszawa = Publ. Działu Geod. Wyższej i Astron. Geod. Zg. PAN No. 12, 273 pp. (1969). – Spherical astronomy of the intertropical zone; Determination of the azimuth of terrestrial direction; Determination of the latitude; Determination of time; Joint methods and approximate method; Reduction of the geographical coordinates and of the azimuth on the astronomical-geodetic points; Supplement.

003.002 **Problems of Astrometry.**
A. K. Korol' (Editor).
Mezhvedomstvennyj Respublikanskij Sbornik. Ser. Astrometriya i Astrofizika No. 2, Akademiya Nauk Ukrainskoj SSR, Glav. Astron. Obs. Izdatel'stvo "Naukova Dumka", Kiev. 153 pp. Price 68 Kop. (1969). In Russian. – The papers included are abstracted in their subject categories.

003.003 **Annuaire 1970** du Bureau des Longitudes. Encyclopédie Physique et Spatiales. Gauthier–Villars, Editeur, Paris. 14 + 959 + A24 + B14 + C4 + D10 + E106 pp. (1969). – Contents: Part 1, Ephémérides astronomiques; Part 2, Géodésie, Géophysique; Part 3, Astrophysique (Interprétation des spectres stellaires; Etoiles doubles visuelles. Masses des étoiles; Etoiles doubles spectroscopiques; Photométrie et étoiles variables; Novae et supernovae galactiques; Physique et évolution des étoiles; L'espace interstellaire; La radioastronomie; Rayonnement cosmique); Part 4, Electromagnétisme; particules; radiations; Part 5, Géographie mondiale; Part 6, Supplément pour l'année 1971; Part 7, Notices.

003.004 **A Long-Range Program in Space Astronomy.** Position Paper of the Astronomy Missions Board, July 1969. R. O. Doyle (Editor).
NASA SP–213. Superintendent of Documents, U.S. Government Printing Office, Washington, D.C. 11 + 365 pp. Price $ 1.50 (1969). – Contents: I.) The unique contribution of space research to the major problems of astronomy and astrophysics, II.) Reports on the subdisciplines of astronomy, III.) AMB long-range plan for space astronomy, IV.) Supporting research and technology, V.) Ground-based astronomy in an integrated national program, VI.) The role of man in space astronomy.

003.005 **Light Scattering in the Atmosphere. Part 2.**
A. I. Ivanov, G. Sh. Livshits, V. E. Pavlov, B. T. Tashenov, Ya. A. Tejfel'.
Izdatel'stvo "Nauka" Kazakhskoj SSR, Alma-Ata. 116 pp. Price 80 Kop. = Trudy Astrofiz. Inst. Alma-Ata, Vol. 10 (1968). In Russian.

003.006 **Cosmic Electrodynamics.** J. H. Piddington.
John Wiley & Sons, Inc., New York – London. 7 + 305 pp. Price 175/– (1969). – Contents: Principles of cosmic electrodynamics; Cosmic plasmas; Electrodynamic effects of universal occurrence; The sun and solar activity; The interplanetary medium and geomagnetic cavity; The earth's magnetosphere and tail; Geomagnetic disturbance and related effects; Theory of the earth's radiation belt, aurora and ionospheric currents; Planets, satellites and comets; Stars and the interstellar medium; Galactic forms and activity; Radio galaxies, quasars and the universe.

003.007 **Physics of Stars and Stellar Systems.** Vol. 2 of a Course in Astrophysics and Stellar Astronomy.

A. A. Mikhailov (Editor).
Translated from Russian.
Israel Program for Scientific Translations, Jerusalem. 8 + 718 pp. Price $ 23.00 (1969).
Contents: Absolute stellar magnitudes (O. A. Mel'nikov); Stellar masses (O. A. Mel'nikov); Visual binaries (A. N. Deich); Spectroscopic binaries (V. A. Krat); Photometric binaries (eclipsing variables) (V. A. Krat); Variables and novae: Introduction (V. G. Gorbatskii); Cepheids (O. A. Mel'nikov); Long-period, semiregular, and irregular high-luminosity variables (V. G. Gorbatskii); Low- and medium-luminosity variables (V. G. Gorbatskii); Stars with variable spectra (B. A. Vorontsov-Vel'yaminov); Novae, supernovae and nova-like stars (B. A. Vorontsov-Vel'yaminov); Gaseous nebulae (B. A. Vorontsov-Vel'yaminov); Dust nebulae (B. A. Vorontsov-Vel'yaminov); Stellar photospheres (V. V. Sobolev); Stellar atmospheres (V. V. Sobolev); Gaseous nebulae (V. V. Sobolev); Stars with bright spectral lines (V. V. Sobolev); Stellar statistics, galactic structure (V. V. Sobolev); Stellar motions (A. N. Deich, T. A. Agekyan); Stellar dynamics (T. A. Agekyan); Star clusters (T. A. Agekyan); Extragalactic astronomy (T. A. Agekyan).

003.008 **Pulsating Stars 2, a "Nature" reprint.**
Introduction by T. Gold.
Macmillan and Co. Ltd, London. 12 + 116 pp. Price 63/– (1969).

003.009 **Astrometricheskie Issledovaniya.**
Izdatel'stvo "Fan", Uzbekskoj SSR, Akademiya Nauk Uzbekskoj SSR, Astronomicheskij Institut, Tashkent. 106 pp. Price 1 Rbl. (1969). In Russian.

003.010 **Earth Photographs from Gemini VI Through XII.**
NASA SP–171, with a foreword by G. E. Mueller.
Scientific and Technical Information Division, Office of Technology Utilization – National Aeronautics and Space Administration, Washington, D. C. – For sale by the Superintendent of Documents, U. S. Government Printing Office, Washington, D. C. 10 + 327 pp. Price $ 8.00 (1968). – Review in Spaceflight, Vol. 11, 442, 1969 (*L. J. Carter*).

003.011 **Ellipsoidal Figures of Equilibrium.**
S. Chandrasekhar.
Yale University Press, New Haven – London. 11 + 253 pp. Price $10, 90/– respectively (1969).
Contents: Historical introduction; The virial equations of the various orders; The potentials of the homogeneous and heterogeneous ellipsoids; Dirichlet's problem and Dedekind's theorem; The Maclaurin spheroids; The Jacobi and the Dedekind ellipsoids; The Riemann ellipsoids; The Roche ellipsoids.

003.012 **Great Radiobursts of the Sun.**
N. Cimakhovich, edited by Ya. Ikaunieks.
Academy of Sciences of the Latvian S.S.R. – Radioastrophysical Observatory. Trudy Observatorii (Transactions of the Observatory), Riga, Vol. 11. Publishing Office : "Zinātne", Riga. 67 + 63 (Catalogue) pp. Price 1 Rbl. 74 Kop. (1968). In Russian.

003.013 **General Catalogue of Variable Stars. Volume 1.**
Constellations Andromeda – Grus.
B. V. Kukarkin, P. N. Kholopov, Yu. N. Efremov, N. P. Kukarkina, N. E. Kurochkin, G. I. Medvedeva, N. B. Perova, V. P. Fedorovich, M. S. Frolov.
Astronomical Council of the Academy of Sciences in the

USSR — Sternberg State Astronomical Institute of the Moscow State University, Moscow. A121 + 474 pp. (1969).
In Russian and English.
The third edition containing information on 20437 variable stars discovered and designated till 1968.

003.014 The Moon.   Z. Kopal.
D. Reidel Publishing Company, Dordrecht—Holland. 16 + 525 pp. Price f 95.00 (1969). — Contents: Motion of the moon and dynamics of the Earth—Moon system; Internal constitution of the lunar globe; Topography of the moon; Radiation of the moon.

003.015 Annual Review of Astronomy and Astrophysics.
L. Goldberg, D. Layzer, J. G. Philips (Editors).
Annual Reviews, Inc., Palo Alto, California. 7 + 717 pp. Price $ 9.00 (1969). — Review in Sky Telescope, Vol. 38, 256 (1969).

003.016 Plasma Physics — Vol. 2, Weakly Ionized Gases.
J. L. Delcroix.
John Wiley & Sons, Inc., New York. 122 pp. Price $ 9.95 (1969). — Review in IEEE Spectrum, Vol. 6, 110 (1969).

003.017 Structures Technology for Large Radio and Radar Telescope Systems.
J. W. Mar, H. Liebowitz (Editors).
M. I. T. Press, Cambridge, Mass. 538 pp. Price $ 30.00 (1969). — Review in Sky Telescope, Vol. 38, 340 (1969).

003.018 Vistas in Astronomy Vol. 11.
A. Beer (Editor).
Pergamon Press, Oxford—New York. 8 + 275 pp. Price 150s., $ 20.00 respectively (1969). — Reviews in Orion Schaffhausen, Band 14, 136; 1969 *(E. Antonini);* Planet. Space Sci. Vol. 17, 2033; 1969 *(L. V. Morrison);* Sky Telescope, Vol. 38, 255, 408 - 409; 1969 *(T. Page).*

003.019 Exploration of the Universe.
G. Abell.
Holt, Rinehart, and Winston, New York; Faber & Faber, Ltd., London. Second edition. 722 pp. Price $ 13.50 (1969). — Reviews in Hemel en Dampkring, Vol. 67, 321; 1969 *(G. P. Können);* Sky Telescope, Vol. 38, 183 (1969).

003.020 Exploration of the Universe. Brief Edition.
G. Abell.
Holt, Rinehart, and Winston, New York. 483 pp. Price $ 10.95 (1969). — Review in Sky Telescope, Vol. 38, 340 (1969).

003.021 Atlas des spectres dans le proche infrarouge de Vénus, Mars, Jupiter et Saturne.
J. Connes, P. Connes, J.-P. Maillard.
Editions du Centre Nationale de la Recherche Scientifique, Paris. 471 pp. (1969).

003.022 The Earth in Space.
A. H. Cook, P. F. Gaskell (Editors).
Blackwell Scientific Publications, Oxford. 248 pp. Price 95s. (1968). — Reviews in Journ. Atmosph. Terr. Phys. Vol. 31, 1135; 1969 *(J. A. Ratcliffe);* Space Sci. Rev. Vol. 9, 872; 1969 *(J. Veldkamp).*

003.023 Atom, Man, and the Universe. The Long Chain of Complications.   H. Alfvén.
Translated from the Swedish edition. W. H. Freeman, San Francisco. 8 + 112 pp. Price $ 3.50, 32s. respectively (1969). Reviews in Nature, Vol. 223, 872; 1969 *(D. W. Hughes);* Sky Telescope, Vol. 38, 111, 333 - 334; 1969 *(G. Reaves);* Space-flight, Vol. 11, 443 - 444; 1969 *(A. E. Slater).*

003.024 Problems of Modern Cosmogony.
V. A. Ambartsumyan, L. V. Mirzoyan, G. S. Saakyan, S. K. Vsekhsvyatskij, V. V. Kazyutinskij.
Izdatel'stvo "Nauka", Moskva. 351 pp. Price 1 Rbl. 74 Kop. (1969).   In Russian. — Contents: I. Instationäre Objekte im Universum und ihre Bedeutung für die Kosmogonie *(V. A. Ambartsumyan).* II. Stellarkosmogonie *(L. V. Mirzoyan).* III. Theorie der überdichten Himmelskörper *(G. S. Saakyan).* IV. Kosmogonie des Sonnensystems *(S. K. Vsekhsvyatskij).* V. Der gegenwärtige Stand der kosmogonischen Theorie *(V. V. Kazyutinskij).*

003.025 The Complete Nautical Astronomer.
C. H. Cotter.
American Elsevier Publishing Company, New York. 336 pp. Price $ 11.75 (1969). — Review in Sky Telescope, Vol. 38, 183 (1969).

003.026 Sundials. A Simplified Approach by Means of the Equatorial Dial.   F. W. Cousins.
Illustrations by M. Chandler.
Baker, London. 247 pp. Price 126s. (1969). — Review in Nature, Vol. 225, 665 - 666; 1970 *(F. A. B. Ward).*

003.027 Invisible Astronomy.   C. A. Ronan.
Eyre and Spottiswoode, London. 16 + 173 pp. Price 50s. (1969). — Reviews in Journ. British Astron. Ass., Vol. 80, 77; 1969 *(H. Miles);* Nature, Vol. 224, 979 - 980; 1969 *(N. K. Reay).*

003.028 Plasma Diagnostics.
W. Lochte-Holtgreven (Editor).
North—Holland Publishing Co., Amsterdam. 946 pp. Price Dfl. 125.00, 292s. respectively (1968). — Review in Space Sci. Rev. Vol. 9, 874; 1969 *(R. Mewe).*

003.029 Seeing and the Eye.   G. H. Begbie.
Natural History Press, Garden City, N. Y. 227 pp. Price $ 5.95 (1969). — Review in Sky Telescope, Vol. 38, 255 - 256, 409 - 411; 1969 *(C. Rosen).*

003.030 The New Space Encyclopaedia.
M. T. Bizony (Editor).
Artemis Press, Horsham, Sussex. 317 pp. Price 65s. (1969). Review in Journ. British Astron. Ass. Vol. 79, 500 - 501; 1969 *(C. A. Ronan).*

003.031 Frontiers of Space.
P. Bono, K. Gatland.
Macmillan Company, New York. 248 pp. Price $ 3.95 (1969). — Review in Sky Telescope, Vol. 38, 340 (1969).

003.032 Man in Space.   H. Brinton.
A. and C. Black, London. 64 pp. Price 12s. 6d. (1969). — Review in Journ. British Astron. Ass. Vol. 79, 415; 1969 *(H. Miles).*

003.033 Talks About the Cosmos and Hypotheses.
V. A. Bronshtén.
"Nauka", Moskva. 240 pp. Price 41 Kop. (1968).   In Russian. — Review in Referativ. Zhurn. 51. Astron., 11.51.53 (1969).

003.034 Les Observatoires Spatiaux.
J.-C. Pecker.
Presses Universitaires de France, Paris. 180 pp. Price F 25.00 (1969).

003.035 L'Astronomie Expérimentale.
J.-C. Pecker.
Presses Universitaires de France, Paris. 155 pp. Price F 20.00

(1969). – Reviews in Sky Telescope, Vol. 38, 255 (1969); Space Sci. Rev. Vol. 10, 314; 1969 *(E. A. Müller)*.

003.036 **Edmond Halley: Genius in Eclipse.**
C. A. Ronan.
Doubleday, Garden City, N. Y. 12 + 252 pp. Price $ 5.95 (1969). – Review in Sky Telescope, Vol. 38, 183 (1969).

003.037 **Sundial.** P. Příhoda.
Štefánikova hvězdárna, Praha. 32 + 8 pp. Price 6.80 Kčs. In Czech. – Review in Říše hvězd, Vol. 50, 159 (1969).

003.038 **Columbia, hier spricht Adler! Der Report der ersten Mondlandung.**
J. von Puttkamer.
Verlag Chemier, Weinheim/Bergstr. 208 pp. Price DM 14.80 (1969). – Review in Umschau, Vol. 69, 854; 1969 *(W. Petri)*.

003.039 **Apollo 8 – Aufbruch ins All.**
J. von Puttkamer.
Heyne Sachbuch Nr. 130, München. 140 pp. Price DM 2.80 (1969). – Review in Weltraumfahrt, 20. Jahrgang, 116 (1969).

003.040 **Star Trackers and Systems Design.**
G. Quasius, F. McCanless.
Macmillan and Co., Ltd., London. 280 pp. Price 84s. Review in Weltraumfahrt, 20. Jahrgang, 158; 1969 *(K. J. Schwenzfeger)*.

003.041 **Lunar Observer's Manual.**
C. L. Ricker.
403 W. Park St., Marquette, Michigan 49855. 30 pp. (1969). – Review in Strolling Astronomer, Vol. 21, 216; 1969 *(W. H. Haas)*.

003.042 **Winds and Turbulence, in Stratosphere, Mesosphere, and Ionosphere.**
K. Rawer (Editor).
North–Holland Publishing Co., Amsterdam; John Wiley and Sons, Inc., New York. 421 pp. Price $ 18.50 (1968). – Review in Strolling Astronomer, Vol. 21, 210; 1969 *(G. W. Rippen)*.

003.043 **Strahlendes Weltall.** H. Rohr.
Rascher-Verlag, Zürich/Schweiz. 85 pp. Price sFr. 28.80 (1969). – Reviews in Orion Schaffhausen, Band 14, 134 - 135; 1969 *(F. Egger);* Sky Telescope, Vol. 38, 416 (1969).

003.044 **L'Astronomie.** P. Rousseau. 2nd édition.
443 pp. (1968). – Review in L'Astronomie, 83e année, 328 - 329 (1969).

003.045 **The Calculation of the Phase Dependence of the Total Planetary Brightness.**
A. A. Rubashevsky, E. G. Yanovitsky.
Izdatel'stvo "Naukova dumka", Kiev. Akademiya Nauk Ukrainskoj SSR. Glavnaya Astronomicheskaya Observatoriya. 100 pp. Price 35 Kop. (1969). In Russian.

003.046 **Mapa Měsíce. (Lunar Chart, 1:10000000).**
A. Rükl.
Kartografické nakladatelství, Praha. 39 × 45 cm². Price 5 Kčs. (1969). – Review in Říše hvězd, Vol. 50, 222 - 223 (1969).

003.047 **The Invasion of the Moon 1969.** P. Ryan.
Penguin Books, London. 190 pp. Price 5s. (1969). Review in Journ. British Astron. Ass., Vol. 80, 78; 1969 *(G. E. Satterthwaite)*.

003.048 **Twentieth Century Discovery.**
I. Asimov.
Doubleday, New York. 178 pp. Price $ 4.95 (1969). – Review in Sky Telescope, Vol. 38, 256 (1969).

003.049 **Radiotelescopes.**
W. N. Christiansen, J. A. Högbom.
At the University Press, Cambridge. Cambridge Monographs on Physics. 10 + 231 pp. Price 90s., $ 14.50 respectively (1969).

003.050 **Beobachtungen des Sternhimmels.**
M. M. Dagaev.
"Nauka", Moskva. 124 pp. Price 24 Kop. (1969). In Russian.

003.051 **Ionospheric Radio Waves.**
K. Davies.
Blaisdell, Massachusetts, Toronto– London. 17 + 460 pp. Price $ 13.50 (1969). – Review in Nature, Vol. 224, 89 - 90; 1969 *(K. G. Budden)*.

003.052 **The Bowl of Night.**
F. P. Dickson.
M. I. T. Press, Cambridge, Mass. 228 pp. Price $ 2.95 (1968). Review in Sky Telescope, Vol. 38, 339 (1969).

003.053 **Bharatiya Jyotish Sastra (History of Indian Astronomy).** S. B. Dikshit.
English translation by R. V. Vaidya.
Part I: History of Astronomy during the Vedic and Vedanga Periods.
Printed by the General Manager, Government of India Press Calcutta, and published by the Manager of Publications Civil Lines. Delhi. 34 + 147 pp. Price 29s. 2d., $ 4.50 respectively (1969).

003.054 **Mechanics in Sixteenth-Century Italy: Selections from Tartaglia, Benedetti, Guido Ubaldo & Galilei.**
S. Drake, I. E. Drabkin.
The University of Wisconsin Press, Madison, Wisc. 428 pp. Price $ 12.50 (1968). – Review in Phys. Today, Vol. 22, No. 12, p. 71, 73; 1969 *(R. S. Shanklamd)*.

003.055 **Galactic Nebulae and Interstellar Matter.**
J. Dufay.
Dover Publications, New York. 352 pp. Price $ 3.00 (1968). Review in Observatory, Vol. 89, 242 - 243; 1969 *(M. V. Penston)*.

003.056 **Lectures in High-Energy Astrophysics.**
H. Ögelman, J. R. Wayland (Editors).
National Aeronautics and Space Administration, NASA Special Publ. 199. 165 pp. Price $ 3.00 (1969). – Review in Sky Telescope, Vol. 38, 183 (1969).

003.057 **Physik der festen Erde.**
L. Egyed.
Akadémiai Kiadó, Budapest. 368 pp. Price DM 60.00 (1969). – Review in Journ. Inst. Navigation, Vol. 22, 331; 1969 *(A. H. Cook)*.

003.058 **Concepts of the Universe.**
P. W. Hodge.
McGraw-Hill Book Company, New York–London. 9 + 125 pp. Price $ 3.50 (1969). – Contents: What is the universe? The size of the universe; The expansion of the universe; Models of the universe; The universe of galaxies; Galaxies in explosion. – Review in Sky Telescope, Vol. 38, 256, 411 - 412; 1969 *(E. H. Cherrington)*.

003.059 **Yearbook of Astronomy 1970.**

P. Moore (Editor).
Sidgwich and Jackson, London. 208 pp. Price 30s. (1969).
Review in Journ. British Astron. Ass., Vol. 80, 79; 1969
(C. A. Ronan).

003.060 **The Development of Astronomical Thought.**
P. Moore.
Oliver and Boyd, Ltd., Edinburgh. 119 pp. Price 7s. 6d.
(1969). — Review in Sky Telescope, Vol. 38, 340 (1969).

003.061 **Moon Flight Atlas.**
P. Moore.
Rand McNally, New York; George Philip & Son, London.
48 pp. Price $ 5.95, 36s. respectively (1969). — Review in
Journ. British Astron. Ass., Vol. 80, 77; 1969 *(H. Miles).*

003.062 **Vom Erdkern zur Magnetosphäre.**
H. Murawski (Editor).
Umschau-Verlag, Frankfurt/Main. 330 pp. Price DM 21.80
(1968). — Review in SuW, Vol. 8, 251; 1969 *(D. Lemke).*

003.063 **Universe, Earth, and Atom: The Story of Physics.**
A. E. Nourse.
Harper & Row, New York. 688 pp. Price $ 10.00 (1969).

003.064 **Telescopes: How to Make Them and Use Them.**
T. Page, L. W. Page.
Macmillan Co., London. 14 + 338 pp (1969). — Review in
Phys. Abstr. Vol. 72, No. 49597 (1969).

003.065 **Beyond the Milky Way. Galaxies, Quasars and the
New Cosmology.**
T. Page, L. W. Page (Editors).
Macmillan, New York; Collier-Macmillan, London. 16 + 336
pp. Price $ 7.95 (1969).

003.066 **Apollo on the Moon.**
H. S. F. Cooper, Jr.
Dial, New York. 12 + 148 pp. Price $ 4.50 (1969).

003.067 **Let's Explore Outer Space.**
S. Engelbrektson, P. Greenleaf.
Sentinel Books Publishers, Inc., New York. 128 pp. Price
$ 1.25 (1969). — Reviews in Science, Vol. 166, 429 (1969);
Sky Telescope, Vol. 38, 184 (1969).

003.068 **The Velocity of Light.**
L. Essen, K. D. Froome.
Academic Press, New York. 200 pp. Price 30s. (1969). — Review in Journ. British Interplanet. Soc., Vol. 22, 460 (1969).

003.069 **Astronomia Elemental.**
A. Feinstein.
Editorial Kapelusz, Buenos Aires. 260 pp. Price $ 3.00
(1969). — Review in Sky Telescope, Vol. 38, 256 (1969).

003.070 **Shadow Bands.**
R. L. Feldman.
2203 Plank Rd., Fredericksburg, Va. 48 pp. Price $ 3.20
(1969). — Review in Sky Telescope, Vol. 38, 416 (1969).

003.071 **The World of Mars.**
V. A. Firsoff.
Oliver and Boyd Ltd., Edinburgh. 8 + 128 pp. Price 7s. 6d.
(1969). — Reviews in Irish Astron. Journ., Vol. 9, 95 - 97;
1969 *(E. Öpik);* Journ. British Astron. Ass., Vol. 80, 79;
1969 *(P. Moore);* Sky Telescope, Vol. 38, 339 (1969).

003.072 **Figur und Dimensionen des Mondes nach astronomischen Beobachtungen.**
I. V. Gavrilov.

"Naukova Dumka", Kiev. 150 pp. Price 85 Kop. (1969).
In Russian. — Contents. 1. Das Problem der Erforschung der
Mondfigur. 2. Astronomische Beobachtungen des Mondes und
ihre Bearbeitung. 3. Basispunkte auf dem Mond. 4. Die Figur
der sichtbaren Seite des Mondes. 5. Hypsometrische Charak-
teristiken der Mondoberfläche. 6. Die Randzone des Mondes.

003.073 **Außerirdische Zivilisationen. Probleme der interstellaren Kommunikation.**
L. M. Gindilis, S. A. Kaplan, N. S. Kardashev, B. N. Panov-
kin, B. V. Sukhotin, G. M. Khovanov.
"Nauka", Moskva. 438 pp. Price 86 Kop. (1969). In Russian.

003.074 **The Constants of the Physical Libration of the
Moon.** A. A. Gorynya.
Izdatel'stvo "Naukova Dumka". Akademiya Nauk Ukrainskoj
SSR, Glavnaya Astronomicheskaya Observatoriya, Kiev.
275 pp. Price 92 Kop. (1969). In Russian.

003.075 **Grundzüge der Ausgleichsrechnung nach der Metho-
de der kleinsten Quadrate nebst Anwendung in der
Geodäsie.** W. Großmann. 3. Edition.
Springer-Verlag, Berlin—Heidelberg—New York. 425 pp.
Price DM 66.00 (1969). — Review in Österr. Zeitschr. Ver-
messungswesen, 57. Jahrgang, 96 - 97; 1969 *(F. Ackerl).*

003.076 **Atlas of Solar Magnetic Fields.**
R. Howard, V. Bumba, S. F. Smith.
Carnegie Institution, Washington, D. C. — Review in Irish
Astron. Journ.,Vol. 9, 45 - 46 (1969).

003.077 **Radiative Transfer and Spectra of Celestial Bodies.**
V. V. Ivanov.
Izdatel'stvo "Nauka", Glavnaya Redaktsiya Fiziko-Matema-
ticheskoj Literatury, Moskva. 472 pp. Price 2 Rbl. 24 Kop.
(1969). In Russian.

003.078 **Sun and Ionosphere.**
**Short-wave Radiation of the Sun and Its Influence
on the Ionosphere.** G. S. Ivanov-Kholodnyj,
G. M. Nikol'skij.
Izdatel'stvo "Nauka". Glavnaya Redaktsiya Fiziko-Matema-
ticheskoj Literatury, Moskva. 456 pp. Price 2 Rbl. 79 Kop.
(1969). In Russian.

003.079 **Stars, Planets and Life: The Evolution of the
Cosmos.** R. Jastrow.
Heinemann, London. 12 + 177 pp. Price 30s. (1968). — Review in Observatory, Vol. 89, 241; 1969 *(B. Pagel).*

003.080 **Observe and understand the Stars.**
A. R. Johnson.
Astronomical League Book Service, R. D., Geneva, N. Y.
40 pp. Price $ 1.00 (1969). — Review in Sky Telescope,
Vol. 38, 255 (1969).

003.081 **Albert Einstein.** P. Jordan.
Verlag Huber, Frauenfeld. 302 pp. Price DM 24.80
(1969).

003.082 **Fragen der atmosphärischen Optik.**
I. G. Kolchinskij (Editor).
Respublikanskij Mezhvedomstvennyj Sbornik. Ser. Astro-
metriya i Astrofiz. No. 5, Akad. Nauk Ukrain. SSR, Glav.
Astron. Obs. Izdatel'stvo "Naukova Dumka", Kiev. 133 pp.
Price 85 Kop. (1969). In Russian. — The papers included
are abstracted in their subject categories.

003.083 **Radiation in the Atmosphere.**
K. Y. Kondratyev.
Academic Press, New York. 915 pp. Price 364s. (1969). — Re-

views in Journ. British Interplanet. Soc.,Vol. 22, 460 (1969); Space Sci. Rev. Vol. 10, 314, 1969 *(M. Nicolet)*.

003.084   Physics of Comets.
V. P. Konopleva (Editor).
Respublikanskij Mezhvedomstvennyj Sbornik. Ser. Astrometriya i Astrofiz. No. 4, Akad. Nauk Ukrain. SSR, Glav. Astron. Obs. Izdatel'stvo "Naukova Dumka", Kiev. 224 pp. Price 98 Kop. (1969).   In Russian.

003.085   Declinations of Bright and Faint Fundamental Stars in a Uniform System.
A. K. Korol'.
Izdatel'stvo "Naukova Dumka". Akademiya Nauk Ukrainskoj SSR, Glavnaya Astronomicheskaya Observatoriya, Kiev. 234 pp. Price 80 Kop. (1969).   In Russian.

003.086   Nuclear and Relativistic Astrophysics and Nuclidic Cosmochemistry: 1963 - 1967. Vol. 1.
B. Kuchowicz.
Nuclear Energy Information Center, Warsaw. Review Report No. 34, 365 pp. (1969).

003.087   Historisch-astronomische Untersuchungen.
Vypusk (No.) 10.
P. G. Kulikovskij.
Izdatel'stvo "Nauka", Moskva. 352 pp. Price 1 Rbl. 72 Kop. (1969).   In Russian.

003.088   High-Energy Astrophysics.
T. C. Weekes.
Chapman and Hall, Limited, London. 11 + 209 pp. Price 60s., $ 9.50 respectively (1969). – Reviews in Nature, Vol. 224, 824 (1969); Sky Telescope, Vol. 38, 339 (1969).

003.089   Progress in Aeronautical Sciences, Vol. 9.
D. Küchemann (Editor).
Pergamon Press, London–New York. 471 pp. Price £ 8. (1968). – Review in Journ. British Interplanet. Soc.,Vol. 22, 150 - 151; 1969 *(W. F. Hilton)*.

003.090   Advances in Geophysics, Vol. 13.
H. E. Landsberg, J. Van Mieghem (Editors).
Academic Press, New York. 10 + 270 pp. Price $ 14.50 (1969).

003.091   To the Moon.
Section 1, The Story in Sound.
Section 2, The Story in Pictures and Text.
M. Kapp, W. R. Young.
Time–Life Records, New York. 192 pp. Price $ 24.95 (1969).

003.092   Nautical Calculations Explained.
J. Klinkert, G. W. White.
Routledge and Kegan Paul, London. 17 + 597 pp. Price £ 5 5s. (1969). – Review in Journ. Inst. Navigation, Vol. 22, 527 - 528; 1969 *(C. H. Cotter)*.

003.093   The Planets.   P. Lauber.
Random House, New York. 8 + 136 pp. Price cloth $ 3.50, paper $ 1.50 (1969).

003.094   Questions récentes d'astrophysique théorique.
P. Ledoux.
Université Libre de Bruxelles, Bruxelles. 273 pp. (1969).

003.095   Thin-Film Optical Filters.
H. A. MacLeod.
American Elsevier, New York. 332 pp. Price $ 22.00 (1969). – Review in Sky Telescope, Vol. 38, 339 (1969).

003.096   Properties of Matter under Unusual Conditions.
H. Mark, S. Fernbach.
John Wiley and Sons, Ltd., New York–Sydney. 389 pp. Price 182s.(1969). – Review in Phys. Blätter, 25. Jahrgang, 429; 1969 *(H. Rechenberg)*.

003.097   Early Solar Physics.
A. J. Meadows.
Pergamon Press, London. 320 pp. Price $ 7.00 (1969). Contents: Ideas of the Sun in the Mid-Nineteenth Century, The New Astronomy (1850 - 1900) – The New Era in Solar Physics.

003.098   Catalog of Emission Lines in Astrophysical Objects.
A. Meinel, A. Aveni, M. W. Stockton.
Optical Sciences Center, University of Arizona, Tucson, Ariz. 198 pp. Price $ 4.00 (1969).

003.099   Tables of Light Trajectories in the Terrestrial Atmosphere.   F. Link, L. Neužil.
Hermann, Paris. 21 + 175 pp. Price F 40.00 (1969).

003.100   Der Mond.   F. Link.
Verständliche Wissenschaft, Band 101. Springer-Verlag, Berlin–Heidelberg–New York. 7 + 94 pp. Price DM 7.80 (1969).

003.101   The Rush Toward the Stars.
T. S. Logsdown.
Brown, Dubuque, Iowa. 6 + 170 pp. Price $ 3.95 (1969).

003.102   A Field Guide to the Stars and Planets.
D. H. Menzel.
Peterson Field Guide Series, Vol. 15, reprint of the 1964 edition. Mifflin, Boston. 16 + 400 pp. Price $ 4.95.

003.103   Light and Colour in the Landscape.
M. Minnaert.   Translation from Dutch into Russian. "Nauka", Moskva. 360 pp. Price 1 Rbl. 32 Kop. (1969). Review in Referativ. Zhurn. 51. Astron., 1.51.69 (1970).

003.104   Dynamics.   T. R. Kane.
Holt, Rinehart and Winston, Inc., New York. Price $ 12.50 (1968).

003.105   Das Schicksal des Sonnensystems.
Populäre Überblicke über Himmelsmechanik.
V. G. Demin.
"Nauka", Moskva. 256 pp. Price 44 Kop. (1969). In Russian.

003.106   The Old Moon and the New.
V. A. Firsoff.
Sidgwick and Jackson, London. 264 pp. + 20 plates. Price 90s., $ 14.50 respectively (1969).

003.107   Abriss der Astronomie II.
H.-H. Voigt.
Bibliographisches Institut, Mannheim. BI Hochschulskripten No. 819/819a. 9 + 282 pp. Price DM 9.80 (1969).

003.108   Introduction to Lunar Physics.
V. N. Zharkov, V. L. Pan'kov, A. A. Kalachnikov, A. I. Osnach.
Izdatel'stvo "Nauka", Moskva. 312 pp. Price 1 Rbl. 37 Kop. (1969).   In Russian. – Review in Referativ. Zhurn. 51. Astron., 11.51.580 (1969).

003.109   The Problem of the Tunguska Catastrophe 1908.
A. V. Zolotov.
"Nauka i tekhn.", Minsk. 204 pp. Price 94 Kop. (1969).

In Russian. – Review in Referativ. Zhurn. 51. Astron., 1.51.378 (1970).

003.110 **Unser Mond.** H. Haber.
Deutsche Verlags-Anstalt, Stuttgart. 128 pp.
Price DM 16.80 (1969). – Reviews in Umschau, Vol. 69, 854; 1969 *(W. Petri)*; VdS Nachrichtenblatt, 18. Jahrgang, 169 - 170; 1969 *(H. B. Brenske)*.

003.111 **Cosmic Rays.** Results of Researches on International Geophysical Projects. Articles No. 10.
S. N. Vernov, L. I. Dorman (Editors).
Publishing House "Nauka", Moscow. 206 pp. Price 1 Rbl. 08 Kop. (1969). In Russian.

003.112 **We Reach the Moon.** J. N. Wilford.
Bantam, New York. 12 + 332 pp. Price $ 1.25 (1969).

003.113 **Elementary Wave Optics.** R. H. Webb.
Academic Press, New York – London. 12 + 268 pp.
Price 107s. (1969). – Review in Nature, Vol. 224, 197; 1969 *(S. Tolansky)*.

003.114 **Constant of Gravitation and the Earth's Mass.**
M. U. Sagitov.
"Nauka", Moskva. 188 pp. Price 83 Kop. (1969).
In Russian.

003.115 **Ebene und Sphärische Trigonometrie mit Anwendungen auf Kartographie, Geodäsie und Astronomie.** R. Sigl.
Akademische Verlagsgesellschaft, Frankfurt am Main. 482 pp.
Price DM 75.00 (1969).

003.116 **Celestial Mechanics, Part 1.** S. Sternberg.
W. A. Benjamin, New York. 12 + 158 pp. Price $ 12.50 cloth, paper $ 3.95 (1969).

003.117 **Atlas zur Himmelskunde.** K. Schaifers.
Bibliographisches Institut, Mannheim – Wien – Zürich. 8 pp. + 28 Starcharts + 12 photographs. Price DM 17.80 (1969).

003.118 **Atlas der Planeten.**
V. de Callatay, A. Dollfus, W. Jahn.
Verlag Goldmann, München. 167 pp. Price öS 607.50 (1969).

003.119 **The Atmosphere of Jupiter.** V. G. Tejfel'.
Izdatel'stvo "Nauka", Moskva. 183 pp. Price 69 Kop. (1969). In Russian. – Contents: Die chemische Zusammensetzung; Die Temperaturverhältnisse des Planeten; Der Aufbau seiner Atmosphäre; Die Wolkendecke des Jupiter; Zirkulation und aktive Prozesse in der Atmosphäre.

003.120 **Relativität und Kosmos. Raum und Zeit in Physik, Astronomie und Kosmologie.** H.-J. Treder.
Akademie-Verlag, Berlin; Pergamon Press, Oxford; Vieweg & Sohn, Braunschweig. 119 pp. Preis DM 6.80 (1968). – Review in Sterne, 45. Jahrgang, 174 - 175; 1969 *(H. Lambrecht)*.

003.121 **Shock Waves Produced by Motion in the Atmosphere of Large Meteoritic Bodies.**
M. A. Tsikulin.
"Nauka", Moskva. 87 pp. Price 45 Kop. (1969). In Russian.

003.122 **The Attractive Universe: Gravity and the Shape of Space.** E. G. Valens.
World Publishing Company, Cleveland, Ohio. 187 pp.

Price $ 5.95 (1969). – Review in Sky Telescope, Vol. 38, 416 (1969).

003.123 **Astronomical Problems.** An Introductory Course in Astronomy. B. A. Vorontsov-Vel'yaminov.
Translated by P. M. Rabbitt, A. Beer, J. B. Hutchings (Editors). Pergamon Press, Oxford, London and New York.
11 + 314 pp. Price 80s., $ 10.00 respectively (1969). – Review in Nature, Vol. 224, 823; 1969 *(A. J. Meadows)*.

003.124 **Essay on the Universe.**
B. A. Vorontsov-Vel'yaminov.
"Nauka", Moskva. 725 pp. Price 1 Rbl. 52 Kop. (1969).
In Russian.

003.125 **The Influence of Solar Flares on the Tropospheric Circulation.** C. J. E. Schuurmans.
The Hague. 122 pp. Price Dfl. 19.75 (1969).

003.126 **The Evolution of the Preplanetary Cloud and the Origin of the Earth and Planets.**
"Nauka", Moskva. 244 pp. Price 1 Rbl. 2 Kop. (1969).
In Russian. – Review in Referativ. Zhurn. 51. Astron., 12.51.727 (1969).

003.127 **Erwachende Wissenschaft II. Die Anfänge der Astronomie.** B. L. van der Waerden.
Birkhäuser Verlag, Basel – Stuttgart. 316 pp. Price DM 38.00 (1968). – Review in Umschau, Vol. 69, 709; 1969 *(W. Petri)*.

003.128 **Astrofysikens Grunder.** A. Wallenquist.
Svenska Bokförlaget, Stockholm. 342 pp. Price Sv. kr. 46:50 (1968). – Review in Astron. Tidssk., Arg. 2, 192; 1969 *(C. Schalén)*.

003.129 **Spectrum Maanatlas.** P. Moore.
Oosthoek Uitgevers-Mij., Utrecht. 48 pp. Price f 15.00. (1969). – Review in Hemel en Dampkring, Vol. 67, 396; 1969 *(A. G. Jansen)*.

003.130 **Der Mond.** R. Meißner.
Suhrkamp–Verlag, Frankfurt, 200 pp. Price DM 12.00 (1969).

003.131 **Untersuchung der Fehler und Analyse der Ergebnisse astrometrischer Beobachtungen.**
Respublikanskij Mezhvedomstvennyj Sbornik. Ser. Astrometriya i Astrofiz. No. 7, Akad. Nauk Ukrain. SSR, Glav. Astron. Obs. Izdatel'stvo "Naukova Dumka", Kiev. 112 pp. Price 96 Kop. (1969). In Russian. – The papers are abstracted in their subject categories.

003.132 **The Riddle of Gravitation.** P. Bergmann.
Translation from English into Russian.
"Nauka", Moskva. 215 pp. Price 61 Kop. (1969).

003.133 **The Universe. From the Flat Earth to Quasars.**
A. Asimov. Translation from English into Russian.
"Mir", Moskva. 350 pp. Price 1 Rbl. 8 Kop. (1969).

003.134 **Comets, Sun and Interplanetary Space.**
Mezhvedomstvennyj Nauchnyj Sbornik. Problemy kosmicheskoj fiziki, No. 4. Izdatel'stvo Kievskogo Universiteta, Kiev. 158 pp. Price 84 Kop. (1969). In Russian.

003.135 **Astrometric Research.**
"Fan", Tashkent. AN Uzb. SSR. Astronomichesk. Int–t. 107 pp. Price 1 Rbl. (1969). In Russian. – Review in Referativ. Zhurn. 51. Astron., 12.51.105 (1969).

003.136 Radioastronomical Telescopes, Equipment, and Observations.
"Nauka", Moskva. 215 pp. Price 1 Rbl. 15 Kop. (1969). In Russian.

003.137 Physical Characteristics of Comets.
S. K. Vsekhsvyatskii.
Israel Program for Scientific Translations (NASA TTF–80; TT 62–1103), Jerusalem. 596 pp. Price $ 3.00 (1969). −

003.138 RR Lyrae Stars. V. P. Tsesevich.
Translated from the Russian. Israel Program for Scientific Translations, Jerusalem. 357 pp. Price $ 3.00. (1969). − Review in Sky Telescope, Vol. 38, 339 - 340 (1969).

003.139 Astronomische Navigation. W. Stein.
Verlag Klasing & Co., Bielefeld − Berlin. 208 pp. (1969).

003.140 Surveyor Program Results, 1969.
National Aeronautics and Space Administration. NASA Special Publication 184. Available from Superindendent of Documents, U. S. Government Printing Office, Washington, D. C. 425 pp. Price $ 4.75 (1969). − Review in Sky Telescope, Vol. 38, 416 (1969).

003.141 Infinity and the Universe.
V. V. Kazyutinskij, G. I. Naan, M. E. Omel'yanovskij, et al. (Editors).
"Mysl'", Moskva. 325 pp. Price 1 Rbl. 54 Kop. (1969). In Russian.

003.142 De nova et nullius aevi memoria prius visa stella iam pridem anno a nato Christo 1572 mense novembri primium conspecta. Tycho Brahé.
Culture et Civilisation, Bruxelles. Photoprint of the original volume from 1573. 110 pp. Price 35 F. − Review in L'Astronomie, 83ᵉ année, 379 (1969).

003.143 Astronomiae instauratae mechanica.
Tycho Brahé.
Culture et Civilisation, Bruxelles. Photoprint of the original volume from 1602. 108 pp. Price 45F. − Review in L'Astronomie, 83ᵉ année, 379 (1969).

003.144 Astronomiae instauratae progymnasmata.
Tycho Brahé.
Culture et Civilisation, Bruxelles. Photoprint of the original volume from 1610. 886 pp. Price 260 F. − Review in L'Astronomie, 83ᵉ année, 379 (1969).

003.145 Significant Achievements in Space Science 1967.
National Aeronautics and Space Administration, NASA SP-167. Superintendent of Documents, U.S. Government Printing Office, Washington, D.C. 558 pp. Price $ 2.50 (1968). − Review in Sky Telescope, Vol. 38, 184 (1969).

003.146 The Planet Venus: Past, Present, and Future.
American Philosophical Society, Philadelphia. 50 pp. Price $ 1.00 (1969). − Review in Sky Telescope, Vol. 38, 111 (1969).

003.147 Astronautics and Aeronautics, 1967: Chronology on Science, Technology, and Policy.
National Aeronautics and Space Administration, Special Publ. SP-4008. Superintendent of Documents, U.S. Government Printing Office, Washington, D.C. 487 pp. Price $ 2.25 (1968). − Reviews in Sky Telescope, Vol. 38, 184 (1969); Spaceflight, Vol. 11, 411 - 412; 1969 (L. S. Butcher).

003.148 Kosmos-Maruzen-Mondglobus.
Frankh'sche Verlagshandlung (Kosmos-Verlag), Stuttgart. Price DM 78.00 (1969).

003.149 Projekt Apollo − Bilddokumentation in Farbe.
W. Büdeler.
Bertelsmann Sachbuchverlag, Gütersloh. 192 pp. Price DM 24.00 (1969). − Review in Umschau, Vol. 69, 854; 1969 (W. Petri).

003.150 A l'Aussaut de la Lune.
J. Tiziou.
Editions Stock, Paris. 220 pp. (1969). − Review in L'Astronomie, 83ᵉ année, 378 - 379 (1969).

003.151 Living in Space. The Astronaut and his Environment. M. R. Sharpe.
Doubleday, Garden City, N. Y. 192 pp. Price $ 5.95, $ 2.45 respectively (1969). − Review in Sky Telescope, Vol. 38, 256 (1969).

003.152 Galilée, penseur libre.
R. Zouckermann, with a preface by P. Couderc.
Les Editions rationalistes, Paris. 330 pp. (1968). − Review in L'Astronomie, 83ᵉ année, 418 (1969).

003.153 Planetary Exploration, 1968 - 1975.
National Academy of Sciences, Washington, D. C. (Available from the Space Science Board, Washington). 8 + 49 pp. (1968). − Review in Icarus, Vol. 11, 274; 1969 (J. W. Findlay).

003.154 Physics of the Earth in Space: A Program of Research, 1968 - 1975.
National Academy of Sciences − National Research Council, Washington, D. C. 5 + 109 pp. (1968). − Review in Icarus, Vol. 11, 273 ; 1969 (S. Chapman).

003.155 El Mundo de los Planetas. W. D. Heintz.
Translated from the German by L. R. de Togores. Ediciones Iberoamericanas, Madrid. 300 pp. Price 100 pesetas (1969).

003.156 Die Welt der Planeten. W. D. Heintz.
2nd, revised German edition. W. Goldmann Verlag, München. 185 pp. Price DM 2.50 (1969).

003.157 Catalogue of B- and V-magnitudes of 12000 Stars.
V. I. Voroshilov, N. B. Kalandadze, L. N. Kolesnik, Eh. P. Polishchuk, G. L. Fedorochenko.
Izdatel'stvo "Naukova Dumka". Akademiya Nauk Ukrainskoj SSR, Glavnaya Astronomicheskaya Observatoriya, Kiev. 256 pp. Price 2 Rbl. 46 Kop. (1969). In Russian.

004 History of Astronomy, Chronology

004.001 The search for the nebulae – VIII.
K. G. Jones.
Journ. British Astron. Ass. Vol. 79, 357 - 370 (1969).

004.002 The search for the nebulae – IX.   K. G. Jones.
Journ. British Astron. Ass. Vol. 79, 450 - 459
(1969).

004.003 Bicentenary of the discovery of the 'Wilson effect'.
W. M. Baxter.
Journ. British Astron. Ass. Vol. 79, 398 (1969). – Historical
section British Astron. Ass., report.

004.004 Ancient Khorezm monument of the IVth century
B. C. Koi-Krylgan-Kala from the point of view of
the history of astronomy.   M. G. Vorobjeva, M. M. Roz-
hanskaya, I. N. Vesselovsky.
Istoriko-Astron. Issled. Vyp. (No.) 10, p. 15 - 34 (1969).
In Russian.

004.005 On the possible astronomical purpose of one of the
places found in Metsamor (Armenia).
E. S. Parsamian, K. A. Mkrtchian.
Istoriko-Astron. Issled. Vyp. (No.) 10, p. 35 - 37 (1969).
In Russian.

004.006 Egyptian decans.   I. N. Vesselovsky.
Istoriko-Astron. Issled. Vyp. (No.) 10, p. 39 - 62
(1969).   In Russian.

004.007 Astronomical work "Canon Mas'uda" by Al Biruni.
B. A. Rozenfeld, M. M. Rozhanskaya.
Istoriko-Astron. Issled. Vyp. (No.) 10, p. 63 - 95 (1969).
In Russian.

004.008 Galileo Galilei's scientific heritage in Russia.
V. L. Tchenakal.
Istoriko-Astron. Issled. Vyp. (No.) 10, p. 97 - 112 (1969).
In Russian.

004.009 Investigations at the Tashkent Observatory in the
domain of Galileo's law of the free fall of bodies
and Italian astronomer's work there.   V. P. Shcheglov.
Istoriko-Astron. Issled. Vyp. (No.) 10, p. 113 - 119 (1969).
In Russian.

004.010 First works on astrophysics at the Petersburg Aca-
demy of Sciences in the XVIIIth century.
N. I. Nevskaya.
Istoriko-Astron. Issled. Vyp. (No.) 10, p. 121 - 157 (1969).
In Russian.

004.011 Contributions to theoretical astronomy by Kiev
astronomers.   P. V. Piaskovsky.
Istoriko-Astron. Issled. Vyp. (No.) 10, p. 199 - 218 (1969).
In Russian.

004.012 Memoirs of the Novorossijsk University and the
Odessa Astronomical Observatory.
N. M. Stoyko-Radilenko.
Istoriko-Astron. Issled. Vyp. (No.) 10, p. 245 - 250 (1969).
In Russian.

004.013 Astrolabe at the Moscow Museum of the Culture
of oriental peoples.
S. V. Smirnov.
Istoriko-Astron. Issled. Vyp. (No.) 10, p. 311 - 330 (1969).

In Russian.

004.014 Some ancient astronomical instruments of the
Astronomical Institute of the Uzbek Academy of
Sciences.   V. P. Shcheglov.
Istoriko-Astron. Issled. Vyp. (No.) 10, p. 331 - 337 (1969).
In Russian.

004.015 Ouloug Beg (1394 - 1449).   J. Kovalevsky.
L'Astronomie, 83ᵉ année, 277 - 281 (1969).

004.016 English henge cathedrals.   L. B. Borst.
Nature, Vol. 224, 335 - 342 (1969).
Is there an architectural relationship between Christian
sanctuaries and henge monuments?

004.017 Die Gregorianische Kalenderreform im Urteil zeit-
genössischer Astronomen.
P. Aufgebauer.
Sterne, 45. Jahrgang, 118 - 121 (1969).

004.018 Foucault's pendulum.   D. Sher.
Journ. Roy. Astron. Soc. Canada, Vol. 63, 227 -
228 (1969).
A descriptive explanation for the motion of the plane
of oscillation of a Foucault pendulum is presented, in terms
which would be suitable for beginning astronomy students.

004.019 Astronomy at the Vilnius University in the second
half of the 18th and at the beginning of the 19th
century.   S. P. Matulajtite.
Liet TSR Mokslų Akad. darbei, Tr. AN Lit. SSR, A, No. 1
(29), p. 69 - 84 (1969).   In Russian.-Abstr. in Referativ.
Zhurn. 51. Astron., 9.51.9 (1969).

004.020 Wie es dazu kam, daß ich den Einsteinturm errichte-
te.   E. Finlay-Freundlich.
Phys. Blätter, 25. Jahrgang, 538 - 541 (1969).
Schon 1911 mit Einstein bekanntgeworden, bemühte
sich Freundlich frühzeitig um experimentelle Beweise der
Allgemeinen Relativitätstheorie. Er unternahm im Sommer
1914 eine Sonnenfinsternisexpedition nach Rußland und ge-
riet durch den Ausbruch des Ersten Weltkrieges in Internie-
rung. Nach Kriegsende errichtete Freundlich bei Potsdam den
Einsteinturm, eine astronomische Beobachtungsstätte, die
speziell dem Nachweis der Einsteinschen Theorie dienen soll-
te.

004.021 Die Rudolphinischen Tafeln von Johannes Kepler –
Mathematische und astronomische Grundlagen.
V. Bialas.
Sitzungsber. Bayer. Akad. Wiss. Math.-Nat. Kl.,Jahrgang 1968,
p. 17* - 20* (1969). – Abstr.

004.022 L'astronomie à travers les siècles. IX. – Les yeux
de verre des astronomes.
E.-H. Geneslay.
L'Astronomie, 83ᵉ année, 381 - 398 (1969).

004.023 L'histoire du cadran solaire.   L. Janin.
La Suisse Horlogère, Rev. Internationale de l'Horlogerie, Vol. 84, No. 4, p. 93 - 101 (1969).

004.024 The lunar observatories of Megalithic man.
A. Thom.
Vistas in Astronomy, Vol. 11, 1 - 29 (1969).
Surveys of a number of Megalithic lunar sites are given,

with horizon profiles carefully determined. It is shown that these contain definite indicators of the rising and setting of the moon at its solstices, to an accuracy limited only by uncertainties in atmospheric refraction. From the profiles the values of the inclination of the lunar orbit and of its small periodic perturbation are deduced and found to agree very closely with the values given by modern astronomers. The difficulties faced by the erectors are made clear and an attempt is made to explain how the accuracy may have been obtained.

004.025 Galileo's contribution to astronomy.
W. Hartner.
Vistas in Astronomy, Vol. 11, 31 - 43 (1969).

The chief aim of the present paper is to destroy the legend of Galileo's early Copernican convictions. This legend is based above all on his letter to Kepler of 4th August 1597, in which he claims to have subscribed "for many years past" to the teaching of Copernicus. On the evidence of his writings and lecture notes, however, it cannot be doubted that Galileo became a convinced Copernican only after the telescopic discoveries.

004.026 Development of astronomy and geophysics at the university of Leningrad (At the 150th anniversary).
V. A. Dombrovskij, G. V. Molochnov.
Zemlya i Vselennaya, No. 5, p. 64 - 68 (1969).   In Russian.

004.027 Some lunar auxiliary tables and related texts from the late Babylonian period.   A. Aaboe.
Kon. Danske Vidensk. Selsk. Mat.-Fys. Medd., Vol. 36, No. 12, 44 pp. (1968). – See Phys. Abstr. Vol. 72, No. 41673 (1969).

004.028 The fate of old telescopes.   J. N. McKie.
Journ. British Astron. Ass., Vol. 80, 48 - 50 (1969).

004.029 An astronomical anniversary: The transit of Venus, 1769 June 3.   B. Hetherington.
Journ. British Astron. Ass., Vol. 80, 52 - 53 (1969).

004.030 Die Mondtheorie des Claudius Ptolemäus.
J. W. Ekrutt.
Sterne, 45. Jahrgang, 191 - 198 (1969).

004.031 The Astronomical Observatory of the University of Tartu (Yuryev, Dorpat), 1805 – 1948. An essay on its history.   G. Zhelnin.
Publ. Tartu Astrofiz. Obs. Vol. 37, 5 - 169 (1969).
In Russian.

004.032 The astronomical instruments of H. M. King George III presented to Armagh Observatory.
E. M. Lindsay.
Irish Astron. Journ., Vol. 9, 57 - 68 (1969).

004.033 La navigation Arabe de jadis.
H. Grosset-Grange.
Navigation (Paris), Vol. 17, 227 - 237 (1969).

004.034 The method of lunar distances and technical advance.   S. Moskowitz.
25. Annual Meeting American Inst. Navigation, New York 1969, 26 pp. (1969).

004.035 History of the Leander McCormick Observatory, circa 1883 to 1928.   C. P. Olivier.
Publ. Leander McCormick Obs. Univ. Virginia, *Charlottesville*, Vol. 11, (Part 26), 203 - 209 (1967).

004.036 Astro-archaeological table for the vertical position of Castor and Pollux (α and β Geminorum).
W. Gleissberg.
Publ. Astron. Inst. Univ. Frankfurt (Main), No. 28, 18 pp. (1969).

This paper may be considered as contribution to that branch of science which was recently named "Astro-archaeology" by Hawkins (1968). It concerns the astronomical explanation of the orientation of antique temples.

004.037 De verandering van het sterrenkundig wereldbeeld in de 16e en 17e eeuw.   F. P. Dijk.
Hemel en Dampkring, Vol. 67, 371 - 375 (1969).

Bharatiya Jyotish Sastra (History of Indian Astronomy).   See Abstr. 003.053.

Historisch-astronomische Untersuchungen.
See Abstr. 003.087.

005 Biography

005.001 **Der Astronom Christian Mayer. Zu seinem 250. Geburtstag.** E. Kollnig-Schattschneider.
SuW, Vol. 8, 190 - 194 (1969).

005.002 **C. C. Reissig, Professor of astronomy and mathematics.** Z. K. Novokshanova (Sokolovskaya).
Istoriko-Astron. Issled. Vyp. (No.) 10, p. 159 - 183 (1969).
In Russian.

005.003 **I. I. Khodzko – the first investigator of the Caucasus from the mathematical point of view, new data on his biography.** F. A. Shibanov.
Istoriko-Astron. Issled. Vyp. (No.) 10, p. 185 - 198 (1969).
In Russian.

005.004 **M. P. Dichenko (1863 - 1932).** N. A. Tchernega.
Istoriko-Astron. Issled. Vyp. (No.) 10, p. 219 - 227 (1969).
In Russian.

005.005 **Women-astronomers of the Pulkovo Observatory: M. V. Zhilova and I. N. Lehmann-Balanovskaya.** M. N. Neujmina.
Istoriko-Astron. Issled. Vyp. (No.) 10, p. 229 - 239 (1969).
In Russian.

005.006 **E. Ya. Perepelkin (1906 - 1937).** M. N. Gnevyshev.
Istoriko-Astron. Issled. Vyp. (No.) 10, p. 241 - 244 (1969).
In Russian.

005.007 **L. Euler's letters to Wettstein.** T. N. Klado (Ed.).
Istoriko-Astron. Issled. Vyp. (No.) 10, p. 253 - 284 (1969).
In Russian.

005.008 **L. Euler's unknown letters to T. Mayer.** Yu. Kopelevich, E. Forbes (Ed.).
Istoriko-Astron. Issled. Vyp. (No.) 10, p. 285 - 310 (1969).
In Russian.

005.009 **Romain Rolland's letter to M. N. Neujmina.** P. G. Kulikovskij (Ed.).
Istoriko-Astron. Issled. Vyp. (No.) 10, p. 339 - 343 (1969).
In Russian.

005.010 **Russell Porter and the Canadian north.** I. Halliday.
Journ. Roy. Astron. Soc. Canada, Vol. 63, 270 - 271 (1969).

005.011 **John Pond: Sixth Astronomer Royal.** J. Ashbrook.
Sky Telescope, Vol. 38, 224 - 225 (1969).

005.012 **James Nasmyth's telescopes and his observations.** J. Ashbrook.
Sky Telescope, Vol. 38, 380 - 381 (1969).

005.013 **Some recollections by contemporaries of Bernhard Schmidt.** P. Müürsepp, E. G. Forbes.
Journ. British Astron. Ass., Vol. 80, 30 - 36 (1969).

005.014 **A research worker of Leningrad on planets.** V. A. Bronshtehn.
Zemlya i Vselennaya, No. 5, p. 70 (1969). In Russian.

005.015 **W. Struve and astronomy.** M. Jõeveer.
Publ. Tartu Astrofiz. Obs. Vol. 37, 170 - 182 (1969). In Russian.

005.016 **W. Struve and geodesy.** G. Zhelnin, L. Vallner.
Publ. Tartu Astrofiz. Obs. Vol. 37, 183 - 201 (1969). In Russian.

005.017 **Life and activities of Prof. T. Rootsmäe.** G. Zhelnin.
Publ. Tartu Astrofiz. Obs. Vol. 37, 202 - 211 (1969).
In Russian.

005.018 **Life and activites of Ernst Öpik.** M. Jõeveer.
Publ. Tartu Astrofiz. Obs. Vol. 37, 212 - 226 (1969). In Russian.

005.019 **Edward Charles Pickering.** D. Eksinger.
Vasiona, Vol. 17, 68 - 70 (1969). In Serbo-Croatian.

005.020 **Leslie C. Peltier – Streiflichter auf Leben und Schaffen eines Amateurastronomen.** A. Oberstatter.
SuW, Vol. 8, 291 - 293 (1969).

005.021 **Th. Brorsen (1819 - 1895).** Říše hvězd, Vol. 50, 234 (1969). In Czech.

005.022 **K. Mayer (1719 - 1783).** B. Pitrun.
Říše hvězd, Vol. 50, 216 - 217 (1969). In Czech.

005.023 **Franz Meyer – zum Gedenken des 100. Geburtstages.** W. Bischoff.
Jenaer Rundschau (Jena Review), 13. Jahrgang, p. 353 - 355 (1968).

005.024 **100th birthday of Nikolaj Nikolaevich Evdokimov.** N. P. Barabashov, K. N. Kuz'menko, V. Kh. Pluzhnikov.
Vestn. Khar'kov. Univ. No. 34, (Ser. Astron. No. 4), p. 3 - 8 (1969). In Russian.

005.025 **My meetings with N. N. Evdokimov.** V. G. Fesenkov.
Vestn. Khar'kov. Univ. No. 34, (Ser. Astron. No. 4), p. 9 - 11 (1969). In Russian.

005.026 **Wilhelm Struve.** D. Eksinger.
Vasiona, Vol. 17, 19 - 21 (1969). In Serbo-Croatian.

Lockyer: Editor, civil servant and man of science.
Nature, Vol. 224, 453 - 456 (1969).

Albert Einstein. See Abstr. 003.081.

006 Personal Notes

V. A. Ambartsumyan, 60th anniversary of birth.
A. B. Severnyi, V. V. Sobolev.
Uspekhi fiz. nauk, Vol. 96, 181 - 183 (1968).   In Russian.

A. Behr, director of the Observatory in Hamburg-Bergedorf.
Astron. Tidssk., Årg. 2, 140 (1969).

R. P. Cesco is named director of the La Plata Observatory.
Publ. Astron. Soc. Pacific, Vol. 81, 456 (1969).

H. Elsässer, Director of Max-Planck-Institut für Astronomie in Heidelberg.
Astron. Nachr. Vol. 291, 224 (1969).

A. H. Jarrett, is named Director of the Boyden Observatory.
Inform. Bull. Southern Hemisphere, No. 14, p. 52 (1969).

P. E. Kustaanheimo, director of the Observatory in Helsingfors.
Astron. Tidssk., Årg. 2, 140 (1969).

H. H. Nininger, award of the Leonard Medal.
Meteoritics, Vol. 4, 147 (1969).

O. Obůrka, 60th birthday.   J. Klapka.
Pokroky, Vol. 14, 238 - 239 (1969).   In Czech.

E. J. Öpik, award of the Leonard Medal.
Meteoritics, Vol. 4, 146 (1969).

D.-H. Sadler received "La septième médaille de l'A.D.I.O.N."
L'Astronomie, 83ᵉ année, 297 - 298 (1969).

R. H. Stoy, Director of the Royal Observatory at Edinburgh, England.
Inform. Bull. Southern Hemisphere, No. 14, p. 53 (1969).

007 Obituaries

L. Beneš, died on 1968, Nov. 3.   E. Buchar.
Stud. Geophys. Geod.,Vol. 13, 334 - 335 (1969).   In Czech.

B. C. Browne, 1911 April 29 - 1968 Aug. 14.
E. C. Bullard.
Quarterly Journ. Roy. Astron. Soc. Vol. 10, 336 - 341 (1969).

G. Burkhardt, died 1969 June 23.
Mitt. Astron. Ges. No. 27, p. 236 (1969).

R. R. S. Cox, 1898 Jan. 11 - 1969 Jan. 1.
T. R. Kaiser.
Quarterly Journ. Roy. Astron. Soc. Vol. 1, 282 (1969).

M.Davidson, 1880 April 6 - 1968 June 25.
W. H. Steavenson.
Quarterly Journ. Roy. Astron. Soc. Vol. 10, 283 - 284 (1969).

G. Demetrescu, 1885 Jan. 22 – 1969 July 15.
C. Drâmbă.
Stud. Cerc. Astron., Vol. 14, 83 - 85 (1969).   In Rumanian.

A. J. Deutsch, died 1969 Nov. 11.
Publ. Astron. Soc. Pacific, Vol. 81, 923 (1969).

G. Gamow, 1904 - 1968.
E. Schatzman.
Astronaut. Acta, Vol. 14, 690 (1969).

Ya. Ya. Ikaunieks, 1912 April 28 - 1969 April 27.
Astron. Tsirk. No. 522, p. 7 - 8 (1969).   In Russian.

H. Kallmann-Bijl, 1908 - 1968.
J. Kaplan.
Astronaut. Acta, Vol. 14, 690 - 691 (1969).

H. Kallmann-Bijl, 1908 Sept. 18 – 1968 Nov. 7.
Space Research IX, Proc. Tokyo 1968, p. VII (1969).

I. A. Khvostikov, died 1969 Aug. 7.
Astron. Tsirk. No. 530, p. 7 - 8 (1969).   In Russian.

I. A. Khvostikov, 1910 - 1969.
Zemlya i Vselennaya, No. 6, p. 53 - 55 (1969).   In Russian.

F. Koebcke, 1909 Oct. 14 – 1969 Febr. 4.
H. Hurnik, J. Witkowski.
Postępy Astron., Vol. 17, 305 - 308 (1969).   In Polish.

F. Koebcke, 1909 Oct 14 – 1969 Febr. 4.
Urania Kraków, Vol. 40, 344 - 345 (1969).   In Polish.

A. Kohlschütter, died 1969 May 28.
Mitt. Astron. Ges. No. 27, p. 236 (1969).

A. König, died 1969 April 24.
Mitt. Astron. Ges. No. 27, p. 236 (1969).

J. J. Kubikowski, 1927 Oct. 9 – 1968 Nov. 11.
Urania Kraków, Vol. 40, 344 (1969).   In Polish.

E. Kühne, died 1969 July 30.
Mitt. Astron. Ges. No. 27, p. 236 (1969).

L. Landová-Štychová, died on 1969, Aug. 31.
Říše hvězd, Vol. 50, 235 (1969).   In Czech.

E. V. Lavrent'eva, 1893 July 26 - 1969 April 18.
E. I. Obrezkova, N. A. Popov, O. V. Chuprunova.
Astron. Tsirk. No. 514, p. 6 - 8 (1969).   In Russian.

C. Lönnqvist, died 1969 April 12.
Astron. Tidssk., Årg. 2, 140 (1969).

O. Mathias died 1969 Febr. 4.
Astron. Nachr. Vol. 291, 224 (1969).

O. Mathias, died 1969 Febr. 4.
Mitt. Astron. Ges. No. 27, p. 236 (1969).

R. W. Michie, 1931 - 1969 March 27.
Phys. Today, Vol. 22, No. 8, p. 103 (1969).

W. Münch, died 1969 July 7.
Mitt. Astron. Ges. No. 27, p. 236 (1969).

M. Mündler died 1969 March 13.
Astron. Nachr. Vol. 291, 224 (1969).

M. Mündler, died 1969 March 13.
Mitt. Astron. Ges. No. 27, p. 236 (1969).

R. J. Northcott, 1913 - 1969 July 29.
J. F. Heard.
Journ. Roy. Astron. Soc. Canada, Vol. 63, 225 - 226 (1969).

R. J. Northcott, 1913 - 1969 July 29.
Sky Telescope, Vol. 38, 211 (1969).

P. I. Popov, 1881 - 1969.
M. M. Dagaev.
Zemlya i Vselennaya, No. 4, p. 55 - 56 (1969). In Russian.

V. M. Slipher, died 1969 Nov. 8.
Publ. Astron. Soc. Pacific, Vol. 81, 922 - 923 (1969).

V. M. Slipher died 1969 Nov. 8.
Science, Vol. 166, 1608 (1969).

E. Szeligiewicz, died 1969 Sept. 26.
A. Mazur.
Urania Kraków, Vol. 40, 290 - 292 (1969).   In Polish.

S. Torrisi, died 1969 October 20.
G. Godoli.
Mem. Soc. Astron. Italiana, Nuova Serie, Vol. 40, 607 (1969).

K. Watanabe, 1911 - 1969 August.
Phys. Today, Vol. 22, No. 11, p. 107.

M. H. Wrubel, died 1968 October 26.
Astron. Tidssk., Årg. 2, 140 (1969).

Reports, communications and publications of observatories and astronomical institutes are recorded in this section; included are numbered series of reprints. Whenever possible, the numbers of the abstracts refering to the publications are given. Observatories and institutes are listed in alphabetical order of their towns. In some cases observatory publications do not give the name of the town. The following list which gives names and towns of some institutions may serve as an aid in such cases.

Algonquin Radio Observatory	Lake Traverse, Ontario, Canada
Allegheny Observatory	Pittsburgh, Pennsylvania
Arthur J. Dyer Observatory	Nashville, Tennessy
Bosscha Observatory	Lembang, Indonesia
Boyden Observatory	Bloemfontein, South Africa
Bureau International de l'Heure	Paris, France
Cajigal Observatory	Caracas, Venezuela
California Institute of Technology	Pasadena, California
Cape of Good Hope	Cape Town, South Africa
Carter Observatory	Wellington, New Zealand
Cavendish Laboratory	Cambridge, England
Ceskoslovenská Akademie Ved Astronomický Ustav	Praha, Czechoslovakia
Chamberlin Observatory, University of Denver	Denver, Colorado
Commonwealth Observatory	Canberra, Australia
David Dunlap Observatory, University of Toronto	Richmond Hill, Ontario
Dearborn Observatory	Evanston, Illinois
Department of Astronomy and Observatory, Univ. California	Los Angeles, California
Department of Astronomy, University of Texas	Austin, Texas
Division Radiophysics, C.S.I.R.O. University Grounds	Sydney, New South Wales
Dominion Astrophysical Observatory	Victoria, British Columbia
Dominion Observatory	Ottawa, Ontario
Dominion Radio Astrophysical Observatory	Penticton, British Columbia
Dudley Observatory	Albany, New York
Dunsink Observatory	Dublin, Ireland
Enhelhardt Observatory	Kazan, R.S.F.S.R.
European Southern Observatory	Hamburg, West Germany
Florida State University Radio Observatory	Tallahassee, Florida
Flower and Cook Observatories, University of Pennsylvania	Philadelphia, Pennsylvania
Four College Observatories	Amherst, Massachusetts
Fraunhofer Institut	Freiburg, West Germany
Georgetown Observatory	Washington, D.C.
Goddard Space Flight Center	Greenbelt, Maryland
Goethe Link Observatory, University of Indiana	Bloomington, Indiana
Griffith Observatory	Los Angeles, California
Harvard College Observatory	Cambridge, Massachusetts
High Altitude Observatory University of Colorado	Boulder, Colorado
Institut of Theoretical Astrophysics, Blindern	Oslo, Norway
Inter-American Observatory	Cerro Tololo, Chile
International Latitude Observatory	Mizusawa, Japan
Kandilli Observatory	Istanbul, Turkey

Kapteyn Astronomical Laboratory	Groningen, Netherlands
Karl Schwarzschild Observatorium	Tautenburg, German Democratic Republic
Kenneth Mees Observatory	Rochester, New York
Kwasan Observatory	Kyoto, Japan
Leander McCormick Observatory University of Virginia	Charlottesville, Virginia
Lee Observatory	Beirut, Lebanon
Leuschner Observatory	Berkeley, California
Lick Observatory	Mount Hamilton, (Santa Cruz), California
Lindheimer Astronomical Research Center	Evanston, Illinois
Lockheed Solar Observatory	Saugus, California
Lohrmann-Institut für Geodätische Astronomie	Dresden, German Democratic Republic
Louisiana State University Observatory	Baton Rouge, Louisiana
Lowell Observatory	Flagstaff, Arizona
Lunar and Planetary Laboratory	Tucson, Arizona
Max-Planck-Institut für Physik und Astrophysik	München, West Germany
McDonald Observatory	Fort Davis, Texas
McMath Hulbert Observatory	Pontiac, Michigan
Molonglo Radio Observatory, University of Sydney	Sydney, New South Wales
Mount Cuba Observatory	Wilmington, Delaware
Mullard Radio Astronomy Observatory	Cambridge, England
Narrabri Observatory, University of Sydney	Sydney, New South Wales
National Bureau of Standards	Washington, D.C.
National Observatory, USA	Kitt Peak, Arizona
National Radio Astronomy Observatory	Green Bank, West Virginia
Naval Research Laboratory, USA	Washington, D.C.
New Mexico State University Observatory	Las Cruces, Mexico
Nizamiah Observatory	Hyderabad, India
Nuffield Radio Astronomy Laboratories, Jodrell Bank, University of Manchester	Manchester, England
Observatoire Royal de Belgique	Uccle, Belgium
Observatorio de Cartuja	Granada, Spain
Observatorio del Ebro	Tortosa. Spain
Observatorio Fabra	Barcelona, Spain
Observatory, University of Michigan	Ann Arbor, Michigan
Ohio State University Radio Observatory	Columbus, Ohio
Ole Roemer-Observatoriet	Aarhus, Denmark
Perkins Observatory, Ohio State and Wesleyan Universities	Delaware, Ohio
Purple Mountain Observatory	Nanking, China
Radcliffe Observatory	Pretoria, South Africa
Radiophysics Laboratory, C.S.I.R.O.	Sydney, New South Wales
Remeis-Sternwarte	Bamberg, West Germany
Rensselaer Observatory	Troy, New York
Republic Observatory	Johannesburg, South Africa
Royal Radar Establishment, Radio Astronomy Division	Malvern, England
Rutherford Observatory, Columbia University	New York, New York
Sagamore Hill Radio Observatory	Bedford, Massachusetts
Saint-Michel, l'Observatoire	Haute Provence, France

Sternberg Observatory   **Moscow, R.S.F.S.R.**
Smithsonian Astrophysical
  Observatory   **Cambridge, Massachusetts**
Specola Astronomica Vaticana   **Castel Gandolfo, Italy**
Specola di Padova   **Asiago, Italy**
Sproul Observatory   **Swarthmore, Pennsylvania**
Steward Observatory,
  University of Arizona   **Tucson, Arizona**
United States Naval Observatory   **Washington, D.C.**

University of Florida,
  Radio Observatory   **Gainesville, Florida**
University of Illinois Observatory **Urbana, Illinois**
Uttar Pradesh State Observatory **Naini Tal, India**
Van Vleck Observatory   **Middletown, Connecticut**
Warner and Swasey Observatory **Cleveland, Ohio**
Washburn Observatory   **Madison, Wisconsin**
Yale University Observatory   **New Haven, Connecticut**
Yerkes Observatory   **Williams Bay, Wisconsin**

008.001   Aarhus

Meddelelser fra Ole Roemer-Observatoriet i Aarhus, Nos. 44 (M. Rudkjøbing, 01.131.007), 45 (P. Gammelgaard, 01.113.001), 46 (M. Rudkjøbing, 02.131.021).

008.002   Alger

L'Observatoire d'Alger en 1968.
T. Weimer, Y. Mentalecheta.
Ann. Obs. Astron. d'Alger, Vol. 3, (Fasc. 1), 1 - 2 (1969).

Université d'Alger. Annales de l'Observatoire d'Alger, Tome 3, Fasc. 1 (A. Ghezloun, M. Benhocine, A. Fresneau, A. Marouf, 02.044.021).

008.003   Alma Ata

Akademiya Nauk Kazakhskoj SSR. **Trudy Astrofizicheskogo Instituta,** *Alma Ata,* Vol. 10 (A. I. Ivanov, G. Sh. Livshits, V. E. Pavlov, B. T. Tashenov, Ya. A. Teifel', 02.003.005); Vol. 12 (F. A. Tsitsin, 02.151.005; G. M. Idlis, 02.151.006; I. L. Genkin, 02.151.007; O. V. Chumak, 02.153.010; I. L. Genkin, L. M. Genkina, 02.158.019; I. L. Genkin, L. M. Genkina, 02.158.020; L. M. Genkina, 02.158.021; L. M. Genkina, 02.158.022; R. H. Gainullina, 02.160.006; E. K. Denisyuk, O. A. Tumakova, 02.158.023); Vol. 13 (V. G. Teifel, 02.099.062; A. N. Aksenov, 02.099.063; Z. N. Grigorjeva, N. V. Priboeva, 02.099.064; L. P. Sorokina, 02.099.065; V. F. Kartashoff, V. G. Teifel, A. A. Usoltzeva, 02.099.066; A. N. Aksenov, 02.099.067; S. V. Karjagina, V. E. Moshajeva, 02.082.129; T. P. Toropova, 02.082.130; T. P. Toropova, 02.082.131; N. M. Ibraimov, T. P. Toropova, G. A. Kharitonova, 02.082.132; T. P. Toropova, V. V. Keguleekhes, K. M. Salamachin, 02.082.133; T. P. Toropova, S. O. Obasheva, 02.082.134; T. P. Toropova, L. L. Slobodkina, 02.082.135; J. I. Rudnev, L. A. Sataeva, 02.082.136; J. I. Rudnev, L. A. Sataeva, 02.034.077; J. I. Rudnev, L. A. Sataeva, 02.034.078; P. N. Boiko, L. A. Sataeva, G. A. Kharitonova, 02.034.079; J. I. Rudnev, L. A. Sataeva, 02.036.017; J. I. Rudnev, T. P. Toropova, 02.034.080); Vol. 14 (D. A. Rozhkovsky, 02.132.041; L. A. Pawlowa, 02.132.042; D. A. Rozhkovsky, 02.022.112; D. A. Rozhkovsky, 02.132.043; I. D. Kupo, 02.122.166; Z. N. Chumak, 02.114.105; D. A. Rozhkovsky, M. I. Musorin, 02.034.081; A. W. Kurchakov, 02.132.044; K. G. Dzhakusheva, V. S. Matjagin, E. G. Michelkin, 02.021.016; E. G. Michelkin, 02.031.028; E. G. Michelkin, 02.022.113); Vol. 15 (S. Obashev, 02.009.017; S. O. Obashev, 02.074.076; S. O. Obashev, E. J. Vilkovisky, A. S. Zubtzov, 02.074.077; S. O. Obashev, 02.073.085; S. O. Obashev, 02.073.086; S. O. Obashev, 02.073.087; N. N. Morozov, S. O. Obashev, 02.072.087; A. S. Zubtzov, S. O. Obashev, 02.073.088).

008.004   Ann Arbor

Publications of the Observatory of the University of Michigan, Vol. 9, No. 9 (L. H. Aller, J. Jugaku, 02.114.050).

008.005   Arcetri

Laboratorio di spettroscopia XUV in Arcetri.
A. M. Cantù, G. G. Noci.
Atti XII Riunione Soc. Astron. Italiana, L'Aquila 1968, p. 47 - 49 (1969). — Abstract SAI.

008.006   Armagh

Armagh Observatory, Leaflet Nos. 89 (A. D. Andrews, AJB 68, 12404), 90 (A. D. Andrews, AJB 68, 12403), 91 (A. D. Andrews, P. Corvan, B. Hardy, P. Johnston, W. Johnston, J. Perrott, 01.123.025), 92 (A. D. Andrews, P. F. Chugainov, R. E. Gershberg, V. S. Oskanian, 01.122.087), 93 (A. D. Andrews, J. Perrott, 01.122.094), 94 (A. D. Andrews, 01.122.098), 95 (E. J. Öpik, 02.065.010), 97 (A. D. Andrews, P. F. Chugainov, 02.122.153).

Contributions from the Armagh Observatory, Nos. 63 (E. J. Öpik, AJB 68, 9463), 64 (H. H. R. Grossie, E. J. Öpik, 01.152.002), 65 (E. J. Öpik, 02.094.043), 66 (A. D. Andrews, 02.122.052).

Contributions from the Armagh Observatory, Quarto Series, No. 3 (A. D. Andrews, T. W. Rackham, P. A. Wayman, 01.105.035).

008.007   Asiago

Contributi dell' Osservatorio Astrofisico dell' Università di Padova in Asiago, Nos. 206 (S. Taffara, AJB 68, 10717), 207 (C. Barbieri, L. A. Erculiani, AJB 68, 13418), 208 (F. Bertola, AJB 68, 145162), 209 (A. Mammano, L. Rosino, AJB 68, 104155), 210 (A. Martini, AJB 68, 104157), 211 (R. Barbon, 01.153.010), 212 (A. Martini, 01.122.050), 213 (R. Margoni, R. Stagni, 01.154.009), 214 (P. L. Bernacca, F. Bertola, 01.158.059), 215 (R. Viotti, 01.132.026), 216 (L. Rosino, R. Stagni, 01.100.012), 217 (G. Colombo, F. Franklin, 01.100.013), 218 (R. Barbon, 01.158.060), 219 (F. Bertola, AJB 68, 145163), 220 (F. Bertola, AJB 68, 145164), 221 (L. Rosino, G. Chincarini, A. Mammano, 01.124.102), 222 (R. Margoni, M. Perinotto, E. Nasi, 02.119.005), 223 (M. Perinotto, G. Chincarini, 01.103.100), 224 (R. Barbon, AJB 68, 145271), 225 (A. Mammano, A. Martini, 02.121.036), 226 (F. Bertola, S.

D'Odorico, W. K. Ford, Jr., V. C. Rubin, 02.158.009), 227
(G. Chincarini, L. Rosino, 02.124.104), 228 (R. Barbon, A.
Mammano, L. Rosino, 02.124.103), 229 (A. Mammano, A.
Martini, 02.122.077), 230 (H. Arp, F. Bertola, 01.158.066).

008.008   Athen

**Astronomical Institute, National Observatory of
Athens.** − Annual report 1967.
D. Kotsakis.
Annual Rep. Astron. Inst. Greece 1967, p. 3 - 5 (1968).

**Astronomical Institute, National Observatory of
Athens.** − Annual report 1968.
D. Kotsakis.
Annual Rep. Astron. Inst. Greece 1968, p. 3 - 5 (1969).

**Research Center for Astronomy and Applied
Mathematics, Academy of Athens.** − Annual report 1967.
J. Xanthakis.
Annual Rep. Astron. Inst. Greece 1967, p. 6 - 7 (1968).

**Research Center for Astronomy and Applied
Mathematics, Academy of Athens.** − Annual report 1968.
J. Xanthakis.
Annual Rep. Astron. Inst. Greece 1968, p. 6 (1969).

**Contributions from the Research Center for Astro-
nomy and Applied Mathematics, Academy of Athens,**
Series I (Astronomy), Nos. 16 (P. G. Alexiou, C. P. Poulakos,
02.072.096), 18 (J. Xanthakis, AJB 66, 6598).

Department of Astronomy, University of Athens.
Annual report 1967.
D. Kotsakis.
Annual Rep. Astron. Inst. Greece 1967, p. 8 - 9 (1968).

Department of Astronomy, University of Athens.
Annual report 1968.
D. Kotsakis.
Annual Rep. Astron. Inst. Greece 1968, p. 7 - 8 (1969).

**Publications of the Laboratory of Astronomy,
University of Athens, Greece,** Series II, Nos. 22 (D. Kotsakis,
M. Zikides, E. Sarris, 02.042.041); 23 (G. Antonacopoulos,
02.021.017).

**Department of Astronomy, Technical University
of Athens.** − Annual report 1967.
J. Argyrakos.
Annual Rep. Astron. Inst. Greece 1967, p. 13 (1968).

**Department of Astronomy, Technical University
of Athens.** − Annual report 1968.
J. Argyrakos.
Annual Rep. Astron. Inst. Greece 1968, p. 14 (1969).

008.009   Auckland

Auckland: Auckland Observatory.
Inform. Bull. Southern Hemisphere, No. 14, p. 20 (1969).

008.010   Babelsberg

Sternwarte Babelsberg, Institut für relativistische

und extragalaktische Forschung.   H.-J. Treder.
Monatsber. Deutsch. Akad. Wiss. Berlin, Band 11, 766 - 768
(1969). − Jahresbericht 1968.

**Deutsche Akademie der Wissenschaften zu Berlin.
Sternwarte Babelsberg, Institut für relativistische und extragalaktische Forschung, Mitteilungen.** Neue Folge, Nos. 17
(G. Dautcourt, 01.066.007), 18 (H.-J. Treder, 01.162.026),
19 (H.-J. Treder, 02.066.043), 20 (H.-J. Treder, 02.162.054),
21 (G. Dautcourt, 02.162.092), 22 (G. Dautcourt,
02.162.093), 23 (E. Kreisel, 02.066.042).

008.011   Bamberg

**Veröffentlichungen der Remeis-Sternwarte Bamberg,
Astronomisches Institut der Universität Erlangen−Nürnberg,**
Vol. 8, Nos. 80 (H. Mauder, U. Köhler, 01.121.002), 81 (H.
Bauernfeind, 02.121.095), 82 (W. Strohmeier, I. Patterson,
01.123.026), 83 (R. Knigge, U. Köhler, 02.121.096), 84
(H. Bauernfeind, 02.123.052), 85 (H. Bauernfeind,
02.122.177), 86 (W. Strohmeier, H. Bauernfeind, 01.123.031).

008.012   Beirut

**Lee Observatory, American University of Beirut,
Lebanon, Monthly Bulletin,** Astronomical Section, 1969
May − July (02.075.028).

008.013   Beograd

Bulletin de l'Observatoire de Beograd, Vol. 27,
No. 1 (P. M. Djurković, G. M. Popović, D. J. Zulević,
02.118.003; G. M. Popović, 02.118.004, 02.118.005; D. J.
Zulević, 02.118.006, 02.118.007; S. Mali, 02.118.008; G. M.
Popović, 02.118.009; I. Semeniuk, 02.121.002).

008.014   Berlin

**Heinrich-Hertz-Institut für solar-terrestrische
Physik** der Deutschen Akademie der Wissenschaften zu
Berlin.   F. W. Jäger, H. Daene.
Monatsber. Deutsch. Akad. Wiss. Berlin, Band 11, 758 - 762
(1969). − Jahresbericht 1968.

**Heinrich-Hertz-Institut, Solare Beobachtungsergeb-
nisse.** Deutsche Akademie der Wissenschaften zu Berlin, Zentralinstitut für Solar-Terrestrische Physik, Berlin−Adlershof.
HHI Solar Data, Vol. 20, 1969 May − December (02.075.025).

**Heinrich-Hertz-Institut, Solare Beobachtungsergeb-
nisse.** Deutsche Akademie der Wissenschaften zu Berlin, Zentralinstitut für Solar-Terrestrische Physik, Berlin−Adlershof.
HHI Supplement Series of Solar Data, Vol. 1, Nos. 5 (A.
Böhme, A. Krüger, H. Künzel, 02.077.049), 6 (A. Böhme,
02.077.050).

008.015   Besançon

Palmarès du 67me concours chronométrique.

Ann. Françaises Chronométrie Micromécanique, 4. année, p. 85 - 90 (1969).

Annales de l'Observatoire de Besançon, Université de Besançon, Vol. 8, Fasc. 1 (G. Hilaire, AJB 67, 4325; C. Froeschle, 02.151.068; L. Arbey, 02.041.044).

008.016 Bloemfontein

Bloemfontein, Mazelspoort: Boyden Observatory. A. H. Jarrett. Inform. Bull. Southern Hemisphere, No. 14, p. 22 - 23 (1969).

008.017 Bologna

Osservatorio Astronomico Universitario di Bologna. Notizie e Rassegne, Nos. 29 - 37 (F. S. Delli Santi, E. Nasi, 02.077.052).

Pubblicazioni dell'Osservatorio Astronomico, Universitario di Bologna, Vol. 9, Nos. 15 (C. Bartolini, P. Battistini, E. Nasi, 02.154.016), 17 (U. Dall'Olmo, 02.099.072), 19 (F. S. Delli Santi, M. G. Pioli, 01.077.027), 20 (V. Castellani, A. Renzini, 01.064.011), 21 (V. Castellani, P. Giannone, A. Renzini, AJB 68, 5426), 22 (V. Castellani, P. Giannone, A. Renzini, 01.154.005); Vol. 10, Nos. 1 (C. Bartolini, P. Battistini, C. Delli Ponti, A. Guarnieri, 01.124.104), 2 (F. S. Delli Santi, 02.141.220), 3 (C. Bartolini, P. Battistini, 02.121.097), 4 (V. Castellani, P. Giannone, A. Renzini, 01.065.029).

Laboratorio Nazionale di Radioastronomia, Istituto di Fisica "A. Righi", Università degli Studi, Bologna (Italia), Contributions, Nos. 56 (02.141.201), 57 (02.141.202), 60 (02.113.052).

008.018 Bordeaux

Rapport, présenté au Conseil de l'Université, année scolaire 1967 - 1968. P. Sémirot. Obs. Univ. Bordeaux, 16 pp. (1968).

Rapport, présenté au Conseil de l'Université, année scolaire 1968 - 1969. P. Sémirot. Obs. Univ. Bordeaux, 18 pp. (1969).

Publications de l'Observatoire de l'Université de Bordeaux (Floirac), Nouvelle Série, Nos. 32 (G. Soulié, L. Pourteau, 01.041.001), 34 (Y. Requième, AJB 66, 2350; P. Mianes, AJB 66, 134162; C. Bardin, M. Chopinet, R. Duflot-Augarde, AJB 67, 145102; M. Chopinet, R. Duflot-Augarde, AJB 68, 13218), 35 (Y. Requième, AJB 68, 2363).

008.019 Borowiec

Polish Academy of Sciences, Astronomical Latitude Station, Borowiec, Circulars, Nos. 109 - 111 (02.044.044, 02.045.031).

008.020 Brno

Astronomical Institute of the University – Brno, Czechoslovakia. Publications, Nos. 8 (B. Onderlička, M. Vetešník, 02.122.171), 13 (Z. Mikulášek, J. E. Purkyně, 02.155.001).

008.021 Bruxelles

Université Libre de Bruxelles. Institut d'Astronomie et d'Astrophysique. Série A, Nos. 10 (M. Arnould, 02.065.107), 11 (R. Coutrez, 02.021.023), 12 (C. Brihaye, G. Reidemeister, 02.022.122).

Université Libre de Bruxelles. Institut d'Astronomie et d'Astrophysique. Série B, Nos. 10 (R. Hendrickx, AJB 66, 6430), 11 (R. Hendrickx, AJB 66, 6429), 12a (M. Arnould, AJB 67, 5402), 12b (M. Arnould, AJB 67, 5403), 12c (M. Arnould, AJB 68, 5407).

008.022 Bucarest

Bucharest Observatory Solar Station. – Report from Solar Institute. C. Popovici. Solar Physics, Vol. 9, 494 - 495 (1969).

008.023 Byurakan

Chronik. Soobshch. Byurakan. Obs. No. 40, p. 108 - 117 (1969). In Russian.

Byurakan Astrophysical Observatory, Armenia, USSR, Reprint No. 32 (V. A. Ambarzumyan, L. V. Mirsoyan, 02.061.007).

Soobshcheniya Byurakanskoj Observatorii, No. 40 (A. T. Kalloghlian, 02.160.002; A. T. Kalloghlian, 02.158.013; L. V. Mirzoyan, E. S. Parsamian, N. L. Kalloghlian, 02.122.010; H. S. Badalian, L. K. Erastova, 02.122.011; H. M. Tovmassian, R. G. Mnatsakanian, 02.160.003; H. M. Tovmassian, 02.158.014; R. A. Vardanian, 02.122.012; V. A. Sanamian, 02.033.004; V. Yu. Terebizh, 02.022.072; V. V. Papoyan, D. M. Sedrakian, E. V. Chubarian, 02.126.003; G. S. Sahakian, M. A. Mnatsakanian, 02.065.008).

008.024 Cambridge, Mass.

Smithsonian Astrophysical Observatory. F. L. Whipple. Smithsonian Year 1968, p. 445 - 489 (1969). – Annual report for the year ended June 30, 1968.

Smithsonian Institution. Astrophysical Observatory. Research in Space Science. SAO Special Reports, Nos. 271 (L. Sehnal, 02.052.006), 302 (S. E. Hamid, 02.042.013), 303 (E. H. Avrett, R. Loeser, 02.064.020), 304 (A. F. Cook, F. A. Franklin, 02.100.005).

008.025 Cape Town

Annals of the Cape Observatory, Vol. 23 (R. H. Stoy, 02.041.015).

Observatory, Cape Province: Royal Observatory at the Cape of Good Hope.
Inform. Bull. Southern Hemisphere, No. 14, p. 26 - 27 (1969).

008.026 Castel Gandolfo

The Vatican Observatory.
P. J. Treanor.
Specola Vaticana, Città del Vaticano. 40 pp. (1969).

Ricerche Astronomiche. Specola Vaticana, Città del Vaticano, Vol. 7, Nos. 18 (W. J. Miller, 02.122.104), 19 (F. C. Bertiau, M. F. McCarthy, 02.114.046).

Specola Vaticana, Miscellanea Astronomica, No. 118 (E. W. Salpeter, 02.022.115).

008.027 Catania

Relazione annuale (per il periodo 1967 novembre 1 - 1968 dicembre 31). G. Godoli.
Oss. Astrofis. Catania Pubbl., Nuova Serie, No. 139, 12 pp. (1969).

Solar research at the Catania Astrophysical Observatory. − Report from Solar Institute. G. Godoli.
Solar Physics, Vol. 9, 246 - 249 (1969).

Osservatorio Astrofisico di Catania, Pubblicazioni, Nuova Serie, Nos. 117 (O. Morgante, AJB 68, 66100), 119 (G. Godoli, B. C. Fossi, AJB 68, 6640), 120 (V. Bumba, G. Godoli, AJB 68, 6622), 121 (G. Godoli, L. Paternò, 02.082.119), 122 (G. Godoli, B. C. Fossi, 02.073.081), 123 (G. Godoli, F. Mazzucconi, S. Nagasawa, 02.073.082), 124 (S. Cristaldi, M. Narbone, M. Rodonò, AJB 68, 12422), 125 (E. Balli, AJB 68, 6511), 126 (N. Dogan, 02.072.083), 127 (C. Blanco, F. Catalano, AJB 68, 10702), 128 (S. Cristaldi, L. Paternò, AJB 68, 2319), 129 (F. Affronti, C. Blanco, 02.082.120), 130 (02.075.021), 136 (M. G. Fracastoro, 02.072.082), 137 (S. Catalano, M. Rodonò, AJB 68, 12210) 139 (G. Godoli, 02.008.027).

008.028 Cerro Tololo

Kitt Peak National Observatory, Tucson, Arizona, and Cerro Tololo Inter-American Observatory, La Serena, Chile. N. U. Mayall.
Bull. American Astron. Soc. Vol. 1, 298 - 331 (1969). − Report 1968 − 1969.

Cerro Tololo Inter-American Observatory, Contributions Nos. 26 (V. Blanco, W. Kunkel, W. A. Hiltner, G. Chodil, H. Mark, R. Rodrigues, F. Seward, C. D. Swift, AJB 68, 13106), 27 (V. Blanco, W. Kunkel, W. A. Hiltner, G. Lynga, H. Bradt, G. Clark, S. Naranan, S. Rappaport, G. Spada, AJB 68, 13506), 28 (V. Blanco, W. Kunkel, W. A. Hiltner, AJB 68, 13507), 29 (N. Sanduleak, AJB 68, 145256), 30 (R. D. McClure, S. van den Bergh, AJB 68, 10343), 31

(A. U. Landolt, AJB 68, 12461), 32 (W. E. Kunkel, 01.122.111), 33 (G. W. Wares, L. H. Aller, AJB 68, 145265), 34 (I. Epstein, AJB 68, 14121), 35 (C. B. Stephenson, N. Sanduleak, R. E. Schild, AJB 68, 124100), 36 (N. Sanduleak, A. G. D. Philip, AJB 68, 14315), 37 (A. G. D. Philip, N. Sanduleak, 01.155.007), 38 (A. Gutierrez-Moreno, H. Moreno, J. Stock, AJB 68, 7327), 39 (A. Gutierrez-Moreno, H. Moreno, J. Stock, AJB 68, 10440), 40 (G. Chodil, H. Mark, R. Rodrigues, F. D. Seward, C. D. Swift, I. Turiel, W. A. Hiltner, G. Wallerstein, E. J. Mannery, AJB 68, 13517), 41 (S. van den Bergh, AJB 68, 13105), 42 (A. Feinstein, O. E. Ferrer, AJB 68, 14123), 43 (S. van den Bergh, G. L. Hagen, AJB 68, 145239), 44 (W. A. Hiltner, C. B. Stephenson, N. Sanduleak, AJB 68, 10450), 45 (N. Sanduleak, AJB 68, 145258), 46 (N. Sanduleak, A. G. D. Philip, AJB 68, 10487), 48 (S. Tapia, 01.122.040), 49 (M. Isakson, 02.097.014), 50 (M. F. Walker, V. M. Blanco, W. E. Kunkel, 01.159.001), 51 (G. A. H. Walker, S. C. Morris, AJB 68, 145263), 52 (A. U. Landolt, 01.118.007), 53 (N. Sanduleak, 01.122.020), 54 (W. A. Hiltner, R. F. Garrison, R. E. Schild, 02.114.004), 55 (A. Feinstein, 01.153.014), 56 (R. E. Schild, W. A. Hiltner, N. Sanduleak, 01.152.004), 57 (H. A. Abt, C. P. Jewsbury, 01.153.024), 58 (W. E. Kunkel, AJB 68, 12458), 59 (C. Sturch, 01.155.002), 60 (W. E. Kunkel, AJB 68, 12457), 61 (D. J. MacConnell, C. L. Perry, 01.152.006), 62 (J. A. Westphal, G. Neugebauer, 01.113.014), 63 (A. Slettebak, P. C. Keenan, R. K. Brundage, 01.114.070), 64 (G. Alcaino, 01.154.011), 65 (H. Albers, 01.114.048), 67 (H. Mark, R. E. Price, R. Rodrigues, F. D. Seward, C. D. Swift, W. A. Hiltner, 01.142.054), 68 (A. G. D. Philip, 02.115.003), 69 (W. E. Kunkel, 01.122.052), 70 (S. Demers, 01.158.048), 71 (H. A. Abt, W. W. Morgan, 02.153.002), 72 (R. E. Schild, 02.122.035), 73 (A. U. Landolt, 02.122.040), 74 (A. U. Landolt, 02.113.011), 76 (S. Demers, 02.154.001), 78 (C. L. Perry, G. Hill, 02.153.006), 80 (N. Sanduleak, 02.159.001).

008.029 Charlottesville

Publications of the Leander McCormick Observatory of the University of Virginia, Vol. 11, Part 26 (C. P. Olivier, 02.004.035).

008.030 Christchurch

Christchurch: West Melton Observatory.
Inform. Bull. Southern Hemisphere, No. 14, p. 20 (1969).

008.031 Cincinnati

University of Cincinnati. Publications of the Cincinnati Observatory, No. 23 (P. Herget, 02.099.057; P. Herget, 02.103.126).

008.032 Cordoba

Córdoba: Observatorio Astronomico. (Astronomical Observatory, National University of Córdoba).
J. L. Sérsic, G. Carranza.
Inform. Bull. Southern Hemisphere, No. 14, p. 8 - 9 (1969).

008.033 Cracow

Cracow Observatory, Reprints Nos. 76 (J. M. Kreiner, AJB 68, 12245), 77 (M. Winiarski, AJB 68, 83325), 78 (J. M. Kreiner, J. Włudarska, 01.124.103), 79 (P. Flin, 01.121.027), 80 (E. Rybka, 02.113.002), 81 (M. Kurpińska, 02.096.003).

008.034 Dresden

Lohrmann-Institut für Geodätische Astronomie der Technischen Universität Dresden. H.-U. Sandig. Monatsber. Deutsch. Akad. Wiss. Berlin, Band 11, 774 - 776 (1969). – Jahresbericht 1968.

Technische Universität Dresden, Lohrmann Observatorium, Zirkular, Nos. 37 - 42 (02.045.018).

008.035 Dublin

Communications of the Dublin Institute for Advanced Studies, Series C. Dunsink Observatory Publications, Vol. 1, No. 5 (T. Kiang, 02.158.012).

Dunsink Observatory, Reprints Nos. 46 (J. H. Reid, 02.073.089), 47 (I. Elliott, AJB 66, 2118), 48 (P. A. Wayman, AJB 67, 14165), 49 (P. A. Wayman, AJB 67, 10110), 50 (P. A. Wayman, AJB 67, 456), 51 (P. A. Wayman, AJB 67, 14238), 52 (T. Kiang, AJB 67, 14536), 53 (P. A. Wayman, AJB 68, 2172), 54 (P. A. Wayman, 02.159.011), 55 (S. M. P. McKenna, AJB 67, 254), 56 (P. A. Wayman, 01.032.042), 57 (I. Elliott, 01.071.006), 58 (T. Kiang, W. C. Saslaw, 01.160.002), 59 (C. J. Butler, P. A. Wayman, 01.113.022).

008.036 Dunedin

Dunedin: University of Otago, Physics Department. P. J. Edwards. Inform. Bull. Southern Hemisphere, No. 14, p. 21 (1969).

Dunedin: Invermay Radio Observatory. (Observatory of the University of Otago). P. J. Edwards. Inform. Bull. Southern Hemisphere, No. 14, p. 20 (1969).

008.037 Edinburgh

Report of the Astronomer Royal for Scotland for the year ending 31st March 1969. H. A. Brück. The Royal Observatory, Edinburgh, 11 pp. (1969).

Publications of the Royal Observatory, Edinburgh, Vol. 6, No. 10 (W. B. Samson, 02.114.071).

Communications from the Royal Observatory, Edinburgh, Nos. 49 (R. D. Wolstencroft, 01.074.047), 59 (C. M. Humphries, 02.034.102), 60 (G. C. Sudbury, 02.034.103), 61 (J. W. Campbell, 02.034.104), 65 (B. N. G. Guthrie, 01.061.016), 66 (M. T. Brück, K. Nandy, H. Seddon, 01.131.112), 67 (V. C. Reddish, N. C. Wickramasinghe, 01.065.028), 68 (V. C. Reddish, 01.065.027), 69 (M. T. Brück, K. Nandy, G. Caprioli, F. Smriglio, 01.113.027), 71 (H. Seddon, 01.125.009), 72 (R. D. Wolstencroft, J. G. Ireland, K. Nandy, H. Seddon, 01.131.102).

008.038 Fort Davis

The University of Texas. Contributions from the McDonald Observatory, Fort Davis, Texas, No. 420 (G. de Vaucouleurs, 02.034.105).

008.039 Frankfurt

Veröffentlichungen des Astronomischen Instituts der Universität Frankfurt (Main) [Publications of the Astronomical Institute of the University of Frankfurt (Main)], Nos. 27 (W. Gleissberg, 01.072.045), 28 (W. Gleissberg, 02.004.036).

008.040 Frascati

Consiglio Nazionale delle Ricerche (Italia). Laboratorio di Astrofisica, Frascati (Roma), Contributi, Nos. 38 (V. Castellani, P. Giannone, A. Renzini, 01.065.029), 39 (V. Castellani, P. Giannone, A. Renzini, 01.154.005), 40 (P. Maffei, A. Martini, 01.122.072), 41 (R. Viotti, 01.122.067), 42 (L. Gratton, 02.142.057).

Consiglio Nazionale delle Ricerche (Italia). Laboratorio di Astrofisica, Frascati (Roma), Contributi, Nos. B1 (F. Occhionero, AJB 66, 5143), B2 (F. Occhionero, AJB 67, 54102), B3 (F. Occhionero, AJB 68, 54136), B4 (F. Occhionero, AJB 68, 5481), B5 (F. Occhionero, AJB 68, 55102), B6 (B. Bertotti, A. Cavaliere, F. Pacini, 01.141.026), B7 (A. Cavaliere, F. Pacini, G. Setti, 01.141.039), B8 (R. Funiciello, M. Fulchignoni, 02.105.183), B9 (A. Martini, 02.122.098).

008.041 Freiburg

Fraunhofer Institut. Map of the Sun. 1969 July 1 – December 31 (02.075.015).

Mitteilungen aus dem Fraunhofer Institut, *Freiburg,* Nos. 84 (A. Bruzek, 02.073.074), 88 (P. N. Brandt, 01.071.033), 89 (R. Göhring, 02.072.001), 90 (W. Mattig, 02.072.004), 91 (A. Bruzek, 02.073.043), 92 (U. Grossmann-Doerth, 02.082.035), 93 (J. P. Mehltretter, 02.072.051), 94 (F.-L. Deubner, 02.071.057).

008.042 Gainesville

Rosemary Hill Observatory, Department of Physics and Astronomy, University of Florida, Gainesville, Florida. Contribution No. 1 (E. E. Clark, AJB 68, 145275).

008.043 Genève

Publications de l'Observatoire de Genève, Série A, Fasc. 76 (L. Martinet, 02.114.096; G. Goy, A. Maeder, 02.113.054; G. Goy, A. Maeder, 02.113.055; M. Golay, E. Peytremann, A. Maeder, 02.114.097; B. Hauck, 02.115.017).

008.044 Green Bank

National Radio Astronomy Observatory, *Green Bank*, Reprints, Series A, Nos. 109 (N. Albaugh, K. H. Wesseling, 02.033.048), 110 (G. L. Verschuur, 01.157.007), 111 (D. H. Staelin, 01.141.225), 112 (D. H. Staelin, 02.022.121), 113 (W. R. Burns, S. S. Yao, 02.021.022), 114 (R. M. Hjellming, C. P. Gordon, K. J. Gordon, 01.131.055), 115 (W. R. Burns, B. G. Clark, 01.141.176), 116 (G. L. Verschuur, 01.131.060), 117 (G. L. Verschuur, 01.131.075), 118 (B. E. Turner, 02.131.001), 119 (J. W. Erkes, J. R. Dickel, 02.125.002), 120 (K. J. Gordon, 02.161.001), 121 (E. Churchwell, M. Felli, P. G. Mezger, 02.131.002), 122 (G. L. Verschuur, 02.131.012), 123 (M. S. Roberts, 02.158.029), 124 (R. M. Hjellming, 02.143.005), 125 (G. L. Verschuur, 02.131.015), 126 (B. E. Turner, 02.131.035), 127 (J. R. Dickel, 02.125.008), 128 (D. K. Milne, T. L. Wilson, F. F. Gardner, P. G. Mezger, 02.131.033), 129 (R. H. Rubin, 02.132.008), 130 (M. H. Andrews, R. M. Hjellming, 02.131.043), 131 (R. M. Hjellming, E. Churchwell, 02.132.011).

National Radio Astronomy Observatory, *Green Bank*, Reprints, Series B, Nos. 137 (P. Palmer, B. Zuckerman, H. Penfield, A. E. Lilley, P. G. Mezger, 01.132.050), 138 (G. L. Verschuur, 01.156.007), 139 (T. J. Sejnowski, R. M. Hjellming, 01.132.051), 140 (P. Palmer, B. Zuckerman, D. Buhl, L. E. Snyder, 01.131.101), 141 (D. H. Staelin, E. C. Reifenstein III, 01.141.197), 142 (B. E. Turner, 02.131.005), 143 (E. C. Reifenstein III, W. D. Brundage, D. H. Staelin, 01.141.198), 144 (R. H. Rubin, B. E. Turner, 02.131.011), 145 (K. I. Kellermann, I. I. K. Pauliny-Toth, P. J. S. Williams, 02.141.012), 146 (R. M. Hjellming, M. H. Andrews, T. J. Sejnowski, 02.131.031), 147 (A. A. Penzias, J. Schraml, R. W. Wilson, 02.066.005), 148 (K. W. Riegel, M. C. Jennings, 02.131.030), 149 (W. E. Howard III, H. Hvatum, 02.141.056), 150 (B. Zuckerman, P. Palmer, L. E. Snyder, D. Buhl, 02.131.036), 151 (R. H. Rubin, 02.131.041), 152 (S. J. Goldstein, Jr., D. D. MacDonald, 02.131.040), 153 (M. R. Kundu, 02.131.023), 154 (D. Buhl, L. E. Snyder, P. R. Schwartz, A. H. Barrett, 02.155.006), 155 (M. H. Cohen, A. T. Moffet, D. Shaffer, B. G. Clark, K. I. Kellermann, D. L. Jauncey, S. Gulkis, 02.158.057).

008.045 Greenbelt

Goddard Space Flight Center, Greenbelt. NASA Technical Note, TN D–5284 (P. Musen, 02.042.027).

008.046 Greenwich

Royal Observatory Annals, (Joint Publications of the Royal Greenwich Observatory, Herstmonceux, Royal Observatory, Cape of Good Hope), No. 3 (R. v. d. R. Woolley, 02.041.017).

Royal Observatory Bulletins, (Joint Publications of the Royal Greenwich Observatory, Herstmonceux, Royal Observatory, Cape of Good Hope), Nos. 152 (A. L. T. Powell, 02.064.047), 154 (R. M. Catchpole, B. E. J. Pagel, A. L. T. Powell, 02.114.012), 155 (D. H. P. Jones, C. M. Haslam, 02.112.009), 156 (R. Woolley, A. S. Asaad, M. P. Candy, M. J. Penston, 02.112.019), 157 (L. S. T. Symms, 02. 118.040).

Geomagnetic Bulletins of the Institute of Geological Sciences, Royal Greenwich Observatory, Herstmonceux

Castle, No. 1 (02.084.244).

008.047 Groningen

Nederlandse Vereniging voor Weer- en Sterrenkunde. Observations of Variable Stars. Report (Kapteyn Astronomical Laboratory, Groningen – Netherlands), No. 16 (02.123.016).

008.048 Hamburg

European Southern Observatory. Rapport annuel 1968. O. Heckmann. Hamburg-Bergedorf, 29 pp. (1969).

Addresses held during the Inauguration of the European Southern Observatory at La Silla on 25 March 1969. O. Heckmann, J. Sahade, J. Bannier. ESO Bull. No. 6, 48 pp. (1969).

The European Southern Observatory. A. Muller. Inform. Bull. Southern Hemisphere, No. 14, p. 3 - 7 (1969).

Deutsches Hydrographisches Institut, Hamburg. Zeitsignalaufnahmen, Astronomische Zeit- und Breitenbestimmungen, 1969 January – June (02.044.026).

008.049 Hartebeesthoek

C.S.I.R., National Institute for Telecommunications Research. Inform. Bull. Southern Hemisphere, No. 14, p. 24 (1969).

008.050 Heidelberg

Veröffentlichungen des Astronomischen Rechen-Instituts, Heidelberg, No. 22 (W. Gliese, 02.041.018).

008.051 Helsinki

Publications of the Finnish Geodetic Institute, *Helsinki*, No. 66 (J. Kakkuri, 02.046.021).

008.052 Houston

Solar physics at the NASA Manned Spacecraft Center. – Report from Solar Institute. D. E. Robbins, J. H. Reid. Solar Physics, Vol. 10, 502 - 510 (1969).

008.053 Ioannina

Department of Astronomy, University of Ioannina. – Annual report 1967. S. N. Svolopoulos.

Annual Rep. Astron. Inst. Greece 1967, p. 16 (1968).

Department of Astronomy, University of Ioannina. – Annual report 1968.
S. N. Svolopoulos.
Annual Rep. Astron. Inst. Greece 1968, p. 12 - 13 (1969).

Aristotelian University of Thessaloniki, Ioannina Campus. **Laboratory of Astronomy, Ioannina, Greece. Contributions,** No. 3 (S. N. Svolopoulos, 02.115.005).

008.054 Izmir

The Solar Department at Izmiran. – Report from Solar Institute. E. I. Mogilevsky.
Solar Physics, Vol. 10, 231 - 234 (1969).

008.055 Jena

Universitäts-Sternwarte der Friedrich-Schiller-Universität Jena. H. Lambrecht.
Monatsber. Deutsch. Akad. Wiss. Berlin, Band 11, 772 - 774 (1969). – Jahresbericht 1968.

Mitteilungen der Universitäts-Sternwarte zu Jena, Nos. 85 (J. Dorschner, C. Friedemann, AJB 68, 12426), 86 (R. Schielicke, 02.034.083), 87 (K.-H. Schmidt, S. van den Bergh, 01.155.003), 88 (W. Pfau, 02.115.006), 89 (C. Friedemann, 02.131.018), 90 (J. Dorschner, C. Friedemann, W. Pfau, 02.124.100).

008.056 Johannesburg

Johannesburg: Republic Observatory.
J. Hers.
Inform. Bull. Southern Hemisphere, No. 14, p. 24 - 25 (1969).

South African Council for Scientific and Industrial Research. **Republic Observatory, Johannesburg. Circulars,** Vol. 7, No. 128 (J. A. Bruwer, B. M. F. Armstrong, 02.098.008, 02.103.102, 02.103.103; G. F. G. Knipe, 02.118.011; J. L. Newburg, 02.118.012; W. S. Finsen, 02.118.013; J. L. Newburg, 02.118.014; W. S. Finsen, 02.118.015; J. L. Newburg, 02.118.016, 02.118.017, 02.118.018; W. S. Finsen, 02.118.019, 02.118.020; G. F. G. Knipe, 02.118.021, 02.118.022, 02.123.012, 02. 123.013, 02.118.023, 02.121.052).

008.057 Kazan

Izvestiya Astronomicheskoj Observatorii im. Ehngel'gardta, *Kazan'*, No. 36 (N. I. Nefed'eva, 02.082.045; S. G. Valeev, 02.094.072; E. A. Vorob'ev, 02.042.014; I. A. Urasina, 02.045.007; Yu. G. Yusupov, 02.045.008; R. A. Bozula, 02.121.026; I. A. Dubyago, S. S. Tokhtas'ev, 02.125.011).

Trudy Kazanskoj Gorodskoj Astronomicheskoj Observatorii, *Kazan'*, No. 35 (N. A. Sakhibullin, 02.114.049; N. A. Sakhibullin, 02.133.020; E. E. Belyaeva, 02.064.034; M. I. Lavrov, 02.121.042; M. I. Lavrov, N. V. Lavrova, 02.121.043; Sh. T. Khabibullin, 02.094.083; Yu. A. Chikanov,

02.094.084; L. E. Nikonova, 02.102.024; L. E. Nikonova, 02.103.100; K. P. Matsukov, 02.103.108).

008.058 Kitt Peak

Kitt Peak National Observatory, Tucson, Arizona, and Cerro Tololo Inter-American Observatory, La Serena, Chile. N. U. Mayall.
Bull. American Astron. Soc. Vol. 1, 298 - 331 (1969). – Report 1968 – 1969.

Kitt Peak National Observatory, Contributions, Nos. 168 (C. R. Lynds, AJB 68, 134155), 169 (C. R. Lynds, AJB 68, 134154), 217 (W. C. Livingston, 02.071.087), 241 (M. J. S. Belton, D. M. Hunten, R. M. Goody, 02.093.037), 242 (D. M. Hunten, 02.097.063), 244 (R. J. W. Henry, M. B. McElroy, 02.091.043), 284 (C. R. Lynds, 02.141.200), 326 (R. Lynds, S. P. Maran, D. E. Trumbo, AJB 68, 134351), 328 (S. P. Maran, A. A. Penzias, J. Schraml, AJB 68, 134356), 348 (D. O. M. Jones, A. A. Hoag, AJB 68, 14139), 349 (A. U. Landolt, AJB 68, 12251), 350 (A. U. Landolt, AJB 68, 12612), 351 (A. U. Landolt, AJB 68, 12462), 352 (C. P. Gordon, AJB 68, 10507), 353 (E. F. Milone, AJB 68, 12263), 354 (R. A. R. Parker, 01.132.008), 355 (M. W. Werner, J. L. Pipher, Y. Terzian, J. R. Houck, 01.132.009), 356 (H. M. Dyck, AJB 68, 12428), 357 (R. J. W. Henry, AJB 68, 1679), 358 (A. Dalgarno, M. B. McElroy, M. H. Rees, J. C. G. Walker, 02.083.045), 359 (H. M. Johnson, J. C. Golson, AJB 68, 13554), 360 (R. J. W. Henry, R. E. Williams, 01.022.032), 361 (J. W. Chamberlain, 01.091.006), 362 (D. M. Hunten, AJB 68, 8122), 363 (A. U. Landolt, AJB 68, 12461), 364 (A. G. D. Philip, AJB 68, 10346), 365 (G. S. Mumford, 01.122.031), 366 (R. D. McClure, S. van den Bergh, AJB 68, 14562), 367 (A. K. Pierce, AJB 68, 66114), 368 (R. C. Henry, 01.114.025), 369 (H. A. Abt, 01.114.031), 370 (J. Bahng, 01.113.003), 372 (G. A. Chapman, N. R. Sheeley, Jr., AJB 68, 6417), 373 (M. B. McElroy, 01.093.007), 374 (L. Wallace, 01.093.081), 375 (D. S. Hall, 02.121.035), 376 (J. A. Graham, 01.126.016), 378 (R. D. McClure, 01.158.013), 379 (R. J. W. Henry, P. G. Burke, A. L. Sinfailam, 02.022.114), 380 (K. Janes, R. Lynds, 01.141.052), 381 (M. S. Snowden, R. H. Koch, 01.121.028), 383 (L. Binnendijk, 01.121.014), 384 (L. Binnendijk, 01.121.015), 385 (M. J. Price, 01.064.040), 386 (H. M. Johnson, J. C. Golson, 01.122.008), 387 (M. B. McElroy, D. F. Strobel, 01.093.008), 388 (F. C. Gillett, W. A. Stein, 01.133.008), 389 (A. G. D. Philip, 01.115.003), 391 (H.-Y. Chiu, V. Canuto, 01.141.081), 392 (R. C. Anderson, R. M. Fike, 02.051.035), 393 (D. B. Wood, 01.113.012), 394 (H. A. Abt, R. J. Dukes, W. B. Weaver, 02.119.004), 395 (D. L. Crawford, J. V. Barnes, 01.153.015), 396 (P. Léna, 01.072.031), 397 (R. Lynds, S. P. Maran, D. E. Trumbo, 01.141.057), 398 (R. C. Canfield, 02.071.008), 399 (C. L. Perry, 01.151.028), 400 (A. U. Landolt, 01.122.027), 401 (A. Stockton, 01.141.086), 402 (C. R. O'Dell, 01.153.011), 403 (Y. Kondo, G. E. McCluskey, Jr., 01.119.007), 404 (M. J. S. Belton, 02.099.005), 405 (G. S. Mumford, 02.121.038), 406 (M. B. McElroy, D. M. Hunten, 01.093.082), 407 (H.-Y. Chiu, V. Canuto, L. Fassio-Canuto, 01.141.022), 408 (W. C. Livingston, 01.080.006), 409 (L. Wallace, A. L. Broadfoot, 01.082.065), 410 (W. A. Stein, F. C. Gillett, 01.132.020), 411 (E. J. Weber, 01.074.037), 412 (A. G. D. Philip, 01.114.104), 413 (D. E. Shemansky, 02.022.071), 414 (M. Breger, 01.122.025), 415 (G. S. Mumford, W. Krzeminski, 02.121.011), 417 (M. Breger, 01.153.006), 418 (P. W. Hodge, R. W. Michie, 01.158.050), 419 (C. L. Perry, 01.113.026), 420 (H. A. Abt, S. G. Levy, 01.119.009), 421 (E. B. Jenkins, D. C. Morton, A. V. Sweigart, 02.093.006), 422 (A. L. Broadfoot, S. P. Maran,

02.022.069), 423 (D. E. Shemansky, N. P. Carleton, 02.022.070), 425 (G. Gonczi, F. Roddier, 02.071.007), 427 (D. M. Peterson, S. E. Strom, 02.064.016), 428 (R. M. Humphrys, 02.121.015), 429 (H. A. Abt, G. H. Smith, 02.112.003), 432 (D. L. Crawford, J. V. Barnes, 02.153.004), 434 (D. DuPuy, J. Schmitt, R. McClure, S. van den Bergh, R. Racine, 01.122.081), 442 (R. Lynds, 02.141.015), 443 (F. E. Stuart, 02.114.021), 445 (H. A. Abt, S. G. Levy, 02.119.002).

008.059 Kodaikanal

Kodaikanal Observatory, Kodaikanal. – Report for the year ending 1968 December 31.   M. K. V. Bappu. Quarterly Journ. Roy. Astron. Soc. Vol. 10, 317 - 323 (1969).

Kodaikanal Observatory, Bulletin No. 169 (M. K. V. Bappu, 02.075.006).

Kodaikanal Observatory. Reprints Nos. 43 (N. S. Roy, 02.033.047), 44 (C. V. Sastry, AJB 68, 8689), 45 (K. R. Sivaraman, 01.072.002), 46 (M. K. V. Bappu, N. Raghavan, 01.122.002).

008.060 Krim

Izvestiya Krymskoj Astrofizicheskoj Observatorii, Akademiya Nauk SSR, Tom (Vol.) 39.

008.061 Kyoto

Contributions from the Institute of Astrophysics and Kwasan Observatory, University of Kyoto, Nos. 175 (M. Matsumoto, AJB 68, 51179), 176 (M. Tadokoro, AJB 68, 145110), 177 (J. Tsujita, AJB 68, 51199), 178 (M. Matsumoto, 02.063.030), 179 (J. Kubota, 01.073.036), 180 (M. Matsumoto, 02.063.003), 181 (J. Tsujita, 02.063.004), 183 (S. Miyamoto, AJB 68, 83191).

008.062 Lake Traverse

Algonquin Radio Observatory, Lake Traverse, Ont., Canada. – Report from Solar Institute.   A. E. Covington. Solar Physics, Vol. 9, 242 - 245 (1969).

008.063 La Plata

La Plata: Observatorio Astronomico. (Astronomical Observatory, National University of La Plata). Inform. Bull. Southern Hemisphere, No. 14, p. 10 (1969).

Separata Astronomica, Observatorio Astronómico, La Plata, Argentina, N. 84 (C. Jaschek, 01.013.003), 85 (A. Feinstein, 01.153.014), 86 (L. Houziaux, A. Ringuelet-Kaswalder, AJB 68, 10452), 87 (M. Jaschek, M. L. Aguilar, 01.114.044), 88 (A. Feinstein, O. E. Ferrer, AJB 68, 14123), 89 (J. Sahade, 02.121.033).

008.064 Leiden

Broederstroom, Transvaal: Leiden Southern Station.

C. J. van Houten. Inform. Bull. Southern Hemisphere, No. 14, p. 23 (1969).

008.065 Lembang

Lembang, Java: Bosscha Observatory. J. Ibrahim. Inform. Bull. Southern Hemisphere, No. 14, p. 19 (1969).

008.066 Leningrad

Ephemerides of minor planets for 1970 (S. G. Makover, 02.098.015).

Trudy Astronomicheskoj Observatorii, (Transactions of the Astronomical Observatory), Vol. 26.

008.067 Lisbonne

Bulletin de l'Observatoire Astronomique de Lisbonne (Tapada), No. 16 (A. P. Botelheiro, A. Baptista dos Santos, 02.041.033).

008.068 Los Angeles

University of California, Los Angeles. Astronomical Papers, Vol. 8, Nos. 1 (H. W. Epps, AJB 68, 10602), 2 (J. Ross, L. Aller, AJB 68, 64112), 3 (L. H. Aller, S. J. Czyzak, AJB 68, 13203), 4 (L. H. Aller, G. Duffner, M. Dworetsky, D. Gudehus, S. Kilston, D. Leckrone, J. Montgomery, J. Oliver, E. Zimmerman, 02.106.031), 5 (J. B. Kaler, S. J. Czyzak, L. H. Aller, AJB 68, 13254), 6 (S. J. Peale, AJB 68, 9539), 7 (E. K. L. Upton, S. J. Little, M. M. Dworetsky, AJB 68, 54103), 8 (S. C. Wolff, L. V. Kuhi, D. Hayes, AJB 68, 104112), 9 (D. M. Popper, AJB 68, 12282), 10 (L. H. Aller, 01.132.044), 11 (L. H. Aller, 01.064.044), 12 (W. J. Kaufmann, III, 01.066.024), 14 (L. H. Aller, AJB 68, 13204), 15 (L. F. Smith, AJB 68, 14435).

008.069 Lund

The branch station of Lund Observatory. See Abstr. 032.017.

Meddelande från Lunds Observatorium, Ser. I, Nos. 241 (J. O. Stenflo, 01.034.005), 242 (J. M. Beckers, J. O. Stenflo, 01.034.006), 243 (A. Ardeberg, K. Särg, S. Wramdemark, 02.113.017), 244 (B. A. Lindblad, 02.105.190), 245 (G. Larsson-Leander, 02.121.024), 246 (B. Karlsson, 02.114.109).

008.070 Lvov

Tsirkulyar. Astronomicheskaya Observatoriya, L'vov. No. 43 (I. A. Klimishin, 02.064.014; I. A. Klimishin, 02.132.006; M. B. Girnyak, 02.123.009; V. V. Golovatyj, 02.123.010; A. T. Dul'tsev, 02.123.011; I. V. Shpychka, 02.122.037; I. V. Shpychka, 02.122.038; B. T. Babij, Yu. V. Fridel', 02.071.017; T. L. Mandrykina, 02.072.015; P. A.

Olijnyk, 02.072.016; G. G. Krajnyuk, A. A. Logvinenko, 02.033.006; T. V. Rad'o, I. I. Terebushko, 02.031.008; A. T. Dul'tsev, 02.096.004).

008.071 Madrid

Boletin Astronómico del Observatorio de Madrid, Instituto Geografico y Catastral, Seccion 2ª, Astronomia, Vol. 7, No. 3 (E. Gullón, 02.061.031; R. Carrasco, 02.098.019), No. 4 (E. Gullón, 02.075.007).

008.072 Manchester

Astronomical Contributions from the University of Manchester, Series II, Jodrell Bank Reprints, Nos. 373 (D. Williams, R. D. Davies, AJB 68, 6108), 375 (G. N. Taylor, R. D. S. Earnshaw, 02.083.056), 376 (A. G. Lyne, F. G. Smith, AJB 68, 134352), 379 (R. G. Conway, P. P. Kronberg, 01.141.002), 383 (F. G. Smith, AJB 68, 13344), 384 (B. H. Bland, AJB 68, 10526), 385 (A. G. Lyne, B. J. Rickett, AJB 68, 134354), 386 (R. D. Davies, 02.141.191), 388 (01.008.076), 389 (F. G. Smith, A. D. Bray, R. A. Porter, W. S. Torbitt, J. V. Jelley, 02.143.071), 390 (F. G. Smith, AJB 68, 134389), 391 (P. W. Horton, R. G. Conway, E. J. Daintree, 01.141.123), 392 (R. D. Davies, R. S. Booth, A. J. Wilson, AJB 68, 14468), 393 (B. J. Rickett, 01.141.010), 394 (J. G. Davies, G. C. Hunt, F. G. Smith, 01.141.005), 395 (P. C. Gregory, 01.084.005), 396 (R. R. Clark, F. G. Smith, 01.141.028), 397 (J. S. Beale, R. D. Davies, 01.158.003), 398 (R. D. Davies, J. E. B. Ponsonby, L. Pointon, G. de Jager, 01.033.005), 399 (B. Lovell, 01.122.051).

008.073 Manila

Solar work at Manila Observatory. – Report from Solar Institute.   J. J. Hennessey. Solar Physics, Vol. 9, 496 - 501 (1969).

008.074 Milano

Astronomical Observatory of Milan, Circular, Nos. 28 (E. Proverbio, 02.044.045), 29 (E. Proverbio, 02.045.032).

Contributi dell'Osservatorio Astronomico di Milano–Merate, Nuova Serie, Nos. 294 (A. Masani, P. Borghese, A. Ferrari, R. Gallino, AJB 68, 5547), 295 (M. Missana, AJB 68, 44101), 296 (F. Zagar, 02.044.001), 297 (E. Proverbio, 02.035.002), 298 (F. Chlistovsky, C. de Concini, 02.035.001), 299 (P. Broglia, G. Guerrero, 02.034.106), 300 (P. Galeotti, L. E. Pasinetti, 01.121.012), 301 (F. Chlistovsky, E. Proverbio, 02.035.003), 302 (E. Proverbio, 02.035.005), 303 (E. Proverbio, F. Chlistovsky, 02.035.007), 304 (E. Proverbio, F. Chlistovsky, 02.035.009), 305 (E. Proverbio, F. Carta, 02.044.002), 306 (E. Proverbio, A. Pensa, 02.044.003), 307 (M. Fracassini, L. E. Pasinetti, AJB 68, 9512), 308 (E. Proverbio, 02.044.004), 309 (M. Missana, 02.044.005), 310 (F. Chlistovsky, 02.035.012), 311 (E. Proverbio, F. Chlistovsky, 02.035.014), 312 (C. Castagnoli, A. Ferrari, R. Riganti, A. Masani, A. Martini, AJB 68, 13417), 313 (M. Fracassini, L. E. Pasinetti, AJB 68,

10506), 314 (M. Fracassini, L. E. Pasinetti, AJB 68, 12221), 315 (G. de Mottoni, 02.032.071).

008.075 Minneapolis

University of Minnesota, Minneapolis, Minnesota, Separate prints (W. J. Luyten, 02.112.017; W. J. Luyten, 02.112.018).

008.076 Mizusawa

Annual report of the geophysical observations made at the International Latitude Observatory of Mizusawa for the year 1967.   T. Okuda. Published by the International Latitude Observatory of Mizusawa, Japan. 22 pp. (1968).

Annual report of the meteorological observations made at the International Latitude Observatory of Mizusawa for the year 1968.   T. Okuda. Published by the International Latitude Observatory of Mizusawa, Japan. 50 pp. (1969).

Bulletins. Time Service of the Mizusawa Observatory, Vol. XII No. 1 - 12, 1967 (02.044.030).

Monthly Notes of the International Polar Motion Service, 1969 Nos. 5 - 10 (02.045.017).

Proceedings of the International Latitude Observatory of Mizusawa. No. 9 (T. Goto, 02.045.023; S. Goto, H. Okawa, H. Kitago, 02.32.059; K. Takahasi, E. Onodera, N. Kikuchi, 02.082.158; C. Sugawa, 02.045.024; M. Ooe, S. Abe, 02.032.060; T. Hara, K. Horiai, 02.035.034; I. Okamoto, G. Murakami, 02.044.028; G. Murakami, 02.044.029; Y. Goto, 02.045.025; S. Goto, T. Goto, 02.045.026; C. Sugawa, K. Hurukawa, H. Okawa, H. Kitago, 02.045.027; C. Sugawa, 02.081.028; G. Teleki, 02.04P.032)..

Publications of the International Latitude Observatory of Mizusawa, Vol. 6, No. 2 (K. Hurukawa, 02.045.019; S. Yumi, H. Ishii, K. Sato, 02.045.020; K. Yokoyama, 02.041.031; S. Takagi, 02.081.027; S. Takagi, G. Murakami, 02.045.021; T. Okuda, 02.045.022).

008.077 Mons

Faculté des Sciences de Mons, Département d'Astrophysique, Separate print (L. Houziaux, 02.041.039).

008.078 Moskva

Soobshcheniya Gosudarstvennogo Astronomicheskogo Instituta im. P. K. Shternberga. Izdatel'stvo Moskovskogo Universiteta. Nos. 158 (F. A. Tsitsin, 02.162.063; V. E. Yakimov, 02.154.015; P. N. Kholopov, 02.120.006; A. S. Sharov, 02.153.035; N. E. Kurochkin, G. A. Starikova, 02.118.034; G. A. Ponomareva, 02.032.043; O. D. Dokuchaeva, 02.036.011), 161 (A. M. Cherepaschuk, 02.121.007), 164 (N. D. Moiseev, 02.042.042; A. A. Orlov, 02.042.043).

008.079 **Mount Hamilton**

The University of California. **Contributions from the Lick Observatory**, Santa Cruz (Mount Hamilton), California, Nos. 205 (G. W. Preston, AJB 67, 10713; G. W. Preston, C. Sturch, AJB 67, 10714), 248 (M. F. Walker, G. E. Kron, AJB 67, 10375; M. F. Walker, AJB 67, 145143), 249 (E. J. Wampler, AJB 67, 104223), 250 (P. S. Conti, S. E. Strom, AJB 68, 14117), 251 (T. D. Kinman, E. Lamla, T. Ciurla, E. Harlan, C. A. Wirtanen, AJB 68, 134133), 252 (P. S. Conti, AJB 68, 10424), 253 (R. P. Kraft, M.-H. Demoulin, AJB 67, 13552), 254 (G. H. Herbig, AJB 68, 13116), 255 (E. J. Wampler, AJB 68, 134259), 256 (G. W. Preston, L. R. Cathey, AJB 68, 104165), 257 (K. Stępień, AJB 68, 13589), 258 (K. Stępień, AJB 68, 104175), 259 (L. R. Cathey, J. E. Hayes, AJB 68, 13514), 260 (G. H. Herbig, A. A. Boyarchuk, AJB 68, 12440), 261 (G. H. Herbig, AJB 68, 104141), 262 (M. F. Walker, G. Chincarini, AJB 68, 124110), 263 (K. Stępień, AJB 68, 12298), 264 (G. H. Herbig, 02.131.132), 265 (J. J. Monaghan, AJB 68, 5475), 267 (K. Stępień, AJB 68, 10716), 268 (P. S. Conti, S. E. Strom, AJB 68, 10425), 269 (M. F. Walker, G. E. Kron, AJB 68, 132102), 270 (G. H. Herbig, R. R. Zappala, AJB 68, 12442), 272 (P. S. Conti, G. Wallerstein, 01.114.013), 274 (J. J. Monaghan, AJB 68, 5476), 275 (R. L. Milton, P. S. Conti, AJB 68, 10476), 276 (G. H. Herbig, 01.122.069), 277 (P. S. Conti, 01.116.007) 278 (G. W. Preston, K. Stępień, AJB 68, 12490), 279 (J. J. Monaghan, AJB 68, 5477), 280 (S. J. Czyzak, M. F. Walker, L. H. Aller, 01.133.024), 281 (M. F. Walker, 01.153.004), 283 (A. R. Klemola, 01.124.105), 284 (R. L. Sears, A. E. Whitford, 01.153.007), 286 (J. S. Miller, AJB 68, 13334), 287 (G. H. Herbig, 02.122.179), 288 (E. J. Wampler, I. I. Papiashvili, AJB 68, 2382), 289 (L. M. Hobbs, 02.131.007), 290 (D. M. Pyper, 02.114.017), 291 (J. G. Bolton, T. D. Kinman, J. V. Wall, AJB 68, 13424), 293 (P. S. Conti, 01.114.037), 294 (R. P. Kraft, J. S. Miller, 01.142.024), 297 (E. K. Conklin, H. T. Howard, J. S. Miller, E. J. Wampler, 01.141.110), 303 (E. J. Wampler, J. D. Scargle, J. S. Miller, 02.141.014).

University of California, **Lick Observatory Bulletin**, Nos. 597 (R. P. Kraft, L. V. Kuhi, P. S. Kuhi, AJB 68, 10333), 599 (P. S. Conti, AJB 68, 11310), 600 (M. F. Walker, AJB 68, 145129; E. J. Wampler, AJB 68, 145131; T. D. Kinman, AJB 68, 145195), 601 (M. F. Walker, AJB 68, 145235), 604 (J. S. Miller, E. J. Wampler, 01.141.039).

008.080 **Mount Stromlo**

Mount Stromlo and Siding Spring Observatories. – Report for year ending 1968 December 31. O. J. Eggen. Quarterly Journ. Roy. Astron. Soc. Vol. 10, 324 - 334 (1969).

008.081 **Mount Wilson**

From Mount Wilson and Palomar Observatories. Publ. Astron. Soc. Pacific, Vol. 81, 453 - 454 (1969).

008.082 **München**

Max-Planck-Institut für Physik und Astrophysik, München, Separate prints (D. Pfirch, K. Schindler, 02.022.018; F. Meyer, H. U. Schmidt, 02.072.089).

008.083 **Naini Tal**

Uttar Pradesh State Observatory, Naini Tal. Reprints Nos. 24 (P. P. Saxena, AJB 68, 7461), 25 (J. B. Srivastava, C. D. Kandpal, AJB 68, 12297), 26 (M. M. Pant, V. P. Gaur, M. C. Pande, AJB 68, 12485).

008.084 **Nashville**

The Arthur J. Dyer Observatory, Vanderbilt University, Nashville, Tennessee, **Reprints,** Nos. 39 (D. S. Hall, AJB 67, 12548), 40 (J. E. Hayes, A. M. Heiser, AJB 68, 12329), 41 (R. C. Barnes, D. S. Hall, R. H. Hardie, AJB 68, 12201), 42 (C. M. Snell, A. M. Heiser, AJB 68, 7335), 43 (A. M. Heiser, AJB 68, 10325), 44 (D. S. Hall, AJB 68, 12225).

008.085 **Neuchâtel**

Rapport annuel du Directeur sur l'exercice 1968 et Rapport sur le Concours chronométrique 1968. J. Bonanomi, G. Udriet. Observatoire Cantonal de Neuchâtel. 36 pp. (1969).

Observatoire de Neuchâtel. Bulletin, Série B, 1969 Mai – Octobre (02.044.042), Série D, 1969 Avril – Octobre (02.044.043).

008.086 **Ottawa**

Dominion Observatory, Ottawa, Ontario. M. W. Grey. Journ. Roy. Astron. Soc. Canada, Vol. 63, 314 - 315 (1969).

Contributions from the Dominion Observatory, Ottawa, Vol. 4, No. 29 (C. S. Beals, 02.105.198; I. Halliday, 02.105.199; J. T. Wilson, 02.105.200); Vol. 8, Nos. 23 (P. B. Robertson, M. R. Dence, M. A. Vos, 02.105.159), 24 (T. E. Bunch, A. J. Cohen, M. R. Dence, 02.105.160), 25 (M. R. Dence, M. J. S. Innes, P. B. Robertson, 02.105.161), 26 (M. R. Dence, 02.105.162).

Contributions from the Dominion Observatory, Ottawa, Nos. 249 (P. B. Robertson, 02.105.029), 252 (P. Andrieux, J. F. Clark, 02.105.201), 255 (C. H. Costain, J. D. Lacey, R. S. Roger, 02.033.057), 262 (R. S. Roger, C. H. Costain, J. D. Lacey, 01.141.133), 264 (F. C. Taylor, M. R. Dence, 01.105.024), I. Halliday, 01.101.008), 270 (J. J. Labrecque, 01.031.009).

Publications of the Dominion Observatory, Ottawa, Vol. 25, No. 13 (C. H. Costain, J. D. Lacey, R. S. Roger, 02.033.009).

008.087 **Oulu**

Aarne Karjaleinen Observatory, University of Oulu, Finland. Publications Nos. 13 (T. Jaakkola, 02.092.006), 14 (T. Pikkarainen, 02.094.235).

008.088 Oxford

Department of Astrophysics,University of Oxford. Report for the year ending 1968 December 31. D. E. Blackwell. Quarterly Journ. Roy. Astron. Soc. Vol. 10, 256 - 260 (1969).

Communications from the University Observatory, Oxford, Nos. 105 (J. V. Peach, AJB 68, 64105), 106 (D. L. Lambert, E. A. Mallia, AJB 68, 6483), 107 (M. F. Ingham, AJB 68, 7429), 108 (M. F. Ingham, R. F. Jameson, AJB 68, 7430), 109 (D. L. Lambert, B. E. J. Pagel, AJB 68, 5337), 110 (B. Warner, AJB 68, 16129), 111 (D. L. Lambert, E. A. Mallia, B. Warner, 01.071.001), 112 (M. S. Hockey, 01.116.001).

008.089 Paris

Bureau International de l'Heure, Circulaires B/C Nos.161 - 164 (02.045.016).

Bureau International de l'Heure, Circulaires, D34 - D37 (02.044.024).

Publications de l'Observatoire de Paris. Satellites Artificiels, Fasc. 22 (P. Muller, 02.055.006), 23 (P. Muller, 02.055.007), 24 (P. Muller, 02.055.008).

Publications de l'Observatoire de Paris. Separate print, (P. Muller, C. Meyer, 02.118.025).

008.090 Pereyra Iraola

Pereyra Iraola: Instituto Argentino de Radioastronomia. Inform. Bull. Southern Hemisphere, No. 14, p. 10 (1969).

008.091 Perth

Perth Observatory. – Report for the year ending 1969 June 30. B. J. Harris. Quarterly Journ. Roy. Astron. Soc. Vol. 10, 335 (1969).

008.092 Porto

Observatorio Astronómico da Universidade do Porto, Monte da Virgem – Vila Nova de Gaia (Portugal). **Publiçacões do Observatorio Astronómico da Faculdade de Ciências do Porto,** Nos. 22 (M. Barros, J. Osório, R. A. Vieira, 02.045.029), 23 (R. S. de Sousa Nunes, D. Appelt, 02.032.066).

008.093 Porto Alegre

Porto Alegre: Instituto de Astronomia da Universidade do Rio Grande do Sul. (Astronomical Institute of the Federal University of Rio Grande do Sul). J. C. Haertel. Inform. Bull. Southern Hemisphere, No. 14, p. 12 (1969).

008.094 Potsdam

Institut für Sternphysik der Deutschen Akademie der Wissenschaften zu Berlin; **Astrophysikalisches Observatorium Potsdam und Sternwarte Sonneberg.** J. Wempe. Monatsber. Deutsch. Akad. Wiss. Berlin, Band 11, 762 - 766 (1969). – Jahresbericht 1968.

Mitteilungen des Astrophysikalischen Observatoriums Potsdam, Nos. 129 (H. Domke, 02.063.018), 130 (E. Gerth, 02.021.022), 131 (E. Gerth, 02.021.023), 132 (G. Scholz, 02.074.007), 133 (G. Scholz, 02.074.008).

Mitteilungen des Astrophysikalischen Observatoriums Potsdam, Nos. 314 - 315 (H. Daene, 02.077.024), 317 - 318 (H. Daene, 02.077.051).

Geodätisches Institut der Deutschen Akademie der Wissenschaften zu Berlin. H. Kautzleben. Monatsber. Deutsch. Akad. Wiss. Berlin, Band 11, 770 - 772 (1969). – Jahresbericht 1968.

Arbeiten aus dem Geodätischen Institut Potsdam (Deutsche Akademie der Wissenschaften zu Berlin), Nos. 25 (E. Büschmann, 02.044.023), 27 (V. Kroitzsch, 02.035.031).

Mitteilungen des Geodätischen Instituts Potsdam, No. 111 (R. Stecher, 02.035.032).

008.095 Praha

Académie Tchécoslovaque des Sciences, Institut Astronomique, **Station de l'Heure à Prague,** Série 5, Nos. 1 - 3 (02.044.025).

Astronomical Institute of the Technical University, Praha – ČSSR. Publikace, Nos. 30 (E. Buchar, AJB 68, 2296), 31 (J. Kabeláč, 02.046.007), 32 (J. Kabeláč, 02.055.011).

Memoirs and Observations of the Czechoslovak Astronomical Society of the Czechoslovak Academy of Sciences, *Praha,* Nos. 12 (Z. Kvíz, F. Žďárský, 02.104.047), 13 (B. Onderlička, M. Vetešník, 02.122.171).

008.096 Preston

Jeremiah Horrocks and Wilfred Hall Observatories, Preston. – Report for the year ending 1968 December 31. V. Barocas. Quarterly Journ. Roy. Astron. Soc. Vol. 10, 261 - 262 (1969).

008.097 Pretoria

Pretoria: Radcliffe Observatory. P. J. Andrews. Inform. Bull. Southern Hemisphere, No. 14, p. 28 - 31 (1969).

Radcliffe Observatory, Pretoria. – Report for the year ending 1969 March 31. A. D. Thackeray. Quarterly Journ. Roy. Astron. Soc. Vol. 10, 263 - 268 (1969).

Communications from the Radcliffe Observatory, Pretoria, No. 99 (A. D. Thackeray, B. Emerson, 01.119.001).

Radcliffe Observatory, *Pretoria*, Reprints Nos. 67 (A. D. Thackeray, AJB 68, 4127), 68 (A. D. Thackeray, AJB 68, 12662), 69 (D. H. P. Jones, 01.121.006), 70 (A. D. Thackeray, 01.122.022), 71 (A. D. Thackeray, 01.122.066), 72 (A. D. Thackeray, 01.124.004).

008.098 Pulkovo

Trudy Glavnoj Astronomicheskoj Observatorii v Pulkove (R. S. Gnevysheva, 02.075.001, 02.075.008).

008.099 Pulsnitz

Veröffentlichungen der Sternwarte Pulsnitz (Sachsen), Nos. 5 (J. Classen, 02.094.234), 6 (J. Classen, 02.105.192).

008.100 Richmond Hill

David Dunlap Observatory, University of Toronto, Richmond Hill, Ontario. D. A. MacRae. Journ. Roy. Astron. Soc. Canada, Vol. 63, 217 - 218 (1969).

Communications from the David Dunlap Observatory, University of Toronto, Richmond Hill, Ontario, Canada, Nos. 209 (L. D. Braun, J. L. Yen, AJB 68, 13433), 210 (W. A. Hiltner, R. F. Garrison, R. E. Schild, 02.114.004), 211 (R. C. Roeder, R. T. Verreault, AJB 68, 134210), 213 (I. J. Sackmann, S. P. S. Anand, 01.065.010), 214 (S. P. S. Anand, O. V. Dubas, AJB 68, 5404), 216 (D. L. DuPuy, AJB 68, 14529), 217 (J. D. Fernie, AJB 68, 12319), 218 (H. B. Sawyer Hogg, 02.120.010), 219 (E. R. Seaquist, AJB 68, 12647), 220 (J. D. Fernie, 01.113.037), 221 (D. Crampton, J. D. Fernie, 01.151.010), 222 (J. D. Fernie, 01.122.029), 226 (J. L. Yen, B. Zuckerman, P. Palmer, H. Penfield, 01.131.046), 227 (R. H. Chambers, R. C. Roeder, 01.162.014), 228 (M. J. Clement, 01.065.061), 229 (J. A. Roberts, G. G. Fahlman, 01.141.114), 230 (D. DuPuy, J. Schmitt, R. McClure, S. van den Bergh, R. Racine, 01.122.081).

Publications of the David Dunlap Observatory, University of Toronto, Richmond Hill, Toronto, Canada, Vol. 3, No. 1 (C. M. Coutts, H. S. Hogg, 02.154.017).

008.101 Rio de Janeiro

Contribuiçoĕs do Observatório do Valongo, Universidade Federal do Rio de Janeiro, Série I, Nos. 4 - 8 (J. F. Caria Caldeira, 02.075.027).

Contribuiçoĕs do Observatório do Valongo, Universidade Federal do Rio de Janeiro, Série II, Nos. 6 - 8 (J. A. Buarque de Nazareth, 02.099.071).

Contribuiçoĕs do Observatório do Valongo, Universidade Federal do Rio de Janeiro, Série III, Nos. 5 - 9 (H. de Souza, L. E. da Silva Machado, 02.096.018).

Rio de Janeiro: Observatorio Nacional. (National Observatory). L. Muniz Barreto. Inform. Bull. Southern Hemisphere, No. 14, p. 12 - 14 (1969).

008.102 Rochester

C. E. Kenneth Mees Observatory, University of Rochester, Rochester, N. Y., **Reprints**, Nos. 19 (G. S. Kutter, M. P. Savedoff, 01.065.060), 20 (P. Murdin, 02.131.009).

008.103 Roma

Osservatorio Astronomico di Roma, Monte Mario — Monte Porzio — Stazione Astrofisica sul Gran Sasso. **Contributi Scientifici**, Ser. III, Nos. 77 (G. Caprioli, A. Palma, AJB 68, 3103), 78 (P. Giannone, M. A. Giannuzzi, AJB 68, 5527), 79 (P. Giannone, N. Virgopia, AJB 68, 5528), 80 (M. Cimino, A. Cacciani, N. Sopranzi, AJB 68, 2314), 81 (T. Fortini, M. Torelli, 02.078.029), 82 (A. Palma, F. Smriglio, 02.044.007), 83 (G. Caprioli, A. Orlando, A. P. Massangioli, 02.035.015), 84 (G. Caprioli, M. Mattei, 02.034.002), 85 (F. Smriglio, 02.034.001), 86 (P. Giannone, M. A. Giannuzzi, AJB 68, 5526), 87 (V. Castellani, P. Giannone, A. Renzini, AJB 68, 5426), 88 (P. Giannone, M. A. Giannuzzi, 01.118.003), 89 (V. Castellani, P. Giannone, A. Renzini, 01.154.005).

Monthly Bulletin, Osservatorio Astronomico di Roma, Nos. 139 - 144 (M. Cimino, 02.075.013).

Photographic Journal of the Sun. Osservatorio Astronomico di Roma, Nos. 20 - 25 (M. Cimino, 02.075.014).

008.104 Sacramento Peak

Sacramento Peak Observatory. Air Force Cambridge Research Laboratories, Sunspot, New Mexico, Separate prints (Y. Öhman, 02.073.091; G. W. Simon, AJB 67 6495; F. Q. Orrall, AJB 68, 6207; J. M. Beckers, 02.072.096; O. R. White, G. W. Simon, AJB 68, 64141; C. L. Hyder, AJB 68, 6656; K. Saito, C. L. Hyder, AJB 68, 6207; J. M. Pasachoff, R. W. Noyes, J. M. Beckers, AJB 68, 66113; R. C. Altrock, AJB 68, 6404; J. M. Beckers, AJB 68, 64148; C. L. Hyder, H. A. Mauter, R. L. Shutt, AJB 68, 6209; G. W. Simon, N. O. Weiss, AJB 68, 64160; I. Elliott, 01.071.006; J. M. Beckers, J. O. Stenflo, 01.034.006; R. C. Altrock, 01.071.020; J. M. Beckers, E. H. Schröter, 01.072.013; S. Musman, 01.071.030; R. C. Altrock, 01.071.034; J. M. Beckers, P. E. Tallant, 01.072.033; C. L. Hyder, 01.080.003; R. L. Parnell, J. M. Beckers, 02.071.021; J. M. Beckers, R. L. Parnell, 02.071.022; J. M. Beckers, 01.061.026).

008.105 San Fernando

Instituto y Observatorio de Marina, San Fernando, Separate print (M. López Palacios, 02.158.025).

008.106 Santiago

Santiago: Departamento de Astronomia, Universidad de Chile. (Astronomy Department, University of Chile). Inform. Bull. Southern Hemisphere, No. 14, p. 16 - 18 (1969). — A) National Astronomical Observatory, Cerro Calán; B) Cerro El Roble Astronomical Station; C) Maipú Radioastronomical Observatory.

Departamento de Astronomia, Universidad de Chile, Facultad de Ciencias Fisicas y Matematicas. Observatorio Astronomico Nacional Cerro Calan, Santiago de Chile, **Publicaciones,** Nos. 5 (J. Stock, S. Tapia, 02.113.008; E. Mendoza V., H. Moreno, J. Stock, 02.082.034; A. Gutiérrez-Moreno, H. Moreno, 02.152.005), 7 (M. Isakson, 02.097.014; E. E. Mendoza V., 02.113.007).

Universidad de Chile. Departamento de Astronomia, *Santiago.* **Separata** 3 (C. Anguita, G. Carrasco, P. Loyola, V. N. Šiškina, M. S. Zverev, 02.041.043), 4 (J. Stock, 02.113.043).

Catholic Univ. Santiago de Chile, Institute of Astrophysics, Separate print (F. Gun-Bayer, 02.085.008).

008.107 Sao José

São José dos Campos: Observatorio Astronomico do Instituto Tecnologico de Aeronautica. (Astronomical Observatory of the Aeronautical Technical Institute). Inform. Bull. Southern Hemisphere, No. 14, p. 14 - 15 (1969).

008.108 Sao Paulo

São Paulo: Instituto Astronomico e Geofisico. (Astronomical and Geophysical Institute). Inform. Bull. Southern Hemisphere, No. 14, 15 (1969).

São Paulo: Centro de Radioastronomia e Astrofisica da Universidade Mackenzie. (Center of Radioastronomy and Astrophysics, Mackenzie University). Inform. Bull. Southern Hemisphere, No. 14, p. 15 (1969).

008.109 Sendai

Sendai Astronomiaj Raportoj, Nos. 107 (Y. Shibata, 01.065.021), 108 (K. Suda, 02.065.068), 109 (S. Kikuchi, AJB 68, 4150), 110 (K. Takakubo, AJB 68, 13352), 111 (M. Takeuti, 01.064.010).

008.110 Skalnaté Pleso

Contributions of the Astronomical Observatory Skalnaté Pleso, Vol. 4 (L. Pajdušáková, 02.072.006; J. Štohl, 02.104.004; J. Štohl, 02.104.005; R. Bajcár, 02.114.008; R. Bajcár, I. Bajcárová, 02.065.007).

008.111 Sonneberg

Institut für Sternphysik, der Deutschen Akademie der Wissenschaften zu Berlin; **Astrophysikalisches Observatorium Potsdam und Sternwarte Sonneberg.** J. Wempe. Monatsber. Deutsch. Akad. Wiss. Berlin, Band 11, 762 - 766 (1969). – Jahresbericht 1968.

Mitteilungen über Veränderliche Sterne, *Sonneberg,* Band 5; Heft 5 (W. Wenzel, 02.122.170; I. Meinunger, 02.123.048; H.-J. Blasberg, 02.123.049; W. Götz, 02.113.057; G. A. Richter, 02.123.050; E. Splittgerber, 02.123.051; H. Geßner, 02.123.056).

008.112 St. Andrews

Communications from the University Observatory, St. Andrews, No. 5 (I. G. van Breda, 01.031.016).

University Observatory, St. Andrews, Reprints, Nos. 31 (P. W. Hill, 01.112.008), 32 (P. W. Hill, 01.122.107), 33 (P. W. Hill, 01.114.100), 34 (R. W. Hilditch, 02.121.060).

008.113 Sydney

Sydney: Sydney Observatory. Inform. Bull. Southern Hemisphere, No. 14, p. 11 (1969).

Division of Radiophysics, C.S.I.R.O., Epping, New South Wales, Australia. Separate prints (J. W. Brooks, 01.033.036; R. A. Batchelor, J. W. Brooks, M. W. Sinclair, 01.033.035; R. T. Stewart, B. Hardwick, 01.077.038; K. Kai, 01.077.039; D. J. McLean, 01.077.040; K. Kai, 01.077.041; N. R. Labrum, 01.077.042; B. J. Robinson, W. M. Goss, R. N. Manchester, 01.131.103; R. N. Manchester, W. M. Goss. B. J. Robinson, 01.131.104; V. Radhakrishnan, J. D. Murray, 01.131.105; O. B. Slee, 01.141.205; J. P. Wild, 01.073.061; J. P. Wild, 02.074.084; P. M. McCulloch, P. A. Hamilton, M. M. Komesaroff, D. J. Cooke, 01.141.206; D. K. Milne, 02.141.081; R. X. McGee, R. A. Batchelor, J. W. Brooks, M. W. Sinclair, 02.132.016; R. M. Price, 02.157.005; T. W. Cole, 02.033.053; R. A. Batchelor, 02.033.054; D. J. Cole, 02.033.055; B. MacA. Thomas, H. C. Minnett, Vu The Bao, 02.033.056).

Radiophysics Laboratory, C.S.I.R.O. Epping, New South Wales, Australia, Separate prints (P. G. Mezger, B. J. Robinson, AJB 68, 133116; J. G. Bolton, 01.141.093; E. K. Bigg, 01.099.012; V. Radhakrishnan, D. J. Cooke, M. M. Komesaroff, D. Morris, 01.141.015; W. M. Goss, B. J. Robinson, R. N. Manchester, 01.132.059; V. Radhakrishnan, R. N. Manchester, 01.141.068; J. G. Bolton, J. V. Wall, 01.141.156; V. Radhakrishnan, D. J. Cooke, 01.141.158; R. N. Manchester, W. M. Goss, B. J. Robinson, 02.131.016; D. K. Milne, T. L. Wilson, F. F. Gardner, P. G. Mezger, 02.131.033; K. V. Sheridan, 02.033.052).

008.114 Tartu

The Astronomical Observatory of the University of Tartu (Yuryev, Dorpat), 1805 - 1948. An essay on its history. See Abstr. 004.031.

Das neue Tartuer Astrophysikalische Wilhelm-Struve-Observatorium. P. Müürsepp. Sterne, 45. Jahrgang, 161 - 165 (1969).

Eesti NSV Teaduste Akadeemia, W. Struve nimelise, Tartu Astrofüüsika Observatooriumi, Publikatsioonid, Köide (Vol.) 37.

Tartu Astronoomia Observatorium, Teated Nos. 20 (T. Feklistova, 02.074.055), 21 (T. Kipper, 02.064.054), 22 (J. I. Einasto, 02.158.035), 23 (I. B. Pustylnick, 02.064.060; I. B. Pustylnick, 02.121.070).

008.115 Tautenburg

Karl-Schwarzschild-Observatorium Tautenburg

der Deutschen Akademie der Wissenschaften zu Berlin.
N. Richter.
Monatsber. Deutsch. Akad. Wiss. Berlin, Band 11, 769 - 770
(1969). – Jahresbericht 1968.

008.116 Teide

Observatorio Astronomico del Teide, Tenerife
«Islas Canarias». Publicación Nos. 1 (J. M. Torroja, F. Sánchez, 01.082.074), 2 (J. Casanovas, 01.082.073).

008.117 Thessaloniki

Astronomical Department, University of Thessaloniki. – Annual report 1967.
G. Contopoulos.
Annual Rep. Astron. Inst. Greece 1967, p. 10 - 12 (1968).

Astronomical Department, University of Thessaloniki. – Annual report 1968.
Annual Rep. Astron. Inst. Greece 1968, p. 9 - 11 (1969).

Department of Geodetic Astronomy, University of Thessaloniki. – Annual report 1967.
L. N. Mavridis.
Annual Rep. Astron. Inst. Greece 1967, p. 14 - 15 (1968).

Department of Geodetic Astronomy, University of Thessaloniki. – Annual report 1968.
L. N. Mavridis.
Annual Rep. Astron. Inst. Greece 1968, p. 15 - 16 (1969).

008.118 Tokyo

Annals of the Tokyo Astronomical Observatory, University of Tokyo, Second Series, Vol. 11, No. 3 (S. Nishimura, K. Ichimura, K. Osawa, 02.122.001; K. Osawa, K. Ichimura, M. Shimizu, 02.122.002; S. Nishimura, K. Ichimura, K. Osawa, K. Nariai, 02.122.003; S. Nishimura, E. Watanabe, 02.122.004).

Bulletin of Solar Phenomena, Tokyo Astronomical observatory, Vol. 20, No. 5 (02.074.006); Vol. 21, Nos. 1 - 2 (02.075.026).

Contributions from the Department of Astronomy, University of Tokyo, Nos. 102 (Y. Fujita, 02.114.113), 103 (Y. Yamashita, AJB 68, 12615), 104 (S. Kikuchi, AJB 68, 104147), 105 (S. Kato, AJB 68, 4346), 106 (G.-i. Hori, AJB 68, 4342), 107 (K. Tanaka, AJB 68, 66152), 108 (W. Unno, 01.065.031), 109 (Y. Fujita, 02.114.114), 110 (M.-a. Kondo, 02.162.003), 111 (M. Hirai, 02.114.001), 112 (Y. Yamashita, 02.114.002).

Data Report of Hydrographic Observations. Series of Astronomy and Geodesy, Maritime Safety Agency, Tokyo, Japan, No. 4 (02.096.014).

Time and Latitude Bulletins, Tokyo Astronomical Observatory, Vol. 43, Nos. 1 - 7 (02.044.027).

Tokyo Astronomical Bulletin, Tokyo Astronomical Observatory, Second Series, Nos. 195 (02.142.065), 196 (G. Ishida, M. Kondo, S. Nishimura, K. Osawa, K. Ichimura).

Tokyo Astronomical Observatory, Reprints Nos. 357 (Y. Kozai, 02.042.002), 358 (Y. Uchida, 02.073.014), 359 (M. Fujimoto, M. Miyamoto, 02.131.013), 360 (N. Kaigu, M. Morimoto, 02.131.014), 361 (Y. Shiomi, 01.077.002), 362 (M. Saito, 02.151.023), 363 (Y. Kozai, 02.042.025), 364 (T. Takakura, 02.076.042), 365 (T. Takakura, 02.077.044).

008.119 Tonantzintla

Boletin de los Observatorios de Tonantzintla y Tacubaya, Universidad Nacional Autonoma de Mexico, Vol. 5, No. 31 (M. Peimbert, R. Costero, 02.132.039; G. Haro, E. Chavira, 02.122.144; E. Parsamian, E. Chavira, 02.122.145; G. Haro, E. Parsamian, 02.122.146; G. Haro, E. Parsamian, 02.122.147; L. Carrasco, M. E. Méndez, 02.132.040; E. E. Mendoza, V., 02.113.056).

008.120 Torino

Contributi dell'Osservatorio Astronomico di Torino, (Pino Torinese), Nos. 48 (F. Job, T. Tamburini, 01.118.020), 49 (C. Egidi, N. Missana, F. Mussino, 02.044.006).

Osservatorio Astronomico di Torino, Studi Monografici, (Pino Torinese), No. 6 (M. G. Fracastoro, 02.091.048).

008.121 Trieste

Vita e attività dell'Osservatorio di Trieste nel 1969.
M. Hack.
Pubbl. Oss. Astron. Trieste, No. 407, p. 2 - 5 (1969).

Pubblicazione Osservatorio Astronomico di Trieste, Nos. 383 (A. Abrami, P. Zlobec, AJB 68, 63), 388 (C. Aydin, AJB 68, 104118), 389 (M. Hack, R. Stalio, 01.114.050), 392 (M. Hack, P. Stenner, 01.122.023), 393 (B. Cester, 01.121.031), 394 (R. Faraggiana, 01.114.066), 395 (P. Zlobec, 02.075.030, 02.075.031), 397 (G. Sedmak, 02.113.012), 398 (P. Zlobec, 02.075.030, 02.075.031), 399 (P. Zlobec, 02.072.097), 400 (R. Faraggiana, M. Hack, 01.122.065), 401 (A. Abrami, 02.008.127), 407 (02.047.037), 408 (P. Zlobec, 02.075.030, 02.075.031).

Solar radio astronomy research at the Trieste Astronomical Observatory. – Report from Solar Institute.
A. Abrami.
Solar Physics, Vol. 9, 502 - 505 (1969).

008.122 Tucson

Communications of the Lunar and Planetary Laboratory, Tucson, The University of Arizona, Vol. 6, Nos. 100 (G. P. Kuiper, F. F. Forbes, 02.093.020), 101 (G. P. Kuiper, 02.093.021), 102/I (G. P. Kuiper, J. W. Fountain, S. M. Larson, 02.093.022), 102/II (W. K. Hartmann, 02.093.023), 102/III (J. Fountain, S. Larson, 02.093.024), 103 (D. P. Cruikshank, A. B. Binder, 02.099.036), 104 (R. I. Mitchell, 02.097.052), Vol. 8, Part 2, Nos. 133 (R. Fryer, C. Titulaer, 02.094.006), 134 (C. Titulaer, 02.094.007)), 135 (S. F. Pellicori, 02.094.008), 136 (W. K. Hartmann, 02.094.009), 137 (W. K. Hartmann, 02.098.003), 138 (F. J. Low, B. J. Smith, 02.122.009), 139 (H. L. Johnson, 02.114.007), 140 (S. F.

Pellicori, 02.034.004), 141 (F. F. Forbes, R. I. Mitchell, 02.113.001), Vol. 9, Part 1, Nos. 160 (L. A. Bijl, G. P. Kuiper, D. P. Cruikshank, 02.071.010), 161(L. A. Bijl, G. P. Kuiper, D. P. Cruikshank, 02.071.011), 162 (G. P. Kuiper, A. B. Thomson, L. A. Bijl, D. C. Benner, 02.071.012), Part 2, Nos. 163 (L. A. Bijl, G. P. Kuiper, D. P. Cruikshank, 02.071.084), 164 (L. A. Bijl, G. P. Kuiper, D. P. Cruikshank, 02.071.085), 165 (L. A. Bijl, G. P. Kuiper, D. P. Cruikshank, 02.071.086).

008.123 Uccle

Bulletin Astronomique, Observatoire Royal de Belgique, Vol. 6, No. 8 (S. Arend, H. Debehogne, G. Roland, 02.098.021; H. Debehogne, G. Roland, 02.103.127; J. Denoyelle, 02.098.022; C. Delys, R. Gonze, 02.077.060; A. G. Velghe, 02.095.006; J. Dommanget, 02.118.039).

Observatoire Royal de Belgique (Koninklijke Sterrenwacht van Belgie), **Communications** (Mededelingen), Série B, Nos. 2(C. Delys, R. Gonze, AJB 68, 63), 35 (P. Paquet, AJB 68, 4160), 37 (J. Dommanget, 01.117.011).

008.124 Uppsala

Uppsala Astronomiska Observatorium, Meddelande Nos. 162 (J.-E. Solheim, R. Stabell, 01.131.054), 164 (L. Häggkvist, T. Oja, 01.113.019), 165 (S. Jaidee, G. Lyngå, 02.113.058), 166 (E. Holmberg, 02.158.091), 167 (L. Häggkvist, T. Oja, 02.113.018), 168 (C. Roslund, 02.113.016), 169 (C. Roslund, 02.122.172).

008.125 Victoria

Dominion Astrophysical Observatory Victoria, B. C. Report for the period 1968 January 1 to 1969 March 31. K. O. Wright. Quarterly Journ. Roy. Astron. Soc. Vol. 10, 269 - 281 (1969).

Dominion Astrophysical Observatory, Victoria, B.C. A. H. Batten. Journ. Roy. Astron. Soc. Canada, Vol. 63, 267 - 268 (1969).

Contributions from the Dominion Astrophysical Observatory, Victoria, B.C., Nos. 127 (K. O. Wright, S. J. Larson, 02.121.037), 128 (J. B. Hutchings, 02.114.045), 129 (J. B. Hutchings, 02.124.102), 131 (D. Crampton, J. D. Fernie, 01.151.010), 132 (J. B. Hutchings, 01.064.051), 134 (G. Hill, C. L. Perry, 02.153.006), 136 (G. A. H. Walker, S. C. Morris, P. F. Younger, 01.122.030), 140 (G. A. H. Walker, J. B. Hutchings, P. F. Younger, 02.113.035), 141 (G. Hill, C. L. Perry, 02.153.011).

Publications of the Dominion Astrophysical Observatory, Victoria, B.C., Vol. 13, Nos. 11 (K. O. Wright, K. H. Hesse, 02.121.091), 12 (G. Hill, 02.119.021).

008.126 Vilnius

Astronomijos Observatorijos, Biuletenis (Bulletin of the Vilnius Astronomical Observatory), No. 25 (R. Bartkus, 02.065.082; R. Bartkus, 02.065.083; G. Kakaras, 02.117.035).

008.127 Warszawa

Astronomical Observatory of the Warsaw University. W. Zonn. Postępy Astron., Vol. 17, 397 - 398 (1969). In Polish.

Publications of the Astronomical Observatory of the Warsaw University, Vol. 15 (M. Karpowicz, K. Rudnicki, 02.125.007).

Publikacje Działu Geodezji Wyższej i Astronomii, Geodezyjnej Zg. PAN. No. 12 (L. Cichowicz, 02.003.001).

Warsaw University Observatory and Astronomical Institute, Polish Academy of Sciences, Reprints Nos. 276 (S. Grzędzielski, 02.064.005), 277 (S. M. Ruciński, 02.117.027), 278 (R. Brukalska, S. M. Ruciński, J. Smak, K. Stępień, 02.121.049), 279 (J. Smak, 02.121.050), 280 (R. Brukalska, 02.161.011).

008.128 Washington

Publications of the United States Naval Observatory, *Washington,* Second Series, Vol. 19, Part II (A. N. Adams, D. K. Scott, 02.041.016), Vol. 22, Part I (R. L. Walker, Jr., 02.118.035).

United States Naval Observatory, *Washington,* Circular, No. 126 (J. S. Duncombe, 02.079.105).

008.129 Waterloo

Contributions of the University of Waterloo Observatory, Nos. 1 (M. P. FitzGerald, J. F. Heard, AJB 68, 615), 2 (M. P. FitzGerald, 02.115.009), 4 (M. P. FitzGerald, N. Houk, 02.114.099).

008.130 Wellington

Report of the Carter Observatory Board for the year ended 1969, March 31. M. A. F. Barnett, I. L. Thomsen. Astron. Bull. Carter Obs. Wellington, No. 71, 6 pp. (1969).

008.131 Wien

Annalen der Universitäts-Sternwarte Wien, Band 28, Nos. 3 (M. G. Firneis, 02.021.008), 4 (P. L. Fischer, 02.122.100), 5 (J. Meurers, F. Prochazka, 02.112.008).

Mitteilungen der Universitäts-Sternwarte Wien, Band 14, Nos. 7 (J. Hopmann, AJB 68, 83112), 8 (J. Hopmann, 02.094.227), 9 (K. D. Rakosch, AJB 68, 10714), 10 (J. Hopmann, 02.118.038), 11 (J. Hopmann, 02.094.228), 12 (J. Meurers, 02.021.018).

Das Leopold Figl-Observatorium der Universitäts-Sternwarte Wien auf dem Mitterschöpfl, eröffnet am 25. September 1969, with a paper by J. Meurers on "Die Entstehung des Leopold Figl-Observatoriums für Astrophysik der Universitäts-Sternwarte Wien." Editor: Vertrieb Optischer Erzeugnisse, Wien. 27 pp. (1969).

Leopold Figl — Observatorium eröffnet.
Sternenbote, 12. Jahrgang, p. 142 - 155 (1969).

Das Leopold-Figl-Observatorium.
SuW, Vol. 8, 285 (1969).

008.132   Wroclaw

Wroclaw University Astronomical Observatory. —
Report from Solar Institute.   J. Mergentaler.
Solar Physics, Vol. 10, 229 - 230 (1969).

Contributions from the Wrocław Astronomical
Observatory, No. 16 (S. Wierzbiński, 02.117.040).

Wrocław Astronomical Observatory. Reprints
Nos. 76 (J. Mergentaler, 02.075.022), 77 (B. Szczodrowska,
01.032.052), 78 (B. Grabowski, 01.122.010), 79 (M. Jakimiec, 01.072.008), 80 (A. Stankiewicz, 02.072.011), 81
(M. A. Abramowicz, 02.141.022), 82 (Z. Kordylewski,
02.032.055).

008.133   Zürich

Tätigkeitsbericht der Eidgenössischen Sternwarte
Zürich für das Jahr 1968.   M. Waldmeier.
Zürich, 8 pp. (1969).

Astronomische Mitteilungen der Eidgenössischen
Sternwarte Zürich, Nos. 287 (M. Waldmeier, S. E. Weber,
02.079.103), 288 (M. Waldmeier, 02.072.088), 289 (J. Dürst,
02.032.070), 290 (M. Waldmeier, S. E. Weber, 02.074.079),
291 (M. Waldmeier, 02.074.080), 292 (M. Waldmeier,
02.074.081).

Publikationen der Eidgenössischen Sternwarte
Zürich, Band 13, Heft 3 (M. Waldmeier, 02.075.016).

Quarterly Bulletin on Solar Activity (Zürich),
Nos. 163, 164 (M. Waldmeier, R. Michard, J. G. Bastiaans,
A. D. Fokker, 02.075.017).

009 Notes on Observatories, Planetaria, and Exhibitions

009.001 Eröffnung der neuen Schul- und Volkssternwarte in Aalen. H.-U. Keller.
SuW, Vol. 8, 210 (1969).

009.002 Die Archenhold-Sternwarte im 20. Jahr der Deutschen Demokratischen Republik.
D. Wattenberg.
Blick in das Weltall, Archenhold-Sternw. Berlin-Treptow, No. 10, p. 83 - 90 (1969).

009.003 Die Eröffnung des Planetariums Longines im Verkehrshaus der Schweiz, Luzern.
N. Hasler-Gloor.
Orion, Band 14, 130 (1969).

009.004 The H. R. MacMillan Planetarium, Vancouver, B.C. D. A. Rodger.
Journ. Roy. Astron. Soc. Canada, Vol. 63, 268 (1969).

009.005 Astronomie im Kleinplanetarium. K. Kockel.
Sterne, 45. Jahrgang, 171 - 174 (1969).

009.006 Modernization at two eastern planetariums.
R. S. Knapp, S. S. Ross.
Sky Telescope, Vol. 38, 382 - 384 (1969).

009.007 Planetarium programs in the United States.
F. M. Branley.
Bull. American Astron. Soc., Vol. 1, 336 - 337 (1969). − Abstract AAS.

009.008 The observatory "Yunost' " in Ufa.
G. D. Tuev.
Zemlya i Vselennaya, No. 4, p. 79 - 82 (1969). In Russian.

009.009 The first director of the planetarium in Moscow.
B. A. Maksi Machev.
Zemlya i Vselennaya, No. 5, p. 60 - 63 (1969). In Russian.

009.010 Cosmic exhibition in Vienna.
V. I. Kuz'min.
Zemlya i Vselennaya, No. 6, p. 70 -75 (1969). In Russian.

009.011 Institute of Astronomy of the Polish Academy of Sciences. W. Iwanowska, S. Piotrowski.

Postępy Astron., Vol. 17, 287 - 290 (1969). In Polish.

009.012 Astronomical Institute of the Wrocław University.
J. Mergentaler.
Postępy Astron., Vol. 17, 291 - 294 (1969). In Polish.

009.013 The Chair of the Geodetical Astronomy of the Warsaw Polytechnic. W. Opalski.
Postępy Astron., Vol. 17, 294 - 296 (1969). In Polish.

009.014 5 Jahre Volks- und Schulsternwarte „Juri Gagarin" Eilenburg. E. Otto.
Sterne, 45. Jahrgang, 201 - 203 (1969).

009.015 Winnipeg (Manitoba Museum of man and nature) Planetarium. B. F. Shinn.
Journ. Roy. Astron. Soc. Canada, Vol. 63, 316 - 317 (1969).

009.016 New University of Western Ontario Observatory opened. W. Wehlau.
Journ. Roy. Astron. Soc. Canada, Vol. 63, 318 - 319 (1969).

009.017 Alpine solar station. S. Obashev.
Trudy Astrofiz. Inst. Alma-Ata, Vol. 15, 3 - 5 (1969). In Russian.
Note on the Alpine solar station of the Academy of Sciences of the Kazakh SSR, 28 km distant from Alma-Ata at a height of 3100 m.

009.018 Das Max-Planck-Institut für Astronomie − gegenwärtiger Stand des Aufbaus. H. Elsässer.
SuW, Vol. 8, 281 - 284 (1969).

009.019 People observatory in Valašské meziříčí.
B. Maleček.
Říše hvězd, Vol. 50, 141 - 143 (1969). In Czech.

009.020 Das neue Universal-Großplanetarium aus Jena.
H. Letsch.
Jenaer Rundschau (Jena Review), 13. Jahrgang, p. 345 - 349 (1968).

009.021 Opbouw van een sterrenwacht in Spanje.
T. Pieraerts.
Hemel en Dampkring, Vol. 67, 390 - 394 (1969). − Report on the construction of a Dutch station in Spain.

010 Societies, Associations, Organizations

010.001 American Association of Variable Star Observers (AAVSO)

Scientific highlights from the AAVSO spring meeting. M. E. Baldwin.
Sky Telescope, Vol. 38, 89 (1969).

010.002 American Astronomical Society (AAS)

American astronomers report.
Sky Telescope, Vol. 38, 302 - 304 (1969).
Highlights of the 130th meeting of the American Astronomical Society at Albany, New York, August 11 - 14, 1969: More computer studies of galactic evolution; Minor planets from comets? ; Education in astronomy; Stellar radii and absolute magnitudes; Light variations of quasars.

010.003 Association of Lunar and Planetary Observers

Announcements.
Strolling Astronomer, Vol. 21, 178 - 179, 214 - 216 (1969).

010.004 Astronomical Society of Australia (ASA)

Astronomical Society of Australia, 3rd. annual meeting – University of Sydney, 4 - 6 December 1968.
Inform. Bull. Southern Hemisphere, No. 14, p. 44 - 47 (1969).

010.005 Astronomical Society of Czechoslovakia.

No publication received.

010.006 Astronomical Society of the Pacific

Activities of the Society.
Publ. Astron. Soc. Pacific, Vol. 81, 295 - 296, 924 (1969).

Activities of the Society: Minutes of the meeting of the directors, May 2, 1969.
Publ. Astron. Soc. Pacific, Vol. 81, 457 - 461 (1969).

Minutes of the 80th annual meeting of the Astronomical Society of the Pacific, May 2, 1969.
Publ. Astron. Soc. Pacific, Vol. 81, 462 - 469 (1969).

Annual report of the treasurer for the year ended December 31, 1968.
Publ. Astron. Soc. Pacific, Vol. 81, 470 - 474 (1969).

Acitivities of the Society: Flagstaff meeting of the Astronomical Society of the Pacific. H. M. Johnson.
Publ. Astron. Soc. Pacific, Vol. 81, 703 - 706 (1969).

010.007 Astronomical Society of Southern Africa

Notices.
Monthly Notes Astron. Soc. Southern Africa, Vol. 28, 67, 79, 91, 104, 115 (1969).

Election of Council for 1969/70.
Monthly Notes Astron. Soc. Southern Africa, Vol. 28, 67 (1969).

Report of Council for Session 1968/69.
Monthly Notes Astron. Soc. Southern Africa, Vol. 28, 69 (1969).

Section reports.
Monthly Notes Astron. Soc. Southern Africa, Vol. 28, 89 - 90 (1969).

Centre reports for 1968 - 1969.
Monthly Notes Astron. Soc. Southern Africa, Vol. 28, 92 - 94 (1969).

010.008 Astronomical Society of Victoria (ASV)

Section directors' reports.
Journ. Astron. Soc. Victoria, Vol. 22, 60 - 61 (1969).
Auroral Section, B. Tregaskis; Lunar and Planetary Section, B. Adcock; Nova Search Section, D. F. Ward.

The A.S.V. exhibition, 1969.
Journ. Astron. Soc. Victoria, Vol. 22, 74 - 80 (1969).

Society notes.
Journ. Astron. Soc. Victoria, Vol. 22, 59 - 60, 84, 96 - 97 (1969).

010.009 Astronomical Society of Western Australia (ASWA)

Reports of proceedings – 203rd – 208th ordinary meeting.
Journ. Astron. Soc. Western Australia, Vol. 20, June; Vol. 21, August – December (1969).

Report of proceedings – Annual general meeting, 1969 July 14.
Journ. Astron. Soc. Western Australia, Vol. 21, July (1969).

010.010 Astronomische Gesellschaft (AG)

Versammlung der Astronomischen Gesellschaft in Mannheim 15. - 19. September 1969. K. Schaifers.
Mitt. Astron. Ges. No. 27, p. 7 - 8 (1969).

Ansprache des Vorsitzenden bei der Eröffnung der Versammlung in Mannheim am 16. September 1969.
R. Kippenhahn.
Mitt. Astron. Ges. No. 27, p. 9 - 14 (1969).

Mitteilungen No. 27.

Aus der Tagung der Astronomischen Gesellschaft.
SuW, Vol. 8, 223 (1969).

010.011 Astronomisk Selskab Kobenhavn

No publication received.

010.012 British Astronomical Association

Meetings of the Association.

Journ. British Astron. Ass., Vol. 79, 340 - 348, Vol. 80, 3 - 6, 10 - 11 (1969).

Lunar section: Report of meeting.
R. C. Maddison.
Journ. British Astron. Ass. Vol. 79, 388 - 390 (1969).

Meteor section meeting.
K. B. Hindley.
Journ. British Astron. Ass. Vol. 79, 391 - 397 (1969).

Report of the Council on work during the session 1968 July 1 to 1969 June 30 to be presented to members of the association at the annual general meeting, 1969 October 29.
Journ. British Astron. Ass. Vol. 79, 421 - 449 (1969).

The annual general meeting of the Association, held on 1969 October 29.
G. E. Taylor, N. J. Goodman, E. P. Duggan.
Journ. British Astron. Ass., Vol. 80, 7 - 11 (1969).

Historical Section. E. A. Beet.
Journ. British Astron. Ass., Vol. 80, 54 (1969).

Solar Section: Observational reports (1969).
W. M. Baxter.
Journ. British Astron. Ass., Vol. 80, 54 (1969).

Meteor Section: Report of Section meeting.
I. F. Philpott, K. B. Hindley.
Journ. British Astron. Ass., Vol. 80, 62 - 65 (1969).

010.013   British Interplanetary Society

Society news.
Spaceflight, Vol. 11, 253 - 254, 294 - 295, 334 - 336, 373, 407 (1969).

010.014   Committee on Space Research (COSPAR)

No publication received.

010.015   European Space Research Organization (ESRO)

No publication received.

010.016   International Astronautical Federation (IAF)

No publication received.

010.017   International Astronomical Union (IAU)

The meeting of the National Committee for Canada of the IAU at Vancouver, May 2 – 3, 1969.
J. R. Auman.
Journ. Roy. Astron. Soc. Canada, Vol. 63, 207 - 214 (1969).
This report includes abstracts of the contributed papers.

010.018   Meteoritical Society

Report of the president of the Meteoritical Society 1967 – 1968. C. B. Moore.
Meteoritics, Vol. 4, 143 - 144 (1969).

The 1968 meeting of the Meteoritical Society. Abstracts of papers.
Meteoritics, Vol. 4, 149 - 215 (1969).

010.019   Nederlandse Vereniging voor Weer-en Sterrenkunde

Verenigingsnieuws.
Hemel en Dampkring, Vol. 67, 258, 301 - 303 (1969).

010.020   Polskie Towarzystwo Astronomiczne (PTA)

No publication received.

010.021   Polskie Towarzystwo Miłośników Astronomii.

Kronika PTMA.
Urania Kraków, Vol. 40, 209 - 212, 282, 340 - 343 (1969).
In Polish.

010.022   Royal Astronomical Society (RAS)

Meetings of the Society.
Observatory, Vol. 89, 129 - 135, 135 - 143, 157 - 163 (1969).

The thirteenth Herstmonceux Conference.
Observatory, Vol. 89, 163 - 178 (1969). – 1969 April 9 and 10.

Reports of Meetings. Annual general meeting 1969 March 14.
Quarterly Journ. Roy. Astron. Soc. Vol. 10, 171 - 174 (1969).

Report of the council to the one hundred and forty-ninth annual general meeting of the Society.
Quarterly Journ. Roy. Astron. Soc. Vol. 10, 175 - 181 (1969).

Report of the honorary auditors for the year 1968.
W. B. Somerville, E. L. G. Bowell.
Quarterly Journ. Roy. Astron. Soc. Vol. 10, 183 (1969).

Treasurer's accounts for the year ending 1968 december 31.
Quarterly Journ. Roy. Astron. Soc. Vol. 10, 184 - 193 (1969).

Reports of meetings.
Quarterly Journ. Roy. Astron. Soc. Vol. 10, 289 - 290, 291 - 292 (1969).

010.023   Royal Astronomical Society of Canada (RAS Canada)

The Royal Astronomical Society of Canada 1968.
M. M. Thomson.
Journ. Roy. Astron. Soc. Canada, Vol. 63, 177 - 188 (1969).

The meeting of the National Committee for Canada of the IAU at London, September 4 – 6, 1969.
V. Gaizauskas.
Journ. Roy. Astron. Soc. Canada, Vol. 63, 309 - 313 (1969).
This report includes abstracts of the contributed papers.

010.024   Royal Astronomical Society of New Zealand (RAS New Zealand)

Royal Astronomical Society of New Zealand,

Variable Star Section.
Inform. Bull. Southern Hemisphere, No. 14, p. 21 (1969).

010.025 Schweizerische Astronomische Gesellschaft (SAG)

**Aus der SAG und den angeschlossenen Gesell-
schaften.**
Orion Schaffhausen, Vol. 14, 110 - 111, 137 - 138, 163 - 168
(1969).

Die ausserordentliche Generalversammlung der SAG.
N. Hasler-Gloor.
Orion Schaffhausen, Vol. 14, 163 - 164 (1969).

Bericht des Generalsekretärs der SAG.
H. Rohr.
Orion Schaffhausen, Vol. 14, 164 - 165 (1969).

Det Schweiziske Astronomiske Selskab (S. A. G.).
P. Darnell.
Astron. Tidsskr., Årg. 2, 133 - 136 (1969).

010.026 Sociedad Astronómica de México

No publication received.

010.027 Società Astronomica Italiana (SAI)

**Verbale di scrutinio delle schede di votazione per
la elezione del consiglio direttivo della Società Astronomica
Italiana per il biennio ottobre 1969 – settembre 1971.**
G. Tagliaferri, M. Rodonò, G. Guerrero.
Mem. Soc. Astron. Italiana, Nuova Serie, Vol. 40, 613 - 614
(1969).

**Discorso de apertura della XII riunione della
Società Astronomica Italiana tenuta a l'Aquila.**
Atti XII Riunione Soc. Astron. Italiana, L'Aquila 1968, p. 9 -
12 (1969).

010.028 Société Astronomique de France

Les séances de la Société.
L'Astronomie, 83ᵉ année, 307 - 310, 324 - 326, 373 - 376,
457 - 459 (1969).

Assemblée générale du 16. juin 1969.
A. Hamon.
L'Astronomie 83ᵉ année, 452 - 456 (1969).

La vie de la Société Astronomique de France.
A. Hamon.
L'Astronomie, 83ᵉ année, 459 - 462 (1969).

010.029 Société Astronomique "R. Bosković"

Annual meeting of the Astronomical Society
"R. Bosković".
Vasiona, Vol. 17, 70 - 72 (1969).   In Serbo-Croatian.

010.030 Société Chronométrique de France

No publication received.

**010.031 Société Belge d'Astronomie, de Météorologie, et
de Physique du Globe**

Réunions mensuelles.   E. Hoge, M. Bauduin.
Ciel et Terre, Vol. 85, 252 - 253, 314 - 343 (1969).

Assemblée générale statutaire du 22 mars 1969.
M. Bauduin.
Ciel et Terre, Vol. 85, 406 - 409 (1969).

010.032 Svenska Astronomiska Sällskapet

No publication received.

**010.033 VAGO (Astronomical–Geodetical Society of the
UdSSR)**

**Plenum of the Central Council of VAGO in
Swerdlowsk.**   E. K. Straut.
Zemlya i Vselennaya, No. 4, p. 62 - 63 (1969).   In Russian.

010.034 Vereiniging voor Sterrenkunde, België

No publication received.

010.035 Argentine Astronomical Association

No publication received.

010.036 IUAA – Internationale Union der Astro-Amateure.
R. A. Naef, H. Rohr.
Orion Schaffhausen, Vol. 14, 112 (1969).

**010.037 International Association of Geodesy. East–Euro-
pean Sub–Commission for Satellite Geodesy.**
**Report on the activity for the period from April 1968 to
June 1969.**   W. Dobaczewska.
Bull. Géod., Nouvelle Série, No. 94, p. 343 - 345 (1969).

010.038 International Union of the Amateur Astronomers.
K. Ziołkowski.
Urania Kraków, Vol. 40, 227 - 234 (1969).   In Polish.

011 Reports on Colloquia, Congresses, Meetings, Symposia, and Expeditions

011.001 **Report from Rome: X-rays and Gamma rays.**
G. S. Mumford.
Sky Telescope, Vol. 38, 96 - 98 (1969). − 1969 May.

011.002 | **Solar-terrestrial relationships.**
D. J. Schove.
Journ. British Astron. Ass. Vol. 79, 384 (1969). − Brussels, 1968 Sept.

011.003 **The European space conference.**
ESRO/ELDO Bull. No. 6, p. 15 (1969). − Report of three meetings in 1969.

011.004 **Chronik. Plenum der Kommission "Veränderliche Sterne" des Astronomischen Rates der Akademie der Wissenschaften der UdSSR.** Yu. N. Efremov.
Astron. Tsirk. No. 513, p. 7 - 8 (1969). In Russian.

011.005 **Symposium commun des commissions de la nouvelle compensation de la triangulation européenne et du réseau européen sur satellites artificiels.**
Bull. Géod. Nouvelle Série, No. 93, p. 213 - 233 (1969). − Paris, 1969 February 24 − March 1.

011.006 **All-Union conference on physics of comets.**
V. I. Ivanchuk.
Astrometriya i Astrofiz., *Kiev,* No. 4, p. 215 - 222 (1969). In Russian. − Kiev, 1966 Oct. 26 - 29.

011.007 **L'école internationale d'été sur le traitement optique d'information.**
G. Ceppatelli, A. Righini.
Mem. Soc. Astron. Italiana, Nuova Serie, Vol. 40, 359 - 360 (1969). − Orsay, France, 1969 July 7 - 12.

011.008 **La scuola estiva 1969 dell'ESRO.**
R. Barletti, G. Poletto.
Mem. Soc. Astron. Italiana, Nuova Serie, Vol. 40, 361 (1969). − Interlaken, Swisse, 1969 Aug. 18 - Sept. 5.

011.009 **Eight days in Kiev.** D. Cruikshank.
Icarus, Vol. 10, 448 - 452 (1969).

011.010 **A summary of "Symposium on the Moon and Planets" Kiev, USSR, October 15 - 22, 1968.**
A. J. Kliore.
Icarus, Vol. 10, 453 - 456 (1969).

011.011 **Vienna planetarium conference.** G. Lovi.
Sky Telescope, Vol. 38, 236 - 239 (1969).

011.012 **Highlights from the Denver convention.**
Sky Telescope, Vol. 38, 240 - 242 (1969).

011.013 **Astrometrische Konferenz der UdSSR in Pulkowo.**
H.-U. Sandig.
Sterne, 45. Jahrgang, 132 - 133 (1969). − 1969 June 2 - 5.

011.014 **Space astronomy: AAAS symposium 28 December 1969, Boston.** M. D. Papagiannis.
Science, Vol. 166, 775 - 776 (1969).

011.015 **Whither lunar and planetary exploration in the 1970's. − AAAS symposium, December 1969, Boston.** D. G. Rea.

Science, Vol. 166, 1184 - 1185 (1969).

011.016 **Fifth symposium on solar physics and magnetohydrodynamics, Potsdam, 30. Sept. − 3. Oct. 1968.**
É. I. Mogilevskij.
Vestn. AN SSSR, No. 3, p. 108 - 109 (1969). In Russian.
Abstr. in Referativ. Zhurn. 51. Astron., 8.51.26 (1969).

011.017 **Study of solar flares. (Symposium at Capri).**
A. B. Severnyj.
Vestn. AN SSSR, No. 11, p. 110 - 112 (1968). In Russian.
Abstr. in Referativ. Zhurn. 51. Astron., 8.51.27 (1969).
1968 June 10 - 12.

011.018 **Problems of planetary astrophysics. (Symposium at Kiev).** L. N. Bondarenko.
Vestn. AN SSSR, No. 3, p. 121 - 125 (1969). In Russian.
Abstr. in Referativ. Zhurn. 51. Astron., 8.51.29 (1969).
1968 Oct. 15 - 21.

011.019 **Cosmic rays − new results and problems. (Conference at Tashkent).** G. B. Zhdanov.
Vestn. AN SSSR, No. 1, p. 113 - 115 (1969). In Russian.
Abstr. in Referativ. Zhurn. 51. Astron., 8.51.31 (1969).
1968 Oct. 7 - 16.

011.020 **Physics of stars, nebulae and galaxies. (Symposium at Byurakan).** M. A. Arakelyan, V. V. Ivanov, G. M. Tovmasyan.
Vestn. AN SSSR, No. 3, p. 125 - 127 (1969). In Russian.
Abstr. in Referativ. Zhurn. 51. Astron., 8.51.32 (1969).
1968 Sept. 16 - 20.

011.021 **Problems of solar-terrestrial physics (Symposium at Crimea).** I. A. Zhulin, V. N. Obridko.
Vestn. AN SSSR, No. 4, p. 95 - 98 (1969). In Russian. −
Abstr. in Referativ. Zhurn. 51. Astron., 10.51.23 (1969).

011.022 **Highlights from the San Diego convention.**
Sky Telescope, Vol. 38, 294 - 298 (1969).

011.023 **Conference on laboratory exercises.**
S. S. Ross.
Sky Telescope, Vol. 38, 304 - 305 (1969).

011.024 **Variable star observers meet.**
C. B. Ford.
Sky Telescope, Vol. 38, 397 (1969).

011.025 **Physics of the magnetosphere (Symposium in Washington).**
S. N. Vernov, K. I. Gringauz, I. A. Zhulin.
Vestn. AN SSSR, No. 3, p. 96 - 100 (1969). In Russian.
Abstr. in Referativ. Zhurn. 62.Issled. kosm. prostranstv., 9.62.16 (1969). − 1968, Sept. 3 - 13.

011.026 **Due colloqui dell'U. A. I. sulle stelle binarie.**
M. G. Fracastoro.
Mem. Soc. Astron. Italiana, Nuova Serie, Vol. 40, 593 - 595 (1969). − Letter.

011.027 **Colloque international de chronométrie.**
E. Proverbio.
Mem. Soc. Astron. Italiana, Nuova Serie, Vol. 40, 597 - 599 (1969). − Parigi, 15 - 20 settembre 1969. − Letter.

011.028 **Die Spiralstruktur unserer Milchstrasse.**
G. A. Tammann.
Orion Schaffhausen, Vol. 14, 143 - 146 (1969). − Report on
IAU Symposium No. 38, Basel, 1969 Aug. 29 − Sept. 4.

011.029 **The COSPAR meetings in Prague.** C. Sagan.
Ícarus, Vol. 11, 268 - 272 (1969). − 12th Plenary
Meeting, 1969 May 11 - 24.

011.030 **Fünfte Konferenz über Kometen.**
Kometn. Tsirk. *Kiev,* No. 89 (1969). In Russian.
Kiev, 1969 Oct. 6 - 9.

011.031 **Impressions from the 12th meeting of the International Committee for Exploration of Cosmic Space.**
I. A. Khvostikov.
Zemlya i Vselennaya, No. 6, p. 50 - 53 (1969). In Russian.

011.032 **International symposium on electromagnetic distance measurement and atmospheric refraction.**
M. R. Richards.
Bull. Géod., Nouvelle Série, No. 94, p. 327 - 335 (1969).
Boulder, Colorado 23rd − 27th June 1969.

011.033 **Meetings in Varna (Bulgaria). 22 − 30.VI.1969. Pertaining to artificial earth satellites.**
J. Łatka.
Bull. Géod., Nouvelle Série, No. 94, p. 337 - 341 (1969).
1. Meeting of the East−European Sub−Commission for Satellite Geodesy of the IAG; 2. Symposium of the Commission for Scientific Studies by Means of Optical Tracking of Satellites.

011.034 **Relazione sul convegno internazionale sulla gravitazione e relatività generale Tbilisi (URSS) 8 − 12**
settembre 1968. V. de Sabbata.
Atti XII Riunione Soc. Astron. Italiana, L'Aquila 1968,
p. 78 - 84 (1969). − Abstract SAI.

011.035 **Dritter Internationaler Planetariumsleiter-Kongress in Wien.** A. Kunert.
SuW, Vol. 8, 246 (1969).

011.036 **Ein Symposium über Himmelsmechanik in Brasilien.**
J. Schubart.
SuW, Vol. 8, 288 (1969).

011.037 **XII plenary meeting of COSPAR.** J. Bouška.
Říše hvězd, Vol. 50, 150 - 154 (1969). In Czech.

011.038 **XII plenary meeting of COSPAR.** J. Bouška.
Vesmír, Vol. 48, 251 (1969). In Czech.

011.039 **Lunar research on the XII COSPAR meeting.**
M. Eliáš.
Říše hvězd, Vol. 50, 213 - 216 (1969). In Czech.

011.040 **Symposium on the Magellanic Clouds.**
Inform. Bull. Southern Hemisphere, No. 14, p.
47 - 48 (1969). − 23 - 29 March 1969 in the E.S.O. Headquarters, Santiago, Chile.

011.041 **XIIth Plenary COSPAR meeting.**
K. Ziołkowski.
Urania Kraków, Vol. 40, 335 - 340 (1969). In Polish.

Astronomical get-together.
Nature, Vol. 224, 214 - 215 (1969).
Report on the October 1969 meeting of the Royal Astronomical Society.

012 Proceedings of Colloquia, Congresses, Meetings, and Symposia

012.001 International Conference on Statistical Mechanics,
Kyoto, September 9–14, 1968.
Journ. Phys. Soc. Japan, Vol. 26, Suppl., 321 pp. (1969).
Review in Bull. Astron. Inst. Czechoslovakia, Vol. 20, 228 -
229, 1969 (*P. Andrle*).

**012.002 Proceedings of the Colloquium on the Problems of
the Time Determination, Keeping and Synchroni-**
zation, Milan, April 23 - 24, 1968.
E. Proverbio (Editor).
Astronomical Observatory, Brera–Milan. Industrie Lito-Tipo-
grafiche Mario Ponzio, Pavia. 350 pp. (1968). – The papers
of this colloquium are abstracted in their subject categories.

012.003 Geochemistry of tektites; papers presented at the
Third International Tektite Symposium, sponsored
by Corning Glass Works and The Smithsonian Institution,
Corning, New York, April 16–18, 1969.
S. R. Taylor, A. A. Levinson (Editors).
Geochim. Cosmochim. Acta, Vol. 33, 1011 - 1147 (1969).

012.004 A Symposium on Jupiter and the Outer Planets.
Icarus, Vol. 10, 353 - 427 (1969). – Dallas, Texas,
1968 Dec. 29 - 30. – The papers included are abstracted in
their subject categories.

012.005 A Symposium on Icarus.
Icarus, Vol. 10, 429 - 456 (1969). – Austin, Texas,
1968 December 10. – The papers included are abstracted in
their subject categories.

012.006 Moon and Planets II. A Session of the Joint Open
Meeting of Working Groups I, II and V of the Tenth
Plenary Meeting of COSPAR, organized by the Committee on
Space Research – COSPAR and the Royal Society.
A. Dollfus (Editor).
North-Holland Publishing Company, Amsterdam. IX + 196 pp.
Price $ 10.00 (1968). – London, 1967 July 26 - 27.

012.007 Non-Periodic Phenomena in Variable Stars.
L. Detre (Editor).
Academic Press, Budapest. 12 + 490 pp. Price $ 26.60 (1969).
– IVth Colloquium on Variable Stars, International Astronomical Union, Commission 27 and 42, held in Budapest, Hun-
gary, 5 - 9 September 1968. – The papers are abstracted in
their subject-categories.

012.008 Mass Loss from Stars. Proceedings of the Second
Trieste Colloquium on Astrophysics, 12 - 17 September, 1968.   M. Hack (Editor).
D. Reidel Publishing Company, Dordrecht–Holland. 12 +
345 pp. Price $ 18.20 (1969).

012.009 Interstellar Ionized Hydrogen.
Y. Terzian (Editor), with concluding remarks by
D. Osterbrock, G. Westerhout.
W. A. Benjamin, Inc., New York – Amsterdam. 9 + 774 pp.
Price $ 15.00 (1968). – The papers of this symposium
(Charlottesville, 1967 Dec. 8, 11) are abstracted in their sub-
ject categories.

**012.010 Abstracts of papers presented at the Conference of
Junior Scientists of the Moscow University.**
Mosk. universitet, Moskva. 281 pp. Price 58 Kop. (1968).
In Russian. – Review in Referativ. Zhurn. 51. Astron.,
9.51.52 (1969).

012.011 Physics of the Magnetosphere, based upon the
Proceedings of the Conference, held at Boston
College, June 19 - 28, 1967.
R. L. Carovillano, J. F. McClay, H. R. Radoski.
D. Reidel Publishing Company, Dordrecht – Holland. 10 +
686 pp. Price $ 36.40 (1968). – The papers included are ab-
stracted in their subject categories.

**012.012 Meteorite Research. Proceedings of a Symposium
on Meteorite Research, held in Vienna, Austria,**
7 - 13 August 1968.
P. M. Millman (Editor).
D. Reidel Publishing Company, Dordrecht – Holland. 15 +
941 pp. Price ƒ 160.00 (1969). – The papers included are abstracted in their subject categories.

**012.013 Publikationen des Internationalen Seminars zur Er-
forschung der Physik des interplanetaren Raumes
mit Hilfe der kosmischen Strahlung.**
Akad. Nauk SSSR, Fiziko-tekhn. in-t, Leningrad. 290 pp.
Price 50 Kop. (1969).   In Russian.

**012.014 Cosmic Ray Studies, in relation to recent develop-
ments in Astronomy & Astrophysics.**
R. R. Daniel, P. J. Lavakare, S. Ramadurai (Editors).
Tata Institute of Fundamental Research, Bombay, India.
6 + 374 pp. (1969). – Proceedings of a Colloquium held in
Bombay, November 11 - 16, 1968. – The papers included are
abstracted in their subject categories.

012.015 Solar Flares and Space Research.
Proceedings of the Symposium held on the occasion
of the Eleventh Plenary Meeting of the Committee on Space
Research, Tokyo, Japan – 9 - 11 May 1968.
C. de Jager, Z. Svestka (Editors), with a summary by
E. N. Parker.
North-Holland Publishing Company, Amsterdam. 9 + 419 pp.
Price Dfl. 72.00 (1969).

**012.016 Atti della XII Riunione della Società Astronomica
Italiana.**
L. U. Japadre Editore, L'Aquila = Quaderni della Universita
degli Studi dell'Aquila. 127 pp. (1969).

**012.017 Papers from AAS/AIAA Astrodynamics Specialist
Conference, 1968.**
AIAA Journ., Vol. 7, 993 - 1063 (1969).

012.018 A Discussion on Infrared Astronomy,
held 1 and 2 May 1966 in London.
Phil.Trans. Royal Soc. London, Ser. A, Vol. 264 (No. 1150),
109 - 320 (1969).

012.019 A discussion on cosmic X-ray astronomy.
Organized by H. Massey, R. L. F. Boyd, E. A.
Stewardson, R. Wilson.
Proc. Roy. Soc. London, Series A, Vol. 313, 299 - 402 (1969).

012.020 Space Research IX. Proceedings of open meetings
of working groups of the Eleventh Plenary Meeting
of COSPAR, Tokyo, 9 - 21 May 1968.
K. S. W. Champion, P. A. Smith, R. L. Smith-Rose (Editors).
North-Holland Publishing Company, Amsterdam. 17 + 770 pp.
Price $ 34.00, Dfl. 120.00 respectively (1969).

012.021 Study of the rotation of artificial satellites based

on photometric data (SPIN program).
Byull. Stantsij Optichesk. Nablyud. Iskusstv. Sputnikov Zemli
No. 54, 49 pp., with opening addresses by B. E. Melnik and
A. G. Massevitch, and concluding remarks by A. G. Masse-
vitch, p. 7, 8, 47 (1969).   In Russian. – Colloquium at
Kishinev, 1968 Sept. 25 - 28.

012.022 **Seminar on geodetic reduction of data from obser-
vations of artificial satellites in Tashkent, 1968
November 23 – 25.**
Byull. Stantsij Optichesk. Nablyud. Iskusstv. Sputnikov
Zemli No. 55, 109 pp., with opening addresses by V. P.
Shcheglov, p. 7 and A. G. Massevitch, p. 8 (1969).
In Russian.

012.023 **Rotation of the Earth and Determination of Time.**
Izdatel'stvo "Nauka", Moskva. Akademiya Nauk
SSSR, Astronomicheskij Sovet. 152 pp. Price 62 Kop. (1969)
In Russian. – Berichte der Konferenz zur Erforschung der
ungleichförmigen Erdrotation in Riga, 1965, Juni 8 - 10.

012.024 **Publications of the 6th annual winter school of
USSR on cosmic physics, 18 March - 1 April 1969,
Part I.**
Apatity. 280 pp. Price 1 Rbl. 50 Kop. (1969).   In Russian.
Review in Referativ. Zhurn. 62. Issled. kosm. prostranstv.,
2.62.24 (1970).

012.025 **Publications of the 6th annual winter school of
USSR on cosmic physics, 18 March - 1 April 1969,
Part II.**
Apatity. 166 pp. Price 1 Rbl. 8 Kop. (1969).   In Russian.
Review in Referativ. Zhurn. 62. Issled. kosm. prostranstv.,
2.62.25 (1970).

012.026 **Proceedings of the international symposium on the
investigation of physical properties of the interplanetary medium by means of cosmic rays.**
Leningrad. AN SSSR. Fiziko-tekhn. in-t. 290 pp. Price 50 Kop.
(1969).   In Russian.

012.027 **Research in Physics and Chemistry.** Third Lunar
International Laboratory Symposium.
F. J. Malina (Editor).
Pergamon Press, Oxford. 145 pp. Price $ 13.50 (1969).
Belgrade, 28 Sept. 1967. – Review in Space Sci. Rev. Vol.
10, 316; 1969 (*G. Fielder*).

012.028 **Relativistic Plasmas.** The Coral Gables Conference.
O. Buneman, W. B. Pardo (Editors).
W. A. Benjamin, Inc., New York. Price $ 13.50 (1969).

012.029 **Extra-Terrestrial Matter.** Proceedings of a Con-
ference, Argonne, Ill., March 1968.
C. A. Randall, Jr. (Editor).
Northern Illinois University Press, Dekalb. 20 + 332 pp.
Price $ 12.50 (1969).

012.030 **Solar-Terrestrial Physics: Terrestrial Aspects.**
Proceedings of a Joint IQSY/COSPAR Symposium,
London, July 1967, Part 1, 2.   A. C. Stickland (Editor).
M.I.T. Press, Cambridge, Mass. 10 + 414 pp., 12 + 468 pp.
Price $ 19.50, $ 22.50 respectively (1969).

012.031 **Atmospheric Emission.** Proceedings of NATO's
Advanced Study Institute, Norway 1968.
B. M. McCormac, A. Omholt (Editors).
Van Nostrand Reinhold, Princeton. 563 pp. Price $ 25.95
(1969). – Review in Sky Telescope, Vol. 38, 255 (1969).

013 Reports on Astronomy in Various Countries and Particular Fields, International Cooperation

013.001 Amateur astronomy in Australia.
J. T. Richards.
Southern Stars, Vol. 23, 58 - 60 (1969).

013.002 Uzbek scientist's research work (25th anniversary of the Academy of Sciences of the Uzbek SSR).
A. S. Sadykov.
Vestn. AN SSSR, No. 2, p. 65 - 70 (1969).   In Russian.

013.003 Das erste internationale astronomische Jugend-lager. R. Baggenstos.
Orion Schaffhausen, Vol. 14, 146 - 147 (1969).

013.004 Les progrès récents de l'astronomie.
J. Rösch.
L'Astronomie, 83ᵉ année, 462 - 463 (1969).

013.005 Optics in the United Kingdom.
R. W. Ditchburn.
Applied Optics, Vol. 8, 1939 - 1942 (1969).
Optics is interpreted to include X-ray optics, electronic optics, and short wave radiooptics as well as the more conventional visible, UV, and IR optics. Recent work in Britain on X-ray optics (applied to molecular biology), on scanning electron microscopy, and in radioastronomy (discovery of pulsars) is mentioned.

013.006 International astronomical youth camp.
A. Glendinning.
Journ. British Astron. Ass., Vol. 80, 53 (1969).

013.007 Current astrophysics in Australia. L. J. Gleeson.
Journ. Astron. Soc. Victoria, Vol. 22, 55 - 59 (1969).

013.008 20 years astronomical science in the German Democratic Republic. O. Günther.
Astron. in der Schule, 6. Jahrgang, 118 - 124 (1969).

013.009 Projekt „JOSO" (Joint Solar Observatory) – Ein kooperatives internationales Sonnen-Observatorium.
K. O. Kiepenheuer.
Mitt. Astron. Ges. No. 27, p. 160 (1969). – Abstract AG.

What future for planetary science in Britain?
Nature, Vol. 224, 744 - 745 (1969).

014 Teaching in Astronomy

014.001 Mathematics, mechanics, astronomy at the Petersburg – Leningrad University.
V. I. Smirnov, N. N. Polyakhov, K. F. Ogorodnikov.
Vestn. Leningr. un–ta, No. 1, p. 5 - 28 (1969). In Russian. – Abstr. in Referativ. Zhurn. 51. Astron., 10.51.9 (1969).

014.002 Auckland Observatory's child-parent astronomy classes. L. A. C. Warner.
Sky Telescope, Vol. 38, 310 - 311 (1969).

014.003 On the career development and education of astronomers in the United States. R. Berendzen.
Bull. American Astron. Soc., Vol. 1, 334 - 335 (1969). – Abstract AAS.

014.004 The elementary astronomy course for nonscience majors. R. C. Bless.
Bull. American Astron. Soc., Vol. 1, 335 (1969). – Abstract AAS.

014.005 The astronomical community and manpower needs.
R. Fleischer, H. Lane.
Bull. American Astron. Soc., Vol. 1, 343 (1969). – Abstract AAS.

014.006 Trends in graduate education in astronomy.
B. F. Peery, Jr.
Bull. American Astron. Soc., Vol. 1, 358 (1969). – Abstr. AAS.

014.007 The role of astronomers in elementary and secondary education. P. H. Vanek.
Bull. American Astron. Soc., Vol. 1, 366 (1969). – Abstr. AAS.

014.008 Ein Schulmodell zur Nachbildung der Lichtkurven von W Ursae Maioris-Sternen.
K. Locher.
Orion Schaffhausen, Vol. 14, 158 - 159 (1969).

014.009 Training of astronomers during studies at university.
N. P. Grushinskij.
Zemlya i Vselennaya, No. 4, p. 76 -78 (1969).   In Russian. German translation by W. Petri, SuW, Vol. 8, 289 - 290 (1969).

014.010 Elements of astronomy and cosmonautics at the secondary school. E. P. Levitan.
Zemlya i Vselennaya, No. 5, p. 72 - 76 (1969).   In Russian.

014.011 School observatory in the village.
L. P. Grebenev.
Zemlya i Vselennaya, No. 5, p. 83 - 85 (1969).   In Russian.

014.012 A voice in the discussion on the curriculum in astronomy in the schools. K. Rudnicki.
Postępy Astron., Vol. 17, 405 - 408 (1969).   In Polish.

014.013 Astronomin i skolan.
R. M. West.
Astron. Tidssk., Årg. 2, 132 (1969).

014.014 Astronomin i skolan.
K. Lindner.
Astron. Tidssk., Årg. 2, 178 - 183 (1969). – Education in astronomy at popular observatories in the German Democratic Republic.

014.015 Some ideas about the scientific investigation tasks

in the sphere of methodics of astronomy instruction in the secondary school.   E. P. Levitan.
Astron. in der Schule, 6. Jahrgang, 98 - 102 (1969).

014.016 Experiences in accomplishing pupils' observations.
         I. Berger.

Astron. in der Schule, 6. Jahrgang, 105 - 110 (1969).

014.017 On the position and the function of astronomy
         within the system of natural sciences at the
secondary school.   O. Mader.
Astron. in der Schule, 6. Jahrgang, 124 - 128 (1969).

015 Miscellanea

015.001 Sind wir allein im Kosmos?   H. Elsässer.
         SuW, Vol. 8, 185 - 189 (1969).

015.002 Paléo-astronomie.   J. Kovalevsky.
         L'Astronomie, 83ᵉ année, 411 - 413 (1969).

015.003 Cosmical philately.   V. V. Polonskij.
         Zemlya i Vselennaya, No. 5, p. 94 - 95 (1969).
In Russian.

015.004 Extraterrestrial civilizations and kybernetics.
         B. N. Panovkin.
Zemlya i Vselennaya, No. 6, p. 2 - 5 (1969).   In Russian.

015.005 Comments on cosmic physics.
         E. J. Öpik.
Irish Astron. Journ., Vol. 9, 81 - 87 (1969).

015.006 Reflections on astronomy, its correlation with
         physics and technology, and its influence on con-
temporary culture.   I. S. Shklovskij.
Vopr. filosofii, No. 5, p. 52 - 62, 187 (1969).   In Russian.

Abstr. in Referativ. Zhurn. 51. Astron., 1.51.2 (1970).

015.007 Die Abstände der Planeten und ihre Entsprechungen
         zu Strukturen der Primzahlfolge und der Mikrophy-
sik.   T. Landscheidt.
Nachr. Olbers–Gesellschaft Bremen, No. 75, p. 3 - 24 (1969).

015.008 Was können Gedankenexperimente nützen?
         E. Verhülsdonk.
SuW, Vol. 8, 229 - 231 (1969).

015.009 La formidable aventura.   R. Compte Porta.
         El Universo, Vol. 23, (No. 89), 101 - 108 (1969).
Popular article.

015.010 La posibilidad de vida extraterrestre.
         C. M. Varsavsky.
Revista Astron., Vol. 41, (No. 169), 5 - 9 (1969). – Popular
article.

         Extraterrestrial optical microscopy.
See Abstr. 034.030.

Applied Mathematics, Physics

021 Mathematics, Computing, Machine Programs

021.001 Approximate spectral analysis by least-squares fit.
P. Vaníček.
Astrophys. Space Sci. Vol. 4, 387 - 391 (1969).

An approximate method of spectral analysis called 'successive spectral analysis' based upon the mean-quadratic approximation of an empirical function by generalised trigonometric polynomial with both unknown frequencies and coefficients is developed. A few quotations describing some properties of the method as well as one of the possible methods for numerical solution are given.

021.002 The IBM 1620 computer at Dunsink Observatory.
C. J. Butler, P. A. Wayman.
Irish Astron. Journ. Vol. 9, 1 - 5 (1969).

021.003 On the choice of the degree of smoothing observational data. Ja. S. Jatskiv.
Astrometriya i Astrofiz., *Kiev*, No. 2, p. 84 - 91 (1969).
In Russian.

When Whittaker's method of smoothing is applied to some observational data the primary task is to choose an appropriate degree of smoothing E. The author proposed a new method of choosing E based on the estimational inspection of the power spectrum of the time series composed of the data under examination. This method was tested on the model of latitude variation.

021.004 A programming system for analytical series expansions on a computer.
R. Broucke, K. Garthwaite.
Celestial Mechanics, Vol. 1, 271 - 284 (1969).

A programming system is described for the manipulation of Poisson series on a computer. The general structure is described together with the most important individual subroutines. The system exists in two versions, one with 3 angular and 3 polynomial variables and the other with twice as many variables. The programming has been done first in FORTRAN IV, but the most crucial subroutines have been rewritten in machine language for more efficiency. The storage mechanism is such that obsolete series can be erased from the main memory in order to make room for new series to be created.

021.005 Die Mathematisierung der Wissenschaften.
K. Strubecker.
Phys. Blätter, 25. Jahrgang, 488 - 495 (1969).

021.006 Lamé functions of the first kind generated by computer. H. G. Walter.
IBM Federal Systems Division, Federal Systems Center, Cambridge, Mass., Final Rep. Task No. 0216, 4 + 21 + A36 pp. (1969).

The definitions of the Lamé functions and the ellipsoidal coordinates inherent to them are introduced. Proceeding from Lamé's differential equation the four classes of Lamé functions of the first kind are generated by computer with the aid of formula manipulation techniques. For this purpose algebraic expressions for the coefficients in the Lamé polynomials are constructed by virtue of recurrence formulae and presented in tabular as well as machine readable form for further processing.

021.007 Random number generators and their application to astronomical statistical problems. J. Meurers.
Astron. Astrophys. Vol. 3, 354 - 363 (1969). In German.

Random number generators represent a new method for constituting and studying accidental events. In the following investigation, a "mixed" random generator is applied to the study of astronomical phenomena. By associating the members of the random series in various ways, it is possible to construct artificial two-dimensional star fields etc. It is shown that in artificial two-dimensional fields, holes, chains and groups of stars occur as temporary chance phenomena. Chains with continuously decreasing magnitude are also possible. Finally, Markarian's chain of galaxies in the Virgo cluster is investigated to determine whether it could be an accidental phenomenon. A statistical comparison between the real chain and artificial point chains shows that the former could have been produced by chance.

021.008 Die Brauchbarkeit von Zufallszahlengebern als Grundlage stellarstatistischer Modelle.
M. G. Firneis.
Ann. Univ.-Sternw. Wien, Vol. 28, (No. 3), 91 - 135 (1969).

Emerging from stellar statistical problems some methods for the generation of pseudo random numbers are discussed. After an exposition of the structure of random number generators the three generators GROBRN, GAS 1 and RANDOM are described functionally. These generators are submitted to the three following tests: pseudodice-, iterated chi square- and d^2-test. The generator RANDOM is recognized to be particularly suitable and is recommended for further research in stellar statistics.

021.009 Integrationstheorie von Krylow-Bogoljubow und gestörte Keplerbewegung. W. Flury.
Diss. No. 4358 der Eidgenössischen Technischen Hochschule, Zürich. Juris Druck + Verlag, Zürich. 76 pp. (1969).

Motion of a satellite in the gravity field of the earth; perturbated Kepler motion; perturbations in the elements; analytic perturbations due to earth flattening according to the method of Krylov-Bogoljubov.

021.010 Le mouvement Keplerien et les oscillateurs harmoniques. C. A. Burdet.
Journ. reine angewandte Math., Band 238, 71 - 84 (1969).

Le premier paragraphe est consacré à une brève présentation de la méthode. Puis, une nouvelle transformation du temps t fournit un autre oscillateur harmonique dont l'avantage principal est de permettre d'éliminer également la singularité $1/r$ qui apparait avec certaines forces perturbatrices (celles due à l'applatissement de la terre notamment).

021.011 Über einen Fall der Verwendung von Potenzreihen in der Himmelsmechanik.
Yu. V. Plakhov.
Izv. vyssh. uchebn. zavedenij. Geod. i aehrofotos'emka, No. 5, p. 101 - 103 (1968). In Russian. − Abstr. in Referativ. Zhurn. 62. Issled. kosm. prostranstv., 8.62.145 (1969).

021.012 A contribution to the problem of smoothing observational data. J. Vondrák.

Bull. Astron. Inst. Czechoslovakia, Vol. 20, 349 - 355 (1969).

The paper describes a general method of graduation of observational data needing neither equidistant argument nor equal weights of observed values of the variable. The method starts from Whittaker's method based on probability and can be used in any general case when necessary. As the method is very laborious it is advisable to use it only by means of a high-speed computer. A practical example of smoothing latitude variations is also given.

021.013 **Investigations in determining astronomic latitudes and their computer programs.**
L. B. Bourquin.
Report IR-68-21, Naval Oceanographic Office, Geodetic Studies Branch, Washington, D. C., 193 pp. (1968). — See Phys. Abstr. Vol. 72, No. 27110 (1969).

021.014 **Inversione di matrici usate nella teoria delle orbite.**
P. Sconzo.
Atti XII Riunione Soc. Astron. Italiana, L'Aquila 1968, p. 23 - 24 (1969). — Abstract SAI.

021.015 **Monograph in calculations of astronomy.**
K. Hukuda.
Mem. Japan Astron. Study Ass., Vol. 3, (No. 12), 167 - 176 (1969). In Japanese.

021.016 **A program for the computer BESM-3M for the treatment of polarization observations received with D. A. Rozhkovski's polarograph.**
K. G. Dzhakusheva, V. S. Matjagin, E. G. Michelkin.
Trudy Astrofiz. Inst. Alma-Ata, Vol. 14, 76 - 95 (1969). In Russian.

A program for the determination of the linear and elliptical polarization parameters of radiation of extended cosmical objects with the computer BESM-3M is given.

021.017 **Laplace coefficients and their derivatives.**
G. Antonacopoulos.
Publ. Lab. Astron. Univ. Athen, Ser. II, No. 23, 70 pp. (1969).

A computer program is given in ALGOL, for the computation and tabulation of Laplace coefficients and their derivatives. The tables for the values of the ratio of semimajor axis of the planetary system are also given.

021.018 **Random number generators and random phenomena in star-fields.** J. Meurers.
Mitt. Univ.-Sternw. Wien, Vol. 14 (No. 12), 179 - 194 (1969). In German.

Random generators represent a new method for constituting and studying accidental events. It is possible on the basis of such generators to construct a two-dimensional distribution of points by combining two following numbers of the generator as rectangular coordinates of a point in a two-dimensional field. This may be interpreted as an artificial star field.

021.019 **Clipping loss in the one-bit autocorrelation spectral line receiver.** W. R. Burns, S. S. Yao.
Radio Science, Vol. 4, No. 5, p. 431 - 436 = National Radio Astron. Obs., Green Bank, Repr. Ser. A, No. 113 (1969).

The one-bit or polarity coincidence correlation method has become a common technique in radio astrophysical spectral line measurement. An error analysis of the one-bit autocorrelation method of spectral estimation is presented, and the variance on the autocorrelation estimate and spectral estimate is given.

021.020 **Applications astronomiques et astrophysiques de la statistique.** R. Coutrez.
Rev. Belge Statistique et de Recherche opérationnelle, Vol. 7, Nos. 3 - 4, 15 pp. = Univ. Libre Bruxelles, Inst. d'Astron. d'Astrophys. , Sér. A, No. 11 (1967).

021.021 **Random Number Generators und ihre Bedeutung für die Beurteilung systematischer Effekte in Sternfeldern.** J. Meurers.
Mitt. Astron. Ges. No. 27, p. 129 - 131 (1969). — Abstract AG.

021.022 **Über Verfahren zur Erfassung, Darstellung und Auswertung von Häufigkeitsverteilungen. Teil I. Theoretische Einführung, mechanische und elektronische Verfahren.** E. Gerth.
Feingerätetechnik, Band 18, Heft 4, 7 pp. = Mitt. Astrophys. Obs. Potsdam No. 130 (1969).

021.023 **Über Verfahren zur Erfassung, Darstellung und Auswertung von Häufigkeitsverteilungen. Teil II. Oszillographische Verfahren.** E. Gerth.
Feingerätetechnik, Band 18, Heft 5, 7 pp. = Mitt. Astrophys. Obs. Potsdam No. 131 (1969).

Grundzüge der Ausgleichsrechnung nach der Methode der kleinsten Quadrate nebst Anwendung in der Geodäsie.
See Abstr. 003.075.

Machine program for the computer M20 for the determination of orbits and their perturbations.
See Abstr. 042.014.

Computer simulation of galactic evolution.
See Abstr. 151.040.

022 Physical Papers Related to Astronomy and Astrophysics

022.001   Electron impact excitation of CN.
O. H. Crawford, A. C. Allison, A. Dalgarno.
Astron. Astrophys. Vol. 2, 451 - 452 (1969).

Close-coupling calculations are presented of the cross-sections for the 0–0, 0–1 and 0–2 rotational transitions in CN induced by impact of low energy electrons.

022.002   An experimental survey of the low energy electron scattering spectrum of nitrogen.
A. J. Williams III, J. P. Doering.
Planet. Space Sci. Vol. 17, 1527 - 1537 (1969).

The electron energy loss spectrum of nitrogen has been studied at incident energies from 9 to 50 eV with an electron spectrometer. The results suggest that the observed intensities of atmospheric N_2 emissions in the aurora are entirely compatible with low energy electron impact excitation mechanisms.

022.003   On the Doppler frequency shift of light using rotating mirrors.   H. Nieuwenhuijzen.
Bull. Astron. Inst. Netherlands, Vol. 20, 300 - 308 = Commun. Astrophys. Lab. Obs. Utrecht (1969).

Moving mirrors can be used to obtain a Doppler frequency shift of a light signal. An important application is the tuning of a laser signal.

022.004   The γ'-system of the TiO molecule.
J. G. Phillips.
Astrophys. Journ. Vol. 157, 449 - 458 (1969).

A rotational analysis of the (0,0) band of the γ'-system of the TiO molecule shows that the system is produced by a transition from a $^3\Pi$ state to the ground state $(X\,^3\Delta)$ of the molecule.

022.005   Intercombination oscillator strengths in the helium sequence.   G. W. F. Drake, A. Dalgarno.
Astrophys. Journ. Vol. 157, 459 - 462 (1969).

The electric dipole oscillator strengths and transition probabilities of the spin-forbidden transitions among the 1^1S, 2^1S, 2^1P, and 2^3P levels of the heliumlike ions from He I to Ne IX are evaluated to high accuracy by purely theoretical methods.

022.006   Effective Gaunt factors g_{eff} for excitation of positive ions by electron collisions in a simplified Coulomb-Born approximation.   M. Blaha.
Astrophys. Journ. Vol. 157, 473 - 477 (1969).

Effective Gaunt factors g_{eff} have been calculated for threshold excitation of positive ions by electrons for transitions in fifteen isoelectronic sequences. A simplified Coulomb-Born approximation is used, the validity of which is given by the inequality (8). Results are presented as a function of Z/k, where $1/2\,k^2$ is the excitation energy in atomic units.

022.007   Comments on quasars and quarks.
B. J. Skutnik.
Phys. Rev., Second Series, Vol. 181, 2145 - 2146 (1969).

In this paper we bring out several points in connection with the recent paper by Huang and Edwards. First, extra-terrestrial spectroscopic searches for quarks have been carried out since 1966 by several groups. Secondly, because of the crudeness of the available data, the criteria used for the identifications were not sufficient and thus it should be stressed that the assignments are quite tentative. Thirdly, order-of-magnitude estimates of relative abundances of the quarked species were not made, based on the identifications. Finally, several of the transition lines identified as λ_{Q1} or λ_{Q2} should have had related normal species λ_{Q0} in the visible region, yet

the related λ_{Q0}'s were not present except in one case.

022.008   Threshold photoneutron cross section for Mg^{26} and a source of stellar neutrons.
B. L. Berman, R. L. Van Hemert, C. D. Bowman.
Phys. Rev. Letters, Vol. 23, 386 - 389 (1969).

The differential photoneutron cross section for Mg^{26} at 135° has been measured as a function of photon energy from 10 keV to 1.5 MeV above threshold by the threshold photoneutron technique. Several prominent resonances have observed, including one located at 54.3 keV above threshold. The existence of this resonance in the Mg^{26} compound system might provide, through the reaction $Ne^{22}\,(\alpha,n)\,Mg^{25}$, the primary production mechanism for neutrons in stars.

022.009   New state of ferromagnetism in degenerate electron gas and magnetic fields in collapsed bodies.
H. J. Lee, V. Canuto, H.-Y. Chiu, C. Chiuderi.
Phys. Rev. Letters, Vol. 23, 390 - 393 (1969).

A new state of "ferro" magnetism in a degenerate electron gas is found and shown to be stable. This magnetism is the sum of all microscopic magnetic moments associated with electrons in their respective Landau levels while the Landau levels of the system are in turn maintained by this macroscopic magnetization. The maximum field in the Landau orbital ferromagnetism state is 10^7 G for white–dwarf densities and 10^{12} G for neutron–star densities.

022.010   Spin down problem of rotating stratified fluid in thermally insulated circular cylinders.
T. Sakurai.
Journ. Fluid Mechanics, Vol. 37, 689 - 699 (1969).

A response of viscous heat-conducting compressible fluid to an abrupt change of angular velocity of a containing thermally insulated circular cylinder under the existence of stable distribution of the temperature is investigated within the framework of the Boussinesq approximation for a time duration of the order of the homogeneous-fluid spin down time in order to resolve the Holton-Pedlosky controversy. The best way to resolve the solar spin down controversy is to investigate the unsteady rotational motion of the model solar interior.

022.011   Classical calculation of inverse bremsstrahlung cross sections in screened potentials.
N. Gyldén, B. Einarsson.
Journ. Quant. Spectrosc. Radiat. Transfer, Vol. 9, 1117 - 1131 (1969).

The Thomas-Fermi potential has been used for calculating atomic cross sections for continuous emission and absorption of radiation. The emission cross section follows a simple scaling law in the atomic number Z. Absorption cross sections are given as a correction factor γ to Kramers' formula. Results are reproduced for $Z = 10, 26, 47$ and 82 for a number of densities and temperatures. Values of γ far below unity are found for low temperatures. At high temperatures, γ may exceed unity. This result is believed to be caused by the inapplicability of Kramers' approximation in those cases.

022.012   Population inversions in ions of astrophysical interest.   H. A. Smith.
Astrophys. Journ. Vol. 158, 371 - 383 (1969).

Simple atomic configurations are examined theoretically for possible population inversions under conditions of astrophysical interest and under the assumption of excitation by electron collisions. Under a range of density and temperature conditions covering those found in planetary nebulae, these ions are found to have transitions that fall into two categories:

those inverted at all low electron densities (extending up to $N_e \approx 10^{10}$ cm^{-3}); and those inverted over a short range of densities ($\Delta N_e \approx 10^2$ cm^{-3}).

022.013 Absolute transition probabilities for some electronic states of CS, SO and S$_2$. W. H. Smith.
Journ. Quant. Spectrosc. Radiative Transfer, Vol. 9, 1191 - 1199 (1969).

Radiative lifetimes of upper states of CS, SO, and S$_2$ have been measured via a phase shift method using modulated electron beam excitation. Lifetimes between 16.2 nsec for SO($B\,^3\Sigma^- - X\,^3\Sigma^-$, $v' = 2$) and 339 nsec for CS($A\,^1\Pi - X\,^1\Sigma$, $v' = 1$) were determined. Substantial lifetime variations are found for some vibrational sequences. Using published Franck-Condon factors, $A_{v'v''}$ values are derived, and used to obtain absolute oscillator strengths or electric transition moments.

022.014 Determination of the gravitational constant G.
R. D. Rose, H. M. Parker, R. A. Lowry, A. R. Kuhlthau, J. W. Beams.
Phys. Rev. Letters, Vol. 23, 655 - 658 (1969).

A new method for measuring the gravitational constant G is described. Preliminary measurements give $G = (6.674 \pm 0.012) \times 10^{-11}$ N m^2/kg^2 where the 0.012 represents 3 standard deviations. Furthermore there is reason to believe that with certain modifications of the apparatus and use of improved metrology techniques an increase in precision of at least one and probably two orders of magnitude will be obtained.

022.015 Quarks – Urbausteine der Materie? V. Linke
Umschau, Vol. 69, 612 - 615 (1969).

022.016 Lichtgeschwindigkeitsmessungen mit Kerrzelle und KDP-Lichtmodulator.
A. Karolus, J. Helmberger.
Deutsche Geod. Kommission Bayer. Akad. Wiss. Reihe A, Heft No. 65, 33 pp. (1969).

Die beschriebenen Messungen der Lichtgeschwindigkeit mit einer Kerrzelle und mit einem KDP-Lichtmodulator verfolgten zwei Ziele: Sie sollten nachweisen, daß der Einfluß der Phasenstruktur bei früheren Messungen in ausreichendem Maße berücksichtigt worden ist, daß das Meßergebnis also nicht vom Modulator abhängt. Vergleichbare Messungen mit verschiedenen Lichtmodulatoren sollten Angaben über die Leistungsfähigkeit und über Vor- und Nachteile dieser Bauelemente liefern, wenn man sie zur Helligkeitsmodulation von Laserlicht im hier gewählten Meßaufbau einsetzt.

022.017 Energy levels, oscillator strengths and forbidden transition probabilities in the Sc II isoelectronic sequence. B. Warner, R. C. Kirkpatrick.
Monthly Notices, Roy. Astron. Soc. Vol. 144, 397 - 410 (1969).

Empirical Slater parameters are derived from a study of energy levels in the isoelectronic sequence Sc II–Ni IX. Positions for some unobserved levels are predicted. Electric dipole oscillator strengths have been calculated for all possible transitions. Electric quadrupole and magnetic dipole transitions between and within the $3d^2$ and $3d4s$ configurations are given.

022.018 Adiabatische Invarianz des Wirkungsintegrals für die Bewegung in nicht-regulären Kraftfeldern.
D. Pfirsch, K. Schindler.
Sitz.-Ber. Bayer. Akad. Wiss., Math.-Naturwiss. Kl. 1968, p. 31 - 43 (1969).

Es ist das Ziel dieser Arbeit, die adiabatische Invarianz des Wirkungsintegrals für die Bewegung in nicht-regulären Kraftfeldern auf einfache und anschauliche Weise herzuleiten und zu verallgemeinern. Dabei wird insbesondere ein Sachverhalt, der für die Theorie der kosmischen Strahlung von Bedeutung ist und der bisher nur durch numerische Rechnungen

gewonnen werden konnte, auf eine allgemeine theoretische Aussage zurückgeführt.

022.019 Energy levels in the Debye field.
D. Schlüter, S. K. Tsoi.
Astron. Astrophys. Vol. 3, 147 - 149 (1969).

The energy levels of an electron in a Debye field have been repeatedly investigated by various authors owing to their significance for the calculation of the ionization potential and the partition function in a plasma. Whereas according to Bonch-Bruevich and Glasko the number of energy levels is finite, Rouse maintains that there are at least as many as a "pure" Coulomb field would yield, viz. infinite. By numerical integration of the Schrödinger equation we have been able to locate the energy levels and confirmed the results of Bonch-Bruevich and Glasko.

022.020 Far-ultraviolet absorption spectra with auto-ionized levels of beryllium and magnesium.
G. Mehlman-Balloffet, J. M. Esteva.
Astrophys. Journ. Vol. 157, 945 - 956 (1969).

This paper describes new features in the beryllium and magnesium absorption spectra from 2000 to 500 Å. The observations have been made using two 3-electrode vacuum sparks, one serving as a background source, the other as a source of the absorbing metallic vapor. New auto-ionized line series have been identified and associated with transitions from the ground state to two-electron excitation states. Most of these resonances, owing to the interaction of the excited states with the photo-ionization continuum, present a shape which can fit into the general profile formulated by Fano's theory.

022.021 Measurements of the Stark broadening of Hγ.
R. D. Bengtson, M. H. Miller, W. D. Davis, J. R. Greig.
Astrophys. Journ. Vol. 157, 957 - 961 (1969).

The Stark broadening of Hγ has been investigated in three independent experiments covering the temperature range $1 \leq kT_e \leq 2$ eV and electron-density range from 2×10^{16} to 2×10^{17} cm^{-3}.

022.022 Calculated electron and ion Stark broadening of the allowed and forbidden $2P-nL$ ($n \geq 5$, $L = 1, 2, \ldots$, $n - 1$) triplet and singlet transitions in neutral helium.
H. A. Gieske, H. R. Griem.
Astrophys. Journ. Vol. 157, 963 - 981 (1969).

Profiles of the He I (2^3P-5^3D)-line at 4026.2 Å inclusive of the forbidden components at 4045.1 Å (2^3P-5^3P), 4025.5 Å (2^3P-5^3F), and 4025.3 Å (2^3P-5^3G), are derived, by use of the quasi-static approximation to account for the perturbing ions and by use of the impact approximation as generalized for overlapping lines to describe the electron effects. Tabulated profiles of the line absorption coefficient are presented.

022.023 Additional shock-tube measurements of absolute Cr I gf-values. S. J. Wolnik, R. O. Berthel, E. H. Carnevale, G. W. Wares.
Astrophys. Journ. Vol. 157, 983 - 995 (1969).

Absolute gf-values for eighty-six Cr I lines have been determined from emission of shock-heated gas mixtures. The present work is continuation of a previous paper, which reported gf-values for forty-one Cr I and two Cr II lines.

022.024 Nontransferable correlation effects and multiplet oscillator strengths for electric dipole transitions in atoms with results on C II, N I, N II, N III, O II, O III, O IV, F II, and Ne II. P. Westhaus, O. Sinanoğlu.
Astrophys. Journ. Vol. 157, 997 - 1005 (1969).

Wave functions which include selected, state-dependent

correlation effects are used to evaluate multiplet oscillator strengths for a number of electric dipole transitions in the first-row atoms and ions.

022.025   **The effects of electron and radiation density on dielectronic recombination.**
A. Burgess, H. P. Summers.
Astrophys. Journ. Vol. 157, 1007 - 1021 (1969).

The statistical-equilibrium populations of excited states of He, O^{+2}, O^{+6}, Ca, Fe^{+7} and Fe^{+14} for a wide range of electron density and temperature have been calculated, including the effects of dielectronic recombination as well as radiative and three-body recombination, together with collisional and radiative transitions between all states. We define, and results are given for, the corresponding overall recombination coefficient, for which we propose the name collisional-dielectronic recombination coefficient.

022.026   **Excitation of the auroral green line of atomic oxygen ($^1S \rightarrow {}^1D$) by $N_2(A^3\Sigma_u{}^+)$.**
J. A. Meyer, D. W. Setser, D. H. Stedman.
Astrophys. Journ. Vol. 157, 1023 - 1025 (1969). – Note.

022.027   **On Compton scattering on relativistic polarized electrons.**   V. P. Gavrilov, I. A. Nagorskaya, V. A. Khoze.
Izv. Akad. Nauk Arm. SSR (Fiz), Vol. 4, Vyp. (No.) 3, p. 137 - 141 (1969).   In Russian.

The polarization properties of γ-quanta, produced in Compton scattering of laser photons by relativistic polarized electrons are discussed. The expressions are derived for the polarization parameters of photons, emitted in the scattering of intense wave by polarized electrons.

022.028   **Theory of atomic structure including electron correlation. III. Calculations of multiplet oscillator strengths and comparisons with experiments for CII, NI, NII, NIII, OII, OIII, OIV, FII, NeII, and NaIII.**
P. Westhaus, O. Sinanoğlu.
Phys. Rev., Second Series, Vol. 183, 56 - 68 (1969).

022.029   **Calculation of bremsstrahlung cross sections with Sommerfeld-Maue eigenfunctions.**
G. Elwert, E. Haug.
Phys. Rev., Second Series, Vol. 183, 90 - 105 (1969).

022.030   **Sublimation of ice water at low temperatures.**
E. A. Kajmakov, V. I. Sharkov.
Astrometriya i Astrofiz., *Kiev*, No. 4, p. 209 (1969).
In Russian. – Abstract.

022.031   **Behavior of dust particles during sublimation of the ice in the system ice water—dust.**
E. A. Kajmakov, V. I. Sharkov.
Astrometriya i Astrofiz., *Kiev*, No. 4, p. 210 (1969).
In Russian. – Abstract.

022.032   **Emission spectra of nitrogen, oxygen, and air ($\lambda\lambda$ 7000 – 11500 Å) excited by fast electrons.**
Ya. M. Fogel', A. G. Koval', V. T. Koppe, V. V. Gritsyna.
Astrometriya i Astrofiz., *Kiev*, No. 4, p. 211 (1969).
In Russian. – Abstract.

022.033   **Derivation of the blackbody radiation spectrum without quantum assumptions.**   T. H. Boyer.
Phys. Rev., Second Series, Vol. 182, 1374 - 1383 (1969).

The Planck radiation law for the blackbody radiation spectrum is derived without the formalism of quantum theory. The hypotheses assume (a) the existence, at the absolute zero of temperature, of classical homogeneous fluctuating electromagnetic radiation with a Lorentz-invariant spectrum;

(b) that classical electrodynamics holds for a dipole oscillator; (c) that a free particle in equilibrium with blackbody radiation has the classical mean kinetic energy $1/2\ kT$ per degree of freedom.

022.034   **Developments in the theory of synchrotron radiation and its reabsorption.**
V. L. Ginzburg, S. I. Syrovatskii.
Annual Rev. Astron. Astrophys. Vol. 7, 375 - 420 (1969).

022.035   **Spectrum of the two-photon emission from the metastable state of singly ionized helium.**
C. J. Artura, N. Tolk, R. Novick.
Astrophys. Journ. *(Letters)*, Vol. 157, L181 - L186 (1969).

Verification has been made of the theoretically predicted spectral distribution of the two-photon emission from the metastable $2\,^2S_{1/2}$ state of singly ionized helium by means of a broad-band spectroscopic coincidence counting technique.

022.036   **Determination of the complex index of refraction of rocks and minerals.**
W. G. Egan, J. F. Becker.
Applied Optics, Vol. 8, 720 - 721 (1969).

022.037   **$3s$ – $3p$ transitions of Ne VII and Ne VIII.**
W. D. Johnston III, H.-J. Kunze.
Astrophys. Journ. Vol. 157, 1469 - 1470 (1969).

The transitions $3s$ – $3p$ of Ne VII and Ne VIII have been identified in a high-temperature plasma produced in a theta-pinch discharge.

022.038   **Production rate of ^{10}Be from oxygen spallation.**
P. S. Goel.
Nature, Vol. 223, 1263 - 1264 (1969).

An analysis of data on cosmogenic Be^{10} in chondritic meteorites permits a reliable estimate of Be^{10} production cross section in high energy spallation of oxygen nuclei. The cross section is (1.8 ± 0.5) mb. This is in good agreement with earlier radiochemical measurements but disagrees with the more recent work of the Orsay group.

022.039   **On the dependence of cross sections of the collision excitation of hydrogen atoms on principal quantum numbers of lower and upper levels.**   I. A. Krinberg.
Astron. Zhurn. Akad. Nauk SSSR, Vol. 46, 993 - 997 (1969).
In Russian. English translation in Soviet Astron. AJ, Vol. 13, No. 5.

022.040   **On the negative reabsorption of synchrotron radiation.**   V. N. Sazonov.
Astron. Zhurn. Akad. Nauk SSSR, Vol. 46, 1016 - 1018 (1969).   In Russian.   English translation in Soviet Astron. AJ, Vol. 13, No. 5.

It is shown that the negative reabsorption (increase) of synchrotron radiation is possible not only in the case when relativistic electrons in the magnetic field are dipped into non-relativistic plasma but also in the case when non-relativistic plasma is absent completely.

022.041   **Spektrale Emission bei der Bildung negativer Ionen.**
H.-P. Popp.
Umschau, Vol. 69, 770 (1969).

022.042   **Effects of strong collisions and lower-state broadening on the H_γ profile.**
M. E. Bacon, D. F. Edwards.
Journ. Quant. Spectrosc. Radiative Transfer, Vol. 9, 951 - 958 (1969).

Calculations have been made of the Stark-broadened H_γ profile which take into account the effect of strong collisions in the electron perturbation and the effect of electron impact broadening of the lower levels. Lower level broadening is shown to be insignificant for this line with the dominant effect being the strong collision term. The agreement between the calculated half-widths and the experimental half-widths for the H_γ line is within experimental error.

022.043 Electronic strengths of the first positive system of N_2 and of the Meinel system of N_2^+ - II.
E. B. Kuprianova, V. N. Kolesnikov, N. N. Sobolev.
Journ. Quant. Spectrosc. Radiative Transfer, Vol. 9, 1025 - 1032 (1969).

The electronic band strengths of the first positive system of N_2 and of the Meinel system of N_2^+ have been calculated on the basis of previous absolute intensity measurements (on very pure nitrogen) in a spectral range from 0.5 to 1.1 μ at $T = 6900°$K and $P = 1$ atm.

022.044 On the evolution of a plasma interacting with radiation. W. Kröll.
Journ. Quant. Spectrosc. Radiative Transfer, Vol. 9, 1331 - 1341 (1969).

One of the basic problems in the theory of irreversible thermodynamics is the formulation of general evolution criteria. A universal criterion of this kind for macroscopic systems with time independent boundary conditions has been derived by Glansdorff and Prigogine. The derivation refers to transport processes in matter like diffusion, heat conduction and chemical reactions. Radiative transfer has not been included. In this investigation, we use statistical mechanics to extend the above evolution criterion and the principle of minimum entropy production to partially ionized, collision-dominated plasmas interacting with radiation. We consider a plasma with collisional – and radiative excitation – and ionization phenomena not restricted to local equilibrium.

022.045 Photoelectron spectra and partial photoionization cross-sections for carbon dioxide.
J. L. Bahr, A. J. Blake, J. H. Carver, V. Kumar.
Journ. Quant. Spectrosc. Radiative Transfer, Vol. 9, 1359 - 1364 (1969).

A photoelectron spectrometer has been used to determine partial photoionization cross-sections for carbon dioxide for monochromatic incident radiation throughout the wavelength range 584 - 720 Å.

022.046 A radiation standard for the vacuum ultraviolet.
J. C. Morris, R. L. Garrison.
Journ. Quant. Spectrosc. Radiative Transfer, Vol. 9, 1407 - 1418 (1969).

A constricted dc argon arc to which nitrogen or oxygen has been added has been calibrated as a radiation standard for the vacuum ultraviolet. The calibration data are given for the strongest atomic lines in the wavelength range of 1800 - 1000 Å. Operation for calibrated output of continuum or blackbody radiation down to 600 Å is given for a helium or neon arc using hydrogen or argon as the radiating gas. The source has been found to be stable and reproducible, thereby allowing use of the comparison method to measure the number of photons incident on the entrance slit of a spectrometer.

022.047 Nonstationary radiation field in infinite media.
D. I. Nagirner.
Astrofizika, Vol. 5, 31 - 53 (1969). In Russian.
English translation in Astrophysics, Vol. 5, No. 1 (1969).

The nonstationary resonance line radiation field in stationary infinite and semi-infinite media is considered. The scattering is assumed to be isotropic and either monochromatic or completely frequency redistributed. The time lag due to the finite velocity of light is neglected. The solutions of the basic integral equations, and the intensity of the emergent radiation are found explicitly.

022.048 Oscillator strengths in complex atoms: Application to N IV. H. Nussbaumer.
Monthly Notices, Roy. Astron. Soc., Vol. 145, 141 - 150 (1969).

N IV is responsible for some prominent features in Wolf Rayet stars. There is however a curious absence of $\lambda 5820$ $2p3p\ ^3P - 2p3d\ ^3P^0$. Neither has this line been found in the laboratory. This work was undertaken with a view to explaining that absence, but the methods employed and the corresponding computer program may be used for the calculation of oscillator strengths of any complex atom. It is based on a program for the calculation of atomic structure allowing for configuration interaction. The present paper gives a first description of the use of this program in conjunction with a program for the calculation of radiative transition probabilities. Results are obtained for the four-electron system of N IV, since this system is of interest for the interpretation of the spectra of Wolf Rayet stars.

022.049 Physical model of hydrodynamic turbulence.
E. N. Parker.
Phys. Fluids, Vol. 12, 1592 - 1604 (1969).

A model for statistically isotropic homogeneous turbulence in an incompressible fluid is constructed, representing the turbulence as a superposition of individual vortex sheets. Each vortex sheet moves in the velocity field of the other sheets which have larger scale. Each sheet is stretched out, and intensified, until obliterated by viscosity at high wavenumber. The model is sufficiently explicit so as to give information on the correlation between different Fourier components of the turbulence. The calculations give the probability of finding a particular value of the vorticity or velocity at a given position and time in the turbulent field.

022.050 The Voigt analog of an Elsasser band.
S. A. Golden.
Journ. Quant. Spectrosc. Radiat. Transfer, Vol. 9, 1067 - 1081 (1969).

An algebraic expression for the spectral absorption coefficient of the Voigt analog of an Elsasser band (i.e. an infinite array of equally spaced, equally intense spectral lines all having a Voigt profile with the same Lorentz and Doppler half-widths) has been obtained.

022.051 Self-reversed profiles of lines broadened by the Stark effect. B. Ya'akobi.
Journ. Quant. Spectrosc. Radiat. Transfer, Vol. 9, 1097 - 1103 (1969).

A simple method is described for employing a self-reversed Stark-broadened spectral line for estimating the average electron concentration in both the regions where the line is emitted and where it is absorbed. The former, by fitting calculated Stark profiles to the wings of the observed line; the latter, by solving the radiative transfer equation assuming a simplified model.

022.052 Calculated transition array for the configurations $3d^2$ - $3d4p$ in Ti III. H. Mendlowitz.
Astrophys. Journ. Vol. 158, 385 - 388 (1969).

Line strengths have been calculated in intermediate coupling for transitions between the configurations $3d^2$ and $3d4p$ in Ti III. The results support the conclusion by Bahcall and others that an earlier identification of Ti III in the quasi-stellar object PKS 0237 – 23 must be rejected.

022.053 Lifetimes of some Fe I states by beam-foil spectroscopy. W. Whaling, R. B. King, M. Martinez-Garcia.
Astrophys. Journ. Vol. 158, 389 - 398 (1969).

The light emitted by 500-keV Fe atoms excited by passage through a thin carbon foil has been analyzed spectroscopically to identify fifty known transitions in Fe I. The light decay downstream from the foil has been measured to find the lifetime of six levels in Fe I between 48 and 57 kK. Our lifetimes are longer than those computed by summing the transition probabilities of Corliss and Tech by a factor that increases from 4.5 at 48 kK to 21 at 57 kK. The dependence of this factor on excitation energy raises doubt about the level population assumed by Corliss and Tech in interpreting the relative line intensities from the arc source.

022.054 **New spectra of the CH molecule.**
G. Herzberg, J. W. C. Johns.
Astrophys. Journ. Vol. 158, 399 - 418 (1969).

Absorption spectra of CH and CD molecules have been studied in the ultraviolet and vacuum ultraviolet by means of the flash photolysis of diazomethane. The three known systems $A - X$, $B - X$, and $C - X$ are fairly strong in absorption. Of these the last two, $B - X$ and $C - X$, have been extended. In particular, in the $B - X$ system diffuseness of the higher rotational lines has been observed. A number of new electronic transitions have been found in the vacuum ultraviolet, including a Rydberg series starting at 1370 Å and yielding an ionization potential of 10.64 eV. At longer wavelengths new electronic transitions occur at 1690, 1560, 1540, and 3007 Å.

022.055 **Cross-sections for destruction of ^{6}Li and ^{7}Li by low-energy protons.**   J. Audouze, H. Reeves.
Astrophys. Journ. Vol. 158, 419 - 421 (1969).

In this note, we give S Gamow factors based on experimental determination of cross-sections for destruction of ^{6}Li and ^{7}Li by low-energy protons. Indeed, accurate determinations of these cross-sections are very important in several domains in astrophysics.

022.056 **Spin change in collisions of hydrogen atoms.**
A. C. Allison, A. Dalgarno.
Astrophys. Journ. Vol. 158, 423 - 425 (1969).

Accurate calculations are presented of the cross-sections for spin change in the collision of a pair of hydrogen atoms.

022.057 **Zur Physik der letzten 25 Jahre.**
H. Rechenberg.
Phys. Blätter, 25. Jahrgang, 481 - 487 (1969).

022.058 **Spin-forbidden resonance multiplets in light elements.**
B. Edlén, H. P. Palenius, K. Bockasten, R. Hallin, J. Bromander.
Solar Physics, Vol. 9, 432 - 438 (1969).

We present new laboratory data on some multiplets in nitrogen, oxygen and fluorine, and discuss the Z-dependence of their wave-numbers. These multiplets are very faint in laboratory light sources, but can become prominent in astrophysical sources of low density. Our results confirm the solar identifications of the nitrogen and oxygen multiplets made by Burton et al. Predicted positions of the corresponding multiplets in neon are given.

022.059 **Resonances in collision strengths for excitation of [O II] and [O III].**
W. Eissner, P. de A. P. Martins, H. Nussbaumer, H. E. Saraph, M. J. Seaton.
Monthly Notices, Roy. Astron. Soc., Vol. 146, 63 - 70 (1969).

In an earlier paper the calculations of resonance structures in the collision strengths are described. In the present paper these results are presented in a form convenient for astrophysical applications.

022.060 **Magnetic-quadrupole radiation and solar coronal de-excitation.**   R. H. Garstang.
Publ. Astron. Soc. Pacific, Vol. 81, 488 - 495 (1969).

Transition probabilities have been computed for a number of spectral lines due to magnetic-quadrupole radiation. Three such lines, in Fe IX, Fe XVII, and Fe XXV, appear to be of importance in the de-excitation of excited atoms in the solar corona.

022.061 **Broadening of hydrogen lines in a neutral medium.**
M. C. Lortet, E. Roueff.
Astron. Astrophys. Vol. 3, 462 - 467 (1969).

The effects of dispersion van der Waals interaction between hydrogen atoms are taken into account for hydrogen lines, in addition to the resonance interaction first introduced by Cayrel et al. (1960) in a neutral medium. Although the widths relative to each interaction are not simply additive the van der Waals interaction is seen to give a contribution for the Balmer lines and even to be preponderant for Hδ. Besides, we question the validity of the $1/r$ expansion for the interatomic potential for Hγ and Hδ lines, since the range of the interaction is less than the dimensions of the atom in its correspondent state.

022.062 **Ultraviolet wavelengths and oscillator strengths for $3d-nf$ transitions in the helium isoelectronic sequence.**   R. T. Brown.
Astrophys. Journ. Vol. 158, 829 - 837 (1969).

This paper reports the results of variational calculations of energies and wave functions for the three lowest ^1F and ^3F states of the helium isoelectronic sequence through S XV, and oscillator strengths for transitions to these states from the 3^1D and 3^3D states.

022.063 **Diamagnetic Zeeman effect and magnetic configuration mixing in long spectral series of Ba I.**
W. R. S. Garton, F. S. Tomkins.
Astrophys. Journ. Vol. 158, 839 - 845 (1969).

Diamagnetic shifts and pronounced effects of mixing between the magnetic sublevels of configurations of differing l and n have been observed in the principal series of Ba I, photographed in absorption to about the $n = 75$ member with a 30-foot spectrograph. An unexpected feature of the spectra is the presence of almost equally spaced absorption maxima, in σ-polarization, which extend across the zero-field series limit to well inside the corresponding ionization continuum, with an interval of about 1.5 times the σ-splitting.

022.064 **Investigation of the rotating and oscillating distributions of molecules excited by different elementary processes, and the importance for astrophysics.**
G. N. Polyakova, Ya. M. Fogel.
Problemy kosmich. fiz. No. 4, p. 103 - 111 (1969).
In Russian.

022.065 **Infra-red emission spectra of some molecular gases excited by fast electrons.**
A. G. Koval, V. T. Koppe, Ya. M. Fogel.
Problemy kosmich. fiz. No. 4, p. 112 - 133 (1969).
In Russian.

Infra-red emission spectra (spectral region 7000 – 11500 Å) of the molecules N_2, O_2, CO, CO_2, NO, CH_4, NH_3, H_2O excited by electrons with an energy of 13 keV were investigated. The nitrogen spectrum was also investigated at the energy of 150 eV. The spectra excited by fast electrons were compared with those excited by slow electrons and protons.

022.066 **Electron impact cross sections for CO_2.**
D. J. Strickland, A. E. S. Green.
Journ. Geophys. Res. Vol. 74, 6415 - 6424 (1969).

022.067 **Multiple "Compton" losses of energy of high-speed electrons and their possible role in various cosmic objects.**   G. G. Getmantsev, Yu. V. Tokarev.

Izv. vyssh. uchebn. zavedenij. Radiofizika, Vol. 12, 624 - 625 (1969). In Russian. – Abstr. in Referativ. Zhurn. 51. Astron., 12.51.309 (1969).

022.068 A study on the statistical equilibrium of the ion A XIV level-population in coronal conditions.
J. Lexa.
Bull. Astron. Inst. Czechoslovakia, Vol. 20, 373 - 380 (1969).

Transition probability coefficients for transitions between energy levels of the ion A XIV have been computed. Beside permitted spontaneous and collisional transitions also spontaneous, radiative and collisional transitions between levels of the fundamental term $2p^2 P^0$ have been taken into consideration. For the afore-mentioned energy levels of the ion A XIV a system of equations of statistical equilibrium have been solved.

022.069 Electronic transition moment for the N_2 Vegard–Kaplan bands.   A. L. Broadfoot, S. P. Maran.
Journ. Chemical Physics, Vol. 51, 678 - 681 = Contr. Kitt Peak National Obs. No. 422 (1969).

Measurements of relative band intensities for the N_2 Vegard–Kaplan bands with $v' = 0$ are used to find a curve that represents the smooth variation of the electronic transition moment with internuclear separation.

022.070 Lifetime of the N_2 Vegard–Kaplan system.   D. E. Shemansky, N. P. Carleton.
Journ. Chemical Physics, Vol. 51, 682 - 685 = Contr. Kitt Peak National Obs. No. 423 (1969).

The results of an experiment by Carleton and Oldenberg, designed for the measurement of the relative lifetimes of the $N_2 B^3 \Pi_g$ and $A^3 \Sigma_u^+$ states, have been reanalyzed.

022.071 N_2 Vegard–Kaplan system in absorption.   D. E. Shemansky.
Journ. Chemical Physics, Vol. 51, 689 - 700 = Contr. Kitt Peak National Obs. No. 413 (1969).

Seven bands $(6,0 - 12,0)$, of the N_2 Vegard–Kaplan system have been observed in the absorption spectrum of 12 m atm of gas. The measuring instrument was a 2-m scanning spectrometer with a resolution of about 30000. Pulse-counting techniques, coupled with the use of a theoretical model based on a measured collision broadening coefficient, have allowed fairly accurate estimates of the transition probabilities of the observed bands.

022.072 The number of scattering of a photon in an inhomogeneous medium.   V. Yu.Terebizh.
Soobshch. Byurakan. Obs. No. 40, p. 76 - 85 (1969). In Russian.

The mean number of scattering of a photon in an inhomogeneous medium is determined. It is assumed that the scattering albedo is an arbitrary function of the position in the medium.

022.073 Effect of a magnetic field and Coriolis forces on Rayleigh–Taylor's instability.   J. M. Gandhi.
Canadian Journ. Phys., Vol. 47, 1621 - 1635 (1969).

We present variational principles which characterize the solution of the equilibrium of a plane horizontal layer of an incompressible, electrically conducting fluid of electrical conductivity σ e.m.u., of magnetic permeability K, having a variable density $\rho(z)$ in the vertical z direction, which is also the direction of gravity having acceleration g, and of viscosity $\mu(z)$ and which is rotating at Ω radians per second about the vertical axis in the presence of a horizontal magnetic field for two cases.

022.074 Spectrum of the hydroxyl radical.   C. Carlone, F. W. Dalby.

Canadian Journ. Phys., Vol. 47, 1945 - 1957 (1969).

We have investigated the $B^2 \Sigma^+ \to A^2 \Sigma^+$ and $C^2 \Sigma^+ \to A^2 \Sigma^+$ systems of OH and OD at high resolution.

022.075 The $n^3P_2 - 1^1S_0$ magnetic-quadrupole transitions of the helium sequence.   G. W. F. Drake.
Astrophys. Journ., Vol. 158, 1199 - 1203 (1969).

Accurate variational calculations are presented for the magnetic-quadrupole decay rates of the heliumlike ions He I to Ne IX.

022.076 Dielectronic recombination.   B. W. Shore.
Astrophys. Journ., Vol. 158, 1205 - 1218 (1969).

This paper derives the rate coefficient for dielectronic recombination (inverse auto-ionization) by using results from the quantum theory of resonance-collision processes, with attention to coupling schemes, degeneracy, and overlapping resonances; numerical results, derived by using screened hydrogenic wave functions, are presented for several configurations.

022.077 Ba I absorption-line series at high resolution.   W. R. S. Garton, F. S. Tomkins.
Astrophys. Journ., Vol. 158, 1219 - 1230 (1969).

The Ba I ultraviolet absorption spectrum has been re-examined at high dispersion by use of the second and third orders of the Argonne 30-foot spectrograph. The principal series has been extended to $n = 75$. Configuration mixing and auto-ionization effects, revealed by the high-dispersion spectra, are discussed qualitatively.

022.078 C V spectra near the $1s–2p$ line of C VI.   U. Feldman, L. Cohen.
Astrophys. Journ. (*Letters*), Vol. 158, L169 - L170 (1969).

022.079 Precision constants of the cyanogen-red system with a perturbation analysis.   T. Fay, I. Marenin, W. van Citters.
Bull. American Astron. Soc., Vol. 1, 342 (1969). – Abstract AAS.

022.080 New vibrational and equilibrium constants for the C_2 Phillips system.   H. R. Johnson, I. R. Marenin.
Bull. American Astron. Soc., Vol. 1, 349 (1969). – Abstract AAS.

022.081 Extension of shock-tube measurements of absolute Cr I gf values to higher excitation potentials.
G. W. Wares, S. J. Wolnick, R. O. Berthel.
Bull. American Astron. Soc., Vol. 1, 367 - 368 (1969). – Abstract AAS.

022.082 Redetermination of the absolute oscillator-strength of the Fe I resonance line λ = 3720 Å by optical double resonance.   R. Wagner, E. W. Otten.
Zeitschr. Physik, Vol. 220, 349 - 361 (1969).   In German.

022.083 Formulas and graphs for a quantitative analysis of the radiation of forbidden lines of emission objects.
A. A. Boyarchuk, R. E. Gershberg, N. V. Godovnikov, V. I. Pronik.
Izv. Krymskoj Astrofiz. Obs. Vol. 39, 147 - 162 (1969). In Russian.

Formulas and graphs necessary for a quantitative analysis of the radiation of forbidden lines of emission objects are given. The results of an earlier paper are precised and graphs are given for a quick determination of the physical conditions in a radiative medium by Seaton's method of "intersection of curves". A common scheme is considered for computer calculations of the ratios of line intensities and absolute line luminosities , considering the dependence of these values on electron temperature, electron density, temperature of ionizing radiation and dilution coefficient.

022.084 **Differential term displacements in spectra of Fe I, Ti I, Ni I.** V. K. Prokofiev, G. A. Terez.
Izv. Krymskoj Astrofiz. Obs. Vol. 39, 170 - 185 (1969). In Russian.

Term displacements in the spectra of Fe I, Ti I and Ni I with the change from atmospheric pressure to a lower one are considered. The sensitivity about the displacements of the terms of higher multiplicity, determined by Babcock, is confirmed.

022.085 **Oscillator strengths for Ti I spectra.**
V. K. Prokofiev, T. A. Ratobylskaya.
Izv. Krymskoj Astrofiz. Obs. Vol. 39, 186 - 231 (1969). In Russian.

Measurements and calculations of relative values of oscillator strengths in Ti I spectra available in the literature are revised critically. A summary of recommended values for 737 lines, belonging to 255 multiplets, is compiled. A formula of the transition from relative values given in King's scale to absolute ones is determined.

022.086 **Diagrams for the application of the Fowler-Milne method for some neon lines.**
A. Petrakiev, T. Vörös.
Applied Optics, Vol. 8, 2152 - 2153 (1969).

The Fowler-Milne method can be used for the determination of temperature of an axially or spherically symmetrical plasma in local thermodynamic equilibrium. This paper gives the calculated and normalized intensity-temperature curve for twelve lines of Ne.

022.087 **Interferometrically measured thorium lines between 2747 and 4572 Å.**
D. Goorvitch, F. P. J. Valero, A. L. Clúa.
Journ. Optical Soc. America, Vol. 59, 971 - 975 (1969).

Interferometrically measured wavelengths are given for 278 thorium lines in the range 2747–4572 Å.

022.088 **Beam-foil excitation of multiply ionized neon.**
A. Denis, J. Desesquelles, M. Dufay.
Journ. Optical Soc. America, Vol. 59, 976 - 980 (1969).

The beam-foil technique has been used to obtain an excited ionized neon beam. Spectroscopic investigations in the wavelength range 2000–6000 Å have revealed many new lines due to multiply ionized neon.

022.089 **Arc measurement of some Ar II optical transition probabilities.** J. B. Shumaker, Jr.,
C. H. Popenoe.
Journ. Optical Soc. America, Vol. 59, 980 - 985 (1969).

022.090 **Transition probabilities for prominent Ar I lines in the near infrared.** W. L. Wiese, J. M. Bridges,
R. L. Kornblith, D. E. Kelleher.
Journ. Optical Soc. America, Vol. 59, 1206 - 1212 (1969).

Relative transition probabilities of 81 infrared Ar I lines in the wavelength range from 9000–24000 Å have been measured.

022.091 **Lifetime and transition probabilities of $np^4(n+1)p$ states of Ne II, Ar II, and Kr II.**
S. H. Koozekanani, G. L. Trusty.
Journ. Optical Soc. America, Vol. 59, 1281 - 1284 (1969).

Lifetime as well as transition probabilities of the first p-excited states of neon⁺, argon⁺, and krypton⁺ have been calculated.

022.092 **Transition probabilities and g values for neon I.**
R. Mehlhorn.
Journ. Optical Soc. America, Vol. 59, 1453 - 1454 (1969).

Measured g values were used to obtain improved eigenvectors for the $2p^5 3p$ configuration of Ne I.

022.093 **Gravitation of the vacuum?** I. D. Novikov.
Zemlya i Vselennaya, No. 5, p. 36 - 41 (1969). In Russian.

022.094 **The effect of electron screening of thermonuclear reactions under high densities.**
V. V. Porfiriev, Yu. N. Redcoborody.
Astrofizika, Vol. 5, 393 - 413 (1969). In Russian. – Engl. translation in Astrophysics, Vol. 5, No. 3.

The effect of electron screening of the Coulomb field of a nucleus is considered. It causes an increase of the thermonuclear fusion rate. The effective potential is derived by means of the self-consistent field method based on the Hartree-Fock approximation. The influence of both reacting nuclei on the distribution of the screening space charge of electrons is taken into account. The screening contribution to interaction energy of colliding nuclei depends essentially on the self-energy of the screening electron cloud. The results differ greatly from the results of other authors.

022.095 **On the equilibrium states of a system of gravitating particles.**
G. S. Bisnovaty-Kogan, Ya. B. Zeldovich.
Astrofizika, Vol. 5, 425 - 431 (1969). In Russian. – Engl. translation in Astrophysics, Vol. 5, No. 3.

The self-consistent solutions of the kinetic equation in the proper gravitational field, which depend on the integrals of motion: energy and angular momentum, are obtained. For the sphere and cylinder the solutions are obtained with arbitrary degree of anisotropy in the velocity space. The solutions for the sphere in the isotropic case are the non-collisional analogy of polytropic solutions. The degenerated solutions with elliptic orbits and gravitational potential $\Phi = \alpha + \beta r^2$ are obtained for sphere, cylinder, and disk.

022.096 **Optical radio-frequency double resonance in molecules: The OH radical.**
K. R. German, R. N. Zare.
Phys. Rev. Letters, Vol. 23, 1207 - 1209 (1969).

022.097 **Stark broadening of two ionized-helium lines by collective electric fields in a laboratory plasma.**
H. R. Griem, H.-J. Kunze.
Phys. Rev. Letters, Vol. 23, 1279 - 1281 (1969).

022.098 **Analysis of some results of quark searches.**
R. K. Adair, H. Kasha.
Phys. Rev. Letter, Vol. 23, 1355 - 1358 (1969).

The interpretation of the results of Cairns, McCusker, Peak, and Woolcott, indicating a discovery of quarks in the cores of very energetic extensive air showers, is shown to be extremely difficult to reconcile with the results of other negative experiments. Alternative explanations of their results are then suggested.

022.099 **Magnetic susceptibility of neutron matter.**
J. W. Clark.
Phys. Rev. Letters, Vol. 23, 1463 - 1466 (1969).

The magnetic susceptibility of a neutron gas at zero temperature, an idealization of neutron-star matter, is estimated for the realistic, soft-core nucleon-nucleon potential of Reid.

022.100 **Hyperfine structure in the molecular ion H_2^+.**

K. B. Jefferts.
Phys. Rev. Letters, Vol. 23, 1476 - 1478 (1969).
30 transitions between hyperfine levels of H_2^+ have been observed.

022.101 Intensity measurements and rotational intensity distribution for the oxygen A-band.
J. H. Miller, R. W. Boese, L. P. Giver.
Journ. Quant. Spectrosc. Radiat. Transfer, Vol. 9, 1507 - 1517 (1969).

022.102 Transition probabilities for the $B^1\Sigma_u^+ - X^1\Sigma_g^+$ band system of H_2. A. C. Allison, A. Dalgarno.
Journ. Quant. Spectrosc. Radiat. Transfer, Vol. 9, 1543 - 1551 (1969).

The electronic dipole moment functions of Browne are used in calculations of the individual radiative transition probabilities for all the bands of the Lyman system of molecular hydrogen.

022.103 Emission cross section of N II lines produced by electron impact on nitrogen. B. N. Srivastava.
Journ. Quant. Spectrosc. Radiat. Transfer, Vol. 9, 1639 - 1641 (1969).

Cross sections have been measured for the emission of the N II lines when excited by electron impact on N_2 in the energy range 150 eV to 4 keV.

022.104 Zur Rotation einer axialsymmetrischen zähen Flüssigkeit um eine feste Achse. E. Schmutzer.
Wiss. Zeitschr. Friedrich-Schiller-Univ. Jena, Jahrgang 18, 187 - 193 (1969).

The axisymmetrical rotation of a viscous fluid is generally investigated on the basis of non-relativistic mechanics of continua and gravitation theory. Especially the balance of energy and angular momentum are studied. Maclaurin's spheroids and Jacobi's ellipsoids appear as special cases. New calculations are represented to the non-rigid rotation of homogeneous fluids. Some numerical tables apply to the problem of the shape of planets.

022.105 Nitric oxide gamma band emission rate factor.
J. B. Pearce.
Journ. Quant. Spectrosc. Radiat. Transfer, Vol. 9, 1593 - 1602 (1969).

The molecular fluorescence emission rate factors for the strong bands of the nitric oxide gamma system have been calculated using a high resolution solar spectrum and individual rotational line strengths. The results show that the intensity of the earth's day airglow, when viewed with resolution insufficient to resolve the details within a band, does not vary significantly with the temperature of the atmosphere.

022.106 Theoretical intensities of recombination lines.
L. Goldberg.
Report NASA–CR–96160, Harvard Coll. Obs., Cambridge, Mass., 42 pp. (1968). – See Phys. Abstr. Vol. 72, No. 24761 (1969).

022.107 Approximate estimate of the transition probabilities $2s^2\ ^1S-2s2p^{1,3}P$ and $3s^2\ ^1S-3s3p^{1,3}P$ in the isoelectronic coronal sequences. I. A. A. Nikitin.
Trudy Astron. Obs. *Leningrad*, Vol. 26 (= Uchenye Zapiski Leningr. Un-ta No. 347 = Seriya Matem. Nauk No. 44), p. 20 - 24 (1969). In Russian.

Semi-empirical methods are used to estimate the transition probabilities of both allowed and forbidden transitions $^1S-^{1,3}P$ for the isoelectronic sequences BeI, BII, CIII ... and MgI, AlII, SiIII ... These transition probabilities appear in some astrophysical problems.

022.108 The HEOS ion cloud experiment in the earth's magnetosphere. E. Rieger.
ESO Bull. No. 7, p. 35 - 43 (1969).

022.109 Effet du champ gravitationnel sur la mecanique ondulatoire. M. Missana.
Atti XII Riunione Soc. Astron. Italiana, L'Aquila 1968, p. 120 - 124 (1969). – Abstract SAI.

022.110 Magnetic dipole rotation spectrum of oxygen.
H. A. Gebbie, W. J. Burroughs, G. R. Bird.
Proc. Roy. Soc. London, Ser. A, Vol. 310, (No. 1503), 579 - 590 (1969).

The identification of the magnetic dipole rotation spectrum of oxygen in atmospheric studies has been confirmed by laboratory measurements in the spectral range 12 to 65 cm^{-1}, using interferometric techniques combined with long absorption paths and pressures of 2 to 3 atmospheres.

022.111 Concerning the energy independence of the velocity of light. B. C. Brown.
Nature. Vol. 224, 1189 (1969).

This letter discusses briefly various experimental and theoretical statements from the recent literature about the postulate that the velocity of light is independent of energy.

022.112 The scattering of quanta as a consequence of accidental events. D. A. Rozhkovsky.
Trudy Astrofiz. Inst. Alma-Ata, Vol. 14, 18 - 21 (1969). In Russian.

022.113 On the radiation of a charge in a magnetic field.
E. G. Michelkin.
Trudy Astrofiz. Inst. Alma-Ata, Vol. 14, 100 - 112 (1969). In Russian.

Transformation properties of a radiation field of relativistic charged particles and different representations of electromagnetic fluxes of energy are considered. On this basis the critical analysis of some exact classical solutions of the problem of radiation of a charge moving in a homogeneous magnetic field is given.

022.114 Scattering of electrons by C, N, O, N^+, O^+, and O^{++}. R. J. W. Henry, P. G. Burke, A.-L. Sinfailam.
Phys. Rev., Second Series, Vol. 178, 218 - 224 = Contr. Kitt Peak National Obs. No. 379 (1969).

022.115 Gli spettri molecolari. E. W. Salpeter.
La Metallurgia Italiana 1969, No. 8, p. 339 - 346 = Specola Vaticana, Miscellanea Astron. No. 118 (1969).

022.116 Line-broadening theory for positive ions.
O. Bely.
Phys. Rev., Second Series, Vol. 185, 79 - 82 (1969).

It is shown that the impact-approximation line width of positive ion lines is generally a continuous function of the energy of the perturbing electrons, when the width is averaged over resonances. Special cases are also discussed.

022.117 Übergangsstrahlung als mögliche sekundäre Standardlichtquelle für das VUV.
W. Böhm.
Mitt. Astron. Ges. No. 27, p. 190 - 192 (1969). – Conference paper.

022.118 Kleine Lichtquellen als Standards für Labor, Teleskop und Weltraumforschung.
H. Römer.
Mitt. Astron. Ges. No. 27, p. 192 - 198 (1969). – Conference paper.

022.119 **Zur Bremsung von H-Atomen in kaltem, atomarem Wasserstoffgas.**
J. Schäfer, E. Trefftz.
Mitt. Astron. Ges. No. 27, p. 226 (1969). – Abstract AG.

022.120 **Über die Genauigkeit von Maschinenprogrammen zur Berechnung von Oszillatorstärken in Coulomb-Approximation.**   H. Friedrich, K. Katterbach, E. Trefftz.
Mitt. Astron. Ges. No. 27, p. 226 - 229 (1969). – Conference paper.

022.121 **Passive remote sensing at microwave wavelengths.**
D. H. Staelin.
Proc. IEEE, Vol. 57, 427 - 439 = National Radio Astron. Obs., Green Bank, Repr. Ser. A, No. 112 (1969).

022.122 **P_4 deformation in an independent-particle model of light nuclei.**   C. Brihaye, G. Reidemeister.
Nuclear Physics, Ser. A, Vol. 100, 65 - 73 = Univ. Libre Bruxelles, Inst. d'Astron. d'Astrophys., Sér. A, No. 12 (1969).

The Velocity of Light.
See Abstr. 003.068.

How important is steady state cosmology to classical and quantum electrodynamics?
See Abstr. 162.031.

Instruments and Astronomical Techniques

031 Optics, Methods of Observation and Reduction

031.001 **Giant mirror blanks poured for Chile and Australia.**
Sky Telescope, Vol. 38, 140 - 143 (1969).

031.002 **Telescope eyepieces.**
H. E. Dall.
Journ. British Astron. Ass. Vol. 79, 349 - 356 (1969).

031.003 **Star finding with an altazimuth mounted telescope.**
F. V. Davies.
Journ. British Astron. Ass. Vol. 79, 467 - 471 (1969).

031.004 **Fotometria en el Observatorio de La Plata.**
A. Feinstein.
Revista Astron. Vol. 40, No. 166, p. 7 - 11 (1968).

031.005 **Teleskopspiegel im Leichtbau.**
A. Hoffmann.
Orion Schaffhausen, Vol. 14, 103 - 105 (1969).

031.006 **Mirror correction of the 125-cm reflector.**
K. A. Voronkov, E. A. Dibay.
Astron. Tsirk. No. 506, p. 5 - 6 (1969).   In Russian.

031.007 **The investigation of the objective of the AVR-2 refractor.**   V. F. Sincheschool, V. N. Sincheschool.
Astrometriya i Astrofiz., *Kiev*, No. 2, p. 135 - 149 (1969). In Russian.
The paper deals with the investigation of the objective of the AVR-2 refractor ($D = 200$ *mm*, $F = 3019$ *mm*) installed at the Poltava Observatory. Spherical aberration, astigmatism, curvature of the field, coma and distortion were investigated.

031.008 **Untersuchung der optischen Systeme der Photokameras NAFA 3c 25-c (No. 714010) und UFIS3-25-2" (No. 42).**   T. V. Rad'o, I. I. Terebushko.
Tsirk. L'vov. Astron. Obs. No. 43, p. 53 - 54 (1969). In Russian.

031.009 **La vision dans les instruments astronomiques et l'observation physique des surfaces planétaires.**
J. Dragesco.
L'Astronomie, 83ᵉ année, 355 - 365 (1969).

031.010 **Binokulare Himmelsbeobachtungen.**
M. Zeller.
Orion, Band 14, 128 - 129 (1969).

031.011 **Mirror blank testing by real-time holographic interferometry.**   W. van Deelen, P. Nisenson.
Applied Optics, Vol. 8, 951 - 955 (1969).
This paper describes an application of real-time holographic interferometry to the testing of unworked mirror blanks. The thermal test of a 70-cm diam, fused silica, eggcrate mirror blank and the mechanical test of 28-cm mirror blank are included.

031.012 **On the possibility of calculating the errors, caused by the deformation of the image of planets.**
V. N. Dudinov.
Astron. Zhurn. Akad. Nauk SSSR, Vol. 46, 1064 - 1073

(1969).   In Russian.   English translation in Soviet Astron. AJ, Vol. 13, No. 5.
The problem of optical reduction of photometric data is discussed. The deformed image of a planet in a telescope can usually be described as a convolution of the true image with distribution of brightness on the image of a point source. However, because of photometric errors and the difficulties in measuring the distribution function the deformation cannot be completely eliminated.

031.013 **A 40-cm welded-segment lightweight aluminum alloy telescope mirror.**   F. F. Forbes.
Applied Optics, Vol. 8, 1361 - 1363 (1969).
A 40-cm aluminum alloy, nickel plated, mirror has been made by welding together six individually cast segments. The mirror is shown to be stable to within one wave over a temperature of 58°C primarily due to the use of the aluminum alloy Tenzaloy. The weldment as well as the alloy itself are now known to be capable of essentially complete stress relief by annealing. The possibilities of extending the welding technique to considerably larger telescope systems is also discussed.

031.014 **Annular aperture diffracted energy distribution for an extended source.**
I. L. Goldberg, A. W. McCulloch.
Applied Optics, Vol. 8, 1451 - 1458 (1969).
The annular aperture diffracted energy distribution in the image plane for an extended incoherent source has been calculated and tabulated for source sizes up to five times the size of the Airy disk and for annular aperture ratios up to 0.5. It is shown that for a scanning instrument an increase in the instantaneous field of view from one to two times the Rayleigh limit degrades the effective resolution by a factor significantly less than two.

031.015 **Corrector systems for Cassegrain telescopes.**
R. N. Wilson.
Applied Optics, Vol. 8, 1924 - 1925 (1969). – Letter.

031.016 **1969–1970 Guide to Scientific Instruments.**
Compiled by E. J. Scherago.
Science, Vol. 165A, No. 3899A, 154pp. (1969).

031.017 **The aberrations of a prism diagonal.**
F. J. Eastman, Jr.
Sky Telescope, Vol. 38, 261 - 262 (1969).

031.018 **A note on curved spiders.**   C. H. Werenskiold.
Sky Telescope, Vol. 38, 262 - 263 (1969).

031.019 **A technique for recording phase-resolved spectra of regularly-varying faint light sources.**
J. R. Powell, D. E. Trumbo, C. R. Lynds.
Publ. Astron. Soc. Pacific, Vol. 81, 601 - 607 = Contr. Kitt Peak National Obs. No. 441 (1969).
A technique was developed for recording phase-resolved spectra of regularly-varying faint light sources. The shortcomings of this technique were investigated. The method was used to record phase-resolved spectra of the Crab nebula pulsar.

031.020 The spot diagrams of Schmidt camera.
 S. Kawai, T. Kogure.
Mem. Fac. Sci. Kyoto Univ., Ser. Phys., Astrophys., Geophys., Chemistry, Vol. 33, 95 - 119 (1968).

The quality of images for some Schmidt cameras, is examined with the aid of the spot diagrams. Particular attention is paid on the color aberration over the wide wave-length range $\lambda\lambda$ 3500 - 8500 Å.

031.021 Principal difficulties of the application of equidensity method for astrometry and photoastrometry.
L. V. Zhukov.
Astron. Zhurn. Akad. Nauk SSSR, Vol. 46, 889 - 894 (1969). In Russian. English translation in Soviet Astron. AJ, Vol. 13, No. 4.

A possibility of the application of equidensity method for photometric and photoastrometric stellar investigations is discussed. From microphotometrical sections of stellar images on plates of a normal astrograph some principal difficulties are pointed out. Disregarding these difficulties, considerable and serious systematic errors can be introduced.

031.022 On a quasi-absolute method of reduction of differential meridian observations.
M. S. Zverev.
Astron. Zhurn. Akad. Nauk SSSR, Vol. 46, 1290 - 1302 (1969). In Russian. English translation in Soviet Astron. AJ, Vol. 13, No. 6.

By reducing to a quasi-instrumental system, differential meridian observations yield results nearly independent on the fundamental catalogue system. The method is described in detail for right ascensions; it is used at present for the reduction of the right ascensions of the SRS, BS, DS, and FK4 stars observed during 1963 – 1967 by the Pulkovo astronomers at Cerro-Calan (Chile) with the Repsold meridian circle. A similar method can be applied for the reduction of differential observations of declinations. The results in the quasi-instrumental system can be used for an improvement of the coordinate system of the fundamental catalogue together with the results of absolute observations.

031.023 Optik für Astro-Amateure.
 E. Wiedemann.
Orion Schaffhausen, Vol. 14, 147 - 151 (1969).

031.024 On the methods of multi-colour polarization observations of the stellar radiation.
Yu. S. Efimov, N. S. Polosukhina, N. M. Shakhovskoy.
Izv. Krymskoj Astrofiz. Obs. Vol. 39, 3 - 10 (1969). In Russian.

Methods and apparatus for multi-colour polarization observations of variable stars are described. The dependence of the instrumental polarization on the wavelength is determined. The accuracy of observations ($\sim$0.1%) is estimated.

The obtained results are compared with the data of other authors. The possibility of the application of narrow-band interference filters for precise polarization measurements is shown.

031.025 Lens objectives free from the secondary spectrum for a wide spectral region.   G. M. Popov.
Izv. Krymskoj Astrofiz. Obs. Vol. 39, 236 - 244 (1969). In Russian.

Lens systems consisting of three lenses and having been corrected with high perfection for chromatic aberration within a wide region of the spectrum (from 3800 Å to 7600 Å) are considered.

031.026 Performance characteristics of a high-dispersion Schmidt–prism combination.   P. J. Treanor.
Vistas in Astronomy, Vol. 11, 147 - 159 (1969).

The high dispersion Schmidt–prism combination of the Vatican Observatory is described, and the problem of obtaining spectra of high quality in long exposures and to faint limiting magnitudes is investigated.

031.027 Einige neuere Untersuchungen zur Kohärenz und zu Schwankungserscheinungen des Lichtes.
E. Wolf.
Jenaer Rundschau (Jena Review), 14. Jahrgang, 315 - 323 (1969).

031.028 Modification of Fesenkov's method for the complete determination of polarization.
E. G. Michelkin.
Trudy Astrofiz. Inst. Alma-Ata, Vol. 14, 96 - 99 (1969). In Russian.

The simplest modification of Fesenkov's method for the complete determination of polarization is given, and working formulas for the calculation of the elliptical polarization parameters are derived.

031.029 Density measurement with radio wave occultation techniques.   B. B. Lusignan.
Space Research IX, Proc. Tokyo 1968, p. 603 - 609 (1969).

031.030 A photoelectric method for registration of transits of stars and its application at the Astronomical Institute of the Academy of Sciences of the Uzbek SSR.
T. Nuraliev.
Astrometr. Issled. p. 50 - 66 (1969).   In Russian.

Observing made easy.
Nature, Vol. 224, 1152 (1969). – News notes.

Color differentiation by computer processing.
See Abstr. 094.077.

032 Astronomical Instruments

032.001 **L'eliminazione dell'errore dell'inclinazione in uno strumento dei passaggi a cannocchiale spezzato.**
J. O. Fleckenstein.
Proc. Colloquium on Problems of Time Determination, Keeping and Synchronization, (Milan 1968), p. 253 - 254 (1968).

L'utilisation des bains de mercure a determiner les temps des passages des étoiles méridiennes à l'aide des observations directes et réfléchies des étoiles permet une augmentation essentielle de la précision des observations visuelles.

032.002 **Der 60-Zoll-Spiegel des Leopold-Figl-Observatoriums für Astrophysik der Universitäts-Sternwarte Wien.** J. Meurers.
SuW, Vol. 8, 195 - 198 (1969).

032.003 **Der Bau eines 10″-Reflektors.** N. Vorstädt.
SuW, Vol. 8, 216 - 217 (1969).

032.004 **Steward 90-inch telescope dedicated.**
Sky Telescope, Vol. 38, 164 - 165 (1969).

032.005 **A very sturdy 10-inch Newtonian reflector.** N. Condoluci.
Sky Telescope, Vol. 38, 189, 191 (1969).

032.006 **Fabrication of a Wright telescope.** T. J. Waineo.
Sky Telescope, Vol. 38, 112 - 118 (1969).

032.007 **Investigation of the time-systems of the satellite-cameras IGN No. 24, IGN No. 25 and BC4-BE2 No. 308 including longwave time-signal receivers T 75 A and E 390.** K. Nottarp. In German.
Deutsche Geod. Kommission Bayer. Akad. Wiss. Reihe B, Heft No. 169, 33 pp. (1969).

The functional blocs of mentioned time-systems are analysed and the different alternations are described. The measured delay times and the statistical uncertainties of definition are given and the arrangements used for the measurements are explained.

032.008 **Beitrag zur Entwicklungsgeschichte des Theodolits.** M. Engelsberger.
Deutsche Geod. Kommission Bayer. Akad. Wiss. Reihe C, Heft No. 134, 98 pp. (1969).

032.009 **Lunar television camera.** E. L. Svensson.
Spaceflight, Vol. 11, 304 - 307 (1969).

032.010 **Deux réalisations pour l'astrographe amateur.** Y. Grandjean.
Orion Schaffhausen, Vol. 14, 85 - 87 (1969).

032.011 **Über den Parswert der Libelle des Bamberg-Zenitteleskops aus Untersuchungen an zwei Niveauprüfern.** E. I. Obrezkova.
Astrometriya i Astrofiz., *Kiev*, No. 2, p. 129 - 134 (1969). In Russian.

From investigation of the Talcott levels of the Bamberg zenith-telescope the author has revealed a considerable dependence of the value of a division on the length of the bubble. This result has been confirmed by measurements with two examinations.

032.012 **The vertical mirror, its potential applications to theodolites and two star image stopping micro-**

meters. L. A. Kivioja.
Bull. Géod. Nouvelle Série, No. 93, p. 263 - 275 (1969).

The application of the described instrumentation is potentially useful in determination of astronomical longitude, changes in longitude differences with time, or the possible east—west component of continental drift, right ascension differences between stars, and the rotation rate of the earth.

032.013 **Le grand télescope astrométrique à cinq ans.** J. Kovalevsky.
L'Astronomie, 83ᵉ année, 369 - 372 (1969).

032.014 **Spiegelteleskop mit sphärischen Flächen und verkürzter Schnittweite.** E. Wiedemann.
Orion Schaffhausen, Band 14, 127 - 128 (1969).

032.015 **Das Protuberanzen-Instrument der Sternwarte Calina.** J. Schaedler.
Orion, Band 14, 131 - 133 (1969).

032.016 **Design procedure for Ritchey-Chrétien corrector.** B. J. Howell.
Applied Optics, Vol. 8, 685 - 695 (1969).

A four-lens corrector system to remove residual aberrations in a Ritchey-Chrétien telescope was selected as the means of studying the relative effectiveness of two design methods: a ray deviation error function and third order aberration theory.

032.017 **The branch station of Lund Observatory.** N. Hansson, C. Schalén.
Ark. Astron. Vol. 5, 197 - 208 (1969).

Section I contains a description of the station on Jävan, 18 km from Lund, and the new 61 cm Cassegrain-Nasmyth reflector. Section II is devoted to a description of the photoelectric photometer attached to the Cassegrain focus of the reflector.

032.018 **A rocket telescope spectrometer with high precision pointing control.**
M. Bottema, W. G. Fastie, H. W. Moos.
Applied Optics, Vol. 8, 1821 - 1826 (1969).

One second of arc pointing accuracy has been achieved by servocontrolling the secondary mirror of a Dall-Kirkham telescope flown in an Aerobee 150 rocket. The primary mirror is weight-relieved, mounted at its nodal line and can resolve 2 arc sec. An objective LiF prism mounted near the focal plane provides a low-resolution far uv spectrum suitable for studying planetary atmospheres. Solar blind photomultiplier tubes with pulse counting electronics provide a dark current background of less than 1 count/sec. Spectra of Venus, Jupiter and η Ursae Majoris (UMa) were obtained in a flight from White Sands, New Mexico, on 5 December 1967. Further flights are planned with the recovered package.

032.019 **Progress report on a high-reflectance Coude telescope.** E. H. Richardson.
Bull. American Astron. Soc. Vol. 1, 259 (1969). – Abstr. AAS.

032.020 **An ultraviolet image converter and folded all-reflecting Schmidt telescope for ultraviolet astronomy.**
J. D. Wray, F. G. O'Callaghan.
Bull. American Astron. Soc. Vol. 1, 268 (1969). – Abstr. AAS.

032.021 The drive system for the McDonald Observatory 107-inch telescope. D. M. Edison, E. J. Rhodes.
Bull. American Astron. Soc. Vol. 1, 240 (1969). – Abstr. AAS.

032.022 An amateur's torque-tube mount. H. Link
Sky Telescope, Vol. 33, 258 - 260 (1969).

032.023 Optical systems for spectroscopic telescopes.
T. Dunham, Jr.
Proc. Astron. Soc. Australia, Vol. 1, 291 - 293 (1969). – Contribution ASA meeting.

032.024 The new Soviet astronomical instruments in Chile.
L. A. Panaiotov, K. N. Tavastsherna.
Vestn. AN SSSR, No. 12, p. 78 - 84 (1968). In Russian.
Abstr. in Referativ. Zhurn. 51. Astron., 8.51.147 (1969).

032.025 Photo album of Kitt Peak's 158-inch telescope building.
Sky Telescope, Vol. 38, 284 - 289 (1969).

032.026 Some fine telescopes are exhibited at Stellafane.
D. Milon.
Sky Telescope, Vol. 38, 342 - 349 (1969).

032.027 Sur l'extension des applications de l'astrolabe de Danjon. S. Debarbat.
Thesis, Sci. Math., Paris. Centre de Documentation du Centre National de la Recherche Scientifique, Paris. No. 3295, 82 pp. (1969).

032.028 Sacramento Peak's new solar telescope.
R. B. Dunn.
Sky Telescope, Vol. 38, 368 - 375 (1969).

032.029 A new three-mirror off-axis amateur telescope.
R. A. Buchroeder.
Sky Telescope, Vol. 38, 418 - 423 (1969).

032.030 Far-ultraviolet solar observatory.
R. N. Watts, Jr.
Sky Telescope, Vol. 38, 390 - 391 (1969).

032.031 The effects of Čerenkov light pulses on a stellar intensity interferometer.
R. H. Brown, J. Davis, L. R. Allen.
Monthly Notices, Roy. Astron. Soc., Vol. 146, 399 - 409 (1969).
Observations of Čerenkov light pulses due to cosmic rays have been made using the large reflectors of the stellar intensity interferometer at Narrabri Observatory. The rate of arrival of correlated pulses in the two reflectors has been measured as a function of the pulse height, zenith angle, separation and relative alignment of the two reflectors. The results are used to calculate the unwanted correlation due to Čerenkov light in the interferometer at Narrabri, and it is found to be negligibly small compared with the correlation due to the faintest stars in the programme.

032.032 Transit circles today.
R. H. Tucker.
Quarterly Journ. Roy. Astron. Soc. Vol. 10, 223 - 232 (1969).
A brief account of how the Transit Circle has managed to survive for 250 years is followed by a survey of modern technical developments that are being applied to this venerable instrument. The various observational programmes now in progress are summarized, and the paper concludes with a few thoughts on future prospects.

032.033 A method for rotating exactly through 180° a meri- dian telescope. I. Rusu.
Stud. Cerc. Astron. Vol. 14, 23 - 27 (1969).

032.034 A method for determining the displacements of the graduations of a divided circle with respect to two diametrical microscopes. I. Rusu.
Stud. Cerc. Astron. Vol. 14, 29 - 34 (1969).

032.035 Das Studium des Niveaus des Zeiss-Passagengeräts.
M. Ştefănescu.
Stud. Cerc. Astron. Vol. 14, 63 - 67 (1969). In Rumanian.

032.036 Comparative assessment of aberrations originating in telescope mirrors from the edge support.
G. Schwesinger.
Astron. Journ., Vol. 74, 1243 - 1254 (1969).
Telescope mirrors suffer the least loss of quality by elastic flexure if the supporting forces are balanced against each other in an optimum manner. There are unresolved differences of opinion whether favorable support conditions for large mirrors should be synthesized from requiring minimum rms values of wave aberrations or of angular ray deviations. Results are presented of a comparative theoretical study for mirrors on radial supports.

032.037 The Pulkovo large transit instrument for absolute determinations of stellar right ascensions mounted in Chile. Yu. A. Beljaev, V. M. Vasiljev, R. Peralta, A. I. Pljugina, Yu. S. Streletsky, K. N. Tavastsherna, R. Tapija.
Astron. Zhurn. Akad. Nauk SSSR, Vol. 46, 919 - 922 (1969). In Russian. English translation in Soviet Astron. AJ, Vol. 13, No. 4.
At Cerro-Calan Observatory the mounting of a new large transit instrument for absolute determinations of right ascensions of stars is completed. The instrument is designed and made at Pulkovo. It retains the construction of the classical type of such instruments and has sufficiently large optical force. It is supplied with a number of automatic and semi-automatic equipments.

032.038 An interferometer with crossed rays.
E. S. Kulagin.
Astron. Zhurn. Akad. Nauk SSSR, Vol. 46, 1310 - 1316 (1969). In Russian. English translation in Soviet Astron. AJ, Vol. 13, No. 6.
Results of the test of a new stellar interferometer, an interferometer with crossed rays, intended first of all for measurements of close binaries, are given.

032.039 A double beam interferometer for the middle infrared. R. Hanel, M. Forman, T. Meilleur, R. Westcott, J. Pritchard.
Applied Optics, Vol. 8, 2059 - 2065 (1969).
A versatile, double beam Michelson interferometer for the middle ir has been constructed.

032.040 A rocket-borne liquid helium-cooled infrared telescope. I: Dewar and optics.
D. P. McNutt, K. Shivanandan, P. D. Feldman.
Applied Optics, Vol. 8, 2199 - 2204 (1969).
A telescope for rocket-borne IR astronomy is described. It consists of a 166-mm Cassegrainian telescope cooled to liquid helium temperature and operated with a total radiation chopper. Details of the optical, cryogenic, and electronic designs are given.

032.041 A rocket-borne liquid helium-cooled infrared telescope. II: Photoconductive detectors.
P. D. Feldman, D. P. McNutt.
Applied Optics, Vol. 8, 2205 - 2210 (1969).
In the absence of a thermal radiation background, it is

possible to obtain a very high responsivity with far IR extrinsic photoconductive detectors.

032.042 Image formation in X-ray telescopes.
J. F. Pastor, R. Tillen.
Journ. Optical Soc. America, Vol. 59, 1518 - 1519 (1969).
Abstract Meeting Optical Soc. America.

032.043 Analysis of the photometric field correction of the 500-mm Maksutov telescope of the southern station of Sternberg Astronomical Institute.
G. A. Ponomareva.
Soobshch. Gos. Astron. Inst. Shternberga, No. 158, p. 52 - 57 (1969).   In Russian.

032.044 Determination of some instrumental characteristics of the Poltava Observatory zenith-telescope ZTL-180.   N. A. Popov, N. I. Panchenko, A. P. Tsapova.
Astrometriya i Astrofiz., *Kiev*, No. 7, p. 67 - 74 (1969).
In Russian.
   The article deals with the basic instrumental characteristics of the Poltava Observatory zenith-telescope ZTL-180: the values of division of the Talcott levels, the scale value of the micrometer screw, the distances between horizontal hatchings on the micrometer glass plate, and others. As a result of the investigation, the authors came to the conclusion that the temperature coefficient of these distances is very small.

032.045 The theory of determining the flexure of the tube with application of autocollimation.
A. S. Kharin.
Astrometriya i Astrofiz., *Kiev*, No. 7, p. 74 - 81 (1969).
In Russian.
   The author considers theoretically the dependence between the angle of rotation of the flat mirror rigidly fastened to the objective casing and the bend angle of the objective end of the tube. On the basis of this dependence, the autocollimation angle is represented by a linear function of the bend angles of both ends of the tube. It enables the differential flexure of the tube to be determined by means of measuring the autocollimation angle by a flat mirror fastened before the objective glass and the bend angle of the eye-piece of the tube by a collimator placed in the tube.

032.046 Investigation of the pivots of the Odessa meridian circle.   M. Yu. Volyanskaya, A. M. Stafeev.
Astrometriya i Astrofiz., *Kiev*, No. 7, p. 83 - 85 (1969).
In Russian.
   The article presents the results concerning the investigation of the pivots of the meridian circle installed at the Odessa Astronomical Observatory. The investigation was carried out by the authors in October 1967. This work was done in two positions of the instrument using the contact method by means of a level with a probe and a photomicroscope instead of a micrometer. The irregularities of the pivot radii are found in intervals of $2^\circ5$. The values of collimation corrections Δc are presented.

032.047 Chromatic curve of the refractor "AVR-2" object glass.   B. F. Sincheskul, V. N. Sincheskul.
Astrometriya i Astrofiz., *Kiev*, No. 7, p. 85 - 91 (1969).
In Russian.
   The authors present the results of the investigation of chromatic aberrations of the refractor "AVR-2" object glass of the Poltava Observatory. The chromatic aberration of position was investigated by three different methods: a) according to the Hartmann scheme with application of a narrow-band interference filter; b) by photos taken with a diffraction grating; c) by photos with an objective prism. The combination of these methods enables to carry out the investigation

in a wide range of wave length as well as to compare the results obtained by three methods different in principle.

032.048 On telescopes for amateurs of astronomy. VI.
N. N. Mikhel'son.
Zemlya i Vselennaya, No. 4, p. 83 - 88 (1969).   In Russian.

032.049 First Soviet astrograph on the southern sphere.
L. A. Panaiotov.
Zemlya i Vselennaya, No. 6, p. 62 - 67 (1969).   In Russian.

032.050 Operating characteristics of the stratoscope. II. Balloon-borne telescope.
D. J. McCarthy.
IEEE Trans. Aerospace Electronic Systems, Vol. AES-5, No. 2, 323 - 329 (1969).

032.051 On improving the effectiveness of large telescopes.
J. G. Baker.
IEEE Trans. Aerospace Electronic Systems, Vol. AES-5, No. 2, 261 - 272 (1969).

032.052 A gamma ray telescope utilizing large area wire spark chambers.   R. W. Ross, C. H. Ehrmann, C. E. Fichtel, D. A. Kniffen, H. B. Ogelman.
IEEE Trans. Nuclear Sci. Vol. S-16, 304 - 308 (1969). – See Phys. Abstr. Vol. 72, No. 38128 (1969).

032.053 Applications of intensity interferometry in physics and astronomy.   R. Q. Twiss.
Optica Acta, Vol. 16, 423 - 451 (1969).

032.054 Effect of non-monochromaticity of light on the visibility of fringes in stellar interferometers.
R. S. Sirohi.
Optik, Vol. 28, 585 - 591 (1969).
   In the present communication the author has discussed the effect of non-monochromaticity of light on the visibility of fringes in a stellar interferometer, when it is used to find the angular diameter of disc objects. It is shown that the peak of the visibility curve diminishes and minimum broadens with increasing departure from monochromaticity.

032.055 A triaxial universal instrument.
Z. Kordylewski.
Acta Astron. Vol. 19, 307 - 321 (1969).
   An astrometric instrument with three axes possessing divided circles considerably facilitates the tracking of the path of an artificial satellite with the cross-wire, of the eyepiece; this renders possible to attain a greater frequency of determination of the positions of a rapidly moving object. A method of observation with such an instrument is given and formulae are presented for transforming the measured angles into horizontal coordinates. A simple method of determination of the eight "instrumental errors" with the help of a level and from observations of Polaris is described.

032.056 The 61 cm photometric telescope of the Bochum University at La Silla.
T. Schmidt-Kaler, J. Dachs.
ESO Bull. No. 5, p. 15 - 18 (1969).

032.057 Large and very large telescopes. Projects and considerations.   K. Bahner.
ESO Bull. No. 5, p. 19 - 34 (1969).
   Large telescopes: Data for existing and planned instruments; Very large telescopes: Design topics (optical systems, mirror support, accuracy of optics, mounting importance of very large telescopes).

032.058 A new automatic camera for satellite tracking.

A. M. Lozinsky, G. A. Leikin.
Space Research IX, Proc. Tokyo 1968, p. 4 - 5 (1969).

A description of a new Soviet camera for tracking artificial celestial bodies is given. The camera has been mounted at the Zvenigorod Station of the Astronomical Council.

032.059 **On the scale values of the floating zenith telescope.** S. Goto, H. Okawa, H. Kitago.
Proc. International Latitude Obs. Mizusawa, No. 9, p. 15 - 24 (1969). In Japanese.

032.060 **On the micrometer constant of the visual zenith telescope determined from the greatest elongation** observations of circum-polar stars at Mizusawa. M. Ooe, S. Abe.
Proc. International Latitude Obs. Mizusawa, No. 9, p. 50 - 58 (1969). In Japanese.

032.061 **Konstruktionsprinzipien eines leistungsfähigen, tragbaren Refraktors für den Sternfreund.** H. Wichmann.
SuW, Vol. 8, 294, 296 - 297 (1969).

032.062 **Comments on the 150-inch Anglo-Australian telescope.** H. Wehner.
Journ. Astron. Soc. Victoria, Vol. 22, 90 - 94 (1969).

032.063 **Über die Stabilität des Azimuts eines Passagen-instruments.** G. K. Gorel', N. S. Kalikhevich.
Rotation of the Earth and Determination of Time, Conference Riga 1965, p. 86 - 91 (1969). In Russian.

032.064 **Instrumentelle Fehler des Prismenastrolab der Firma ORL.** L. N. Nadeev.
Rotation of the Earth and Determination of Time, Conference Riga 1965, p. 91 - 96 (1969). In Russian.

032.065 **Investigation of the pivots of the Tashkent meridian circle.** O. S. Tursunov.
Astrometr. Issled. p. 83 - 105 (1969). In Russian.

032.066 **Digital recording of the circle at the Oporto university mirror transit circle.**
R. S. de Sousa Nunes, D. Appelt.

Anais Faculdade Ciências do Porto, Vol. 51, Fasc. 3, 11 pp. = Publ. Obs. Astron. Porto No. 23 (1968).

The mirror transit circle at Oporto was designed in 1953 and since its installation it is being continuously improved both in its mechanical and data readout aspects. Now a system is being developped to perform the automatic digital readout of the circle position. This paper gives a brief description of this new system and its operating.

032.067 **Beobachtungen am Passage-Instrument 100/1000 des VEB Carl Zeiss.** S. Wächter.
Jenaer Rundschau (Jena Review), 13. Jahrgang, p. 337 - 340 (1968).

032.068 **Kuppelsteuerung in den 2-Meter-PCC-Teleskopan-lagen.** A. Jensch, M. Steinbach.
Jenaer Rundschau (Jena Review), 13. Jahrgang, p. 341 - 344 (1968).

032.069 **How to construct an amateur telescope? (VIII).** L. Newelski.
Urania Kraków, Vol. 40, 328 - 332 (1969). In Polish.

032.070 **Kinematographie der Chromosphäre.** H. Dürst.
Astron. Mitt. Eidg. Sternw. Zürich, No. 289, 14 pp. (1969).

The automatic solar patrol telescope of the Swiss Federal Observatory is described. An Arriflex-35 mm-cine-camera has been mounted on a Zeiss-Coudé-refractor. The telescope is equipped with a photoelectric guider, a Halle H-alpha filter, and a device for turning the dome corresponding to the daily motion of the telescope.

032.071 **Il telescopio a specchio metallico di 137 cm dell'Osservatorio di Merate.**
G. de Mottoni.
Contr. Oss. Astron. Milano–Merate, Nuova Ser., No. 315, 25 pp. (1969).

New solar telescope.
Nature, Vol. 224, 204 - 205 (1969).
Solar telescope at Sacramento Peak Observatory.

Telescopes: How to Make Them und Use Them.
See Abstr. 003.064.

033 Radio Telescopes and Equipment

033.001 **Studies of the focal region of a spherical reflector: Stationary phase evaluation.** G. Hyde.
IEEE Trans. Antennas Propagation, Vol. AP-16, 646 - 656 (1968).
A study of focal fields. Calculated fields are compared with measurement.—*RAB*

033.002 **The pointing calibration of the Haystack antenna.** M. L. Meeks, J. A. Ball, A. B. Hull.
IEEE Trans. Antennas Propagation, Vol. AP-16, 746 - 751 (1968).
Optical astronomy techniques used to calibrate antenna by observing radio sources at 15.25 GHz. Rms residual pointing errors after calibration were 2.90 millidegrees in azimuth and 3.45 millidegrees in elevation.—*RAB*

033.003 **The determination of antenna parameters by the use of extraterrestrial radio sources.**
D. A. Guidice, J. P. Castelli.
U. S. Air Force Cambridge, Res. Lab. Office Aerospace Res. Ionosph. Phys. Lab., Phys. Sci. Res. Papers, No. 357. AFCRL-68-0231, 38 pp. (1968).
Discusses the technique of determining the radiation pattern, effective area, aperture efficiency and gain of large aerials by means of known radio sources, the sun and the moon.—*BMT*

033.004 **On an application of the phase-switching method.** V. A. Sanamian.
Soobshch. Byurakan. Obs. No. 40, p. 69 - 75 (1969).
In Russian.
By means of the phase-switching method it is possible to increase the accuracy of the positions of radio sources.

033.005 **Reduction of radio telescope flux density recordings.** J. R. Smith.
Journ. British Astron. Ass. Vol. 79, 371 - 374 (1969).

033.006 **Zur Bestimmung der Signalverzögerung eines Radioempfängers.** G. G. Krajnyuk, A. A. Logvinenko.
Tsirk. L'vov. Astron. Obs. No. 43, p. 50 - 52 (1969).
In Russian.

033.007 **Synthetic-aperture radio telescopes.** G. W. Swenson, Jr.
Annual Rev. Astron. Astrophys. Vol. 7, 353 - 374 (1969).

033.008 **Buckland Park aerial array.**
B. H. Briggs, W. G. Elford, D. G. Felgate, M. G. Golley, D. E. Rossiter, J. W. Smith.
Nature, Vol. 223, 1321 - 1325 (1969).
An aerial array 1 km in diameter has been constructed near Adelaide, South Australia. It will be used for observations of ionospheric drifts and meteors, and for other experiments in ionospheric physics.

033.009 **The 22.25 MHz radio telescope at the Dominion Radio Astrophysical Observatory.**
C. H. Costain, J. D. Lacey, R. S. Roger.
Publ. Dominion Obs. Ottawa, Vol. 25, 323 - 335 (1969).
Radio telescopes of the T and Cross configurations are compared and their relative merits discussed. A large T array built at the Dominion Radio Astrophysical Observatory for radio astronomical observations at 22.25 MHz (λ = 13.5 m) is described. It consists of 624 full-wave dipoles above a reflecting screen 65,000 m² in area. The dipoles are arranged in an east-west section of dimensions 96λ × 2.5λ and in a north-south section 32.5λ × 4λ. The instrument has a pencil-beam response of 1°1 × 1°7 at the zenith. Simultaneous observations at five adjacent declinations are made with a time-sharing technique. Observations commenced in 1965 and will provide flux density measures for 400 to 500 radio sources down to a limiting flux density of 30 × 10^{-26} Wm⁻² Hz⁻¹. A map of the galactic background radiation from the sky north of –20° declination is being prepared.

033.010 **Focus broadening by astigmatism of large microwave parabolic antennas.** G. Feix.
Applied Optics, Vol. 8, 1631 - 1634 (1969).
The primary astigmatism of a parabolic antenna of an aperture of $10^3 \lambda$ and of f/D = 0.43 was investigated. The measured astigmatic aberration is caused predominantly by large span surface deformations. In the case described, the focus of a Cassegrain system of 11.85 magnification is broadened up to 153 mm at the secondary and up to 2 mm at the primary focus. This aberration effect accounts for 5% - 6% of the total intensity.

033.011 **Astronomy : Tight budget gains stranglehold on radio facilities.** R. W. Holcomb.
Science, Vol. 166, 984 - 986 (1969).

033.012 **Techniques for pulsar observation.**
G. R. A. Ellis.
Proc. Astron. Soc. Australia, Vol. 1, 289 - 290 (1969). – Contribution ASA meeting.

033.013 **A new radio-wave technique in X-ray astronomy.**
P. J. Edwards.
Proc. Astron. Soc. Australia, Vol. 1, 290 - 291 (1969). – Contribution ASA meeting.

033.014 **Impedance of a thin cylinder antenna in isotropic plasma.** V. Ya. Eidman.
Izv. vyssh. uchebn. zavedenij. Radiofizika, Vol. 12, 36 - 43 (1969). In Russian. – Abstr. in Referativ. Zhurn. 51. Astron., 8.51.459 (1969).

033.015 **A possibility of synthetizing a multi-element radio interferometer with autonomous reception for investigating images of radio sources.** V. A. Alekseev.
Izv. vyssh. uchebn. zavedenij. Radiofizika, Vol. 12, 487 - 490 (1969). In Russian. – Abstr. in Referativ. Zhurn. 51. Astron., 11.51.443 (1969).

033.016 **Some problems of aerial tolerance theory and evaluation of reflex aerial parameters.**
É. Ya. Bervalds.
Latv. PSR Zinatnu Akad. vēstis, Izv. AN Latv. SSR, No. 5, p. 34 - 40 (1969). In Russian. – Abstr. in Referativ. Zhurn. 51. Astron., 11.51.449 (1969).

033.017 **On the application of the compensation method for spectral measurements.**
É. V. Borodzich, Yu. S. Rusinov, R. L. Sorochenko.
Trudy Fiz. in–t. AN SSSR, Vol. 47, 21 - 24 (1969). In Russian. – Abstr. in Referativ. Zhurn. 51. Astron., 12.51.414 (1969).

033.018 **Two-frequency radio interferometer with independent heterodynes.**
V. A. Alekseev, V. D. Krotikov, V. N. Nikonov, V. S. Troits-

kij.
Izv. vyssh. uchebn. zavedenij. Radiofizika, Vol. 12, 644 - 650 (1969). In Russian. – Abstr. in Referativ. Zhurn. 51. Astron., 12.51.491 (1969).

033.019 On diminishing the influence of the electric length instabilities of transmission lines of heterodyne signals in the system of radio interferometer with a high angular resolution. V. A. Alekseev, V. D. Krotikov.
Izv. vyssh. uchebn. zavedenij. Radiofizika, Vol. 12, 651 - 654 (1969). In Russian. – Abstr. in Referativ. Zhurn. 51. Astron., 12.51.493 (1969).

033.020 Radio interference with retranslation. V. V. Balinov, V. V. Vitkevich.
Trudy Fiz. in–t. AN SSSR, Vol. 47, 163 - 172 (1969). In Russian. – Abstr. in Referativ. Zhurn. 51. Astron., 12.51.494 (1969).

033.021 Aspect determination of the electric axis of the cross–type radio telescope (DKR – 1000) east – west arm by statistical reduction of observations of many discrete sources. V. V. Vitkevich, V. N. Kozhukhov.
Trudy Fiz. in–t. AN SSSR, Vol. 47, 160 - 162 (1969). In Russian. – Abstr. in Referativ. Zhurn. 51. Astron., 12.51.495 (1969).

033.022 Principal parameters of the "east–west" aerial feed of the FIAN wide band cross-type radio telescope. Yu. P. Ilyasov.
Trudy Fiz. in–t. AN SSSR, Vol. 47, 173 - 182 (1969). In Russian. – Abstr. in Referativ. Zhurn. 51. Astron., 12.51.496 (1969).

033.023 On the statics of a parabolic mirror with multiple support mounting. P. D. Kalachev.
Trudy Fiz. in–t. AN SSSR, Vol. 47, 36 - 54 (1969). In Russian. – Abstr. in Referativ. Zhurn. 51. Astron., 12.51.497 (1969).

033.024 The rigidity estimation of parabolic mirror cantilever mounting. P. D. Kalachev.
Trudy Fiz. in–t. AN SSSR, Vol. 47, 55 - 61 (1969). In Russian. – Abstr. in Referativ. Zhurn. 51. Astron., 12.51.498 (1969).

033.025 The multiple support mountings with radial symmetry. P. D. Kalachev.
Trudy Fiz. in–t. AN SSSR, Vol. 47, 77 - 84 (1969). In Russian. – Abstr. in Referativ. Zhurn. 51. Astron., 12.51.499 (1969).

033.026 Calculated elastic deformations of a 7.5-m parabolic mirror model.
P. D. Kalachev, V. P. Nazarov, V. Ya. Chashnikov, A. A. Parshchikov.
Trudy Fiz. in–t. AN SSSR, Vol. 47, 62 - 76 (1969). In Russian. – Abstr. in Referativ. Zhurn. 51. Astron., 12.51.500 (1969).

033.027 The "null" radio spectrometer for observations of the galactic radio line of excited hydrogen.
V. P. Bibinova, É. V. Borodzich, R. L. Sorochenko, I. V. Shavlovskij.
Trudy Fiz. in–t. AN SSSR, Vol. 47, 137 - 143 (1969). In Russian. – Abstr. in Referativ. Zhurn. 51. Astron., 12.51.503 (1969).

033.028 The "null" spectral radiometer at 5.2 cm with symmetrical reception.
V. M. Gudnov, I. M. Goryachev, V. A. Kolbasov, G. S. Mise-

zhnikov, R. L. Sorochenko, B. V. Sestroretskij, V. B. Shtejnshlejger.
Trudy Fiz. in–t. AN SSSR, Vol. 47, 5 - 20 (1969). In Russian. – Abstr. in Referativ. Zhurn. 51. Astron., 12.51.506 (1969).

033.029 The 8-cm radiometer with a quantum paramagnetic amplifier.
L. I. Matveenko, G. S. Misezhnikov, M. M. Mukhina, V. B. Shtejnshlejger.
Trudy Fiz. in–t. AN SSSR, Vol. 47, 25 - 35 (1969). In Russian. – Abstr. in Referativ. Zhurn. 51. Astron., 12.51.507 (1969).

033.030 1.6- and 3.3-cm radiometers with parametric amplifiers for the RT-22 radio telescope.
V. P. Bibinova, A. D. Kuz'min, M. T. Levchenko, V. I. Pushkarev, A. E. Salomonovich, I. V. Shavlovskij.
Trudy Fiz. in–t. AN SSSR, Vol. 47, 149 - 152 (1969). In Russian. – Abstr. in Referativ. Zhurn. 51. Astron., 12.51.508 (1969).

033.031 A 21-cm parametric amplifier for radio astronomical investigations.
I. I. Berulis, B. Z. Kanevskij, E. A. Spangenberg, I. A. Strukov.
Trudy Fiz. in–t. AN SSSR, Vol. 47, 153 - 159 (1969). In Russian. – Abstr. in Referativ. Zhurn. 51. Astron., 12.51.509 (1969).

033.032 Instrumentation and methods for radio astronomical measurements of solar wind velocity.
I. A. Alekseev, V. V. Vitkevich, V. I. Vlasov, Yu. P. Ilyasov, S. M. Kutuzov, M. M. Tyaptin.
Trudy Fiz. in–t. AN SSSR, Vol. 47, 183 - 200 (1969). In Russian. – Abstr. in Referativ. Zhurn. 51. Astron., 12.51.510 (1969).

033.033 Phase principle for measuring antenna temperature. A. J. Rainal.
Proc. IEEE, Vol. 57, 1678 - 1680 (1969).

A phase principle for measuring the temperature of a radio or radar antenna is described. The phase principle makes use of phase information exclusively and is therefore insensitive to receiver gain fluctuations. Under certain conditions the potential accuracy of the measurement is somewhat better than the potential accuracy of the corresponding measurement resulting from a balanced Dicke radiometer.

033.034 A method for detecting weak radio sources in the presence of stronger sources in observations made with the Cambridge One-mile Telescope.
A. C. Neville, M. D. Windram, S. Kenderdine.
Observatory, Vol. 89, 186 - 192 (1969).

A technique is described which has been used in the analysis of observations with the Cambridge One-mile Telescope to remove powerful sources from maps without appreciably affecting nearby weaker ones. These can then be studied with much higher accuracy than would otherwise be possible.

033.035 S-band Y-junction stripline circulator with triangular ferrite. R. C. Kumar.
Electronic Engineering, Vol. 41, (No. 492), 214 - 215 (1969).

033.036 The Italian Cross radiotelescope. III. Operation of the telescope.
A. Braccesi, M. Ceccarelli, G. Colla, R. Fanti, A. Ficarra, G. Gelato, G. Grueff, G. Sinigaglia.
Nuovo Cimento, Vol. 62B, 13 - 19 (1969).

033.037 The radio telescope for millimetre waves at the

Simeis Observatory (USSR).
M. Laffineur, S. Koutchmy.
L'Onde Electr., Vol. 49, (No. 503), 246 - 249 (1969).
In French.

Brief description of 22-metre radio telescope in Crimea. Capable of operating at wavelengths down to 2 mm. – DNC

033.038   A novel circular array antenna.
F. H. Cleveland, N. P. Kernweis, P. R. Franchi.
U. S. Air Force Cambridge Res. Labs., Office Aerospace Res. Microwave Phys. Lab. Instrumentation Papers No. 152, AFCRL-68-0582, 17 pp. (1968).

Antenna uses 103 horizontal north-south dipoles arranged in a semi-random pattern around a circle of 2000 feet diameter. A 3 degree beam with reasonably low sidelobes in the frequency range 5 to 7.5 MHz is obtained. – BFC

033.039   A 36 GHz travelling wave maser for use in radio astronomy.   P. N. Swanson.
Penn. State Univ. Coll. Sci. Dep. Astron. Radio Astron. Obs. Sci. Rep. No. 017. NASA Grant NGL 39-009-015 (8)., 89 pp. (1969).

033.040   Computer-design of diode-using microwave components, and a computer-dimensioned, X-band parametric amplifier.   W. J. Getsinger, A. H. Kessler.
Microwave Journ., Vol. 12, 119 - 123 (1969).

Discusses the extension of equivalent circuit techniques to active microwave components using diodes. A first approximation to complete computer design of a complicated, diode-using microwave component was achieved by creating two computer programs; one dimensions an X-band parametric amplifier and the other predicts the details of its performance. *MWS*

033.041   On two-idler parametric amplifiers.
A. R. Kerr.
IEEE Trans. Microwave Theory and Techn., Vol. MTT-17, No. 1, p. 39 - 40 (1969).

033.042   Microwave power transistors.
W. C. Lee.
Microwave Journ., Vol. 12, 51 - 65 (1969).

033.043   Pulsar facility.
B. F. Burke, R. M. Price, M. S. Ewing.
Massachusetts Inst. Technol. Res. Lab. Electronics, Quarterly Progr. Rep. No. 93, p. 25 (1969).

Progress report on the 16 × 50-ft dish pulsar facility. *MMK*

033.044   Very long baseline interferometry.
B. F. Burke, I. I. Shapiro, H. F. Hinteregger, A. R. Whitney, C. A. Knight.
Massachusetts Inst. Technol. Res. Lab. Electronics, Quarterly Progr. Rep. No. 93, p. 23 - 24 (1969).

Preliminary report on V. L. B. experiments at X and L band in which the phase information is retained. Possible applications include testing gravitational bending prediction of relativity, measuring continental drift and proper motions of quasars. – *MMK*

033.045   A high-resolution swept-frequency reflectometer.
D. L. Hollway, P. I. Somlo.
IEEE Trans. Microwave Theory and Techn., Vol. MTT-17, No. 4, p. 185 - 188 (1969).

033.046   Radiometro solare automatizzato a stato solido a 239 MHz.   G. Sedmak.
Atti XII Riunione Soc. Astron. Italiana, L'Aquila 1968, p. 32 - 33 (1969). – Abstract SAI.

033.047   Dual input null networks utilizing RC ladders.
N. S. Roy.
Indian Journ. Pure Applied Phys., Vol. 6, 647 - 648 = Kodaikanal Obs. Repr. No. 43 (1968).

Two RC ladder networks are presented each of which, with an auxiliary input channel, shows transmission zero at a frequency that can be varied simply and widely by varying the ratio of the two inputs.

033.048   A novel way of beam switching, particularly suitable at millimeter wavelengths.
N. Albaugh, K. H. Wesseling.
IEEE Trans. Antennas Propagation, Vol. AP-17, No. 1, p. 98 - 100 (1969).

033.049   An interferometer with retranslation for the meter-range.   G. I. Dobysh.
Trudy Fiz. in-ta AN SSSR, Vol. 47, 201 - 211 (1969). In Russian. – Abstr. in Referativ. Zhurn. 51. Astron., 2.51.101 (1970).

033.050   Radar receiver with logarithmic IFA for radar observations of meteor trains.
S. N. Yudin, I. A. Delov.
Vestn. Khar'kovsk. politekhn. in-ta, No. 36 (84), p. 44 - 47 (1969).   In Russian. – Abstr. in Referativ. Zhurn. 51. Astron., 2.51.107 (1970).

033.051   Bericht über den Bau des 100-m-Radioteleskops des Max-Planck-Instituts für Radioastronomie Bonn.
O. Hachenberg.
Mitt. Astron. Ges. No. 27, p. 31 - 38 (1969). – Report on the construction of the 100 m radiotelescope at Bonn.

033.052   Use of atomic frequency standards for phase calibration of large aerial arrays.
K. V. Sheridan.
Electronics Letters, Vol. 5, No. 16, 2 pp. = Separate print Radiophys. Lab. C.S.I.R.O., Sydney, Australia (1969).

The use of two extremely stable radio-frequency oscillators, one stationary and the other mobile, is described for measuring relative phase difference between widely separated elements of large aerial arrays. Results of measurements made at 80 MHz on the 96-aerial Culgoora radioheliograph, using an atomic frequency standard for the mobile unit, demonstrate that the method is suitable for phase-calibration purposes.

033.053   Electro-optic spectrograph for radio astronomy.   T. W. Cole.
Optics Technology, Vol. 1, No. 1, p. 31 - 35 = Separate print Div. Radiophys. C.S.I.R.O., Sydney (1968).

This paper describes an experimental investigation of a type of spectrograph that employs electro-optic processing and its application, for the first time, to radio astronomy.

033.054   Transistor R.F. and I.F. amplifiers for radio astronomy applications.   R. A. Batchelor.
Proc. Institution Radio Electronics Engineers Australia, Vol. 30, No. 4, p. 99 - 105 = Separate print Div. Radiophys. C.S.I.R.O., Sydney (1969).

The paper describes transistor r.f. and i.f. amplifiers which have been designed for use on the 210-foot antenna of the Australian National Radio Astronomy Observatory. General criteria for obtaining low noise temperature and high stability are outlined and a number of specific designs are described, including an i.f. amplifier with 200 MHz bandwidth and 140° noise temperature.

033.055   A simple integrated circuit driver for a stepping motor.   D. J. Cole.
Proc. Institution Radio Electronics Engineers Australia, Vol.

30, No. 5, p. 153 - 154 = Separate print Div. Radiophys. C.S.I.R.O., Sydney (1969).

A reversible stepping motor drive circuit employing integrated circuits is described. It may be used up to the frequency limit of the stepping motor.

033.056 **Fields in the focal region of a spherical reflector.**
B. MacA. Thomas, H. C. Minnett, Vu The Bao.
IEEE Trans. Antennas Propagation, Vol. AP-17, No. 2, p. 229 - 232 = Separate print Div. Radiophys. C.S.I.R.O., Sydney (1969).

An earlier analysis of the field structure and energy flow near the axis of any circularly symmetric focusing reflector is applied to the case of a spherical reflector.

033.057 **Large 22-MHz array for radio astronomy.**
C. H. Costain, J. D. Lacey, R. S. Roger.
IEEE Trans. Antennas Propagation, Vol. AP-17, 162 - 169 = Contr. Dominion Obs. Ottawa, No. 255 (1969).

The relative merits of T- and cross-type radio telescopes are discussed. A large array used for radio astronomical studies at a frequency of 22.25 MHz (λ = 13.5 m) is described.

More telescopes urged.
Nature, Vol. 223, 988 (1969).

Below three millimetres.
Nature, Vol. 224, 300 - 301 (1969).

News notes on a 25 m dish aerial at Chilbolton, Hampshire.

Structures Technology for Large Radio and Radar Telescope Systems. See Abstr. 003.017.

Radiotelescopes.
See Abstr. 003.049.

Interferometric observations with a baseline of 127 kilometres — I. See Abstr. 141.104.

034 Astronomical Accessories

034.001 **A new principle for the contacts of an impersonal micrometer.** F. Smriglio.
Proc. Colloquium on Problems of Time Determination, Keeping and Synchronization, (Milan 1968), p. 229 - 234 (1968). In Italian.

034.002 **Development of an electronic cronograph with punched cards output for the use at a passage instrument.** G. Caprioli, M. Mattei.
Proc. Colloquium on Problems of Time Determination, Keeping and Synchronization, (Milan 1968), p. 241 - 246 (1968). In Italian.

034.003 **Auxiliary equipment of an astronomical photometer.** R. Rijf, J. Tinbergen, T. Walraven.
Bull. Astron. Inst. Netherlands, Vol. 20, 279 - 299 = Commun. Obs. Leiden (1969).
This paper discusses an instrumental system for efficient photoelectric photometry of large numbers of stars; Leiden Observatory's 36-inch light-collector and the simultaneous five-colour photometer designed by Walraven are used as a concrete example. Simultaneous multi-band photometry can yield colours more accurate than the magnitudes contributing to them; this leads to the extra requirement that the response of the measurement system to undetected accidental disturbances be "grey". After a brief description of the telescope and photometer, the design is dicussed in more detail.

034.004 **Notes on liquid-filter cells.**
S. F. Pellicori.
Commun. Lunar Planet. Lab. Vol. 8 (No. 140), 97 - 98 (1969).
Design and use of a leakproof liquid-filter cell is described.

034.005 **Planetary spectroscopy with the 107-inch telescope.** R. G. Tull.
Sky Telescope, Vol. 38, 156 - 160 (1969).

034.006 **A homebuilt machine for scanning plates.**
H. Vehrenberg.
Sky Telescope, Vol. 38, 186 - 188 (1969).

034.007 **A Harvard-Smithsonian tube sensitometer.**
D. W. Latham.
American Astron. Soc. Photo-Bull. No. 1, p. 3 - 7 (1969).
The paper describes a tube sensitometer designed for calibrating astronomical spectrograms.

034.008 **Aluminized filters for solar photography.**
S. W. Mathers, G. A. J. Ferris.
Journ. British Astron. Ass. Vol. 79, 376 - 380 (1969).

034.009 **Uno spettrografo per gioco.**
U. Dall'Olmo.
Coelum, Vol. 37, 220 - 221 (1969).

034.010 **Fadenkreuzokulare und ihre Beleuchtungseinrichtungen. 1. Teil.** H. G. Ziegler.
Orion Schaffhausen, Vol. 14, 88 - 93 (1969).

034.011 **Polarimeter for celestial X rays.**
H. W. Schnopper, K. Kalata.
Astron. Journ. Vol. 74, 854 - 858 (1969).
The properties of a polarimeter using reflection at a Bragg angle of 45° are discussed and applied to the design of a rocket payload for determining the polarization of X rays from celestial sources. Small detector areas are accomplished through the use of torroidally bent crystals to focus the X rays. This device will detect polarizations of as small as a few percent from sources such as Sco XR-1 and the Crab nebula. With slight modifications, this device can also be used to obtain polarization images of X-ray sources, and also to obtain a spectrum of polarization as a function of energy.

034.012 **The TV system used as a guiding set at the prime focus.** V. M. Ljuty, V. A. Malarev, G. A. Tambovski, L. V. Turkina.
Astron. Tsirk. No. 521, p. 1 - 3 (1969). In Russian.

034.013 **Electronic device for measurement of photographic recordings of circle readings.**
V. M. Ivakin, E. M. Tilk.
Astrometriya i Astrofiz.,*Kiev*, No. 2, p. 150 - 152 (1969). In Russian.

034.014 **Accidental errors of solar magnetographs.** O. V. Nikonov, E. S. Nikonova.
Solnechnye Dannye Byull. No. 4, p. 97 - 100 (1969). In Russian.
Accidental errors of the canals of longitudinal and transversal components of the magnetic field in a solar magnetograph were evaluated. The obtained results are in good agreement with the experimental data.

034.015 **A digital photo-electric photometer.**
C. Rowe.
Southern Stars, Vol. 23, 63 - 66 (1969).

034.016 **Quarzglas für Mond-Laser-Reflektor.**
P. Bäumler.
Laser, Vol. 1, No. 3, p. 29 - 31 (1969).

034.017 **On polarimetry in solar active regions. I: The new Locarno polarimeter; observing procedures.**
E. Wiehr.
Solar Physics, Vol. 9, 225 - 234 (1969).
The miscentering by the Doppler compensator of the Locarno polarimeter is investigated in detail. It is shown that the linear polarization is strongly falsified by this effect which also occurs at the Crimean and Izmiran polarimeters. The new design for the exit slits of the Locarno polarimeter is described. It avoids the ambiguities in the determination of the magnetic field vector that always occur when using two exit slits. A new simple electronic setup avoids most of the difficulties which are usually involved in eliminating instrumental polarization and compensating intensity fluctuations. The observational techniques for solar polarimetry at the Locarno observatory are described.

034.018 **Interferometry and grating spectroscopy: An introductory survey.** P. Jacquinot.
Applied Optics, Vol. 8, 497 - 499 (1969).

034.019 **Diffraction grating spectroscopy.**
F. Kneubühl.
Applied Optics, Vol. 8, 505 - 519 (1969).
The present state of grating spectroscopy is reviewed with special emphasis on the far infrared. The review includes the discussion of the properties of diffraction gratings, the intensity distribution among different orders of echelette gratings, Wood anomalies, infrared and submillimeterwave filters, detectors, grating spectrometers with thermal sources, rules for the construction of far infrared and submillimeterwave grating spectrometers, diffraction at the monochromator slits, compa-

rative performance of interferometers and grating spectrometers, and spectroscopy of far infrared laser emissions. Extensive references are presented.

034.020 Aberrations of Fabry-Pérot interferometers when used as filters. J. V. Ramsay.
Applied Optics, Vol. 8, 569 - 574 (1969).

The spectral and optical performances of Fabry-Pérot interferometers having nonuniformities in spacing are discussed, and the finesse F of an interferometer with a circular aperture and a parabolic nonuniformity in spacing and that of one with a rectangular aperture and a linear nonuniformity are given.

034.021 Open multipliers in the soft X-ray region.
U. Mayer, M. Mozer, M. v. Reinhardt.
Applied Optics, Vol. 8, 617 - 625 (1969).

The Bendix magnetic multiplier M 306, and the Mullard channel multipliers B 400 A and B 200 B were operated in the wavelength range 1.5 Å to 44 Å. Pulse height distributions, gain, and counting rate characteristics were investigated.

034.022 Photomultiplier reception of satellite beacon flashes.
D. G. Abby, T. E. Wirtanen.
Applied Optics, Vol. 8, 627 - 631 (1969).

Experiments have been performed on the electrooptical detection of flashes from satellite-borne beacons for the purpose of establishing the time of flash at the observing site, measurement of received pulse shape, and relative measurement of received energy. Initial observations have been made of the beacons carried by the geodetic satellite GEOS-B. Time of flash has been obtained to a precision of 0.1 msec.

034.023 Magnetically focused electronographic image converters for space astronomy applications.
G. R. Carruthers.
Applied Optics, Vol. 8, 633 - 638 (1969).

A series of magnetically focused, electronographic image converters has been developed for use in far-UV photography and spectroscopy. An all-reflecting objective spectrograph using this system with Schwarzschild optics has been constructed for obtaining stellar spectra in the 1000 – 1400-Å wavelength range, as well as a similar system based on Schmidt type optics for stellar spectroscopy in the 1230 – 2000-Å wavelength range. Also, smaller Schmidt type instruments have been developed for direct stellar photography in the far UV.

034.024 Infrared radiometer for the 1969 Mariner mission to Mars. S. C. Chase, Jr.
Applied Optics, Vol. 8, 639 - 643 (1969).

A two-channel radiometer has been developed to measure the equivalent blackbody surface temperatures of the equatorial and colder south polar regions of Mars. Two spectral channels, $8 – 12\,\mu$ and $18 – 25\,\mu$, cover the temperature range of 140 K to 325 K.

034.025 Stellar photometric data for various photocathode materials. F. F. Forbes, R. I. Mitchell.
Applied Optics, Vol. 8, 718 - 720 (1969).

034.026 Objective transmission gratings for large Schmidt telescopes. J. Strong, F. Zwicky.
Applied Optics, Vol. 8, 1021 - 1022 (1969).

Several lamellar gratings of 18-in. aperture, with 300 diffraction elements per inch, have been made for determining stellar spectra with wide and angle telescopes. Central orders are missing at λ4800, and weak at adjacent wavelengths. The two, equal first orders are about 1^m weaker than spectra by a prism or echelette grating. Dispersion is linear.

034.027 Synchronous three colour stellar photometry at the Catania Astrophysical Observatory.
S. Cristaldi, L. Paternò.
Non-Periodic Phenomena in Variable Stars, IAU Colloquium, Budapest, 1968, p. 51 - 56 (1969).

A synchronous three colour stellar photometer using a single photomultiplier has been constructed at Catania. In this communication the characteristics and the efficiency of this photometer are briefly described. At present the instrument is used for simultaneous UBV photometry of flare stars. A graph of simultaneous measurements in the UBV system of a flare of EV Lac is shown. A more detailed description of the instrument had been published elsewhere (Cristaldi, Paternò 1968).

034.028 Vizier en hoekmeetinstrument.
A. H. Boerdijk.
Hemel en Dampkring, Vol. 67, 336 - 342 (1969).

034.029 Sampling. G. M. Hotz.
Applied Optics, Vol. 8, 1329 - 1339 (1969).

Sampling requirements for prospective landed extraterrestrial geological and biological instruments performing on-site analyses are satisfied by three general classes of samplers: bulk, selective particulate, and hard rock.

034.030 Extraterrestrial optical microscopy.
G. A. Soffen.
Applied Optics, Vol. 8, 1341 - 1347 (1969).

The microscope is an important tool in detecting and identifying extraterrestrial life. The interpretation of the data may be the most difficult aspect of the experiment, but an unequivocal result is certainly possible. From an instrumental point of view, it is a complex instrument to design maximally without requiring a human operator. Three instruments have been investigated to some extent.

034.031 Wavelength errors in spectrographs. I: The effect of surface irregularities in camera mirrors.
R. G. Tull.
Applied Optics, Vol. 8, 1635 - 1638 (1969).

High-dispersion astronomical coudé spectrographs frequently give lower-precision radial velocities than expected. This is explained in part in terms of the variable spectral shifts along the plate produced by a camera mirror with zonal imperfections. For a mirror with a sinusoidal zonal irregularity, the standard deviation in the radial velocity depends inversely on the angular dispersion and size of the grating or prism and directly on the amplitude of the surface fluctuation.

034.032 Interferometry and the Doppler effect.
D. Malacara, I. Rizo, A. Morales.
Applied Optics, Vol. 8, 1746 - 1747 (1969).

034.033 Spectrometer slit programmer with no moving parts. B. Krakow, S. A. Dolin.
Applied Optics, Vol. 8, 1827 - 1829 (1969).

An optical system with no moving mechanical parts has been installed in a rapid scanning spectrometer to program the spectral slit width as a function of wavelength. A set of contoured slits and baffles maintains a good balance between signal and resolution and keeps the dynamic range of the background signal within reasonable limits.

034.034 A grille spectrometer for measurements near 14μ.
B. A. Tinsley.
Applied Optics, Vol. 8, 1831 - 1835 (1969).

A new design of the grille spectrometer allows its use for measurements of thermal ir emissions. The grilles are made of self-supporting etched metal foil, and the curved chessboard pattern allows optical chopping of one spectral element at a time by oscillation of the exit grille.

034.035 A pressure-scanned Fabry-Pérot interferometer for the Kitt Peak National Observatory. M. G. Smith.
Bull. American Astron. Soc. Vol. 1, 262 (1969). – Abstr. AAS.

034.036 Progress report on a 40-channel magnetograph.
R. S. Aikens, L. A. Doe, W. C. Livingston, C. D. Slaughter.
Bull. American Astron. Soc. Vol. 1, 270 (1969). – Abstr. AAS.

034.037 Observations of the sun and laboratory sources with a three-meter XUV spectrograph.
W. E. Behring, L. Cohen, K. Saffer, U. Feldman.
Bull. American Astron. Soc. Vol. 1, 272 - 273 (1969). – Abstr. AAS.

034.038 Computer display of magnetograph observations.
E. N. Frazier.
Bull. American Astron. Soc. Vol. 1, 276 (1969). – Abstr. AAS.

034.039 A filter magnetograph. H. E. Ramsey.
Bull. American Astron. Soc. Vol. 1, 291 (1969). – Abstr. AAS.

034.040 The flare videometer at the manned spacecraft center, Houston. J. H. Reid.
Bull. American Astron. Soc. Vol. 1, 291 (1969). – Abstr. AAS.

034.041 Dynamical errors of solar magnetographs.
O. V. Nikonov, E. S. Nikonova.
Solnechnye Dannye 1969 Byull. No. 5, p. 86 - 90 (1969). In Russian.

034.042 On the influence of the width of the intermediate slit of a double monochromator on the profiles of the Fraunhofer lines. E. A. Gurtovenko.
Solnechnye Dannye 1969 Byull. No. 6, p. 91 - 99 (1969). In Russian.

It is shown that the variations in the width (within the limits of 0.15 - 1.07 mm) of the intermediate slit of the Golosseyevo solar double monochromator do not influence appreciably the distant wings of the line. They affect insignificantly closer wings and give rise to essential changes of the central intensity of the line.

034.043 Determination of the instrumental profile of the Coudé spectrograph of the Ondřejov two-meter telescope. C. Veth.
Bull. Astron. Inst. Netherlands, Vol. 20, 312 - 316 (1969).

A helium-neon gas laser has been used to determine the instrumental profile of a Coudé spectrograph for stellar spectra. A comparison of the theoretical profile with the experimental profile shows the important influence of the characteristics of the photographic plate.

034.044 Utilisation de réflexions sur des surfaces coniques pour analyser sans rotation une lumière polarisée.
C. Aimé.
Comptes Rendus Acad. Sci. Paris, Sér. B, Vol. 269, 964 - 967 (1969).

Réalisation en laboratoire d'un polarimètre utilisant des réflexions sur deux surfaces coniques dans le domaine visible. L'analyse de la lumière se fait sans rotation. Une étude théorique est faite en vue de mesurer la polarisation interstellaire en lumière ultraviolette.

034.045 An integrating isodensitometer.
A. V. Hewitt.
Publ. Astron. Soc. Pacific, Vol. 81, 541 - 542 (1969). Abstract ASP.

034.046 A filter for photography of comet continua.
F. D. Miller.
Publ. Astron. Soc. Pacific, Vol. 81, 594 - 600 (1969).

Criteria adopted in planning an interference filter for photography of the dust component of comet heads and an application of the filter to comet Honda (1968 c) are described.

034.047 The use of Fabry-Pérot filters for spectral scanning.
M. D. Waterworth.
Proc. Astron. Soc. Australia, Vol. 1, 293 - 294 (1969). – Contribution ASA meeting.

034.048 Some photoelectric tracking systems for telescope guiding. Kkhong Din' Khong, V. G. Vafiadi.
Vestn. Belorussk. un–ta, Ser. 1, No. 1, p. 89 - 92 (1969). In Russian. – Abstr. in Referativ. Zhurn. 51. Astron., 10.51.805 (1969).

034.049 Vibrational mode behavior of rotating cruciform gravitational gradient sensors. C. C. Bell, J. R. Morris, J. M. Richardson, R. L. Forward.
Journ. Applied Phys., Vol. 39, 3193 - 3200 (1968). – See Phys. Ber., Vol. 48, No. 2 – 3346 (1969).

034.050 Use of highly reflecting crystals for spectroscopy and polarimetry in X-ray astronomy.
J. R. P. Angel, M. Weisskopf.
Bull. American Astron. Soc., Vol. 1, 333 (1969). – Abstract AAS.

034.051 Multiplex grating spectrometer.
J. A. Decker, Jr., M. Harwitt.
Bull. American Astron. Soc., Vol. 1, 339 (1969). – Abstract AAS.

034.052 The Kron electronic camera.
M. R. Lewis, G. E. Kron.
Bull. American Astron. Soc., Vol. 1, 352 - 353 (1969). – Abstract AAS.

034.053 Calcul d'un spectromètre interférentiel Fabry – Pérot intégral astronomique.
S. I. Sheglova.
Astron. Zhurn. Akad. Nauk SSSR, Vol. 46, 885 - 888 (1969). In Russian. English translation in Soviet Astron. AJ, Vol. 13, No. 4.

Formulas and nomograms for the construction of an astronomical interference spectrometer of Fabry–Pérot type are given. It is provided for obtaining stellar and planetary spectra with high resolving power.

034.054 L'interferometro Fabry-Pérot per l'analisi fotometrica dei profili di righe solari.
A. Righini.
Mem. Soc. Astron. Italiana, Nuova Serie, Vol. 40, 475 - 487 (1969).

Photometric measurements of the instrumental profile of an interferometer Fabry-Pérot, the evaluation of the performances when this instrument is used as scanning element and comparison with those of a conventional grating spectrometer are given.

034.055 Photomètre pour photométrie visuelle directe.
Y. Grandjean.

Orion Schaffhausen, Vol. 14, 141 - 142 (1969).

034.056 A rocket-borne photoelectric spectrophotometer using convergent beam dispersion to observe far ultraviolet stellar spectra. G. C. Sudbury.
Applied Optics, Vol. 8, 2013 - 2017 (1969).

The advantages of the Monk-Gillieson dispersion system using a plane grating off axis in the convergent beam from a paraboloidal collecting mirror have been applied to obtain low resolution photoelectric spectra of over forty bright stars in the 1500–3000 Å region. Techniques of construction, alignment, calibration, and dc and pulse counting output data handling are described. The flight performance, in unstabilized Skylark rockets, is discussed.

034.057 A two-channel scanning spectrophotometer for use in studies of collision broadening and shift in optical spectra. A. D. Petford, G. Smith.
Applied Optics, Vol. 8, 2025 - 2028 (1969).

A two-channel scanning spectrophotometer is described which has been used, in conjunction with a high resolution grating spectrograph, to make measurements of collision broadening and shift in optical absorption spectra to an accuracy of 0.001 cm^{-1}.

034.058 Far infrared band-pass filters in the 400–16 cm^{-1} spectral region. S. P. Varma, K. D. Möller.
Applied Optics, Vol. 8, 2151 - 2152 (1969).

034.059 A spectral line discriminator interferometer: an analysis. C. T. Foskett, J. M. Weinberg.
Applied Optics, Vol. 8, 2185 - 2190 (1969).

This work considers the Michelson spectral line discriminator interferometer (SLDI), a field compensated instrument, as a linear device. The ability of the instrument to measure spectral discontinuities in the presence of strong background continua, the harmonic distribution of the signal, and the stability of the signal are examined.

034.060 Reflectance of semitransparent platinum films on various substrates in the vacuum ultraviolet.
G. Hass, J. B. Ramsey, W. R. Hunter.
Applied Optics, Vol. 8, 2255 - 2259 (1969).

The reflectance of semitransparent Pt films deposited on various substrates at close to room temperature was measured at λ = 584 Å and λ = 736 Å and in the wavelength region 1000–2000 Å and was compared with that of opaque films of Pt.

034.061 The improved solar magnetograph of the High Altitude Observatory.
R. H. Lee, J. W. Harvey, E. Tandberg-Hanssen.
Applied Optics, Vol. 8, 2370 - 2372 (1969).

034.062 A small lamellar grating interferometer for the very far-infrared. R. C. Milward.
Infrared Physics, Vol. 9, 59 - 74 (1969).

This paper describes the construction and performance of a lamellar grating interferometer designed for routine spectroscopy in the 3–80 cm^{-1} spectral region.

034.063 Photoexposimeter for photographic registration of circle readings. A. M. Stafeev, I. N. Nabokov.
Astrometriya i Astrofiz., *Kiev*, No. 7, p. 81 - 83 (1969).
In Russian.

A photoelectrical apparatus is suggested to be used for choosing the exposure time during photographing the meridian circle limb. The authors made such an apparatus and called it photoexposimeter. A description of this apparatus and its design are presented.

034.064 The sensitivity distribution in the field of the FKT-type image tube. M. K. Babadjanjanz.
Trudy Astron. Obs. *Leningrad,* Vol. 26 (= Uchenye Zapiski Leningr. Un-ta No. 347 = Seriya Matem. Nauk No. 44), p. 84 - 87 (1969). In Russian.

The sensitivity distribution of FKT-type tubes is measured. It is shown that these tubes can be used for photometric research.

034.065 Control systems of telescopes.
N. N. Mikhel'son.
Zemlya i Vselennaya, No. 5, p. 51 - 54 (1969). In Russian.

034.066 Accessories of telescopes.
N. N. Mikhel'son.
Zemlya i Vselennaya, No. 6, p. 46 - 49 (1969). In Russian.

034.067 A rocket borne scintillation spectrometer for observing cosmic X rays.
J. Harri, M. McGee, A. Toor.
Rev. Sci. Instruments, Vol. 40, 703 - 708 (1969).

034.068 Design of a zenith mirror. M. Drodofsky.
Optik, Vol. 29, 45 - 51 (1969). In German.

A device eliminating the drawbacks of an exposed mercury surface is suggested for determining the direction of the plumb line in astrometrical instruments. Hints are given on dimensioning for a residual error of 0.01″.

034.069 The polarimeter of the 1 m photometric telescope. A. Behr.
ESO Bull. No. 5, p. 9 - 13 (1969).

The principle of the polarimeter rests on an idea of Hiltner, who proposed to split the light of the star by means of a Wollaston prism into two beams of light polarized in planes of vibration perpendicular to each other and being observed simultaneously. In this way the effects of atmospheric scintillation should be minimized, and changes in extinction during a set of measurements compensated.

034.070 Flexions d'un spectrographe.
A. Baranne, E. Maurice.
ESO Bull. No. 7, p. 5 - 10 (1969).

034.071 Le spectrographe Cassegrain du télescope de 1.52 mètre. A. Baranne, E. Maurice, L. Prévot.
ESO Bull. No. 7, p. 11 - 18 (1969).

034.072 Un tipo di polarimetro privo di elementi ruotanti.
A. Cacciani.
Atti XII Riunione Soc. Astron. Italiana, L'Aquila 1968, p. 20 - 22 (1969). – Abstract SAI.

034.073 Mercury cadmium telluride as a 1–20 μm wavelength infrared detector for space applications.
R. D. Packard.
AIAA Journ., Vol. 7, 1570 - 1573 (1969).

Mercury cadmium telluride offers attractive possibilities for fast, elevated-operating-temperature infrared detection from 1–20 μm wavelengths, particularly in the eight to fourteen region for which it was developed. It is considered in this report for use in satellites for terrestrial, atmospheric, and ocean mapping, infrared astronomy, and optical communications.

034.074 Photometric error analysis. IX: Optimum use of photomultipliers. A. T. Young.
Applied Optics, Vol. 8, 2431 - 2447 (1969).

034.075 Up–down photometers for auroral profile studies.

K. A. Dick, W. G. Fastie.
Applied Optics, Vol. 8, 2457 - 2460 (1969).
An ultraviolet rocket-borne photometer system currently under development is described which alternately measures zenith and nadir intensities at rocket altitude.

034.076   The calibration of sky photometers.
F. Rössler, D. Bock.
Optik, Vol. 28, 299 - 310 (1969).   In German.
Systematic tests have been performed to improve the accuracy in the calibration of sky photometers. Measurement by indirect illumination and with the aid of neutral filters is the only method which corresponds to the conditions in the sky and which in itself is free from discrepancies. In special cases, the technique of direct calibration cannot be used. To achieve sufficient illuminance, a reflectorlamp with a condenser must be employed.

034.077   On the exact definition of wavelength with a Fabry–Pérot étalon.
J. I. Rudnev, L. A. Sataeva.
Trudy Astrofiz. Inst. Alma-Ata, Vol. 13, 98 - 102 (1969). In Russian.

034.078   Application of a Fabry–Pérot étalon for investigation of temperature dependence of the wavelength
of ruby laser radiation.   J. I. Rudnev, L. A. Sataeva.
Trudy Astrofiz. Inst. Alma-Ata, Vol. 13, 103 - 108 (1969). In Russian.

034.079   Dependence of the wavelength of ruby laser radiation on temperature.
P. N. Boiko, L. A. Sataeva, G. A. Kharitonova.
Trudy Astrofiz. Inst. Alma-Ata, Vol. 13, 109 - 111 (1969). In Russian.

034.080   On the visual observation of infrared laser emission.
J. I. Rudnev, T. P. Toropova.
Trudy Astrofiz. Inst. Alma-Ata, Vol. 13, 115 - 116 (1969). In Russian.

034.081   A double-beam microphotometer.
D. A. Rozhkovsky, M. I. Musorin.
Trudy Astrofiz. Inst. Alma-Ata, Vol. 14, 54 - 64 (1969). In Russian.
A double-beam microphotometer for simultaneous measurements of two photographs of the same object is described.

034.082   High power zone plates as image forming systems for soft X-rays.   G. Schmahl, D. Rudolph.
Optik, Vol. 29, 577 - 585 (1969).   In German.
A method is described to make zone plates with high zone numbers ($N \geq 10^3$) using laser interferences in combination with photoresist layers. By focussing soft X-rays with such zone plates spherical aberration occurs. This can be corrected if one uses optics with special spherical aberration for producing the zone plates. Applications for high power zone plates (X-ray microscope, X-ray telescope) are discussed.

034.083   Meßwertlocher mit Handeingabemöglichkeit für ein Digitalvoltmeter.   R. Schielicke.
Radio, Fernsehen, Elektronik, Band 17, H. 19, 595 - 597 = Mitt. Univ.-Sternw. Jena No. 86 (1968).
Beim Auswerten photographischer Himmelsaufnahmen zur Bestimmung von Sternhelligkeiten mit einem Irisblendenphotometer wird wegen des hohen Informationsgehaltes eine möglichst große Meßgeschwindigkeit gefordert. Durch den nachträglichen Anbau einer digitalen Meßwertausgabe an ein Sartorius-Irisblendenphotometer konnte die Meßgeschwindigkeit von etwa 90 auf 350 Sterne je Stunde gesteigert werden.

034.084   Ein Plangitterspektrograph im Cassegrain-Fokus des 2m-Universal-Spiegel-Teleskopes.   H. Artus.
Jenaer Rundschau (Jena Review), 14. Jahrgang, 346 - 352 (1969).

034.085   Probleme der "Lichtelektrischen Photometer".
H. Ziegler.
BAV Rundbrief, 18. Jahrgang, 21 - 25 (1969).

034.086   A sky-suppressing automatic trailing device.
C. R. O'Dell.
Publ. Astron. Soc. Pacific, Vol. 81, 854 - 860 (1969).
An automatic trailing device for stellar spectroscopy is described. This system, employing a rocking plate behind the entrance slit, permits one to eliminate most of the sky background, increases observing efficiency, improves the homogeneity of spectra, and the stellar image remains fixed.

034.087   Versuche zur Verkleinerung des periodischen Fehlers der Mikrometerschraube eines Libellen-
prüfers.   S. S. Tovchigrechko.
Rotation of the Earth and Determination of Time, Conference Riga 1965, p. 117 - 119 (1969).   In Russian.

034.088   Trägheitseigenschaften eines schmalbandigen Verstärkers.   A. I. Yazev.
Rotation of the Earth and Determination of Time, Conference Riga 1965, p. 119 - 124 (1969).   In Russian.

034.089   Photoelektrische Transistoranlage zur Registrierung von Sterndurchgängen.   M. I. Il'kiv.
Rotation of the Earth and Determination of Time, Conference Riga 1965, p. 125 - 129 (1969).   In Russian.

034.090   Elektronische Rechenanlage zur Bestimmung der Zeitmittel von Sterndurchgängen.
B. I. Brodskij.
Rotation of the Earth and Determination of Time, Conference Riga 1965, p. 129 - 134 (1969).   In Russian.

034.091   A gamma ray telescope utilizing large area wire spark chambers.
R. W. Ross, C. H. Ehrmann, C. E. Fichtel, D. A. Kniffen, H. B. Ögelman.
IEEE Trans. Nuclear Sci., Vol. NS-16, No. 1, p. 304 - 308 (1969).

034.092   Onboard calibration system for γ ray spectrometers in space satellites.
E. L. Chupp, P. J. Lavakare, A. A. Sarkady.
IEEE Trans. Nuclear Sci., Vol. NS-16, No. 1, p. 309 - 313 (1969).

034.093   A current-to-frequency converter for astronomical photometry.   D. J. Taylor.
Rev. Sci. Instruments, Vol. 40, 559 - 562 (1969).

034.094   An amplitude discriminator with following threshold.   V. V. Lizogub, V. A. Nechitajlenko.
Vestn. Khar'kovsk. politekhn. in-ta, No. 36 (84), p. 27 - 31 (1969).   In Russian. – Abstr. in Referativ. Zhurn. 51. Astron., 2.51.106 (1970).

034.095   The position contact micrometer of the astronomical universal instrument.
A. V. Gozhij, V. A. Ovchinnikov.
Geod., kartogr. i aehrofotos''emka. Mezhved. resp. nauchnotekhn. sb., Vyp. (No.) 9, p. 3 - 12 (1969).   In Russian. Abstr. in Referativ. Zhurn. 51. Astron., 2.51.194 (1970).

034.096   The Surveyor VII laser pointing experiment.
J. W. Young.
Strolling Astronomer, Vol. 21, 176 - 178 (1969).

034.097   Eine empfindliche Kamera zur Beobachtung licht-
schwacher, flächenhafter astronomischer Objekte.
J. Loidl.
Mitt. Astron. Ges. No. 27, p. 199 (1969). – Abstract AG.

034.098   Ein lichtelektrisches Abtastphotometer.
E. Hög.
Mitt. Astron. Ges. No. 27, p. 199 (1969). – Abstract AG.

034.099   Die Verwendung von Ringzonenplatten zur Ab-
bildung der Röntgensonne.
G. Elwert, J. Feitzinger.
Mitt. Astron. Ges. No. 27, p. 199 - 200 (1969). – Abstract
AG.

034.100   Ein Spektroheliograph für ultraweiche Röntgen-
strahlung.
H. Bräuninger, H. J. Einighammer, H. H. Fink, U. Mayer.
Mitt. Astron. Ges. No. 27, p. 200 - 201 (1969). – Abstract
AG.

034.101   Astronomische Anwendungsmöglichkeiten holo-
graphisch hergestellter Beugungsgitter.
G. Schmahl, D. Rudolph.
Mitt. Astron. Ges. No. 27, p. 201 - 203 (1969). – Abstract
AG.

034.102   Photo-electric calibration of an ultraviolet
Schmidt camera.   C. M. Humphries.
Commun. Roy. Obs. Edinburgh, No. 59 [Reprinted from
Symposium on "Calibration Methods in the Ultraviolet and
X-ray regions of the Spectrum", Munich, 1968] p. 199 - 204
(1968).
   The absolute intensity calibration of photographic
recording stellar ultraviolet instruments using in-flight photoelectric techniques is discussed. In particular, the calibration
of a dual sounding rocket experiment consisting of an objective prism Schmidt camera and an ultraviolet photometer is
described.

034.103   An on-board calibration unit for the Liège-Edin-
burgh ultraviolet stellar photometers in the TD1
satellite.   G. C. Sudbury.
Commun. Roy. Obs. Edinburgh, No. 60 [Reprinted from
Symposium on "Calibration Methods in the Ultraviolet and
X-ray Regions of the Spectrum", Munich, 1968] p. 191 -

198 (1968).
   The subject of this paper is essentially an applied secondary calibration problem, how to transfer an instrument which
has been calibrated by other means into orbit with an assur-
ance that its calibration is reasonably intact.

034.104   The absolute calibration of rocket-borne stellar
photometers in the region 2000 – 3000 Å.
J. W. Campbell.
Commun. Roy. Obs. Edinburgh, No. 61 [Reprinted from
Symposium on "Calibration Methods in the Ultraviolet and
X-ray Regions of the Spectrum", Munich, 1968] p. 183 -
190 (1968).
   In order to compare the results of stellar surveys over a
period of time it is necessary that reliable laboratory calibration standards are maintained. This paper describes the
laboratory procedures adopted by the author in calibrating
four stellar photometers recently flown in the ESRO programme. Such procedures include the measurement of spectral response, quantum efficiency and cathode uniformity for
a variety of "solar blind" photomultipliers.

034.105   Radial velocities of extended sources by a slitless
scanning technique.
G. de Vaucouleurs.
Contr. McDonald Obs., Fort Davis, No. 420 [Reprinted from
I.A.U. Symposium, No. 30, p. 91 - 101], 11 pp. (1967).

034.106   Calibrazione di due spettrosensitometri.
P. Broglia, G. Guerrero.
Contr. Oss. Astron. Milano–Merate, Nuova Ser., No. 299,
8 pp. (1968).
   By means of photoelectric scanning two spectrosensito-
meters, the one with rotating sector, the other with linear
variation of intensity are calibrated. In relation to their use
the intermittence effect for some photographic emulsions
is studied.

   The vertical mirror, its potential application to
theodolites and two star image stopping micrometers.
See Abstr. 032.012.

   An ultraviolet image converter and folded all-reflecting Schmidt telescope for ultraviolet astronomy.
See Abstr. 032.020.

   Infrared photometry. Theoretical limits for signal
to noise ratio in the general astronomical case in the μ range.
See Abstr. 113.012.

035 Clocks and Frequency Standards

035.001 Methods and experiences for the utilization and synchronization of time signals and standard frequencies. F. Chlistovsky, C. de Concini.
Proc. Colloquium on Problems of Time Determination, Keeping and Synchronization, (Milan 1968), p. 23 - 30 (1968). In Italian.

In the first part some fields of technical and scientific applications which require use of time signals and standard frequencies are briefly treated. The necessary relative precision in these cases go from 10^{-7} to 10^{-13}. The more and more pressing requirements of precision of time signals and standard frequency make urgent and actual the problem of synchronization (of time and frequency). In the second part experiences of synchronization, some of these currently used, other in experimental phase are related.

035.002 Phase and time variations in VLF synchronization. E. Proverbio.
Proc. Colloquium on Problems of Time Determination, Keeping and Synchronization, (Milan 1968), p. 31 - 40 (1968). In Italian.

035.003 Statistical analysis of radio time signal propagation in band 9. F. Chlistovsky, E. Proverbio.
Proc. Colloquium on Problems of Time Determination, Keeping and Synchronization, (Milan 1968), p. 41 - 47 (1968).

035.004 Standard time and frequency comparison accuracy using HF ground-waves. S. Leschiutta.
Proc. Colloquium on Problems of Time Determination, Keeping and Synchronization, (Milan 1968), p. 49 - 58 (1968). In Italian.

035.005 Propagation velocity variations of radio time signals and related problems. E. Proverbio.
Proc. Colloquium on Problems of Time Determination, Keeping and Synchronization, (Milan, 1968), p. 59 - 67 (1968). In Italian.

035.006 Phase comparisons of myriametric waves for distances of about 1000 km. S. Leschiutta.
Proc. Colloquium on Problems of Time Determination, Keeping and Synchronization, (Milan 1968), p. 69 - 90 (1968). In Italian.

035.007 Results of phase comparisons of VLF standard frequencies over mean distances.
E. Proverbio, F. Chlistovsky.
Proc. Colloquium on Problems of Time Determination, Keeping and Synchronization, (Milan 1968), p. 91 - 109 (1968). In Italian.

035.008 Long term keeping of time scales. S. Leschiutta.
Proc. Colloquium on Problems of Time Determination, Keeping and Synchronization, (Milan 1968), p. 111 - 132 (1968). In Italian.

Two points concerning the problem of maintaining remotely located clocks in synchronism are here investigated. First the timekeeping accuracy using the VLF phase comparison data is obtained, second the criteria to be used in order to maintain a time scale are given.

035.009 Coherent system for travel time of radio time signals. E. Proverbio, F. Chlistovsky.
Proc. Colloquium on Problems of Time Determination, Keeping and Synchronization, (Milan 1968), p. 133 - 137 (1968).

035.010 Time transport over great distances. H. Brandenberger, G. Cauderay.
Proc. Colloquium on Problems of Time Determination, Keeping and Synchronization, (Milan 1968), p. 139 - 142 (1968). In Italian.

035.011 VLF narrow band time signals. C. Egidi.
Proc. Colloquium on Problems of Time Determination, Keeping and Synchronization, (Milan 1968), p. 143 - 160 (1968). In Italian.

035.012 Semiautomatic devices in time keeping and synchronization at the Brera Observatory.
F. Chlistovsky.
Proc. Colloquium on Problems of Time Determination, Keeping and Synchronization, (Milan 1968), p. 273 - 280 (1968). In Italian.

035.013 Development of the IEN facilities for frequency and time calibrations.
E. Angelotti, C. Egidi, G. Giachino, S. Leschiutta, G. Roncalli.
Proc. Colloquium on Problems of Time Determination, Keeping and Synchronization, (Milan 1968), p. 281 - 299 (1968). In Italian.

035.014 Developments and prospects of time signals and standard frequencies transmission at the Brera Observatory. E. Proverbio, F. Chlistovsky.
Proc. Colloquium on Problems of Time Determination, Keeping and Synchronization, (Milan 1968), p. 301 - 312 (1968). In Italian.

035.015 Time and frequency standard telephonic connection between I. S. P. T. and the Astronomical Observatory in Rome.
G. Caprioli, A. Orlando, A. Porreca Massangioli.
Proc. Colloquium on Problems of Time Determination, Keeping and Synchronization, (Milan 1968) p. 313 - 320 (1968). In Italian.

035.016 Synchronization of standard clocks by time signal emissions.
S. Leschiutta, A. Orlando, A. P. Massangioli.
Proc. Colloquium on Problems of Time Determination, Keeping and Synchronization, (Milan 1968), p. 321 - 329 (1968). In Italian.

035.017 Experiences with a caesium resonator. S. Leschiutta.
Proc. Colloquium on Problems of Time Determination, Keeping and Synchronization, (Milan 1968), p. 331 - 345 (1968). In Italian.

035.018 A portable reference clock for phase and time comparisons between standard signals of frequency and time. G. Giachino.
Proc. Colloquium on Problems of Time Determination, Keeping and Synchronization, (Milan 1968), p. 347 - 350 (1968). In Italian.

035.019 Centenary of the New Zealand time-service. E. A. Beet, G. A. Eiby.
Journ. British Astron. Ass. Vol. 79, 489 - 490 (1969).

– Historical section report British Astron. Ass.

035.020 Chronometrie. G. Glaser.
Phys. Blätter, 25. Jahrgang, 437 - 447 (1969).

035.021 Sternzeituhr für den Amateur, III.
E. Wiedemann.
Orion Schaffhausen, Vol. 14, 157 - 158 (1969).

035.022 The atomic clock meets the needs of modern science and technology.
La Suisse Horlogère, Rev. Internationale de l'Horlogerie, Vol. 84, No. 3, p. 35 - 38 (1969).

035.023 On smoothing clock corrections.
A. N. Kuryanova.
Astrometriya i Astrofiz., *Kiev,* No. 7, p. 61 - 65 (1969). In Russian.

The expression of the linear transformation operator, the action of which on the observed sequence of clock corrections is equivalent to that of parabolic smoothing adopted by the Time Service of the USSR, is found. The transfer function of the above transformation is derived. The fluctuations with periods from 35 to 270 days are shown to contribute to both deviations of the observed corrections from the smoothed curve and to the smoothed curve itself.

035.024 Daily variations of the frequency of a very accurate radio frequency. D. S. Sadeh, B. D. Au.
Nature, Vol. 224, 1291 - 1293 (1969).

The rate of two cesium clocks, one in Washington D. C. and the other in Cape Fear North Carolina, was compared. An apparent decrease in the rate of the North Carolina clock is evident at sunrise and an increase at sun-set. A smaller decrease is apparent at moon rise. Such an effect cannot be caused by the influence of the height of the ionosphere on the carrier wave and not on known environmental effects.

035.025 A re-evaluation of the NRC long cesium beam frequency standard.
A. G. Mungall, R. Bailey, H. Daams, D. Morris.
Metrologia, Vol. 4, 165 - 168 (1968). – See Phys. Abstr. Vol. 73, No. 2793 (1970).

035.026 Mass-frequency effect on v.l.f. and portable clock comparisons of atomic frequency standards.
A. G. Mungall, H. Daams, R. Bailey, D. Morris.
Metrologia, Vol. 5, No. 1, p. 31 - 32 (1969). – See Phys. Abstr. Vol. 73, No. 2794 (1970).

035.027 Study of some anomalous deviations of the clock corrections from the system of standard time.
A. V. Shirjaev, M. P. Mishchenko.
Trudy Astron. Obs. *Leningrad,* Vol. 26 (= Uchenye Zapiski Leningr. Un-ta No. 347 = Seriya Matem. Nauk No. 44), p. 125 - 133 (1969). In Russian.

The analysis of data obtained with the photoelectric transit instrument of Leningrad shows that the precision of the determination of time depends on the thermal deformations of the instrument, the humidity of the atmosphere and the method of the determination of the most probable value of the azimuth of the instrument.

035.028 Régimes transitoires et non linéarités dans les oscillations des quartz. M. Gagnepain.
Ann. Françaises Chronométrie Micromécanique, 4. année, p. 25 - 48 (1969).

035.029 Générateurs d'impulsions horaires.
L. Genoux, M. Sauzeat.
Ann. Françaises Chronométrie Micromécanique, 4. année, p. 69 - 71 (1969).

Au cours des années précédentes les appareils de distribution horaire ont suivi l'évolution constante de la technologie, et, aujourd'hui, nous nous proposons de décrire un appareil réalisé à l'aide de circuits intégrés.

035.030 Compte-rendu d'activité du Service Chronométrique du 1er avril 1968 au 31 mars 1969.
A. Remond.
Ann. Françaises Chronométrie Micromécanique, 4. année, p. 81 - 83 (1969).

035.031 Über die Ursache der Alterungserscheinung an Schwingquarzen. V. Kroitzsch.
Arbeiten Geod. Inst. Potsdam (Deutsche Akad. Wiss. Berlin), No. 27, 109 pp. (1969).

The resonance frequency of quartz resonators changes with time and due to some other influences. The possible causes of quartz resonator aging are analysed in this paper.

035.032 Betrachtungen zum Präzisionsfrequenzvergleich.
R. Stecher.
Nachrichtentechnik, 19. Jahrgang, 272 - 276 = Mitt. Geod. Inst. Potsdam No. 111 (1969).

035.033 Note on atomic time keeping at the National Research Council.
A. G. Mungall, H. Daams, R. Bailey.
Metrologia, Vol. 5, No. 3, p. 73 - 76 (1969).

035.034 The integrated time of the VLF radio waves received at Mizusawa. T. Hara, K. Horiai.
Proc. International Latitude Obs. Mizusawa, No. 9, p. 59 - 67 (1969). In Japanese.

035.035 Analyse des Uhrganges mit Hilfe der Korrelationsfunktion. K. A. Shtejns.
Rotation of the Earth and Determination of Time, Conference Riga 1965, p. 113 - 116 (1969). In Russian.

035.036 Über einen Frequenzteiler. E. M. Vinnikov.
Rotation of the Earth and Determination of Time, Conference Riga 1965, p. 134 - 139 (1969). In Russian.

035.037 Vervollkommnung der Uhren "Rohde und Schwartz" der Produktion des Jahres 1957.
N. A. Tel'pukhovskij, A. M. Moroz.
Rotation of the Earth and Determination of Time, Conference Riga 1965, p. 139 - 145 (1969). In Russian.

035.038 The photoelectric device for the time service of the Charkov Astronomical Observatory and the National Research Institute for Metrology in Charkov.
A. D. Egorov.
Vestn. Khar'kov. Univ. No. 34, (Ser. Astron. No. 4), p. 66 - 70 (1969). In Russian.

035.039 Improvement to the National Physical Laboratory atomic clock. L. Essen, D. S. Sutcliffe.
Nature, Vol. 223, 602 - 603 (1969).

The measurement of time.
See Abstr. 044.017.

036 Photographic Auxiliaries

036.001 **Filmmaterial zur Protuberanzenphotographie.**
B. Wedel.
SuW, Vol. 8, 215 (1969).

036.002 **Report on a new emulsion.** B. J. Bok.
American Astron. Soc. Photo-Bull. No. 1, p. 8 - 9 (1969).
Kodak special plate, type 098-01 seems decidedly superior to Kodak spectroscopic plate, type 103a-E on two counts: First, for the same exposure conditions and with identical processing, the 098-01 emulsion records fainter nebulosity than does the 103a-E emulsion; and second, the 098-01 emulsion is sixtenths of a magnitude faster for the recording of faint stars.

036.003 **Cutting plates for astronomical uses.**
W. C. Miller.
American Astron. Soc. Photo-Bull. No. 1, p. 10 - 12 (1969).
Basic principles of plate cutting with wheel and diamond hand cutters are discussed, and recommendations are given for cutting plates that meet the high standards of accuracy and reliability required by the astronomer. Brief comments regarding diamond plate cutting machines are included.

036.004 **Recent research in hypersensitization.**
D. Morrison.
American Astron. Soc. Photo-Bull. No. 1, p. 12 - 13 (1969).

036.005 **Enlarged star charts from prints or plates.**
W. C. Miller.
American Astron. Soc. Photo-Bull. No. 1, p. 13 - 14 (1969).
Preparation of accurate finding charts for identification of objects at the focal plane of large telescopes is a troublesome and time-consuming task. When the charts must be copied to an enlarged scale, accuracy is lost unless photographic methods are employed. This article describes a copy camera designed for making $5\times$ enlargements of prints or plates.

036.006 **Color photographs of the night sky are made by refrigerating the film.** C. L. Stong.
Sci. American, Vol. 221, No. 2, p. 124 - 129 (1969).

036.007 **Chilled-emulsion photography at Fernbank Observatory.** P. H. Knappenberger.
Bull. American Astron. Soc. Vol. 1, 249 (1969). – Abstr. AAS.

036.008 **Color astrophotography at the Flagstaff Station.**
J. W. Christy.
Publ. Astron. Soc. Pacific, Vol. 81, 532 - 533 (1969).
Abstract ASP.

036.009 **A theoretical investigation of focal stellar images in the photographic emulsion and application to photographic photometry.** A. F. J. Moffat.
Astron. Astrophys. Vol. 3, 455 - 461 (1969).
Two-dimensional intensity profiles of focal stellar images in photographic emulsions were derived by convolving gaussian seeing distributions with diffraction profiles and emulsion scattering functions. The results predict too low an intensity for large radial distances from the centre of the image, contradicting the assumption that seeing is gaussian. An analytic formula representing observed image profiles combined with an expression for the characteristic curve led to calibration curves for iris photometers.

036.010 **The comparison of the most high-sensitive plates**
of various firms. I. I. Breido, K. P. Ermoshina.
Astron. Zhurn. Akad. Nauk SSSR, Vol. 46, 916 - 918 (1969).
In Russian. English translation in Soviet Astron. AJ, Vol. 13, No. 4.
The ORWO plates NP 27 and Ilford plates HPS are compared by their general and spectral sensitivity, resolving power, and granulation. It turns out that by all pointed parameters the ORWO plates NP 27 are most preferable.

036.011 **Investigation of the films made by the V. V. Kujbishev chemical factory.**
O. D. Dokuchaeva.
Soobshch. Gos. Astron. Inst. Shternberga, No. 158, p. 58 - 66 (1969). In Russian.
The results of the investigation of some sensitometric parameters of the films A-500, A-600 and A-700 are given. The data obtained concern the development factors after processing with 3 different developers, the failure of the reciprocity law and the different ways of storing the films.

036.012 **Laboratory test for revealing small-contrast objects by photographic technique.**
V. A. Dombrovsky, V. G. Khristich.
Trudy Astron. Obs. *Leningrad,* Vol. 26 (= Uchenye Zapiski Leningr. Un-ta No. 347 = Seriya Matem. Nauk No. 44), p. 63 - 79 (1969). In Russian.
The problem of detecting small-contrast objects by a photographic method is considered. Measurements of the gradients, the noise of the granularity and the large-scale variations of density for several emulsions are made. Recommendations for optimal conditions of the detection of the small-contrast objects are given. Various methods of obtaining the composite photographs for the purpose of reducing the noise-to-gradient ratio are analysed.

036.013 **Hypersensitization of I-920 and I-1030 infra-red films.** T. E. Derviz.
Trudy Astron. Obs. *Leningrad,* Vol. 26 (= Uchenye Zapiski Leningr. Un-ta No. 347 = Seriya Matem. Nauk No. 44), p. 80 - 83 (1969). In Russian.
Results of laboratory and astronomical tests of hypersensitization for the I-920 and I-1030 films are given. It is shown that sensitivity increases by the factor 25 for the I-920 and by the factor 7 for the I-1030 films. These films can be used for stellar photometry in the infra-red region of the spectrum.

036.014 **The investigation of the aerophoto film deformation.**
V. P. Bolshakov, E. I. Vingisaar.
Trudy Astron. Obs. *Leningrad,* Vol. 26 (= Uchenye Zapiski Leningr. Un-ta No. 347 = Seriya Matem. Nauk No. 44), p. 144 - 154 (1969). In Russian.
The magnitude and direction of the shift of the aerophoto film emulsion layer and backing are studied. Experimentally obtained data are studied by statistical methods. It is concluded that for precise astrophotographic work photoplates and not aerophoto films must be used. The deformations of the latter, including permanent ones, were significant.

036.015 **Results of comparative testings of the plates Eastman Kodak 103aO, ORWO ZU-2 and ZU-1,**
and the astronomical film A-500.
I. I. Breido, O. M. Michailova.
Astron. Tsirk. No. 534, p. 4 - 6 (1969). In Russian.

036.016 **Application of photographic materials in astronomy.**
A. A. Hoag, W. C. Miller.

Applied Optics, Vol. 8, 2417 - 2430 (1969).

A list of emulsions used for astronomical photography in the United States is presented together with descriptions of techniques of hypersensitization, exposure, processing, and image evaluation applied in their use.

036.017 Construction of the characteristic curve of photographic materials for using a pulsed light source.
J. I. Rudnev, L. A. Sataeva.
Trudy Astrofiz. Inst. Alma-Ata, Vol. 13, 112 - 114 (1969). In Russian.

036.018 A history of the application of photography to astronomy. V. Juska.

Journ. Astron. Soc. Victoria, Vol. 22, 86 - 87 (1969).

036.019 Simultaneous multicolor planetary photography.
C. F. Capen.
Strolling Astronomer, Vol. 21, 173 - 176 (1969).

036.020 Metodos de fotografia planetaria.
H. G. Marraco.
Revista Astron., Vol. 41, (No. 169), 17 - 25 (1969). – Popular article.

036.021 Über die Herstellungstechnik photographischer Äquidensiten. W. Högner.
Jenaer Rundschau (Jena Review), 14. Jahrgang, 340 - 345 (1969).

Positional Astronomy. Celestial Mechanics

041 Positional Astronomy, Star Catalogues and Atlases

041.001 Results of measuring the positions of the moon by the method of equal altitudes. J. Vondrák.
Bull. Astron. Inst. Czechoslovakia, Vol. 20, 223 - 226 (1969).
The paper gives a survey of the differences in observed and calculated zenith distances of the bright lunar limb. The measurements were performed at the Geodetical Observatory Pecný with a small circumzenithal in 1965 − 1968. On the basis of these measurements a relation is derived between the Atomic Time $A3$ and the Ephemeris Time and also a relation between the $TU2$ time and the Ephemeris Time. The mean deviation of the moon in latitude and the correction of the ephemeris radius of the moon together with the personal error of the observer were calculated for each year as secondary quantities.

041.002 Variation law of random errors of the stellar passages and Albrecht relation.
C. de Concini, S. Mancuso, E. Proverbio.
Proc. Colloquium on Problems of Time Determination, Keeping and Synchronization, (Milan 1968), p. 195 - 217 (1968). In Italian.
The representation of observational errors in meridian transits is discussed. The hypothesis of a random distribution of the observational errors is proved to be not valid neither from experimental nor from theoretical point of view. By means of three samples each of which contains about 150 stars the statistical dependence of observational errors on magnitude or contacts of the micrometer is studied.

041.003 L'astronomie méridienne. R. Dejaiffe.
Ciel et Terre, Vol. 85, 193 - 235 (1969).
Review: Le problème fondamental de l'astronomie de position; Les principes généraux de l'astronomie méridienne; Les principaux instruments d'astronomie méridienne à l'Observatoire Royal de Belgique; Les catalogues d'étoiles.

041.004 Genauigkeitsuntersuchung zur photogrammetrischen Sternkoordinatenbestimmung durch geschlossene Blockausgleichung. H. Ebner.
Deutsche Geod. Kommission Bayer. Akad. Wiss. Reihe C, Heft No. 141, 91 pp. (1969).
Die vorliegende Arbeit untersucht die Genauigkeit einer Methode, welche eine simultane Koordinatenbestimmung von Sternen ermöglicht, die gleichmäßig über den gesamten Himmel verteilt sind. Durch einzelne photogrammetrische Bildbündel wird dabei ein geschlossener Block gebildet und ausgeglichen. Allen behandelten Blöcken wird das gleiche stochastische Modell zugrundegelegt, aber verschiedene funktionelle Modelle, die zum Teil Korrekturen der Aufnahmebündel erlauben.

041.005 Het gebruik van sterren-atlassen.
A. Mak.
Hemel en Dampkring, Vol. 67, 233 - 237 (1969).

041.006 Nogmaals: Drievoudige conjuncties.
J. Meeus.
Hemel en Dampkring, Vol. 67, 272 - 282 (1969).

041.007 Sur un effet systématique dans les observations méridiennes en ascension droite.
L. Arbey.
Comptes Rendus, Acad. Sci. Paris, Sér. B, Vol. 269, 388 - 390 (1969).
L'instant du passage au méridien de l'image d'une étoile est, en général, affecté systématiquement par une erreur sur la valeur observée du tour de la vis d'ascension droite. On évalue cet effet et l'on fournit la correction correspondante.

041.008 Systematic errors Δa_δ in the FK4 fundamental catalogue as deduced from astrolabe and meridian observations in the southern hemisphere.
C. Anguita, F. Noel.
Astron. Journ. Vol. 74, 954 - 957 (1969).
Investigations of the Δa_δ error in the FK4, based on astrolabe and meridian observations made at Cerro Calán, Santiago de Chile, gave similar results. This indicates that the most important features are not due to the instrumental systems. The results are compared with the results obtained from astrolabe observations of FK4 stars, made at Tananarive, Madagascar. This comparison shows that the Δa_δ curves obtained at Cerro Calán are really due to a Δa_δ error in the FK4.

041.009 Corrections to right ascensions of the FK4 from observations with the meridian circle in Chile.
C. Anguita, M. S. Zverev, G. Carrasco, D. D. Polojentsev.
Astron. Tsirk. No. 506, p. 6 - 8 (1969). In Russian.

041.010 Orientation of FK3 and FK4 from meridian observations during the period 1963 - 1964.
D. P. Duma, A. I. Gamjanina.
Astrometriya i Astrofiz., *Kiev*, No. 2, p. 36 - 40 (1969). In Russian.
The equinox and equator differences FK4 −FK3 have been obtained from meridian observations of the sun for the period 1963 - 1964. The results are +0$\overset{s}{.}$003 and +0$\overset{''}{.}$13 correspondingly. The differences of that type should be taken into account when discussing the equinox corrections obtained for different catalogues. The authors point out some difficulties in determination of equinox and equator corrections by means of observations reduced to different catalogues and different epochs.

041.011 A comparison of the system of the Pulkovo and Golosejevo vertical circles. E. M. Nenakhova.
Astrometriya i Astrofiz., *Kiev*, No. 2, p. 41 - 45 (1969). In Russian.
Absolute declinations of the fundamental bright and faint stars were determined at Pulkovo and Golosejevo with large vertical circles. The author has compared these declinations. The results are given in tables and in figures. The paper deals also with the accuracy of the comparison of two catalogues with the intermediate catalogue. The differences "intermediate comparison minus direct comparison" are given in tables.

041.012 Positional observations of Mars by means of local desensitization of plates. R. F. Meshkova.
Astrometriya i Astrofiz., *Kiev*, No. 2, p. 46 - 51 (1969). In Russian.

At the Main Astronomical Observatory of the Ukrainian Academy of Sciences photographic observations of Mars were made using the method of local desensitization of plates. 32 positions of the planet have been obtained. The mean differences between the observed and ephemeris positions are given.

041.013 On the influence of some instrumental errors on the $\Delta\delta_\delta$ errors in star positions. A. M. Stafeyev.
Astrometriya i Astrofiz., *Kiev*, No. 2, p. 124 - 128 (1969). In Russian.

Inclination of the film plane to the optical axis of the photomicroscope affects measurement of photographic recording of circle readings. The author deduced formulae for estimating this effect. He has designed and made a device for adjustment of the photomicroscope capable of giving an accuracy of ±5'. Description of the device is also given.

041.014 Erratum: A study of the positions and proper motions of 83 stars in the region of Sco X-1.
S. Sofia, H. Eichhorn, G. Gatewood.
Astron. Journ. Vol. 74, 1053 (1969).
See Astron. Astrophys. Abstr. Vol. 1, 041.002.

041.015 Second Cape Catalogue for 1950.0. Observations of the Sun, Moon and Planets 1936 - 1959.
Prepared under the direction of R. H. Stoy.
Ann. Cape Obs. Vol. 23, 50 + 405 pp. (1968). — Contents: Introduction (Observing programme, the instrument and general procedure, pivot errors, right ascensions, declinations, proper motions, comparison with other catalogues, observations of the sun, moon and planets, analysis of the daytime observations); The catalogue (Main section, polar section); Observations of the sun, moon, and planets.

041.016 Results of observations made with the six-inch transit circle 1956 - 1962. Observations of the sun, moon, and planets; catalog of 2554 stars for 1950.0; corrections to FK4, GC, and N30.
A. N. Adams, D. K. Scott.
Publ. United States Naval Obs., Washington, Second Ser., Vol. 19, Part II, p. 295 - 435 (1968).

041.017 First Greenwich Catalogue of Stars for 1950.0.
Prepared under the direction of R. v. d. R. Woolley.
Roy. Obs. Ann. No. 3 [Science Research Council, London], 204 pp. Price £3, 16s. 0d. (1969). — The catalogue contains three parts: Observed R. A. and Dec. of 1399 fundamental stars; 6173 stars in Henry Draper Catalogue, in declination zone 0° to +24°, brighter than $8.^m00$; 337 circumpolar stars observed in both culminations.

041.018 Catalogue of nearby stars, edition 1969.
W. Gliese.
Veröff. Astron. Rechen-Inst. Heidelberg No. 22, [Verlag G. Braun, Karlsruhe. Price DM 25.00], 117 pp. (1969).

The catalogue contains 1529 single stars and systems with altogether 1890 components.

041.019 An atlas of the southern Milky Way.
J. D. Wray, B. E. Westerlund.
Bull. American Astron. Soc. Vol. 1, 268 - 269 (1969). — Abstr. AAS.

041.020 Utilisation des matrices de rotation dans les formules de passage au méridien.
E. Marcus.
Stud. Cerc. Astron. Vol. 14, 17 - 22 (1969). In Rumanian.

Des formules rigoureuses qui experiment l'ascension droite d'une étoile au moment du passage au méridien sont déduites au moyen de la rotation d'un trièdre de référence.

Les formules classiques de Bessel, Mayer, Hansen résultent comme des cas limites de ces relations générales.

041.021 Orientation of fundamental catalogue systems from observations of major and minor planets.
D. P. Duma.
Astrometriya i Astrofiz., *Kiev*, No. 7, p. 3 - 39 (1969). In Russian.

Conditional equations for determining the fundamental catalogue zero-points from observations of major and minor planets are derived. The conditions on separating corrections to equinox and equator from corrections to the orbital elements of the planets and the earth are discussed. It is concluded that the determination of corrections to the catalogue zero-points from observations of minor planets loses its advantage because of the complex correlations between the coefficients of the equations. The use of stationary artificial satellites is suggested as a possible method for the solution of the problem.

041.022 A Messier Album. J. H. Mallas, E. Kreimer.
Sky Telescope, Vol. 38, 86 - 87, 163, 235, 239, 299, 396 (1969).

041.023 Comparison of the accuracy of the photographic position of an object with the NAFA-3c/25 camera on a film and on plates. S. S. Smirnov.
Trudy Astron. Obs. *Leningrad*, Vol. 26 (= Uchenye Zapiski Leningr. Un-ta No. 347 = Seriya Matem. Nauk No. 44), p. 134 - 139 (1969). In Russian.

The mean-square errors of the determination of coordinates of basic stars on the photographs taken with the NAFA-3c/25 camera are found. Near the optical center the errors are 30–40% greater when a film is used as compared with those for plates.

041.024 L'influence de l'erreur de centrage sur les mesures des plaques photographiques.
H. Debehogne.
Acad. Roy. Belgique, Bull. Cl. Sci., 5e Série, Vol. 54, 1434 - 1438 (1968).

The centering error is here determinated exactly by means of logarithmic derivation.

041.025 Etude des effets d'une erreur intervenant dans la détermination par moindres carrés des constantes de plaque au moyen d'un système de référence en réseau.
H. Debehogne.
Acad. Roy. Belgique, Bull. Cl. Sci., 5e Série, Vol. 55, 415 - 422 (1969).

Our purpose is to find which of several possible formulas for the transformation of measured to standard coordinates will give the most accurate results and how many reference stars are required to keep certain systematic errors below a given limit. They are determined theoretically sometimes by the covariance matrix of the plate constants.

041.026 La réduction des positions astrométriques et la méthode de l'erreur locale. J. Dommanget.
Acad. Roy. Belgique, Bull. Cl. Sci., 5e Série, Vol. 55, 549 - 550 (1969).

041.027 Réduction des positions astronomiques. Méthode de l'erreur locale. H. Debehogne.
Acad. Roy. Belgique, Bull. Cl. Sci., 5e Série, Vol. 55, 551 - 552 (1969).

041.028 Determinazione del passo del micrometro mediante coppie scalari e studio dei moti propri del catalogo zenitale di Washington. E. Proverbio.
Atti XII Riunione Soc. Astron. Italiana, L'Aquila 1968,

p. 105 - 108 (1969). – Abstract SAI.

041.029 On the determination of a systematic error in the observations of star coordinates.
A. M. Stafeev, L. L. Vagushenko.
Astron. Tsirk. No. 534, p. 7 - 8 (1969).   In Russian.

041.030 Stellar atlas containing for both hemispheres all stars brighter than $8^m 25$ with designation of variable and binary stars, stellar clusters and nebulae.
A. A. Mikhajlov. Third Edition.
"Nauka", Leningrad. 60 pp. Price 2 Rbl. 40 Kop. (1969).
In Russian.

041.031 Provisional corrections to the star places derived from the observations with the Danjon astrolabe at Mizusawa.   K. Yokoyama.
Publ. International Latitude Observatory of Mizusawa, Vol. 6, 185 - 204 (1968).
 The results of the observations with the Danjon astrolabe at Mizusawa are analysed. No systematic errors with magnitude and spectral type are found in the internal adjustments (l'accord interne) derived from the arithmetic mean of residuals. Corrections to the right ascension and declination of the observed FK4 stars are calculated.

041.032 The influence of refraction in fundamental astrometry.   G. Teleki.
Proc. International Latitude Obs. Mizusawa, No. 9, p. 222 - 233 (1969).   In Japanese and English.
 Problems connected with astronomical refraction can be placed in three groups: problems connected with pure (normal) refraction, problems connected with anomalous refraction of systematical character and problems connected with anomalous refraction of accidental character. This paper points out the size of refractional influences, indicating their importance for astronomical measuring, especially for fundamental astrometry.

041.033 Résultats définitifs des observations faites par photographie d'étoiles zénithales à l'instrument des passages établie dans le I.^{e} vertical.
A. Baptista dos Santos, with an introduction by A. P. Botelheiro.
Bull. Obs. Astron. Lisbonne (Tapada), No. 16, 14 pp. (1968).
 Valeurs déterminées d'un tour du micromètre de mesure et valeur moyenne définitive; Corrections définitives aux déclinaisons des étoiles zénithales photographiées; Latitude.

041.034 Katalog der Rektaszensionen der Z1-Sterne.
V. Eh. Brandt.
Rotation of the Earth and Determination of Time, Conference Riga 1965, p. 35 - 61 (1969).   In Russian.

041.035 Katalog der Rektaszensionen von 372 Sternen.
A. I. Yazev.
Rotation of the Earth and Determination of Time, Conference Riga 1965, p. 61 - 76 (1969).   In Russian.

041.036 Die Helligkeitsgleichung in photoelektrischen Beobachtungen von Sterndurchgängen.
L. F. Roze.
Rotation of the Earth and Determination of Time, Conference Riga 1965, p. 77 - 79 (1969).   In Russian.

041.037 Die Eliminierung des Einflusses der Lagerung eines Passageninstruments auf die Genauigkeit der Beob-
achtungen.   L. A. Solov'eva.
Rotation of the Earth and Determination of Time, Conference Riga 1965, p. 82 - 85 (1969).   In Russian.

041.038 Differential catalogue of the right ascensions of 102 stars of the Tashkent broad zenith-zone.
Eh. A. Sanakulov.
Astrometr. Issled. p. 67 - 82 (1969).   In Russian.

041.039 Correspondance des numéros d'étoiles de la BD, CD et CPD avec les numéros S. A. O. et HD pour les étoiles du catalogue S. A. O. Zone - 0° à - 15°.
L. Houziaux.
Separate print Faculté des Sciences de Mons, Département d'Astrophysique. 23 pp. (1969).

041.040 Right ascensions of the major planets with the meridian circle of the Charkov Astronomical Observatory 1966 – 1967.
K. N. Derkach, N. G. Zuev.
Vestn. Khar'kov. Univ. No. 34, (Ser. Astron. No. 4), p. 79 - 81 (1969).   In Russian.

041.041 Bericht über die Hamburger Meridiankreis-Expedition in Australien.   A. Behr.
Mitt. Astron. Ges. No. 27, p. 126 (1969). – Abstract AG.

041.042 Neue Zonenkataloge mit Eigenbewegungen.
W. Dieckvoss.
Mitt. Astron. Ges. No. 27, p. 127 (1969). – Abstract AG.

041.043 Meridian observations at Santiago, Chile.
C. Anguita, G. Carrasco, P. Loyola, V. N. Šiškina, M. S. Zverev.
Univ. de Chile, Dep. Astron., Santiago, Separata 3 [Reprinted from "Highlights of Astronomy" I.A.U. (L. Perek, Ed.), p. 292 - 296], 5 pp. (1968).

041.044 Analyse de la série chronométrique définissant l'instant du passage d'une étoile au méridien.
L. Arbey.
Ann. Obs. Besançon, Vol. 8, 19 - 22 (1969).

041.045 Smithsonian Astrophysical Observatory. Star Atlas of Reference Stars and Nonstellar Objects.
Prepared by the Staff of the SAO. Foreword by J. Ashbrook.
Editor: Sky and Telescope. 12 pp. + 152 charts, Price 180/– (1969).

041.046 Atlas Stellarum 1950.0.   H. Vehrenberg.
Treugesell–Verlag KG, Düsseldorf. Deliveries 7 and 8. 60 charts (1969).

Declinations of Bright and Faint Fundamental Stars in a Uniform System.   See Abstr. 003.085.

Atlas zur Himmelskunde.
See Abstr. 003.117.

On a quasi-absolute method of reduction of differential meridian observations.   See Abstr. 031.022.

The relation between Z term and catalogue errors.
See Abstr. 045.014.

Ascensioni rette del pianeta Saturno 1967, 1968.
See Abstr. 100.007.

042 Celestial Mechanics

042.001 Literal developments in the analytical theory of the moon. J. Chapront, L. Mangeney.
Astron. Astrophys. Vol. 2, 425 - 445 (1969). In French.

The authors describe the analytical method they have used to construct a completely literal solution of the equations of motion for the main problem of the lunar theory. The method was so chosen as to lead to a formula fit to manipulate lengthy analytical developments, in a process of successive approximations. In this work it is always kept in mind that those developments are to be handled on an electronic computer. (The disturbing function R; Systems derived from Lagrange's equations; Analysis of the equations; First type of integration in variable $z = \xi \exp i \eta$; Second type of integration; Study of some couples of variables (x, y); Determination of ν; Definite choice of variables; Method of integration; Study of the convergency of the solution).

042.002 Periodic solutions of the third sorte for the restricted problem of three bodies. Y. Kozai.
Proc. Japan Acad. Vol. 45, 394 - 398 = Tokyo Astron. Obs. Repr. No. 357 (1969).

Five periodic solutions with moderate eccentricities and high inclinations for the three-dimensional restricted problem of three bodies are found for cases of 3 : 2, 2 : 1, and 4 : 1 of the mean motions by expanding the disturbing functions by use of a high-speed computer. The expansion with respect to the inclination is made by Tisserand's polynomials and that to the eccentricity is made by Newcomb operators up to the tenth power. The periodic solutions found here correspond to orbits, for which secular and long-periodic perturbations of orbital elements except for the mean anomaly vanish. The existence of such periodic orbits are verified by numerical integration method for a case that the disturbing mass is 0.001.

042.003 The precession and nutation of deformable bodies, II. Z. Kopal.
Astrophys. Space Sci. Vol. 4, 330 - 364 (1969).

In a previous paper of this series the Eulerian equations have been set up which govern the precession and nutation of selfgravitating bodies of viscous fluid in inertial coordinates which are at rest in space. In order to facilitate their solution, in the present investigation we shall transform these equations to the rotating body-axes; and shall explicitly evaluate all their coefficients arising as a result of second-harmonic dynamical tides. Following the introductory section 1 which contains a mathematical statement of the problem, the requisite transformation of coordinates will be outlined in section 2, and applied to the equations of motion in section 5. The corresponding moments and products of inertia appropriate for self-gravitating configurations of arbitrary internal structure will be formulated in section 4; while the deformation terms arising from second-harmonic dynamical tides raised on centrally-condensed configurations will be evaluated in sections 3 and 6. The concluding section 7 will then contain a specification of the components of the disturbing force.

042.004 Second-order stability of dynamical systems with two degrees of freedom. J. D. Hadjidemetriou.
Astron. Journ. Vol. 74, 789 - 795 (1969).

It is proved that symmetric periodic orbits in the plane-restricted three-body problem and similar Hamiltonian systems, which are stable in the linear sense, are stable to second-order terms in the initial displacements provided $2\pi/a\omega \neq \nu/n$ (where a is a characteristic exponent, ω the period, $\nu = 1$, 2, or 3, and n any integer). One can find time-dependent integrals of the second order valid near the periodic orbit.

042.005 Families of symmetric periodic orbits of the restricted three body problem, when the perturbing mass is small. P. Guillaume.
Astron. Astrophys. Vol. 3, 57 - 76 (1969).

The object of this paper is to study the families of symmetric periodic orbits of the restricted three body problem, by analytical continuation from the two body problem. The solution of the equations of symmetry to the first order allows: 1) a good comparison with numerous numerical results obtained during the last few years; 2) a qualitative description of the characteristics of symmetric periodic orbits for a mass ratio $\neq 0$.

042.006 Disturbing function and analytical solution of the problem of the motion of a satellite. A. Challe, J. J. Laclaverie.
Astron. Astrophys. Vol. 3, 15 - 28 (1969). In French.

The disturbing function R due to the development in spherical harmonics of the Earth potential is considered. It is shown that it can be expressed as the real part of a complex function R which can be written as a multiple Fourier series of three independent angular quantities ω, M and $\Omega - T - \lambda_{nq}$. The coefficients can be written in the form of products of functions $A_{n,q}^k (I)$ and $B_{n,p}^k (e)$ where I is the inclination and e, the eccentricity. Two functions $\Phi (p, q, k)$ and $\Psi (p, q, k)$ are constructed with these functions A and B. The disturbing function R and its partial derivatives respective to the six elliptic elements can be simply expressed when using Ω, Ψ and their partial derivatives. From these expressions, the formal first order solution of the problem of the motion of an artificial earth satellite is fully derived. A similar approach is used in order to compute the first order perturbations of the motion of a satellite as disturbed by another satellite or the sun.

042.007 On the periodical orbits emanating from L_4 in the resonance case 1/4. A. Deprit, J. Henrard.
Astron. Astrophys. Vol. 3, 88 - 93 (1969). In French.

Pour la résonance 1/4 entre les fréquences fondamentales à l'équilibre, le point L_4 constitue un centre naturel de libration.

042.008 The precession and nutation of deformable bodies, III. Z. Kopal.
Astrophys. Space Sci. Vol. 4, 427 - 458 (1969).

The Eulerian differential equations which govern the precession and nutation of self-gravitating fluid components of close binary systems deformed by dynamical tides are simplified and linearized by certain assumptions. The solution of these equations — giving the periods of precession and nutation, as well as the rate of nodal regression which is synchronized with precession — are expressed in terms of the physical properties of the respective system and of its constituent components; the concluding section contains a discussion of the results, in which the differences between the precession and nutation of rigid and fluid bodies are pointed out.

042.009 Sur un nouveau processus d'élimination des termes périodiques par la méthode de von Zeipel pour les théories du premier et du second ordre. J. Meffroy.
Comptes Rendus, Acad. Sci. Paris, Sér. A, Vol. 269, 309 - 312 (1969).

Contrairement à la procédure habituellement adoptée, on élimine les termes à longue période avant les termes à courte période. Cette élimination n'est possible que si l'ensemble $\hat{F}_1$ des termes à longue période de l'hamiltonien F qui sont de degré 1 par rapport aux masses est identiquement nul.

042.010 The influence of external forces on the inertial acceleration in the relative motion of two bodies.
R. A. Sahakian.
Akad. Nauk Armyanskoj SSR, Byurakanskaya Astrofiz. Obs.
47 pp. (1969). In Russian.

For a two-body system the additional inertial acceleration resulting from the curvilinear motion of its center of mass is computed. The influence of this quantity on tidal phenomena is discussed.

042.011 Les perturbations séculaires en mécanique céleste.
Séminaires d'été du Bureau des Longitudes, La Coûme. 143 pp. (1968).

The methods of Lindstedt, Gylden, Lagrange, Von Zeipel, Hill-Brown, Krylov, Bogoliubov and Mitropolsky are reviewed and some applications are discussed.

042.012 On the influence of a finite speed of gravitation in celestial mechanics. J. M. J. Kooy, J. Berghuis.
Astronaut. Acta, Vol. 15, 37 - 48 (1969).

In the equations of celestial mechanics the speed of propagation of the gravitational field has always been considered as infinite. However, it seems probable that this speed, although very large, is finite. In a former article, which appeared in this journal, the motion of a binary star system was considered on the assumption that the speed of the cosmical background, would be equal to the speed of light, In this article the motion of the same binary star system has been considered in relation to a reference system at rest in the metric cosmical field, assuming the gravitational speed in relation to this reference system equal to the speed of light.

042.013 Second-order planetary theory. Part I: Outline of the method. S. E. Hamid.
SAO Cambridge, Mass. Special Rep. No. 302, 6 + 83 pp. (1969).

The analytical procedure for computing second-order perturbations in rectangular coordinates, according to Brouwer's theory of planetary motion, is given. Single- and double-harmonic analyses and the multiplication of Fourier series with numerical coefficients are used in the computations. In the series multiplication, a variable tolerance is considered, enabling us to avoid the difficulties arising from a small divisor. Also presented is an example computing that part of the second-order perturbation of Mars containing the masses of Jupiter and Saturn. The analytical solution of this perturbation is compared with the numerical integration of the differential equations defining this perturbation. The numerical integration covered the interval from 0 to 40000 days. The comparison shows an agreement within 1×10^{-9}.

042.014 Machine program for the computer M20 for the determination of orbits and their perturbations.
E. A. Vorob'ev.
Izv. Astron. Obs. Ehngel'gardta, *Kazan'*, No. 36, p. 185 - 199 (1968). In Russian.

042.015 Internal resonances at an equilibrium.
A. Deprit, J. Henrard.
Bull. American Astron. Soc. Vol. 1, 239 (1969). — Abstr. AAS.

042.016 Accurate ephemerides for planets and moon.
C. Oesterwinter, C. J. Cohen.
Bull. American Astron. Soc. Vol. 1, 255 - 256 (1969). — Abstr. AAS.

042.017 Families of periodic orbits continued in regularizing coordinates. A. Deprit, J. Palmore.

Celestial Mechanics, Vol. 1, 150 - 162 (1969).

A predictor-corrector algorithm is proposed for continuing analytically families of periodic orbits beyond collision trajectories in the restricted problem of three bodies. It is based on Hill's equation for normal variations in Thiele's regularizing coordinates.

042.018 Expansion formulae in canonical transformations depending on a small parameter.
A. A. Kamel.
Celestial Mechanics, Vol. 1, 190 - 199 (1969).

The theory of perturbation based on Lie transforms is considered. Deprit's equation is reduced to a form which enables us to generate simplified general recursion formulae. These expansions are then modified to speed up the implementation of such perturbation theory in the computerized symbolic manipulation.

042.019 Note on a statement in A. Wintner's 'Analytical foundations of celestial mechanics.'
H. J. Sperling.
Celestial Mechanics, Vol. 1, 210 - 212 (1969).

042.020 The collision singularity in a perturbed two-body problem. H. J. Sperling.
Celestial Mechanics, Vol. 1, 213 - 221 (1969).

It is shown that, in the neighborhood of a collision singularity, the motion in a perturbed two-body problem has the same basic properties as the motion in the neighborhood of a collision in the unperturbed two-body problem.

042.021 Birkhoff's normalization.
A. Deprit, J. Henrard, J. F. Price, A. Rom.
Celestial Mechanics, Vol. 1, 222 - 251 (1969).

Birkhoff's normalizing canonical transformation at an equilibrium of elliptic type with no internal resonance can be built explicitly and recursively, without partial inversions or substitutions, by means of Lie transforms. Invariant sections and ordinary families of periodic orbits for truncated normalized systems are analyzed in detail.

042.022 Sur une solution analytique des équations canoniques de Hori résultant de l'élimination des termes à courte période dans une théorie planétaire du premier ordre.
J. Meffroy.
Comptes Rendus Acad. Sci. Paris, Sér. A, Vol. 269, 1047 - 1050 (1969).

Lorsqu'on réduit la partie principale F_{1p} de la fonction perturbatrice à ses termes de degrés 0, 1, 2 par rapport aux variables canoniques de Hori X_u, Y_u, P_u, Q_u; $u = 1, 2, ..., n$, la résolution des équations canoniques de Hori résultant de l'élimination des termes à courte période de F_{1p} se ramène à celle de deux systèmes d'équations différentielles linéaires et à coefficients constant, l'un en X_u, Y_u; l'autre en P_u, Q_u. On cherche pour chacun d'eux une solution périodique sans terme séculaire.

042.023 The capture of comets by Jupiter. O. Havnes.
Astrophys. Space Sci. Vol. 5, 272 - 282 (1969).

The capture of comets with parabolic orbits by Jupiter is investigated. The influence of the gravitational force of the sun on the cometary orbit during the passage of Jupiter's sphere of influence is taken into account. A comparison of the present results with previous calculations demonstrate the importance of the solar perturbations. It is also shown that captures of comets with parabolic orbits and repeated close passages to Jupiter cannot explain all of the observed cometary orbits found in the family of Jupiter.

042.024 The Roche coordinates and their use in hydrodynamics or celestial mechanics. Z. Kopal.

Astrophys. Space Sci. Vol. 5, 360 - 384 (1969).

The aim of the present paper will be to introduce a new system of curvilinear coordinates – hereafter referred to as Roche coordinates – in which spheres of constant radius are replaced by equipotential surfaces of a rotating gravitational dipole (which consists of two discrete points of finite mass, revolving around their common center of gravity); while the remaining coordinates are orthogonal to the equipotentials. It will be shown that the use of such coordinates offers a new method of approach to the solution of certain problems of particle dynamics (such as, for instance, the construction of certain types of trajectories in the restricted problem of three bodies); as well as of the hydrodynamics of gas streams in close binary systems, in which the equipotential surfaces of their components distorted by axial rotation and mutual tidal interaction constitute essential boundary conditions.

042.025 Stationary and periodic solutions for restricted problem of three bodies in three-dimensional space.
Y. Kozai.
Publ. Astron. Soc. Japan, Vol. 21, 267 - 287 = Tokyo Astron. Obs. Repr. No. 363 (1969).

Stationary and periodic solutions are found for commensurable cases of the mean motions for the restricted problem of three bodies by expanding the disturbing functions by use of a high-speed digital computer. For the stationary solutions secular and long-periodic perturbations vanish except for the mean anomaly and the longitude of the ascending node which moves secularly. Closed periodic solutions fixed in space are generated from stationary solutions, for which the secular perturbations for the node also vanish. The periodic orbits found here are almost polar with moderate eccentricities.

042.026 Analytical methods of computing perturbations of coordinates of the planets.
L. K. Babadzhanyants.
Vestn. Leningr. un–ta, No. 7, p. 121 - 132 (1969). In Russian. Abstr. in Referativ. Zhurn. 51. Astron. 11.51.89 (1969).

042.027 Planetary effects in the motion of natural satellites.
P. Musen.
NASA Technical Note, NASA TN D-5284, 5 + 21 pp. (1969).

Addition of the direct and indirect planetary perturbations to the author's 1963 modification of Hansen's theory is proposed. The method described can determine the Saturnian effects upon fhe motion of the outer Jovian satellites. The expansion of the disturbing function and of its derivatives is reduced to a form convenient for programming.

042.028 Perturbation of libration points of the restricted three-body problem due to gravitational and radiative influence of a fourth body. Existence of a periodic solution in the vicinity of the libration points.   V. Matas.
Bull. Astron. Inst. Czechoslovakia, Vol. 20, 322 - 326 (1969).

In this paper, equations of the motion of infinitesimal body M_4 for generalized Huang's model of restricted four-body problem are derived. Especially, motion of the body M_4 is studied in the vicinity of the libration points of a system of bodies M_2, M_3 having finite, non-vanishing masses; the perturbation is realized by gravitational and radiative influence of body M_1. Existence of a periodic solution for body M_4 is proved when (I) the motion of M_1, M_2, M_3 takes place according to Huang's (1960) model and (II) the perturbation mentioned above is taken into account simultaneously. In the proof, Poincaré's small parameter method is applied and some results of Cronin et al. (1968) are formally used.

042.029 Resonance in the restricted problem of three bodies.   G. E. O. Giacaglia.
Astron. Journ., Vol. 74, 1254 - 1261 (1969).

A formal solution of the motion of an asteroid whose mean motion is approximately rational with that of Jupiter is developed. The series obtained are extended formally to an arbitrary degree of approximation in powers of the small parameter of the problem: the ratio of the mass of Jupiter to that of the sun. Specific numerical applications have been and will be published separately.

042.030 Energy changes during close encounters.
E. Everhart.
Bull. American Astron. Soc., Vol. 1, 342 (1969). – Abstract AAS.

042.031 Second-order planetary theory: Outline of the method.   S. E. Hamid.
Bull. American Astron. Soc., Vol. 1, 346 - 347 (1969). – Abstract AAS.

042.032 A preliminary special perturbation theory for the moon.   D. B. Holdridge, J. D. Mulholland.
Bull. American Astron. Soc., Vol. 1, 348 - 349 (1969). – Abstract AAS.

042.033 Some periodic orbits in the elliptic restricted problem of three bodies.
P. J. Shelus, S. S. Kumar.
Bull. American Astron. Soc., Vol. 1, 361 - 362 (1969). – Abstract AAS.

042.034 Fundamental problems of celestial mechanics.
G. A. Chebotarev.
Astron. Zhurn. Akad. Nauk SSSR, Vol. 46, 1274 - 1278 (1969).   In Russian. English translation in Soviet Astron. AJ, Vol. 13, No. 6.

A short review of fundamental problems of celestial mechanics in its historical development is given. The principal aim of celestial mechanics consists in constructing a dynamical model of the solar system. The mathematical basis is the three body problem.

042.035 Some generalized problems of celestial mechanics.
G. N. Duboshin.
Astron. Zhurn. Akad. Nauk SSSR, Vol. 46, 1279 - 1289 (1969).   In Russian. English translation in Soviet Astron. AJ, Vol. 13, No. 6.

Some problems of the motion of a system of material points, each of them acting on any point with a force proportional to the product of masses of both points with a certain function of their mutual distance, are considered. Besides, Newton's third law of motion, the law of action and reaction, is not assumed to hold.

042.036 Parametric representation of resonance in the restricted problem.   G. E. O. Giacaglia.
Mem. Soc. Astron. Italiana, Nuova Serie, Vol. 40, 499 - 515 (1969).

The parametric representation of deep resonance motion introduced by Poincaré (1893) as Bohlin's methods and discussed in a recent paper by Garfinkel (1966) is extended to the Hamiltonian pertinent to the restricted problem of three bodies. We analyse perturbations of orbits far from close approaches to the primaries whose mass ratio is supposed to be small.

042.037 The regularization of the restricted n–body problem.
B. Szczodrowska.
Postępy Astron., Vol. 17, 375 - 386 (1969).   In Polish.

In this paper we discussed the means of removing the singularity resulting by the collisions and the "integral Jacobi" in the restricted n-body problem. We examined the peculiar cases of the configurations for primary bodies too.

042.038 On the redundant constant of integration of Hill's planetary method.   L. K. Babadzhanjanz.
Astron. Tsirk. No. 539, p. 4 - 5 (1969).   In Russian.

042.039 The expansion of the disturbing function of the oblateness factor in series up to the ninth power of eccentricity.   D. Z. Koenov.
Dokl. AN Tadzh. SSR, Vol. 6, No. 2, p. 17 - 19 (1969). In Russian. — Abstr. in Referativ. Zhurn. 51. Astron., 1.51.132 (1970).

042.040 Two-body perturbation matrix.
D. F. Crawford.
AIAA Journ., Vol. 7, 1163 - 1164 (1969).

042.041 Contraction of orbits under the influence of resisting medium in the plane restricted three-body problem.   D. Kotsakis, M. Zikides, E. Sarris.
Publ. Lab. Astron. Univ. Athen, Ser. II, No. 22, 6 pp. (1969). In Greek.
In the present paper we study the contraction of orbits under the influence of resisting medium in the plane restricted three-body problem. The study of the contraction of the orbits is made by calculating the changes of a, e, and T of the osculating elements.

042.042 Some simplified basic schemes in celestial mechanics obtained by averaging the restricted problem of three bodies.   N. D. Moiseev.
Soobshch. Gos. Astron. Inst. Shternberga, No. 164, p. 3 - 10 (1969).   In Russian.

042.043 Improvement of the elements of an intermediate orbit.   A. A. Orlov.
Soobshch. Gos. Astron. Inst. Shternberga, No. 164, p. 11 - 35 (1969).   In Russian.

042.044 Singularities of the n–body problem. I.
H. Pollard, D. G. Saari.
Arch. Rational Mechanics and Analysis, Vol. 30, 263 - 269 (1968). — See Bull. Signal., Vol. 30, Section 120, No. 9269 (1969).

Über einen Fall der Verwendung von Potenzreihen in der Himmelsmechanik.   See Abstr. 021.011.

Gravitationally consistent planetary ephemerides based on meridian circle, radar, and Mariner observations. See Abstr. 047.014.

A note on the evaluation of the latitude of the moon.   See Abstr. 047.019.

Long-range variations of orbits with arbitrary inclination and eccentricity.   See Abstr. 052.020.

Stability of periodic orbits in the elliptic, restricted three-body problem.   See Abstr. 052.030.

Stability of the triangular points in the elliptic restricted problem of three bodies.
See Abstr. 052.033.

Comparison of the Newtonian and general relativistic orbits of a point mass in an inverse-square law force field. See Abstr. 052.034.

Planetary orbits for a moving sun.
See Abstr. 066.047.

Corrections to the improved lunar ephemeris.
See Abstr. 094.159.

The sphere of predominance of the sun relative to other stars.   See Abstr. 102.044.

043 Astronomical Constants

043.001 Time variation of the fundamental constants of physics. J. O'Hanlon, K.-K. Tam.
Progress Theoret. Phys. Japan, Vol. 41, 1596 - 1598 (1969). – Letter.

043.002 **Algunas constantes astronómicas fundamentales. Elementos orbitales y físicos del sistema solar. Tablas de conversión de tiempo.** E. Marin, J. A. Pardi.
Revista Astron. Vol. 41, No. 168, 24 pp. (1969).

043.003 **Evaluation of the gravitational effect of adjacent masses in determining the gravitational constant by the method of torsion oscillations.** M. U. Sagitov.
Vestn. Mosk. un–ta. Fiz., Astron., No. 3, p. 87 - 92 (1969). In Russian. – Abstr. in Referativ. Zhurn. 51. Astron., 11.51.658 (1969).

043.004 **Astronomical constants from observations of the inner planets and Icarus.**
G. W. Null, J. H. Lieske.
Bull. American Astron. Soc., Vol. 1, 356 - 357 (1969). Abstr. AAS.

043.005 **The present situation of the problem of the determination of the gravitation constant and the earth's mass.** M. U. Sagitov.
Astron. Zhurn. Akad. Nauk SSSR, Vol. 46, 907 - 915 (1969). In Russian. English translation in Soviet Astron. AJ, Vol. 13, No. 4.

Various determinations of the gravitation constant are reviewed. The Cavendish gravitation constant, which influences the precise knowledge of the mass and the average density of the earth, is known particularly with low precision. An increase of precision is possible by means of a more precise theory and technique of measurements.

043.006 **Free diurnal nutation of the earth based on observations at Pulkovo from 1915 to 1928.**
Ya. S. Yatskiv.
Astrometriya i Astrofiz., *Kiev*, No. 7, p. 49 - 60 (1969). In Russian.

The author dicusses the possibility of studying the free diurnal nutation of the earth using common latitude observations. He gives the results of determination of the parameters of this nutation based on the observations at Pulkovo from 1915 to 1928. The period of the free diurnal nutation thus obtained is in good agreement with the theoretical value ($23^h 56^m 54^s$ sidereal time) calculated by M. S. Molodensky for his second model of the earth. The essential changes of the amplitude and phase of the nutation are revealed. The average value of the amplitude for the series of observations under consideration equals to $0.''010$.

043.007 **The new system of astronomical constants.** K. A. Kulikov.
Izdatel'stvo "Nauka", Moskva. 91 pp. Price 28 Kop. (1969). In Russian.

043.008 **Improvement of astronomical constants and ephemerides from Pioneer radio-tracking data.**
J. D. Anderson, D. E. Hilt.
AIAA Journ., Vol. 7, 1048 - 1054 (1969).

Some preliminary, weighted, least-squares fits to two-way coherent S-band Doppler data from Pioneers VI and VII are tabulated in this paper. The principal results are that values are recommended for the geocentric gravitational constant (GE), the earth-moon mass ratio μ^{-1}, and the gravitational constant (GM) for the moon. These are $GE = 398601.5 \pm 0.4$ km^3/sec^2; $GM = 4902.75 \pm 0.12$ km^3/sec^2; $\mu^{-1} = 81.3016 \pm 0.0020$. The uncertainties associated with the values are believed to represent realistic standard errors.

044 Time, Rotation of the Earth

044.001 Researches and prospects on the time problems.
F. Zagar.
Proc. Colloquium on Problems of Time Determination, Keeping and Synchronization, (Milan 1968), p. 13 - 22 (1968). In Italian.

After expounding the reasons which have determined the present dualistic concept of unit and time scale, the method and the modern applications of time determination, time-keeping and synchronisation are discussed.

044.002 On two new periods in the rate of earth's rotation.
E. Proverbio, F. Carta.
Proc. Colloquium on Problems of Time Determination, Keeping and Synchronization, (Milan 1968), p. 163 - 170 (1968).

The study of the short term variations of the rotational time scale for the period 1955.5 - 1967.5 points out the existence of two preferential periods around 3.3 and 5.3 years, independent of the particular form attributed to long term periods of the rotation itself.

044.003 Irregular variations of the earth's rotation and the solar activity. E. Proverbio, A. Pensa.
Proc. Colloquium on Problems of Time Determination, Keeping and Synchronization, (Milan 1968), p. 171 - 176 (1968).

044.004 Long period effects of earth tides on the earth's rotation. E. Proverbio.
Proc. Colloquium on Problems of Time Determination, Keeping and Synchronization, (Milan 1968), p. 177 - 180 (1968).

044.005 Time measurements, discriminating experiences among the theories of the gravitational field.
M. Missana.
Proc. Colloquium on Problems of Time Determination, Keeping and Synchronization, (Milan 1968), p. 181 - 193 (1968). In Italian.

044.006 Photographic time determination with the transit instrument.
C. Egidi, N. Missana, F. Mussino.
Proc. Colloquium on Problems of Time Determination, Keeping and Synchronization, (Milan 1968), p. 219 - 227 (1968). In Italian.

Preliminary results of a photographic method for the stars transit time determination, applied to one of two instruments of Astronomic Observatory of Pino Torinese, are given. The used apparatus is a camera without the optics and with the shutter replaced by the IEN standard frequency signals, sent to the Observatory by a direct telephone line. The apparatus has been made chiefly in order to obtain some information on photographic and mechanical problems which arise in such devices and to obtain a first accuracy estimation. Using the gathered data, a final construction is planned.

044.007 Alcuni criteri di peso delle equazioni nelle determinazioni di tempo e di azimut strumentale secondo il metodo di Mayer. A. Palma, F. Smriglio.
Proc. Colloquium on Problems of Time Determination, Keeping and Synchronization, (Milan 1968), p. 235 - 240 (1968).

Criteria are given for the weighting determinations of time and of the azimuth of the instrument.

044.008 Prime vertical time and latitude determinations.

S. Leone.
Proc. Colloquium on Problems of Time Determination, Keeping and Synchronization, (Milan 1968), p. 255 - 261 (1968). In Italian.

044.009 Analysis of time observations in the meridian by the method of Doellen. C. de Concini.
Proc. Colloquium on Problems of Time Determination, Keeping and Synchronization, (Milan 1968), p. 265 - 271 (1968). In Italian.

044.010 Accidental and systematic errors of time observations and time systems. E. Buschmann.
Bull. Géod. Nouvelle Série, No. 93, p. 277 - 282 (1969).

The results of astronomical time observations contain many kinds of accidental and systematic errors. Their causes are very different. Important research is done by many authors to analyse these errors or at least to eliminate some of them. The results are very different and contradictory. In some cases this may be caused by using small and inhomogeneous series of observations. The following results have been got by analysing an extensive material of observations made at the time service station of the Geodetic Institute Potsdam during the period from 1957.5 to 1963.5.

044.011 Analisi critica degli errori personali relativi nelle osservazioni meridiane di tempo e di longitudine.
S. Mancuso, E. Proverbio.
Mem. Soc. Astron. Italiana, Nuova Serie, Vol. 40, 325 - 340 (1969).

The different physical and astronomical methods employing for determining the personal equation in time observations are statistically examined and is shown that the direct astronomical method supplied equivalent performances than physical methods and better accuracy as regards other astronomical methods. The statistical analysis points out that personal variations represent one fourth of the whole mean square error in time determinations.

044.012 Secular accelerations of the earth and moon.
R. R. Newton.
Science, Vol. 166, 825 - 831 (1969).

The acceleration of the earth's spin, which is obviously important to astronomy, has also been used extensively to provide information about important geophysical processes. This article will deal with components having a time scale of centuries or more. The average acceleration over an interval of several centuries or longer is usually called the secular acceleration. Paleontology, satellites, and ancient astronomy yield accelerations that geophysics cannot yet explain.

044.013 Average acceleration of the earth's rotation and the viscosity of the deep mantle.
R. H. Dicke.
Journ. Geophys. Res. Vol. 74, 5895 - 5902 (1969).

The acceleration of the earth's rotation averaged over the past 3000 years has been restudied using a new analysis of the ancient eclipses. The relaxation time for a second-order harmonic distortion of the earth has been calculated to fall in the range 870 – 1600 years. This short relaxation time seems to require a viscosity for the deep mantle of $\sim 10^{22}$ poises, permitting the convective transport of heat from the deep interior of the earth. If the viscosity is this low, deep convective currents are required to support the excess equatorial bulge.

044.014 Variations saisonnières de la rotation de la terre

déterminées par l'observation visuelle à la lunette Zeiss à l'Observatoire de Bucarest.   G. Oprescu.
Stud. Cerc. Astron. Vol. 14, 49 - 51 (1969).   In Rumanian.

En utilisant les observations de passage effectuées visuellement à l'Observatoire de Bucarest avec la lunette Zeiss 100/1000 mm, nous avons étudié les variations saisonnières de la rotation de la Terre rapportées à l'échelle du temps local.

044.015 On the systems of coordinates used in the study of the earth's rotation.   N. T. Mironov,
E. P. Fedorov.
Astron. Zhurn. Akad. Nauk SSSR, Vol. 46, 1303 - 1309 (1969).   In Russian. English translation in Soviet Astron. AJ, Vol. 13, No. 6.

For the study of the rotation of the earth a rotating frame of reference fixed to the earth may be constructed by attaching its axes to the zeniths of several selected observatories. The equatorial plane of this system is so situated that its angular distances from the zeniths are always as near as possible to some adopted constants. Its motions relative to the system of the principal axes of inertia can be derived from the analysis of the polar motion.

044.016 Scale di tempo coordinato ed integrato e precisione nella conservazione del tempo.
E. Proverbio, G. Chlistovsky.
Mem. Soc. Astron. Italiana, Nuova Serie, Vol. 40, 435 - 446 (1969).

Two scales of physical coordinated time UTC_{MI} and integrated time UTC_{INT} established at the Brera Observatory of Milan are studied. The deviations between the two scales, analysed by statistical methods, can be interpreted as short and long period variations caused by radiowave ionospheric propagation, and as irregularities depending by comparison methods.

044.017 The measurement of time.   L. Essen.
Vistas in Astronomy, Vol. 11, 45 - 67 (1969). – Review article.

044.018 Zum Problem künftiger Zeitskalen.
G. Becker.
PTB Mitt., 79. Jahrgang, 441 - 446 (1969).

After a presentation of the problems of time scales now in use, international activities in the field of time scales within the work of CIPM, CCIR and URSI are reported. Advantages and disadvantages of the 1 s SAT system under discussion by an "International Working Party" of the CCIR are described.

044.019 Rotation de la terre, phénomènes géophysiques et activité du soleil.   A. Stoyko, N. Stoyko.
Acad. Roy. Belgique, Bull. Cl. Sci., 5ᵉ Série, Vol. 55, 279 - 285 (1969).

L'étude de la période 1900 – 1963 a montré l'existence de fortes corrélations entre les phénomènes géophysiques et solaires.

044.020 Temps des éphémérides, temps atomique, temps rotationnel et leur comparaison.
A. Stoyko.
Acad. Roy. Belgique, Bull. Cl. Sci., 5ᵉ Série, Vol. 55, 286 - 296 (1969).

Les définitions de la seconde des éphémérides, de la seconde atomique et de la seconde rotationnelle permettent de faire la comparaison de trois échelles de temps. Deux premières échelles définissent les temps uniformes concomittants et permettent de déceler certaines régularités à longue période dans la variation de rotation de la terre.

044.021 Résultats des observations faites à Alger avec

l'astrolabe impersonnel A. Danjon OPL 8. Temps et latitude 1968.   A. Ghezloun, M. Benhocine, A. Fresneau, A. Marouf.
Ann. Obs. Astrom. d'Alger, Vol. 3, (Fasc. 1), 3 - 25 (1969).

044.022 Comparaisons de temps à grande distance dans le domaine de la microseconde.   B. Guinot.
Ann. Françaises Chronométrie Micromécanique, 4. année, p. 13 - 15 (1969).

Les recherches destinées à améliorer les comparaisons d'horloges sont très actives. Nous passerons en revue les différentes méthodes expérimentées ou déjà utilisées couramment.

044.023 Zufällige und systematische Fehler in geodätisch-astronomischen Zeitbestimmungen.
E. Buschmann.
Arbeiten Geod. Inst. Potsdam (Deutsche Akad. Wiss. Berlin), No. 25, 131 pp. (1969).

The results of about 3000 time observations are used for an investigation on sources of errors. It is found that instrumental errors predominate.

044.024 Temps universel et coordonnées du pôle; temps d'émission des signaux horaires.
Bureau International de l'Heure, Paris, Circ. D34 – D37 (1969).

Circular D of the BIH provides users with the current results relative to universal time, the time of emission of time-signals, the coordination of time maintained by the laboratories.

044.025 Détermination astronomique de l'heure et heures demi-définitives de réception des signaux horaires.
Acad. Tchécoslov. Sci., Inst. Astron., Station de l'Heure, Prague, Sér. 5, Nos. 1 - 3 (1969). – 1969 January – June.

044.026 Astronomische Zeit- und Breitenbestimmungen. Empfangszeiten von Zeitsignalen.
Edited by Deutsches Hydrographisches Institut, Hamburg. 1969 January – June. (1969).

044.027 Time and latitude.
Time and Latitude Bull., Tokyo Astron. Obs., Vol. 43, Nos. 1 – 7, January – July, (1969).

Coordinates of the instantaneous pole on the earth; Corrections for the longitude and the seasonal variations; Astronomical observations made with the PZT; Times of emission of radio time signals on UT2; Times of arrival and frequency deviations of GBR signal as received.

044.028 On the selection of stars for time and latitude observations with the photographic zenith tube at the International Latitude Observatory of Mizusawa.
I. Okamoto, G. Murakami.
Proc. International Latitude Obs. Mizusawa, No. 9, p. 68 - 98 (1969).   In Japanese.

A list of 452 stars selected from the Smithsonian Star Catalogue is given. Arguments are presented for the suitability of these stars for observations of time and latitude.

044.029 On the corrections for the positions of the stars of the Mizusawa PZT by modified Blaser's method.
G. Murakami.
Proc. International Latitude Obs. Mizusawa, No. 9, p. 99 - 105 (1969).   In Japanese.

The group corrections to the PZT star system have been derived by using modified Blaser's method.

044.030 Time Service of the Mizusawa Observatory.
Bulletins, Vol. 12, No. 1 - 12, 1967.

Edited by the International Latitude Observatory of Mizusawa, Mizusawa-Shi, Iwate-Ken, Japan. 46 pp. (1969).

This bulletin contains the results of time service and the astronomical observations made at the Mizusawa Observatory during the period beginning with January 1967 to December 1967.

044.031 On the problem of accurate time transmission by using the meteoric propagation of radio frequency waves. A. N. Smirnov.
Vestn. Khar'kovsk. politekhn. in-ta, No. 36 (84), p. 25 - 26 (1969). In Russian. – Abstr. in Referativ. Zhurn. 51. Astron., 2.51.151 (1970).

044.032 Schwankungen der Geschwindigkeit der Erdrotation und horizontale Bewegungen der Erdkruste. N. N. Pavlov.
Rotation of the Earth and Determination of Time, Conference Riga 1965, p. 5 - 14 (1969). In Russian.

044.033 Änderungen der Rotationsgeschwindigkeit der Erde aus einer Analyse astronomischer Beobachtungen unter Verwendung von Atom- und Quarzuhren. D. Yu. Belotserkovskij.
Rotation of the Earth and Determination of Time, Conference Riga 1965, p. 14 - 25 (1969). In Russian.

044.034 Der Einfluß der halbtägigen lunaren Gezeitenwelle auf die astronomische Bestimmung der Uhrkorrektion. L. A. Solov'eva, T. K. Nikol'skaya.
Rotation of the Earth and Determination of Time, Conference Riga 1965, p. 26 - 34 (1969). In Russian.

044.035 Über einen möglichen Fehler bei der Zeitbestimmung aus astronomischen Beobachtungen. A. V. Shiryaev, M. P. Mishchenko.
Rotation of the Earth and Determination of Time, Conference Riga 1965, p. 79 - 82 (1969). In Russian.

044.036 Programm der Zeitbestimmung im Astronomischen Observatorium der Leningrader Staatsuniversität und im Nikolajewer Observatorium aus Sternen, die symmetrisch zum Zenit liegen. V. N. Pyshnenko.
Rotation of the Earth and Determination of Time, Conference Riga 1965, p. 96 - 100 (1969). In Russian.

044.037 Beobachtungen von Zinger-Paaren mit dem Danjon-Astrolab. L. N. Nadeev.
Rotation of the Earth and Determination of Time, Conference Riga 1965, p. 100 - 102 (1969). In Russian.

044.038 Arbeiten zur Erhöhung der Genauigkeit der Beobachtungen des Nikolajewer Zeitdienstes von 1960 bis 1964. M. I. Il'kiv, N. S. Kalikhevich.
Rotation of the Earth and Determination of Time, Conference Riga 1965, p. 102 - 105 (1969). In Russian.

044.039 Über die Auswahl der Gewichte bei der Berechnung der Korrektionen der Standardzeit nach Beobachtungsergebnissen einzelner Observatorien. N. S. Blinov.
Rotation of the Earth and Determination of Time, Conference Riga 1965, p. 106 - 113 (1969). In Russian.

044.040 Corrections to Czechoslovak time signals. V. Ptáček.
Říše hvězd, Vol. 50, 140, 157, 183, 220, 238 (1969). – 1969 April – September.

044.041 Erdrotation und Gezeitenreibung. P. Brosche, J. Sündermann.
Mitt. Astron. Ges. No. 27, p. 131 - 132 (1969). – Abstract AG.

044.042 Détermination astronomique de l'heure et de la latitude.
Obs. Neuchâtel, Bull. (B), 1969 Mai – Octobre (1969).

044.043 L'heure astronomique définitive de l'Observatoire de Neuchâtel.
Obs. Neuchâtel, Bull. (D), 1969 Avril – Octobre (1969).

044.044 Results of the determination of time.
Polish Acad. Sci. Astron. Latitude Station Borowiec, Circ. Nos. 109 - 111 (1969). – 1969 January – September.

044.045 Time service. E. Proverbio.
Astron. Obs. Milan, Circ. No. 28, 35 pp. (1969).

045 Latitude Determination, Polar Motion

045.001   The results of observations at the Gorki latitude station.   L. D. Kovbasjuk, S. G. Kulagin.
Astron. Tsirk. No. 510, p. 4 - 6 (1969).   In Russian.

045.002   Nutation with the period of 18.6 years from the data of the International Latitude Service.
V. K. Tarady.
Astrometriya i Astrofiz., *Kiev*, No. 2, p. 7 - 26 (1969). In Russian.

For deriving the coefficients of the main terms of nutation in obliquity, N, and longitude, M, the author utilized latitude observations at Mizusawa, Carloforte and Ukiah from 1900 to 1942. The results are as follows: $N = 9''1970 \pm 0''0010$, $M = 6''8476 \pm 0''0010$.

045.003   Motion of the earth's pole during the period 1960 - 1965.   A. A. Korsun'.
Astrometriya i Astrofiz., *Kiev*, No. 2, p. 27 - 31 (1969). In Russian.

The author discusses the problem of estimating the weights of latitude observations which should be taken into account when deriving the polar motion. The method was used for determination of the coordinates of the pole from latitude observations at 31 observatories during the period 1960 - 1965.

045.004   The latitude variations during the night in observations with the Danjon astrolabe at Poltava.
A. A. Slavinskaya, N. I. Panchenko.
Astrometriya i Astrofiz., *Kiev*, No. 2, p. 32 - 35 (1969). In Russian.

The latitude variations during the night obtained from the observations with the Danjon astrolabe at Poltava from 1961.0 to 1964.4 are discussed. It has been found that the latitude always decreases from evening to midnight and then increases.

045.005   Pearson's distribution of type VII in the errors of latitude observations.   J. V. June.
Astrometriya i Astrofiz., *Kiev*, No. 2, p. 101 - 115 (1969). In Russian.

Concurrent observations with two zenith telescopes have been carried out at the Poltava Observatory. Differences between latitudes obtained with the two telescopes are discussed. It has been found that actual distribution of these differences better follows the Pearson's law of type VII than the normal law which may be explained as due to the fact that the precision of observation is unstable because of variability of the observational conditions.

045.006   Determination of deflection of vertical with a small instrument.   O. Mathisen.
Bull. Géod. Nouvelle Série, No. 93, p. 283 - 286 (1969).

045.007   Analysis of the deflections of the vertical due to tides observed at Engelhardt Observatory, 1961 — 1966.   I. A. Urasina.
Izv. Astron. Obs. Ehngel'gardta, *Kazan*, No. 36, p. 200 - 228 (1968).   In Russian.

045.008   Smoothing of latitude observations at Engelhardt Observatory, 1957 - 1965.   Yu. G. Yusupov.
Izv. Astron. Obs. Ehngel'gardta, *Kazan'*, No. 36, p. 229 - 239 (1968).   In Russian.

045.009   Comparison between the Upper Palaeozoic and Me- sozoic palaeomagnetic poles for South America, Africa and Australia.**   K. M. Creer, B. J. J. Embleton, D. A. Valencio.
Earth Planet. Sci. Letters, Vol. 7, 288 - 292 (1969).

When South America, Africa and Australia are placed in their pre-drift relative positions a common polar wandering curve can be constructed through their upper Carboniferous and Lower and Upper Permian poles. The Triassic poles for South America and Africa are coincident but the position of that for Australia is significantly different. Maps have been drawn showing the latitudes of these three continents during this interval of time.

045.010   Les positions succesives du pôle d'inertie de la terre dans l'intervalle août 1959 - décembre 1962.
C. Drâmbă, G. Stănilă.
Stud. Cerc. Astron. Vol. 14, 9 - 15 (1969).

045.011   Annual report of the International Polar Motion Service for the year 1967.   S. Yumi.
Published for the International Council of Scientific Unions by Central Bureau of the International Polar Motion Service, Mizusawa. 4 + 160 pp. (1969).

Results of the latitude observations in 1967 made with 43 instruments at the 37 stations, inclusive of the five ILS stations are described in this volume.

045.012   General analysis of the results of latitude observations with the Danjon astrolabe at Poltava during 1961.0 - 1964.4.   A. A. Slavinskaya.
Astrometriya i Astrofiz., *Kiev*, No. 7, p. 39 - 49 (1969). In Russian.

For reduction of observations with the Danjon astrolabe at Poltava during the period of 1961.0 - 2964.4 moments of star transits were calculated using FK3 postions, and 20.''47 as the constant of annual aberration. The corrections to be applied to latitudes for transition to the new aberration constant, 20.''496, are given. To decrease the systematic errors latitudes were smoothed by the chain method. The closing error proved to be 0.''137. The group corrections are given for transition from the FK3 to the FK4 system. 124 normal latitude values were smoothed by an analytical method. The spectrum of errors was investigated by means of analysing deviations of the normal latitudes from the smoothed ones. The value $S_c = 0.''000110$ was taken as a characteristic of the spectrum. The amplitude and phase of the annual z-term proved to be unstable. It was found that the latitude decreases from morning till midnight and then increases again.

045.013   Discussion of the results of the latitude observations at Pulkovo with the ZTF-135 according an expanded program during 1955 - 1961.3.   L. D. Kostina.
Astrometriya i Astrofiz., *Kiev*, No. 7, p. 95 - 106 (1969). In Russian.

The paper deals with the analysis of the six-year latitude observations with the Freiberg zenith-telescope ZTF-135. The observations were executed according to an extended program. The abundant material was thoroughly analysed for a correction to the constant of aberration, for corrections to the scale values and its temperature coefficients, the declinations and proper motions of the centres of the Talcott pairs. Erros of star declinations and their proper motions taken from different catalogues were considered in detail. The accuracy of the latitude observations at Pulkovo with the ZTF-135 during 1955.0 - 1961.3 is estimated. The curve of latitude variations and the value of the z-term are presented.

045.014 The relations between Z term and catalogue errors.
Y. Wako.
Acad. Roy. Belgique, Bull. Cl. Sci., 5ᵉ Série, Vol. 55, 537 - 548 (1969).

045.015 The coordinates of the earth's pole (1968.0 – 1969.6) referred to the mean pole.
N. I. Panchenko, I. P. Ogorodnik.
Astron. Tsirk. No. 537, p. 3 - 5 (1969).   In Russian.

045.016 Coordonnées du pôle instantané rapportées à l'origine conventionnelle internationale et corrections de longitude TU 1 – TU 0, à 0h TU.
Bureau International de l'Heure, Paris, Circ B/C, Nos. 161 - 164 (1969). – Valeurs interpolées et extrapolées.

045.017 Monthly Notes of the International Polar Motion Service.
IPMS Monthly Notes, International Latitude Obs. Mizusawa (Japan). 1969, Nos. 5 (July) – 10 (December), p. 47 - 98 (1969). – Announces the values of latitude observed at the collaborating stations during 1969 May until October.

045.018 Breitenbestimmungen.
Techn. Univ. Dresden, Lohrmann-Obs. Zirk. Nos. 37 - 42 (1969). – 1969 January – December.

045.019 Secular variation of latitude derived from the FZT observations at Mizusawa.
K. Hurukawa.
Publ. International Latitude Observatory of Mizusawa, Vol. 6, 103 - 159 (1968).
In supplying the unpublished data for the period from 1950 to 1954, all the data reduced by the standardized uniform method in the GC system are given. Errors, both internal and external, in the observed values of latitude are examined in some details. After removing the terms of Chandler and the annual from the reduced data, the secular variation of latitude at Mizusawa referred to the FK4 system is derived.

045.020 Assumed deformation of the earth deduced from the observations of the polar motion.
S. Yumi, H. Ishii, K. Sato.
Publ. International Latitude Observatory of Mizusawa, Vol. 6, 161 - 184 (1968).
From the analyses of the residual latitudes at the IPMS stations from 1962 to 1967, a local trend in latitude variation for each station other than by the polar motion was found. A gradual deformation of the earth or a variation of geopotential surface may be responsible for the effect. Coordinates of the pole calculated from the results of the IPMS stations were compared with those obtained from the five ILS stations and the systematic difference between them was discussed from the viewpoint of the deformable earth or of the variable geopotential surface.

045.021 Numerical analysis of the Oppolzer terms.
S. Takagi, G. Murakami.
Publ. International Latitude Observatory of Mizusawa, Vol. 6, 225 - 230 (1968).
The effect of the Oppolzer terms attains the value of about $\pm 0\rlap{.}{''}02$ in the polar motion and about ± 2.0 ms in the rotational velocity. The value $0\rlap{.}{''}02$ might offer us a new clue in the analysis of the z term.

045.022 Local non-polar variation of latitude deduced from the ILS data for the period 1933 – 1965.
T. Okuda.
Publ. International Latitude Observatory of Mizusawa, Vol. 6, 231 - 243 (1968).
Using the differences between the common z terms obtained from five ILS stations, we have made a study of the local non-polar variation of latitude at each station based on the data for the period from 1933.0 to 1966.0. The results show marked local variations of latitude with a period of approximately 19 years; the phase is just the same for Mizusawa and Gaithersburg while the phase is opposite for Ukaih and Kitab. The local non-polar variations of latitude suggest that a global periodic change of the earth's potential surface should be taken into consideration for the determination of astronomical position.

045.023 On the relation between the night error of latitude observations and the meso-scale pressure distribution, (III). T. Goto.
Proc. International Latitude Obs. Mizusawa, No. 9, p. 1 - 14 (1969).   In Japanese.
In the preceding notes, the relation between individual observed values of latitude and the meso-scale perturbations of air cells on one night were kinematically investigated relating to the inclinations of airstrata of equal density. In this note, the relation between daily mean latitudes and the meso-scale pressure distribution without perturbing cells during latitude observations has been investigated.

045.024 On the Chandlerian periodicity in the variation of atmospheric pressure at Mizusawa. C. Sugawa.
Proc. International Latitude Obs. Mizusawa, No. 9, p. 35 - 37 (1969).   In Japanese.
The Chandlerian periodicity in the variation of barometric pressure at the International Latitude Observatory of Mizusawa was examined by using the periodogram analysis for the trial periods from 13 to 16 months.

045.025 On the wind effect in latitude observations with the floating zenith telescope. Y. Goto.
Proc. International Latitude Obs. Mizusawa, No. 9, p. 119 - 124 (1969).   In Japanese.

045.026 On the effect of the anomalous refraction due to the inclinations of air layers upon the observed latitudes. S. Goto, T. Goto.
Proc. International Latitude Obs. Mizusawa, No. 9, p. 125 - 134 (1969).   In Japanese.

045.027 1962, 1963 data on latitude observations with the floating zenith telescope.
C. Sugawa, K. Hurukawa, H. Okawa, H. Kitago.
Proc. International Latitude Obs. Mizusawa, No. 9, p. 143 - 190 (1969).   In Japanese.
Work at the Floating Zenith Telescope (FZT) is described. The system difference between the ILS system and the proper FZT system was deduced as $+0\rlap{.}{''}059$. Thus, we could first obtain the system difference between the ILS and the proper FZT systems from the mixed programme in the years 1962 - 1963.

045.028 The motion of the earth's poles and astronomical and geodetic work. I. D. Zhongolovitch.
Byull. Stantsij Optichesk. Nablyud. Iskusstv. Sputnikov Zemli No. 55, p. 38 - 43 (1969).   In Russian.
This paper deals with a detailed derivation of formulas necessary for taking into account the influence of the polar motion on latitude, longitude, and azimuth determinations, as well as on the geocentric rectangular coordinates of surface points. The formulas obtained for the latter case differ from those published formerly by G. Veis.

045.029 Results of observations of latitude at Monte da Virgem. M. Barros, J. Osório, R. A. Vieira.
Anais Faculdade Ciências do Porto, Vol. 51, Fasc. 1, 2, 30 pp. = Publ. Obs. Astron. Porto No. 22 (1968).

In the paper the instrumental constants and the values of individual latitudes are given, and it is discussed the evaluation of the mean latitude of the Observatory as well as the evaluation of a set of corrections to the mean declinations of the star pairs observed.

045.030 **Following the pole.** G. Cecchini, S. Yumi.
Southern Stars, Vol. 23, 88 - 89 (1969).

045.031 **Results of the latitude determination.**
Polish Acad. Sci. Astron. Latitude Station Borowiec, Circ. Nos. 109 - 111 (1969). – 1969 January – September.

045.032 **Provisional results of latitude observations by zenith stars program (1966 – 1968).**
E. Proverbio.
Astron. Obs. Milan, Circ. No. 29, 24 pp. (1969).

Résultats définitifs des observations faites par photographie d'étoiles zénithales à l'instrument des passages établie dans le I.ᵉ vertical. See Abstr. 041.033.

Prime vertical time and latitude determinations. See Abstr. 044.008.

Analisi critica degli errori personali relativi nelle osservazioni meridiane di tempo e di longitudine. See Abstr. 044.011.

Résultats des observations faites à Alger avec l'astrolabe impersonnel A. Danjon OPL 8. Temps et latitude 1968. See Abstr. 044.021.

Astronomische Zeit- und Breitenbestimmungen. Empfangszeiten von Zeitsignalen. See Abstr. 044.026.

Time and latitude. See Abstr. 044.027.

On the selection of stars for time and latitude observations with the photographic zenith tube at the International Latitude Observatory of Mizusawa. See Abstr. 044.028.

On the corrections for the positions of the stars of the Mizusawa PZT by modified Blaser's method. See Abstr. 044.029.

046 Geodetic Astronomy, Navigation

046.001 Transformationen zwischen ellipsoidischen Koordinatensystemen. S. Heitz.
Deutsche Geod. Kommission Bayer. Akad. Wiss. Reihe A, Heft No. 64, 31 pp. (1969).

First the author gives general equations for the transformation of the ellipsoidal longitudes, latitudes and heights between two systems basing on different reference ellipsoids in an arbitrary relative orientation. Then he deals with the determination of the six elements of relative orientation, especially in consideration to the transformation of geoid representations.

046.002 Die Vorbereitung der Satellitentriangulation beim Deutschen Geodätischen Forschungsinstitut und bei der Bayerischen Kommission für die Internationale Erdmessung. H. Deker, E. Wolf.
Veröff. Bayer. Kommission Internationale Erdmessung Bayer. Akad. Wiss. Astron.-Geod. Arbeiten, Heft No. 25, 82 pp. (1969).

Es werden zwei Stationssysteme (BE1 und BE2) beschrieben, die zur photographischen Beobachtung von Erdsatelliten am Deutschen Geodätischen Forschungsinstitut München zum Zwecke der Satellitentriangulation entwickelt wurden. Anschließend werden Formeln zur Berechnung der Azimute und Höhen als Einstellwerte für eine Kamera zur Satellitenphotographie entwickelt. *U. Güntzel-Lingner*

046.003 Bestimmung des Azimutes Braunschweig – Wesendorf – München – Hohenpeißenberg aus Simultanbeobachtungen der Satelliten Echo I und Echo II. K. Kaniuth.
Veröff. Bayer. Kommission Internationale Erdmessung Bayer. Akad. Wiss. Astron.-Geod. Arbeiten, Heft No. 26, 35 pp. (1969).

In this paper the Arnold formula system for the determination of azimuths from simultaneous satellite observations is extended to the derivation of the systematic time error of a camera, then it is applied to the determination of the azimuth of the geodetic line between the observation stations Braunschweig – Wesendorf and München – Hohenpeissenberg. The evaluation of the photos was carried out according to a purely "photogrammetric" method and by means of a "fictitious" satellite point resulting from a two-dimensional curve fit. From four pairs of simultaneous observations of Echo I and II the azimuth results with a mean square error of ± 0.22″ and the difference to the geodetic azimuth computed from the coordinates of the triangulation net is 0.60″.

046.004 A contribution to the mathematical foundation of physical geodesy. T. Krarup.
Geod. Inst. Copenhagen, Medd. No. 44, 80 pp. (1969).

046.005 Errors of the azimuth derived from observations of circumpolar stars in digression. A. P. Tzapova.
Astrometriya i Astrofiz., *Kiev*, No. 2, p. 92 - 100 (1969).
In Russian.

The paper deals with the accuracy of the azimuth of two marks (northern and southern) determined from observation of circumpolar stars in greatest elongation. The observation was carried out with the transit instrument of the Poltava Observatory in 1961 - 1963.

046.006 Liaison Nice – Beyrouth à l'aide des observations du satellite D1A (Diapason).
J. Kovalevsky, F. Barlier, I. Stellmacher.
Bull. Géod. Nouvelle Série, No. 93, p. 235 - 242 (1969).
La réduction des observations avait pour but d'étudier tous les problèmes liés à l'application des mesures Doppler à la géodésie et, en définitive, de tenter une liaison géodésique entre Nice et Beyrouth.

046.007 Determination of astronomical equatorial coordinates of spatial network side direction by the astrophotographic method. J. Kabeláč.
Bull. Géod. Nouvelle Série, No. 93, p. 255 - 261 (1969).

The determination of astronomical equatorial coordinates of the direction of a spatial network side as well as some special conclusions concerned with this problem are described.

046.008 The Schuler pendulum and inertial navigation. F. C. Bell.
Journ. Inst. Navigation, Vol. 22, 516 (1969).

046.009 I satelliti artificiali nell'impiego geodetico e cartografico. G. Birardi.
L'Universo, Anno 49, 605 - 620 (1969). – Review article on the use of artificial satellites in geodesy and cartography.

046.010 Moderne physikalische Orientierungsmethoden. K. Peters.
Österreich. Zeitschr. Vermessungswesen, 57. Jahrgang, 84 - 93 (1969).

Instruments and methods for rapid orientation in geodesy are described.

046.011 Evaluation of the deflection of the vertical due to the topographic masses and the underlying compensating masses which should be present according to the isostatic hypotheses. Parte II, III.
S. Ballarin.
Boll. Geod. Sci. Affini, Vol. 28, 133 - 147, 193 - 214 (1969).
In Italian.

046.012 Geodesy of artificial satellites. Star selection, computation and observation of their plate coordinates.
G. Birardi.
Boll. Geod. Sci. Affini, Vol. 28, 217 - 235 (1969).
In Italian.

046.013 Compensation of the relative personal errors during the national campaign of longitudes, 1967 - 1968.
C. de Concini, E. Proverbio.
Boll. Geod. Sci. Affini, Vol. 28, 237 - 260 (1969).
In Italian.

In this paper the personal equations of the observers partaking to the national enterprise for re-determining the astronomical longitude, have been obtained with the method of simultaneous observations. After the analysis of the variation of the personal equations, with the stellar magnitudes the observed values of personal equations suitably weighted have been submitted to general compensation following two different methods. Besides the systematic personal-instrumental corrections reducing all personal equations to the principal instrument of the Milan Observatory are determined.

046.014 Sphärische Integralformeln in der Geodäsie. E. Ecker.
Deutsche Geod. Kommission, Bayer. Akad. Wiss. Reihe C, No. 142, 3 + 54 pp. (1969).

046.015 Determination of longitude of the Main Astronomical Observatory (MAO) of the Ukrainian Academy of Sciences (Golosejevo). N. A. Vasilenko.
Astrometriya i Astrofiz., *Kiev*, No. 7, p. 91 - 95 (1969).

In Russian.

The longitude of the Golosejevo Observatory has been determined. The local time was determined from observation of star pairs on equal altitudes. The methods of observation and reduction are described. The following values of longitude are obtained: for the observation pillar "Golosejevo" $\lambda° = -2^h 02^m 0^s.191 \pm 0^s.016$, for the centre of the vertical circle $\lambda° = -2^h 01^m 59^s.931 \pm 0^s.016$.

046.016 La definition précise du système géocentrique des coordonnées. J. B. Zieliński.
Geodezja Kartografia, Vol. 18, 207 - 216 (1969). In Polish.

046.017 Longitude without time. B. Ortlepp.
Navigation (US), Vol. 16, 29 - 31 (1969).

046.018 Time-difference position–determination systems for aerospace and terrestrial applications.
J. E. Gaffney.
Navigation (US), Vol. 16, 182 - 190 (1969).

046.019 Meßverfahren zur optischen Ortung und Nachführung von Satelliten mittels Laser. H. Helbig.
Ortung und Navigation IV/69, p. 13 - 21 (1969).

046.020 Star/horizon measurement for onboard spacecraft navigation. J. A. Hand.
National Space Meeting on Space Navigation – Present and Future, Houston 1969, 23 pp. (1969).

046.021 Errors in the reduction of photographic plates for the stellar triangulation. J. Kakkuri.
Publ. Finnish Geod. Inst. No. 66 [Reprinted from Photogrammetric Journ. Finland, Vol. 3, No. 2], 14 pp. (1969).

The modified Turner's method, which has been discussed in this article, is suitable for the reduction of stellar plates. The only condition is that the altitude over the horizon from which the stellar photograph is taken is not less than 10 degrees. No systematical errors exist. Accuracy obtained is $\pm 0''.8$ in the right ascension and $\pm 0''.7$ in the declination. In the stellar photographs taken from an altitude lower than 10 degrees, the displacements for astronomical refraction are so strong that the effect cannot be corrected by the third order terms.

046.022 General notes on possibilities of using data obtained from observations of artificial satellites for elaboration of ground geodetic nets.
W. Dobaczewska.
Byull. Stantsij Optichesk. Nablyud. Iskusstv. Sputnikov Zemli No. 55, p. 19 - 22 (1969). In Russian.

046.023 Proper ways of the joint reduction of ground and space triangulations. L. P. Pellinen.
Byull. Stantsij Optichesk. Nablyud. Iskusstv. Sputnikov Zemli No. 55, p. 23 - 30 (1969). In Russian.

046.024 Construction of a coordinate system related to the mass centre and the rotation axis of the earth.
A. A. Izotov.
Byull. Stantsij Optichesk. Nablyud. Iskusstv. Sputnikov Zemli No. 55, p. 31 - 38 (1969). In Russian.

046.025 The role of satellite geodesy for further development of continental astronomical-geodetic nets.
M. Shedlikh.
Byull. Stantsij Optichesk. Nablyud. Iskusstv. Sputnikov Zemli No. 55, p. 73 - 81 (1969). In Russian.

Methods of Geodetic Astronomy for the Intertropical Zone. See Abstr. 003.001.

047 Ephemerides, Almanacs, Calendars

047.001 Astronomische Grundlagen für den Kalender 1971.
Edited by Astronomisches Rechen-Institut in Heidelberg. Verlag G. Braun GmbH., Karlsruhe. 85 pp. + Appendix. Price DM 22.50 (1969).

047.002 The Nautical Almanac for the Year 1970.
Issued by Her Majesty's Nautical Almanac Office, London; and Nautical Almanac Office United States Naval Observatory, Washington. Her Majesty's Stationery Office, London. 276 + 35 pp. Price £1 0s. 0d. (1969).

047.003 Efemerides astronomicas y manual del aficionado para el año 1969.
Revista Astron. Vol. 40, No. 167, 32 pp. (1968).

047.004 1970 Nautical Almanac. Pub. No. 681.
Published by Hydrographic Office of Japan, Tokyo. 3 + 466 pp. (1969).

047.005 1970 Abridged Nautical Almanac. Pub. No. 683.
Published by Hydrographic Office of Japan, Tokyo. 2 + 239 pp. (1969).

047.006 Efemérides Astronómicas año 1970, publicadas de orden de la Superioridad por el Instituto y Observatorio de Marina de San Fernando (Cádiz).
Observatorio de Marina, San Fernando. XIII + 7A + 535 pp. Price 200 pesetas (1969).

047.007 Ephémérides Nautiques pour l'an 1970. Ouvrage publié par le Bureau des Longitudes spécialement à l'usage des marins.
Gauthier-Villars, Editeur, Paris. 473 pp. (1969).

047.008 The Air Almanac 1970, January - April.
Her Majesty's Stationery Office, London; United States Naval Observatory, Washington. 242 + A82 + F4 pp. Price £ 1 10s. 0d. (1969).

047.009 Astronomical phenomena for the year 1972.
Issued by the Nautical Almanac Office, United States Naval Observatory.
U. S. Government Printing Office, Washington, D. C. 71 pp. Price 55 cents (1968).

047.010 Almanacco Astronomico della Rivista Coelum per l'anno 1970.
Compiled by E. Nasi, A. Betti, L. Dall'Olio, L. D'Ascanio. Coelum Suppl. Vol. 37, Fasc. 9 - 10 [Osservatorio Astronomico Universitario,Bologna]. 28 + 40 pp. Price L. 2000 (1969).

047.011 Efemérides Astronómicas para o ano de 1970.
Edited by Osservatório Astronómico da Universidade de Coimbra.
Imprensa de Coimbra, Limitada, Coimbra. 13 + 236 pp. (1969).

047.012 The Indian Ephemeris and Nautical Almanac for the Year 1970.
Office of preparation: Nautical Almanac Unit, Regional Meteorological Centre, Alipore, Calcutta. Printed by the General Manager, Government of India Press, Calcutta. 32 + 466pp. Price Rs. 14.00, 32s. 8d., $5 4 cents, respectively (1969).

047.013 Connaissance des Temps ou des Mouvements Célestes pour l'an 1970, à l'usage des astronomes et des navigateurs, publiée par le Bureau des Longitudes.
Gauthier-Villars Editeurs, Paris. 42 + 643 pp. (1969).

047.014 Gravitationally consistent planetary ephemerides based on meridian circle, radar, and Mariner observations. W. G. Melbourne, D. A. O'Handley, R. Reed.
Bull. American Astron. Soc. Vol. 1, 253 (1969). – Abstr. AAS.

047.015 Anuario del Observatorio Astronómico de Madrid para 1970.
Published by Instituto Geográfico y Catastral, Madrid. 443 pp. Price 75 pesetas (1969).

047.016 Apparent Places of Fundamental Stars 1972, containing the 1535 Stars in the Fourth Fundamental Catalogue (FK4).
Edited by Astronomisches Rechen-Institut, Heidelberg. To be purchased from Verlag G. Braun, Karlsruhe. 44 + 510 pp. Price DM 36.00 (1969).

047.017 Der Sternenhimmel 1970.
Kleines astronomisches Jahrbuch für Sternfreunde. R. A. Naef (Editor).
Verlag Sauerländer, Aarau. 30. Jahrgang, 182 pp. Price Fr. 15.00 (1969).

047.018 Kalender für Sternfreunde 1970.
P. Ahnert (Editor).
Johann Ambrosius Barth, Leipzig. 200 pp. Price DM 4.00 (1969).

047.019 A note on the evaluation of the latitude of the moon. W. J. Eckert, T. C. Van Flandern, G. A. Wilkins.
Monthly Notices, Roy. Astron. Soc., Vol. 146, 473 - 478 (1969).

Investigations of a discrepancy of about $0\rlap{.}{''}034 \sin (F\text{-}2D)$ between two methods for the evaluation of the fundamental lunar ephemeris have revealed a numerical error in a coefficient used in the reverse transformation by Eckert, Walker and Eckert of Brown's series for the latitude of the moon. As a consequence, the series given by them for the differential correction of the Improved Lunar Ephemeris requires amendment. The investigations have also shown that the interpretation of Brown's formula for the latitude that was used in the Improved Lunar Ephemeris does not give the best representation of Brown's original series.

047.020 Annuaire 1970 du Bureau des Longitudes. Encyclopédie Physique et Spatiales. Gauthier–Villars, Editeur, Paris. 14 + 959 + A24 + B14 + C4 + D10 + E106 pp. (1969).

047.021 Graphische Zeittafel des Himmels, Januar bis Juni 1970. N. Hasler-Gloor.
Orion Schaffhausen, Vol. 14, 153 - 155 (1969).

047.022 Japanese Ephemeris 1971.
Compiled by Astronomical Division, Hydrographic Department, Tokyo, Japan. Pub. No. 684. 6 + 426 pp. (1969).

047.023 Polaris Almanac for Azimuth Surveying 1970.
Published by Hydrographic Department, Tokyo, Japan. 9 pp. (1969).

047.024 Annuaire de l'Observatoire Royal de Belgique [Jaarboek van de Koninklijke Sterrenwacht van België] 1970.
Imprimerie Hayez, Bruxelles. 137ᵉ année (jaargang). 225 pp. (1969).

047.025   **Himmelskalender 1970.**
Ein astronomisches Jahrbuch für Österreich.
H. Mucke, K. Mayrhofer (Editors).
Verlag H. Mucke, Wien. 79 pp. Price öS 25.00 (1969).

047.026   **Rocżnik Astronomiczny na Rok 1970.**
Prepared under the supervision of J. Radecki.
Instytut Geodezji i Kartografii, Państwowe przedsiębiorstwo wydawnictw kartograficznych, Warszawa. Vol. 25, 113 pp. Price zł 55.- (1969).

047.027   **Anuarul Observatorului din București – 1970.**
Editura Academiei Republicii Socialiste România.
222 pp. Price Lei 25 (1969).

047.028   **Almanaque Nautico y Aeronautico para el año 1970.**
Republica Argentina. Armada Argentina, Servicio de Hidrografia Naval, Buenos Aires. 382 pp. Price $ 600m/arg (1969).

047.029   **Supplemento al Almanaque Nautico y Aeronautico para el año 1970. Sol, Planetas y Estrellas.**
Republica Argentina. Armada Argentina, Servicio de Hidrografia Naval, Buenos Aires, 8 + 133 pp. Price $ 200 m/arg (1969).

047.030   **Astronomischer Kalender des Observatoriums in Sofia für das Jahr 1970.**   N. Bonev (Editor).
Izdatelstvo na B'lgarskata Akademiya na Naukite, Sofiya. 92 pp. Price 0.80 Lv. (1969).   In Bulgarian.

047.031   **The Nautical Almanac for the year 1971.**
Her Majesty's Nautical Almanac Office, London;
Nautical Almanac Office, United States Navel Observatory, Washington. 276 + 35 pp. Price £1 0s. 0d. (1969).

047.032   **Astronomical Yearbook 1970.**
Published by the Astronomical Society of Victoria, Melbourne. 32 pp. Price 60c. (1969).

047.033   **Visibility of the planets, 1970.**   L. P. Lee.
Southern Stars, Vol. 23, 90 - 91 (1969).

047.034   **The American Ephemeris and Nautical Almanac for the year 1971.**
Issued by Nautical Almanac Office, United States Naval Observatory, Washington; Her Majesty's Nautical Almanac Office, Royal Greenwich Observatory, London. U.S. Government Printing Office, Washington. 17 + 520 pp. Price $ 6.25 (1969).

047.035   **Das Himmelsjahr 1970.**
Sonne, Mond und Sterne im Jahr 1970.
Compiled by M. Gerstenberger.
Kosmos-Verlag, Franckh'sche Verlagshandlung, Stuttgart. 110 pp. Price DM 5.80 (1969).

047.036   **The Air Almanac 1970, May – August.**
Her Majesty's Stationery Office, London; United States Naval Observatory, Washington. 248 pp. + A82 + F4 pp. Price £1 10s. 0d. (1969).

047.037   **Annuario Astronomico 1970.**
Pubbl. Oss. Astron. Trieste, No. 407, 77 pp.
(1969).

047.038   **Standard dates for ephemerides in 1970.**
IAU Circ. No. 2174 (1969).

Accurate ephemerides for planets and moon.
See Abstr. 091.021.

Corrections to the improved lunar ephemeris.
See Abstr. 094.159.

Ephemerides of minor planets for 1970.
See Abstr. 098.015.

Space Research

051 Extraterrestric Research, Spaceflight Related to Astronomy

051.001 **Die erfolgreich gestarteten künstlichen Erdsatelliten und Raumsonden. (1.1. bis 15.6.1969).**
U. Güntzel-Lingner.
SuW, Vol. 8, 205 - 209 (1969).

051.002 **Space report.**
Spaceflight, Vol. 11, 269 - 274 (1969).
Barium cloud from HEOS; Aurorae after six months; Moon-earth interaction; Meteor weather Satellite (*1969-29A*); OAO triumph; Water on Mars; Mars landing radar; Apollo 8 inspection.

051.003 **Space report.**
Spaceflight, Vol. 11, 326 - 333 (1969).
Results from Venus (*concerning the spaceprobes Venus 5 and 6*); Ocean waves on moon? More lunar mascons; Active lunar volcano; Crop conditions by satellite; Skylark probes hot stars.

051.004 **Space report.**
Spaceflight, Vol. 11, 348 - 357 (1969).
Photographing Mars; OAO-2 "repaired" in orbit; German-US sun probe; Radio astronomy satellite; Solar radiation measured; Ballon satellite descends (*Echo 2 = 1964-4A*); The earth's poles; Asteroid "Geographer".

051.005 **Astronautica.** R. Migliavacca.
Coelum, Vol. 37, 205 - 219 (1969).

051.006 **Kurzberichte.**
Weltraumfahrt, 20. Jahrgang, p. 150 - 155 (1969).
Bemanntes Mondauto; Pegasus 3 verglüht (*1965-60A*); Studie für erste Merkur-Sonde abgeschlossen; OSO 6 funktioniert einwandfrei (*1969-68A*).

051.007 **Space report.**
Spaceflight, Vol. 11, 390 - 396 (1969).
Laser light on moon; Seventh IMP (*= Explorer 41 = 1969-53A*); Mars in focus.

051.008 **Les satellites artificiels de l'année 1968.**
J. Thurnheer.
Orion, Band 14, 118 - 122 (1969).

051.009 **Astronautique 1968.** J. Meeus.
Ciel et Terre, Vol. 85, 257 - 307 (1969).

051.010 **Relativistic interstellar spaceflight.**
J. F. Fishback.
Astronaut. Acta, Vol. 15, 25 - 35 (1969).

051.011 **Belastung von Satelliten durch die Korpuskular-strahlung im Weltraum.** K. Wohlleben.
Umschau, Vol. 69, 694 - 695 (1969).

051.012 **Light scattering by manned spacecraft atmospheres. II. Large particles.** N. S. Kovar, R. P. Kovar.
Bull. American Astron. Soc. Vol. 1, 250 (1969). – Abstr. AAS.

051.013 **Space notes.** R. N. Watts, Jr.
Sky Telescope, Vol. 38, 230 - 231 (1969).
Another successful OSO (*1969–68A*); Two Soviet moon probes (*Luna 15, Zond 7*); European satellite plans.

051.014 **Effects of secondary electron emission on electron trap measurements in the magnetosphere and solar wind.** E. C. Whipple, Jr., L. W. Parker.
Journ. Geophys. Res. Vol. 74, 5763 - 5774 (1969).
The theory developed earlier for the behavior of an electron trap mounted on a charged spacecraft is extended to include the contributions of secondary electrons emitted from the spacecraft surfaces.

051.015 **Space report.**
Spaceflight, Vol. 11, 424 - 431 (1969). – Men to Mars; Findings on the moon; Moon exploration crews; Age of the moon; Soviet weather satellites; Research on meteors; Evidence of gravity waves; Tektite discovery; Book of Mars.

051.016 **The Orbiting Geophysical Observatory (OGO).**
P. J. Parker.
Spaceflight, Vol. 11, 436 - 438 (1969).

051.017 **Cometary probes.** Rh. Lüst.
Space Sci. Rev. Vol. 10, 217 - 229 (1969).
The studies carried out in the U.S.A. and in Europe to investigate the possibilities and the scientific merit of a cometary probe are surveyed and reviewed. The scientific objectives of such a mission are given and the conditions are stated which a feasible comet must fulfill. Further, proposals of the different groups for the instrumentation of the probe are discussed. Though not all the groups propose the same comets for a first mission, due to different selection criteria, they agree that a mission to a 'new', non-periodic comet is out of consideration at present, and that a mission to a periodic comet, though possible in principle, presents various difficulties with respect to orbit accuracy and energy requirements. It is emphasized that a mission to comet Halley in 1986 would be of special value.

051.018 **Ein photographisch-photometrisches Verfahren zur Bestimmung der Periode der Helligkeitsände-rungen künstlicher Himmelskörper.**
Balekh Bishara Bagkhos, B. E. Tumanyan.
Uch. zap. Erevansk. un-t. Estestv. n. No. 1 (110), p. 85 - 91 (1969). In Russian. – Abstr. in Referativ. Zhurn. 62. Issled. kosm. prostranstv., 11.62.145 (1969).

051.019 **Report to the consultative group on the potentially harmful effects of space experiments from the Panel on Planetary Quarantine.**
Icarus, Vol. 11, 221 - 224 (1969).

051.020 **A penetration criterion for double-walled structures subject to meteoroid impact.**
J. P. D. Wilkinson.
AIAA Journ., Vol. 7, 1937 - 1943 (1969).
A penetration criterion is developed for double-walled

structures subject to hypervelocity impact. The penetration criterion is applied to the problem of calculating the optimum protection requirements for given spacecraft applications during interplanetary flight. Here, a Monte Carlo method is used to account for the observed distribution of meteoroid properties.

051.021   **The first men on the moon.**   V. V. Mikhajlov.
Zemlya i Vselennaya, No. 5, p. 4 - 5 (1969).
In Russian.

051.022   **Problems and achievements of cosmonautics.**
V. G. Fesenkov.
Zemlya i Vselennaya, No. 5, p. 6 - 14 (1969).   In Russian.

051.023   **The Pioneer 8 cosmic dust experiment.**
O. E. Berg, F. F. Richardson.
Rev. Sci. Instruments, Vol. 40, 1333 - 1337 (1969).

051.024   **A general relativity test using two or more solar satellites.**   J. G. Laframboise, M. Sachs.
Astronaut. Acta, Vol. 15, 65 - 66 (1969).
A method is proposed for placing artificial satellites in short period, highly eccentric solar orbits of arbitrary inclination to the sun's equator. This would permit accurate measurement of the sun's gravitational quadrupole moment and resolve present ambiguities in using Mercury's perihelion precession to test competing formulations of general relativity.

051.025   **La velocità di fuga.**   G. Mannino.
Coelum, Vol. 37, 253 - 255 (1969).

051.026   **The application of the gradient method of V. V. Kavraisky for the analysis of celestial fixes.**
A. L. Abramenko.
Trudy Astron. Obs. *Leningrad,* Vol. 26 (= Uchenye Zapiski Leningr. Un-ta No. 347 = Seriya Matem. Nauk No. 44), p. 140 - 143 (1969).   In Russian.
In accordance with the ideas of the method of V. V. Kavraisky a generalized solution of the problem of the correction of the coordinates of a space vehicle and the board time from celestial fixes is proposed. It is shown that the method can be used to make choice of the optimal version of the solution of the problem.

051.027   **Solar activity and supersonic flight.**
J. H. Reid.
Irish Astron. Journ., Vol. 9, 69 - 77 (1969).

051.028   **Künstliche Erdsatelliten und Raumsonden: Situationsbericht.**
Weltraumfahrt, Jahrgang 20, 110, 148, 184, 186 (1969).
1969 April 16 – October 31.

051.029   **United Kingdom space science in the 1970s.**
R. L. F. Boyd.
Proc. Roy. Soc. London, Ser. A, Vol. 308, (No. 1493), 145 - 156 (1968).
The current status of space science; space astronomy; astronomical satellites (stellar ultraviolet, cosmic X-ray astronomy, solar space astronomy).

051.030   **An introduction to the Ariel III satellite project.**
A. C. Ladd, J. F. Smith.

Proc. Roy. Soc. London, Ser. A, Vol. 311 (No. 1507), 479 - 487 (1969).

051.031   **De første resultater fra OAO (Orbiting Astronomical Observatory).**   P. E. Nissen.
Astron. Tidssk., Årg. 2, 190 - 191 (1969).

051.032   **A spacecraft-based navigation instrument for outer planet missions.**   T. C. Duxburg.
AIAA Paper 69–902, 8 pp. (1969).

051.033   **A simplified technique for estimating the navigation accuracy of interplanetary spacecraft.**
L. Kingsland, Jr., W. E. Bollman.
AIAA Paper 69–899, 7 pp. (1969).

051.034   **Results and aims of the international SPIN program.**
V. M. Grigorevsky.
Byull. StantsijOptichesk. Nablyud. Iskusstv. Sputnikov Zemli No. 54, p. 9 - 13 (1969).   In Russian.
Main problems, which can be solved with satellite photometry, and some data obtained with this method are presented.

051.035   **Measurements of the pressure and thermal environment in an aerobee 150 sounding rocket.**
R. C. Anderson, R. M. Fike.
Journ. Spacecraft and Rockets, Vol. 6, (No. 2), 214 - 215 = Contr. Kitt Peak National Obs. No. 392 (1969).
This note discusses a study of the heat transfer between an aerodynamically heated skin and the barrel of a telescope in a sounding rocket.

051.036   **The sixth Orbiting Solar Observatory.**
Journ. Astron. Soc. Victoria, Vol. 22, 100 - 101 (1969).

051.037   **Supplement to Data Catalog of satellite experiments (NSSDC 69-01).**
Data Catalog, National Space Science Data Center, NSSDC 69-17 (NASA, Goddard Space Flight Center, Greenbelt, Maryland), 25 pp. (1969).

051.038   **Novedades astronomicas algunas relaciones actuales entre la astronomia y la cosmonautica.**
C. J. Lavagnino.
Revista Astron., Vol. 41, (No. 169), 11 - 15 (1969).

051.039   **Cooperation of socialistic countries in astronautics.**
J. Kožešník.
Vesmír, Vol. 48, 355 (1969).   In Czech.

051.040   **Astronautics in the year 1968.**   J. Bouška.
Říše hvězd, Vol. 50, 121 - 129 (1969).   In Czech.

051.041   **Astronautics in the year 1969.**   P. Toufar.
Vesmír, Vol. 48, 356 - 359 (1969).   In Czech.

Apollo 11 success brings astronomy down to earth.
Phys. Today, Vol. 22, No. 9, p. 65, 67 (1969).

Calculated dose rates in Jupiter's van Allen belts.
See Abstr. 099.048.

Photoelectric photometry from a space vehicle.
See Abstr. 113.027.

052 Astrodynamics and Navigation of Space Vehicles

052.001 Perturbations of existing resonant satellites.
C. A. Wagner, B. C. Douglas.
Planet. Space Sci. Vol. 17, 1505 - 1517 (1969).
This paper presents an analysis of the resonant perturbations to the 20th degree and order on 83 existing satellites. This analysis reveals that terms of order (m) 2 and 12 - 14 are well represented by existing resonant satellites. Only a few resonances for the other orders are available.

052.002 On the energy integral for satellites.
A. Bjerhammar.
Tellus, Vol. 21, 1 - 9 (1969).
The energy integral can be used for studing the gravity field of the earth from satellite orbits. With known satellite velocities in the orbit we can compute the potential in an extremely simple way. Polar satellites give full coverage all over the earth and permit a high accuracy. Non-polar satellites require a small correction for the rotation of the earth.

052.003 Luni-solar perturbations of an earth satellite.
A. E. Roy.
Astrophys. Space Sci. Vol. 4, 375 - 386 (1969).
Luni-solar perturbations of the orbit of an artificial earth satellite are given by modifying the analytical theory of an artificial lunar satellite derived by the author in recent papers. Expressions for the first order changes, both secular and periodic, in the elements of the geocentric Keplerian orbit of the earth satellite are given, the moon's geocentric orbit, including solar perturbations in it, being found by using Brown's lunar theory. The effects of sun and moon on the satellite orbit are described to a high order of accuracy so that the theory may be used for distant earth satellites.

052.004 The differential correction of close-earth satellite orbits. Part II: The differential correction orbit programme (DCOP).
H. G. Walter, I. M. Wales, S. Pallaschke.
Sci. Rep. European Space Research Organization, ESRO SR-8 (ESOC), V + 53 pp. (1968).
Whilst part I of this report, "The differential correction of close-earth satellite orbits", ESRO SR-7 (ESDAC), deals with the theory and principles, Part II is intended as an operations manual for the differential correction orbit program (DCOP).

052.005 Determination of the perturbations from Jupiter and the sun in the motion of an artificial satellite of Jupiter's IV satellite (Kallisto). N. B. Batueva.
Astron. Tsirk. No. 510, p. 1 - 3 (1969). In Russian.

052.006 The motion of a charged satellite in the earth's magnetic field. L. Sehnal.
SAO Cambridge, Mass. Special Rep. No. 271, 3 + 15 pp. (1969).
The perturbations of the orbital elements of a charged artificial earth satellite caused by the earth's magnetic field are studied. A rough estimate of the size of the disturbing effects gives them a very small value. A detailed computation is made in the case of the changes of the inclination of the satellite's orbit.

052.007 Response to Garfinkel's comments on 'How critical is the critical inclination?'
A. G. Lubowe.
Celestial Mechanics, Vol. 1, 143 (1969). – See Astron. Astrophys. Abstr. Vol. 1, 052.020 and 052.021.

052.008 Tesseral resonance effects on satellite orbits.
G. S. Gedeon.
Celestial Mechanics, Vol. 1, 167 - 189 (1969).
Resonance effects on satellite orbits due to tesseral harmonics in the potential field have been studied by many authors. Most of these studies have been restricted to nearly circular 24-hour orbits and to the deep resonance regime, where there is exact commensurability between earth rotation and orbit period. Resonance effects have also been noted, however, on eccentric synchronous and subsynchronous orbits and on orbits with far from commensurate periods. These have received much less attention; the object of this paper is to study the whole spectrum of orbits with respect to resonance effects.

052.009 Motion of near-polar satellites.
V. S. Ural'skaya.
Vestn. Mosk. un-ta. Fiz., Astron., No. 2, p. 38 - 46 (1969). In Russian. – Abstr. in Referativ. Zhurn. 51. Astron., 8.51.99 (1969).

052.010 Perturbation of the orbital elements of a synchronous satellite moving in the earth's noncentral gravitational field. S. G. Zhuravlev.
Vestn. Mosk. un-ta. Fiz., Astron., No. 2, p. 105 - 110 (1969). In Russian.
The first-order secular and periodic perturbations of the Keplerian orbital elements are found for a triaxial earth. A numerical example is given.

052.011 The earth's shadowing effects in the short-periodic perturbations of satellite orbits.
P. Lála, L. Sehnal.
Bull. Astron. Inst. Czechoslovakia, Vol. 20, 327 - 330 (1969).
The authors investigate the short-periodic perturbations of the orbits of an artificial satellite, during one revolution of the satellite around the earth, caused by the direct solar radiation pressure. Special interest is paid to the study of the influence of the earth's shadow, which is considered as a special mathematical function in the equations for the perturbations. This semi-analytical theory is applied to the perturbations of the semimajor axis and eccentricity. A comparison is made with the values of perturbations obtained in a previous paper in which the effect of the shadow was neglected.

052.012 Die Drehung der Bahnebene eines Satelliten.
H. Scharn.
Zeitschr. Angew. Math. Mech. Vol. 48, 405 - 413 (1968). – See Phys. Ber., Vol. 48, No. 3 – 3358 (1969).

052.013 Säkulare Effekte der Bahnentwicklung unter dem Einfluß des Strahlungsdrucks. M. L. Lidov.
Kosmich. Issled. Vol. 7, 467 - 484 (1969). In Russian.

052.014 Geometrische Lösung für die Impulsübertragung zwischen nahen, fast kreisförmigen Bahnen.
E. I. Bushuev, A. A. Krasovskij.
Kosmich. Issled. Vol. 7, 485 - 489 (1969). In Russian.

052.015 Methodische Fehler bei vereinfachter Voraussage der Umlaufperiode eines künstlichen Erdsatelliten nach Bahnkorrektur.
I. V. Aleksakhin, R. V. Bodnarchuk, E. I. Bushuev, A. A. Krasovskij.
Kosmich. Issled. Vol. 7, 490 - 497 (1969). In Russian.

052.016 Bearbeitung von Reihenbeobachtungen zur Lösung

eines Problems der kosmischen Navigation.
V. M. Rudakov.
Kosmich. Issled. Vol. 7, 498 - 504 (1969). In Russian.

Verf. diskutiert die Lösung zur Bestimmung der Keplerschen Bahnparameter nach der Methode der größten Wahrscheinlichkeit für den Fall von Reihenbeobachtungen. Eine Ableitung der entsprechenden Gleichung und ihre Lösung werden gegeben. Das Problem wird aus Reihenmessungen des Winkels "Navigationsstern – Raumflugkörper – zentraler Planet" gelöst.

052.017 Bestimmung der Orientierung künstlicher Erdsatelliten aus einem vorgegebenen System von Messungen. V. V. Golubkov, I. G. Khatskevich.
Kosmich. Issled. Vol. 7, 510 - 521 (1969). In Russian.

052.018 Untersuchung der relativen Bewegung von Satelliten in verallgemeinerten Parametern.
V. B. Sokolov, O. I. Ivashchenko.
Kosmich. Issled. Vol. 7, 667 - 675 (1969). In Russian.

052.019 Die Bewegung eines Raumflugkörpers im normalen Gravitationsfeld der Erde unter der Einwirkung zusätzlicher Kräfte. V. K. Kajsin.
Kosmich. Issled. Vol. 7, 686 - 693 (1969). In Russian.

052.020 Long-range variations of orbits with arbitrary inclination and eccentricity. Y. Kozai.
Vistas in Astronomy, Vol. 11, 103 - 117 (1969).

Secular and long-periodic perturbations for satellites and asteroids with arbitrary inclination and eccentricity are discussed by analyzing the energy integral.

052.021 Astrodynamic peculiarities of the flight Earth–Moon–Earth. Yu. A. Ryabow.
Zemlya i Vselennaya, No. 4, p. 12 - 16 (1969). In Russian.

052.022 Optimal construction of the transfer ellipse using complete rotations in the equatorial plane of an axially symmetric planet. V. S. Novoselov.
Trudy Astron. Obs. *Leningrad,* Vol. 26 (= Uchenye Zapiski Leningr. Un-ta No. 347 = Seriya Matem. Nauk No. 44), p. 97 - 113 (1969). In Russian.

The optimal transfer ellipse is obtained. The eccentricities and given motions in the equatorial plane of the orbits of departure and arrival are taken into consideration.

052.023 Use of the moon's attraction to accelerate a spacecraft to hyperbolic flight.
H. Hiller.
Journ. British Interplanet. Soc., Vol. 22, 60 - 74 (1969).

A study has been made of the trajectories required for a spacecraft spiralling out to the vicinity of the moon under the action of microthrust and then utilising the moon's gravitational attraction to accelerate it to hyperbolic speed relative to the earth, i.e. into a heliocentric orbit.

052.024 The perturbed motion of a stationary artificial earth satellite during short time intervals.
M. A. Vashkov'yak.
Kosmich. Issled. Vol. 7, 841 - 851 (1969). In Russian.

052.025 A statistic method for the determination of the plane of an orbit. Yu. N. Zybin.
Kosmich. Issled. Vol. 7, 852 - 856 (1969). In Russian.

052.026 A survey of impulsive trajectories.
F. W. Gobetz, J. R. Doll.
AIAA Journ., Vol. 7, 801 - 834 (1969). – Review article.

052.027 Numerical analysis of the asymptotic solution for

earth-to-moon trajectories.
J. Kevorkian, G. Brachet.
AIAA Journ., Vol. 7, 885 - 889 (1969).

The theoretical results of a high-order asymptotic solution for the motion of a particle from earth to moon are summarized for the idealized case of the restricted three-body problem. Various definite integrals arising in the theory are evaluated, then used to calculate the elements during close passage to the moon for a set of 108 orbits in the actual earth-moon system. The initial conditions defining these 108 orbits then are used to integrate numerically the equations of motion for the trajectories from earth to moon. Prediction of the orbit during close passage to the moon is chosen as a basis for assessing the accuracy of the theoretical results, and it is shown that the largest errors are less than 4%.

052.028 Approximate finite-thrust trajectory optimization.
F. T. Johnson.
AIAA Journ., Vol. 7, 993 - 997 (1969).

A method which rapidly computes optimum finite-thrust space trajectories is presented. The approach is based on the approximation of the state time history by a polynomial. Differentiation and integration are performed in closed form and the problem is reduced to one of ordinary calculus. The technique is applied to constant power, variable-thrust trajectories, and results are presented for earth-Mars transfers. Apparently the method is an order of magnitude faster than standard techniques. The method is not restricted to low-thrust levels or to transfers between nearby orbits.

052.029 A new method for constructing periodic orbits in nonlinear dynamical systems.
A. Bennett, J. Palmore.
AIAA Journ., Vol. 7, 998 - 1002 (1969).

An iterative method for constructing periodic orbits in nonlinear dynamical systems is developed. The method is a modification of the Generalized Newton-Raphson technique. Application of the method to numerically continuing natural and isoperiodic families in the restricted three-body problem was implemented on a high-speed computer and the results show broad regions of convergence, lack of sensitivity, and strong convergence properties of the method.

052.030 Stability of periodic orbits in the elliptic, restricted three-body problem. R. Broucke.
AIAA Journ., Vol. 7, 1003 - 1009 (1969).

A systematic study has been made of periodic orbits in the two-dimensional, elliptic, restricted three-body problem. All ranges of eccentricities, from 0 to 1, and mass-ratios, from 0 to 1/2, have been investigated. Eleven hundred periodic orbits have been obtained. It is concluded that the elliptic problem behaves in a way which is completely different from the circular problem. The main difference is in the stability properties of the periodic orbits. The stability of the periodic orbits has been determined by numerically integrating the variational equations with a recurrent power series method. The results are in contrast with the circular problem. The elliptic, restricted three-body problem can be considered as the prototype of all nonintegrable, nonconservative Hamiltonian systems, and in this paper, probably for the first time, a classification of the multipliers is given for these systems.

052.031 Trajectory optimization using regularized variables.
B. D. Tapley, V. Szebehely, J. M. Lewallen.
AIAA Journ., Vol. 7, 1010 - 1017 (1969).

In this investigation, regularized equations for the optimal trajectory ot a space vehicle with continuous thrust are obtained. The computational characteristics of the regularized equations are compared with the characteristics of the unregularized equations using a perturbation type numerical optimization method. The comparison is made for a three-dimensional, low-thrust, Earth-Jupiter rendezvous transfer. The comparison indicates that, when the regularized equations are used, a significant reduction in computing time is obtained. Furthermore, for the values considered in this study, the convergence of the regularized equations is much less sensitive to errors in the guesses for the unknown boundary conditions.

052.032 Rapid analysis of moon-to-earth trajectories.
J. E. Lancaster, J. C. Walker, F. I. Mann.
AIAA Journ., Vol. 7, 1017 - 1023 (1969).

Analytical and numerical techniques are applied to transearth mission analysis to illustrate the effects of parameters defining the transearth trajectory. Approximate analytical solutions for transearth trajectories (previously derived by the method of matched asymptotic expansions) are extended to a mixed boundary-value problem. Relations are derived showing the dependence of the hyperbolic elements near the moon on specified earth-entry conditions, and on three arbitrary parameters. Only order-unity results are given explicitly, and the method for determining higher-order corrections is outlined.

052.033 Stability of the triangular points in the elliptic restricted problem of three bodies.
K. T. Alfriend, R. H. Rand.
AIAA Journ., Vol. 7, 1024 - 1028 (1969).

The two variable expansion method is used to study the stability of infinitesimal motions about the triangular libration points in the elliptic restricted problem of three bodies. This perturbation technique entails replacing the independent variable (here f, the true anomaly of the smaller primary) with two new independent variables. The results of the study are analytical expressions for the transition curves bounding regions of stability in the $\mu - e$ plane, accurate to $0(e^3)$. For small e, these expressions are seen to compare favorably with the numerical analysis of Danby.

052.034 Comparison of the Newtonian and general relativistic orbits of a point mass in an inverse-square law force field. H. Lass, C. B. Solloway.
AIAA Journ., Vol. 7, 1029 - 1031 (1969).

The averaging method of Krylov-Bogoliubov is applied to the motion of a particle that moves along a geodesic due to the Schwarzschild line element of general relativity theory. The approximate analytic solutions are compared with the solutions for the motion of a particle in a Newtonian force field of a point mass.

052.035 Nonlinear attitude motion near resonance.
D. L. Hitzl.
AIAA Journ., Vol. 7, 1039 - 1047 (1969).

The roll-yaw attitude motion of a spinning symmetric satellite in an elliptic orbit has been investigated with particular emphasis on the behavior near resonance. Resonance will occur when there is a low-order commensurability between the frequencies of the two normal modes of the attitude motion or between these attitude frequencies and the orbital mean motion. Assuming certain types of initial conditions, sufficient conditions for the interchange of energy to be small are derived. Finally, the determination of stable and unstable periodic attitude motions is outlined.

052.036 Bounds on the librations of parametrically resonant satellites. P. W. Likins, G. M. Wrout.
AIAA Journ., Vol. 7, 1134 - 1139 (1969).

The objective of this study is the determination and portrayal of the bounds of rotational motion of satellites previously shown to exhibit parametric resonance when performing small librations in the orbital plane. Attention is restricted to the behavior of rigid satellites in circular orbit in an inverse square field, ignoring translation/rotation coupling

and truncating the gravitational potential expansion in terms of the ratio (satellite dimension/orbital radius) to retain only first-approximation torque expressions.

052.037 Adaptive control for Mars entry based on sensitivity analysis. C. N. Shen, P. J. Cefola.
AIAA Journ., Vol. 7, 1145 - 1150 (1969).

The Mars entry guidance problem is complicated by the great uncertainty which exists in the Martian atmospheric parameters. This paper suggests a guidance scheme that will produce a given terminal condition whatever the atmosphere encountered on Mars entry. The approach is analytical. Sensitivity analysis is applied to the entry dynamics in order to compute the effects of both density parameter deviations and control changes. The required sensitivity coefficients are obtained by simultaneous numerical solution of the first-order sensitivity equations with the equations of motion.

052.038 Integrals for impulsive orbit transfer from Noether's theorem. H. G. Moyer.
AIAA Journ., Vol. 7, 1232 - 1235 (1969).

The problem considered is the minimum impulse transfer of a vehicle between noncoplanar Kepler orbits. The transfer time, the number of impulses, and the points of departure and arrival on the terminal orbits are all unspecified. The objectives are to derive constants of the motion and to discuss their properties. These integrals are composed of orbital elements and the Lagrange multipliers that are used in the optimal control treatment.

052.039 Recurrent nature of Lagrange multipliers for optimal low-thrust Earth-Jupiter transfers.
J. D. Hart, W. T. Fowler, J. M. Lewallen.
AIAA Journ., Vol. 7, 1357 - 1358 (1969).

In the study reported here, the initial Lagrange multipliers and mission durations for a series of low-thrust Earth-Jupiter transfers are determined and examined for repetitive behavior. A three-dimensional solar system model is used.

052.040 Determination of orbits of planetary artificial satellites and planetary gravitational fields.
C. C. H. Tang, C. L. Greer.
AIAA Journ., Vol. 7, 1469 - 1476 (1969).

In this paper, the planetary satellite velocity component, in the direction from the planet center to earth center instead of that from the satellite to an observation station, is computed in a mathematical model. The least-squares estimation criterion obtained this way is relatively simple for the case of a planet assumed at infinite distance from the center of a non-rotating earth. The resulting simplification in the mathematical model facilitates comparison studies of numerical methods. For the actual case of finite distance between the earth and a planet, the least-squares estimation criterion thus obtained is slightly different from that of the conventional approach.

052.041 Vector-matrix second-order sensitivity equation with application to Mars entry.
P. J. Cefola, C. N. Shen.
AIAA Journ., Vol. 7, 1633 - 1635 (1969).

052.042 Genauigkeitsbestimmung der Bahnelemente und Positionen von Satelliten in Abhängigkeit von den Beobachtungen. H. G. Walter.
Ortung und Navigation I/69, p. 1 - 7 (1969).

052.043 Outline of a general orbit determination method.
M. Schneider.
Space Research IX, Proc. Tokyo 1968, p. 37 - 40 (1969).

A general orbit determination method allowing the determination of the earth's gravitational field and correction of station coordinates is discussed. No orbital theory and no reduction of satellite observations due to nongravitational terms of the acting force is necessary. Furthermore there is no major indeterminateness in the proposed method.

052.044   First-order theory of orbital transfer for geodetic satellite missions.   F. M. Calabria, A. Vallone. Space Research IX, Proc. Tokyo 1968, p. 41 - 48 (1969).

Integrationstheorie von Krylow-Bogoljubow und gestörte Keplerbewegung.   See Abstr. 021.009.

053 Lunar and Planetary Probes and Satellites

053.001 **The first men on the moon.** R. Hillenbrand.
Sky Telescope, Vol. 38, 144 - 149 (1969).

053.002 **Apollo 10 – The last rehearsal.**
P. J. Parker.
Spaceflight, Vol. 11, 275 - 278, 290 (1969).

053.003 **Apollo and beyond.** G. E. Mueller.
Spaceflight, Vol. 11, 298 - 303 (1969).
The author, one of the principal architects of the Apollo programme, looked forward to the Apollo 11 mission and surveyed future prospects.

053.004 **Man on the moon.** P. J. Parker.
Spaceflight, Vol. 11, 313 - 317, 338 - 341 (1969).

053.005 **Preliminary results from the Soviet Venus probes.**
H. Miles.
Journ. British Astron. Ass. Vol. 79, 494 - 495 (1969).
– Artificial Satellite Section report British Astron. Ass.

053.006 **Apollo 11. Die Landung auf dem Mond.**
W. Büdeler.
Weltraumfahrt, 20. Jahrgang, p. 125 - 135 (1969).

053.007 **Russia's moon programme.**
P. J. Parker.
Spaceflight, Vol. 11, 378 - 383, 395 (1969). – Review on Russian lunar probe launchings from Luna 1 to Zond 7.

053.008 **Objects on the moon–1.**
Compiled by G. Falworth.
Spaceflight, Vol. 11, 384 - 385 (1969).

053.009 **The Apollo spacecraft: Guidance and navigation.**
D. Baker.
Spaceflight, Vol. 11, 386 - 389 (1969).

053.010 **Apollo 11 – Die ersten Menschen auf dem Mond.**
N. Hasler-Gloor.
Orion, Band 14, 115 - 118 (1969).

053.011 **The 1969 missions to Mars.** R. J. Fryer.
Journ. British Interplanet. Soc. Vol. 22, 212 - 222 (1969).
This paper describes the principal design features of the two 1969 Mariner (Mariner 6 and 7) probes to Mars and the experimental observations which it is hoped to undertake.

053.012 **The use of Brown's lunar theory in lunar satellite perturbations by sun and earth.** A. E. Roy.
The Moon, Vol. 1, 143 (1969). – Abstract.

053.013 **Apollo lunar module engine exhaust products.**
B. R. Simoneit, A. L. Burlingame, D. A. Flory,
I. D. Smith.
Science, Vol. 166, 733 - 738 (1969).
The return of lunar samples by the Apollo lunar landing missions offers an opportunity for the study of extraterrestrial material free of the ambiguity surrounding meteorite analysis caused by unknown contamination histories. The nature of the Apollo program is such, however, that it will be impossible during early missions to return lunar samples that are completely free of significant amounts of contamination.

053.014 **The influence of Jupiter's oblateness on the motion of an artificial satellite of one of the Jupiter's**

Galilean satellites. N. B. Batueva.
Vestn. Leningr. un–ta, No. 7, p. 133 - 144 (1969). In Russian. – Abstr. in Referativ. Zhurn. 51. Astron., 9.51.102; 62. Issled. kosm. prostranstv., 9.62.188 (1969).

053.015 **Surveyor 7 lunar mission.**
L. D. Jaffe, R. H. Steinbacher.
Journ. Geophys. Res. Vol. 74, 6702 - 6705 (1969).
The spacecraft Surveyor 7 landed on the outer rim of the crater Tycho on January 10, 1968. Presented is a brief report on the lunar mission.

053.016 **Apollo 11: Les premiers hommes sur la lune.**
G. Bodifee, J. Meeus.
Ciel et Terre, Vol. 85, 376 - 391 (1969).

053.017 **The flight of the Soviet automatic station "Zond 7".**
Zemlya i Vselennaya, No. 6, p. 6 - 7 (1969).
In Russian.

053.018 **Optical tracking of Apollo 12.**
W. G. Grimwood.
Monthly Notes Astron. Soc. Southern Africa, Vol. 28, 134 (1969).

053.019 **Apollo 12. Die zweite Landung auf dem Mond.**
W. Büdeler.
Weltraumfahrt, Jahrgang 20, 170 - 175 (1969).

053.020 **Die Ergebnisse der sowjetischen Venussonden V und VI.** H. Zimmer.
Weltraumfahrt, Jahrgang 20, 178 - 180 (1969). – News notes.

053.021 **Nya rymdsondobservationer av planeterna Venus och Mars.** Å. Wallenquist.
Astron. Tidsskr., Årg. 2, 141 - 150 (1969).
Popular article on planetary probes.

053.022 **Mariner Mars 1969 flight design and mission analysis.** C. E. Kohlhase.
Journ. Spacecraft and Rockets, Vol. 6, 537 - 544 (1969).

053.023 **A new method for optical tracking of space probes.**
P. P. Dobronravin, V. M. Mojzerin, V. K. Prokofiev,
N. S. Chernykh.
Space Research IX, Proc. Tokyo 1968, p. 1 - 3 (1969).
A new method has been developed for optical position determination of distant space probes using a TV system attached to the 2.6 meter reflecting telescope of the Crimean Astrophysical Observatory. Observations of the space probes Luna 11, Luna 12, Luna 13 and Luna 14 as well as the satellite Molnija 1 have shown that this method allows highly accurate positions of the probes to be obtained very quickly.

053.024 **The Lunar Orbiter program.**
W. E. Brunk.
Space Research IX, Proc. Tokyo 1968, p. 625 - 656 (1969).

053.025 **Lunar Explorer 35.** N. F. Ness.
Space Research IX, Proc. Tokyo 1968, p. 678 - 703 (1969).

053.026 **1969 Mariner Mars launch.** M. Grün.
Říše hvězd, Vol. 50, 171 - 175 (1969). In Czech.

053.027 **Apollo 11 and the Soviet lunar programme.**
A. Blagonravov.

Spaceflight, Vol. 11, 414 - 416 (1969).

053.028 **The scientific program of the Pioneer lunar expedi-**
tion.   S. R. Brzostkiewicz.
Urania Kraków, Vol. 40, 268 - 274 (1969).   In Polish.

Zeitplan für den Flug Apollo 11. (16. bis 24. Juli
1969).

Weltraumfahrt, Jahrgang 20, 93 - 95 (1969).

The first four Lunar Orbiter photographic missions.
See Abstr. 094.064.

Radar observations of Mars.
See Abstr. 097.051.

054 Artificial Earth Satellites

054.001 **Das Astronomische Satelliten-Observatorium OAO-2.** D. Lemke.
SuW, Vol. 8, 199 - 202 (1969).

054.002 **The latest biosatellite.** R. N. Watts, Jr.
Sky Telescope, Vol. 38, 80 - 81 (1969). – Concerning Bios 3 (=1969-56A).

054.003 **Last of the OGO's.** P. J. Parker.
Spaceflight, Vol. 11, 363 - 365 (1969).
Concerning research program of OGO 6 = 1969-51A.

054.004 **HEOS-1: Scientific aims and experiments.** B. G. Taylor.
ESRO/ELDO Bull. Suppl. August, p. 14 - 25 (1969).

054.005 **The HEOS-1 satellite.** G. H. Booth.
ESRO/ELDO Bull. Suppl. August, p. 26 - 30 (1969).

054.006 **La chute de la fusée de Cosmos 253 (1968-102B).** R. Futaully.
L'Astronomie, 83ᵉ année, 366 - 368 (1969).

054.007 **A determination of the orbit of Secor 6 rocket (1966 – 51A).** D. W. Scott.
Journ. British Interplanet. Soc. Vol. 22, 123 - 140 (1969).
36 sets of orbital parameters for 1966-51A have been obtained using optical and radar observations made between June and October 1966. The results should be useful in studies of the density and rotational speed of the upper atmosphere.

054.008 **The ESRO Large Astronomical Satellite (LAS) project. – The observatory in orbit.** D. Marsh.
Journ. British Interplanet. Soc. Vol. 22, 189 - 201 (1969).
The present paper describes the operation of the complete observatory in orbit, and the design definition of the spacecraft and its interfaces with the scientific package.

054.009 **Ground operations for the Large Astronomical Satellite (LAS) project.** G. L. Reijns.
Journ. British Interplanet. Soc. Vol. 22, 202 - 211 (1969).
The paper discusses the ground operations and hardware aspects of the ground data system, which comprises a 1700 MHz telemetry, a PCM telecommand system and ground processing and display equipment.

054.010 **A (relatively) low altitude 24-hour satellite.** A. R. Collar, J. W. Flower.
Journ. British Interplanet. Soc. Vol. 22, 442 - 457 (1969).
A twin-satellite system is suggested whereby an outer satellite, situated beyond the 24-hour orbit radius, is joined by an enormously long, light, cord to an inner satellite within the 24-hour orbit, the whole system being arranged to have a period of 24-hours.

054.011 **Satellite digest.** G. Falworth.
Spaceflight, Vol. 11, 238, 279 - 280, 321 - 322, 323, 358, 359, 397 - 399, 432 (1969). – Listing of all known artificial earth satellites on a month-by-month basis.

054.012 **The determination of the approximate orbit of an artificial earth satellite using little separated optical observations.** V. V. Gavrilov.
Trudy Astron. Obs. *Leningrad,* Vol. 26 (= Uchenye Zapiski Leningr. Un-ta No. 347 = Seriya Matem. Nauk No. 44), p. 114 - 124 (1969). In Russian.

The problem of determination of the approximate orbit of an earth satellite from little separated optical observations at two known stations is considered. This method is applied mainly in the case when during a small time interval many observations were made at one station, while at the other only one observation was made.

054.013 **Analyse de quelques termes correctifs appliqués à réduction des observations photographiques de satellites artificiels de la Terre.** J. Łatka.
Geodezja Kartografia, Vol. 18, 135 - 144 (1969). In Polish.

054.014 **Détermination des coordonnées topocentriques sphériques des satellites artificiels de la Terre à la base d'observations photographiques.** S. Domaradzki.
Geodezja Kartografia, Vol. 18, 283 - 298 (1969). In Polish.
Cet article décrit une méthode de réduction des observations satellitaires programmées pour la machine à calculer électronique GIER.

054.015 **Azur 1 (1969-97A), Deutschlands erster künstlicher Erdsatellit.**
Weltraumfahrt, Jahrgang 20, 165 - 169 (1969).
Report on the successful launching of the first German artificial earth satellite. Orbital elements are given and the research projects described.

054.016 **Use of pulsed laser tracking system for tracking satellites.**
J. L. Hughes, O. Randva, B. W. Sweeney, P. C. Stedman, L. D. T. Moyle.
Journ. Sci. Instruments, Ser. 2, Vol. E2, 456 (1969).

054.017 **Aussichten der Nutzung von Satelliten für See– und Luftfahrt.** H. C. Freiesleben.
Der Seewart, Vol. 30, 225 - 229 (1969).

054.018 **The effect of solar-radiation pressure on determination of the semimajor axis in satellite-orbit computation.** J. W. Slowey.
Space Research IX, Proc. Tokyo 1968, p. 76 - 82 (1969).

054.019 **Possibilities for obtaining orbital elements from observations of satellites during small time intervals.**
I. D. Zhongolovitch.
Byull. Stantsij Optichesk. Nablyud. Iskusstv. Sputnikov Zemli No. 54, p. 14 - 18 (1969). In Russian.
Tracking data obtained from reduction of simultaneous observations of the satellite Midas 4 are used to show the possibility of determining osculating orbital elements of artificial satellites from observations carried out during a short period of time.

054.020 **A device for electrophotometric observations of artificial satellites entering the earth's shadow.**
P. N. Boiko, V. S. Matiagin.
Byull. Stantsij Optichesk. Nablyud. Iskusstv. Sputnikov Zemli No. 54, p. 19 (1969). In Russian.

054.021 **Data on the deceleration of the satellite 1965-11-4 obtained from simultaneous observations based on the INTEROBS and SPIN programs.**
V. M. Grigorevsky, T. V. Kassimenko, I. M. Panich, V. A. Vorobjeva.
Byull. Stantsij Optichesk. Nablyud. Iskusstv. Sputnikov Zemli No. 54, p. 30 (1969). In Russian.

054.022 **Influence of the magnetic field of the earth on the motion of artificial satellites.**
G. M. Shmelev, V. M. Grigorevsky, N. S. Shmeleva.
Byull. Stantsij Optichesk. Nablyud. Iskusstv. Sputnikov Zemli No. 54, p. 35 - 38 (1969).   In Russian.

054.023 **Application of the vector formula for the great circle in space triangulation.**
A. A. Kisselev.
Byull. Stantsij Optichesk. Nablyud. Iskusstv. Sputnikov Zemli No. 55, p. 86 - 91 (1969).   In Russian.

054.024 **Construction of space triangulation nets on the basis of directions and distances.**
B. M. Klenitsky.
Byull. Stantsij Optichesk. Nablyud. Iskusstv. Sputnikov Zemli No. 55, p. 92 - 99 (1969).   In Russian.

054.025 **Use of satellite data with time errors in space geodesy.**   Yu. V. Batrakov.
Byull. Stantsij Optichesk. Nablyud. Iskusstv. Sputnikov Zemli No. 55, p. 99 - 106 (1969).   In Russian.

054.026 **Artificial satellites and economical advantages.**
S. Grzędzielski.
Urania Kraków, Vol. 40, 197 - 200, 234 - 237, 274 - 277 (1969).   In Polish.

054.027 **Some second order inequalities in the motion of distant artificial earth satellites.**
V. P. Dolgachev.
Vestn. Mosk. un-ta. Fiz., Astron., No. 5, p. 74 - 80 (1969). In Russian. – Abstr. in Referativ. Zhurn. 62. Issled. kosm. prostranstv., 2.62.286 (1970).

054.028 **Kunstmanen.**   J. Meeus.
     Hemel en Dampkring, Vol. 67, 283 - 285 (1969).
1969 January – April.

ESRO–Ib.
European Space Research Organization Separate print, 31 pp. Neuilly/Seine (1969).
     Satellite ESRO–1B (= 1969-83A), launched 1969 Oct. 1, carries 8 experiments to study ionospheric and auroral phenomena particularly over the northern polar regions in darkness in winter. The scientific missions, the satellite, and the tracking, control and data handling are described.

Erstes deutsches Satelliten-Experiment erfolgreich.
Umschau, Vol. 69, 591 (1969).
– Concerning HEOS-A-1 (= 1968 - 109 A).

055 Observations of Earth Satellites, Lunar and Planetary Probes

055.001 Data Catalog of satellite and rocket experiments.
National Space Science Data Center, Data Catalog, NSSDC 69-01, NASA, Goddard Space Flight Center, Greenbelt, Md. VIII + 171 pp. (1969).

The purpose of this Data Catalog of Satellite and Rocket Experiments is to announce the availability of reduced experimental space science data, to describe these data, and to inform potential data users of the services provided by the National Space Science Data Center (NSSDC). This catalog has previously been published semiannually, in January and July. The volume of data is now such that the Data Center is considering either an annual or biennial edition with supplements as necessary.

055.002 Flow and use of information at the National Space Science Data Center.
N. Karlow, J. I. Vette.
National Space Science Data Center, NSSDC 69-02, NASA, Goddard Space Flight Center, Greenbelt, Md. VI + 41 pp. (1969).

An integrative, overall view of the flow and use of information at the National Space Science Data Center (NSSDC) is presented by examining the tasks facing the NSSDC, present resources, and the flow of information. The information system used to handle this flow is then discussed in terms of its four main subsystems: The Automated Internal Management file, the Machine-Oriented Data System, the Technical Reference File, and the Request Accounting Status and History file.

055.003 An information retrieval system for photographic data. B. I. Blum.
National Space Science Data Center, NSSDC 69-09, NASA, Goddard Space Flight Center, Greenbelt, Md. V + 15 pp. (1969).

An introduction of the Extra-terrestrial Photographic Information Center (EPIC) maintained at the National Space Science Data Center (NSSDC) is presented. The keystone of EPIC is an information retrieval system for the identification of extra-terrestrial photographs resulting from satellites and manned space explorations. The concepts and operation of EPIC are described. An appendix contains a description of the contents of a file devoted to information about Lunar Orbiter photographs.

055.004 La station de guidage radioélectrique ELDO de Gove. R. Alexis.
ESRO/ELDO Bull. No. 6, p. 4 - 8 (1969).

Le système de guidage est basé sur l'emploi d'une station sol qui détermine la trajectoire du véhicule pendant le vol propulsé du 3ème étage (c'est la fonction localisation), puis calcule les ordres de guidage pour les lui transmettre (c'est la fonction guidage).

055.005 Daytime ground-satellite laser ranging experiments.
R. L. Iliff, G. Hadgigeorge.
Applied Optics, Vol. 8, 1742 - 1743 (1969).

055.006 Observations faites à Strasbourg, (11.11.1961 - 31.05.1964: Echo et Divers).
P. Muller.
Publ. Obs. Paris, Satellites Artificiels, [Centre National d'Etudes Spatiales], Fasc. 22, 129 pp. (1969).

The observations include 4031 measurements obtained during 511 transits.

055.007 Observations faites à Strasbourg, (1.06.1964 -
31.12.1965: Echo et Divers).
P. Muller.
Publ. Obs. Paris, Satellites Artificiels, [Centre National d'Etudes Spatiales], Fasc. 23, 129 pp. (1969).

The observations include 3072 measurements obtained during 740 transits.

055.008 Observations faites à Besançon, 1 juillet 1963 - 31 décembre 1963: Divers, 1 janvier 1964 - 14 avril 1964: Echo et Divers. P. Muller.
Publ. Obs. Paris, Satellites Artificiels, [Centre National d'Etudes Spatiales], Fasc. 24, 122 pp. (1969).

The observations include 3977 measurements obtained during 215 transits.

055.009 Photometry from Apollo tracking.
C. A. Lundquist.
The Moon, Vol. 1, 143 - 144 (1969). – Abstract.

055.010 Photographische Beobachtung künstlicher Erdsatelliten ohne Hilfe registrierender Zeiteinrichtungen.
R. Rajchl.
Bull. Astron. Inst. Czechoslovakia, Vol. 20, 331 - 343 (1969).

Es wird eine neue Methode der photographischen Beobachtung künstlicher Erdsatelliten beschrieben. Sie setzt sich zum Ziel, die hier vorkommende doppelte Problematik – die der Positions- und der Zeitbestimmung – mit Hilfe eines Indikators, nämlich des Negativs, zu lösen. Zu diesem Zweck braucht man zwei Negative, die im gleichen Augenblick mit zwei achsenparallelen Kameras gleicher optischer Parameter aufgenommen werden, wobei die Verschlüsse mit einer gewissen Phasenverschiebung arbeiten: Wenn sich eine Kamera öffnet, schließt sich die andere. Die theoretischen Erwägungen stützen sich auf Differenzen zwischen den aufeinanderfolgenden Endpunkten der unterbrochenen Spur.

055.011 Influence of the magnitude of the evaluated region, of the objective distortion and of the transformation methods in evaluation of a satellite photograph.
J. Kabeláč.
Bull. Astron. Inst. Czechoslovakia, Vol. 20, 344 - 349 (1969).

Photographic cameras of different types are being used for optical observations of artificial satellites of the earth. The interrupted path of a satellite and of stars are taken on either photographic plates or films. By means of the latter, equatorial coordinates of unknown points of the satellite should be determined by using known reference stars on the picture.

055.012 La rifrazione parallattica dei satelliti terrestri.
A. Vassallo.
Mem. Soc. Astron. Italiana, Nuova Serie, Vol. 40, 585 - 587 (1969). – Letter.

055.013 Berechnung einer Gewichtskoeffizientenmatrix für die Zwecke der europäischen Satellitentriangulation. M. Näbauer.
Sitzungsber. Bayer. Akad. Wiss. Math.-Nat. Kl., Jahrgang 1968, p. 27*, 141 - 155 (1969).

055.014 Simultaneous tracking of the Pageos satellite with small cameras placed at large distances.
A. G. Massevitch, S. K. Tatevian, N. N. Kovalenko.
Space Research IX, Proc. Tokyo 1968, p. 6 - 14 (1969).

Results of the reduction of simultaneous observations of the balloon-satellite Pageos obtained by stations in the USSR, East Europe and Africa during cooperative observations in autumn 1966 are discussed.

055.015 Premières réductions de l'experience française sur satellite D 1 A.
J. Kovalevsky, F. Barlier, I. Stellmacher.
Space Research IX, Proc. Tokyo 1968, p. 29 - 36 (1969).

Toutes les observations Doppler du satellite D 1 A lancé en 1966 ont été exploitées dans plusieurs conditions différentes. On a étudié l'effet du groupement des observations sur la détermination de la position de la station du Liban par rapport à celle de Nice.

055.016 Dispersion des positions des éclairs de GEOS; étude des causes. P. Muller.
Space Research IX, Proc. Tokyo 1968, p. 49 - 52 (1969).

La réduction d'un cliché de satellite fournit d'abord un critère de la précision sur la position d'une étoile inconnue, laquelle ne supporte que les erreurs des catalogues et celles des mesures faites sur la plaque. Les positions du satellite comportent en outre l'effet des écarts sur les temps, de ceux du guidage, enfin de la turbulence atmosphérique. Ces divers effets peuvent être étudiés et, dans une certaine mesure, séparés grâce à des lissages des positions successives sur un même cliché en fonction du temps.

055.017 L'utilisation du «cercle de simultanéité» dans la triangulation cosmique. C. Popovici.
Stud. Cerc. Astron., Vol. 14, 87 - 90 (1969).

L'auteur expose une méthode d'utilisation du «cercle de simultanéité» approché pour le calcul de la direction dans l'espace de la droite joignant deux stations d'observation de satellites en connaissant le temps à ces deux stations. Il donne de même un exemple numérique de la méthode proposée.

055.018 Preliminary determination of the spatial direction Riga–Cairo using simultaneous observations of the satellite Pageos A. H. Alexandrescu.
Stud. Cerc. Astron., Vol. 14, 193 - 196 (1969).
In Rumanian.

The author uses 14 pairs of simultaneous photographic observations of the satellite Pageos A for the determination of the space direction Riga–Cairo. The method used was that of the "simultaneity circle".

055.019 Horizontal coordinates (August 1966).
Rezul'taty Nablyud. Sovet. Iskusstv. Sputnikov Zemli No. 110, 55 pp. (1968). In Russian. – Concerning Cosmos 54 rocket = 1965-11 D.

055.020 Horizontal coordinates (September – December 1966).
Rezul'taty Nablyud. Sovet. Iskusstv. Sputnikov Zemli No. 111, 46 pp. (1968). In Russian. – Concerning Cosmos 54 rocket = 1965-11 D.

055.021 Results of reduction of data obtained from observations based on the SPIN program and some problems referred to observational techniques.
I. M. Panich, V. A. Vorobjeva, V. M. Grigorevsky.
Byull. Stantsij Optichesk. Nablyud. Iskusstv. Sputnikov Zemli No. 54, p. 40 - 46 (1969). In Russian.

055.022 On the precision of data obtained from photographic tracking of artificial satellites.
A. M. Losinsky.
Byull. Stantsij Optichesk. Nablyud. Iskusstv. Sputnikov Zemli No. 55, p. 9 - 11 (1969). In Russian.

055.023 Requirements to the accuracy of optical observations of artificial satellites for geodetic purposes.
Yu. V. Batrakov.
Byull. Stantsij Optichesk. Nablyud. Iskusstv. Sputnikov Zemli No. 55, p. 11 - 18 (1969). In Russian.

055.024 Apollo 12.
J. Dommanget, H. Debehogne, G. Roland.
IAU Circ. No. 2194 (1969).

Theoretical Astrophysics

061 General Theoretical Problems of Astrophysics, Gravitational Instability, Neutrino Astronomy, X Ray- and Gamma Ray-Astronomy, Frequency and Origin of Elements, etc.

061.001 Contrastreaming instability in gravitating fluids with thermal effects.
M. Aggarwal, S. P. Talwar.
Publ. Astron. Soc. Japan, Vol. 21, 176 - 184 (1969).

The problem of stability of contrastreaming self-gravitating streams is investigated taking account of the thermal conduction and radiative effects. Conditions of stability are derived. It is found that thermal conductivity has a destabilising influence in that it shortens the range of stable wavelengths of perturbations and increases the growth rates of the monotonically unstable perturbations for slow interpenetration speeds. Further a finite heat conductivity enlarges the range and the growth rates of the overstable wave numbers for fast interpenetration speeds. The combined effect of thermal conductivity and a heat-loss function is also discussed.

061.002 What next with solar neutrinos ?
J. N. Bahcall.
Phys. Rev. Letters, Vol. 23, 251 - 254 (1969).

The capture rate of solar neutrinos is estimated for a number of targets that have been suggested previously as possible detectors of solar neutrinos. It is shown that the most important feasible experiment to be carried out in the future employs ^{7}Li as a detector.

061.003 Weak interaction of photons and Photocoulomb neutrinos. P. R. Chaudhuri.
Progr. Theor. Phys. Japan, Vol. 42, 116 - 120 (1969).

Taking into account that photons can interact weakly with neutrinos, the cross section for the Photocoulomb neutrinos $Z + \gamma \rightarrow Z + \nu + \bar{\nu}$ has been calculated. From this, the rate of stellar energy loss due to neutrino pair emission has been calculated and compared with the result obtained on the basis of the current-current coupling theory.

061.004 On gravitational instability of the interstellar gas. B. Bertotti, A. Cavaliere.
Astrophys. Space Sci. Vol. 5, 78 - 91 (1969).

In the galaxy, Jean's critical length for the interstellar gas is appreciably smaller than the critical length for the stars, a necessary condition for the gravitational instability of the former to have a local character. An accurate discussion of the orders of magnitude involved leads to the establishment of a well defined limiting procedure and to simplified equations in which the effect of stars occurs only through the equilibrium, but disappears from the perturbations. The equations are spatially local, but their coefficients are time-dependent, in that they describe the evolution of a small wave packet dragged along by the supersonic gas motion. They have been solved in several interesting cases by the introduction of an effective, time-dependent wave vector, which describes the deformation of a wave profile due to the velocity gradients. The ordinary Jeans' instability is recovered only when the velocity gradient is a skew tensor; otherwise we find a stabilizing effect in accelerated and sheared flows, a destabilizing effect in a decelerated flow. Possible connections of this mo-

del with the observed turbulent structure of the interstellar gas are discussed.

061.005 Nuclear chronologies for the galaxy.
G. J. Wasserburg, D. N. Schramm, J. C. Huneke.
Astrophys. Journ. *(Letters)*, Vol. 157, L91 - L96 (1969).

The ratios U^{235}/U^{238}, Th^{232}/U^{238}, Pu^{244}/U^{238}, and I^{129}/I^{127} have been used to obtain the time evolution of r-process nuclei. Using $Pu^{244}/U^{238} = \frac{1}{30}$, the solutions all have a large amount of initial production, with a duration of from 0 to 10^{10} years, followed by a relatively quiescent period ($\sim 3 \times 10^9$ years) terminated by a nucleosynthetic event that possibly initiated the separation of the solar system. For values of Pu^{244} less than $\frac{1}{60}$, models of uniform synthesis terminated by a sharp nucleosynthetic event are possible.

061.006 Role of the creation e in the stellar nucleosynthesis. A. A. Joukoff.
Astron. Astrophys. Vol. 3, 186 - 196 (1969). In French.

We develop the bound-state beta decay theory (β^- emission towards a bound electronic state of the formed atom), for any nuclear transition, based on the electron capture theory. We calculate the relative rates of bound-state beta decay compared with ordinary β^- emission for allowed transitions, first forbidden in parity and first unique forbidden transitions which generally are predominant in astrophysical applications. We also study the synthesis of 8 elements rich in protons (Kr^{80}, Mo^{94}, Cd^{108}, Te^{120}, Xe^{126}, Ba^{132}, Gd^{152} and Er^{164}) by branchings of the s capture chains of neutrons at the stage of carbon burning.

061.007 Astrophysik.
V. A. Ambartsumyan, L. V. Mirsoyan.
Otdel'nyj ottisk iz sbornika "Akademiya nauk Armyanskoj SSR za 25 let". Izdatel'stvo AN Arm. SSR. Erevan, 1968, p. 19 - 42 = Byurakan Astrophys. Obs. Repr. No. 32 (1968). In Russian.

061.008 Concerning the origin of superheavy elements.
E. E. Berlovich, Yu. N. Novikov.
JETP Letters, Vol. 9, 445 - 448 (1969). [Translated from ZhETF Pis. Red. 9, No. 12, p. 708 - 712 (1969). In Russian.]

Paths of r-process for heavy nuclei are calculated at various values of the neutron binding energy, and fission barriers of nuclei are indicated along the r-process path.

061.009 Lithium and beryllium in stars.
G. Wallerstein, P. S. Conti.
Annual Rev. Astron. Astrophys. Vol. 7, 99 - 120 (1969).

061.010 Implications of the far-infrared background measurements. R. V. Wagoner.
Nature, Vol. 224, 481 - 484 (1969).

Some consequences of far-infrared background radiation originating beyond the solar system are derived. The most likely source, narrow emission features produced by our galaxy, can also lead to a flux from all other galaxies similar to that of the 2.7 K microwave background.

061.011 Direct production of ^{56}Fe in silicon quasi-equilibria and the problem of ^{58}Ni.
D. D. Clayton, S. E. Woosley.
Astrophys. Journ. Vol. 157, 1381 - 1389 (1969).

We consider two questions of importance to the science of nucleosynthesis: (1) Can ^{56}Fe, rather than ^{56}Ni, have been synthesized in its observed ratio to ^{28}Si within a silicon-burning quasi-equilibrium? (2) What accounts for the large abundance of ^{58}Ni?

061.012 The instability of the congruent Darwin ellipsoids.
S. Chandrasekhar.
Astrophys. Journ. Vol. 157, 1419 - 1434 (1969).

In this paper a class of synchronous coupled oscillations of the congruent Darwin ellipsoids is considered; and it is shown that two of the five modes of oscillation belonging to this class excite instabilities along the entire Darwin sequence.

061.013 Focusing of the emission by the gravitational field.
A. V. Bjalko.
Astron. Zhurn. Akad. Nauk SSSR, Vol. 46, 998 - 1002 (1969). In Russian. English translation in Soviet Astron. AJ, Vol. 13, No. 5.

A focusing of the light near gravitational bodies is investigated. Intensity distributions of the emission, occurring near a spherical-symmetric body in the approximation of geometric and wave optics are obtained. The possibility of the observation of the effect is discussed. It is shown that observations of a distant radio source through the galactic center are most favourable for discovery.

061.014 The influence of a poloidal magnetic field on convection in spherical shells.
D. L. Moss, R. J. Tayler.
Monthly Notices, Roy. Astron. Soc., Vol. 145, 217 - 240 (1969).

The influence of a poloidal magnetic field on the onset of convection in spherical shells of a compressible fluid is investigated. It is shown that, provided the change in the gravitational field produced by the convective motions can be neglected, a separate stability criterion can be found on each field line. The changes of the gravitational field are a destabilizing influence but they are likely to be unimportant if both the layer thickness and the pressure scale height are small compared to the local gravitational Jeans' length. It is pointed out that if linear theory predicts that one part of the shell is stable whilst another part is unstable, this property may not persist when fully developed convection occurs in the unstable region.

061.015 Solar neutrinos.
J. N. Bahcall, N. A. Bahcall, R. K. Ulrich.
Bull. American Astron. Soc. Vol. 1, 272 (1969). – Abstr. AAS.

061.016 On the formation of energy spectra in synchrotron sources. D. B. Melrose.
Astrophys. Space Sci. Vol. 5, 131 - 149 (1969).

The observed energy spectra in synchrotron sources are power laws, $N(E) = KE^{-\gamma}$, with the distribution in γ peaked around 2.5. Contrary to the results of Kardashev (1962), it is shown that statistical acceleration when coupled with synchrotron losses lead naturally to the formation of flat stationary spectra with $\gamma \leqslant 1$. The time evolution of an initial spectrum towards the stationary spectrum is investigated. It is suggested that the initially flat spectra with $\gamma \simeq 1$ to 1.5 observed in some variable sources result from an incomplete

approach to the stationary spectrum, and that in sources with constant acceleration spectra with $\gamma \simeq 2$ are to be expected.

061.017 Radioisotopes and the history of nucleosynthesis in the galaxy. C. M. Hohenberg.
Science, Vol. 166, 212 - 215 (1969).

A new model for galactic r-process element synthesis is proposed. From modern astronomical observations and recent data on the abundances of natural radioisotopes (both present and extinct), it seems likely that the bulk of r-process synthesis occurred early in galactic history, followed by a period of relatively little production. A "last minute" addition of r-process material is necessary to explain the abundances of short-lived radioisotopes. A galactic age between 8 and 9 billion years is inferred, with the initial "prompt" synthesis accounting for 81 - 89 percent of the total r-process material ever produced, a "last minute" synthesis contributing 11 - 13 percent, and a "continuous" synthesis providing 0 - 8 percent. In this model the time interval between the collapse of the solar nebula and the formation of solid bodies appears to be 176 - 179 million years.

061.018 The mass–radius relation for cold spheres of low mass. H. S. Zapolsky, E. E. Salpeter.
Astrophys. Journ. Vol. 158, 809 - 813 (1969).

The relationship between mass and radius for zero-temperature spheres is determined for each of a number of chemical elements by using a previously derived equation of state and numerical integration. The maximum radius of a cold sphere is thus found as a function of chemical composition, and a semi-empirical formula for the mass–radius curve is derived.

061.019 Measuring the rate of nucleosynthesis with a gamma-ray detector. D. D. Clayton, J. Silk.
Astrophys. Journ. (*Letters*), Vol. 158, L43 - L48 (1969).

The gamma-ray lines emitted when ^{56}Ni decays to ^{56}Fe are shown to provide a photon flux which is a significant fraction of the diffuse background near 1 MeV. Successful measurement of the line profiles can reveal both the present and past rates of nucleosynthesis in the universe.

061.020 Low-energy gamma radiation in the atmosphere at midlatitudes. R. C. Haymes, S. W. Glenn, G. J. Fishman, F. R. Harnden, Jr.
Journ. Geophys. Res. Vol. 74, 5792 - 5805 (1969).

Balloon-borne experiments with directional scintillators that measured the spectrum of gamma radiation with energies between 30 and 570 kev in the atmosphere up to 130.000 feet were conducted during 1967 and 1968. These data are of interest in connection with understanding the origin of atmospheric photons and with estimating planetary and stellar gamma albedos, as well as effecting improvements in gamma-ray astronomy.

061.021 On the neutrino luminosity of superdense stars.
G. V. Shishkin.
Dokl. AN BSSR, Vol. 13, 227 - 230 (1969). In Russian. – Abstr. in Referativ. Zhurn., 51. Astron., 10.51.302 (1969).

061.022 Gravitational instability of a finitely conducting, infinitely extending layer of finite thickness, surrounded by a non-conducting matter.
K. M. Srivastava, R. C. Sharma.
Acta Astron. Vol. 19, 291 - 300 (1969).

We have studied the gravitational instability of a finitely conducting incompressible fluid layer of finite thickness in the presence of the surrounding non-conducting medium. The magnetic field is directed parallel to the interfaces. The symmetric and asymmetric perturbations have been considered separately. The dispersion relation is obtained and examined under various limiting cases of electrical resistivity; in particular, for the case of small conductivity an approximate formula for the frequency of oscillation is obtained.

061.023 Magnetothermal instability in a rotating gravitating fluid. M. Aggarwal, S. P. Talwar.
Monthly Notices, Roy. Astron. Soc., Vol. 146, 235 - 242 (1969).

The problem of incipient fragmentation of interstellar matter to form condensations is investigated taking into account thermal and radiative effects. A uniform rotation and magnetic field are incorporated in a viscous, electrically conducting fluid. It is found that the classical Jeans' result regarding the size of initial break-up is considerably modified due to a heat-loss function in an inviscid and ideally conducting fluid. Instability criteria are obtained in special cases involving different relative orientations of the magnetic field, rotation and the propagation vector.

061.024 Laboratory investigation of the nuclear reactions in the $p - p$ and CNO cycles.
B. Kuchowicz.
Postępy Astron., Vol. 17, 297 - 302 (1969).   In Polish.

061.025 Gravitational stability of cylinders in an aligned magnetic field. R. A. Wentzell.
Canadian Journ. Phys., Vol. 47, 1349 - 1353 (1969).

An examination has been made of the gravitational instability of an infinite, perfectly-conducting cylinder, maintained by its intrinsic gravitational force, with an axial magnetic field, and surrounded by fluid of a different density. The magnetic field will suppress any axisymmetric instability due to axial flow if the Alfvén velocity is greater than the streaming velocity.

061.026 The enhancement of the 3 ^{4}He $\rightarrow$ ^{12}C reaction rate in dense matter by inelastic-scattering processes.
J. W. Truran, B.-Z. Kozlovsky.
Astrophys. Journ., Vol. 158, 1021 - 1032 (1969).

The rates for the de-excitation of the 7.644-MeV (0$^+$) and 9.638-MeV (3$^-$) excited states of ^{12}C by neutron, proton, and α-particle inelastic scattering are calculated as a function of temperature and density. The implications of the results for astrophysical problems are discussed.

061.027 Abundances of the elements.
A. G. W. Cameron.
Cosmic Ray Studies, Bombay 1968, p. 1 - 24 (1969).

061.028 Nucleosynthesis. W. A. Fowler.
Cosmic Ray Studies, Bombay 1968, p. 25 - 44 (1969).

061.029 Low exposures in the s process.
J. G. Peters, D. D. Clayton.
Bull. American Astron. Soc., Vol. 1, 359 (1969). – Abstr. AAS.

061.030 Equilibrium of rotating degenerate fluid cylinders.
S. P. Talwar, M. Aggarwal.
Publ. Astron. Soc. Japan, Vol. 21, 337 - 349 (1969).

Equilibrium configurations of uniformly rotating completely degenerate gravitating cylinders are considered.

061.031 Las temperaturas en astrofisica. E. Gullon.
Bol. Astron. Obs. Madrid, Vol. 7, No. 3, p. 5 - 27 (1969). – Review article.

061.032 Electron bremsstrahlung in intense magnetic fields.
V. Canuto, H.-Y. Chiu, L. Fassio-Canuto.
Phys. Rev., Second Series, Vol. 185, 1607 - 1613 (1969).

In this paper we present detailed calculations of the most important radiation process in ponderable matter in the presence of an intense magnetic field: the electron bremsstrahlung. We have obtained the emission rate and the absorption coefficient for fields much less than 10^{13} G. This calculation is valid in the large quantum number limit.

061.033 Quantized synchrotron radiation in intense magnetic fields. H.-Y. Chiu, L. Fassio-Canuto.
Phys. Rev., Second Series, Vol. 185, 1614 - 1618 (1969).

In this paper, we compute the radiation rate from de-excitation of electrons in quantized orbits in magnetic fields. Such radiation in the classical limit is known as synchrotron radiation. We stress the radiation properties of electrons in magnetic orbits of low quantum numbers.

061.034 Gravitational instability of an infinite isothermal stratified medium under uniform rotation using the principle of exchange of stabilities.
P. L. Bhatnagar, R. J. Isaac.
Proc. National Inst. Sci. India, Ser. A., Vol. 34, 169 - 178 (1968). – See Phys. Abstr. Vol. 72, No. 26992 (1969).

061.035 The californium hypothesis.
T. E. Margrave, Jr.
Journ. Washington Acad. Sci., Vol. 59, No. 4 - 5, p. 70 - 76 (1969).

061.036 Remarks on quark stars.
D. Ivanenko, D. F. Kurdgelaidze.
Indian Journ. Pure Applied Phys., Vol. 7, 585 - 586 (1969).

061.037 Etude qualitative des solutions de l'équation différentielle d'Emden-Fowler.
M. Lefranc, J. Mawhin.
Acad. Roy. Belgique, Bull. Cl. Sci., 5e Série, Vol. 55, 763 - 770 (1969).

Qualitative methods of the theory of ordinary differential equations are used to study the singularities of Emden-Fowler's equation in the (u, v)-plane. The phase-portrait of the trajectories is obtained by the method of isoclines for a number of values of the polytropic index.

061.038 Produzione di coppie $\nu-\bar{\nu}$ nello scattering e$^-$ e$^-$.
P. Cazzola, A. Saggion.
Atti XII Riunione Soc. Astron. Italiana, L'Aquila 1968, p. 109 - 110 (1969). – Abstract SAI.

061.039 Consistent analysis of gamma-ray astronomy experiments. D. Hearn.
Nuclear Instrument., Vol. 70, 200 - 204 (1969). – See Phys. Ber., Vol. 48, No. 11–3319 (1969).

061.040 On the thermal instability of a rotating-fluid sphere containing heat sources. P. H. Roberts.
Phil. Trans. Roy. Soc. London, Ser. A, Vol. 263, (No. 1136), 93 - 117 (1968).

The theory of marginal convection in a uniformly rotating, self-gravitating, fluid sphere, of uniform density and containing a uniform distribution of heat sources, is developed to embrace modes of convection which are asymmetric with respect to the axis of rotation.

061.041 Photon-photon scattering and photon-scalar particle scattering via gravitational interaction (one-graviton exchange) and comparison of the processes between classical (general relativistic) theory and the quantum linearized field theory.
D. Boccaletti, V. de Sabbata, C. Gualdi, P. Fortini.
Nuovo Cimento, Vol. 60B, 320 - 330, Vol. 64B, 320 (1969).

061.042 **Electron-electron neutrino bremsstrahlung. I, II.**
P. Cazzola, A. Saggion.
Nuovo Cimento, Vol. 63A, 354 - 366, 367 - 376 (1969).

061.043 **Influence of finite Larmor radius and finite conductivity on the gravitational instability of a plasma.**
P. K. Bhatia.
Nuovo Cimento, Vol. 59B, 228 - 235 (1969).

Object of the present note is to study the effect of finite Larmor radius, finite conductivity and rotation on the gravitational instability of interstellar and interplanetary plasmas. The effect of Hall current has been included because of its importance in the dynamics of interstellar matter.

061.044 **Weak interaction of photons and the process**
$\gamma + \gamma \rightarrow \gamma + \nu + \bar{\nu}$.   P. R. Chaudhuri.
Nuovo Cimento, Vol. 64A, 528 - 532 (1969).

From the point of view that photons can interact weakly with neutrinos the cross section for the process has been calculated. The rate of stellar energy loss due to neutrino pair emission and neutrino luminosity of the star in this process has been calculated and the result is compared with that obtained on the basis of the current–current coupling theory.

061.045 **Weak interaction of photons and neutrino – electron scattering.**   P. R. Chaudhuri, P. Bandyopadhyay.
Nuovo Cimento, Vol. 64A, 658 - 668 (1969).

The rate for neutrino-electron scattering has been discussed for non-degenerate and degenerate gases in a previous paper. In the present note the neutrino opacity of stars due to the neutrino-electron scattering according to photon-neutrino coupling theory of weak interactions is investigated.

061.046 **Neutrino and the theory of gravitation.**
B. Kuchowicz.
Fortschritte Physik, Vol. 17, 517 - 534 (1969). – Review article.

061.047 **On the possibility of a hydrodynamic model with finite radius of the region of frequent collisions.**
M. V. Konyukov.
Trudy Fiz. in-ta AN SSSR, Vol. 47, 96 - 105 (1969).
In Russian. – Abstr. in Referativ. Zhurn. 51. Astron., 2.51.223 (1970).

061.048 **Isotopic composition of cosmic importance.**
D. D. Clayton.
Nature, Vol. 224, 56 - 57 (1969).

062 Magneto-Hydrodynamics, Plasma

062.001   A Rossby-wave dynamo for the sun, I.
P. A. Gilman.
Solar Physics, Vol. 8, 316 - 330 (1969).

There is increasing interest in the possible existence of large horizontally flowing eddies or 'Rossby waves' in the sun's convection zone and photosphere. We present here and in part II a mathematical model which shows that flows of this type, driven by an assumed latitudinal temperature gradient, can act as hydromagnetic dynamos to induce magnetic fields that periodically reverse. In this part, we discuss the assumptions for the model, review earlier linear analyses that demonstrate the ability of Rossby waves to induce solar-like magnetic fields, and finally derive the non-linear equations that govern the model.

062.002   Tables for computing dielectronic recombination rates from Burgess' general formula.   V. Letfus.
Bull. Astron. Inst. Czechoslovakia, Vol. 20, 159 - 163 (1969).

The general formula for computing dielectronic recombination, as derived by Burgess (1965), is arranged for use in tabular form. Some conditions relevant for the computation in practice are given.

062.003   Radiation fields of energetic electrons in helical orbits within a magnetoactive plasma.
K. Sakurai, T. Ogawa.
Planet. Space Sci. Vol. 17, 1449 - 1458 (1969).

The radiation fields produced by an energetic electron which is moving in helical orbit in the magnetoactive plasma are derived by solving Maxwell's equations exactly. Those fields consist of three components which are characterized by different wave frequencies related to the anisotropic nature of the medium. The result, obtained in this paper, seems to be useful in investigating the suppression of the low frequencies of solar radio type IV bursts and galactic emission, and suggests that the explanation for the influence of ambient plasmas on the gyro-synchrotron radiation from energetic electrons must be modified.

062.004   The equilibrium and stability of magnetopolytropes.
S. P. S. Anand.
Astrophys. Space Sci. Vol. 4, 255 - 274 (1969).

The theory of the oscillations of axisymmetric gaseous configurations with a prevalent magnetic field is presented. The virial tensor method is used to obtain the nine second harmonic modes of oscillations of the system. It is found that out of the nine modes, three are neutral, four are non-radial, and two are coupled. For the Prendergast spherical model it is found that one of the coupled modes is radial and the other non-radial. Both the radial and the non-radial modes obtained in this case agree with the corresponding formulae obtained by Chandrasekhar and Limber (1954) and Woltjer (1962). The equilibrium structure of gaseous polytropes with toroidal magnetic fields is also investigated in detail for several values of the polytropic index n.

062.005   Monte-Carlo study of Stark fields in dense plasmas.
A. Gervat, D. Rossignol-Guzzi.
Astron. Astrophys. Vol. 3, 5 - 14 (1969).

Using the Monte-Carlo model of a plasma in thermodynamic equilibrium, proposed by Brush, Sahlin and Teller, the ionic electric field is studied in order to determine the probability of a given electric field in the plasma. This study may be used in astrophysics to obtain Stark fields influence on opacity. The results obtained are compared with the work of Mayer and Broyles.

062.006   Material coefficients and transfer of polarized radio radiation in a plasma.   J. Heyvaerts.
Astrophys. Space Sci. Vol. 5, 36 - 52 (1969).

By a perturbation and diagram resummation method, a transport equation for the transverse field polarization matrix is established. This equation is then transformed into an equation for the Stokes parameters of the radiation. The equation takes the usual form of a transfer equation; the absorption and emission coefficients are matrix, the elements of which are given as a function of the dissipative part of the microcurrent correlation tensor and conductivity tensor. Finally this equation is expressed as a system for the intensities of the proper modes. The equations of the system are usually coupled.

062.007   An axisymmetric magnetic field with differential rotations in a spherical fluid shell. II.
Y. Nakagawa.
Astrophys. Journ. Vol. 157, 881 - 886 (1969).

The differential rotation in a spherical shell of an incompressible inviscid fluid of infinite electrical conductivity is examined when the magnetic field and the fluid motions are both purely toroidal. It is shown that a steady-state solution is possible, of the Alfvén velocity associated with the toroidal magnetic field is comparable with the velocity of rotation. By considering a particular set of solutions which satisfy the observed solar differential rotation on the outer surface and a constant rotation on the inner surface, it is shown that the toroidal magnetic field must be of the order of 10^4 gauss for the sun.

062.008   Zur Dynamotheorie stellarer und planetarer Magnetfelder. II. Berechnung planetenähnlicher Gleichfeldgeneratoren.   M. Steenbeck, F. Krause.
Astron. Nachr., Vol. 291, 271 - 286 (1969).

In this second part we investigate models on conditions similar to those in the planets, especially the earth. Therefore differential rotation is not taken into account; only induction actions of non-reflectionsymmetric turbulence are regarded. The models show dynamo maintenance of stationary fields. As a surprising result wer find that the critical values for field maintenance are nearly the same for the antisymmetric fields (dipol-type) and the symmetric fields (quadrupol-type). A quantitative comparison with the data of the earth core shows a sufficient agreement. – The conditions for dynamo action of the other planets are discussed and some suggestions for future work in this field are given.

062.009   Zur Dynamotheorie stellarer und planetarer Magnetfelder. III. Über die Lösung der Eigenwertprobleme und die Berechnung der Feldgrößen.
F. Krause, H. Hiller.
Astron. Nachr., Vol. 291, 287 - 294 (1969).

The mathematical analysis and the technic of electronic computing of the eigenvalue problems are described, which occur in the dynamo theory of stellar and planetary magnetic fields developed by M. Steenbeck and F. Krause. The eigenvalue-problem is solved by integrating the initial-value problem with fixed parameters. Than the parameters are changed until there is the possibility of solving the boundary conditions. If the eigenvalues are determined the system is integrated once more and the fields are calculated.

062.010   Possible mechnism for the acceleration of ions in some astrophysical phenomena.   M. Friedman.
Phys. Rev., Second Series, Vol. 182, 1408 - 1414 (1969).

A model is suggested for the acceleration of ions, based

on Petschek's theory of magnetic field annihilation. It is shown that ions can gain high energies in this model, and that energy distributions similar to those observed can be obtained. It is also shown that this model prevents electrons from gaining high energy, consistent with cosmic-ray observations. It is suggested that if this model is correct, we can obtain, from the observed ion energy distribution, a relation between the (average) plasma density and the (average) magnetic field in regions where the ions are accelerated.

062.011 Amplification of weak magnetic fields in turbulent flow. E. N. Parker.
Astrophys. Journ. Vol. 157, 1119 - 1127 (1969).

The problem of a weak magnetic field carried in a turbulent-velocity field is considered under the circumstance that the velocity field is a prescribed random function of space and time. It is assumed that the time over which the velocity field is correlated with itself is very short. In this approximation the theory of random functions is used to show that the velocity field amplifies the magnetic field at all wave-numbers. Thus the calculation establishes that the large-scale components of a magnetic field in a turbulent-velocity field grow without limit as long as the velocity field is maintained. The unlimited amplification is of interest in astrophysical problems.

062.012 Growth rate of the relativistic fire-hose instability. P. D. Noerdlinger, A. K.-M. Yui.
Astrophys. Journ. Vol. 157, 1147 - 1156 (1969).

The growth rate of the fire-hose instability in a relativistic plasma with anisotropic pressure is given as a function of wavenumber, degree of anisotropy, magnetic field, density of the cold background, and mean relativistic energy factor $\langle\gamma\rangle = \langle E/Mc^2\rangle$. The growth rates indicate that the instability should be important in liberating cosmic rays at the outer boundary of the galactic halo. If Smith's recent measurement of the magnetic field in the galactic disk, $B_0 < 2 \times 10^{-7}$ gauss, is typical in the disk, the instability would also be important in isotropizing cosmic rays there.

062.013 Ionization equilibrium and radiative cooling of a low-density plasma. D. P. Cox, W. H. Tucker.
Astrophys. Journ. Vol. 157, 1157 - 1167 (1969).

The results of calculations of the ionization equilibrium and rate of radiative cooling of a high-temperature low-density plasma are presented. The elements H, He, C, N, O, Ne, Mg, Si, and S are considered, and the temperature range is taken to be $10^3 - 10^8$ °K. The effect of the process of dielectronic recombination, radiative losses due to bremsstrahlung, recombination radiation, and collision-induced line emission are considered.

062.014 The excitation and spectra of plasma turbulence in objects with large density of the electromagnetic emission. E. N. Krivorutsky, V. N. Tsytovich.
Astron. Zhurn. Akad. Nauk SSSR, Vol. 46, 1003 - 1015 (1969). In Russian. English translation in Soviet Astron. AJ, Vol. 13, No. 5.

Effects of excitation of various types of turbulent pulsations in plasma, in objects with high density of the electromagnetic radiation, such as quasars, supernova shells and so on, are considered. Spectra of turbulence are obtained. Some estimations of the generated power are given.

062.015 Radiative transport quantities for a hydrogen plasma. D. A. Mandell, R. D. Cess.
Journ. Quant. Spectrosc. Radiative Transfer, Vol. 9, 981 - 994 (1969).

The object of this investigation is to evaluate various radiative transport quantities for a hydrogen plasma. These transport quantities pertain to the analysis of radiative energy transfer in nonisothermal gases, and they include the Planck mean, modified Planck mean, linear Planck mean, and Rosseland mean absorption coefficients, in addition to the modified emissivity and its derivative which appear in the formulation for the linearized radiative flux. Twenty one spectral lines are included in the calculations, and the line profiles are taken to be Stark profiles. Results are presented for temperatures from 10000°K to 20000°K and electron densities of 10^{16}, 10^{17}, and 10^{18} cm^{-3}.

062.016 The scattering of radiation in the spectral lines of a turbulent plasma. S. A. Kaplan, V. N. Tsitovich.
Astrofizika, Vol. 5, 21 - 30 (1969). In Russian. English translation in Astrophysics, Vol. 5, No. 1 (1969).

The formation of satellites and the broadening of spectral lines during radiation transfer in a turbulent plasma are considered. A mechanism on radiation pumping in the spectral lines is proposed which can in principal give a "maser-effect" on natural conditions.

062.017 Spherical and cylindrical nonlinear waves. M. A. Gintsburg.
Astron. Zhurn. Akad. Nauk SSSR, Vol. 46, 1019 - 1028 (1969). In Russian. English translation in Soviet Astron. AJ, Vol. 13, No. 5.

Nonlinear evolution of spherical, cylindrical and plane plasma inhomogeneities is considered. The author solves numerically and by series expansion some problems. The main new effects discovered are: 1) density and velocity pulse formation in expanding plasma, 2) velocity and field sign reversal, oscillations, 3) particle acceleration in expanding magnetized plasma.

062.018 Propagation of coherent magnetohydrodynamic waves. C. H. Liu.
Phys. Fluids, Vol. 12, 1642 - 1647 (1969).

The propagation properties of coherent magnetohydrodynamic waves in the presence of plasma density fluctuations are investigated. A perturbation technique borrowed from quantum field theory (renormalization procedure) is used to solve the stochastic wave equation. Effective propagation and attenuation constants are derived for propagation of plane coherent magnetohydrodynamic waves. It is found that, in general, the phase velocity will be less than the Alfvén velocity and there will be attenuation (or damping) of the waves due to the random mixing of phase.

062.019 Electron temperature effects in fast magnetohydrodynamic shock waves. B. P. Leonard.
Phys. Fluids, Vol. 12, 1816 - 1829 (1969).

The interaction between magnetic and electron temperature effects is analyzed in the simplest nontrivial case of fast magnetohydrodynamic shock waves in a slightly ionized plasma.

062.020 Electron temperature distribution across a shock wave in a partially ionized gas. S. S. R. Murty.
Phys. Fluids, Vol. 12, 1830 - 1832 (1969).

The electron temperature structure in a weakly ionized plasma is studied allowing the degree of ionization to vary across the shock wave. The values of the elctron temperature and the downstream equilibrium temperature obtained with variable ionization are less than those for frozen ionization. The electron temperature rises sharply behind the shock for variable ionization while a gradual increase is predicted by frozen ionization.

062.021 Solutions of the linearized equations of magnetohydrodynamics in nonhomogeneous magnetic fields. E. Infeld.

Phys. Fluids, Vol. 12, 1845 - 1848 (1969).

Earlier results, concerning the conditions when the solutions of the linearized equations of magnetohydrodynamics are exact, are extended. Only a partial answer to the question is still possible for varying initial fields. An exact solution is given as an illustration. This solution might be useful as a model for investigating the effect of currents on Alfvén waves in the radiation belts.

062.022 Stochastic acceleration in the presence of streaming. D. E. Hall.
Bull. American Astron. Soc. Vol. 1, 243 (1969). – Abstr. AAS.

062.023 Plasma wave propagation in the neighborhood of a magnetic neutral point. R. L. Harkness, Jr.
Solar Physics, Vol. 10, 154 - 167 (1969).

A linearized magnetohydrodynamic formalism is used to examine the propagation in two dimensions of transverse waves in a plasma in which is embedded a curl-free magnetic field. Only waves of frequency less than the ion cyclotron frequency are considered. It is shown that a disturbance propagates through the medium with a group velocity that decreases from the speed of light at large distances from the neutral point to zero at the neutral point, and that the amplitudes of the velocities associated with the disturbance diverge there.

062.024 Propagation of gravity waves in a non-isentropic rotating, hydromagnetic medium. G. L. Kalra.
Publ. Astron. Soc. Japan, Vol. 21, 263 - 266 (1969).

A dispersion relation governing the propagation of hydromagnetic gravity waves in a rotating, non-isentropic medium is derived. It is found that a convectively stable medium remains stable and that the effect of a constant entropy gradient is most pronounced in a direction normal to the gravity field. The effect of a small coriolis force is to reduce the phase speed of the Alfvén mode. The speed of the entropy mode is, however, increased due to the presence of rotation.

062.025 Gyrosynchrotron emission and absorption in a magnetoactive plasma. R. Ramaty.
Astrophys. Journ. Vol. 158, 753 - 770 (1969).

The intensity, spectrum, and polarization of gyrosynchrotron radiation from an arbitrary distribution of electrons in a magnetoactive plasma are calculated. These calculations are based on detailed theories of gyrosynchrotron emission and absorption as well as on the equations of radiation transfer in the limit of large Faraday rotation. We investigate the influence of reabsorption and the influence of a cold plasma on the spectrum and polarization of the radiation.

062.026 Diffusion of strong inhomogeneities in a weakly ionized magnetoactive plasma.
B. N. Gershman, G. I. Grigor'ev.
Izv. vyssh. uchebn. zavedenij. Radiofizika, Vol. 12, 20 - 25 (1969).   In Russian. – Abstr. in Referativ. Zhurn. 51. Astron., 9.51.366 (1969).

062.027 On nonlinear interaction of intensive plasma waves.
V. N. Tsytovich, A. S. Chikhachev.
Izv. vyssh. uchebn. zavedenij. Radiofizika, Vol. 12, 26 - 35 (1969).   In Russian. – Abstr. in Referativ. Zhurn. 51. Astron., 9.51.368 (1969).

062.028 Magnetic turbulence in shocks.
C. F. Kennel, H. E. Petschek.
Physics of the Magnetosphere, Boston College 1967, p. 485 - 513 (1968). – Invited research paper.

062.029 Polarization transfer of electromagnetic waves in a
magnetized plasma. P. C. W. Fung.
Astrophys. Space Sci. Vol. 5, 448 - 458 (1969).

In this investigation, the polarization transfer equations in terms of the Stokes parameters are derived for electromagnetic waves propagating in an arbitrary direction in an inhomogeneous magnetized plasma. This system of transfer equations is then solved analytically in the case when the magnetized plasma is homogeneous. For simplicity in presentation, the source term in the equation of transfer has been omitted. Transitting to the special case of quasi-longitudinal propagation, the results obtained here are shown to be in agreement to that derived by Zheleznyakov earlier.

062.030 Zur Hydrodynamik des mittleren Geschwindigkeitsfeldes in einem turbulenten Medium.
F. Krause.
Monatsber. Deutsch. Akad. Wiss. Berlin, Band 11, 188 - 194 (1969).

062.031 Zur Elektrodynamik in turbulenten, Coriolis–Kräften unterworfenen leitenden Medien.
K.-H. Rädler.
Monatsber. Deutsch. Akad. Wiss. Berlin, Band 11, 194 - 201 (1969).

062.032 Hydromagnetic oscillations of a differentially rotating spherical shell. S. K. Trehan, Y. Nakagawa.
Astrophys. Journ., Vol.158, 1075 - 1080 (1969).

The axisymmetric hydromagnetic oscillations of an incompressible inviscid fluid of infinite electrical conductivity in a differentially rotating spherical shell are considered when the prevailing magnetic and velocity fields are purely toroidal. It is shown that stable oscillations result with the observed form of surface differential rotation of the sun and with the assumption that the core rotates with a constant angular velocity. The effect of differential rotation is shown to increase the period of oscillations.

062.033 The connection of Stokes' parameters of synchrotron radiation with a tensor of Maxwellian stresses in the emitting region. M. M. Molodensky.
Astron. Zhurn. Akad. Nauk SSSR, Vol. 46, 797 - 799 (1969). In Russian. English translation in Soviet Astron. AJ, Vol. 13, No. 4.

062.034 Sopra alcune generalizzazioni di un teorema di Poincaré in magnetofluidodinamica.
T. Zeuli.
Atti Accad. Nazionale Lincei, Rend. Cl. Sci. fis. mat. nat., Serie Ottava, Vol. 46, 561 - 568 (1969).

Noted bounds are given in magnetohydrodynamics, for the angular velocity of a rotating and in relative equilibrium fluid mass, from which follow some results given by Poincaré and Armellini.

062.035 Über eine neue Möglichkeit eines Dynamomechanismus in turbulenten leitenden Medien.
K.-H. Rädler.
Monatsber. Deutsch. Akad. Wiss. Berlin, Band 11, 272 - 279 (1969).

062.036 Intensities of the hydrogen lines emitted by an optically thick plasma. V. P. Grinin.
Astrofizika, Vol. 5, 371 - 381 (1969).   In Russian. – Engl. translation in Astrophysics, Vol. 5, No. 3.

The problem of the radiation of the hydrogen plasma, partially opaque in the Balmer lines, is considered. It is assumed that the source of excitation is radiative recombination. The redistribution of the radiation between the lines as well as the frequency redistribution within the lines themselves is taken into consideration. Relative intensities and equivalent widths of the Balmer lines are calculated for a set of electron temperatures and optical thickness in Hα.

062.037 Exact motion of a charged particle in an arbitrary plane wave propagating along a constant homogeneous magnetic field. E. Leer.
Phys. Fluids, Vol. 12, 2206 - 2210 (1969).

062.038 Enhanced bremsstrahlung from plasmas with relativistic electron tails. K. Papadopoulos.
Phys. Fluids, Vol. 12, 2185 - 2193 (1969).

The bremsstrahlung emitted from thermal plasmas which coexist with a flux of relativistic electrons is calculated. The emission is found to be greatly enhanced at the fundamental and second harmonic of the electron plasma frequency. Some possible astrophysical applications of the theory are discussed.

062.039 Relativistic magnetohydrodynamics of a gravitating fluid. K.-K. Tam, J. O'Hanlon.
Nuovo Cimento, Vol. 62B, 351 - 359 (1969).

062.040 Solution of a class of magnetohydrodynamic problems with strengthening of the magnetic field.
A. V. Getling.
Dokl. Akad. Nauk SSSR, Ser. Mat. Fiz., Vol. 187, 301 - 304 (1969).   In Russian.

Subphotospheric convective layers may be considered as responsible in a large measure for the action of the solar dynamo. Magnetic configurations created by hydromagnetic activity of convective cells under the effect of Coriolis forces are investigated. The calculations are performed for a simple model based on a vertical ring. It is shown that different situa-

tions with similar velocity fields can easily be solved on the basis of the theory given in this paper. These cases are important for the investigation of the solar magnetic cycle and of the solar wind.   *B. Onderlička*

062.041 Kinetic theory of weakly ionized time-varying magnetoplasmas. W. Stiller, G. Vojta.
Physica, Vol. 43, 216 - 228 (1969).

Using Boltzmann's equation a general method is developed for the calculation of the electron distribution function f of a weakly ionized plasma in external homogeneous time-dependent magnetic fields and additional electromagnetic fields. Only elastic collisions are considered. In the field of astrophysics and geophysics this theory can be useful for the understanding of cosmic rays, magnetic stars and other phenomena.

062.042 Unified classical-path treatment of Stark-broadening in plasmas.
E. W. Smith, J. Cooper, C. R. Vidal.
Phys. Rev., Second Series, Vol. 185, 140 - 151 (1969).

A theoretical treatment of spectral line broadening in plasmas is developed using classical-path methods. This treatment unifies certain aspects of the familar impact, one-electron, and relaxation theories to produce results which are valid from the line center to the far line wings where electrons may behave quasistatistically. Calculations of the Lyman α line of hydrogen are used to illustrate the theory.

Plasma Physics – Vol. 2, Weakly Ionized Gases.
See Abstr. 003.016.

063 Radiative Transfer

063.001 The effects of polarization on diffusion descriptions of radiative transfer. G. C. Pomraning.
Astron. Astrophys. Vol. 2, 419 - 424 (1969).

The effect of a proper treatment of polarization in the equation of transfer is investigated, with particular emphasis on energy transport. It is found that for a large class of radiative transfer problems the neglect of the effects of polarization introduces a small error. The problems in this class are those which can be described by one of several diffusion equations, such as results from the widely used Eddington approximation. It is also found that the error associated with the neglect of polarization effects is comparable in magnitude to the error introduced due to the neglect of the anisotropy of the Rayleigh (Thomson) phase function. This suggests that if polarization effects are neglected, as is frequently the case in actual calculations of energy transport, it is not inconsistent to neglect the anisotropy of Rayleigh scattering as well. This observation has practical consequences since the equation of transfer is significantly easier to solve if the scattering phase function can be assumed isotropic.

063.002 Non-coherent scattering – V. Thermalization distances and their distribution function.
G. B. Rybicki, D. G. Hummer.
Monthly Notices, Roy. Astron. Soc., Vol. 144, 313 - 323 (1969).

The distribution function for thermalization distances is derived for an infinite atmosphere with a plane source. Precise definitions of the thermalization length are discussed from the point of view of representing the distribution by a single characteristic length; of these a definition in terms of the median of the distribution seems to be most useful. The distribution of longest flights is derived and shown to provide a good approximation to the distribution of thermalization lengths at large distances from the source. Extensive numerical illustrations are provided.

063.003 Diffuse reflection and transmission of time-dependent collimated light by a finite inhomogeneous atmosphere. M. Matsumoto.
Publ. Astron. Soc. Japan, Vol. 21, 1 - 14 (1969).

We consider the diffuse reflection and transmission of time-dependent collimated light, incident at an instant at a point on the boundary surface, by a finite inhomogeneous atmosphere. The principles of invariance for the non-stationary and three-dimensional radiation field are formulated. With the aid of these principles, we obtain the functional equations for the S- and T-functions which govern the diffuse reflection and transmission, allowing for two time scales t_1, the duration of temporal capture, and t_2, the mean free time.

063.004 The invariant imbedding equation for the dissipation function of an infinite inhomogeneous cylindrical shell. J. Tsujita.
Publ. Astron. Soc. Japan, Vol. 21, 15 - 20 (1969).

In a non-emitting and isotropically-scattering inhomogeneous infinite cylindrical shell region, surrounding an absorbing core, a functional relation of the dissipation function for uniform incident radiation is derived, by the invariant imbedding particle counting method.

063.005 Chandrasekhar's X- and Y-functions for isotropic scattering in thick slabs. B. E. Clancy.
Australian Journ. Phys. Vol. 22, 317 - 326 (1969).

Approximate forms are developed for the X- and Y-functions of isotropic scattering of Chandrasekhar in terms of the well-tabulated H-function. The approximations that are asymptotically correct for slabs of large thickness are compared with available tabulated values.

063.006 The treatment of resonance scattering of polarized radiation in weak magnetic fields by the Monte Carlo technique. L. L. House, L. C. Cohen.
Astrophys. Journ. Vol. 157, 261 - 274 (1969).

The problem of resonance scattering in a finite medium which is permeated by a uniform magnetic field is described. A parallel beam of unpolarized photons is assumed incident upon the lower surface of the scattering medium, the direction of propagation being perpendicular to the magnetic field. Through inclusion of the Hanle effect which influences the phase coherency of the scattering process, it has been possible to treat the case of weak fields for the normal Zeeman triplet. This is in contrast to most studies of the transport of polarized radiation in which a strong field is assumed, giving rise to well-separated Zeeman components. By the Monte Carlo process "photons" are traced through the scattering medium by using the probabilistic laws of interaction, and, by counting the "photons" as they exit, line profiles and the degree of polarization are determined.

063.007 Probabilistic model for radiative transfer problems for spherical shell medium.
T. K. Leong, K. K. Sen.
Publ. Astron. Soc. Japan, Vol. 21, 167 - 175 (1969).

A probabilistic model for transmission functions for solving transfer problems in spherical shell medium surrounding (a) an emitting black core and (b) a point source have been proposed. The integro-differential equations for transmission functions are deduced and the expressions for emergent intensities are obtained. The integro-differential equations for transmission functions are found to be the same as those obtained by Bellman, Kagiwada, Kalaba, and Ueno by the invariant imbedding method.

063.008 A new computational method for the X and Y functions of radiative transfer. H. Cohen.
Journ. Quant. Spectrosc. Radiative Transfer, Vol. 9, 931 - 942 (1969).

A new method of calculating the X and Y functions of radiative transfer is presented. This new method involves the solution of an integral equation rather than coupled differential equations. A numerical comparison with previously published data shows this method to be quite accurate. The advantages of this technique over the method of coupled differential equations are discussed.

063.009 The extended Eddington approximation with anisotropic scattering. G. C. Pomraning.
Journ. Quant. Spectrosc. Radiative Transfer, Vol. 9, 1011 - 1016 (1969).

The modified Eddington approximation proposed earlier for isotropic scattering is formulated for a general scattering law. Numerical results are given for the special case of Thomson scattering.

063.010 An exactly soluble problem of radiative transfer without redistribution in frequency in an inhomogeneous atmosphere. M. Lecar.
Journ. Quant. Spectrosc. Radiative Transfer, Vol. 9, 1017 - 1024 (1969).

The transfer of radiation by coherent scattering and gray absorption is described by an integral equation. For the case when the medium is one dimensional, so that the kernel is an exponential, and homogeneous (ϵ constant), the resolvent was given by Milne. The resolvents are given here for a class of prescribed variations of ϵ.

063.011 Radiative transfer in a nongray gas between parallel walls. H.-C. Kung, M. Sibulkin.

Journ. Quant. Spectrosc. Radiative Transfer, Vol. 9, 1447 - 1461 (1969).

The problem of radiative transfer in a gas bounded by parallel walls is examined for the case in which the spectral absorption coefficient of the gas is approximated by a two-level model. Solutions for the heat flux and temperature distribution have been obtained by numerical integration of the exact equations of transfer and by series solutions of a differential approximation to the transfer equations. The results obtained are compared with each other and with the appropriate limiting gray-gas results. It is found that, when a high temperature gas is approximated by a two-level absorption coefficient, the average temperature of the gas is higher than when a gray-gas approximation is used.

063.012 Diffuse reflection and transmission of light by an atmosphere with anisotropic scattering. V. V. Sobolev.

Astrofizika, Vol. 5, 5 - 20 (1969). In Russian.
English translation in Astrophysics, Vol. 5, No. 1 (1969).

A former investigation of the linear integral equations for the reflection and transmission coefficients $\rho(\eta, \zeta)$ and $\sigma(\eta, \zeta)$ and for the auxiliary functions $\varphi_i(\eta)$ and $\Psi_i(\eta)$ is continued. The complementary relations for these quantities are found. A general method is given which enables one to obtain the functions $\varphi_i(\eta)$ and $\Psi_i(\eta)$ in terms of Chandrasekhar's functions $X(\eta)$ and $Y(\eta)$. As an example the case of the two-term scattering indicatrix is considered. Formulae are given for the albedo of the atmosphere and for the illuminance of the surface.

063.013 On some asymptotic formulae in the theory of anisotropic light scattering. A. K. Kolesov, V. V. Sobolev.

Astrofizika, Vol. 5, 175 - 186 (1969). In Russian.
English translation in Astrophysics, Vol. 5, No. 2 (1969).

The problem of anisotropic light scattering in a semi-infinite medium is considered. Particle albedo λ is assumed to be close to unity. The asymptotic formula is found containing the zeroth and the first terms of the expansion of the function $H(\eta)$ in powers of $(1 - \lambda)^{1/2}$.

063.014 Non-coherent scattering – VI. Solutions of the transfer problem with a frequency-dependent source function. D. G. Hummer.

Monthly Notices, Roy. Astron. Soc., Vol. 145, 95 - 120 (1969).

A generalized discrete-ordinate method is used to obtain accurate numerical solutions of the line transfer problem in which the scattering is described by a redistribution function. Extensive results are obtained and discussed for the cases of pure Doppler broadening and of Doppler and natural broadening combined. It is shown that, in the latter case, the intensity of radiation emerging from a semi-infinite isothermal atmosphere approaches that for coherent scattering in the line wings instead of approaching the value of the Planck function.

063.015 Theory of radiative transfer in inhomogeneous atmospheres. I. Perturbation method. A. L. Fymat, K. D. Abhyankar.

Astrophys. Journ. Vol. 158, 315 - 324 (1969).

It it is assumed that the albedo for single scattering $\Omega(\tau)$ differs from a constant value Ω_0 by a small amount throughout the atmosphere, the nonlinear singular integral equations for the X- and Y-functions of Chandrasekhar and the X^* and Y^* functions of Ueno, which describe the transfer of radiation

through an inhomogeneous plane-parallel atmosphere of arbitrary stratification, are linearized by use of a perturbation technique.

063.016 Theory of radiative transfer in inhomogeneous atmospheres. II. Application of the perturbation method to a semi-infinite atmosphere. A. L. Fymat, K. D. Abhyankar.

Astrophys. Journ. Vol. 158, 325 - 335 (1969).

The nonlinear singular integral equation for the H-function of an inhomogeneous plane-parallel atmosphere of arbitrary stratification is linearized by using the perturbation method developed in paper I. The N-solution of this equation is given. The iteration procedure for computing the fractional perturbation in H is also described. For illustrating the method, H-functions for semi-infinite homogeneous atmospheres with $\Omega = 0.300, 0.700, 0.925$, and 0.975 are computed from the H-function for an atmosphere with $\Omega = 0.900$.

063.017 An exact solution in the theory of line formation. C. E. Siewert, M. N. Özişik.

Monthly Notices, Roy. Astron. Soc., Vol. 146, 351 - 360 (1969).

The singular eigenfunction expansion technique is used to solve rigorously the equation of radiative transfer describing the interlocking-doublet model in the theory of line scattering.

063.018 Zur Theorie der Linienentstehung bei NLTE unter Einwirkung eines Magnetfeldes. H. Domke.

Monatsber. Deutsch. Akad. Wiss. Berlin, Band 11, 269 - 272 (1969). – Research note.

063.019 Line formation in a magnetic field. D. E. Rees.

Solar Physics, Vol. 10, 268 - 282 (1969).

The effect of noncoherent scattering is examined for an absorption line formed in a uniform magnetic field. It is shown that the Stokes parameters of the line radiation may be computed by using the line source function in the absence of a magnetic field as a first approximation for that in the presence of a magnetic field.

063.020 Diffuse reflection and transmission of light by an atmosphere with arbitrary scattering indicatrix. V. V. Sobolev.

Astron. Zhurn. Akad. Nauk SSSR, Vol. 46, 1137 - 1148 (1969). In Russian. English translation in Soviet Astron. AJ, Vol. 13, No. 6.

The reflection and transmission functions are expanded in terms of the cosines of the multiples of the azimuth.

063.021 On the equation of radiative transfer in dispersive anisotropic media. S. Enomé.

Publ. Astron. Soc. Japan, Vol. 21, 367 - 369 (1969).

Zheleznyakov (1969) has obtained a transfer equation in a dispersive magneto-active plasma from the kinetic equation in the phase space of spatial and wave vectors. His transfer equation is incorrect owing to a false relation between energy density in the plasma space and the specific intensity.

063.022 Diffuse reflection of a searchlight beam by slab, cylindrical, and spherical media. A. Uesugi, J. Tsujita.

Publ. Astron. Soc. Japan, Vol. 21, 370 - 383 (1969).

With the aid of Chandrasekhar's (1950) principle of invariance in connection with the equation of transfer, the functional equations for the scattering function are obtained in slab, cylindrical, and spherical geometries, when a narrow pencil of radiation is incident on the outer surface. In the case of a constant conical flux of radiation uniformly incident on the outer boundary, the result reduces to that given by several authors.

063.023 Anisotropic light scattering in an atmosphere of finite optical thickness. V. V. Sobolev.
Astrofizika, Vol. 5, 343 - 358 (1969). In Russian. – Engl. translation in Astrophysics, Vol. 5, No. 3.

Light scattering is considered in an atmosphere of optical thickness τ_0. Scattering indicatrix x (γ) and particle albedo λ are arbitrary. The atmosphere is assumed to be illuminated by parallel rays. The angle of incidence of the rays is arc cos ζ. The quantity under consideration is $I (\tau, \eta, \zeta, \varphi)$, i.e. the intensity of scattered radiation at the optical depth τ in the direction characterized by the angle with the normal arc cos η and the azimuth φ. The intensity is expanded in terms of cos $m\varphi$.

063.024 The polychromatic light scattering in a semi-infinite medium. V. Yu. Terebizh.
Astrofizika, Vol. 5, 359 - 370 (1969). In Russian. – Engl. translation in Astrophysics, Vol. 5, No. 3.

A pure scattering of radiation in a plane-parallel layer of gas illuminated by an external radiation is considered. It is assumed that: a) the gas consists of three-level atoms, b) the optical thickness is infinitely large for all frequencies, c) the absorption coefficient in the lines has a rectangular contour. The solution of the problem is found in the Schwarzschild-Schuster approximation. The physical interpretation of the results is given.

063.025 On the solution $H_1(\mu)$ of Ambarzumian–Chandrasekhar H-equation.
K. D. Abhyankar, A. L. Fymat.
Journ. Quant. Spectrosc. Radiat. Transfer, Vol. 9, 1563 - 1566 (1969).

It is shown in a simple manner why the H-equation admits two solutions and how they are related to each other. A justification for rejecting the solution $H_1(\mu)$ is given without referring to the auxiliary equation for the source function.

063.026 The normalized on-the-spot approximation for line transfer problems.
D. van Blerkom, D. G. Hummer.
Journ. Quant. Spectrosc. Radiat. Transfer, Vol. 9, 1567 - 1571 (1969).

A simple approximation for line transfer problems is presented which is useful when photons scatter only a few times before being destroyed. Comparison is made to solutions of high accuracy.

063.027 The Monte Carlo technique applied to radiative transfer. L. L. House, L. W. Avery.
Journ. Quant. Spectrosc. Radiat. Transfer, Vol. 9, 1579 - 1591 (1969).

A general review is given of the application of the Monte Carlo method to the numerical calculations of non-equilibrium radiation transport. With the aid of a general flow chart, the various steps required in such Monte Carlo calculations are detailed.

063.028 Radiative transfer in a semi-infinite medium with a three-term scattering indicatrix.
A. K. Kolesov, V. V. Sobolev.
Trudy Astron. Obs. Leningrad, Vol. 26 (= Uchenye Zapiski Leningr. Un-ta No. 347 = Seriya Matem. Nauk No. 44), p. 3 - 19 (1969). In Russian.

Radiative transfer in a semi-infinite medium with a three-term scattering indicatrix and an arbitrary particle albedo is considered. The formulas are given for the intensity of radiation in three problems, namely: 1) diffuse reflection, 2) radiation field in the deep layers, 3) the diffuse transmission.

063.029 The transport equation of radiative transfer with isotropic scattering. The solution of the auxiliary equation by a Green's function method.
G. E. Hunt.
Proc. Cambridge Phil. Soc. (Math. Phys. Sci.), Vol. 65, 199 - 208 (1969).

The kernel of the integral equation for the source function in a three-dimensional homogeneous atmosphere possesses the properties of a Green's function. These properties are used to transform the integral equation into a singular integral equation for the kernel. The particular case of a homogeneous plane parallel atmosphere is discussed and a solution to the kernel equation is obtained at all points of the atmosphere.

063.030 Duhamel's principle in the nonstationary radiation field. M. Matsumoto.
Journ. Math. Analysis Applications, Vol. 21, 445 - 457 (1968).

In the present paper, we discuss Duhamel's principle for time-dependent radiation field in an inhomogeneous atmosphere of finite or semi-infinite thickness. We assume the medium in the atmosphere is independent on time, but the incident radiation is nonstationary.

Radiative Transfer and Spectra of Celestial Bodies.
See Abstr. 003.077.

Transfer of resonance line radiation from a point source in the half-space. See Abstr. 064.023.

On the nonlinear nonstationary problem of radiation transfer in a spectral line. See Abstr. 064.024.

064 Stellar Atmospheres , Stellar Envelopes

064.001 **Theoretical line intensities. VIII. Comment on the effectively thin approximation.** R. G. Athay.
Astrophys. Journ. Vol. 157, 281 - 285 (1969).

The effectively thin approximation for the formation of spectral lines is discussed from the standpoints of photon-degradation processes and the random walk of scattered photons. It is shown that for some lines the optical thickness of the effectively thin layer is fixed by atomic rate coefficients and is independent of the model atmosphere.

064.002 **Why the temperature rise does not occur in radiative equilibrium in stellar chromospheres of dominant H⁻ opacity.** S. D. Jordan.
Astrophys. Journ. Vol. 157, 465 - 467 (1969).

It is demonstrated that the temperature inversion in the low solar chromosphere is very unlikely to result from the influence of the photospheric radiation field on the low-density gas. Therefore, dissipation of mechanical energy is necessary to produce the rise. Since this effect follows directly from the dominance of the continuum opacity by the H⁻ ion, the same conclusion follows for all stellar chromospheres where H⁻ dominates the continuum opacity.

064.003 **The effect of electron scattering on curves of growth.** J. D. Rosendhal.
Astrophys. Space Sci. Vol. 4, 419 - 426 = Contr. Kitt Peak National Obs. No. 416 (1969).

A new exact solution of the macroscopic line transfer equation including electron scattering terms has been obtained for a Milne-Eddington Model atmosphere, and curves of growth based on this solution have been calculated. The results indicate that for lines formed by scattering there is a systematic change in the appropriate theoretical curve of growth as electron scattering becomes an increasingly important source of continuous opacity. In a case where electron scattering is the dominant opacity source, the abundance necessary to produce a given line strength may be decreased by a factor of 2–5 and the derived velocity parameter decreased by 20–30% due to the shifts in the theoretical curves.

064.004 **Neutral helium lines and departures from LTE in hot stars.** H. R. Johnson, A. I. Poland.
Journ. Quant. Spectrosc. Radiat. Transfer, Vol. 9, 1151 - 1165 = Publ. Goethe Link Obs., Indiana Univ. No. 94 (1969).

In order to assess the importance of departures from thermodynamic equilibrium among the populations of the bound states of He I in explaining the observations, we have calculated departures for several model helium atoms, including some with levels through $n = 20$, in model atmospheres of 20000, 30000 and 40000 °K. The coupled equations of radiative transfer and steady state were solved for all significant bound-bound transitions and for all bound-free continua. Results for various atomic models are presented and the influence of the model is discussed.

064.005 **Asymptotic solutions of the solar (stellar) wind equations for large distances from the central star.** S. Grzędzielski.
Acta Astron. Vol. 19, 189 - 198 (1969).

The asymptotic behaviour of solar (stellar) wind solutions at large distances r from the central star is discussed. The equations include the effects of rotation and of the anisotropy in thermal conductivity due to the presence of a (weak) magnetic field. The nature of the singularity at $r = \infty$ is discussed. Examples of numerical integrations are shown which indicate that for a fixed critical point the temperature $T \to 0$ for $r \to \infty$ only for solutions corresponding to one set of numerical values for the parameters F (total mass flow), E (total energy flux) and Ω (angular velocity of the central star). Analytical discussion of such solutions is also presented. It is shown that the conduction flux may not vanish in infinity even though the thermal conductivity vanishes for $r \to \infty$.

064.006 **Differential fine analysis Sirius versus Vega.** U. K. Gehlich.
Astron. Astrophys. Vol. 3, 169 - 178 (1969).

Differential chemical abundances for Sirius and Vega are determined by means of model atmosphere analysis. The parameters of the model atmospheres are found to be $T_{eff} = 9700°$ K, $\log g = 4.3$, $\xi_t = 2.0$ km/s for Sirius and $T_{eff} = 9200°$ K, $\log g = 4.0$, $\xi_t = 3.5$ km/s for Vega. Nearly all metals are more abundant in Sirius than in Vega. However, Sirius does not meet all Am-star criteria. Remaining uncertainties in the abundance differences are mainly due to the errors of the measured equivalent widths of the lines in the spectrum of Vega.

064.007 **The atmosphere of the helium-carbon star BD+10°2179.** K. Hunger, D. Klinglesmith.
Astrophys. Journ. Vol. 157, 721 - 735 (1969).

A fine analysis of the helium-carbon star BD+10°2179 has been performed by using a grid of constant flux models. Opacities considered are: H I, H⁻, He I, He II, He⁻, C II, and electron scattering. The C II opacities are taken from quantum-defect calculations of Peach. The parameters of the models are: $T_{eff} = 16000°$, $18000°$ K; $\log g = 2.5$, 3.0; X (mass fraction of hydrogen) $= 10^{-3}$; Z_c (mass fraction of carbon) $= 0.035$, 0.152, 0.52. Abundances of thirteen elements are derived and compared with previous analyses. The mass of BD+10°2179 is estimated to $\geq 0.5 M_\odot$.

064.008 **The structure and kinematics of the envelopes of Be stars: Pleione's shell episode, 1938 - 1954.** D. N. Limber.
Astrophys. Journ. Vol. 157, 785 - 797 (1969).

An attempt is made to interpret Pleione's 1938 - 1954 shell episode in terms of relatively simple model for the structure and evolution of the circumstellar envelope. It is found that the observed variations with time of the intensities, the radial velocities, and the profiles of the shell lines for the higher Balmer lines can be readily understood in terms of a model in which (1) the support of the envelope in the star's equatorial plane is centrifugal, while its support perpendicular to the equatorial plane is thermal, (2) the outward component of velocity, assumed to be independent of time, increases essentially monotonically outward from a value of a few kilometers per second near the star's equator to a value of 60 km sec⁻¹ or more at large distances from the star, and (3) the rate at which mass moves into the base of the envelope from the star's equator rises monotonically over a period of several years from zero to a maximum value, and then drops essentially to zero over a much shorter period of time.

064.009 **Model atmospheres of late-type stars.** J. R. Auman, Jr.
Astrophys. Journ. Vol. 157, 799 - 826 (1969).

Model atmospheres with solar abundances have been calculated with effective temperatures between 2000° and 4000° K and with gravities corresponding to dwarf, giant, and supergiant stars. The opacity due to water vapor was included. Convection was included by using the Böhm-Vitense mixing-length theory with the mixing length equal to the pressure-scale height. In order to test the effects of varying the composition, two atmospheres were calculated with $T_e = 4000°$ K and log

$g = 2.0$, with the solar metal abundances reduced by factors of 10 and 100. The emitted flux changed only slightly. In addition, two atmospheres were calculated with $T_e = 3000°$ K and $\log g = 1.0$, with the carbon abundance varied. The strengths of the H_2O features vary much less than the H_2O abundance.

064.010 The excitation of nonspherical waves in differentially rotating stellar convective envelopes.
S. Kato.
Astrophys. Journ. Vol. 157, 827 - 834 (1969).

A possible mechanism which excites nearly horizontal nonspherical waves in differentially rotating stellar convective envelopes is examined. It is shown that such waves can be excited either by the dynamical instability due to the horizontal-velocity shear of differential rotations or by the overstable thermal instability in the convective envelope; the latter results from the cooperative interaction of the superadiabatic temperature stratification and thermal conduction. The excitation by the differential rotation is similar to that of Rossby waves in meteorology. A numerical example is considered for the sun, and it is suggested that the sinusoidal deformation of quiescent prominences could be explained in terms of the excitation of such waves.

064.011 Transfer of radiation in circumstellar dust envelopes. I. Extreme cases. S.-S. Huang.
Astrophys. Journ. Vol. 157, 835 - 842 (1969).

Transfer of radiation in circumstellar dust clouds has been treated for two simplified cases: (1) the thin cloud and (2) the completely opaque cloud. The condition and the consequences of these two cases are examined. For the thin cloud the problem is reducible to a single integral equation that links the emergent infrared spectrum with the dust distribution around the star, and for the completely opaque cloud the problem resembles that of extended stellar atmospheres.

064.012 Transfer of radiation in circumstellar dust envelopes. II. Intermediate case. S.-S. Huang.
Astrophys. Journ. Vol. 157, 843 - 855 (1969).

The gradual conversion of optical radiation to infrared radiation in the circumstellar dust envelope with spherical symmetry is disccued on the basis that the dust particles not only absorb optical radiation but also scatter it. Scattering is treated as isotropic and coherent, while the temperature at any point in the dust envelope is assumed to be in equilibrium with the combined infrared radiation and the absorbed portion of optical radiation at the point. If the opacity is due to the geometrical cross-sections of dust grains, it has been shown that the problem can be expressed by two independent integro-differential equations. Two cases — one of uniform distribution of dust in the envelope and the other of decreasing distribution outward — have been numerically carried out. Finally, the formal solution is given for the radiation field in the spherical shell of a finite thickness.

064.013 Models for envelopes of stars of 4 $M_\odot$ and 20 $M_\odot$.
E. Ergma.
Nauchn. Informatsii, Vyp. (No.) 12, 95 pp. (1969).
In Russian and English.

Models for stellar envelopes on stages of evolution from the main sequence towards the red giants have been computed for stars with $M = 4 M_\odot$ (X = 0.602, Z = 0.044) and $M = 20 M_\odot$ (X = 0.70, Z = 0.04). The models have been computed for different values of the luminosity $L/L_\odot$ and the effective temperature.

064.014 Über die Explosion einer Sternhülle durch eine Stoßwelle. I. A. Klimishin.
Tsirk. L'vov. Astron. Obs. No. 43, p. 3 - 7 (1969).
In Russian.

064.015 Vibrational stability of pure helium stars surrounded by pure hydrogen envelopes.
R. van der Borght.
Australian Journ. Phys. Vol. 22, 497 - 503 (1969).

This paper investigates the influence of a pure hydrogen envelope on the vibrational stability of pure helium stars and establishes the limiting mass as a function of the fractional mass of such an envelope. It is found that an envelope with a mass fraction of the order of 10^{-4} will make the star vibrationally stable, even when the total mass of the star is well above the limiting mass for pure helium stars.

064.016 Departures from LTE in the hydrogen lines of late B stars. D. M. Peterson, S. E. Strom.
Astrophys. Journ. Vol. 157, 1341 - 1361 = Contr. Kitt Peak National Obs. No. 427 (1969).

Departures from LTE for atomic hydrogen have been computed under the assumption of detailed balance in the hydrogen lines for a number of models of B star atmospheres. This assumption has been found to be valid not only for the physical conditions encountered in B stars of all luminosities but also for a much wider range of temperatures and luminosities. Photoelectric and photographic observations of these Balmer lines have been obtained for a number of B stars. The observations suggest that Hα and Hγ are indeed affected by departures from LTE, and that the observed effects are close to those predicted by the models reported here.

064.017 The effects of line shifts on the temperature distribution in stellar atmospheres. D. Mihalas.
Astrophys. Journ. Vol. 157, 1363 - 1367 (1969).

Simplified models are constructed to estimate the radiative effects upon the temperature structure of a stellar atmosphere due to an abrupt change in the line spectrum at some depth in the atmosphere. The physical motivation is that such changes can occur in the presence of velocity fields or shock fronts in the atmosphere.

064.018 Variations rapide, probablement non périodiques d'enveloppes d'étoiles Be.
A. M. Deplace, R. Herman, A. Peton.
Non-Periodic Phenomena in Variable Stars, IAU Colloquium, Budapest, 1968, p. 223 - 226 (1969).

064.019 Radiation-hydrodynamic phenomena in the atmospheres of luminous stars. R. W. Hillendahl.
Astrophys. Letters, Vol. 4, 179 - 181 (1969).

A physical mechanism which can result in the generation of extended stellar atmospheres is discussed. The process involves the unloading of stellar material following the arrival of a shock wave at the edge of a star. A non-gray transport type radiation-hydrodynamics code was used to obtain quantitative results for a cepheid atmosphere. The computed line profiles compare favorably with those observed for supergiants. The unloading mechanism provides an explanation for the low pressure gradients (low g) in supergiant atmospheres and illustrates the importance of hydrodynamic flow in determining the temperature and density profiles in the atmospheres of luminous stars. The additional acceleration occurring in the process provides a suggested mechanism for mass loss.

064.020 Formation of line and continuous spectra. I. Source-function calculations. E. H. Avrett, R. Loeser.
SAO Cambridge, Mass. Special Rep. No. 303, 9 + 98 pp. (1969).

We present in full detail a method for solving the combined equations of statistical equilibrium and radiative transfer for line and continuum transitions. The method can be used to determine the variation of atomic number densities with depth in a stellar atmosphere. Local thermodynamic equilibrium is not assumed. In this paper, we discuss only the basic source-function calculations. In subsequent papers, we shall discuss related iterative procedures and the general determination of a theoretical spectrum.

064.021 Radiative transfer within a stellar absorption line. The contribution curves of fine-analysis methods.
F. N. Edmonds, Jr.
Journ. Quant. Spectrosc. Radiative Transfer, Vol. 9, 1427 - 1446 (1969).

Three methods of fine analysis of stellar spectral lines, the formal-solution, the Planckian-gradient, and the weight-saturation, are formally equivalent under the condition of pure absorption or LTE. Comparisons of these methods have been carried out for various model atmospheres of Procyon and the sun and for four representative spectral lines. These comparisons were made between curves of growth, line profiles, and related contribution curves. The principal result is that the contribution curves for the three methods and for residual flux are quite different.

064.022 The curve of growth for an expanding atmosphere.
M. A. Arakelian.
Astrofizika, Vol. 5, 75 - 81 (1969).   In Russian.
English translation in Astrophysics, Vol. 5, No. 1 (1969).

The curves of growth for an expanding atmosphere in the case of pure absorption are calculated. The velocity of expansion is supposed to be constant and the intensity of the continuous radiation is accepted to be a linear function of $\cos \varphi$ (φ is the angular distance from the center to the limb of the star).

064.023 Transfer of resonance line radiation from a point source in the half-space.
Y. Y. Abramov, A. P. Napartovich.
Astrofizika, Vol. 5, 187 - 202 (1969).   In Russian.
English translation in Astrophysics, Vol. 5, No. 2 (1969).

The case is considered when the point source is situated on the boundary of the half-space. The assumption of complete redistribution is used. The continuous absorption is neglected. Quite simple expressions are found for the asymptotic behaviour of the excitation degree of the atoms at large distances from the source.

064.024 On the nonlinear nonstationary problem of radiation transfer in a spectral line.
R. S. Vardanian, N. B. Yengibarian.
Astrofizika, Vol. 5, 203 - 211 (1969).   In Russian.
English translation in Astrophysics, Vol. 5, No. 2 (1969).

The nonlinear nonstationary problem of radiation transfer in plane-parallel layers or in a one-dimensional medium, consisting of two-level atoms is considered. The cases of monochromatic scattering and complete redistribution of frequencies are discussed. The solution of this problem is obtained as power series of t.

064.025 The Rayleigh scattering cross-sections of He, C, N and O.   S. P. Tarafdar, M. S. Vardya.
Monthly Notices, Roy. Astron. Soc., Vol. 145, 171 - 180 (1969).

The Rayleigh scattering cross-sections of He, C, N and O have been computed by quantum defect method. Fairly good agreement has been obtained between the results of dipole length and dipole velocity formalisms.

064.026 Radiative transfer effects in Cepheid atmospheres.
C. G. Davis, Jr., J. Bendt.
Bull. American Astron. Soc. Vol. 1, 238 (1969). − Abstr. AAS.

064.027 Hydrodynamic atmospheres of a Cepheid model.

C. F. Keller, Jr.
Bull. American Astron. Soc. Vol. 1, 247 (1969). − Abstr. AAS.

064.028 Microturbulence in A-type supergiants.
J. D. Rosendhal.
Bull. American Astron. Soc. Vol. 1, 260 (1969). − Abstr. AAS.

064.029 Line formation in multidimensional media.
H. P. Jones, A. Skumanich.
Bull. American Astron. Soc. Vol. 1, 281 - 282 (1969). − Abstr. AAS.

064.030 Mass loss from stars: A review.
A. J. Deutsch.
Mass Loss from Stars, Trieste 1968, p. 1 - 14 (1969).

064.031 Microturbulence in stellar atmospheres. Methodic remark.   E. A. Gussmann.
Mass Loss from Stars, Trieste 1968, p. 72 - 74 (1969).

064.032 Mass loss from coronae and its effect upon stellar rotation.   K. Nariai.
Mass Loss from Stars, Trieste 1968, p. 122 - 131 (1969).

The acoustic energy-generation rate from the convective zone was calculated for various models. Results show that chromosphere and corona can be expected around stars with temperature lower than 8000 K at the main sequence, and lower than 6500 K at $\log g = 2$. When a star is rotating rapidly, mass loss from its corona is large, and can be an effective mechanism of braking the stellar rotation. If this mechanism is effective, we can explain the slow rotation of stars later than F2 to be the result of the loss of the angular momentum through a stellar wind that is effective in their main sequence phase. Stars with mass $M > 1.5\,M_\odot$ lose mass through a stellar wind during their contraction phase.

064.033 Coronae around helium stars and X-ray sources.
K. Nariai.
Mass Loss from Stars, Trieste 1968, p. 132 - 134 (1969).

Calculations of the acoustic energy generation for helium-rich composition show that the maximum acoustic energy generation is located around 12000 K at $\log g = 4$ and 15000 K at $\log g = 6$. The author's suggestion in his last paper that a helium star v Sgr may have a corona seems to be justified.

064.034 Über die Möglichkeit einer Anwendung der Methode Mustels zur Berechnung von Photosphärenmodellen von A5 – F5-Sternen.   E. E. Belyaeva.
Trudy Kazan. Gorod. Astron. Obs. No. 35, p. 44 - 59 (1968). In Russian.

Die Formel von Mustel zur Berechnung von Photosphärenmodellen reiner Wassterstoffsterne wird jetzt auch auf gleichartige Berechnungen von A5 – F5-Sternen mit unterschiedlicher chemischer Zusammensetzung angewandt. Verf. erläutert ausführlich die hierzu notwendigen Modifikationen seiner Methode.

064.035 Nucleosynthesis in neutron star atmospheres.
L. C. Rosen.
Astrophys. Space Sci. Vol. 5, 150 - 170 (1969).

The composition of neutron star atmospheres is calculated as a function of time including effects of diffusion, cooling and thermonuclear reactions. A seven-component nuclear reaction network which includes He^4, C^{12}, O^{16}, Ne^{20}, Mg^{24}, Si^{28} and Fe^{56} is utilized. Neutron star models with different initial nuclear abundances are compared as to subsequent nucleosynthesis. It is found that the final abundances are independent of original composition assuming He^4 as the major initial constituent. The final composition of the atmosphere is predominantly Fe^{56}. Mass loss from an evolving neutron star is examined as a possible source of cosmic rays. It is found that a neutron star contributes only Fe^{56} significantly to the cosmic-ray spectrum.

064.036 The formation and growth of carbon particles in the atmospheres of cool carbon stars. J. D. Fix.
Monthly Notices, Roy. Astron. Soc., Vol. 146, 37 - 49 (1969).

A system of equations is developed which describes the formation and growth of solid particles in an adiabatic gas undergoing sinusoidal time variations of temperature. The system of equations is solved throughout a cycle of temperature variation for a gas having an initial temperature, pressure and composition representative of the atmosphere of a cool carbon star and having a period and amplitude of temperature variation typical of long period variables. The results show that the particles form and grow to nearly their maximum size during a relatively small fraction of the period of temperature variation.

064.037 The formation and growth of carbon particles in the atmospheres of Mira variables. J. D. Fix.
Monthly Notices, Roy. Astron. Soc., Vol. 146, 51 - 55 (1969).

In light of recent evidence that the oxygen to carbon ratio in Mira variables may be only slightly greater than 1.0, molecular dissociative equilibrium calculations are made using an O/C ratio of 1.03, O/H ratio of 10^{-4} and Si/H ratio of 4×10^{-5} to determine the region of the P, T plane in which carbon is super-saturated and particle formation is possible. A system of equations describing the formation and growth of carbon particles throughout the temperature cycle of a variable star is solved for a gas undergoing temperature variations characteristic of those occurring in the atmospheres of Mira variables.

064.038 Theoretical studies on stellar stability. II. Undisturbed convective nongrey atmospheres.
W. Unno.
Publ. Astron. Soc. Japan, Vol. 21, 240 - 262 (1969).

A method for the construction of convective nongrey atmospheres is developed on the same theoretical basis as employed in paper I so that the construction of undisturbed model and the stability analysis can be made with consistent scheme of calculation. Improvement on the current mixing-length theory of Vitense is made. Non-locality of convective flow and nongrey radiative transfer are taken into account.

064.039 Contribution functions. H. Ruhm.
Astron. Astrophys. Vol. 3, 277 - 280 = Mitt. Landessternw. Heidelberg-Königstuhl No. 161 (1969).

Gussmann's criticism of Elste's contribution functions is discussed. It is shown that Gussmann's interpretation of his contribution functions should be revised.

064.040 Model atmospheres of the early spectral type stars.
S. Ruciński.
Postępy Astron., Vol. 17, 201 - 216 (1969). In Polish.

The contemporary methods of constructing the non-gray model atmospheres of early-type stars without assuming local thermodynamic equilibrium (non-LTE models) are described. Feautrier's method for the numerical solution of the transfer equation in case of a complicated source function (for example with non-coherent scattering included) is given.

064.041 The effective temperatures of the O stars.
D. C. Morton.
Astrophys. Journ. Vol. 158, 629 - 640 (1969).

Effective temperatures of O-type stars imbedded in diffuse nebulae are derived from measurements of Hα and radio fluxes from the nebulae and the apparent magnitudes of the stars. Accurate model atmospheres, with ultraviolet line blanketing where appropriate, are used for the theoretical relation between effective temperature and the ratio of Lyman-continuum fluxes to visual stellar fluxes. Although there is considerable scatter in the results, an average temperature of 48000°K is found for spectral type O5, 40000°K for O6, and 35000°K for O7.

064.042 Non-LTE model atmospheres. III. A complete-linearization method.
L. H. Auer, D. Mihalas.
Astrophys. Journ. Vol. 158, 641 - 655 (1969).

In this paper we present a method of solving transfer problems subject to constraints of radiative, hydrostatic, and statistical equilibrium, by using a complete-linearization technique. The form of the equations assures that the coupling among all physical variables is fully accounted for self-consistently to first order at each stage of the calculation; convergence is global and quadratic. The constraint equations are included in a particularly transparent way, which allows easy generalization and elaboration. Sample models of pure-hydrogen atmospheres are presented.

064.043 Relativistic stellar-wind torques. F. C. Michel.
Astrophys. Journ. Vol. 158, 727 - 738 (1969).

Several authors have calculated the torque on the sun owing to the solar wind. We extend those treatments to encompass both special- and general-relativistic effects. A direct parallel treatment is feasible, and the major correction is, as would be expected, from the increased effective mass density of a relativistic fluid.

064.044 High-dispersion spectroscopy and the interpretation of stellar atmospheres. L. H. Aller.
Proc. Astron. Soc. Australia, Vol. 1, 264 - 268 (1969). − Review article.

064.045 Observational results concerning departures from LTE in late B stars. M. A. Smith, S. E. Strom.
Astrophys. Journ., Vol. 158, 1161 - 1165 (1969).

Measures of the size of the Balmer and Paschen discontinuities have been obtained for a number of B stars by use of an intermediate-band photometric system described in this paper. The ratio of observed Paschen-to-Balmer discontinuities, ϕ, has been shown to be a sensitive indicator of departures from LTE. The ϕ-values obtained from the observations are compared with the predictions of recently computed non-LTE models. It is concluded that the departures are negligible for stars less luminous than class III but are significant for supergiants.

064.046 Opacity-probability distribution functions for carbon monoxide at elevated temperatures.
V. G. Kunde.
Astrophys. Journ., Vol. 158, 1167 - 1176 (1969).

It is the purpose of this paper to determine opacity-probability distribution functions for CO for inclusion in calculations of model atmospheres. Opacity-probability distribution functions have been determined from computed theoretical spectra of $^{12}C^{16}O$ at temperatures of 1680°, 2016°, 2520°, and 3360°K and at turbulent velocities of 0, 2, and 8 km sec^{-1}.

064.047 Van der Held curves of growth.
A. L. T. Powell.
Roy. Obs. Bull. Greenwich−Cape, No. 152, p. E 385 - E 408 (1969).

Thirty-one van der Held theoretical curves of growth have been computed for a range of the damping constant from 0.0005 to 0.5 in equal logarithmic intervals. The curves are normalized with a total Doppler broadening parameter of 1.69 km/s, which corresponds approximately to the solar curve of growth derived by Cowley and Cowley.

064.048 Intrinsic polarization in nongray atmospheres.
G. W. Collins, II.
Bull. American Astron. Soc., Vol. 1, 338 (1969). – Abstract AAS.

064.049 A preliminary atmospheric analysis of θ Ursae Majoris. J. C. Evans, L. W. Schroeder.
Bull. American Astron. Soc., Vol. 1, 341 (1969). – Abstract AAS.

064.050 Model atmospheres for degenerate stars.
S. S. Kumar, R. J. Doyle.
Bull. American Astron. Soc., Vol. 1, 351 (1969). – Abstract AAS.

064.051 Envelope models of hydrogen-deficient stars.
K. Nariai.
Bull. American Astron. Soc., Vol. 1, 355 (1969). – Abstract AAS.

064.052 Mass loss: A hydrodynamic envelope for stellar models. D. W. Schuerman, M. P. Savedoff.
Bull. American Astron. Soc. Vol. 1, 360 - 361 (1969). – Abstract AAS.

064.053 Meridional circulation and gravitational settling.
R. C. Smith.
Observatory, Vol. 89, 208 - 209 (1969). – Letter.

064.054 Relative abundances of elements in the atmospheres of 41 Cyg, ν Her and σ Boo.
T. Kipper.
Izv. Akad. Nauk Estonskoj SSR, Vol. 18, (Fiz., Mat. 1969 No. 1), 65 - 78 = Tartu Astron. Obs. Teated No. 21 (1969). In Russian.

Using an approximate model atmosphere for F stars theoretical curves of growth were calculated with a "Minsk 2" computer. The Planckian method, developed by Mugglestone, was modified. By means of these curves of growth and equivalent widths, obtained from high dispersion spectra, the relative element abundances for the atmospheres of 41 Cyg, ν Her, and σ Boo were derived and compared with solar abundances.

064.055 Model atmospheres for RR Lyrae stars.
S. E. Strom.
Report NASA–CR–95982, Smithsonian Astrophys. Obs., Cambridge, Mass. 12 pp. (1968). – See Phys. Abstr. Vol. 72, No. 24776 (1969).

064.056 Hydrogen molecules and the Hayashi effect.
M. V. Penston.
Phil. Trans., Ser. A, Vol. 264 (No. 1150), 235 - 240 (1969).

064.057 La stratification des couches à émissions discrètes dans les atmosphères des étoiles variables à longue période. J.-P. Swings.
Acad. Roy. Belgique, Bull. Cl. Sci., 5e Série, Vol. 55, 553 - 559 (1969).

The behavior of [Fe II] lines and AlO bands in long period variables is examined and discussed from the point of view of stratified models of the emitting layers of these objects.

064.058 Righe di emissione del ferro ionizzato in oggetti celesti. R. Viotti.
Atti XII Riunione Soc. Astron. Italiana, L'Aquila 1968, p. 45 - 46 (1969). – Abstract SAI.

064.059 Soluzione dell'equazione del trasporto relativa ad una riga e informazione contenuta in un profilo.
G. G. Noci.
Atti XII Riunione Soc. Astron. Italiana, L'Aquila 1968, p. 85 - 86 (1969). – Abstract SAI.

064.060 Models of the F–K stars with extended atmospheres.
I. B. Pustylnick.
Tartu Astron. Obs. Teated No. 23, p. 3 - 34 (1969).
In Russian.

A brief description of the computational procedure and numerical results of calculations of models of extended atmospheres for the F–K stars are given.

064.061 Die Atmosphären einiger K-Hauptsequenzsterne.
P. Strohbach.
Mitt. Astron. Ges. No. 27, p. 230 - 232 (1969). – Conference paper.

064.062 Modellatmosphären für heliumreiche Weiße Zwerge.
I. Bues.
Mitt. Astron. Ges. No. 27, p. 232 - 235 (1969). – Conference paper.

An approximative model of two late-type stars.
See Abstr. 065.007.

Progrès récents dans le calcul de l'opacité stellaire.
See Abstr. 065.086.

Revised solar iron abundance and its influence on the photosperic model. See Abstr. 071.003.

Resonance-broadening absorption in the wings of Lyman alpha. See Abstr. 071.014.

The carbon, nitrogen, and oxygen abundances in four K giants. See Abstr. 114.014.

Theoretical continuous and line spectra of stars in a close binary system. See Abstr. 117.030.

065 Stellar Structure, Stellar Evolution, Stellar Nucleosynthesis

065.001 Rotational perturbation of a radial oscillation in a gaseous star. R. Simon.
Astron. Astrophys. Vol. 2, 390 - 397 (1969).

The perturbation method is applied to the problem of the oscillations of a gaseous star rotating around a fixed z axis according to a general law of the type $\Omega = \Omega(r, \theta)$. This rotation is assumed to be slow and the analysis includes all the effects in Ω and Ω^2, and particularly the rotational distortion of the stationary motion. The role of the trivial modes of oscillation of the zero order spherical star is emphasized, since these transversal and divergence-free modes which have a vanishing σ frequency appear to be important in the Ω^2 terms. The particular application to a radial mode is considered in detail and numerical results are given for a rigid rotation, in the case of the fundamental radial oscillation of the homogeneous and of the standard models.

065.002 Models for main-sequence stars.
J. Horn, S. Kříž, M. Plavec.
Bull. Astron. Inst. Czechoslovakia, Vol. 20, 193 - 201 (1969).

Evolutionary model sequences are computed for stars of 2, 3, 4, 5, 6, 7, and 9 solar masses, covering the phase of hydrogen burning in the star's core. Relations between various characteristics for the zero-age models are given, as well as for the top of the main-sequence expansion. Conditions are studied under which the star, if it happens to be the primary component of a binary system, becomes dynamically unstable during the main-sequence phase. Initial conditions for the process of mass exchange are derived.

065.003 Isentropic stars in general relativity.
A. Kovetz.
Astrophys. Space Sci. Vol. 4, 365 - 369 (1969).

In an investigation of the evolution of homogeneous, isentropic stars through stages of diminishing entropy, Rakavy and Shaviv (1968) have recently found that stars of mass less than M_c (Chandrasekhar's limiting mass for white dwarfs) evolve into white dwarfs, while stars of mass greater than M_c approach a (singular) state of minimum entropy. An elementary explanation of these results is given and qualitative effects of general relativity are discussed. It is found that stars which are lighter than the Oppenheimer and Volkoff (1939) limit become white dwarfs, while heavier stars must become dynamically unstable at a finite stage in their evolution.

065.004 Carbon-burning nucleosynthesis at constant temperature. W. D. Arnett, J. W. Truran.
Astrophys. Journ. Vol. 157, 339 - 365 (1969).

Synthesis of elements during thermonuclear burning of carbon is examined at a series of temperatures (T_9 = 0.6, 0.8, 1.0, 1.2, and 1.4) and for several initial compositions. Recent experimental results for the ($^{12}C + ^{12}C$)-reaction are used. A general method for numerical solution of nuclear-reaction networks is described. At the higher temperatures in the range now thought to be appropriate for carbon burning in stars, $T_9 \gtrsim 1$, the Na/Mg ratio is similar to that of the sun. At all temperatures considerable production of ^{20}Ne occurs. By using the results of the reaction-network calculations, a simple analytic expression for energy generation during carbon burning is constructed. Neutron exposures during carbon burning are presented for investigation of the s-process.

065.005 Experimental investigation of the stellar nuclear reaction $^{12}C + ^{12}C$ at low energies.
J. R. Patterson, H. Winkler, C. S. Zaidins.
Astrophys. Journ. Vol. 157, 367 - 373 (1969).

The reactions have been studied to provide reliable predictions of their cross-sections in the energy region of interest in stellar carbon burning. Many excited-state transitions have been observed for the α-particles and protons. Measurements for these particles have been made in the center-of-mass energy range 3.23–8.75 MeV. The relatively small yield of the $^{12}C(^{12}C, n)^{23}Mg$ reaction has been investigated from 4.25 to 6.25 MeV. The new results for these reactions lead to lower cross-sections than previously estimated for the region of astrophysical interest.

065.006 The effect of auto-ionization lines on the opacity of stellar interiors. W. D. Watson.
Astrophys. Journ. Vol. 157, 375 - 387 (1969).

The contribution to the Rosseland opacity from auto-ionization lines is computed for an element mixture consisting of X = 0.74, Y = 0.24, and Z = 0.02. This contribution is found to be of roughly the same importance as that from normal lines at temperatures greater than about 5×10^5 °K. Under conditions that occur within stellar interiors, the contribution from auto-ionization lines often represents an increase of one-third to one-half in the total Rosseland opacity. The increase can be as great as 100 per cent at some temperatures and densities. Computations are performed at twenty temperature-density points. At these points, the following Rosseland opacities are presented: continuous, continuous plus normal line, continuous plus auto-ionization line, and total radiative.

065.007 An approximative model of two late-type stars.
R. Bajcár, I. Bajcárová.
Contr. Astron. Obs. Skalnaté Pleso, Vol. 4, 93 - 103 (1969).

Models of two late-type stars are studied by the analytical method. One of the stars (HD 37160 spectrum K0 III) is a high-velocity star, the other (HD 188512 spectrum G8 IV) is a standard star. The atmospheric curves of both stars are determined from quantities given in the foregoing paper, under certain simplifying conditions (assumed absorption coefficient and gas pressure). These curves are compared with values given by de Jager and Neven, and with the values for the sun.

065.008 Stellar configuration of degenerate electron gas.
G. S. Sahakian, M. A. Mnatsakanian.
Soobshch. Byurakan. Obs. No. 40, p. 98 - 107 (1969).
In Russian.

The parameters of spherically symmetric static stellar configurations consisting of atomic nuclei and degenerate electron gas are computed. The equations of Newton's generalized gravitation theory are used. The mass and radius dependencies from the central density consist of two branches. The first represents the known ordinary white dwarf configurations, the second one corresponds to stellar bodies with masses of $5 \times 10^8 \lesssim M/M_\odot \lesssim 10^{15}$ and radii of $1.4 \times 10^5 \lesssim R \lesssim 1.3 \times 10^9$ km.

065.009 How are neutron stars formed?
A. G. W. Cameron.
Comments Astrophys. Space Phys. Vol. 1, 172 - 177 (1969).

065.010 Stellar interiors: The source of life and death.
E. J. Öpik.
Irish Astron. Journ. Vol. 9, 15 - 31 (1969).

Basis of lecture given to the Armagh Centre of the Irish Astronomical Society on November 19, 1968.

065.011 Particle diffusion in degenerate, ionized gases.
L. C. Rosen.
Astrophys. Space Sci. Vol. 5, 92 - 102 (1969).

A perturbation treatment is used to find the distribution functions from the Boltzmann transport equation for a two-component ionized, degenerate gas. General expressions are found for the diffusion coefficient and relative diffusion velocity. Tables are given for specific values of spin and mass ratio.

065.012 **Treatment of pulsating white dwarfs including general relativistic effects.**
J. M. Cohen, A. Lapidus, A. G. W. Cameron.
Astrophys. Space Sci. Vol. 5, 113 - 125 (1969).

In this paper, pulsating white dwarfs are treated via general relativity. Numerical integration of Einstein's equations was used to find equilibrium white dwarfs models and the fundamental periods of small oscillations about these equilibrium models. In these calculations account was taken of coulomb, Thomas-Fermi, and exchange interactions as well as ion zero point energies. It is shown that general relativity makes not just a quantitative difference in the results but a qualitative difference; pure C^{12} models which are stable in Newtonian mechanics can be unstable against collapse (at a central density of 3×10^{10} g/cm^3) when general relativity is taken into account. The collapsing model may become a neutron star or may continue towards the Schwarzschild radius. More realistic white dwarf models with carbon burning products at the center, also were studied.

065.013 **Minimal and maximal values of the central pressure and temperature in convectively stable stars.**
A. Kovetz.
Monthly Notices, Roy. Astron. Soc. Vol. 144, 459 - 460 (1969).

For the convective stability of a star it is necessary that the specific entropy should not decrease outwards. Subject to this condition the minimal value of the central pressure in a star of mass M and radius R, and the maximal values of central pressure and temperature in a star of mass M and central density ρ_c, are obtained when the configurations are isentropic.

065.014 **The effect of rotation on the pulsation periods of degenerate white dwarf models.**   J. R. Gribbin.
Monthly Notices, Roy. Astron. Soc. Vol. 144, 549 - 552 (1969).

It is shown that for models of degenerate stars the fundamental pulsation periods in the presence of solid body rotation can be found to high accuracy (error <2 per cent) without solving the pulsation equation for a rotating body. The appropriate periods are given for models obeying both the Hamada–Salpeter and Chandrasekhar equations of state.

065.015 **Hydroxyl and water masers in protostars.**
M. M. Litvak.
Science, Vol. 165, 855 - 861 (1969).

Properties of these maser molecules are related to the earliest evolution of a star.

065.016 **Electric field in neutron stars.**
M. S. Bhatia, S. Bonazzola, G. Szamosi.
Astron. Astrophys. Vol. 3, 206 - 209 (1969).

The general-relativistic conditions for simultaneous hydrostatic and chemical equilibrium in a spherical neutron star are given. The critical mass does not change appreciably but a small outward radial electric field appears in the star as a result of the displacement of electrons from the core of the star.

065.017 **Fast evolution towards the white dwarfs.**
W. Deinzer, C. J. Hansen.
Astron. Astrophys. Vol. 3, 214 - 221 (1969).

Four evolutionary sequences of stellar models have been investigated numerically to improve upon an earlier attempt by Deinzer to explain some puzzling features in the evolution of the central stars of planetary nebulae. The present models consist of highly contracted, semi-degenerate, isothermal cores containing a large fraction of the stellar mass, and radiative envelopes of very small mass content which are capable of large contractions within short time-scales.

065.018 **General relativistic neutron star models.**
H. Heintzmann.
Astron. Astrophys. Vol. 3, 243 - 246 (1969).

Two topics of general relativity are reconsidered with the help of an exact solution of Einstein's field equations. We give an upper limit for the number of baryons of a stable neutron star and for the mass-defect due to gravitational binding. We show further that stable hyperon stars may exist with mean densities a hundred times greater than nuclear density. These can only be formed by catastrophic processes.

065.019 **The effect of excited nuclear states on stellar reaction rates.**   N. A. Bahcall, W. A. Fowler.
Astrophys. Journ. Vol. 157, 645 - 657 (1969).

The effect of excited states of target nuclei on stellar reaction rates is discussed. Two specific examples, ^{19}F(p, a) ^{16}O and ^{19}F$(p, \gamma)^{20}$Ne, are treated in detail using the available experimental information.

065.020 **Endoergic (p, n) and (a, n) reactions and their reverse reaction rates.**
N. A. Bahcall, W. A. Fowler.
Astrophys. Journ. Vol. 157, 659 - 672 (1969).

Experimental data on cross-sections for endoergic (p, n) and (a, n) reactions near threshold are used to calculate, for twelve nuclei, the corresponding reaction rates in the temperature range $10^8 - 10^{10}$ °K. Phenomenological expressions for average endoergic (p, n) and (a, n) cross-sections and rates are also derived. The reverse exoergic neutron rates are calculated from the experimental (p, n) or (a, n) data, using the reciprocity theorem, and are expected to be valid in the temperature range from 0° to about 10^{10} °K. Cross-sections for thermal-neutron reactions are deduced from these rates and compared where possible (three cases) with measured values. Calculations of cross-sections and rates for low-energy neutrons have also been carried out using a black-nucleus model.

065.021 **An explanantion for the blue sequence of variable stars.**   R. Stothers, N. R. Simon.
Astrophys. Journ. Vol. 157, 673 - 681 (1969).

Nuclear energy of the stellar core is found to be the direct source of energy fed into radial pulsations in two classes of blue variable stars. The μ-mechanism (reversal of the gradient of mean molecular weight, which reduces the central condensation) or the β-mechanism (high radiation pressure) can explain the basic variability of β Cephei stars if they are burning hydrogen in their cores, have accreted a helium-rich envelope, and are the luminous secondaries of close binary systems.

065.022 **Methods for the calculation of stationary spherical symmetric stars and their evolution.**
G. Ruben.
Nauchn. Informatsii, Vyp. (No.) 14, 139 pp. (1969).
In Russian.

In the present paper the foundations for the study of stationary spherical symmetric stars and of their evolution are given together with simple approximations to the auxiliary quantities and the boundary conditions. The main methods for the calculation of the corresponding models, i. e. the fitting method and the difference method are described and compared. Different variants of these methods are discussed. Programs written in Algol on the base of the derived formulas and algorithms for both methods are published together with the control values necessary for testing them. Evolutionary sequences and the different ways of comparing them with observations are discussed.

065.023 Nuclear Q values in a dense stellar plasma.
P. B. Shaw.
Phys. Rev., Second Series, Vol. 182, 1369 - 1373 (1969).

Corrections to nuclear Q values in a dense stellar plasma are calculated by means of a modified ionic cluster expansion in which the electrons are treated as a dielectric medium. Corrections beyond the weak-screening (Debye) approximation are given in tabular form for purposes of interpolation. A comparison between the Debye correction and the double-cluster correction is made for the chemical potential of a one-component plasma; the latter correction is ~20% of the former for a plasma parameter as large as unity.

065.024 Observational studies relating to star formation. II.
A. F. Aveni, J. H. Hunter, Jr.
Astron. Journ. Vol. 74, 1021 - 1023, 1058 - 1059 (1969).

Observational evidence is presented in support of the view that star formation may occur in small, primary condensations having masses $\lesssim 100$ M$_\odot$. The small aggregate containing the T Tauri-like star BM And is discussed in some detail, and other examples are cited of similar groups of early-type stars.

065.025 Observational clues to the evolution of M giant stars. A. M. Boesgaard.
Publ. Astron. Soc. Pacific, Vol. 81, 365 - 373 (1969).

High dispersion (4–8 Å/mm) spectrograms are used to search for incipient traces of S-star characteristics in a homogeneous sample of M giants. Even at this dispersion there are very few signs that M-star atmospheres are enriched by the elements found in S stars. It is concluded that if M stars evolve into S stars, they do so on a very short time scale.

065.026 Neutrino production by two-photon annihilation.
J. A. Campbell.
Astrophys. Journ. *(Letters)*, Vol. 157, L177 - L179 (1969).

A rate of loss of energy in stellar evolution through two-photon annihilation into neutrino-anti-neutrino pairs is given. The estimate is valid only if the weak interaction is mediated by an intermediate-vector boson. The effects of various assumptions about the mass of the intermediate-vector boson and about the size of the cutoff for the weak interaction are discussed.

065.027 Explosive nucleosynthesis in stars.
W. D. Arnett.
Astrophys. Journ. Vol. 157, 1369 - 1380 (1969).

Conditions for explosive nucleosynthesis in stars are derived for carbon burning and silicon burning. Numerical calculations for explosive carbon burning predict abundances of relevant isotopes which agree remarkably well with those observed in the solar system and in stars. Implications of these results for the theory of the late stages of stellar evolution and for the enrichment of the galaxy in heavy elements are discussed.

065.028 Entropy of a relativistic electron gas.
R. F. Tooper.
Astrophys. Journ. Vol. 157, 1391 - 1393 (1969).

An expression for the entropy of a relativistic electron-proton gas is derived, and limiting forms of this expression for the nonrelativistic and extremely relativistic cases are presented.

065.029 Thermonuclear pion production and high temperature stars. D. F. Falla, R. G. Sale.
Nature, Vol. 223, 725 - 726 (1969).

Thermonuclear pion production rates in hydrogen at temperatures 5×10^{10} °K and 10^{11} °K are computed, using a nucleon isobar model; the relative probability of muon neutrino production from pion decay is also found. These results are applied to the case of a star with an interior of high temperature and density, and the rate of energy loss through neutrino emission is estimated. It is concluded that this particular neutrino production process has a negligible effect on the thermal stability of the star.

065.030 Cepheid pulsation – I. Numerical technique and test calculations. R. S. Stobie.
Monthly Notices, Roy. Astron. Soc. Vol. 144, 461 - 484 (1969).

This paper describes the numerical technique developed to compute the non-linear, radial pulsation of a spherical star. The models constructed consist of a radiative envelope down to a temperature of 10^6 °K corresponding in radius to the outer four-fifths of the star. The only ionization zones considered as contributing substantially to the excitation are those of hydrogen and helium. Results of test calculations are reported to illustrate the sensitivity of the full amplitude motion to the free parameters in the construction of a model.

065.031 Cepheid pulsation – II. Models fitted to evolutionary tracks. R. S. Stobie.
Monthly Notices, Roy. Astron. Soc. Vol. 144, 485 - 510 (1969).

The results of a survey of classical cepheid models are presented. The masses and luminosities of the models were chosen to correspond to where evolutionary tracks of 5 - 9 $M_\odot$ cross the cepheid strip. The effect of helium content on the position of the high T_e boundary of the instability strip is examined and it is found that for every 15 per cent increase in helium (by mass) the strip shifts by ~600°K to higher T_e. Three different modes of pulsation are investigated for instability – fundamental, first and second modes. The existence of second mode instability is confirmed for three stars all of which have a period less than two days. The prescence of secondary bumps in the light and velocity curves was a prominent feature of many of the models. However, in no case did the bump occur (for a given period) at the phase expected from a comparison with observation. The only way in which the phase of the bump can be brought into agreement with observation is for the mass, at a given luminosity, to be reduced by a factor ~2 compared to the evolutionary calculations.

065.032 Cepheid pulsation – III. Models fitted to a new mass-luminosity relation. R. S. Stobie.
Monthly Notices, Roy. Astron. Soc. Vol. 144, 511 - 535 (1969).

A new mass-luminosity relation for classical cepheids was chosen so that models, which exhibited a prominent secondary bump in the light and velocity curves, agreed well with the observed period-phase of bump relation. As the high T_e edge of the strip is sensitive to the helium content, comparison with observation indicated a helium content of 45 per cent by mass though this value depends critically on the assumed $(B-V) - T_e$ transformation. The models which were examined covered a period range of 1 – 40 days. It was found that in general cepheids of period greater than 7 days pulsate in the fundamental, those of period between 2 and 7 days pulsate in the first mode and those of period less than 2 days pulsate in the second mode. The existence of a period-radius-mass relation from the models automatically implies the existence of a period-luminosity-colour relation provided that classical cepheids obey some mass-luminosity relation. This period-luminosity-colour relation was derived separately for the fundamental and the first mode pulsators.

065.033 Statistical and physical interpretation of non-perio-dic phenomena in variable stars.
L. Detre.
Non-Periodic Phenomena in Variable Stars, IAU Colloquium, Budapest, 1968, p. 3 - 19 (1969). — Introductory report.

065.034 A note on the solution of Eddington's equation.
S. Africk.
Astrophys. Letters, Vol. 4, 153 - 154 (1969).

By considering the eigenfunctions of Eddington's equation to be functions of the differential moment of inertia, one can construct a set of functions which *a priori* satisfy the orthonormality condition on the eigenfunctions. A particular set of these functions is exhibited and used to calculate the pulsation modes for a white dwarf.

065.035 On the gravitational stability of a gaseous sphere.
E. E. Shnol.
Astron. Zhurn. Akad. Nauk SSSR, Vol. 46, 970 - 977 (1969). In Russian. English translation in Soviet Astron. AJ, Vol. 13, No. 5.

A classical problem on the equilibrium stability of a gaseous sphere in a proper gravitational field is considered. It is shown that for the stability relative to any perturbations the fulfilment of the following conditions is necessary and sufficient: a) stability relative to radial perturbations, b) absence of convection.

065.036 Numerical calculations of the dynamics of a collap-sing proto-star. R. B. Larson.
Monthly Notices, Roy. Astron. Soc., Vol. 145, 271 - 295 (1969).

Numerical calculations of the dynamics of a spherically collapsing proto-star of one solar mass have been made for various initial conditions. Calculations have also been made for masses of $2M_\odot$ and $5M_\odot$. In all cases the collapse is found to be extremely non-homologous and is such that a very small part of the cloud's mass at the centre reaches stellar densities and stops collapsing before most of the cloud has had time to collapse very far. During most of this time the stellar core is completely obscured by the dust in the infalling cloud, the absorbed radiation reappearing in the infra-red as thermal emission from the dust grains. The resulting star is almost a conventional Hayashi pre-main sequence model, but it appears rather low on the Hayashi track. For masses much greater than about $2M_\odot$ the convective Hayashi phase does not exist at all.

065.037 The emitted spectrum of a proto-star.
R. B. Larson.
Monthly Notices, Roy. Astron. Soc., Vol. 145, 297 - 308 (1969).

The spectral energy distribution emitted by a collapsing proto-star has been calculated on the basis of some simple approximations for the radiative transfer in the extended spherical 'atmosphere' of the proto-star. Results for the evolution in spectral appearance are presented for several of the collapsing proto-star calculations described in a previous paper (Larson 1969). Comparison is made with some observations of infra-red objects in Orion and with recent infra-red observations of T Tauri stars.

065.038 Star formation in clouds of solid hydrogen grains — II. Some properties of model galaxies.
V. C. Reddish.
Monthly Notices, Roy. Astron. Soc., Vol. 145, 357 - 366 (1969).

It has been pointed out that delay in the negative feedback in the loop 'grain cooling → hydrogen freezing → star formation → grain heating' may cause overshoot in the rate of star formation. It is now shown that this will produce the observed properties of elliptical galaxies. It is also shown that the negative feedback in the evolution of a galaxy, due to the dependence of stellar mass on grain abundance described in (I), results in a high degree of uniformity among galaxies in the apparent average heavy element abundance of dwarf stars.

065.039 Interaction between low and high frequency stellar pulsations. Y. V. Vandakurov, Y. B. Zeldovich.
Astrofizika, Vol. 5, 235 - 241 (1969). In Russian.
English translation in Astrophysics, Vol. 5, No. 2 (1969).

The interaction between two adiabatic pulsations is considered. One of them is radial and has a frequency near zero, while the frequency of the second is much greater. They will be called briefly as Hr and br pulsations. The stabilizing influence of the br pulsation on the Hr pulsation is shown. The stability criterion is derived for the case of radial br pulsation. The equilibrium equation for the case of br pulsation and the excitation of the Hr pulsation are discussed.

065.040 Propagation of shock waves in a polytrope with a toroidal magnetic field. I. Simplified solution of differential equations. N. K. Sinha.
Australian Journ. Phys. Vol. 22, 589 - 603 (1969).

The propagation of an initially spherical shock wave in a polytrope with a magnetic field has been studied. The model chosen for the purpose was that of a polytrope with a toroidal magnetic field given previously by Sinha. Butler's method has been extended to transform the set of governing partial differential equations into a set of ordinary differential equations involving derivatives in the direction of propagation of the shock element at any point.

065.041 Propagation of shock waves in a polytrope with a toroidal magnetic field. II. Solution of complete differential equations. N. K. Sinha.
Australian Journ. Phys. Vol. 22, 605 - 612 (1969).

The differential equations for the shock parameters along shock rays in the case of propagation of a spherically developed shock wave in a polytrope with a toroidal magnetic field, obtained in part I, have been integrated numerically for a particular set of initial values. The results are compared with the corresponding results in part I obtained by neglecting certain small terms and it is found that the effect of this omission is not significant. This substantiates the results and justifies the simplification made in part I.

065.042 Nonradial pulsation of general-relativistic stellar models. III. Analytic and numerical results for neutron stars. K. S. Thorne.
Astrophys. Journ. Vol. 158, 1 - 16 (1969).

Numerical techniques are developed for calculating the properties of the nonradial, quasi-normal pulsations of fully relativistic stellar models. These techniques are then applied to calculate, for four realistic neutron-star models, the following characteristics of the lowest few quadrupole p-type modes: the pulsation periods ($\sim 10^{-3}$ sec), the gravitational-radiation damping times (~ 1 sec), the pulsation energies ($\sim 10^{52}$ ergs), the power emitted in gravitational waves ($\sim 10^{52}$ ergs sec^{-1}), and the shapes of the fluid motions.

065.043 The effect of collective interactions on the electron-scattering opacity of stellar interiors.
W. D. Watson.
Astrophys. Journ. Vol. 158, 303 - 313 (1969).

The effect of collective interactions on the electron-scattering opacity is calculated for conditions prevailing in stellar interiors. Use of the Debye-Hückel radial distribution function and neglect of collisions are shown to be adequate approximations. Collective effects decrease the electron-scattering opacity. They are important when the product of the Debye length and the photon wavenumber is near unity or smaller. The estimated influence on the total opacity in some stellar-interior regions is: (i) at the core of the sun the opacity is reduced by approximately 10 percent, (ii) the opacity within the hydrogen-burning shell of low-mass red giants is decreased by 5 - 20 percent at many stages of evolution, and (iii) during much of the evolution, a reduction of 5 - 15 percent is expected in the opacity of the core and the hydrogen-burning shell of a typical horizontal-branch star.

065.044 Radio observations of pre-protostars.
V. A. Hughes.
Bull. American Astron. Soc. Vol. 1, 244 (1969). — Abstr. AAS.

065.045 Nonlinear Cepheid pulsation calculations.
D. S. King, J. P. Cox.
Bull. American Astron. Soc. Vol. 1, 248 (1969). — Abstr. AAS.

065.046 Stellar evolution from main sequence to white dwarf or supernova.   B. Paczynski.
Bull. American Astron. Soc. Vol. 1, 256 (1969). — Abstr. AAS.

065.047 An analytical model for the evolution of shell-burning stars.   S. Refsdal, A. Weigert.
Bull. American Astron. Soc. Vol. 1, 258 - 259 (1969). — Abstr. AAS.

065.048 The final evolutionary stages of a low-mass star.
W. K. Rose, R. Smith.
Bull. American Astron. Soc. Vol. 1, 260 (1969). — Abstr. AAS.

065.049 URCA shells in white-dwarf stars.
S. Tsuruta, A. G. W. Cameron.
Bull. American Astron. Soc. Vol. 1, 263 (1969). — Abstr. AAS.

065.050 Recent studies of models for unstable stars.
W. K. Rose.
Mass Loss from Stars, Trieste 1968, p. 77 - 88 (1969).

Recent studies of stellar models have uncovered instabilities associated with helium and hydrogen shell burning. The present paper describes some calculations of stellar models that include these unstable stages and discusses some of their possible consequences. The influence of neutrino emission becomes very important during the helium shell-burning stage for stars more massive than $1 M_\odot$. Calculations are described that illustrate the influence of neutrino emission on stellar evolution up to the onset of carbon burning.

065.051 Mass loss from fast rotating stars.
L. Nobili, L. Secco.
Mass Loss from Stars, Trieste 1968, p. 135 - 138 (1969).

Evolutionary models of a rotationally unstable $7 M_\odot$ star are computed; preliminary results during the H-burning stages are obtained and discussed.

065.052 Angular momentum considerations and mass loss.
I. P. Williams.
Mass Loss from Stars, Trieste 1968, p. 139 - 145 (1969).

The conservation of energy and angular momentum in contracting star that is ejecting material is considered. It is found that in cases likely to be of interest, either an upper limit to the mass loss exists when the material escapes to infinity, or, when the primary aim is to remove angular momentum, a lower limit to the rate of mass loss exists.

065.053 Vulcanism and seismicity in neutron stars.
F. J. Dyson.
Comments Astrophys. Space Phys. Vol. 1, 198 - 206 (1969).

The purpose of the present comment is to discuss the consequences that follow, if one takes seriously the notion that a neutron star behaves more like the earth than like a normal star. The most conspicuous phenomena associated with the earth's mantle are vulcanism and seismicity. Do these phenomena have analogs in neutron stars? If so, is there any way to decide observationally whether such phenomena actually occur?

065.054 Thermal generation of magnetic fields with radiation pressure in rotating stars.
S. Kato, Y. Nakagawa.
Astrophys. Space Sci. Vol. 5, 171 - 179 (1969).

It has been suggested by Biermann that in rotating stars the electron partial pressure could generate a toroidal magnetic field of a considerable strength. However, Mestel and Roxburgh have shown recently that the generation of such a toroidal magnetic field could almost completely be suppressed when a weak primodial poloidal magnetic field exists in the star. In this paper it is shown that a toroidal magnetic field of a moderate strength could be generated even in the presence of a primodial poloidal magnetic field, if the effect of radiation pressure is taken into consideration. This considered mechanism is effective for moderately massive stars, and numerical estimate indicates that in A type stars a toroidal magnetic field of the order of a thousand gauss can be generated near the surface within the time scale of the evolution of the star.

065.055 Effects of hyperons on the vibrations of neutron stars.   W. D. Langer, A. G. W. Cameron.
Astrophys. Space Sci. Vol. 5, 213 - 253 (1969).

The effects of hyperons and resonance particles on the vibrations of neutron stars are calculated. Vibrating neutron stars can store large amounts of energy in their vibrations; the interaction of the vibrations with the atmosphere would produce electromagnetic radiation. If any process damps out the vibrations rapidly on an astronomical time scale (~ 1000 years) then vibrating neutron stars are not likely to be found. Previous work indicates that radiation by a neutrino URCA process ($N + N \rightarrow P + N + e^- + \bar{\nu}_e$) does not rapidly damp many of the neutron star models. Some neutron stars are predicted to contain massive baryons; here we study thermal damping by nonequilibrium reactions involving these baryons.

065.056 An equation of state at subnuclear densities.
W. D. Langer, L. C. Rosen, J. M. Cohen, A. G. W. Cameron.
Astrophys. Space Sci. Vol. 5, 259 - 271 (1969).

An equation of state for cold matter above white dwarf densities is evaluated. The gas is considered to be a mixture of degenerate neutrons, protons and electrons combined with nuclei of one type (that is only one A and Z value). We derive the equilibrium equations for the mixture and calculate the number densities as well as the A and Z of the nucleus. Finally we calculate an equation of state, which smoothly goes over to that of a neutron, proton electron gas mixture at a density of $\sim 5 \times 10^{13}$ g/cm^3.

065.057 Production of proton-rich heavy shielded elements in the stars.
A. Agnese, M. La Camera, A. Wataghin.
Monthly Notices, Roy. Astron. Soc., Vol. 146, 57 - 62 (1969).

For explanation of the observed abundances of proton-rich elements in stars Scotto, Paoli and Wataghin have proposed a mechanism taking into account the β-processes because of the presence of electrons and positrons in thermodynamical equilibrium at high temperatures. The mechanism analysed is found to be efficient for the production in the required abundance for: Er^{164}, Se^{74}, Cd^{108}, Sn^{114}, Xe^{126}, W^{180}, Hg^{196}, and partially efficient for Pd^{102}.

065.058 Neutral hydrogen in H II regions and regions of star formation. K. W. Riegel.
Interstellar Ionized Hydrogen, Charlottesville 1967, p. 61 - 85 (1968).
This report will be a review of our knowledge of the relationship between neutral hydrogen and young regions of star formation.

065.059 The formation of stars. S. von Hoerner.
Interstellar Ionized Hydrogen, Charlottesville 1967, p. 101 - 169 (1968).
The rate of star formation as a function of mass. The formation rate as a function of time. Star formation in clusters and associations. The process of star formation. The angular momentum problem.

065.060 The effects of star formation and evolution on the evolution of H II regions. R. M. Hjellming.
Interstellar Ionized Hydrogen, Charlottesville 1967, p. 171 - 188 (1968).

065.061 Gravitational separation of the elements and turbulent transport. E. Schatzman.
Astron. Astrophys. Vol. 3, 331 - 346 (1969).
The equation of diffusion is solved in a particular case, taking into account the transport due to gravitational separation. When the convective zone is shallow (F0 stars), the diffusion coefficient is not very small and the surface of the star could experience depletion of heavy and light elements, in the star life time, by a large factor. When the convective zone is deep, (like in the sun), depletion takes place very slowly. The amount of depletion which would take place in F0—F2 stars is so large that the question of a diffusion barrier is raised. Stellar spin-down, or meridional circulation can produce the appropriate turbulent velocity field. Even very slowly rotating stars can build a diffusion barrier sufficient to prevent gravitational separation.

065.062 Radiative and conductive opacities for twenty three stellar mixtures. A. N. Cox, J. N. Stewart.
Nauchn. Informatsii, Vyp. (No.) 15, 103 pp. (1969).
In English and Russian.
Detailed tables of Rosseland mean opacities, without bound-bound absorptions, with bound-bound absorptions, and finally including the effects of electron conduction, are presented for twenty three stellar mixtures. Brief descriptions are given for changes in the methods since earlier publications.

065.063 Measurements of cross-sections for nuclear reactions related to the anomalous composition of 3 Centauri A. R. J. Griffiths, S. A. Harbison.
Astrophys. Journ. Vol. 158, 711 - 718 (1969).
Measurements of the reactions $^3He(p,d)2p$, $^3He(p, pn)2p$, $^4He(p, d)^3He$, and $^4He(p, pn)^3He$ at energies of 30 and 50 MeV show that these processes are able to account for the generation of anomalous abundances of helium isotopes found in the peculiar star 3 Cen A. This supports the model involving spallation by high-energy particles as suggested by Wallerstein and by Novikov and Syunyaev.

065.064 Slowly rotating relativistic stars. III. Static criterion for stability. J. B. Hartle, K. S. Thorne.
Astrophys. Journ. Vol. 158, 719 - 726 (1969).
A method is given for analyzing the stabilizing effect of a slow and rigid rotation on the radial pulsations of relativistic stellar models. This method is restricted to stellar models with a one-parameter equation of state, white dwarfs and neutron stars. The method is also applicable to hot, isentropic stellar models. It is the generalization to slowly rotating stars of the "static stability analysis," which is now used widely for nonrotating stellar models.

065.065 Thermal and dynamical evolution of gas clouds of various masses.
T. Hattori, T. Nakano, C. Hayashi.
Progr. Theor. Phys. Japan, Vol. 42, 781 - 798 (1969).
The evolution of gas clouds of masses, 10^4, 10^2, 1 and $10^{-2}M_\odot$, with the population I composition is investigated both for transparent and opaque stages by comparing the rates of cooling, heating, contraction and expansion.

065.066 The formation of stars with particular application to temporary stars and quasars. I. P. Williams.
Monthly Notices, Roy. Astron. Soc., Vol. 146, 339 - 350 (1969).
It is assumed that all gas clouds have tendencies to fragment into unstable cloudlets (or floccules as McCrea has termed them). Formulae giving the average mass of the stars that form out of a contracting gas cloud in terms of various parameters are then derived. When numerical values lying within reasonable ranges are inserted, it is shown that a gas cloud of average stellar cluster mass, 10^{36} g to 10^{38} g, will form stars with a mass inside the usual stellar mass range. An initial cloud of mass 10^8 $M_\odot$ forms objects of mass slightly above the stellar range. These are shown to explode at a rate of about 30 per year as is required in order to explain the quasar phenomenon.

065.067 The hydrodynamics of the helium flash.
A. C. Edwards.
Monthly Notices, Roy. Astron. Soc., Vol. 146, 445 - 472 (1969).
Time-dependent numerical calculations, based on the conservation equations in difference equation form, are presented for a $1M_\odot$ Pop II star evolving dynamically through its helium flash. With the heuristic method used to calculate time-dependent convection it is found that the convective core of the star explodes. The evolution begins with an oscillation of the core of period 60 seconds imposed on the general expansion. The core cannot expand fast enough and slowly heats up. A composition gradient is formed in the convective core opposed to convection. Finally the core explodes. Throughout the calculation the total energy is conserved to 1 part in 10^4.

065.068 A method to treat the convective regions in the evolving stellar interior. K. Suda.
Sci. Rep. Tôhoku Univ., First Series, Vol. 52, 10 - 18 (1969).
A method is described for treating generally the convective regions in the automatic computation of stellar evolution. A generalization is done in the scope of mixing length theory and the necessary formulations are provided for use in the method proposed by Henyey, Forbes and Gould (1964).

065.069 On the unimportance of bound electrons in influencing nuclear-reaction rates. I. Iben, Jr.
Astrophys. Journ., Vol. 158, 1033 - 1035 (1969).
Screening by bound electrons has a negligible effect on thermonuclear-reaction rates in stellar interiors.

065.070 Advanced phases of evolution in massive red supergiants. R. Stothers, C.-w. Chin.
Astrophys. Journ., Vol. 158, 1039 - 1057 (1969).
Very general models have been constructed for the advanced phases of evolution in massive stars of 15, 30, and

60 $M_\odot$. Independent of the surface condition and the specification of the central energy source, the models may be specified by the two quantities E and T_e. Introduction of the physical boundary conditions shows that the models represent evolutionary sequences of convective red supergiants. The phases of carbon, neon, and oxygen burning in the stellar core are considered, with and without the inclusion of the hypothetical electron-neutrino processes. Three tests of the hypothesis of electron-neutrino interaction have been made by comparing our theoretical results with the relevant observational data on the statistics of M supergiants, N and S stars, and yellow supergiants.

065.071 The influence of metal content on the evolution of stars of five solar masses. B. M. Schlesinger.
Astrophys. Journ., Vol. 158, 1059 - 1074 (1969).

Evolutionary tracks have been computed for stars of 5 $M_\odot$ with two choices of chemical composition: $X = 0.67$, $Z = 0.03$, and $X = 0.64928$, $Z = 0.06$. The tracks extend from the main sequence through the entire evolution to the blue in the H–R diagram during core helium burning. The H/He ratio is the same for both compositions. Physical properties of the two stars are compared over the evolutionary tracks. From a study of clusters in the galaxy and in the Magellanic Clouds in the light of the new theoretical results, it is clear that the composition of at least parts of the Small Cloud is different from that of the galaxy. Definite conclusions cannot be drawn regarding the composition of the Large Cloud.

065.072 A comparison of static and hydrodynamic Cepheid models. C. F. Keller, J. P. Mutschlecner.
Bull. American Astron. Soc., Vol. 1, 350 (1969). – Abstr. AAS.

065.073 The nature of the β Cephei phenomenon. J. R. Percy.
Bull. American Astron. Soc., Vol. 1, 358 - 359 (1969). Abstr. AAS.

065.074 New horizontal-branch models. R. T. Rood.
Bull. American Astron. Soc., Vol. 1, 360 (1969). – Abstr. AAS.

065.075 The effect of collective interactions on the electron-scattering opacity of stellar interiors. W. D. Watson.
Bull. American Astron. Soc., Vol. 1, 368 (1969). – Abstr. AAS.

065.076 On the mechanism of X-rays of a neutron star. G. S. Bisnovaty-Kogan, A. M. Fridman.
Astron. Zhurn. Akad. Nauk SSSR, Vol. 46, 721 - 724 (1969). In Russian. English translation in Soviet Astron. AJ, Vol. 13, No. 4.

It is shown in the case of gas accretion by a neutron star, that the magnetic field in a stream and the dipole field near the star can be $\sim 10^8$ Gauß. A collisionless shock wave, in which electrons may have a temperature of 10^{11} °K and radiate by a synchrotron mechanism in the X-region, is generated.

065.077 L'evoluzione delle stelle. II. - Le stelle doppie. P. Giannone, M. A. Giannuzzi.
Coelum, Vol. 37, 256 - 270 (1969).

065.078 The braking of stellar rotation in the pre-main-sequence phase. I. Okamoto.
Publ. Astron. Soc. Japan, Vol. 21, 350 - 366 (1969).

The loss of angular momentum from stars in the pre-main-sequence phase is discussed. The internal structure of stars during the wholly convective phase is approximated by a rotating polytrope with an index of 1.5, and a modified form of the Schatzman-type mechanism is used as the means of the loss of angular momentum. For stars of 0.4, 0.6, 1.0, and 2.0 solar masses, the rotation equation is numerically integrated, and the change of angular momentum, rotational velocity, etc., with evolution is calculated. Stellar rotation from the star formation to the main sequence is discussed on the basis of the results of the calculations. Also, with respect to the braking of stellar rotation, the occurrence of planetary systems is discussed.

065.079 Stellar evolution. I. Iben.
Zemlya i Vselennaya, No. 4, p. 19 - 26 (1969). In Russian.

065.080 The outer regions and the internal constitution of the stars. P. Ledoux.
Zemlya i Vselennaya, No. 6, p. 14 - 20 (1969). In Russian.

065.081 The ellipticity of slowly rotating configurations. Z. F. Seidov.
Astrofizika, Vol. 5, 503 - 505 (1969). In Russian. – Engl. translation in Astrophysics, Vol. 5, No. 3.

Upper and lower bounds for m/σ, where m is the ratio of centrifugal force to the gravity force at the equator and σ is the ellipticity of configuration, are set up.

065.082 Dynamic action of radiation from a massive protostar. R. Bartkus.
Bull. Vilnius Astron. Obs. No. 25, p. 3 - 22 (1969). In Russian.

The early evolution of a massive star is assumed to go through the convective and subsequent radiative phases. A protostar is adopted to be surrounded by interstellar gas (originally unionized hydrogen and dust in the normal proportion). The cases M = 11, 20, 30 $M_\odot$, and N_H = 10, 100, 1000 cm^{-3} are considered. The protostars are shown to "blow away" the dust by photon flux up to the distances of the order of 0.2 pc. The gases reach velocities V_g during the dust flight. As a result of the strong viscous drag from protons the dust grains stop in gases when hydrogen becomes ionized. The formed transparent (dust-free) zone expands until radiation and gas pressures are balanced.

065.083 Dynamic action of radiation from a massive star. R. Bartkus.
Bull. Vilnius Astron. Obs. No. 25, p. 23 - 38 (1969). In Russian.

The evolution of an H II region around a massive star born in a uniform density neutral gas is considered. The H I shell is adopted to be formed at t = 0 after the star reached the main sequence. The subsequent propagation of the shock and ionization fronts is calculated in dimensionless variables assuming a mean gas density for the compressed region and a mean pressure for the H II region. Models of evolution are tabulated for the various parameters. Then the motion of interstellar grains is examined.

065.084 Random independent splitting model for the mass spectrum of protostars and interstellar clouds. A. H. Marcus.
Report NASA–CR–95938, Bellcomm. Inc., Washington, D.C., 35 pp. (1968). – See Phys. Abstr. Vol. 72, No. 24779 (1969).

065.085 Pulsations of massive stars. N. R. Simon.
Thesis, Univ. Yeshiva. Univ. Microfilms, Ann Arbor, Mi., 140 pp. (1968). – See Phys. Abstr. Vol. 72, No. 24803 (1969).

065.086 Progrès récents dans le calcul de l'opacité stellaire. A. N. Cox.

Acad. Roy. Belgique, Bull. Cl. Sci., 5ᵉ Série, Vol. 55, 771 - 782 (1969).

065.087 **Dynamics of a collapsing protostar.**
R. B. Larson.
Thesis, California Inst. Technology. Univ. Microfilms, Ann. Arbor, Mi., 267 pp. (1968). − See Phys. Abstr. Vol. 72, No. 29008 (1969).

065.088 **Non-linear theory of pulsation.**
A. Zytkow.
Postępy Astron., Vol. 17, 241 - 265 (1969). In Polish.
A short account of Christy's results on non-linear calculations of stellar pulsations is given.

065.089 **Probleme enger Doppelsterne.**
A. Weigert.
Sterne, 45. Jahrgang, 177 - 191 (1969).

065.090 **Contrazione gravitazionale di un modello di stella di 2 $M_\odot$.** P. Giannone, A. Vignato.
Atti XII Riunione Soc. Astron. Italiana, L'Aquila 1968, p. 42 (1969). − Abstract SAI.

065.091 **Analisi della evoluzione di "remnants" di giganti rosse di ammassi globulari.**
V. Castellani, P. Giannone, A. Renzini.
Atti XII Riunione Soc. Astron. Italiana, L'Aquila 1968, p.44 (1969). − Abstract SAI.

065.092 **The proton–proton cycle in stars in the temperature range $10^7 - 5 \times 10^7 \,°K$.**
V. A. Dergachev, G. E. Kocharov.
Cosmic Rays No. 10, Moscow, p. 177 - 196 (1969). In Russian.
The problem of the proton–proton cycle in the interior of stars is investigated. The concentrations are calculated of the particles participating in the $p - p$ cycle, dependent on the hydrogen and helium content. The reaction times of the $p - p$ cycle are determined. The dependence of the quantity of energy released on the temperature and the chemical composition of stellar matter is computed.

065.093 **Rotational properties of neutron star models.**
J. M. Cohen, A. G. W. Cameron.
Nature, Vol. 224, 566 - 567 (1969).
It is shown that a rigidly rotating neutron star (with rotational period and fractional change of period corresponding to those of the Crab pulsar) must have a minimum mass $\sim 0.5\,M_\odot$ in order to supply 10^{38}ergs/sec to the Crab nebula. Stable models are given for which induced rotation of inertial frames at the center is more than 70 percent of the angular velocity of the star.

065.094 **Superconducting state of neutron stars.**
N. Itoh.
Progr. Theor. Phys. Japan, Vol. 42, 1478 - 1479 (1969).
The possibility of the existence of the superconducting state of neutron stars was first pointed out by Migdal in 1959. The aim of this letter is to give more exact results.

065.095 **Relation between the viscosity of a plasma and the possible generation of a constant magnetic field in a star.** D. Cattani.
Nuovo Cimento Lettere, Prima Ser., Vol. 1, 343 - 345 (1969).
The contribution of the viscosity of a plasma to the explanation of stellar magnetic fields is considered.

065.096 **Stellar models and evolution of stars.**
H. Zimmermann.
Astron. in der Schule, 6. Jahrgang, 89 - 94 (1969). − Popular article.

065.097 **Champ scalaire et configurations d'équilibre de grosses masses.** L. Bel, B. Léauté.
Ann. Inst. Henri Poincaré, Sér. A, Vol. 10, No. 3, p. 317 - 348 (1969).

065.098 **The evolution of a vibrationally unstable 130 $M_\odot$ main sequence star.** I. Appenzeller.
Mitt. Astron. Ges. No. 27, p. 161 (1969). − Abstract AG.

065.099 **Nonlinear periodic pulsations of stars.**
N. H. Baker, K. v. Sengbusch.
Mitt. Astron. Ges. No. 27, p. 162 - 167 (1969). − Conference paper.

065.100 **Zum Verständnis von Sternmodellen mit Schalenquellen.** S. Refsdal, A. Weigert.
Mitt. Astron. Ges. No. 27, p. 167 - 168 (1969). − Abstract AG.

065.101 **A simple method for the solution of the stellar structure equations including rotation and tidal forces.** R. Kippenhahn, H.-C. Thomas.
Mitt. Astron. Ges. No. 27, p. 168 - 169 (1969). − Abstract AG.

065.102 **Evolution of a rotating star of nine solar masses.** R. Kippenhahn, E. Meyer-Hofmeister, H.-C. Thomas.
Mitt. Astron. Ges. No. 27, p. 169 - 171 (1969). − Abstract AG.

065.103 **Der Einfluß des entarteten Elektronengases auf die Kontraktionsphasen von Sternen.**
U. Vogel.
Mitt. Astron. Ges. No. 27, p. 176 (1969). − Abstract AG.

065.104 **Neutron starquakes and pulsar periods.**
M. Ruderman.
Nature, Vol. 223, 597 - 598 (1969).

065.105 **Superfluidity in neutron stars.**
G. Baym, C. Pethick, D. Pines.
Nature, Vol. 224, 673 - 674 (1969).
Here were discuss certain general features of superfluid states in the interior of neutron stars and the extent to which they influence the properties of the stars.

065.106 **Electrical conductivity of neutron star matter.**
G. Baym, C. Pethick, D. Pines.
Nature, Vol. 224, 674 - 675 (1969).

065.107 **Importance of the photo-beta process for the synthesis of "p" elements in stellar conditions.**
M. Arnould.
Nuclear Physics, Ser. A, Vol. 100, 657 - 672 = Univ. Libre Bruxelles, Inst. d'Astron. d'Astrophys., Sér. A, No. 10 (1967).
In this work, we compare position capture and photo-beta disintegration probabilities in several stellar conditions.

Source of stellar neutrons.
Nature, Vol. 223, 889 (1969).

Magnetic susceptibility of neutron matter.
See Abstr. 022.099.

The instability of the congruent Darwin ellipsoids.
See Abstr. 061.012.

Models for envelopes of stars of 4 $M_\odot$ and 20 $M_\odot$.
See Abstr. 064.013.

Vibrational stability of pure helium stars surrounded by pure hydrogen envelopes.　See Abstr. 064.015.

Radiative transfer effects in Cepheid atmospheres. See Abstr. 064.026.

Slowly rotating relativistic polytropic models. See Abstr. 066.021.

Some general relativitistic inequalities for a star in hydrostatic equilibrium.　See Abstr. 066.022.

Nonradial pulsation of general-relativistic stellar models. IV. The weak-field limit.　See Abstr. 066.050.

Quantitative Spektralklassifikation und ihre Anwendung auf Probleme der Entwicklung der Sterne und der Milchstraße.　See Abstr. 114.112.

Scanner abundance studies. I. An investigation of supermetallicity in late-type evolved stars. See Abstr. 114.115.

Hydrodynamic theory of stellar rotation. II. Stationarity of the rotation and meridional circulation. See Abstr. 116.018.

Magnetic stars. Part II. Theory. See Abstr. 116.021.

Comparison of main-sequence binaries with theoretical models.　See Abstr. 117.001.

The expected fraction of evolved close binaries among main-sequence stars of spectral type earlier than A5. See Abstr. 117.026.

New evidence for the oblique-rotator model for a^2 Canum Venaticorum.　See Abstr. 122.036.

Some integral characteristics of the rotating white dwarfs and neutron stars.　See Abstr. 126.003.

A search at 15 GHz for compact H II regions in regions of possible star formation.　See Abstr. 131.002.

The accretion of interstellar matter by a neutron star with magnetic field.　See Abstr. 131.028.

The collapse of interstellar gas clouds. – IV. Models of collapse and a theory of star formation. See Abstr. 131.071.

Magnetic field decay in a neutron star and the distribution of pulsar periods.　See Abstr. 141.050.

Rotating neutron stars, pulsars and cosmic X-ray sources.　See Abstr. 141.089.

Large-scale shock formation in spiral galaxies and its implications on star formation.　See Abstr. 151.013.

Dynamics of self-gravitating gaseous spheres – II. Collapses of gas spheres with cooling and the behaviour of polytropic gas spheres.　See Abstr. 151.017.

Dynamics of self-gravitating gaseous spheres – III. Analytical results in the free-fall of isothermal cases. See Abstr. 151.025.

Distribution of Wolf-Rayet stars in the Galaxy, and their evolution.　See Abstr. 155.001.

066 Relativistic Astrophysics (without Cosmology), Background Radiation, Gravitation Theory

066.001 Small-scale variations in the cosmic microwave background. G. Dautcourt.
Monthly Notices, Roy. Astron. Soc., Vol. 144, 255 - 278 (1969).

A discussion is given of some processes which produce fluctuations in the $3°K$ radiation on a small angular scale ($\leq 1°$). Apart from fluctuations caused by a primordial spectrum of spatial inhomogeneities in the matter and radiation densities, effects of scattering on the gravitational fields of clusters of galaxies as well as scattering by gravitational radiation of extremely large coherence or wave-lengths are considered. While scattering by static gravitational field perturbations may be neglected in most cases, scattering by gravitational radiation with large wave-lengths may produce variations on an observable level. The measurements by Conklin & Bracewell provide upper limits on the equivalent mass density of a gravitational radiation background for coherence lengths $L > 1$ Mpc. The fluctuations may be strongly affected by electron scattering in ionized intergalactic matter.

066.002 An iterative scheme of solution for the problem of relativistic incompressible fluid sphere.
S. J. Wilson.
Publ. Astron. Soc. Japan, Vol. 21, 21 - 24 (1969).

Buchdahl and Land investigated the interior solutions of an incompressible fluid sphere defining incompressibility such that the speed of sound wave is equal to the speed of light. They derived a second-order nonlinear differential equation and obtained approximate solutions. The present paper describes another scheme to investigate the same problem. An integral equation is derived, an iterative scheme is suggested and the consequences are investigated.

066.003 On relativistic statistical thermodynamics.
S. Nakajima.
Progress Theoret. Phys. Japan, Vol. 41, 1450 - 1460 (1969).

It is shown that, on the basis of the principle of maximum entropy, statistical thermodynamics can be constructed in as general and unambiguous manner for relativistic systems as for non-relativistic systems.

066.004 Limits on the density and temperature of massive spheres in general relativity. A. Kovetz.
Astrophys. Journ. Vol. 157, 335 - 337 (1969).

A comparison of the maximal central pressure provided by gravity at the center of a star with the minimal pressure predicted on the exclusion principle brings out the existence of a critical mass. The central density in any star with a mass less than the critical mass is limited to two or more well-determined intervals, in agreement with known numerical results (e.g., Harrison et al. 1965). Similar limitations are obtained for the central temperature. No such limits exist for stars of supercritical mass.

066.005 Observational constraints on a discrete-source model to explain the microwave background.
A. A. Penzias, J. Schraml, R. W. Wilson.
Astrophys. Journ. *(Letters),* Vol. 157, L49 - L51 (1969).

We present an experimental limit to the small-scale anisotropy of the microwave background at 3.5 mm. This upper limit corresponds to an rms variation in antenna temperature of $0.024°K$ observed with an effective beam size of 1.4×10^{-3} square degree. From this result we conclude that the discrete-source models presented in a recent paper by Wolfe and Bur-

bidge require a source density at least as great as that for all observed galaxies in order to explain the microwave background.

066.006 The origin of the X-ray background.
M. S. Longair, R. A. Sunyaev.
Astrophys. Letters, Vol. 4, 65 - 70 (1969).

A source model is described in which it is possible to account for the details of the observed X-ray background spectrum for energies $\epsilon > 100$ eV. As a result of inverse Compton scattering, relativistic electrons originating in the nuclei of intense sources of infrared radiation produce (i) hard X-rays by scattering the infrared quanta in the nuclei and (ii) soft X-rays by scattering photons of the relict radiation once the electrons escape into the intergalactic medium.

066.007 Fluktuationen in der 3 °K-Strahlung.
G. Dautcourt.
Monatsber. Deutsch. Akad. Wiss. Berlin, Vol. 11, 163 - 164 = Sternw. Babelsberg, Inst. Relativistische und Extragalaktische Forschung, Mitt. Neue Folge, No. 17 (1969).

066.008 Kann die allgemeine Relativitätstheorie heute als bestätigt gelten? H. Dehnen.
Phys. Blätter, 25. Jahrgang, 400 - 407 (1969).

066.009 Is gravitatiestraling eindelijk ontdekt?
H. Rosenberg.
Hemel en Dampkring, Vol. 67, 297 - 299 (1969).

066.010 Observational possibility of detection of the relativistic pericentrum motion of natural members of the solar system. M. Pańków.
Acta Astron. Vol. 19, 237 - 239 (1969).

The search for natural members of the solar system, other than Mercury and Icarus, suitable for investigation of the relativistic pericentrum motion leads to a negative result.

066.011 A satellite observation of the relativistic Doppler shift. R. E. Jenkins.
Astron. Journ. Vol. 74, 960 - 963 (1969).

Relativity predicts a periodic variation in the observed frequency of satellite radio signals as the satellite moves from perigee to apogee in an eccentric orbit. This variation has been observed using the NASA sponsored satellite 1956—89A (GEOS-1). The observation was made over a four-day span using data obtained in 1966 by the TRANET Doppler system.

066.012 Yilmaz' theory of gravitation and some modifications. B. O. J. Tupper, C. Page.
Journ. Phys. A, General Phys. Ser. 2, Vol. 2, 521 - 528 (1969).

An apparent error in the formulation of Yilmaz' theory of gravitation is corrected. Yilmaz' field equations are unaffected but the corrected theory suggests alternative field equations.

066.013 On the so-called gravitational darkening.
I. B. Pustilnik.
Astron. Tsirk. No. 524, p. 1 - 3 (1969). In Russian.

066.014 Stationary axially symmetric generalizations of the Weyl solutions in general relativity.
R. B. Hoffman.
Phys. Rev., Second Series, Vol. 182, 1361 - 1368 (1969).

066.015 Physics of massive objects.
R. V. Wagoner.
Annual Rev. Astron. Astrophys. Vol. 7, 553 - 576 (1969).

066.016 Hydrostatic equilibrium in rotating relativistic stars, I.   F. Occhionero.
Mem. Soc. Astron. Italiana, Nuova Serie, Vol. 40, 317- 324 (1969).

In general relativity, a mass current generates a gravitational vector potential, which acts "magnetically" on mass currents. This has no equivalent in Newtonian gravitation, where there is only one scalar potential. Therefore in the presence of mass currents, great carefulness is required when concepts typical of Newton's theory are extended into Einstein's. This paper deals with the self-gravitating core of a relativistic star; a forthcoming paper will deal with the non-gravitating external envelope.

066.017 More tests of general relativity.
G. H. Keswani.
Nature, Vol. 223, 1264 - 1265 (1969).

Experiments using sensitive, stable and portable atomic clocks, now available, are suggested to verify the exterior and interior (Schwarzschild) metrics near the surface of the earth. One clock (C_A) remains at the base laboratory while the second clock (C_B) is taken to another location for $3 \times 10^7 s$ (~ one year) and brought back for comparison with C_A. If the location is 3 km higher on a mountain, C_B will gain $10 \mu s$ and if it is 2 km deep in a mine it will loose $7 \mu s$. If C_B is taken from geocentric latitude 30° to 45°, to a location with the same gravitational potential, it will gain $9 \mu s$. This latter experiment is like the "turn-table" experiments made to determine frequency-shift due to differences in angular motion. The result now follows from the Schwarzschild metric, the earth being used as a "turn-table".

066.018 The gravitational field of a disk.
T. Morgan, L. Morgan.
Phys.Rev., Second Series, Vol. 183, 1097 - 1101 (1969).

The general solution of the static field equations of general relativity is given for a disk of "counter-rotating" dust particles. Bardeen has pointed out that a family of such disks can have arbitrarily large red-shifts without violating the velocity condition. However, it is shown that their red-shift cannot exceed 1.9015 before their binding energy becomes negative. This work suggests that the largest gravitational red-shift to which counter-rotating dust disks can give rise is of order of magnitude 1.

066.019 Spheres with two density distributions in general relativity.   M. C. Durgapal, G. L. Gehlot.
Phys.Rev., Second Series, Vol. 183, 1102 - 1104 (1969).

Exact internal solutions have been found in a massive sphere with two different density distributions. The density is a minimum at the surface and increases as we move towards the center; but the density in the central region is constant at a certain maximum value. The restrictions imposed by pressure-density relations limit the solutions to certain values of n for various ratios of core radius and external radius. n is the parameter expressing the densities in the two regions and the solutions of the field equations.

066.020 Sur le caractère non-linéaire des ondes gravifiques.
J. Carstoiu.
Comptes Rendus Acad. Sci. Paris, Sér. A, Vol. 269, 826 - 828 (1969).

066.021 Slowly rotating relativistic polytropic models.
V. V. Papoyan, D. M. Sedrakian, E. V. Chubarian.
Astrofizika, Vol. 5, 97 - 111 (1969).   In Russian.
English translation in Astrophysics, Vol. 5, No. 1 (1969).

The slow rotation of relativistic polytropic models is studied. Only the linear term is taken into account. The distribution for the moment of inertia and the nondiagonal components of metric is derived in case of $n = 1; 1.5; 2; 2.5; 3$. In connection with the Hoyle and Fowler hypothesis, a rotating polytrope with $n = 3$ is specially discussed. It is found that two configurations correspond to the same barion number. In the case of transition between these states, $10^{58} - 10^{60}$ erg of energy are radiated. A part of the radiated energy is spent on the increase of the rotation energy (30%). During the transition matter is thrown out from the equator of the star. The proposed model is probably a satisfactory model of quasars.

066.022 Some general relativistic inequalities for a star in hydrostatic equilibrium.   J. N. Islam.
Monthly Notices, Roy. Astron. Soc., Vol. 145, 21 - 29 (1969).

A minimal theorem for the pressure inside a star in hydrostatic equilibrium is obtained in general relativity that is the analogue of a well-known classical theorem. An improvement is made on some limiting inequalities obtained by Buchdahl (1959) related to the gravitational potential energy.

066.023 On Salmona's paper on the model of static configurations.   R. M. Avakian, M. A. Mnatsakanian.
Astrofizika, Vol. 5, 169 - 171 (1969).   In Russian.
English translation in Astrophysics, Vol. 5, No. 1 (1969).

The incorrectness of Salmona's analysis of the problem of static configurations by Brans-Dicke theory is shown.

066.024 Conservation laws in general relativity and in the post-Newtonian approximations.
S. Chandrasekhar.
Astrophys. Journ. Vol. 158, 45 - 54 (1969).

It is shown how the exact conservation laws of general relativity, expressed in terms of the symmetric energy-momentum complex of Landau and Lifshitz, can be used to determine the various conserved quantities in the different post-Newtonian approximations. Particular attention is given to the conserved energy which emerges as the integral over the whole of space of the difference between the (0,0)-component of the Landau-Lifshitz complex and the energy of the conserved mass present. The method is illustrated in the framework of the first post-Newtonian equations of hydrodynamics.

066.025 The second post-Newtonian equations of hydrodynamics in general relativity.
S. Chandrasekhar, Y. Nutku.
Astrophys. Journ. Vol. 158, 55 - 79 (1969).

In this paper the equations of hydrodynamics governing a perfect fluid in the second post-Newtonian approximation to general relativity are derived; in this approximation all terms inclusive of $O(c^{-4})$ are retained consistently with Einstein's field equations. And the equation $T^{ij}_{;j} = 0$ governing the energy-momentum tensor is also derived to $O(c^{-4})$.

066.026 Post-Newtonian n-body equations of the Brans-Dicke theory.   F. B. Estabrook.
Astrophys. Journ. Vol. 158, 81 - 83 (1969).

The post-Newtonian equations of the Brans-Dicke scalar-tensor theory are derived, for the case of n gravitating point masses. They are a set of coupled second-order differential equations for the accelerations of the point masses, which prove to be derivable from a classical velocity-dependent Lagrangian.

066.027 Structure of a rotating polytrope in the post-Newtonian approximation of general relativity.
W.-Y. Chau.
Astrophys. Journ. Vol. 158, 85 - 89 (1969).

A variational approach based on the work of Roberts and of Krefetz is used to examine the structure of a fast-rotating polytrope with symmetry about the axis of rotation and under stationary conditions in the post-Newtonian approximation of general relativity. The effects are evaluated by means of a simple trial function.

066.028　The scattering of photons by a highly relativistic star.　W. J. Kaufmann, R. M. Crutcher, S. Pierce.
Bull. American Astron. Soc. Vol. 1, 246 - 247 (1969). −
Abstr. AAS.

066.029　On the Brans solution in the scalar-tensor theory of gravitation.　H. Nariai.
Progr. Theoret. Phys. Vol. 42, 742 - 744 (1969). − Letter.

066.030　The astronomical significance of mass loss by gravitational radiation.
G. B. Field, M. J. Rees, D. W. Sciama.
Comments Astrophys. Space Phys. Vol. 1, 187 - 193 (1969).

066.031　Observational tests of gravitational theories.
J. R. Percy.
Journ. Roy. Astron. Soc. Canada, Vol. 63, 265 - 266 (1969).

066.032　Velocity of gravitational waves.　W. W. Salisbury.
Nature, Vol. 224, 782 - 783 (1969).
General relativity theory indicates that between spinning masses there exists an extra gravitation-like force that may be called a spin-spin coupling. Measurement of this force now seems possible. Such a measurement can be interpreted to give a value for the velocity of propagation of the gravitational force, a fundamental physical constant. Such a measurement will possibly also provide a definitive test of the general theory of relativity. Support for the design of appropriate apparatus for this experiment has been made available by the Smithsonian Institution.

066.033　Electromagnetic radiation from relativistic oscillators.　R. V. Wagoner.
Astrophys. Journ. Vol. 158, 739 - 751 (1969).
The oscillatory motion of a charged particle in a linear force field is considered in the limit of a highly relativistic maximum velocity. Expressions are derived for the flux of electromagnetic radiation per unit frequency received from such particles, and various limiting cases are considered in detail. Several distinctive features of this process are then discussed.

066.034　Uniformly rotating disks in general relativity.
J. M. Bardeen, R. V. Wagoner.
Astrophys. Journ. (*Letters*), Vol. 158, L65 - L69 (1969).
General relativistic corrections to the structure of an infinitesimally thin, uniformly rotating disk have been calculated. The binding energy is found to increase monotonically with redshift to a value of about 40 percent of the rest mass energy in the limit of infinite redshift. Other interesting properties of such configurations are discussed.

066.035　On an unified description of gravitation and electromagnetism.　B. A. Arbuzov.
Zh. ehksperim. i teor. fiz. Vol. 56, 1046 - 1056 (1969).
In Russian. − Abstr. in Referativ. Zhurn. 51. Astron., 9.51.710 (1969).

066.036　On the nature of singularities in the general solutions of the gravitational equations.
V. A. Belinskij, I. M. Khalatnikov.
Zh. ehksperim. i teor. fiz. Vol. 56, 1700 - 1712 (1969).
In Russian. − Abstr. in Referativ. Zhurn. 51. Astron. 9.51.711 (1969).

066.037　Superstar stability and the problem of galaxy for-

mation.　É. Ya. Vil'kovskij.
Vestn. AN Kaz. SSR, No. 4, p. 44 - 49 (1969). In Russian.
Abstr. in Referativ. Zhurn. 51. Astron., 10.51.764 (1969).

066.038　On the density of the spatial gravitational field.
N. P. Suvorov.
Uch. zap. Mosk. obl. ped. in−t, Vol. 165, 183 - 203 (1969).
In Russian. − Abstr. in Referativ. Zhurn. 51. Astron., 10.51.773 (1969).

066.039　The equations of motion in general relativity.
G. P. Vypov.
Mekhan. tverd. tela. Resp. mezhved. sb., vyp. (No.) 1, p. 153 - 157 (1969). In Russian. − Abstr. in Referativ. Zhurn. 51. Astron., 10.51.787 (1969).

066.040　The principle of the minimum action, Lagrange's and Hamilton's equations in physical space of general relativity.　G. P. Vypov.
Mekhan. tverd. tela. Resp. mezhved. sb., vyp. (No.) 1, p. 157 - 162 (1969). In Russian. − Abstr. in Referativ. Zhurn. 51. Astron., 10.51.788 (1969).

066.041　Zur kugelsymmetrischen Vakuumlösung der Trederschen Gravitationstheorie.
H.-H. v. Borzeszkowski.
Ann. Physik, 7. Folge, Band 22, 326 - 330 (1969).
Spherically symmetric vacuum solutions of Treder's gravitation theory are investigated and the problem of the gravitational collapse in this theory is discussed.

066.042　Die statischen Gleichgewichtszustände von kugelsymmetrischen Massen mit beliebiger Baryonenzahl in der Tetraden-Theorie von Treder.　E. Kreisel.
Ann. Physik, 7. Folge, Band 23, 180 - 191 = Sternw. Babelsberg, Inst. Relativistische und Extragalaktische Forschung, Mitt. Neue Folge, No. 23 (1969).
Die Gleichgewichtszustände von statischen, kugelsymmetrischen Materieverteilungen in der Tetraden-Theorie von Treder werden untersucht. Beim Anschluß an die Vakuumlösungen erweisen sich zwei Konstanten als notwendig. Eine der Konstanten entspricht der üblichen Schwarzschild−Masse, die zweite weicht bei hohen zentralen Dichten wesentlich von ihr ab. Mit wachsender zentraler Dichte nimmt die Baryonenzahl im Gegensatz zur Einsteinschen Theorie unbeschränkt zu.

066.043　Äquivalenzprinzip und Abschirmung der Schwerkraft.　H.-J. Treder.
Monatsber. Deutsch. Akad. Wiss. Berlin, Band 11, 207 - 217 = Sternw. Babelsberg, Inst. Relativistische und Extragalaktische Forschung, Mitt. Neue Folge, No. 19 (1969).

066.044　A conjecture regarding quantized gravitational theory and elementary particles.　P. Harris.
Canadian Journ. Phys., Vol. 47, 1884 - 1885 (1969).
The gravitational action integral is evaluated for the case of a sperical potential surrounding a spherical mass shell particle. The gravitational action integral is quantized in the Bohr approximation, and a relation is found which connects the particle mass and radius to the Hubble constant.

066.045　Charged spheres in general relativity.
M. C. Faulkes.
Canadian Journ. Phys., Vol. 47, 1989 - 1994 (1969).
The Einstein−Maxwell equations for a spherically symmetric distribution of charged matter are studied.

066.046　Equation of state of neutron gas and the critical mass of neutron stars.
M. Binder, R. H. Pierce, M. Razavy.
Canadian Journ. Phys., Vol. 47, 2101 - 2114 (1969).

The equation of state of the neutron gas is calculated by a variational method, using the Hamada–Johnson potential for the two-nucleon interaction. The energy-density relation determined in this way is valid for low and moderate densities. This result is extrapolated to higher densities and is joined smoothly to the relativistic energy-density relation for free particles. The critical mass of neutron stars is obtained by integrating the general relativistic equations for hydrostatic equilibrium using the equation of state. The critical mass is found to be less than 0.7 of the mass of the sun.

066.047   Planetary orbits for a moving sun.
P. Rastall.
Canadian Journ. Phys., Vol. 47, 2161 - 2164 (1969).

The scalar theory of gravitation is known to be in agreement with observed planetary motions if the sun is assumed to be stationary with respect to the preferred coordinate systems of the theory. We now assume that the sun is moving, and we find that, unless its speed is improbably small, there are observable effects on the planetary orbits. The difficulty can be overcome if one assumes that the Newtonian charts are determined by the distribution of matter.

066.048   Exact solution of a static charged sphere in general relativity.   S. J. Wilson.
Canadian Journ. Phys., Vol. 47, 2400 - 2404 (1969).

066.049   The energy-momentum complex in the Brans–Dicke theory.   Y. Nutku.
Astrophys. Journ., Vol. 158, 991 - 996 (1969).

A symmetric energy-momentum complex, similar to the Landau–Lifshitz complex in general relativity, is obtained for the Brans–Dicke theory. An evaluation of this complex shows that the conservation laws in the post-Newtonian approximations of this theory follow the same scheme as in general relativity and may be obtained in a similar way.

066.050   Nonradial pulsation of general-relativistic stellar models. IV. The weak-field limit.   K. S. Thorne.
Astrophys. Journ., Vol. 158, 997 - 1019 (1969).

In previous papers of this series techniques were developed for calculating the properties of the nonradial, quasi-normal pulsations of fully relativistic stellar models. In this paper those techniques are reexamined in the weak-field, nearly Newtonian limit, and are found to predict an emission of gravitational radiation at a rate which agrees with the standard weak-field approximation to general relativity. The present analysis also exhibits explicitly the radiation-reaction forces inside the star and shows that they sap pulsation energy from the star at the same rate as the waves carry it off.

066.051   Gravitational radiation from dense star clusters.
G. Greenstein.
Astrophys. Journ. (*Letters*), Vol. 158, L145 - L149 (1969).

The pulsed gravitational radiation that Weber has recently observed almost certainly does not arise from binary encounters between stars in a nonrelativistic star cluster. The relativistic case is quite unlikely but is not explicitly ruled out. Relativistic clusters evolve extremely rapidly by star-star collisions and loss of energy by gravitational radiation; only very massive clusters are able to withstand these effects.

066.052   Uniformly rotating disks in general relativity.
J. M. Bardeen, R. V. Wagoner.
Bull. American Astron. Soc., Vol. 1, 333 - 334 (1969). – Abstract AAS.

066.053   Relativistically covariant equations of coupling between emission and matter.
V. S. Imshennik, Yu. I. Morozov.
Astron. Zhurn. Akad. Nauk SSSR, Vol. 46, 800 - 809 (1969).

In Russian. English translation in Soviet Astron. AJ, Vol. 13, No. 4.

In a medium moving with relativistic velocities, it is very important that the non-equilibrium radiation should be taken into account. Such a physical system in the general case is described by equations of radiative hydrodynamics and transfer equation, formulated in the relativistic covariant form. Momentary radiation equations, taking into account, besides an absorption, photon scattering on free electrons, are deduced in the paper.

066.054   Non-static fluid spheres in general relativity.
M. C. Faulkes.
Progr. Theor. Phys. Japan, Vol. 42, 1139 - 1142 (1969).

An exact solution for a spherically symmetric distribution of inhomogeneous matter in an empty background is derived and investigated. The solution serves as a simple model for gravitational collapse.

066.055   Electric fields in rotating, magnetic, relativistic stars.
F. Occhionero, M. Demianski.
Phys. Rev. Letters, Vol. 23, 1128 - 1130 (1969).

All the current models for pulsars call for very large magnetic fields in rotating neutron stars. Since general relativistic effects are important in the latter, electromagnetism too must be framed consistently (which has been overlooked so far). It is shown that in a rotating neutron star even a uniform magnetic field, static in the corotating frame, implies in the same frame the static electric field which is crucial to the pulsar emission theories.

066.056   Spherically symmetric static solutions of Einstein's equations.   C. Leibovitz.
Phys. Rev., Second Series, Vol. 185, 1664 - 1670 (1969).

A formal general solution of Einstein's equations in the static case containing an arbitrary function of r is obtained. A necessary and sufficient condition that the arbitrary function must satisfy in order that the solution be physically meaningful in the neighborhood of the center is established. A mapping from Newtonian solutions is indicated. The case of infinite pressure at the center is considered. New solutions are given as examples.

066.057   Gravitational waves in general relativity. X. Asymptotic expansions for the Einstein-Maxwell field.
M. G. J. van der Burg.
Proc. Roy. Soc. London, Ser. A, Vol. 310 (No. 1501), 221 - 230 (1969).

066.058   Covariant black-body and the 3°K radiation field.
M. Alexanian.
Nuovo Cimento Lettere, Prima Ser., Vol. 1, 75 - 78 (1969).

066.059   Ferromagnetic transition in superdense matter and neutron stars.
D. H. Brownell, Jr., J. Callaway.
Nuovo Cimento, Vol. 60B, 169 - 188 (1969).

The conditions for dense nuclear matter to undergo a ferromagnetic transition are estimated, and shown to be applicable to neutron stars.

066.060   The hard-sphere Fermi gas and ferromagnetism in neutron stars.   M. J. Rice.
Phys. Letters, Vol. 29A, 637 - 638 (1969).

066.061   Loss of mass and rotational momentum of a star during collapse.
O. H. Guseinov, F. K. Kasumov.
Astron. Tsirk. No. 539, p. 6 - 7 (1969).   In Russian.

066.062   On possible experimental observation of neutrinos

from collapsing stars.
G. V. Domogatsky, G. G. Zatsenin.
Cosmic Rays No. 10, Moscow, p. 139 - 141 (1969).
In Russian.

The possibility of recording positrons or electrons from the reactions $\bar{\nu} + p \rightarrow n + e^+$ or $\nu + z \rightarrow (z + 1) + e^-$, induced by antineutrino – neutrino fluxes from collapsing stars of our Galaxy, is examined.

066.063 Reception of gravitational radiation of extra-terrestrial origin. V. B. Braginskii,
Ya. B. Zel'dovich, V. N. Rudenko.
JETP Letters, Vol. 10, 280 - 283 (1969). [Translated from ZhETF Pis. Red. 10, No. 9, 437 - 441 (1969). In Russian].

Some remarks are made on the experiment of Weber, who recorded signals, that might be due to gravitational radiation of cosmic origin.

066.064 Statical gravitational fields in second approxima-tion. J. L. Synge.
Proc. Roy. Irish Acad., Ser. A, Vol. 67, No. 5, 47 - 66 (1969).
See Phys. Ber., Vol. 48, No. 10–259 (1969).

066.065 Gravitationswellen. W. Hofmann.
SuW, Vol. 8, 234 - 235 (1969).

066.066 On Schwarzschild's metric in synchronous frame of reference. K. P. Staniukovich.
Dokl. Akad. Nauk SSSR, Ser. Mat. Fiz., Vol. 187, 75 - 78 (1969). In Russian.

Transformations removing the singularity in the standard Schwarzschild metric and leading to a synchronous frame of reference, as well as inverse transformations are investigated. It is pointed out that the transformed metrics must turn into Minkowski metric in the limiting case when the gravitational field is not taken into account. In a first approximation a metric is derived which may prove convenient for quantization of an interval. *B. Onderlička*

066.067 Precession of Schiff's proposed gyroscope in an arbitrary force field. R. F. O'Connell.
Nuovo Cimento Lettere, Prima Ser., Vol. 1, 933 - 935 (1969).

Schiff's experiment is of great interest because of the possibility of testing Einstein's general relativity and for a decision between Einstein and Brans–Dicke theories of gravitation.

066.068 Remarks on quark stars.
D. Ivanenko, D. F. Kurdgelaidze.
Nuovo Cimento Lettere, Vol. 2, 13 - 16 (1969).

Some time ago we have drawn attention to the necessity of admitting the existence of a quarkian core in the interior of hypothetical superstars of very great masses. The authors propose to investigate some properties of a hypothetical quarkian substance.

066.069 Effect of attractive nuclear forces on the onset of ferromagnetism in neutron star matter.
J. W. Clark, N.-C. Chao.
Nuovo Cimento Lettere, Prima Ser., Vol. 2, 185 - 188 (1969).

066.070 Solution of the Dirac equation in orthogonal electric and magnetic fields. V. Canuto, C. Chiuderi.
Nuovo Cimento Lettere, Prima Ser., Vol. 2, 223 - 227 (1969).

The exact solution is derived in the case of constant electric and magnetic field orthogonal to each other. Such process is important in neutron stars or white dwarfs with intense magnetic fields.

066.071 Remarks on the black–body radiation at very high temperature. C. Bouchiat.

Nuovo Cimento Lettere, Prima Ser., Vol. 2, 243 - 248 (1969).

066.072 Gravitational coupling of negative matter.
A. Inomata, D. Peak.
Nuovo Cimento, Vol. 63B, 132 - 142 (1969).

The relevance of the sign of mass in general relativity is examined by analysing a simple model universe in which Dirac matter is distributed uniformly. Mass reversal, converting a source of positive matter into one of negative matter, gives rise to a concomitant change in sign of the gravitational coupling.

066.073 A relativistic effect in pulsating fluid spheres.
F. de Felice.
Nuovo Cimento, Vol. 63B, 649 - 660 (1969).

The purpose of this note is to describe a relativistic effect of energy accumulation which arises when spherically symmetric configurations are enclosed in a sufficiently small volume. Conditions for a release of this energy are analysed and possible applications are suggested.

066.074 Isotropic solutions of the Einstein–Liouville equations. J. Ehlers, P. Geren, R. K. Sachs.
Journ. Math. Phys., Vol. 9, 1344 - 1349 (1968).

066.075 Champ vectoriel et configurations d'équilibre de grosses masses en relativité générale.
D. Gerbal, B. Léauté.
Ann. Inst. Henri Poincaré, Sér. A, Vol. 10, 349 - 357 (1969).

066.076 Motion in a Schwarzschild field. I. Precession of a moving gyroscope. L. Parker.
American Journ. Phys., Vol. 37, 309 - 312 (1969).

066.077 Motion in a Schwarzschild field. II. Deflection of light. L. Parker.
American Journ. Phys., Vol. 37, 313 - 314 (1969).

066.078 First-order approximation to a spherically symmetric solution of Einstein's equation. R. Stabell.
Journ. Math. Phys., Vol. 10, 735 - 739 (1968).

066.079 Relativity theory and astronomy. A. Zięba.
Urania Kraków, Vol. 40, 292 - 303, 322 - 328 (1969). In Polish.

066.080 Disturbances at the passage of a collapsing sphere under a Schwarzschild's sphere.
I. D. Novikov.
Zhurn. ehksperim. i teor. fiz.,Vol. 57, 949 - 951 (1969). In Russian. – Abstr. in Referativ. Zhurn. 51. Astron., 2.51.231 (1970).

066.081 Discussion of possible sources of gravitational radiation. P. Kafka.
Mitt. Astron. Ges. No. 27, p. 134 - 139 (1969). – Conference paper.

066.082 Fluctuations in the microwave background radiation. M. S. Longair, R. A. Sunyaev.
Nature, Vol. 223, 719 - 721 (1969).

Fluctuations in the microwave background radiation are expected as a result of perturbations in the primaeval plasma from which large-scale structures such as galaxies and clusters of galaxies were formed. Here we compare the magnitudes of the temperature fluctuations $\Delta T/T$ expected as a result of "primaeval" perturbations with several other possible sources of fluctuations, particularly discrete radio sources.

066.083 Distortions of the background radiation spectrum.
R. A. Sunyaev, Ya. B. Zeldovich.

Nature, Vol. 223, 721 - 722 (1969).

The discovery of cosmic blackbody radiation with a temperature of 2.7 K is widely considered as powerful evidence for the hot, big-bang model of the universe. But if departures from the equilibrium spectrum were discovered, what would their meaning be? We believe that one need not abandon the hot, big-bang theory. There are natural processes which can lead to pronounced distortions of the spectrum.

How real are gravitational waves?
Nature, Vol. 224, 411 (1969).

Die kosmische 3 K-Hintergrundstrahlung.
Phys. Blätter, 25. Jahrgang, 370 - 372 (1969). – News notes.

Weber reports 1660-Hz gravitational waves from outer space.
Phys. Today, Vol. 22, No. 8, p. 61 - 62 (1969).

Gravitational waves detected.
Sky Telescope, Vol. 38, 71, 81 (1969).

Eine neue Gravitationstheorie.
Umschau, Vol. 69, 665 (1969).
– Comments to the gravitation theory by R. H. Dicke.

New state of ferromagnetism in degenerate electron gas and magnetic fields in collapsed bodies.
See Abstr. 022.009.

The present situation of the problem of the determination of the gravitation constant and the earth's mass.
See Abstr. 043.005.

A general relativity test using two or more solar satellites.　See Abstr. 051.024.

Focusing of the emission by the gravitational field.
See Abstr. 061.013.

Neutrino and the theory of gravitation.
See Abstr. 061.046.

Relativistic stellar-wind torques.

See Abstr. 064.043.

Isentropic stars in general relativity.
See Abstr. 065.003.

General relativistic neutron star models.
See Abstr. 065.018.

Nonradial pulsation of general-relativistic stellar models. III. Analytic and numerical results for neutron stars.
See Abstr. 065.042.

Slowly rotating relativistic stars. III. Static criterion for stability.　See Abstr. 065.064.

Rotational properties of neutron star models.
See Abstr. 065.093.

Superconducting state of neutron stars.
See Abstr. 065.094.

Weiße Zwerge, Neutronensterne und der Endzustand der Materie.　See Abstr. 126.010.

Leakage electrons from normal galaxies: The diffuse cosmic X-ray source.　See Abstr. 143.011.

Upper limit to radiation of mass energy derived from expansion of galaxy.　See Abstr. 151.054.

Is the galaxy losing mass on a time scale of a billion years?　See Abstr. 155.014.

The interaction of matter and radiation in a hot-model universe.　See Abstr. 162.004.

Finite-range gravitation.
See Abstr. 162.020.

Zur Anisotropie der kosmischen Mikrowellen-Hintergrundsstrahlung.　See Abstr. 162.092.

Magnetic fields and highly condensed objects.
See Abstr. 162.094.

Sun

071 Solar Photosphere, Spectrum

071.001 **Observation of profiles and asymmetry of Fraun-hofer lines.** R. Boyer.
Astron. Astrophys. Vol. 2, 375 - 380 (1969). In French.

Nous avons observé par des méthodes spectrophotoélectriques les profiles de quatre raies du titane neutre et déterminé le profil moyen de chacune d'elles. Connaissant les intensités centrales relatives moyennes et les courbes de dissymétrie moyennes, nous avons cherché à tenir compte de l'influence du profil instrumental sur les quantités observées. Nous avons été conduits à un profil instrumental symétrique, à ailes étendues, dont l'influence est peu marquée sur la forme globale des courbes de dissymétrie. Les observations révèlent d'autre part un important effet de fluctuation temporelle de la forme et de la dissymétrie des profils.

071.002 **Atmospheric transmission and solar spectroscopy in the submillimeter wavelength range.**
Y. Biraud, J. Gay, J. P. Verdet, Y. Zéau.
Astron. Astrophys. Vol. 2, 413 - 418 (1969). In French.

We have studied from the new high altitude Gornergrat Observatory, the spectrum of the solar radiation between 300μ and 1 mm wavelengths. From this we deduced, by means of a continuation of the Bouguer's line, that the atmospheric transmission was about 30 per cent in the 300μ window. We also observed the presence of a new absorption at $26\ cm^{-1}$ noticed by Gebbie et al. (1968), and we have shown that it had an extra-atmospheric and probably solar origin proved either by means of a Bouguer's continuation or by the comparison of sun and moon spectra taken with the same resolution. Finally we identified several lines due to the atmospheric absorption of O_2, O_3, N_2O and perhaps D_2O, NO_2 appearing in both nocturnal and diurnal spectra.

071.003 **Revised solar iron abundance and its influence on the photospheric model.**
T. Garz, H. Holweger, M. Kock, J. Richter.
Astron. Astrophys. Vol. 2, 446 - 450 (1969). In German.

Using the new f-values of Garz and Kock (1969), an analysis of 26 photospheric Fe I lines is carried out yielding a solar iron abundance $\log \epsilon(Fe) = 7.60$, $\log \epsilon(H) = 12$. This new value − 10 times greater than recently determined photospheric values − is a consequence of the revised scale of f-values and agrees with coronal abundance determinations. As a by-product we find from the dependence of abundance on equivalent width, that in the solar model of Holweger (1967) the microturbulence has to be decreased by 0.7 km/s, which leads to 1.0 km/s at $\tau_{5000} \sim 0.1$. In order to obtain the same central intensities of Fraunhofer lines the macroturbulence must be increased to about 1.6 km/s. In solar-type stars, the increase of $\epsilon(Fe)$ by a factor of 10 leads to an increase of the electron pressure and to a corresponding decrease of the gas pressure.

071.004 **Inversion of the intensity integral for the Doppler core of strong solar absorption lines.**
G. Worrall, A. M. Wilson.
Astron. Astrophys. Vol. 2, 458 - 468 (1969).

It is shown that intensity integral, considered as an equation for the profile in the Doppler core of strong solar absorption lines, can be inverted to obtain the line source function.

Numerical methods for performing the inversion in the general case in which the absorption coefficient (Doppler width) varies with depth are described.

071.005 **An analysis of the solar sodium D lines.**
A. M. Wilson, G. Worrall.
Astron. Astrophys. Vol. 2, 469 - 476 (1969).

Waddell's observations of the D lines in the solar spectrum have been analyzed by a direct inversion of the line profiles. The principal conclusion is that the line source function in the Doppler core is significantly frequency dependent.

071.006 **The differences between quiet and active regions measured by spectroheliograms in the neutral helium resonance lines.**
A. G. Hearn, R. W. Noyes, G. L. Withbroe.
Monthly Notices, Roy. Astron. Soc., Vol. 144, 351 - 357 (1969).

Spectroheliograms of the 537 and 584 Å lines of neutral helium were obtained in November 1967 by the Harvard spectroheliometer on OSO-IV. The increased intensities of these lines in active regions cannot be explained by an increased electron temperature. Calculations show that the variation of the ratio of the intensity of the 537 Å line to the 584 Å line as a function of the intensity of the 584 Å line is consistent with the layers emitting these lines having a higher electron density in the active regions. The calculations require the layer emitting the neutral helium lines in a quiet region to have an electron temperature of 32000°K and an electron density of 4.5×10^{10} cm^{-3}. The error in this electron density may be a factor of 3. The active regions that have been observed require an increase in the electron density of up to 2.5 times that of a quiet region.

071.007 **A Fourier spectrum analysis of long samples of solar line oscillations.** G. Gonczi, F. Roddier.
Solar Physics, Vol. 8, 255 - 259 = Kitt Peak National Obs. Contr. No. 425 (1969).

Sequences of the oscillations of solar lines up to 2 hours 20 min long have been recorded at the same point on the sun. The power spectra show several peaks separated by 0.85×10^{-3} cps on average from each other. A sharp main peak at 3.3×10^{-3} cps (300 sec period) is almost always present. These results suggest that the lifetime of the phase of the oscillation is much longer than that of the amplitude and is likely to exceed one hour. We actually observe the modulation of a wave in smaller wave trains about 11 min long and 20 min apart (average values). Observations with low spatial resolution also suggest that the area of coherence is much greater for the phase than for the amplitude.

071.008 **Reversals of selected Ce II solar lines.**
R. C. Canfield.
Astrophys. Journ. Vol. 157, 425 - 437 = Contr. Kitt. Peak National Obs. No. 398 (1969).

Observations of six Ce II lines in the solar spectrum show that the lines reverse from absorption to emission on the disk, inside the limb. Furthermore, the position of this reversal varies with wavelength. The absorption-line profiles favor a non-thermal velocity field that is anisotropic. The emission-line profiles require a horizontal non-thermal velocity of 2.0 ± 0.2 km sec^{-1}, averaged over heights $0 < h < 400$ km. The equivalent widths and central intensities of the absorption lines require a cerium abundance log N_{Ce} = 1.4 ± 0.3 (log N_H = 12.0), and favor the Bilderberg temperature distribution over that of Holweger.

071.009 Helium abundance determination from solar-model photospheres. C. A. Rouse.
Astron. Astrophys. Vol. 3, 122 - 125 (1969).

Preliminary results from a new method for determining the helium abundance of the sun by the theoretical prediction of line and continuum radiation from solar-model photospheres are given. Using a solar-model photosphere (SMP) with assumed He/H = 0.14, the predicted residual intensities in the wings of the resonance line of Ca I (λ 4226.73 Å) and the K-line of Ca II (λ 3933.67 Å) and the predicted continuum intensities at λ = 4000 Å and 5000 Å are in good agreement with observations. On the other hand, a SMP with an assumed He/H = 0.05 predicts line profiles and continuum intensities significantly lower than observed.

071.010 Arizona-NASA atlas of infrared solar spectrum, report IV.
L. A. Bijl, G. P. Kuiper, D. P. Cruikshank.
Commun. Lunar Planet. Lab. Vol. 9 (No. 160), 1 - 27 (1969).

This paper is a continuation of earlier papers covering the interval $\lambda\lambda$ 13138–14707 Å. Part of the spectrum of the 1.4 μ H_2O band taken with the 4-meter spectrometer in flight is included to show the absorption in the spectrometer itself. For purposes of further identification, laboratory spectra of the 1.4 μ H_2O band are given in the addendum.

071.011 Arizona-NASA atlas of infrared solar spectrum, report V.
L. A. Bijl, G. P. Kuiper, D. P. Cruikshank.
Commun. Lunar Planet. Lab. Vol. 9 (No. 161), 29 - 51 (1969).

In this report Charts 23 - 32 of the *Atlas* are given, containing the solar spectrum $\lambda\lambda$ 12187–17731 Å, obtained from the NASA CV-990 Jet at an altitude of 39 000 ft with the LPL 4-meter spectrometer. A 600-lines/mm grating was used, blazed at 1.6 μ. The Michigan Atlas spectra and an LPL laboratory spectrum of the 2 ν_3 (1.6 μ) CH_4 band are included for comparison.

071.012 Arizona-NASA atlas of infrared solar spectrum, report VI.
G. P. Kuiper, A. B. Thomson, L. A. Bijl, D. C. Benner.
Commun. Lunar Planet. Lab. Vol. 9 (No. 162), 53 - 63 (1969).

In this paper we reproduce the solar spectrum as recorded with the LPL B-spectrometer on the NASA CV-990, using the 1-μ grating (1200 lines/mm). The resolution is about 4 times lower than in the corresponding records obtained with the LPL 4-m spectrometer.

071.013 Contribution à l'étude de la sélection des images solaires. P. Gaujard.
Comptes Rendus Acad. Sci. Paris, Sér. B, Vol. 269, 228 - 231 (1969).

L'étude, la réalisation et l'exploitation d'un sélecteur d'images solaires par analyse du contraste de la photosphère, a permis d'obtenir de nombreux résultats. Parmi eux, la détermination de vitesses radiales montre l'intérêt de la sélection qui permet une amélioration sensible du rendement des installations solaires.

071.014 Resonance-broadening absorption in the wings of Lyman alpha.
K. Sando, R. O. Doyle, A. Dalgarno.
Astrophys. Journ. *(Letters)*, Vol. 157, L143 - L145 (1969).

It is pointed out that the conventional formula descri-

bing absorption in the wings of Lyman alpha seriously overestimates the magnitude of the absorption at long wavelengths, and it seems improbable that the resonance-broadened wings of Lyman alpha are a significant source of opacity at wavelengths in excess of 2000 Å.

071.015 Near-limb solar brightness distribution at 1216 Å and 1300 Å. J. E. Blamont, C. Malique.
Astron. Astrophys. Vol. 3, 135 - 146 (1969). In French.

Two photometers were placed on board of a sounding rocket for the observation in Argentina of the total solar eclipse of November 12, 1966. Measurements were started at 220 km of altitude during the ascent phase, and ended at 123 km of altitude during the descending phase for a total time duration of 296 s. The distributions of brightness were determined with a great relative accuracy at the solar limb and into the chromosphere up to an altitude of 12000 km at the wavelength of 1216 Å and 1300 Å. The geometrical limb of the solar disk is formed at these wavelengths at the altitude of 2490 km in the chromosphere. An increase of brightness, or bright ring, is found at the limb. At 1216 Å the maximum of brightness is 4600 km far from the geometric limb as defined by a simultaneous infra-red observation; the bright ring is 10000 km wide. At 1300 Å the bright ring coincides with the limb defined in the infra-red, and is 4300 km wide.

071.016 Photoelectric eclipse observation of the continuum at the extreme solar limb.
S. R. Weart, J. E. Faller.
Astrophys. Journ. Vol. 157, 887 - 901 (1969).

The total solar eclipse of November 12, 1966 was observed with a photoelectric spectrometer. We discuss limb-darkening curves for regions of the continuum 1.5 Å wide around 5278 and 6404 Å. Our curves are for fairly quiet regions of the sun near both second and third contacts. To get surface brightness from our data, we differentiated with respect to the position of the lunar limb; the roughness of the moon was taken into account. On the disk, our surface-brightness results agree with those of previous investigation. But at the extreme limb our curves continue to fall with decreasing scale height (as low as 50 km) out to about 550 km above the height zero that is conventional in eclipse work. Here there is a sharp break in the curves, and the surface brightness falls very slowly for the next 1000 km. This break and plateau are not predicted by existing models of the sun. We argue that the low scale heights can be explained only if hydrogen ionization is negligible, and hence that the temperature of the atmosphere is under 5000° K out to 550 km; above, the temperature is probably over 6000° K.

071.017 Zur Interpretation der Profile schwacher Fraunhofer-Linien. B. T. Babij, Yu. V. Fridel'.
Tsirk. L'vov. Astron. Obs. No. 43, p. 38 - 42 (1969).
In Russian.

071.018 A determination of the velocities of large- and small-scale motions by the method of line-width correlation. G. D. Poljakova.
Solnechnye Dannye Byull. No. 4, p. 106 - 111 (1969).
In Russian.

The velocities of large- and small-scale motions are determined by the method of line-width correlation. The influence of photographic factors on the values of the velocity of large-scale motions is considered. The determined velocity values of large- and small-scale motions are compared with those obtained by the above mentioned method and with the data of photoelectric measurements.

071.019 The photospheric abundance of iron.
G. L. Withbroe.
Solar Physics, Vol. 9, 19 - 30 (1969).

The center-to-limb variation of equivalent widths of 198 FeI lines in the spectral region 5500 to 7000 Å was studied with five photospheric models. The gf-values of Corliss and Warner (1964) were used in the analysis. The photospheric iron abundance was found to vary with excitation potential. This can be explained by a systematic error in the gf-values of high excitation lines and an error of 250 to 500 K in the temperature of the arcs used for measuring the gf-values. Departures from LTE in the solar FeI lines are also a possibility. The adopted photospheric abundance of iron, $\log(N_{Fe}/N_H)$ is –5.2.

071.020 Comments on forbidden sulphur I lines in the solar spectrum. J. W. Swensson.
Solar Physics, Vol. 9, 31 - 34 (1969). – Research note.

071.021 The interpretation of velocity filtergrams. I. The effective depth of line formation.
R. L. Parnell, J. M. Beckers.
Solar Physics, Vol. 9, 35 - 38 (1969).

The paper describes a numerical experiment in which the effect of an assumed velocity distribution in the solar atmosphere on the intensity difference between a blue- and a red-wing filtergrams is derived. This results in the effective optical depth at which the velocity is measured. It is shown that this τ_{eff} strongly depends on the assumed velocity distribution.

071.022 The interpretation of velocity filtergrams. II: The velocity and intensity field of the central solar disk. J. M. Beckers, R. L. Parnell.
Solar Physics, Vol. 9, 39 - 50 (1969).

From simultaneous filtergrams obtained in the blue and red wings of a Fraunhofer line we analyzed the velocity and intensity field at the center of the solar disk. Results are as follows: (a) Cross correlation between velocity and intensity is 0.6. It increases somewhat when long wavelengths (> 5000 km) are eliminated. (b) An rms velocity and intensity variation of 0.45 km/sec and 3.9 %.

071.023 Abundances of iron and some other elements in the sun and in meteorites.
T. Garz, M. Kock, J. Richter, B. Baschek, H. Holweger, A. Unsöld.
Nature, Vol. 223, 1254 - 1255 (1969).

New measurements of f-values of FeI lines by Garz and Kock lead to a revised photospheric iron abundance $\log \epsilon$ (Fe) = 7.60, normalized to $\log \epsilon$ (H) = 12. There is now agreement between the iron abundance obtained for the solar corona and that of the photosphere. Furthermore the anomaly that the meteoritic abundance of iron in the type I carbonaceous chondrites was about ten times greater than in the sun is removed. The hypothesis of Urey about a heterogeneous origin of the carbonaceous chondrites becomes unnecessary. The hypothesis about partial irradiation in an early state of the solar system is not supported by recent determinations of the isotopic ratio $^{12}C/^{13}C$ and Li abundance.

071.024 The difference of turbulent velocities in the active and undisturbed photosphere.
O. G. Badaljan, M. A. Livshitz.
Astron. Zhurn. Akad. Nauk. SSSR, Vol. 46, 1035 - 1039 (1969). In Russian. English translation in Soviet Astron. AJ, Vol. 13, No. 5.

A comparison of turbulent velocities in the active and undisturbed photosphere was made. For lines with relatively deep formation levels in photospheric faculae, velocities are approximately by 0.20 km/sec larger than in the undisturbed photosphere.

071.025 Sur le temps de relaxation des perturbations en

température dans la photosphère.
B. Schmieder.
Comptes Rendus Acad. Sci. Paris, Sér. B, Vol. 269, 935 - 937 (1969).

F. N. Edmonds, R. Michard et R. Servajean ont mis en évidence deux composantes du champ microscopique des vitesses: une oscillatoire de période 5 mn et une « convective » de basse fréquence. Nous étudions les variations de densité et de température associées à la première composante.

071.026 Observations of the 5000-Å water vapor band.
J. W. Brault.
Bull. American Astron. Soc. Vol. 1, 234 - 235 (1969). – Abstr. AAS.

071.027 The solar wavelength program – Kitt Peak National Observatory.
J. W. Brault, J. B. Breckinridge, A. K. Pierce.
Bull. American Astron. Soc. Vol. 1, 235 (1969). – Abstr. AAS.

071.028 Brightness fluctuations from waves in the Sun's atmosphere. M. Y. Cha, F. Q. Orrall.
Bull. American Astron. Soc. Vol. 1, 235 - 236 (1969). – Abstr. AAS.

071.029 Some solar line profiles obtained from small regions of the disk with a high-speed spectral scanner.
J. G. Kirk.
Bull. American Astron. Soc. Vol. 1, 248 (1969). – Abstr. AAS.

071.030 The variation of the central intensities of Fraunhofer lines during the solar cycle. W. E. Mitchell, Jr.
Bull. American Astron. Soc. Vol. 1, 253 - 254 (1969). – Abstr. AAS.

071.031 Photospheric and chromospheric magnetic fields and the brightness of faculae and flocculi.
T. T. Tsap.
Izv. Krymskoj Astrofiz. Obs. Vol. 39, 265 - 275 (1969). In Russian.

Photospheric and chromospheric longitudinal magnetic fields and the brightness of faculae and flocculi are compared. The connection between the brightness distributions at different depths in active regions on the sun is investigated. The obtained results conform well with the fact that distributions of the magnetic fields in the photosphere and chromosphere are similar.

071.032 Small-scale structures as produced by electric currents. H. Alfvén.
Bull. American Astron. Soc. Vol. 1, 270 (1969). – Abstr. AAS.

071.033 Source functions of infrared Fraunhofer lines from equivalent widths. R. C. Altrock.
Bull. American Astron. Soc. Vol. 1, 270 (1969). – Abstr. AAS.

071.034 Computed profiles for solar Mg b and Na D lines.
R. G. Athay, R. C. Canfield.
Bull. American Astron. Soc. Vol. 1, 272 (1969). – Abstr. AAS.

071.035 A new technique to obtain solar-velocity maps directly in one spectroheliogram.
A. Bhatnagar, J. O. Stenflo.
Bull. American Astron. Soc. Vol. 1, 273 (1969). – Abstr. AAS.

071.036 **The temperature minimum from far-infrared measurements.**
J. A. Eddy, P. J. Lena, R. M. MacQueen.
Bull. American Astron. Soc. Vol. 1, 275 (1969). – Abstr. AAS.

071.037 **Solar line formation in a magnetic field.**
J. C. Evans.
Bull. American Astron. Soç. Vol. 1, 276 (1969). – Abstr. AAS.

071.038 **Solar site selection based on time-lapse photography of granulation.** V. Gaizauskas.
Bull. American Astron. Soc. Vol. 1, 276 - 277 (1969). – Abstr. AAS.

071.039 **High-resolution solar spectra between 1.2 and 24 microns.** D. N. B. Hall.
Bull. American Astron. Soc. Vol. 1, 277 - 278 (1969). – Abstr. AAS.

071.040 **The difference between the spectra of granular and intergranular regions.**
R. Howard, A. Bhatnagar.
Bull. American Astron. Soc. Vol. 1, 279 - 280 (1969). – Abstr. AAS.

071.041 **Have recent f-value measurements changed the observed photospheric solar iron abundance?**
M. Huber.
Bull. American Astron. Soc. Vol. 1, 280 (1969). – Abstr. AAS.

071.042 **Observations and interpretation of the solar Lyman continuum.** R. W. Noyes, W. Kalkofen.
Bull. American Astron. Soc. Vol. 1, 288 (1969). – Abstr. AAS.

071.043 **Measurements in the solar spectrum between 1400 and 1875 Å with a rocket-borne spectrometer.**
W. H. Parkinson, E. M. Reeves.
Bull. American Astron. Soc. Vol. 1, 288 - 289 (1969). – Abstr. AAS.

071.044 **Analysis of velocity filtergrams.**
R. Parnell, J. M. Beckers.
Bull. American Astron. Soc. Vol. 1, 289 (1969). – Abstr. AAS.

071.045 **K-line profiles of solar fine structure.**
J. M. Pasachoff.
Bull. American Astron. Soc. Vol. 1, 289 (1969). – Abstr. AAS.

071.046 **Comparisons of photospheric magnetograms and H-alpha filtergrams.** D. M. Rust, S. F. Smith.
Bull. American Astron. Soc. Vol. 1, 292 (1969). – Abstr. AAS.

071.047 **Spectroheliograms in Fe II λ4924.**
N. R. Sheeley, Jr., O. Engvold.
Bull. American Astron. Soc. Vol. 1, 292 - 293 (1969). – Abstr. AAS.

071.048 **The acoustic properties of the solar surface.**
R. K. Ulrich.
Bull. American Astron. Soc. Vol. 1, 294 (1969). – Abstr. AAS.

071.049 **Temperature fluctuations and convection in the solar photosphere.** P. R. Wilson.

Bull. American Astron. Soc. Vol. 1, 296 (1969). – Abstr. AAS.

071.050 **On the diffusion of solar magnetic fields. II.**
V. V. Kassinsky.
Solnechnye Dannye 1969 Byull. No. 5, p. 82 - 85 (1969). In Russian.

Some aspects of diffusion of magnetic fields in the solar photosphere are considered with allowance for convective transport. It is shown that in the model of hierarchy of diffusion mechanisms the fine structure of the field ($\gtrsim$ 1000 km) complies with the equation of continuity. The consequence of this fact is the fine structure conservation. A time variation of the field at the borders of supergranulation cells is discussed.

071.051 **On the influence of sound waves upon the form and shift of faint Fraunhofer lines in the solar photosphere.** B. T. Babij, A. D. Altman.
Solnechnye Dannye 1969 Byull. No. 5, p. 103 - 109 (1969). In Russian.

The problem of possible reasons of asymmetry and shift of photospheric Fraunhofer lines is discussed. It is shown that sound waves can essentially influence the peculiarities of the line profiles.

071.052 **Determination of the velocities of large- and small-scale motions using the method of correlations of central depths and equivalent widths of the lines.**
G. D. Poljakova.
Solnechnye Dannye 1969 Byull. No. 5, p. 110 - 114 (1969). In Russian.

071.053 **Equator-pole effect in the central intensities of some strong solar Fraunhofer lines.**
M. Burger, J. Houtgast.
Solar Physics, Vol. 9, 296 - 302 (1969).

Uncorrected central intensities of the Fraunhofer lines NaD_1 and D_2, Mgb_1 and b_2, and $H\beta$ have been determined at a number of points along the S–W limb of the sun, from equator to pole, using photographic spectra taken at the Locarno station of the Göttingen Observatory. No significant pole-equator variation could be found in excess of the observational errors which are of the order of 2%.

071.054 **Temperature fluctuations in the solar photosphere. II: The mean limb-darkening and the second maximum.** P. R. Wilson.
Solar Physics, Vol. 9, 303 - 314 (1969).

Two models, model 1 exhibiting a single temperature fluctuation maximum and model 2 which has two temperature fluctuation maxima, were put forward as worthy of further investigation. The theoretical mean limb-darkening for these models is compared with the observed limb-darkening. Neither is satisfactory and several modifications are discussed. Models of the first type can be made to fit these data only by making adjustments which appear to be inconsistent with convection as an explanation of the temperature fluctuations. Further, the agreement with the fluctuation data is now less satisfactory. However, a modified model of the second type is developed which is consistent with the convection hypothesis, which is in good agreement with the mean limb-darkening and is in qualitative agreement with the fluctuation data.

071.055 **More on granulation models.**
T. E. Margrave, Jr., T. L. Swihart.
Solar Physics, Vol. 9, 315 - 316 (1969). – Research note.

071.056 **Solar velocity fields: 5-min oscillations and supergranulation.**
A. S. Tanenbaum, J. M. Wilcox, E. N. Frazier, R. Howard.
Solar Physics, Vol. 9, 328 - 342 (1969).

One dimensional magnetograph scans have been used to study the 5-min photospheric velocity oscillations and the supergranulation. The oscillations in wing brightness lead the oscillations in velocity by less than 90° in the photosphere, and about 90° in the chromosphere, suggesting that they are traveling waves at lower levels and standing waves at higher levels. Downward flows have been observed to be coincident with the chromospheric network confirming the hypothesis that material is flowing downward at supergranular boundaries.

071.057 Some properties of velocity fields in the solar photosphere. II: The spatial distribution of the oscillatory field. F.-L. Deubner.
Solar Physics, Vol. 9, 343 - 346 = Mitt. Fraunhofer Inst. Freiburg, No. 94 (1969). — Research note.

071.058 The evolution of the photospheric network. N. R. Sheeley, Jr.
Solar Physics, Vol. 9, 347 - 357 = Contr. Kitt Peak National Obs. No. 453 (1969).

A time-lapse sequence of spectroheliograms in the bandhead of CN at λ3883 reveals the following behavior of the photospheric network with time: (1) There is a steady flow of bright 'points' ($\simeq$ 1000 km in diameter) laterally outward from sunspots at speeds on the order of 1 km × sec^{-1}. (2) Spatial changes in the network pattern seem to take place by means of the shifting of network fragments laterally on the solar surface. (3) Occasionally 'new' network, not resulting from the lateral motion of bright features from either previously existing network or sunspots, appears on the solar surface. These observed changes of the photospheric network with time are interpreted as formation and motions of photospheric magnetic fields. It is suggested that these motions reflect the presence of both short-lived small-scale and long-lived large-scale photospheric currents such as one might expect from the granulation and the supergranulation.

071.059 Spectroheliograms in the Mg II line at 2795.5 Å. K. Fredga.
Solar Physics, Vol. 9, 358 - 371 (1969).

A rocket-borne spectroheliograph designed to take monochromatic pictures of the sun in the Mg II line at 2795.5 Å was successfully launched from White Sands Missile Range on May 20, 1968. Pinhole photographs in the soft X-ray region were also secured in the same flight. The Mg II and X-ray photographs are compared with simultaneous Ca II and Hα spectroheliograms, radiomaps at 9.1 cm and 3.3 cm, a Fraunhofer map and a magnetogram and some preliminary results are discussed.

071.060 The profiles of Fraunhofer lines in the presence of Zeeman splitting. I: The Zeeman triplet.
J. M. Beckers.
Solar Physics, Vol. 9, 372 - 386 (1969).

For the case of pure absorption lines (LTE) a method is described which enables the general computation of Zeeman-split line profiles. The magnetic field vector, the Doppler shift and the line absorption coefficient is permitted to vary arbitrarily with optical depth. Elliptical birefringence (e.g., Faraday rotation) of the solar atmosphere is taken into account. Some numerical examples are given and some interesting behaviors of the line profiles are discussed.

071.061 Magnetic 'knots' in the solar photosphere. H. I. Abdusamatov, V. A. Krat.
Solar Physics, Vol. 9, 420 - 422 (1969). — Research note.

071.062 Clarification in the identification of certain lines in the infrared solar spectrum.
E. F. Montgomery.

Solar Physics, Vol. 10, 60 - 62 (1969).

Forty-two unidentified spectral lines in Mohler's (1955) table of infrared solar spectrum wavelengths are shown to be solar, and three of these tentatively have been identified. One line, listed by Mohler as solar or telluric, is shown to be telluric.

071.063 On the instrumental profile of the Utrecht Photometric Atlas. M. Minnaert.
Monthly Notices, Roy. Astron. Soc., Vol. 146, 91 - 92 (1969).

071.064 Fabry-Perot interferograms of the solar Mg II doublet and XUV solar images obtained during a stabilized Skylark rocket flight.
B. Bates, D. J. Bradley, C. D. McKeith, N. E. McKeith, W. M. Burton, H. J. B. Paxton, D. B. Shenton, R. Wilson.
Nature, Vol. 224, 161 - 163 (1969).

A total of 500 useful profiles of both the 2795.53 and 2802.7 Å resonance lines of Mg II were obtained, with a spectral resolution of 0.03 Å, using an optically contacted Fabry–Perot interferometer which was internally mounted in an echelle spectrograph. The spectral resolution element of 0.03 Å corresponded to a spatial resolution of 5 sec of arc along the spectrograph slit. Final solar image stabilisation on the latter was achieved by servo controlling the main telescope collector mirror, to an indicated accuracy of 16 and 5 arc secs peak to peak in the directions along and perpendicular to the slit respectively. A sample interferogram and reduced profiles are presented along with solar images in the extreme ultraviolet obtained using a multiple aperture pinhole camera.

071.065 On the abundance of iron in the solar photosphere. J. B. Rogerson, Jr.
Astrophys. Journ. Vol. 158, 797 - 802 (1969).

An infrared supermultiplet of lines of neutral iron has been used for an abundance study of the sun, since their formation deep in the photosphere and their high level of excitation should make them less subject to deviations from local thermodynamic equilibrium. The resulting abundance is log A(Fe) = 6.85 on a scale where log A(H) = 12.00. No evidence is found that non-LTE effects may explain the difference between photospheric and coronal abundance determinations. A mean microturbulent velocity of 3.4 km sec^{-1} is required to explain all the equivalent-width data.

071.066 On the spectrum of granular and intergranular regions. R. Howard, A. Bhatnagar.
Solar Physics, Vol. 10, 245 - 253 (1969).

A very high quality wiggly-line spectrogram was analyzed by making high-resolution spectral scans of numerous small solar features. An attempt from the line profiles to detect a magnetic field difference between the granular and intergranular regions, resulted in a field increase of 20 ± 15 G in the darker regions of the granular field. The present investigation is an attempt to obtain quantitative estimates of the difference in the magnetic field and turbulent velocities and variations in line profile in the granular and intergranular regions.

071.067 Studies of granular velocities. I: Granular Doppler shifts and convective motion.
W. Mattig, J. P. Mehltretter, A. Nesis.
Solar Physics, Vol. 10, 254 - 261 = Mitt. Fraunhofer Inst. Freiburg, No. 95 (1969).

Size of the elements and the influence of atmospheric seeing; Observations of granular Doppler shifts.

071.068 The profiles of Fraunhofer lines in the presence of Zeeman splitting. II: Zeeman multiplets for dipole and quadrupole radiation. J. M. Beckers.
Solar Physics, Vol. 10, 262 - 267 (1969).

The radiative transfer equations (LTE) in the four Stokes parameters are derived for the general case of a Zeeman multiplet for both electric and magnetic dipole as well as for electric quadrupole radiation.

071.069 Observational evidence for quantization in photospheric magnetic flux.
W. Livingston, J. Harvey.
Solar Physics, Vol. 10, 294 - 296 = Contr. Kitt Peak National Obs. No. 501 (1969).

Observations are presented which suggest that away from sunspots photospheric magnetic flux is quantized. Assuming the elemental area of a magnetic region to be 1 (arc-sec)^2 the elemental field strength is 525 G.

071.070 Multi-channel magnetograph observations. I: Comparison with spectroheliograms.
E. N. Frazier, P. H. Scherrer.
Solar Physics, Vol. 10, 297 - 310 = Contr. Kitt Peak National Obs. No. 499 (1969).

A new technique for displaying magnetograph observations is presented and applied to the 12-channel magnetograph at Kitt Peak National Observatory. Using the data from a raster scan, a digital 'spectroheliogram' is constructed on the face of a cathode ray tube and photographed. This enables one to recognize patterns in magnetograph data as easily as with conventional photographs. Comparisons with simultaneous spectroheliograms show no qualitative differences and indicate that the magnetograph is quite capable of studying morphology of individual solar features.

071.071 The forbidden line [Ca II] λ 7323 in the Fraunhofer spectrum. D. L. Lambert, E. A. Mallia.
Solar Physics, Vol. 10, 311 - 314 (1969).

New observations of the [Ca II] λ 7323 Fraunhofer line are reported. The blending H_2O line was weak at the time of observation. Accurate estimates of the centre-limb variation of the equivalent width of the [Ca II] transition are obtained and shown to be consistent with the calcium abundance log N(Ca) = 6.33.

071.072 The abundance of cadmium in the solar atmosphere.
Ø. Hauge.
Solar Physics, Vol. 10, 315 - 318 (1969).

The solar spectrum at 3261 Å has been studied using the spectrograph at the Oslo Solar Observatory. From analysis of this wavelength region and recent results at 5085 Å, a solar cadmium abundance log N_{Cd} = 1.86 ± 0.15 is obtained.

071.073 Identification of the SiH lines in the solar disk spectrum. A. J. Sauval.
Solar Physics, Vol. 10, 319 - 329 (1969).

A new investigation of the presence of SiH lines in the solar disk spectrum has been performed. It may be concluded that molecular absorption lines of SiH are present in the disk spectrum with maximum equivalent widths of about 2mÅ. A value of the oscillator strength of SiH has been derived $(f_{00} = 0.0008 ± 0.0004)$.

071.074 Measurements in the solar spectrum between 1400 and 1875 Å with a rocket-borne spectrometer.
W. H. Parkinson, E. M. Reeves.
Solar Physics, Vol. 10, 342 - 347 (1969).

Intensity measurements in the solar continuum throughout the wavelength range 1400 Å − 1875 Å are reported for a rocket flight from White Sands, New Mexico. These intensities are approximately a factor of 3 lower than other published estimates and suggest a solar temperature minimum of 4400 K or lower which is in agreement with infrared observations.

071.075 Astrophysical damping constants for neutral iron.
C. R. Cowley, G. H. Elste, R. H. Allen.
Astrophys. Journ., Vol. 158, 1177 - 1181 (1969).

Empirical damping constants are derived for twenty-six lines of neutral iron by using the solar spectrum. Eight of these lines have also been measured by Kusch, and the agreement between the two sets of measurements is satisfactory. We suggest that these damping constants may be used to obtain abundances and to strengthen spectroscopic determinations of surface gravity.

071.076 Neutral neon in the visible solar spectrum.
R. D. Dietz, F. Q. Orrall.
Astrophys. Journ., Vol. 158, 1239 - 1242 (1969).

A faint emission line at λ6402.2 occasionally appears in the spectrum of bright solar prominences. Calculations support the suggestion that the line is due to Ne I.

071.077 The revised solar abundance of iron and the solar neutrino flux. W. D. Watson.
Astrophys. Journ. (*Letters*), Vol. 158, L189 - L191 (1969).

The influence on the solar-core opacity of the revised photospheric abundance of iron is investigated. Within the inner 0.3 $M_\odot$ of current models for the present sun, the Rosseland opacity is found to be increased by about 25 percent as a result of bound-free absorption by the iron. This change is estimated to double the predicted value for the neutrino flux in the ^{37}Cl experiment relative to that obtained by using otherwise "best values" for all physical parameters.

071.078 Excitation temperatures of Fe I in the sun and A stars and systematic errors in the *gf* scale.
G. Elste.
Bull. American Astron. Soc., Vol. 1, 340 - 341 (1969). − Abstract AAS.

071.079 Review of early photographic observation of solar granulation. T. E. Margrave, Jr.
Journ. Washington Acad. Sci., Vol. 58, No. 4, p. 75 - 79 (1968).

071.080 High resolution spectra of the sun.
L. Delbouille, G. Roland.
Phil. Trans., Ser. A, Vol. 264 (No. 1150), 171 - 182 (1969).

071.081 Infrared solar spectrum as observed from balloons.
D. G. Murcray, F. H. Murcray, W. J. Williams, T. G. Kyle.
Phil. Trans., Ser. A, Vol. 264 (No. 1150), 183 - 194 (1969).

071.082 The solar continuum in the far-infrared and millimetre regions. R. W. Noyes.
Phil. Trans., Ser. A, Vol. 264 (No. 1150), 205 - 208 (1969).

071.083 Solar extreme limb spectrum of the rare earth cerium. R. C. Canfield.
Thesis, Univ. of Colorado. Univ. Microfilms, Ann Arbor, Mi., 153 pp. (1968). − See Phys. Abstr. Vol. 72, No. 29089 (1969).

071.084 Arizona-NASA atlas of the infrared solar spectrum, Report VII.
L. A. Bijl, G. P. Kuiper, D. P. Cruikshank.
Commun. Lunar Planet. Lab. Vol. 9, (No. 163), 65 - 92 (1969).

This paper is a continuation of Commun. LPL 161, covering the interval λλ 17731–21492 Å. For purposes of identification a laboratory spectrum of the 1.8 μ water-vapor band is given and Courtoy's CO_2 spectrum is reproduced for the interval λλ 19374–20930 Å.

071.085 Arizona-NASA atlas of the infrared solar spectrum,

Report VIII.

L. A. Bijl, G. P. Kuiper, D. P. Cruikshank.

Commun. Lunar Planet. Lab. Vol. 9, (No. 164), 93 - 120 (1969).

In this paper we reproduce the solar spectrum λλ 21492–25583 Å, as obtained from the NASA CV-990 Jet at high altitude. The paper is a continuation of Commun. LPL Nos. 161 and 163. Included are laboratory spectra of the 2.20, 2.32, and 2.37 μ bands of methane and of the wings of the water-vapor bands at 2.7 μ. The 8–6 and 9–7 solar CO bands are found to be present in the spectra.

071.086 Arizona-NASA atlas of the infrared solar spectrum, Report IX.

L. A. Bijl, G. P. Kuiper, D. P. Cruikshank.

Commun. Lunar Planet. Lab. Vol. 9, (No. 165), 121 - 153 (1969).

In this paper we give the solar spectrum λλ 25583–30920 Å as obtained from the NASA CV-990 Jet. A laboratory spectrum of the 2.7 μ H_2O bands is included; Courtoy's laboratory spectrum of the 2.7 μ CO_2 bands is given.

071.087 On correlations between brightness, velocity, and magnetic fields in the solar photosphere.

W. C. Livingston.

Proc. Fifth Berkeley Symposium on Mathematical Statistics and Probability, Vol. 3, 61 - 72 = Contr. Kitt Peak National Obs. No. 217 (1967).

071.088 The use of television systems for observations of the solar granulation.

M. B. Kerimbekov, Ch. A. Éfendiev.

Izv. AN Azerb. SSR, Ser. fiz.-tekhn. i matem. n. No. 1, p. 72 - 76 (1969).   In Russian. – Abstr. in Referativ. Zhurn. 51. Astron., 1.51.112 (1970).

071.089 Variation of the infrared solar spectrum between 700 cm^{-1} and 2240 cm^{-1} with altitude.

D. G. Murcray, F. H. Murcray, W. J. Williams, T. G. Kyle, A. Goldman.

Applied Optics, Vol. 8, 2519 - 2536 (1969).

A grating spectrometer with a Ge : Cu detector was flown on three balloon flights. Spectra in the 4 – 14.3-μ region were obtained at various altitudes from the ground through 30 km with a resolution considerably better than that achieved in previous flights. Some of the spectra were obtained over long paths at float altitude, during the sunset. Data from these flights are presented with a discussion of the significant features of the observed absorptions.

071.090 Solar photosphere charts.   L. Schmied.

Říše hvězd, Vol. 50, 141, 158, 220 (1969). –

Rotations Nos. 1542 - 1547.

071.091 Photoelektrische Registrierungen des photosphärischen Netzwerkes.

F.-L. Deubner.

Mitt. Astron. Ges. No. 27, p. 204 - 205 (1969). – Abstract AG.

Spin-forbidden resonance resonance resonance multiplets in light elements.   See Abstr. 022.058.

On polarimetry in solar active regions. I: The new Locarno polarimeter; observing procedures. See Abstr. 034.017.

On the influence of the width of the intermediate slit of a double monochromator on the profiles of the Fraunhofer lines.   See Abstr. 034.042.

L'interferometro Fabry-Pérot per l'analisi fotometrica dei profili di righe solari.   See Abstr. 034.054.

Solution of a class of magnetohydrodynamic problems with strengthening of the magnetic field. See Abstr. 062.040.

On the stability of sunspots inside the supergranulation net.   See Abstr. 072.021.

Observation of hydrogen fluoride in sunspots and the determination of the solar fluorine abundance. See Abstr. 072.027.

A mechanism for the acceleration of solar chromospheric spicules.   See Abstr. 073.014.

Recent eclipse data and the solar limb. See Abstr. 073.017.

On some flare-sensitive high photospheric and low chromospheric lines.   See Abstr. 073.051.

Revision of the ultraviolet solar spectrum in the range 3650 – 3000 Å.   See Abstr. 076.007.

On the diffusion of solar magnetic fields. I. See Abstr. 080.006.

Magnetograph measurements with temperature-sensitive lines.   See Abstr. 080.031.

Line statistics for solar type stars. See Abstr. 114.032.

072 Sunspots, Faculae, Solar Activity

072.001 On the pseudo-π-component in sunspot spectra.
R. Göhring.
Solar Physics, Vol. 8, 271 - 274 = Mitt. Fraunhofer Inst. Freiburg No. 89 (1969).
The visibility, contrast and displacement of the pseudo-π-component of a Zeeman triplet, explained as due to saturation, is computed for the triplet at 5250 Å (Fe 1). Examples of the computational results are given.

072.002 A model for the penumbra of sunspots.
O. K. Moe, P. Maltby.
Solar Physics, Vol. 8, 275 - 283 (1969).
A penumbra model in hydrostatic equilibrium is presented. The model accounts for the continuum observations as well as the observations of Fraunhofer lines in the penumbra. The uncertainty in the model in deeper layers is discussed. It is shown that the penumbra is probably not in strict radiative equilibrium.

072.003 On the position of sunspots in the core of Hα relative to the continuum. O. Engvold.
Solar Physics, Vol. 8, 284 - 290 (1969).
The relative position of sunspots as observed in the core of Hα and in the continuum has been studied in 316 spectra of 84 different sunspots. We find that chromospheric features surrounding sunspots may produce apparent shifts of the spots in the core and in the wing of Hα. In addition a shift directed towards the limb is found. This shift is found to be a height effect. The difference in height between the levels of the Hα core and the continuum varies from 2300 km to 1000 km for different sunspots.

072.004 The geometrical height-scale and the pressure equilibrium in the sunspot umbra. W. Mattig.
Solar Physics, Vol. 8, 291 - 309 = Mitt. Fraunhofer Inst. Freiburg, No. 90 (1969).
Spectra of spots very near to the solar limb are used to determine the hight difference between the levels of formation of the continuum and the line cores of 60 medium-strong Fraunhofer lines. For all lines (with Rowland Intensity < 10), this difference is < 1″ (= 725 km) and well correlated with the Rowland intensity. The line absorption coefficient is calculated for some lines with known oscillator strength. This gives a possibility to deduce a value for the scale hight of the umbra, which is found to be about 100 km, thus being equal to the photospheric scale height. Pure hydrostatic equilibrium exists, therefore, in the umbra, and vertical magnetic forces are negligible. The horizontal pressure equilibrium is discussed. The magnetic field is confirmed to be force-free in higher layers (chromosphere). The pressure difference umbra-photosphere increases towards deeper layers.

072.005 Some notes on Babcock's theory of solar activity.
M. Kopecký.
Bull. Astron. Inst. Czechoslovakia, Vol. 20, 172 - 176 (1969).
The change in the initial magnetic field of the Sun from cycle to cycle can explain, on the basis of Babcock's theory, the different shape of the course of the mean heliographic latitude of spots in the different 11-year cycles. This would also lead to changes in the magnetic field of the corona. It is concluded that the process of amplification of the magnetic field by the differential rotation of the Sun begins 0.85 years before the beginning of the 11-year cycle.

072.006 The east-west asymmetry of sunspots.
L. Pajdušáková.
Contr. Astron. Obs. Skalnaté Pleso, Vol. 4, 6 - 24 (1969).
The east-west asymmetry of sunspots observed at Skalnaté Pleso is not systematically positive: it shows a marked dependence on the 11-year cycle, being positive with increasing and high activity, and negative when activity declines or is low. A similar dependence was also reported by other stations. The number of groups in the Zurich material has been almost systematically higher in the west since 1933, in other words, the asymmetry has been negative. Negative asymmetry in the number of groups and the spot area over relatively longer periods has also been reported by other stations in the last decades. This inconsistent result of different stations suggests the conclusion that there may be further reasons for both positive and negative asymmetry. The material obtained at Skalnaté Pleso, as an independent station, was examined for the effect of meteorological conditions, and the result was as follows: 1. Asymmetry is positive for most years with above-average precipitation, negative for those with below-average precipitation. 2. Groups of the types D, E and F observed for at least 13 days have better atmospheric observational conditions before than after the central-meridian passage.

072.007 Über Temperatur und Dichte in chromosphärischen Fackeln, ermittelt aus Profilen starker Fraunhofer-linien. G. Bachmann.
Astron. Nachr. Vol. 291, 131 - 153 (1969).
Spectrograms of faculae with the Fraunhofer lines Hα, Hβ, H, and K were used for the determination of the excitation temperature, the kinetic temperature and the turbulence velocity. For comparison, the same quantities for undisturbed regions of the solar atmosphere were calculated by means of already existing limb-darkening measurements. The residual intensities of the four lines were used to calculate the excitation temperatures at the height of formation of the line centres. From the line profiles of the two hydrogen lines Doppler widths were determined by the multiplet method (Goldberg) extended on lines with equal excitation temperatures. The Doppler width of the Ca II lines could be calculated from the half width of the central absorption line by the relation $\Delta\lambda_{1/2} = \Delta\lambda_D$. The combination of the Doppler widths of the hydrogen line Hα with those of the Ca II line K yields under certain conditions kinetic temperature and turbulence velocity in the corresponding range of height.

072.008 Another contribution to the question of the Joule dissipation of magnetic fields in the solar atmosphere. M. Kopecký.
Bull. Astron. Inst. Czechoslovakia, Vol. 20, 296 - 300 (1969).
If the magnetic field dissipates in the elements of its fine structure, it may happen, to a great extent, during the lifetime of a spot group even though the dissipation is determined by electric currents flowing in regions of high electric conductivity. In certain conditions, the Joule dissipation may therefore play an important role in the life of magnetic fields in the solar atmosphere. In this connection there also arises the question of the possibility of a continuous formation of magnetic fields in the solar atmosphere.

072.009 Planetary influences on sunspots.
G. A. J. Ferris.
Journ. British Astron. Ass. Vol. 79, 385 - 388 (1969).

072.010 Hemispherical inequality in the distribution of sunspots. J. Friends.
Journ. British Astron. Ass. Vol. 79, 472 - 474 (1969).

072.011 The transfer of energy in sunspots. II. The axisymmetric semi-empirical models.

A. Stankiewicz.
Acta Astron. Vol. 19, 199 - 223 (1969).

Energy transport by radiation in the sunspot region was considered. The model computations were performed for the hydrostatic and magnetohydrostatic conditions of equilibrium. Basing on a detailed analysis of the results obtained it has been suggested that the sub-photospheric flux of energy must be reduced in the sunspot. It is shown that the reduced flux can be estimated for any given magnetic field in sunspots. This result follows from the analysis of the models only, and the theory of the convective or turbulent motion in the magnetic field in deeper layers has not been used. This independent estimate of the reduced flux may be important for the theory of inhibitating of convection in the presence of a magnetic field.

072.012 **Kinetic temperature of the upper layers of faculae.**
V. V. Polonsky.
Astron. Tsirk. No. 509, p. 6 - 7 (1969).   In Russian.

072.013 **On the length of the solar activity cycle.**
A. R. Kolomietz.
Astron. Tsirk. No. 509, p. 7 - 8 (1969).   In Russian.

072.014 **On the supergranulation structure of sunspots.**
Eh. P. Surkov.
Astron. Tsirk. No. 519, p. 6 - 8 (1969).   In Russian.

072.015 **Der Index a für die Jahre 1953 - 1965.**
T. L. Mandrykina.
Tsirk. L'vov. Astron. Obs. No. 43, p. 43 - 45 (1969).
In Russian.

072.016 **Der Einfluß des Magnetfeldes auf die Zunahme der Äquivalentbreiten der Absorptionslinien von Sonnenflecken.**   P. A. Olijnyk.
Tsirk. L'vov. Astron. Obs. No. 43, p. 46 - 49 (1969).
In Russian.

072.017 **On the value of the turbulence velocity and spectral lines' magnetic intensification in a sunspot.**
I. Sattarov.
Solnechnye Dannye Byull. No. 2, p. 88 - 92 (1969).
In Russian.

The value of the turbulent velocity (ξ) is determined by the method of line-width correlation. Altogether 24 spectral lines unaffected and slightly affected by the magnetic field of a sunspot are used for this purpose. The line profiles are corrected for the instrumental profile and magnetic broadening. The ξ_t value derived from the comparison of the observed curve with the theoretical one is equal to 2.5 km/sec. The scattering in the data on the sunspot curve of growth is studied.

072.018 **Relation between three consecutive 11-year solar cycles.**   A. D. Bonov.
Solnechnye Dannye Byull. No. 2, p. 93 - 95 (1969).
In Russian.

Relations between the maximum value of Wolf's numbers in a seperate 11-year cycle and the sum of these values in two neighbouring 11-year cycles are considered. These relations are expressed by formulas for odd and even 11-year cycles correspondingly.

072.019 **Non-uniform distribution of sunspots and their groups over an area.**   P. V. Florensky.
Solnechnye Dannye Byull. No. 2, p. 96 - 99 (1969).
In Russian.

On the base of studying the data for 1967 it is shown that the existence of sunspots and their groups with the area corresponding to the sizes of "spotted supergranules" is most probable. An extremely non-uniform distribution in their area

was detected. The formulas, describing the areas of the sunspots and their groups more frequently observed are given.

072.020 **Some peculiarities of the H and K Ca II lines in the light bridges of sunspots.**
R. B. Teplitskaya, Z. B. Korobova.
Solnechnye Dannye Byull. No. 3, p. 102 - 110 (1969).
In Russian.

Fifteen spectrograms of sunspots with light bridges have been investigated. The emission-core width and separation of the emission peaks of the H and K Ca II lines in the spectra of light bridges are compared with those in plages and umbrae. Some peculiarities of the chromosphere above sunspots are found to remain above light bridges too.

072.021 **On the stability of sunspots inside the supergranulation net.**   V. V. Kassinsky.
Solnechnye Dannye Byull. No. 3, p. 111 - 115 (1969).
In Russian.

On the basis of the energetic principle the transposition instability of magnetic field tubes in the photosphere is considered. In the case of hydrostatic equilibrium with surroundings the tubes are stable. Supergranulation leads to "discreteness" of areas in the photosphere, which disturbs hydrostatic equilibrium. In such a situation young spots in the intersupergranular space and the developed spots occupying one elementary cell of supergranulation will be stable.

072.022 **On the active longitudes of the solar background magnetic fields.**   Y. I. Vitinsky.
Solnechnye Dannye Byull. No. 4, p. 88 - 97 (1969).
In Russian.

The active longitudes of the background magnetic regions of the northern and southern polarity were detected by the method of isolines. It is shown that a degree of concentration in these active longitudes corresponds to that for the least active formations of the sun. The active longitudes of the background magnetic regions coincide in practice with the active longitudes of sunspots and flares in the northern hemisphere of the sun. It is shown that differential rotation does not influence essentially on detecting the active longitudes of the background magnetic regions.

072.023 **Über den Zusammenhang zwischen Penumbraformen und magnetischer Polaritätsverteilung in Sonnenfleckengruppen.**   H. Künzel.
Astron. Nachr., Vol. 291, 265 - 270 (1969).

The characteristics of six typical penumbra forms are described. It was found that the penumbra of a spot is influenced by the polarity, the magnetic field strength, and the size of neighbouring spots. Probably, the form and the structure of the penumbra is a copy of the lines of magnetic force. A test showed that in 93 per cent of 600 examples the form of the penumbra is significantly correlated with the distribution of the polarity of spots. Experiments with circular magnets gave figures of the lines of force which resemble the typical penumbra forms of spots with the same distribution of the polarities. The results of the test show that the characteristics of the penumbra forms give the possibility to make a statement as to the relative distribution of the polarities in the spot groups without measuring the real magnetic data. In this way it is possible to find the boundary between the north- and the south-polarity in the spot groups.

072.024 **On the chromospheric velocity field in sunspot regions.**   E. Haugen.
Solar Physics, Vol. 9, 88 - 101 (1969).

The wavelength shifts of approximately 8000 absorption elements in the Hα-line from spectra of 66 different sunspot regions have been measured. The average velocity field in the chromosphere close to sunspots is determined. Inside

15000 km from the spot's penumbral rim the average velocity vector is directed towards the spot and downwards in the chromosphere; the angle with the horizontal direction is on the average equal to 20°. The magnitude of the average velocity vector shows a maximum of 6.8 ± 1.2 km/sec just outside the penumbral rim and decreases quickly with increasing distance from the spot. Outside 15000 km from the penumbral rim the average velocity vector is small ($\simeq$ 0.7 km/sec) and directed nearly vertically outwards from the sun. No significant tangential component of the average velocity field is found.

072.025 Sunspot observation through water clouds.
D. Deirmendjian.
Applied Optics, Vol. 8, 833 (1969).

The author gives a theoretical explanation of a naked eye observation of a sunspot through a layer of coastal stratus cloud.

072.026 Der 20. Sonnenfleckenzyklus zwischen Minimum und Maximum. W. Schulze.
Sterne, 45. Jahrgang, 107 - 112 (1969).

072.027 Observation of hydrogen fluoride in sunspots and the determination of the solar fluorine abundance.
D. N. B. Hall, R. W. Noyes.
Astrophys. Letters, Vol. 4, 143 - 148 (1969).

Observations of the spectrum of sunspots in the near infrared have revealed the presence of lines of the fundamental vibration-rotation band of the molecule HF. The excellent agreement between observed and laboratory wavelengths, as well as agreement between the observed and the predicted variation in equivalent width for spots of different temperatures, make the identification secure. Model calculations carried out for two spots differing in temperature by 400°K yield a solar fluorine abundance $\log A_F$ = 4.56 ± 0.33, on a scale where $\log A_H$ = 12.00.

072.028 The electromagnetic acceleration of charged particles in nonstationary magnetic fields of sunspots.
P. E. Kolpakov.
Astron. Zhurn. Akad. Nauk SSSR, Vol. 46, 1047 - 1056 (1969). In Russian. English translation in Soviet Astron. AJ, Vol. 13, No. 5.

Conditions and possibilities of the acceleration of non-relativistic charged particles in magnetic fields of sunspots changing (increasing) with time are considered. Energy losses of accelerated particles are taken into account and conditions are found, on which electrons and protons of a chromospheric-coronal plasma can be accelerated.

072.029 Magnetic field observations for the sunspot C.M.P. 1966 September 19. M. G. Adam.
Monthly Notices, Roy. Astron. Soc., Vol. 145, 1 - 20 (1969).

Magnetic vector directions have been measured at 119 points over the field of a large, long-lived sunspot on 1966 September 19, when the spot was closely at central meridian passage. Further developments in Treanor's method of determining field directions enable the directions to be obtained independently for the violet and red sigma-components in the Zeeman patterns of the absorption lines. Reliability of the determinations is indicated by agreement between these two results, and for such agreeing determinations the field configuration shows a simple axially symmetric arrangement which is also supported by the field strength measurements. A discussion of the accuracy obtainable in field direction measurements, by Treanor's and by other methods, shows the greater potentiality of Treanor's procedure. The discussion leads also to the conclusion that apparent anomalies in the present results, at least in the sunspot umbra, lie within the uncertainty of the observations. For some of the penumbral regions, however, the apparent inconsistencies in the deduced magnetic field directions may indicate a genuine solar effect.

072.030 Vertical velocities in and surrounding a sunspot group. E. V. P. Smith.
Bull. American Astron. Soc. Vol. 1, 262 (1969). – Abstr. AAS.

072.031 On the correlation between sunspot structure and solar flares. S. M. Adler.
Bull. American Astron. Soc. Vol. 1, 270 (1969). – Abstr. AAS.

072.032 Magnetic development of an active region.
K. L. Angle, W. C. Livingston.
Bull. American Astron. Soc. Vol. 1, 271 (1969). – Abstr. AAS.

072.033 Chromospheric inhomogeneities in sunspot umbrae.
J. M. Beckers, P. E. Tallant.
Bull. American Astron. Soc. Vol. 1, 272 (1969). – Abstr. AAS.

072.034 Photometry of a white-light penumbral brightening.
G. Emerson, J. M. Malville.
Bull. American Astron. Soc. Vol. 1, 276 (1969). – Abstr. AAS.

072.035 A model study of persistent magnetic regions.
R. B. Leighton.
Bull. American Astron. Soc. Vol. 1, 284 (1969). – Abstr. AAS.

072.036 A model for the penumbra of sunspots.
P. Maltby, O. K. Moe.
Bull. American Astron. Soc. Vol. 1, 285 - 286 (1969). – Abstr. AAS.

072.037 The birth of active regions.
S. Weart, H. Zirin, B.-Z. Kozlovsky.
Bull. American Astron. Soc. Vol. 1, 295 (1969). – Abstr. AAS.

072.038 On the magnetic field in pores.
N. O. Weiss, G. W. Simon.
Bull. American Astron. Soc. Vol. 1, 295 (1969). – Abstr. AAS.

072.039 Large-scale aspects of solar activity. J. M. Wilcox.
Bull. American Astron. Soc. Vol. 1, 296 (1969). – Abstr. AAS.

072.040 Circular components of the π-component of the Fe I 6302 Å triplet and vertical velocity in sunspot umbrae. E. P. Surkov.
Solnechnye Dannye 1969 Byull. No. 5, p. 91 - 98 (1969). In Russian.

Additional data on the Severny effect of displacement of circular components in the π-component of the Fe I 6302 Å triplet are given. The correctness of the interpretation of this effect as the Faraday effect is shown. A vertical motion in sunspot umbrae directed along the radius to the center of the sun is detected. The center-limb variations of this motion are considered.

072.041 Some comments on a forecast of main characteristics of the 20th cycle of solar activity.
Y. L. Vitinsky.
Solnechnye Dannye 1969 Byull. No. 5, p. 99 - 102 (1969). In Russian.

More precise values of the epoch of maximum (1969.2 –

1969.5) and mean-year maximum of Wolf numbers (W_M = 105 - 110) were derived for the 20th cycle of solar activity. A forecast of quarterly Wolf numbers for 1969 – 1972 is given.

072.042 Continuous spectra of umbra and penumbra of sunspots. V. I. Makarov.
Solnechnye Dannye 1969 Byull. No. 6, p. 85 - 90 (1969). In Russian.

The analysis of the continuous spectrum of sunspots in the region of 4000 – 7000 Å was made using the spectrograms obtained at Pulkovo with the four-cameras spectrograph. A non-uniform distribution of the intensity in sunspot umbrae was taken into account. A dependence $I_N / I_\psi (\lambda)$ was derived for the "cold" and "hot" elements in the sunspot umbrae.

072.043 Time variations of the sunspot contrast from observations of the chromosphere. E. P. Surkov.
Solnechnye Dannye 1969 Byull. No. 6, p. 100 - 103 (1969). In Russian.

During flares of importance 3f, 2n and Ib the essential variations of the sunspot contrast were observed with a birefringent filter. The importance of taking into account the physical conditions in such an active sunspot for forecasts and investigations of the mechanism of flares is stated.

072.044 On the distribution of solar activity centers. G. S. Minassjants, S. O. Obashev.
Solnechnye Dannye 1969 Byull. No. 6, p. 108 - 112 (1969). In Russian.

For nine longitudinal intervals the latitude-time distribution of flares with importance ≥ 2 during the 19. cycle is considered. The active longitudes for cycle 19 were detected for flares with importance ≥ 2 and sunspot groups with $S_{max} >$ 500 millionths of solar hemisphere.

072.045 On the law of limb darkening of umbra and penumbra of sunspots. V. I. Makarov.
Solnechnye Dannye 1969 Byull. No. 7, p. 72 - 78 (1969). In Russian.

An investigation is made of the law of limb darkening for umbra and penumbra of a stable sunspot with the size of $24''$ at $\lambda_{eff} = 4100$ Å which was living during about 4 rotations of the sun. It is shown that up to $\rho = 0.80$ this law is the same as for the photosphere.

072.046 Dependence of thermodynamical variables on the distribution of the magnetic field in sunspots. Y. B. Ponomarenko.
Solnechnye Dannye 1969 Byull. No. 7, p. 78 - 81 (1969). In Russian.

Thermodynamical variables are expressed in terms of distribution of the magnetic field of an asymmetric sunspot in hydrostatic equilibrium. It was shown that the azimuthal component of the field increases the pressure and density in the sunspot.

072.047 Magnetic intensification of absorption lines in sunspots derived with respect to stratification of the sunspot atmosphere. R. B. Teplitskaya, V. D. Turchina.
Solnechnye Dannye 1969 Byull. No. 7, p. 82 - 86 (1969). In Russian.

The tables are given for the determination of the magnetic intensification of the Fe and Ti lines with moderate intensity. The measurements were made with respect to stratification of the sunspot atmosphere. It is shown, that magnetic intensification influences the equivalent widths of the lines.

072.048 On convective lifting of the magnetic field of sunspots in the sub-photospheric layer. V. E. Merkulenko.

Solnechnye Dannye 1969 Byull. No. 7, p. 86 - 94 (1969). In Russian.

The dynamics of a magnetic loop lifting in the convective zone under action of magnetic and hydrostatic forces is investigated on the base of a variational method. It is shown that a sum of these forces remains positive at all stages of tube lifting when the original field topology is force free.

072.049 The magnetic field in the region of the sunspot group N 268 (1967). M. M. Mamadazimov.
Solnechnye Dannye 1969 Byull. No. 7, p. 94 - 98 (1969). In Russian.

The distribution of the longitudinal component of a magnetic field in the circular leader of the sunspot group has been investigated. It is shown that the magnetic lines of force in the penumbra are directed nearly to radial direction. In the bright formation in a penumbra a decrease of the longitudinal magnetic field intensitiy takes place. The matter in a penumbra moves mainly along the lines of force. The magnetic field of the leader extends nearly to the following sunspot. The magnetic lines of force in the leader penumbra are supposed to be concentrated in this direction.

072.050 Checking up the Waldmeier empirical formulas. D. Nohonoi.
Solnechnye Dannye 1969 Byull. No. 7, p. 99 - 100 (1969). In Russian.

A more precise formula is given for the connection of the growth of the 11-year cycle with the maximum Wolf numbers.

072.051 On π-components in Zeeman-split lines of the umbra spectrum. J. P. Mehltretter.
Solar Physics, Vol. 9, 387 - 390 = Mitt. Fraunhofer Inst. Freiburg No. 93 (1969). – Research note.

072.052 The mean temperature gradient in the umbra. P. R. Wilson.
Solar Physics, Vol. 9, 391 - 393 (1969). – Research note.

072.053 On H_2O in sunspots. H. Wöhl.
Solar Physics, Vol. 9, 394 - 396 (1969). – Research note.

072.054 Sunspot motion statistics for 1965–67. H. E. Coffey, P. A. Gilman.
Solar Physics, Vol. 9, 423 - 426 (1969). – Research note.

072.055 Gas-pressure and pressure-stratification in the sunspot. H. Ruhm.
Solar Physics, Vol. 10, 104 - 111 = Mitt. Landessternw. Heidelberg-Königstuhl No. 162 (1969).

By a curve of growth analysis, using photospheric and umbral equivalent widths, published by Fricke and Elsässer, and absolute oscillator-strengths, it has been found, that the gas pressure in a sunspot is lower than in the corresponding layers of the photosphere. Contribution curves have been computed to get more accurate information to which layers of the spot the results relate, which were derived by comparing on the one hand the damping parts and on the other hand the linear parts of the photospheric and umbral curves of growth. From Mattig's observations of a sunspot near the sun's limb, a scale-height of about 300 km is derived. This indicates a deviation from the hydrostatic pressure stratification.

072.056 On the absence of the (0, 0) C_2 Swan band from sunspot spectra. D. Branch.
Solar Physics, Vol. 10, 112 - 114 (1969). – Research note.

072.057 On a relation between the indices of solar activity in the photosphere and the corona.

J. Xanthakis.
Solar Physics, Vol. 10, 168 - 177 (1969).

In the first part a new index of solar activity in the photosphere is introduced i.e. the areas index I_a defined by the total areas of the sunspots and faculae corrected for foreshortening. In the second part a relation is proposed between the intensity of solar radio-emission in the frequency range 1000 to 10000 MHz and the index of solar flares. In the third part a relation is proposed between the mean intensity of the coronal line 5303 Å and the number of proton flares.

072.058 **A rapid scanning low noise spectrometer for study of sunspot spectra.**
D. E. Blackwell, E. A. Mallia, A. D. Petford.
Monthly Notices, Roy. Astron. Soc., Vol. 146, 93 - 99 (1969).

A rapid scanning high resolution photoelectric spectrometer designed for the study of sunspot spectra is described. The spectrometer employs an on-line computer for signal integration, and to reject scans that are contaminated by false light due to 'seeing'. The apparatus is installed at the Gornergrat, Switzerland, at a height of 3090 metres. An example of its performance is given.

072.059 **Magnetic flux-trapping experiment with a moving conductor.** J. Hovorka.
Science, Vol. 166, 877 - 878 (1969).

An aluminum conductor moving into and out of a magnetic field of 75 gauss traps within itself for varying lengths of time a detectable fraction of the encountered flux, which subsequently decays. A time constant of about 0.005 second, which is the order of magnitude predicted by classical electrodynamics, is measured. The result is of interest in connection with the "frozen-in field" concept of Babcock's sunspot model.

072.060 **On the magnetic splitting of molecular lines in sunspot spectra.** H. Wöhl.
Astron. Astrophys. Vol. 3, 378 - 379 (1969).

In sunspot spectra, MgH lines show Zeeman splitting which can be used for magnetic field measurements.

072.061 **Spectrometrical and spectrophotometrical studies of a sunspot. II.** I. Sattarov.
Vestn. Leningr. un–ta, No. 1, p. 129 - 139 (1969). In Russian.
Abstr. in Referativ. Zhurn. 51. Astron., 10.51.594 (1969).

072.062 **Moleküle in Sonnenflecken.**
H. Wöhl.
Umschau, Vol. 69, 845 (1969). – News notes.

072.063 **The intensity, velocity and magnetic structure of a sunspot region. V: On the gradients of temperature and pressure in sunspots.** A. Wittmann, E. H. Schröter.
Solar Physics, Vol. 10, 357 - 369 (1969).

White light photographs in 3 continuous wavelengths were used to obtain intensity profiles of two large sunspots for various positions on the disc. The intensity profiles were corrected for the influence of scattered light, seeing, and the differences in line absorption between photosphere and spot. From the center-to-limb variation of the intensity, the temperature distribution of both umbra and penumbra was derived. A homogeneous model was obtained for the case of a force-free magnetic field, i.e. for hydrostatic equilibrium (HE). Comparison of the model scale height with its value as derived from the Wilson effect allows one to check the validity of assuming HE.

072.064 **Structure of a sunspot. V: What is the Wilson effect?**
P. R. Wilson, P. S. McIntosh.
Solar Physics, Vol. 10, 370 - 383 (1969).

From enlargements of patrol photographs of the disk passage of a sunspot intensity profiles across the spot are obtained at several positions near the disk-center and at each limb. It is found that these profiles show asymmetric features near each limb. Conventionally the Wilson effect is described as the extreme foreshortening and eventual disappearance of the disk-side penumbra and, recently, Suzuki has referred to this as the occultation of the penumbra by the photosphere. We find no evidence at all for the disappearance of the disk-side penumbra at the limb in this spot.

072.065 **The intensity, velocity and magnetic structure of a sunspot region. IV: Properties of a unipolar sunspot.**
J. M. Beckers, E. H. Schröter.
Solar Physics, Vol. 10, 384 - 403 (1969).

From an investigation of spectra in a magnetically sensitive (λ 6173, g = 2.5) and insensitive line (λ 5576, g = 0), we derive the variation of the magnetic field strength with the distance from the sunspot-center, and the variation of the zenith angle of the magnetic field. Also described are details about the variation of the magnetic regions of the penumbra and motions in umbral dots.

072.066 **On the properties of umbral dots.**
P. R. Wilson.
Solar Physics, Vol. 10, 404 - 415 (1969).

On the basis of a three-dimensional radiative transfer analysis of several models it is shown that bright structures in sunspot umbrae which have horizontal diameters of 300 km or less cannot extend more than 300 km down into the umbra. A model having a diameter of 200 km is shown to be consistent with the available observations but these are not sufficiently precise to warrant any strong claim for the validity of this model.

072.067 **Magnetic and turbulent effects on sunspot curves of growth.** J. C. Evans, L. A. Dreiling.
Bull. American Astron. Soc., Vol. 1, 341 (1969). – Abstr. AAS.

072.068 **Isotopic abundance of magnesium in the sun.**
C. K. Kumar.
Bull. American Astron. Soc., Vol. 1, 351 (1969). – Abstr. AAS.

072.069 **A magnetostatic model of a sunspot with "twisted" fields.** H. S. Yun.
Bull. American Astron. Soc., Vol. 1, 370 (1969). – Abstr. AAS.

072.070 **Erratum: On the magnetic splitting of molecular lines in sunspot spectra. [Astron. Astrophys.,**
Vol. 3, 378 - 379 (1969)]. H. Wöhl.
Astron. Astrophys. Vol. 3, 487 (1969). – See Abstract 072.060.

072.071 **Structure and development of a sunspot group on July 4 - 8, 1966.** N. V. Steshenko.
Izv. Krymskoj Astrofiz. Obs. Vol. 39, 245 - 252 (1969). In Russian.

The structural peculiarities of a sunspot group in the process of its growth are investigated from photoheliograms obtained with high resolution. The group is characterized by the complex structure and rapid change of the umbrae of sunspots. The comparison of the position of penumbra filaments with maps of the transversal magnetic field, obtained by A. B. Severny and T. T. Tsap, shows on the whole a good coincidence in the orientation of filaments and the field. It is an additional confirmation of the fact that the investigation of the fine structure of spots from direct photographs of the sun can give additional information on the structure and evolution of a strong magnetic field.

072.072 The determination of the magnetic field strength from different absorption lines.
M. J. Guseynov.
Izv. Krymskoj Astrofiz. Obs. Vol. 39, 253 - 264 (1969).
In Russian.
From spectrograms, obtained with the echelette spectrograph of the tower solar telescope of the Crimean Astrophysical Observatory, the absolute values of the magnetic field strength H are determined from different absorption lines. It is found that the magnetic field strength changes from one line to another and depends both on the Rowland intensity and on the excitation potential of the lower level of the line.

072.073 On the multiplicative character of fluctuations in Wolf numbers.
L. V. Zhukov, Y. S. Muzalevsky.
Solnechnye Dannye 1969 Byull. No. 8, p. 88 - 97 (1969).
In Russian.
It is shown that the fluctuations treated as a non-stationary centered stochastic process are a multiplicative process. A synchronous diagram $W_\phi - W_c$ is plotted. A directly proportional dependence of the mean amplitude of fluctuations on the corresponding smoothed value of the Wolf numbers is detected. It is argued that the anomalous excess of relatively large fluctuations at small W_c can be due to classical determining of series of the solar activity index. The results obtained are of certain importance for forecasts.

072.074 Variations in the position of the axis of a bipolar group of sunspots in dependence on the group
evolution.   K. F. Kuleshova.
Solnechnye Dannye 1969 Byull. No. 8, p. 97 - 100 (1969).
In Russian.
Variations of the inclination angle of the axis of a bipolar sunspot towards the heliographical parallel are considered. It was found that with the development of a group the inclination angle of the axis increases in the northern hemisphere and decreases in the southern one.

072.075 On the recent phases of development of the activity centers.   Y. I. Vitinsky.
Solnechnye Dannye 1969 Byull. No. 8, p. 100 - 105 (1969).
In Russian.
The peculiarities of the development of activity centers after a decay of flocculi for the periods 1961 and 1963 – 1965 are considered.

072.076 The Wilson effect and the transparency of sunspot models.   E. Jensen, R. Brahde, P. Ofstad.
Solar Physics, Vol. 9, 397 - 419 (1969).
Hydrostatic models of sunspot penumbra and umbra are evaluated using Bode's tables of monochromatic absorption coefficients and T–τ relations given by Makita and Morimoto (1960) and by Zwaan (1965). These models are placed side by side to simulate a complete sunspot corresponding to an area of 480×10^{-6} of a hemisphere. Intensity profiles are evaluated for aspect angles up to 85° and compared to observations. The gas-pressure at the zero-level in the geometrical depth (z = 0) corresponding to optical depth, $\tau = 10^{-3}$, appear as adjustable parameters. By varying the pressures in the umbra and penumbra the transparencies are changing and the effects on the Wilson-effect and other limb-effects are studied. Taking curvature into account the best fit was found for $P_o^p = 3200$ and $P_o^u = 800$ (cgs) and a depression of 400 km in the umbra zero-level relative to the photosphere.

072.077 Theoretical sunspot models.   Hong Sik Yun.
Thesis, Univ. of Indiana. Univ. Microfilms, Ann Arbor, Mi., 168 pp. (1968). – See Phys. Abstr. Vol. 72, No.

21974 (1969).

072.078 The solar cycles.
Yu. I. Vitinsky.
Astron. Vestn. Vol. 3, 121 - 134 (1969).   In Russian.
A review of modern ideas on the main peculiarities of the solar cycles is given. Peculiar attention is paid to the consideration of the 11- and the 80–90-year cycles of the solar activity and their principal differences. In conclusion, theories of the origin of the solar cycles are considered and a critical analysis of them is given.

072.079 Fast changes of sunspot equilibrium conditions and solar flares.   Y. D. Žugžda.
Solar Flares and Space Research, Tokyo 1968, p. 368 - 373 (1969).
Oscillatory convection, transport of energy and the structure of sunspots are considered. The possibility of fast changes in the sunspot equilibrium conditions is investigated at various stages of spot evolution. It is suggested that these fast changes may be a cause of magnetic reconstruction and origin of solar flares.

072.080 On long-term forecasts of solar activity.
V. Bumba, R. Howard.
Solar Flares and Space Research, Tokyo 1968, p. 387 - 396 (1969).
A study of the Mount Wilson magnetic field synoptic chart material divided into latitude zones and a comparison of these data with sunspot groups show us a striking regularity in the background field pattern and a relation between concentrations of active regions and concentrations of background fields over certain longitude zones.

072.081 Physics of the sunspots, Part I.
J. Jakimiec.
Postępy Astron., Vol. 17, 315 - 335 (1969).   In Polish.
Recent investigations of the structure of the solar atmosphere in sunspots based on the spectrophotometric measurements, are reviewed.

072.082 Alcune considerazioni sulle macchie solari negli anni dal 1943 al 1967.
M. G. Fracastoro.
Oss. Astrofis. Catania Pubbl., Nuova Serie, No. 136, 16 pp. (1968).

072.083 Research on the 27 day recurrence of sunspot groups.   N. Dogan.
Oss. Astrofis. Catania Pubbl., Nuova Serie, No. 126, 23 pp. (1968).
In this research the distribution of sunspot groups on the solar surface has been studied for the period 1849 - 1964 considering geophysical rotation. It has been found that (1) during the minimum phase the spot distribution is less uniform; (2) according to results reported by different authors, preferred longitudes exist; (3) the same longitudinal belts seem to be involved.

072.084 Variazione della distribuzione spettrale tra 0.5 e 16 Å al variare dell'attività solare.
M. Landini, B. B. C. Fossi, G. L. Tagliaferri.
Atti XII Riunione Soc. Astron. Italiana, L'Aquila 1968, p. 55 - 57 (1969). – Abstract SAI.

072.085 Preliminary results of investigations of the motion of matter across the penumbra of a sunspot.
H. I. Abdussamatov.
Solnechnye Dannye 1969 Byull. No. 9, p. 101 - 105 (1969).
In Russian.
The motion of matter across the penumbra of a sunspot

from the lines Fe I λ 6574.254 Å and H α (by the method of escalation) is studied. It is shown that the Evershed effect appears to represent the streams of matter, which flow out of the sunspot umbra along separate dark filaments of the penumbra in the photosphere and which flow along separate dark filaments in the chromosphere. The flowing of matter (on the photospheric level) takes place only within the penumbra and ends sharply at its external border. Radial velocities of 10 - 20 km/sec (in the H α line) are observable in separate dark filaments of the penumbra or at its very border.

072.086 On some peculiarities of the solar cycle.
O. B. Vassiljev, Y. I. Vitinsky.
Solnechnye Dannye 1969 Byull. No. 9, p. 105 - 116 (1969). In Russian.

A spectral analysis of various indices of sunspots was made for the whole solar disk and separately for the northern and southern hemispheres over 1874 – 1961. The 22-year cycle has not been detected for these indices. The main periodic components of the spectra of sunspot areas for the northern and southern hemispheres were found to coincide in practice. The sunspot cycles are more sharply defined for the Wolf numbers than for the areas; they are also more pronounced for the whole solar disk than for its separate hemispheres.

072.087 On the longitudinal distribution of sunspot groups in the last 11-year cycle of solar activity.
N. N. Morozov, S. O. Obashev.
Trudy Astrofiz. Inst. Alma-Ata, Vol. 15, 38 - 45 (1969). In Russian.

Using the method of correlation functions, it is established that the period in the longitudinal distribution of sunspot groups does not remain constant during the cycle and is not always similar in the different hemispheres of the sun. The value of the period is, evidently, correlated with the mean characteristic linear size of active regions, which undergoes essential changes possibly dependent on the phase of the cycle. In the different hemispheres active regions are often displaced with respect to one another.

072.088 Die Sonnenaktivität im Jahre 1968.
M. Waldmeier.
Vierteljahresschrift Naturforsch. Ges. Zürich, 114. Jahrgang, p. 223 - 243 = Astron. Mitt. Eidg. Sternw. Zürich, No. 288 (1969).

The present paper gives the frequency numbers of sunspots, photospheric faculae and prominences as well as the intensity of the coronal line 5303 Å and of the solar radio emission at the wavelength of 10.7 cm, all characterizing the solar activity in the year 1968.

072.089 Magnetisch ausgerichtete Strömungen zwischen Sonnenflecken.
F. Meyer, H. U. Schmidt.
Zeitschr. Angew. Math. Mechanik, Vol. 48, 218 - 221 (1968).

072.090 Double-K_2 emission line observed in sunspots and in prominences. O. Engvold, W. Livingston.
Publ. Astron. Soc. Pacific, Vol. 81, 795 - 803 = Contr. Kitt Peak National Obs. No. 484 (1969).

The narrow- and double-K_2 emission lines frequently observed in spectra of sunspots are most likely emitted by quiescent-type prominences pointing into the spots.

072.091 Decrease in the number of solar flares and sunspots near the central meridian. M. Kopecký.
Říše hvězd, Vol. 50, 211 - 212 (1969). In Czech.

072.092 New opinions on the periodicity of sunspots.
M. Kopecký.
Vesmír, Vol. 48, 349 (1969). In Czech.

072.093 Photoelektrische Magnetfeldmessungen an einem regulären Sonnenfleck mit besonders dunkler Umbra. F.-L. Deubner, R. Göhring.
Mitt. Astron. Ges. No. 27, p. 205 - 206 (1969). – Abstract AG.

072.094 Moleküle in Sonnenflecken.
H. Wöhl.
Mitt. Astron. Ges. No. 27, p. 206 - 207 (1969). – Conference paper.

072.095 Fleckenkontrast als Funktion der Fleckenfläche; einige Eigenschaften der Granulation in unmittelbarer Fleckennähe. M. Roßbach, E. H. Schröter.
Mitt. Astron. Ges. No. 27, p. 208 - 209 (1969). – Conference paper.

072.096 The microstructure of sunspots.
J. M. Beckers.
Separate print Sacramento Peak Obs., Sunspot, New Mexico, 14 pp. (1969).

072.097 Misure di aree delle macchie solari dal 13 novembre 1967 al 31 dicembre 1968. P. Zlobec.
Pubbl. Oss. Astron. Trieste, No. 399 (Osservazioni Solari, No. 14 Suppl.), 46 pp. (1969).

The solar sunspot areas observed from November 13th 1967 to December 31st 1968 are given. The photographic method and the reduction system are explained. Finally the results are compared with those given by other authors.

Pioneering solar activity.
Nature, Vol. 223, 881 - 882 (1969).

Photospheric and chromospheric magnetic fields and the brightness of faculae and flocculi.
See Abstr. 071.031.

Solar soft X-rays and solar activity. I: Relationships between reported flares and radio bursts, and X-ray bursts.
See Abstr. 076.001.

The emission of solar X-rays in the 0.5–3 Å wavelength range and its relation to the magnetic configuration of active centers. See Abstr. 076.002.

11-year modulation of cosmic ray intensity and distribution of spots in heliographic latitude.
See Abstr. 078.020.

073   Solar Chromosphere, Flares, Prominences

073.001 ` **Solar flare optical, neutron and gamma-ray emission.**   R. E. Lingenfelter.
Solar Physics, Vol. 8, 341 - 347 (1969).

It has previously been suggested that the energy for the optical emission of solar flares was provided by ionization losses of accelerated particles in the flares. We show that nuclear interaction of these particles would also produce fluxes of secondary neutrons and gamma-rays detectable at the earth. A comparison of the expected intensities of these secondaries with the present upper limit intensities during solar flares shows that such an origin from the optical emission energy is consistent with the measured limits.

073.002   **On a possible proton origin for type V continuum radiation from a solar flare.**
M. Friedman, S. M. Hamberger.
Solar Physics, Vol. 8, 398 - 399 (1969). − Research note.

073.003   **Comment on the note by Friedman and Hamberger.**   Z. Švestka.
Solar Physics, Vol. 8, 400 (1969). − Research note. See Abstr. 073.002.

073.004   **Development and spatial structure of proton flares near the limb and coronal phenomena. II. Flare of 26. IX. 1963 and its emission.**   L. Krivský.
Bull. Astron. Inst. Czechoslovakia, Vol. 20, 163 - 171 (1969).

From a series of photographs of a proton flare of the ascending-tube type of 26. IX. 1963 in H-alpha, some further typical properties of proton flares were demonstrated. The length and the rate of elongation of chromospheric flare ribbons, the width and the height of the flare channel and the rates of its growth, the length of the channel according to the peak emission ribbon and the width of this ribbon were investigated. The wave and the particle emissions of the flare are briefly described. To conclude, the spatial dimensions of some of the stages and formations of the flare channel, significant for the release of energy in the form of accelerated particles and of X-emission, are given.

073.005   **Flight time of solar fast particles from flares to the earth. Supplement II.**   L. Křivský.
Bull. Astron. Inst. Czechoslovakia, Vol. 20, 293 - 296 (1969).

Ejections of solar fast particles from flares that produce the PCA (polar cap absorption) and GLE (ground level effect − increase of cosmic rays) effects are considered to originate in flares with a Y-shaped phase of development at the time of their rapid blaze up. Thus we know the exact time of the ejection and escape of the particles, coincident with the first top of radio emission on cm and dm range and with peaks of very hard X-emission. Determined is the time lag of the beginning of PCA effect or of the beginning of increase of cosmic rays after the occurrence of the Y-shaped phase of the flare. With cosmic rays the delay time ranges between 3 and 85 minutes ($+8^m$ for light), with particles producing the PCA-effect it ranges from 10 to 870 minutes ($+8^m$ for light).

073.006   **The yellow line of helium in prominences.**   N. A. Jakovkin, M. Y. Zeldina.
Solnechnye Dannye Byull. No. 4, p. 82 - 88 (1969). In Russian.

The mechanism of helium excitation in a prominence is suggested. The short-wave solar emission ionizes helium. Recombinations and resonance scattering of photospheric radiation generate the population of triplet levels in conformity with observations. The kinetic temperature of the prominence matter in this case remains low, equal to 7000° on average.

073.007   **Boundary conditions on model solar chromospheres.**   R. G. Athay.
Solar Physics, Vol. 9, 51 - 55 (1969).

It is shown that the gas pressure in the corona provides a rather stringent boundary condition on model chromospheres and limits the thickness of those regions of the chromosphere in which the temperature is less than 10000 K to less than or about 2000 km.

073.008   **On the chromospheric observations at the 1962 eclipse.**   W. Henze, Jr.
Solar Physics, Vol. 9, 56 - 64 (1969).

A discussion is given of slitless spectrograms of the chromosphere obtained by an expedition of the High Altitude Observatory, Sacramento Peak Observatory, and National Bureau of Standards at the eclipse of 4 - 5 February 1962. The data which are considered consist of previously published line intensities plus continuum data presented here for the first time. The data reduction procedure is briefly reviewed and a source of error introduced during the reduction is described. The error can possibly affect many of the reported line intensities. Comparison of the 1962 observations with the HAO data from the 1952 eclipse indicates that the previously suggested factor-of-two decrease in the earlier data should be modified to a wavelength-dependent correction: a decrease by a factor of three or four at the high Balmer lines, a decrease by a factor of two at λ 4700 A, no change at Hα, and an increase by a factor of about 1.5 at the Paschen lines.

073.009   **Analysis of the chromospheric hydrogen spectrum at the 1962 eclipse.**   W. Henze, Jr.
Solar Physics, Vol. 9, 65 - 76 (1969).

Slitless spectrograms of the chromosphere obtained during the eclipse of 4 - 5 February 1962 have been analyzed to obtain the decrements of the level populations of hydrogen, the self-absorption in the Balmer lines, and parameters useful in construction of models of the low chromosphere. The chromospheric continuum was generally underexposed; the absence of observed continuum in the visible region of the spectrum made it impossible to derive a unique model from the 1962 data alone. However, the high Balmer line data and new theoretical solutions of the statistical equilibrium equations for hydrogen combined with corrected 1952 observations at λ 4700 A are compatible with a model having approximately the same temperature and neutral hydrogen structure as the 1952 model by Pottasch and Thomas but half the electron density.

073.010   **Periodic structures in quiescent prominences.**   Y. Nakagawa, J. M. Malville.
Solar Physics, Vol. 9, 102 - 115 (1969).

The observed structural periodicities in quiescent prominences and filaments are examined in terms of the instability of a plasma supported by a magnetic field against gravity. It is suggested that the spacing of arch-like structures may be identified with the most unstable wavelength of the interface between the prominence and the supporting magnetic field. The results of analysis further suggest that the observed spacing of periodic structures corresponds to the supporting magnetic field which lies at an angle 90° to 60° with respect to the long axis of the prominence.

073.011   **On the localization, size and structure of the regions of the X-ray flares on the sun.**
I. L. Beigman, Yu. I. Grineva, S. L. Mandel'stam, L. A. Vain-

stein, I. A. Žitnik.
Solar Physics, Vol. 9, 160 - 165 (1969).

With the use of X-ray heliographs carried by the satellites 'Cosmos-166' and 'Cosmos-230' the height of an X-ray flare was found to be about 20 - 25000 km. The regions of the X-ray flares possess a filamentary structure which, during the development of the flares, shows spatial changings with speeds up to 10^7 cm/sec.

073.012 Mechanisms of solar flares. P. A. Sweet.
Annual Rev. Astron. Astrophys. Vol. 7, 149 - 176 (1969).

073.013 On the measurement of the polarization of radiation of a prominence in the Ca I λ 4227 line.
G. M. Nikolsky, Ts. S. Khetsuriani.
Astron. Zhurn. Akad. Nauk SSSR, Vol. 46, 1040 - 1046 (1969). In Russian. English translation in Soviet Astron. AJ, Vol. 13, No. 5.

By means of polarization photographs of the spectrum of a quiet prominence in the Ca I λ 4227 line the degree of polarization P ≈ 4% was estimated. The expected degree of polarization in the case of photoexcitation λ 4227 amounts to about 20%. The halfwidth of the λ 4227 line corresponds to nonthermal velocity 4 km/sec, a radial velocity ≤ 2 km/sec.

073.014 A mechanism for the acceleration of solar chromospheric spicules. Y. Uchida.
Publ. Astron. Soc. Japan, Vol. 21, 128 - 140 (1969).

Observed magnetic polarity distribution in the quiet photosphere suggests that magnetically neutral lines and points occur along the supergranulation boundary. The time scale of the magnetic field annihilation in these magnetically neutral regions is estimated to be short enough if Petschek's mechanism is applicable, when these long lasting magnetic configurations are further disturbed, for example, by the appearance of new granulation or observed oscillation. The plasma with magnetic lines of force thus disconnected from their photospheric root is pushed upwards by "melon seed" effect in the existing gradient in the field strength between the photosphere and the corona. The motion of the jet, or the train of such plasma bubbles, guided in the magnetic tube of force with varying cross-section is treated under the action of "melon seed" force and gravity, and it is shown that the flow has a de Lavals transition point at which an originally subsonic flow transforms itself into a supersonic one as in Parker's solar wind theory. The resulting velocity, the reaching height and the density in the jet are in good agreement with observation.

073.015 White-light solar flares. K. Najita, F. Orrall.
Bull. American Astron. Soc. Vol. 1, 254 - 255 (1969). – Abstr. AAS.

073.016 On the stability of solar filaments. U. Anzer.
Bull. American Astron. Soc. Vol. 1, 271 - 272 (1969). – Abstr. AAS.

073.017 Recent eclipse data and the solar limb.
K. B. Gebbie, S. R. Weart, R. N. Thomas.
Bull. American Astron. Soc. Vol. 1, 277 (1969). – Abstr. AAS.

073.018 The oblique shock of the proton flare of 7 July 1966. E. W. Greenstadt.
Bull. American Astron. Soc. Vol. 1, 277 (1969). – Abstr. AAS.

073.019 On the velocity field in the chromosphere close to sunspots. E. Haugen.
Bull. American Astron. Soc. Vol. 1, 278 - 279 (1969). – Abstr. AAS.

073.020 Line-width analysis of the chromosphere and quiescent prominences. T. Hirayama.
Bull. American Astron. Soc. Vol. 1, 279 (1969). – Abstr. AAS.

073.021 A solar flare disturbance as observed in the interplanetary medium.
J. Hirshberg, S. J. Bame, A. J. Hundhausen.
Bull. American Astron. Soc. Vol. 1, 279 (1969). – Abstr. AAS.

073.022 The response of the chromosphere to rapidly falling material and solar flares, plages, etc.
C. L. Hyder, Y. Nakagawa.
Bull. American Astron. Soc. Vol. 1, 281 (1969). – Abstr. AAS.

073.023 Optical and radio 2B flare of 8 August 1968.
T. J. Janssens, K. P. White III.
Bull. American Astron. Soc. Vol. 1, 281 (1969). – Abstr. AAS.

073.024 Study of 2B flare by filtergrams – H alpha.
T. J. Janssens.
Bull. American Astron. Soc. Vol. 1, 281 (1969). – Abstr. AAS.

073.025 Why the temperature rise does not occur in radiative equilibrium in low solar chromosphere.
S. D. Jordan.
Bull. American Astron. Soc. Vol. 1, 282 (1969). – Abstr. AAS.

073.026 Halo structure of solar proton events.
S. W. Kahler, R. P. Lin, E. C. Roelof.
Bull. American Astron. Soc. Vol. 1, 283 (1969). – Abstr. AAS.

073.027 The emission of ~40-keV electrons by the sun.
R. P. Lin.
Bull. American Astron. Soc. Vol. 1, 284 (1969). – Abstr. AAS.

073.028 A recalibration of solar millimeter brightness temperatures based upon lunar observations.
J. L. Linsky.
Bull. American Astron. Soc. Vol. 1, 284 - 285 (1969). – Abstr. AAS.

073.029 A critical evaluation of temperature determinations in the low solar chromosphere as obtained from the ultraviolet continuum data. J. L. Linsky, D. R. Brown.
Bull. American Astron. Soc. Vol. 1, 285 (1969). – Abstr. AAS.

073.030 On the periodic structure of quiescent prominence.
J. M. Malville, Y. Nakagawa.
Bull. American Astron. Soc. Vol. 1, 286 (1969). – Abstr. AAS.

073.031 Flare forecasting and solar patrol observations.
P. S. McIntosh.
Bull. American Astron. Soc. Vol. 1, 286 (1969). – Abstr. AAS.

073.032 Magnetogravitational instability. F. C. Michel.
Bull. American Astron. Soc. Vol. 1, 287 (1969). – Abstr. AAS.

073.033 The transport of mechanical energy in the chromosphere. R. W. Milkey.

Bull. American Astron. Soc. Vol. 1, 287 (1969). − Abstr. AAS.

073.034 Two-dimensional observations of solar oscillating regions. S. A. Musman, D. M. Rust.
Bull. American Astron. Soc. Vol. 1, 287 (1969). − Abstr. AAS. ˎ

073.035 Transverse motion of chromospheric mottles. C. Sawyer.
Bull. American Astron. Soc. Vol. 1, 292 (1969). − Abstr. AAS.

073.036 A mechanism for the buildup of flare energy. J. O. Stenflo.
Bull. American Astron. Soc. Vol. 1, 293 (1969). − Abstr. AAS.

073.037 Observations of magnetic fields in prominences using hydrogen, helium, and metal lines.
E. Tandberg-Hanssen.
Bull. American Astron. Soc. Vol. 1, 293 - 294 (1969). − Abstr. AAS.

073.038 Analysis of high-resolution X-ray photographs. I. An importance 1N flare.
G. S. Vaiana, T. Zehnpfennig.
Bull. American Astron. Soc. Vol. 1, 294 (1969). − Abstr. AAS.

073.039 A statistical analysis of flare events and its implications concerning the sun's internal rotation.
G. Van Hoven, P. A. Sturrock, P. Switzer.
Bull. American Astron. Soc. Vol. 1, 294 - 295 (1969). − Abstr. AAS.

073.040 An example of radio and optical homologous flares.
K. P. White III, T. J. Janssens.
Bull. American Astron. Soc. Vol. 1, 295 (1969). − Abstr. AAS.

073.041 High-resolution flare observations. H. Zirin.
Bull. American Astron. Soc. Vol. 1, 296 - 297 (1969). − Abstr. AAS.

073.042 Round-the-clock photoheliography.
H. Zirin, J. D. Bohlin, S. Weart, U. Feldman.
Bull. American Astron. Soc. Vol. 1, 297 (1969). − Abstr. AAS.

073.043 The coarse structure of the solar atmosphere.
M. Simon, H. Zirin.
Solar Physics, Vol. 9, 317 - 327 (1969).
Observations of the quiet sun at wavelengths from 3 Å to 75 cm show (with two exceptions: the O VI line at 1032 Å and possibly the continuum at 1.2 mm) either no limb brightening or less than had been supposed. On the other hand, the brightness temperature is observed to increase with wavelength in the millimeter and centimeter range. If this increase is due to greater visibility of hot overlying material, that material ought to be evident at the limb at shorter wavelengths, resulting in limb brightening. The only possible explanation for the absence of limb brightening at almost all wavelengths is that the emitting surface is rough at all wavelengths, with a scale of roughness approximately equal to the scale height at each temperature.

073.044 Observations of rotational motion in prominences.
Y. Öhman.
Solar Physics, Vol. 9, 427 - 431 = Contr. Kitt Peak National Obs. No. 454 (1969).

The writer reports some recent observations carried out with the McMath Solar Telescope, which can be interpreted as due to rotational motions in prominences.

073.045 Proton Flare Project, 1966. Summary of the August/September particle events in the McMath region 8461.   Z. Švestka, P. Simon.
Solar Physics, Vol. 10, 3 - 59 (1969).
The paper summarizes observations of solar and space phenomena related to the McMath region number 8461 which passed over the solar disk during the 1966 Proton Flare Project period, from August 21 to September 4, and produced two important solar particle events on August 28 and September 2. The most important results are reviewed and interpretation of some of them is suggested.

073.046 High-resolution photography of the solar chromosphere. VI: Properties of the bright mottles.
R. J. Bray.
Solar Physics, Vol. 10, 63 - 70 (1969).
Bright chromospheric mottles observed at the Hα line centre are found to have sizes ranging from 1450 to 4400 km and lifetimes of about 11 min. They occur in close juxtaposition to dark mottles which, at intermediate heliocentric angles ($\approx 60°$), are found to be displaced towards the limb relative to the associated bright mottles. The magnitude of the displacement indicates a height difference of 4300 km. In conjunction with height measurements of bright mottles beyond the limb (Loughhead, 1969), this implies that bright and dark mottles are phenomena of the lower ($\lesssim 3300$ km) and upper ($\approx 5000 - 7600$ km) chromosphere respectively.

073.047 High-resolution photography of the solar chromosphere. VII: Structure of the low chromosphere.
R. E. Loughhead.
Solar Physics, Vol. 10, 71 - 78 (1969).
High-resolution observations of the low chromosphere beyond the limb at the centre of the Hα line reveal the existence of two types of fine structure whose presence has hitherto been unobserved: (1) small bright features which show a close correspondence in properties with the bright mottles on the disk and are unequivocally identifiable as the latter seen beyond the limb; and (2) a narrow dark band lying immediately above the photospheric limb whose physical significance is not yet clear. The discovery and identification of the bright mottles beyond the limb, taken in conjunction with recent observations of Bray (1969), lend further weight to the view that the chromospheric spicules are to be identified with the dark mottles seen on the disk.

073.048 Ca II resonance lines in non-homogeneous chromospheres.   H. A. Beebe, H. R. Johnson.
Solar Physics, Vol. 10, 79 - 87 = Publ. Goethe Link Obs., Indiana Univ., No. 95 (1969).
Profiles of the K line of Ca II are computed for a two component solar chromosphere, chosen to simulate with a simple geometry the chromospheric supergranular network. The line source function and the optical depth are obtained from a self-consistent treatment of the steady state and radiative transfer equations, with complete redistribution assumed for scattering in the line. The atomic model consists of two bound levels and a continuum. It is found that a 4600 K minimum can lead to the successful theoretical prediction of the observed limb darkening and 4300 K radiation temperature of the K 1 feature only when very large values of turbulent velocity are assumed to exist in the cell region.

073.049 A solar spicule model based upon calcium II K line radiative transfer studies.
L. W. Avery, L. L. House.
Solar Physics, Vol. 10, 88 - 103 (1969).

Monte Carlo radiative transfer techniques are used to develop a height-dependent spicule model based upon a more realistic configuration than has hitherto been considered. The spicule is represented by a uniform cylinder, of finite length, standing vertically upon a plane chromosphere. The observed, limb-darkened, anisotropic chromospheric flux incident upon the cylinder is incorporated into the transfer calculations. The resulting model is characterized by a random, line broadening velocity of 20 km/sec, with electron temperature increasing from 6×10^3 K at the base to about 1.5×10^4 K at 11500 km above the solar surface. The corresponding values of electron density are 8×10^{11} cm^{-3} and 4×10^{10} cm^{-3}.

073.050 **No evidence of any solar limb brightening in the range of 3.5 mm–2 cm.** A. Tlamicha.
Solar Physics, Vol. 10, 150 - 153 (1969). – Research note.

073.051 **On some flare-sensitive high photospheric and low chromospheric lines.** Y. Öhman.
Solar Physics, Vol. 10, 178 - 183 = Contr. Kitt Peak National Obs. No. 481 (1969).

During a stay at the Kitt Peak National Observatory the writer has tried to find an influence of flare radiation on the high photospheric and low chromospheric lines of the area occupied by the flare. Observations have been made in the Hα region and in the region of the H and K lines. When flare emission is present in sunspots some of the faint (molecular) lines seem to be weakened. When a flare appears near the solar limb some of the Evershed-type (chromospheric) lines are strongly influenced.

073.052 **Neutron monitor and Pioneer 6 and 7 studies of the January 28, 1967 solar flare event.**
R. P. Bukata, P. T. Gronstal, R. A. R. Palmeira, K. G. McCracken, U. R. Rao.
Solar Physics, Vol. 10, 198 - 211 (1969).

A discussion of the January 28, 1967 solar flare event is presented. High energy data from several neutron monitor stations are supplemented by low energy data from the interplanetary space probes Pioneers 6 and 7. A study of the data obtained from these three observation stations widely separated in solar azimuth has provided information on the location of the responsible flare, on the particle diffusion across the interplanetary magnetic field lines, the spectral exponent of the rigidity spectrum, and on the occurrence of a low energy solar injection prior to the high energy event.

073.053 **Solar flare Alpha to proton ratio changes following interplanetary disturbances.**
L. J. Lanzerotti, M. F. Robbins.
Solar Physics, Vol. 10, 212 - 218 (1969).

A discussion is presented on the half hour averaged low energy solar alpha to solar proton flux ratios observed following the three large solar flares of May 23, 1967. One of the large changes observed in the particle ratios (following a sudden commencement (SC) storm observed on the earth) is interpreted as due to a source effect. The second large change, again observed following an SC, is observed in the equal velocity and equal rigidity ratios and not in the equal energy/charge ratios. This observation suggests that electric fields in an interplanetary disturbance may be the cause of the modulations.

073.054 **Selective excitation in the spectrum of two solar flares.** J. M. Marlborough, C. R. Cowley, A. P. Cowley.
Publ. Astron. Soc. Pacific, Vol. 81, 543 - 544 (1969). Abstract ASP.

073.055 **Selective excitation in the flare of 1958 August 7.** C. Cowley, J. M. Marlborough.

Astrophys. Journ. Vol. 158, 803 - 807 (1969).

The Fe I line λ4063 is enhanced in a region of the flare of 1958 August 7. The enhancement is interpreted as a selective-excitation effect similar to that observed in late-type dwarf stars. The appearance of this selective excitation is used to obtain information on the temperature and electron density in the region of the flare where the line arises.

073.056 **Microwave pulse trains observed before and during a solar flare.** T. J. Janssens, K. P. White III.
Astrophys. Journ. (*Letters*), Vol. 158, L127 - L128 (1969).

073.057 **Length of neutral lines as an indicator of flare productivity.** M. T. C. de Peralta, D. E. Billings.
Journ. Geophys. Res. Vol. 74, 5822 - 5823 (1969). – Letter.

073.058 **Aspects of N–S asymmetry of solar flares associated with geophysical effects of the electromagnetic and corpuscular radiation.** G. Mariş, V. Dinulescu.
Stud. Cerc. Astron. Vol. 14, 45 - 48 (1969). In Rumanian.

A difference appears between the flares followed by electromagnetic radiation effects (SWF) and the flares producing geocorpuscular effects (PCA and Geomagnetic Storms).

073.059 **Far infrared measurement of the solar minimum temperature.** J. A. Eddy, P. J. Léna, R. M. MacQueen.
Solar Physics, Vol. 10, 330 - 341 (1969).

Radiometric measurement of the brightness temperature of the mean solar disk has been made in the wavelength range from 238 μ to 312 μ (42.1 cm^{-1} to 32.1 cm^{-1}) using a Michelson interferometer of resolution 0.25 cm^{-1}, carried on the NASA research aircraft at altitude 12.6 km. A mean temperature 4370 ± 260 K is obtained.

073.060 **On the energy dissipation of fast hydromagnetic shock waves in the solar chromosphere.**
R. Mäckle.
Solar Physics, Vol. 10, 348 - 356 (1969).

MHD equations including dissipation terms are applied to study the most important irreversible processes occurring in fast hydromagnetic shock waves under the conditions of the outer solar atmosphere. The atmosphere is assumed to be permeated by a nearly horizontal, uniform magnetic field, the magnitude and inclination angle of which being parameters of the analysis. Numerical examples, corresponding to situations which might occur in the upper chromosphere, are computed in order to demonstrate the procedure.

073.061 **Isophotal photometry and morphological changes in the flares.** M. Dizer.
Solar Physics, Vol. 10, 416 - 428 (1969).

We report measurements made on the brightness in Hα of all parts of the flare photographed through a birefringent filter centered on Hα, using a scanning isodensitometer. From obtained isophotes of the flares we derived some information on the morphological changes in the flare and estimated the total energy in Hα of the flare.

073.062 **Selective excitation in two solar flares.** C. R. Cowley, J. M. Marlborough.
Bull. American Astron. Soc., Vol. 1, 339 (1969). – Abstract AAS.

073.063 **An empirical model of the lower chromosphere based upon millimeter data calibrated by lunar observations.** J. L. Linsky.
Bull. American Astron. Soc., Vol. 1, 372 (1969). – Abstr. AAS.

073.064 **Chromospheric flares and phenomena in the upper**

layer of an active region. II. Yu. M. Slonim.
Astron. Zhurn. Akad. Nauk SSSR, Vol. 46, 697 - 714 (1969).
In Russian. English translation in Soviet Astron. AJ, Vol. 13,
No. 4.

A connection of flares and fast processes with stationary
filaments of an active region is considered. A physical classification of fast processes, based on the evolution of the magnetic field of a filament, is substantiated.

073.065 **Peculiarities of motions in flocculi and outside of
them.** S. I. Gopasjuk, T. T. Tsap.
Astron. Zhurn. Akad. Nauk SSSR, Vol. 46, 923 - 925 (1969).
In Russian. English translation in Soviet Astron. AJ, Vol. 13,
No. 4.

A connection between velocities and the brightnesses in
active and undisturbed regions of the sun is investigated. Radial velocities and brightnesses are measured with help of a
magnetograph in the lines FeI λ 5250 Å, CaI λ 6103 Å, Hα,
Hβ and K$_3$CaII. It is found that the brightest parts of knots
in active regions as well as bright knots, forming a chromospheric network of the quiet sun, are situated on a zero line
of radial velocities or lie near to it.

073.066 **Excitation of helium in solar prominences.**
N. N. Morozhenko.
Astron. Zhurn. Akad. Nauk SSSR, Vol. 46, 1184 - 1189
(1969). In Russian. English translation in Soviet Astron.
AJ, Vol. 13, No. 6.

Comparing the observed and calculated values of the population of the 2^3 P level of helium yields the preliminary conclusion that in quiescent prominences helium can be excited
in places of hydrogen glow, the metastable helium level 2^3 S
having been conditioned by an ionized recombination
mechanism in the case of ionization of helium atoms by ultraviolet solar radiation with $\lambda \leqslant 504$ Å.

073.067 **The magnetic field of a solar prominence.**
V. A. Kotov.
Izv. Krymskoj Astrofiz. Obs. Vol. 39, 276 - 278 (1969).
In Russian.

Results of the measurement of the longitudinal magnetic field in a prominence in the Hα line made with the solar
magnetograph of the Crimean Astrophysical Observatory are
given. The magnetic field of separate elements is $\sim$100 Gauß.

073.068 **Study of spectroheliograms in metal lines and the
structure of the chromosphere.** E. E. Dubov.
Izv. Krymskoj Astrofiz. Obs. Vol. 39, 279 - 294 (1969).
In Russian.

Distributions of bright and dark parts of spectroheliograms, obtained in metal lines (Fe I 4202 Å, 4384 Å, Ca I
4227 Å, Sr II 4078 Å, Na I 5890 Å and K$_3$ Ca II) at different
distances from line centres, are compared. It is concluded that
these brightness distributions do not contradict the ideas on
the structure of the chromosphere, developed by the author
in a former paper. The brightness distributions on spectroheliograms in metal lines reflect, mainly, the structure of velocity fields in the photosphere and the distribution of photospheric faculae in active regions.

073.069 **The turbulence in quiescent prominences and shock
waves in the upper atmosphere.** E. E. Dubov.
Izv. Krymskoj Astrofiz. Obs. Vol. 39, 295 - 298 (1969).
In Russian.

It is shown that shock waves penetrating into a prominence may support the turbulence in quiescent prominences.

073.070 **On the effect of the cylindric form of a spicule
upon the appearance of spectral line profiles.**
S. G. Mamedov, E. S. Orudzhev.
Solnechnye Dannye 1969 Byull. No. 8, p. 78 - 80 (1969).

In Russian.

It is shown that in the first approximation the cylindric
form of a spicule does not influence the appearance of a line
profile. This effect is to be taken into account in precise
measurements.

073.071 **Solar flares; properties and problems.**
C. de Jager.
Solar Flares and Space Research, Tokyo 1968, p. 1 - 15
(1969). — Introductory review.

073.072 **The optical flare.** Z. Švestka.
Solar Flares and Space Research, Tokyo 1968,
p. 16 - 37 (1969).

A review of flare observations in the optical spectral
region and their interpretation.

073.073 **Solar flares and magnetic fields.**
A. B. Severny.
Solar Flares and Space Research, Tokyo 1968, p. 38 - 60
(1969).

Examples and illustrations based on the examination of
the magnetic field in flaring active regions are presented.

073.074 **Flare associated optical phenomena.**
A. Bruzek.
Solar Flares and Space Research, Tokyo 1968, p. 61 - 77 =
Mitt. Fraunhofer Inst. Freiburg, No. 84 (1969).

A survey is given of the various types of visible, mainly
dynamic phenomena and effects associated with the occurrence of flares in the solar atmosphere.

073.075 **Theoretical aspects of the flare phenomenon.**
H. U. Schmidt.
Solar Flares and Space Research, Tokyo 1968, p. 331 - 345
(1969). — Review paper.

073.076 **On the mechanism of solar flares.**
S. I. Syrovatskii.
Solar Flares and Space Research, Tokyo 1968, p. 346 - 355
(1969).

The process of collisionless (dynamic) dissipation of
magnetic energy near the zero lines of the magnetic field is
supposed to be the base of the solar flare mechanism.

073.077 **A possible mechanism for solar flares.**
H. Elliot.
Solar Flares and Space Research, Tokyo 1968, p. 356 - 362
(1969).

073.078 **The nature of solar flares.**
D. H. Menzel.
Solar Flares and Space Research, Tokyo 1968, p. 363 - 367
(1969).

073.079 **Laboratory experiments on solar flare model by
laser.** C. Yamanaka, T. Yamanaka, Y. Izawa,
T. Sasaki, N. Tsuchimori, M. Onishi.
Solar Flares and Space Research, Tokyo 1968, p. 374 - 383
(1969).

073.080 **Flare forecasting.**
P. Simon, M. J. Martres, J.-P. Legrand.
Solar Flares and Space Research, Tokyo 1968, p. 405 - 411
(1969).

073.081 **The Ca plage of the active region No. 18521
(25 N 184^5) of the proton flares of August, 28th
and September 2nd, 1966.** G. Godoli, B. C. Fossi.
Oss. Astrofis. Catania Pubbl., Nuova Serie, No. 122, 10 pp.
(1968).

The evolution curves show that the plage size had a secondary minimum during the third rotation. During the fifth rotation (the rotation of the proton flare) a nearby plage became connected with the main plage.

073.082 The limb activity on September 4, 1966 of the active region No. 18521 (25 N 184°) of the proton flares of August 28th and September 2nd.
G. Godoli, F. Mazzucconi, S. Nagasawa.
Oss. Astrofis. Catania Pubbl., Nuova Serie, No. 123, 14 pp. (1968).
On the basis of observations made at 15 observatories, a detailed description of the limb activity is given.

073.083 Correlazione brillamenti–bursts in funzione del tipo di macchie.   F. D. Chiuderi, F. Mazzucconi.
Atti XII Riunione Soc. Astron. Italiana, L'Aquila 1968, p. 58 - 60 (1969). – Abstract SAI.

073.084 Magnetic flares on the sun.
V. F. Tshystjakov.
Astron. Tsirk. No. 533, p. 6 - 8 (1969).   In Russian.

073.085 On the main eruptive phase of solar flares.
S. O. Obashev.
Trudy Astrofiz. Inst. Alma-Ata, Vol. 15, 16 - 21 (1969). In Russian.
The theory of the point flare in a heterogeneous atmosphere is used for solar flares. The energy of the solar flares in the main eruptive phase is estimated to be about 3×10^{25} erg, and the form of the shock wave front for different moments of time is obtained. It can be concluded that the solar flares are generated in the lower layers of the chromosphere.

073.086 Eruptions in the solar atmosphere and the structure of the chromosphere.   S. O. Obashev.
Trudy Astrofiz. Inst. Alma-Ata, Vol. 15, 22 - 26 (1969). In Russian.
Some physical parameters of solar eruptions observed on 1957 Oct. 23 and 1960 April 2 with the photosphere–chromosphere telescope AFR–2 using the IPF have been considered. The total number of hydrogen atoms enclosed in the eruption of 1957 Oct. 23 equals 10^{41}. Comparing the gas mass contained in the eruption with the mean chromospherical mass, the author comes to the conclusion that eruptions are deep-layer formations, developing in the photosphere or the lower chromosphere. The kinetic energy of the eruption is about 10^{31} erg. That is approximately equal to the energy of a flare. It is possible that flares and eruptions are the same phenomenon, but have a different appearance in dependence on the configuration of the magnetic field in active regions. In the line wings of Hα in an active region a dark halo was discovered, the size of which is by ten times larger than that of the structural elements of the chromosphere.

073.087 On a possible relation between different parameters of solar flares.   S. O. Obashev.
Trudy Astrofiz. Inst. Alma-Ata, Vol. 15, 27 - 37 (1969). In Russian.
Some relations and quantitative estimates of the parameters of solar flares are obtained from simple physical assumptions that in solar flares a certain part of energy is transformed into radiation. In particular the formula for changing of a flare area is deduced, and the relation between the values S_m and T is found. The total energy of a flare is estimated and an energy classification is proposed.

073.088 A catalogue of solar flares.
A. S. Zubtzov, S. O. Obashev.

Trudy Astrofiz. Inst. Alma-Ata, Vol. 15, 46 - 60 (1969). In Russian.
A catalogue of the solar flares observed at the Alpine solar station of the Academy of Sciences of the Kazakh SSR during the IGY–IQSY (October 1957 - June 1962) is given. This catalogue contains the main flare characteristics: the time of beginning and end of the appearance of the flare, the flare class of importance, and the coordinates of the flare. On the graphs the time of patrol of the solar flares is given.

073.089 Solar flares and magnetic fields.
J. H. Reid.
Dunsink Obs. Repr. No. 46, [Reprinted from "Magnetism and Cosmos", Oliver & Boyd Ltd, Edinburgh, p. 233 - 239], 7 pp. (1967).

073.090 The infall-impact mechanism and solar flares.
C. L. Hyder.
Nobel Symposium 9 ["Mass Motions in Solar Flares and Related Phenomena", Almqvist & Wiksell, Stockholm], p. 57 - 65 = Separate print from Sacramento Peak Obs. Sunspot, New Mexico (1968). – Conference paper.

Near-limb solar brightness distribution at 1216 Å and 1300 Å.   See Abstr. 071.015.

Photoelectric eclipse observation of the continuum at the extreme solar limb.   See Abstr. 071.016.

Brightness fluctuations from waves in the Sun's atmosphere.   See Abstr. 071.028.

Photospheric and chromospheric magnetic fields and the brightness of faculae and flocculi.
See Abstr. 071.031.

Solar velocity fields: 5-min oscillations and supergranulation.   See Abstr. 071.056.

On the correlation between sunspot structure and solar flares.   See Abstr. 072.031.

Magnetic development of an active region.
See Abstr. 072.032.

Time variations of the sunspot contrast from observations of the chromosphere.   See Abstr. 072.043.

Fast changes of sunspot equilibrium conditions and solar flares.   See Abstr. 072.079.

Double-K_2 emission line observed in sunspots and in prominences.   See Abstr. 072.090.

Solar wind disturbances associated with flares.
See Abstr. 074.071.

Chromospheric flare enhancements in the extreme ultraviolet and their relationship with solar abundances.
See Abstr. 076.014.

Energetic X-ray and extreme-ultraviolet flashes of solar flares.   See Abstr. 076.033.

Solar EUV enhancements associated with flares.
See Abstr. 076.034.

X-ray observations of solar flares.
See Abstr. 076.035.

074   Solar Corona, Solar Wind

074.001   Thermal continuum radiation from coronal plasmas at soft X-ray wavelengths.   J. L. Culhane.
Monthly Notices, Roy. Astron. Soc., Vol. 144, 375 - 389 (1969).

The continuous spectra, arising from the free-free and free-bound transitions of electrons in coronal plasmas, are calculated for wavelengths in the range 1 Å to 30 Å and at temperatures in the range $0.8 \ 10^6$ °K to $100.0 \ 10^6$ °K. The effect of variations in the element abundances is investigated. Estimates of the continuum flux from the solar corona are presented and the observed line to continuum ratios discussed.

074.002   Corotating structure in the solar wind.
R. L. Carovillano, G. L. Siscoe.
Solar Physics, Vol. 8, 401 - 414 (1969).

The hydrodynamic equations which describe the radial solar wind expansion are linearized and specialized to treat corotating perturbations. Approximate solutions are found which are time stationary in the corotating reference frame. The solutions predict the behavior of corotating structures for a given boundary condition close to the sun. In particular, the structure resulting from the interaction of fast and slow streams is described. Comparison with sector structure data shows reasonable qualitative and quantitative agreement.

074.003   On the north-south asymmetry in the solar wind.
G. L. Siscoe, P. J. Coleman, Jr.
Solar Physics, Vol. 8, 415 - 421 (1969).

The orientations of tangential discontinuities seen by Mariner 4 are interpreted as implying a sector dependent asymmetry in the north-south component of the solar-wind flow. In two sectors, fast solar wind streams had a southward motion relative to slow streams, in one sector the reverse obtained, and in the remaining sector the asymmetry was not clearly defined. We interpret this as being due to greater pressure in the north hemisphere in two sectors and greater pressure in the south hemisphere in one. It is possible this asymmetry could produce a small average southward magnetic field component.

074.004   Hydromagnetic shocks in the solar wind.
K. W. Ogilvie, L. F. Burlaga.
Solar Physics, Vol. 8, 422 - 434 (1969).

The Rankine-Hugoniot relations are applied to shock-like discontinuities measured by both magnetic field and plasma instruments on the satellite Explorer 34 between May 30, 1967 and Jan. 11, 1968. Shock normals were either determined from the magnetic field observations, or from the times of occurrence of the discontinuity at Explorers 33, 34 amd 35. The Rankine-Hugoniot relations are obeyed to the accuracy of the observations, and the values of shock velocities, density ratios, and Mach numbers indicate that at 1 AU the typical interplanetary shock is not strong, although all the events studied caused geomagnetic impulses.

074.005   Helium abundance in the solar wind.
K. W. Ogilvie, T. D. Wilkerson.
Solar Physics, Vol. 8, 435 - 449 (1969).

Observations of hydrogen and helium ions in the solar wind have been carried out by the Goddard Space Flight Center – University of Maryland plasma instrument on Explorer 34. These ions are completely separated by means of electrostatic and magnetic fields. The average value of the ratio number densities is $0.051 \pm .02$, derived from over 3000 h of measurement. Variations about this value from about 0.01 up to greater than 0.15 occur, and there are more high values than can be explained by random variation. A tentative association

with some geomagnetic storms is suggested.

074.006   Solar electron corona.
Tokyo Astron. Obs. Bull. Solar Phenomena, Vol. 20, (No. 5), 105 - 112 (1968).

074.007   Untersuchungen über die Struktur der Sonnenkorona aus photographischen Polarisationsmessungen während der totalen Sonnenbedeckung am 15.2.1961.
G. Scholz.
Astron. Nachr. Vol. 291, 187 - 210 = Mitt. Astrophys. Obs. Potsdam No. 132 (1969).

The corona was photographed at the solar eclipse of 1961 February 15 with a quadruple camera. Six plates were obtained (three in the blue and three in the red spectral region) each with four corona images, three through polaroids and one without polaroid. The photometry has provided the intensities, the excess of red intensity to blue intensity, the amount and direction of polarisation. These data allow the separation of K- and F-coronas and the calculation of the electron densities. The results lead to an oblateness of the F-corona or to the assumption that the F-corona is polarized. The reason may be a partial alignment of the interplanetary dust particles. The intensity of the F-corona and the excess of red to blue intensity was computed theoretically by the aid of the Fraunhofer-Airy theory of the diffraction of light. These results were compared with the observations. We obtain 10^{-1} to 5×10^{-2} cm as an upper limit for the radii of the particles and no dependence of the distributions of the particles with increasing distance from the sun.

074.008   Polarimetrische Untersuchungen in einem diffusen Koronastrahl und die Form der nördlichen Polarstrahlen am 15.2.1961.   G. Scholz.
Astron. Nachr. Vol. 291, 211 - 216 = Mitt. Astrophys. Obs. Potsdam No. 133 (1969).

The corona was photographed at the solar eclipse of 1961 February 15 with a quadruple camera. Three plates were obtained, each with four corona images, three through polaroids and one without polaroid. The photometry has provided the intensities, the amount and direction of polarisation, and the electron densities for a diffuse coronal ray and an intermediary region. We obtain from geometrical examinations on the northern polar rays no connection between the direction of the polar rays along the sun's limb and the lines of force of a dipole field or a magnetic field with a simple structure.

074.009   Analysis of a method of wide-slit photometry of coronal lines.   V. Rušín, M. Rybanský.
Bull. Astron. Inst. Czechoslovakia, Vol. 20, 300 - 303 (1969).

The analysis of a method used in photometry of emission lines of the corona is presented in this paper. It is shown that the method is accompanied with a systematic error. Causes of the occurence of this error, the magnitude of the latter in dependence on different factors are ascertained and a suggestion is presented how to eliminate the error by varying the observation method.

074.010   Discovering of 1″ thin streams in the solar corona, 22 September 1968.   V. I. Ivanchuk, S. K. Vsekhsvjatsky, N. I. Dzubenko, G. A. Rubo.
Astron. Tsirk. No. 504, p. 1 - 3 (1969).   In Russian.

074.011   The electron density in thin streams of the solar corona in 1968.   N. I. Dzubenko, V. I. Ivanchuk, G. A. Rubo.

Astron. Tsirk. No. 504, p. 3 - 6 (1969). In Russian.

074.012 Photometry of the solar corona on 22 Sept. 1968 by the equidensity method.
A. T. Nesmjanovich, Yu. A. Khomenko, O. S. Popov.
Astron. Tsirk. No. 525, p. 1 - 3 (1969). In Russian.

074.013 Structure of the solar corona on 22 September 1968.
A. T. Nesmjanovich.
Astron. Tsirk. No. 525, p. 3 - 6 (1969). In Russian.

074.014 L'observation photographique pondérée de la couronne solaire. Eclipse totale du 22 septembre 1968 à Yourgamish (Sibérie). M. Laffineur.
L'Astronomie, 83ᵉ année, 337 - 353 (1969).

074.015 Interferometric investigation of the red and green coronal lines during the total solar eclipse of May 30, 1965. A. B. Delone, E. A. Makarova.
Solar Physics, Vol. 9, 116 - 130 (1969).

Fabry-Pérot interferometric observations of the corona were carried out. The 6374 Å line shows radial velocities between 10 and 70 km sec^{-1}, both positive and negative. Most profiles of the 6374 Å line are not Gaussian. The widths of the lines indicate unacceptably high temperatures, and thus suggest turbulent velocities, which appear to be of the same order as the line displacement velocities. Arguments are put forward that the corona consists mainly of individual non-turbulent knots with relative velocities similar to the measured ones.

074.016 Magnetic fields and the structure of the solar corona. I: Methods of calculating coronal fields.
M. D. Altschuler, G. Newkirk, Jr.
Solar Physics, Vol. 9, 131 - 149 (1969).

Several different mathematical methods are described which use the observed line-of-sight component of the photospheric magnetic field to determine the magnetic field of the solar corona in the current-free (or potential-field) approximation. Discussed are (1) a monopole method, (2) a Legendre polynomial expansion assuming knowledge of the radial photospheric magnetic field, (3) a Legendre polynomial expansion obtained from the line-of-sight photospheric field by a least-mean-square technique, (4) solar wind simulation by zero-potential surfaces in the corona, (5) corrections for the missing flux due to magnetograph saturation. We conclude that the field given by a Legendre polynomial is a rigorous and self-consistent solution with respect to the available data.

074.017 Coronal densities and magnetic fields from K-coronameter and type IV radio burst data.
J. D. Bohlin, M. Simon.
Solar Physics, Vol. 9, 183 - 193 (1969).

From K-coronameter data we have obtained an electron density profile above the active region responsible for the type IV burst observed on 14 September 1966. If the observed frequency cutoff in the burst's spectrum is caused by the Razin effect, then the coronal electron density may be derived from the intensity variation in the burst as it propagates outwards from the sun. We show that the electron density profiles obtained from K-coronameter data and from the radio data form a continuous distribution. We conclude that the cutoff is due to the Razin effect, and that radiation in the burst is due to relativistic electrons having a steep inverse power-law energy distribution. From the electron density profile derived from the radio data, we find that the coronal magnetic field was 0.26 G at $r/R_\odot = 2.2$.

074.018 Solar wind tail and the anisotropic production of fast hydrogen atoms. P. W. Blum, H. J. Fahr.

Nature, Vol. 223, 936 - 937 (1969).

The solar system possesses a peculiar velocity of probably 10 - 40 km/sec relative to the surrounding interstellar neutral gas. The supersonic solar wind is slowed down to subsonic velocities at the so-called magnetic shock front which is nearly spherical and is mainly determined by the interstellar magnetic field. Between the shock front and interstellar space is a transition region where the density of the hot turbulent solar protons decreases due to charge exchanges with neutral interstellar hydrogen. It is shown by considerations of particle continuity that the outer boundary of the transition region is highly aspherical: compressed at its front side and drawn out to a tail at its back side.

074.019 The acceleration of the solar wind and the heating of the coronal plasma above active regions.
G. S. Bisnovaty-Kogan, I. M. Gordon.
Astrophys. Letters, Vol. 4, 149 - 152 (1969).

Analysis of the spectra of strong radio echoes from the sun reveals a rapid rise in the velocity of the coronal plasma above plage regions. This velocity sometimes reaches 120 km/sec at a height of only $1.7 R_\odot$. In the framework of the new ideas on the formation of the reflected signals, these data provide observational grounds for the development of a theory of the solar wind based on the results of the radar explorations of the sun. On the assumption that the acceleration of the subsonic flow is determined by the heating caused by the dissipation of plasma turbulence, the rise of temperature and the input of energy in the region of the acceleration has been computed.

074.020 Rates of excitation, ionization and recombination of ions in the plasma of the solar corona.
I. Beigman, L. Vainstein, A. Vinogradov.
Astron. Zhurn. Akad. Nauk SSSR, Vol. 46, 985 - 992 (1969). In Russian. English translation in Soviet Astron. AJ, Vol. 13, No. 5.

The rates of excitation, ionization, photorecombination and dielectronic recombination of highly ionized atoms in a plasma with Maxwellian velocity distribution are considered. By means of the rates of these processes the ionization equilibria were calculated for the most important ions which provide for the (3 – 35) Å X-ray emission from the solar corona.

074.021 On the nature of the inhomogeneous structure of the circumsolar plasma.
I. S. Bajkov, N. A. Lotova.
Astron. Zhurn. Akad. Nauk SSSR, Vol. 46, 1057 - 1063 (1969). In Russian. English translation in Soviet Astron. AJ, Vol. 13, No. 5.

Various types of instabilities, possible in the conditions of the circumsolar plasma, are analysed. Drift, slipping instabilities, and instabilities, caused by the anisotropic distribution of temperature are considered. It is shown that the small-scale structure, observed by the methods of radioastronomy, can be connected with various types of instability in different regions of the circumsolar plasma.

074.022 Thermal state and effective collision frequency in the solar wind plasma. A. Nishida.
Journ. Geophys. Res. Vol. 74, 5155 - 5157 (1969). – Letter.

074.023 Correction to paper by Milo A. Schield, 'Pressure balance between solar wind and magnetosphere'.
M. A. Schield.
Journ. Geophys. Res. Vol. 74, 5189 - 5190 (1969). – Letter.

074.024 Coronal structure from polarization measurements.
W. N. Arnquist.
Bull. American Astron. Soc. Vol. 1, 232 (1969). – Abstr. AAS.

074.025 Preliminary determination of solar wind velocities from ionic comet tail orientations without the assumption of coplanarity. J. C. Brandt.
Bull. American Astron. Soc. Vol. 1, 234 (1969). – Abstr. AAS.

074.026 Ne I in the sun and in RS Ophiuchi.
R. D. Dietz, F. Q. Orrall.
Bull. American Astron. Soc. Vol. 1, 240 (1969). – Abstr. AAS.

074.027 Differential rotation of the solar electron corona.
R. T. Hansen, S. F. Hansen, H. G. Loomis.
Bull. American Astron. Soc. Vol. 1, 243 - 244 (1969). – Abstr. AAS.

074.028 Radar evidence of plasma that refracts or reflects 38-MHz radio energy at three or more solar radii from the sun, and the solar-wind implications. J. C. James.
Bull. American Astron. Soc. Vol. 1, 245 - 246 (1969). – Abstr. AAS.

074.029 Models for coronal condensations.
J. T. Jefferies, F. Q. Orrall, J. B. Zirker.
Bull. American Astron. Soc. Vol. 1, 246 (1969). – Abstr. AAS.

074.030 Infrared observations of the solar corona.
K. E. Kissell, P. L. Byard.
Bull. American Astron. Soc. Vol. 1, 248 (1969). – Abstr. AAS.

074.031 Coronal polarization and structure at the total solar eclipse of 22 September 1968.
K. H. Schatten, D. H. Menzel, J. M. Pasachoff.
Bull. American Astron. Soc. Vol. 1, 261 (1969). – Abstr. AAS.

074.032 Electron density and magnetic field distribution in coronal region responsible for type-IV radio bursts of 14 September 1966. M. Simon, J. D. Bohlin.
Bull. American Astron. Soc. Vol. 1, 261 - 262 (1969). – Abstr. AAS.

074.033 Solar coronal streamers: Spatial locations and associated solar phenomena.
J. D. Bohlin, R. T. Hansen, G. Newkirk, Jr..
Bull. American Astron. Soc. Vol. 1, 273 - 274 (1969). – Abstr. AAS.

074.034 The excitation of the coronal lines of Fe XIII and Ca XV. R. A. Chevalier, D. L. Lambert.
Bull. American Astron. Soc. Vol. 1, 274 - 275 (1969). – Abstr. AAS.

074.035 Brightness variations of the white-light corona during the years 1964 - 1967.
R. T. Hansen, C. J. Garcia, S. F. Hansen, H. G. Loomis.
Bull. American Astron. Soc. Vol. 1, 278 (1969). – Abstr. AAS.

074.036 Differential rotation of the solar electron corona.
R. T. Hansen, S. F. Hansen, H. G. Loomis.
Bull. American Astron. Soc. Vol. 1, 278 (1969). – Abstr. AAS.

074.037 Coronal streamers photographed on 27 and 29 April 1968.
M. J. Koomen, R. T. Seal, R. Tousey.
Bull. American Astron. Soc. Vol. 1, 283 (1969). · Abstr. AAS.

074.038 An investigation of the solar wind by interplanetary scintillation. L. T. Little, R. D. Ekers.
Bull. American Astron. Soc. Vol. 1, 285 (1969). – Abstr. AAS.

074.039 Magnetic fields and the structure of the solar corona. G. Newkirk, M. Altschuler.
Bull. American Astron. Soc. Vol. 1, 288 (1969). – Abstr. AAS.

074.040 Coronal streamer configurations with energy transport. G. W. Pneuman, R. A. Kopp.
Bull. American Astron. Soc. Vol. 1, 289 - 290 (1969). – Abstr. AAS.

074.041 Observations of a coronal streamer at 1 a.u.
K. H. Schatten, J. M. Wilcox.
Bull. American Astron. Soc. Vol. 1, 292 (1969). – Abstr. AAS.

074.042 Radiation-driven waves and the structure of the sun's outer temperature profile.
A. J. Skalafuris.
Bull. American Astron. Soc. Vol. 1, 293 (1969). – Abstr. AAS.

074.043 Solar-wind model including the effects of rotation, magnetic fields, and anisotropic heat conduction.
S. Grzędzielski.
Mass Loss from Stars, Trieste 1968, p. 110 - 121 (1969).
A time-independent solar-wind model is considered in the case of spherical symmetry and of radial magnetic field at the sun's surface. The model leads to azimuthal velocity at earth between 0.6 and 2.7 km/sec, to radial velocity at earth between 350 and 500 km/sec, and to angular momentum loss of 5×10^{18} cm^2/sec per unit mass of gas leaving the solar equator.

074.044 On the identification of Ar X and Ar XIV in the solar corona and the origin of the unidentified coronal lines. B. Edlén.
Solar Physics, Vol. 9, 439 - 445 (1969).
A study of the Z-dependence of the ^2P intervals of $2s^2 2p$ and $2s^2 2p^5$, aided by recent observational results, confirms the identification in the coronal spectrum of $\lambda 4412$ with Ar XIV, and of $\lambda 5533.4$ with Ar X. It is further shown that transitions from metastable levels in the configurations $3s^2 3p^k 3d$, with $k = 3$, 4 and 5, of Fe XI, X, IX, and Ni XIII, XII, XI can well account for the remaining unidentified coronal lines.

074.045 Meridional (north–south) motions of the solar wind. G. L. Siscoe, L. T. Finley.
Solar Physics, Vol. 9, 452 - 466 (1969).
The steady state hydrodynamic equations which describe the solar wind flow are linearized and used to study the spatial behavior of zonal pressure perturbations. Such perturbations produce meridional (north–south) motions in the solar wind. A simplified problem involving a north–south magnetic field asymmetry is also treated. The emphasis of the paper is to determine what pressure perturbations are required at the inner boundary (0.1 AU) to produce at earth north–south deviations from radial flow of 1° to 3°.

074.046 Tangential discontinuities in the solar wind.
L. F. Burlaga, N. F. Ness.
Solar Physics, Vol. 9, 467 - 477 (1969).
This paper considers six discontinuity surfaces which were observed by magnetometers on 3 spacecrafts in the solar wind. It is shown that the actual surface orientations, determined from the measured time delays and solar wind speed,

are consistent with the theoretical orientations. The plasma and magnetic field data for the discontinuities are consistent with the pressure balance condition, and the magnetic field vectors in the associated current sheets are parallel to the discontinuity surface, as required theoretically. The 6 discontinuity surfaces extended without much distortion over ~0.002 AU.

074.047 The excitation of the forbidden coronal lines. I: Fe XIII λλ10747, 10798, and 3388.
R. A. Chevalier, D. L. Lambert.
Solar Physics, Vol. 10, 115 - 134 (1969).

The excitation of the lowest $(3s^2 3p^2)$ configuration of Fe XIII is discussed for the range of density and temperatures experienced in the solar corona. The principal features are the introduction of proton collisions as an important mechanism for exciting the $3p^{2\, 3}P$ levels, the use of improved electron collision strengths and a detailed discussion of the influence of the excited configurations. The predicted intensity ratios are shown to be consistent with available observations with the single exception of an eclipse measurement of the ratio of the intensities of the infrared lines.

074.048 Differential rotation of the solar electron corona.
R. T. Hansen, S. F. Hansen, H. G. Loomis.
Solar Physics, Vol. 10, 135 - 149 (1969).

Autocorrelation analyses of K-coronameter observations made at Haleakala and Mauna Loa, Hawaii, during 1964 – 1967 have established average yearly rotation rates of coronal features as a function of latitude and height above the limb. At low latitudes the corona was found to rotate at the same rate as sunspots but at higher latitudes was consistently faster than the underlying photosphere. There were differences as large as 3 – 4 % in the rate at specific latitudes from year to year and between the two hemispheres. In 1967 a nearly constant rotation was found for heights ranging from 1.125 to 2.0 R_0. For 1966 there was a more complicated pattern of height dependence, with the rate generally decreasing with height at low latitudes and increasing at high latitudes.

074.049 A model of the magnetized solar wind.
I. H. Urch.
Solar Physics, Vol. 10, 219 - 228 (1969).

A steady state, inviscid, single fluid model of the solar wind in the equatorial plane is developed using magneto-hydrodynamics and including the heat equation with thermal conduction but no non-thermal heating (i.e. a conduction model). The effects of solar rotation and magnetic field are included enabling both radial and azimuthal components of the velocity and magnetic fields to be found in a conduction model for the first time. Under these assumptions the solar wind temperature, density, velocity and magnetic fields are calculated and compared with observed values.

074.050 Some morphological particularities of the solar corona on 22 September 1968. S. Koutchmy.
Astrophys. Letters, Vol. 4, 215 - 220 (1969).

Three coronal streamers of different kinds are considered. None of these streamers seems to have developed autonomously. The magnetostatic model not being able to account for this, closer connections with the solar activity are suggested.

074.051 The effect of galactic cosmic rays upon the dynamics of the solar wind.
S. F. Sousk, A. M. Lenchek,
Astrophys. Journ. Vol. 158, 781 - 795 (1969).

The scattering of galactic cosmic rays by the fluctuations in the magnetic field embedded in the solar wind transfers energy and momentum to the cosmic-ray gas. The hydrodynamic equations of the wind, including both these effects, are integrated. It is found that the diffusion coefficient K must vary with heliocentric distance in such a way that, at distances greater than ~ 4 a.u., K increases to a value very much larger than its value at 1 a.u. With such models of K (r) the cosmic-ray gas then exerts only a small influence on the wind.

074.052 Realization of solutions of Parker's type.
M. V. Konyukov.
Problemy kosmich. fiz. No. 4, p. 42 - 53 (1969). In Russian.

Approximate solutions of the equations for a one-dimensional stationary spherical flow are constructed. Parker's solutions cannot be used for a description of plasma flows from the sun. A new model is proposed.

074.053 Nonthermal heating in the quiet solar wind.
A. J. Hundhausen.
Journ. Geophys. Res. Vol. 74, 5810 - 5813 (1969). – Letter.

074.054 The heating of the solar corona. M. Kuperus.
Space Sci. Rev. Vol. 9, 713 - 739 (1969).

A discussion is given of the present state of the theory of the heating of the solar corona by shock waves. The heating of the outer layers by dissipation of shock waves is found to be sufficient to account for the observed radiative and corpuscular energy losses. Much emphasis is laid on the competitive role played by the four fundamental processes of energy transfer: mechanical heating, radiation, heat conduction and convection of energy in establishing the equilibrium structure of the corona. The atmosphere may be divided in several regions according to the pre-dominance of one of the energy processes mentioned above.

074.055 The probabilities of some forbidden coronal lines in the vacuum ultraviolet region.
T. Feklistova.
Izv. AN Est. SSR. Fiz., Matem., Vol. 18, No. 1, p. 57 - 64 = Tartu Astron. Obs. Teated No. 20 (1969). In Russian.

The transition probabilities of some forbidden coronal lines in the vacuum ultraviolet region are calculated. These are transitions for the ions of C, N, O, F, Ne, Na.

074.056 Solar wind interactions and the magnetosphere.
A. J. Dessler.
Physics of the Magnetosphere, Boston College 1967, p. 65 - 105 (1968). – Tutorial lecture.

074.057 Observations of the solar wind, bow shock and magnetosheath by the Vela satellites.
I. B. Strong.
Physics of the Magnetosphere, Boston College 1967, p. 376 - 391 (1968). – Invited research paper.

074.058 Radio evidence of directive shock-wave propagation in the solar corona. K. Kai.
Solar Physics, Vol. 10, 460 - 464 (1969).

Radioheliograph observations at 80 MHz are reported of a flare-associated event in which two type II bursts occur in four different sources. On the opposite side of the flare centre, outside the shock cone, there was a stable bipolar source. Strong magnetic fields in this source may have acted as a 'magnetic wall' to the shock wave and inhibited its propagation in this direction.

074.059 Interaction of the solar wind with planetary atmospheres. R. A. Elco.
Journ. Geophys. Res. Vol. 74, 5073 - 5082 (1969).

Neither Venus nor the moon have a significant dipole magnetic field, and their atmospheres are exposed to the solar wind and the interplanetary magnetic field. A model for the supersonic collisionless hydromagnetic reacting flow of the solar wind into a planetary atmosphere is developed to determine the limits on the position of a bow shock in the planets ionosphere. The position and existence of the shock is dependent on the UV optical depth of the atmosphere, the strength of the solar wind, the mass of the atmospheric atoms or molecules and the scale height of the atmosphere. Recombination sets a lower limit on the optical depth at which a shock can form. In the case of Venus (CO_2 atmosphere) the shock forms for $10^{-9} < \tau < 2 \times 10^{-4}$, i.e., above the usual Chapman layer.

074.060 Plasma instabilities associated with heat conduction in the solar wind and their consequences.
D. W. Forslund.
Bull. American Astron. Soc., Vol. 1, 343 (1969). – Abstract AAS.

074.061 Energy source for the solar corona.
J. F. Hashemi.
Bull. American Astron. Soc., Vol. 1, 347 (1969). – Abstract AAS.

074.062 A solar-wind model with two temperatures, viscosity, rotation, and magnetic field.
C. L. Wolff, J. C. Brandt, R. G. Southwick.
Bull. American Astron. Soc., Vol. 1, 369 (1969). – Abstract AAS.

074.063 Temperature determination in the corona from a time profile of radio bursts of type III.
V. V. Fomichev, I. M. Chertok.
Astron. Zhurn. Akad. Nauk SSSR, Vol. 46, 1319 - 1321 (1969). In Russian. English translation in Soviet Astron. AJ, Vol. 13, No. 6.

It is paid attention to the necessity of taking into account the extent of the region of the radioemission generation at a fixed frequency in the case of temperature determination in the corona from a time profile of a type III burst. To take this into consideration is very important in particular at high frequencies (> 150 – 200 MHz).

074.064 The solar wind. T. G. Cowling.
Observatory, Vol. 89, 217 - 224 (1969). – The Halley lecture for 1969, delivered in Oxford on May 8.

074.065 Apollo 11 solar wind composition experiment: First results. F. Bühler, P. Eberhardt, J. Geiss, J. Meister, P. Signer.
Science, Vol. 166, 1502 - 1503 (1969).

The helium-4 solar wind flux during the Apollo 11 lunar surface excursion was $(6.3 \pm 1.2) \times 10^6$ atoms per square centimeter per second. The solar wind direction and energy are essentially not perturbed by the moon. Evidence for a lunar solar wind albedo was found.

074.066 The structure of the magnetic fields in dark coronal arcs. V. E. Merkulenko.
Solnechnye Dannye 1969 Byull. No. 8, p. 81 - 87 (1969). In Russian.

The structure of the field in dark coronal arcs is analysed. The problem of equilibrium conditions in a magnetic loop with the ends fixed on the plane is considered. It is shown that the system can be in equilibrium if the strength of the toroidal component of the magnetic field is equal to 0.25 times the strength of the poloidal component and magnetic force is balanced by hydrostatic pressure.

074.067 Photometry of the solar corona on September 22, 1968. N. S. Shilova.
Solnechnye Dannye 1969 Byull. No. 8, p. 106 - 112 (1969). In Russian.

The main results of a reduction of two photographs of the solar corona are given. The photographs were taken during the total eclipse on September 22, 1968 in the region of Yurgamysh by astrophysicists of the IZMIRAN.

074.068 Infrared observations of the outer solar corona. R. M. MacQueen.
Thesis, Univ. Johns Hopkins. Univ. Microfilms, Ann Arbor, Mi., 144 pp. (1968). – See Phys. Abstr. Vol. 72, No. 21982 (1969).

074.069 Collecting a sample of solar wind: an experimental study of its capture in metal films.
D. Lal, W. F. Libby, G. Wetherill, J. Leventhal.
Journ. Applied Phys., Vol. 40, 3257 - 3267 (1969).

074.070 Isophotes of the solar corona on September 22, 1968. N. A. Nesterko, V. E. Solovjev, E. A. Shilov.
Astron. Vestn. Vol. 3, 170 - 172 (1969). In Russian.

074.071 Solar wind disturbances associated with flares.
J. M. Wilcox.
Solar Flares and Space Research, Tokyo 1968, p. 294 - 309 (1969).

The structure of the quiet solar wind and interplanetary magnetic field is reviewed to provide background and perspective for the discussion of solar wind disturbances associated with flares.

074.072 The solar (stellar) wind as an one-dimensional flow. Part I. S. Grzędzielski.
Postępy Astron., Vol. 17, 337 - 346 (1969). In Polish.

The solar wind phenomenon is discussed in terms of an one-dimensional, stationary gas flow (de Laval nozzle problem). The influence of rotation and of the magnetic field upon the solar wind model with heat conduction is also discussed in the case when spherical symmetry applies to the ecliptic plane.

074.073 Are extremely thin streams in the solar corona jets of fast particles?
V. I. Ivanchuk.
Astron. Tsirk. No. 537, p. 5 - 8 (1969). In Russian.

074.074 On the distribution of perturbations by chromospheric eruptions in the solar wind.
V. P. Korobejnikov, Yu. M. Nikolaev.
Kosmich. Issled. Vol. 7, 891 - 894 (1969). In Russian.

074.075 About a peculiarity of connection between solar wind velocity and geomagnetic activity.
S. M. Mansurov, L. G. Mansurova.
Geomagn. Aeronom. Vol. 9, 693 - 696 (1969). In Russian.

074.076 Some results of comparisons of coronal data.
S. O. Obashev.
Trudy Astrofiz. Inst. Alma-Ata, Vol. 15, 6 - 9 (1969). In Russian.

By comparisons of the data of different coronal stations using the same type of instruments and equal methods for observations and treatment of data received, the author comes to the conclusion that it is impossible to expect total identity of the obtained data. The differences, on the one hand, are due to the nature of the corona itself, and on the other, to arbitrary observational and treatment errors.

074.077 An influence of electric fields on the profiles of coronal lines.
S. O. Obashev, E. J. Vilkovisky, A. S. Zubtzov.
Trudy Astrofiz. Inst. Alma-Ata, Vol. 15, 10 - 15 (1969). In Russian.

The consideration of the influence of weak electric

fields allows to obtain the profiles of the coronal lines matching with observed ones. The kinetic temperature appears to be less than that obtained from the observed half-width.

074.078   Solar wind observations with satellite ESRO HEOS-1 in December 1968.
A. Bonetti, G. Moreno, S. Cantarano, A. Egidi, R. Marconero, F. Palutan, G. Pizzella.
Nuovo Cimento, Vol. 64B, 307 - 323 (1969).
    Preliminary results from the solar wind experiment are presented. The properties of the solar wind during a period near solar maximum are similar to the general properties during previous years.

074.079   Die Korona bei der Sonnenfinsternis vom 22. September 1968. I. Die Form der Korona.
M. Waldmeier, S. E. Weber.
Astron. Mitt. Eidg. Sternw. Zürich, No. 290, 13 pp. (1969).

074.080   Beobachtungen der Sonnenkorona in den Jahren 1965 und 1966.   M. Waldmeier.
Astron. Mitt. Eidg. Sternw. Zürich, No. 291, 24 pp. (1969).

074.081   Beobachtungen der Sonnenkorona in den Jahren 1967 und 1968.   M. Waldmeier.
Astron. Mitt. Eidg. Sternw. Zürich, No. 292, 22 pp. (1969).

074.082   On the theory of the solar wind.
I. M. Dagkesamanskaya, M. V. Konyukov.
Trudy Fiz. in-ta AN SSSR, Vol. 47, 85 - 95 (1969).
In Russian. – Abstr. in Referativ. Zhurn. 51. Astron., 2.51.428 (1970).

074.083   Koronale Expansionsphänomene.
A. Bruzek.
Mitt. Astron. Ges. No. 27, p. 211 - 212 (1969). – Abstract AG.

074.084   Radio evidence of instabilities and shock waves in

the solar corona.   J. P. Wild.
Separate print Div. Radiophys. C.S.I.R.O., Sydney [Reprinted from Proceedings of Conference on "Plasma Instabilities in Astrophysics", held at Asilomar, California, Oct. 1968 (D. A. Tidman, D. G. Wentzel, Editors)], p. 119 - 138 (1969).

A study on the statistical equilibrium of the ion A XIV level-population in coronal conditions.
See Abstr. 022.068.

Asymptotic solutions of the solar (stellar) wind equations for large distances from the central star.
See Abstr. 064.005.

On a relation between the indices of solar activity in the photosphere and the corona.   See Abstr. 072.057.

On the solar corpuscular radiation obtained from measurements with satellites and rockets. The four-stream model of the solar corona near the minimum of activity.
See Abstr. 078.005.

Cosmic ray acceleration by corpuscular streams and the characteristics of the solar wind.   See Abstr. 078.031.

Die Expedition zur Beobachtung der totalen Sonnenfinsternis vom 22. September 1968.
See Abstr. 079.103.

Note on the solar wind-induced drag on comets.
See Abstr. 102.019.

Interplanetary shock waves. I. Gross structure.
See Abstr. 106.003.

Investigation of the solar wind with help of solar and galactic cosmic rays.
See Abstr. 143.031.

075   Solar Patrol

075.001   Katalog der Sonnentätigkeit für das Jahr 1966.
R. S. Gnevysheva.
Trudy Glav. Astron. Obs. v Pulkove, 100 pp. Price 70 Kop. (1969). In Russian.

075.002   Magnetfelder von Sonnenflecken.
Prilozheniya k Byulletenyu "Solnechnye Dannye", 1969 No. 5 - 10.   In Russian.

075.003   Nombres relatifs définitifs de Wolf pour l'année 1968.   M. Waldmeier.
L'Astronomie, 83e année, 306 (1969).

075.004   Die Sonnentätigkeit im zweiten Halbjahr 1968.
R. Müller.
Sterne, 45. Jahrgang, 113 - 115 (1969).

075.005   Definitive Sonnenflecken-Relativzahlen für 1968.
M. Waldmeier.
Sterne, 45. Jahrgang, 121 (1969).

075.006   Summary of prominence observations, magnetic observations, and ionospheric observations for the

second half of 1962.   M. K. V. Bappu.
Kodaikanal Obs. Bull. No. 169, 313 pp. (1968).

075.007   Actividad solar en 1968.   E. Gullón.
Bol. Astron. Obs. Madrid, Vol. 7, No. 4, 218 pp. (1969). – I. Números relativos de Wolf; II. Estadística de manchas y superficie de las mismas; III. Fáculas cromosféricas brillantes; IV. Filamentos de hidrógeno; V. Protuberancias; VI. Gráficas de actividad solar.

075.008   Katalog der Sonnentätigkeit für 1967.
R. S. Gnevysheva.
Trudy Glav. Astron. Obs. v Pulkove, 140 pp. Price 1 Rbl. 33 Kop. (1969).   In Russian.

075.009   Reports on the progress of astronomy: Solar activity (sunspots, prominences).
P. S. Laurie, M. K. V. Bappu.
Quarterly Journ. Roy. Astron. Soc. Vol. 10, 239 - 240 (1969).

075.010   Indices of geomagnetic activity of the observatories Hartland, Eskdalemuir, Lerwick.
Journ. Atmosph. Terr. Phys. Vol. 31, 1229, 1231, 1297, 1371,

1451 (1969). – 1969 April – August.

075.011 **Solar activity and geomagnetic storms 1968.**
P. S. Laurie, B. R. Leaton.
Observatory, Vol. 89, 215 - 216 (1969).

075.012 **L'activité solaire.** M.-J. Martres.
L'Astronomie, 83ᵉ année, 327, 354, 409 - 410,
470 - 471 (1969). – Rotations Nos. 1542 - 1546.

075.013 **Solar phenomena.** M. Cimino (Editor).
Oss. Astron. Roma, Monthly Bull. Nos. 139 - 144
(1969). – 1969 July – December: Daily total areas of sunspot-groups; Heliographic position, classification and area of
sunspot-groups; Longitudinal sunspot magnetic fields; Hours
of K-line cinematographic patrol; Hours of H_α cinematographic patrol; Sudden cosmic noise absorption S.C.N.A. and
sudden enhancement of atmospherics S.E.A.

075.014 **Daily Hα chromosphere pictures, daily K_{232} chro-**
mosphere pictures, daily white light photosphere
pictures. M. Cimino (Editor).
Photographic Journal of the Sun. Oss. Astron. Roma, N. 20 -
25 (1969). – 1969 May 21 – Oct. 31.

075.015 **Map of the Sun.**
Edited by Fraunhofer Institut, Freiburg. – 1969
July 1 – December 31.

075.016 **Heliographic maps of the photosphere for the year**
1968. M. Waldmeier.
Publ. Sternw. Zürich, Vol. 13, (No. 3), 61 - 91 (1969).

The present publication gives heliographic maps of the
photosphere and evolution tables of sunspot-groups for the
year 1968. Maps and tables are based on daily drawings of
spots and faculae using a projected solar image with a diameter of 25 cm.

075.017 **Sunspots; Eruptions chromosphériques brillantes;**
Intensité de la couronne solaire; Solar radio emis-
sions.
M. Waldmeier, R. Michard, J. G. Bastiaans, A. D. Fokker.
Quarterly Bull. Solar Activity (published by Eidgen. Sternw.
Zürich), Nos. 163, 164, p. 197 - 398 (1969). – Observations
of the co-operating observatories are given.

075.018 **Sunspot numbers.**
Sky Telescope, Vol. 38, 127, 200, 273, 357,
428 (1969).

075.019 **Fenomeni solari.** F. Mazzucconi, S. Delli Santi,
M. L. Sturiale, A. Abrami.
Coelum, Vol. 37, 182 - 190, 236 - 243, 288 - 293 (1969). –
1969 March – August.

075.020 **Osservatorio Magnetico de L'Aquila. Bolletino**
magnetico. F. Molina.
Coelum, Vol. 37, 193 - 197, 246 - 247, 296 - 297 (1969). –
1969 Jan. - July.

075.021 **Solar observations made at Catania Astrophysical**
Observatory during 1967.
R. Campisi Cristaldi, O. Morgante, L. Paternò, M. L. Sturiale,
G. Celeani, C. D'Arrigo, G. Domina, G. Patti, S. Rifici,
G. Sapienza, S. Torrisi.
Oss. Astrofis. Catania Pubbl., Nuova Serie, No. 130, 143 pp.
(1968).

075.022 **Sunspots in the year 1967.**
J. Mergentaler.
Acta Geophys. Polonica, Vol. 17, (No. 1), 93 - 96 = Wrocław
Astron. Obs. Repr. No. 76.

075.023 **Tageskarten der Sonne und geophysikalische**
Schaubilder.
Solnechnye Dannye 1969 Byull. No. 2, p. 1 - 84, No. 3, p. 1 -
96, No. 4, p. 1 - 75, No. 5, p. 1 - 81, No. 6, p. 1 - 76, No. 7,
p. 1 - 71, No. 8, p. 1 - 77, No. 9, p. 1 - 87 (1969). In Russian.

075.024 **Solar and solar system activity.**
R. J. J. Langton, J. R. Smith.
Journ. British Astron. Ass. Vol. 79, 399 - 402, 491 - 493,
Vol. 80, 66 - 69 (1969). – 1969 March – Aug.

075.025 **Solare Beobachtungsergebnisse (Solar Data).**
E. A. Lauter, H. Daene, F. W. Jäger, F. Fürstenberg,
H. Künzel, D. Scholz, W. Dittmar.
Zentralinstitut für Solar-Terrestrische Physik, (Heinrich-Hertz-
Inst.), Deutsche Akad. Wiss. Berlin, HHI Solar Data, Vol. 20,
May – December (1969). – Solar radio emission; Sunspot
magnetic data.

075.026 **Solar phenomena.**
Tokyo Astron. Obs. Bull. Solar Phenomena, Vol.
21, Nos. 1 - 2, p. 1 - 51 (1969). – Sunspots; Evolution table
of sunspot groups; Map of sunspots; Hα flocculi and Hα dark
filaments; Solar flares; Hours of Hα patrol; Prominences and
filaments; Intensity of coronal emission line 5303 Å; Solar
radio emission (1969 January – June).

075.027 **Manchas e grupos de manchas do sol. Sunspots and**
sunspots-groups. J. F. Caria Caldeira.
Contr. Obs. Valongo, Univ. Federal Rio de Janeiro, Ser. I,
Nos. 4 - 8 (1969).

075.028 **Solar photospheric observations.**
F. Bruin, H. Hourani, T. Assaf.
Lee Obs. American Univ. Beirut, Monthly Bull., Astron.
Section, 1969 May – July (1969).

Sunspot relative numbers; Heliographic mean position
and classification of the sunspot groups; Number of facular
zones.

075.029 **Zonnevlekkengetallen.**
Hemel en Dampkring, Vol. 67, 296, 325, 362,
384 (1969). – 1969 April – September.

075.030 **Visual observations of the solar photosphere.**
P. Zlobec.
Pubbl. Oss. Astron. Trieste, No. 395 (Osservazioni Solari,
No. 13), p. 3 - 12; No. 398 (Osservazioni Solari, No. 14), p.
3 - 18; No. 408 (Osservazioni Solari, No. 15), p. 3 - 16 (1969).
1969 January - September.

075.031 **Measures of the solar flux at 239 MHz.**
P. Zlobec.
Pubbl. Oss. Astron. Trieste, No. 395 (Osservazioni Solari,
No. 13), p. 14 - 21; No. 398 (Osservazioni Solari, No. 14),
p. 20 - 27; No. 408 (Osservazioni Solari, No. 15), p. 18 - 25
(1969). – 1969 January - September.

075.032 **Provisional sunspot-numbers.**
Yamamoto Circ. Nos. 1701, 1702, 1705, 1706,
1708, 1710 (1969). In Japanese. – 1969 May – November.

076   Solar UV, X Rays, Gamma Radiation

076.001   **Solar soft X-rays and solar activity. I: Relationships between reported flares and radio bursts, and X-ray bursts.**   R. G. Teske, R. J. Thomas.
Solar Physics, Vol. 8, 348 - 368 (1969).

Soft solar X-rays were observed from OSO-III. An analysis of the X-ray enhancements associated with 165 solar flares revealed that there is a tendency for a weak soft X-ray enhancement to precede the cm-λ burst and Hα flare. The peak soft X-ray flux follows the cm-λ peak by about 4 min, on the average. Additionally, it was found that flare-rich active centers tend to produce flares which are stronger X-ray and cm-λ emitters than are flares which take place in flare-poor active centers.

076.002   **The emission of solar X-rays in the 0.5–3 Å wavelength range and its relation to the magnetic configuration of active centers.**   G. Chambe.
Solar Physics, Vol. 8, 369 - 375 (1969).

The slowly varying component of solar X-rays in the 0.5–3 Å wavelength range has been studied using data obtained by the satellite Explorer 30 (Solrad 8). The intensity of these X-rays is poorly correlated with the centimeter radio flux, contrary to the good correlation found in the spectral bands 1–8, 8–16 and 44–60 Å. On the other hand the 0.5–3 Å X-ray intensity is often connected to the development of a specific magnetic configuration in the sun spot group which may thus be associated with the X-ray producing active center.

076.003   **Further investigations of solar X-ray sources using D-layer ionization behavior during eclipses.**
D. D. Meisel.
Solar Physics, Vol. 8, 477 - 490 (1969).

Radio absorption records obtained in or near the zone of totality at two solar eclipses (May 30, 1965 and July 20, 1963) have been examined in detail. It is concluded that all major radio absorption changes during an eclipse are ionization controlled and occur in the $D-E$ layer. Corrections for the ultraviolet sensitivity of the region below 150 kilometers have been applied, so the results indicate effects attributable to the X-ray flux alone. The residual curves clearly show a 'threshold' effect similar to that described by Rastogi et al. (1956) and later by Schmidt and Sharp (1965). Arguments are presented for interpreting this effect in terms of limb configurations involving small, hot X-ray sources similar to that described previously (Meisel, 1968). Once again it is necessary to postulate that grazing incidence reflections from the lunar limb occur. Source positions have been derived from intersections of lunar arcs as seen from different geographic locations.

076.004   **The hard solar X-ray spectrum observed from the third Orbiting Solar Observatory.**
H. S. Hudson, L. E. Peterson, D. A. Schwartz.
Astrophys. Journ. Vol. 157, 389 - 415 (1969).

The hard solar X-ray scintillation-counter telescope on the OSO-III satellite covers the energy range 7.7–210 keV with 15-sec time resolution, and six logarithmically spaced energy channels. Approximately ten bursts per day were detected during the interval March 9–March 23, 1967. About once per day a burst of peak energy flux greater than 1.6×10^{-5} erg (cm^2 sec)$^{-1}$ was observed. Although many variations were observed, the typical event had an e-folding rise time of 86 sec and a decay time of 458 sec. The bursts occurred in correlation with almost all listed flares and subflares (88 per cent), microwave bursts (92 per cent), and SID's (100 per cent). Numerous bursts were also detected without these ac-

companying phenomena. The correlation with type III radio bursts, although still positive, is not as good (31 per cent), a fact which suggests that coronal disturbances are not an inevitable consequence of the process which produces X-ray emission. The X-ray spectrum is appreciably non-thermal in the initial phase of the burst and thermal in the decay phase, with an effective temperature often exceeding 50×10^6 °K. The average peak temperature of subflares exceeds 10×10^6 °K, while that of importance 1 or greater exceeds 14×10^6 °K.

076.005   **Rocket observations of profiles of solar ultraviolet emission lines.**
E. C. Bruner, Jr., W. A. Rense.
Astrophys. Journ. Vol. 157, 417 - 424 (1969).

High-resolution rocket spectra have been photographed of solar hydrogen Ly a, Si III at 1206.52 Å, and O I at 1302–1306 Å; the profiles were corrected for instrumental broadening. The hydrogen Ly a profile can be accounted for by a non-LTE theory except in the wings. There is some evidence of solar self-reversal in the O I triplet, but none in the Si III line. Effects of plage areas are present for all lines.

076.006   **Solar-cycle variation of extreme ultraviolet radiation.**
L. A. Hall, J. E. Higgins, C. W. Chagnon, H. E. Hinteregger.
Journ. Geophys. Res. Vol. 74, 4181 - 4183 (1969). – Letter.

076.007   **Revision of the ultraviolet solar spectrum in the range 3650 – 3000 Å.**
W. E. Mitchell, Jr., O. C. Mohler.
Astrophys. Journ. Suppl. Series, Vol. 18, 379 - 427 (1969).

This paper tabulates the wavelengths of 1987 features in the range of the solar spectrum 3650 – 3000 Å. These features come out of the detailed examination and intercomparison of several sets of high-dispersion photoelectric records and the Second Revised Rowland. Most of the listed features (a) are new (551), (b) agree sufficiently well in wavelength to be considered to confirm features of the Second Revised Rowland (557), or (c) are features of the Second Revised Rowland whose existence is called into question (237). Small relocations of wavelength are indicated for 172 features of the Second Revised Rowland. Of the new lines, equivalent widths have been measured for 198 which suffer the least blending. Their strengths range from 16.6 down to 0.2 mÅ and average 3.2 mÅ. A total of 726 identifications are proposed for 476 of the new or relocated lines. A total of 1911 comments are given as to their visibility and appearance on the photoelectric records of these lines and a number of the weakest features of the Second Revised Rowland.

076.008   **Correlation of solar microwave and soft X-ray radiation. 1. The solar cycle and slowly varying components.**   C. D. Wende.
Journ. Geophys. Res. Vol. 74, 4649 - 4660 (1969).

The integral solar X-ray flux between 2 and 12 Å was measured by the spacecrafts Injun 1, Injun 3, Explorer 33, and Explorer 35. The integral flux between 2 and 9 Å was observed by the Mariner 5 spacecraft. The data from Injuns 1 and 3 and Explorers 33 and 35, when averaged in monthly intervals, show a long-term variation in activity of at least a factor of 5 between 1961 and 1969. A slowly varying component that tracks the 10-cm radio flux is observed in the X-ray flux. This X-ray flux variation, about a factor of 10, correlates with the appearance of major active regions on the solar disk. This slowly varying component correlates well with radio fluxes at frequencies greater than about 1 GHz and less well with radio fluxes at lower frequencies. The X-ray spectrum, obtained by comparing the 2 and 9 Å flux with the 2 and 12 Å flux, hardens during periods of high solar activity and may harden during flares, but it does not soften.

076.009 Observations of two components in energetic solar X-ray bursts. S. R. Kane.
Astrophys. Journ. *(Letters)*, Vol. 157, L139 - L142 (1969).

Measurements made with the solar X-ray detector aboard the OGO-5 satellite show that some energetic ($\gtrsim$ 9.6 keV) solar X-ray bursts consist of two components, viz., (1) an impulsive component that reaches its peak early in the event approximately in coincidence with the peak in the microwave burst and has a photon spectrum consistent with a power law in energy and (2) a slower component that attains its maximum later in the event and has a photon spectrum steeper than that for the impulsive component. The measurements also show that in some X-ray bursts only the slower component is observable. The impulsive component is attributed to the bremsstrahlung emission from electrons with a nonthermal energy distribution.

076.010 Cosmic X-ray bremsstrahlung associated with suprathermal protons. E. Boldt, P. Serlemitsos.
Astrophys. Journ. Vol. 157, 557 - 562 (1969).

Copious X-ray production is shown to attend the collisions of suprathermal ($\gtrsim$ MeV) protons with ambient electrons. Such suprathermal protons have been detected in association with solar-flare events and cosmic rays. The X-ray emission to be expected for the well-observed solar-proton event of September 28, 1961 agrees with measurements. It is demonstrated that suprathermal cosmic-ray protons propagating through the interstellar gas could be responsible for a detectable component of the diffuse sky background of X-rays. The possible role of suprathermal protons in producing the X-rays emitted by the Crab Nebula is examined and found to be inconclusive.

076.011 Interpretation of XUV spectroheliograms. H. Zirin.
Solar Physics, Vol. 9, 77 - 87 (1969).

Some parameters of chromospheric structure are drawn from recently published XUV spectroheliograms. The HeII emission above the limb arises from the small amount of He⁺ still existing at $10^{6\,\circ}$. The larger amounts of He⁺ in the cooler corona at the poles explain the polar cap absorption in λ 304. The flat distribution of emission in OIV and OV, with a sharp spike at the limb, is caused by the rough structure of the chromosphere and the variable excitation in the emitting spicules. The intensity of the NeVII lines shows that the transition zone between chromosphere and corona is very sharp.

076.012 X rays from the sun. W. M. Neupert.
Annual Rev. Astron. Astrophys. Vol. 7, 121 - 148 (1969).

076.013 Interpretation of solar helium-like ion line intensities. A. H. Gabriel, C. Jordan.
Monthly Notices, Roy. Astron. Soc., Vol. 145, 241 - 248 (1969).

Recent identification of the $2^3S \rightarrow 1^1S$ line from helium-like ions in the solar soft X-ray spectrum, followed by calculation of its transition probability, enables an analysis of the observations to be carried out, based on intensities of the three lines $2^1P \rightarrow 1^1S$, $2^3P \rightarrow 1^1S$, and $2^3S \rightarrow 1^1S$. The relative collision rates to the excited levels and the electron densities in the emitting regions have been determined, subject to the limitation of available observations.

076.014 Chromospheric flare enhancements in the extreme ultraviolet and their relationship with solar abundances. B. C. Bowers.
Nature, Vol. 224, 352 - 353 (1969).

Increases in the extreme ultraviolet during chromospheric flares, observed by Hinteregger and Hall from OSO III, are found to vary with solar abundance. Some deviations are observed but with one exception these are attributed to blending. The exception is an oxygen line at 629.7 Å which is observed at the short wavelength limit of the spectrograph and is therefore suspect.

076.015 Some relations between occurrence of soft solar X-rays, solar flares, and cm-λ bursts.
R. G. Teske, R. J. Thomas.
Bull. American Astron. Soc. Vol. 1, 263 (1969). – Abstr. AAS.

076.016 Simultaneous X-ray and electron emission from the sun. H. S. Hudson, R. P. Lin.
Bull. American Astron. Soc. Vol. 1, 280 (1969). – Abstr. AAS.

076.017 De-occultation X-ray event of 2 December 1967.
H. Hudson, D. McKenzie, H. Zirin, W. Ingham.
Bull. American Astron. Soc. Vol. 1, 280 - 281 (1969). – Abstr. AAS.

076.018 Results from the AS and E X-ray telescope on OSO-IV. A. S. Krieger, G. S. Vaiana.
Bull. American Astron. Soc. Vol. 1, 283 - 284 (1969). – Abstr. AAS.

076.019 Lunar occultation of small X-ray plages during the 22 September 1968 solar eclipse. D. D. Meisel.
Bull. American Astron. Soc. Vol. 1, 286 - 287 (1969). – Abstr. AAS.

076.020 New solar EUV emission lines observed during flares by OSO-III. W. M. Neupert.
Bull. American Astron. Soc. Vol. 1, 287 (1969). – Abstr. AAS.

076.021 XUV solar features observed on 22 September 1968. J. D. Purcell, R. Tousey.
Bull. American Astron. Soc. Vol. 1, 290 (1969). – Abstr. AAS.

076.022 Analysis of high-resolution solar X-ray photographs. II. Solar active regions.
W. P. Reidy, L. VanSpeybroeck.
Bull. American Astron. Soc. Vol. 1, 291 (1969). – Abstr. AAS.

076.023 Brightness temperatures in the solar ultraviolet continuum: 1450 - 2080 Å.
K. Widing, J. D. Purcell.
Bull. American Astron. Soc. Vol. 1, 295 - 296 (1969). – Abstr. AAS.

076.024 OSO-IV observations of UV limb brightening. G. L. Withbroe.
Bull. American Astron. Soc. Vol. 1, 296 (1969). – Abstr. AAS.

076.025 A study of the solar soft X-ray spectrum.
J. L. Culhane, P. W. Sanford, M. L. Shaw, K. J. H. Phillips, A. P. Willmore, P. J. Bowen, K. A. Pounds, D. G. Smith.
Monthly Notices, Roy. Astron. Soc., Vol. 145, 435 - 455 (1969).

A proportional counter X-ray spectrometer, sensitive in the wavelength ranges 1 - 20 Å and 44 - 60 Å, was placed in orbit on the Orbiting Solar Observatory –4. The instrument began to acquire data on 1967 October 23. The spectrometer and its operation are briefly described. Its performance and the method of data analyses are also discussed. Preliminary studies have been made of the slowly varying component of solar X-radiation, the nature of the X-ray active regions in the solar corona and the characteristics of impulsive X-ray events.

076.026 De-occultation X-ray events of 2 December, 1967.
H. Zirin, W. Ingham, H. Hudson, D. McKenzie.
Solar Physics, Vol. 9, 269 - 277 (1969).

A flare rising from behind the solar limb was recorded simultaneously by the UCSD X-ray detector on OSO-III (7.7 – 200 keV) and the Caltech photoheliograph. Spectra suggest that the material was already heated to 27000000° and that the increase in flux was due to the de-occultation. Comparison of the deduced volume with the bremsstrahlung formula gives a density of about 10^{10} for the 27000000° component of the flare.

076.027 Some relationships between solar X-ray bursts and SPA's produced on VLF propagation in the lower ionosphere. P. Kaufmann, M. H. Paes de Barros.
Solar Physics, Vol. 9, 478 - 486 (1969).

This paper discusses SPA's measured at long VLF propagation paths in the lower ionosphere and their association with solar X-ray bursts observed by USNRL satellites in the 0–3 Å, 0–8 Å and 8–20 Å bands.

076.028 Identification of two solar X-ray sources at the 22 September 1968 total eclipse.
D. D. Meisel.
Solar Physics, Vol. 9, 487 - 493 (1969).

Study of nearly 200 records of radio signal strength indicates that a major portion of the ambient D layer absorption on 22 September 1968 can be attributed to two strong X-ray sources. The hottest source was located on the eastern limb and showed evidence of cooling from $T_{wien} = (3.6 \pm 0.3) \times 10^6$ K to $T_{wien} = (2.7 \pm 0.2) \times 10^6$ K over a $2^1/_2$ h period. The other source was located near the western limb and showed no appreciable temperature change during the eclipse period.

076.029 An outstanding Lyman-alpha event.
B. C. Fossi, G. Poletto, G. L. Tagliaferri.
Solar Physics, Vol. 10, 196 - 197 (1969). – Research note.

076.030 Temporal variations of solar Lyman alpha.
R. R. Meier.
Journ. Geophys. Res. Vol. 74, 6487 - 6490 (1969).

Geocoronal Lyman α was observed by the OGO 4 spacecraft from August through December 1967. The emission rate at a fixed orientation with respect to the sun was found to have short-term fluctuations of less than ±5% superimposed on a monthly (or 27-day) variation of as much as ±15%. These phenomena are attributed to variability of the Lyman-α flux at the center of the solar emission line.

076.031 Iowa catalog of solar X-ray flux (2 – 12 Å).
J. F. Drake, Sr., J. Gibson, J. A. van Allen.
Solar Physics, Vol. 10, 433 - 459 (1969).

The absolute X-ray flux from the whole disc of the sun in the wavelength range 2 to 12 Å has been observed for a prolonged period by University of Iowa equipment on the earth-orbiting satellite Explorer 33 and the moon-orbiting satellite Explorer 35. A comprehensive catalog of the flux F (2 – 12 Å) is being produced. The observational technique and the scheme of reducing data are described herein. A catalog of tabular and graphical data with a time resolution of either 81.8 or 163.6 sec has been completed for the following periods: From Explorer 33: 2 July 1966 to 27 July 1967;

from Explorer 35: 26 July 1967 to 18 September 1968, and made available through that agency to interested workers in solar and ionospheric physics.

076.032 Rapid fine structure in a burst of hard solar X-rays observed by OSO-5.   K. J. Frost.
Astrophys. Journ. (*Letters*), Vol. 158, L159 - L163 (1969).

A number of bursts of hard solar X-rays have been observed in nine channels simultaneously covering the energy range from 14 to 250 keV. The spectra of these bursts were sampled for 0.2 sec every 1.8 sec. One such burst, of 9-min duration, with an intensity-time profile having a complex, quasi-periodic structure, is presented. A mechanism based on the repetitive production of monoenergetic electrons is qualitatively shown to explain the time structure of the X-ray burst.

076.033 Energetic X-ray and extreme-ultraviolet flashes of solar flares. R. F. Donnelly.
Astrophys. Journ. (*Letters*), Vol. 158, L165 - L167 (1969).

Extreme-ultraviolet flashes from solar flares and X-rays of energies ≥ 50 keV exhibit similar, concurrent impulsive fine time structure. The extreme-ultraviolet enhancement observed by means of sudden frequency deviations is compared with hard-X-ray observations for the Hα class 2B flare of 1816 U.T. 1968 August 8, when the ratio of the enhancement of energy flux in the wavelength range 1 - 1030 Å to the hard-X-ray energy flux was about 10^4.

076.034 Solar EUV enhancements associated with flares.
L. A. Hall, H. E. Hinteregger.
Solar Flares and Space Research, Tokyo 1968, p. 81 - 86 (1969).

An extreme ultraviolet spectrometer aboard the OSO-III satellite has been used to measure temporal variations of solar extreme ultraviolet in the wavelength range from 1300 Å to 260 Å over a six month period.

076.035 X-ray observations of solar flares.
H. Friedman.
Solar Flares and Space Research, Tokyo 1968, p. 87 - 94 (1969). – Review paper.

076.036 Observations of the solar flare soft X-ray spectrum and comparison with centimetric radio bursts.
W. M. Neupert, M. Swartz, W. A. White, R. M. Young.
Solar Flares and Space Research, Tokyo 1968, p. 95 - 101 (1969).

Solar X-ray and extreme ultraviolet emission spectra between 1 Å and 400 Å have been obtained over a period of greater than ten months by the OSO-III satellite. The observational data are discussed.

076.037 Enhancement of the solar X-ray spectrum below 25 Å during solar flares.
A. B. C. Walker, Jr., H. R. Rugge.
Solar Flares and Space Research, Tokyo 1968, p. 102 - 112 (1969).

The scanning Bragg crystal spectrometer flown aboard the satellite 1966-111 B observed the solar X-ray spectrum between 8 Å and 25 Å shortly after the occurrence of a class 3 flare on 13 February 1967.

076.038 The time structure of solar X-ray bursts above 7.7 keV.   H. S. Hudson, L. E. Peterson,
D. A. Schwartz.
Solar Flares and Space Research, Tokyo 1968, p. 113 - 120 (1969).

We discuss the processes which cause the solar X-ray emission.

076.039   Measurements of solar X-ray emission from the
OGO-IV spacecraft.
R. W. Kreplin, D. M. Horan, T. A. Chubb, H. Friedman.
Solar Flares and Space Research, Tokyo 1968, p. 121 - 130
(1969).

076.040   Observations of solar X-ray activity with a propor-
tional counter spectrometer on OSO-IV.
J. L. Culhane, P. W. Sanford, M. L. Shaw, K. A. Pounds,
D. G. Smith.
Solar Flares and Space Research, Tokyo 1968, p. 131 - 133
(1969). — Abstract.

076.041   Sudden increase in high energy Gamma-ray intensity
observed at balloon altitude.
I. Kondo, F. Nagase.
Solar Flares and Space Research, Tokyo 1968, p. 134 - 143
(1969).
The relation between a flare and a sudden increase in
high energy gamma-ray intensity is discussed.

076.042   Interpretation of time characteristics of solar X-ray
bursts.   T. Takakura.
Solar Flares and Space Research, Tokyo 1968, p. 144 - 150
(1969).

076.043   Theory of hard X-ray bursts.
R. Snijders.
Solar Flares and Space Research, Tokyo 1968, p. 151 - 154
(1969). — Note.

076.044   An investigation of solar X-ray and radio emissions
and their relationship to ionospheric phenomena.
A. Krüger, J. Taubenheim, G. Entzian.
Solar Flares and Space Research, Tokyo 1968, p. 181 - 193
(1969).
The development of the solar X-ray spectrum in dependence on the type of associated spot group is investigated.

076.045   Mitte-Rand-Variation von Intensität und Linien-
profil solarer UV-Linien.
G. Elwert, R. Mäckle.
Mitt. Astron. Ges. No. 27, p. 211 (1969). — Abstract AG.

Observations of the sun and laboratory sources
with a three-meter XUV spectrograph.
See Abstr. 034.037.

Observations and interpretation of the solar Lyman
continuum.   See Abstr. 071.042.

Fabry–Perot interferograms of the solar Mg II
doublet and XUV solar images obtained during a stabilized
Skylark rocket flight.   See Abstr. 071.064.

Solar flare optical, neutron and gamma-ray emission.   See Abstr. 073.001.

On the localization, size and structure of the regions of the X-ray flares on the sun.
See Abstr. 073.011.

Analysis of high-resolution X-ray photographs. I.
An importance 1N flare.   See Abstr. 073.038.

Thermal continuum radiation from coronal plasmas
at soft X-ray wavelengths.   See Abstr. 074.001.

The probabilities of some forbidden coronal lines
in the vacuum ultraviolet region.   See Abstr. 074.055.

The relation of sudden frequency deviations to the
spectrum and other characteristics of solar microwave bursts.
See Abstr. 077.009.

Correlation of solar microwave and soft X-ray
radiation. 2. The burst component.
See Abstr. 077.034.

077 Solar Radio Radiation

077.001 On the nature of the high-frequency cutoffs of type IV solar radio bursts. R. Ramaty.
Astrophys. Letters, Vol. 4, 43 - 45 (1969).

The high-frequency cutoff of synchrotron radiation resulting from electrons with an anisotropic pitch-angle distribution is investigated. It is shown that this effect could be responsible for the high-frequency cutoffs of type IV bursts at decimeter, meter and decameter wavelengths.

077.002 Observation of solar radio emission by the 22-m radio telescope at the Crimean Astrophysical Observatory at 2.25 mm and 8.15 mm wavelengths.
V. A. Efanov, A. G. Kislyakov, I. G. Moiseev, A. I. Naumov.
Solar Physics, Vol. 8, 331 - 340 (1969).

The results are given of observation of solar radio emission of the S-component at 8.15 mm-λ and 2.25 mm-λ made with the 22 m radio telescope of the Crimean Astrophysical Observatory. Solar radio images are obtained at both wavelengths. The data are presented of radio emission intensity and brightness temperatures of 10 sources of the S-component as well as the result of a flare observed. The sources of the S-component appear to be opaque at millimetre wavelengths.

077.003 Spectral characteristics of medium-sized solar radio events. A. D. Fokker.
Solar Physics, Vol. 8, 376 - 387 (1969).

Peak intensities at different frequencies, as reported by several solar radio patrol stations, are used to study the spectrum at the time of maximum intensity of medium-sized solar radio events that cover both the centimetric and metric frequency bands. Two types of spectrum can be distinguished: a V-type of spectrum, where the straight lines, that can be drawn to represent the centimetric and the metric branches, meet each other at a frequency somewhere in the decimetric frequency range and a Jump-type of spectrum, where a discontinuity occurs somewhere in the low-frequency part of the decimetric spectrum. The aspect of the radio response at 600 MHz may have a character which is more 'centimetric' or more 'metric'.

077.004 Type III radio bursts in the outer corona.
J. K. Alexander, H. H. Malitson, R. G. Stone.
Solar Physics, Vol. 8, 388 - 397 (1969).

Type III solar radio bursts observed from 3.0 to 0.45 MHz with the ATS-II satellite over the period April-October 1967 have been analyzed to derive two alternative models of active region streamers in the outer solar corona. Assuming that the bursts correspond to radiation near the electron plasma frequency, 'pressure equilibrium' arguments lead to streamer model I in which the streamer electron temperature derived from collision damping time falls off much more rapidly than in the 'average' corona and the electron density is as much as 25 times the average coronal density at heights of 10 to 50 solar radii ($R_\odot$). In model II the streamer electron temperature is assumed to equal the average coronal temperature, giving a density enhancement which decreases from a factor of 10 close to the sun to less than a factor of two at large distances ($> \frac{1}{4}$ AU).

077.005 Some features of the solar microwave emission and their connection with geomagnetic activity. III: Sunspots and geomagnetic activity. J. Roosen.
Solar Physics, Vol. 8, 450 - 463 (1969).

The distinction between faint and bright sources of the slowly varying component of the solar microwave emission (paper II) is applied to statistics of the geomagnetic activity. Superposed epoch diagrams (recurrent disturbances excluded)

show that the increased number of disturbed days after the CMP of sunspot groups can almost exclusively be ascribed to the spot groups associated with faint sources. The variation of the disturbance amplitude with the heliographic latitude of the spot groups is discussed. A tentative model for the solar-wind enhancement associated with a faint source is presented.

077.006 A contribution to the model of the slowly varying component of the sun derived from observations at 8.6 mm. G. Feix.
Astrophys. Journ. Vol. 157, 903 - 912 (1969).

Various authors have measured the degree of polarization of the slowly varying component of the sun and have derived values of 20 - 30 percent at λ = 3.2 cm, < 10 percent at λ = 7.5 cm, and < 2 percent at λ = 21 cm. We have found at λ = 8.6 mm a degree of polarization of ≈ 2 percent originating from extended plage areas. The observations were carried out by means of a pencil-beam antenna approximately $3\overset{.}{.}6 \times 3\overset{.}{.}1$ in size. Measured degree of polarization is a lower limit only because of the large beam width of the antenna which averages out left-circularly polarized and right-circularly polarized components. The data obtained are analyzed in accordance with the magneto-ionic theory; an electron temperature of 8500° K is found for layers below 10000 km.

077.007 Strip-scans of the S-component of the solar radio emission published in the "Solar Data".
V. N. Ikhsanova.
Solnechnye Dannye Bull. No. 2, p. 85 - 87 (1969).
In Russian.

Conditions for obtaining the S-component strip-scans of solar radio emission and a method of their reduction are given.

077.008 On the dependence of specific characteristics of solar radio bursts on the phase of the solar cycle.
A. Krüger, F. Fürstenberg.
Solnechnye Dannye Byull. No. 3, p. 97 - 101 (1969).

On the basis of extensive data obtained from observations which had been made during the last solar cycle at the Heinrich-Hertz Institute, the morphological and spectral properties of the bursts in cm- and dm-ranges were investigated in dependence on the phase of the solar cycle. The relations with the other phenomena of solar activity were discussed. The results were compared with tendencies of the development of the current solar cycle.

077.009 The relation of sudden frequency deviations to the spectrum and other characteristics of solar microwave bursts. F. M. Strauss, M. D. Papagiannis, J. Aarons.
Journ. Atmosph. Terr. Phys. Vol. 31, 1241 - 1249 (1969).

Sudden Frequency Deviation (SFD) events were correlated with solar radio bursts observed at five different frequencies (606, 1415, 2695, 4995 and 8800 MHz). The best single frequency correlation was found at 4995 MHz. Using the spectral classification of AFCRL for microwave bursts, we have obtained the highest correlation for bursts with a spectrum showing an intensity maximum in the observing frequency range (606 – 8800 MHz) and the lowest correlation for bursts with intensity spectra decreasing with frequency. The correlation increased with the intensity of the bursts for all spectral classes.

077.010 Unpolarized impulsive solar bursts observed at 7 GHz. P. Kaufmann.
Solar Physics, Vol. 9, 166 - 172 (1969).

The occurrence of very faintly polarized, or unpolarized impulsive bursts observed at 7 GHz is discussed. It appears that some of them show a peculiar spectral peak somewhere between 5 GHz and 7 GHz. Possible interpretations are suggested, emphasizing the need to associate to the burst the state of polarization of the S-component in which it occurred.

077.011 A magnetohydrodynamic approach for interpreting solar polarization bursts at 7 GHz.
O. T. Matsuura.
Solar Physics, Vol. 9, 173 - 182 (1969).

A magnetohydrodynamic (MHD) approach is presented that appears to be comprehensive for the interpretation of the recently discovered microwave solar events, in which only the degree of circular polarization changes, without any increase in the output of the total solar flux. On the basis of this explanation experimental evidence is suggested for Alfvén waves, in relation to the velocity fields in the solar chromosphere.

077.012 On the origin of type II and IV radio sources during flares observed by a radioheliograph on 80 MHz. L. Krivský.
Solar Physics, Vol. 9, 194 - 197 (1969). − Research note.

077.013 Flare-time sudden enhancements of low frequency field strength and associated meter wave solar radio bursts. S. K. Alurkar, R. V. Bhonsle.
Solar Physics, Vol. 9, 198 - 204 (1969).

A number of meter wavelength solar radio bursts of spectral type-III have been observed by means of a solar radio spectroscope (40 - 240 MHz) simultaneously with sudden enhancements of low frequency (164 KHz) field strength (SES's) of Radio Tashkent which are known to take place due to the enhancements of D-layer ionization caused by flare-time solar X-rays. The association between the solar X-ray flares as detected by the SES's and the type-III meter-wave solar bursts is discussed. It is found that the association of SES's and meter wave solar bursts, which implies the ejection of flare-time electrons towards the photosphere as well as corona, is about 72%.

077.014 Absolute calibration method and technique of the daily patrol of the solar flux density at 1470 MHz. J. Priese.
Solar Physics, Vol. 9, 235 - 240 (1969).

In the frequency range from 1 to 2 GHz discrepancies exist between the values of the solar flux density observed at several stations. In order to clarify the situation a new absolute calibration has been carried out by the HHI at 1470 MHz. The equipment used for this purpose consists of an electromagnetic horn with calculated gain and of a receiver, the temperature scale of which has been calibrated by comparison with the noise of a heated absorber. The noise temperatures of the absorber have been identified with the mean thermodynamic temperatures of the absorbing material. The calibration procedure of the receiver and the observation method of the sun's radiation are described.

077.015 Pioneer 6: Measurement of transient Faraday rotation phenomena observed during solar occultation.
G. S. Levy, T. Sato, B. L. Seidel, C. T. Stelzried, J. E. Ohlson, W. V. T. Rusch.
Science, Vol. 166, 596 - 598 (1969).

Pioneer 6 was occulted by the sun in the last half of November 1968. During the period in which the spacecraft was occulted by the solar corona, the S-band telemetry carrier underwent Faraday rotation as a result of this anisotropic plasma. The measurement of these phenomena indicated that Faraday rotation on the order of 40 degrees occurred. The duration of each phenomenon was approximately 2 hours. These phenomena appear to be correlated with observations

of solar radio bursts with wavelengths in the dekametric region.

077.016 Superior conjunction of Pioneer 6.
R. M. Goldstein.
Science, Vol. 166, 598 - 601 (1969).

Spectrograms of the radio signals from Pioneer 6 were taken as the spacecraft was occulted by the sun. The spectral bandwidths increased slowly at first, then very rapidly at 1 degree from the sun. In addition, six solar "events" produced marked increases of bandwidth lasting for several hours. The received signal power seemed unaffected by the solar corona.

077.017 Satellite observations of solar radio bursts to kilometric wavelengths.
H. H. Malitson, J. Fainberg, R. G. Stone, J. K. Alexander.
Bull. American Astron. Soc. Vol. 1, 252 - 253 (1969). − Abstr. AAS.

077.018 Summary of the first three months of solar monitoring at 3.3 mm. F. I. Shimabukuro.
Bull. American Astron. Soc. Vol. 1, 261 (1969). − Abstr. AAS.

077.019 Statistical considerations of centimeter-wavelength solar-burst directivity.
J. P. Castelli, W. R. Barron.
Bull. American Astron. Soc. Vol. 1, 274 (1969). − Abstr. AAS.

077.020 Solar radio spectra of type-IV bursts.
E. G. Howard.
Bull. American Astron. Soc. Vol. 1, 279 (1969). − Abstr. AAS.

077.021 Gyrosynchrotron radiation and solar microwave bursts. R. Ramaty.
Bull. American Astron. Soc. Vol. 1, 290 - 291 (1969). − Abstr. AAS.

077.022 Time dependence of Razin spectra in type-IV solar radio bursts. M. Simon.
Bull. American Astron. Soc. Vol. 1, 293 (1969). − Abstr. AAS.

077.023 Computer processing of solar radio maps and television images of the sun.
C. K. Sourk, T. J. Janssens.
Bull. American Astron. Soc. Vol. 1, 293 (1969). − Abstr. AAS.

077.024 Messungen der solaren Radiostrahlung.
H. Daene.
Mitt. Astrophys. Obs. Potsdam, No. 314 - 315, p. 109 - 204 (1969). − 1962 January − 1963 December.

077.025 Investigation of faint solar bursts at λ = 4.0 cm.
G. B. Gelfreich, N. P. Stasjuk.
Solnechnye Dannye 1969 Byull. No. 7, p. 100 - 107 (1969). In Russian.

Faint solar bursts were observed using a radio interferometer with small base. The connection of very faint radio bursts (∼1% of the total flux) with other manifestations of solar flares was investigated. The increase of burst numbers with a decrease of their amplitude can be found up to the amplitude of 2 - 3%; the bursts with an amplitude less than 1% are observed very rarely. Altogether 66% of faint bursts are not connected with flares, whereas 70% of flares do not give rise to bursts. The analogous non-conformity takes place also in the bursts of the meter wavelength range.

077.026 **Polarization of solar radio bursts at 3.2 cm and its connection with the magnetic field at photospheric level.** V. P. Nefedjev, L. E. Treskova.
Solnechnye Dannye 1969 Byull. No. 7, p. 107 - 113 (1969). In Russian.

27 microwave bursts observed simultaneously with chromospheric flares are investigated. A comparison of the direction of the magnetic fields in the Hα bright points with the sign of microwave burst polarization allows the conclusion that a microwave burst is generated in the region with a strong magnetic field of sunspots.

077.027 **Radio spectra and related observations of a solar active region in July 1968.**
J. M. Pasachoff, J. P. Castelli.
Bull. American Astron. Soc. Vol. 1, 289 (1969). – Abstr. AAS.

077.028 **Pencilbeam observation of solar bursts at 36 GHz.** G. Feix.
Solar Physics, Vol. 9, 265 - 268 (1969).

From a burst survey at 36 GHz, the diameter of the burst core was always found less than 1'. Several limb bursts with a remarkable flash intensity have been observed. Comparison of corresponding bursts on the disk exhibit in general a recognizable post-increase phase which seems to become faint at the solar limb.

077.029 **The radio-emission and Ca-brightness of two outstanding active regions during their lifetime.**
G. Feix.
Solar Physics, Vol. 10, 184 - 187 (1969). – Research note.

077.030 **The mm wave outbursts of November 1 and 2, 1968.** M. Anastassiades, C. Macris.
Solar Physics, Vol. 10, 188 - 195 (1969).

During the period October 22 to November 4, 1968 an important active region ($\varphi = S15°$, $L = 173°$) on the solar disk presented particular activity. The November 1 and 2 flares, of importance 2b, produced with a difference in time of almost 24 hours, were associated with radio bursts recorded at Athens in both the cm and mm bands. The events are discussed in detail.

077.031 **Spectrum and polarization of solar radio bursts on a 10-millisecond time scale.**
J. W. Warwick, G. A. Dulk.
Astrophys. Journ. (*Letters*), Vol. 158, L123 - L125 (1969).

We have recently observed several common types of solar radio bursts between 24 and 37 MHz, with swept-frequency spectral resolution of 60 kHz, polarization (linear, right circular, and left circular), and time structure (on a scale of about 10 msec). We find that many bursts exhibit an unexpectedly complex time, frequency, and polarization structure.

077.032 **Solar microwave burst from behind-the-limb proton flare.** V. L. Badillo, J. E. Salcedo.
Nature, Vol. 224, 503 - 504 (1969).

On 30 March 1969 Manila Observatory recorded an eruptive prominence on the west limb at 0302 U.T. The EPL came from McMath region 9994 located 18°N and 15° behind the west limb. The peak flux densities (in units of 10^{-22} Wm^{-2} Hz^{-1}) of the radio burst are among the largest recorded in the cm range: 44,000 flux units at 8800 MHz; 39,000 for 4995; 35,000 for 2695; 9,200 for 1415 and 4,800 for 606.

077.033 **The frequency splitting of transient solar radio bursts.** G. R. A. Ellis.
Proc. Astron. Soc. Australia, Vol. 1, 273 - 274 (1969). – Contribution ASA meeting.

077.034 **Correlation of solar microwave and soft X-ray radiation. 2. The burst component.**
C. D. Wende.
Journ. Geophys. Res. Vol. 74, 6471 - 6481 (1969).

This paper is limited to a study of the quasi-thermal bursts. A flare model is developed in which the peak temperature and angular size of the flare can be determined from the correlation of the X-ray and microwave fluxes.

077.035 **Study of the slowly varying component of the solar radio emission.** G. B. Gel'frejkh.
Vestn. AN SSSR, No. 4, p. 46 - 54 (1969). In Russian. – Abstr. in Referativ. Zhurn. 51. Astron., 10.51.447 (1969).

077.036 **The spectral features of the S-component of the solar radio emission.**
A. T. Nesmyanovich, Y. A. Homenko.
Problemy kosmich. fiz. No. 4, p. 13 - 27 (1969). In Russian.

A statistical study of flux density measurements of solar radio emission at Toyakawa (Japan) is given. The B component does not depend on the phase of the 11-year solar cycle. The spectra of B- and S-components are obtained and their connections with flocculi, sunspots and solar activity are discussed.

077.037 **Observations on the time structure of solar radio bursts at a wavelength of 12 m.**
Ch. V. Sastry.
Solar Physics, Vol. 10, 429 - 432 (1969).

077.038 **Observation of the magnetic structure of a type IV solar radio outburst.** J. P. Wild.
Solar Physics, Vol. 9, 260 - 264 (1969).

A continuous record of the 80 MHz image and polarization of a type IV solar outburst has been made with the Culgoora radioheliograph from which the magnetic structure of the event can be directly inferred. The radio-emitting arch appears to lie above an eruptive prominence seen in Hα. The stationary part is seen later as a separate highly polarized source on the disk above the projected position of the flare that had previously triggered the prominence activity.

077.039 **Regions of enhanced emission on the sun, observed at 1.2-mm wavelength.** J. E. Beckman.
Bull. American Astron. Soc., Vol. 1, 334 (1969). – Abstract AAS.

077.040 **High time-resolution observations of fine structure concurrent with type-III bursts.**
K. W. Philip.
Bull. American Astron. Soc., Vol. 1, 359 - 360 (1969). Abstr. AAS.

077.041 **On the role of the compression and expansion of the generative region in the synchrotron radio emission.**
S. I. Gopasyuk, N. N. Erjushev, Y. I. Neshpor.
Izv. Krymskoj Astrofiz. Obs. Vol. 39, 299 - 316 (1969). In Russian.

The observed data on bursts of the eruptive type during solar flares in the centimeter region are analysed. It is shown that the explanation of the decreasing phase of these bursts by ionization losses only makes difficulties. The temporal variation of the synchrotron flux density of relativistic electrons with the power spectrum, allowing for ionization losses, is discussed. It is supposed that the generative region of this emission goes at first through a homogeneous compression and after that through the expansion.

077.042 **A catalogue of unusual phenomena in solar radio emission at the frequency of 208 MHz during IGY and IQSY.**

I. G. Moiseev, L. I. Yurovskaya, Yu. F. Yurovsky.
Izv. Krymskoj Astrofiz. Obs. Vol. 39, 325 - 537 (1969).
In Russian.

The catalogue contains 622 bursts of the solar radiation (the frequency is 208 MHz), recorded at the Crimean Astrophysical Observatory of the USSR Academy of Sciences during the IGY and IQSY from July 1957 to December 1959. Bursts are represented as copies from original recordings of the general solar flux density.

077.043 **Solar radio emission during the International Geophysical Year.** S. F. Smerd.
Ann. International Geophys. Year, Vol. 34, 1 - 357 (1969).

077.044 **Flare-associated radio bursts.**
T. Takakura.
Solar Flares and Space Research, Tokyo 1968, p. 165 - 180 (1969).

The solar radio measurements based on ground observations and the observations of X-rays and solar particles in space combine to a better understanding of the flare-associated phenomena.

077.045 **Spectral considerations of microwave solar bursts.**
J. P. Castelli, J. Aarons, G. A. Michael, C. Jones, H. C. Ko.
Solar Flares and Space Research, Tokyo 1968, p. 194 - 201 (1969).

The correlation of single-frequency centimeter radio bursts with flares and associated active regions investigated by others is extended in the present paper.

077.046 **Ricerche sulla natura dei due outbursts solari osservati a Trieste il 13 gennaio 1968.**
A. Abrami.
Atti XII Riunione Soc. Astron. Italiana, L'Aquila 1968, p. 97 - 99 (1969). – Abstract SAI.

077.047 **Measurements of the spectral slope of the solar radio emission at 3.3 cm during the eclipse of 22 September, 1968.** I. F. Belov, M. M. Kobrin, A. I. Korshunov, B. V. Timofejev.
Astron. Tsirk. No. 529, p. 5 - 6 (1969). In Russian.

077.048 **Periodic fluctuations of the fluxes of local radio sources of the sun.** G. B. Gelfreikh, O. G. Derevjanko, A. N. Korzhavin, N. P. Stasjuk.
Solnechnye Dannye 1969 Byull. No. 9, p. 88 - 94 (1969). In Russian.

Fluctuations of the solar emission flux were investigated using a radio interferometer with the small base at 4 cm. Construction of autocorrelation functions allows to detect fluctuations with periods of 10 – 15 min and more, and amplitudes up to 0.5 – 0.7% of the undisturbed sun. The periodic fluctuations of the flux are connected with separate local sources. These fluctuations are supposed to be due to proper oscillations of a coronal condensation, i.e. a condensed plasma.

077.049 **Report on the solar type IV radio burst activity observed in July 1968.**
A. Böhme, A. Krüger, H. Künzel.
Heinrich-Hertz-Inst. für Solar-Terrestrische Physik, Deutsche Akad. Wiss. Berlin, HHI Suppl. Ser. Solar Data, Vol. 1, (No. 5), 79 - 99 (1969).

In this paper the observations of the HHI are collected with main emphasis to the radio phenomena happened on the sun during that period.

077.050 **Remarks concerning the classification of the type IV bursts during the second PFP period 1966.**
A. Böhme.

Heinrich-Hertz-Inst. für Solar-Terrestrische Physik, Deutsche Akad. Wiss. Berlin, HHI Suppl. Ser. Solar Data, Vol. 1, (No. 6), 101 - 124 (1969).

077.051 **Messungen der solaren Radiostrahlung.**
H. Daene.
Mitt. Astrophys. Obs. Potsdam, No. 317 - 318, p. 61 - 151 (1969). – 1966 January – 1967 December.

077.052 **327 MHz solar radio observations.**
F. S. Delli Santi, E. Nasi.
Oss. Astron. Univ. Bologna, Notizie Rassegne, Nos. 29 - 37 (1969). – 1968 March – 1969 August.

077.053 **Observations of the quasi-periodic components in the 8-mm solar radio emission.**
M. S. Durasova, M. M. Kobrin, Ja. B. Losovsky, A. K. Chandaev, O. I. Yudin.
Astron. Tsirk. No. 531, p. 1 - 3 (1969). In Russian.

077.054 **On the correlation of the quasi-periodic components in fluctuations of the solar radio emission at 9100 and 9300 MHz.**
I. F. Belov, M. M. Kobrin, A. I. Korshunov, B. V. Timofejev, A. K. Chandaev.
Astron. Tsirk. No. 531, p. 3 - 5 (1969). In Russian.

077.055 **On the method of reduction of the solar radio emission scannings on the large Pulkovo radio-telescope at wavelengths shorter than the critical one.**
V. G. Nagnibeda.
Vestn. Leningr. un-ta, No. 13, p. 150 - 158 (1969). In Russian. – Abstr. in Referativ. Zhurn. 51. Astron., 2.51.97 (1970).

077.056 **Karten der solaren Radiostrahlung bei 8.6 mm Wellenlänge.** E. Fürst.
Mitt. Astron. Ges. No. 27, p. 213 - 215 (1969). – Conference paper.

077.057 **Untersuchungen solarer Bursts im Mikrowellengebiet anhand von Intensitäts- und Polarisationsmessungen bei 17 GHz.** W. Wassenberg.
Mitt. Astron. Ges. No. 27, p. 216 - 218 (1969). – Conference paper.

077.058 **Das Abklingen von Energie und Strahlungsleistung im solaren Plasma infolge von Synchrotronstrahlungsverlusten.** W. Hirth.
Mitt. Astron. Ges. No. 27, p. 218 - 221 (1969). – Conference paper.

077.059 **Coherent synchrotron deceleration and the emission of type II and type III solar radio bursts.**
A. F. Kuckes, R. N. Sudan.
Nature, Vol. 223, 1048 - 1049 (1969).

077.060 **Observations radioélectriques solaires faites sur 600 MHz en 1967 au laboratoire de radioastronomie de Humain-Rochefort.** C. Delys, R. Gonze.
Bull. Astron. Obs. Roy. Belgique, Vol. 6 (No. 8), 329 - 341 (1969).

Great Radiobursts of the Sun.
See Abstr. 003.012.

Radiation fields of energetic electrons in helical orbits within a magnetoactive plasma. See Abstr. 062.003.

The coarse structure of the solar atmosphere.
See Abstr. 073.043.

Temperature determination in the corona from a time profile of radio bursts of type III.
See Abstr. 074.063.

Correlation of solar microwave and soft X-ray radiation. 1. The solar cycle and slowly varying components.
See Abstr. 076.008.

Observations of the solar flare soft X-ray spectrum and comparison with centimetric radio bursts.
See Abstr. 076.036.

Results of observing the partial solar eclipse on September 22, 1968 at 3.15 cm.
See Abstr. 079.103.

078 Solar Cosmic Radiation

078.001 Relative abundance of iron-group nuclei in solar cosmic rays.
D. L. Bertsch, C. E. Fichtel, D. V. Reames.
Astrophys. Journ. *(Letters)*, Vol. 157, L53 - L56 (1969).

The abundance of the iron-group nuclei relative to oxygen in a solar cosmic-ray event has been determined for the first time in the event of September 2, 1966; it was found to be $(1.1 \pm 0.3) \times 10^{-2}$ above 24.5 MeV nucleon^{-1}. This ratio is consistent with the solar value determined spectroscopically but is over an order of magnitude smaller than the galactic cosmic-ray ratio. This result is in agreement with the concept already evolving from measurements on other nuclei that the relative abundances of solar cosmic rays reflect those of the solar photosphere for multicharged nuclei with approximately the same nuclear charge-to-mass ratio.

078.002 Increase in cosmic-ray cutoffs at high latitudes during magnetospheric substorms. J. R. Barcus.
Journ. Geophys. Res. Vol. 74, 4694 - 4700 (1969).

Balloon observations obtained at Byrd Station, Antarctica ($L \approx 7$), are presented of time variations in the nuclear γ-ray flux produced by proton bombardment of the terrestrial atmosphere during the January 28 and February 2, 1967 solar proton events. The observations reveal evidence for transient increases in the proton cutoff during the local evening and morning hours, and these changes are associated with the occurrence of modest magnetospheric substorms.

078.003 Neutrinos from the sun. J. N. Bahcall.
Sci. American, Vol. 221, No. 1, p. 29 - 37 (1969).

A giant trap has been set deep underground to catch a few of the neutrinos that theory predicts should be pouring out of the sun. Their capture would prove that the sun runs on thermonuclear power.

078.004 Remarks on the solar corpuscular radiation based on space data. A. S. Dvoryashin.
Astrometriya i Astrofiz., *Kiev*, No. 4, p. 7 - 39 (1969).
In Russian.

Data on corpuscular radiation obtained from geophysical observations and space experiments are considered. The existence of the stable four-stream structure that causes four consequences of geomagnetic disturbances has been established. Every stream carries a magnetic field mainly of one polarity. The four-stream model of the interplanetary magnetic field with definite boundaries which exists during some revolutions of the sun may be constructed.

078.005 On the solar corpuscular radiation obtained from measurements with satellites and rockets. The four-stream model of the solar corona near the minimum of activity. A. S. Dvoryashin.
Astrometriya i Astrofiz., *Kiev*, No. 4, p. 40 - 53 (1969).
In Russian.

The four-stream model of the solar corona was constructed for a period near the minimum of solar activity using the data from satellite and rocket observations on the plasma and field. The peculiarities of the magnetic disturbances of the earth's magnetic field may be explained by the complex structure of coronal rays.

078.006 Access of solar protons into the polar cap: A persistent north-south asymmetry.
L. C. Evans, E. C. Stone.
Journ. Geophys. Res. Vol. 74, 5127 - 5131 (1969).

Before the magnetic storm sudden commencement during the November 2, 1967, solar particle event, the access of 1.2- to 40-MeV protons into the high-latitude portion of the northern polar region was delayed by $\sim$ 20 hours. At the same time the access delay for 10- to 40-MeV protons was $\lesssim$ 1 hour in the southern polar region and at middle northern latitudes. The implications of the north-south asymmetry are discussed.

078.007 Electrons and protons in long-lived streams of energetic solar particles. K. A. Anderson.
Bull. American Astron. Soc. Vol. 1, 270 - 271 (1969). – Abstr. AAS.

078.008 Iron nuclei in solar cosmic radiation.
D. L. Bertsch, C. E. Fichtel, D. V. Reames.
Bull. American Astron. Soc. Vol. 1, 273 (1969). – Abstr. AAS.

078.009 Solar circumstances at the time of the cosmic-ray increase on 28 January 1967.
H. W. Dodson, E. R. Hedeman.
Bull. American Astron. Soc. Vol. 1, 275 (1969). – Abstr. AAS.

078.010 Solar cosmic-ray activity in 1968. A. J. Masley.
Bull. American Astron. Soc. Vol. 1, 286 (1969). – Abstr. AAS.

078.011 Observation of a transient anisotropic increase during the onset stage of a Forbush decrease.
R. A. R. Palmeira, R. P. Bukata, P. T. Gronstal.
Planet. Space Sci. Vol. 17, 1913 - 1918 (1969). – Research note.

078.012 Solar circumstances at the time of the cosmic ray increase on January 28, 1967.
H. W. Dodson, E. R. Hedeman.
Solar Physics, Vol. 9, 278 - 295 (1969).

Solar circumstances have been evaluated for January 28, 1967, the date of an observed ground level enhancement of cosmic rays which was not preceded by observation of a suitably great Hα flare. On the visible solar hemisphere, a bright subflare at S23° E19° occurred in appropriate time association with the cosmic ray event, and was accompanied by weak X-ray enhancement and radio frequency emission. If this flare, alone, or in combination with other minor flares observed on the visible hemisphere on January 28 was the source of the energetic cosmic rays recorded on that date, then current thinking regarding the characteristics of cosmic ray flares must be modified.

078.013 Confinement of solar flare cosmic rays to sectors of the corotating solar magnetic field.
J. Feit.
Journ. Geophys. Res. Vol. 74, 5579 - 5589 (1969).

A method is presented for solving the problem of confining solar flare particles to definite sectors of the solar magnetic field by the use of Laplace transform coordinates.

078.014 Streaming of solar cosmic rays.
K. G. McCracken, I. D. Palmer.
Proc. Astron. Soc. Australia, Vol. 1, 276 - 278 (1969). – Contribution ASA meeting.

078.015 Advance warning of proton emission from the sun.
J. B. Blizard.
Bull. American Astron. Soc., Vol. 1, 336 (1969). – Abstract AAS.

078.016 **Solare Protonen niedriger Energie.**
V. N. Lutsenko, N. F. Pisarenko.
Kosmich. Issled. Vol. 7, 711 - 730 (1969). In Russian.

078.017 **Modeling of solar cosmic ray events based on recent observations.** M. B. Baker, R. E. Santina, A. J. Masley.
AIAA Journ., Vol. 7, 2105 - 2110 (1969).
In order to assess the magnitude of a solar cosmic ray event after the first particles have been detected, solar cosmic ray events models have been developed which relate a spectral parameter to a dose parameter. Spectrometer data obtained early in an event would be used to make projections of the total expected event dose and other dose parameters. Two models were considered.

078.018 **Solar cosmic radiation.**
L. I. Miroshnichenko.
Zemlya i Vselennaya, No. 6, p. 31 - 37 (1969). In Russian.

078.019 **A method of solar protons spectrum determination.**
Dj. Heristchi, E. Barouch, P. Masse.
Nuclear Instruments and Methods, Vol. 71, 353 - 357 (1969).
See Phys. Abstr. Vol. 72, No. 45685 (1969).

078.020 **11-year modulation of cosmic ray intensity and distribution of spots in heliographic latitude.**
Yu. I. Stozhkov, T. N. Charakhchyan.
Geomagn. Aeronom. Vol. 9, 803 - 808 (1969). In Russian.

078.021 **Solar cosmic-ray flares deduced from the developing patterns of SID's.** K. Sakurai.
Solar Flares and Space Research, Tokyo 1968, p. 155 - 162 (1969).
Solar cosmic-ray flares are usually associated with sudden ionospheric disturbances (SID) such as SWF, SEA and SFD.

078.022 **High energy particle events associated with solar flares.** K. G. McCracken.
Solar Flares and Space Research, Tokyo 1968, p. 202 - 214 (1969). – Review paper.

078.023 **Observations of solar protons aboard OV3-3 and ATS-1.** J. B. Blake, G. A. Paulikas, S. C. Freden.
Solar Flares and Space Research, Tokyo 1968, p. 258 - 266 (1969).

078.024 **Measurement of cosmic ray anisotropy of solar origin by Explorer 34 satellite.**
U. R. Rao, F. R. Allum, W. C. Bartley, R. A. R. Palmeira, J. A. Harries, K. G. McCracken.
Solar Flares and Space Research, Tokyo 1968, p. 267 - 276 (1969).

078.025 **The charge composition of solar cosmic rays and solar abundances.**
D. V. Reames, C. E. Fichtel.
Solar Flares and Space Research, Tokyo 1968, p. 277 - 278 (1969). – Note.

078.026 **The 28 January 1967 solar cosmic ray event and related complications.** A. J. Masley.
Solar Flares and Space Research, Tokyo 1968, p. 279 - 283 (1969).

078.027 **The 23 and 28 May 1967 solar cosmic ray events.**
A. D. Goedeke, A. J. Masley.
Solar Flares and Space Research, Tokyo 1968, p. 284 - 293 (1969).

078.028 **Solar corpuscular streams and the general circulation of the earth's atmosphere.**
E. R. Mustel.
Astron. Tsirk. No. 530, p. 1 - 5 (1969). In Russian.

078.029 **Le regioni solari sedi di eventi protoni negli anni 1965, 1966, 1967.** T. Fortini, M. Torelli.
Oss. Astron. Roma, Contr. Sci., Ser. III, No. 81, 7 pp. (1968).

078.030 **Observations of the solar particle event of 7 July 1966 with University of Iowa detectors.**
T. P. Armstrong, S. Krimigis, J. A. van Allen.
Ann. International Quiet Sun Year, Vol. 3, 329 - 336 (1969).

078.031 **Cosmic ray acceleration by corpuscular streams and the characteristics of the solar wind.**
L. I. Dorman, N. S. Kaminer, T. V. Kebuladze.
Geomagn. Aeronom. Vol. 9, 617 - 624 (1969). In Russian.
The authors investigate the connection between intensity increases of the cosmic radiation before Forbush-decrease and the characteristics of the plasma and the magnetic field of the interplanetary medium. The amplitude of the increase depends on the velocity of the corpuscular stream, on the velocity of the undisturbed solar wind, and on the heliographic longitude of the chromospheric eruption. Estimates of the connection of these factors with the amplitude of the intensity increase are given.

078.032 **Propagation of protons with energies of 1.5 MeV, generated during solar flares.**
S. N. Vernov, E. V. Gorchakov, G. A. Timofeev.
Geomagn. Aeronom. Vol. 9, 961 - 967 (1969). In Russian.

078.033 **Measurements of solar protons with the satellite Molniya 1 on May 25, 1967.**
S. N. Vernov, I. N. Senchuro, M. V. Teltzov, P. I. Shavrin.
Geomagn. Aeronom. Vol. 9, 968 - 971 (1969). In Russian.

078.034 **Statistical analysis of intensity increases of cosmic rays before Forbush-effects.**
L. I. Dorman, N. S. Kaminer, T. V. Kebuladze.
Geomagn. Aeronom. Vol. 9, 1069 - 1071 (1969). In Russian. – Brief information.

078.035 **Some features of the energy spectrum of solar cosmic rays.** L. I. Miroshnichenko.
Cosmic Rays No. 10, Moscow, p. 50 - 55 (1969). In Russian.
Some features of the energy spectrum of solar cosmic rays are analysed. The role of various factors which determine the form of the spectrum near the earth's orbit (drift, anti-Fermi deceleration, convective transfer, diffusion, prolonged emission from the acceleration region) are examined. A number of flares are taken as an example to show that experimental data deviate from the exponential law of the rigidity spectrum of solar particles suggested by Freier and Webber. Probable causes of this disagreement are discussed.

078.036 **The structure of magnetic heterogeneity flows from the sun according to the data on recurrent disturbances in cosmic radiation and magnetic activity.**
L. I. Dorman, A. A. Luzov, G. V. Khrustselevskaya.
Cosmic Rays No. 10, Moscow, p. 156 - 161 (1969).
In Russian.
A detailed study of the 27-day recurrent events in cosmic radiation intensity and geomagnetic activity as observed during the period from July 11, 1957, to December 31, 1958 was made.

078.037 **A study of solar and cosmic radiation from the Venus 4 space probe.**

S. N. Vernov, A. E. Chudakov, P. V. Vakulov, E. V. Gorchakov, P. P. Ignatiev, N. N. Kontor, S. N. Kuznetsov, Yu. I. Lo-
gachev, G. P. Lyubimov, A. G. Nikolaev, N. V. Pereslegina.
Space Research IX, Proc. Tokyo 1968, p. 203 - 214 (1969).

**Flight time of solar fast particles from flares to the
earth. Supplement II.**   See Abstr. 073.005.

**The geomagnetic and cosmic-ray storm of May 25/
26, 1967.**   See Abstr. 084.201.

079 Solar Eclipses

079.001 A shadow-band experiment.
   R. D. Burgess, M. E. Hults.
Sky Telescope, Vol. 38, 95 (1969).

**Further investigations of solar X-ray sources using
D-layer ionization behavior during eclipses.**
See Abstr. 076.003.

**A new analysis of eclipse effect in the equatorial
F-region.**   See Abstr. 083.018.

079.100 Solar eclipse, 1962 February 5

**The flash spectrum observed at the total eclipse of
February 5, 1962.**   H. Kurokawa, S. Tominaga,
J. Kubota, I. Kawaguchi.
Publ. Astron. Soc. Japan, Vol. 21, 141 - 166 (1969).
   At the total solar eclipse of February 5, 1962, the flash
spectrum was observed at Lae, New Guinea. Logarithmic intensities of about 130 emission lines were obtained at several
positions on the east and west limbs of the sun in the wavelength range from $\lambda 5850$ Å to $\lambda 6563$ Å. A comparison of our
intensities with those previously published is made. Two types
of abnormal intensity gradients of the Hα and D3 lines were
found in the active region very near the west limb and the observed spectral features are described in some detail. A comparison is given between the flash spectrograms and the
white-light or the monochromatic images of the solar disk
before the eclipse.

**On the chromospheric observations at the 1962
eclipse.**   See Abstr. 073.008.

**Analysis of the chromospheric hydrogen spectrum
at the 1962 eclipse.**   See Abstr. 073.009.

079.101 Solar eclipse, 1969 September 11

September's solar eclipse.
Sky Telescope, Vol. 38, 161 - 162 (1969).

Observations of the September solar eclipse.
Sky Telescope, Vol. 38, 351 - 355 (1969).

079.102 Solar eclipse, 1966 May 20

**Electron content measurements by beacon S-66 sa-
tellite during the May 20, 1966, solar eclipse.**
M. Anastassiadis, D. Matsoukas.
Journ. Atmosph. Terr. Phys. Vol. 31, 1217 - 1222 (1969).
   During annular solar eclipse of May 20, 1966, over the
Northern part of Africa and the Southern part of Europe, it
was possible for the first time, to investigate the effect of an
eclipse on the total electron content of the ionosphere, by

using the satellite technique.

**Air temperature variations during a recent solar
eclipse.**   K. J. H. Phillips.
Journ. British Astron. Ass. Vol. 79, 460 - 466 (1969).

079.103 Solar eclipse, 1968 September 22

**Die Expedition zur Beobachtung der totalen Son-
nenfinsternis vom 22. September 1968.**
M. Waldmeier, S. E. Weber.
Astron. Mitt. Eidg. Sternw. Zürich, No. 287, 19 pp. (1969).

**Observations of the total solar eclipse of 1968,
September 22.**   V. V. Sachanov, S. B. Alexandrov,
V. A. Sachanova, V. N. Emeljanov.
Astron. Tsirk. No. 520, p. 7 - 8 (1969).   In Russian.

**The solar eclipse of 22 September 1968: Predic-
tion and observations.**   K. H. Schatten.
Bull. American Astron. Soc. Vol. 1, 292 (1969). – Abstr.
AAS.

**Radio observations of the partial solar eclipse of
1968 September 22.**   L. D. J. Harris.
Journ. British Astron. Ass., Vol. 80, 46 - 47 (1969).

**Ergebnisse der Expedition zur Beobachtung der
totalen Sonnenfinsternis vom 22. September 1968.**
M. Waldmeier.
Mitt. Astron. Ges. No. 27, p. 212 - 213 (1969). – Abstract
AG.

**Preliminary report on photographing the solar co-
rona in 5303 Å with a polaroid and a Fabry-Pérot interferometer during the total solar eclipse of September 22, 1968.**
A. Delone, E. Makarova.
Solar Physics, Vol. 9, 446 - 447 (1969). – Research note.

**Results of observing the partial solar eclipse on
September 22, 1968 at 3.15 cm.**
N. N. Erjushev, L. I. Tsvetkov.
Solnechnye Dannye 1969 Byull. No. 9, p. 95 - 101 (1969).
In Russian.
   The results of the partial solar eclipse observed on Sept.
22, 1968 at the wavelength of 3.15 cm are given. It has been
found that near the central meridian the radio source with
circularly polarized radiation is located, in practice, just
above a sunspot. For a spot group distant from the central
meridian the polarized radio sources are displaced in the plane
of the disk from the sunspots to the limb. The height of the
sources above the photosphere in this case is equal to more
than 10^4 km. The extension of the radio diameter above the
photosphere is equal to $(16-20) \times 10^3$ km. The fine structure of the polarized radio sources was discovered. Some details have a size less than 5″.

Die Korona bei der Sonnenfinsternis vom
22. September 1968. I. Die Form der Korona.
See Abstr. 074.079.

Measurements of the spectral slope of the solar
radio emission at 3.3 cm during the eclipse of 22 September,
1968.   See Abstr. 077.047.

079.104   Solar eclipse, 1965 May 30

Interferometric investigation of the red and green
coronal lines during the total solar eclipse of May 30, 1965.
See Abstr. 074.015.

079.105   Solar eclipse, 1970 March 7

Solar eclipse plans for March 7, 1970 by the
Astronomical League for U.S.A. stations and for Mexico.
R. C. Maag.
Strolling Astronomer, Vol. 21, 204 - 209 (1969).

079.106   Solar eclipse, 1970 August 31

Annular solar eclipse of 31 August - 1 September
1970.   J. S. Duncombe.
U.S. Naval Obs. Washington, Circ. No. 126, 21 pp. (1969).

080   Solar Figure, Internal Constitution, Rotation, Miscellanea

080.001   The Zeeman effect for weak magnetic fields.
J. O. Stenflo.
Solar Physics, Vol. 8, 260 - 263 (1969).

The polarization of a normal Zeeman triplet is discussed for the case in which the lifetime τ of the excited state of the atom is comparable to or shorter than the period of Larmor precession.

080.002   About the influence of inhomogeneities of magnetic fields on line contours and magnetographic measurements.   J. Staude.
Solar Physics, Vol. 8, 264 - 270 (1969).

The discrepancies between theoretical and experimental calibration curves for solar magnetographs (Severny, 1967) may be explained by horizontal inhomogeneities of the observed but not resolved magnetic field region. Using the Unno solution of the equations of transfer simple two-stream models have been constructed. For the more complicated case of a depth dependence of the magnetic field vector it is shown assuming pure absorption and permitting arbitrary variations of the magnetic field vector and the atmosphere model with depth that a solution of the equations of transfer may be found by iteration.

080.003   General magnetic field and rotation of the outer layers of the sun.   F. Unz, K. Walter.
Solar Physics, Vol. 8, 310 - 315 = Mitt. Astron. Inst. Univ. Tübingen, No. 117 (1969).

We investigate how the rotation of the outer layers of the sun will be influenced by a variable general magnetic field. Applying the resulting formulae to the spectroscopic observations of the velocity of rotation at the solar limb in middle and high latitudes, the variation of the rotational velocity during the cycle 1901 - 1912 as found by Newall and by Halm can be made to agree with modern views on the general magnetic field.

080.004   On the structure of the outer solar convection zone.   S. Mizuno, M. Nishida.
Publ. Astron. Soc. Japan, Vol. 21, 121 - 127 (1969).

The effect of the change in values of the parameters included in the mixing length theory on the structure of the outer solar convection zone is examined. The models of the solar convection zone are constructed, also, on the assumption adopted by Böhm and Stückl (1967), and are compared with their results.

080.005   An outline of solar physics.   J. H. Reid.
Irish Astron. Journ. Vol. 9, 6 - 11 (1969).

080.006   On the diffusion of solar magnetic fields. I.
V. V. Kassinsky.
Solnechnye Dannye Byull. No. 4, p. 76 - 81 (1969).
In Russian.

The diffusion of solar magnetic fields within the frame of the Brownian motion is considered. The extension of magnetic structures is discussed from the point of view of hierarchy of different scale diffusion mechanisms. An usual ohmic damping is the final stage. The conclusion on relative stability of the fine structure of the magnetic field ($\gtrsim$ 1000 km) inspite of the effect of transient fields (supergranulation and granulation) is drawn.

080.007   On variations of the rotation velocity of the sun with heliographical latitude and atmospherical depth.   Y. A. Solonsky.
Solnechnye Dannye Byull. No. 4, p. 100 - 105 (1969).
In Russian.

The dependence of the axial solar rotation velocity on different heliographical latitudes 0°, 5°, 10°, ... 80° and different depths in the atmosphere is investigated. A slight decrease of velocity with an increase of the depth near 25° latitude is noticed.

080.008   A Rossby-wave dynamo for the Sun, II.
P. A. Gilman.
Solar Physics, Vol. 9, 3 - 18 (1969).

We analyze in detail the dynamo action from a typical Rossby wave motion and compare it with the solar cycle. The field reversal process is similar in some respects to that put forth by Babcock. Toroidal fields are dragged up by vertical motions in the Rossby waves to form large-scale vertical fields, whose polarities alternate with longitude roughly like bipolar magnetic regions. Vertical fields of preferentially one polarity are carried toward the pole by the meridional motion in the wave to form an axisymmetric poloidal field. This poloidal field is then stretched out by the differential rotation into a new toroidal field of the opposite sign from the original. The poloidal field changes sign when the toroidal and bipolar region like fields are maximum, and vice versa.

080.009   The rotation of the solar atmosphere.
E. J. Weber.
Solar Physics, Vol. 9, 150 - 159 = Contr. Kitt Peak National Obs. No. 439 (1969).

A model of the solar atmosphere is presented in which we discuss the conservation of angular momentum for the two basic states in which the solar gas can be: namely, either confined by closed field lines or outflowing along open magnetic field lines.

080.010   Neutrino's van de zon.   R. J. Rutten.
Hemel en Dampkring, Vol. 67, 322 - 325 (1969).

080.011   Determining the rotation of the sun.
W. C. Livingston.
Astron. Soc. Pacific, Leaflet No. 484, 8 pp. (1969).

080.012   Solar irradiance measurements from a research aircraft.
M. P. Thekaekara, R. Kruger, C. H. Duncan.
Applied Optics, Vol. 8, 1713 - 1732 (1969).

Measurements of the solar constant and solar spectrum were made from a research aircraft flying at 11.58 km, above almost all of the highly variable and absorbing constituents of the atmosphere. A wide range of solar zenith angles was covered during six flights for over 14 h of observation. A new value of the solar constant, 135.1 mW cm^{-2}, has been derived, as well as a revised solar spectral irradiance curve for zero air mass.

080.013   A rapidly rotating core and solar neutrinos.
R. K. Ulrich.
Astrophys. Journ. Vol. 158, 427 (1969).

Solar models have been constructed with the hydrostatic equation perturbed by the centripetal acceleration from a rapidly rotating core. These models show that the observed low counting rate of solar neutrinos cannot be due to this cause.

080.014   Hydromagnetic solar spin down.
E. R. Benton, P. A. Gilman, D. Loper.
Bull. American Astron. Soc. Vol. 1, 273 (1969). – Abstr. AAS.

080.015 **Solar differential rotation and oblateness.**
A. Clark, Jr., J. H. Thomas, P. A. Clark.
Bull. American Astron. Soc. Vol. 1, 275 (1969). — Abstr.
AAS.

080.016 **Polar magnetic fields of the sun from 1960–68.**
R. Howard.
Bull. American Astron. Soc. Vol. 1, 280 (1969). — Abstr.
AAS.

080.017 **Solar rotation 1966–68.** W. C. Livingston.
Bull. American Astron. Soc. Vol. 1, 285 (1969). —
Abstr. AAS.

080.018 **H_3^+, an important missing continuous opacity source in the sun.** J. L. Linsky.
Bull. American Astron. Soc. Vol. 1, 251 (1969). — Abstr.
AAS.

080.019 **Calibration of solar magnetograms obtained with narrow-band birefringent filters.**
G. E. Brueckner, M. Hagyard.
Bull. American Astron. Soc. Vol. 1, 274 (1969). — Abstr.
AAS.

080.020 **A new solar model atmosphere.**
O. J. Gingerich.
Bull. American Astron. Soc. Vol. 1, 277 (1969). — Abstr.
AAS.

080.021 **Interaction of giant convection cells with the differential rotation.**
R. Davies-Jones, P. Gilman.
Bull. American Astron. Soc. Vol. 1, 282 (1969). — Abstr.
AAS.

080.022 **Excitation of the Rossby-type waves in solar atmosphere.** S. Kato, Y. Nakagawa.
Bull. American Astron. Soc. Vol. 1, 283 (1969). — Abstr.
AAS.

080.023 **Semi-empirical line blanketing for solar model atmospheres.** J. P. Mutschlecner, C. F. Keller.
Bull. American Astron. Soc. Vol. 1, 287 (1969). — Abstr.
AAS.

080.024 **The angular velocity of the solar gas.**
E. J. Weber.
Bull. American Astron. Soc. Vol. 1, 295 (1969). — Abstr.
AAS.

080.025 **Some statistical properties of the fine structure of the solar magnetic field and the influence of the resolving power on results of field measurements.**
V. M. Grigorjev.
Solnechnye Dannye 1969 Byull. No. 6, p. 77 - 85 (1969).
In Russian.

Two schemes of the field structure based on the statistical properties of the solar magnetic field distribution which permit to estimate the influence of the resolving power on the measurement of the longitudinal and transversal field components are considered. It is shown, that with the variation of the resolving power from 4″ to 18″ the average longitudinal component of the field strength is decreased by 1.5 times more than the transversal component.

080.026 **Variations of solar brightness depending on sunspot numbers.** G. A. Zakharova.
Solnechnye Dannye 1969 Byull. No. 6, p. 104 - 108 (1969).
In Russian.

From a comparison of variations in solar brightness with the number of sunspots it is shown that these variations caused by spot activity of the sun are real.

080.027 **On the differential rotation with height in the solar atmosphere.** W. C. Livingston.
Solar Physics, Vol. 9, 448 - 451 = Contr. Kitt Peak National Obs. No. 450 (1969).

Spectroscopic measurements of solar rotation having good height discrimination show no change in angular velocity through the photosphere layers but an increase of 8% for the Hα chromosphere (epoch 1968.9). Spectroscopic results in general are compared with measures made with tracers, i.e. sunspots, filaments, etc., and it is seen that the spectroscopic method always shows increased differential rotation with height, while tracers indicate none. A westward flowing wind is proposed that increases in velocity with height, but produces negligible movement to magnetic regions associated with tracers.

080.028 **Buoyancy and solar spin-down.**
J. L. Modisette, J. E. Novotny.
Science, Vol. 166, 872 - 874 (1969).

The transition of a cylinder of water with a vertical density gradient between convective and viscous spin-down occurs near the point at which the maximum buoyancy force equals the radial pressure gradient that is driving the convection. A similar analysis applied to the sun shows the maximum buoyancy force to be 1500 times the convective force. A rapidly rotating solar interior would not be damped by large-scale convection.

080.029 **Is the sun a magnetic rotator?** A. Severny.
Nature, Vol. 224, 53 - 54 (1969).

The photoelectric measurements of the magnetic field of the whole solar disk seen as a star during 1968 showed the periodicity in sign and strength of the field with the period twice as short as the period of solar rotation as if the sun behaved itself like rotating magnetic quadrupole. This implies the exsistence of the four-sector structure of the interplanetary magnetic field. The variations of the total field are rather in antiphase with the variations of total magnetic flux from sunspots.

080.030 **The early despinning of the sun.**
K. Schwartz, G. Schubert.
Astrophys. Space Sci. Vol. 5, 444 - 447 (1969).

It is shown that if the sun passed through a T Tauri stage, then a mass loss of only 15% would be sufficient to despin the sun to an angular velocity of $0(10^{-5} \text{rad/sec})$ at 10^7 years without the additional braking effect of an enhanced magnetic field. Thus the present sun could have a core rotating at most ten times faster than its surface.

080.031 **Magnetograph measurements with temperature-sensitive lines.** J. Harvey, W. Livingston.
Solar Physics, Vol. 10, 283 - 293 = Contr. Kitt Peak National Obs. No. 500 (1969).

Certain discrepancies between theoretical and empirical calibrations of magnetograph response are resolved by recognizing the existence of line profile changes in magnetic regions. Many of the photospheric lines commonly used for magnetic field measurements weaken greatly in magnetic regions outside of sunspots. Unless due account is made of the line profile change, the magnetograph measurements underestimate magnetic flux and field strengths.

080.032 **The maintenance of solar differential rotation by two-dimensional turbulence.** G. H. Nickel.
Solar Physics, Vol. 10, 472 - 475 (1969).

A numerical model has been made to test the theory that solar differential rotation is maintained by the 'Countergradient transport' of energy peculiar to two-dimensional turbulence. The results of one problem are presented, indicating that this model can represent the observed large-scale nature of the sun's surface.

080.033 The solar differential rotation and 'Rossby-type' waves.   S. Kato, Y. Nakagawa.
Solar Physics, Vol. 10, 476 - 493 (1969).
It has been suggested that the solar differential rotation might be maintained by nearly horizontal non-spherical convective circulation called the Rossby-type waves. In this paper, such Rossby-type waves which could be excited in the upper solar convection zone are considered, and the possibility of maintenance of the solar differential rotation by such waves is examined. A numerical estimate, in terms of the rate of conversion of the kinetic energy of such wave motions into the mean rotational motion, indicates this possibility.

080.034 A possible connection between N–S and E–W solar asymmetries.   G. Godoli, G. Poletto.
Solar Physics, Vol. 10, 494 - 495 (1969). – Research note.

080.035 The sun as a variable star.
R. Albrecht, H. M. Maitzen, K. D. Rakos.
Astron. Astrophys. Vol. 3, 236 - 242 (1969).
Small periodical variations of the solar B magnitude could be established. The siderial period of the light curve varies between 26.6 and 29.0 days as a simple function of the solar cycle. The light variation coincides with the variation of the general solar magnetic field. The spectrum variation in the far ultraviolet changes in the same way as found at the Ap magnetic variable stars. From the light variations and the variations of the maximum appearence of the sunspots during the same time interval, it seems that it would be necessary to postulate a solar subsurface source for the sunspots and a magnetic field with the siderial rotational period of 27.6 days.

080.036 Interior structure of the sun.   C. A. Rouse.
Nature, Vol. 224, 1009 - 1010 (1969).
Three arguments are given, in addition to the negative solar neutrino results, why the interior structure of the present sun derived from contemporary solar evolution calculations is propably very significantly different from the structure of the real sun. It is concluded that the structure of the sun is more complex than believed heretofore and more accurate models for the present sun must use real gas effects in the equation of state and opacity calculations, as well as consider the gravitational separation of high–Z elements and/or other possible sources of stellar energy production.

080.037 Solar neutrinos.
W. A. Fowler.
Cosmic Ray Studies, Bombay 1968, p. 245 - 258 (1969).

080.038 A theoretical solar model with metal-line blanketing.   D. F. Carbon.
Bull. American Astron. Soc., Vol. 1, 337 (1969). – Abstract AAS.

080.039 Über elektrische Sonnenschwingungen und ihre Wirkung.   W. O. Schumann.
Sitzungsber. Bayer. Akad. Wiss. Math.-Nat. Kl., Jahrgang 1968, p. 8* - 9* (1969). – Abstr.

080.040 The depths of the formation of some absorption lines in the solar atmosphere.   V. G. Buslavsky.
Izv. Krymskoj Astrofiz. Obs. Vol. 39, 317 - 324 (1969).
In Russian.
The effective depths of formation of the absorption lines ($\lambda\lambda$4808, 5250, 5302 (Fe I), 6103 (Ca I), 4554 (Ba II)) in the solar atmosphere used for the recording of the magnetic field

with the double magnetograph of the Crimean Astrophysical Observatory are determined.

080.041 Solar standard.
P. Dejonc.
Journ. Optical Soc. America, Vol. 59, 1536 (1969).
Abstract Meeting Optical Soc. America.

080.042 A large scale pattern in the solar magnetic field.
J. M. Wilcox, R. Howard.
Solar Flares and Space Research, Tokyo 1968, p. 327 - 328 (1969). – Abstract.

080.043 Neutrino spectroscopy of the sun.
G. T. Zatsepin, V. A. Kuzmin.
Cosmic Rays No. 10, Moscow, p. 75 - 77 (1969). In Russian.
A programme of solar neutrino spectroscopy is discussed; two sets of detectors are to be used.

080.044 On thermometry of the sun's interior in neutrino detection experiments.   V. A. Kuzmin.
Cosmic Rays No. 10, Moscow, p. 81 - 83 (1969). In Russian.
The possibilities of thermometry of the solar interior with radiochemical methods for solar neutrino detection are considered and the relevant uncertainties are discussed. It is examined whether the Davis experiment can be interpreted in terms of the upper limit of the solar central temperature.

080.045 Variation in the gravitational constant and the neutrino luminosity of the sun.   V. A. Kuzmin.
Cosmic Rays No. 10, Moscow, p. 84 - 86 (1969). In Russian.
The influence of the time variation of the gravitational constant, $G \sim t^{-n}$, upon the evolution and the structure of the interior regions of the sun is discussed. The structure of the central region of the sun and the B^8-, and N^{13}-, O^{15}-neutrino flux intensities prove to be quite sensitive to the value of G during the evolution.

080.046 Splitting of the deuteron by solar neutrinos.
Yu. S. Kopysov, V. A. Kuzmin.
Cosmic Rays No. 10, Moscow, p. 90 - 91 (1969). In Russian.
The cross sections of the process $D^2 + \nu \rightarrow H^1 + H^1 + e^-$ are calculated for neutrino energies 4 MeV $\leqslant E_\nu \leqslant$ 20 MeV. Integral cross sections for B^8 - and He^3 p-solar neutrinos have been obtained. The possibilities of using deuterium for solar neutrino detection are examined, and the characteristics of a deuterium scintillation detector are discussed.

080.047 Solar neutrino detection by means of recording electrons in reactions $\nu + z \rightarrow (z + 1) + e^-$.
G. V. Domogatsky, R. A. Eramzhen.
Cosmic Rays No. 10, Moscow, p. 132 - 135 (1969).
In Russian.
Light nuclei of Li^7, Be^9, B^{11} and a number of other nuclei as possible solar neutrino detectors are examined.

080.048 The future of the sun.   I. W. Roxburgh.
Adv. Sci. London, Vol. 25, (No. 125), 289 - 297 (1969).

080.049 Die heutige Erforschung der Sonne und die Kosmosforschung der Gegenwart.
M. Schwarzschild.
Universitas, Vol. 24, 457 - 468 (1969).

080.050 Planetary influences on the sun.   L. Křivský.
Říše hvězd, Vol. 50, 161 - 165 (1969).   In Czech.

080.051 Über Probleme, Aufgaben und Arbeitsmittel der Sonnenforschung.   F. W. Jäger.
Jenaer Rundschau (Jena Review), 13. Jahrgang, p. 326 - 329

(1968).

080.052 **Differentielle Rotation als Folge anisotroper turbulenter Viskosität.** H. Köhler.
Mitt. Astron. Ges. No. 27, p. 172 - 176 (1969). − Conference paper.

080.053 **Die Sonne − ein veränderlicher Stern.** K. D. Rakosch.
Mitt. Astron. Ges. No. 27, p. 177 - 178 (1969). − Abstract AG.

080.054 **Inhomogene Sonnenmodelle und Geschwindigkeitsfluktuationen der Granulation.**
W. Mattig, J. P. Mehltretter.
Mitt. Astron. Ges. No. 27, p. 204 (1969). − Abstract AG.

080.055 **Vergleich verschiedener Lösungen der Unnoschen Gleichungen im homogenen Magnetfeld.**
R. Göhring.
Mitt. Astron. Ges. No. 27, p. 209 - 210 (1969). − Abstract AG.

080.056 **Information on shock waves deduced from empirical solar models and a theoretical continuation into the transition region.** P. Ulmschneider.
Mitt. Astron. Ges. No. 27, p. 210 - 211 (1969). − Abstract AG.

Spin down problem of rotating stratified fluid in thermally insulated circular cylinders.
See Abstr. 022.010.

A Rossby-wave dynamo for the sun, I.
See Abstr. 062.001.

An axisymmetric magnetic field with differential rotations in a spherical fluid shell. II.
See Abstr. 062.007.

Analysis of velocity filtergrams.
See Abstr. 071.044.

On the diffusion of solar magnetic fields. II.
See Abstr. 071.050.

The revised solar abundance of iron and the solar neutrino flux. See Abstr. 071.077.

A statistical analysis of flare events and its implications concerning the sun's internal rotation.
See Abstr. 073.039.

Earth

081 Figure, Composition, and Gravity of the Earth

081.001 **Solution of the geodetic boundary value problem for a reference ellipsoid.** K.-R. Koch.
Journ. Geophys. Res. Vol. 74, 3796 - 3803 (1969).

The solution of the geodetic boundary value problem for a reference ellipsoid is given with a relative error of the order of the square of the earth's flattening. The solution is obtained by representing the disturbing potential at the earth's surface as the potential of a simple layer. If the earth's topography is neglected in the derived solution, a new solution of Stokes' problem for a reference ellipsoid is obtained.

081.002 **Inhomogeneous accumulation of the earth from the primitive solar nebula.**
K. K. Turekian, S. P. Clark, Jr.
Earth Planet. Sci. Letters, Vol. 6, 346 - 348 (1969).

To avoid the difficulties posed by Ringwood's model of accumulation and subsequent planetary processing of homogeneous material from the solar nebula, the case of the acceleration of the earth and planets in a non-homogeneous manner is considered. Some earlier models are reviewed and it is proposed that a combination of parts of these models can be made to provide a unified model for the accumulation and structure of the earth and other planets.

081.003 **Statistical models for the vertical deflection from gravity-anomaly models.**
L. Shaw, I. Paul, P. Henrikson.
Journ. Geophys. Res. Vol. 74, 4259 - 4265 (1969).

Statistical models for vertical deflections in the form of power spectral densities and autocorrelation functions are derived from theoretical gravity-anomaly models by means of the Vening Meinesz equations. Details are given for homogeneous and isotropic gravity-anomaly models described by exponential and Bessel autocorrelation functions. The use of these models for evaluation of mean-squared output errors in an inertial navigation system is described.

081.004 **Accuracy of geoid heights from modified Stokes kernels.** L. Wong, R. Gore.
Geophys. Journ. Roy. Astron. Soc. Vol. 18, 81 - 91 (1969).

The dependence of the r.m.s. geoid height error on the degree of the first term in the zonal harmonics expansion of the kernel in Stokes's integration formula is examined. It is shown that kernels with the lower degree terms removed have some advantage over the conventional kernel when a significant error in the zeroth term of the gravity anomaly expansion is present. Numerical estimates of r.m.s. geoid height error vs integration cap size are obtained for several kernels.

081.005 **Messung der Schwerkraft mit Gravimetern und Satelliten.** E. Groten.
Umschau, Vol. 69, 585 - 586 (1969).

081.006 **Zur Bestimmung des Gravitationsfeldes der Erde aus Satellitenbeobachtungen.** C. Reigber.
Deutsche Geod. Kommission Bayer. Akad. Wiss. Reihe C, Heft No. 137, 129 pp. (1969).

Im Anschluß an eine gründliche Beschreibung des Gravitationsfeldes der Erde und einer Diskussion der Bahnbestimmungsmethoden für Erdsatelliten entwickelt Verf. eine neue Methode zur Bestimmung der Parameter des Gravitationsfeldes, wobei er von den diesbezüglichen Untersuchungen von M. Schneider ausgeht. *U. Güntzel-Lingner*

081.007 **Ein- und zweiparametrige Gleichgewichtsfiguren.** K. Ledersteger.
Nachr. Karten- und Vermessungswesen, Inst. Angew. Geodäsie, Frankfurt, Sonderheft, p. 79 - 93 (1969).

081.008 **Continental drift?** W. R. Cook.
Phys. Today, Vol. 22, No. 8, p. 11, 13, with a reply by D. L. Turcotte, E. R. Oxburgh, p. 13, 15 (1969).

081.009 **Altitude extension of the three anomalous gravity components.** L. de Witte.
Bull. Géod. Nouvelle Série, No. 93, p. 287 - 305 (1969).

081.010 **Irdischer und lunarer Vulkanismus – Ergebnisse einer Reise nach Island.** W. Sandner.
Orion, Band 14, 123 - 126 (1969).

081.011 **Age of the separation of South America and Africa.** D. A. Valencio, J. F. Vilas.
Nature, Vol. 223, 1353 - 1354 (1969).

081.012 **Ammonite biostratigraphy, continental drift and oscillatory transgressions.** R. A. Reyment.
Nature, Vol. 224, 137 - 140 (1969).

An appraisal of continental drift in the South Atlantic area from the palaeontological viewpoint.

081.013 **Comments on the earth's figure.** S. J. Paddack.
Nature, Vol. 224, 254 - 255 (1969).

081.014 **Topological inconsistency of continental drift on the present-sized earth.** R. Meservey.
Science, Vol. 166, 609 - 611 (1969).

Certain continents have in the past moved with respect to each other in a manner clearly implied by sea-floor spreading and other data. However, the resulting collective motion of all the continents was apparently not topologically possible on the present-sized earth. An expanding earth might resolve this difficulty.

081.015 **Direct measurements of the earth's gravitational potential using a satellite pair.** M. Wolff.
Journ. Geophys. Res. Vol. 74, 5295 - 5300 (1969).

It is possible to easily measure variations of intensity of the earth's gravity field by orbiting two geometrically identical satellites spaced about 200 kilometers apart and equipped to measure their relative velocity. The velocity difference is related to the space gradient of the potential field, which is a totally different concept of measurement, compared with present methods that measure the long-term integrated effects of gravity on orbit elements. It is expected that one month of satellite time would yield one set of measurements of the gravity anomalies of the entire earth with sensitivity at least 1 mgal ten times better than present world surveys.

081.016 **Search for seismic signals at pulsar frequencies.**

R. A. Wiggins, F. Press.
Journ. Geophys. Res. Vol. 74, 5351 - 5352 (1969).

This paper reports on an effort to detect seismic signals at pulsar frequencies by means of the 525-element large aperture seismic array (LASA) in Montana. No correlation between pulsar frequencies and spectral peaks was found.

081.017 Lead isotopes, lunar capture and mantle evolution.
T. J. Ulrych.
Nature, Vol. 224, 766 - 768 (1969).

Lead isotope data from young mantle derived volcanics suggest that the mantle is a two-stage system. The second stage was formed at the time of the global anorthosite event, and both events may have been caused by capture of the moon by the earth.

081.018 Error analyses of resonant orbits for geodesy.
B. C. Douglas, C. F. Martin, R. G. Williamson, C. A. Wagner.
Celestial Mechanics, Vol. 1, 252 - 270 (1969).

This study concerns determining the realistic accuracy to which certain geopotential coefficients can be recovered by observing satellites on resonant orbits.

081.019 Oceanic sediment volumes and continental drift.
J. Gilluly.
Science, Vol. 166, 992 - 994 (1969).

The volume of sediment off the Atlantic Coast of the United States is at least six times as great as that off the Pacific Coast. This disparity is readily accounted for if the continent is drifting westward and has overrun large volumes of sediment on a former Benioff zone. Such an overrunning is also consonant with many features of the geology of the western United States.

081.020 Tidal friction with latitude-dependent amplitude and phase angle. W. M. Kaula.
Astron. Journ. Vol. 74, 1108 - 1114 = Publ. Inst. Geophys. Planet. Phys., Univ. California, Los Angeles No. 744 (1969).

Tidal-disturbing functions were developed in which the amplitude factor κ and lag angle ϵ are expressed as sums of zonal spherical harmonics. In regard to the current evolution of the moon's orbit, the existence of a second-degree harmonic in the lag angle could make a significant contribution to energy transfer to the moon. If the moon formed in an inclined orbit, tidal friction could increase the inclination further if there was a commensurability between the earth's rotation and the moon's revolution. Latitudinal variations in tidal properties would have appreciable effects on close-satellite orbits.

081.021 General solution of the problem of hydrostatic equilibrium of the earth. M. A. Khan.
Geophys. Journ. Roy. Astron. Soc., Vol. 18, 177 - 188 (1969).

081.022 The free air geoid for Australia.
R. S. Mather.
Geophys. Journ. Roy. Astron. Soc., Vol. 18, 499 - 516 (1969).

The solutions obtained for the free air geoid are compared with the astrogeodetic determination of the geoid on the Australian Geodetic Datum by Fischer and Slutsky and the accuracy of the comparisons is estimated.

081.023 The normal modes of a rotating, elliptical earth – II. Near-resonance multiplet coupling.
F. A. Dahlen.
Geophys. Journ. Roy. Astron. Soc., Vol. 18, 397 - 436 (1969).

This paper uses Rayleigh's variational principle to specify the selection rules for strong mode coupling and to examine the nature and effects of this mode coupling.

081.024 Sea-floor spreading, continental drift and lithosphe- re sinking with an asthenosphere at melting point.
L. Lliboutry.
Journ. Geophys. Res. Vol. 74, 6525 - 6540 (1969).

081.025 La nouvelle conception de la dérive des continents.
A. de Vuyst.
Ciel et Terre, Vol. 85, 345 - 375 (1969).

081.026 Interglacial high sea levels and the control of Greenland ice by the precession of the equinoxes.
C. Emiliani.
Science, Vol. 166, 1503 - 1504 (1969).

The precession of the equinoxes appears to control the occurrence of high sea levels by partial or even total melting of the Greenland ice cap during interglacial ages.

081.027 Equations of motion for the earth tides.
S. Takagi.
Publ. International Latitude Observatory of Mizusawa, Vol. 6, 205 - 223 (1968).

Recent development of geophysics makes it necessary to construct a more detailed theory of the earth tides which is based on a model more similar to the actual earth.

081.028 On the triaxiality of the earth deduced from Chandler ellipse. C. Sugawa.
Proc. International Latitude Obs. Mizusawa, No. 9, p. 191 - 211 (1969). In Japanese.

The inequality of A and B, the principal moments of inertia in the equatorial plane of the earth, is indicated by free nutation (the Chandler term in elliptic).

081.029 The Indian Ocean and ancient Gondwanaland. Appendix: The Asian continent and the sub- oceanic ridges. A. Niini.
Ann. Acad. Sci. Fennicae, *Helsinki, Finland*. Series A, III. Geol.–Geogr. 103, 36 pp. (1969).

081.030 Determination of the figure and of the gravitational field of the earth from observations of artificial satellites. K. Arnold.
Byull. Stantsij Optichesk. Nablyud. Iskusstv. Sputnikov Zemli No. 55, p. 43 - 58 (1969). In Russian.

081.031 Joint leveling of gravimetric and satellite data for determination of the gravitational field of the earth.
L. P. Pellinen.
Byull. Stantsij Optichesk. Nablyud. Iskusstv. Sputnikov Zemli No. 55, p. 58 - 68 (1969). In Russian.

081.032 On the errors of determining the earth's potential from measurements of gravity on the known sur- face. V. G. Shkodrov.
Byull. Stantsij Optichesk. Nablyud. Iskusstv. Sputnikov Zemli No. 55, p. 68 - 73 (1969). In Russian.

081.033 On the determination of the geopotential from perturbations in satellite orbits using the orthogonality of harmonics. M. Burša.
Stud. Geophys. Geod., Vol. 13, 359 - 372 (1969). In Czech.

081.034 Potential of the geoidal surface, the scale factor for lengths and earth's figure parameters from sa- tellite observations. M. Burša.
Stud. Geophys. Geod., Vol. 13, 337 - 358 (1969).

081.035 Expansion of a formula determining the earth's figure in a Taylor's series. M. I. Marych.
Geod., kartogr. i aehrofotos"emka. Mezhved. resp. nauchno-tekhn. sb., Vyp. (No.) 9, p. 29 - 32 (1969). In Russian. – Abstr. in Referativ. Zhurn. 52. Geod. Aehrofot., 2.52.58 (1970).

082 The Earth's Atmosphere including Refraction, Scintillation, Extinction, Airglow, Site Testing

082.001 Air density at heights of 140 - 180 km, from analysis of the orbit of 1968-59A.
D. G. King-Hele, D. M. C. Walker.
Planet. Space Sci. Vol. 17, 1539 - 1556 (1969).

Separate profiles of air density vs. height are obtained for July—August and for September—October 1968, with the latter giving values of density about 10 per cent higher. This increase in density reveals the existence of the semi-annual variation in density, well known at greater heights: the maximum density in October 1968 was about 20 per cent higher than the minimum in July, for heights near 170 km.

082.002 Turbulence spectra in stable and convective layers in the free atmosphere. L. O. Myrup.
Tellus, Vol. 21, 341 - 354 (1969).

082.003 Zones of the visibility of a noctilucent cloud.
G. Dietze.
Tellus, Vol. 21, 436 - 442 (1969).

For a noctilucent cloud (NLC) with its typical features the brightness contrast k of such a cloud against the sky is computed as observed from any point within a circle of more than 800 km. This contrast is compared with the contrast threshold ϵ of the eye. The quotient $|k| : \epsilon$ is a measure of the degree of visibility. Its distribution over the earth's surface in the vicinity of an NLC at 5 different times during twilight reveals zones of differential visibility.

082.004 Far-infrared nightsky emission above 120 kilometers. J. R. Houck, M. Harwit.
Astrophys. Journ. *(Letters)*, Vol. 157, L45 - L48 (1969).

We have verified the existence of a flux of 5 (+5, −2.5) × 10^{-9} W cm^{-2} sterad^{-1} in the spectral range from 0.4 to 1.3 mm, at an altitude above 120 km. This result is discussed in the context of other recent investigations.

082.005 A theory of thermospheric dynamics—I. Diurnal and solar cycle variations. H. Volland.
Planet. Space Sci. Vol. 17, 1581 - 1597 (1969).

It is shown that the diurnal density and temperature variation within the thermosphere is generated not only by solar EUV heat input but also by a tidal wave from the lower atmosphere which penetrates into the thermosphere and predominates the diurnal variations below 250 km altitude. Assuming a tidal wave from below which is not dependent on solar activity and assuming solar EUV heat input to be proportional to the solar activity factor $F_{10.7}$, the observed diurnal density variation and its dependence on solar activity can be explained and reproduced quantitatively with the help of a two-dimensional model between 100 and 400 km height.

082.006 Equatorial measurements of the [OI] 5577 Å emission of the dayglow with a rocket photometer.
B. S. Dandekar.
Planet. Space Sci. Vol. 17, 1609 - 1618 (1969).

Observations of the [OI] 5577 Å emission of the dayglow have been obtained with a rocket photometer from the equatorial launch site at Natal, Brazil. Our observations show that the contribution to the [OI] 5577 Å emission of the dayglow comes from three different ranges of altitudes.

082.007 Hydroxyl emission of the upper atmosphere—II. Effects of a sunlit atmosphere. N. N. Shefov.
Planet. Space Sci. Vol. 17, 1629 - 1639 (1969).

On the basis of the observations made at Zvenigorod the measured average seasonal variations of the population of the fifth vibrational level and the vibrational temperature of OH molecules seasonal variations of the populations of the levels with $V' = 1$ to $V' = 9$, and the total energy of the hydroxyl emission, have been obtained.

082.008 Predawn enhancement of 6300 Å airglow in conjugate regions.
E. H. Carman, G. J. Hatzopoulos, M. P. Heeran.
Planet. Space Sci. Vol. 17, 1677 - 1679 (1969). – Research note.

082.009 Absorption of a grazing ray calculated by ray tracing. K. H. Lloyd.
Planet. Space Sci. Vol. 17, 1683 - 1687 (1969).–Research note.

082.010 Interpretation of OV1-5 spectrometric observations.
R. G. Breene, Jr., J. S. Swant, L. P. Marcotte.
Journ. Quant. Spectrosc. Radiative Transfer, Vol. 9, 1239 - 1249(1969).

Scattering and absorption by water or ice particles in a cloud deck are treated by Mie theory. Gaseous absorption is analyzed by using familiar emissivity theory. Three spectrometric scans are interpreted as examples. The method for N_2O matrix element estimation is discussed.

082.011 The simultaneous investigation of attenuation and emission by the earth's atmosphere at wavelengths from 4 centimeters to 8 millimeters.
G. G. Haroules, W. E. Brown, III.
Journ. Geophys. Res. Vol. 74, 4453 - 4471 (1969).

Daily solar measurements of atmospheric absorption and atmospheric emission were conducted simultaneously for a period of nine months during 1967 and 1968. Nighttime observations of emission followed daytime observations of absorption and emission. The observational frequencies of 8, 15, 19, and 35 GHz were used to determine the frequency correlation between meteorological phenomena and the atmospheric parameters associated with the fine structure characteristics of absorption and emission.

082.012 Temperature shape parameter of the thermosphere determined from probe data. G. R. Swenson.
Journ. Geophys. Res. Vol. 74, 4074 - 4078 (1969).

The static diffusion model of the upper atmosphere developed by L. G. Jacchia in 1965 uses an exospheric temperature dependence to arrive at a temperature shape parameter that is used to generate the thermospheric temperature profile. Contained here is the analysis of shape parameters deduced from eighteen thermosphere probe flights. Findings show a dependence of the shape parameter on the hour angle, indicating the thermospheric temperature profile to be more dependent upon the dynamics of heating and cooling than on the exospheric temperature.

082.013 Predissociation in the Schumann-Runge band system of O_2: Laboratory measurements and atmospheric effects. R. D. Hudson, V. L. Carter, E. L. Breig.
Journ. Geophys. Res. Vol. 74, 4079 - 4086 (1969).

082.014 An on-board technique for estimating the effect of water vapor in radio occultation measurements of atmospheric density. W. G. Tank.
Journ. Geophys. Res. Vol. 74, 4147 - 4156 (1969).

S-band phase path measured between two earth satellites—one in an occultation orbit relative to the other—determines the integrated radio refractivity of the neutral earth atmosphere. Atmospheric density is recoverable from such determinations provided an independent estimate of the contribution of water vapor to integrated refractivity is obtained. Transmission measurements in the vicinity of the 22.235-GHz resonant water vapor absorption line provide this estimate.

082.015 Interferometric measurements of the 6300 Å Doppler temperature during a magnetic storm.
P. B. Hays, A. F. Nagy, R. G. Roble.
Journ. Geophys. Res. Vol. 74, 4162 - 4168 (1969).
The 6300 Å Doppler broadened emission line of atomic oxygen $[O(^1D) - O(^3P)]$ was measured with a 6-inch, high resolution Fabry-Perot interferometer during the magnetic storm period of October 30 to November 2, 1968.

082.016 Latitudinal variations in the neutral atmospheric density. G. P. Newton, D. T. Pelz.
Journ. Geophys. Res. Vol. 74, 4169 - 4174 (1969).

082.017 Escaping photoelectrons and dayglow.
S. S. Prasad.
Journ. Geophys. Res. Vol. 74, 4772 - 4776 (1969).
Theoretical calculations were made for the photoelectron excitation rates of $\lambda 1304$ and $\lambda 1356$ radiations of O I. In the lower region, about 225 km and below, the excitation rates were practically the same whether the photoelectrons were localized or escaping. But the rates in the two cases differ rapidly as the altitude increases. At about 300 km the excitation rates with escaping photoelectrons are lower by a factor of about 1.5 to 2.0 relative to those with localized electrons. The theoretical results were compared with the corresponding observational data by scaling them to a common (experimental) value at about 175 km. It was found that the altitude dependence in the escaping photoelectron case was in better accord with the experiment.

082.018 Far infrared nightglow emission from atomic oxygen. P. D. Feldman, D. P. McNutt.
Journ. Geophys. Res. Vol. 74, 4791 - 4793 (1969).
We have observed nightglow emission in a spectral band from 50μ to 125μ by using a rocket-borne liquid helium-cooled telescope. The observed zenith intensity as a function of altitude is consistent with the assumption that the emission is due to the $^3P_1 - {}^3P_2$ fine structure transition of O I at 63μ. At 120 km, the zenith intensity is 1.7 megarayleighs or 5.5×10^{-2} erg sec^{-2} (column), and the derived value for the atomic oxygen density is 4×10^{10} cm^{-3}.

082.019 Discussion of paper by J. B. Pearce, 'Rocket measurement of nitric oxide between 60 and 96 kilometers'. G. C. Tisone.
Journ. Geophys. Res. Vol. 74, 4803 - 4804 (1969).

082.020 Simulation von physikalisch-chemischen Vorgängen in der höheren Atmosphäre. W. Groth.
Umschau, Vol. 69, 622 (1969).

082.021 Observations of H_a -nightglow at Abastumani during summer 1968. L. M. Fishkova, P. V. Sheglov.
Astron. Tsirk. No. 505, p. 5 - 8 (1969).   In Russian.

082.022 Diurnal and long-term variations of the H_a -nightglow. Z. V. Karjagina, V. E. Mozjaeva, P. V. Sheglov.
Astron. Tsirk. No. 516, p. 4 - 6 (1969).   In Russian.

082.023 Ha-nightglow measurements at Abastumani in the second half of 1968. L. M. Fishkova, P. V. Sheglov.

Astron. Tsirk. No. 516, p. 6 - 7 (1969).   In Russian.

082.024 Simultaneous observations of Ha-nightglow in the direction of the celestial pole at Abastumani and Alma-Ata. Z. V. Karjagina, V. E. Mozjaeva, L. M. Fishkova.
Astron. Tsirk. No. 516, p. 7 - 8 (1969).   In Russian.

082.025 A simple wind-integrating device for site testing.
P. V. Sheglov.
Astron. Tsirk. No. 517, p. 5 - 7 (1969).   In Russian.

082.026 Results of double-beam site testing at Mt. Sanglock during 6 months. N. A. Abramenko, A. V. Bagrov, Yu. F. Nikitin, G. V. Novikova, S. B. Novikov, P. V. Sheglov.
Astron. Tsirk. No. 518, p. 4 - 6 (1969).   In Russian.

082.027 Quelques réflexions sur la conception moderne du choix des sites. P. V. Sheglov.
Astron. Tsirk. No. 518, p. 6 - 8 (1969).   In Russian.

082.028 Tables of variation of refraction due to small temperature differences. A. K. Korol, R. N. Koval.
Astrometriya i Astrofiz., *Kiev*, No. 2, p. 116 - 123 (1969).
In Russian.
The tables give the values of additional refraction due to temperature differences between two volumes of air devided by a horizontal or vertical plane.

082.029 Climatic factors and telescope output.
D. S. Evans.
Astron. Astrophys. Vol. 3, 247 - 251 (1969).
It is contended that telescope output is a function not only of the average clear time available for observations but also of the average length of a clear interval. An extreme case is quoted where a ratio of about 3 : 1 in the former produces an efficiency ratio of the order of 50 : 1. A formula for the average productivity of a telescope per clear interval is derived and used in various examples to illustrate how management decisions may be guided. The analysis may also apply to other large facilities subject to random withdrawal from service which are used in different modes by a number of workers.

082.030 A theory of thermospheric dynamics—II. Geomagnetic activity effect, 27-day variation and semiannual variation. H. Volland.
Planet. Space Sci. Vol. 17, 1709 - 1724 (1969).
In this paper we shall deal with the geomagnetic activity effect, with the 27-day variation and with the semiannual variation of the thermospheric density. All three effects are nearly independent of local time which means that the energy input into the thermosphere occurs on a global scale. Therefore, we can treat these effects theoretically by a one dimensional model in which horizontal exchange of mass, momentum and energy is neglected.

082.031 $O(^1D)$ quenching in the ionospheric F-region.
V. L. Peterson, T. E. Vanzandt.
Planet. Space Sci. Vol. 17, 1725 - 1736 (1969).
The collisional quenching rate of $O(^1D)$ is determined by comparing observed 6300 Å nightglow intensities with intensities calculated from simultaneously observed F-region electron density profiles. We obtain a quenching frequency which, when extrapolated to the reference height of 120 km, is 20 sec^{-1}, with an accuracy of about a factor of two. The corresponding rate coefficient, assuming N_2 is the quencher, is $5-10 \times 10^{-11}$ cm^3 sec^{-1}, depending upon the model atmosphere used.

082.032　**A theoretical study of the 27-day variation of upper-atmospheric temperature.**　G. E. Thomas, B. K. Ching.
Planet. Space Sci. Vol. 17, 1737 - 1747 (1969).

082.033　**Utilisation d'observations d'amateurs pour l'étude de la haute atmosphère en corrélation avec l'activité solaire.**　F. Barlier.
L'Astronomie, 83e année, 282 - 286 (1969).

082.034　**Variations of the atmospheric extinction in the ultraviolet.**　E. Mendoza V., H. Moreno, J. Stock.
Publ. Dep. Astron. Univ. Chile, Obs. Astron. Nacional, Cerro Calan, Santiago, Vol. 1 (No. 5), 67 - 71 (1968).

Spectrophotometric observations obtained at Cerro Tololo show that the atmospheric extinction in the region of the ultraviolet ozone band ($\lambda\lambda$ 3100 - 3500 Å) is significantly and systematically variable during the night. Photometric observations obtained simultaneously at Cerro La Silla show a correlation of the atmospheric extinction at both sites.

082.035　**On astronomical seeing: The single Schlieren model.**　U. Grossmann-Doerth.
Solar Physics, Vol. 9, 210 - 224 (1969).

A model of astronomical seeing with particular view to solar observations is developed which assumes the atmospheric disturbances to consist of individual turbulence elements called Schlieren. A quantitative account is given of each image motion, image blurring and scintillation as function of Schlieren properties and telescope parameters. The theory permits to explain the observational results under conditions of good seeing; furthermore, it provides a basis for the discussion of the physical phenomena in the atmosphere that cause image deterioration.

082.036　**Analisi micrometeorologica della torre solare di Arcetri. Risultati preliminari.**　A. Righini.
Mem. Soc. Astron. Italiana, Nuova Serie, Vol. 40, 355 - 357 (1969).

082.037　**On the accuracy of Chapman function approximations.**　W. Swider, Jr., M. E. Gardner.
Applied Optics, Vol. 8, 725 (1969).

082.038　**Photometric error analysis. VIII. The temporal power spectrum of scintillation.**　A. T. Young.
Applied Optics, Vol. 8, 869 - 885 (1969).

Previous theoretical work on the scintillation spectra of stars is extended to include rectangular as well as annular apertures. The theory is then generalized to encompass planetary scintillation and the effects of diffraction, atmospheric dispersion, and seeing. The results are quite accurate for apertures larger than 3 in. (7.6 cm) or 4 in. (10.2 cm). Observations of planetary scintillation show that the telescopic scintillation is produced throughout the atmosphere rather than primarily in a thin layer; the turbulent scale height is usually near 8 km and can be found with an accuracy of about 1 km. The twinkling seen with the naked eye arises within 1 km of the ground and is more closely related to telescopic seeing than to telescopic scintillation.

082.039　**Molecular radiation of the upper atmosphere in the $3 - 8$-μ spectral region.**　M. N. Markov.
Applied Optics, Vol. 8, 887 - 891 (1969).

The upper atmosphere infrared radiation has been investigated by rockets and satellites. The upper atmosphere radiation has a layer structure, its maximum intensity being observed at altitudes of 150 km, 280 km, 430 km, and 500 km. The radiation energy density in the layers reaches 10^{-4} erg/cm^3 sec, increasing to 10^{-3} erg/cm^3 sec during a magnetic storm. The character of spectral distribution corresponds to radiation of the rotation-vibration bands of diatomic molecules, apparently NO, OH, and CO.

082.040　**Features of tropospheric and stratospheric dust.**　L. Elterman, R. Wexler, D. T. Chang.
Applied Optics, Vol. 8, 893 - 903 (1969).

A statistical treatment of data comprising 119 aerosol attenuation profiles provides information on aerosol attenuation coefficients, optical mixing ratios, stratospheric dust of volcanic origin and its abatement. These in turn permit examination of relative aerosol concentrations or layer features, aerosol scale height, and the determination of atmospheric attenuation parameters to 50 km.

082.041　**On the theory of oxygen red-line emission.**　T. F. Tuan.
Astrophys. Journ. Vol. 157, 1449 - 1454 (1969).

A general time-dependent formula has been derived for the total intensity at zenith of the red oxygen line at 6300 Å of the night airglow. The formula takes specifically into account the variations in the shape of the profiles of electron density as a function of time; it reduces to the well-known Barbier formula when the distribution of electron density is shape-invariant.

082.042　**Calculations of atmospheric transmittance from 1.7 to 20 μ.**　T. G. Kyle.
Journ. Quant. Spectrosc. Radiative Transfer, Vol. 9, 1477 - 1488 (1969).

A method for performing line by line calculations of atmospheric transmittance along variable pressure and variable mixing ratio paths is described. This method has been used to calculate the atmospheric transmittance for a vertical path through the atmosphere from a pressure level of 200 millibars and with a resolution of 0.5 cm^{-1} for the interval from 1.7 to 20 μ.

082.043　**Strengths, widths, and shapes of the oxygen lines near 13,100 cm^{-1} (7620 Å).**
D. E. Burch, D. A. Gryvnak.
Applied Optics, Vol. 8, 1493 - 1499 (1969).

082.044　**Atmospheric transmittance in the $590 - 750$ cm^{-1} interval.**
T. G. Kyle, J. N. Brooks, D. G. Murcray, W. J. Williams.
Applied Optics, Vol. 8, 1926 - 1927 (1969). — Letter.

082.045　**Astronomical refraction. I.**　A. I. Nefed'eva.
Izv. Astron. Obs. Ehngel'gardta, *Kazan'*, No. 36, p. 3 - 18 (1968).　In Russian.
Contents: History of refraction from its discovery until the 19. century; publications on refraction in the 19. century; refraction theory by Harzer; refraction theory on the basis of a polytrope model of the atmosphere.

082.046　**Atmospheric density above 158 kilometers inferred from magnetron and drag data from the satellite OV1-15 (1968-059 A).**　V. L. Carter, B. K. Ching, D. D. Elliott.
Journ. Geophys. Res. Vol. 74, 5083 - 5091 (1969).

This paper reports the first results of the cold-cathode gage measurement of density during times of relatively moderate solar and magnetic activity and compares the perigee density with that determined from drag.

082.047　**Vibrational populations of O_2 ($A\ ^3\Sigma_u{}^+$) and synthetic spectra of the Herzberg bands in the night airglow.**　V. Degen.
Journ. Geophys. Res. Vol. 74, 5145 - 5154 (1969).

This paper presents, first, the relative populations of the vibrational levels $0 < v' \leq 10$, determined by matching calculatcd synthetic spectra of the Herzberg bands to the various published nightglow spectrograms, and second, as an aid to nightglow identification, the vibrationally identified 'pure' Herzberg synthetic spectra at the various resolutions in the literature and as might be encountered in future observations.

082.048 Dissociation of water vapor and evolution of oxygen in the terrestrial atmosphere.
R. T. Brinkmann.
Journ. Geophys. Res. Vol. 74, 5355 - 5368 (1969).

Previous studies of the photodissociation of water vapor and the resulting evolution of oxygen in the earth's atmosphere have led to the conclusion that, over most of geologic time, the atmospheric oxygen abundance has been $\lesssim 10^{-3}$ times the present atmospheric level.

082.049 Ice-nucleating properties of dust collected above 80 kilometers. N. R. Gokhale, J. Goold, Jr.
Journ. Geophys. Res. Vol. 74, 5374 - 5378 (1969).

A specially constructed apparatus was used to compare the ice-nucleating properties of dust particles collected above 80 km with those of soil and AgI particles. The dust particles were found to be relatively inactive in ice nucleation when compared with AgI and soil particles of terrestrial origin.

082.050 Distribution of water vapor in the stratosphere as derived from setting sun absorption data.
D. G. Murcray, T. G. Kyle, W. J. Williams.
Journ. Geophys. Res. Vol. 74, 5369 - 5373 (1969).

The variation of the infrared solar spectrum in the region around 6.3μ during sunset over New Mexico was observed by a balloon-borne spectrometer system. Analysis of the variation of the water vapor absorption lines in this region with the increased high-altitude optical path has shown that the water vapor mixing ratio in the region from 25 to 30 km is between 2.5×10^{-6} and 3.0×10^{-6} g/g.

082.051 Tables of refraction corrections for observations of objects in the earth's atmosphere.
I. G. Kolchinsky, A. N. Kuryanova, E. B. Shmelkina.
Astrometriya i Astrofiz., *Kiev*, No. 5, p. 7 - 46 (1969).
In Russian.

Tables of refraction angles for atmospheric objects are presented. The calculations are given for the conditions of a standard atmosphere.

082.052 Observations of astronomical refraction at great zenith distances made at Goloseyevo (Kiev suburb).
N. A. Vasilenko.
Astrometriya i Astrofiz., *Kiev*, No. 5, p. 47 - 61 (1969).
In Russian.

The results of observations of astronomical refraction near the horizon made with the $2''$ universal instrument are given. The maximum deviation of the measured refraction from the calculated on the basis of the Pulkovo tables reaches $1'$ at a zenith distance of $89°.5$.

082.053 The atmospheric dispersion and magnitude equation for the 400 mm astrograph of the Main Astronomical Observatory of the Ukrainian Academy of Sciences.
A. B. Onegina.
Astrometriya i Astrofiz., *Kiev*. No. 5, p. 62 - 71 (1969).
In Russian.

Results of the determination of atmospheric dispersion and magnitude equation on the plates taken with the 400 mm astrograph are given. The obtained value of atmospheric dispersion constant is $0''.17 \pm 0''.03$.

082.054 A study of the star image motion by the method of traces made with the 200 mm reflector (AZT-7) in 1963. G. V. Moroz.

Astrometriya i Astrofiz., *Kiev*, No. 5, p. 72 - 87 (1969).
In Russian.

Star image motion was investigated in 1963 in Goloseyevo by the method of traces with the 200 mm Maksutov reflector (AZT-7). The results of measuring 267 traces obtained during 20 nights show that the r.m.s. value of the oscillations is proportional to $(\sec z)^{0.5}$.

082.055 Dependence of star image motion on meteorological conditions. G. V. Moroz.
Astrometriya i Astrofiz., *Kiev*, No. 5, p. 88 - 99 (1969).
In Russian.

Meteorological conditions and their influence on the dependence between the value of image motions and zenith distances are studied. It was established that under antycyclone conditions the image motion is nearly constant at all zenith distances and decreases slightly in zenith.

082.056 Some results of observations of spectra of starlight scintillation carried out with the 200 mm Maksutov reflector AZT-7 at Goloseyevo (Kiev suburb).
E. M. Diamant, I. G. Kolchinsky, Yu. K. Filippov.
Astrometriya i Astrofiz., *Kiev*, No. 5, p. 100 - 117 (1969).
In Russian.

082.057 On the astroclimate of two points in the Transcarpathian and Odessa regions of the Ukrainian SSR.
L. R. Lisina, E. S. Kheylo.
Astrometriya i Astrofiz., *Kiev*, No. 5, p. 118 - 126 (1969).
In Russian.

082.058 Investigation of the scintillation amplitude components by traces of artificial satellites and stars and the bead structure of star traces.
I. V. Shvalagin, Ya. M. Sinichenko.
Astrometriya i Astrofiz., *Kiev*, No. 5, p. 127 - 131 (1969).
In Russian.

The paper shows that image motion determined by the traces of stars and artificial satellites and obtained roughly at one and the same time is approximately equal.

082.059 The pre-dawn enhancement of the 6300 Å [O I] line in the nightglow. M. F. Ingham.
Monthly Notices, Roy. Astron. Soc., Vol. 145, 389 - 400 (1969).

Measurements of the emission rate at 6300 Å in the spectrum of the nightglow have been made which show in detail the onset of the pre-dawn enhancement. The results are analysed to (i) fix the time of onset of the enhancement, (ii) determine the form of the curve of emission rate vs. time near onset and (iii) display short-term variations in the emission rate.

082.060 Transformations of the forms of tropospheric circulation in connection with the macrostructure of the solar wind. R. V. Smirnov.
Dokl. Akad. Nauk. SSSR, Ser. Mat. Fiz., Vol. 187, 1278 - 1281 (1969). In Russian.

082.061 Atmospheric penetration of ultra-violet and visible solar radiations during twilight periods.
L. Thomas, M. R. Bowman.
Journ. Atmosph. Terr. Phys. Vol. 31, 1311 - 1322 (1969).

A numerical study is made of the penetration of solar radiation of wavelength 100 - 600 mm (1000 - 6000 Å) to heights below 100 km during the pre-sunrise period. Account is taken of absorption by molecular oxygen and ozone, Rayleigh scattering, and atmospheric refraction; the effects of changes in the height distribution of ozone are considered. The results of the computations are related to observations of the D-region and of the airglow during the pre-sunrise period.

082.062 The excitation of characteristic X-radiation in an atmosphere. S. G. Tomlin.
Journ. Atmosph. Terr. Phys. Vol. 31, 1323 - 1332 (1969).

It is shown how the intensity of characteristic X-radiation generated in an atmosphere could be calculated given sufficient information about the spatial and energy distribution of the irradiating electrons in it. In the absence of such data detailed discussions of idealized models are presented. In an exponential nitrogen atmosphere the intensity reaches values of the order of 10^{-3} quanta $cm^{-2} sec^{-1}$ for each electron incident normally per sec per cm^2.

082.063 Observations of a magnetic conjugate effect in the OI 6300 Å airglow at Saskatoon.
L. L. Cogger, G. G. Shepherd.
Planet. Space Sci. Vol. 17, 1857 - 1865 (1969).

A variation in the OI 6300 Å nightglow emission intensity at Saskatoon (60.5°N magnetic latitude) has been found to follow the changes in the solar depression angle at the magnetic conjugate region. Observations made with a multichannel Fabry-Perot spectrometer during January–March, 1967 are described and discussed.

082.064 Upper-atmosphere winds and their interpretation –II. Turbulence in the lower E-region.
D. Layzer, J. F. Bedinger.
Planet. Space Sci. Vol. 17, 1891 - 1911 (1969).

This paper examines observational evidence bearing on the question of turbulence on the lower E-region. New observational material, mainly high-resolution photographs of artificial vapor trails, is presented.

082.065 Sources of the Lyman-α emission in the night sky. M. F. Ingham.
Monthly Notices, Roy. Astron. Soc., Vol. 145, 401 - 404 (1969).

Observations of Lyman-α in the night sky have been made with a Skylark rocket and indicate both a terrestrial and a galactic source for the emission.

082.066 Discussion of letter by Robert T. Bennett, 'Latitude dependence of 6300 Å (O I) twilight airglow enhancement'. M. R. Torr, D. G. Torr.
Journ. Geophys. Res. Vol. 74, 5187 (1969). – Letter.

082.067 Mass spectrometric investigation of the thermosphere at high latitudes. U. von Zahn, J. Gross.
Journ. Geophys. Res. Vol. 74, 4055 - 4063 (1969).

The number densities for N_2, O_2, O, and Ar in the 115–155 km range were measured by means of a monopole spectrometer flown aboard a Black Brant 3 rocket launched at the Churchill Research Range on December 12, 1966, at 1320 CST.

082.068 Sodium distribution in the terrestrial upper atmosphere. R. J. Moffett.
Nature, Vol. 224, 1097 (1969).

Observations of the twilight glow and dayglow have shown that there is a thin layer of free sodium atoms at a height of about 90 km.

082.069 Atmospheric tides. R. S. Lindzen, S. Chapman.
Space Sci. Rev. Vol. 10, 3 - 188 (1969).
Review article. – Introductory and historical; The solar daily atmospheric oscillations as revealed by meteorological data; The lunar atmospheric tide as revealed by meteorological data; Quantitative theory of atmospheric tides and thermal tides.

082.070 Atmospheric absorption anomalies in the ultraviolet near an altitude of 50 kilometers.
A. J. Krueger.

Science, Vol. 166, 998 - 1000 (1969).

Rocket-borne radiometer determinations of ozone distributions by absorption of ultraviolet sunlight show anomalous effects near 3000 angstroms.

082.071 Revised profiles of air density at heights of 130 - 180 km, from the orbits of 1968–59A and B.
D. G. King-Hele, D. M. C. Walker.
Planet. Space Sci. Vol. 17, 2027 - 2029 (1969). – Research note.

082.072 Intensity and polarization of the sky twilight. N. B. Dyvary.
Problemy kosmich. fiz. No. 4, p. 139 - 151 (1969).
In Russian.

The results of a calculation of intensity, polarization and polarization angle of the twilight are given.

082.073 Resolution of the difference between atmospheric density measurements from Explorer 17 satellite by density gage and drag techniques. G. P. Newton.
Journ. Geophys. Res. Vol. 74, 6409 - 6414 (1969).

082.074 Deuterium in the earth's exosphere.
E. C. Bruner, Jr., T. E. Wilson.
Journ. Geophys. Res. Vol. 74, 6491 - 6493 (1969).

Signal averaging techniques, applied to high resolution spectra of the solar Lyman-alpha line, have revealed a weak absorption feature in the short wavelength wing of the line. The position and shape of this feature suggest that it be identified as the Lyman-alpha transition in deuterium.

082.075 Structure in the fast spectra of atmospheric neutrons. J. W. Wilson, J. J. Lambiotte, T. Foelsche.
Journ. Geophys. Res. Vol. 74, 6494 - 6496 (1969).

The results of a Monte Carlo transport calculation of fast neutron spectra produced by galactic cosmic-ray protons in the atmosphere are presented. These results show that definite structure in the fast neutron spectra should be observed when these measurements are made.

082.076 The influence of astronomical refraction in the Zinger method. V. N. Baranov.
Izv. vyssh. uchebn. zavedenij. Geod. i aehrofotos"emka, No. 4, p. 97 - 101 (1968). In Russian. – Abstr. in Referativ. Zhurn. 51. Astron., 11.51.111 (1969).

082.077 On the absorption band 0.43μ in the solar spectrum and in the spectra of scattered radiation of the sky.
Yu. S. Georgievskij.
Izv. AN SSSR. Fiz. atmosf i okeana, Vol. 5, 298 - 299 (1969). In Russian.

082.078 Remote sensing of winds and atmospheric turbulence by cross-correlation of passive optical signals.
A. J. Montgomery.
Space Sci. Rev. Vol. 10, 291 - 313 (1969).

Review paper including a report on measurements of scattered sunlight fluctuations and on the design of photometers for scattered sunlight measurements.

082.079 Local variations of astronomical refraction.
G. Teleki.
Stud. Cerc. Astron. Vol. 14, 3 - 7 (1969).

082.080 Some observations of twilight He (1.083) emission.
A. W. Harrison.
Canadian Journ. Phys., Vol. 47, 1377 - 1380 (1969).

The 1.083μ resonance line of He I has been observed in the evening twilight airglow spectrum over the period February 1968 to February 1969.

082.081 Submillimeter telluric absorption at Mauna Kea Observatory.   I. G. Nolt, T. Z. Martin, W. M. Sinton.
Bull. American Astron. Soc., Vol. 1, 356 (1969). – Abstr. AAS.

082.082 Die Helligkeiten des nächtlichen Luftleuchtens während des Sonnenfleckenminimums nach Messungen in Südwestafrika. II. Diskussion der Ergebnisse und Vergleich mit Ionosphärenbeobachtungen.   J. Dachs.
Beiträge Phys. Atmosphäre, Vol. 41, 184 - 215 = Mitt. Astron. Inst. Univ. Tübingen, No. 104 (1968).

082.083 Les variations de la densité atmosphérique vers 150 km déduites de l'étude du freinage des satellites.   J. Vercheval.
Ann. Géophys. Vol. 25, 405 - 417 (1969).
The variations in air density at altitudes near 150 km have been deduced from orbital data of the satellite 1966–101G during the period from November 1966 till April 1967.

082.084 The semi-annual variation in the upper atmosphere: A review.   G. E. Cook.
Ann. Géophys. Vol. 25, 451 - 469 (1969).

082.085 Atmospheric densities at a height of 275 km derived from drag data of Explorer 32.   D. Schaefer, C. Wulf-Mathies.
Ann. Géophys. Vol. 25, 471 - 474 (1969).

082.086 Influence of interstellar matter on the density of atmospheric hydrogen.   H. J. Fahr.
Ann. Géophys. Vol. 25, 475 - 478 = Mitt. Astron. Inst. Univ. Bonn, No. 109 (1969).
The influence of a flux of neutral interstellar matter on the density of atmospheric hydrogen is investigated. It is found that the hydrogen density varies in phase with the incoming flux of interstellar hydrogen atoms which itself shows an annual variation with a possible semiannual component.

082.087 Mesure de la répartition verticale de l'hydrogène atomique entre 200 et 500 kilomètres d'altitude le 14 juin 1966.   J.-A. Quessette.
Ann. Géophys. Vol. 25, 479 - 484 (1969).

082.088 Sur l'utilité des phénomènes crépusculaires pour l'exploration de la haute atmosphère.   F. Link.
Ann. Géophys. Vol. 25, 551 - 553 (1969).
The usefulness of twilight measurements for the exploration of the upper atmosphere is demonstrated, and the possibilities of meteoric influences are briefly discussed.

082.089 Mesures crépusculaires au niveau de 25 mb.   I. Zacharov.
Ann. Géophys. Vol. 25, 555 - 561 (1969).

082.090 Polarisation du ciel crépusculaire au niveau de 25 mb.   L. Neužil.
Ann. Géophys. Vol. 25, 563 - 565 (1969).
Raw data from a series of measurements in the visible part of the spectrum (4500 Å and 5300 Å) obtained in 1968.

082.091 Antarctic twilight observations. 1. Search for metallic emission lines.   M. Gadsden.
Ann. Géophys. Vol. 25, 667 - 677 (1969).

082.092 Antarctic twilight observations. 2. Sodium emission at 90°S.   M. Gadsden.
Ann. Géophys. Vol. 25, 721 - 730 (1969).

082.093 Die Veränderlichkeit der Intensität der Korpuskularstrahlung in der Hochatmosphäre in niedrigen Breiten.   V. F. Tulinov, Yu. N. Moiseev, I. G. Shapiro.
Kosmich. Issled. Vol. 7, 613 - 615 (1969).   In Russian. – Brief information.

082.094 Determination of extraterrestrial solar spectral irradiance from a research aircraft.   J. C. Arvesen, R. N. Griffin, Jr., B. D. Pearson, Jr.
Applied Optics, Vol. 8, 2215 - 2232 (1969).
Results are presented of an experiment to determine extraterrestrial solar spectral irradiance at the earth's mean solar distance within the 300–2500 mm wavelength region.

082.095 An atlas of air absorptions from 4000 cm^{-1} to 250 cm^{-1}.   D. J. Lovell.
Applied Optics, Vol. 8, 2368 - 2369 (1969).

082.096 Presence of HNO_3 in the upper atmosphere.   D. G. Murcray, T. G. Kyle, F. H. Murcray, W. J. Williams.
Journ. Optical Soc. America, Vol. 59, 1131 - 1134 (1969).
Observations of absorption of solar radiation by atmospheric nitric acid obtained on different balloon flights are presented. This absorption occurs in three different wavelength intervals. The use of the very long paths, occurring near sunset, for enhancing weak absorption is discussed.

082.097 Observations of wavelength dependence in stellar scintillation.   J. J. Burke.
Journ. Optical Soc. America, Vol. 59, 1513 (1969).
Abstract Meeting Optical Soc. America.

082.098 The dependence of parameters of telluric lines on the air mass and on the height of the observation place.
N. I. Kozhevnikov, M. A. Kljakotko, G. A. Porfirjeva, G. F. Sitnik.
Astron. Vestn. Vol. 2, 114 - 119 (1968).   In Russian.
The dependence of the half-width and depth of a line on the conditions of an observation is considered. An absorption line contour is calculated for an isothermic atmosphere with density changing according to the barometric law. Two cases are considered: 1) the line absorption coefficient is determined by collision broadening, and 2) the line absorption coefficient is determined by Doppler broadening.

082.099 An optical edge of noctilucent clouds.   N. I. Novozhilov.
Astron. Vestn. Vol. 2, 158 - 160 (1968).   In Russian.
A visible edge of noctilucent clouds is usually not real but only an apparent, optical edge, as far as it corresponds more often not to an edge of a cloudy field but to the earth's terminator on a cloudy field.

082.100 Statistic data on the appearance of noctilucent clouds during the IQSY (1964 - 1965).   Ch. I. Villmann.
Astron. Vestn. Vol. 2, 161 - 170 (1968).   In Russian.

082.101 Experiences with observations of noctilucent clouds.   E. I. Kreem.
Astron. Vestn. Vol. 2, 178 - 182 (1968).   In Russian.

082.102 Phenomena connected with turbidity of the earth's atmosphere.   V. M. Chernov.
Astron. Vestn. Vol. 2, 185 - 188 (1968).   In Russian.

082.103 The determination of positions of artificial noctilucent clouds or meteor trains by the method of transformation of coordinates.   D. B. Uvarov.

Astron. Vestn. Vol. 3, 50 - 56 (1969).   In Russian.
The problem of the determination of coordinates of a diffuse object of the type of noctilucent clouds or meteor trains by coupling of spherical coordinates with measured ones is considered. The coefficients of the corresponding equations are corrected for atmospheric refraction.

082.104 **Patrol of noctilucent clouds in Novosibirsk 1963 - 1967.**
S. S. Vojnov, E. I. Malkov.
Astron. Vestn. Vol. 3, 59 - 61 (1969).   In Russian.

082.105 **Observations of noctilucent clouds in Gorki.**
V. A. Bakanov, E. G. Demidovich, A. P. Poroshin.
Astron. Vestn. Vol. 3, 62 - 63 (1969).   In Russian.
1966 - 1967.

082.106 **Astroclimatic observations in Bodaibo.**
Yu. P. Gorin.
Astron. Vestn. Vol. 3, 159 - 162 (1969).   In Russian.

082.107 **Preliminary results of the instability of the earth's atmosphere in Novosibirsk-Akademgorodok in 1965 - 1966.**   A. D. Belkin, S. S. Voinov.
Yu. M. Fedorov.
Astron. Vestn. Vol. 3, 162 - 166 (1969).   In Russian.

082.108 **The visibility of the penumbra of the earth during lunar eclipses.**   V. M. Chernov.
Astron. Vestn. Vol. 3, 166 - 169 (1969). In Russian.

082.109 **Observations of noctilucent clouds in southern Kazakhstan in September 1968.**
V. V. Chichmar'.
Astron. Vestn. Vol. 3, 169 - 170 (1969).   In Russian.

082.110 **Les variations de la densité atmosphérique dans la thermosphère.**   J. Vercheval.
Acad. Roy. Belgique, Bull. Cl. Sci., 5e Série, Vol. 55, 821 - 841 (1969).
An analysis of the various types of density variations in the mean thermosphere is made by using the orbital data of the satellites 1966–101 G, Secor 6 and ERS 16 obtained between November 1966 and June 1967.

082.111 **Tropospheric and stratospheric concentration of aerosol obtained from daytime measurements of sky polarization in the zenith.**   F. Unz.
Beiträge Phys. Atmosph., Vol. 42, 1 - 35 = Mitt. Astron. Inst. Univ. Tübingen, No. 114 (1969).

082.112 **Morphological investigation of noctilucent clouds on July 3 - 4, 1967.**   I. N. Grishin.
Astron. Vestn. Vol. 3, 108 - 113 (1969).   In Russian.

082.113 **Distribution of excited molecules $O_2(A^3 \Sigma_u^+)$ in the upper atmosphere.**   M. N. Vlasov.
Geomagn. Aeronom. Vol. 9, 894 - 898 (1969). In Russian.

082.114 **Estimation of ionospheric reaction coefficients coordinating the equatorial N(h)-profile and the model of the neutral atmosphere of Jacchia.**
L. P. Goncharov, L. A. Tshepkin.
Geomagn. Aeronom. Vol. 9, 929 - 931 (1969).   In Russian.
Brief information.

082.115 **Concentration of excited molecules $O_2(^1\Delta g)$ in the upper atmosphere.**   M. N. Vlasov.
Geomagn. Aeronom. Vol. 9, 940 - 942 (1969).   In Russian.
Brief information.

082.116 **Atmospheric extinction.**   A. W. J. Cousins.
Monthly Notes Astron. Soc. Southern Africa,
Vol. 28, 105 - 106 (1969). – Note.

082.117 **Astronomical observing conditions in Northern Chile.**   J. Stock.
ESO Bull. No. 5, p. 35 - 40 (1969).

082.118 **Meteorological observations on La Silla in 1968.**
A. B. Muller.
ESO Bull. No. 7, p. 19 - 33 (1969).

082.119 **Analisi del seeing solare nella zona di Catania, mediante misuratore di microfluttuazioni termiche in vicinanza del suolo.**   G. Godoli, L. Paternò.
Atti 17. Convegno Ass. Geofis. Italiana, Roma 1968, 8 pp. = Oss. Astrofis. Catania Pubbl., Nuova Serie; No. 121 (1968).
A short description is given of the construction and operation at the Catania Astrophysical Observatory of an apparatus for recording microthermal fluctuation in order to determine seeing conditions for solar observations.

082.120 **Condizioni meteorologiche a Serra La Nave nel 1967 carattere del gradiente pluviometrico verticale dell'Etna.**   F. Affronti, C. Blanco.
Oss. Astrofis. Catania Pubbl., Nuova Serie, No. 129, 18 pp. (1968).

082.121 **Osservazioni meteorologiche e preparazione della campagna di seeing per l'Osservatorio Astrofisico Nazionale.**   G. Mannino, O. Bendinelli.
Atti XII Riunione Soc. Astron. Italiana, L'Aquila 1968, p. 112 (1969). – Abstract SAI.

082.122 **On the influence of corpuscular streams upon the atmospheric circulation of the earth. I. The character of atmospheric pressure variations for different seasons.**
E. R. Mustel.
Astron. Zhurn. Akad. Nauk SSSR, Vol. 46, 1169 - 1183 (1969). In Russian. English translation in Soviet Astron. AJ, Vol. 13, No. 6.
The principal purpose is to find out the distribution of signs of the ground level atmospheric pressure variations over the surface of the northern hemisphere of the earth after the arrival of a corpuscular stream.

082.123 **On the accuracy of Hα-nightglow measurements and its intensity variations.**
Z. V. Karjaguina, V. E. Mozjaeva, P. V. Sheglov.
Astron. Tsirk. No. 532, p. 7 - 8 (1969).   In Russian.

082.124 **Spatial distribution of single-charged particles and nuclei of helium above the atmosphere in the equatorial region.**   R. N. Basilova, N. L. Grigorov, I. A. Savenko.
Kosmich. Issled. Vol. 7, 895 - 899 (1969).   In Russian.

082.125 **The temperature of the emission [OI] $\lambda = 6300$ Å by observations in Zvenigorod.**
G. A. Nasirov.
Geomagn. Aeronom. Vol. 9, 762 - 763 (1969).
In Russian. – Brief information.

082.126 **Hydrogen and helium emissions in the upper atmosphere.**   N. N. Shefov.
Geomagn. Aeronom. Vol. 9, 1048 - 1052 (1969).   In Russian.

082.127 **Some results of regular measurements of the intensity of X-ray photons in the atmosphere.**
G. A. Bazilevskaya, A. N. Kvashnin, A. F. Krasotkin,

A. K. Pankratov, A. N. Charakhch'yan.
Geomagn. Aeronom. Vol. 9, 1071 - 1073 (1969).
In Russian. — Brief information.

082.128   Effects of changes in complex part of the refractive
index on polarization of light scattered from haze
and clouds.   G. N. Plass, G. W. Kattawar.
Applied Optics, Vol. 8, 2489 - 2495 (1969).
The polarization and radiance of the reflected and
transmitted radiation is calculated for a continental haze
model and for a nimbostratus cloud model.

082.129   Results of airglow observations according to the
IQSY program.
S. V. Karjagina, V. E. Moshajeva.
Trudy Astrofiz. Inst. Alma-Ata, Vol. 13, 43 - 54 (1969).
In Russian.
The paper presents results of spectrographic observations
of forbidden OI, Na, Hα, OH lines during the IQSY. The inten-
sities of all emissions are obtained in absolute units. The rota-
tional temperatures of OH bands (6.1), (9.3), and (5.2) are
determined.

082.130   About some properties of atmospheric aerosols.
T. P. Toropova.
Trudy Astrofiz. Inst. Alma-Ata, Vol. 13, 55 - 62 (1969).
In Russian.
From measurements of some optical properties of the
atmosphere a conclusion on the distribution of particles of
atmospherical aerosols according to size is drawn.

082.131   On the wavelength dependence of the aerosol
scattering indicatrix in the atmosphere.
T. P. Toropova.
Trudy Astrofiz. Inst. Alma-Ata, Vol. 13, 63 - 66 (1969).
In Russian.
A wavelength-dependent definition of the form of the
aerosol scattering indicatrix from data of distribution of the
sky's brightness in any relative units is suggested.

082.132   About some optical atmospheric properties at
heights of 750 and 1450 m above sea level.
N. M. Ibraimov, T. P. Toropova, G. A. Kharitonova.
Trudy Astrofiz. Inst. Alma-Ata, Vol. 13, 67 - 72 (1969).
In Russian.
In this paper some data on optical properties of the
earth's atmosphere at heights of 750 and 1450 m above sea
level are given. It is shown that small particles have a considerable influence on the extinction of light in the smoke of cities.

082.133   Some data on the scattering function in the bottom
layer of the atmosphere.
T. P. Toropova, V. V. Keguleekhes, K. M. Salamachin.
Trudy Astrofiz. Inst. Alma-Ata, Vol. 13, 73 - 77 (1969).
In Russian.

082.134   Some data about polydisperse mediums with re-
fractive index 1.50 and 1.25.
T. P. Toropova, S. O. Obasheva.
Trudy Astrofiz. Inst. Alma-Ata, Vol. 13, 78 - 81 (1969).
In Russian.

082.135   Experimental data about extinction of coherent
radiation in a turbid medium.
T. P. Toropova, L. L. Slobodkina.
Trudy Astrofiz. Inst. Alma-Ata, Vol. 13, 82 - 90 (1969).
In Russian.

082.136   On using a Fabry–Pérot étalon for investigating the
absorption spectra of the atmosphere.
J. I. Rudnev, L. A. Sataeva.

Trudy Astrofiz. Inst. Alma-Ata, Vol. 13, 91 - 97 (1969).
In Russian.

082.137   Determination of geophysical parameters from
thermal radio emission measurements on the artificial earth satellite "Cosmos-243".
A. E. Basharinov, A. S. Gurvich, S. T. Egorov.
Dokl. Akad. Nauk SSSR, Ser. Mat. Fiz., Vol. 188, 1273 -
1276 (1969).   In Russian.
Thermal radio emission of the earth was measured on
Cosmos 243 at λ 8.5, 3.4, 1.35, and 0.8 cm. Water vapour and
hydrometeor content over oceans as well as surface temperature profiles on oceans and continents have been derived and
discussed.   B. Onderlička

082.138   Inhomogeneous semi-infinite atmosphere in the
case of pure scattering.   E. G. Ianovitskii.
Dokl. Akad. Nauk SSSR, Ser. Mat. Fiz., Vol. 189, 74 - 76
(1969).   In Russian.
General formulae for the brightness coefficient of diffu-
sely reflected radiation and for Lambert's albedo are derived
for an inhomogeneous plane semi-infinite atmosphere with a
nonspherical indicatrix depending on optical depth. The results
are specialized for pure scattering, and two cases are discussed:
a) the indicatrix is of the form $1 + x_1 \cos \gamma$, x_1 depending
on optical depth, b) two layers, the indicatrix being spherical
in the upper one and of the form just mentioned, but with
constant x_1, in the lower one.   B. Onderlička

082.139   The nature of noctilucent clouds.
G. Witt.
Space Research IX, Proc. Tokyo 1968, p. 157 - 169 (1969).

082.140   Nucleation and growth of noctilucent cloud
particles.   E. Hesstvedt.
Space Research IX, Proc. Tokyo 1968, p. 170 - 174 (1969).

082.141   A condensation model of noctilucent cloud formation.   A. D. Christie.
Space Research IX, Proc. Tokyo 1968, p. 175 - 182 (1969).

082.142   Extraterrestrial origin of noctilucent clouds.
R. K. Soberman.
Space Research IX, Proc. Tokyo 1968, p. 183 - 189 (1969).

082.143   On the determination of night sky brightness from
a space vehicle.   N. A. Dimov, A. B. Severny.
Space Research IX, Proc. Tokyo 1968, p. 228 (1969).

082.144   The effective electron-loss coefficient at heights
of 200 to 400 km during increasing solar activity
(1965 – 1966).   N. M. Shutte, I. A. Knorin.
Space Research IX, Proc. Tokyo 1968, p. 267 - 272 (1969).

082.145   Ionospheric effects of the faster rotation of the
upper atmosphere.   N. Matuura.
Space Research IX, Proc. Tokyo 1968, p. 273 - 278 (1969).

082.146   Upper atmosphere parameters obtained from recent
falling sphere measurements at Eglin, Florida.
A. C. Faire, K. S. W. Champion.
Space Research IX, Proc. Tokyo 1968, p. 343 - 353 (1969).

082.147   Consequences of fine structure in the vertical tempe-
rature profile on radiative transfer in the mesosphere.
S. R. Drayson, E. S. Epstein.
Space Research IX, Proc. Tokyo 1968, p. 376 - 384 (1969).

082.148   Review of atmospheric structure in the region
30 – 100 km.   G. V. Groves.
Space Research IX, Proc. Tokyo 1968, p. 449 - 458 (1969).

082.149 **Review of the properties of the lower thermosphere.**
K. S. W. Champion.
Space Research IX, Proc. Tokyo 1968, p. 459 - 477 (1969).

082.150 **The neutral atmosphere above 200 km: A progress report.** L. G. Jacchia.
Space Research IX, Proc. Tokyo 1968, p. 478 - 486 (1969).
Recent results of observations and analyses of atmospheric data in the region above 200 km have uncovered several shortcomings in existing atmospheric models. These shortcomings are reviewed in detail, and suggestions are made how to overcome them in the construction of future models.

082.151 **Temperature and density of the thermosphere in 1966 – 1967.** M. Ya. Marov, A. M. Alpherov.
Space Research IX, Proc. Tokyo 1968, p. 487 - 498 (1969).

082.152 **Thermospheric densities and temperatures from EUV absorption measurements by OSO-III.**
H. E. Hinteregger, L. A. Hall.
Space Research IX, Proc. Tokyo 1968, p. 519 - 529 (1969).

082.153 **Changes in the lower exosphere since solar minimum.** G. M. Keating.
Space Research IX, Proc. Tokyo 1968, p. 534 - 546 (1969).

082.154 **Equatorial atmospheric density obtained from San Marco II satellite between 200 and 350 km.**
L. Broglio.
Space Research IX, Proc. Tokyo 1968, p. 547 - 554 (1969).

082.155 **Angular characteristics of the reflectance of the earth-atmosphere system as obtained from a synchronous satellite.** E. Raschke.
Space Research IX, Proc. Tokyo 1968, p. 580 - 585 (1969).

082.156 **Der Jahresgang der Nachmittagsmaxima der terrestrischen Szintillation.**
D. Paperlein.
Optik, Vol. 30, 93 - 95 (1969).

082.157 **On atmospheric propagation of a laser beam.**
D. H. Höhn.
Optik, Vol. 30, 161 - 170, 234 - 256 = Mitt. Sternw. Univ. Tübingen No. 118 (1969). In German.
The results of experimental investigations of the propagation of a laser beam through the lower atmosphere are compared with the results of theoretical optics in a turbulent medium.

082.158 **On the atmospheric frictional height at Mizusawa.**
K. Takahasi, E. Onodera, N. Kikuchi.
Proc. International Latitude Obs. Mizusawa, No. 9, p. 25 - 29 (1969). In Japanese.

082.159 **Determination of atmospheric density from observations of meteors and artificial luminescent clouds.**
L. A. Andreeva, L. A. Katasev.
Byull. Stantsij Optichesk. Nablyud. Iskusstv. Sputnikov Zemli No. 54, p. 20 - 29 (1969). In Russian.
Experimental data on the atmospheric density in the meteor zone and results of studies of atmospheric winds obtained from observations of the drift of meteor trains are discussed together with data based on determinations of the atmospheric density from observations of artificial luminescent clouds.

082.160 **Main source of thermal fluctuations of the upper atmosphere within time intervals of 1 - 30 days.**
V. E. Tchertoprud.
Byull. Stantsij Optichesk. Nablyud. Iskusstv. Sputnikov Zemli No. 54, p. 31 - 35 (1969). In Russian.

082.161 **Determination of the parameters of the atmosphere from the deceleration of a satellite taking into account its orientation.**
V. V. Beletsky, G. I. Zmievskaya, M. Ya. Marov.
Byull. Stantsij Optichesk. Nablyud. Iskusstv. Sputnikov Zemli No. 54, p. 39 (1969). In Russian.

082.162 **Astronomical observing conditions in Chile, II.**
G. Carrasco, J. Stock.
Inform. Bull. Southern Hemisphere, No. 14, p. 42 - 43 (1969).
Night-time cloud statistics for Central Chile and Northern Central Chile are reported.

082.163 **On the quality of planetary images obtained with the AZT-8 telescope of the Charkov Astronomical Observatory.** V. N. Dudinov.
Vestn. Khar'kov. Univ. No. 34, (Ser. Astron. No. 4), p. 39 - 49 (1969). In Russian.

082.164 **Die Helligkeit des Nachthimmels bei der Deklination –19° im Wellenlängenbereich von 3900 bis 7100 Å.** J. Dachs.
Mitt. Astron. Ges. No. 27, p. 158 (1969). – Abstract AG.

082.165 **De polarisatie van het hemellicht.**
G. P. Können.
Hemel en Dampkring, Vol. 67, 386 - 387 (1969).

Light Scattering in the Atmosphere. Part 2.
See Abstr. 003.005.

Fragen der atmosphärischen Optik.
See Abstr. 003.082.

Radiation in the Atmosphere.
See Abstr. 003.083.

The influence of refraction in fundamental astrometry. See Abstr. 041.032.

A determination of the orbit of Secor 6 rocket (1966 – 51A). See Abstr. 054.007.

Atmospheric transmission and solar spectroscopy in the submillimeter wavelength range. See Abstr. 071.002.

Solar site selection based on time-lapse photography of granulation. See Abstr. 071.038.

Studies of granular velocities. I: Granular Doppler shifts and convective motion. See Abstr. 071.067.

Determining the parameters of small-scale turbulence of the atmosphere from combined observations of meteors. See Abstr. 104.045.

The atmospheric extinction in photoelectric photometry. See Abstr. 113.043.

083 Ionosphere

083.001 New theory of mid-latitude sporadic E.
D. Layzer.
Nature, Vol. 223, 794 - 797 (1969).
Unified explanations have been derived for the main statistical and structural characteristics of sporadic E.

083.002 The electron content of the low latitude ionosphere.
J. E. Titheridge, W. D. Smith.
Planet. Space Sci. Vol. 17, 1667 - 1676 (1969).

083.003 Diurnal and seasonal variation of atmospheric ion composition; correlation with solar zenith angle.
H. C. Brinton, R. A. Pickett, H. A. Taylor, Jr.
Journ. Geophys. Res. Vol. 74, 4064 - 4073 (1969).
A global measurement of the diurnal and seasonal variations of the primary atmospheric ions in the earth's topside ionosphere between 280 and 2700 km has been made using the ion mass spectrometer on Explorer 32. Evidence of direct correlation between solar zenith angle and the low and mid-latitude distributions of thermal O^+, N^+, He^+, and H^+ was obtained as the satellite orbit plane traversed one complete diurnal cycle between June 11 and October 5, 1966.

083.004 Measurement and interpretation of power spectrums of ionospheric scintillation at a sub-auroral location.
T. J. Elkins, M. D. Papagiannis.
Journ. Geophys. Res. Vol. 74, 4105 - 4115 (1969).
A study of the power density spectrum of ionospheric scintillation of a radio star and satellites, at a subauroral location, has revealed the following: (1) Scintillations generally display a 'pink' noise spectrum, with almost uniform spectral density at frequencies below ~0.01 Hz, and decreasing spectral density at frequencies above this value. (2) At frequencies $f \gtrsim 0.01$ Hz, the power spectrum varies as f^{-n}, with n being typically 2.7 (i.e., a decrease of 8 db per octave). (3) The width of the spectrum displays a diurnal variation such that, on the average, higher scintillation frequencies are observed at night than during the day.

083.005 Correlation of sudden frequency deviations with solar microwave bursts.
S. Basu, S. R. Chowdhury.
Journ. Geophys. Res. Vol. 74, 4175 - 4177 (1969).
The paper presents some investigations on the correlation of sudden frequency deviations of HF radio waves propagated through the ionosphere and solar microwave bursts at 606, 2695, and 8800 MHz.

083.006 Ionization rates due to the attenuation of 1 - 100 Å nonflare solar X rays in the terrestrial atmosphere.
W. Swider, Jr.
Rev. Geophys. Vol. 7, 573 - 594 (1969).
Solar fluxes and absorption cross sections are reviewed for 1 - 100 Å X rays. Review of the cross sections is extended to 300 Å to more completely validate the cross sections selected at shorter wavelengths. Convenient lists of the solar flux and appropriate cross sections are provided, and simple accurate formulas are given for them. Numerical rates for q, the total ion-pair production rate, are tabulated for three nonflare solar conditions at various altitudes and solar zenith angles. Production rates due to other ionization sources important at these altitudes and angles are also tabulated.

083.007 Ionospheric electron content at midlatitudes near the minimum of the solar cycle.
R. G. Merrill, R. S. Lawrence.
Journ. Geophys. Res. Vol. 74, 4661 - 4666(1969).

083.008 Sporadic-E ionization over Ahmedabad through the half solar cycle 1954 - 1957.
K. M. Kotadia.
Journ. Atmosph. Terr. Phys. Vol. 31, 1137 - 1146 (1969).
In this paper, characteristic variations of sporadic-E-layer over Ahmedabad are given for the half solar cycle (1953 - 1957) of increasing sunspot activity. Data of years prior to the I. G. Y. have been reexamined.

083.009 Electron content measurements at Hyderabad using beacon satellite transmissions.
C. R. Rao, E. B. Rao, G. C. Subbaraju.
Journ. Atmosph. Terr. Phys. Vol. 31, 1197 - 1203 (1969).
Measurements of electron content carried out at Hyderabad, a low latitude station, are presented for the period October 1964 – February 1967. The diurnal and seasonal variations are discussed and results compared with those from other stations.

083.010 The conductivity of lower ionosphere deduced from sudden enhancements of strength (SES) of
v.l.f. transmissions.   M. Yamashita.
Journ. Atmosph. Terr. Phys. Vol. 31, 1049 - 1057 (1969).

083.011 Daytime ion composition in the height range of 150 - 1500 km for solar minimum condition.
V. P. Bhatnagar.
Journ. Atmosph. Terr. Phys. Vol. 31, 1059 - 1076 (1969).
In this paper an electron and ion composition model of the ionosphere is developed for the height range 150 - 1500 km for day-time at moderate latitudes, on the basis of given reference data. The model will specifically give the vertical distribution of atomic oxygen, helium and atomic hydrogen ions, electron density, mean ionic mass and plasma scale height. These distributions would normally hold for a dip value of 60°, a solar zenith of 60° and minimum solar activity.

083.012 Perturbations in a non-linear F-region at night.
C. H. Cummack.
Journ. Atmosph. Terr. Phys. Vol. 31, 1107 - 1118 (1969).
The effects of non-linear recombination and variable drift in the night-time F-region are examined. It is found that close agreement with observations is possible and a wide variety of ionograms may be synthesised.

083.013 The hysteresis variation in $F2$-layer parameters.
M. S. V. Gopal Rao, R. S. Rao.
Journ. Atmosph. Terr. Phys. Vol. 31, 1119 - 1125 (1969).
The variations in $f_0 F2$ and $M(3000)F2$ with sunspot number for the 19th solar cycle are studied to assess the difference between the rising and falling parts of the solar cycle. The results are discussed in the light of a hysteresis effect observed in the variation of magnetic activity with the sunspot number.

083.014 Drift and diffusion of filament ion clouds in the earth's ionosphere.   M. Giles, G. Martelli.
Planet. Space Sci. Vol. 17, 1693 - 1707 (1969).
The behaviour of a needle shaped perturbation ('filament ion cloud') in the plasma of the ionospheric $F1$-layer in the presence of a neutral wind and a uniform electric field having a component perpendicular to the geomagnetic field has been investigated.

083.015 Some short-term auroral breakup effects on sporadic-E.   B. J. Cheney.
Australian Journ. Phys. Vol. 22, 549 - 553 (1969). – Short

communication.

083.016 **Effect on the lower ionosphere of X-rays from Scorpius XR-1.**
S. Ananthakrishnan, K. R. Ramanathan.
Nature, Vol. 223, 488 - 489 (1969).

The records of field strengths of radiowaves on 164 MHz transmitted from Tashkent received at Ahmedabad from 1960 onward have shown that there is a pronounced minimum in field strength at night and that the time of this minimum nearly coincides with the time of transit of the X-ray star Sco X-1 across the meridian of 71°E the midpoint between Tashkent and Ahmedabad. This is attributed to the ionization in the lower D region by X-rays from Sco X-1.

083.017 **Ionospheric aerodynamics.**
A. V. Gurevich, L. P. Pitaevskii, V. V. Smirnova.
Space Sci. Rev. Vol. 9, 805 - 871 (1969).

Contents: Basic equations and law of similarity; Flow past a semi-plane; Flow of plasma, containing a mixture of ions, past a semi-plane; Flow past a wedge; Flow past the rounded edge of a body; Flow past a plate; Flow past a cylinder; Flow past a disc; Influence of an electric field on the motion of the ions; Flow past highly elongated bodies; Plasma instability in the trail of a moving body; The potential of the body's surface; Numerical calculations taking account of a finite Debye radius; Experimental investigations of the disturbed zone in the vicinity of rockets and satellites in the ionosphere; Scattering of radiowaves by a disturbed zone in the vicinity of a layer with $\epsilon = 0$.

083.018 **A new analysis of eclipse effects in the equatorial F-region.** N. J. Skinner.
Journ. Atmosph. Terr. Phys. Vol. 31, 1333 - 1344 (1969).

F-region electron density measurements obtained during the solar eclipse of 2 October, 1959 at Ibadan and Maiduguri are re-examined. A method has been developed to analyse the data in which the transport term in the electron density continuity equation is retained, and its effect is found to be important.

083.019 **Messung elektrischer Felder in der Ionosphäre mit künstlichen Plasmawolken.** G. Haerendel.
Naturwissenschaften, 56. Jahrgang, 545 - 552 (1969).

083.020 **Origin of the term ionosphere.**
G. W. Gardiner.
Nature, Vol. 224, 1096 (1969).

083.021 **Lunar oscillations in the D-region absorption at Singapore.** S. C. Chakravarty, R. G. Rastogi.
Nature, Vol. 223, 939 - 940 (1969).

The letter describes the lunar oscillations in the D-region deduced from the ananlysis of the absorption of vertical pulsed radio waves at Singapore. The amplitude of the lunar semi-monthly tide in the D-months is found to be more than that in the E or J-months following closely the variation observed in the absorption tide at Colombo. The lunar tides in the D and F-regions for Singapore are opposite in phase possibly because the transition region for the D-region tides lies closer to the magnetic equator than that for the F-region tides.

083.022 **Search for the effect of stellar X-ray on the night-time lower ionosphere.**
B. Burgess, T. B. Jones.
Nature, Vol. 224, 680 - 681 (1969).

An examination of VLF phase delay propagation data has been made in order to determine whether stellar X-radiation contributes significantly to the night-time level of ionization in the lower ionosphere. It is concluded that X-ray

effects are probably being swamped by ionization variations which appear to originate from density changes at the 85 - 90 km level.

083.023 **Secular variation in F-region response to sunspot number.** L. M. Muggleton.
Journ. Atmosph. Terr. Phys. Vol. 31, 1413 - 1419 (1969).

A long-term oscillatory variation has been detected in the relationship between F-region ionization and visible solar activity. The dominant component appears to be a cosine term of period equal to four sunspot-cycles. The last crossing of the zero line appears to have occurred during the 1947 - 48 sunspot maximum, making that epoch of importance to statistical methods designed to remove this secular variation from ionospheric studies.

083.024 **The measurement of a lunar tide in the phase height of the E-region.** E. C. Butcher, K. Weekes.
Journ. Atmosph. Terr. Phys. Vol. 31, 1421 - 1434 (1969).

The measurements of the lunar tidal influence on the group height of radio waves reflected in the E-region made by Appleton and Weekes and others have given rise to problems of interpretation. In this paper we describe an attempt to determine the corresponding lunar influence on the phase height of reflection in the hope that this would give some help in the interpretation.

083.025 **An intermediate ionospheric layer at high latitudes.**
A. G. French.
Journ. Atmosph. Terr. Phys. Vol. 31, 1435 - 1438 (1969).

The behaviour of an intermediate ionospheric layer at high latitudes, its relationship to geomagnetic and auroral activity, and its continuity with the F-region are discussed briefly.

083.026 **On the field alignment of small ionospheric irregularities.** J. E. Titheridge.
Journ. Atmosph. Terr. Phys. Vol. 31, 1439 - 1444 (1969).

All transits of the satellites BeB and BeC observed over a period of 3 yr were examined to find cases when the angle between the ray path and the magnetic field in the ionosphere became less than 10°. The amplitude and polarization scintillations occurring at these times were compared with the scintillations occurring, on the same transits, when the ray path made an angle of 30° to the magnetic field.

083.027 **Whistlers and VLF emissions.**
R. A. Helliwell.
Physics of the Magnetosphere, Boston College 1967, p. 106 - 146 (1968). Tutorial lecture.

The following topics are treated: (1) a historical background of whistlers and related phenomena; (2) the observational facts, from both the ground and satellites; (3) the application of these measurements to electron density mapping and techniques for the detection of ions using whistlers; (4) the nature of electromagnetic emissions from the magnetosphere and ideas relating to their origin.

083.028 **Energy transfer to and through ionospheric electrons.** N. P. Carleton.
Physics of the Magnetosphere, Boston College 1967, p. 556 - 562 (1968). – Invited research paper.

083.029 **Uniform stratified layers in the equatorial E-region.**
U. R. Rao, S. S. Degaonkar, Banshidhar, R. M. Patel.
Nature, Vol. 224, 676 - 677 (1969).

The paper describes the results on the measurements of electron density in the E layer over Thumba Equatorial Rocket Launching Station, India, obtained using a rocket-borne High Frequency Capacitance Probe, on February 26, 1969. The results show a sharp decrease in the electron density beyond a peak at 110 km, indicating a valley in the ionization. Beyond this altitude, a wave-like structure having a vertical wavelength of 8 km is indicated. The vertical stratification in the E-region with a characteristic uniform spacing of about 8 km between 120 - 140 km is shown to be consistent with the internal gravity wave mechanism.

083.030 Spread–F echoes at Ahmedabad over a solar cycle.
R. G. Rastogi, P. P. Kulkarni.
Ann. Géophys. Vol. 25, 577 - 587 (1969).

083.031 Über die Rolle der L_α-Strahlung bei der Ionisation der unteren Ionosphäre. V. F. Tulinov.
Kosmich. Issled. Vol. 7, 575 - 579 (1969).   In Russian.

083.032 Korpuskularstrahlung als Ionisationsquelle der unteren Ionosphäre zur Nachtzeit.
V. F. Tulinov.
Kosmich. Issled. Vol. 7, 738 - 740 (1969).   In Russian.

083.033 Global electron density distributions from topside soundings.   K. L. Chan, L. Colin.
Proc. IEEE, Vol. 57, 990 - 1004 (1969).

083.034 The high-latitude ionosphere.
D. H. Jelly, L. E. Petrie.
Proc. IEEE, Vol. 57, 1005 - 1012 (1969).

083.035 Ionospheric irregularities observed by topside sounders.   W. Calvert, J. M. Warnock.
Proc. IEEE, Vol. 57, 1019 - 1025 (1969).

083.036 The topside ionosphere during geomagnetic storms.
E. S. Warren.
Proc. IEEE, Vol. 57, 1029 - 1036 (1969).

083.037 A review of the theories concerning the equatorial F2 region ionosphere.   R. A. Goldberg.
Proc. IEEE, Vol. 57, 1119 - 1126 (1969).

083.038 The middle-latitude F region during some severe ionospheric storms.   R. B. Norton.
Proc. IEEE, Vol. 57, 1147 - 1149 (1969).

083.039 Electron densities less than 100 electron cm^{-3} in the topside ionosphere.
P. L. Timleck, G. L. Nelms.
Proc. IEEE, Vol. 57, 1164 - 1171 (1969).

083.040 Ionospheric effect of X-rays from Scorpius XR-1.
I. G. Poppoff, R. C. Whitten.
Nature, Vol. 224, 1187 - 1188 (1969).
      The effect of X-rays from Scorpius XR-1 on the night-time terrestrial atmosphere is compared with the effect of other night-time ionization sources. The ionization rate due to Lyman-alpha was found to be greater than that due to Sco XR-1; the relative rates, however, are not as important as the time required for the candidate sources to build-up a detectable concentration of free electrons. It is concluded that it is not possible to justify the hypothesis that astronomical X-ray sources can be detected by their effects on the terrestrial ionosphere.

083.041 The influence of Forbush-effects on the state of the cosmic layer in the low ionosphere.
P. Velinov, L. I. Dorman, G. Nestorov.
Geomagn. Aeronom. Vol. 9, 813 - 817 (1969).   In Russian.

083.042 About some effects of radiowave dispersion in the ionosphere. I.
L. M. Erukhimov, V. Yu. Trakhtengertz.

Geomagn. Aeronom. Vol. 9, 834 - 841 (1969).   In Russian.

083.043 About cyclic variations of anomalous ionization of the F2-layer in the region of the geomagnetic pole.   A. S. Besprozvannaya, L. A. Yudovitch.
Geomagn. Aeronom. Vol. 9, 856 - 859 (1969).   In Russian.

083.044 Influence of the ionosphere on radio-astronomical observations with artificial earth satellites.
V. A. Danilin, A. N. Kazantzev.
Geomagn. Aeronom. Vol. 9, 922 - 923 (1969).   In Russian.
Brief information.

083.045 The effect of oxygen cooling on ionospheric electron temperatures.   A. Dalgarno, M. B. McElroy, M. H. Rees, J. C. G. Walker.
Planet. Space Sci. Vol. 16, 1371 - 1380 = Contr. Kitt Peak National Obs. No. 358 (1968).
      There are discrepancies between ionospheric electron temperatures derived from Thomson scatter data and electron temperatures predicted from the solar ultra-violet heat source. The inclusion of electron cooling by excitation of the fine-structure levels of atomic oxygen removes the discrepancy throughout the day at all altitudes above 320 km.

083.046 Polar cap ionospheric response to solar cosmic ray events observed by Mariners 2 and 4.
H. L. Stolov.
Solar Flares and Space Research, Tokyo 1968, p. 310 - 318 (1969).

083.047 Anomalous features of the electron density in the topside ionosphere.
J. Sayers, J. W. G. Wilson, B. Loftus.
Proc. Roy. Soc. London, Ser. A, Vol. 311 (No. 1507), 501 - 516 (1969).

083.048 A dynamic model of the F-region of the ionosphere, including temperature variations.
N. N. Klimov, V. M. Polyakov, G. I. Gershengorn, G. M. Kuznetzova.
Geomagn. Aeronom. Vol. 9, 655 - 660 (1969).   In Russian.

083.049 Distribution of electron concentration in the outer part of the F2 region under calm and disturbed conditions.   M. N. Fatkullin, A. D. Legenka.
Geomagn. Aeronom. Vol. 9, 743 - 746 (1969).
In Russian. – Brief information.

083.050 Horizontal ionospheric drifts in the period of low solar activity.   E. S. Kazimirovsky.
Geomagn. Aeronom. Vol. 9, 1041 - 1047 (1969).   In Russian.

083.051 The dependence of the characteristics of the sporadic E-layer of the equatorial type on solar activity.
V. N. Pogrebnoi.
Geomagn. Aeronom. Vol. 9, 1095 - 1096 (1969).
In Russian. – Brief information.

083.052 Dynamic characteristics and the structure of the sporadic E-layer in the course of aurorae.
G. A. Zherebtsov, V. A. Kurilov.
Geomagn. Aeronom. Vol. 9, 1099 - 1101 (1969).
In Russian. – Brief information.

083.053 Absorption of cosmic radio radiation.
N. A. Kochenova, M. D. Fligel.
Geomagn. Aeronom. Vol. 9, 1101 - 1103 (1969).
In Russian. – Brief information.

083.054 Rocket observation of the ionosphere in twilight

conditions.   K. Hirao, H. Oya, T. Tohmatsu,
T. Ogawa.
Space Research IX, Proc. Tokyo 1968, p. 262 - 266 (1969).

**083.055 Lower thermosphere ions in the nighttime auroral
zone.**   G. Horiuchi.
Space Research IX, Proc. Tokyo 1968, p. 433 - 441 (1969).

**083.056 Changes in the total electron content and slab
thickness of the ionosphere during a magnetic
storm in June 1965.**   G. N. Taylor, R. D. S. Earnshaw.
Journ. Atmosph. Terr. Phys., Vol. 31, 211 - 216 = Astron.
Contr. Univ. Manchester, Ser. II, Jodrell Bank Repr. No. 375
(1969).

083.057 Mesures ionosphériques.
Edited by "Division des Prévisions Ionosphériques
du C.N.E.T.".
Bull. Mesures Ionosph. (BMI), Vol. 8, Nos. 4 - 9 [Dakar,
Djibuti, Paris–Saclay, Tahiti, Tananarive] (1969). – 1966
Avril – Septembre.

**Some relationships between solar X-ray bursts and
SPA's produced on VLF propagation in the lower ionosphere.**
See Abstr. 076.027.

**Consequences of very small planetary magnetic
moments.**   See Abstr. 084.222.

084 Aurorae, Geomagnetic Field, Radiation Belts

Aurorae

084.001 Conjugate and closely-spaced observations of auroral radio absorption—I. Seasonal and diurnal behaviour. J. K. Hargreaves.
Planet. Space Sci. Vol. 17, 1459 - 1484 (1969).

084.002 Conjugate and closely-spaced observations of auroral radio absorption—II. Correlation properties. J. K. Hargreaves.
Planet. Space Sci. Vol. 17, 1485 - 1495 (1969).

084.003 Auroral morphological similarities at two magnetically conjugate stations: Buckles Bay and Kotzebue. F. R. Bond.
Australian Journ. Phys. Vol. 22, 421 - 433 (1969).

This paper gives a preliminary study of auroral displays photographed by all-sky cameras during November 1962 to March 1963 at the approximately magnetically conjugate stations, Buckles Bay, Macquarie Island, and Kotzebue, Alaska. This information suggests conformity with the concept that electron precipitation guided along magnetic field lines is a major factor in the morphological pattern of the aurora, and that the visible aurora may portray in miniature the pattern of deformation in the magnetosphere of the magnetic field lines connecting the conjugate phenomena.

084.004 A physical model of a radio aurora event. D. R. McDiarmid, A. G. McNamara.
Canadian Journ. Phys. Vol. 47, 1271 - 1281 (1969).

A physical model is proposed to relate and explain the data obtained during a radio aurora event which occurred during the International Geophysical Year. One of Boström's electrojet models and the Farley two-stream instability form the basis of this model. The echoes obtained during the event were found to originate in the plasma bands associated with the visual forms and the orientation of the electric field was found to be east—west, which is in agreement with that predicted by the magnetospheric convection model of Brice.

084.005 Bursts of VLF-emission simultaneous with sharp dip of absorption and SC in geomagnetic H. L. Harang.
Planet. Space Sci. Vol. 17, 1565 - 1572 (1969).

084.006 Morphology of the pulsating aurora. G. J. Kvifte, H. Pettersen.
Planet. Space Sci. Vol. 17, 1599 - 1607 (1969).

Observations of pulsating aurora were made from Tromsø during the winter 1967—68. The data which were obtained from 4 photometers pointing in different directions in the meridian plane, have been analysed to investigate the diurnal variation of the occurrence of pulsating aurora. Also the dependence of latitude and magnetic activity has been investigated. The results indicate that the pulsating aurora mainly occurs in a spiral shaped region being situated in the southern part of the morning sector of the auroral oval.

084.007 The auroral orientation curves for the IQSY. G. Gustafsson, Y. I. Feldstein, N. F. Shevnina.
Planet. Space Sci. Vol. 17, 1657 - 1666 (1969).

The auroral orientation curves for IQSY have been evaluated from the all-sky camera data obtained at 13 stations. Comparing the orientation curves from IGY and IQSY it has been found that the dawn-dusk asymmetry which was pronounced in the IGY curves has disappeared in those of IQSY.

On the other hand the day-night asymmetry remained unchanged between the two periods of time.

084.008 The occurrence of the NII 4176 Å emission line in connection with type-B auroras. J.-C. Gérard, O. E. Harang.
Planet. Space Sci. Vol. 17, 1680 - 1682 (1969). —Research note.

084.009 Electric acceleration of auroral particles. J. W. Chamberlain.
Rev. Geophys. Vol. 7, 461 - 482 = Contr. Kitt Peak National Obs. No. 448 (1969).

Measurements of the distribution of auroral particles with energy and pitch angle have not given much of a clue to the precipitating mechanism, except to indicate that it occurs locally, in the magnetosphere. The available data, including observations of periodicities in auroral bombardment, hint that at least two different mechanisms are operative. With the aid of a simplified model for an accelerating electric field, the kind of dependences on angle and energy to be expected are illustrated.

084.010 Satellite observations of the average properties of auroral particle precipitations: Latitudinal variations. R. D. Sharp, D. L. Carr, R. G. Johnson.
Journ. Geophys. Res. Vol. 74, 4618 - 4630 (1969).

084.011 Satellite measurements of auroral ultraviolet and 3914-Å radiation. E. G. Joki, J. E. Evans.
Journ. Geophys. Res. Vol. 74, 4677 - 4686 (1969).

Auroral ultraviolet radiation in the 1230- to 1840-Å region was observed on many satellite crossings of the auroral zones and was seen to occur with charged-particle precipitation. At those times when the auroral 3914-Å intensity was sufficiently great to permit its observation despite the high moonlight background, good spatial correlation with the ultraviolet radiation was found. At the peaks of the auroral forms the intensities of the 3914-Å radiation and of detected ultraviolet were comparable.

084.012 Characteristics of polar cap auroras. R. H. Eather, S.-I. Akasofu.
Journ. Geophys. Res. Vol. 74, 4794 - 4798 (1969). – Letter.

084.013 Auroral absorption of HF radio waves in the ionosphere: A review of results from the first decade of riometry. J. K. Hargreaves.
Proc. IEEE, Vol. 57, 1348 - 1373 (1969).

In the decade since the International Geophysical Year, information about the ionospheric absorption of radio waves during periods of auroral and magnetic disturbances has been greatly increased by the use of the riometer technique.

084.014 Comparison of auroral light emission and electron density. M. Jespersen, B. Landmark, K. Maseide.
Journ. Atmosph. Terr. Phys. Vol. 31, 1251 - 1257 (1969).

Simultaneous measurements of the auroral light emission and the electron density have been made during three rocket flights into auroral glow. The optical measurements were made on the (0–1) first negative band of N_2^+ at 4278 Å, and the volume emission rate vs. altitude has been obtained. The electron density measurements were made by a propagation experiment based upon the Faraday rotation technique.

084.015 N₂ Vegard-Kaplan band system in aurora. Rotational temperatures and vibrational populations.
M. Ahmed.
Journ. Atmosph. Terr. Phys. Vol. 31, 1259 - 1271 (1969).
  The rotational structure of the Vegard-Kaplan bands of N_2 has been successfully synthesized. The rotational temperatures of the Vegard-Kaplan bands of N_2 determined from the auroral spectra are in the range, 900 – 2200°K.

084.016 Two-station auroral infrasonic wave observations.
C. R. Wilson.
Planet. Space Sci. Vol. 17, 1817 - 1847 (1969).

084.017 VLF observations of auroral beams as sources of a class of emissions.   R. L. Smith.
Nature, Vol. 224, 351 - 352 (1969).

084.018 Röntgenstrahlungseinbrüche in die Polarlichtzone. Ein Teilaspekt der Auswirkung magnetosphärischer Dynamik.   G. Pfotzer.
Naturwissenschaften, 56. Jahrgang, 477 - 485 (1969).

084.019 Rocket measurements of H beta production in a hydrogen aurora.   J. R. Miller, G. G. Shepherd.
Journ. Geophys. Res. Vol. 74, 4987 - 4997 (1969).
  The Hβ vertical emission rate profile, obtained with a two-channel photometer during a rocket flight, is combined with proton measurements above 30 keV to obtain the proton energy spectrum below 30 keV. The derived energy spectrum cuts off sharply below about 25 keV and may have an additional low-energy component with an energy of a few keV.

084.020 Short-period auroral pulsations in λ6300 O I.
R. H. Eather.
Journ. Geophys. Res. Vol. 74, 4998 - 5004 (1969).
  λ6300 O I pulsations with quasi-periods of 2 to ~20 seconds have been observed in pulsating auroras. The percentage modulation was only 0.03 – 0.9%, compared with modulations of up to 60% in λ5577 O I and λ4278 N_2^+.

084.021 Ionization and excitation of nitrogen by protons and hydrogen atoms in the energy range 1 – 25 keV.
R. J. McNeal, D. C. Clark.
Journ. Geophys. Res. Vol. 74, 5065 - 5072 (1969).
  Recent measurements of the energy spectra of protons and hydrogen atoms in breakup auroras have found significant fluxes at energies near 1 keV. Atomic beam measurements are reported of ionization and charge-capture cross sections for proton bombardment of N_2, of ionization and stripping cross sections for hydrogen atom bombardment of N_2, and of the relative cross sections for emission of N_2^+ 391.4-nm radiation in collisions of protons and hydrogen atoms with N_2 in the energy range 1 – 25 keV.

084.022 Electric and magnetic field measurements near an auroral electrojet.   W. E. Potter, L. J. Cahill, Jr.
Journ. Geophys. Res. Vol. 74, 5159 - 5160 (1969). – Letter.

084.023 The Edmonton survey of the aurora borealis.
E. R. Milton.
Journ. Roy. Astron. Soc. Canada, Vol. 63, 238 - 250 (1969).
  An attempt is made to correlate solar activity with three aspects of auroral activity by analysis of the observations made over nine years by a group of amateur observers in Alberta.

084.024 The role of post-breakup pulsating aurora in conjugate point absorption increases.   L. A. Hajkowicz.
Journ. Atmosph. Terr. Phys. Vol. 31, 1365 - 1370 (1969).
  On the basis of the available yearly riometer, photome-

ter, magnetometer and all-sky camera records from Macquarie Island and the northern conjugate area, it is evident that the typical development of auroral substorms is similar at the auroral zone conjugate stations.

084.025 Electron density and temperature in the vicinity of the 29 September 1967 middle latitude red arc.
R. B. Norton, J. A. Findlay.
Planet. Space Sci. Vol. 17, 1867 - 1877 (1969).
  Electron temperature and density data were obtained as the Alouette satellites passed over the middle latitude red arc that was observed by the Fritz Peak and Richland airglow observatories on September 29, 1967. These data indicate that the electron temperature is enhanced and that the electron density is depressed along the magnetic field lines that intersect the red arc.

084.026 Conjugate and closely-spaced observations of auroral radio absorption – III. On the influence of the interplanetary magnetic field.   J. K. Hargreaves.
Planet. Space Sci. Vol. 17, 1919 - 1922 (1969). – Research note.

084.027 Plasma acoustic propagation in the ionosphere.
R. S. Iyengar.
Nature, Vol. 224, 1191 - 1192 (1969).
  We present here a discussion of how a new implication of the plasma acoustic theory is related to some optical observations of auroras.

084.028 Transient emissions on the wavelength of helium I, 5876 Å recorded during auroral break-up.
W. Stoffregen.
Planet. Space Sci. Vol. 17, 1927 - 1935 (1969).
  The records on the wavelength of 5876 Å shown in the present paper, which because of the high instrumental resolution are believed to be helium emissions, are difficult to explain in terms of a kind of helium glow. The high intensity observed, the sudden appearance and the short duration of the He I emission is believed to be related to an increased content of α-particles in the solar wind due to solar flare activity.

084.029 Excitation of the $A^2\pi_u$ state of N_2^+ by electrons.
F. R. Simpson, J. W. McConkey.
Planet. Space Sci. Vol. 17, 1941 - 1948 (1969).
  Absolute values of the emission cross sections for five vibrational bands in the Meinel system of N_2^+ excited by electron impact are presented.

084.030 Spatial and temporal relations between auroral emission and cosmic noise absorption.
G. Gustafsson.
Planet. Space Sci. Vol. 17, 1961 - 1975 (1969).
  Auroral emission and absorption of cosmic radio noise recorded by the use of a new 60 m long rotatable antenna support were studied with the aim of elucidating energy spectrum and distribution of particles responsible for auroral phenomena and cosmic noise absorption. Inter-phenomenal spatial and temporal relations were correlated and interpreted with the aid of certain simplifying assumptions.

084.031 Secondary electrons in aurora.
M. H. Rees, A. I. Stewart, J. C. G. Walker.
Planet. Space Sci. Vol. 17, 1997 - 2008 (1969).
  The production spectrum of secondary auroral electrons is derived. Consideration of energy loss processes leads to the flux of secondary electrons, the measurable quantity.

084.032 Relationship between a monochromatic auroral arc of 6300 A and a visible aurora.
T. Ichikawa, T. Old, J. S. Kim.

Journ. Geophys. Res. Vol. 74, 5819 - 5821 (1969).−Letter.

084.033 Auroral Lyman-alpha observations.
    M. A. Clark, P. H. Metzger.
Journ. Geophys. Res. Vol. 74, 6257 - 6265 (1969).
    Observations have been made of Lyman-alpha auroral emissions during January and February 1967. The observations were made with a narrow band sky-scanning photometer mounted on an earth-oriented satellite in polar orbit.

084.034 Penetration of auroral electrons into the atmosphere.
M. Walt, W. M. Macdonald, W. E. Francis.
Physics of the Magnetosphere, Boston College 1967, p. 534 - 555 (1968). − Invited research paper.

084.035 A widely seen September aurora.
    S. W. Schultz, Jr.
Sky Telescope, Vol. 38, 431 (1969).

084.036 Auroral electrons.   M. H. Rees.
    Space Sci. Rev. Vol. 10, 413 - 441 (1969).
    Satellite and rocket measurements of auroral electrons (which have been made since Brown's (1966) and Pfister's (1967) papers have appeared) are reviewed, and the salient characteristics of auroral electrons which emerge from all types of measurements are summarized. Effects of the atmosphere on the energy distribution of electron fluxes are discussed. Ionization rates associated with typical fluxes are derived. Observable effects produced in the atmosphere and the fate of auroral electrons are briefly described.

084.037 Rocket measurements of plasma densities and temperatures in visual aurora.   A. G. McNamara.
Canadian Journ. Phys., Vol. 47, 1913 - 1927 (1969).

084.038 Low-energy bremsstrahlung X-ray spectra from a stable auroral arc.
B. G. Wilson, A. J. Baxter, D. W. Green.
Canadian Journ. Phys., Vol. 47, 2427 - 2430 (1969).
    During a rocket experiment launched to investigate cosmic X-rays, the directional features and spectral characteristics of X rays from an auroral arc have been determined in the 1.6 to 10 keV energy range. The spectrum was best represented by a power law of slope −3.365 ± 0.07.

084.039 Auroral activity during 1968.
    J. Paton.
Observatory, Vol. 89, 246 - 247 (1969).

084.040 Bremsstrahlung of hard electrons during intensive auroras.   A. M. Novikov, Yu. G. Shafer.
Geomagn. Aeronom. Vol. 9, 880 - 886 (1969).   In Russian.

084.041 The width of band-formed auroras.
    Yu. A. Nadubovitch.
Geomagn. Aeronom. Vol. 9, 887 - 893 (1969).   In Russian.

084.042 Peculiarities of fragmental formations of strip-like forms of auroras.   Yu. A. Nadubovitch.
Geomagn. Aeronom. Vol. 9, 938 - 940 (1969).   In Russian.
Brief information.

084.043 PCA in July, August, and September 1966 by observations in the auroral zone.
V. V. Belikovitch, E. A. Benedictov, Z. Tz. Rapoport.
Geomagn. Aeronom. Vol. 9, 666 - 673 (1969).   In Russian.

084.044 Influence of a ring current on the location of the auroral absorption zone.
E. M. Zhulina.
Geomagn. Aeronom. Vol. 9, 751 - 752 (1969).
In Russian. − Brief information.

084.045 An analytical presentation of the equatorial boundary of the oval zone of auroras.
G. V. Starkov.
Geomagn. Aeronom. Vol. 9, 759 - 760 (1969).
In Russian. − Brief information.

084.046 The electron density profile of auroral layers as observed with ESRO rockets at Kiruna.
K. G. Jacobs, R. Kist, K. Rawer.
Space Research IX, Proc. Tokyo 1968, p. 246 - 255 (1969).

084.047 Résultats recents sur l'étude du rayonnement X auroral.   F. Cambou, G. Maral, J.-P. Treilhou.
Space Research IX, Proc. Tokyo 1968, p. 317 - 326 (1969).

$O(^1D)$ quenching in the ionospheric F-region.
See Abstr. 082.031.

Geomagnetic Field

084.201 **The geomagnetic and cosmic-ray storm of May 25/ 26, 1967.**
S.-I. Akasofu, P. D. Perreault, S. Yoshida.
Solar Physics, Vol. 8, 464 - 476 (1969).
A series of geomagnetic disturbances and cosmic ray variations caused by the McMath plage region 8818 in the latter half of May 1967 were examined. The systematic changes of the geomagnetic disturbances were observed as the relative location between the responsible flares and the earth changed during the half solar rotation period.

084.202 **Long time-scale magnetodynamic noise in the geomagnetic tail.** A. Hruška, J. Hrušková.
Planet. Space Sci. Vol. 17, 1497 - 1504 (1969).
Hourly ranges of the magnetic field in the earth's magnetotail have been determined from 5.46-min field averages measured by the IMP 1 satellite. The amplitude of fluctuations parallel to the local average field B decreases with increasing distance from the neutral sheet. The amplitude of fluctuations perpendicular to B decreases with increasing distance from the earth.

084.203 **A paleomagnetic study of secular variation in New Zealand.** A. Cox.
Earth Planet. Sci. Letters, Vol. 6, 257 - 267 (1969).
Ancient secular variation in New Zealand was determined from paleomagnetic measurements on 22 volcanic formations with ages of less than 0.68 m.y. The angular standard deviation from the field of an axial dipole is 13.2° with 95% confidence limits between 10.9° and 16.7°. The angular standard deviation of the corresponding virtual geomagnetic poles is 19.6° with confidence limits between 16.2° and 24.7°. These values are larger than those predicted by most models for secular variation. No difference was detected between the angular secular variation in New Zealand and that at the same latitude in North America.

084.204 **VLF electric and magnetic fields observed in the auroral zone with the Javelin 8.46 sounding rocket.**
D. A. Gurnett, S. R. Mosier.
Journ. Geophys. Res. Vol. 74, 3979 - 3991 (1969).

084.205 **Transverse wave propagation and instabilities within the magnetosphere.**
H. H. Bird, G. Schmidt.
Journ. Geophys. Res. Vol. 74, 3993 - 4002 (1969).

084.206 **A study of polar magnetic substorms. 2. Three-dimensional current system.**
C.-I. Meng, S.-I. Akasofu.
Journ. Geophys. Res. Vol. 74, 4035 - 4053 (1969).

084.207 **Radial motion resulting from pitch-angle scattering of trapped electrons in the distorted geomagnetic field.** C.-G. Fälthammar, M. Walt.
Journ. Geophys. Res. Vol. 74, 4184 - 4186 (1969). – Letter.

084.208 **Measurement and interpretation of magnetic anomaly components in low latitude.**
C. E. Adegbohungbe.
Journ. Geophys. Res. Vol. 74, 4233 - 4245 (1969).
A method of magnetic field survey that allows for the absolute measurement of all components in low latitude is described, and a computer method for presenting the results is adopted.

084.209 **International geomagnetic reference field 1965.0.**
Journ. Geophys. Res. Vol. 74, 4407 - 4408 (1969). Letter.

084.210 **High-frequency magnetic fluctuations associated with the earth's bow shock.**
J. V. Olson, R. E. Holzer, E. J. Smith.
Journ. Geophys. Res. Vol. 74, 4601 - 4617 (1969).

084.211 **Experimental verification of drift-shell splitting in the distorted magnetosphere.**
K. A. Pfitzer, T. W. Lezniak, J. R. Winckler.
Journ. Geophys. Res. Vol. 74, 4687 - 4693 (1969).

084.212 **Characteristics of the diurnally varying electron flux near the polar cap.**
M. H. Israel, R. E. Vogt.
Journ. Geophys. Res. Vol. 74, 4714 - 4720 (1969).

084.213 **The daily variation of trajectory-derived high-latitude cutoff rigidities in a model magnetosphere.**
D. F. Smart, M. A. Shea, R. Gall.
Journ. Geophys. Res. Vol. 74, 4731 - 4738 (1969).

084.214 **On the cause of geomagnetic bays.**
G. L. Siscoe, W. D. Cummings.
Planet. Space Sci. Vol. 17, 1795 - 1802 (1969).

084.215 **Spatial variations of the magnetosheath magnetic field.** K. W. Behannon, D. H. Fairfield.
Planet. Space Sci. Vol. 17, 1803 - 1816 (1969).
Results of Explorer measurements indicate that the magnetosheath field is several times the strength of the simultaneously measured interplanetary field in the sunward magnetosheath. This magnetosheath to interplanetary magnitude ratio decreases with distance from the subsolar point to values which are frequently less than unity at distances beyond $30 R_E$ and away from the bow shock. This ratio also displays a dawn-dusk asymmetry which is dependent on the interplanetary field orientation.

084.216 **Geomagnetic micropulsations and the location of the magnetospheric boundary.**
N. D'Angelo.
Planet. Space Sci. Vol. 17, 1849 - 1850 (1969). – Research note.

084.217 **Multipole analysis. II. Geomagnetic secular variation.** R. W. James.
Australian Journ. Phys. Vol. 22, 481 - 495 (1969).
The method of multipole analysis described in part I is applied to the earth's magnetic field for various epochs between 1845 and 1965, allowing the geomagnetic secular variation to be illustrated by time trends in the multipole parameters. The rates of change of the multipole parameters are used to separate the secular variation into non-drifting, meridional drifting, and longitudinal drifting components, which are discussed in detail for the epoch 1965.

084.218 **The influence of a uniform magnetic field on thermal convection in an enclosed fluid core.**
A. Kovetz.
Phys. Earth Planet. Interiors, Vol. 2, 88 - 92 (1969).

084.219 **Daily variation of the geomagnetic field and the deformation of the magnetosphere.**
V. Sarabhai, K. N. Nair.
Nature, Vol. 223, 603 - 605 (1969).
The range Δ H, of the daily variation of the horizontal component of the geomagnetic field at a low latitude station outside the influence of the equatorial electrojet is largely caused by the weakening of the ambient field on the night time due to magnetospheric drift currents. There exists a

good relation between the observed kinetic energy density of the solar wind and $(\Delta H)^2$. It is suggested that effects of the magnetospheric currents developing from the tail of the magnetosphere dominate over the effects due to ionospheric currents in producing ΔH.

084.220 Particle substorms observed at the geostationary orbit. R. L. Arnoldy, K. W. Chan.
Journ. Geophys. Res. Vol. 74, 5019 - 5028 (1969).

084.221 Conjugate point problems. T. Oguti.
Space Sci. Rev. Vol. 9, 745 - 804 (1969).
Conjugate research of upper atmospheric phenomena has developed rapidly after the IGY, when a number of observatories were established all over the world. Since then, many investigators have reported on the conjugate behaviour of upper atmospheric phenomena such as auroral displays and polar magnetic disturbances. In this review article the conjugate point problems are described with some proposals for future conjugate research.

084.222 Consequences of very small planetary magnetic moments. B. M. McCormac, J. E. Evans.
Nature, Vol. 223, 1255 (1969).
Atmospheric erosion by the solar wind when earth's magnetic dipole is near zero during reversal does not appear to be large but aeronomic modifications may be significant. Ions above the peak of the ionosphere should be removed by the solar wind.

084.223 A seasonal effect in the mid-latitude slab thickness variations during magnetic disturbances.
M. Mendillo, M. D. Papagiannis, J. A. Klobuchar.
Journ. Atmosph. Terr. Phys. Vol. 31, 1359 - 1364 (1969).
By monitoring the v.h.f. signals from geostationary satellites, it has been possible to study the diurnal changes in total electron content during individual magnetic storms.

084.224 On the interpretation of some observed features of geomagnetic pulsations. A. Hruška.
Planet. Space Sci. Vol. 17, 1937 - 1940 (1969).
The evidence for a latitude-period relation for geomagnetic micropulsations is re-examined and the physical meaning of various experimental results is discussed. Coupling between the various components of the magnetic field fluctuations is also investigated.

084.225 The distribution of stably trapped magnetospheric plasma as determined from Pi2 micropulsations and the auroral electrojet index, *AE*. M. W. Haurwitz.
Planet. Space Sci. Vol. 17, 2017 - 2023 (1969).

084.226 On electrical fields in the earth's magnetosphere. B. A. Tverskoi.
Dokl. Akad. Nauk SSSR, Ser. Mat. Fiz., Vol. 188, 575 - 578 (1969). In Russian.

084.227 The "pulse" of the earth's magnetosphere. V. A. Troitskaya.
Priroda No. 12, p. 9 - 17 (1969). In Russian.

084.228 The occasional reversal of the geomagnetic field. E. N. Parker.
Astrophys. Journ. Vol. 158, 815 - 827 (1969).
It is presently believed that the geomagnetic field is generated by the combined effects of the cyclonic convective motions and the nonuniform rotation of the liquid core of earth. Fossil magnetism shows that the field remains steady for periods of the order of $10^6 - 10^7$ years, and then, at apparently random times, reverses abruptly, thereafter remaining steady for another $10^6 - 10^7$ years, etc. We demonstrate that a statistical fluctuation in the distribution of the fifteen to twenty cyclonic convective cells in the core produces an abrupt reversal of the geomagnetic field.

084.229 Drift mirror instability in the magnetosphere: Particle and field oscillations and electron heating.
L. J. Lanzerotti, A. Hasegawa, C. G. MacLennan.
Journ. Geophys. Res. Vol. 74, 5565 - 5578 (1969).

084.230 Observations of the geomagnetic tail at 500 earth radii by Pioneer 8. F. Mariani, N. F. Ness.
Journ. Geophys. Res. Vol. 74, 5633 - 5641 (1969).
In January 1968, Pioneer 8 passed through the extended geomagnetic tail region at $470 - 580\ R_E$. The magnetic field observations suggest detection of the geomagnetic tail; the observations are similar in characteristics to those observed by Pioneer 7 at $1000\ R_E$, but they indicate a higher percentage of time was spent in the region of an extended, aberrated tail.

084.231 The shape of the tilted magnetopause. W. P. Olson.
Journ. Geophys. Res. Vol. 74, 5642 - 5651 (1969).
A model for the determination of the shape of the magnetopause is developed that permits the inclusion of cases where the solar wind is directed obliquely toward the geomagnetic dipole axis.

084.232 Daily variation and secular variation of the geomagnetic field from shipboard observations in the Gulf of Aden. R. B. Whitmarsh, M. T. Jones.
Geophys. Journ. Roy. Astron. Soc., Vol. 18, 477 - 488 (1969).

084.233 Effects of solar flare duration on a double shock pair at 1 AU. A. J. Hundhausen, R. A. Gentry.
Journ. Geophys. Res. Vol. 74, 6229 - 6237 (1969).
Numerical solutions of the time-dependent equations of motion for spherically symmetric flow are obtained to study the propagation and modification of the forward-reverse shock pair in the region between the sun and 1 AU.

084.234 Two-dimensional Chapman-Ferraro problem with neutral sheet. 3. Implied magnetospheric flows and their time dependence. G. Atkinson, T. Unti.
Journ. Geophys. Res. Vol. 74, 6275 - 6280 (1969).
In a magnetosphere with a magnetic tail and a perfect neutral sheet, the steady-state field line configuration is found to depend on two variables: the magnetic flux in the tail and the solar wind momentum flux. Thus, changes in solar wind momentum flux, or transport of magnetic field lines into and from the tail cause a large-scale magnetospheric flow toward the new equilibrium configuration. The solutions to the two-dimensional case are used to study the flows.

084.235 Resonant compressional waves in the geomagnetic tail. G. L. Siscoe.
Journ. Geophys. Res. Vol. 74, 6482 - 6486 (1969).
Irregularities in the solar wind can excite compression waves in the geomagnetic tail that have their maximum amplitude in the plasma sheet. Details such as the exact frequency and spatial distribution of the waves depend on the over-all geometry of the tail. Estimates of these properties are obtained by use of a single-layered, two-dimensional model.

084.236 Dynamical properties of the magnetosphere. E. N. Parker.
Physics of the Magnetosphere, Boston College 1967, p. 3 - 64 (1968). – Tutorial lecture.

084.237 Particle description of the magnetosphere. J. A. van Allen.
Physics of the Magnetosphere, Boston College 1967, p. 147 -

217 (1968). – Tutorial lecture.

084.238 Waves and particles in the magnetosphere.
J. W. Dungey.
Physics of the Magnetosphere, Boston College 1967, p. 218 - 259 (1968). – Tutorial lecture.

084.239 Inflation of the inner magnetosphere.
L. J. Cahill, Jr.
Physics of the Magnetosphere, Boston College 1967, p. 263 - 270 (1968). – Invited research paper.

084.240 Magnetic energy relationships in the magnetosphere.
R. L. Carovillano, J. J. Maguire.
Physics of the Magnetosphere, Boston College 1967, p. 290 - 300 (1968). – Invited research paper.

084.241 External aerodynamics of the magnetosphere.
J. R. Spreiter, A. Y. Alksne, A. L. Summers.
Physics of the Magnetosphere, Boston College 1967, p. 301 - 375 (1968). – Invited research paper.

084.242 Review and interpretation of particle measurements made by the Vela satellites in the magnetotail.
E. W. Hones, Jr.
Physics of the Magnetosphere, Boston College 1967, p. 392 - 408 (1968). – Invited research paper.

084.243 Satellite studies of the earth's magnetic tail.
K. W. Behannon, N. F. Ness.
Physics of the Magnetosphere, Boston College 1967, p. 409 - 434 (1968). – Invited research paper.

084.244 Review of Ames Research Center plasma-probe results from Pioneers 6 and 7.
J. H. Wolfe, D. D. McKibbin.
Physics of the Magnetosphere, Boston College 1967, p. 435 - 460 (1968). – Invited research paper.

084.245 The geomagnetic tail: Topology, reconnection and interaction with the moon.
C. P. Sonett, D. S. Colburn, R. G. Currie, J. D. Mihalov.
Physics of the Magnetosphere, Boston College 1967, p. 461 - 484 (1968). – Invited research paper.

084.246 Magnetospheric and high latitude ionospheric disturbance phenomena. N. Brice.
Physics of the Magnetosphere, Boston College 1967, p. 563 - 585 (1968). – Invited research paper.

084.247 Particle dynamics at the synchronous orbit.
J. W. Freeman, Jr., J. J. Maguire.
Physics of the Magnetosphere, Boston College 1967, p. 586 - 604 (1968). – Invited research paper.

084.248 Shock and magnetopause boundary observations with IMP-2. J. H. Binsack.
Physics of the Magnetosphere, Boston College 1967, p. 605 - 621 (1968). – Invited research paper.

084.249 Low-energy electrons in the magnetosphere as observed by OGO-1 and OGO-3.
V. M. Vasyliunas.
Physics of the Magnetosphere, Boston College 1967, p. 622 - 640 (1968). – Invited research paper.

084.250 Summary of experimental results from M.I.T. detector on IMP-1. S. Olbert.
Physics of the Magnetosphere, Boston College 1967, p. 641 - 659 (1968). – Invited research paper.

084.251 Charged particle diffusion by violation of the third adiabatic invariant.
T. J. Birmingham, T. G. Northrop, C.-G. Fälthammar.
Physics of the Magnetosphere, Boston College 1967, p. 660 - 677 (1968). – Invited research paper.

084.252 Steady state charge neutral models of the magnetopause. W. Alpers.
Astrophys. Space Sci. Vol. 5, 425 - 437 (1969).
Plane models of the magnetopause are investigated under the assumption that ionospheric electrons are able to short-circuit electric fields (exact charge neutrality). Using the Vlasov theory a general method is presented for constructing distribution functions that lead to given magnetic field and tangential bulk velocity profiles. As an example we describe the magnetic field transition in terms of error functions and obtain particle distributions in explicit form, including bulk velocities.

084.253 Some problems of the earth's magnetosphere physics. A. I. Ershkovich, G. A. Skuridin, V. P. Shalimov.
Space Sci. Rev. Vol. 10, 262 - 290 (1969).
The solar wind flow around the magnetosphere and quantitative models of the magnetosphere; Energetic charged particles in the earth's magnetosphere and ring current problem; Mechanisms of transfer of the solar wind energy into the magnetosphere.

084.254 Geomagnetic pulsations. T. Saito.
Space Sci. Rev. Vol. 10, 319 - 412 (1969). – Review article.

084.255 British world magnetic charts. S. R. C. Malin.
Quarterly Journ. Roy. Astron. Soc. Vol. 10, 309 - 316 (1969).

084.256 Geomagnetic disturbances at conjugate points and some problems of the influence of solar plasma upon the magnetosphere. M. S. Bobrov.
Astron. Zhurn. Akad. Nauk SSSR, Vol. 46, 862 - 871 (1969). In Russian. English translation in Soviet Astron. AJ, Vol. 13, No. 4.
The magnetic disturbances, registered at the conjugate observatories Mirny and Murchison Bay (at the polar caps), are studied in detail with respect to the similarity of their forms. As to the sources of the disturbances the influence of chaotic motions of turbulent plasma and of flare plasma invasions is discussed.

084.257 Study of Čerenkov radiation from electrons in the magnetosphere and ionosphere – II.
R. N. Singh, R. P. Singh.
Ann. Géophys. Vol. 25, 629 - 638 (1969).

084.258 Study of Doppler-shifted cyclotron radiation from energetic electrons in the magnetosphere – III.
R. P. Singh, R. N. Singh.
Ann. Géophys. Vol. 25, 639 - 645 (1969).

084.259 Über ein ebenes Modell der Magnetosphäre.
Yu. S. Sigov.
Kosmich. Issled. Vol. 7, 731 - 737 (1969). In Russian.

084.260 Die magnetische durch tangentiales Abreißen des Sonnenwindes hervorgerufene Störung in der Magnetosphäre. N. V. Mikerina.
Kosmich. Issled. Vol. 7, 794 - 795 (1969). In Russian. – Brief information.

084.261 **About plasma movement within the magnetosphere near the axis of symmetry.** M. V. Samokhin.
Geomagn. Aeronom. Vol. 9, 798 - 802 (1969). In Russian.

084.262 **Non-linear dynamics of the irregularity of fast electrons in the magnetosphere.**
A. V. Gurevitch, E. E. Tzedilina.
Geomagn. Aeronom. Vol. 9, 818 - 827 (1969). In Russian.

084.263 **The geomagnetic storm families for 1957 – 1964. III. A connection of geomagnetic activity with different forms of solar activity.** V. I. Afanasieva.
Geomagn. Aeronom. Vol. 9, 899 - 902 (1969). In Russian.

084.264 **Non-stationarity and the time of passing of the geo-effective stream of interplanetary plasma responsible for the magnetospheric storm on April 17 - 19, 1965.** K. G. Ivanov.
Geomagn. Aeronom. Vol. 9, 913 - 914 (1969). In Russian.
Brief information.

084.265 **Effects associated with the sector boundary crossing on 8 July 1966.** Z. Švestka.
Solar Flares and Space Research, Tokyo 1968, p. 319 - 326 (1969).

Thirty hours after the proton flare of 7 July 1966, the earth and nearby satellites crossed a sector boundary of the interplanetary magnetic field. The boundary crossing caused a short-lived geomagnetic disturbance.

084.266 **On the proper oscillations of the earth's magnetic tail.** A. I. Ershkovich.
Kosmich. Issled. Vol. 7, 944 - 947 (1969). In Russian.

084.267 **Dynamics of irregularities of fast electrons and ions in the magnetosphere of the earth. II.**
A. V. Gurevitch, E. E. Tzedilina.
Geomagn. Aeronom. Vol. 9, 642 - 649 (1969). In Russian.

084.268 **Families of geomagnetic storms for 1957 – 1964. II. 27-day recurrence of geomagnetic storms with sudden and gradual commencements.**
V. I. Afanasieva.
Geomagn. Aeronom. Vol. 9, 697 - 699 (1969). In Russian.

084.269 **About variations of the magnetic field of the earth.** D. Zidarov.
Geomagn. Aeronom. Vol. 9, 1060 - 1066 (1969). In Russian.

084.270 **The geomagnetic storm families for 1957 – 1964. V. The character of development of activity within storm families.** V. I. Afanasieva.
Geomagn. Aeronom. Vol. 9, 1116 - 1117 (1969). In Russian. – Brief information.

084.271 **Reversals of the earth's magnetic field.** E. Bullard.
Phil. Trans. Roy. Soc. London, Ser. A, Vol. 263, (No. 1143), 481 - 524 (1968). – The Bakerian Lecture, 1967. – Reversals of the earth's magnetic field can be studied in the magnetization of lavas and sediments on land, in the magnetization of deep sea cores and in the magnetic pattern on the ocean floor. The dynamo theory of the earth's magnetic field may be able to account for reversals as an instability in the dynamo, but only models with a finite number of degrees of freedom have been investigated. Spreading of the ocean floor is believed to be associated with convective motions in the upper mantle, although there are difficulties connected with the equality of the oceanic and continental heat flows. There is some evidence for the extinction of radiolaria at times of reversal of the magnetic field; it has been suggested that this is due to the effect of the field on cosmic rays but this appears impossible. If the extinctions are due to the reversals, the mechanism is unknown.

084.272 **Drift mirror instability in the magnetosphere.** A. Hasegawa.
Phys. Fluids, Vol. 12, 2642 - 2650 (1969).

084.273 **Changes of the earth's magnetic moment and radiocarbon dating.** V. Bucha.
Nature, Vol. 224, 681 - 683 (1969).

Zur Dynamotheorie stellarer und planetarer Magnetfelder. II. Berechnung planetenähnlicher Gleichfeldgeneratoren. See Abstr. 062.008.

Solar wind interactions and the magnetosphere. See Abstr. 074.056.

Observations of the solar wind, bow shock and magnetosheath by the Vela satellites. See Abstr. 074.057.

About a peculiarity of connection between solar wind velocity and geomagnetic activity. See Abstr. 074.075.

Some features of the solar microwave emission and their connection with geomagnetic activity. III: Sunspots and geomagnetic activity. See Abstr. 077.005.

Observation of interplanetary field lines in the magnetotail. See Abstr. 106.001.

New evidence of the connection between magnetic fields of cosmic space and the earth. See Abstr. 106.026.

A connection of the variations of the azimuthal (eastern) component of the interplanetary electric field with variations of the geomagnetic field in the polar regions. See Abstr. 106.027.

Radiation Belts

084.401 **Simultaneous observations of 5- to 15-second period modulated energetic electron fluxes at the synchronous altitude and the auroral zone.**
G. K. Parks, J. R. Winckler.
Journ. Geophys. Res. Vol. 74, 4003 - 4017 (1969).

084.402 **Evaluation of the CRAND source for 10- to 50-MeV trapped protons.**
T. A. Farley, A. D. Tomassian, M. C. Chapman.
Journ. Geophys. Res. Vol. 74, 4721 - 4730 (1969).
Proton intensities in two energy channels covering the range from 10 to 50 MeV, measured in November 1965, with directional detectors on the OV1 2(1965-78A) spacecraft, are reported and discussed.

084.403 **Intensity correlations and substorm electron drift effects in the outer radiation belt measured with the OGO 3 and ATS 1 satellites.**
K. A. Pfitzer, J. R. Winckler.
Journ. Geophys. Res. Vol. 74, 5005 - 5018 (1969).

084.404 **Source of high-energy protons trapped on low L shells.** W. L. Imhof, J. B. Reagan.
Journ. Geophys. Res. Vol. 74, 5054 - 5064 (1969).

084.405 **Initial observations of geomagnetically trapped protons and alpha particles with OGO 4.**
T. A. Fritz, S. M. Krimigis.
Journ. Geophys. Res. Vol. 74, 5132 - 5138 (1969).

084.406 **Large temporal variations of energetic electron intensities at midlatitudes in the outer radiation zone.**
D. M. Yeager, L. A. Frank.
Journ. Geophys. Res. Vol. 74, 5697 - 5708 (1969).
A thorough search of Explorers 12 and 14 observations of electron (40 kev $\lesssim E \lesssim$ 2 Mev) intensities within the outer radiation zone during the period 1961 - 1963 has provided evidences of several catastrophic rapid decreases and recoveries of these electron intensities within periods ~ several minutes at L values ~5 to 6.

084.407 **Relation of solar proton latitude profiles to outer radiation zone electron measurements.**
I. B. McDiarmid, J. R. Burrows.
Journ. Geophys. Res. Vol. 74, 6239 - 6246 (1969).
The present study was undertaken in an attempt to relate some of the structure observed in low-energy solar proton latitude profiles to features observed in latitude profiles of outer radiation zone electrons ($E > 35$ keV). A secondary objective was to examine as a function of local time a relatively large number of proton latitude profiles and attempt to give a description of the average properties of these profiles in the region of the cutoffs.

084.408 **Some observations on energetic electrons in the outer Van Allen zone during auroral substorms in relation to open and closed field lines.**
C. S. R. Rao.
Journ. Geophys. Res. Vol. 74, 6513 - 6517 (1969). – Letter.

084.409 **Particles and fields: Significant achievements, 2.**
A. W. Schardt, A. G. Opp.
Rev. Geophys. Vol. 7, 799 - 849 (1969).
This paper is a sequel to previous summaries (Hess et al., 1965; Schardt and Opp, 1967) and covers primarily the areas of research in which major new information has been developed since 1965. Results published prior to July 1967 have generally been included. It has been decided, however, to omit the interplanetary medium and solar cosmic rays from this summary. Instead, emphasis has been placed on the terrestrial magnetosphere.

084.410 **Recent observations of low-energy charged particles in the earth's magnetosphere.** L. A. Frank.
Physics of the Magnetosphere, Boston College 1967, p. 271 - 289 (1968). – Invited research paper.
Several recent observations of low-energy proton and electron intensities within the energy range ~ 100 eV to 50 keV in the earth's radiation zones with a sensitive array of electrostatic analyzers borne on the earth-satellite OGO 3 during mid-1966 are summarized.

084.411 **Cyclotron- and bounce–resonance scattering of electrons trapped in the earth's magnetic field.**
C. S. Roberts.
Physics of the Magnetosphere, Boston College 1967, p. 514 - 533 (1968). – Invited research paper.

084.412 **Untersuchung der Protonenspektren des inneren Strahlungsgürtels an Bord des Satelliten "Kosmos 137".** I. A. Savenko, O. I. Savun, P. I. Shavrin.
Kosmich. Issled. Vol. 7, 553 - 558 (1969). In Russian.

085 Solar-Terrestrial Relations

085.001 The IQSY 27-day recurrence sequence for 1964 solar and terrestrial activity. M. C. Ballario.
Mem. Soc. Astron. Italiana, Nuova Serie, Vol. 40, 271 - 294 (1969).

The solar and terrestrial phenomena observed during the 27-day recurrence sequence chosen by the *IQSY* Committee for the year 1964 have been examined. Plages, plage-groups, coronal condensations, filaments and flares have been taken into consideration.

085.002 Geophysical mechanism for C^{14} variations. D. C. Grey.
Journ. Geophys. Res. Vol. 74, 6333 - 6340 (1969).

Simple models are employed to discuss the effects of the modulation of the cosmic-ray flux in the earth's atmosphere by the solar wind and by geomagnetic field variations on the variations of C^{14} in the atmosphere. It is shown that a good predictive model of short-term variations in atmospheric C^{14} can be based on sunspot numbers, whereas long-term variations appear to correlate well with paleomagnetic variations.

085.003 VLF phase disturbances, HF absorption, and solar protons in the events of August 28 and September 2, 1966. T. A. Potemra, A. J. Zmuda, C. R. Haave, B. W. Shaw.
Journ. Geophys. Res. Vol. 74, 6444 - 6458 (1969).

The VLF phase disturbances and HF absorption observed during the proton events of August 28 and September 2, 1966, were quantitatively connected to the proton intensities, energies, and cutoff latitudes observed by satellite 1963-38C, at 1100 km in a polar orbit.

085.004 Unusual or neglected optical phenomena in the landscape. M. Minnaert.
Zemlya i Vselennaya, No. 5, p. 43 - 49 (1969). In Russian.

085.005 The terrestrial efficiency of corpuscular sources depending on their heliographic latitude.
A. T. Nesmjanovich, V. V. Chmil.
Astron. Vestn. Vol. 3, 151 - 156 (1969). In Russian.

The terrestrial efficiency of corpuscular sources depending on their position on the solar disk is investigated. Sources of effective corpuscular streams are connected with active regions, characterized by high flare activity, the high-latitude regions are responsible for weaker perturbations. Active regions without flares, apparently, do not accelerate the solar plasma.

085.006 Polar cap absorption and ground level effects. B. Hultqvist.
Solar Flares and Space Research, Tokyo 1968, p. 215 - 257 (1969).

A review is given of solar energetic particle effects as observed in the earth's upper atmosphere and at ground level.

085.007 On forecasts of interplanetary and geophysical conditions. R. Howard, V. Bumba.
Solar Flares and Space Research, Tokyo 1968, p. 397 - 404 (1969).

The low-latitude, weak background magnetic-field pattern in the photosphere shows a distinct tendency to repeat itself over intervals of many months. During times of high solar activity, the recurrent geophysical disturbances appear to be correlated with a relatively weak background field pattern, and during intervals of low solar activity these recurrent disturbances are correlated with relatively strong "streams" of photospheric fields caused by active regions.

085.008 Some discoveries in seismology and geomagnetism. F. Gun-Bayer.
Separate print from Inst. Astrophys. Catholic Univ. Santiago de Chile, 16 pp. (1969).

Discussed are the daily period of earthquakes, the influence of solar activity cycles, and effects produced by electromagnetic forces.

085.009 Cosmic data. Monthly review.
In-t zemn. magn., ionosfery i rasprostr. radiovoln. AN SSSR, avg., No. 8, 45 pp. 1968 (1969). In Russian.
Abstr. in Referativ. Zhurn. 51. Astron., 2.51.484 (1970).

085.010 On scales and forms of the influence of the solar activity. M. Kh. Bajdal.
Trudy Kazakhsk. n.-i. gidrometeorol. in-ta, Vyp. (No.) 38, p. 19 - 21 (1969). In Russian. – Abstr. in Referativ. Zhurn. 51. Astron., 2.51.486 (1970).

Utilisation d'observations d'amateurs pour l'étude de la haute atmosphère en corrélation avec l'activité solaire.
See Abstr. 082.033.

Ionization rates due to the attenuation of 1 – 100 Å nonflare solar X rays in the terrestrial atmosphere.
See Abstr. 083.006.

Planetary System

091 Physics of the Planetary System (Planetary Atmospheres, Figure, Interior, Magnetic Fields, Rotation, etc.)

091.001 Greenhouse effect in a finite anisotropically scattering atmosphere. J. K. Shultis, H. G. Kaper.
Astron. Astrophys. Vol. 3, 110 - 121 (1969).

An exact analytical solution is given for the steady-state distribution of visible and infrared radiation in a planetary atmosphere. It is assumed that the atmosphere is illuminated from outside by visible light which provides the sole source of atmospheric heating. The atmosphere may have any finite thickness and any degree of anisotropic scattering is allowed. The solution, with either diffuse or mirror reflection of the visible radiation and infrared emission at the surface of the planet, is expressed as a linear combination of known solutions to various monoenergetic albedo problems and an infinite medium Green's function. Once the visible and infrared intensities have been found, the atmospheric temperature profile can be readily calculated.

091.002 Components as nonunique functions of total magnetic field; a note on extraterrestrial magnetic prospecting. J. F. Claerbout.
Journ. Geophys. Res. Vol. 74, 4187 - 4188 (1969). – Letter.

091.003 Remarques sur le rayonnement diffusé par une atmosphère planétaire épaisse suivant la loi de Rayleigh. J. Lenoble.
Comptes Rendus Acad. Sci. Paris, Sér. B, Vol. 269, 232 - 234 (1969).

On a étudié dans le cas de la diffusion Rayleigh, la variation de la luminance et du taux de polarisation dans une raie d'absorption observée dans le rayonnement diffusé au centre du disque planétaire. On a montré que pour calculer la luminance on pouvait, avec une bonne précision, négliger l'effet polarisant de la diffusion.

091.004 Permissible modes of non-thermal convection in planetary mantles. D. G. Ashworth.
Astrophys. Space Sci. Vol. 5, 71 - 75 (1969).

Boundary conditions are imposed upon the solutions of the conservation equations for non-thermal convective motion in a self-gravitating, homogeneous, non-rotating sphere of radius R, consisting of a core extending to a fraction η of the radius of the sphere and with a viscous mantle overlying the core. It is shown that convective modes are permissible in the mantle only for certain values of n and η.

091.005 Kernreaktionen der Nukleogenese der Xenonisotope in der Materie des Sonnensystems.
A. K. Lavrukhina.
Meteoritika, No. 29, p. 9 - 29 (1969). In Russian.

Für Materie, aus der sich die Meteorite gebildet haben, werden Kernreaktionen, die zur Synthese von Xenonisotopen führen, diskutiert. Der Prozeß des Neutroneneinfangs ist dabei vorherrschend. Für Meteorite verschiedenen Typs werden die Isotopenzusammensetzungen analysiert. Sie weisen durch den unterschiedlichen Anteil der vier Xenonkomponenten eine große Vielfalt auf. Die Frage des X^{129}–Exzesses wird gesondert behandelt. *D. Krahn*

091.006 Heating of a planetary thermosphere: Effects of non-linear flow conductance terms and of convective terms. R. J. Moffett.
Planet. Space Sci. Vol. 17, 1850 - 1851 (1969). – Research note.

091.007 Shaking a planet. G. Eiby.
Southern Stars, Vol. 23, 67 - 75 (1969).

091.008 The interiors of the planets. K. E. Bullen.
Annual Rev. Astron. Astrophys. Vol. 7, 177 - 200 (1969).

091.009 Radar studies of planetary surfaces.
J. V. Evans.
Annual Rev. Astron. Astrophys. Vol. 7, 201 - 248 (1969).

091.010 Motions of planetary atmospheres.
R. Goody.
Annual Rev. Astron. Astrophys. Vol. 7, 303 - 352 (1969).

091.011 Densities of the terrestrial planets.
W. H. McCrea.
Nature, Vol. 224, 28 - 29 (1969).

Recent determinations of the masses and radii of the planets may be interpreted as showing that the systems Earth-Moon-Mars and Mercury-Venus could have resulted from the break-up of two unstable planetary bodies of identical chemical composition. Thereby various features of planetary evolution would be elucidated.

091.012 Résultats récents sur les planètes Mercure, Vénus et Mars, obtenus par les observations astronomiques au sol. A. Dollfus.
Moon and Planets II, London 1967, p. 1 - 25 (1968).

Le présent mémoire, préparé en juillet 1967, résume quelques-uns des derniers résultats obtenus récemment sur les planètes Mercure, Vénus et Mars. On se limite à l'étude des connaissances établies part les méthodes d'observation classique de l'astronomie optique, radio et radar.

091.013 The evolution of terrestrial-type planets.
D. L. Anderson.
Applied Optics, Vol. 8, 1271 - 1277 (1969).

The present internal configuration of the earth shows that it is an extremely differentiated body, and the ages of surface rocks indicate that this differentiation took place early in the hirstory of the earth. The gravitational energy associated with the formation of the earth and the thermal energy of core formation are the primary sources of energy for the melting and subsequent gravitational separation of the earth into a core, a crust, a chemically inhomogeneous mantle, and an atmosphere and hydrosphere. The larger the planet the more likely it is to be extensively differentiated and outgassed.

091.014 Planetary ultraviolet spectroscopy.
C. A. Barth.

Applied Optics, Vol. 8, 1295 - 1304 (1969).

Ultraviolet spectroscopy can determine if the atoms and molecules basic to life are present in a planetary atmosphere: in particular, molecular nitrogen and the photodissociation products of water vapor. Ultraviolet spectrometer experiments can also tell if the atmosphere has been changed by the presence of life such as the production of oxygen from photosynthesis. Local regions of photosynthetic oxygen production will produce ozone which is detectable by uv spectroscopy.

091.015 Photogrammetry with surface-based images.
R. M. Batson.
Applied Optics, Vol. 8, 1315 - 1322 (1969).

Stereoscopic pictures returned by surface-based imaging systems can be used to reconstruct the topography of landing sites on Mars and other planets. Large surface relief with respect to distance and the large scale variation inherent in surface-based pictures produce problems in stereoscopic measurement very different from those presented by high altitude photography.

091.016 Distribution of slopes on a cratered planetary surface: Theory and preliminary applications.
A. H. Marcus.
Journ. Geophys. Res. Vol. 74, 5253 - 5267 (1969).

The distribution of slopes over any finite span on a surface excavated by primary impact craters is derived from a representation of the surface as a 'moving average' of impact events. The cumulants are always positive, and they are large for typical mare crater densities. In some cases the distribution can be approximated by a rapidly convergent Gram-Charlier type A series. Observations of slopes in Mare Cognitum are in good agreement with theory, if the validity of the photoclinometric data and the model can be accepted.

091.017 Simple model for a rotating neutral planetary exosphere.
K. M. Hagenbuch, R. E. Hartle.
Phys. Fluids, Vol. 12, 1551 - 1559 (1969).

The model neutral exosphere of Öpik and Singer for a nonrotating planet is generalized by permitting the corresponding barosphere to rotate uniformly at an angular velocity which may or may not be equal to that of the planet. For this case the velocity-distribution function, satisfying the collisionless Boltzmann equation, is constructed. Then, the density is determined from the distribution and compared with the corresponding result for a nonrotating planet obtained by Öpik and Singer. In addition, the radial and azimuthal fluxes are derived. Based on the result for the azimuthal flux, the point at which exospheric corotation can be said to have broken down is indicated for several conditions. It is shown that in all cases, for a given radius, the density at the equator exceeds the density at the pole. For example, a model terrestrial neutral exosphere of hydrogen, helium, and oxygen has density ratios between the pole and the equator of 0.984, 0.869, and 0.530, respectively, at $r = 2R$.

091.018 Absorption-line formation in a scattering planetary atmosphere: A test of van de Hulst's similarity relations.
J. E. Hansen.
Astrophys. Journ. Vol. 158, 337 - 349 (1969).

Van de Hulst's similarity relations, which reduce the problem of anisotropic scattering in a homogeneous atmosphere to one of isotropic scattering by scaling the optical thickness and the single-scattering albedo, are tested for line formation in clouds and hazes. The relations are shown to give good approximations for a useful range of scattering angles when k (the first characteristic exponent occurring in the solution of the transfer equation in unbounded media) is the basis for the scaling relations. It is indicated that the density of cloud particles on Venus is about 6 times greater than the value suggested by the synthetic-spectra calculations of Belton, Hunten,

and Goody for isotropic scattering, if it is assumed only that the cloud particles are at least $\sim 1\mu$ in radius. This implies that the density of cloud particles on Venus is comparable with that of cirrus clouds on earth.

091.019 Planetary surfaces of equilibrium of Helmert's type.
P. Sconzo.
Bull. American Astron. Soc. Vol. 1, 261 (1969). – Abstr. AAS.

091.020 Evaporation from rotating planets.
J. A. Burke.
Monthly Notices, Roy. Astron. Soc., Vol. 145, 487 - 492 (1969).

Evaporation rates from a rapidly rotating planet are considerably augmented over those from a stationary planet. Simply regarding the rotation as diminishing the surface gravity may greatly underestimate the evaporation rate. Evaporating material carries off an angular momentum per unit mass larger than that of the remaining material.

091.021 Accurate ephemerides for planets and moon.
C. Oesterwinter, C. J. Cohen.
The Moon, Vol. 1, 121 (1969). – Abstract.

091.022 Distribution of slopes on a cratered planetary surface.
A. H. Marcus.
The Moon, Vol. 1, 133 (1969). – Abstract.

091.023 Convection in planetary interiors.– The effects of rotation.
D. G. Ashworth.
Astrophys. Space Sci. Vol. 5, 289 - 299 (1969).

The conservation equations of mass, energy and momentum are applied to the problem of thermal and non thermal convective motions inside a homogeneous, compressible fluid sphere of uniform viscosity which is rotating with a constant angular velocity about the z-axis. The resulting equations are manipulated into a form which should be suitable for solution.

091.024 Speculations on mass loss by meteoroid impact and formation of the planets.
A. H. Marcus.
Icarus, Vol. 11, 76 - 87 (1969).

The ratio between mass of a meteoroidal projectile and the mass of ejecta from the impact of the projectile which escapes the target planet is computed from extrapolations of available laboratory hypervelocity impact data. If primitive protoplanets have grown by accretion, the meteoroids and planetesimals they accumulated could not have had impact speeds much in excess of the protoplanetary escape speed. The similarities in sizes of the major satellites can be explained by the presence of a high mass loss ratio for escape speeds between 2 and 4 km/sec. At present, asteroids and lunar-sized objects are probably undergoing net erosion by impacts.

091.025 Critique of "The resonant structure of the solar system" by A. M. Molchanov.
G. E. Backus.
Icarus, Vol. 11, 88 - 92 (1969).

A. M. Molchanov's eight resonance relations among the orbital frequencies of the nine planets in the solar system are satisfied not exactly but with a relative error of 4.5×10^{-4}. It is shown that with such a large error, those relations can be the result of chance. Such relations could be expected among any nine numbers chosen at random.

091.026 A comment on "The resonant structure of the solar system", by A. M. Molchanov.
M. Hénon.
Icarus, Vol. 11, 93 - 94 (1969).

091.027 Resonances in complex systems: A reply to critiques. A. M. Molchanov.
Icarus, Vol. 11, 95 - 103 (1969).

The conclusions reached by Backus and Hénon, that resonance relations in the solar system of the sort proposed by Molchanov are a result of chance, is based on a very crude statistical model. A more accurate model gives a value $P \sim 10^{-10}$ for the probability of chance formation of systems similar to the solar system.

091.028 The reality of resonances in the solar system. A. M. Molchanov.
Icarus, Vol. 11, 104 - 110 (1969).

The object of this paper is to attempt a quantitative evaluation of the probability of a given resonant structure. It is shown that the formation of a "good" resonant structure by chance is not very likely, and that the random probability of the resonant structure of the solar system is less than 10^{-10}.

091.029 Kepler and the resonant structure of the solar system. O. Gingerich.
Icarus, Vol. 11, 111 - 113 (1969).

Molchanov's ordering the solar system by a table of resonance relations recalls an earlier attempt by Johannes Kepler (1571 - 1630).

091.030 On the existence of unknown satellites of major planets of the solar system.
S. S. Gamburg.
Astron. Vestn. Vol. 3, 157 - 158 (1969).   In Russian.

Forecasts on the existence of satellites yet undiscovered in the system of major planets, made by the author and Bowell and Wilson, are compared.

091.031 On the origin of planetary rotation. A. V. Artem'ev.
Uch. zap. Gor'kovsk. gos. ped. in-ta, vyp. (No.) 98, p. 98 - 105 (1969). In Russian. – Abstr. in Referativ. Zhurn. 51. Astron., 12.51.728 (1969).

091.032 Are the libration clouds real? R. G. Roosen, C. L. Wolff.
Nature, Vol. 224, 571 (1969).

Attention is called to considerable observational evidence presently in the literature against the existence of dust clouds at the L_4 and L_5 libration points of the earth–moon system. Theoretical difficulties in obtaining a sufficient number of particles with stable orbits to populate the clouds are also cited. As an example that alternative explanations exist, the results from the dust–counting experiments on orbiting artificial satellites are considered. They suggest that clouds of dust of obervable brightness may pass through interplanetary space, causing brightness fluctuations in the zodiacal light and Gegenschein, and an occasional erroneous report of a "libration cloud".

091.033 Photoelectric spectrophotometry of the major planets. J. S. Neff.
Bull. American Astron. Soc., Vol. 1, 355 - 356 (1969). Abstr. AAS.

091.034 A rapid method for computation of absorption spectra in a planetary atmosphere.
A. Uesugi, W. M. Irvine.
Bull. American Astron. Soc., Vol. 1, 366 (1969). – Abstr. AAS.

091.035 Theory of the figure of hydrostatic equilibrium of rotating planets. The third approximation.
V. N. Zharkov, V. P. Trubitsyn.
Astron. Zhurn. Akad. Nauk SSSR, Vol. 46, 1252 - 1263 (1969).   In Russian. English translation in Soviet Astron. AJ, Vol. 13, No. 6.

The figure of equilibrium of a gravitating inhomogeneous rotating liquid, slightly different from a sphere, is investigated. The integral equations are solved for the particular case of constant density and for the case of a gradual density distribution with a precision up to terms of the third order. The theory can be used for the construction of more precise models of Jupiter and Saturn.

091.036 Stability of planetary interiors.
G. Schubert, D. L. Turcotte, E. R. Oxburgh.
Geophys. Journ. Roy. Astron. Soc., Vol. 18, 441 - 460 (1969).

In this paper we examine the effects on stability of a strongly depth-dependent viscosity and the removal of the lower boundary, with and without internal heat generation. In the linear stability analysis a depth-dependent viscosity can take account of any pressure and temperature dependence that the viscosity may have since the required viscosity is only a function of the zero-order temperature and pressure which are functions only of depth. Using temperature profiles obtained from conduction calculations the results of the stability analysis are applied to the interiors of the earth, Mars, Venus and the moon to determine whether thermal convection is occurring.

091.037 La vision dans les instruments astronomiques et l'observation physique des surfaces planétaires.
J. Dragesco.
L'Astronomie, 83e année, 399 - 408, 439 - 447 (1969).

091.038 Über das Eindringen von Antimaterie in das Sonnensystem und in die Erdatmosphäre.
V. A. Bronshtehn, K. P. Stanyukovich.
Kosmich. Issled. Vol. 7, 597 - 601 (1969).   In Russian.

Die Bewegung hypothetischer Antikörper im Sonnensystem und in der Erdatmosphäre wird diskutiert. Bei Zusammenstößen eines Antikörpers mit Atomen des interplanetaren Gases und der Atmosphäre findet nicht nur eine Annihilation, sondern auch eine intensive Verdampfung des Antikörpers statt, deren spezifische Energie um 10 Größenordnungen geringer ist als die der Annihilation. Bei einer Dichte des interplanetaren Gases von 1 Atom/cm^3 kann ein Antikörper mit dem Radius von $r_0 \leqslant 1$ cm das Sonnensystem deshalb nicht durchqueren. Beim Eintritt in die Erdatmosphäre findet eine starke Bremsung des Antikörpers auf Kosten eines reaktiven Impulses der Strahlung statt, die bei der Annihilation entsteht, sowie auch eine intensive Verdampfung, die für Körper von Meteordimensionen in Höhen zwischen 450 und 800 km, und für $r_0 = 1$ m in 200 km Höhe endet. Somit ist jegliche Möglichkeit einer Antimaterie-Natur der Meteore und Kometen, die B. P. Konstantinov und andere in einer früheren Arbeit postulieren, ausgeschlossen.

091.039 Comment on the papers by P. Goldreich and S. Soter; F. F. Fish, Jr.; and W. K. Hartmann and S. M. Larson, related to the mass-angular momentum diagram of planets. P. Brosche.
Icarus, Vol. 11, 220 (1969). – Note.

091.040 Free bodily vibrations of the terrestrial planets.
B. A. Bolt, J. S. Derr.
Vistas in Astronomy, Vol. 11, 69 - 102 (1969).

A comparison is made of the work in geophysics, astronomy, and atomic physics on free oscillations of spherical models. The mathematical formulation of the eigenvibrations of an elastic sphere is outlined and the geometry of the vibrations and the effects of rotation and ellipticity discussed. Conditions for the generation of planetary free oscillations and their measurement on seismometers, gravimeters, and magnetometers are presented along with a discussion of observations made already on the earth. For the earth, moon, Venus, and Mars, models of velocity and density as functions of depth are defined which satisfy available data on mass, radius, and other physical parameters.

091.041 Weather on other planets.
G. S. Golitsyn.
Zemlya i Vselennaya, No. 6, p. 25 - 30 (1969). In Russian.

091.042 Numerical studies of planetary circulations in a model atmosphere [Final Scientific Report, 1 Aug. 1965–31 Jul. 1967] A. Huss.
Report AFCRL–68–0114, Hebrew Univ., Dep. Meteorology, Jerusalem. 192 pp. (1967). – See Phys. Abstr. Vol. 72, No. 20223 (1969).

091.043 Photoelectrons in planetary atmospheres.
R. J. W. Henry, M. B. McElroy.
Contr. Kitt Peak National Obs. No. 244, [Reprinted from *"Atmospheres of Venus and Mars"*, Gordon and Breach, 1968, p. 251 - 285], 35 pp. (1968).

091.044 Comparable characteristics of the planets Mars, Venus, and earth.
K. I. Gringauz, T. K. Breus.
Kosmich. Issled. Vol. 7, 871 - 890 (1969). In Russian.

091.045 Estimates of parameters of boundary layers in the atmospheres of terrestrial planets.
G. S. Golitsyn.
Izv. AN SSSR. Fiz. atmosf. i okeana, Vol. 5, 775 - 781 (1969). In Russian. – Abstr. in Referativ. Zhurn. 51. Astron., 1.51.277 (1970).

091.046 The spectral brightness of an inhomogeneous spherical atmosphere.
R. Bellman, H. Hagiwada, R. Kalabra, S. Ueno.
Space Research IX, Proc. Tokyo 1968, p. 385 - 391 (1969).

091.047 Über die Verteilung der am häufigsten vorkommenden Atmosphären-Bestandteile auf den Planeten des Sonnensystems. G. G. Kandilarov.
Monatshefte Chemie, Vol. 100, 213 - 223 (1969).
Unter Berücksichtigung der kosmischen Häufigkeitsstufen der am häufigsten vorkommenden Elemente, läßt sich die Verteilung der am häufigsten vorkommenden Atmosphärenbestandteile auf den Planeten des Sonnensystems, sowohl auf ihre Atom- bzw. Molekulargewichte, Siede- bzw. Sublimationstemperaturen, Entfliehensfähigkeiten und andere physikalische und chemische Eigenschaften als auch auf die auf den betreffenden kosmischen Körpern herrschenden Bedingungen zurückführen.

091.048 La terra, il sole e il sistema planetario.
M. G. Fracastoro.
Oss. Astron. Torino, Studi Monografici, No. 6, 44 pp. (1969). Reprinted from Coelum, Vol. 36, 1 - 13, 37 - 46, 119 - 127 (1968); Vol. 37, 49 - 67 (1969).

091.049 Radar surveys of the solar system.
J. V. Evans.
Proc. American Phil. Soc., Vol. 113, 203 - 223 (1969).

091.050 An approximative formula for the brightness distribution over the disk of a planet with thin

atmosphere.   Yu. V. Aleksandrov, V. I. Garazha.
Vestn. Khar'kov. Univ. No. 34, (Ser. Astron. No. 4), p. 31 - 38 (1969). In Russian.

Atlas des spectres dans le proche infrarouge de Vénus, Mars, Jupiter et Saturne.
See Abstr. 003.021.

El Mundo de los Planetas.
See Abstr. 003.155.

Die Welt der Planeten.
See Abstr. 003.156.

Zur Rotation einer axialsymmetrischen zähen Flüssigkeit um eine feste Achse.   See Abstr. 022.104.

Nitric oxide gamma band emission rate factor.
See Abstr. 022.105.

La vision dans les instruments astronomiques et l'observation physique des surfaces planétaires.
See Abstr. 031.009.

On the possibility of calculating the errors, caused by the deformation of the image of planets.
See Abstr. 031.012.

Planetary spectroscopy with the 107-inch telescope.
See Abstr. 034.005.

Zur Dynamotheorie stellarer und planetarer Magnetfelder. II. Berechnung planetenähnlicher Gleichfeldgeneratoren.   See Abstr. 062.008.

Zur Dynamotheorie stellarer und planetarer Magnetfelder. III. Über die Lösung der Eigenwertprobleme und die Berechnung der Feldgrößen.   See Abstr. 062.009.

Interaction of the solar wind with planetary atmospheres.   See Abstr. 074.059.

An on-board technique for estimating the effect of water vapor in radio occultation measurements of atmospheric density.   See Abstr. 082.014.

The moon and the planets.
See Abstr. 094.044.

Proton induced chemical reactions on lunar and planetary surfaces.   See Abstr. 094.145.

Lunar and planetary mass concentrations.
See Abstr. 094.155.

Optical properties and the structure of Jupiter's atmosphere. II. The influence of the multiple scattering in a cloud layer on planetary absorption line profiles.
See Abstr. 099.054.

Isotopic composition of lithium in some meteorites and the role of neutrons in the nucleosynthesis of the light elements in the solar system.   See Abstr. 105.097.

On the possible distribution of mass among celestial bodies.   See Abstr. 115.008.

092 Mercury

092.001 **Spin-orbit resonance of the inner planets.**
P. M. Campbell.
Science, Vol. 165, 930 (1969).

092.002 **On the figure of the planet Mercury.**
H.-S. Liu.
Celestial Mechanics, Vol. 1, 144 - 149 (1969).

The figure of Mercury is estimated in terms of an isostatic form of equilibrium which tends to be controlled by the situation near perihelion passage at the 3 : 2 resonance spin rate. The ratios of the principal moments of inertia for Mercury are : (1) $(C - A)/C \geqslant 7 \times 10^{-5}$; (2) $(C - B)/C \geqslant 5 \times 10^{-5}$ and (3) $(B - A)/C \geqslant 2 \times 10^{-5}$. The thermal effect on Mercury's figure during solidification forces Mercury's rotation to be trapped in the 3 : 2 resonance lock as its spin rate is being slowed by tidal effects. It is shown that the process of trapping of Mercury has been naturally affected by the instantaneous solidification of Mercury into a shape with two thermal bulges, and that the two permanent thermal bulges stabilize the planet's rotation.

092.003 **The microwave spectrum of Mercury.**
M. J. Klein, D. Morrison.
Bull. American Astron. Soc., Vol. 1, 350 (1969). — Abstr. AAS.

092.004 **A review of the environment of Mercury, Venus, and Mars.** K. M. Foreman, E. Nowatzki, F. R. Pomilla, J. Reichman, J. H. Scanlon.
Report RM-420, Grumman Aircraft Engineering Corp., Research Dep., Bethpage, N. Y., 245 pp. (1968). — See Phys. Abstr. Vol. 72, No. 27086 (1969).

092.005 **Laboratory astronomy: A geometric experiment to determine the orbit of Mercury.**
R. B. Herr.
American Journ. Phys., Vol. 37, 74 - 81 (1969).

092.006 **A re-examiniation of Danjon's observations of the planet Mercury.** T. Jaakkola.
Ann. Acad. Sci. Fennicae, Series A, VI. Physica 301, 11 pp. = Aarne Karjalainen Obs. Univ. Oulu, Finland, Publ. No. 13 (1969).

Danjon's measurements on Mercury's integrated brightness are re-examined. Its phase curve turns to be asymmetrical with respect to the zero phase. This phenomenon may partially be explained by two bright spots on the opposite sides of the planet but a residual asymmetry remains unexplained.

092.007 **Mercury in 1965.** R. G. Hodgson.
Strolling Astronomer, Vol. 21, 210 - 214 (1969).

092.008 **On the internal structures of Mercury and Venus.**
R. A. Lyttleton.
Astrophys. Space Sci. Vol. 5, 18 - 35 (1969).

Recent radar measures of the radius and mass of Mercury imply a composition for the planet containing about 60% iron. Either Mercury is a highly exceptional object among terrestrial planets, or all measures to date of the planet involve substantial systematic error. Independent checking of the radius and mass of Mercury by some entirely different means has become of the greatest importance. The recent radar and other determinations of the solid radius of Venus imply an internal structure similar to that of the earth. It has seemed worthwhile to calculate the internal structures of these two planets by existing theoretical means in order to ascertain what the measures imply regarding their possible compositions as compared with the earth.

Lunar and planetary mass concentrations.
See Abstr. 094.015.

Radiative heat transfer in the lunar and Mercurian surfaces. See Abstr. 094.202.

The surface structures of the Moon and Mercury derived from integrated photometry.
See Abstr. 094.235.

093 Venus

093.001 **Photodissociation of molecular hydrogen on Venus.**
A. Dalgarno, A. C. Allison.
Journ. Geophys. Res. Vol. 74, 4178 - 4180 (1969). –Letter.

093.002 **Zur Bestimmung der visuellen Venus-Dichotomie.**
H. Jungblut.
VdS Nachrichtenblatt, Vol. 18, 120 - 123 (1969).

093.003 **Venusonderzoek van nabij.** F. Israel.
Hemel en Dampkring, Vol. 67, 259 - 271 (1969).

093.004 **Venus: Mapping the surface reflectivity by radar
interferometry.** A. E. E. Rogers, R. P. Ingalls.
Science, Vol. 165, 797 - 799 (1969).
The surface reflectivity of Venus obtained by radar
interferometry at a wavelength of 3.8 centimeters has been
mapped for a region extending approximately from $-80°$ to
$0°$ in longitude and from $-50°$ to $+40°$ in latitude. The map
is free from the twofold range-Doppler ambiguity because
the interferometer fringe pattern makes possible the separation of two points of equal range and Doppler shift. The map
presents many new features and clearly delineates features
already observed. Most notably, the map shows large circular
regions of significantly lower reflectivity than their surroundings.

093.005 **TV pictures of ultra-violet markings on Venus.**
V. V. Prokofieva, S. I. Usliber.
Astron. Tsirk. No. 518, p. 1 - 2 (1969).   In Russian.

093.006 **Rocket spectra of Venus and Jupiter from 2000 to
3000 Å.**   E. B. Jenkins, D. C. Morton,
A. V. Sweigart.
Astrophys. Journ. Vol. 157, 913 - 924 = Contr. Kitt Peak
National Obs. No. 421 (1969).
A rocket-borne spectrograph simultaneously photographed spectra of Venus from 2100 to 3070 Å and of Jupiter
from 2400 to 3000 Å while the two planets were only $1°8$
apart on June 9, 1967. Gyrostabilization along the dispersion
direction allowed the f/2 objective-grating camera to provide
four exposures having resolutions of about 1 Å. Divisions by
the solar spectrum yielded planetary reflectivities which were
devoid of any especially strong absorption or emission features. Possible weak absorptions by the atmosphere of Venus
may exist near 2174 and 2450 Å, and broad variations in albedo seem apparent between 2470 and 2650 Å. An overall
depression in albedo at 2200 Å could be caused by 5×10^{-2}cm
atm of COS. The following approximate upper limits of abundance for the upper atmosphere of Venus were provided by
a lack of detectable absorptions: 5×10^{-4} cm atm O_3, $9 \times
10^{-3}$ cm atm NH_3, 10^{-3} cm atm SO_2, 6×10^{-2} cm atm NO_2,
10^{-1} cm atm NO, and 5×10^{-1} cm atm C_3O_2. The low abundance of ozone suggests that oxygen may be nearly absent on
Venus.

093.007 **Venus: The next phase of planetary exploration.**
D. M. Hunten, R. M. Goody.
Science, Vol. 165, 1317 - 1323 (1969).
The atmosphere and clouds of Venus are ripe for direct
exploration by means of entry probes. Taking into account
the observations of Venus 4 and Mariner 5 the following subjects are treated: Greenhouse model, infrared opacity, convection, cloud layer, dust clouds, planetary circulation, upper atmosphere.

093.008 **Polar temperature of Venus.**
W. A. Gale, A. C. E. Sinclair.

Science, Vol. 165, 1356 - 1357 (1969).
The presence of substantial polar cooling of Venus, as
derived from microwave interferometry at 10.6 cm wavelength,
is shown to be open to doubt. Other microwave measurements
give little evidence for significant poleward variation in temperature on the planet.

093.009 **Die Untersuchungen des Schröter-Effektes in der
Sowjetunion.**   W. A. Bronshten.
Sterne, 45. Jahrgang, 103 - 107 (1969).

093.010 **Effect of temperature on the strength and composition of the upper lithosphere of Venus.**
R. F. Mueller.
Nature, Vol. 224, 354 - 356 (1969).
Application of the kinetic theory of viscosity indicates
that the high temperatures on Venus should lead to a mechanical weakening of the lithosphere and a corresponding reduction of surface relief. These effects should also be accompanied by a more thorough chemical differentiation of the
upper lithosphere than on earth. As a result there should be
a greater upward concentration of granitic rocks and radioactive elements in the planetary crust. However a partially
molten lithosphere seems to be excluded.

093.011 **The mass and dynamical oblateness of Venus.**
J. D. Anderson, L. Efron.
Bull. American Astron. Soc. Vol. 1, 231 - 232 (1969). –
Abstr. AAS.

093.012 **The Mariner V dual-frequency occultation measure-
ments of the neutral atmosphere of Venus.**
G. Fjeldbo, V. R. Eshleman.
Bull. American Astron. Soc. Vol. 1, 241 (1969). – Abstr.
AAS.

093.013 **On the phase curve of Venus.**   J. F. Potter.
Bull. American Astron. Soc. Vol. 1, 258 (1969). –
Abstr. AAS.

093.014 **The ionospheres of Venus and Mars. I. Mariner IV
and preliminary ideas.**   R. Goody.
Comments Astrophys. Space Phys. Vol. 1, 194 - 197 (1969).
The Mariner-IV Mars flyby in July 1965 yielded a profile of electron density as a function of height, and Mariner V
in October 1967 gave both daytime and nighttime profiles
for Venus. These new data have led to a major reorientation
of our ideas about ionospheric physics. The author would like
to review some of these developments in a historical perspective.

093.015 **Interpretation of high-resolution spectra of Venus.
I. The $2\nu_3$ band of $^{13}C^{16}O^{18}O$ at 2.21 microns.**
L. G. Young.
Icarus, Vol. 11, 66 - 75 (1969).
The high-resolution spectra of Venus, obtained by Janine
and Pierre Connes in the region of 2 microns have been used
to obtain a curve of growth for the $2\nu_3$ band of the $^{13}C^{16}O^{18}O$
isotope of carbon dioxide. Several methods of data reduction
are compared; the rotational temperature found from this
band, using the curve of growth, is $245° \pm 3°K$ (standard deviation). Values for the effective pressure for line formation
and the absorber amount above the clouds are also derived,
but are dependent on the strengths of the lines.

093.016 **Solar cosmic ray effects in the lower ionosphere of
Venus.**   R. R. Brown.

Planet. Space Sci. Vol. 17, 1923 - 1926 (1969).

The similarity of atomic parameters for the CO_2 atmosphere of Venus and that of the earth is used to calculate the ionization and optical emission rate in the upper atmosphere of Venus resulting from a major solar cosmic ray event. The possibility of as much as 10 per cent of N_2 in the atmospheric composition of Venus does not change these effects appreciably.

093.017   **Thermal structure of the ionosphere of Venus.**
R. C. Whitten.
Journ. Geophys. Res. Vol. 74, 5623 - 5628 (1969).

There are two possible heat sources for the sunlit ionosphere of Venus, photoionization of the neutral species, mainly CO_2, and influx from the solar wind. It is shown that the electron temperature should be substantially higher (at least ~500°K) than the temperature of the neutral atmosphere at altitudes above 250 km but that the ion temperature probably does not greatly exceed that of the neutral species. The significance of the solar wind as a heat source is also discussed.

093.018   **Estimate of radiogenic He⁴ and Ar⁴⁰ concentration in the Cytherean atmosphere.**
W. C. Knudsen, A. D. Anderson.
Journ. Geophys. Res. Vol. 74, 5629 - 5632 (1969).

We present a computation for the abundance of He⁴ in the lower atmosphere of Venus, which when extrapolated into the upper atmosphere leads to essentially the same He⁴ concentration as that needed to explain the ionospheric results. The approach used to compute the He⁴ content of the atmosphere also can be used to compute the Ar⁴⁰ content.

093.019   **Absorption of radio waves by water vapour in the atmospheres of Venus and Mars.**
A. A. Viktorova, A. P. Naumov.
Izv. vyssh. uchebn. zavedenij. Radiofizika, Vol. 12, 621 - 624 (1969). In Russian. — Abstr. in Referativ. Zhurn. 51. Astron., 11.51.560 (1969).

093.020   **High altitude spectra from NASA CV-990 Jet. II. Water vapor on Venus.**
G. P. Kuiper, F. F. Forbes, D. L. Steinmetz, R. I. Mitchell, U. Fink.
Commun. Lunar Planet. Lab. Vol. 6, (No. 100), 209 - 228 (1969).

Venus and moon spectra obtained with a new interferometer (resolution 8 cm⁻¹) during two flights with the NASA CV-990 Jet on November 27 and 28, 1967, are presented. The ratio spectrum Venus/moon is also derived which eliminates remaining weak telluric absorptions as well as solar absorption lines of any strength. The ratio spectrum shows Venus essentially as if illuminated in white light, allowing a direct comparison with laboratory spectra. A lunar spectrum with increased precision was obtained on 5 - 6 May 1968, allowing the computation of an improved ratio spectrum for Venus. Finally, ground-based Venus and moon spectra, and their ratio, are added for increased precision of the Venus spectrum outside the telluric H_2O bands.

093.021   **Identification of the Venus cloud layers.**
G. P. Kuiper.
Commun. Lunar Planet. Lab. Vol. 6, (No. 101), 229 - 250 (1969).

Observations establish the existence of two well-separated cloud layers on Venus, which are described in the paper. A compilation of published photometric data from 0.2 - 4.0 μ and new spectrophotometric data, compared to new laboratory measures, show that the chief constituent of the yellow haze layer is incompletely-hydrated $FeCl_2$. The published H_2O measurements by Venera 4 - 6 for the deeper layers are discussed but found to be incompatible with well-established

results. It is concluded that Venus has a halide meteorology, compared to a water meteorology on earth, and an ammonia meteorology in Jupiter and Saturn. The near-absence of water on Venus must be a basic planetary property, apparently resulting from a protoplanet temperature being substantially higher than that of proto earth, which caused H_2O to be in the vapor phase and lost with the inert gases. This must have been caused by Venus forming later in the solar development as well as closer to the Sun.

093.022   **Venus photographs. Part I: Photographs of Venus taken with the 82-inch telescope at McDonald Observatory, 1950-56.**   G. P. Kuiper, J. W. Fountain, S. M. Larson.
Commun. Lunar Planet. Lab. Vol. 6, (No. 102/I), 251 - 262 (1969).

In this communication are collected three series of Venus photographs, taken in our continuing planetary program: one at the McDonald Observatory 82-inch telescope (discussed in Part I); one with the Steward Observatory 36-inch telescope (Part II); and one with the 61-inch NASA telescope at the Catalina Observatory (Part III).

093.023   **Venus photographs. Part II: Multicolor photography of Venus, 1962.**   W. K. Hartmann.
Commun. Lunar Planet. Lab. Vol. 6, (No. 102/II), 262 - 264 (1969).

093.024   **Venus photographs: Part III: UV photographs of Venus taken with the 61-inch NASA telescope, 1967.**   J. Fountain, S. Larson.
Commun. Lunar Planet. Lab. Vol. 6, (No. 102/III), 264 - 274 (1969).

093.025   **The atmosphere of Venus.**
M. Ya. Marov.
Vestn. AN SSSR, No. 5, p. 72 - 81 (1969).   In Russian.

093.026   **Identification of the Venus cloud layers.**
G. P. Kuiper, G. T. Sill.
Bull. American Astron. Soc., Vol. 1, 351 (1969). — Abstr. AAS.

093.027   **The surface temperature distribution on Venus.**
A. C. E. Sinclair, J. P. Basart, D. Buhl, W. A. Gale, M. Liwshitz.
Bull. American Astron. Soc., Vol. 1, 362 (1969). — Abstr. AAS.

093.028   **The evolution of water vapor in the atmosphere of Venus.**   L. L. Smith, S. H. Gross.
Bull. American Astron. Soc., Vol. 1, 363 (1969). — Abstr. AAS.

093.029   **Ergebnisse zur Dynamik der Venusatmosphäre aus Radialgeschwindigkeitsmessungen mit der automatischen interplanetaren Station "Venus 4".**
V. K. Kerzhanovich, V. M. Gotlib, N. V. Chetyrkin, B. N. Andreev.
Kosmich. Issled. Vol. 7, 592 - 596 (1969).   In Russian.

Vorgelegt werden Radialgeschwindigkeitsmessungen der automatischen Station "Venus 4" während der Zeit, in der diese in der Venusatmosphäre herabsank. Methodik und Ergebnisse der Bestimmung der vertikalen und horizontalen Strömungen während des Herabsinkens werden mitgeteilt. Die Meßfehler werden angegeben.

093.030   **Über die Natur des Magnetfeldes in der Umgebung der Venus.**
Sh. Sh. Dolginov, E. G. Eroshenko, L. Davis.
Kosmich. Issled. Vol. 7, 747 - 752 (1969).   In Russian.

093.031 High dispersion spectroscopic observation of Venus. IV: The weak carbon dioxide band at 7883 Å.
L. D. Gray, R. A. Schorn, E. Barker.
Applied Optics, Vol. 8, 2087 - 2093 (1969).

The average rotational temperature of the Cytherean atmosphere above the cloud tops was found to be $T_{rot} = 244 \pm 10$ K based on twelve plates of the 7883-Å CO_2 band. If the temperatures found from the 7820-Å band on the same plate are averaged with the temperatures found from the 7883-Å band, we obtain a temperature of $T_{rot} = 245 \pm 6$ K. The observations of Venus were made between April and December of 1967. Laboratory measurements of the 7883-Å band are lacking, but we infer that the band is half as strong as the 7820-Å band.

093.032 Infrared transmission properties of CO, HCl, and SO_2 and their significance for the greenhouse effect on Venus.
I. J. Eberstein, B. N. Khare, J. B. Pollack.
Icarus, Vol. 11, 159 - 170 (1969).

Low resolution (20 cm^{-1}) transmission measurements were obtained for strong infrared absorption features of CO, HCl, and SO_2. It was concluded that none of the above gases are significant for the strong greenhouse effect on Venus, either because they are not present in sufficient amounts, or because they produce opacity in spectral regions where carbon dioxide is already quite opaque.

093.033 High surface temperature on Venus: Evaluation of the greenhouse explanation. G. Ohring.
Icarus, Vol. 11, 171 - 179 (1969).

In the present study, we use the recent Mariner 5 and Venera 4 observations of temperature, pressure, and atmospheric composition to evaluate the greenhouse effect on Venus with the use of a nongray radiative model.

093.034 Report on the elongation of Venus: 1969 January.
J. H. Robinson.
Journ. British Astron. Ass., Vol. 80, 55 - 61 (1969).

093.035 Once again on the radius of Venus.
D. Ya. Martynov.
Astron. Vestn. Vol. 3, 82 - 84 (1969). In Russian.

It is shown that the mean error of the height above the surface of Venus (H = 113 km) at its occultation of Regulus is not less than ± 10 km, the error in the usual determination of this radius is about ±(25 – 30) km.

093.036 Observations of the dichotomy of Venus.
V. G. Lozitsky.
Astron. Vestn. Vol. 3, 114 - 117 (1969). In Russian.

093.037 Quantitative spectroscopy of Venus in the region 8.000 – 11.000 Å.
M. J. S. Belton, D. M. Hunten, R. M. Goody.
Contr. Kitt Peak National Obs. No. 241, [Reprinted from *"Atmospheres of Venus and Mars,"* Gordon and Breach, 1968, p. 69 - 97], 29 pp. (1968).

093.038 On the luminosity of the night sky of Venus.
N. B. Ibragimov.
Astron. Tsirk. No. 533, p. 3 (1969). In Russian.

093.039 Absorption of radiowaves in the ionosphere of Venus. A. N. Kazantsev, V. A. Danilin.
Kosmich. Issled. Vol. 7, 900 - 904 (1969). In Russian.

093.040 Refraction of radiowaves and the field-strength in the atmosphere of Venus.
D. S. Lukin, Yu. G. Spiridonov, V. A. Shkol'nikov.

Kosmich. Issled. Vol. 7, 905 - 910 (1969). In Russian.

093.041 The distribution of radio brightness over the disc of Venus at 8 mm-wavelength.
V. J. Golnev, Iu. N. Pariiskii, P. A. Fridman, O. N. Shivris.
Dokl. Akad. Nauk SSSR, Ser. Mat. Fiz., Vol. 188, 297 - 299 (1969). In Russian.

Observations at 8 mm have been made in Pulkovo 1967, Sept. 6 - 22 by means of the large radio telescope with a resolution $15'' \times 4'$. The brightness distribution on the disc of Venus obtained from 10 scannings agrees well with the hypothesis that millimetre waves are absorbed by atmospheric layers near the surface with an adiabatic gradient ~10°/km. The observations disagree with the supposition that strong absorption of 8 mm radiation takes place in the cloud layer as well as with the hypothesis of strong winds in the Venus atmosphere. There is an asymmetry in the brightness distribution, the east (partially sunlit) part being 12 - 18°K warmer. Observations of phase variation of the brightness temperature of Venus in the centimetre band are in progress in Pulkovo.
B. Onderlička

093.042 Venus 1969. L. Lebedev, S. Nikitin.
Bild der Wiss., 1969, No. 12, 1213 - 1219 (1969).

Report on the results of space probes Venus 5 and 6. Chemical composition of the Venus atmosphere: CO_2 93 - 97%, N_2 + rare gases 2 - 5%, water vapour 4 - 11 mg/litre (at heights of atmospheric pressure 0.6 atm), O_2 below 0.4%. Extrapolated surface values from Venus 5 : 530°C, 140 atm; from Venus 6: 400°C, 60 atm. *B. Onderlička*

093.043 The atmosphere of Venus from recent investigations.
A. D. Kuzmin.
Space Research IX, Proc. Tokyo 1968, p. 704 - 711 (1969).

On the basis of Venera 4 measurements, a model of the atmosphere of Venus has been constructed up to a height of 100 km. Data from optical and infrared ground measurements were utilized. The spectra of the planet's radio emission have been calculated. It is shown that radio astronomical measurements are in good agreement with the Venera 4 atmosphere model.

093.044 Structure of the atmosphere of Venus derived from Mariner V S-band measurements.
A. Kliore, D. L. Cain, G. S. Levy, G. Fjeldbo, S. I. Rasool.
Space Research IX, Proc. Tokyo 1968, p. 712 - 729 (1969).

093.045 Temperature and density of the Venus atmosphere according to measurements obtained by Venera 4.
V. V. Mikhnevitch, V. A. Sokolov.
Space Research IX, Proc. Tokyo 1968, p. 730 - 744 (1969).

093.046 The atmosphere of the planet Venus from data of the Soviet space probe Venera 4.
V. S. Avduevsky, M. Ya. Marov, M. K. Rozhdestvensky.
Space Research IX, Proc. Tokyo 1968, p. 745 - 759 (1969).

093.047 Induced magnetosphere of Venus.
F. S. Johnson, J. E. Midgley.
Space Research IX, Proc. Tokyo 1968, p. 760 - 763 (1969).

093.048 Research results and speculations about the Venus' surface and atmosphere. K.-H. Remane.
Astron. in der Schule, 6. Jahrgang, 95 - 98 (1969). – Popular article.

093.049 Effect of cloud scattering on line formation in the atmosphere of Venus. J. F. Potter.
Journ. Atmosph. Sci., Vol. 26, 511 - 517 (1969).

093.050 A radar view of the surface of Venus.

R. M. Goldstein.
Proc. American Phil. Soc., Vol. 113, 224 - 228 (1969).

093.051   **Venus through a spectroscope.**
          P. Swings.
Proc. American Phil. Soc., Vol. 113, 229 - 246 (1969).

093.052   **Some results of a photometry of Venus based on
          observations in 1964.**
O. M. Starodubtseva.
Vestn. Khar'kov. Univ. No. 34, (Ser. Astron. No. 4), p. 56 -
65 (1969).   In Russian.

093.053   **Spectrophotometric comparisons of some regions
          of Venus.**   O. M. Starodubtseva.
Vestn. Khar'kov. Univ. No. 34, (Ser. Astron. No. 4), p. 91 -
102 (1969).   In Russian.

Spin-orbit resonance of the inner planets.

See Abstr. 092.001.

**A review of the environment of Mercury, Venus,
and Mars.**   See Abstr. 092.004.

On the internal structures of Mercury and Venus.
See Abstr. 092.008.

**Solar cycle variation of exospheric temperatures on
Mars and Venus: A prediction for Mariner 6 and 7.**
See Abstr. 097.002.

Exospheric temperatures on Mars and Venus.
See Abstr. 097.071.

**Empirical determination of heating efficiencies in
the Mars and Venus atmospheres.**   See Abstr. 097.072.

**Interplanetary plasma densities deduced from radar
observations of Venus.**   See Abstr. 106.005.

094 Moon

094.001 Internal constitution of the Moon: Is the lunar interior chemically homogeneous?
Y. Nakamura, G. V. Latham.
Journ. Geophys. Res. Vol. 74, 3771 - 3780 (1969).

A procedure for constructing lunar models has been developed in which pressure, temperature, and compositional effects are taken into account. Pertinent lattice-dynamical relations are also incorporated. By using this procedure, a series of lunar models have been constructed based on the latest measurements on the physical properties of rock-forming minerals. Chemically homogeneous models thus constructed are found to have too high a density at the surface and probably too large a moment of inertia. We conclude that (1) there must be a concentration of lighter material near the surface of the Moon, and (2) the deep interior of the Moon is more likely to be chemically heterogeneous than to be homogeneous throughout although the possibility of chemical homogeneity cannot be ruled out judging from the current range of uncertainty of the moment of inertia of the Moon.

094.002 Lunar thermal anomalies and internal heating.
J. M. Saari.
Astrophys. Space Sci. Vol. 4, 275 - 284 (1969).

The evidence for the existence of the so-called linear anomaly recently reported on the western margin of Mare Humorum is examined critically. Whether the anomalous cooling of Mare Humorum considered as a whole can be caused by internal heating is rejected on the basis of (1) the measured temperature differential compared to the environs observed during the lunar day, (2) the required temperature gradient, and (3) energy considerations compared to terrestrial heat flow. It is concluded that the observed hot spots are unlikely to be caused by internal heating.

094.003 Comments on 'Lunar thermal anomalies and internal heating' by J. M. Saari.
G. R. Hunt, J. W. Salisbury, R. K. Vincent.
Astrophys. Space Sci. Vol. 4, 370 - 372 (1969).

094.004 Moon dust and coal ash.
A. B. Hart, E. Raask.
Nature, Vol. 223, 762 - 763 (1969). – Letter.

094.005 Some comments on the paper 'Moments of inertia and gravity field of the moon' by C. L. Goudas.
P. Melchior.
Astrophys. Space Sci. Vol. 4, 417 - 418 (1969). – Research note. – See AJB 64 Abstr. 8368.

094.006 Crater statistics near the Flamsteed P ring.
R. Fryer, C. Titulaer.
Commun. Lunar Planet. Lab. Vol. 8 (No. 133), 51 - 61 (1969).

Previous work on the population curves and crater distribution in the vicinity of Surveyor 1 is reviewed and extended. A total-population curve is derived for craters ranging in size from centimeters to kilometers. Comparison is made with Ranger data covering an equivalent diameter range. The floor of the low ring Flamsteed P is found to be unusually young, younger than the surrounding mare; this is consistent with the recent hypotheses of O'Keefe and Fielder which regard the ring as a recent extrusive structure.

094.007 Crater overlap on the near-side of the moon.
C. Titulaer.
Commun. Lunar Planet. Lab. Vol. 8 (No. 134), 63 - 72 (1969).

An investigation of the overlap of craters on the front side of the moon shows a well-defined anomalous area around Tycho. This Tycho Association of craters may have originated as a result of the collision of a cometary shower, the nucleus of which formed Tycho itself. Alternative explanations are also considered. No full explanation is possible without additional research.

094.008 Polarization–albedo relationship for selected lunar regions. S. F. Pellicori.
Commun. Lunar Planet. Lab. Vol. 8 (No. 135), 73 - 74 (1969). – Reprinted from Nature, Vol. 221, 162 (1969).

094.009 Terrestrial, lunar and interplanetary rock fragmentation (synopsis). W. K. Hartmann.
Commun. Lunar Planet. Lab. Vol. 8 (No. 136), 75 - 79 (1969).

In a table collected data on mass distributions of fragmented rocks are presented. Mass distributions are typically power-law functions with exponent b; the b-values (slopes on log-log plots) increase as the samples are exposed to greater grinding and crushing or greater energy expenditure per particle. Plots of mass distributions can be used to interpret terrestrial and extraterrestrial rock samples.

094.010 Electrostatic potential distribution of the sunlit lunar surface. W. D. Grobman, J. L. Blank.
Journ. Geophys. Res. Vol. 74, 3943 - 3951 (1969).

The steady-state lunar surface charge and potential distributions are determined by the condition that the net current to a small surface area vanish, where the dominant currents are due to photoemission of electrons and collection of solar wind particles. The lunar crust and photoelectron cloud are too resistive to carry a significant flux. A calculation similar to one used in collisionless electrostatic probe theory shows that the current from the solar wind is predominantly due to the electrons, is independent of potential, and is weakly dependent upon the polar angle θ measured from the moon-sun line.

094.011 Differences between proposed Apollo sites. 1. Synthesis. B. C. Murray, A. F. H. Goetz, H. H. Kieffer, T. B. McCord.
Journ. Geophys. Res. Vol. 74, 4382 - 4384 (1969).

Recent observations of the spectral reflectivity and emissivity of the five prime Apollo landing sites are evaluated in the context of similar observations of other localities on the moon and of data returned from unmanned lunar probes. We conclude that those five sites differ significantly only in minor constituents and/or relative valence states and that those differences are more modest than the differences that characterize mare regions generally. Recommendations of priorities for the five prime Apollo sites are made based on their uniqueness for sample return.

094.012 Differences between proposed Apollo sites. 2. Visible and infrared reflectivity evidence.
T. B. McCord, T. V. Johnson, H. H. Kieffer.
Journ. Geophys. Res. Vol. 74, 4385 - 4388 (1969).

The relative spectral reflectivity, 0.40 to 1.10μ, was measured for 10- to 18-km-diameter lunar areas centered on the five prime Apollo landing sites. Though the spectral reflectivities of the Apollo sites appear typical of the lunar maria in general, significant differences were found among the five sites. The reflectivity differences are attributed to compositional and/or mineralogical differences in the lunar surface materials.

094.013 Differences between proposed Apollo sites. 3. Far infrared emissivity evidence.

A. F. H. Goetz, L. A. Soderblom.
Journ. Geophys. Res. Vol. 74, 4389 - 4394 (1969).

Infrared emissivity comparison spectra of nine areas on the lunar surface, each 40 km in diameter, indicate that the majority of the lunar surface, including the five Apollo sites, has a constant Si-O ratio so far as present infrared techniques are able to detect. However, an anomaly in the $8.2-9\,\mu$ region of the emissivity spectrum of the crater Plato is interpreted as evidence of a significantly different Si-O ratio in the mineral assemblage exposed on that surface.

094.014 **Relative spectral reflectivity 0.4 - 1μ of selected areas of the lunar surface.**
T. B. McCord, T. V. Johnson.
Journ. Geophys. Res. Vol. 74, 4395 - 4401 (1969).

The relative spectral reflectivity, 0.72 to 1.10μ, was measured for several areas of the lunar surface. A double-beam photoelectric filter photometer was used with the 24- and 60-inch (0.61 and 1.52 meter) telescopes on Mount Wilson. The infrared relative spectral curves were combined with curves for the visible portion of the spectrum obtained in an earlier study. The combined curves have considerable structure, which differs with lunar area and is correlated with lunar morphology. The series of curves cannot be derived from a linear combination of any two end-member curves. The spectral structure does not seem to correlate with relative age of the lunar areas. The position and structure of spectral features present in the curves are similar to absorption bands found in the reflectance spectrum for common terrestrial silicate materials. These results suggest that compositional and mineralogical differences on the lunar surface are responsible for the relative spectral reflectivity differences.

094.015 **Lunar and planetary mass concentrations.**
B. T. O'Leary, M. J. Campbell, C. Sagan.
Science, Vol. 165, 651 - 657 (1969).

Mascons beneath large circular basins may explain dynamical asymmetries in the moon, Mars, and Mercury.

094.016 **Lunar maria: Structure and evolution.**
W. G. van Dorn.
Science, Vol. 165, 693 - 695 (1969).

The lunar maria are considered to have evolved as homologous, transient, gravity-wave systems from large impact craters on a crustal layer 50 kilometers thick, fluidized from beneath by prompt, shock-induced melting inside an initially hot moon.

094.017 **Mondbeobachtungen während des Apollo 11-Fluges.**
J. Classen.
VdS Nachrichtenblatt, Vol. 18, 124 - 125 (1969).

094.018 **The Apollo 11 experiments.** J. Ashbrook.
Sky Telescope, Vol. 38, 149 - 151, 163 (1969).

094.019 **Time-dependent lunar electric and magnetic fields induced by a spatially varying interplanetary magnetic field.** K. Schwartz, G. Schubert.
Journ. Geophys. Res. Vol. 74, 4777 - 4780 (1969).

The response of a homogeneous conducting moon to variations and irregularities in the interplanetary magnetic field is determined. It is found that the time-dependent lunar electric and magnetic fields are forced by the oscillatory magnetic field and the oscillatory motional electric field in the solar wind plasma. In the limit of a highly conducting plasma surrounding the moon and for length scales of the interplanetary magnetic field irregularities much larger than the moon's radius, the magnetic field fluctuations in the lunar interior are the sum of the variations in the driving interplanetary magnetic field plus induced magnetic field fluctuations whose magnitude is proportional to the product of the magnitude of

the interplanetary magnetic field fluctuations and the lunar magnetic Reynolds number. This induced field is toroidal about an axis in the direction of the forcing motional electric field.

094.020 **Fluidization on the moon (?).**
J. D. Murray, E. A. Spiegel, J. Theys.
Comments Astrophys. Space Phys. Vol. 1, 165 - 171 (1969).

094.021 **Geophysics of the moon.**
H. C. Urey, G. J. F. MacDonald.
Science Journ. Vol. 5, No. 5, p. 60 - 65 (1969).

With its low density and mysterious surface features, the moon's evolution and the nature of its interior have long been the source of speculation. Now physical data are resolving many of the disputes.

094.022 **Die wissenschaftliche Untersuchung der Mondproben.** H. Wänke.
Umschau, Vol. 69, 580 - 581 (1969).

094.023 **Pictures from Apollo 10.**
Sky Telescope, Vol. 38, 90 - 94 (1969).

094.024 **Feldlabor im Mare Tranquillitatis. Apollo 11-Besatzung richtet auf dem Mond eine erste Versuchsstation ein.**
Weltraumfahrt, Jahrgang 20, 96 - 99 (1969).

094.025 **Massenkonzentrationen im Mondinneren.**
R. Engel.
Weltraumfahrt, Jahrgang 20, 99 - 102 (1969).

094.026 **The cinder lake moon simulation.**
R. J. Fryer.
Spaceflight, Vol. 11, 281 - 283 (1969).

094.027 **The mechanism of origin of the lunar seas.**
A. L. Sukhanov.
Spaceflight, Vol. 11, 283 - 286 (1969).

094.028 **Where they landed.** R. J. Fryer.
Spaceflight, Vol. 11, 318 - 320, 336 (1969).

094.029 **Charts for the moon.** E. Burgess.
Spaceflight, Vol. 11, 359 (1969).

094.030 **Observemos los satelites naturales de la tierra.**
H. G. Marraco.
Revista Astron. Vol. 40, No. 166, p. 17 - 20 (1968).

094.031 **Strength-density relations in particulate silicates of complex shape and their possible lunar significance.** L. D. Jaffe.
Science, Vol. 165, 1121 - 1123 (1969).

Some terrestrial particulate silicate rocks with complex particle shapes have internal friction angles over $45°$ and cohesion of about 0.1 newton per square centimeter at bulk densities of 0.6 to 0.8 g cm^{-3}. Mechanical and other properties of the lunar surface layer, observed with spacecraft, may be consistent with a low bulk density and complex reentrant shapes for the fine particles.

094.032 **Application of a statistical surface model to planetary radar astronomy.** A. H. Marcus.
Journ. Geophys. Res. Vol. 74, 4958 - 4962 (1969).

A statistical model of the distribution of slopes and elevations on a cratered planetary surface is applied to scattering of radio waves by the moon and Venus. It is found that the quasi-specular component of backscattered power can be fit by any symmetric stable distribution for elevation differences whose characteristic exponent is between 1 and 2 (crater diameter population index between 2 and 3). The lack of dependence on wavelength of the roughness parameter for decimeter and meter wavelengths, however, suggests that surface elevation differences have nearly a Cauchy distribution, not a Gaussian distribution. The increase of the roughness parameter with wavelength at millimeter and centimeter wavelengths suggests that pebbles and blocks are the major component of small-scale surface roughness.

094.033 On the distribution of brightness across the lunar disc at full moon. A. V. Morozjenko, E. G. Yanovitsky.
Astron. Tsirk. No. 524, p. 5 - 7 (1969). In Russian.

094.034 Determination of parameters of the lunar rotation by a method independent of the moon's profile. A. A. Gorynya.
Astrometriya i Astrofiz., *Kiev*, No. 2, p. 52 - 57 (1969). In Russian.

From 89 photographs of the moon taken in 1950 - 1955 and 300 visual observations made in 1962 - 1966 the parameters are obtained. The values are I = 1°33′18″ ± 25″, f = 0.88 ± 0.14.

094.035 Estimation of the constants of the moon's physical libration from Schlüter's observations with the heliometer at Königsberg (1841 - 1843). L. N. Kizjun.
Astrometriya i Astrofiz., *Kiev*, No. 2, p. 58 - 76 (1969). In Russian.

The paper deals with the reduction of Schlüter's heliometer observations for deriving the constants of the physical libration of the moon. The classical method was used and allowance was made for limb irregularities according to Watts and Hain, using the model "C" proposed by A. A. Jakovkin.

094.036 Determination of the constants of the moon's rotation from distances between Mösting A and limb craters. V. S. Kisljuk.
Astrometriya i Astrofiz., *Kiev*, No. 2, p. 77 - 83 (1969). In Russian.

The following values of constants *I* and *f* have been derived from 275 distances between Mösting A and 18 limb craters, measured on 26 photographs of the moon: I = 1°32′ 47″ ± 24″, f = 0.82 ± 0.05.

094.037 Preliminary examination of lunar samples from Apollo 11.
Science, Vol. 165, 1211 - 1227 (1969).

A physical, chemical, mineralogical, and biological analysis of 22 kilograms of lunar rocks and fines is given.

094.038 Early temperature history of the moon. H. C. Urey.
Science, Vol. 165, 1275 (1969).

The observed lunar data lend support to a model of the moon accumulated at low temperatures, melted at the surface to some depth, possibly 50 km, and then slowly solidified and bombarded briefly by objects of terrestrial composition.

094.039 Die Sichtweite auf dem Mond. F. Fleig.
Orion, Band 14, 133 (1969).

094.040 Apollo 11 observations of a remarkable glazing phenomenon on the lunar surface. T. Gold.
Science, Vol. 165, 1345 - 1349 (1969).

Some glazing is apparently due to radiation heating; it suggests a giant solar outburst in geologically recent times.

094.041 Bed forms in base-surge deposits: Lunar implications. R. V. Fisher, A. C. Waters.
Science, Vol. 165, 1349 - 1352 (1969).

Undulating dunelike deposits of surface debris, widespread over parts of the lunar landscape, are similar in form but greater in size than base-surge deposits found in many maar volcanoes and tuff rings on earth. The bed forms of base-surge deposits develop by the interaction of the bed materials with those in the current passing overhead. Therefore the "patterned ground" produced differs from that formed by ballistic fallout.

094.042 Some results of the global investigation of the lunar surface. Yu. N. Lipsky.
Astronaut. Acta, Vol. 14, 639 - 651 (1969).

The global exploration of the lunar surface, according to information obtained by Soviet automatic stations "Luna-3" and "Zond-3", fully confirms that the sea areas on the visible and observe hemisphere are distributed very asymmetrically, as deduced from the first photographs of "Luna-3" in 1959. It was established that approximately 80% of the whole surface of the moon is coated with a monolithic material crust. The sea areas, which are depressions filled with lava, are concentrated mainly on the lunar surface facing the earth. The formations found on the other side, which are termed thalassoids form, as it were, a single generically connected sequence with the lunar seas, in which the thalassoids are the oldest formations and the sea formations are the most recent.

094.043 The moon's surface. E. J. Öpik.
Annual Rev. Astron. Astrophys. Vol. 7, 473 - 526 (1969).

094.044 The moon and the planets.
Nature, Vol. 223, 1026 - 1029 (1969).

News notes concerning the formation of craters, water on the moon, filling the craters, moon experiments, probes to Mars and Venus, looking to Jupiter.

094.045 Mascons, marid and sinuous rilles. – A postulated igneous origin. J. Kane, G. Carucci, B. Turner, J. McEntee.
Nature, Vol. 224, 164 (1969).

Mascons may be subsurface magma chambers or batholiths which would be denser than the surrounding proto-lunar particulate matter from which they formed. Maria may be huge caldera-type collapse features filled with extruded lava or tephra from the underlying magma. Sinuous rilles may be channels formed by tephra flows.

094.046 Free oscillations of new lunar models. J. S. Derr.
Phys. Earth Planet. Interiors, Vol. 2, 61 - 68 (1969).

Twenty new moon models are developed, based on a moment of inertia of 0.39 as determined from the Lunar Orbiters, and on composition and phase change studies of the corresponding pressure region, the upper 150 km, of the earth. Periods of free oscillations and surface wave dispersion are computed and compared to those of homogeneous, self-compression, heavy core, and density reversal models found previously. The mechanical spectra and dispersion curves are sufficiently different that observations of free oscillations on the moon will be able to choose among most models.

094.047 The moon's photometric function near zero phase angle from Apollo 8 photography. H. A. Pohn, H. W. Radin, R. L. Wildey.
Astrophys. Journ. *(Letters)*, Vol. 157, L193 - 195 (1969).

A preliminary evaluation of the moon's photometric function near zero phase has been obtained from Apollo 8 closeup photography. The results indicate that the lunar reflected brightness is 19 percent higher at zero phase angle than at 1°.5 phase angle.

094.048 Selenodetic implications of mascons.
W. M. Boyce.
Science, Vol. 164, 1189 - 1190 (1969).

The acceleration of a lunar satellite due to a point mass ("mascon") directly below it is expanded as a series in 1/r (r its distance from the center of the moon) and compared with a truncated spherical harmonic representation of lunar gravity. For Apollo-type lunar orbits, mascons within 100 km of the lunar surface, and truncations at up to the fifteenth degree, the truncated series fails to include the majority of the mascon's effect. It is concluded that "mascon parameters" must be used in addition to spherical harmonics to represent gravity near the lunar surface.

094.049 Het onderzoek van de maan na de Apollo 11.
F. P. Israel.
Hemel en Dampkring, Vol. 67, 320 - 321 (1969).

094.050 Waarneming van de krater Arzachel op 5 april en 3 juli 1968.   Backer.
Hemel en Dampkring, Vol. 67, 332 - 335 (1969).

094.051 The structure and the supposed composition of rocks from floor of young lunar craters.
V. V. Novikov.
Astron. Zhurn. Akad. Nauk SSSR, Vol. 46, 1115 - 1118 (1969). In Russian. English translation in Soviet Astron. AJ, Vol. 13, No. 5

Morphological analogies of the Kamchatka volcanic region for some lunar formations are given. The relative brightness distribution over the solar light spectrum in the spectral region up to 2.5 μ for volcanic covers and for some lunar parts is compared on the basis of results, obtained by the author during the Kamchatka expedition of 1967. It is concluded on the presence of lava covers on the floor of young lunar craters which are more sour than basaltic slags with regard to their composition.

094.052 On the distribution of potential and gravity on the physical surface of the moon.
N. A. Chujkova.
Astron. Zhurn. Akad. Nauk SSSR, Vol. 46, 1119 - 1123 (1969). In Russian. English translation in Soviet Astron. AJ, Vol. 13, No. 5.

It is possible to find the distribution of potential and gravity on the physical surface of the moon if both the external gravitational lunar potential and the equation of the lunar physical surface are known. Spherical harmonics appear in all those orders, which are present in the external potential and in the lunar relief, as well as of all differential and summary orders.

094.053 Krater Kopernikus und seine Umgebung.
R. Mühlfeld.
Umschau, Vol. 69, 729 - 731 (1969).

094.054 Maria des Mondes in neuer Sicht.
W. G. van Dorn.
Umschau, Vol. 69, 738 (1969).

094.055 Mondproben von Apollo 11. – Vorläufige Ergebnisse.   H. Wänke.
Umschau, Vol. 69, 776 - 777 (1969).

094.056 Lunar radius from radar measurements.
A. Shapiro, E. A. Uliana, B. S. Yaplee, S. H. Knowles.
Moon and Planets II, London 1967, p. 34 - 46 (1968).

The distance to the moon was measured by radar from 11 August 1966 to 22 December 1966 with a measurement accuracy of ± 100 m. Over the same period, the range measurements of Lunar Orbiter I and II provided predicted earth – moon distances to within ± 100 m. It was thus possible to apply the full radar measurement accuracy to the determination of the lunar radius facing the earth in about 230 lunar areas. The variation of the radar values of the lunar radius agree with the local topography, but differences of 500 m to 1 km are found in widely separated areas. The lunar radius in the subearth region varies between 1736.28 km near Mösting to 1739.58 km near Herschel, with a mean value for all areas of 1737.81 km.

094.057 The spectral albedo of the moon's surface in the mid-ultraviolet according to data from the Zond-3 space probe.   A. I. Lebedinsky, V. A. Krasnopolsky, M. U. Aganina.
Moon and Planets II, London, 1967, p. 47 - 54 (1968).

Fourteen spectra of the moon's surface have been obtained in the 1900 – 2750 Å wavelength range with 14 Å resolution using an ultraviolet spectrophotometer flown on the Zond-3 space probe. Examination of the spectra has shown that the moon's mean albedo in the above spectral range is about 1 – 1.5%. A sharp increase in brightness in the 2420 – 2470 Å range was found in all the spectrograms which is likely to be connected with the moon's surface luminescence.

094.058 Infrared spectrophotometry of lunar surface from Zond-3 space probe.   A. I. Lebedinsky, G. A. Leikin, V. I. Tulupov, A. G. Fomichev, B. V. Khlopov, T. E. Shvidkovskaya, G. I. Shuster, D. N. Glovatsky.
Moon and Planets II, London 1967, p. 55 - 63 (1968).

The back-scattered radiation from the lunar surface in the 3 – 4μ wavelength range was measured from the Zond-3 space probe using a diffraction spectrophotometer.

094.059 Comparative study of the lunar surface structure at the landing sites of automatic stations.
A. I. Lebedinsky, G. A. Leikin.
Moon and Planets II, London 1967, p. 64 - 69 (1968).

The comparative surface characteristics of the landing sites of Luna 9 and Luna 13 automatic stations are discussed. The two stations were located in small craters. The results of the statistical calculation of the number of stones of various size, according to data from these two stations, are set forth and compared with the published results from Surveyor 1. The lunar landscape was similar at both landing sites.

094.060 Determination of the physical and mechanical properties of the lunar surface layer by means of Luna 13 automatic station.   I. I. Cherkasov, V. M. Vakhnin, A. L. Kemurjian, L. N. Mikhailov, V. V. Mikheyev, A. A. Musatov, M. I. Smorodinov, V. V. Shvarev.
Moon and Planets II, London 1967, p. 70 - 76 (1968).

The paper outlines the main design features of the instruments for direct determination of the mechanical properties and the density of the surface layer of the lunar soil from the automatic lunar station Luna 13. A description is given of the instrument calibration on materials simulating lunar soils in conformity with the existing hypotheses. The effect of reduced gravitation and vacuum on the instrument readings under lunar conditions is discussed.

094.061 Gamma investigation of the moon and composition of the lunar rocks.   A. P. Vinogradov, Yu. A. Surkov, G. M. Chernov, F. F. Kirnozov, G. B. Nazarkina.
Moon and Planets II, London 1967, p. 77 - 90 (1968).

As a result of measurements made by Luna 10, it has been found that the main part (not less than 90%) of gamma rays of lunar rocks is due to the interaction of cosmic rays with rocks. Another part of gamma rays belongs to the natural radioactive elements contained in the lunar rocks. The content of the natural radioactive elements in lunar rocks corresponds to earth-type rocks of basic and ultrabasic compositions. Lunar rocks with a natural radioactive content the same as in earth rocks of acid composition (granite type), were discovered nowhere on the lunar surface.

094.062 Scientific results of the Surveyor 1 lunar landing.
L. D. Jaffe.
Moon and Planets II, London 1967, p. 91 - 118 (1968).

Surveyor 1 landed in a mare area, about 100 m from a crater 170 m in diameter. The surface at this site consists mostly of discrete particles smaller than 100 microns. The bearing strength of this material is about 3×10^5 dyne/cm^2, the cohesion is between 10^2 and 10^5 dyne/cm^2. The granular layer is at least 1 m deep. Rocks in sizes up to a meter or more are abundant, many of them concentrated along crater rims. Some of these rock concentrations are apparently associated with local increases in radar cross-section at 2-cm wavelength.

094.063 A summary of Surveyor III science results.
S. E. Dwornik.
Moon and Planets II, London 1967, p. 119 - 144 (1968).

Surveyor III, NASA's second spacecraft to make a soft landing on the moon, came to rest at 2.94°S and 23.34°W on the Oceanus Procellarum on 20 April 1967. The spacecraft, which operated for 14 earth days, transmitted 6315 television pictures, some of which were in color. The pictures included views not only of the immediate landing vicinity, but also of a lunar eclipse and the planet earth. Data on the bearing strength of the surface and the cohesiveness of the soil, as well as thermal and electrical properties, were also obtained. A soil-mechanics surface-sampler device was included on Surveyor III to dig trenches and otherwise manipulate the lunar surface or move objects found there. The device also provided data on the surface bearing strength.

094.064 The first four Lunar Orbiter photographic missions.
L. R. Scherer.
Moon and Planets II, London 1967, p. 145 - 177 (1968).

The Lunar Orbiter Program consists of a series of five automated spacecraft whose prime objective is to perform photographic missions from close-in orbits about the moon. Other objectives are to provide information on the lunar gravitational field and on the radiation and micrometeoroid flux levels. The first four missions were completed by June 1967. All four were successful. The first three missions were used to photograph the more promising areas along the equatorial belt in the search for candidate man landing sites. A total of eight such sites have been selected on the basis of this photography.

094.065 An analysis of the lunar gravitational field as obtained from Lunar Orbiter tracking data.
R. H. Tolson, J. P. Gapcynski.
Moon and Planets II, London 1967, p. 178 - 186 (1968).

A set of coefficients, defining the lunar gravitational field has been determined from analysis of the tracking data from the US Lunar Orbiters I, III and IV. The gravity potential is represented by the usual series of spherical harmonics and a set of coefficients through degree and order five are presented. The resulting gravity set is compared with a smaller set obtained from Luna 10 and the agreement is good. The lunar gravitational constant is also determined and the result compares favorably with recently determined values from other space missions.

094.066 The lunar crescent and earthshine observed at 2° solar elongation.
M. J. Koomen, R. Tousey, R. T. Seal, Jr.
Moon and Planets II, London 1967, p. 187 - 195 (1968).

Using a rocket-borne pair of externally occulted coronagraphs, launched when the moon was close to the sun, photographs of the moon's earthshine and crescent were obtained at 2° elongation on 12 November 1966, and of the earthshine at 1.2° on 9 May 1967.

094.067 Laser beam directed at the lunar retro-reflector array: Observations of the first returns.
J. Faller, I. Winer, W. Carrion, T. S. Johnson, P. Spadin, L. Robinson, E. J. Wampler, D. Wieber.
Science, Vol. 166, 99 - 102 (1969).

On 1 August between 10:15 and 12:50 universal time, with the Lick Observatory 120-inch (304-cm) telescope and a laser operating at 6943 angstroms, return signals from an optical retro-reflector array placed on the moon by the Apollo 11 astronauts were successfully detected. After the return signal was first detected it continued to appear with the expected time delay for the remainder of the night. The observed range is in excellent agreement with the predicted ephemeris.

094.068 Moon: Infrared studies of surface composition.
D. P. Cruikshank.
Science, Vol. 166, 215 - 218 (1969).

Infrared reflectance studies of small lunar regions reveal several absorption bands which match those of ferrous iron in laboratory spectra of olivines and orthopyroxenes. The craters Kepler and Aristarchus exhibit absorption bands suggestive of orthopyroxene, whereas the background mare material shows a band probably due to olivine.

094.069 Alpha-particle emissivity of the moon: An observed upper limit.
R. S. Yeh, J. A. Van Allen.
Science, Vol. 166, 370 - 372 (1969).

Measurements made by the moon-orbiting spacecraft Explorer 35 during 1967 - 1968 show that it is unlikely that the alpha-particle emissivity of the moon is greater than 0.064 per square centimeter per second per steradian and exceedingly unlikely that it is greater than 0.128, these values being respectively 0.1 and 0.2 of the provisional estimates made by Kraner and al. in 1966. This result implies that the abundance of uranium-238 in the outer crust (approximately a few meters thick) of the moon is much less than that typical of the earth's lithosphere, though it is consistent with the abundance of uranium-238 in terrestrial basalt or in chondritic meteorites.

094.070 Gravity: First measurement on the lunar surface.
R. L. Nance.
Science, Vol. 166, 384 - 385 (1969).

The gravity at the landing site of the first lunar-landing mission has been determined to be 162,821.680 milligals from data telemetered to earth by the lunar module on the lunar surface. The gravity was measured with a pulsed integrating pendulous accelerometer. These measurements were used to compute the gravity anomaly and radius at the landing site.

094.071 Pyroxene gabbro (anorthosite association): Similarity to Surveyor V lunar analysis.
E. Olsen.
Science, Vol. 166, 401 - 402 (1969).

Two typical analyses from the Adirondack Mountains are close to the Surveyor data for all oxides except Na_2O. The observed plagioclase compositions in these gabbros range from An 42 to An 50.

094.072 Ephemeris time and the accuracy of selenodetic control systems.
S. G. Valeev.
Izv. Astron. Obs. Ehngel'gardta, *Kazan'*, No. 36, p. 169 - 184 (1968). In Russian.

094.073 **Lunar thermal anomalies: Magnetic phase transitions on the lunar surface?**   D. R. Waldbaum.
Science, Vol. 166, 531 - 532 (1969). – Research note.

094.074 **Fluidization phenomena and possible implications for the origin of lunar craters.**   A. A. Mills.
Nature, Vol. 224, 863 - 866 (1969).
The experimental production of craters is described. Craters resembling those of the moon and Mars are produced on a fluidized bed.

094.075 **An analysis of the distribution of boulders in the vicinity of small lunar craters.**
W. S. Cameron, G. J. Coyle.
Bull. American Astron. Soc. Vol. 1, 235 (1969). – Abstr. AAS.

094.076 **The nature of the lunar mascons.**
J. J. Gilvarry, P. M. Muller, W. L. Sjogren.
Bull. American Astron. Soc. Vol. 1, 241 - 242 (1969). – Abstr. AAS.

094.077 **Color differentiation by computer processing.**   A. F. H. Goetz, F. C. Billinsley, J. N. Lindsley.
Bull. American Astron. Soc. Vol. 1, 242 (1969). – Abstr. AAS.

094.078 **Periodicity of lunar spectral anomalies.**
J. Green.
Bull. American Astron. Soc. Vol. 1, 243 (1969). – Abstr. AAS.

094.079 **AFCRL lunar laser ground station status.**
M. S. Hunt, F. F. Forbes.
Bull. American Astron. Soc. Vol. 1, 244 - 245 (1969). – Abstr. AAS.

094.080 **Washington meridian observations of the moon 1925 - 1968.**   B. L. Klock, D. K. Scott.
Bull. American Astron. Soc. Vol. 1, 249 (1969). – Abstr. AAS.

094.081 **Preliminary results from tests of Van Flandern's corrections to the lunar elements.**
J. D. Mulholland, N. A. Mottinger, C. J. Vegos, F. B. Winn.
Bull. American Astron. Soc. Vol. 1, 254 (1969). – Abstr. AAS.

094.082 **On the use of gravity-field models to postulate hidden maria on the far side of the moon.**
M. R. Warner.
Bull. American Astron. Soc. Vol. 1, 265 (1969). – Abstr. AAS.

094.083 **Zur Existenz der freien physischen Libration des Mondes.**   Sh. T. Khabibullin.
Trudy Kazan. Gorod. Astron. Obs. No. 35, p. 110 - 115 (1968).   In Russian.

094.084 **Koeffizienten in der trigonometrischen Reihenentwicklung für die Komponenten der physischen Mondlibration.**   Yu. A. Chikanov.
Trudy Kazan. Gorod. Astron. Obs. No. 35, p. 116 - 156 (1968).   In Russian.

094.085 **Fortschritte auf dem Gebiete der Mondforschung.**   J. Hoppe.
Sterne, 45. Jahrgang, 148 - 152 (1969).

094.086 **Unipolar induction in the moon and a lunar limb shock mechanism.**   K. Schwartz, C. P. Sonett, D. S. Colburn.
The Moon, Vol. 1, 7 - 30 (1969).
The unipolar induction mechanism is employed to calculate electric field profiles in the interior of a chemically homogeneous moon possessing a steep radial thermal gradient characteristic of long-term radioactive heating. The thermal models used are those of Fricker, Reynolds, and Summers. From the magnetic field, the magnetic back pressure upon the solar wind is found. The results indicate that a hot moon can yield sufficient current flow so that the magnetic back pressure is observable as a vestigial limb shock wave using an activation energy of about $^2/_3$ eV together with a conductivity coefficient of about 10^3 mhos/m.

094.087 **A quantitative evaluation of the uniformity of the light scattering properties of the lunar surface.**
M. T. Jones.
The Moon, Vol. 1, 31 - 58 (1969).
A catalogue was compiled in the author's previous paper (Jones, 1969) of the relative brightness of 199 lunar features as observed at phase angles of $2°.1$, $17°.5$, $32°.5$, $46°.2$, $59°.4$ and $72°.1$ before full moon, in the wavelength interval 5500 to 7000 Å. The present paper is concerned with an interpretation of this data in terms of the uniformity of the photometric function of the lunar surface. If due account is taken of the effect of second order scattering it is found that all the survey points studied scatter light according to the same photometric function independent of the type of terrain on which they are located.

094.088 **The earliest maps of the moon.**   Z. Kopal.
The Moon, Vol. 1, 59 - 66 (1969).
The aim of the present paper is to give a brief account of the history of lunar mapping in the pre-telescopic era, and that immediately following the discovery of the telescope.

094.089 **A photometric investigation of the lunar crater rays.**   J. van Diggelen.
The Moon, Vol. 1, 67 - 84 (1969).
This investigation deals with accurate photometric data concerning a number of rays of Tycho, Copernicus, Kepler, and Aristarchus. They have been derived from plates taken at the Yerkes Observatory in a night of a total lunar eclipse near phase angle 0°. By comparing the normal albedo with that of the surroundings of the rays we found that they can be interpreted as samples of telescopically unresolved bright patches. The distribution of the brightness along the rays has also been compared with the mass distribution of the ejecta in the rays around terrestrial explosion craters.

094.090 **A theory for the interpretation of lunar surface magnetometer data.**
G. Schubert, K. Schwartz.
The Moon, Vol. 1, 106 - 117 (1969).
The solution to the problem of the motion of the moon relative to spatial irregularities in the interplanetary magnetic field is found. The lunar electrical conductivity is modeled by a two-layer conductivity profile. For the interaction of the moon with the corotating sector structure of the interplanetary magnetic field it is found that the magnetic field in the lunar shell is the superposition of an oscillatory uniform field, an oscillatory dipole field and an oscillatory field that is toroidal about the axis of the motional electric field. With various lunar conductivity models and the theory of this paper, lunar surface magnetometer data can be quantitatively interpreted to yield information on the conductivity and consequently the temperature of the lunar core.

094.091 **An analysis of the distribution of boulders in the vicinity of small lunar craters.**
W. S. Cameron, G. J. Coyle.

The Moon, Vol. 1, 118 (1969). — Abstract.

094.092 **The nature of the lunar mascons.**
J. J. Gilvarry, P. M. Muller, W. L. Sjogren.
The Moon, Vol. 1, 118 - 119 (1969). — Abstract.

094.093 **Periodicity of lunar spectral anomalies.**
J. Green.
The Moon, Vol. 1, 119 (1969). — Abstract.

094.094 **AFCRL lunar laser ground station status.**
M. S. Hunt.
The Moon, Vol. 1, 120 (1969). — Abstract.

094.095 **Washington meridian observations of the moon,**
1925 – 1968. B. L. Klock, D. K. Scott.
The Moon, Vol. 1, 120 - 121 (1969). — Abstract.

094.096 **Preliminary results from tests of Van Flandern's**
corrections to the lunar elements.
J. D. Mulholland, N. A. Mottinger, C. J. Vegos, F. B. Winn.
The Moon, Vol. 1, 121 (1969). — Abstract.

094.097 **On the use of gravity field models to postulate**
hidden maria on the far side of the moon.
M. R. Warner.
The Moon, Vol. 1, 122 (1969). — Abstract.

094.098 **Lunar Explorer 35: Picogram lunar ejecta related**
to the Orionid and Leonid meteor showers, 1968
and **1969.** W. M. Alexander, J. L. Bohn.
The Moon, Vol. 1, 123 (1969). — Abstract.

094.099 **Lunar Explorer 35: Indications of mass limit for**
lunar ejecta resulting from hypervelocity impact of
meteoroids on the lunar surface.
W. M. Alexander, J. L. Bohn.
The Moon, Vol. 1, 123 (1969). — Abstract.

094.100 **The case for recent lunar surface water.**
P. R. Bell.
The Moon, Vol. 1, 123 - 124 (1969). — Abstract.

094.101 **Progress on lunar gravitational potential determi-**
nation by analysis of Lunar Orbiter tracking data.
W. T. Blackshear.
The Moon, Vol. 1, 124 (1969). — Abstract.

094.102 **The effect of rocks and roughness on the anoma-**
lous cooling of the lunar craters. D. Buhl.
The Moon, Vol. 1, 124 - 125 (1969). — Abstract.

094.103 **Some results and implications of new observations**
of lunar backscatter at decameter wavelengths.
A. A. Burns.
The Moon, Vol. 1, 125 (1969). — Abstract.

094.104 **Lunar cratering and erosion from Orbiter V photo-**
graphs.
C. R. Chapman, J. A. Mosher, G. Simmons.
The Moon, Vol. 1, 125 - 126 (1969). — Abstract.

094.105 **Paramagnetic resonance spectra of simulated lunar**
rocks and other silicate rocks: A survey.
A. Chatelain, J. L. Kolopus, R. A. Weeks.
The Moon, Vol. 1, 126 (1969). — Abstract.

094.106 **Attempted observations of meteor impacts on the**
lunar surface.
G. Davidson, O. Shephard, F. Franklin, J. W. Carpenter.
The Moon, Vol. 1, 126 (1969). — Abstract.

094.107 **Free oscillations of new lunar models.**
J. S. Derr.
The Moon, Vol. 1, 127 (1969). — Abstract.

094.108 **The effect of proton irradiation on the spectra of**
silicates. J. P. Dybwad.
The Moon, Vol. 1, 127 (1969). — Abstract.

094.109 **Non-random distribution of major lunar craters.**
W. E. Elston, M. J. Aldrich.
The Moon, Vol. 1, 127 - 128 (1969). — Abstract.

094.110 **Lunar structural implications of lineaments map-**
ped from lunar orbital pictures. C. V. Fulmer.
The Moon, Vol. 1, 128 (1969). — Abstract.

094.111 **Erosion and fragmentation of rocks on the lunar**
surface. D. E. Gault.
The Moon, Vol. 1, 128 (1969). — Abstract.

094.112 **Mass distributions inferred from Lunar Orbiter**
tracking data. P. Gottlieb, P. A. Laing.
The Moon, Vol. 1, 128 - 129 (1969). — Abstract.

094.113 **Remote sensing of sulfur-bearing compounds on**
the moon and Mars as defluidization centers.
J. Green.
The Moon, Vol. 1, 129 (1969). — Abstract.

094.114 **A stochastic model of the distribution of lunar**
thermal anomalies. R. R. Green, L. B. Ronca.
The Moon, Vol. 1, 129 (1969). — Abstract.

094.115 **The distribution of a transient lunar atmosphere.**
J. J. Grossman, T. M. Henderson, S. W. Benson.
The Moon, Vol. 1, 130 (1969). — Abstract.

094.116 **Electrical resistivity and impact testing of a lunar**
analog soil. G. C. Henderson.
The Moon, Vol. 1, 130 (1969). — Abstract.

094.117 **Dynamic modelling of lunar features with a viscoe-**
lastic material. E. A. Kaarsberg.
The Moon, Vol. 1, 130 - 131 (1969). — Abstract.

094.118 **Analysis of gravitational anomalies of the moon.**
M. F. Kane, E. M. Shoemaker.
The Moon, Vol. 1, 131 (1969). — Abstract.

094.119 **Lunar soil temperatures near cm-sized craters.**
J. A. Krupp, D. F. Winter.
The Moon, Vol. 1, 131 - 132 (1969). — Abstract.

094.120 **Shape and internal structure of moon from Lunar**
Orbiter data. D. L. Lamar, J. V. McGann-Lamar.
The Moon, Vol. 1, 132 (1969). — Abstract.

094.121 **The shape of the moon's gravity field as determined**
from the Lunar Orbiter tracking data.
J. Lorell.
The Moon, Vol. 1, 132 (1969). — Abstract.

094.122 **Lunar surface plasma environment and electric**
potential. R. H. Manka, H. R. Anderson.
The Moon, Vol. 1, 132 - 133 (1969). — Abstract.

094.123 **The domes and cones in the Marius Hills region. —**
Evidence for lunar differentiation?
J. F. McGauley.
The Moon, Vol. 1, 133 - 134 (1969). — Abstract.

094.124 **Spectral reflectivity of selected regions of the lunar surface.** T. B. McCord, T. V. Johnson.
The Moon, Vol. 1, 134 (1969). – Abstract.

094.125 **Tidal cycles and seismic causes for lunar events.** B. M. Middlehurst, W. B. Chapman.
The Moon, Vol. 1, 134 - 135 (1969). – Abstract.

094.126 **Photographic evidence for the presence of water cut beaches, benches, arroyos, and rivers on the moon.** P. M. Muller, W. L. Sjogren.
The Moon, Vol. 1, 135 (1969). – Abstract.

094.127 **A new reduction of Lunar Orbiter radio tracking data in describing the lunar near-side gravity field.** P. M. Muller, W. L. Sjogren.
The Moon, Vol. 1, 135 (1969). – Abstract.

094.128 **Compositional differences between the prime Apollo sites suggested by visible and IR evidence.** B. C. Murray, A. F. H. Goetz, T. B. McCord, H. H. Kieffer.
The Moon, Vol. 1, 136 (1969). – Abstract.

094.129 **Electrical conductivity of the moon.** N. F. Ness.
The Moon, Vol. 1, 136 (1969). – Abstract.

094.130 **Detection of interplanetary magnetic field fluctuations stimulated by the lunar wake.** N. F. Ness, K. H. Schatten.
The Moon, Vol. 1, 136 - 137 (1969). – Abstract.

094.131 **Dependence of the lunar wake on solar wind plasma characteristics.** K. W. Ogilvie, N. F. Ness.
The Moon, Vol. 1, 137 (1969). – Abstract.

094.132 **Relative dating of the lunar surface.** L. B. Ronca, R. R. Green.
The Moon, Vol. 1, 137 (1969). – Abstract.

094.133 **The geoid and selenoid.** S. K. Runcorn.
The Moon, Vol. 1, 137 - 138 (1969). – Abstract.

094.134 **A critique of internal heating as an explanation for lunar thermal anomalies.** J. M. Saari.
The Moon, Vol. 1, 138 (1969). – Abstract.

094.135 **The distribution and morphology of lunar sinuous rilles.** G. Schubert, R. E. Lingenfelter, S. J. Peale.
The Moon, Vol. 1, 138 (1969). – Abstract.

094.136 **Evolution of the moon's orbit near the earth.** S. F. Singer.
The Moon, Vol. 1, 138 - 139 (1969). – Abstract.

094.137 **Rümker Hills: A volcanic plateau in the Oceanus Procellarum.** E. I. Smith.
The Moon, Vol. 1, 139 (1969). – Abstract.

094.138 **Radar maps of the moon at 70 centimeter wavelength.** T. W. Thompson.
The Moon, Vol. 1, 139 - 140 (1969). – Abstract.

094.139 **Effects of cellular convection within the moon.** D. L. Turcotte, E. R. Oxburgh, G. Schubert.
The Moon, Vol. 1, 140 (1969). – Abstract.

094.140 **Impact craters and the origin of the lunar maria.** E. H. Walker.
The Moon, Vol. 1, 140 (1969). – Abstract.

094.141 **The solar wind plasma and fields in the lunar wake.** Y. C. Whang.
The Moon, Vol. 1, 140 - 141 (1969). – Abstract.

094.142 **Volcanic materials in the lunar terrae.– Orbiter observations.** D. E. Wilhelms, J. F. McCaulay.
The Moon, Vol. 1, 141 (1969). – Abstract.

094.143 **Mascons as structural relief on a lunar Moho.** D. U. Wise, M. T. Yates.
The Moon, Vol. 1, 141 - 142 (1969). – Abstract.

094.144 **Dynamical determination of mascons on the moon.** L. Wong, G. Buechler, W. D. Downs, R. H. Prislin, W. L. Sjogren, P. Muller, P. Gottlieb.
The Moon, Vol. 1, 142 (1969). – Abstract.

094.145 **Proton induced chemical reactions on lunar and planetary surfaces.** E. J. Zeller, G. Dreschhoff.
The Moon, Vol. 1, 142 (1969). – Abstract.

094.146 **Results on the mass and the gravitational field of the moon as determined from dynamics of lunar satellites.** W. H. Michael, Jr., W. T. Blackshear, J. P. Gapcynski.
The Moon. Vol. 1, 143 (1969). – Abstract.

094.147 **The Apollo 11 laser ranging retro-reflector experiment: Opportunity for new precision in the study of the earth-moon system.** C. O. Alley.
The Moon, Vol. 1, 144 (1969). – Abstract.

094.148 **Lunar Explorer 35: 1967 – 1968 measurements of picogram dust particle flux in selenocentric space.** W. M. Alexander, C. W. Arthur, J. L. Hohn, J. D. Corbin.
The Moon, Vol. 1, 144 - 145 (1969). – Abstract.

094.149 **Lunar surface: Recent spacecraft observations.** L. D. Jaffe.
The Moon, Vol. 1, 145 (1969). – Abstract.

094.150 **Comparison of the chemical composition of lunar surface material determined by radio astronomical observations with the results of chemical analysis obtained by Surveyor.** V. S. Troitski.
The Moon, Vol. 1, 146 (1969). – Abstract.

094.151 **Some data on lunar surface microstructure deduced from studies on the moon's infrared and ultraviolet radiation obtained from 'Zond-3' space probe.** G. A. Leikin, T. E. Shvidkovskaya, V. A. Krasnopolsky.
The Moon, Vol. 1, 146 - 147 (1969). – Abstract.

094.152 **Structure of the lunar surface.** S. Hayakawa, T. Matsumoto, T. Nishimura.
The Moon, Vol. 1, 147 (1969). – Abstract.

094.153 **The measurement of scattering characteristics of local lunar regions from space vehicles.** N. N. Krupenio.
The Moon, Vol. 1, 147 (1969). – Abstract.

094.154 **Mascons on the moon.** H. C. Urey.
The Moon, Vol. 1, 147 - 148 (1969). – Abstract.

094.155 **Lunar and planetary mass concentrations.** B. T. O'Leary, M. J. Campbell, C. Sagan.
The Moon, Vol. 1, 148 - 149 (1969). – Abstract.

094.156 **Lunar gravimetrics.** P. M. Muller, W. L. Sjogren.

The Moon, Vol. 1, 149 (1969). — Abstract.

094.157 The electrical conductivity and internal temperature of the moon.   N. F. Ness.
The Moon, Vol. 1, 149 - 150 (1969). — Abstract.

094.158 On the origin of lunar relief.   P. Feschotte.
The Moon, Vol. 1, 150 (1969). — Abstract.

094.159 Corrections to the improved lunar ephemeris.
T. C. van Flandern.
Celestial Mechanics, Vol. 1, 163 - 166 (1969).

Based primarily upon the formation of new conditional equations using analytical partial derivatives of the moon's mean elements, meridian circle observations of the moon from 1952 – 67 have been examined to determine corrections to the constants of lunar theory and to the fundamental coordinate system (FK4). With certain exceptions, the new corrections are in agreement with those published earlier by the author. Systematic corrections to FK4 are surprisingly large, although in agreement with some other recent determinations. New corrections to the lunar ephemeris, resulting from the discussion, are also presented.

094.160 Internal temperature of the moon.
J. J. Gilvarry.
Nature, Vol. 224, 968 - 970 (1969).

Temperatures past and present are inferred from the creep rate of lunar mascons and compared with the results of other methods.

094.161 A statistical analysis of the reflectance of igneous rocks from 0.2 to 2.65 microns.
H. P. Ross, J. E. M. Adler, G. R. Hunt.
Icarus, Vol. 11, 46 - 54 (1969).

The reflectance spectra of igneous rocks in the 0.2 to 2.65 micron region have been studied to provide information for the design and interpretation of lunar and terrestrial remote sensing experiments. Reflectance spectra from 0.2 to 2.65 microns were obtained for several igneous rocks and minerals. The reflectance of all the rock samples, when crushed, increases with descreasing particle size. A statistical analysis established the correlation between reflectance and the wavelength of energy, the generalized composition, and the particle size of the rock samples.

094.162 Correlation of mineral luminescent phenomena and its selenological implications.
R. T. Greer, J. N. Weber.
Icarus, Vol. 11, 55 - 65 (1969).

The luminescence response for several representative natural terrestrial, extraterrestrial (separated from meteorites), and synthetic silicates has been investigated by studying the interrelationships of crystal host, activator, and impurity as they influence the wavelength and intensity of the emission colors. The observed correlations and contrasts provide a reasonable basis to expect useful information from studying the luminescent response and patterns that can be obtained from lunar surface specimens.

094.163 Maria, mascons, and the history of the moon.
M. von Reinhardt.
Astrophys. Letters, Vol. 4, 225 - 226 (1969).

It is proposed that all the lunar maria originated simultaneously in an impact of large bodies, and that the side on which the impact occurred was turned around by gravitational forces and thus became the front side of the moon.

094.164 Nonexistence of large mascons at Mare Marginis and Mare Orientale.
P. Gottlieb, P. M. Muller, W. L. Sjogren.

Science, Vol. 166, 1145 - 1147 (1969).

The analysis of line-of-sight residual accelerations from Lunar Orbiters 3 and 5 does not show any evidence for large mascons near the lunar limbs. Although unfavorable geometry reduces the acceleration effect due to any mascons near the limb, simulations show that large masses at Mare Orientale and Mare Marginis would produce substantial accelerations, in complete disagreement with the actual Doppler tracking data obtained from a Lunar Orbiter experiment.

094.165 Interpretation of lunar mass concentrations.
W. M. Kaula.
Phys. Earth Planet. Interiors, Vol. 2, 123 - 137 (1969).

The lunar "mascons" appear to be too big to be caused primarily by the infalling bodies which created the ringed maria. The mechanism of mass transfer which appears most consistent with the low level of geologic activity is the generation of excess pressures consequent upon the more rapid cooling of the outer parts of the moon. In addition to mass transfer, there also could have operated mechanisms of densification, the most significant of which appears to be the outgassing of water after the impact, which resulted in the ringed maria not being serpentinized to the same extent as other parts of the moon.

094.166 A preliminary study of lunar specimens brought by "Apollo 11".
Priroda No. 12, p. 50 - 59 (1969).   In Russian.
(Translation from "Science" by G. R. Roshkovan).

094.167 The normal albedo of the Apollo 11 landing site and intrinsic dispersion in the lunar Heiligenschein.
R. L. Wildey, H. A. Pohn.
Astrophys. Journ. (*Letters*), Vol. 158, L129 - L130 (1969).

The normal albedo of the Apollo 11 landing site is 0.0995 from Apollo and earth-based observations. The brightness surge between $g = 1.5°$ and absolute full phase varies by at least 250 percent over the lunar surface.

094.168 Wavelength dependence of polarization. XIX. Comparison of the lunar surface with laboratory samples.   S. F. Pellicori.
Astron. Journ. Vol. 74, 1066 - 1072, 1115 - 1116 (1969).

Photoelectric polarimetry at several wavelength bands between 0.32 and 0.56 μ of samples of "pure" mare and terra and of lavas and chemicals are presented. The lunar particles are somewhat translucent, rather opaque. The wavelength dependence of the polarization of the moon, and by comparison, of Mercury, Mars, and the asteroids is produced by the decrease in translucency of the particles with decreasing wavelength. The maria show relatively higher polarization at short wavelengths than the terrae, suggesting either a greater absorption coefficient or larger particle sizes than on the terrae.

094.169 Electrical properties of rocks and their significance for lunar radar observations.
M. J. Campbell, J. Ulrichs.
Journ. Geophys. Res. Vol. 74, 5867 - 5881 (1969).

In this paper we report measurements of the dielectric constant and loss tangent of a wide variety of terrestrial rocks and minerals at two frequencies, 450 MHz and 35 GHz. These frequencies, corresponding to wavelengths of 67 cm and 8.6 mm respectively, lie near the extremes of the range spanned by current radar and radiometric observations of the moon and planets.

094.170 Meteoroid impacts as sources of seismicity on the moon.   A. McGarr, G. V. Latham, D. E. Gault.
Journ. Geophys. Res. Vol. 74, 5981 - 5994 (1969).

There are two possible sources of natural seismic activity on the moon: (1) 'moon quakes', i.e., seismic energy released by sudden rupture or changes in volume within the moon; and (2) meteoroids colliding with the lunar surface. The objective of the research reported here is to estimate the number and character of seismic signals produced by meteoroid impacts on the moon that will be recorded during the lifetime of the Apollo seismic experiments.

094.171 Origin of the moon from the earth: some new mechanisms and comparisons. D. U. Wise.
Journ. Geophys. Res. Vol. 74, 6034 - 6045 (1969).

Goldreich's calculation of the history of the lunar orbit, with which he argues strongly against fission and against all other lunar theories as well, omits one significant possibility, that earth's equator had been tilted about 10° to the ecliptic prior to the time of fission. If this is true, the inclination of earth's equator and of the lunar orbit to the ecliptic fit nicely into Goldreich's curves. It is concluded that there are fewer outstanding criticisms of the hypothesis of lunar origin by fission than of other hypotheses. The average lunar density and the results of the Surveyor chemical analyses lead us to argue that fission from the earth triggered by the earth core formation is the most probable origin of the moon, considering the data available now. The purposes of this paper are (1) to strengthen the fission alternative by proposing new solutions to the traditional difficulties of angular momentum discrepancy and the history of the lunar orbit, (2) to spread criticism more equitably by applying to MacDonald's own capture theory his arguments against fission on dynamical grounds, and (3) to raise objections to the capture theory that are unique to that theory.

094.172 Relative reflectivity (0.4μ to 1.1μ) of the lunar landing site Apollo 7.
T. V. Johnson, L. A. Soderblom.
Journ. Geophys. Res. Vol. 74, 6046 - 6048 (1969).

The results show that the reflectivity of site Apollo 7 differs significantly from that of Apollo 2.

094.173 Implications of the Surveyor 7 results.
R. A. Phinney, J. A. O'Keefe, J. B. Adams, D. E. Gault, G. P. Kuiper, H. Masursky, R. J. Collins, E. M. Shoemaker.
Journ. Geophys. Res. Vol. 74, 6053 - 6080 (1969).

This paper discusses two unique aspects of the Surveyor 7 data. The a-scattering experiment shows a surface composition that is lower in iron than the mare sites previously analyzed. The television images (Shoemaker et al., 1968; Bird et al., 1968) show a significantly rockier moonscape than seen by earlier Surveyors. A number of additional clues to the origin of the lunar surface and to the mechanisms that determine its current state are contained in these observations.

094.174 Observations of the lunar regolith and the earth from the television camera on Surveyor 7.
E. M. Shoemaker, R. M. Batson, H. E. Holt, E. C. Morris, J. J. Rennilson, E. A. Whitaker.
Journ. Geophys. Res. Vol. 74, 6081 - 6119 (1969).

Location and topography of landing site; Regional geologic setting; Geology of area around spacecraft (geologic units, thickness of the regolith, fragmental debris, disturbances of the surface); Photometric observations of lunar surface; Polarimetric observations of lunar surface (method of polarimetric measurements, polarization of sunlight scattered from the lunar surface, depolarization of earthlight scattered from the lunar surface); Evolution of lunar regolith; Observations of earth.

094.175 Alpha-scattering experiment on Surveyor 7: Comparison with Surveyors 5 and 6.
J. H. Patterson, E. J. Franzgrote, A. L. Turkevich, W. A. Anderson, T. E. Economou, H. E. Griffin, S. L. Grotch, K. P.

Sowinski.
Journ. Geophys. Res. Vol. 74, 6120 - 6148 (1969).

The last three Surveyor missions (5, 6, and 7) included an a-scattering experiment that obtained elemental analyses of surface material at three widely separated locations on the moon.

094.176 Lunar surface mechanical properties.
R. Choate, S. A. Batterson, E. M. Christensen, R. E. Hutton, L. D. Jaffe, R. H. Jones, H. Y. Ko, R. L. Spencer, F. B. Sperling.
Journ. Geophys. Res. Vol. 74, 6149 - 6174 (1969).

Although the lunar surface at the Surveyor 7 highland landing site is somewhat rougher than the surface at previous mare landing sites, many of the physical properties of the soil at the sites are similar. The soil is primarily fine-grained, compressible, and slightly cohesive; only 2.8% of the surface is covered by rocks larger than 5 cm in diameter. The average soil static bearing strength is 0.2 N/cm^2 at 0.2-cm depth and 3.4 N/cm^2 at 4-cm depth.

094.177 Soil mechanics surface sampler.
R. F. Scott, F. I. Roberson.
Journ. Geophys. Res. Vol. 74, 6175 - 6214 (1969).

Subsystem description; Functional and operational description; Mission description; Lunar operations: First lunar day; Lunar operations: Second lunar day; Discussion of tests; Preliminary analyses and results.

094.178 On the apparent motion of the moon.
M. A. Janevich, I. A. Paribok-Aleksandrovich.
Uch. zap. Vologodsk. gos. ped. in-t, Vol. 34, p. 149 - 164 (1968). In Russian. – Abstr. in Referativ. Zhurn. 51. Astron., 8.51.47 (1969).

094.179 First studies of lunar material.
R. N. Watts, Jr.
Sky Telescope, Vol. 38, 312 - 314 (1969).

094.180 Transient lunar phenomena.
K. E. Chilton.
Journ. Roy. Astron. Soc. Canada, Vol. 63, 203 - 205 (1969).

094.181 Observation of lunar retroreflector array.
J. Faller, E. J. Wampler.
IAU Circ. No. 2160 (1969).

094.182 Why is the moon gray? W. F. Libby.
Science, Vol. 166, 1437 - 1438 (1969).

I suggest that the answer lies in the solar wind's bringing in atomic hydrogen to replace that lost by the photolytic decomposition of water vapor.

094.183 The lunar controversy – I. An assessment of the position before Apollo 11. G. J. H. McCall.
Journ. British Astron. Ass., Vol. 80, 19 - 29 (1969). – Popular article.

094.184 Water on the moon.
G. N. Katterfel'd, P. M. Frolov.
Izv. Vses. geogr. o-va, Vol. 101, No. 3, p. 260 - 264 (1969). In Russian. – Abstr. in Referativ. Zhurn. 62. Issled. kosm. prostranstv., 11.62.233 (1969).

094.185 Determination of the albedo of the moon at a wavelength of 6 m. T. Hagfors, J. L. Green, A. Guillén.
Astron. Journ., Vol. 74, 1214 - 1219 (1969).

A method is described whereby the reflectivity of the lunar surface may be determined by measuring the amount of cosmic noise being reflected. The method is applied to data obtained at the Jicamarca Radar Observatory. It is determined that the apparent dielectric constant of the lunar surface material at a wavelength of 6 m is 3.7 ± 0.7. This indicates that relatively loosely packed surface strata must exist to depths of several meters.

094.186 Photometry of lunar libration regions.
J. L. Weinberg, D. E. Beeson, P. B. Hutchison.
Bull. American Astron. Soc., Vol. 1, 368 (1969). – Abstr. AAS.

094.187 Diffuse component of lunar radar echoes.
A. A. Burns.
Journ. Geophys. Res. Vol. 74, 6553 - 6566 (1969).
A simple model consisting of volume backscattering from within the lunar regolith can explain the observed diffuse component of lunar radar echoes. At a wavelength of 68 cm, a good match of the model with the data yields a value of 2.5 – 3 for the relative permittivity of the regolith in good agreement with the quasi-specular backscattering estimates.

094.188 Doppler gravity, a new method.
M. F. Kane.
Journ. Geophys. Res. Vol. 74, 6579 - 6582 (1969).
The method recently applied by P. M. Muller and W. L. Sjogren in mapping gravity anomalies of the moon employed a new principle of gravity measurement: time differentiation of Doppler-tracked velocities.

094.189 Phase changes of spectropolarimetric characteristics of lunar details. L. N. Bondarenko.
Astron. Zhurn. Akad. Nauk SSSR, Vol. 46, 877 - 884 (1969). In Russian. English translation in Soviet Astron. AJ, Vol. 13, No. 4.
Lunar details are grouped according to the type of their spectropolarimetric characteristics P (λ). There is a general tendency of polarization to decrease with increasing wavelengths. The slope of the curves P (λ) grows up to the lunar phase 90°. Some details show abnormal spectropolarimetric curves.

094.190 On the inhomogeneity of properties of lunit in the depth of the surface layer. O. B. Shchuko.
Astron. Zhurn. Akad. Nauk SSSR, Vol. 46, 1264 - 1269 (1969). In Russian. English translation in Soviet Astron. AJ, Vol. 13, No. 6.
An inhomogeneous model of the upper cover of the moon is constructed supposing dependence of thermal parameters (heat conductivity and heat) on temperature. It is shown that the surface temperatures in a lunar midnight and at the moment of the middle of the eclipse, calculated for the considered model, agree well with values of surface temperatures following from radio measurements in the infrared region; the density of the matter of the upper lunar cover remains nearly invariable.

094.191 Selenodetic investigations from positional observations of the moon. S. G. Valeev.
Astron. Zhurn. Akad. Nauk SSSR, Vol. 46, 1270 - 1273 (1969). In Russian. English translation in Soviet Astron. AJ, Vol. 13, No. 6.
A possibility of solving a number of selenodetic problems from large-scale photographs of the moon with stars is considered. Formulas for the transition from equatorial coordinates of the moon to orbital ones, and inversely, with the use of cracovians are given.

094.192 The spectrum of the reflection coefficient of radio waves on the lunar surface with changing lunar material properties with depth.

T. V. Tikhonova, V. S. Troitskii.
Astron. Zhurn. Akad. Nauk SSSR, Vol. 46, 1324 - 1328 (1969). In Russian. English translation in Soviet Astron. AJ, Vol. 13, No. 6.
The power spectra of the reflection coefficient for two models of a lunar surface layer are obtained by solving Riccati's equation on a digital computer. A comparison with an experiment shows that both the models explain sufficiently well the radar spectrum of the reflection coefficient. The thickness of a porous layer is estimated to (200 ± 100) cm.

094.193 Zur Bestimmung der statistischen Merkmale des lunaren Oberflächenreliefs. Yu. S. Tumashev.
Kosmich. Issled. Vol. 7, 547 - 552 (1969). In Russian.
Zur Bearbeitung der Messungen der Flughöhe eines Raumflugkörpers über der Mondoberfläche wird nach der Methode der stochastischen Differentialgleichungen eine Korrelationsfunktion des lunaren Oberflächenreliefs erhalten.

094.194 Untersuchung der physikalisch-mechanischen Eigenschaften des arktischen Tuffs als Analogie zu dem lunaren Oberflächengestein. F. M. Lyakhovitskij.
Kosmich. Issled. Vol. 7, 615 - 617 (1969). In Russian. – Brief information.

094.195 Über eine Möglichkeit zur Typenbestimmung des Mondgesteins mit der massenspektrometrischen Methode.
A. F. Kuz'min, A. Eh. Rafal'son, N. A. Kholin, G. E. Tsigel'man, R. P. Shirshov.
Kosmich. Issled. Vol. 7, 618 - 620 (1969). In Russian. – Brief information.

094.196 Über Möglichkeiten und Vorteile der massenspektrometrischen Analyse des Mondbodens aus der Ionenkomponente der Verdampfungsprodukte durch Elektronenstrahlen.
A. F. Kuz'min, N. A. Kholin, G. E. Tsigel'man.
Kosmich. Issled. Vol. 7, 620 - 622 (1969). In Russian. – Brief information.

094.197 Die Verteilung der Helligkeit aus dem Spektrum des von vulkanischen Bedeckungen reflektierten Sonnenlichtes. Yu. N. Lipskij, V. V. Novikov.
Kosmich. Issled. Vol. 7, 753 - 759 (1969). In Russian.

094.198 Poröse vulkanische Ablagerungen – eine mögliche terrestrische Analogie zu den Mondböden.
I. I. Cherkasov, V. V. Shvarev, G. S. Shtejnberg.
Kosmich. Issled. Vol. 7, 760 - 772 (1969). In Russian.

094.199 Die räumliche Streuindikatrix der lunaren Kontinente. Yu. N. Lipskij, V. V. Shevchenko.
Kosmich. Issled. Vol. 7, 773 - 776 (1969). In Russian.

094.200 Die durch Einwirkung kosmischer Strahlung entstehende Neutronenstrahlung des Mondes.
Yu. N. Gnedin, A. Z. Dolginov, A. I. Tsygan.
Kosmich. Issled. Vol. 7, 777 - 785 (1969). In Russian.

094.201 Ion-induced changes in simulated lunar rocks.
D. P. Cruikshank.
Icarus, Vol. 11, 145 - 154 (1969).
To study the effects of the solar wind flux on the infrared reflectivity of rocks and minerals on the moon, a laboratory apparatus was constructed for irradiation of rock samples with an ion flux having the same energy as the ambient solar wind, but with 10^7 to 10^8 times the flux density. The infrared colorimetric profiles from 0.8 to 2.2μ were measured for several natural and irradiated samples. The relationship of the visual darkening of the rocks on irradiation and the

changes in iron absorption bands are discussed relative to the lunar surface.

094.202 Radiative heat transfer in the lunar and Mercurian surfaces. J. Ulrichs, M. J. Campbell.
Icarus, Vol. 11, 180 - 188 (1969).

In this paper we examine the validity of the usual approximate solutions to a simple model of heat transfer, including radiative transfer, in powders, by comparing them with an exact numerical solution. We find that the approximate solutions can, under some circumstances, adequately describe the planetary observations provided cognizance is taken of the fact that emission is a volume rather than surface effect; neglect of volume emission may have caused errors of as much as 20°K in the interpretation of lunar eclipse observations.

094.203 Astronaut activity on the lunar surface.
Icarus, Vol. 11, 260 - 267 (1969).

Presented are photographs of astronaut activity on the lunar surface supplied by the Manned Spacecraft Center, National Aeronautics and Space Administration.

094.204 The gravitational field of the moon.
W. M. Kaula.
Science, Vol. 166, 1581 - 1588 (1969).

Lunar satellite perturbations indicate striking correlations of gravity anomalies with ancient features.

094.205 Lunar gravitational field as determined from Lunar Orbiter tracking data.
J. P. Gapcynski, W. T. Blackshear, H. R. Compton.
AIAA Journ., Vol. 7, 1905 - 1908 (1969).

This paper presents results that have recently been obtained at the NASA Langley Research Center on the problem of determining the gravitational field of the moon from an analysis of the tracking data from the United States Lunar Orbiter series of spacecraft. A set of coefficients defining the gravitational field through degree and order 7 are presented.

094.206 Determination of the constants of lunar rotation with allowance for the errors of the coordinates
of the crater Mösting A. V. S. Kislyuk.
Astrometriya i Astrofiz., Kiev, No. 7, p. 65 - 67 (1969). In Russian.

The author presents the values of the rotational constants of the moon, the errors in the coordinates of the crater Mösting A with respect to the system of limb craters being taken into account. The derived values of I and f practically do not differ from those presented in a former paper.

094.207 Physics of the lunar surface layer.
V. V. Shevchenko.
Zemlya i Vselennaya, No. 4, p. 57 - 62 (1969). In Russian.

094.208 Lunar landscapes on the volcanoes of Kamchatka.
I. I. Cherkasov, V. V. Shvarev, G. S. Shtejnberg.
Zemlya i Vselennaya, No. 5, p. 26 - 35 (1969). In Russian.

094.209 On the determination of the figure of the moon.
N. L. Makarenko.
Astron. Vestn. Vol. 2, 81 - 89 (1968). In Russian.

A theoretical possibility to determine the lunar figure by means of meridian measurements of the visible lunar diameters is considered.

094.210 Observations of thermal radiation of the moon.
N. P. Yesaulov, N. S. Nikulin.
Report FTD-MT-67-25, Air Force Systems Command, Wright–Patterson AFB, Ohio. 16 pp. (1967). – See Phys. Abstr. Vol. 72, No. 20219 (1969).

094.211 Radar studies of the moon, volume 2 [Final Report].
Report NASA–CR–95218, Massachusetts Inst.
Technology, Lincoln Lab., Lexington, Ma., 104 pp. (1968). – See Phys. Abstr. Vol. 72, No. 20220 (1969).

094.212 Lunar millimeter-wavelength radiation and thermal model. D. E. Clardy.
Thesis, Univ. of Texas. Univ. Microfilms, Ann Arbor, Mi., 97 pp. (1968). – See Phys. Abstr. Vol. 72, No. 27074 (1969).

094.213 The kinetic theory of the solar wind and its interaction with the moon. D. H. Griffel.
Thesis, California Inst. Technology. Univ. Microfilms, Ann Arbor, Mi., 122 pp. (1968). – See Phys. Abstr. Vol. 72, No. 27100 (1969).

094.214 Mid-infrared spectroscopic observations of the moon. G. R. Hunt, J. W. Salisbury.
Phil. Trans., Ser. A, Vol. 264 (No. 1150), 109 - 139 (1969).

094.215 Temperature distribution of the moon.
D. H. Menzel.
Phil. Trans., Ser. A, Vol. 264 (No. 1150), 141 - 144 (1969).

094.216 The origin of the moon. R. A. Lyttleton.
Science Journ. Vol. 5, No. 5, p. 53 - 59 (1969).

094.217 Geophysics of the moon.
H. C. Urey, G. J. F. MacDonald.
Science Journ. Vol. 5, No. 5, p. 60 - 65 (1969).

094.218 Lunar radar scattering at 6 and 12 meter wavelengths. A. A. Burns.
Report NASA–CR–95405, Stanford Univ., Radioscience Lab., 139 pp. (1968). – See Phys. Abstr. Vol. 73, No. 5300 (1970).

094.219 The geometrical figure of the marginal zone of the moon from measurements of its apparent diameters
in Greenwich. N. L. Makarenko.
Astron. Vestn. Vol. 3, 135 - 141 (1969). In Russian.

Parameters and orientation of an ellipsoid, representing a geometrical figure of the marginal belt of the moon, are deduced from the reduction of Greenwich meridional measurements of apparent vertical and horizontal diameters of the moon.

094.220 On the regularity in position and size of craters in crater-chains on the moon.
M. M. Shemyakin.
Astron. Vestn. Vol. 3, 65 - 75 (1969). In Russian.

The author investigates some crater-chains on the moon (about 20 on the visible side, about 30 in the border zones and on the reverse side). As to the position and dimensions of the craters, regularities are found, which can be formulated mathematically.

094.221 Particles of lunar origin. I. Determination of the elements of geocentric orbits of lunar particles in
space. V. P. Orlov.
Astron. Vestn. Vol. 3, 76 - 81 (1969). In Russian.

Elements are deduced of geocentric orbits of particles ejected from the moon by explosions of meteor bodies as functions of initial conditions of the ejection: selenocentric longitude λ_0 and latitude φ_0, initial velocity V_0, azimuth A_0 and zenith distance δ_0 of the direction of ejection. It is shown that the nodes of geocentric orbits of the particles of lunar origin are concentrated near the moon's position at the moment of ejection. From the deduced elements of geocentric orbits the region of the lunar surface, from which a vertical flight of particles to the earth is possible, is determined.

094.222 Moon's spherical astronomy.
B. Kołaczek.
Geodezja Kartografia, Vol. 18, 95 - 134 (1969).   In Polish.
   A short introduction contains the basic information about the moon, its orbit and a description of astronomical phenomenae which can be observed from the surface of the moon. The next part describes the parallaxes related with transformation of the geo-centric or heliocentric coordinates into the seleno-centric or lunar-topocentric ones. It also discusses the phenomenae of the aberrations caused by the motion of the moon and its precession and nutation. The article contains the formulae needed for calculation of the influence of the above mentioned phenomenae upon the coordinates of seleno-equatorial, ecliptic and geo-equatorial systems. The table of the moon constants is given at the end.

094.223 Mondproben in Heidelberg.
J. Zähringer.
Ruperto Carola, Univ. Heidelberg, Vol. 47, 12 - 18 (1969).

094.224 Rilles and water on the moon?
E. J. Öpik.
Irish Astron. Journ., Vol. 9, 79 - 80 (1969).

094.225 A qualitative method of determination of colour differences on the lunar surface.
N. N. Petrova.
Astron. Tsirk. No. 537, p. 1 - 2 (1969).   In Russian.

094.226 Measurement of areas of lunar maria and thalassoids.
J. F. Rodionova.
Astron. Tsirk. No. 538, p. 1 - 3 (1969).   In Russian.

094.227 Relative Höhen auf dem Monde, dritte Reihe mit 182 Beobachtungen,Genauigkeitsuntersuchungen.
J. Hopmann.
Sitzungsber. Österreich. Akad. Wiss., Math.-Naturwiss. Kl., Abt. II, Vol. 176, 343 - 361 (1968) = Mitt. Univ.-Sternw. Wien, Vol. 14, No. 8 (1969).
   This is a continuation of two earlier papers on relative lunar heights. Small objects were preferred during the observations. After the discussion of the internal p.e. of the Vienna measures follows a comparison of 5 independent publications of relative lunar heights. It seems, that the visual micrometrical measures are better than the maps.

094.228 Relative Höhen auf dem Monde. 4. Reihe mit 172 Beobachtungen, Genauigkeitsuntersuchungen.
J. Hopmann.
Sitzungsber. Österreich. Akad. Wiss., Math-Naturwiss. Kl., Abt. II, Vol. 177, 95 - 112 = Mitt. Univ.-Sternw. Wien, Vol. 14, No. 11 (1969).
   The results of micrometrical measures with the 8''-refractor of the Vienna-Observatory are given in the same way as in the three preceding papers. 90% of the objects are on the (astronomical) west-side of the moon. The computed heights are compared with those of J. Schmidt, two maps of the US Army-Map-Service and the charts of the "Aeronautical Chart Information Center" of the US Air Force. So it is possible to give an idea on the accuracy of the five different sets.

094.229 Gross estimates of the conductivity, dielectric constant, and magnetic permeability distributions in the moon.   S. H. Ward.
Radio Sci., Vol. 4, No. 2, 117 - 137 (1969). – See Phys. Ber., Vol. 48, No. 12–3438 (1969).

094.230 Non-diffuse infrared emission from the lunar surface.   J. K. Harrison.
International Journ. Heat Mass Transfer, Vol. 12, 689 - 697 (1969).

094.231 The erosion processes on the lunar surface.
Z. Kopal.
Space Research IX, Proc. Tokyo 1968, p. 657 - 677 (1969).

094.232 Lunar Orbiter photographic data.
Prepared by M. Beeler, K. Michlovitz.
Data Users' Note, National Space Science Data Center, NASA, Goddard Space Flight Center, Greenbelt, Md. NSSDC 69-05, 5 + 37 pp. + 17 charts (1969).
   The primary purpose of this *Data Users' Note* is to announce the availability of Lunar Orbiter 1-5 pictorial data and to aid the investigator in the selection of Lunar Orbiter photographs for study. In addition, this *Note* can give some guidance to the interpretation of the pictures. As background information, the *Note* includes a brief description of the mission objectives and the photographic subsystem.

094.233 Erste Untersuchungsergebnisse an Apollo-11-Mondproben.
SuW, Vol. 8, 255 - 257 (1969).

094.234 Veränderungen auf dem Mond.
J. Classen.
Veröff. Sternw. Pulsnitz, No. 5, 20 pp. (1969).
   Since 1540, flares can be observed on the moon. But it is only since 1958 that the opinion gathered way these flares being due to real processes. The flares are assumed to be processes of luminescence or perhaps lunar gas eruptions. Based on a flare the observation of which succeeded in Pulsnitz, the author recommends to observe the ashy moonlight, too. Furthermore, comparison of brightness between Aristarchus and Kepler is proposed.

094.235 The surface structures of the Moon and Mercury derived from integrated photometry.
T. Pikkarainen.
Ann. Acad. Sci. Fennicae, Series A, VI. Physica 316, 12 pp. = Aarne Karjaleinen Obs. Univ. Oulu, Finland, Publ. No. 14 (1969).
   The integrated brightnesses of the Moon and Mercury are theoretically calculated by using a surface model with parabolical holes, Lommel-Seeliger's reflection law and the phase function of the irradiated basalt powder. The theoretical integrated brightnesses are fitted with observations by using two free parameters which describe the surface structures of the Moon and Mercury. The results show that Mercury's surface is more densely covered with holes but that their depths are smaller on the Moon.

094.236 ALPO selected areas program: Plato.
C. L. Ricker, H. W. Kelsey.
Strolling Astronomer, Vol. 21, 161 - 166 (1969).

094.237 A report and analysis of seventeen recent lunar transient phenomena, addendum.
C. L. Ricker, H. W. Kelsey.
Strolling Astronomer, Vol. 21, 166 - 167 (1969).

094.238 Apollo Mission 10 Photography Indexes.
Prepared under the Direction of the Department of Defense by the Aeronautical Chart and Information Center, United States Air Force for the NASA, Greenbelt. 6 sheets (1969).

094.239 First men on the moon.   J. Bouška.
Říše hvězd, Vol. 50, 185 - 189 (1969).   In Czech.

094.240 Lunar mapping and research in Czechoslovakia.
F. Kadavý.
Říše hvězd, Vol. 50, 201 - 205 (1969).   In Czech.

094.241 **Mare Orientale and other lunar basins.**
P. Příhoda.
Říše hvězd, Vol. 50, 189 - 197 (1969). In Czech.

094.242 **Mare Tranquillitatis, lunar probes Ranger VIII and Surveyor V, and Apollo XI.** A. Rükl.
Vesmír,Vol. 48, 360 - 365 (1969). In Czech.

094.243 **News on lunar "rolling stones".** M. Grün.
Říše hvězd, Vol. 50, 231 - 234 (1969). In Czech.

094.244 **The discussion on lunar craters origin is still going on.** S. R. Brzostkiewicz.
Urania Kraków, Vol. 40, 304 - 309 (1969). In Polish.

094.245 **The inconstant moon.** N. E. Heath.
Southern Stars, Vol. 23, 85 - 87 (1969).

094.246 **The characteristics of the lunar surface relief in the vicinity of the station Luna 13.**
N. P. Barabashov, L. A. Akimov, D. F. Lupishko.
Vestn. Khar'kov. Univ. No. 34, (Ser. Astron. No. 4), p. 12 - 16 (1969). In Russian.

094.247 **The microrelief of surfaces imitating the lunar surface.** N. P. Barabashov, L. A. Akimov.
Vestn. Khar'kov. Univ. No. 34, (Ser. Astron. No. 4), p. 17 - 30 (1969). In Russian.

094.248 **Measurement of the luminescence of the lunar surface in the H and K lines of Ca II.**
V. S. Tsvetkova.
Vestn. Khar'kov. Univ. No. 34, (Ser. Astron. No. 4), p. 50 - 55 (1969). In Russian.

094.249 **Luminous rays and the microrelief of the lunar surface.** N. P. Barabashov.
Vestn. Khar'kov. Univ. No. 34, (Ser. Astron. No. 4), p. 103 - 108 (1969). In Russian.

094.250 **The A. L. P. O. Lunar Section Aristarchus–Herodotus Mapping Project: Final report.**
J. E. Westfall.
Strolling Astronomer, Vol. 21, 181 - 201 (1969).

094.251 **Der derzeitige Stand der Physik des Mondes.**
R. Meissner.
Mitt. Astron. Ges. No. 27, p. 87 - 108 (1969). – Review article.

094.252 **Lunar limb shock wave.**
J. V. Hollweg.
Mitt. Astron. Ges. No. 27, p. 222 - 226 (1969). – Conference paper.

094.253 **Apollo 10 Photography Index.**
Prepared by the Mapping Sciences Laboratory at the NASA Manned Spacecraft Center, Houston, Texas. Published by the National Space Science Data Center, Greenbelt. NSSDC 69-14, 110 pp. (1969).
This index contains supporting information about each photograph taken during the Apollo 10 mission, including those photographs taken of the earth.

094.254 **Origin and history of the moon.**
H. C. Urey.
Bull. Atomic Scientists, Vol. 25, No. 7, p. 46 - 51 (1969).
The following aspects are considered: Time of surface formation, temperature history, gravitational field, composition of the moon, lunar heat balance, potassium uncertainty, surface composition, water on the moon, the moon as a planet, accumulating debris, lunar landing.

094.255 **Lunar retroreflector array.**
D. G. Currie, S. K. Poultney.
IAU Circ. No. 2164 (1969).

094.256 **Particle size distribution of lunar surface material.**
C. C. Mason.
Bull. Geol. Soc. America, Vol. 80, 587 - 594 (1969).

094.257 **The moon as an abode of life?** D. H. Menzel.
Proc. American Phil. Soc., Vol. 113, No. 2, 102 - 126 (1969).

094.258 **Origin of lunar maria.** J. H. Mackin.
Bull. Geol. Soc. America, Vol. 80, 735 - 747 (1969).

Moon and Mars.
Nature, Vol. 223, 1197 (1969).
News notes on the lunar surface material collected by Apollo 11 and the analysis of two hundred photographs of Mars by Mariners 6 and 7.

Moon rocks. First results published.
Nature, Vol. 223, 1305 - 1306 (1969).

Material vom Mond.
Naturwissenschaften, 56. Jahrgang, 487 (1969).

Landing on the moon.
Priroda No. 8, p. 104 - 106 (1969). In Russian.

The moon in colour.
Spaceflight, Vol. 11, 307 (1969).

Apollo pictures.
Spaceflight, Vol. 11, 416 (1969).

Lunar globe.
Spaceflight, Vol. 11, 416 (1969).

The Moon. See Abstr. 003.014.

Mapa Měsíce. (Lunar Chart, 1:10000000).
See Abstr. 003.046.

Figur und Dimensionen des Mondes nach astronomischen Beobachtungen. See Abstr. 003.072.

The Constants of the Physical Libration of the Moon. See Abstr. 003.074.

Introduction to Lunar Physics.
See Abstr. 003.108.

Literal developments in the analytical theory of the moon. See Abstr. 042.001.

Surveyor 7 lunar mission.
See Abstr. 053.015.

Lunar Explorer 35.
See Abstr. 053.025.

Irdischer und lunarer Vulkanismus – Ergebnisse einer Reise nach Island. See Abstr. 081.010.

Lead isotopes, lunar capture and mantle evolution.
See Abstr. 081.017.

Tidal friction with latitude-dependent amplitude

and phase angle.   See Abstr. 081.020.

Distribution of slopes on a cratered planetary surface: Theory and preliminary applications.
See Abstr. 091.016.

Accurate ephemerides for planets and moon.

See Abstr. 091.021.

Vergleichende Charakteristik der elektromagnetischen Parameter der Mondoberfläche, der Meteorite und Tektite.   See Abstr. 105.050.

Field and plasma in the lunar wake.
See Abstr. 106.022.

095   Lunar Eclipses

095.001   **Photoelektrische Helligkeitsmessung der Mond-Halbschattenfinsternis vom 2. April 1969 in Bregenz.**
K. P. Schneider.
SuW, Vol. 8, 214 - 215 (1969).

095.002   **Eclipse de luna del 12/13 de Abril de 1968. Algunos resultados.**   A. J. Camponovo.
Revista Astron. Vol. 40, No. 166, p. 21 - 24 (1968).

095.003   **Photometric brightness asymmetry during a lunar eclipse.**   R. Gerharz.
Arch. Meteorol. Geophys. Biokl., Ser. A, Vol. 18, 221 - 226 (1969).

The changes of lunar brightness during the eclipse of the moon on 13 April 1968 were measured photoelectrically in the visible part of the spectrum. Superposition of the data functions on a common chart revealed a distinct brightness asymmetry. Its probable cause could be a non-uniform diurnal distribution of atmospheric scatterers in the two half-zones of the terrestrial terminator.

095.004   **Die Abplattung des Erdschattens bei Mondfinster-**

nissen.   J. Meeus.
Sterne, 45. Jahrgang, 116 - 117 (1969). − Some remarks by P. Ahnert, p. 117 (1969).

095.005   **Analyse photométrique de la pénombre pendant les éclipses de lune.**   J. Dubois, F. Link.
The Moon. Vol. 1, 85 - 105 (1969).

Photometric analysis of the penumbra during 21 eclipses between 1921 and 1968 based upon the homogeneous observational material reveals some anomalies which may be explained by the lunar luminescence excited by UV-X solar radiations whose sources are located in the low corona and above the K-3 plages. The influence of the terrestrial upper atmosphere is detectable on the border of the umbra.

095.006   **Observations de l'éclipse totale de Lune du 13 avril 1968.**   A. G. Velghe.
Bull. Astron. Obs. Roy. Belgique, Vol. 6, (No. 8), 342 (1969).

The visibility of the penumbra of the earth during lunar eclipses.   See Abstr. 082.108.

096   Lunar Occultations

096.001   Occultation highlights – September-December, 1969.   D. W. Dunham.
Sky Telescope, Vol. 38, 204 (1969).

096.002   Beobachtung von Sternbedeckungen durch den Mond – eine interessante Amateurbeschäftigung.
E. Mayer.
Orion Schaffhausen, Vol. 14, 102 - 103 (1969).

096.003   Occultations of stars by the moon observed at the Cracow Astronomical Observatory in the year 1968.
M. Kurpińska.
Acta Astron. Vol. 19, 241 - 244 (1969).
The paper contains the results of 57 observations of occultations of stars by the moon in the year 1968 made at the old Cracow Observatory, Sniadecki Collegium, and at the new observatory, M. Kopernik Observatory, at Fort Skala.

096.004   Beobachtungen von Sternbedeckungen durch den Mond in Lwow.   A. T. Dul'tsev.
Tsirk. L'vov. Astron. Obs. No. 43, p. 55 (1969).
In Russian.

096.005   L'occultation des Pléiades du 21 décembre 1969.
J. Meeus.
Ciel et Terre, Vol. 85, 248 - 251 (1969).

096.006   L'observation pratique des occultations stellaires.
A. Oberstatter.
L'Astronomie, 83ᵉ année, 299 - 305 (1969).

096.007   Lunar occultation observations of Jupiter at 234 and 405 MHz.   S. Gulkis.
Bull. American Astron. Soc. Vol. 1, 243 (1969). – Abstr. AAS.

096.008   The discovery and measurement of double stars by lunar occultations.
R. E. Nather, D. S. Evans.
Publ. Astron. Soc. Pacific, Vol. 81, 548 (1969). – Abstract ASP.

096.009   Occultations of stars by the moon observed at the Rio de Janeiro National Observatory, Brazil, in the year 1968.   J. J. D. Nogueira, M. Migon.
Acta Astron. Vol. 19, 323 - 325 (1969).
Results are given of 59 observations of occultations of stars by the moon at the Rio de Janeiro National Observatory, Brazil.

096.010   1970 occultation supplement. Predictions for the United States and Canada.
Sky Telescope, Vol. 38, 317 - 321, 326 - 329 (1969).

096.011   Theory of lunar occultation: New methods of estimating brightness distributions and their widths.
K. R. Lang.
Astrophys. Journ., Vol. 158, 1189 - 1198 (1969).

A new method of obtaining an estimated brightness distribution, which is based upon extracing the spatial Fourier components of the brightness distribution from the observed power, is described in this paper. The limit to the resolution attainable by either method is determined by the highest available spatial frequency. This cutoff frequency is shown to be a function of antenna aperture, source width, bandwidth and noise effects. When source width is all that is required, it may be determined from a comparison of the observed diffraction lobes with those of a theoretical brightness distribution.

096.012   A discussion of 1950 - 1968 occultations of stars by the moon.   T. C. Van Flandern.
Bull. American Astron. Soc., Vol. 1, 367 (1969). – Abstr. AAS.

096.013   Quelques occultations rasantes visibles en France (janvier – juin 1970).   J. Meeus.
L'Astronomie, 83ᵉ année, 448 - 451 (1969).

096.014   Occultation observation in 1968.
Data Rep. Hydrographic Observations, Ser. Astron. Geod., *Tokyo*, (Pub. 691) No. 4, p. 1 - 12 (1969).
This is a continuation of the series of the report of occultation observations made by the Hydrographic Department and some cooperators in Japan, and contains the data for 1968.

096.015   Occultations – observing reappearances.
M. J. Hendrie.
Journ. British Astron. Ass., Vol. 80, 70 - 73 (1969).

096.016   Observations of occultations of stars by the moon in Gorki in 1967.
E. G. Demidovich, A. P. Poroshin.
Astron. Vestn. Vol. 2, 258 (1968).   In Russian.

096.017   Grazing occultation of ZC 1925 (Spica) observed at Potgietersrus, South Africa, on 1969, June 24.
B. Armstrong.
Monthly Notes Astron. Soc. Southern Africa, Vol. 28, 97 - 101 (1969).

096.018   Lunar occultations.
H. de Souza, L. E. da Silva Machado.
Contr. Obs. Valongo, Univ. Federal Rio de Janeiro, Sér. III, Nos. 5 - 9 (1969).

096.019   Predictions of grazing occultations.
R. Abileath.
Yamamoto Circ. No. 1701 (1969).   In Japanese.

August occultation of the Pleiades.
Sky Telescope, Vol. 38, 269 - 270 (1969).

Studio preliminare per un programma di occultazioni lunari di radiosorgenti a medicina.
See Abstr. 141.220.

097 Mars

097.001 Windblown dust on Mars.
C. Sagan, J. B. Pollack.
Nature, Vol. 223, 791 - 794 (1969).
The wave of darkening in the Martian springtime, which some say may have a biological explanation, can be explained in terms of windblown dust.

097.002 Solar cycle variation of exospheric temperatures on Mars and Venus: A prediction for Mariner 6 and 7.
R. W. Stewart, J. S. Hogan.
Science, Vol. 165, 386 - 388 (1969).
Calculations have been made to determine the effects of variations in the extreme ultraviolet solar radiation on the upper atmospheres of Mars and Venus. The results indicate that the exospheric temperature on Mars varies from 300 °K to 600 °K during the solar cycle, with a corresponding range on Venus of 450 °K to 850 °K. At the present time, the temperature of the Martian exosphere should be approximately 500 °K.

097.003 Mariner 6 television pictures: First report.
R. B. Leighton, N. H. Horowitz, B. C. Murray, R. P. Sharp, A. G. Herriman, A. T. Young, B. A. Smith, M. E. Davies, C. B. Leovy.
Science, Vol. 165, 684 - 690 (1969).
On 31 July 1969, the more advanced Mariner 6 spacecraft, carrying two television cameras, passed Mars and recorded 75 pictures. A twin spacecraft, Mariner 7, passed Mars on 5 August 1969. This report summarizes the results of a first, qualitative study of the Mariner 6 television pictures, carried out on the uncalibrated data within a few days after receipt on Earth.

097.004 The Martian grid system, continental blocks and continental drift. R. A. Wells.
Geophys. Journ. Roy. Astron. Soc. Vol. 18, 109 - 128 (1969).
Mars possesses two grid systems — a diagonal grid made up of lineaments trending approximately NW—SE and NE—SW, and a meridian-latitude grid composed of lineaments trending approximately N—S and E—W. The distribution of angles between the two families of lineament planes in the diagonal grid implies the existence of a 4-cell asymmetric convection pattern within the Martian mantle. The convection pattern readily explains the distribution of the major dark features, considered to be elevated continental blocks, by a continental drift hypothesis. The equatorial maria are elevated regions composed of E—W overthrust structures, while Mare Acidalium, the only major dark feature in the northern hemisphere, is a ridge-and-trough system comparable to a terrestrial oceanic rise. The Martian maria bear structural resemblances to the lunar highlands and the Martian deserts are similar to the lunar maria. Mars has undergone extensive tectonic development over a geological time-scale.

097.005 Fernaugen beobachten Mars. M. Römer.
Umschau, Vol. 69, 616 - 618 (1969).

097.006 Some highlights of the current apparition of Mars.
Sky Telescope, Vol. 38, 72 - 74 (1969).

097.007 Mariner 7 television pictures: First report.
R. B. Leighton, N. H. Horowitz, B. C. Murray, R. P. Sharp, A. G. Herriman, A. T. Young, B. A. Smith, M. E. Davies, C. B. Leovy.
Science, Vol. 165, 787 - 795 (1969).

097.008 Mars: Water vapor in its atmosphere.

T. Owen, H. P. Mason.
Science, Vol. 165, 893 - 895 (1969).
With the use of newly obtained spectrograms of Mars in the region of the water vapor band at 8200 Å, we have derived a value of 35 ± 15 μ for the amount of precipitable water in a vertical column in the Martian atmosphere.

097.009 Mars: Interpretation of spectral reflectivity of light and dark regions. J. B. Adams, T. B. McCord.
Journ. Geophys. Res. Vol. 74, 4851 - 4856 (1969).
New spectral reflectivity curves for the light and dark areas of Mars and for seasonal changes of the dark regions were modeled in the laboratory. Light areas are more oxidized and are probably composed of finer particles than the dark areas. Curves for both regions indicate a combination of ferric iron oxide and mafic silicate rock such as basalt. Seasonal changes in the spectral reflectivity of Syrtis Major are not compatible with wind transport of materials or with surface chemical changes. Models of local seasonal condensation of ice or moisture or of growth of macroscopic very dark gray vegetation satisfy the spectral data.

097.010 On the determination of relative levels of continental and sea regions of Mars. I. K. Koval.
Astron. Tsirk. No. 506, p. 2 - 5 (1969). In Russian.

097.011 Cloudiness in the Martian atmosphere as possible cause of the canals' invisibility on the pictures taken by Mariner-4. M. S. Bobrov.
Astron. Tsirk. No. 514, p. 3 - 4 (1969). In Russian.

097.012 On the Martian colour-index.
D. F. Lupishko, T. A. Lupishko.
Astron. Tsirk. No. 525, p. 6 - 8 (1969). In Russian.

097.013 Mariner 6: Ultraviolet spectrum of Mars upper atmosphere. C. A. Barth, W. G. Fastie, C. W. Hord, J. B. Pearce, K. K. Kelly, A. I. Stewart, G. E. Thomas, G. P. Anderson, O. F. Raper.
Science, Vol. 165, 1004 - 1005 (1969).
Emission features from ionized carbon dioxide and carbon monoxide were measured in the 1900 to 4300 Å spectral region. The Lyman alpha 1216 Å line of atomic hydrogen and the 1304, 1356, and 2972 Å lines of atomic oxygen were observed.

097.014 Spectrophotometric studies of Mars, Jupiter and Saturn. M. Isakson.
Publ. Dep. Astron. Univ. Chile, Obs. Astron. Nacional, Cerro Calan, Santiago, Vol. 1 (No. 7), 94 - 105 (1968).
A spectrophotometric study of the planets Mars, Jupiter and Saturn has been made. The curves of the relative magnitude of these planets have been compared with those of an average solar-type star and in this way, the albedo relative curves of the planets mentioned have been determined.

097.015 An analysis of the Mariner-4 cratering statistics.
C. R. Chapman, J. B. Pollack, C. Sagan.
Astron. Journ. Vol. 74, 1039 - 1051 (1969).
A catalogue of craters and crater-like objects has been prepared from several sets of contrast-enhanced high-quality positive transparencies of the Mariner-4 photography. Craters were identified and counted by the same procedures used in the compilation of lunar crater catalogues; particular attention was given to crater class and quality. Counts of craters with diameters $D < 20$ km begin to show the effects of incompleteness. Substantial erosion and obliteration of all but the

largest craters has occurred during the history of Mars. The population of impacting objects assumed responsible for craters is taken as having a differential number density varying as $X^{-\beta}$, where X is the diameter of the impacting object. The number of "live" comets and Apollo objects crossing the orbit of Mars is insufficient by more than 2 orders of magnitude to explain the observed number density of craters on Mars. For $\beta = 4$ or 5, agreement between the predicted and observed number densities can be secured with a nearly uniform rate of asteroidal bombardment. The dust produced by impact during the history of Mars is estimated to have depths between 0.1 and several km. For $\beta < 3$, the mean ages of Martian craters are found to be approximately equal to the age of the planet. However, for β significantly larger than 3, different craters will have different mean ages, ranging down to $\lesssim 10^7$ yr for craters smaller than 20 km. Thus, the absence of such signs of running water as river valleys in the Mariner-4 photography is quite irrelevant to the question of the existence of bodies of water in early Martian history.

097.016 The meteorite flux at the surface of Mars.
R. D. Dycus.
Publ. Astron. Soc. Pacific, Vol. 81, 399 - 414 (1969).

Estimates of the average meteorite flux at the Martian surface are presented. These estimates were derived by generating a meteorite flux distribution for the top of the Martian atmosphere, and then adjusting this flux for mass loss and deceleration in the atmosphere. The top-of-the-atmosphere meteorite flux distribution was based on meteorite fall, asteroid, and lunar and Martian crater data. The effects of the Martian atmosphere were evaluated. Results of this preliminary study indicate the Martian surface is afforded varying degrees of protection from impacting meteorites in the mass range 10 grams to 1 metric ton.

097.017 Carbon monoxide in the Martian atmosphere.
L. D. Kaplan, J. Connes, P. Connes.
Astrophys. Journ. *(Letters)*, Vol. 157, L187 - L192 (1969).

Carbon monoxide has been detected in the Martian atmosphere by the appearance, in high-resolution interferometric spectra, of prominent lines of the 2 - 0 band of the principal isotope and weaker lines of other isotopes and of the 3 - 0 band. Quantitative analysis of the spectra results in a CO content of 5.6 cm Amagat, a surface pressure of 5.3 mb, and a CO concentration of 0.8 part per thousand by volume.

097.018 De Mariners 6 en 7 fotograferen Mars.
T. E. de Vries.
Hemel en Dampkring, Vol. 67, 306 - 319 (1969).

097.019 New interpretations of Mayeda's flare on Mars. II.
V. D. Davydov.
Astron. Zhurn. Akad. Nauk SSSR, Vol. 46, 1074 - 1086 (1969). In Russian. English translation in Soviet Astron. AJ, Vol. 13, No. 5.

In the first part of the paper (Astron. Astrophys. Abstr., Vol. 1, 097.039) two versions of the hypothesis of Mayeda's flare are given. A quantitative elaboration and comparison of both the versions of the hypothesis are given in this second part of the paper. Besides Mayeda's flare, other cases of the appearance of a temporary white point on Mars are pointed out in which the hypothesis is applicable.

097.020 The Martian atmosphere from polarization observations. A. V. Morozhenko.
Astron. Zhurn. Akad. Nauk SSSR, Vol. 46, 1087 - 1094 (1969). In Russian. English translation in Soviet Astron. AJ, Vol. 13, No. 5.

Optical parameters of the Martian atmosphere are determined from polarization observations. The light scattering on uniform spherical particles with the refractive index 1.5 and

Junge distribution of particle sizes have been taken into account. It is obtained that the atmospheric pressure near the planet surface is about 11 mbar, the full optical thickness of the atmosphere is 355 mμ is about 0.064, and the Junge distribution parameter is equal to 4.25.

097.021 The diurnal temperature variation in the aerosol-gaseous atmosphere and in the Martian ground.
V. I. Aleshin, T. N. Fedoseeva.
Astron. Zhurn. Akad. Nauk SSSR, Vol. 46, 1095 - 1103 (1969). In Russian. English translation in Soviet Astron. AJ, Vol. 13, No. 5.

The height distribution and diurnal temperature variations in the aerosol-gaseous atmosphere and in the Martian ground are calculated. The models of the atmosphere, in which the pressure near the surface is equal to 5, 10 and 20 mbar, are considered. Each model is characterized by the optical thickness $\tau_\odot$, conditioned by the absorption of the solar radiation in the dust atmosphere. It is shown that the presence even of a weak atmosphere with minimal dust increases the surface temperature of the planet at night by 17°. In the case of considerable dust this increase can reach up to 30°. The temperature variations are discussed in detail.

097.022 Exobiology and the exploration of Mars.
D. G. Rea.
Applied Optics, Vol. 8, 1267 - 1269 (1969).

Our limited understanding of the Martian environment has stimulated much speculation on the possible existence of a Martian biota. This understanding is being steadily improved by ground-based observing techniques. Rapid advances have been made by the Mariner IV flyby and will be made by the Mariner 1969 flybys and the Mariner 1971 orbiters. In situ measurements, including some aimed at detecting manifestations of life, will be carried out on the Viking landers in 1973. We should then be in a position to discuss more quantitatively than at present the various questions related to indigenous Martian life forms.

097.023 Mars through a crystal ball.   J. Lederberg.
Applied Optics, Vol. 8, 1269 - 1270 (1969).

097.024 Mars: Theoretical aspects of meteorology.
C. B. Leovy.
Applied Optics, Vol. 8, 1279 - 1286 (1969).

The Mars atmosphere responds by radiative and convective heating to surface temperature changes in a time-scale of the order of one day. This is about 50 times as fast as the corresponding response on the earth. As a consequence, the mean wind distribution is under strong solar control but may be modified by latent heat release in CO_2 condensation and by the mass flow required to balance such condensation. Unstable eddies, analogous to terrestrial cyclones, are to be expected in all seasons except summer.

097.025 Polarimetry of Mars.   K. L. Coulson.
Applied Optics, Vol. 8, 1287 - 1294 (1969).

A summary of results of observations of the polarization of Mars shows that surface pressure determinations from these data have not yielded satisfactory results in spite of the extensive number of observations available. Computations of atmospheric effects to be expected from various Rayleigh and aerosol models of the atmosphere show that the polarizing effects of realistic aerosol models can vary widely, depending on particle parameters, and that polarization due to Rayleigh scattering by representative models of the Martian atmosphere can only serve to shift the position of the neutral point to smaller phase angles and to shift the polarization curve in the positive direction from its position for only the surface-reflected radiation. The extent to which the surface of Mars is composed of hydrated iron oxides is still an open question. Polarization measurements show that limonite is one, but by no means unique, possibility for a major constituent of the surface.

097.026 Measurements in the atmosphere of Mars.
A. Seiff.
Applied Optics, Vol. 8, 1305 - 1314 (1969).

The detailed definition of the key features of Mars' atmosphere from one or a few entries and landings is a challenging task involving a variety of measurements, taken during entry and after landing, and correlated with observations to be taken from orbiters and flyby missions. The properties of interest include profiles to high altitudes of the atmospheric state properties, composition, and meteorological factors such as winds, clouds, and variations in pressure and temperature. Profiles of the atmosphere are best measured during entry, by techniques which are described.

097.027 Mineralogical investigations of Mars.
J. T. O'Connor.
Applied Optics, Vol. 8, 1323 - 1328 (1969).

A mineralogical study of Mars will offer a great deal of information on the present state of the planet and on the processes which have operated on it during the past. The degree of planetary development, the amount of igneous activity, the development of a secondary atmosphere, the presence of metamorphic or sedimentary processes, and the effects of impacting meteorites will all be recorded in the mineral assemblages. The identification of the mineral phases in the Martian rocks is the fundamental step in conducting such an investigation. The author gives a brief description to obtain mineralogical information of Mars.

097.028 Optical activity for exobiology and the exploration of Mars. B. Halpern.
Applied Optics, Vol. 8, 1349 - 1353 (1969).

The crucial role of a sterically specified informational macromolecule argues for optical activity as an assay for the presence of biogeny on a planet. Recent developments in the measurement of optical activity and their relevance to exobiology are discussed.

097.029 Extraterrestrial life detection.
R. S. Young.
Applied Optics, Vol. 8, 1355 - 1360 (1969).

A fundamental objective of the NASA Planetary Exploration Program is to attempt to cast light on the origin of the planets and the solar system. The basic life attributes discussed are chemistry, morphology, growth, and metabolism. Included are experimental approaches to the acquisition of such data.

097.030 Mariner 6 and 7 television pictures: Preliminary analysis.
R. B. Leighton, N. H. Horowitz, B. C. Murray, R. P. Sharp, A. H. Herriman, A. T. Young, B. A. Smith, M. E. Davies, C. B. Leovy.
Science, Vol. 166, 49 - 67 (1969).

The purpose of this article is to draw together the preliminary television results from the two spacecraft; to present tentative data concerning crater size distributions, wall slopes, and geographic distribution; to discuss evidences of haze or clouds; to describe new, distinctive types of topography seen in the pictures, and to discuss the implications of the results with respect to the present state, past history, and possible biological status of Mars.

097.031 Mariner 1969: Preliminary results of the infrared radiometer experiment.
G. Neugebauer, G. Münch, S. C. Chase, Jr., H. Hatzenbeler, E. Miner, D. Schofield.
Science, Vol. 166, 98 - 99 (1969).

The thermal energy emitted by Mars was measured in the 8- to 12- and 18- to 25-micrometer bands. The minimum temperature derived for the southern polar cap is $150°K$, an indication that the cap is formed by frozen carbon dioxide. No significant temperature fluctuations were detected with a 100-kilometer scale.

097.032 Spectrographic detection of topographic features on Mars. M. J. S. Belton, D. M. Hunten.
Science, Vol. 166, 225 - 227 (1969).

Observations of the Martian carbon dioxide band at 1.05μ made with a three-channel multislit spectrophotometer indicate gross height variations in the vicinity of Syrtis Major and surrounding desert regions. Syrtis Major appears to be very high with essentially no detectable carbon dioxide above it. The data appear to confirm local trends and, in magnitude at least, the large variations of height found in earlier radar observations. A one-to-one correlation of height with albedo is not evident in the results. Elevated areas are found in both desert and dark regions. In several regions dark areas are associated with relatively steep slopes.

097.033 Motion of Phobos. G. A. Wilkins.
Nature, Vol. 224, 789 (1969).

An attempt has been made to explain the secular acceleration of the motion of Phobos, and it is concluded that it is much less than derived by Sharpless in 1945.

097.034 Infrared absorptions near three microns recorded over the polar cap of Mars.
K. C. Herr, G. C. Pimentel.
Science, Vol. 166, 496 - 499 (1969).

During the Mariner 7 flyby of Mars, the infrared spectrometer recorded distinct, sharp absorptions near 3020 and 3300 reciprocal centimeters between $61°S$ and $80°S$, at the edge of the southern polar cap, with maximum optical density near $68°S$ and $341°E$. These bands, which match in frequency the ν_3 bands of methane and ammonia, can be associated with previously unreported spectral features of solid carbon dioxide exceeding 1 millimeter in thickness. Possible reasons for the geographic localization are discussed.

097.035 Mars pictures from Mariners 6 and 7.
Sky Telescope, Vol. 38, 212 - 221 (1969).

097.036 First findings from the Mariner flybys.
Sky Telescope, Vol. 38, 232 - 234 (1969).

097.037 Betrachtungen zur Vulkanaktivität auf dem Planeten Mars sowie synoptische Studien atmosphärischer Trübungserscheinungen. H. Heuseler.
Sterne, 45. Jahrgang, 133 - 148 (1969).

097.038 Statistik von Zentralbergen in Marskratern.
H. Heuseler.
Sterne, 45. Jahrgang, 152 - 155 (1969).

097.039 Über die Farbe des Planeten Mars.
G. G. Kandilarov.
Sterne, 45. Jahrgang, 155 - 157 (1969).

097.040 High-resolution photometry of a thin planetary atmosphere. A. T. Young.
Icarus, Vol. 11, 1 - 23 (1969).

High-resolution photometry near the limb of a planet can be inverted to yield the scale height and optical depth of the atmosphere if the scattering phase function is known. The method is applied to Mariner IV observations of Mars, assuming a Rayleigh phase function. After making a large correction for scattered light, we find $h_0 \approx 10$ km and $\tau \approx 0.03$ at $\lambda = 0.6 \mu$. There is some evidence for a dusty layer within a few kilometers of the ground. Better space-craft data are needed to determine the optical properties of the Martian atmosphere.

097.041 Topography and surface features of Mars.
A. B. Binder.
Icarus, Vol. 11, 24 - 35 (1969).

Topographic data obtained by radar ranging along the + 21°5 parallel of Mars have been correlated with surface detail. The data show that the high areas are deserts, the canals occur in broad and frequently deep valleys, and the maria occur in low areas or on slopes. The data confirm the concept that frost is deposited predominantly in the high areas and add support to the concept that the dark areas are biological in nature.

097.042 Martian temperatures and thermal properties.
D. Morrison, C. Sagan, J. B. Pollack.
Icarus, Vol. 11, 36 - 45 (1969).

In this paper the results of a reduction of a total of 31 scans yielding more than 500 independent temperature measurements on the illuminated hemisphere of Mars are discussed. From these brightness temperatures, thermometric temperatures are derived and compared with the predictions of thermal models for the planet.

097.043 Martian topography: Large-scale variations.
R. A. Wells.
Science, Vol. 166, 862 - 865 (1969).

The variation in carbon dioxide abundances detected in the 1.05-micron band over small, discrete areas on Mars indicates that larger-scale topographical differences are present than had previously been believed. Spectroscopic mapping of the surface also indicates no apparent correlation between albedo and height; the results are in good agreement with topographical data derived from the range-gated radar scan along + 21°N. High and low areas are found in both the major equatorial maria and the bright deserts in the northern hemisphere.

097.044 Mars : Is the surface colored by carbon suboxide?
W. T. Plummer, R. K. Carson.
Science, Vol. 166, 1141 - 1142 (1969).

The reflection spectrum of Mars can be well matched from 0.2 through 1.6 microns (and farther) by polymers of carbon suboxide, reflection spectra for which have now been measured. We propose that the reddish color of Mars might be attributed to carbon suboxide, not the commonly considered limonite or other iron-bearing minerals.

097.045 Seasonal behavior of the Martian polar caps.
G. E. Fischbacher, L. J. Martin, W. A. Baum.
Publ. Astron. Soc. Pacific, Vol. 81, 538 (1969). – Abstract ASP.

097.046 A study of cloud motions on Mars.
L. J. Martin.
Publ. Astron. Soc. Pacific, Vol. 81, 545 (1969). – Abstract ASP.

097.047 Molecular hydrogen in the atmosphere of Mars.
M. B. McElroy, D. M. Hunten.
Journ. Geophys. Res. Vol. 74, 5807 - 5809 (1969).

Attention is directed to the destruction of H_2 by reaction with CO_2^+ in the Martian ionosphere. It is shown that nearly all the H_2 flowing into the ionosphere is converted into H and escapes in the latter form.

097.048 Modification of the Martian ionosphere by the solar wind.
P. A. Cloutier, M. B. McElroy, F. C. Michel.
Journ. Geophys. Res. Vol. 74, 6215 - 6228 (1969).

A dynamical model is developed for interaction of solar wind with Martian ionosphere. It is assumed that Mars has a negligible magnetic moment, and a radiative equilibrium model, characterized by an exospheric temperature of 490°K, is adopted for the neutral atmosphere. The computed ionospheric structure is compared with the ionospheric profile observed by Mariner 4 and is found to agree satisfactorily. In particular, the dynamic model predicts a constant scale height of 29 km over an extensive altitude regime, in excellent agreement with the observed value.

097.049 Water on Mars. A. Young.
Spaceflight, Vol. 11, 441 (1969).

097.050 Mariners 6 and 7: Radio occultation measurements of the atmosphere of Mars.
A. Kliore, G. Fjeldbo, B. L. Seidel, S. I. Rasool.
Science, Vol. 166, 1393 - 1397 (1969).

Radio occultation measurements with Mariners 6 and 7 provided refractivity data in the atmosphere of Mars at four points above its surface. For an atmosphere consisting predominantly of carbon dioxide, surface pressures between 6 and 7 millibars are obtained at three of the points of measurement, and 3.8 at the fourth, indicating an elevation of 5 to 6 kilometers. The temperature profile measured by Mariner 6 near the equator in the daytime indicates temperatures in the stratosphere about 100°K warmer than those predicted by theory.

097.051 Radar observations of Mars.
A. Zachs, A. K. Fung.
Space Sci. Rev. Vol. 10, 442 - 454 (1969).

In view of the increasing interest in Mars and the practicability of missions to Mars this paper uses the published data to evaluate the angular behavior of the radar backscattering characteristics of Mars; a required information for the design of radar equipment of spacecrafts. In addition, results of past observations are summarized, analyzed and discussed in terms of a general interpretation of the Martian surface.

097.052 New multicolor filter photometry of Mars.
R. I. Mitchell.
Commun. Lunar Planet. Lab. Vol. 6, (No. 104), 289 - 294 (1969).

New photometry of Mars for the wavelength region 0.33 - 5.0μ with broad and medium-narrow filters is presented. The observations show a variation of 10 percent in whole-disk reflectivity in the 1μ region. The reflectivity of Mars is found to decrease between 2.2 and 3.5μ. The integrated albedo is found to be 0.31 ± .03 pe. The apparent 5μ brightness temperature is 243°K for the 1967 opposition period.

097.053 Area scanning of Mars during 1969.
P. B. Boyce.
Bull. American Astron. Soc., Vol. 1, 336 (1969). – Abstract AAS.

097.054 The significance of Mie scattering in the Martian atmosphere.
W. G. Egan, K. M. Foreman, J. Mead.
Bull. American Astron. Soc., Vol. 1, 340 (1969). – Abstract AAS.

097.055 Photometry of Mars near the time of opposition.
R. E. Murphy.
Bull. American Astron. Soc., Vol. 1, 355 (1969). – Abstract AAS.

097.056 A solution for the mass and dynamical oblateness of Mars using Mariner-IV Doppler data.
G. W. Null.

Bull. American Astron. Soc., Vol. 1, 356 (1969). — Abstract AAS.

097.057 On the phase curve of Mars and the brightness distribution on its disk. N. Barabashev.
Astron. Zhurn. Akad. Nauk SSSR, Vol. 46, 1247 - 1251 (1969). In Russian. English translation in Soviet Astron. AJ, Vol. 13, No. 6.
The method of determination of the smoothing factor of the planet from the curve of integrated brightness is given. The necessity of visual observations of Mars, simultaneous with an integral photometry, is indicated for calculating the state of the atmosphere and surface of the planet.

097.058 Some speculations on adsorption and desorption of CO_2 in Martian bright areas. B. W. Davis.
Icarus, Vol. 11, 155 - 158 (1969).
Conjectures are made on the importance of CO_2 adsorption in Martian bright areas. Reasonable assumptions relative to temperature, the partial pressure of CO_2, and the adsorptive properties of Martian dust permit computation of 4.4×10^{-6} moles/cc of bulk as a conservative estimate for the amount desorbed per unit bulk volume due to solar warming of bright area dust.

097.059 First pictures from Mariner VI and VII.
Icarus, Vol. 11, 225 - 259 (1969).
The Mariner VI and VII photographs of Mars which were released in mid–August 1969 by NASA's Office of Public Information are presented here.

097.060 Multispectral imaging of Mars from a lander. V. Klemas, E. L. G. Odell, F. Huck.
Journ. Optical Soc. America, Vol. 59, 1519 (1969). Abstract Meeting Optical Soc. America.

097.061 The lower ionosphere of Mars. A. C. Aikin.
Report NASA–TM–X–63237, Goddard Space Flight Center, Greenbelt, Md., 29 pp. (1968). — See Phys. Abstr. Vol. 72, No. 20224 (1969).

097.062 Seasonal and latitudinal variations of the average surface temperature and vertical temperature profile on Mars. G. Ohring, J. Mariano.
Journ. Atmosph. Sci. Vol. 25, 673 - 681 (1968). — See Phys. Abstr. Vol. 72, No. 24861 (1969).

097.063 The ionosphere and upper atmosphere of Mars. D. M. Hunten.
Contr. Kitt Peak National Obs. No. 242, [Reprinted from *"Atmospheres of Venus and Mars"*, Gordon and Breach, 1968, p. 147 - 180], 34 pp. (1968).

097.064 The coordinates of Deimos during the Mars opposition 1967. D. K. Karimova, P. N. Kholopov.
Astron. Tsirk. No. 530, p. 5 - 7 (1969). In Russian.

097.065 On the NO_2 band (6200 Å) in the Martian atmosphere. N. B. Ibragimov.
Astron. Tsirk. No. 533, p. 3 - 4 (1969). In Russian.

097.066 Ein neues Dunkelgebiet entsteht auf dem Mars. H. Heuseler.
SuW, Vol. 8, 240 - 243 (1969).

097.067 The parhelic halo on Mars. Brightness of the parhelic halo and quantity of ice crystals in the Martian atmosphere. V. D. Davydov.
Dokl. Akad. Nauk SSSR, Ser. Mat. Fiz., Vol. 189, 70 - 73 (1969). In Russian.
Occasionally observed bright spots of more or less short

duration on Mars may be halo phenomena. The brightness of a parhelic halo on the side faces of hexagonal plates with vertical main axes for the sun near horizon has been calculated for two cases: without and with diffraction on the crystals. The condition of visibility of the halo from the earth leads to minimum masses of the crystals in the line of sight 2.0×10^{-3} and 6.7×10^{-4} g/cm² for the two cases mentioned. These values make the halo explanation quite probable if a H_2O content of $10 - 20\mu$ in the Martian atmosphere is supposed (R.A. Schorn et al., 1967). Brighter halo forms than the parhelic halo give even more favourable conditions of visibility.
B. Onderlička

097.068 Mariner 6 und 7 – Betrachtungen ausgewählter Photographien.
H. Heuseler, E. Mädlow.
SuW, Vol. 8, 267 - 272 (1969).

097.069 The high level blue layer and blue clearing in the atmosphere of the planet Mars. B. S. Adcock.
Journ. Astron. Soc. Victoria, Vol. 22, 62 - 65 (1969).

097.070 Geology of Mars, and probes Mariner VI and VII. R. Valach.
Vesmír, Vol. 48, 371 - 373 (1969). In Czech.

097.071 Exospheric temperatures on Mars and Venus. J. S. Hogan, R. W. Stewart.
Journ. Atmosph. Sci., Vol. 26, 332 - 333 (1969).

097.072 Empirical determination of heating efficiencies in the Mars and Venus atmospheres. R. W. Stewart, J. S. Hogan.
Journ. Atmosph. Sci., Vol. 26, 330 - 332 (1969).

097.073 The 1969 Mariner mission to Mars. C. E. Kohlhase, H. W. Norris, J. A. Stallkamp, H. M. Schurmeier.
Astronaut. Aeronaut. U. S. A., Vol. 7, No. 7, p. 80 - 96 (1969).

097.074 Integral spectrophotometry of Mars. T. A. Lupishko.
Vestn. Khar'kov. Univ. No. 34, (Ser. Astron. No. 4), p. 82 - 90 (1969). In Russian.

Mariners 6 and 7 answer some questions, and raise some others, in trip around Mars.
IEEE Spectrum, Vol. 6, No. 10, p. 16 - 17 (1969).

Mars still mysterious.
Nature, Vol. 223, 561 - 562 (1969). — News notes.

Keep feet off Mars.
Nature, Vol. 223, 657 - 658 (1969). — News notes.

Moon and Mars.
Nature, Vol. 223, 1197 (1969).

Planets: Contour map of Mars.
Nature, Vol. 224, 843 (1969). — News notes.

The World of Mars.
See Abstr. 003.071.

Positional observations of Mars by means of local desensitization of plates. See Abstr. 041.012.

Photogrammetry with surface-based images.
See Abstr. 091.015.

A review of the environment of Mercury, Venus, and Mars.   See Abstr. 092.004.

The ionospheres of Venus and Mars. I. Mariner IV and preliminary ideas.   See Abstr. 093.014.

Absorption of radio waves by water vapour in the atmospheres of Venus and Mars.   See Abstr. 093.019.

Lunar and planetary mass concentrations. See Abstr. 094.015.

Remote sensing of sulfur-bearing compounds on the moon and Mars as defluidization centers. See Abstr. 094.113.

098 Minor Planets

098.001 **Resonant asteroids in the Kirkwood gaps and statistical explanations of the gaps.** F. Schweizer.
Astron. Journ. Vol. 74, 779 - 788 (1969).

If many of the asteroids nearly commensurable with Jupiter were to regularly cross the Kirkwood gaps, these could be explained as statistically underpopulated regions in the distribution of osculating mean motions n_0. In that case no gaps should appear in the distribution of time-averaged mean motions $\bar{n}$, the average being taken over a long-period perturbation. The $\bar{n}$ distribution of 185 asteroids around the Hecuba gaps shows a similar gap shape as the n_0 distribution and the statistical explanation in this form must be abandoned, but seems to be correct in Brown's modified form. Three asteroids recognized by Schubart to be resonant in the Hecuba gap are confirmed, and (887) Alinda is found to be resonant in the Hestia gap. Asteroids forming 13 pairs and one triplet recognizable by very similar $\bar{n}$, mean eccentricities, and inclinations, are suggested to be fragments of larger asteroids.

098.002 **Minor planets. III. Lightcurves of a Trojan asteroid.** J. L. Dunlap, T. Gehrels.
Astron. Journ. Vol. 74, 796 - 803 (1969).

Lightcurves for 624 Hektor were obtained with various telescopes. Hektor is strongly elongated and its rotational axis is close to the plane of the ecliptic. The shape probably is that of a cylinder with rounded ends, the total length being about 110 km and the diameter 40 km. The reflectivity and the UBV colors are uniform over the surface. The rotation period has been determined.

098.003 **Growth of asteroids and planetesimals by accretion.** W. K. Hartmann.
Commun. Lunar Planet. Lab. Vol. 8 (No. 137), 81 - 85 (1969). — Reprinted from Astrophys. Journ. Vol. 152, 337 - 342 (1968).

098.004 **Statistical arguments for asteroidal jet streams.** L. Danielsson.
Astrophys. Space Sci. Vol. 5, 53 - 58 (1969).

The distribution of asteroid orbits in space is studied and a detailed investigation for Hirayama's Flora family is presented. Some tendencies to (5-dimensional) clustering are found by means of comparing actual with random distribution simulated on a computer.

098.005 **Große Annäherung des Planetoiden (1620) Geographos an die Erde im August/September 1969.** R. A. Naef.
Orion Schaffhausen, Vol. 14, 107 - 108 (1969).

098.006 **Determination of the position of the minor planet Icarus from photographic observations in Alma-Ata.** N. K. Sumzina.
Astron. Tsirk. No. 510, p. 3 - 4 (1969). In Russian.

098.007 **Erratum: The minor planet (1221) Amor.** [Astron. Astrophys. Vol. 2, 173 - 181 (1969)]. J. Schubart.
Astron. Astrophys. Vol. 3, 256 (1969). — See Astron. Astrophys. Abstr. Vol. 1, 098.017.

098.008 **Photographic observations of minor planets.** J. A. Bruwer, B. M. F. Armstrong.
Republic Obs. Johannesburg, Circ. Vol. 7 (No. 128), 174 - 176 (1969).

Observations at the Republic Observatory Annexe, Hartbeespoort, with the Franklin-Adams Star Camera.

098.009 **Radar observations of Icarus.** R. M. Goldstein.
Icarus, Vol. 10, 430 - 431 (1969).

Recent radar and optical observations of Icarus indicate that the central latitudes have a smooth surface, whereas the higher latitudes are considerably rougher. According to the data, the radius of Icarus is $\geqslant 490$ m, and the radar reflectivity $\leqslant 0.13$.

098.010 **Radar observations of Icarus.** G. H. Pettengill, I. I. Shapiro, M. E. Ash, R. P. Ingalls, L. P. Rainville, W. B. Smith, M. L. Stone.
Icarus, Vol. 10, 432 - 435 (1969).

The minor planet Icarus was observed with the 3.8-cm Haystack radar from 12 to 15 June 1968. Weak but significant echoes were detected indicating a radar cross section of about 0.1 km², a radius of about 1 km, and an effective reflectivity of about 0.05. These results have an uncertainty of about a factor of 2.

098.011 **Photometric determination of the rotation period of 1566 Icarus.** E. Miner, J. Young.
Icarus, Vol. 10, 436 - 440 (1969).

Measurements made of Icarus on the nights of 1968 June 19 UT and June 20 UT indicate a rotation period of $2^h 16^m1 \pm 0^m3$ (estimated error). The amplitude of the brightness variation amounted to 0^m056. On a previous night (June 16 UT), UBV colors were found to be $U - B = 0^m45 \pm 0^m05$ (RMS) and $B - V = 0^m78 \pm 0^m01$ (RMS). All light-curve data were obtained with a pulse-counting photometer mounted at the Cassegrain focus of the 60-cm telescope at JPL's Table Mountain Observatory.

098.012 **Observations of Icarus: 1968.** J. Veverka, W. Liller.
Icarus, Vol. 10, 441 - 444 (1969).

Icarus is a nearly spherical object that rotates once every 2.25 (± 0.05) hr. From the degree of polarization of the reflected light as a function of phase angle, we estimate that the reflectivity of the surface (in the visual) does not exceed 0.20. Combining this estimate with the inferred absolute magnitude $g_\nu = 17.05$, we obtain a lower limit on the radius of 750 m.

098.013 **Angular momentum of Icarus.** W. K. Hartmann.
Icarus, Vol. 10, 445 - 446 (1969).

Inclusion of Icarus in an angular momentum density–mass diagram supports a 1967 prediction that asteroidal bodies smaller than about 10^{17} gm will depart from the planetary rotation law described earlier. Icarus' asteroidal nature and origin by fragmentation is thereby suggested.

098.014 **Observed positions of minor planets: Atami (1139), Holmia (378), Bavaria (301), and 1931 TT₁ (1624).** H. K. Herglotz, J. W. McCarter.
Astron. Journ. Vol. 74, 1052 (1969).

Four minor planets were photographed when near their opposition during the months of November 1967 through April 1968. Positions were determined relative to coordinates of identified stars from the SAO Star Catalog, with a precision of one second of arc or better.

098.015 **Ephemerides of minor planets for 1970.** Editor: Institut Teoreticheskoj Astronomii, Akademii Nauk SSR, under the editorship of S. G. Makover. Izdatel'stvo "Nauka", Leningrad. 163 pp. Price 1 Rbl. 81 Kop. (1969). In Russian and English.

Contents: Introduction, p. 3 - 8; Information on new elements, p. 9 - 11; Elements, p. 12 - 43; Opposition dates, p. 44 - 51; Ephemerides, p. 52 - 148; Ephemeris of Icarus, p. 149; Ephemerides of bright planets, p. 150 - 161; Status of observations of minor planets, p. 162.

098.016 Asteroid families and "jet streams".
J. R. Arnold.
Astron. Journ., Vol. 74, 1235 - 1242 (1969).

The asteroid families of Hirayama and Brouwer have been reexamined. The original families 1 - 9 are confirmed, while most of the later families are extensively revised (and renumbered). Eleven entirely new families are reported. The jet stream Flora A described by Alfvén is confirmed and extended, and several new streams are added. The search for regularities and resonances in the mean family elements has been unsuccessful. The origin of families by collisional breakup appears best to account for the data.

098.017 A photometric research on the minor planet
12 Victoria. P. Tempesti, R. Burchi.
Mem. Soc. Astron. Italiana, Nuova Serie, Vol. 40, 415 - 432 (1969).

Photoelectric observations in V and B light of the minor planet Victoria carried out during the 1068 opposition show a regular light variation with amplitude up to $0.^m 33$, period $0.^d 36060$ and double maximum. The absolute magnitude is $V (1,0) = 7.^m 55$ and the phase coefficient $\Phi = 0.030$ mag/degree. Finally a mean radius of 45 km has been derived by these photometric parameters.

098.018 Orbite dei pianetini Victoria (12), Egeria (13)
ed Urania (30). A. Kranjc.
Mem. Soc. Astron. Italiana, Nuova Serie, Vol. 40, 589 - 592 (1969). – Letter.

098.019 Posiciones exactas de pequeños planetas obtenidas
en el Observatorio Astronomico de Madrid.
R. Carrasco.
Bol. Astron. Obs. Madrid, Vol. 7, No. 3, p. 28 - 31 (1969).

098.020 Secular motion of resonant asteroids.
G. E. O. Giacaglia.
Report NASA–CR–96742, Smithsonian Astrophys. Obs., Cambridge, Mass., 78 pp. (1968). – See Phys. Abstr. Vol. 72, No. 24850 (1969).

098.021 Observations photographiques de petites planètes
et de Pluton, effectuées à l'astrographe double
de 40 cm au cours de l'année 1967.
S. Arend, H. Debehogne, G. Roland.
Bull. Astron. Obs. Roy. Belgique, Vol. 6, (No. 8), 322 - 327 (1969).

098.022 Observations photographiques de petites planètes,
effectuées au triplet Zeiss de 30 cm au cours du
1er semestre de l'année 1967. J. Denoyelle.
Bull. Astron. Obs. Roy. Belgique, Vol. 6, (No. 8), 328 (1969).

098.023 Some indications concerning the quality of the
observations of minor planets carried out at the
Bucharest Observatory. C. Cristescu, V. I. Vlăsceanu, I. Ghețu, G. Bocșa.
Stud. Cerc. Astron., Vol. 14, 123 - 171 (1969).

Photographic positions of the following minor planets observed from 1954 to 1965 are reported: 1 Ceres, 2 Pallas, 3 Juno, 4 Vesta, 6 Hebe, 7 Iris, 11 Parthenope, 18 Melpomene, 39 Laetitia, 40 Harmonia.

098.024 (1566) Icarus.
Stud. Cerc. Astron., Vol. 14, 197 (1969). – Note.

098.025 The close approach of minor planet 1620 Geographos in 1969. R. G. Hodgson.
Strolling Astronomer, Vol. 21, 145 - 146 (1969).

098.026 About the observation of minor planets.
J. Židů.
Říše hvězd, Vol. 50, 165 - 170 (1969). In Czech.

098.027 (1620) Geographos.
R. J. Reichert, P. C. Tiffany, S. Herrick.
British Astron. Ass. Circ. No. 510 (1969).

098.028 (1620) Geographos. T. Seki, H. Debehogne.
IAU Circ. No. 2166 (1969).

098.029 (1620) Geographos.
B. Milet, T. Seki, H. Debehogne.
IAU Circ. No. 2168 (1969).

098.030 (1620) Geographos.
S. Herrick, P. A. Thompson, P. C. Tiffany.
IAU Circ. No. 2171 (1969).

098.031 (1620) Geographos.
B. Mintz, H. Debehogne, B. Milet, P. Wild.
IAU Circ. 2172 (1969).

098.032 (887) Alinda.
B. Mintz, P. Wild.
IAU Circ. No. 2173 (1969).

098.033 Minor planets. B. Harris, M. P. Candy, D. Gans.
IAU Circ. No. 2175 (1969).

098.034 (1620) Geographos. G. Soulié, L. Pourteau.
IAU Circ. No. 2193 (1969).

098.035 (1620) Geographos.
S. Herrick.
Yamamoto Circ. No. 1699 (1969). In Japanese.

098.036 Geographos (1620).
Yamamoto Circ. No. 1704 (1969). In Japanese.

Radarwaarnemingen van Icarus.
Hemel en Dampkring, Vol. 67, 250 (1969). – Short report.

Icaro.
Revista Astron. Vol. 40, No. 166, p. 25 (1968).

Bulletin of the Astronomical Institutes of the
Netherlands 1921 – 1969. Index of minor planets.
See Abstr. 002.015.

The mass of Jupiter and the motion of four minor
planets. See Abstr. 099.003.

Sur les masses et les densités de Io et Europe, les
lacunes joviennes et l'origine des comètes.
See Abstr. 099.011.

On the relationship between comets and minor
planets. See Abstr. 102.039.

099 Jupiter

099.001 Determination of the mass of Jupiter using the motion of its ninth satellite. A. Bec.
Astron. Astrophys. Vol. 2, 381 - 387 (1969). In French.

The equations of motion of the ninth satellite are integrated numerically, the accuracy being chosen in such a way that the differences between the computed positions and the observed ones may be attributed to a lack of precision on the initial values used to start the integration, namely the six elements and the mass of Jupiter. The integration was carried out according to Cowell's method. To improve the initial conditions and the mass of Jupiter we have used the "method of variations": seven numerical integrations were necessary to obtain the variations of the computed positions when a variation of one unit is given to each of the seven initial values (elements and mass). The equations of condition system was solved by the method of least squares. The following result was found for the mass of Jupiter: $1/m = 1047.386 \pm 0.041$. This result was compared with the latest results obtained. Moreover the list of the observations of the ninth satellite used in this paper was given.

099.002 A report on observations of Jupiter at frequencies between 3 and 8 mc/s.
F. R. Zabriskie, W. A. Solomon.
Pennsylvania State Univ. Coll. Sci. Dep. Astron. Radio Astron. Obs. Sci. Rep. No. 016, 62 pp. (1968).

Description of equipment, procedure used to analyse the records and catalogue of storms recorded.—GD

099.003 The mass of Jupiter and the motion of four minor planets. W. J. Klepczynski.
Astron. Journ. Vol. 74, 774 - 775 (1969).

An analysis of the observations of four minor planets suggested by G. W. Hill as being especially suitable for the determination of the mass of Jupiter yields the corresponding reciprocal masses with their mean error: 10 Hygiea, 1047.351 ± 0.006; 24 Themis, 1047.359 ± 0.010; 31 Euphrosyne, 1047.372 ± 0.006; 52 Europa, 1047.337 ± 0.027. A simultaneous solution for the mass of Jupiter utilizing the combined observations of all four minor planets yields the value 1047.360 ± 0.004.

099.004 Faraday effect on Jupiter's radio bursts.
G. D. Parker, G. A. Dulk, J. W. Warwick.
Astrophys. Journ. Vol. 157, 439 - 448 (1969).

From Faraday fringes present on swept-frequency records of Jupiter's decametric radio bursts the orientation of the polarization ellipse at the source can be determined. At Jupiter the ellipse has an orientation independent of frequency over the range 15−35 MHz, and its major axis has an average position angle of −25° ± 15° from the magnetic dipole axis for early source emissions. Apparently radiation is generated in one elliptically polarized magnetoionic mode. This ellipse undergoes no observable Faraday rotation in passing through the Jovian magnetosphere. The inferred upper limit to Jupiter's magnetospheric electron density is 10 cm^{-3}.

099.005 An estimate of the abundance and the rotational temperature of CH$_4$ on Jupiter.
M. J. S. Belton.
Astrophys. Journ. Vol. 157, 469 - 472 = Contr. Kitt Peak National Obs. No. 404 (1969).

An analysis of the $3\nu_3$ band of CH$_4$ yields a rotational temperature of 163° ± 20° K on Jupiter. The observed distribution of absorption in the lines $R(1)−R(6)$ is in good agreement with prediction. The abundance of CH$_4$ is found to be 30 m-atm. The effects of line saturation are taken into account in a simplified way, and both the reflecting layer and scattering models for line formation are discussed.

099.006 Observations of localized 5-micron radiation from Jupiter. J. A. Westphal.
Astrophys. Journ. (Letters), Vol. 157, L63 - L64 (1969).

Localized thermal emission from Jupiter has been observed at 5μ. During April and May 1969 peak brightness temperatures of at least 310° K were associated with narrow elongate regions between approximately 5° and 20°N. latitude.

099.007 Jupiter's decametric rotation period.
F. F. Donivan, T. D. Carr.
Astrophys. Journ. (Letters), Vol. 157, L65 - L68 (1969).

The mean value of Jupiter's rotation period over an assumed 11.86-year drift cycle was found to be $9^h 55^m 29\overset{s}{.}73 \pm 0\overset{s}{.}04$. The position of source A in a central-meridian-longitude system based on the new rotation period displayed a correlation with the Jovicentric declination of the earth distinctly higher than its anticorrelation with the sunspot number, substantiating an earlier suggestion that the observed drift is a geometrical effect due to Jupiter's changing aspect in its orbit around the sun. A model consisting of a single curved emission sheet fixed to the magnetic field can be made to account for the observed drifts of sources A and B.

099.008 The internal powers and effective temperatures of Jupiter and Saturn.
H. H. Aumann, C. M. Gillespie, Jr., F. J. Low.
Astrophys. Journ. (Letters), Vol. 157, L69 - L72 (1969).

The total power emitted by Jupiter and Saturn has been measured by observing the planets from a jet aircraft at 15-km altitude with a telescope system open from 1.5 to 350μ. The two planets were found to radiate 2.7 and 2.4 times the amount of power they receive from the sun, respectively. These new results put observational restraints on models for the internal structures and atmospheres of the two planets.

099.009 A 3-month oscillation in the longitude of Jupiter's Red Spot. H. G. Solberg, Jr.
Planet. Space Sci. Vol. 17, 1573 - 1580 (1969).

Photographic measurements of Jupiter's Red Spot show that it oscillated semi-regularly between 1963 and 1968; the oscillation had a mean period of 90.0 days and a mean amplitude of 0°.8. No comparable oscillation has ever been observed in the Jovian atmosphere.

099.010 Neuere Beobachtungen des Planeten Jupiter.
W. Grüner.
VdS Nachrichtenblatt, Vol. 18, 127 - 128 (1969).

099.011 Sur les masses et les densités de Io et Europe, les lacunes joviennnes et l'origine des comètes.
A. Dauvillier.
Comptes Rendus, Acad. Sci. Paris, Sér. B, Vol. 269, 478 - 480 (1969).

Les masses et les densités anormales de Io et Europe montrent que ces satellites ont intercepté la majeure partie des noyaux cométaires captés par Jupiter. La discussion numérique permet d'estimer la masse de matière soustraite aux Anneaux d'astéroïdes par Jupiter et apparaissant comme lacunes joviennes.

099.012 Io. G. Comello, H. Pietersma.
Hemel en Dampkring, Vol. 67, 254 (1969).

099.013 The determination of Jupiter's mass correction by

observations of 10 Hygiea from 1932 - 1967.
N. S. Chernykh.
Astron. Tsirk. No. 526, p. 5 - 6 (1969).   In Russian.

099.014   The 2.8 – 14-micron spectrum of Jupiter.
F. C. Gillett, F. J. Low, W. A. Stein.
Astrophys. Journ. Vol. 157, 925 - 934 (1969).

Observations of Jupiter have been made from 2.8 to 14 μ with $\Delta\lambda/\lambda \approx 0.02$. The most pronounced features of this spectrum are: (1) strong absorption in the range 3 – 4 μ, (2) high brightness temperature in the range 4.6 – 5.1 μ, and (3) a brightness temperature that is relatively constant in the range 9 – 12 μ. A preliminary analysis indicates the following: (1) Most of the features of the 2.8 – 14-μ spectrum of Jupiter can be accounted for on the basis of absorption by NH_3, CH_4, and H_2. Upper limits to abundances of minor constituents are consistent with completely reduced solar abundance, except for SiH_4. (2) The temperature decreases with height to a minimum ($T_{min} \leq 115°$ K), increasing again above this level to $T \geq 145°$ K. A model is suggested to maintain this temperature profile. (3) Ammonia is probably saturated around the 125° K level, and the abundance of H_2 above the 125° K level is $\omega \approx 12$ km atm.

099.015   Studies of methane absorption in the Jovian atmosphere. I. Rotational temperature from the $3\nu_3$ band.
J. S. Margolis, K. Fox.
Astrophys. Journ. Vol. 157, 935 - 943 (1969).

The rotational temperature of methane in the Jovian atmosphere has been deduced from Walker and Hayes's equivalent widths of lines in the R-branch of the $3\nu_3$ band at 9050 cm^{-1}. Line positions and relative intensities from high-resolution laboratory spectra were used to calculate curves of growth for the J-manifolds. The fine-structure components were assumed to have Lorentz shape, corresponding to a simple reflecting-layer model, with half-width, γ, ranging from 0.03 to 0.10 cm^{-1}. Saturation was found to be important in obtaining correct line strengths. The analyses of the data were satisfactory, but the fact that the computed temperature was sensitive to the value of γ indicates the need for a measurement of the line width in this spectral region.

099.016   Présentation de la planète Jupiter 1967 - 1968.
J. Dragesco.
L'Astronomie, 83e année, 311 - 324 (1969).

099.017   The spectra of Jupiter and Saturn in the photographic infrared.   T. Owen.
Icarus, Vol. 10, 355 - 364 = Contr. Kitt Peak National Obs. No. 433 = Contr. McDonald Obs. No. 445 (1969).

Recent studies of the near-infrared spectra of Jupiter and Saturn have led to new evaluations of the relative abundances of atmospheric constituents and some indication of the distribution of these constituents with altitude. A preliminary model for the atmospheric structure of Jupiter that accommodates observations made at a variety of wavelengths is discussed. The model suggests that UV observations only reach a level near the tropopause, while in the near-infrared the lower bound of the observable atmosphere is set by a dense cloud layer whose upper boundary has a temperature of 225°K. Between this level and the tropopause there is a region of ammonia crystal clouds.

099.018   The clouds of Jupiter and the $NH_3 - H_2O$ and $NH_3 - H_2S$ systems.   J. S. Lewis.
Icarus, Vol. 10, 365 - 378 (1969).

The structure and composition of clouds made up of ammonia- and water-bearing condensates are investigated for two principal compositional models for the atmosphere of Jupiter.

099.019   Far-ultraviolet spectroscopy of Jupiter.
E. B. Jenkins.
Icarus, Vol. 10, 379 - 385 (1969).

The spectrum of Jupiter from 2400 to 3000 Å was photographed at a resolution of 1 Å by a rocket-borne spectrograph on June 9, 1967. The derived reflectivity revealed a definite absence of any particularly strong features longward of 2650 Å. A slight reduction of reflectivity was registered at 2600 Å, but the match with a puzzling absorption feature seen in earlier, low-resolution observations was not very precise. This drop, whose existence may be subject to doubt, appeared to be caused by a broad, diffuse absorption rather than a profusion of narrow, strong features.

099.020   Organic synthesis in a simulated Jovian atmosphere.
F. Woeller, C. Ponnamperuma.
Icarus, Vol. 10, 386 - 392 (1969).

Reactions which may occur in the Jovian atmosphere were simulated by passing an electrical discharge through a mixture of methane and ammonia. Analysis of the volatile fraction revealed the presence of several precursors of biologically important compounds. The nonvolatile fraction consisted of an orange-red polymer. This result may provide an explanation for the appearance of the red spot on the planet Jupiter.

099.021   Observability of spectroscopically active compounds in the atmosphere of Jupiter.   J. S. Lewis.
Icarus, Vol. 10, 393 - 409 (1969).

The abundances of several hundred volatile compounds have been calculated at several different levels in the atmosphere of Jupiter. Complete chemical equilibrium has been assumed, and a solar-composition, adiabatic-equilibrium model of the atmospheric composition and structure is used throughout. The principal results are that only H_2, CH_4, and NH_3 are predicted to be observable with present techniques. A simple solar-composition adiabatic model for the atmosphere of Jupiter is found to be completely compatible with present observational evidence. It is not, however, possible to conclude that this model is correct: a subadiabatic model or one in which the H_2 abundance is varied by as much as a factor of 10 would agree with the observations equally well, except of course with regard to the H : C ratio. The general conclusions of this paper apply without substantial modification to Saturn, but not necessarily to Uranus or Neptune, since the latter cannot have solar composition.

099.022   The transparency of the Jovian polar zones.
T. Gehrels.
Icarus, Vol. 10, 410 - 411 (1969).

Polarimetric and photometric observations indicate the presence of a thin molecular atmosphere – above the cloud-top – between Jovian latitudes +45° and –45°, and a thick one at the polar zones. The greater optical transparency and the possibility of a subadiabatic condition at the poles are discussed. A marked difference in molecular optical depth between North and South pole was found in 1960 and 1963. Both poles show evidence of thin clouds, presumably cirrus high above the ammonia-crystal cloudtop.

099.023   Jupiter's Red Spot in 1967 - 68.
H. G. Solberg, Jr.
Icarus, Vol. 10, 412 - 416 (1969).

Photographic observations of Jupiter's Red Spot between 12 September 1967 and 17 July 1968 are reported. As the activity in the south component of the South Equatorial Belt (SEBs) decreased, the Red Spot became larger and darker, finally regaining the prominence it had during early 1966.

099.024   The magnetosphere of Jupiter.

T. D. Carr, S. Gulkis.
Annual Rev. Astron. Astrophys. Vol. 7, 577 - 618 (1969).

099.025 Photographic photometry of Jupiter's belts and zones during 1967-68 apparition.
U. dall'Olmo.
Mem. Soc. Astron. Italiana, Nuova Serie, Vol. 40, 341 - 343 (1969).

The results of photometric measures of Jupiter's belts and zones, made on 13 photographs taken during the 1967-68 apparition at the newtonian focus of a 12″ reflector, are reported in this paper.

099.026 Visibility of Jupiter's Red Spot. E. Argyle.
Nature, Vol. 223, 485 (1969).

A recent attempt to establish a relationship between the visibility of Jupiter's Red Spot and solar activity violates the rule that the method of analysis should, wherever possible, be decided upon before the data have been examined for the features sought in the analysis.

099.027 Hα auroral activity on Jupiter.
J. H. Hunter.
Nature, Vol. 223, 388 - 389 (1969).

Observations have been made of what may be Hα auroral activity from the planet Jupiter. The experiment was carried-out at the U. S. Naval Research Laboratory using a 16-inch telescope equipped with an image orthicon camera with an S-20 photocathode. The Hα feature was isolated by placing in the optical path various combinations of Fabry-Pérot etalons centered upon Hα. Further observations are planned during the next several Jovian oppositions.

099.028 Io-related polarization characteristics of the Jovian decameter emission. T. C. Green, W. M. Sherrill.
Astrophys. Journ. Vol. 158, 351 - 363 (1969).

Measurements of the polarization of the decameter radiation from Jupiter at 15 – 24 MHz were made during the 1966 - 1967 apparition and combined with data from the 1963 - 1964 apparitions for analysis. Polarization data analyzed in terms of Io departure angle from superior geocentric conjunction showed axial-ratio profiles consistent with the Dulk model of the Jovian source and also with previously observed polarization characteristics reported as a function of system III longitude. Comparison of the Southwest Research Institute (SwRI) data with cyclotron-model predictions and the recently reported data from Florida State University were made.

099.029 A new examination of the nonthermal microwave spectrum of Jupiter. G. L. Berge.
Bull. American Astron. Soc. Vol. 1, 233 - 234 (1969). – Abstr. AAS.

099.030 The microwave spectrum of Jupiter.
J. R. Dickel, J. J. Degioanni, G. C. Goodman.
Bull. American Astron. Soc. Vol. 1, 240 (1969). – Abstr. AAS.

099.031 Narrow-band colorimetry of some of the satellites of Jupiter and Saturn.
T. B. McCord, T. V. Johnson.
Bull. American Astron. Soc. Vol. 1, 252 (1969). – Abstr. AAS.

099.032 Helligkeitsschwankung von Jupitermond IV.
J. Classen.
Sterne, 45. Jahrgang, 169 - 171 (1969).

099.033 Refutation of Bigg's evidence for new Jovian satellites. J. N. Douglas, F. A. Bozyan.
Astrophys. Letters, Vol. 4, 227 - 228 (1969).

The periodic terms near 20 hr 40 min in the Jupiter decameter radiation time series found by Bigg and attributed by him to the influence of hitherto undiscovered Jovian satellites are shown to be artificial.

099.034 Circularly polarized synchrotron radiation in dipolar magnetic fields with application to the planet Jupiter. L. J. Gleeson, M. P. C. Legg, K. C. Westfold.
Proc. Astron. Soc. Australia, Vol. 1, 274 - 276 (1969). – Contribution ASA meeting.

099.035 The influence of zonal harmonics on the motion of the fifth satellite of Jupiter.
V. N. Kiryushenkov.
Vestn. Mosk. un-ta. Fiz., Astron., No. 2, p. 80 - 87 (1969). In Russian. – Abstr. in Referativ. Zhurn. 51. Astron., 8.51.101 (1969).

099.036 The infrared spectrum of Jupiter, 0.95 – 1.60 microns, with laboratory calibrations.
D. P. Cruikshank, A. B. Binder.
Commun. Lunar Planet. Lab. Vol. 6, (No. 103), 275 - 288 (1969).

New spectra of Jupiter in the region 0.95 - 1.60 microns are presented with calibrations from laboratory studies of the bands of CH_4 and NH_3. Estimates of the quantities of these gases above the effective reflecting layer in the Jupiter atmosphere are given for each of several individual bands. The presence of C_2H_2, H_2S, and HCN is investigated with the aid of composite spectra of these gases in addition to CH_4 and NH_3. We find as upper limits 8 cm-atm for C_2H_2, 50 cm-atm for H_2S, and 10 cm-atm for HCN.

099.037 Rationale and procedures for a Jovian rank and colour report. P. K. Mackal.
Journ. British Astron. Ass., Vol. 80, 37 - 39 (1969).

099.038 Studies of methane absorption in the Jovian atmosphere. II. Abundance from the $3\nu_3$ band.
J. S. Margolis, K. Fox.
Astrophys. Journ., Vol. 158, 1183 - 1188 (1969).

ηN (η = air mass, N = abundance) for methane in the Jovian atmosphere has been deduced from Walker and Hayes's equivalent widths of lines in the R-branch of the $3\nu_3$ band at 1.1μ. Laboratory curves of growth were used to obtain the line intensity for each J-manifold. Saturation effects were found to be important, and ηN was determined to be between 86 and 185 m-atm, depending on the value used for the line half-width.

099.039 New circular-polarization measurements of Jupiter's decimeter radiation.
B. Gary, S. Gulkis.
Astrophys. Journ. (*Letters*), Vol. 158, L193 - L195 (1969).

Circular-polarization measurements of Jupiter at a wavelength of 13 cm were made during 1969 April and May with the 210-foot radio telescope at Goldstone, California. An upper limit to the net degree of circular polarization in the integrated emission is 1 percent over a limited longitude-range.

099.040 Determination of the mass of Jupiter from a study of the motion of 57 Mnemosyne.
A. D. Fiala.
Bull. American Astron. Soc., Vol. 1, 342 (1969). – Abstract AAS.

099.041 A search for an anomalous brightening of Io after eclipse. O. G. Franz, R. L. Millis, T. V. Pettauer.
Bull. American Astron. Soc., Vol. 1, 344 (1969). – Abstract AAS.

099.042 Jupiter circular polarization measurements.
B. Gary, S. Gulkis, M. Klein, B. Meredith.
Bull. American Astron. Soc., Vol. 1, 344 (1969). – Abstract AAS.

099.043 Motion of the Red Spot of Jupiter.
R. Smoluchowski.
Bull. American Astron. Soc., Vol. 1, 363 (1969). – Abstract AAS.

099.044 On some properties of the Jupiter atmosphere. III.
L. S. Galkin.
Izv. Krymskoj Astrofiz. Obs. Vol. 39, 232 - 235 (1969). In Russian.

The analysis of the change of the width of Jupiter spectra is carried out according to observations of 1964. Changes different from that observed earlier are discovered.

099.045 Satellite shadows on Jupiter as probes of its upper atmosphere. S. O. Kastner.
Icarus, Vol. 11, 208 - 211 (1969).

The photometric profiles of the shadows cast on Jupiter by its large satellites are determined to some extent by the physical properties of the Jovian upper atmosphere. Profiles of Io's shadow are constructed on the assumption of exponentially decreasing mean free path and different path lengths. Variations from the incident intensity distribution occur, which may be useful in the atmospheric analysis.

099.046 Commencement times and durations of Jupiter's radio noise storms.
C. N. Olsson, A. G. Smith, H. I. Register, J. May.
Icarus, Vol. 11, 212 - 217 (1969).

Analysis of 10 years of observations confirms Duncan's conclusion that the commencement times of Jupiter's decametric noise storms define a constant planetary rotation period, the value for which in the present investigation is $9^h 55^m 29^s37$. However, the study indicates that the apparent 12-year oscillation in the longitudes of the centers of the conventional sources does not result from a simple variation in storm length.

099.047 Vorticity, Rossby number, and geostrophy in the atmosphere of Jupiter. S. L. Hess.
Icarus, Vol. 11, 218 - 219 (1969).

Recent observations of spots moving around the Great Red Spot are used to calculate circulation, vorticity and a Rossby number at this scale of atmospheric motions. The results are quite comparable to values applying to large-scale atmospheric systems on earth and indicate that geostrophy obtains on Jupiter.

099.048 Calculated dose rates in Jupiter's van Allen belts.
J. W. Haffner.
AIAA Journ., Vol. 7, 2305 - 2311 (1969).

A series of particle flux and dose rate calculations have been carried out based upon the assumption that the decimetric rf noise emanating from the vicinity of Jupiter is synchrotron radiation from electrons trapped in a dipole magnetic field.

099.049 Circular polarization of Jupiter at 9.26 cm.
E. R. Seaquist.
Nature, Vol. 224, 1011 - 1012 (1969).

The observations were made during two observing periods in April and July 1969 with the 46meter paraboloid of the Algonquin Radio Observatory, Ontario, Canada. An analysis of the data shows that the degree of circular polarization varies with system III longitude of the central meridian. The result leads to the conclusion that there exist magnetic fields in the radiating regions with $B \gtrsim 0.04$ gauss, and that the longitude λ_0^{III} of the north magnetic pole of Jupiter is $204°4 \pm 12°9$.

099.050 A mechanism for Jupiter's equatorial acceleration.
P. J. Gierasch, P. H. Stone.
Journ. Atmosph. Sci. Vol. 25, 1169 - 1170 (1968). – See Phys. Abstr. Vol. 72, No. 21937 (1969).

099.051 The photochemistry of Jupiter above 1000 Å.
J. R. McNesby.
Journ. Atmosph. Sci. Vol. 26, 594 - 599 (1969).

099.052 A report on observations of Jupiter at frequencies between 3 and 8 mc/s.
F. R. Zabriskie, W. A. Solomon.
Pennsylvania State Univ., Coll. Sci. Dep. Astron. Radio. Astron. Obs., Sci. Rep. No. 016, 62 pp. (1968).

Description of equipment, procedure used to analyse the records and catalogue of storms recorded. – GD

099.053 Results of Jupiter observations at Pulkovo at 3.95 cm. N. S. Soboleva, G. M. Timofeeva.
Astron. Zhurn. Akad. Nauk SSSR, Vol. 46, 872 - 876 (1969). In Russian. English translation in Soviet Astron. AJ, Vol. 13, No. 4.

New radio observations of Jupiter at 3.95 cm with resolution of $1'45$ showed that the effective dimension of the radio emission region at this wavelength is $35'' \pm 3''$, and the integral percentage of polarization is $(3 \pm ^{1.5}_{0.5})\%$. A joint reduction of all Pulkovo radio observations of Jupiter showed that 1) the character of radio emission of radiation belts at these wavelengths sharply differs from their emission in decimetre wavelength region, 2) the brightness temperature of the planetary disc (after the emission of radiation belts having been taken into account) is considerably higher than in the infra-red (128°K) and the millimetre region of the spectrum (111°K) and continuously increases to the side of long wavelengths. The spectrum of the brightness temperature of the planetary disc obtained can be used for a more precise definition of the model of the planetary atmosphere.

099.054 Optical properties and the structure of Jupiter's atmosphere. II. The influence of the multiple scattering in a cloud layer on planetary absorption line profiles. V. G. Teifel.
Astron. Vestn. Vol. 3, 85 - 95 (1969). In Russian.

Calculations of the changes in the intensity and form of Lorentz absorption line profiles, appearing in planetary atmospheres with various degrees of multiple scattering in a cloud layer of a planet in the formation of these lines, are carried out on the supposition of scattering isotropy.

099.055 Jupiter from observations of 1966 – 1967.
N. K. Andrianov, L. A. Vishnjakova.
Astron. Vestn. Vol. 3, 117 - 119 (1969). In Russian.

The authors give the results of visual observations of position and intensity of the belts and polar caps of Jupiter.

099.056 Results of micrometer measurements of Jupiter's belts made during 1966 and 1968.
M. Migon, J. J. D. Nogueira.
Monthly Notes Astron. Soc. Southern Africa, Vol. 28, 72 - 75 (1969).

099.057 Ephemerides and residuals of the outer satellites of Jupiter. P. Herget.
Publ. Cincinnati Obs. No. 23, p. 1 - 56 (1968).
Ephemerides and residuals of satellites VIII to XII.

099.058 Chemical composition and internal constitution of Jupiter. E. L. Ruskol.

Kosmich. Issled. Vol. 7, 857 - 870 (1969). In Russian.

099.059 Perturbations of the motion of the fifth satellite of Jupiter by the Galilean satellites.
V. N. Kiryushenkov.
Vestn. Mosk. un-ta. Fiz., Astron., No. 4, p. 65 - 70 (1969).
In Russian. – Abstr. in Referativ. Zhurn. 51. Astron.,
1.51.138 (1970).

099.060 Reobservation of the tenth satellite of Jupiter.
M. L. Charnow.
AIAA Journ., Vol. 7, 1151 - 1152 (1969).

099.061 Die Bestimmung der Jupitermasse aus Kreismikrometer-Beobachtungen des IV. Satelliten.
L. Brandt.
SuW, Vol. 8, 244 - 245 (1969).

099.062 The continuous spectrum and molecular absorption in the Red Spot of Jupiter. V. G. Teifel.
Trudy Astrofiz. Inst. Alma-Ata, Vol. 13, 3 - 12 (1969).
In Russian.
Some spectrograms of the Red Spot and the South Tropical Zone were measured to study the continuous spectrum and the molecular absorption bands CH_4 6190 Å and NH_3 6475 Å. The continuous absorption in the Red Spot increases towards short wavelengths in the region λ 6100 - 4400 Å. Between 3300 and 4000 Å absorption is nearly constant and the relative spectral curve is approximately horizontal. The molecular absorption bands CH_4 and NH_3 in the spectrum of the Red Spot have the same intensity as those in the surrounding areas of the cloudy surface. The temperature of the Red Spot cannot differ from that of the South Tropical Zone by more than 2 - 3°K.

099.063 Investigation of the absorption in the NH_3 band λ 6450 Å in the atmosphere of Jupiter.
A. N. Aksenov.
Trudy Astrofiz. Inst. Alma-Ata, Vol. 13, 13 - 15 (1969).
In Russian.
Spectral observations of Jupiter were made at the Astrophysical Institute in 1965 - 1967. The absorption in the NH_3 band λ 6450 Å was investigated. It was found that the absorption in this band does not undergo remarkable variations in time.

099.064 Spectral and colour contrasts on the disk of Jupiter.
Z. N. Grigorjeva, N. V. Priboeva.
Trudy Astrofiz. Inst. Alma-Ata, Vol. 13, 16 - 21 (1969).
In Russian.
From 18 spectrograms obtained in 1965 with the 72-cm reflector (dispersion 30 Å/mm) the changes of spectral and colour contrasts of details on Jupiter's disk are investigated. The contrasts vary from 0.00 to 0.42.

099.065 Photometric investigations of the atmospheric activity of Jupiter in 1965 - 1966.
L. P. Sorokina.
Trudy Astrofiz. Inst. Alma-Ata, Vol. 13, 22 - 30 (1969).
In Russian.
The activity of Jupiter's atmosphere during 1965 - 1966 is investigated by calculating the photometric activity factor. The quantity of visible dark matter on the disk of the planet is determined by photometric methods. It is shown that the northern hemisphere of the planet is more active than the southern one.

099.066 Experiments on a photographic equidensitometry of Jupiter.
V. F. Kartashoff, V. G. Teifel, A. A. Usoltzeva.
Trudy Astrofiz. Inst. Alma-Ata, Vol. 13, 31 - 38 (1969).
In Russian.
Some experiments on a photographic equidensitometry of the moon and Jupiter are made. The method of determination of the planetary albedo using equidensitometry and measurements of the integral magnitude of the planet was applicated for the determination of the albedo of dark and bright matter of the Jovian cloudy layer in 1962.

099.067 Determination of the convective energy in Jupiter's atmosphere. A. N. Aksenov.
Trudy Astrofiz. Inst. Alma-Ata, Vol. 13, 39 - 42 (1969).
In Russian.
The determination of the convective energy of Jupiter's atmosphere is considered for different thicknesses of the atmosphere and under varying temperature gradients supposing a polytropic model of the planet. The existence of inner energy sources in Jupiter which condition the difference between the observed surface temperature of the cloudy layer and the equilibrium temperature is supposed.

099.068 Bewegung des GRF auf Jupiter. C. Kowalec.
VdS Nachrichtenblatt, 18. Jahrgang, 143 (1969).
Motion of the Great Red Spot.

099.069 Der Grosse Rote Fleck und die W.O.S.-Objekte auf Jupiter. C. Kowalec.
VdS Nachrichtenblatt, 18. Jahrgang, 153 - 154 (1969). –
Motion of the Great Red Spot.

099.070 Jupiter in 1964-65: Rotation periods.
P. W. Budine.
Strolling Astronomer, Vol. 21, 153 - 161 (1969).

099.071 Satellites of Jupiter. J. A. Buarque de Nazareth.
Contr. Obs. Valongo, Univ. Federal Rio de Janeiro,
Ser. II, Nos. 6 - 8 (1969).

099.072 Fotometria fotografica delle bande e delle zone di Giove nel corso delle opposizioni del decennio 1957 – 1967. U. Dall'Olmo.
Pubbl. Oss. Astron. Univ. Bologna, Vol. 9, No. 17, 14 pp. (1968).
Photometric measurements of Jupiter's belts and zones made on 47 photographs, obtained during 1957 – 1967 apparitions with a 12″ Newtonian reflector in Bologna are reported in this paper.

099.073 Détermination de la masse de Jupiter par l'étude du mouvement de son neuvième satellite.
A. Bec.
Thesis 3e Cycle, Spéc. Astron. fondam., Mention Mécanique céleste, Paris. 55 pp. (1968). – See Bull. Signal., Vol. 30, Section 120, No. 9569 (1969).

099.074 Changes in Jupiter's atmosphere. S. Cífka.
Říše hvězd, Vol. 50, 132 - 134 (1969). In Czech.

099.075 Observation of Jupiter during the years 1968 and 1969. M. Druckmüller, Z. Okáč.
Říše hvězd, Vol. 50, 139 - 140 (1969). In Czech.

099.076 Inertial Taylor columns and Jupiter's Great Red Spot. A. P. Ingersoll.
Journ. Atmosph. Sci., Vol. 26, 744 - 752 (1969).

099.077 Some peculiarities of the Jovian atmosphere.
M. F. Khodyachikh.
Vestn. Khar'kov. Univ. No. 34, (Ser. Astron. No. 4), p. 71 - 78 (1969). In Russian.

Ein neuer Jupitermond wird postuliert.

Umschau, Vol. 69, 661 (1969).

The Atmosphere of Jupiter.
See Abstr. 003.119.

The influence of Jupiter's oblateness on the motion of an artificial satellite of one of the Jupiter's Galilean satellites. See Abstr. 053.014.

Rocket spectra of Venus and Jupiter from 2000 to 3000 Å. See Abstr. 093.006.

Spectrophotometric studies of Mars, Jupiter and Saturn. See Abstr. 097.014.

Sur la densité de Saturne et la structure des planètes géantes. See Abstr. 100.001.

100 Saturn

100.001 Sur la densité de Saturne et la structure des planètes géantes. A. Dauvillier.
Comptes Rendus, Acad. Sci. Paris, Sér. B, Vol. 269, 391 - 396 (1969).
Les densités de Jupiter et de Saturne peuvent être calculées à partir du mélange solaire de Russell, en tenant compte de la chaleur interne due à la radioactivité de la lithosphère du noyau central.

100.002 A search for an atmosphere enveloping Saturn's rings. F. A. Franklin, A. F. Cook, II.
Icarus, Vol. 10, 417 - 420 (1969).
Spectrograms of Saturn's brighter satellites taken near superior conjunction and when the earth lay nearly in the ring plane might be expected to show the absorption spectrum of a possible ring atmosphere. Observations on two nights gave negative results, which we have used to set an upper limit of ~100 atoms/cm³ on the density of neutral Na vapor surrounding the ring.

100.003 The microwave spectrum of Saturn. S. Gulkis, T. R. McDonough, H. Craft.
Icarus, Vol. 10, 421 - 427 (1969).
Two atmospheric models which may account for Saturn's microwave spectrum are discussed. The two models considered are (1) thermal emission by a deep atmosphere with a temperature gradient, and (2) free-free emission by an ionosphere. An attempt was made to detect radio emission from Saturn at 70 cm. Saturn was not detected. The upper limit we place on the equivalent brightness temperature at 70 cm is $1250°$ K.

100.004 The erosion of particles in the rings of Saturn. L. W. Bandermann, R. D. Wolstencroft.
Bull. American Astron. Soc. Vol. 1, 233 (1969). – Abstr. AAS.

100.005 The effect of meteoroidal bombardment on Saturn's rings. A. F. Cook, F. A. Franklin.
Smithsonian Astrophys. Obs. Special Rep. No. 304, 6 + 31 pp. (1969).
This study tries to establish whether Saturn's rings are undergoing a loss or a gain of material as a result of impacts with interplanetary particles. Cometary meteoroids probably dominate over interstellar dust at Saturn, and meteoroids from comets like P/comet Halley probably constitute the bulk of the interplanetary dust near Saturn. Most of the spalled fragments from the ring particles travel either to the other node or back to the node of origin and are recovered by the rings. Some fragments are lost into escaping orbits; rather more enter the planet's atmosphere. There is some doubt as to whether the rings are slowly accreting or losing matter, although our results somewhat favor the latter. If net erosion is taking place, its maximum amount in 4×10^9 years is about 3 g cm^{-2}.

100.006 Meteoroidal bombardment of Saturn's rings. A. F. Cook, F. A. Franklin.
Bull. American Astron. Soc., Vol. 1, 338 (1969). – Abstract AAS.

100.007 Ascensioni rette del pianeta Saturno 1967, 1968. S. Mancuso, L. Milano.
Mem. Soc. Astron. Italiana, Nuova Serie, Vol. 40, 433 - 434 (1969).
In this paper we give the observed $O - C$ of Saturn for the years 1967, 1968.

100.008 Bright spot on Saturn. A. W. Heath, R. L. Waterfield.
British Astron. Ass. Circ. No. 514 (1969).

100.009 A revision of the magnitudes of Saturn's satellites. ["The magnitudes of Saturn's satellites", by Patrick Moore: Journ. British Astron. Ass. Vol. 79, 121 - 123 (1969)]. E. J. Öpik.
Irish Astron. Journ., Vol. 9, 92 - 95 (1969).

100.010 Bright spot on Saturn. E. J. Reese.
IAU Circ. No. 2181, 2183 (1969).

Spectrophotometric studies of Mars, Jupiter and Saturn. See Abstr. 097.014.

The internal powers and effective temperatures of Jupiter and Saturn. See Abstr. 099.008.

The spectra of Jupiter and Saturn in the photographic infrared. See Abstr. 099.017.

Narrow-band colorimetry of some of the satellites of Jupiter and Saturn. See Abstr. 099.031.

101   Uranus, Neptune, Pluto, Transplutonian Planet

101.001   Diameter, flattening and optical properties of the upper atmosphere of Neptune as derived from the occultation of BD - 17° 4388.
J. Kovalevsky, F. Link.
Astron. Astrophys. Vol. 2, 398 - 412 (1969).   In French.

The occultation of BD-17° 4388 by Neptune that occurred on April 7, 1968 was observed photometrically in Japan and Australia. Precise photometric curves obtained at Dodaira and Okayama observatories were carefully studied and this paper presents the results of this new reduction (Kovalevsky and Link, 1968). We obtained the apparent dimensions and shape of a certain atmospheric layer at which the ratio of the light flux from the star arriving at the Earth is reduced to one half, and which we call the half-intensity layer. As results of the observations we obtained the position of Neptune with respect to the star, the equatorial radius of the half-intensity layer, corrected for refraction and relativistic effects, at the distance of 30.055 A.U., and the flattening of the half-intensity layer. The diameter is 50450 ± 60 km and the flattening 0.021 ± 0.004.

101.002   The mass of Neptune and the orbit of Uranus.
P. K. Seidelmann, R. L. Duncombe, W. J. Klepczynski.
Astron. Journ. Vol. 74, 776 - 778 (1969).

Analysis of the observations of Uranus from 1781 – 1968 indicates a reciprocal mass of Neptune of 19349 ± 28 (m.e.).

101.003   Observations d'Uranus effectuées à l'astrolabe A. Danjon de l'Observatoire de Sao Paulo.
S. Débarbat, E. Koppe, P. Piazza, A. Postoiev, A. Rabanaque.
Astron. Astrophys. Vol. 3, 126 (1969).

This paper contains results for Uranus observed with the Danjon's astrolabe (Sao Paulo Observatory) between February 22 and May 28, 1968.

101.004   Photographic observations of Neptune and BD –17° 4388 near the time of the occultation in April 1968.
D. W. Dunham.
Astron. Journ. Vol. 74, 958 - 959 (1969).

Photographic observations were made before and after the occultation of 7 April 1968, with the 102-cm reflector at Yale Observatory's Bethany Station. Analysis of these showed that the minimum geocentric separation was 0.''57 ± 0.''05 at $16^h 15^m 7 ± 1^m 5$ U. T. Such observations can be useful in the determination of a planet's radius, especially if an occultation is observed.

101.005   The rotational temperature and the upper limit of pressure in the outer atmosphere of Uranus.
V. G. Teifel, G. A. Kharitonova.
Astron. Zhurn. Akad. Nauk SSSR, Vol. 46, 1104 - 1114 (1969).   In Russian.   English translation in Soviet Astron. AJ, Vol. 13, No. 5.

From a spectrogram of Uranus (dispersion 30.4 Å/mm), the absorption band CH_4 6800 Å and the line $H_2 S$ (0) (4 - 0), induced by pressure, are investigated and some unidentified absorption lines, intensive enough, and absorption details in the region $\lambda\lambda$ 6590 - 6789 Å are found. The rotational temperature T_{rot} = 60° - 68° K and the value of pressure at the effective level $2.5 < P_{eff} < 5$ atm are estimated from 7 lines of the R-branch of the band CH_4 6800 Å.

101.006   The planet Neptune.   G. E. Taylor.
Journ. British Astron. Ass., Vol. 80, 13 - 18 (1969). Presidential address.

101.007   The diameter of Neptune.
J. E. Bixby, T. C. Van Flandern.
Astron. Journ., Vol. 74, 1220 - 1222 (1969).

Fourteen meridian circle observations made before and after the occultation of ZC 2232 by Neptune on 7 April 1968 and observations of the occultation from Japan, New Zealand, and Australia have been used to determine corrections to the ephemeris of the relative positions of the star and Neptune, of +0.^{s}260 ± 0.^{s}001 in right ascension and –0.''46 ± 0.''07 in declination, and a new semidiameter of Neptune of 33.''9 ± 1.''8 at unit distance.

101.008   The spectrum of Uranus in the region 4800 – 7500 Å.   L. S. Galkin, L. A. Bugaenko,
O. I. Bugaenko, A. V. Morozhenko.
Astron. Tsirk. No. 533, p. 1 - 2 (1969).   In Russian.

101.009   Some visual observations of markings on Uranus.
E. W. Cross, Jr.
Strolling Astronomer, Vol. 21, 152 - 153 (1969).

101.010   Uranus in early 1969.   R. G. Hodgson.
Strolling Astronomer, Vol. 21, 167 - 170 (1969).

101.011   Lack of a noticeable methane atmosphere on Triton.
H. Spinrad.
Publ. Astron. Soc. Pacific, Vol. 81, 895 - 896 (1969).

The purpose of this note is to point out the lack of an easily measurable CH_4 band in the spectrum of Triton, Neptune's large satellite.

Observations photographiques de petites planètes et de Pluton.   See Abstr. 098.021.

102 Comets

102.001 Total gas concentration in atmospheres of the short-period comets and impulsive forces upon their nuclei. Z. Sekanina.
Astron. Journ. Vol. 74, 944 - 950 (1969).

The mass-loss rates are calculated for 23 short-period comets with known systematic deviations from the gravitational law in their motion. The Ca_2 amount in the atmospheres of the comets, calculated from photometric data, is compared with the total abundance of gases (mostly nonluminous), derived from dynamical data. The results show that C_2 represents not more than 0.1 per mille of the total gas concentration, on an average. It is suspected that the ratio between the total and C_2 densities might tend to increase secularly. It is concluded that the vast amounts of gases liberated from the nucleus each revolution (at least $10^{12} - 10^{14}$ g) are able to explain the whole dynamical effect observed in the short-period comets and that their nuclei should be $10^{17} - 10^{18}$ g in mass.

102.002 Physische Beobachtungen von Kometen. XVI. M. Beyer.
Astron. Nachr., Vol. 291, 239 - 256 (1969).

The results are given of systematic visual observations of seven bright comets (1966 b, 1966 c, 1966 d, 1967 n, 1968 a, 1968 c, 1968 d) in the years 1966 – 68. As in preceding numbers of this series, the photometric parameters n and H_0 are derived from observations of the total magnitude of the comet's head, and the correlation of magnitude with solar activity is examined. Furthermore, observations are given of the magnitude of the nucleus, the diameter of coma, and direction and structure of the tail. For the first time, photoelectric magnitudes of the comet's head, measured since 1966 by K. Wenske, are included in the results.

102.003 Nachweis und Ergebnisse von Kometen-Beobachtungen aus den Jahren 1921 – 1968. M. Beyer.
Astron. Nachr., Vol. 291, 257 - 264 (1969).

The references and main results of physical observations of 103 comets, obtained and published by the author in the years 1921 – 1968, are compiled (total magnitude of the head; photometric parameters H_0 and n; magnitude of nucleus; diameter of coma; type and direction of the tail; correlation of magnitude with solar activity).

102.004 Gaseous dynamics in the circum-nucleus region of comets. L. M. Shulman.
Astrometriya i Astrofiz., *Kiev*, No. 4, p. 101 - 116 (1969). In Russian.

The flow of particles in a region around the nucleus of a comet's head is considered. In the region mentioned the intermolecular collisions are important and motion of matter is to be described in terms of gaseous dynamics. It is shown that only supersonic flow can take place if outer radiation heating is small. Distribution of brightness is found in the case of an adiabatic flow.

102.005 Physical conditions in the wall-attached layer of the cometary nucleus. L. M. Shulman.
Astrometriya i Astrofiz., *Kiev*, No. 4, p. 117 - 126 (1969). In Russian.

The connection between initial parameters of the gaseous flow (density, velocity, temperature and pressure) and the temperature of the cometary nucleus is found from conservation laws. It is shown that the initial Mach number depends on sublimation and accomodation coefficients only. Initial velocity of flow is somewhat greater than results of calculations.

102.006 Some problems of photoelectric observations of comets. V. Vanysek.
Astrometriya i Astrofiz., *Kiev*, No. 4, p. 127 - 133 (1969). In Russian.

The methods of photoelectric observations of comets with narrow and wide band filters are discussed. Some possible combinations of filters allow to obtain the ratio of intensity of the emission bands of the molecules CN, C_2, C_3 and the natrium lines D_1, D_2 to intensity of the continuous spectrum. A new physical classification of comets depending on the ratio is given.

102.007 Some problems of cometary spectroscopy. V. I. Cherednichenko.
Astrometriya i Astrofiz., *Kiev*, No. 4, p. 134 - 141 (1969). In Russian.

Some unidentified emission in the cometary spectra and their possible interpretations are described. Existence of some paternal molecules and atoms yet undiscovered in the comets has been found. The abundance of hydrogen atoms in the cometary atmospheres is discussed. The process of recombination of negative hydrogen ions with the positive ones is assumed to be a source of the cometary continuous spectra. The theoretic and observed brightness of the continuous spectrum and the position of maximum in the spectral distribution are in good agreement for comet Mrkos 1957 V.

102.008 On the problem of the dust component in cometary atmospheres. M. Z. Markovich.
Astrometriya i Astrofiz., *Kiev*, No. 4, p. 166 - 173 (1969). In Russian.

Some properties of the dust component in cometary atmospheres are considered: dust mass, number of particles in the cometary head, dust concentration near the nucleus, dust capacity, distribution of dust particle sizes. It is assumed that the size distribution is the same as in meteor showers. All values are changing within wide ranges for the different comets.

102.009 Investigations of cometary movements (1963 – 1966) carried out by the Institute of Theoretical Astronomy. N. A. Bokhan.
Astrometriya i Astrofiz., *Kiev*, No. 4, p. 187 - 194 (1969). In Russian.

The review contains short contents of 37 papers.

102.010 Acceleration of cometary ions by arbitrary magnetic fields of the solar wind. A. Z. Dolginov.
Astrometriya i Astrofiz., *Kiev*, No. 4, p. 195 (1969). In Russian. – Abstract.

102.011 Models of cometary motions in Oort's Cloud. I. Zal'kalne.
Astrometriya i Astrofiz., *Kiev*, No. 4, p. 196 (1969). In Russian. – Abstract.

102.012 Ray systems in comet tails. V. P. Tarashchuk.
Astrometriya i Astrofiz., *Kiev*, No. 4, p. 197 (1969). In Russian. – Abstract.

102.013 Temperature of ice nuclei of comets near the sun. M. Z. Markovich, L. N. Tulenkova.
Astrometriya i Astrofiz., *Kiev*, No. 4, p. 198 (1969). In Russian. – Abstract.

102.014 A rule in the mechanical theory of cometary forms. O. V. Dobrovol'skij.
Astrometriya i Astrofiz., *Kiev*, No. 4, p. 199 (1969).
In Russian. – Abstract.

102.015 A new method for calculating cometary tails. O. V. Dobrovol'skij, Kh. Ibadinov.
Astrometriya i Astrofiz., *Kiev*, No. 4, p. 200 (1969).
In Russian. – Abstract.

102.016 A new statistical law in comets. O. V. Dobrovol'skij, R. S. Osherov.
Astrometriya i Astrofiz., *Kiev*, No. 4, p. 201 (1969).
In Russian. – Abstract.

102.017 One of the possible causes for the asymmetry of light curves of comets with respect to their perihelion passage. P. Egibekov.
Astrometriya i Astrofiz., *Kiev*, No. 4, p. 203 (1969).
In Russian. – Abstract.

102.018 On Oort's cometary cloud. S. K. Vsekhsvyatskij.
Astrometriya i Astrofiz., *Kiev*, No. 4, p. 207 - 208 (1969).
In Russian. – Abstract.

102.019 Note on the solar wind-induced drag on comets. D. E. Gonzales.
Solar Physics, Vol. 9, 205 - 209 (1969).

The solar wind-induced drag on magnetically large comets is estimated as follows. As the comet approaches the sun, solar radiation striking the comet surface generates a surrounding neutral atmosphere which is subsequently ionized. The resulting plasma cloud interacts with the solar wind to produce a comet magnetosphere and associated collision-free shock wave. An approximation to the accompanying drag is obtained using the similarity between the comet magnetosphere and that of the earth, and is shown to be much less than the mechanical mass loss force.

102.020 Sudden changes in the brightness of comets before their discovery. E. M. Pittich.
Bull. Astron. Inst. Czechoslovakia, Vol. 20, 251 - 293 (1969).

The problem as to whether and to what extent comets are discovered as a consequence of a sudden increase of their brightness is investigated. The data about the discovery conditions of individual comets (their brightness, the position in space and in the sky, the apparent motion, the phases of the moon, the solar activity, the number of independent discoverers) enable us to draw statistical conclusions about the course of the brightness at the time when the comet was not yet discovered. An analysis of the distribution of the time intervals during which the comets were not discovered despite generally favourable observing conditions, enables to distinguish between comets discovered shortly after a sudden increase in their brightness and the others. The analysis of 537 unpredicted comet discoveries from the period 1750 - 1967 shows that in 8% of the cases the discovery was very probably influenced by the effect under discussion. The results obtained are used for the comparison of some dynamical and physical parameters which might be connected with anomalous changes between the group of comets discovered after sudden increases of brightness and the other comets.

102.021 Die teleskopische Beobachtung der Kometen. I. J. Classen.
Sterne, 45. Jahrgang, 95 - 102 (1969).

102.022 Lifetime of neutral sodium in comets. N. S. Kovar, R. P. Kovar.
Bull. American Astron. Soc. Vol. 1, 250 (1969). – Abstr.

AAS.

102.023 A method of general perturbations for cometary orbits. P. Nacozy.
Bull. American Astron. Soc. Vol. 1, 254 (1969). – Abstr. AAS.

102.024 Komplexes Programm zur Bahnberechnung von Kometen auf Ellipsen mit kleiner Periheldistanz und einer Exzentrizität nahe 1. L. E. Nikonova.
Trudy Kazan. Gorod. Astron. Obs. No. 35, p. 157 - 168 (1968). In Russian.

Das vorgeschlagene Programm wurde für die elektronische Maschine M-20 entwickelt. Ein Beispiel für den Kometen Pereyra (1963 e) wird vorgeführt.

102.025 Die teleskopische Beobachtung der Kometen. II. J. Classen.
Sterne, 45. Jahrgang, 157 - 161 (1969).

102.026 Maximal length of cometary tails and some remarks on their classification. S. K. Vsekhsviatsky.
Problemy kosmich. fiz. No. 4, p. 28 - 41 (1969). In Russian.

For the comets of 1870 - 1967 a catalogue of types and maximal lengths of their tails is in preparation. Some correlations between maximal lengths, types, perihelion distances, absolute magnitudes and the spectra of the comets are studied. The existence of dust tails for comets with small perihelion distances and of molecular tails for bright comets is found. The possibility of composite tails with different structure at different distances from the head is discussed.

102.027 The possibility of maser effects in cometary atmospheres. D. A. Varshalovich.
Problemy kosmich. fiz. No. 4, p. 54 - 57 (1969). In Russian.

Cometary atmospheres have no thermodynamical equilibrium; atom and molecule level populations are not Boltzmanian. The conditions which are necessary for coherent stimulated amplification of the radiation are discussed.

102.028 Distribution of particles in non-ionized cometary tails. Yu. N. Gnedin, A. Z. Dolginov, G. G. Novikov.
Problemy kosmich. fiz. No. 4, p. 58 - 60 (1969). In Russian.

Boltzmann's equation is used for obtaining the distribution function of the particles in non-ionized cometary tails and for the derivation of their surface brightness.

102.029 Physical parameters of regions near cometary nuclei. Yu. N. Gnedin, A. Z. Dolginov, G. G. Novikov.
Problemy kosmich. fiz. No. 4, p. 61 - 63 (1969). In Russian.

On the basis of previous results the isophotes of four comets are derived, and some parameters are determined.

102.030 Recombination processes in comets. V. I. Cherednichenko.
Problemy kosmich. fiz. No. 4, p. 64 - 68 (1969). In Russian.

On the basis of experimental data formulas are derived for computing effective cross-sections of both ion-ion and electron-ion recombinations. The dominating role of solar wind electrons has been proved in forming the green and red lines of atomic oxygen in cometary atmospheres as result of electron-ion recombination.

102.031 A new method for investigating comets' surface brightness. O. V. Dobrovolsky, R. S. Osherov.
Problemy kosmich. fiz. No. 4, p. 69 - 77 (1969). In Russian.

A method is proposed for determining the mean degree of diffusion of a cometary head. The method is applied to a number of comets.

102.032 **A possible mechanism of the growth of dust particles in cometary atmospheres.**   P. Egybekov.
Problemy kosmich. fiz. No. 4, p. 78 - 80 (1969).   In Russian.
A dust-forming mechanism and the processes of collision in the inner parts of the head due to the attracting force of the nucleus have been examined. The distribution of apparent densities in a dust atmosphere is obtained.

102.033 **The possibility of radiolocation of comets in the short-wave range.**   I. S. Vsekhsviatskaya.
Problemy kosmich. fiz. No. 4, p. 81 - 83 (1969).   In Russian.
It is shown that radar echoes of type I tails can be obtained in the short-wave range. This method may be used to study the physical behavior of cometary tails.

102.034 **Some characteristics of the solar wind inferred from the study of sodium emission from cometary nuclei.**
M. K. V. Bappu, K. R. Sivaraman.
Solar Physics, Vol. 10, 496 - 501 (1969).
Seventeen comets, having information on sodium D-line emission during their apparition, have been studied. The heliocentric distances corresponding to the sodium emission commencement or termination epoch are found to have a dependence on the phase of the solar cycle. It is concluded that the spatial properties of the solar wind during a solar maximum and minimum are responsible for the observed dependence.

102.035 **Dynamical and evolutionary aspects of gradual deactivation and disintegration of short-period comets.**   Z. Sekanina.
Astron. Journ., Vol. 74, 1223 - 1234 (1969).
Mass-loss rates averaging 0.1 to 1% of the total mass per revolution have been found for six short-period comets from the nongravitational terms in the equations of motion applied by Marsden. The steep decrease in nongravitational activity noticed for some of the six comets is interpreted as a rapid-loss process of the comets' reservoirs of volatile materials. The comparison of the present values of the half-lives of the comets with the half-lives of their nongravitational activity strongly suggests that nowadays the relative deactivation rates characterizing the decrease in the activity of the comets are several orders of magnitude higher than the corresponding mass-loss rates. Particular attention is given to the secular variations in the nongravitational parameter κ of P/Encke. It is believed that after complete loss of its volatile materials, the deactivated core of the original nucleus of P/Encke will become a minor planet of the Apollo type.

102.036 **Search for solar proximate comets.**
H. C. Courten, D. W. Brown, D. B. Albert, R. W. Genberg.
Bull. American Astron. Soc., Vol. 1, 338 - 339 (1969). – Abstract AAS.

102.037 **A new model of the cometary nucleus and coma.**
A. H. Delsemme, D. C. Miller.
Bull. American Astron. Soc., Vol. 1, 339 (1969). – Abstract AAS.

102.038 **Orientation-dependence of the evolution of long-period comet orbits.**   P. C. Joss, M. Harwit, D. O. Muhleman.
Bull. American Astron. Soc., Vol. 1, 349 - 350 (1969). – Abstract AAS.

102.039 **On the relationship between comets and minor planets.**   B. G. Marsden.
Bull. American Astron. Soc., Vol. 1, 353 - 354 (1969). – Abstract AAS.

102.040 **Dynamical aspects of gradual disactivation and dis**integration of short-period comets.
Z. Sekanina.
Bull. American Astron. Soc., Vol. 1, 361 (1969). – Abstract AAS.

102.041 **Dissociation processes in comets.**
V. Vanysek.
Bull. American Astron. Soc., Vol. 1, 367 (1969). – Abstract AAS.

102.042 **On the mechanism of generation of Alfvén waves in comets.**   Z. M. Ioffe.
Astron. Zhurn. Akad. Nauk SSSR, Vol. 46, 1328 - 1329 (1969).   In Russian. English translation in Soviet Astron. AJ, Vol. 13, No. 6.
The mechanism of development of wave structure in type I cometary tails is investigated. It is shown, that these structures can arise as a result of Kelvin-Helmholtz instability on the boundary between cometary plasma and solar wind.

102.043 **Comet-hunting in southern skies.**
J. C. Bennett.
Monthly Notes Astron. Soc. Southern Africa, Vol. 28, 80 - 88 (1969). – Presidential address 1969.

102.044 **The sphere of predominance of the sun relative to other stars.**   G. Sitarski.
Postępy Astron., Vol. 17, 303 - 304 (1969).   In Polish.

102.045 **On surface and space density in a cometary head.**
M. Z. Markovich, R. S. Osherov.
Dokl. AN Tadzh. SSR, Vol. 12, No. 1, p. 16 - 19 (1969). In Russian. – Abstr. in Referativ. Zhurn. 51. Astron., 1.51.321 (1970).

102.046 **Kometen.**   W. G. Fessenkow.
Bild der Wissenschaft, 6. Jahrgang, No. 2, 156 - 164 (1969).

102.047 **Determination of the lifetime of parent molecules in comets from the brightness outbursts.**
V. Vanysek.
Publ. Astron. Soc. Pacific, Vol. 81, 840 - 847 = Contr. Four College Obs. No. 45 (1969).
The possibility of using the brightness outbursts or flares in comets for determination of the lifetime of parent molecules is discussed. It is shown that monochromatic measurements of the comets' flares carried out photoelectrically may be a very useful method for a determination of the upper limit of the lifetime of CN and C_2 precursors.

102.048 **Motion of comets and non-gravitational forces.**
J. Bouška.
Vesmír, Vol. 48, 292 - 295 (1969).   In Czech.

102.049 **Verdampfung von Kometenstaub.**
W. F. Hübner.
Mitt. Astron. Ges. No. 27, p. 222 (1969). – Abstract AG.

Physics of Comets.
See Abstr. 003.084.

A filter for photography of comet continua.
See Abstr. 034.046.

The capture of comets by Jupiter.
See Abstr. 042.023.

Cometary probes.
See Abstr. 051.017.

103 Comets: Listed Objects

103.001 Comet notes. E. Roemer.
Publ. Astron. Soc. Pacific, Vol. 81, 448 - 450 (1969).

103.002 Comet notes. E. Roemer.
Publ. Astron. Soc. Pacific, Vol. 81, 700 - 701, 907 - 910 (1969). – Concerning the comets 1968 j, 1969 a, b, c; 1908 II, 1958 VI, 1968 g, 1969 a, b, c, d, e.

103.003 Reports on the progress of astronomy: Comets. B. G. Marsden.
Quarterly Journ. Roy. Astron. Soc. Vol. 10, 240 - 255 (1969).

103.004 Photographische Beobachtungen der Helligkeit von Kometen der Jahre 1958 – 1960. V. G. Rijves.
Kometn. Tsirk. *Kiev*, No. 86 (1969). In Russian.

103.005 Roman numeral designations of comets in 1968.
IAU Circ. No. 2196 (1969).

Bulletin of the Astronomical Institutes of the Netherlands 1921 - 1969. Index of comets.
See Abstr. 002.016.

103.100 Comet 1963 V Pereyra

Definitive Bahn des Kometen 1963 e Pereyra.
L. E. Nikonova.
Trudy Kazan. Gorod. Astron. Obs. No. 35, p. 169 - 176 (1968). In Russian.

103.101 Comet 1965 VIII Ikeya-Seki

Spectrum of the tail of comet Ikeya-Seki (1965 f).
Z. V. Karyagina.
Astrometriya i Astrofiz., *Kiev*, No. 4, p. 142 - 147 (1969). In Russian.

The energy distribution in the spectrum of the cometary tail is obtained in absolute units for 1965 Nov. 1 and Nov. 2. Comparison with the energy distribution in the solar spectrum shows that the spectral characteristics of the tail are similar to those of the sun, while the colors of the tail of the comets Arend-Roland (1965 h) and Mrkos (1957 d) are redder than sunlight.

Polarimetric and photometric observations of comet Ikeya-Seki 1965 f. T. A. Polyakova, T. K. Pisareva.
Astrometriya i Astrofiz., *Kiev*, No. 4, p. 148 - 154 (1969). In Russian.

On the density of Na in comet Ikeya-Seki (1965 f).
E. A. Gurtovenko.
Astrometriya i Astrofiz., *Kiev*, No. 4, p. 155 - 158 (1969). In Russian.

To estimate the number of atoms N and the mass of sodium M in the head of comet 1965 f a resonance scattering mechanism and the data of spectral observations at Sacramento Peak are used. Obtained values $N \approx 0.25 \times 10^{32}$ and $M \approx 1 \times 10^9$ g are nearly equal to the corresponding estimations deduced from the emission of other elements and combinations in comets far distant from the sun. This points out that comets consist of heavy elements to a considerably extent. These elements can educe the comet's nucleus at a near distance from the sun only.

Photographic photometry of the tail of comet Ikeya-Seki. E. Eroshevich.
Astrometriya i Astrofiz., *Kiev*, No. 4, p. 159 - 165 (1969). In Russian.

Photometric measurements from observations with a Schmidt camera (D = 170 mm, D : F = 1 : 1) lead to isophotes of head and tail of the comet. According to these the quantity of dust is determined.

Evolution of the tail of comet Ikeya-Seki up to perihelion passage. Kh. Ibadinov.
Astrometriya i Astrofiz., *Kiev*, No. 4, p. 202 (1969). In Russian. – Abstract.

Observations of comet Ikeya-Seki (1965 f) in Dushanbe. A. M. Bakharev.
Astrometriya i Astrofiz., *Kiev*, No. 4, p. 204 (1969). In Russian. – Abstract.

Observations of comet Ikeya-Seki in Alma-Ata.
Sh. N. Sabitov.
Astrometriya i Astrofiz., *Kiev*, No. 4, p. 205 (1969). In Russian. – Abstract.

Investigation of the tail of comet Ikeya-Seki.
N. S. Chernykh.
Astrometriya i Astrofiz., *Kiev*, No. 4, p. 206 (1969). In Russian. – Abstract.

Observations of comet Ikeya-Seki 1965 November with an electron telescope.
P. G. Petrov, K. L. Mench, V. S. Rylov.
Astrometriya i Astrofiz., *Kiev*, No. 4, p. 212 (1969). In Russian. – Abstract.

The emission spectrum of comet Ikeya-Seki 1965-f at perihelion passage. C. D. Slaughter.
Astron. Journ. Vol. 74, 929 - 943 = Contr. Kitt Peak National Obs. No. 447 (1969).

Daylight observations of comet Ikeya-Seki (1965-f) were obtained on 20 and 21 October 1965, at solar distances less than 0.2 a. u. The spectrograms obtained show many atomic and molecular emission lines within the comet head. The plates contain a combination of three spectra: (1) the above-mentioned emission lines, (2) the solar spectrum from the daylight sky, and (3) a Fraunhofer spectrum reflected from the comet. Contamination of the emission-line spectra by the reflected spectra necessitated a procedure for determining an exact representation of the two absorption spectra to permit their effective removal. The resulting emission spectrum, containing lines due to Na I, Ca I, Ca II, Cr I, Co I, Mn I, Fe I, Ni I, Cu I, V I, and CN, is catalogued in an atlas containing 477 entries. The anomalous Cu/Fe abundance found by Preston is confirmed; no evidence is found of the [O I] line at $\lambda 6300.30$ Å. An effective Boltzmann temperature of $4480°$K is derived from the Fe I and Ni I lines, which do not exhibit marked effects of fluorescence. The violet (0,0) CN band at $\lambda 3880$ Å exhibits many intensity deviations when compared to solar or laboratory reference spectra.

Photometry by the equidensity method and the spectra of the comet Ikeya-Seki 1965 f.
I. R. Beytrishvili.
Problemy kosmich. fiz. No. 4, p. 91 - 102 (1969). In Russian.

Comet Ikeya-Seki 1965 f was observed at Abastumani Astrophysical Observatory from October 29 to December 7, 1965. The photometry of the comet was made by the equidensity method. The author comes to the conclusion that during the whole observational period there was noticed a CN emission in the comet's head which caused a symmetrical distribution of intensity in the head.

The tail of comet Ikeya-Seki 1965f.
D. Milon.
Strolling Astronomer, Vol. 21, 146 - 152 (1969).
Analysis of ALPO comets section observations of comet Ikeya-Seki 1965f by 24 observers gives a maximum tail length of 0.7 astronomical units and a radial velocity of the end of the tail of 32 kilometers per second and further reveals an anomalous tail.

Coma diameter of comet Ikeya-Seki 1965f.
D. Milon.
Strolling Astronomer, Vol. 21, 201 - 204 (1969).

103.102   Comet 1968 VI Honda

Comets Honda (1968c and 1968e).
R. R. de Freitas Mourão, I. Mourilhe.
IAU Circ. No. 2167 (1969).

Komet Honda, 1968 c.
N. M. Bronnikova, G. R. Kastel'.
Kometn. Tsirk. Kiev, No. 86 (1969).   In Russian.

Photographic observations of comets.
J. A. Bruwer, B. M. F. Armstrong.
Republic Obs. Johannesburg, Circ. Vol. 7 (No. 128), 176 (1969).
Observations at the Republic Observatory Annexe, Hartbeespoort, with the Franklin-Adams Star Camera.

Poziţiile cometelor Honda (1968 c) şi Bally-Clayton (1968 d) observate in anul 1968.
C. Cristescu, G. Bocşa, V. I. Vlăsceanu.
Stud. Cerc. Astron., Vol. 14, 173 - 176 (1969).
In Rumanian.

103.103   Comet 1968 IX Honda

Comets Honda (1968c and 1968e).
R. R. de Freitas Mourão, I. Mourilhe.
IAU Circ. No. 2167 (1969).

Photographic observations of comets.
See Abstr. 103.102.

103.104   Comet 1967 XIII Encke

Secular variations in the absolute brightness of comet Encke.   Z. Sekanina.
Astrometriya i Astrofiz., Kiev, No. 4, p. 54 - 76 (1969).
In Russian.
The light curves of comet Encke are studied in detail from its 1805 to 1961 returns. The secular decrease of its absolute brightness ranges between 2^m and 4^m per century, consisting with the author's 1964 results. Vsekhsvjatskij's definition of the absolute magnitude, H_{10} has been found in a good approximation of the comet's luminosity as far as observations from long enough interval of heliocentric distances are applied.

103.105   Comet 1957 III Arend-Roland

An analysis of distribution of the surface brightness in the tail of comet 1956 h.   G. K. Nazarchuk.
Astrometriya i Astrofiz., Kiev, No. 4, p. 77 - 100 (1969).
In Russian.
The photometry of large-scale photographs of high quality of the comet Arend-Roland obtained by S. K. Vsekhsvyatsky with the 40 cm telescope at the Crimean Astrophysical Observatory was carried out. For the sake of data on kinematic and physical parameters of streams, a simple model has been used, taking into account the constant acceleration, dissociation of luminous molecules and the expansion of matter. All the parameters were determined by comparison of photometric and calculated results. Anisotropic expansion of matter has been found. It may be explained by the presence of a longitudinal magnetic field.

Erratum: Comet Arend-Roland (1957 III) on 5 May 1957. I. Development and kinematics of the type-I tail.
F. D. Miller.
Astron. Journ. Vol. 74, 845 (1969). – See Astron. Astrophys. Abstr. Vol. 1, 103.112.

On the dust component in the head of comet Arend-Roland.   M. Z. Markovich, R. S. Osherov.
Dokl. AN Tadzh. SSR, Vol. 12, No. 3, p. 13 - 14 (1969). In Russian. – Abstr. in Referativ. Zhurn. 51. Astron., 12.51. 695 (1969).

103.106   Comet 1966 V Kilston

Investigation of the brightness of comet Kilston 1966b.   D. Andrienko, A. A. Demenko.
Problemy kosmich. fiz. No. 4, p. 84 - 87 (1969).   In Russian.
A correlation between the brightness increase of comet Kilston and solar flares is noted.

The integral brightness of comet Kilston 1966b.
A. M. Baharev.
Problemy kosmich. fiz. No. 4, p. 88 - 90 (1969).   In Russian.
Results of visual observations of comet 1966b Kilston made during 1966 Aug. 16 – Oct. 20 are communicated. Integral magnitudes of the head, a description of the comet, the observational conditions, and magnitudes of comparison stars are given.

Photographic observations of comet Kilston (1966 b).   G. A. Garazdo-Lesnykh, V. P. Konopleva.
Astrometriya i Astrofiz., Kiev, No. 4, p. 174 - 186 (1969).
In Russian.
From observations at Kiev and Skalnate Pleso photographic and photovisual magnitudes were determined and reduced to the coma's volume with radius of 25000 km. The changes of the appearance of the comet are described.

103.107   Comet 1959 IV Alcock

Definitive, initial and future orbits of comet 1959 IV Alcock.   G. T. Yanovitskaya.
Astrometriya i Astrofiz., Kiev, No. 4, p. 213 - 214 (1969).
In Russian. – Abstract.

103.108   Comet 1960 III Schaumasse

Komet Schaumasse (1960 III – 1959 h).
K. P. Matsukov.
Trudy Kazan. Gorod. Astron. Obs. No. 35, p. 177 - 185 (1968).   In Russian.

103.109   Comet 1969a Faye

Periodic comet Faye (1969a).
T. Seki.
IAU Circ. No. 2165 (1969).

Periodic comet Faye (1969a).
A. Mrkos, B. Milet, G. van Biesbroeck, Petrovičová.
IAU Circ. No. 2171 (1969).

Periodic comet Faye (1969a).
A. Mrkos, B. Milet, T. Seki.
IAU Circ. No. 2182 (1969).

Periodic comet Faye (1969a).
A. Mrkos, Petrovičová, B. Milet, R. L. Waterfield, T. Seki.
IAU Circ. No. 2192 (1969).

Periodic comet Faye (1969a).
A. Mrkos, B. A. Burnaševa, S. l. Gerasimenko.
IAU Circ. No. 2196 (1969).

P/Faye 1969a.
J. E. Bortle.
British Astron. Ass. Circ. No. 513 (1969).

Der kurzperiodische Komet Faye, 1969 a.
Kometn. Tsirk. *Kiev*, No. 89 (1969).   In Russian.

Periodischer Komet Faye, 1969 a.
Kometn. Tsirk. *Kiev*, No. 92 (1969).   In Russian.

Investigation of the motion of comet Faye.
F. B. Khanina.
Problemy kosmich. fiz. No. 4, p. 152 - 156 (1969).
In Russian.

This paper presents a short survey of work done at the Institute for Theoretical Astronomy, Leningrad, on the orbital evolution of short-period comets, and a brief historical account and principal results on the motion of comet Faye from its discovery to the present time. Suggestions are made for future studies of the motion of comet Faye.

Periodic comet Faye.   J. Bouška.
Říše hvězd, Vol. 50, 155 - 156 (1969).   In Czech.

Periodic comet Faye (1969a).
Sky Telescope, Vol. 38, 426 - 427 (1969).

P/Faye, 1969a.
Yamamoto Circ. Nos. 1699, 1704, 1709 (1969).
In Japanese.

Beobachtungen des Kometen Kohoutek, 1969 b und Faye, 1969 a.   See Abstr. 103.112.

103.110   Comet 1969h Čurjumov-Gerasimenko

Comet Čurjumov (Čurymov?)-Gerasimenko (1969h).   K. I. Čurjumov, S. I. Gerasimenko.
IAU Circ. No. 2179 (1969).

Comet Čurjumov-Gerasimenko (1969h).
D. Ja. Martynov, M. P. Candy, D. Gans, J. E. Bortle.
IAU Circ. No. 2181 (1969).

Comet Čurjumov-Gerasimenko (1969h).
E. Roemer.
IAU Circ. No. 2184 (1969).

Periodic comet Čurjumov-Gerasimenko (1969h).
K. I. Curjumov, S. I. Gerasimenko, M. Ja. Šmakova.
IAU Circ. No. 2186 (1969).

Periodic comet Čurjumov-Gerasimenko (1969h).
S. I. Gerasimenko, B. Milet, N. S. Černyh, C. Scovil, Lipovetskij, S. K. Vsehsvjatskij.
IAU Circ. No. 2187 (1969).

Periodic comet Čurjumov-Gerasimenko (1969h).
A. Mrkos, G. van Biesbroeck.
IAU Circ. No. 2195 (1969).

New comet Churiumov-Gerasimenko 1969 h.
S. K. Vsechsviatsky.
Astron. Tsirk. No. 539, p. 1 (1969).   In Russian.

Ephemeris of comet Churiumov-Gerasimenko 1969 h.
Astron. Tsirk. No. 539, p. 2 (1969).   In Russian.

Comet P/Curjumov-Gerasimenko 1969h.
M. P. Candy, B. G. Marsden.
British Astron. Ass. Circ. No. 514 (1969).

Neuer Komet Čurjumov-Gerasimenko, 1969 h.
M. Ya. Shmakova.
Kometn. Tsirk. *Kiev*, No. 90 (1969).   In Russian.

Komet Čurjumov-Gerasimenko, 1969 h.
Kometn. Tsirk. *Kiev*, No. 91 (1969).   In Russian.

Komet Čurjumov-Gerasimenko, 1969 h.
Kometn. Tsirk. *Kiev*, No. 92 (1969).   In Russian.

Periodischer Komet Čurjumov-Gerasimenko, 1969 h.
Kometn. Tsirk. *Kiev*, No. 93 (1969).   In Russian.

Kurzperiodischer Komet Čurjumov-Gerasimenko, 1969 h.   V. L. Afanas'ev.
Kometn. Tsirk. *Kiev*, No. 94 (1969).   In Russian.

Comet Churjumov-Gerasimenko, 1969h.
Yamamoto Circ. No. 1707 (1969).   In Japanese.

Comet P/Čurjumov-Gerasimenko, 1969h.
Yamamoto Circ. Nos. 1708, 1709, 1710 (1969).   In Japanese.

103.111   Comet 1968 g Comas Solá

Periodic comet Comas Solá (1968 g).
B. Milet.
IAU Circ. No. 2173 (1969).

Periodic comet Comas Solá (1968 g).
S. I. Gerasimenko, K. I. Čurjumov, B. Milet.
IAU Circ. No. 2188 (1969).

Periodic comet Comas Solá (1968 g).
A. Mrkos, T. Seki.
IAU Circ. No. 2195 (1969).

Komet Comas-Solá.
Kometn. Tsirk. *Kiev*, No. 91 (1969).   In Russian.

Periodic comet Comas Sola (1968g).
Sky Telescope, Vol. 38, 426 (1969).

P/Comas Solá, 1968g.
Yamamoto Circ. Nos. 1702, 1709 (1969). In Japanese.

103.112 Comet 1969b Kohoutek

Comet Kohoutek (1969b).
L. Kohoutek, B. Milet.
IAU Circ. No. 2159 (1969).

Comet Kohoutek (1969b).
R. L. Waterfield, S. W. Milbourn.
IAU Circ. No. 2161 (1969).

Comet Kohoutek (1969b).
L. Kohoutek, M. Antal.
IAU Circ. No. 2162 (1969).

Comet Kohoutek (1969b).
B. Milet.
IAU Circ. No. 2163 (1969).

Comet Kohoutek (1969b).
T. Seki, N. S. Černyh, B. A. Burnaševa, B. Milet, H. Morgan,
R. L. Waterfield.
IAU Circ. No. 2166 (1969).

Comet Kohoutek (1969b)
A. Mrkos, Petrovičová, B. Milet, T. Seki, D. Larcome, B. G.
Marsden, R. L. Waterfield.
IAU Circ. No. 2168 (1969).

Comet Kohoutek (1969b).
G. van Biesbroeck, T. Seki, B. Milet.
IAU Circ. No. 2174 (1969).

Comet Kohoutek (1969b).
A. Mrkos, L. I. Černyh, N. S. Černyh.
IAU Circ. No. 2179 (1969).

Comet Kohoutek (1969b).
L. I. Cernyh, L. Kohoutek, A. Mrkos, Petrovičová, E. Roemer,
B. Milet, M. Schreur.
IAU Circ. No. 2191 (1969).

New comet Kohoutek 1969b.
L. Kohoutek.
British Astron. Ass. Circ. No. 510 (1969).

Comet Kohoutek 1969b.
B. G. Marsden, R. L. Waterfield, H. Morgan.
British Astron. Ass. Circ. No. 511 (1969).

Comet Kohoutek 1969b.
B. G. Marsden, R. L. Waterfield.
British Astron. Ass. Circ. No. 512 (1969).

Neuer Komet Kohoutek, 1969 b.
Kometn. Tsirk. *Kiev*, No. 86 (1969). In Russian.

Komet Kohoutek, 1969 b.
N. S. Chernykh, G. R. Kastel'.
Kometn. Tsirk. *Kiev*, No. 87 (1969). In Russian.

Komet Kohoutek, 1969 b.
G. R. Kastel'.
Kometn. Tsirk. *Kiev*, No. 88 (1969). In Russian.

Komet Kohoutek, 1969 b.
M. Shmakova.
Kometn. Tsirk. *Kiev*, No. 89 (1969). In Russian.

Komet Kohoutek, 1969 b.
Kometn. Tsirk. *Kiev*, No. 90 (1969). In Russian.

Komet Kohoutek, 1969 b.
N. S. Chernykh, M. Ya. Shmakova.
Kometn. Tsirk. *Kiev*, No. 91 (1969). In Russian.

Komet Kohoutek, 1969 b.
Kometn. Tsirk. *Kiev*, No. 93 (1969). In Russian.

**Beobachtungen des Kometen Kohoutek, 1969 b
und Faye, 1969 a.**
S. I. Gerasimenko, M. Ya. Shmakova.
Kometn. Tsirk. *Kiev*, No. 94 (1969). In Russian.

On the discovery of comet Kohoutek.
J. Bouška.
Říše hvězd, Vol. 50, 229 - 231 (1969). In Czech.

Comet Kohoutek (1969b).
Sky Telescope, Vol. 38, 427 (1969).

Comet Kohoutek, 1969b.
L. Kohoutek, B. Milet, B. G. Marsden.
Yamamoto Circ. Nos. 1702, 1703, 1706 (1969). In Japanese.

103.113 Comet 1969d Fujikawa

Comet Fujikawa (1969d).
S. Fujikawa, K. Tomita, B. Milet.
IAU Circ. No. 2161 (1969).

Comet Fujikawa (1969d).
K. Tomita, M. Antal.
IAU Circ. No. 2162 (1969).

Comet Fujikawa (1969d).
B. Milet.
IAU Circ. No. 2163 (1969).

Comet Fujikawa (1969d).
T. Seki, B. Milet, I. Hasegawa.
IAU Circ. No. 2165 (1969).

Comet Fujikawa (1969d).
B, Milet, T. Seki, B. G. Marsden.
IAU Circ. No. 2169 (1969).

Comet Fujikawa (1969d).
G. van Biesbroeck, T. Seki, B. Milet, I. Hasegawa.
IAU Circ. No. 2172 (1969).

Comet Fujikawa (1969d).
B. A. Burnaševa, B. Milet, K. Locher.
IAU Circ. No. 2174 (1969).

Comet Fujikawa (1969d).
B. A. Burnaševa, N. S. Černyh, H. K. Raudsaar, A. Mrkos.
IAU Circ. No. 2184 (1969).

Comet Fujikawa (1969d).
T. Seki, B. Milet.
IAU Circ. No. 2188 (1969).

Comet Fujikawa (1969d).
A. Mrkos.
IAU Circ. No. 2190 (1969).

Further observations of comets.
M. Antal.
IAU Circ. No. 2195 (1969).

New comet Fujikawa 1969d.
B. G. Marsden.
British Astron. Ass. Circ. No. 511 (1969).

Comet Fujikawa 1969d.
B. G. Marsden.
British Astron. Ass. Circ. No. 512 (1969).

Neuer Komet Fujikawa-Tomita, 1969 d.
Kometn. Tsirk. *Kiev*, No. 87 (1969). In Russian.

Komet Fujikawa, 1969 d.
B. A. Burnasheva, G. R. Kastel', M. Ya. Shmakova.
Kometn. Tsirk. *Kiev*, No. 88 (1969). In Russian.

Komet Fujikawa, 1969 d.
M. Shmakova.
Kometn. Tsirk. *Kiev*, No. 89 (1969). In Russian.

Komet Fujikawa, 1969 d.
Kometn. Tsirk. *Kiev*, No. 90 (1969). In Russian.

Komet Fujikawa, 1969 d.
Kometn. Tsirk. *Kiev*, No. 91 (1969). In Russian.

Elemente und Ephemeride des Kometen Fujikawa, 1969 d.
Kometn. Tsirk. *Kiev*, No. 94 (1969). In Russian.

Comet Fujikawa (1969d).
Sky Telescope, Vol. 38, 427 (1969).

Comet Fujikawa, 1969d.
Yamamoto Circ. Nos. 1703, 1704, 1705, 1710 (1969).
In Japanese.

Comet Fujikawa, 1969d.
S. Fujikawa.
Yamamoto Circ. No. 1706 (1969).

Comet Fujikawa, 1969d. T. Seki.
Yamamoto Circ. No. 1709 (1969). In Japanese.

103.114 Comet 1969e Honda-Mrkos-Pajdušáková

Periodic comet Honda-Mrkos-Pajdušáková (1969e).
A. Mrkos.
IAU Circ. No. 2162 (1969).

Periodic comet Honda-Mrkos-Pajdušáková (1969e).
T. Seki, P. Wild.
IAU Circ. No. 2166 (1969).

Periodic comet Honda-Mrkos-Pajdušáková (1969e).
A. Mrkos, T. Seki, B. Milet, G. van Biesbroeck.
IAU Circ. No. 2170 (1969).

Periodic comet Honda-Mrkos-Pajdušáková (1969e).
N. Kojima, T. Seki.
IAU Circ. No. 2173 (1969).

Periodic comet Honda-Mrkos-Pajdušáková (1969e).
A. Mrkos, B. Milet, T. Seki.
IAU Circ. No. 2179 (1969).

Periodic comet Honda-Mrkos-Pajdušáková (1969e).
H. Morgan, A. Mrkos, B. Milet, R. L. Waterfield.
IAU Circ. No. 2194 (1969).

Further observations of comets.

M. Antal.
IAU Circ. No. 2195 (1969).

Comet P/Honda-Mrkos-Pajdušáková.
B. G. Marsden.
British Astron. Ass. Circ. No. 510 (1969).

Comet P/Honda-Mrkos-Pajdušáková 1969e.
A. Mrkos.
British Astron. Ass. Circ. No. 511 (1969).

Neuer Komet Mrkos, 1969 e.
Kometn. Tsirk. *Kiev*, No. 87 (1969). In Russian.

Kurzperiodischer Komet Honda-Mrkos-Pajdušáková, 1969 e.
Kometn. Tsirk. *Kiev*, No. 88 (1969). In Russian.

Komet Honda-Mrkos-Pajdušáková, 1969 e.
Kometn. Tsirk. *Kiev*, No. 89 (1969). In Russian.

Periodic comet Honda-Mrkos-Pajdusakova (1969e).
Sky Telescope, Vol. 38, 427 (1969).

P/Honda-Mrkos-Pajdušáková, 1969e.
Yamamoto Circ. Nos. 1699, 1704, 1705 (1969).
In Japanese.

P/Honda-Mrkos-Pajdušáková, 1969e.
A. Mrkos.
Yamamoto Circ. No. 1703 (1969). In Japanese.

103.115 Comet 1962 III Seki-Lines

Comet Seki-Lines (1962 III).
B. G. Marsden.
IAU Circ. No. 2160 (1969).

103.116 Comet 1960 II Burnham

Comet Burnham (1960 II).
N. V. Fatčihin, O. N. Orlova, N. M. Bronnikova.
IAU Circ. No. 2161 (1969).

Komet Burnham, 1959 k.
Kometn. Tsirk. *Kiev*, No. 86 (1969). In Russian.

The tail oscillations of comet Burnham (1960 II) and comet Halley (1835 III).
D. R. L. Jones, A. J. Meadows.
Observatory, Vol. 89, 184 - 185 (1969).

103.117 Comet 1968 I Ikeya-Seki

Comet Ikeya-Seki (1967 n).
E. Roemer, M. Schreur.
IAU Circ. No. 2192 (1969).

Komet Ikeya-Seki, 1967 n.
Kometn. Tsirk. *Kiev*, No. 86 (1969). In Russian.

Comet Ikeya-Seki, 1967n.
Yamamoto Circ. Nos. 1701, 1710 (1969). In Japanese.

Photoelectric measurements of comet Ikeya-Seki 1967n. V. Vanýsek.
Bull. Astron. Inst. Czechoslovakia, Vol. 20, 355 - 362 (1969).
Photoelectric measurements in narrow bands centered

at 0.388, 0.474 and 0.486 μ are presented for "dust rich" comet Ikeya–Seki 1967 n. The relative molecular abundance of CN and C_2 is determined and is about a factor 2 lower than the CN/C_2 abundance in comets with weak continuum. From the continuum intensity an upper limit to the mass of dust in the cometary atmosphere is found to be 3×10^{11} g. The photoelectric scanning across the coma provides an estimate of the average intensity distribution of molecular emission and the lifetime of parent molecules of CN is discussed. The maximum value of velocity $\times$ lifetime of parent molecules is 1.1×10^4 km.

103.118 Comet 1835 III Halley

The tail oscillations of comet Burnham (1960 II) and comet Halley (1835 III).   See Abstr. 103.116.

103.119 Comet 1969c Whipple

Periodic comet Whipple (1969c).
Z. M. Pereyra, J. J. Rodriguez.
IAU Circ. No. 2166 (1969).

Comet P/Whipple 1969c.
Z. M. Pereyra.
British Astron. Ass. Circ. No. 510 (1969).

Beobachtungen von Kometen am Observatorium Pulkovo.
Kometn. Tsirk. *Kiev*, No. 86 (1969).   In Russian.

Periodischer Komet Whipple, 1969 c.
Kometn. Tsirk. *Kiev*, No. 87 (1969).   In Russian.

Kurzperiodischer Komet Whipple, 1969 c.
Kometn. Tsirk. *Kiev*, No. 89 (1969).   In Russian.

P/Whipple, 1969c.
Yamamoto Circ. No. 1702 (1969).   In Japanese.

103.120 Comet 1969g Tago-Sato-Kosaka

Comet Tago-Sato-Kosaka (1969g).
A. Tago, Y. Sato, Kosaka.
IAU Circ. No. 2175 (1969).

Comet Tago-Sato-Kosaka (1969g).
N. Kobayashi, H. Kosai.
IAU Circ. No. 2176 (1969).

Comet Tago-Sato-Kosaka (1969g).
T. Seki, H. Kosai, D. Harwood, I. Wardrop, I. Andruszkiw, M. Antal.
IAU Circ. No. 2177 (1969).

Comet Tago-Sato-Kosaka (1969g).
M. Antal, S. W. Milbourn, Z. Sekanina.
IAU Circ. No. 2178 (1969).

Comet Tago-Sato-Kosaka (1969g).
D. Harwood, I. Wardrop, I. Andruszkiw, I. Nikoloff.
IAU Circ. No. 2180 (1969).

Comet Tago-Sato-Kosaka (1969g).
R. L. Waterfield.
IAU Circ. No. 2183 (1969).

Comet Tago-Sato-Kosaka (1969g).
T. Seki.
IAU Circ. No. 2186 (1969).

Comet Tago-Sato-Kosaka (1969g).
T. Seki, B. Milet, A. Mrkos, Petrovičová, Z. M. Pereyra, B. Oviedo, J. J. Rodriguez, F. Dossin, S. W. Milbourn, B. G. Marsden.
IAU Circ. No. 2189 (1969).

Comet Tago-Sato-Kosaka (1969g).
P. Davison, F. W. Gerber, L. González.
IAU Circ. No. 2195 (1969).

Comet Tago-Sato-Kosaka 1969g.
Astron. Tsirk. No. 533, p. 1 (1969).   In Russian.

New comet Tago-Sato-Kosaka 1969g.
S. W. Milbourn.
British Astron. Ass. Circ. No. 513 (1969).

Comet Tago-Sato-Kosaka 1969g.
S. W. Milbourn.
British Astron. Ass. Circ. No. 514 (1969).

Neuer Komet Tago-Sato-Kosaka, 1969 g.
Kometn. Tsirk. *Kiev*, No. 89 (1969).   In Russian.

Komet Tago-Sato-Kosaka, 1969 g.
Kometn. Tsirk. *Kiev*, No. 90 (1969).   In Russian.

Komet Tago-Sato-Kosaka, 1969 g.
Kometn. Tsirk. *Kiev*, No. 91 (1969).   In Russian.

Komet Tago-Sato-Kosaka, 1969 g.
Kometn. Tsirk. *Kiev*, No. 92 (1969).   In Russian.

Komet Tago-Sato-Kosaka, 1969 g.
Kometn. Tsirk. *Kiev*, No. 94 (1969).   In Russian.

Comet Tago-Sato-Kosaka, 1969g.
A. Tago, Y. Sato, Kosaka.
Yamamoto Circ. No. 1706 (1969).   In Japanese.

Comet Tago-Sato-Kosaka, 1969g.
Yamamoto Circ. No. 1707, 1708, 1710 (1969).   In Japanese.

Comet Tago-Sato-Kosaka, 1969g.
T. Seki, B. G. Marsden.
Yamamoto Circ. No. 1709 (1969).   In Japanese.

103.121 Comet 1969f Slaughter-Burnham

Periodic comet Slaughter-Burnham (1969f).
Z. M. Pereyra.
IAU Circ. Nos. 2167, 2170 (1969).

Periodic comet Slaughter-Burnham (1969f).
E. Roemer.
IAU Circ. No. 2184 (1969).

Periodic comet Slaughter-Burnham (1969f).
B. Schreur, E. Roemer.
IAU Circ. No. 2190 (1969).

Comet P/Slaughter-Burnham 1969f.
Z. M. Pereyra.
British Astron. Ass. Circ. No. 512 (1969).

Kurzperiodischer Komet Slaughter-Burnham.
Kometn. Tsirk. *Kiev*, No. 89 (1969).   In Russian.

Periodischer Komet Slaughter-Burnham, 1969 f.
Kometn. Tsirk. *Kiev*, No. 92 (1969).   In Russian.

P/Slaughter-Burnham, 1969f.
Yamamoto Circ. Nos. 1705, 1706, 1708 (1969).
In Japanese.

103.122   Comet 1968 j Thomas

Comet Thomas (1968 j).
A. Mrkos, Petrovičova.
IAU Circ. No. 2170 (1969).

Komet Thomas, 1968 j.
Kometn. Tsirk. *Kiev*, No. 89 (1969).   In Russian.

Comet Thomas, 1968j.
Yamamoto Circ. No. 1699 (1969).   In Japanese.

103.123   Comet 1963 VI Ashbrook-Jackson

Periodic comet Ashbrook-Jackson.
N. A. Beljaev, M. A. Mersljakova.
IAU Circ. No. 2193 (1969).

Kurzperiodischer Komet Ashbrook-Jackson.
M. A. Merzlyakova, N. A. Belyaev.
Kometn. Tsirk. *Kiev*, No. 91 (1969).   In Russian.

103.124   Comet 1908 II Tempel-Swift

Periodic comet Tempel-Swift.
B. G. Marsden.
IAU Circ. No. 2164 (1969).

Periodic comet Tempel-Swift.
G. A. Tammann.
IAU Circ. No. 2181 (1969).

Periodischer Komet Tempel-Swift.
Kometn. Tsirk. *Kiev*, No. 92 (1969).   In Russian.

103.125   Comet 1964 III Kopff

Periodic comet Kopff.

G. Sitarski.
IAU Circ. No. 2183 (1969).

Periodischer Komet Kopff.
Kometn. Tsirk. *Kiev*, No. 93 (1969).   In Russian.

103.126   Comet 1957 IV Schwassmann-Wachmann 1

Periodic comet Schwassmann-Wachmann 1
B. G. Marsden.
IAU Circ. No. 2185 (1969).

Ephemeride des Kometen Schwassmann-Wachmann 1.
Kometn. Tsirk. *Kiev*, No. 94 (1969).   In Russian.

Ephemerides of comet Schwassmann-Wachmann I.
P. Herget.
Publ. Cincinnati Obs. No. 23, p. 57 - 62 (1968).

103.127   Comet 1967 III Wild

Observations de la comète Wild (1967c), effectuées en 1967 à l'aide de l'astrographe double de 40 cm.
H. Debehogne, G. Roland.
Bull. Astron. Obs. Roy. Belgique, Vol. 6 (No. 8), 327 (1969).

103.128   Comet 1969i Bennett.

Comet Bennett (1969i).
J. C. Bennett.
IAU Circ. No. 2196 (1969).

103.129   Comet 1964 I Pons-Winnecke

Periodic comet Pons-Winnecke.
B. G. Marsden.
IAU Circ. No. 2185 (1969).

103.130   Comet 1963 VII d'Arrest

Periodic comet d'Arrest.
B. G. Marsden.
IAU Circ. No. 2191 (1969).

104 Meteors, Meteor Streams

104.001 The discrimination between cometary and asteroidal meteors. I. The orbital criteria. L . Kresák.
Bull. Astron. Inst. Czechoslovakia, Vol. 20, 177 - 188 (1969).

The differences between the orbits of comets and asteroids are discussed with special emphasis on the limiting cases. Various criteria are considered as a means of discriminating between the meteors of cometary and asteroidal origin. It is shown that for the orbits of $q \simeq 1$ all criteria sharply and consistently delimit the elements of the asteroidal meteors.

104.002 On the connection between fireballs and "new" comets. J. Rajchl.
Bull. Astron. Inst. Czechoslovakia, Vol. 20, 189 - 192 (1969).

Spectral and orbital evidence of a connection between fireballs with $i > 30°$ and $v > 25$ km/s, new comets and carbonaceous meteorites is given and discussed. In this respect also the problem of the origin of some groups of photographic and radar meteors is discussed.

104.003 The reflection of radio waves from irregularly ionized meteor trains. J. Jones.
Planet. Space Sci. Vol. 17, 1519 - 1526 (1969).

A theory of the reflection process of radio waves from irregularly ionized meteor trains has been developed which predicts a scatter in the decay time of underdense radio meteors. The magnitude of this scatter is close to that deduced previously by a numerical method, and is shown to be wavelength dependent. Application of the theory to available data indicates that irregularities in the electron line density with scale lengths between 300 and 500 m are less than 20 per cent of the mean electron line density while those with scale lengths between 500 m and 6 km are less than 43 per cent of the mean.

104.004 Sporadic meteor rates 1944 - 1953. J. Štohl.
Contr. Astron. Obs. Skalnaté Pleso, Vol. 4, 25 - 45 (1969).

The following paper presents 28511 visual observations of sporadic meteors obtained at the Skalnaté Pleso Observatory from 1944 until 1953. The net observational time was 2342 hours. The hourly rates, reduced by the method of personal factors, are listed.

104.005 The diurnal and annual variations of sporadic meteors. J. Štohl.
Contr. Astron. Obs. Skalnaté Pleso, Vol. 4, 46 - 62 (1969).

A number of visual and radar series of observations made in different geographical latitudes are examined for the two-dimensional diurnal-seasonal variation in the frequencies of sporadic meteors. The effect of streams on the determination by different techniques of the variation in the sporadic rate is dealt with. It is seen that the half-yearly phase displacement of the annual activity maximum of sporadic meteors on the southern sky is real. A new method is used to examine the annual variation in the activity of the principal radiant sources of sporadic meteors (model: apex, helion, antihelion, background) in Springhill radar data. The dependence of $f \sim \cos^n z$ of the frequency observed upon the radiant's zenith distance was assumed for the computations; from the three exponent values considered ($n = 1.0, 1.5, 2.0$), the results for $n = 1.0$ were the most consistent.

104.006 Eine ausserordentliche Feuerkugel. H. Suszek.
VdS Nachrichtenblatt, Vol. 18, 128 - 129 (1969).

104.007 Meteor notes – January to June 1969. K. B. Hindley.
Journ. British Astron. Ass. Vol. 79, 475 - 488 (1969).
—Meteor section report British Astron. Ass.

104.008 Le Leonidi. C. Bertaud.
Coelum, Vol. 37, 222 - 226 (1969).

104.009 De grote fotografische augustusaktie der Perseiden. B. Apeldoorn.
Hemel en Dampkring, Vol. 67, 251 - 252 (1969).

104.010 On the height of bright meteors. I. S. Shestaka, E. N. Kramer.
Astron. Tsirk. No. 504, p. 6 - 8 (1969). In Russian.

104.011 Determination of the densities of meteoroids. V. G. Kruchinenko.
Astron. Tsirk. No. 510, p. 6 - 8 (1969). In Russian.

104.012 Feuerkugel am 8. Juni 1969. R. A. Naef.
Orion, Band 14, 134 (1969).

104.013 The discrimination between cometary and asteroidal meteors. II. The orbits and physical characteristics of meteors. L. Kresák.
Bull. Astron. Inst. Czechoslovakia, Vol. 20, 231 - 251 (1969).

Orbital criteria discussed in Part 1 of this paper, are applied to compare the meteors of possible asteroidal origin with those of definite cometary origin. Statistical differences in various physical and dynamical parameters are pointed out. Unlike the cometary meteors, those moving in asteroidal orbits do not exhibit the alignment of the lines of apsides, required by the capture process and by the theory of secular perturbations. No commensurability gaps, and no associations analogous to meteor streams, Hirayama's families, Brouwer's groups or Alfvén's jet streams are found among the asteroidal orbits.

104.014 Southern hemisphere meteor rates. C. S. L. Keay, C. D. Ellyett.
Mem. Roy. Astron. Soc., Vol. 73, 185 - 232 (1969).

Meteor rate data of high quality is presented from a controlled-parameter radar survey of meteor activity in the southern hemisphere. The survey extended continuously for 31 months from February 1963 until August 1965 inclusive. The data have been very thoroughly screened and the meteor rates presented in tabular form cover almost 90 per cent of the survey period. The tables contain a total of 2304333 meteors, an average of 114 per hour. The average diurnal variation in meteor activity is exactly similar to that obtained in an earlier survey in 1960–61 but some changes are evident in the annual variation. The various factors which influence the observed meteor rates are discussed.

104.015 De herfstzwermen Tauriden, Leoniden en Geminiden. B. Apeldoorn.
Hemel en Dampkring, Vol. 67, 344 - 345 (1969).

104.016 Favorable skies aid Perseid watchers. S. S. Ross.
Sky Telescope, Vol. 38, 265 - 268 (1969).

104.017 Dzhaus bolide. A. M. Bakharev.
Izv. AN Tadzh. SSR. Otd. fiz.-matem. i geol.-khim. n. No. 4 (30), p. 19 - 27 (1968). In Russian. – Abstr. in Referativ. Zhurn. 51. Astron., 9.51.669 (1969).

104.018 Photographic networks for fireballs.
R. E. McCrosky, Z. Ceplecha.
Meteorite Research, Vienna 1968, p. 600 - 612 (1969).

Two photographic networks, designed to acquire orbital and trajectory data on meteors and impact points of meteorites occurring during night-time hours, have been in operation in the U.S.A. and Czechoslovakia since 1964. More than 100 meteors brighter than –9 mag have been photographed by more than one station. The major result to date is the fact that bright fireballs occur far more frequently than expected; paradoxically, the meteorite rate, as judged by the determination of terminal masses of the meteoroids, is below our most pessimistic estimate.

104.019 A network for rapid analysis of fireball trajectories.
V. D. Chamberlain.
Meteorite Research, Vienna 1968, p. 613 - 622 (1969).

The purposes of this network are threefold: (1) to recover meteorites shortly after their fall: (2) to obtain detailed information on phenomena accompanying meteorite fall; (3) to obtain data on the orbital elements of meteoroids.

104.020 Perseid meteor spectra photographed in 1969.
J. A. Russell.
Sky Telescope, Vol. 38, 424 - 425 (1969).

104.021 The Orionid meteor shower in 1969.
Sky Telescope, Vol. 38, 428 (1969).

104.022 Telescopic observation of a meteor and its train.
W. A. Feibelman.
Journ. Roy. Astron. Soc. Canada, Vol. 63, 189 - 192 (1969).

Photographic observations of meteors and their trains are fairly well documented by now, but these photographs do not show the micro-structure of the trains. The present note describes a high resolution visual observation obtained with a large refractor at a magnification of about 600 ×.

104.023 Structure and fragmentation of meteoroids.
F. Verniani.
Space Sci. Rev. Vol. 10, 230 - 261 (1969).

A summary of the evidence concerning the common occurrence of fragmentation among both photographic and radio meteors is given first. Then, an attempt is made to examine all the present observational, theoretical and laboratory data on the luminous and ionizing efficiencies of meteors, with the aim of establishing a mass scale. This allows the computation of the bulk density of meteoroids. A porous and fragile structure for most of these particles is suggested. In turn, the crumbly structure and the cometary origin confirm Whipple's theory of comets and meteor production.

104.024 Nature's secrets. J. A. Russell.
Astron. Soc. Pacific, Leaflet No. 486, 8 pp. (1969).

104.025 On the interaction layer in front of a meteor body.
J. Rajchl.
Bull. Astron. Inst. Czechoslovakia, Vol. 20, 363 - 372 (1969).

In this paper a model of an interaction layer reported in (Rajchl, 1968a) is developed in more detail. The main physical processes as ionization, diffusion, excitation, recombination and heat generation are described. The role of the shielding action of the cloud of particles of increased density in the close vicinity ahead of a meteor body and its transition into the shock interaction is investigated with respect to the ionization efficiency in the interaction layer. Connection of this layer with the head echo phenomenon, the forbidden emission at 5577 Å and other meteor phenomena is presented and some further consequencies are discussed.

104.026 The effect of wind shear on the decay constant of
meteor echoes. B. A. McIntosh.
Canadian Journ. Phys., Vol. 47, 1337 - 1341 (1969).

For a meteor trail, rotational motion caused by wind gradients leads to errors in the decay constant of the radar echo. Two cases are assessed. In the first, large wind shears produce gross motion of the reflection point along a trail in which the electron density varies exponentially. Secondly, smaller motions on a short, finite trail are examined. These models cannot account for all of the observed scatter in decay-time measurements.

104.027 Untersuchung von Meteormaterie an Bord von Raumflugkörpern mit Lumineszenzzählern.
T. N. Nazarova, A. K. Rybakov, G. D. Komissarov, Z. V. Vasyukova, I. N. Orlov.
Kosmich. Issled. Vol. 7, 795 - 797 (1969). In Russian. – Brief information.

104.028 Minor meteor showers of November, December, and January. M. S. Rao, P. V. S. R. Rao, P. Ramesh.
Australian Journ. Phys. Vol. 22, 767 - 774 (1969).

Systematic visual observations of 7987 meteors were made during the months November, December, and January in 1961–7. Major shower meteors were identified and the remaining data were analysed for evidence of minor meteor shower activity. By tracing the paths of 894 meteors which appeared to make up 70 groups, common radiant points were found suggesting the existence of minor meteor showers. The majority of the minor shower radiants were found to lie close to the plane of the ecliptic.

104.029 Photometric standardization and calibration of meteor spectrograms. V. A. Smirnov.
Astron. Vestn. Vol. 3, 36 - 44 (1969). In Russian.

A method of the reduction of meteor spectrograms is described. The results of the reduction of three spectrograms are given. The possibility of their application for a quantitative spectral analysis of meteor spectrograms is underlined.

104.030 The effect of accumulation and estimate of brightness of meteors from visual and telescopic observations. I. N. Latyshev.
Astron. Vestn. Vol. 3, 149 - 150 (1969). In Russian.

104.031 A new catalogue of 1344 fireballs by Nielsen.
V. V. Fedynsky.
Astron. Vestn. Vol. 3, 173 - 174 (1969). In Russian. – Review of the 'Catalogue of Bright Meteors', Medd. Ole Roemer Obs. Aarhus, No. 30 (1968).

104.032 Statistics and origin of meteor flares.
E. N. Kramer, A. K. Markina.
Astron. Vestn. Vol. 3, 96 - 105 (1969). In Russian.

From investigations of 318 flare meteors statistics of meteor flares were carried out. It is found that slower and more massive meteor bodies flash more frequently. A number of flares decreases monotonously with the increase of flare durations. The analysis of five mechanisms, by which flares and pulsations of meteor lights can be explained, is given.

104.033 Estimate of the exponent in the law of mass distribution of meteor bodies for the Leonid stream 1967.
Yu. V. Bytsenko, V. N. Donij, R. I. Mojsja, E. I. Fialko.
Astron. Vestn. Vol. 3, 106 - 107 (1969). In Russian.

The exponent in the law of mass distribution of meteor bodies is estimated from the distribution of the duration of fluctuating meteor radio echoes during the action period of the Leonid stream in 1967.

104.034 Meteor radioelectronics and its sections.

E. I. Fialko.
Geomagn. Aeronom. Vol. 9, 942 - 944 (1969).   In Russian.
Brief information.

104.035 Report of meteor observations in 1966.
H. Kobayashi.
Mem. Japan Astron. Study Ass., Vol. 3, (No. 12), 133 - 136 (1969).   In Japanese.

104.036 Electrophotometric registration of meteor spectra.
A. Bekbolotov.
Astron. Tsirk. No. 532, p. 3 - 5 (1969).   In Russian.

104.037 Determination of the attachment rate of electrons from combined photographic and radar observations of meteors.   P. B. Babadzhanov, R. Sh. Bibarsov.
Astron. Tsirk. No. 532, p. 5 - 7 (1969).   In Russian.

104.038 Light curves of 103 meteors.
N. N. Israetskaja, V. I. Mysiy, M. Toktogulov, I. S. Shestaka.
Astron. Tsirk. No. 535, p. 1 - 4 (1969).   In Russian.

104.039 Perihelion distances of sporadic meteor particles.
E. N. Kramer.
Astron. Tsirk. No. 535, p. 4 - 6 (1969).   In Russian.

104.040 On the maximum of the Leonid stream in 1968.
I. S. Astapovich.
Astron. Tsirk. No. 539, p. 2 - 4 (1969).   In Russian.

104.041 Radiowave reflection from meteor trains. II. Diffraction pictures.
V. N. Lebedinetz, A. K. Sosnova.
Geomagn. Aeronom. Vol. 9, 680 - 688 (1969).   In Russian.

104.042 Radar observations of meteor rates at Kharkov on six registration levels.
B. L. Kashcheev, Yu. I. Voloshchuk, B. S. Dudnik, N. V. Novoselova, A. A. Tkachuk.
Vestn. Khar'kovsk. politekhn. in-ta, No. 36 (84), p. 3 - 14 (1969).   In Russian. – Abstr. in Referativ. Zhurn. 51. Astron., 1.51.332 (1970).

104.043 On the characteristics of the sporadic-meteor flow.
N. Carrara, A. Consortini, F. Pasqualetti, L. Ronchi.
Atti Fondaz. Ronchi, Vol. 24, 120 - 152 (1969).

104.044 Measurement of radio-meteor ionization profiles.
J. Jones.
Canadian Journ. Phys., Vol. 47, 1467 - 1473 (1969).
   The theory of a method for the determination of electron line density profiles of underdense radio meteors is developed, and the effect of cosmic noise, winds in the meteor region, and ambipolar diffusion on the measured profiles are estimated.

104.045 Determining the parameters of small-scale turbulence of the atmosphere from combined observations of meteors.   P. B. Babadzhanov, R. Sh. Bibarsov.
Dokl. Akad. Nauk SSSR, Ser. Mat. Fiz., Vol. 189, 67 - 69 (1969).   In Russian.
   Processes leading to a decrease of electron density in a meteor train are discussed. From a critical moment t_2 turbulent diffusion becomes effective. On the basis of combined photographic and radar observations of 9 meteors the following characteristics of small-scale turbulence are calculated: time constant t_2, specific dissipative energy of atmospheric turbulence ω, characteristic dimension l_2, pulsation velocity v_2 and coefficient of atmospheric turbulence D_e. The values of D_e thus found are compared with those derived from observations of steady meteor trains and of artificial clouds. The agreement is satisfactory below 100 km. Further combined meteor observations are desirable.   *B. Onderlička*

104.046 Collection of meteoric dust after the Leonid meteor shower 1965.
C. L. Hemenway, D. S. Hallgren.
Space Research IX, Proc. Tokyo 1968, p. 140 - 146 (1969).

104.047 Visual meteor observation at Úpice, 1964.
Z. Kvíz, F. Žďárský.
Mem. Observations Czechoslovak Astron. Soc., Czechoslovak Acad. Sci., No. 12, 23 pp. (1967).
   The paper gives the results of visual meteor observations, as made at the Úpice Public Observatory during the time of the Perseids activity in August 1964. The observational material gives a number of relatively reliable results. The slope of the luminosity function of the Perseids differs markedly from that of sporadic meteors, the former being smaller. The luminosity function of the Perseids was found to be humpy, just as it was during the Mt. Bezovec expedition in 1961.

104.048 Observations of a very bright bolide.
W. Sędzielowski.
Urania Kraków, Vol. 40, 283 (1969).   In Polish.

104.049 Remarkable display of Geminids observed.
Y. Yabu.
Yamamoto Circ. No. 1710 (1969).   In Japanese.

Un gros bolide au Mexique.
L'Astronomie, 83ᵉ année, 296 (1969).

Few Leonids expected.
Nature, Vol. 224, 639 - 640 (1969).

June Lyrids confirmed.
Sky Telescope, Vol. 38, 271 (1969).

Long-lasting meteor trains.
Sky Telescope, Vol. 38, 379 (1969).

The determination of positions of artificial noctilucent clouds or meteor trains by the method of transformation of coordinates.   See Abstr. 082.103.

Lunar Explorer 35: Picogram lunar ejecta related to the Orionid and Leonid meteor showers, 1968 and 1969.
See Abstr. 094.098.

Lunar Explorer 35: Indications of mass limit for lunar ejecta resulting from hypervelocity impact of meteoroids on the lunar surface.   See Abstr. 094.099.

Astronomical information on meteorite orbits.
See Abstr. 105.130.

On the possible mechanism of meteoric variation of the intensity of cosmic rays.   See Abstr. 143.051.

105 Meteorites, Meteorite Craters

105.001 Porphyrins in meteorites: metal complexes in Orgueil, Murray, Cold Bokkeveld, and Mokoia carbonaceous chondrites. G. W. Hodgson, B. L. Baker.
Geochim. Cosmochim. Acta, Vol. 33, 943 - 958 (1969).

Porphyrin metal complexes were found to be present in samples of the Orgueil, Murray, Cold Bokkeveld, and Mokoia carbonaceous chondrites. Results obtained by absorption spectrophotometry in earlier studies were extended by excitation spectrofluorometry based on analytical demetallation of the porphyrin complexes using methanesulfonic acid. Extensive evaluations were made of the possibilities of contamination for the origin of the meteorite pigments involving terrestrial soils, dusts, and microorganisms. Since the specific biogenic attributes previously assigned to extraterrestrial porphyrins can no longer be accepted, the meteorite porphyrins may be interpreted as representing either prebiotic chemical evolution, or forms of life or diagenesis appreciably different from those responsible for terrestrial porphyrins.

105.002 Isotopic analyses of krypton and xenon in fourteen stone meteorites.
O. Eugster, P. Eberhardt, J. Geiss.
Journ. Geophys. Res. Vol. 74, 3874 - 3896 (1969).

The concentrations and isotopic compositions of Kr and Xe in twelve chondrites and two achondrites have been determined. In addition to the trapped component, Kr and Xe produced by spallation and $(n, \gamma\beta)$ reactions as well as fission and radiogenic Xe have been found. Because of the chemical uniformity of the ordinary chondrites the concentrations of spallation-produced Kr and Xe found in these meteorites correlate with the concentrations of the spallation-produced light rare gases. It is shown that variations in the Kr spallation spectrum are caused by differences in the hardness of the energy spectrum, which are also responsible for the variations in the spallation ratio Ne^{21}/He^3 in ordinary chondrites. Excesses of Kr^{80}, Kr^{82}, and Xe^{128} produced by capture of secondary neutrons in Br and I, respectively, are sensitive indicators for shielding and preatmospheric meteorite sizes. Neutron slowing down densities and minimal preatmospheric radii and masses are calculated on this basis for some of the investigated chondrites. Kr^{81} radiation ages are obtained for three chondrites and for the achondrites.

105.003 Isotopic analyses of barium in meteorites and in terrestrial samples.
O. Eugster, F. Tera, G. J. Wasserburg.
Journ. Geophys. Res. Vol. 74, 3897 - 3908 (1969).

Isotopic composition and concentration of barium in six stone meteorites and the silicate inclusions of two iron meteorites and three terrestrial samples were measured by use of a 'double spike' isotopic dilution technique in order to correct for laboratory fractionation. Any differences between the abundances of the isotopes in meteoritic and terrestrial Ba were found to be less than 0.1% for all isotopes. The per cent abundances of Ba found in this work for the isotopes are given.

105.004 Production of Na^{22} and H^3 in a thick silicate target and its application to meteorites.
B. M. P. Trivedi, P. S. Goel.
Journ. Geophys. Res. Vol. 74, 3909 - 3917 (1969).

A number of glass plates chosen from a stack of thick target having the gross chemical composition of stone meteorites, and exposed to 3 Gev protons, were analyzed for Na^{22} and H^3. The data are used to calculate H^3 and Na^{22} production rates in spherical chondrites exposed to cosmic radiation, assuming that only primary nucleons of energy greater than

1.2 Gev are effective in the nuclide production, and that these can be represented by an average energy of 3 Gev. The calculated production rates of H^3 (450–670 atom min^{-1} kg^{-1}) and Na^{22} (80–110 atom min^{-1} kg^{-1}) are comparable to the observed range of concentrations of these nuclides in stone meteorites.

105.005 Evidence for extraterrestrial life: Identity of sporopollenin with the insoluble organic matter present in the Orgueil and Murray meteorites and also in some terrestrial microfossils. J. Brooks, G. Shaw.
Nature, Vol. 223, 754 - 756 (1969).

105.006 Rubidium-strontium correlation study of moldavites and Ries Crater material.
C. C. Schnetzler, J. A. Philpotts, W. H. Pinson, Jr.
Geochim. Cosmochim. Acta, Vol. 33, 1015 - 1021 (1969).

The Sr^{87}/Sr^{86} values of nine moldavites are essentially constant at 0.722 ± 0.001, although the Rb/Sr ratios vary from 0.77 to 1.2. This very uniform Sr isotopic composition indicates either that the parent material was (a) quite homogeneous or homogenized during the fusion event, or (b) a heterogeneous mixture of two or more phases with different Rb/Sr, perhaps created during the fusion event, in which the isotopic composition of Sr was dominated by that of a Sr-rich component. The Ries materials have distinctly lower Rb/Sr and Sr^{87}/Sr^{86} values than the moldavites.

105.007 Sr isotope patterns within the Southeast Australasian strewn-field.
W. Compston, D. R. Chapman.
Geochim. Cosmochim. Acta, Vol. 33, 1023 - 1036 (1969).

New Rb, Sr and Sr^{87}/Sr^{86} determinations are reported for seventy-three tektites from the Southeast Australasian strewn-field. The specimens represent several of the new chemical types of Chapman and Scheiber (1969) within this field. They were selected for maximum dispersion of Rb/Sr from knowledge of their K and Ca contents. The Rb-Sr data do not distinguish between a lunar or a terrestrial origin. If each of the different chemical types of tektite represent separate differentiated igneous suites from the moon, as proposed by Chapman and Scheiber (1969) then the Rb-Sr isochrons would measure their ages of differentiation.

105.008 Rb-Sr study of impact glass and country rocks from the Henbury meteorite crater field.
W. Compston, S. R. Taylor.
Geochim. Cosmochim. Acta, Vol. 33, 1037 - 1043 (1969).

Rb-Sr data for the Henbury country rock (Winnall Beds) and the laterally equivalent Pertatataka Formation 95 miles ENE of Henbury, fall close to an isochron with age 7.30 m.y. ± 45 m.y. and an Sr^{87}/Sr^{86} initial ratio of 0.724. The scatter about the isochron is interpreted as variation in initial Sr^{87}/Sr^{86} reflecting small changes in Sr provenance. The impact glass data are particularly well-aligned over a 15% range in Rb/Sr, and independently define an isochron of 675 ± 100 m.y.

105.009 Isotopic comparison of lead in tektites with lead in earth materials.
J. M. Wampler, D. H. Smith, A. E. Cameron.
Geochim. Cosmochim. Acta, Vol. 33, 1045 - 1055 (1969).

Lead samples from moldavites and from Ries glass are isotopically similar, except that the Ries glass tends to be slightly enriched in the radiogenic lead isotopes relative to moldavites. The lead in these materials is similar to the lead found in oceanic sediments and falls on the growth curves typical of most terrestrial rock and ore lead samples.

105.010 Apparent K-Ar dates on cores and excess Ar in flanges of australites.
I. McDougall, J. F. Lovering.
Geochim. Cosmochim.. Acta, Vol. 33, 1057 - 1070 (1969).

Potassium and radiogenic argon contents have been measured on 13 australite cores from widely separated regions in Australia. The apparent K−Ar dates average 0.86 ± 0.06 m.y. (standard deviation), approximately 0.15 m.y. older than found by previous workers. The internal consistency of the K−Ar dates on the australite cores, which have potassium contents ranging from 1.61 to 2.18 per cent, suggests that the dates are geologically meaningful, and probably record the time of primary melting.

105.011 Ages of Darwin glass, Macedon glass, and Far Eastern tektites.
R. L. Fleischer, P. B. Price, J. R. M. Viertl, R. T. Woods.
Geochim. Cosmochim. Acta, Vol. 33, 1071 - 1074 (1969).

A group of possibly related glasses from the Far East, including seven samples of Darwin glass and two known samples of Macedon glass, have been dated by the fission track technique. With the exception of a lower age for one Darwin glass sample and one Muong Nong tektite all ages are indistinguishable, suggesting that the Darwin and Macedon impact glasses are related to the Far Eastern tektites.

105.012 New fission track ages of tektites and related glasses.
W. Gentner, D. Storzer, G. A. Wagner.
Geochim. Cosmochim. Acta, Vol. 33, 1075 - 1081 (1969).

Fission track dating is applied to a large number of australites, Muong Nong type tektites, Darwin glasses and Libyan Desert glasses. It is found that these glasses often have lowered fission track ages due to annealing effects. The same specimens have also smaller fission track etch pits. According to the diminishing percentage of etched fossil fission track diameters, lowered fission track ages are corrected. A mean age of 0.7 m.y. is found for eleven australites, six Darwin glasses and seven Muong Nong type tektites (from Laos and Thailand), indicating a related genesis of these glasses.

105.013 Genetic significance of the chemical composition of tektites: A review.
S. R. Taylor, M. Kaye.
Geochim. Cosmochim. Acta, Vol. 33, 1083 - 1100 (1969).

Evidence from internal compositional variations and ratios, selective vaporisation and impact glass analogies, bearing on the relationship of the present composition to that of the parent material indicates that concentration changes during melting have been minor, thus enabling the parent material to be identified. Terrestrial igneous rocks are not suitable source material, but terrestrial sandstones show strong similarities in major and trace element abundances, and exhibit analogous inter-element variations to tektites. A lunar origin of tektites is not compatible with the chemistry of the lunar surface as presently known. A terrestrial origin by impact on sandstones of appropriate composition is consistent with the chemical evidence.

105.014 The radioactivity of the Ivory Coast tektites and the formation of the Bosumtwi crater (Ghana).
L. Rybach, J. A. S. Adams.
Geochim. Cosmochim. Acta, Vol. 33, 1101 - 1102 (1969).

Further evidence for the impact origin of the Ivory Coast tektites and the Bosumtwi crater is presented: uranium, thorium and potassium abundances show very similar patterns in tektites and crater rocks.

105.015 Chemical composition and bulk density of moldavites.   J. Konta, L. Mráz.
Geochim. Cosmochim. Acta, Vol. 33, 1103 - 1111 (1969).

The weight, dimensions, bulk density, abundance of bubbles, abundance of lechatelierite, size frequency of lechatelierite and maximum projection sphericity were determined in thirty five moldavites (29 from the Bohemian and 6 from Moravian localities). Fifteen new chemical analyses of Bohemian moldavites are reported. Chemical composition and size frequency of lechatelierite, were the factors determining the bulk density of moldavites.

105.016 Magnetite bearing spherules in tektites.
B. Kleinmann.
Geochim. Cosmochim. Acta, Vol. 33, 1113 - 1120 (1969).

In the tektites, three kinds of spherules were found: type I, magnetic spherules consisting of magnetite skeletons embedded in a glass matrix with a small nucleus of pure iron; type II, intergrowths of magnetite and wüstite and some interstitial glass; and type III, a single glass spherule which contains finely distributed magnetic skeletons and idiomorphic magnetite crystals partially intergrown with wüstite. The absence of nickel in the metallic phase and the lack of diffusion boundaries around the spherules in tektites is evidence that the spherules are not of meteoritic origin but formed within the tektite glass.

105.017 Petrology of moldavites.   V. E. Barnes.
Geochim. Cosmochim. Acta, Vol. 33, 1121 - 1134 (1969).

105.018 Chemical composition of Ivory Coast microtektites.
B. P. Glass.
Geochim. Cosmochim. Acta, Vol. 33, 1135 - 1147 (1969).

The chemical composition of twenty-eight microscopic glassy objects, found in a deep-sea sediment core taken near the Ivory Coast, has been determined by electron microprobe analysis. Like the Australasian microtektites, these glassy objects can be divided into two main groups. One group is composed of individuals with chemical compositions similar to tektites. These are referred to as "normal" microtektites. The other group is composed of bottle-green glassy objects with low silica contents ($< 53\%$) and high MgO contents (16−21%).

105.019 The composition of the stony meteorites, (IV). Some analytical data on Orgueil, Nogoya, Ornans and Ngawi.   L. H. Ahrens, H. von Michaelis, H. W. Fesq.
Earth Planet. Sci. Letters, Vol. 6, 285 - 288 (1969).

New analytical data on Fe, Si, Mg, Ca, Al, Mn, K, Ti and P in three carbonaceous chondrites (Orgueil, Nogoya and Ornans) and one LL-group chondrite (Ngawi) are presented and compared with other data in the literature. The new data confirm a previous conclusion that the Ca/Al ratio of 1.10 (± a small factor) is typical of all common stony meteorites. A systematic difference between our phosphorus values and others given in the literature has been noted.

105.020 The composition of stony meteorites, (V). Some aspects of the composition of the basaltic achondrites.   L. H. Ahrens, H. von Michaelis.
Earth Planet. Sci. Letters, Vol. 6, 304 - 308 (1969).

Unique compositional features characterize the basaltic achondrites (eucrites and howardites), notably; remarkably uniform Si and almost uniform Fe contents; a constant Ca/Al ratio which is equivalent to that of the chondrites; an almost uniform Na/K ratio also similar in magnitude to that in the chondrites and a seemingly unusually well-developed inverse Ca (or Al)-Mg relationship. These compositional features suggest strongly that the basaltic achondrites could not have been formed by a process of magmatic differentiation and that physical processes played a dominant role.

105.021 Some halogen measurements on achondrites.

G. W. Reed, S. Jovanovic.
Earth Planet. Sci. Letters, Vol. 6, 316 - 320 (1969).

Data are presented on halogen concentration in various classes of achondritic meteorites. The importance of the lability of the halogens to the interpretation of rare gas data is pointed out.

105.022 Determination of an internal ^{87}Rb-^{87}Sr isochron for the Olivenza chondrite.
H. G. Sanz, G. J. Wasserburg.
Earth Planet. Sci. Letters, Vol. 6, 335 - 345 (1969).

Rb-Sr isotopic determinations were made on five single chondrules, ranging from 4 to 27 milligrams, extracted from Olivenza, an olivine-hypersthene chondrite. Several density fractions with a particle size smaller than 37 μ were separated and Rb-Sr analyses made. These data form a well defined linear array on the Sr-Rb evolution diagram and yield an age of 4.63×10^9 years and an initial ^{87}Sr/^{86}Sr = 0.6994. The typical strontium and rubidium blanks were 8×10^{-10} and 4×10^{-11} g, respectively.

105.023 I-Xe dating of silicates from Toluca iron.
E. C. Alexander, Jr., B. Srinivasan, O. K. Manuel.
Earth Planet. Sci. Letters, Vol. 6, 355 - 358 (1969).

A Reynolds type heating experiment on neutron irradiated silicates from Toluca indicates that the high temperature sites began to retain radiogenic ^{129}Xe when the ^{129}I/^{127}I ratio was $(1.4 \pm 0.4) \times 10^{-3}$. This ratio suggests that Toluca silicates formed 65^{+6}_{-8} million years before the chondrites and the Shallowater achondrite.

105.024 Trapped neon in meteorites.—II.
D. C. Black, R. O. Pepin.
Earth Planet. Sci. Letters, Vol. 6, 395 - 405 (1969).

Studies of the structure of trapped meteoritic neon are continued through stepwise heating experiments on carbonaceous chondrites and gas-rich meteorites. Trapped ^{20}Ne/^{22}Ne ratios are observed to vary by more than a factor 4, from extremes of 3.4−4.8 in 1000° C gas fractions from six carbonaceous chondrites to ~ 14 in neon evolved from Kapoeta at very low temperatures. The description of trapped neon in meteorites as a variable mixture of two components with ^{20}Ne/^{22}Ne = 8.2 (Neon-A) and 12.5 (Neon-B) is qualitatively valid for total neon, but it cannot account for the extreme compositions revealed in these individual temperature fractions. At least two other components, with ^{20}Ne/^{22}Ne $\lesssim$ 3.4 and $\gtrsim$ 14, are required.

105.025 Correlation between fission tracks and fission type xenon in meteoritic whitlockite.
G. J. Wasserburg, J. C. Huneke, D. S. Burnett.
Journ. Geophys. Res. Vol. 74, 4221 - 4232 = Contr. 1630 California Inst. Technology, Pasadena (1969).

Whitlockite from the St. Severin chondrite, previously shown to contain excess fission tracks, is here shown to have a large concentration of excess neutron-rich xenon isotopes. The concentration of excess heavy Xe in the whitlockite is about twenty-five times that calculated from the track density. An isotopic spectrum is deduced that is identical to the spectrum calculated previously for excess heavy xenon in the Pasamonte achondrite. These results uniquely associate this xenon spectrum with in situ fission in meteorites. Chemical arguments support the correlation of this with Pu^{244}. Identification of the fissioning nucleus as Pu^{244} gives Pu^{244}/U^{238} $\approx$ 1/30. Neither 'sudden' nor 'uniform' nucleosynthetic models give consistent solutions for Pu^{244}/U^{238} and U^{235}/U^{238}

105.026 Rubidium-strontium age of amphoterite (LL) chondrites. K. Gopalan, G. W. Wetherill.
Journ. Geophys. Res. Vol. 74, 4349 - 4358 (1969).

Eleven falls and three finds from the amphoterite (LL)

group of chondrites have been analyzed for K, Rb, and Sr concentrations and Sr isotopic composition with negligible blank contributions. The samples measured show a sufficiently good spread in their Rb/Sr ratios to define a good isochron with a slope of 0.0654 ± 0.00037 (1σ) and an intercept of 0.7005 ± 0.00015.

105.027 Rotational bursting of small celestial bodies: Effects of radiation pressure. S. J. Paddack.
Journ. Geophys. Res. Vol. 74, 4379 - 4381 (1969).

Solar radiation pressure can cause rotational bursting and eventual elimination from the solar system of small nonmagnetic bodies in space by a sort of windmill effect. A time to bursting of about sixty years was calculated for an idealized body. On the basis of a simple experiment, it is estimated that actual bodies, such as nonmagnetic meteorites and tektites, will reach bursting speed in about 60.000 years.

105.028 Comments on paper by D. E. Fisher and M. F. Swanson, 'Frequency distribution of meteorite-earth collisions'. G. W. Wetherill.
Journ. Geophys. Res. Vol. 74, 4402 - 4405, with a reply by D. E. Fisher, p. 4406 (1969). − Letter.

105.029 La Malbaie structure, Quebec. − A palaeozoic meteorite impact site. P. B. Robertson.
Meteoritics, Vol. 4, 89 - 112 = Contr. Dominion Obs. Ottawa, No. 249 (1968).

Evidence in support of an origin by meteorite impact has been found at La Malbaie structure, 105 km northeast of Quebec City on the north shore of the St. Lawrence River. A detailed description of the crater geology is given. The stratigraphic and tectonic history of the region indicates that the crater originated between Middle Ordovician and late Devonian times.

105.030 Kwantitatief onderzoek van nikkelijzer micrometeorieten. A. Hermans, A. van Waarde, T. van Dijk.
Hemel en Dampkring, Vol. 67, 253 (1969).

105.031 Enstatite: Disorder produced by a megabar shock event. S. S. Pollack, P. S. DeCarli.
Science, Vol. 165, 591 - 592 (1969).

Shocked Bamle enstatite partly transforms to disordered enstatite. Debye-Scherrer patterns of some shocked material are almost identical to those of disordered enstatite from portions of various enstatite achondrites. No disordered single crystals have been found.

105.032 Radioactive isotopes in Hoba West and other iron meteorites. R. H. McCorkell, E. L. Fireman, J. D'Amico, S. O. Thompson.
Meteoritics, Vol. 4, 113 - 122 (1968).

The radioactivities Be^{10}, Al^{26}, Cl^{36}, Mn^{53}, Ni^{59}, and Co^{60} and the rare gas isotopes were measured in a sample taken from the surface of the Hoba meteorite. The spallation-produced radioactivities indicate that the sample was at a depth of 35 to 40 cm when the body was in space. The Ni^{59} activity indicates that the terrestrial age of Hoba is less than 80.000 years. Its Cl^{36} − Ar^{36} exposure age is 263 ± 40 million years. The Cl^{36} and rare-gas isotopes were also measured in some other meteorites.

105.033 Effect of pressure and temperature on the reversal transitions of stishovite.
P. D. Gigl, F. Dachille.
Meteoritics, Vol. 4, 123 - 136 (1968).

The work concerning the persistence of stishovite gives support to the hypothesis that its presence at the earth's surface could only have been the result of meteoritic impact.

This paper explores this thesis further by studying the influence of various pressure-temperature (p – t) conditions on the reversal transitions of stishovite. The parameters of shear stresses and mineralizers were also utilized in a number of experiments.

105.034 The Oshkosh meteorite.   W. F. Read.
Meteoritics, Vol. 4, 137 - 140 (1968).
In the fall of 1961, fragments of an olivine-bronzite chondrite were found about 2 miles NNW of Oshkosh, Wisconsin, the total weight being 144.8 g. This paper fixes the exact location and describes the circumstances of the find.

105.035 Zn^{65}, nickel, copper, and zinc contents of the Bogou iron meteorite.   H. R. Heydegger, A. Turkevich.
Journ. Geophys. Res. Vol. 74, 4949 - 4957 (1969).
The Zn^{65} content of a portion of the Bogou iron meteorite, which fell in Upper Volta on August 14, 1962, has been determined by means of radiochemical techniques. The specific radioactivity of Zn^{65} in Bogou at the time of fall is estimated to be 5.0 ± 1.7 dpm kg^{-1}. It is suggested that this nuclide may have been produced principally by the Ni^{62} (a, n) Zn^{65} reaction induced by solar flare particles. The elemental abundances of Cu and of Zn in the same portion of the Bogou meteorite have been determined by neutron activation analysis to be 154 ± 8 ppm and 67 ± 3 ppm, respectively. The Ni abundance was determined gravimetrically to be $7.0 \pm 0.4\%$.

105.036 Muzzaffarpur meteorite.
D. R. Das Gupta, T. V. Viswanathan, N. R. Sen Gupta, S. Banerjee.
Geochim. Cosmochim. Acta, Vol. 33, 1298 - 1302 (1969).
Petrological, X-ray and chemical analyses of the Muzzaffarpur nickel-rich ataxite show that it consists mainly of kamacite and taenite and minor schreibersite. The relative amount of kamacite in the meteorite increases in the interior of the meteorite. The change in the chemical composition of the meteorite during heating in the laboratory showed that as the temperature increases the meteorite oxidises forming minerals similar to trevorite and hematite at 1000°C.

105.037 Possible meteoritic crater.
V. V. Teljnjuk-Adamchuk.
Astron. Tsirk. No. 507, p. 8 (1969).   In Russian.

105.038 Zur Bestimmung des metallischen und sauren Eisens in Chondriten.
A. A. Yavnel', M. I. D'yakonova.
Meteoritika, No. 29, p. 30 - 35 (1969).   In Russian.

105.039 Der Steinmeteorit von Odessa.
O. A. Kirova, M. I. D'yakonova.
Meteoritika, No. 29, p. 36 - 47 (1969).   In Russian.

105.040 Der Chondrit von Severnyj Kolchim, UdSSR.
O. K. Ivanov.
Meteoritika, No. 29, p. 48 - 56 (1969).   In Russian.

105.041 Morphologie und Zusammensetzung des Meteoriten Gumoshnik, Bulgarien.   D. Dimov.
Meteoritika, No. 29, p. 57 - 67 (1969).   In Russian.

105.042 Die Struktur des nickelhaltigen Eisens und das Sulfid des Meteoriten Santa Catharina, Brasilien.
L. G. Kvasha, V. D. Kolomenskij, I. A. Bud'ko.
Meteoritika, No. 29, p. 68 - 75 (1969).   In Russian.

105.043 Der Eisenmeteorit Burgavli, UdSSR.
V. F. Alyavdin.
Meteoritika, No. 29, p. 76 - 90 (1969).   In Russian.

105.044 Die chemische Zusammensetzung der Chondrite Kaande, Kargonal'e, Dimmitt, Chico und New Almelo aus der Sammlung des Komitees für Meteorite der Akademie der Wissenschaften der UdSSR.
V. Ya. Kharitonova.
Meteoritika, No. 29, p. 91 - 93 (1969).   In Russian.

105.045 Röntgenographische Untersuchung des Maskelynits und eine neue chemische Analyse des Meteoriten Andronishkis.   V. A. Vasil'ev.
Meteoritika, No. 29, p. 94 - 100 (1969).   In Russian.

105.046 Die Häufigkeit von Tonerde und Titanoxyd in Steinmeteoriten.   M. I. D'yakonova.
Meteoritika, No. 29, p. 101 - 103 (1969).   In Russian.
In früher analysierten Meteoriten (20 Chondrite, 3 Achondrite) wurde die Häufigkeit von Aluminium und Titan bestimmt. Das Aluminium wurde vorher von anderen Ionen getrennt. Zur Bestimmung des Titans wurde die kolorimetrische Methode verwandt.   *D. Krahn*

105.047 Die Mikrohärte einiger meteoritischer Mineralien.
G. P. Vdovykin.
Meteoritika, No. 29, p. 104 - 109 (1969).   In Russian.

105.048 Ultra- und Mikroporosität von Eisenmeteoriten.
A. S. Shur, N. T. El'kina, I. A. Yudin.
Meteoritika, No. 29, p. 110 - 115 (1969).   In Russian.

105.049 Magnetische Eigenschaften von Meteoriten einiger Sammlungen der Sowjetunion.
E. G. Gus'kova.
Meteoritika, No. 29, p. 116 - 127 (1969).   In Russian.

105.050 Vergleichende Charakteristik der elektromagnetischen Parameter der Mondoberfläche, der Meteorite und Tektite.   K. N. Alekseeva, E. G. Gus'kova.
Meteoritika, No. 29, p. 128 - 131 (1969).   In Russian.
Verf. messen die magnetische Suszeptibilität von 23 Meteoriten, den elektrischen Widerstand und die Dielektrizitätskonstante von 16 Meteoriten sowie die Dielektrizitätskonstante von 4 Tektiten. Der Zusammenhang dieser Größen mit der Metallhäufigkeit, dem Gehalt an Sulfiden sowie den elastischen Eigenschaften der Meteorite wird diskutiert. Ein Vergleich mit entsprechenden Werten für die Mondoberfläche liefert keine Bestätigung für die Hypothese des lunaren Ursprungs der Tektite.   *D. Krahn*

105.051 Untersuchung von künstlichem Meteorstaub (Kügelchen).   I. A. Yudin.
Meteoritika, No. 29, p. 132 - 141 (1969).   In Russian.

105.052 Physikalische Parameter einiger Tektite aus der Tschechoslowakei und aus Indochina.
K. N. Alekseeva.
Meteoritika, No. 29, p. 142 - 145 (1969).   In Russian.

105.053 Thermolumineszenz von Tektiten und Obsidianen.
V. G. Kashkarova, L. L. Kashkarov.
Meteoritika, No. 29, p. 146 - 151 (1969).   In Russian.

105.054 Über den Fund des Eisenmeteoriten Sejmchan, UdSSR.   V. I. Tsvetkov.
Meteoritika, No. 29, p. 152 - 153 (1969).   In Russian.

105.055 Über die Meteorite Tubil und Abakan, UdSSR.
A. A. Yavnel'.
Meteoritika, No. 29, p. 154 - 156 (1969).   In Russian.

105.056 Korrektur der veröffentlichten Angaben über Ra-

dianten und Bahnen von Meteoriten.
A. N. Simonenko.
Meteoritika, No. 29, p. 157 - 159 (1969).   In Russian.

105.057 Untersuchung der Meteorite Lettlands im 19. Jahrhundert.   I. A. Daube.
Meteoritika, No. 29, p. 160 - 162 (1969).   In Russian.

105.058 Überlegungen im Zusammenhang mit dem neuen Meteoritenfund in Australien.   J. Classen.
Meteoritika, No. 29, p. 163 - 165 (1969).   In Russian.

105.059 Über Reliktstrukturen von Steinmeteoriten in den Ablagerungen des Mesozoikums des mittleren Urals.   I. A. Yudin.
Meteoritika, No. 29, p. 166 - 169 (1969).   In Russian.

105.060 Die anomale mit dem Tungusischen Meteoriten zusammenhängende Dämmerung.   I. T. Zotkin.
Meteoritika, No. 29, p. 170 - 176 (1969).   In Russian.

105.061 Katalog der Meteoritensammlungen der Litauischen SSR.   V. A. Vasil'ev.
Meteoritika, No. 29, p. 177 - 179 (1969).   In Russian.

105.062 Jadeite: Shock-induced formation from oligoclase Ries Crater, Germany.   O. B. James.
Science, Vol. 165, 1005 - 1008 (1969).

105.063 Allende meteorite: Age determination by thermoluminescence.
S. A. Durrani, C. Christodoulides.
Nature, Vol. 223, 1219 - 1221 (1969).
An attempt is made to measure the age of Allende meteorite, which fell in Mexico on February 8, 1969, by the thermoluminescence method.

105.064 Lanthanides in the silicate inclusion of the Woodbine meteorite.   A. Masuda.
Nature, Vol. 224, 164 - 165 (1969).
Silicate inclusion material weighing 235 mg was analysed for lanthanides, and their abundances were determined by the isotope dilution technique.

105.065 Reworking of deep-sea sediments as indicated by the vertical dispersion of the Australasian and Ivory Coast microtektite horizons.   B. P. Glass.
Earth Planet. Sci. Letters, Vol. 6, 409 - 415 = Contr. Lamont Geol. Obs. No. 1360 (1969).
Microtektites (small glassy objects) have been found in deep-sea pelagic sediments associated with both the Australasian and Ivory Coast tektite strewn fields. Because of their wide geographical distribution, age of deposition (0.7 to 1.0 m.y. ago), size (~1 mm down to at least 20 microns in diameter) and ease of recognition, and investigation of the vertical dispersion of the microtektite horizons can tell us much about reworking of pelagic sediments. In the eight cores investigated the microtektites are dispersed through a vertical section of from 35 to 90 cm. This represents an average of 120000 yr of deposition. The distribution of the microtektites within the dispersed zone seems to be related in a general way to the amount of burrowing evident in the core.

105.066 Nitrogen abundances in enstatite chondrites.
C. B. Moore, E. K. Gibson, Jr., K. Keil.
Earth Planet. Sci. Letters, Vol. 6, 457 - 460 (1969).

105.067 Eine neue Kohlenstoff-Modifikation aus dem Nördlinger Ries.   A. El Goresy.
Naturwissenschaften, 56. Jahrgang, 493 - 494 (1969).

105.068 „Peckelsheim". A new bronzite achondrite.
P. Ramdohr, A. El Goresy.
Naturwissenschaften, 56. Jahrgang, 512 (1969).

105.069 Verschleppte Tektite in Liberia.
E. Preuss.
Naturwissenschaften, 56. Jahrgang, 512 (1969).

105.070 Attempts to measure micrometeoroid flux on the OGO 2 and OGO 4 satellites.
C. S. Nilsson, F. W. Wright, D. Wilson.
Journ. Geophys. Res. Vol. 74, 5268 - 5276 (1969).
The micrometeoroid experiments on the OGO 2 and OGO 4 satellites are described. The aim of the OGO 2 experiment was to measure the velocities, masses, and orbits of dust particles in the earth's dust cloud. No orbits were determined. The OGO 4 experiment was modified in an attempt to measure a flux obviously much smaller than previously anticipated. No micrometeoroids capable of penetrating 4000 A of Al have impacted. We find that the flux of micrometeoroids $> 10^{-12}$ g in the neighborhood of the earth is less than 2×10^{-3} particles/m^2 sec 2π ster.

105.071 Initial strontium for a chondrite and the determination of a metamorphism or formation interval.
G. J. Wasserburg, D. A. Papanastassiou, H. G. Sanz.
Earth Planet. Sci. Letters, Vol. 7, 33 - 43 (1969).
A precise Rb-Sr internal isochron was determined for Guareña, an H6 chondrite, yielding an age of $4.56 \pm 0.08 \times 10^9$ years. Rb-poor, Sr-rich phosphate phases (whitlockite and apatite) were obtained resulting in a precise measurement of the initial Sr isotopic composition. It is shown that from precise age and initial ($^{87}Sr/^{86}Sr$)$_1$ measurements we can obtain information on the differential evolution of Rb-Sr systems involving either simple metamorphism of closed systems or multi-stage processes.

105.072 A second tektite fall in Australia.
R. L. Fleischer, P. B. Price, R. T. Woods.
Earth Planet. Sci. Letters, Vol. 7, 51 - 52 (1969).
A chemical subgroup of tektites found in Australia (Chapman's "high sodium" tektites) have fission track ages of approximately 4 m.y. and hence are part of a distinct tektite fall, one that preceded the formation of the well known Australasion strewn field 0.7 m.y. ago.

105.073 Uranium and thorium in tektites.   J. W. Morgan.
Earth Planet. Sci. Letters, Vol. 7, 53 - 63 (1969).
Neutron activation analysis for uranium and thorium are reported for two australites, a javaite, a bediasite and a moldavite. A critical survey is made of published uranium and thorium abundances in tektites, and superior analyses selected. Uranium and thorium abundances in the tektite groups are compared with terrestrial sedimentary and acid igneous rocks. Granophyres match tektites most closely in the abundance of these two elements, but cannot be considered probable parent material for a terrestrial tektite origin because of their limited distribution.

105.074 Isotopic composition of meteoritic thallium.
R. G. Ostic, H. M. El-Badry, T. P. Kohman.
Earth Planet. Sci. Letters, Vol. 7, 72 - 76 (1969).
Thallium contents have been determined as $2.5 \pm 0.5 \times 10^{-6}$ in Ivigtut, Greenland, galena; $7 \pm 2 \times 10^{-10}$ in Plainview chondrite; $1.5 \pm 0.3 \times 10^{-9}$ in Canyon Diablo troilite; and $2.5 \pm 0.5 \times 10^{-10}$ in Canyon Diablo metal.

105.075 Recent advances in the study of fossil tracks in meteorites due to heavy nuclei of the cosmic radiation.
D. Lal.

Space. Sci. Rev. Vol. 9, 623 - 650 (1969). − Review article.

105.076 Allende meteorite: Some major and trace element abundances by neutron activation analysis.
J. W. Morgan, T. V. Rebagay, D. L. Showalter, R. A. Nadkarni, D. E. Gillum, D. M. McKown, W. D. Ehmann.
Nature, Vol. 224, 789 - 791 (1969).

In assuming that the Allende meteorite which belongs to the carbonaceous chondrites will become important in the interlaboratory comparison of analytical techniques, the authors have analysed for fourteen elements in homogenized powders. The element abundances are presented.

105.077 Statistics of meteorite falls. B. Hellyer.
Earth Planet. Sci. Letters, Vol. 7, 148 - 150 (1969).

Data from a large number of observed meteorite falls have been investigated. Correlations have been sought between the ratio of multiple falls to total falls and (a) the mass of the meteorite, (b) the local time of fall and (c) the class of the meteorite. It was found that the probability of fragmentation is strongly dependent on the mass of the body. The ratio of multiple falls is relatively high for carbonaceous chondrites and low for irons but seems independent of all other factors considered.

105.078 Shock, reheating, and the gas retention ages of chondrites. G. J. Taylor, D. Heymann.
Earth Planet. Sci. Letters, Vol. 7, 151 - 161 (1969).

In order to investigate shock and reheating effects in ordinary chondrites, 71 bronzite, 26 hypersthene and 6 amphoterite chondrites were examined by X-ray diffraction and 39 bronzite and 19 hypersthene chondrites were studied metallographically. In addition, the inert gas contents of 11 bronzites and 4 hypersthene chondrites were measured mass spectrometrically. We have found that virtually all chondrites with short gas retention ages (<2 b.y.) are substantially shocked and reheated, whereas those with ages >3 b.y. are not reheated to any appreciable extent, although they may be strongly shocked.

105.079 Silicon concentrations in the metal of iron meteorites. C. M. Wai, J. T. Wasson.
Geochim. Cosmochim. Acta, Vol. 33, 1465 - 1471 (1969).

Electron microprobe determinations of Si in the metal of 19 iron meteorites, including representatives of the nine chemical groups, reveal Si concentrations below our detection limit (about 25 ppm) in all but two cases. The two objects are Tucson, an iron of anomalous composition containing highly reduced silicate inclusions, and Horse Creek, a meteorite which is more appropriately classified with the enstatite chondrites than with the iron meteorites. We conclude that most (probably more than 99%) of the iron meteorites contain less than 30 ppm Si in the metal, and that meteoritic evidence for the presence of Si in the earth's core is very weak. Some details regarding the structures and composition of Tucson are given.

105.080 Cliftonite: A proposed origin, and its bearing on the origin of diamonds in meteorites.
R. Brett, G. T. Higgins.
Geochim. Cosmochim. Acta, Vol. 33, 1473 - 1484 (1969).

Cliftonite, a polycrystalline aggregate of graphite with spherulitic structure and cubic morphology, is known in 14 meteorites. Some workers have considered it to be a pseudomorph after diamond, and have used the proposed diamond ancestry as evidence of a meteoritic parent body of at least lunar dimensions. Careful examination of meteoritic samples indicates that cliftonite forms by precipitation within kamacite. We have also demonstrated that graphite with cubic morphology may be synthesized in a Fe - Ni - C alloy annealed in a vacuum. We therefore suggest that a high pressure origin is unnecessary for meteorites which contain cliftonite, and that these meteorites were formed at low pressures.

105.081 Solar-type xenon: A new isotopic composition of xenon in the Pesyanoe meteorite. K. Marti.
Science, Vol. 166, 1263 - 1265 (1969).

Xenon in the Pesyanoe meteorite is a mixture of several components. Solar-type xenon is a new component deficient in the neutron-rich isotopes as compared to both trapped chondritic and terrestrial atmospheric xenon.

105.082 Radionuclide composition of the Allende meteorite from nondestructive gamma-ray spectrometric analysis.
L. A. Rancitelli, R. W. Perkins, J. A. Cooper, J. H. Kaye, N. A. Wogman.
Science, Vol. 166, 1269 - 1272 (1969).

The concentrations of beryllium-7, sodium-22, aluminum-26, potassium-40, scandium-46, vanadium-48, chromium-51, manganese-54, cobalt-57, cobalt-60, and thorium-232 (thallium-208) have been measured in the Allende meteorite by nondestructive gamma-ray spectrometry. The high cobalt-60 content of the meteorite is indicative of a preatmospheric body with a minimum effective radius of 50 centimeters and a weight of 1650 kilograms; the aluminum-26 activity indicates a minimum exposure age of 3 million years.

105.083 Shock induced thermal metamorphism and mechanical deformations in the Ramsdorf chondrite.
F. Begemann, F. Wlotzka.
Geochim. Cosmochim. Acta, Vol. 33, 1351 - 1370 (1969).

The grey hypersthene chondrite Ramsdorf, which has short and concordant He-U/Th- and Ar-K-gas retention ages, shows signs of a severe reheating, probably caused by shock. These are recrystallization of the silicates, formation of silicate glass, melting and redistribution of metal and troilite. Rapid cooling is indicated by the formation of Ni-rich rims surrounding the metal. The cooling time calculated from the width of these rims shows that proto-Ramsdorf had a radius of less than two meters at that time. The time at which this event happened is discussed.

105.084 "Lunar" craters on the earth. I. T. Zotkin.
Priroda No. 9, p. 95 - 105 (1969). In Russian.

105.085 Mineralogical study of nickel-free iron in stony meteorites. I. A. Yudin.
Mineralog. sb. In–t geol. i geokhimii, Ural'skij fil. AN SSSR, No. 8, p. 137 - 140 (1969). In Russian. − Abstr. in Referativ. Zhurn. 51. Astron., 9.51.676 (1969).

105.086 Noble gases in iron meteorites and fission reactions.
L. K. Levskij, A. N. Murin.
Geokhimiya, No. 4, p. 478 - 483 (1969). In Russian. − Abstr. in Referativ. Zhurn. 51. Astron., 9.51.677 (1969).

105.087 On the mineralogy of meteor dust. I. A. Yudin.
Mineralog. sb. In–t geol. i geokhimii. Ural'skij fil. AN SSSR, No. 8, p. 136 - 137 (1968). In Russian. − Abstr. in Referativ. Zhurn. 51. Astron., 10.51.726 (1969).

105.088 Nuclear abundance rules and the composition of meteorites. H. E. Suess.
Meteorite Research, Vienna 1968, p. 3 - 6 (1969).

This paper deals with the questions of which type of meteorite can be expected to best resemble the composition of primordial matter and to what degree fractionation processes can still be recognized.

105.089 Origin of meteorites and planetary cosmogony.
B. Ju. Levin.

Meteorite Research, Vienna 1968, p. 16 - 30 (1969).

105.090 Recent fossil track studies bearing on extinct Pu²⁴⁴ in meteorites. J. Shirck, M. Hoppe, M. Maurette, R. Walker.
Meteorite Research, Vienna 1968, p. 41 - 50 (1969).

105.091 Prior diagram for chondrites.
A. A. Yavnel.
Meteorite Research, Vienna 1968, p. 51 - 59 (1969).
In Russian.

105.092 Meteorites and the high-temperature origin of terrestrial planets. W. Kiesl, F. Hecht.
Meteorite Research, Vienna 1968, p. 67 - 74 (1969).
In this paper the authors try to relate their own investigations on the chemical composition of meteorites, as well as petrography, mineralogy and age relationships. From these, and from the monistic theory of the origin of the solar system which is today generally admitted, it seems that the planets were condensed or accreted in a period of a fully convective early protosun, which may have had a luminosity of some orders of magnitude higher than the present sun, thus producing very high temperatures in the inner regions of the solar system. A possible mechanism for the origin of carbonaceous chondrites is also given.

105.093 Study of trace element abundance in meteorites by neutron activation. R. Rieder, H. Wänke.
Meteorite Research, Vienna 1968, p. 75 - 86 (1969).
About 20 elements with highly varying chemical and physical properties have been selected, ranging from the alkaline to the noble metals and from indium to tungsten. The measurements were carried out by thermal neutron activation technique.

105.094 Content and isotopic composition of carbon in the light and dark portions of gas-rich chondrites.
F. Begemann, K. Heinzinger.
Meteorite Research, Vienna 1968, p. 87 - 92 (1969).

105.095 Mössbauer investigation of the unequilibrated ordinary chondrites. E. L. Sprenkel-Segel.
Meteorite Research, Vienna 1968, p. 93 - 105 (1969).

105.096 Mössbauer spectroscopy applied to the classification of stone meteorites. W. Herr, B. Skerra.
Meteorite Research, Vienna 1968, p. 106 - 122 (1969).

105.097 Isotopic composition of lithium in some meteorites and the role of neutrons in the nucleosynthesis of the light elements in the solar system.
R. Bernas, E. Gradsztajn, A. Yaniv.
Meteorite Research, Vienna 1968, p. 123 - 131 (1969).

105.098 Determination of boron, lithium, and chlorine in meteorites. M. Quijano-Rico, H. Wänke.
Meteorite Research, Vienna 1968, p. 132 - 145 (1969).

105.099 The Sharps chondrite – New evidence on the origin of chondrules and chondrites.
K. Fredriksson, E. Jarosewich, J. Nelen.
Meteorite Research, Vienna 1968, p. 155 - 165 (1969).

105.100 Fractionation of some abundant lithophile element ratios in chondrites. L. H. Ahrens, H. von Michaelis, A. J. Erlank, J. P. Willis.
Meteorite Research, Vienna 1968, p. 166 - 173 (1969).

105.101 On the formation of chondrules and metal particles by 'shock melting'. F. Wlotzka.

Meteorite Research, Vienna 1968, p. 174 - 184 (1969).

105.102 The formation of chondrules and chondrites and some observations on chondrules from the Tieschitz meteorite. G. Kurat.
Meteorite Research, Vienna 1968, p. 185 - 190 (1969).

105.103 The major-element composition of individual chondrules of the Bjurböle meteorite.
L. S. Walter.
Meteorite Research, Vienna 1968, p. 191 - 205 (1969).

105.104 Genesis of the calcium-rich achondrites in light of rare-earth and barium concentrations.
C. C. Schnetzler, J. A. Philpotts.
Meteorite Research, Vienna 1968, p. 206 - 216 (1969).

105.105 The Leoville, Kansas, meteorite: A polymict breccia of carbonaceous chondrites and achondrite.
K. Keil, G. I. Huss, H. B. Wiik.
Meteorite Research, Vienna 1968, p. 217 (1969). – Abstract.

105.106 Cosmic-radiation-induced radioactivity of the moon and meteorites and the origin of meteorites.
A. K. Lavrukhina, G. K. Ustinova, T. A. Ibraev, R. I. Kuznetsova.
Meteorite Research, Vienna 1968, p. 227 - 245 (1969).
In Russian.
The authors propose an analytical method for calculating the depth distribution of fluxes of primary cosmic radiation and secondary nuclear-active particles in stone meteorites and in the surface layer of large bodies: the planets, the moon and the asteroids.

105.107 Cosmic-ray produced radionuclides and rare gases near the surface of Saint-Séverin meteorite.
K. Marti, J. P. Shedlovsky, R. M. Lindstrom, J. R. Arnold, N. G. Bhandari.
Meteorite Research, Vienna 1968, p. 246 - 266 (1969).

105.108 Cosmic-ray produced radionuclides in the Barwell and Saint-Séverin meteorites.
F. Begemann, R. Rieder, E. Vilcsek, H. Wänke.
Meteorite Research, Vienna 1968, p. 267 - 274 (1969).

105.109 On the energy spectrum of iron-group nuclei as deduced from fossil-track studies in meteoritic minerals. D. Lal, J. C. Lorin, P. Pellas, R. S. Rajan, A. S. Tamhane.
Meteorite Research, Vienna 1968, p. 275 - 285 (1969).
We present here experimental data on the depth variation of fossil-track densities in the hypersthene and oligoclase crystals sampled from several cores taken from the least ablated regions of the meteorite Saint-Séverin. The implications of these results (and of those in the Patwar meteorite) to the extent of atmospheric/preatmospheric ablation, and to the time-averaged flux and energy spectrum of cosmic-ray iron-group nuclei during the last $10 - 50 \times 10^6$ years, are discussed.

105.110 Fossil tracks in meteorites and the chemical abundance and energy spectrum of extremely heavy cosmic rays. M. Maurette, P. Thro, R. Walker, R. Webbink.
Meteorite Research, Vienna 1968, p. 286 - 315 (1969).
It has been previously shown that certain meteorites contain fossil tracks that are produced by slowing down heavy nuclei in the primary cosmic radiation. In this paper we give calculations of the angular distributions and densities of such tracks as a function of the depth and orientation of test samples. Comparison of the theoretical results with previous experimental work in the meteorite Saint-Séverin shows that the cosmic-ray energy spectrum has not changed appreciably in

the last 10^7 years.

105.111 On the flux of low-energy particles in the solar system during the last 10 million years.
B. S. Amin, D. Lal, J. C. Lorin, P. Pellas, R. S. Rajan, A. S. Tamhane, V. S. Venkatavaradan.
Meteorite Research, Vienna 1968, p. 316 - 327 (1969).

The work presented here provides, for the first time, conclusive evidence for the presence of an appreciable flux of low-energy particles ($10 < E < 300$ MeV) in the solar system, at distances of the order of 2 - 3 AU. The experimental work was based on radio-chemical analysis of near surface materials from the amphoterite Saint-Séverin. The implications of this study to the flux of low-energy cosmic radiation in the solar system during prehistoric era, and its origin are discussed.

105.112 Heat generation in meteorites during the early stage of the solar system. H. Reeves, J. Audouze.
Meteorite Research, Vienna 1968, p. 328 - 334 (1969).

105.113 Some stable and long-lived nuclides produced by spallation in meteoritic iron. M. Shima, M. Imamura, H. Matsuda, M. Honda.
Meteorite Research, Vienna 1968, p. 335 - 347 (1969).

The concentrations of cosmic-ray produced stable nuclides of chromium, vanadium, titanium, and calcium, and two long-lived radioactive nuclides, K-40 and Mn-53, were measured in iron meteorites. In the Grant iron meteorite they were determined as a function of the depth of the specimen.

105.114 Chlorine-36 and argon-39 production rates in the metal of stone and stony-iron meteorites.
F. Begemann, E. Vilcsek.
Meteorite Research, Vienna 1968, p. 355 - 362 (1969).

105.115 ^{39}K (n, p)-produced ^{39}Ar in chondrites: New data and their interpretation in terms of size, exposure age and orbital elements. F. Begemann, H. Wänke.
Meteorite Research, Vienna 1968, p. 363 - 371 (1969).

105.116 Radiation ages of chondrites.
G. Spannagel, G. Heusser.
Meteorite Research, Vienna 1968, p. 372 - 386 (1969).

Cosmic-ray-produced activities have been studied in 5 amphoterite-, 7 hypersthene-, and 2 bronzite chondrites. ^{26}Al was determined by non-destructive γ-γ coincidence spectrometry. The technical construction and the background properties of two spectrometers are described. From the ^{26}Al activity, the ^{39}Ar activity, and the rare-gas content measured by mass spectrometry, including all available data from the literature, exposure ages were derived.

105.117 Evaluation of ^{53}Mn by (n, γ) activation, ^{26}Al and special trace elements in meteorites by γ-coincidence techniques. U. Herpers, W. Herr, R. Wölfle.
Meteorite Research, Vienna 1968, p. 387 - 396 (1969).

In a large number of meteorites the cosmic-ray-produced nuclides ^{53}Mn and ^{26}Al were measured by intense neutron bombardment or by 'low-level' γ-coincidence techniques. Conclusions on 'exposure ages' of individual stone meteorites were drawn by comparing our results with others based on rare-gas contents. The fact that ^{53}Mn should also be present in extraterrestrial cosmic dust led us to search for this radionuclide in deep-sea sediments, in order to get data on the cosmic-dust influx rate.

105.118 Beryllium-10 in iron meteorites, their cosmic-ray exposure and terrestrial ages.
C. Chang, H. Wänke.
Meteorite Research, Vienna 1968, p. 397 - 406 (1969).

Cosmic-ray-produced ^{10}Be has been determined in about 40 iron meteorites. Together with the values for ^{36}Cl, and the stable cosmic-ray-produced rare-gas isotopes obtained previously in our laboratory from the same samples, exposure ages, and in particular terrestrial ages, have been calculated for these meteorites.

105.119 Thermal histories of meteorites by the ^{39}Ar-^{40}Ar method. G. Turner.
Meteorite Research, Vienna 1968, p. 407 - 417 (1969).

Small-scale variations in the relative distribution of potassium and radiogenic argon in potassium-bearing minerals may be investigated using neutron activation and inert gas mass spectrometry. The application of this technique, the ^{39}Ar-^{40}Ar method, to meteorites is considered and the results of recent investigations of both low- and high-age hypersthene chondrites discussed in terms of current theories of origin.

105.120 The stony meteorite Krähenberg. Its chemical composition and the Rb-Sr age of the light and dark portions. W. Kempe, O. Müller.
Meteorite Research, Vienna 1968, p. 418 - 428 (1969).

The amphoterite chondrite Krähenberg (type LL5) shows remarkable inclusions of dark material. The boundaries between the light and the dark phase are well defined. In both phases the major constituents were analysed by standard chemical methods. The trace elements In, Ga, Ge, As, and Cs were determined by neutron activation analysis. Remarkable chemical differences do exist between the light and the dark portion.

105.121 K/Ar-age determinations of iron meteorites. V. W. Kaiser, J. Zähringer.
Meteorite Research, Vienna 1968, p. 429 - 443 (1969).

K and ^{40}Ar were determined in 24 iron meteorites by neutron activation. We have also analyzed troilite of Hoba and Odessa, schreibersite of Odessa and Sikhote-Alin, and silicates from Odessa. The techniques for determining the K and ^{40}Ar content were improved. Measurements on Sikhote-Alin led to the result that schreibersite contains water-soluble K compounds.

105.122 Ages of the Ca-rich achondrites.
D. Heymann, E. Mazor, E. Anders.
Meteorite Research, Vienna 1968, p. 444 - 457 (1969).

105.123 The status of isotopic age determinations on iron and stone meteorites.
G. J. Wasserburg, D. S. Burnett.
Meteorite Research, Vienna 1968, p. 467 - 479 (1969).

A discussion of the ages determined by various methods is given. The results of ^{87}Rb$-^{87}$SR, and ^{40}K$-^{40}$Ar ages on the silicate phases of iron meteorites are summarized as well as the internal isochrons on stone meteorites. The data for Kodaikanal are reviewed and the importance of recognizing events of intermediate age in the formation of planetary objects is emphasized. It is shown that it is possible to resolve time events by high precision Sr measurements to 5×10^7 years for chondrites.

105.124 Mineralogy, petrology, and classification of types 3 and 4 carbonaceous chondrites.
W. R. van Schmus.
Meteorite Research, Vienna 1968, p. 480 - 491 (1969).

105.125 Etude minéralogique de la chondrite C III de Lancé. M. C. Michel-Lévy.
Meteorite Research, Vienna 1968, p. 492 - 499 (1969).

105.126 The use of selected-area electron diffraction in meteorite mineralogy. J. F. Kerridge.
Meteorite Research, Vienna 1968, p. 500 - 504 (1969).

105.127 **Genetical interrelations between ureilites and carbonaceous chondrites.** G. Mueller.
Meteorite Research, Vienna 1968, p. 505 - 517 (1969).

105.128 **Gas-chromatographic mass-spectrometric studies in the isoprenoids and other isomeric alkanes in meteorites.** J. Oró, E. Gelpi.
Meteorite Research, Vienna 1968, p. 518 - 523 (1969).

105.129 **Formation of organic compounds in solid bodies by solar and cosmic proton bombardment.** E. J. Zeller, G. Dreschhoff.
Meteorite Research, Vienna 1968, p. 524 - 533 (1969).

105.130 **Astronomical information on meteorite orbits.** P. M. Millman.
Meteorite Research, Vienna 1968, p. 541 - 551 (1969).
Observational data available for fireballs from which meteorites have been recovered are analysed and the normal ranges of values for the orbital elements of this class of object have been derived. A general tendency for these orbits to have small semi-major axes, moderate eccentricities, and low inclinations, is confirmed. Their closest dynamical association is with the group of peculiar small asteroids which have orbits penetrating the zone between Mars and Venus.

105.131 **Meteorite radiants and orbits.** B. Ju. Levin, A. N. Simonenko.
Meteorite Research, Vienna 1968, p. 552 - 558 (1969).

105.132 **Orbital clues to the nature of meteorite parent bodies.** E. Anders, P. J. Mellick.
Meteorite Research, Vienna 1968, p. 559 - 572 (1969).

105.133 **Relationships between orbits and sources of chondritic meteorites.** G. W. Wetherill.
Meteorite Research, Vienna 1968, p. 573 - 589 (1969).
Available data related to a meteorite's orbit are cosmic-ray exposure ages, local time of fall, apparent radiant and, rarely, the orbital elements. Initial orbits associated with satisfactory sources of meteorites should evolve into the observed distribution of these quantities.

105.134 **Magnetic properties of meteorites in the Soviet collection.** E. G. Guskova, V. I. Pochtarev.
Meteorite Research, Vienna 1968, p. 633 - 637 (1969).
In Russian.
The authors give the results of investigations on the magnetic properties – the natural remnant magnetization I_n and the magnetic susceptibility κ – for the Soviet Union's eight collections of meteorites. More than 900 specimens of stone, stony iron and iron meteorites were examined. In investigating the magnetic properties of the different meteorites it was found that the magnetization of meteorites is not affected by the fusion crust or by impact with the earth in the presence of the earth's magnetic field, and that the natural remnant magnetization can only be of extraterrestrial origin.

105.135 **Phase relations in the system Cr-Fe-S.** A. El Goresy, G. Kullerud.
Meteorite Research, Vienna 1968, p. 638 - 656 (1969).
The purpose of the present study was to determine the phase relations in the systems Cr-S and Cr-Fe-S and to correlate the physical-chemical conditions of formation of meteoritic sulfide assemblages of which daubreelite is a constituent with assemblages containing other chromium sulfides but in which daubreelite is absent.

105.136 **Phase métallique des météorites pierreuses. Etude de l'interdépendance de la composition chimique et des caractères optiques.** R. Cayé, R. Giraud, A. Sandrea.
Meteorite Research, Vienna 1968, p. 657 - 668 (1969).
Using an electronic microprobe the authors make a systematic study of the reflecting power and chemical composition of minerals of the metallic phase in stone meteorites, and measure the reflecting power over a given range.

105.137 **Investigations on a quantitative mineralogical characterization of meteorites by modal analysis.** G. Dörfler, H. G. Hiesböck.
Meteorite Research, Vienna 1968, p. 669 - 682 (1969).
The determination of volume-fractions and mean grain size of different phases is well known in mineralogy as modal analysis. This method is applied to the mineralogical characterization of chondrites. The following structural parameters will be given for each meteorite: (a) The volume fractions of the most important phases: (b) The mean particle size of these phases: (c) Their surface area/unit volume; (d) Their grain-size distribution; (e) The proximity values between important phases.

105.138 **The phosphate mineralogy of meteorites.** L. H. Fuchs.
Meteorite Research, Vienna 1968, p. 683 - 695 (1969).

105.139 **A chondritic inclusion of unique type in the Cumberland falls meteorite.** R. A. Binns.
Meteorite Research, Vienna 1968, p. 696 - 704 (1969).

105.140 **Reconstitution de la météorite Saint-Séverin dans l'espace.** Y. Cantelaube, P. Pellas, D. Nordemann, J. Tobailem.
Meteorite Research, Vienna 1968, p. 705 - 713 (1969).

105.141 **The classification of iron meteorites.** J. I. Goldstein.
Meteorite Research, Vienna 1968, p. 721 - 737 (1969).
A genetic classification system that attempts to group iron meteorites which form in the same parent body has been developed. This system uses a comparison of two independent parameters, which have been used to group iron meteorites in the past. One of these is the cooling rate of the parent body during the formation of the Widmanstätten pattern, and the other is the Ga-Ge trace element content.

105.142 **Superior analyses of iron meteorites.** C. B. Moore, C. F. Lewis, D. Nava.
Meteorite Research, Vienna 1968, p. 738 - 748 (1969).

105.143 **Phosphorus in meteoritic nickel-iron.** S. J. B. Reed.
Meteorite Research, Vienna 1968, p. 749 - 762 (1969).
This paper describes the results of a survey of 6 chondrites, 1 pallasite and 54 irons, in which the kamacite was analysed for phosphorus and nickel.

105.144 **The formation of phosphides in iron meteorites.** A. S. Doan, Jr., J. I. Goldstein.
Meteorite Research, Vienna 1968, p. 763 - 779 (1969).

105.145 **Phosphide and carbide inclusions in iron meteorites.** M. F. Comerford.
Meteorite Research, Vienna 1968, p. 780 - 795 (1969).

105.146 **Pre-terrestrial deformation effects in iron meteorites.** H. J. Axon.
Meteorite Research, Vienna 1968, p. 796 - 805 (1969).

105.147 **Dynamically deformed structures in some meteorites.** B. Baldanza, G. Pialli.
Meteorite Research, Vienna 1968, p. 806 - 825 (1969).

105.148 Shock histories of hexahedrites and Ga-Ge group III octahedrites. A. V. Jain, M. E. Lipschutz.
Meteorite Research, Vienna 1968, p. 826 - 837 (1969).

105.149 Rare gases in stony meteorites.
H. W. Müller, J. Zähringer.
Meteorite Research, Vienna 1968, p. 845 - 856 (1969).

105.150 Rare-gas measurements in separate mineral phases of the Otis and Elenovka chondrites.
P. Bochsler, P. Eberhardt, J. Geiss, N. Grögler.
Meteorite Research, Vienna 1968, p. 857 - 874 (1969).

105.151 Short exposure ages of meteorites determined from the spallogenic $^{36}Ar/^{38}Ar$ ratios.
L. E. Nyquist, F. Begemann, J. C. Huneke, P. Signer.
Meteorite Research, Vienna 1968, p. 875 - 886 (1969).

105.152 Spallogenic rare gases in taenite separated from iron meteorites. L. Schultz, J. C. Huneke,
L. E. Nyquist, P. Signer.
Meteorite Research, Vienna 1968, p. 887 - 894 (1969).

105.153 Rare gases in the iron and in the inclusions of the Campo del Cielo meteorite El Taco.
H. Hintenberger, L. Schultz, H. Weber.
Meteorite Research, Vienna 1968, p. 895 - 900 (1969).

105.154 The thermal release of rare gases from separated minerals of the Mócs meteorite.
J. C. Huneke, L. E. Nyquist, H. Funk, V. Köppel, P. Signer.
Meteorite Research, Vienna 1968, p. 901 - 921 (1969).

105.155 Distribution and origin of primordial helium, neon, and argon in the Fayetteville and Kapoeta meteorites. G. H. Megrue.
Meteorite Research, Vienna 1968, p. 922 - 930 (1969).

105.156 Titanium distribution in enstatite chondrites and achondrites, and its bearing on their origin.
K. Keil.
Earth Planet. Sci. Letters, Vol. 7, 243 - 248 (1969).

Titanium is a major constituent of a new sulfide mineral, a titanium-iron sulfide with Ti > Fe, which occurs in the Bustee enstatite achondrite. The element may also combine with nitrogen to form TiN (osbornite), a mineral found in the Bustee and Khor Temiki enstatite achondrites. On the basis of similar bulk titanium contents but strikingly different troilite contents and titanium concentrations in troilite of enstatite chondrites and enstatite achondrites, it is concluded that enstatite achondrites are not directly genetically related to enstatite chondrites.

105.157 Shock-induced planar deformation structures in experimentally shock-loaded olivines and in olivines from chondritic meteorites. W. F. Müller, U. Hornemann.
Earth Planet. Sci. Letters, Vol. 7, 251 - 264 (1969).

Shock wave compression experiments in the peak pressure range $\simeq$ 50 to 430 kb were performed on single olivine crystals and on specimens drilled from an olivine nodule. All shocked specimens display planar deformation structures. Planar structures observed in olivines from 10 chondritic meteorites indicate that these meteorites have been subjected to shock waves during their history.

105.158 Uranium measurements in hypersthene chondrites and their relation to the 600 – 700 million year 'event'. D. E. Fisher.
Earth Planet. Sci. Letters, Vol. 7, 278 - 280 (1969).

In this paper the results of an investigation by homogenized fission track analysis of the U abundance in 19 hyper-

sthene chondrite falls are reported.

105.159 Deformation in rock-forming minerals from Canadian craters.
P. B. Robertson, M. R. Dence, M. A. Vos.
Contr. Dominion Obs. Ottawa, Vol. 8, No. 23, 20 pp. (1968).

Deformation features from twelve Canadian craters are described in terms of a single scheme of progressive changes in strain patterns. The deformation features described here are attributed to shock loading at low to moderate pressures, which experimental results suggest are between 40 kb and 250 kb. Hypervelocity impact is the only natural mechanism known to generate such shocks, indicating a meteoritic origin for all the craters.

105.160 Shock-induced structural disorder in plagioclase and quartz
T. E. Bunch, A. J. Cohen, M. R. Dence.
Contr. Dominion Obs. Ottawa, Vol. 8, No. 24, 10 pp. (1968).

The physical properties of the isotropic pseudomorph of plagioclase, maskelynite, and those of coexisting disordered quartz have been studied in shocked materials from the Sedan nuclear crater ejecta. It is concluded that the natural maskelynite and disordered quartz samples were formed by shock pressures in excess of 150 kb, probably through hypervelocity meteorite impact.

105.161 Recent geological and geophysical studies of Canadian craters.
M. R. Dence, M. J. S. Innes, P. B. Robertson.
Contr. Dominion Obs. Ottawa, Vol. 8, No. 25, 24 pp. (1968).

Field studies were carried out at four craters in Canada in 1965 and early 1966. The presence of shocked materials confirms an impact origin for all four craters, raising the number of meteorite craters on the Canadian Shield to twelve.

105.162 Shock zoning at Canadian craters: Petrography and structural implications. M. R. Dence.
Contr. Dominion Obs. Ottawa. Vol. 8, No. 26, 16 pp. (1968).

The implications for cratering mechanics are considered in terms of two models. Emphasis is placed on deep penetration of the meteorite to form a primary crater, or ejecta void, with sides inclined at approximately 30° to the horizontal.

105.163 Meteorites: An X-ray analysis of deformed kamacite. M. F. Comerford.
Journ. Geophys. Res. Vol. 74, 6675 - 6678 (1969).

An X-ray line-broadening analysis was carried out to study the degree and nature of lattice damage that is present in the kamacite phase of iron meteorites. Single plates of kamacite were removed from three octahedrites, and diffraction-line profiles were measured on a spectrometer for several orders of (110) and (200) reflections.

105.164 Preservation of the iodine-xenon record in meteorites. C. M. Hohenberg, J. H. Reynolds.
Journ. Geophys. Res. Vol. 74, 6679 - 6683 (1969).

105.165 Maximum tektite size as limited by thermal stress and aerodynamic loads. F. J. Centolanzi.
Journ. Geophys. Res. Vol. 74, 6723 - 6736 (1969).

105.166 Chemical investigation of Australasian tektites.
D. R. Chapman, L. C. Scheiber.
Journ. Geophys. Res. Vol. 74, 6737 - 6776 (1969).

105.167 The destruction of tektites by micrometeoroid impact. D. E. Gault, J. A. Wedekind.
Journ. Geophys. Res. Vol. 74, 6780 - 6794 (1969).

Damage to tektites caused by collisions with micrometeoroids has been studied by firing small projectiles at high speed against glass spheres. The results, combined with estimates for the flux of micrometeoroids at 1 AU, indicate that the mean survival time before the complete destruction of tektites in circular heliocentric orbits is of the order of 10^3 and 10^4 years for, respectively, objects 1 to 10 cm in diameter.

105.168 The microtektite data: Implications for the hypothesis of the lunar origin of tektites.
J. A. O'Keefe.
Journ. Geophys. Res. Vol. 74, 6795 - 6804 (1969).

Arguments are presented in favor of the hypothesis that the microtektites must come directly from the moon, as postulated by Chapman, if they are extraterrestrial at all. Microtektites may represent the unwelded component of a lunar ash flow, of which the Muong Nong tektites are the welded components.

105.169 The relationship of nickel and chromium in tektites: New data on the Ivory Coast tektites.
W. H. Pinson, Jr., T. B. Griswold.
Journ. Geophys. Res. Vol. 74, 6811 - 6815 (1969).

105.170 Sculpturing of moldavites and the problem of micromoldavites. R. Rost.
Journ. Geophys. Res. Vol. 74, 6816 - 6824 (1969).

105.171 Correlations between O^{18}/O^{16} ratios and chemical compositions of tektites.
H. P. Taylor, Jr., S. Epstein.
Journ. Geophys. Res. Vol. 74, 6834 - 6844 (1969).

105.172 Inhomogeneities and iron diffusion in a Thailand tektite. A. K. Varshneya, A. R. Cooper.
Journ. Geophys. Res. Vol. 74, 6845 - 6852 (1969).

105.173 Changes of optical properties of various transmitting materials after simulated micrometeoroid exposure. R. L. Bowman, M. J. Mirtich, A. J. Weigand.
Journ. Optical Soc. America, Vol. 59, 1518 (1969).
Abstract Meeting Optical Soc. America.

105.174 When you find a meteorite.
I. T. Zotkin.
Zemlya i Vselennaya, No. 6, p. 77 - 81 (1969).   In Russian.

105.175 Magnetic susceptibility of some synthetic and natural tektites.
W. Ostertag, A. A. Erickson, J. P. Williams.
Journ. Geophys. Res. Vol. 74, 6805 - 6810 (1969).

The synthetic tektites were prepared under conditions which might simulate three possible sources of tektites as described by O'Keefe [1963]: (1) tektites from extraterrestrial sources were thought to be simulated by vacuum melts, (2) tektites from terrestrial sources were thought to be simulated by crucible and arc melts in air, and (3) tektites formed in the earth–moon system were thought to be simulated by melts synthesized in a very oxygen poor atmosphere.

105.176 Experimental study of meteorite ablation.
V. A. Bronshten, Yu. A. Buevich, O. K. Egorov, Yu. I. Portnjagin, M. I. Yakushin.
Astron. Vestn. Vol. 2, 139 - 152 (1968).   In Russian.

For the study of the mechanism of meteorite ablations special experiments were carried out with an electrodeless plasmotron, creating the heat flow $q \sim 0.5~kW/cm^2$ and temperature of plasma stream of $10000°K$. The fusion of three samples of meteorites was investigated: two stony (Elenovka, friable chondrite) and one iron (Sikhote–Alin).

105.177 Size and spacial distribution in meteoritic showers.

M. J. Frost.
Meteoritics, Vol. 4, 217 - 232 (1969).

Size distribution in the Barwell, Bruderheim, Gibeon, Johnstown, Sikhote-Alin and Tenham showers is studied.

105.178 The Needles (California) iron meteorite.
J. T. Wasson, J. Kimberlin.
Meteoritics, Vol. 4, 233 - 239 (1969).

The Needles fine octahedrite, which weighs 45.3 kg, was found in 1962 in the Turtle Mountains, about 50 km SSW of Needles, California. The chemical composition is described.

105.179 The Ladder Creek, Horace, and Tribune meteorites (Greeley County, Kansas).
B. Mason.
Meteoritics, Vol. 4, 240 - 243 (1969). − Results of an analysis of the principal minerals.

105.180 New specimens from the Faucett, Missouri, meteorite locality. W. F. Read.
Meteoritics, Vol. 4, 244 - 250 (1969).

105.181 Pulverization of stony meteorites at high temperatures. V. N. Lebedinets, V. M. Stuchenkov, V. B. Shushkova.
Astron. Vestn. Vol. 3, 142 - 148 (1969).   In Russian.

Results of laboratory experiments on the investigation of pulverization of copper, quartz and stony meteorite-chondrites at high temperatures to $2000°K$ are given. A close dependence of the pulverization coefficient on temperature is obtained for quartz and meteorites. The value of the mass decrease of meteor bodies due to pulverization before the beginning of intensive evaporation is estimated.

105.182 Micrometeors in the cosmic space close to earth from observations with "Kosmos 163".
B. P. Konstantinov, M. M. Bredov, E. P. Mazets, V. N. Panov, R. L. Aptekar', S. V. Golenetskij, Yu. A. Gur'yan, V. N. Il'inskij.
Kosmich. Issled. Vol. 7, 911 - 917 (1969).   In Russian.

105.183 First remarks on the abundance and structure of cosmic spherules in central Italy sediments.
R. Funiciello, M. Fulchignoni.
Geologia Romana, Vol. 8, 117 - 128 = Lab. Astrofis. Frascati, Contr. No. 38 (1969).

105.184 Activation analysis of major elements in normal chondrites.
E. Garcia Agudo, A. H. W. Aten, Jr.
Meteoritics, Vol. 4, 257 (1969).

An instrumental non-destructive method by neutron activation for simultaneous analysis of Si, O, Fe, Mg, and Al, comprising 93% by weight of the principal components of chondrites, has been carried out. In addition to being rapid and precise, this method is specially useful for comparative studies of meteorites, because excepting statistical errors in the measurement of radioactivity, all other causes of error are minimized.

105.185 The iron meteorite Barranca Blanca.
H. J. Axon, D. Faulkner.
Meteoritics, Vol. 4, 257 (1969).

Barranca Blanca is catalogued as a brecciated octahedrite with a nickel content of about eight percent. We have made a metallographic and microprobe examination and the distribution of the metallic phases suggests that the parent structure was an aggregate of sulphides and polycristalline taenite. The plessite is unusual in that it shows internal grain boundaries.

105.186　Meteor physics and the density of particles at satellite and balloon altitudes.
P. W. Hodge, D. E. Brownlee.
Space Research IX, Proc. Tokyo 1968, p. 116 - 119 (1969).

Recent results of satellite measurements of the near-earth space density of small particles are compared with estimates, made from chemical measurements, of the terrestrial influx of cosmic material. The comparison shows that the majority of the particles of suspected cosmic origin, both in the terrestrial atmosphere and in sediments, cannot be micrometeorites as defined by Whipple. If they are extraterrestrial, they must be ablation products from larger bodies.

105.187　Terminal velocities of small particles in the earth's upper atmosphere.　U. Shafrir, G. J. Dittberner.
Space Research IX, Proc. Tokyo 1968, p. 120 - 128 (1969).

105.188　Particle collection results from recent rocket and satellite experiments.　R. A. Skrivanek,
S. A. Chrest, R. F. Carnevale.
Space Research IX, Proc. Tokyo 1968, p. 129 - 139 (1969).

105.189　The measurement of micrometeorite impact fluxes.
R. C. Jennison, J. A. M. McDonnell.
Space Research IX, Proc. Tokyo 1968, p. 155 - 156 (1969).

105.190　Measurement of micrometeorite impacts from a sounding rocket during noctilucent cloud display.
B. A. Lindblad.
Space Research IX, Proc. Tokyo 1968, p. 190 - 197 (1969).

105.191　An optical model for the detection of cosmic dust in the upper atmosphere.
F. Link.
Space Research IX, Proc. Tokyo 1968, p. 198 - 200 (1969).

105.192　Mondvulkanismus und Perlstein als Ursachen der Tektiteschauer.　J. Classen.
Veröff. Sternw. Pulsnitz, No. 6, 15 pp. (1969).

Up to now, all theories have explained only partially the phenomena of tektite generation. All results gathered by observations may be explained by considering the tektites to be pearlstone-type meteorites arising from the interior of the moon, and breaking to balls like earth-borne pearlstones when entering the terrestrial atmosphere. Then the balls fall to the earth in a dense bulk and can be found in the form of tektites.

105.193　Age and development of meteorites.
J. Židů.
Říše hvězd, Vol. 50, 205 - 211 (1969).　In Czech.

105.194　Microtektites.　R. Rost.
Vesmír, Vol. 48, 237 - 239 (1969).　In Czech.

105.195　Once more on the fossil traces of particles in meteorites.　B. Kuchowicz.
Urania Kraków, Vol. 40, 237 - 241 (1969).　In Polish.

105.196　Irdische Meteoritenkrater und Tektite.
W. Gentner.
Mitt. Astron. Ges. No. 27, p. 109 - 123 (1969). – Review article.

105.197　Surveyor alpha-scattering data: Consistency with lunar origin of eucrites and howardites.
M. B. Duke.
Science, Vol. 165, 515 - 517 (1969).

105.198　Theories of the origin of Hudson Bay. Part I. On the possibility of a catastrophic origin for the great arc of eastern Hudson Bay.　C. S. Beals.
Contr. Dominion Obs. Ottawa, Vol. 4, No. 29, p. 1 - 15 (1968).

Local evidence is given for an asteroidal impact origin. Among others a comparison with an asteroidal impact crater is made.

105.199　Theories of the origin of Hudson Bay. Part II. Supporting astronomical evidence from three members of the solar system.　I. Halliday.
Contr. Dominion Obs. Ottawa, Vol. 4, No. 29, p. 15 - 31 (1968).

Comparisons are made with terrestrial impact craters and with craters on the moon and on Mars.

105.200　Theories of the origin of Hudson Bay. Part III. Comparison of the Hudson Bay arc with some other features.　J. T. Wilson.
Contr. Dominion Obs. Ottawa, Vol. 4, No. 29, p. 31 - 49 (1968).

Comparison with terrestrial features.

105.201　Applications des méthodes électriques de prospection à l'étude du cratère d'Holleford.
P. Andrieux, J. F. Clark.
Canadian Journ. Earth Sci., Vol. 6, 1325 - 1337 = Contr. Dominion Obs. Ottawa, No. 252.

Über den hypothetischen Zusammenhang zwischen Mond und Tektiten.
Umschau, Vol. 69, 558 (1969).

Hypothesen-Katalog zur Tungusischen Explosion.
Umschau, Vol. 69, 811 (1969).

The 1968 meeting of the Meteoritical Society. Abstracts of papers.　See Abstr. 010.018.

Production rate of ^{10}Be from oxygen spallation.
See Abstr. 022.038.

Kernreaktionen der Nukleogenese der Xenonisotope in der Materie des Sonnensystems.
See Abstr. 091.005.

The meteorite flux at the surface of Mars.
See Abstr. 097.016.

On the possible distribution of mass among celestial bodies.　See Abstr. 115.008.

106 Interplanetary Matter, Interplanetary Magnetic Field, Zodiacal Light

106.001 Observation of interplanetary field lines in the magnetotail. K. A. Anderson, R. P. Lin.
Journ. Geophys. Res. Vol. 74, 3953 - 3968 (1969).

A method of finding the topology of magnetic field lines in the magnetotail using solar electrons as field line tracers is described. The method depends on the presence of a large absorber such as the moon. Applying this method to spacecraft data we find: (1) Most of the field lines in the magnetotail at $60\,R_E$ geocentric distance are connected to the interplanetary field. (2) On one occasion, field lines of interplanetary character were found at the center of the magnetotail at a geocentric distance of about $60\,R_E$. One interpretation of this observation is that the reconnection region must have been at a geocentric distance less than $60\,R_E$.

106.002 Dependence of the lunar wake on solar wind plasma characteristics. K. W. Ogilvie, N. F. Ness.
Journ. Geophys. Res. Vol. 74, 4123 - 4128 (1969).

Simultaneous measurements in cislunar space of the characteristics of the solar wind plasma as observed by Explorer 34, and the perturbed magnetic field in the lunar wake as detected by Explorer 35, were performed during July-October 1967. The plasma parameter β_i for the ions is found to be more important in determining the magnitude of the umbral positive and penumbral negative anomalies than the direction of the interplanetary magnetic field. A quantitative comparison of these observations with the lunar wake theory of Whang indicates that the solar wind electrons must contribute even more significantly than the ions to the effective β for the plasma. Using the lunar wake as a solar wind shock for the electrons yields $\beta_e \sim 2\beta_i$, so that on the average the temperature of the electrons is indirectly determined to be $T_e/T_i \approx 2$ and the net effective β of the plasma is $\sim 3\beta_i$.

106.003 Interplanetary shock waves. I. Gross structure. C. P. Sonett.
Comments Astrophys. Space Phys. Vol. 1, 178 - 185 (1969).

106.004 Über die Sondierung der die Erde umgebenden kosmischen Wolke mit der optischen Methode. V. G. Fesenkov.
Meteoritika, No. 29, p. 3 - 8 (1969). In Russian.

Verf. gibt Hinweise für das Vorhandensein einer die Erde umgebenden Wolke kosmischen Staubes, deren untere Schichten in die Erdatmosphäre hineinreichen und mittels optischer Methoden entdeckt werden können. Dazu empfiehlt es sich, Beobachtungen in symmetrischen Punkten des Sonnenvertikals von tiefer Dämmerung bis zum totalen Eintritt der Nacht durchzuführen. Eine Bearbeitung derartiger Beobachtungen, die in Alma-Ata und Abastumani durchgeführt worden waren, zeigte, daß in Höhen über 100 km die Gesamtstreuung der Atmosphäre allein durch die staubförmige, d. h. die kosmische Komponente, bei relativ langsamer Abnahme der optischen Dichte bedingt ist. *D. Krahn*

106.005 Interplanetary plasma densities deduced from radar observations of Venus. W. B. Smith, K. L. Bowles, I. I. Shapiro.
Moon and Planets II, London 1967, p. 26 - 33 (1968).

Echo delays of 50 MHz (6 meter) radar signals reflected from Venus near inferior conjunction in January 1966 have enabled the average plasma densities between earth and Venus to be determined for the first time. Comparison of these group delays with an ephemeris produced from observations at far higher radar frequencies shows the average electron densities to vary from about 10 to 20 particles per cubic centimeter. The frequency spectra and delay profiles obtained from the reflections of the 50 MHz radar signals can also be used to study the properties of Venus' surface.

106.006 Discussion of paper by J. L. Blank and W. R. Sill, 'Response of the moon to the time-varying interplanetary magnetic field'. B. D. Fuller, S. H. Ward.
Journ. Geophys. Res. Vol. 74, 5173 - 5174, with a reply of J. L. Blank and W. R. Sill, p. 5175 - 5177 (1969).

106.007 Zodiacal dust particles: Some comments on recent evidence concerning their motion. N. K. Reay.
Nature, Vol. 224, 54 - 55 (1969).

Measurements have been made recently of the Doppler shift in the solar $H\beta$ absorption line scattered from the zodiacal dust cloud. Preliminary interpretation of this data suggests that the dust is in prograde orbits about the sun. A subsequent attempt to refine the interpretation, using a model in which the particles are in highly eccentric orbits, is commented upon. It is shown that a model of the dust cloud in which the effect of radiation pressure is included allows an assessment of the particle size distribution to be made.

106.008 The zodiacal light at the north ecliptic pole. R. D. Wolstencroft, F. E. Roach.
Astrophys. Journ. Vol. 158, 365 - 369 (1969).

A critical evaluation is given of some recent measurements of the zodiacal light, with particular reference to its brightness at the north ecliptic pole.

106.009 Ion cyclotron resonant instability of RH waves propagating at an angle to the interplanetary magnetic field. S. Cuperman, R. W. Landau.
Astrophys. Space Sci. Vol. 5, 333 - 341 (1969).

The anomalous Doppler-shift interaction between positive ions and right-hand (RH) polarized E.M. waves propagating at a small angle to a static magnetic field is investigated. The linear rate of growth of the resulting instability is obtained and compared with the growth rate for the parallel propagation case. For conditions typical of the solar wind at about 1 AU, the rate of growth always decreases with increasing propagation angle. For very large ion pressures ($\beta_\parallel \gg 1$) and temperature anisotropies ($T_\parallel/T_\perp \gg 1$), the rate of growth may increase with increasing propagation angle.

106.010 Heliographic latitude dependence of the dominant polarity of the interplanetary magnetic field. R. L. Rosenberg, P. J. Coleman, Jr.
Journ. Geophys. Res. Vol. 74, 5611 - 5622 (1969).

The measurements of the interplanetary magnetic field taken with the Mariners 2, 4, and 5, and OGO 5 cover several parts of the interval from September 1962 to the present and several paths through the region between 0.7 and 1.5 AU and between ±7.3° in solar equatorial latitude. From an analysis of these measurements we find evidence for a distinct dominant polarity effect in the magnetic field. Our results suggest that over most of a solar cycle, the dominant polarity of the interplanetary field in either the norther or southern hemisphere of interplanetary space is just that of the dipolar component of the sun's field in the same hemisphere.

106.011 Interplanetary magnetic field during the rising part of the solar cycle. J. Hirshberg.
Journ. Geophys. Res. Vol. 74, 5814 - 5818 (1969).

This letter compares the interplanetary magnetic field observed during the rising part of the solar cycle with that observed during solar minimum.

106.012 Solar source of interplanetary magnetic fields.
J. M. Wilcox, A. Severny, D. S. Colburn.
Nature, Vol. 224, 353 - 354 (1969).

A comparison of the measurements of the solar magnetic field as described in (A. Severny, Nature, Vol. 224, 53 - 54 (1969)) with the measurements of the polarity of the interplanetary magnetic field near the earth (made with the aid of the spacecrafts Explorers 33 and 35) showed very close correspondence between both magnetic fields. This points that the solar source of interplanetary field represents an appreciable part of the solar disk having the predominant polarity of one sign.

106.013 Detection of interplanetary magnetic field fluctuations stimulated by the lunar wake.
N. F. Ness, K. H. Schatten.
Journ. Geophys. Res. Vol. 74, 6425 - 6438 (1969).

The analysis of detailed measurements at 5.11-second intervals by the NASA-GSFC magnetic field experiment on lunar Explorer 35 in the vicinity of the moon has revealed the presence of rapid fluctuations up to the instrument bandpass of 5 Hz with amplitudes of several gammas. These disturbances are transmitted both up- and downstream from the penumbra into regions of space directly connected to the penumbra by the magnetic field.

106.014 Magnetic field fluctuations near the moon.
N. A. Krall, D. A. Tidman.
Journ. Geophys. Res. Vol. 74, 6439 - 6443 (1969).

It is proposed that magnetic field fluctuations in the solar wind near the moon are due to ballistic effects, i.e., they derive from electrons recently arrived from regions of plasma turbulence and are not connected with either wave or stability properties in the regions where the fluctuations are observed.

106.015 Interplanetary dust measurements near the earth.
L. W. Bandermann, S. F. Singer.
Rev. Geophys. Vol. 7, 759 - 797 (1969). – Review article: Dust concentration near the earth; Calculation of impact rates; Discussion of satellite data; Time variations in impact rate; Morning-to-evening asymmetry; Appendix A: Effects on the concentration of dust near the earth due to the dispersion in geocentric speeds; Appendix B: Effects on the concentration of dust near the earth and on the accretion rate due to the finite size of the sphere of influence; Appendix C: Concerning dust measurements with large flat detectors; Appendix D: Penetration formulas.

106.016 Cosmic protons interaction with interplanetary plasma.   Yu. M. Nikolaev.
Vestn. Mosk. un-ta, No. 1, p. 113 - 116 (1969).   In Russian. Abstr. in Referativ. Zhurn. 51. Astron., 8.51.333 (1969).

106.017 Observations of cosmic ray variations as a method to study the interplanetary medium.
L. I. Dorman.
Izv. AN SSSR. Ser. fiz., Vol. 32, No. 11, p. 1770 - 1775 (1969). In Russian. – Abstr. in Referativ. Zhurn. 51. Astron., 10.51.240 (1969).

106.018 Interplanetary dust.   S. F. Singer.
Meteorite Research, Vienna 1968, p. 590 - 599 (1969).

The existence of dust in the solar system can be inferred from astronomical observations of the zodiacal light, from impact penetrations on space vehicles, and from chemical and radio-chemical analyses of deep sea sediments. A self-consis-

tent interpretation of these and other data has been made and used to derive some of the major properties of the interplanetary dust particles: their physical properties such as size, shape, density, composition, surface properties; as well as their orbital properties, such as inclination to the ecliptic, and distribution in interplanetary space.

106.019 The radial gradient of interplanetary radiation measured by Mariners 4 and 5.
S. M. Krimigis, D. Venkatesan.
Journ. Geophys. Res. Vol. 74, 4129 - 4145 (1969).

Our study refers to interplanetary protons and alpha particles of energy $E \gtrsim 50$ MeV/nucleon. Comparing data from similar detectors on Mars bound Mariner 4 and earth orbiting IMP-OGO series of spacecraft, we find the following: (a) Differences appear in time and space development of Forbush decreases. We estimate the region of Forbush decrease could be as small as 0.5×0.7 AU. (b) The gradient of particles in the interplanetary medium is $-14.4 \pm 2\%$/AU (i.e. directed towards the sun) during the solar minimum of 1964 - 1965, and possibly larger in magnitude during 1967 (from Mariner 5 measurements). (c) On the basis of (b) particles in interplanetary space must in part be of solar origin. (d) The astrophysical consequences derived by O'Gallagher (1967) regarding the $(R\beta)^{-1}$ dependence of the gradient and the size of modulating region are open to question.

106.020 Physics of the interplanetary plasma and laboratory experiments.   I. M. Podgornyj, R. Z. Sagdeev.
Uspekhi fiz. nauk, Vol. 98, 409 - 440 (1969). In Russian. Abstr. in Reterativ. Zhurn. 62. Issled. kosm. prostranstv., 11.62.222 (1969).

106.021 Radioastronomical investigations of the drift of the inhomogeneous interplanetary plasma.
V. V. Vitkevich, V. I. Vlasov.
Astron. Zhurn. Akad. Nauk SSSR, Vol. 46, 851 - 861 (1969). In Russian. English translation in Soviet Astron. AJ, Vol. 13, No. 4.

Basic results of radioastronomical measurements of characteristics of the inhomogeneous interplanetary plasma from observations 1966 - 1967 are communicated. The reduction of observational data is carried out by the similarity method and by correlative analysis. Values of velocities and directions of the motion of inhomogeneities, their character, shape, orientation in the space are obtained.

106.022 Field and plasma in the lunar wake.
Y. C. Whang.
Phys. Rev., Second Series, Vol. 186, 143 - 150 (1969).

A theory is presented to explain the observed variations of the magnetic field and plasma in the vicinity of the moon. Under the guiding–center approximation, solutions for the plasma flow near the moon are obtained from the kinetic equation. The creation of a plasma cavity in the core region of the lunar shadow disturbs the interplanetary magnetic field.

106.023 Directional discontinuities in the interplanetary magnetic field.   L. F. Burlaga.
Report NASA–TM–X–63243, Goddard Space Flight Center, Greenbelt, Md., 36 pp. (1968). – See Phys. Abstr. Vol. 72, No. 20235 (1969).

106.024 The thermal emission of interplanetary dust cloud models.   C. B. Kaiser.
Thesis, Univ. of Colorado. Univ. Microfilms, Ann Arbor, Mi., 150 pp. (1968). – See Phys. Abstr. Vol. 72, No. 27092 (1969).

106.025 Acceleration of charged particles in the interplanetary plasma.   M. E. Kats, A. K. Yukhimuk.
Ukrain. Fiz. Zhurn. Vol. 14, 1019 - 1022 (1969).   In Rus-

sian. English translation in Ukrainian Phys. Journ.

106.026 New evidence of the connection between magnetic fields of cosmic space and the earth.
S. M. Mansurov.
Geomagn. Aeronom. Vol. 9, 768 - 770 (1969).
In Russian. – Brief information.

106.027 A connection of the variations of the azimuthal (eastern) component of the interplanetary electric field with variations of the geomagnetic field in the polar regions. N. V. Mikerina.
Geomagn. Aeronom. Vol. 9, 770 - 771 (1969).
In Russian. – Brief information.

106.028 Influence of irregularities of the solar wind velocity on the structure of the interplanetary magnetic field. I. I. Alekseev, A. P. Kropotkin, A. R. Shister.
Geomagn. Aeronom. Vol. 9, 1067 - 1068 (1969).
In Russian. – Brief information.

106.029 Acceleration of charged particles in the interplanetary plasma. M. E. Kats, A. K. Yukhimuk.
Ukr. fiz. Zhurn. Vol. 14, 1019 - 1022 (1969). In Russian.
Abstr. in Referativ. Zhurn. 51. Astron., 1.51.232 (1970).

106.030 On the problem of interplanetary plasma magnetosphere interaction simulation.
V. B. Baranov.
Advances Applied Mech. Suppl. No. 5, Part 2, p. 1587 - 1599 (1969). – See Bull. Signal., Vol. 30, Section 120, No. 13851 (1969).

106.031 Some models of the zodiacal cloud.
L. H. Aller, G. Duffner, M. Dworetsky, D. Gudehus, S. Kilston, D. Leckrone, J. Montgomery, J. Oliver, E. Zimmerman.
Univ. California, Los Angeles, Astron. Papers, Vol. 8, No. 4 (1967). – Reprinted from "The Zodiacal Light and the Interplanetary Medium", Proceedings of a Symposium, 1967 January 30 – February 2 (J. L. Weinberg, Editor), NASA SP-150, p. 243 - 256 (1967).

Numerous observational studies have been made of the zodiacal light. Some of these papers have tried to explain the optical properties of the zodiacal light; others have dealt with the dynamics of the cloud. In this investigation, we confine our attention to an examination of certain simplified heliocentric models which attempt to reproduce the observed polarization, brightness, and color distribution in the zodiacal light.

A solar flare disturbance as observed in the interplanetary medium. See Abstr. 073.021.

Interaction of the solar wind with planetary atmospheres. See Abstr. 074.059.

Remarks on the solar corpuscular radiation based on space data. See Abstr. 078.004.

Conjugate and closely-spaced observations of auroral radio absorption – III. On the influence of the interplanetary magnetic field. See Abstr. 084.026.

Time-dependent lunar electric and magnetic fields induced by a spatially varying interplanetary magnetic field. See Abstr. 094.019.

A possible inter-relation between interstellar and interplanetary cosmic dust. See Abstr. 131.130.

Interplanetary scintillations. V. A survey of the northern ecliptic. See Abstr. 141.179.

107   Cosmogony of the Planetary System

107.001   Strategy for scientific exploration of the terrestrial planets.   J. B. Adams, J. E. Conel, J. A. Dunne, F. Fanale, G. B. Holstrom, A. A. Loomis.
Rev. Geophys. Vol. 7, 623 - 661 (1969).

A strategy is presented for exploration of the inner solar system that will lead to knowledge of its origin and history. Several critical points in the evolution of the solar system are identified at which alternate evolutionary paths might have been possible. Experiments are suggested to help reveal the actual course of events at these critical points. The result is a framework relating experiments to be performed on the terrestrial planets and satellites to the major goal of determining the origin and history of the solar system. From this background, the lunar-exploration portion of the suggested over-all program is discussed in further detail. It is concluded that the moon offers a unique opportunity for investigating certain problems of the solar system.

107.002   On the loss of angular momentum from the proto-sun and the formation of the solar system.
I. Okamoto.
Publ. Astron. Soc. Japan, Vol. 21, 25 - 53 (1969).

The loss of angular momentum from the protosun during the wholly convective phase of the Hayashi track is discussed, the internal structure of which is approximated by the rotating polytrope with the index of 1.5. The angular momentum equation is numerically integrated, taking into account the gravitational contraction and the magnetic torque. The Schatzman mechanism is used as the means of loss of angular momentum. The magnetic energy is assumed to change with evolution proportionally to R^s, where R is the radius and s a parameter. The initial state is taken to be the instant when the protosun began the quasi-static contraction, and it is assumed that the protosun was then rotationally unstable or nearly so and had about 58×10^{50} g cm^2/sec as the amount of the angular momentum. It is shown that protosun could lose almost all of its initial momentum during the wholly convective phase, if the initial magnetic field is taken to be 600 gauss and s = –1, for example. Also, the effect of loss of angular momentum from the protosun upon the formation of the solar system is discussed as quantitatively as possible.

107.003   Planetary formation and lunar material.
F. Hoyle.
Science, Vol. 166, 401 (1969).

107.004   Exchange capture as a mechanism for the origin of planetary satellites.   E. H. Walker.
Bull. American Astron. Soc. Vol. 1, 264 - 265 (1969). – Abstr. AAS.

107.005   Origin of the solar system. II.   M. L. White.
Bull. American Astron. Soc. Vol. 1, 266 (1969). – Abstr. AAS.

107.006   Physical conditions in the primitive solar nebula.
A. G. W. Cameron.
Meteorite Research, Vienna 1968, p. 7 - 15 (1969).

Models of the primitive solar nebula have been constructed. A model is required to be in centrifugal equilibrium radially in the plane of the disk and in hydrostatic equilibrium perpendicular to the plane of the disk. The distribution of angular momentum per unit mass in the initial model is that appropriate to a fragment of a collapsing interstellar gas cloud.

107.007   Consolidation and differentiation in the development of the solar system.
V. I. Baranov, K. G. Knorre.
Meteorite Research, Vienna 1968, p. 31 - 40 (1969).
In Russian.

107.008   Planetary masses and distances.
B. W. Pendred, I. P. Williams.
Astrophys. Space Sci. Vol. 5, 420 - 424 (1969). – Research note.

We show that, provided the values for the numerical parameters involved are suitably chosen a planetary system can be produced that is in good agreement with the observed system as far as mass and distance are concerned, provided a particular assumption is made regarding the 'reach' of a condensing planet.

107.009   Chemie der Entstehung des Planetensystems.
E. Anders.
Umschau, Vol. 69, 846 (1969). – News notes.

107.010   The genesis of the solar system.
C. C. Leiby, Jr.
Bull. American Astron. Soc., Vol. 1, 352 (1969). – Abstr. AAS.

107.011   A relativistic collision model of the origin of the solar system.   S. Zaromb.
Bull. American Astron. Soc., Vol. 1, 371 (1969). – Abstr. AAS.

107.012   The evolution of the solar system.
M. M. Woolfson.
Rep. Progr. Phys., Vol. 32, 135 - 185 (1969). – Review article: A general description of the solar system; A review of theories up to 1960; Hoyle nebula theory; Accretion theory; McCrea's floccule theory; Capture theory; Some recent work related to chemical evidence.

107.013   The earliest past of the Earth–Moon system.
H. Gerstenkorn.
Icarus, Vol. 11, 189 - 207 (1969).

Fundamental equations are derived for the secular changes of the elliptic orbit of a satellite under the action of tidal friction. We restrict ourselves to the two-body problem and consider radial tides and "weak" tidal friction. Equations are developed to account for temporal changes of the moment of inertia of the planet caused by changes in the structure and size of its core. Calculations are essentially restricted to $r < 10R$, where r is the distance of the moon from the earth's center and R is the radius of the earth. The eccentricity of the moon's orbit at 10 earth radii is numerically calculated.

107.014   Died out isotopes in the history of the solar system.
G. V. Vojtkevich.
Zemlya i Vselennaya, No. 6, p. 8 - 13 (1969).   In Russian.

107.015   Effect of the high-luminosity stage of the protosun on the composition of planets and meteorites.
A. Miyashiro.
Chemie der Erde, Vol. 27, 252 - 259 (1968). – See Bull. Signal., Vol. 30, Section 120, No. 9307 (1969).

107.016   Cambrian fossils and origin of earth–moon system: Discussion.   M. Pollard.
Bull. Geol. Soc. America, Vol. 80, 729 - 734 (1969).

Nuclear chronologies for the galaxy.
See Abstr. 061.005.

Inhomogeneous accumulation of the earth from the primitive solar nebula.   See Abstr. 081.002.

Densities of the terrestrial planets.
See Abstr. 091.011.

Speculations on mass loss by meteoroid impact and formation of the planets.   See Abstr. 091.024.

Origin of meteorites and planetary cosmogony.
See Abstr. 105.089.

Meteorites and the high-temperature origin of terrestrial planets.   See Abstr. 105.092.

Occurrence of planetary systems in the universe as a problem in stellar astronomy.
See Abstr. 117.034.

Stars

111 Stellar Parallaxes

111.001 **Cepheiden als indicatoren voor de bepaling van afstanden.** G. A. Tammann.
Hemel en Dampkring, Vol. 67, 211 - 226 (1969). — Translation from the German paper in SuW, Vol. 8, 28 - 54 (1969).

111.002 **Trigonometric parallaxes of 17 stars.** K. A. Strand, R. K. Riddle.
Astron. Journ. Vol. 74, 1038 (1969).
Trigonometric parallax determinations are given for 17 stars. The photographic plate material was obtained with the 40-inch Yerkes refractor.

111.003 **De sterren binnen vijf parsec.** F. Israel.
Hemel en Dampkring, Vol. 67, 326 - 331 (1969).

111.004 **Trigonometric parallax determinations for faint stars.** R. K. Riddle, J. Priser, K. A. Strand.
Bull. American Astron. Soc. Vol. 1, 259 (1969). — Abstract AAS.

111.005 **An astrometric study of L726-8.** L. W. Fredrick, P. J. Shelus.
Bull. American Astron. Soc. Vol. 1, 241 (1969). — Abstr. AAS.

Catalogue of nearby stars, edition 1969.
See Abstr. 041.018.

112 Proper Motions, Radial Velocities, Space Motions

112.001 The stars of very large proper motion.
J. H. Anderson.
Sky Telescope, Vol. 38, 76 - 78 (1969).

112.002 Possible period of radial velocity variation of HD 193793. E. A. Vitrichenko, T. S. Galkina, P. N. Kholopov.
Astron. Tsirk. No. 522, p. 6 - 7 (1969). In Russian.

112.003 Accuracies of radial-velocity measurements.
H. A. Abt, G. H. Smith.
Publ. Astron. Soc. Pacific, Vol. 81, 332 - 338 = Contr. Kitt Peak National Obs. No. 429 (1969).

The accuracies of radial velocities measured with current techniques are studied to determine their dependence on dispersion, spectrum width, and line width. It is concluded that (1) illumination or guiding errors still cause a loss of accuracy by a factor of roughly 2; (2) the overall accuracies are much higher than those derived in the past; (3) broadening the spectra above 0.3–0.6 mm produces very little improvement for the increased observing time required; (4) increasing the dispersion is generally the best way to increase accuracy; (5) velocity accuracies are almost independent of line width for rotational velocities above about 100 km/sec.

112.004 Stars with motions of more than 100 km/sec perpendicular to the galactic plane.
O. J. Eggen.
Publ. Astron. Soc. Pacific, Vol. 81, 346 - 358 (1969).

From the space motions of several thousand stars, 40 are found to definitely have velocities perpendicular to the galactic plane greater than 100 km/sec. Most of these stars are of four kinds: CH stars, subdwarfs, short-period cepheids, and horizontal-branch stars. In addition, five giants are included.

112.005 Investigations of a milky way field in Scorpius. VI. Radial velocities of bright O and B stars.
C. Roslund.
Ark. Astron. Vol. 5, 209 - 220 (1969).

Radial velocities have been determined from prism spectrograms with a dispersion of 90 Å/mm at Hγ for twenty-six O and B stars brighter than the ninth magnitude in a milky way field in Scorpius. Twenty-one of the stars have no previously reported determinations of radial velocity.

112.006 Photoelectric radial velocities of four K stars.
R. F. Griffin.
Monthly Notices, Roy. Astron. Soc., Vol. 145, 163 - 170 (1969).

Photoelectric measurements of radial velocities are at present differential rather than absolute, and they therefore require reference stars of known velocities. The velocities of four reference stars have been carefully intercompared photoelectrically; the derived velocities are as follows: HR 152 –32.6, 63 Aur –27.1, 41 Com –14.7, λ Lyr –16.9 km s^{-1}.

112.007 Radial velocities of field horizontal-branch stars. I.
A. G. D. Philip.
Astrophys. Journ. (*Letters*), Vol. 158, L113 - L115 = Contr. Kitt Peak National Obs. No. 498 (1969).

Spectra have been obtained for twelve of the field horizontal-branch stars found by Philip. The Z-velocity dispersion for seven NGP stars is 113 km sec^{-1}, which confirms their membership in population II and shows that the Z-velocity dispersion increases with decreasing brightness.

112.008 Die Verwendbarkeit des Großen Wiener Refraktors für die Bestimmung von Positionen und Eigenbewegungen.(Die absolute Eigenbewegung von NGC 6838).
J. Meurers, F. Prochazka.
Ann. Univ.–Sternw. Wien, Vol. 28, (No. 5), 211 - 239 (1969).

The present paper gives an investigation of the ability of the great Vienna refractor to work in photographic astrometry especially in the field of proper motions of star clusters. These investigations are based on the method of the "absolute determination" of star positions. As a further example and preliminary result are given in an appendix new proper motions of the NGC 6838. In literature this cluster is partly determined as globular partly as open. The new proper motions would rather indicate its open status and are also in agreement with other investigations based on proper motions. But this is not a definite decision concerning the character of this cluster until a new three colour photometry is carried out which goes to fainter magnitudes.

112.009 Radial velocities of stars of spectral types A and F.
D. H. P. Jones, C. M. Haslam.
Roy. Obs. Bull. Greenwich – Cape, No. 155, p. 19 - 34 (1969).

Radial velocities are presented for twenty-seven stars with spectral types between A5 and F6. The plates were exposed with the grating spectrograph on the 36-inch reflector at Herstmonceux, using the f/1.75 camera at 86 Å/mm. The lines measured include some in the near ultraviolet not hitherto used for radial-velocity determination. Fresh wavelengths are derived for all the lines. The accuracy achieved is closely comparable with that obtained by other observatories. Six stars have been intensively observed to investigate short-period variability. One of them, HD 107904 is a δ Scuti variable. Upper variation limits are given for the other five.

112.010 Proper motion data processing.
K. C. Blackwell.
Quarterly Journ. Roy. Astron. Soc. Vol. 10, 233 - 237 (1969).

During more than 200 years meridian observers have recorded vast numbers of accurate positions of stars extending from pole to pole. The electronic computer makes it possible to collect together all the material into one 'bank' so that the proper motion of any one of several hundred thousand stars may be calculated from all available positions to date. This paper sets out the broad outlines of the plan upon which the Meridian Department of the Royal Greenwich Observatory is working in order to achieve such an objective, and the stage reached at present.

112.011 A program for the study of the motions and distribution of intermediate-age stars at the Leander McCormick Observatory. A. Blaauw, P. A. Ianna.
Bull. American Astron. Soc., Vol. 1, 335 (1969). – Abstract AAS.

112.012 M supergiants in the Perseus arm.
R. M. Humphreys.
Bull. American Astron. Soc., Vol. 1, 349 (1969). – Abstract AAS.

Radial velocities of 45 known M supergiants in the Perseus arm were measured from infrared spectra obtained with the 84-inch telescope at Kitt Peak National Observatory.

112.013 Radial velocities of field horizontal-branch stars.
A. G. D. Philip.
Bull. American Astron. Soc., Vol. 1, 359 (1969). – Abstr. AAS.

112.014 **On a method for selection of high-velocity OB stars.**
E. D. Pavlovskaya.
Astron. Zhurn. Akad. Nauk SSSR, Vol. 46, 840 - 850 (1969).
In Russian. English translation in Soviet Astron. AJ, Vol. 13,
No. 4.

The paper is devoted to a critical analysis of the method
for selection of high-velocity OB stars according to space velocities. It is shown that if the precision of the kinematic characteristics of these stars is not high, the method leads to over-
estimating the numbers of high-velocity stars in the direction
of galactic longitudes 90° and 270°.

112.015 **An investigation of high velocity early type stars**
(run-aways). II. E. A. Vitrichenko.
Izv. Krymskoj Astrofiz. Obs. Vol. 39, 63 - 95 (1969).
In Russian.

The program of observations for the most safely picked
out fast OB-stars is compiled on the basis of a given list. The
equivalent line-widths of 13 investigated and 5 standard stars
are determined from spectrograms obtained with the grating
spectrograph ASP-11 of the 122-cm reflector of the Crimean
Observatory (dispersion 15 and 37 Å/mm). No peculiarities
differing fast stars from standard ones were found in the line
spectrum. The two-dimensional classification of investigated
and standard stars is given. The content of helium with respect
to hydrogen, determined by two independent methods, is
found to be equal in fast and standard stars. Velocities of rotation ($v \sin i$) of 13 fast stars are determined. The duplicity of
fast stars was investigated by spectroscopic and photoelectric
methods. The orbital elements of HD 3950 have been improved. The variability of the radial velocity of HD 188439 is not
corroborated. The variability of the radial velocities of HD
167330, HD 167451 and HD 175514 are corroborated.

112.016 **Note on the distribution of small proper motions**
in 97 regions common to Potsdam Photographische
Himmels-Karte and Oxford Astrographic Catalogues of +32°
and +33°. A. N. Goyal, R. S. Khandelwal.
Proc. National Inst. Sci. India, Ser. A, Vol. 35, 434 - 438
(1969).

112.017 **Proper Motion Survey with the forty-eight inch**
Schmidt telescope. XXI. Double stars with common
proper motion. W. J. Luyten.
Separate print Univ. Minnesota, Minneapolis, Minnesota,
29 pp. (1969).

The catalogue which follows gives data for all 1229
double stars found and measured to-date on Palomar Schmidt
plates.

112.018 **A search for faint blue stars. L. Proper motions for**
951 faint blue stars. W. J. Luyten.
Separate print Univ. Minnesota, Minneapolis, Minnesota,
21 pp. (1969).

Herewith are given all proper motions I have derived for
faint blue stars; most of these motions were originally
published in scattered issues of these same publications. The
only data omitted from the present list are those for some
300 faint blue stars found in the region of Praesepe, which
were originally published in No XLI.

112.019 **Motion of A0 stars perpendicular to the galactic**
plane. III. Radial velocities observed at Kottamia.
R. Woolley, A. S. Asaad, M. P. Candy, M. J. Penston.
Roy. Obs. Bull. Greenwich–Cape, No. 156, p. 36 - 50 (1969).

Observations of radial velocities of 64 stars classed as A0
in the Henry Draper Catalogue made with the Cassegrain
spectrograph of the 74-inch telescope of the Helwan Observatory are described. These are analysed together with data from
Wilson's General Catalogue of Stellar Radial Velocities and
Wayman's observations in the South Galactic Cap. It is found
that there is a significant increase in the velocity dispersion
with height above the galactic plane. Comparison with the
theory of Paper II (Woolley and Stewart 1967) is satisfactory.

112.020 **The motions of the A stars at the North Galactic**
Pole. O. J. Eggen.
Publ. Astron. Soc. Pacific, Vol. 81, 741 - 753 (1969).

The extensive photometric and radial velocity material
for A-type stars near the NGP, collected by Perry, is discussed
in light of his difficulty in deriving K_Z. New proper motions
are derived for many of the stars. These proper motions permit a separation of young and old disk objects on the basis of
their (U, V) motion. Perry's conclusion that there are two
types of A stars at the NGP is shown to be the expected
result from the mixture of old and young disk stars in his
sample, which contains objects of types B6 to F0. The young
disk stars extend to only 300 parsecs above the galactic
plane and the old disk population contributes 20 percent of
the stars to this height.

112.021 **Proper motions of four variable stars of U Gemi-**
norum-type and of stars in their neighbourhood.
Sh. Primkulov.
Astrometr. Issled. p. 24 - 49 (1969). In Russian.

Some properties of K-giant field stars.
See Abstr. 115.012.

The frequency of spectroscopic binaries among high-
velocity dwarf stars. See Abstr. 119.002.

New subdwarfs. II. Radial velocities, photometry,
and preliminary space motions for 112 stars with large pro-
per motion. See Abstr. 126.007.

Radial velocities of the gas in H II regions and their
associated stars. See Abstr. 131.009.

113 Stellar Magnitudes, Colors, Photometry

113.001 Stellar photometric data for six different photo-
cathode materials and the silicon detector.
F. F. Forbes, R. I. Mitchell.
Commun. Lunar Planet. Lab. Vol. 8 (No. 141), 99 - 119 (1969).

A photodetector responses to the radiant energy incident on the Earth's atmosphere from the 964 brightest stars north of declination –20° are presented for the S-1, 4, 11, 17, and 20 and Bialkali photocathode materials and for the silicon detector. The computations for these data are based on recent Lunar and Planetary Laboratory 13-color narrow-band filter photometry. Photodetector response date for 57 navigational stars, based on existing UBVRIJ photometry for stars south of declination –20°, are also included.

113.002 The corrected magnitudes and colours of 278 stars
near S. A. 1-139 in the UBV system.
E. Rybka.
Acta Astron. Vol. 19, 229 - 236 (1969).

The v magnitudes and $b - v$ and $u - b$ colours from the Crimean photometry of the 6. magnitude stars near S. A. 1-139 (Nekrasova, Nikonov, Rybka 1965) have been reduced anew to the UBV system. With the aid of these new photometric values the Wroclaw Y magnitudes and CI colours (Rybka 1957) have been expressed in the BV system. The results are given in tables.

113.003 H$_\alpha$ and H$_\beta$ photoelectric photometry for 80 bright
stars.   P. L. Tebbe.
Astron. Journ. Vol. 74, 920 - 924 (1969).

H$_\alpha$ and H$_\beta$ indices are obtained by interference-filter methods at Georgetown College Observatory in Washington, D. C. Straight-line transformations of the β indices are made to the Crawford and Mander (1966, Astron. Journ. Vol. 71, 114) standard system. The technique of using α and β indices together, as described by Abt and Golson (1966, Astrophys. Journ. Vol. 143, 306) is used to obtain a good separation of B-type supergiants from Be stars.

113.004 Double stars in the Vilnius photometric system.
G. Kakaras.
Astron. Tsirk. No. 507, p. 2 - 4 (1969).   In Russian.

113.005 A group of carbon stars with anomalous relation of
colours in the spectrum.   M. V. Dolidze.
Astron. Tsirk. No. 514, p. 4 - 6 (1969).   In Russian.

113.006 Local interstellar reddening.
D. M. Gottlieb, W. L. Upson, II.
Astrophys. Journ. Vol. 157, 611 - 621 (1969).

With the help of the Bright Star Catalogue, an analysis has been made of interstellar reddening as a function of heliocentric distance and galactic latitude and longitude. Intrinsic $B - V$ colors have been determined for spectral types O9–K5 for each luminosity class I–V, and also for types M0–M6 for luminosity class III. Color excess was then plotted as a function of distance for 204 zones, covering the entire sky. At low galactic latitudes ($|b^{II}| \leq 10°$), an average reddening of 0.17 mag kpc^{-1} was determined. Several regions where the reddening occurred predominantly in a narrow distance range were found. At high galactic latitudes ($|b^{II}| > 20°$), color excesses were found to be <0.1 mag in nearly every direction.

113.007 Some characteristics of infrared colors.
E. E. Mendoza V.
Publ. Dep. Astron. Univ. Chile, Obs. Astron. Nacional, Cerro Calan, Santiago, Vol. 1 (No. 7), 106 - 126 (1968).

A catalogue of nearly 500 stars with UBVRIJKL photometry is presented. All the stars in this catalogue have 3.4μ (L-magnitudes) measurements. The data have been used to investigate some effective temperature relationships and color characteristics. Preliminary results indicate that the average effective temperature, derived from both V-R and R-I color indices probably is the best temperature obtainable for the majority of the stars. However, V-R alone seems to yield better results for early type stars with infrared excesses and with mild or no emission characteristics; and R-I alone seems to give better temperature for carbon stars; because the "UV depression" perhaps it also carries over into the visual spectral region, in at least some of them. In the catalogue there are more than 50 stars which have infrared excesses, too large to be explained by interstellar extinction alone. The main causes of these infrared excesses could be for T Tauri-like objects, circumstellar dust clouds; for Of and Be stars, the presence of stellar "shells", for carbon stars, abnormal stellar atmospheric chemical composition; and for B type stars with no emission, either circumstellar dust clouds or "infrared" companions.

113.008 Photometry of electronographic images.
J. Stock, S. Tapia.
Publ. Dep. Astron. Univ. Chile, Obs. Astron. Nacional, Cerro Calan, Santiago, Vol. 1 (No. 5), 65 - 66 (1968).

From theoretical considerations it is concluded that no linear relation may be expected between total grain counts and stellar intensities in electronographic images. However, a linear relation between magnitudes and grain counts can be expected for images with a saturated center.

113.009 uvby photometry of A and F stars.
D. C. Barry.
Publ. Astron. Soc. Pacific, Vol. 81, 339 - 345 (1969).

uvby photometry for 42 stars transformed to the system of the Strömgren-Perry catalog is presented. The A- and F-type MK standards not contained in the catalog are included.

113.010 Polarimetric observations of late-type dwarfs.
R. R. Zappala.
Publ. Astron. Soc. Pacific, Vol. 81, 433 - 437 = Contr. Lick Obs. No. 295 (1969).

Ten dM and dMe stars were observed polarimetrically. None of the stars, including the UV Ceti variable AD Leo, appeared to have intrinsic plane polarization during the period of observation. The mean observed polarization was 0.11 percent. For unpolarized stars the observational error involved would have produced an apparent mean polarization of 0.09 percent.

113.011 UBV observations of selected double systems, II.
A. U. Landolt.
Publ. Astron. Soc. Pacific, Vol. 81, 443 - 446 = Contr. Louisiana State Univ. Obs. No. 22 = Contr. Cerro Tololo Inter-American Obs. No. 74 (1969).

UBV magnitudes and color indices have been obtained for 22 double systems listed in the Bright Star Catalogue for which these data have been lacking.

113.012 Infrared photometry. Theoretical limits for signal
to noise ratio in the general astronomical case in
the μ range.   G. Sedmak.
Mem. Soc. Astron. Italiana, Nuova Serie, Vol. 40, 247 - 260 (1969).

A theoretical analysis was made, in order to obtain the S/N ratio for a semiconductor photovoltaic cell, under general hypotheses in the astronomical case. In conjunction with the analytical S/N expression, some graphs are shown which permit the graphical computation of the S/N ratio, or of inverse problems. Moreover, we show some generally true optimization criteria for the S/N ratio.

113.013 Remarks on linearization of characteristic curves in photographic photometry. M. Margoshes.
Applied Optics, Vol. 8, 818 (1969).

113.014 Comments on several published and unpublished reactions to a paper on photographic characteristic curves. G. de Vaucouleurs.
Applied Optics, Vol. 8, 818 - 819 (1969).

113.015 Photometric standard sequences in Norma $l^{II} = 320°$ - 340°. L. O. Lodén, B. Nordström.
Ark. Astron. Vol. 5, 231 - 239 (1969).
Photoelectric UBV photometry of 290 stars in the Norma regions, performed at the Boyden Observatory, is given primarily in order to provide standard connections sequences for photographic photometry.

113.016 A photoelectric sequence near NGC 5128. C. Roslund.
Ark. Astron. Vol. 5, 249 - 251 (1969).
A magnitude sequence of twenty-four stars ranging in brightness from the seventh to the fifteenth visual magnitude has been set up photoelectrically in the UBV system less than one degree south of the peculiar galaxy NGC 5128 (Centaurus A).

113.017 Photometry with the Jävan reflector. A. Ardeberg, K. Särg, S. Wramdemark.
Ark. Astron. Vol. 5, 297 - 301 (1969).
Transformation formulae to the standard UBV system are given for the Lund Observatory photometric telescope at Jävan. The accuracy and the limiting magnitudes are discussed.

113.018 BV photometry of standard stars. L. Häggkvist, T. Oja.
Ark. Astron. Vol. 5, 303 - 304 (1969).
During the years 1961 - 1967 altogether 34 stars have been used as standards for the photoelectric BV photometry by members of the staff at the Uppsala Observatory. Accurate values of V and $B - V$ for these stars have been derived.

113.019 Narrow- and broad-band photometry of red stars. IV. Population separation in giant stars. O. J. Eggen.
Astrophys. Journ. Vol 158, 225 - 241 (1969).
Narrow-band (102,65,62) and broad-band $(R - I)$ photometry of the M- and K-type giants in the Bright Star Catalogue is used to construct population discriminants for the young disk stars ($\leq 5 \times 10^8$ years), old disk stars (to about 5×10^9 years), and the older halo stars. The $B - V$ scale collapses for stars redder than about $(B - V) = +1.55$ mag, but the $(U - B)$ colors are a smoothly varying function of color temperature (blackbody) between about 6000° and 2000°. The main population discriminants are the space motion, the ultraviolet colors, and the TiO absorption. The small ultraviolet excess known for old disk dwarfs is also found in the subgiants but is very small or absent in the giants. The halo-population subgiants and giants show large values of $\Delta(U - B)$ relative to the young disk stars. The TiO absorption of young and old disk giants is indistinguishable for stars hotter than color temperatures (blackbody) of about 2500°, but the halo stars show no TiO absorption. The old disk giants define a tight sequence in the (M_{bol}, T_e)-plane, and this sequence differs little from that for the young-disk group in the Hyades.

Also, the halo giants join this sequence at color temperatures near 2800°.

113.020 Electrographic stellar photometry. H. D. Ables, G. E. Kron, A. V. Hewitt.
Bull. American Astron. Soc. Vol. 1, 231 (1969). – Abstr. AAS.

113.021 Infrared colors of carbon stars. J. Bahng.
Bull. American Astron. Soc. Vol. 1, 232 - 233 (1969). – Abstr. AAS.

113.022 H-alpha and H-beta photometry of selected bright stars. C. R. Chambliss, P. L. Tebbe.
Bull. American Astron. Soc. Vol. 1, 236 (1969). – Abstr. AAS.

113.023 Far-ultraviolet photometry of stars obtained with the Celescope experiment in OAO-A2. R. J. Davis.
Bull. American Astron. Soc. Vol. 1, 238 - 239 (1969). – Abstr. AAS.

113.024 Ultraviolet intensities of stars observed in Vela by the Celescope experiment on the Orbiting Astronomical Observatory. W. A. Deutschman.
Bull. American Astron. Soc. Vol. 1, 239 (1969). – Abstr. AAS.

113.025 Photometry of A stars at the south galactic pole. A. G. D. Philip.
Bull. American Astron. Soc. Vol. 1, 257 (1969). – Abstr. AAS.

113.026 Mariner V ultraviolet photometer measurements of the Lyman-alpha galactic background. C. A. Barth.
Bull. American Astron. Soc. Vol. 1, 233 (1969). – Abstr. AAS.

113.027 Photoelectric photometry from a space vehicle. A. D. Code.
Publ. Astron. Soc. Pacific, Vol. 81, 475 - 487 (1969).
Public lecture presented at the Flagstaff meeting of the Astronomical Society of the Pacific, on June 19, 1969.

113.028 Photometric calibration of direct plates. R. Racine.
Publ. Astron. Soc. Pacific, Vol. 81, 549 - 550 (1969). Abstract ASP.

113.029 Interference filters for standard magnitude-color systems with arbitrary sensors. R. L. Wildey.
Publ. Astron. Soc. Pacific, Vol. 81, 688 - 690 (1969). – Note.

113.030 Calibration of photographic plates for the photometry of extended objects. V. G. Rijves.
Problemy kosmich. fiz. No. 4, p. 134 - 138 (1969). In Russian.
Some formulae are proposed for the region of small and average density of characteristic curves. As standards weak or intrafocal images of stars of known magnitudes are proposed.

113.031 Photometric calibration of direct plates. R. Racine.
Astron. Journ. Vol. 74, 1073 - 1078, 1117 (1969).
It is shown that if certain strict conditions are satisfied, photographic techniques can be efficiently used to set up accurate magnitude scales without recourse to photoelectric ob-

servations. Of all such techniques used in the past only the "auxiliary-prism" method is judged satisfactory. Its application and adaptation to large reflectors are described in detail. Tested at the Mount Wilson 60-inch telescope, the method is shown capable of an accuracy equal to or better than the accuracy obtained from calibrations using local photoelectric standards.

113.032 On the relation between total absorption and color excess. V. I. Voroshilov, E. P. Polishchuk.
Dopovidi AN USSR B, No. 5, p. 410 - 412 (1969). In Ukrainian. – Abstr. in Referativ. Zhurn. 51. Astron., 11.51.361 (1969).

113.033 Ultraviolet photometry from a spacecraft.
A. D. Code, T. E. Houck, J. F. McNall, R. C. Bless, C. F. Lillie.
Sky Telescope, Vol. 38, 290 - 293 (1969).

113.034 Far-ultraviolet photometry of Orion stars.
G. R. Carruthers.
Astrophys. Space Sci. Vol. 5, 387 - 402 (1969).

Far-ultraviolet photometric data for early type stars in Orion, in the 1050 - 1180 and 1230 - 1350 Å wavelength ranges, were obtained in an Aerobee rocket flight on 30 January 1969. The results corrected for interstellar extinction, appear in good agreement with model atmospheres in the case of main-sequence stars. Bright giant and supergiant stars, however, appear to be up to one magnitude fainter than main-sequence stars of similar spectral class in the 1050 - 1180 Å range.

113.035 Photometric standards for the southern hemisphere.
B. J. Bok, P. F. Bok.
Astron. Journ., Vol. 74, 1125 - 1130 = Contr. Cerro Tololo Inter-American Obs. No. 85 (1969).

About ten years ago, the authors published a paper (Bok and Bok 1960) with several standard sequences for southern hemisphere work. Additional photoelectric measurements have been made since then, first from Australia (Mount Stromlo and Siding Spring Observatories), and more recently from Chile (Cerro Tololo Inter-American Observatory). Since so much effort is now being expended on research of the Southern Milky Way and of the Magellanic Clouds, it may be helpful to bring some of the results together in one place.

113.036 On the *uvby* photometric system.
S. Matsushima.
Astrophys. Journ., Vol. 158, 1137 - 1149 (1969).

A model-atmosphere analysis has been carried out for *uvby* photometric data for a large sample of A– and early F– type main-sequence stars. The normalization constants for the theoretical color indices in the four-color system are determined on the basis of photoelectric scanner observations of twenty A-type stars. Correction factors for absorption lines derived from various methods are found to be in good agreement, and mutually consistent values for the effective temperatures are predicted by different color indices.

113.037 On the spectral energy distribution of the barium stars. H. E. Bond, J. S. Neff.
Astrophys. Journ., Vol. 158, 1235 - 1237 = Contr. Louisiana State Univ. Obs., Baton Rouge, No. 27 = Contr. Kitt Peak National Obs. No. 457 (1969).

Intermediate-band photometry reveals that the Ba II stars show a broad ultraviolet absorption similar to, but weaker than, that seen in the N-type carbon stars.

113.038 The unusual infrared object IRC + 10216.
E. E. Becklin, J. A. Frogel, A. R. Hyland, J. Kristian, G. Neugebauer.

Astrophys. Journ. (*Letters*), Vol. 158, L133 - L137 (1969).
IRC + 10216 is an extended object located out of the galactic plane in an unreddened region. At 5μ it is the brightest source observed outside the solar system; at 2.2μ it varies by as much as 2 mag, with a time scale on the order of 600 days. Its energy distribution resembles that of a $650°$K blackbody, and no spectral features have been observed in the wavelength range from 1.5 to 14μ. The object is interpreted as being consistent with a galactic source surrounded by an optically thick dust shell.

113.039 Photometric classification of high-latitude blue stars. J. A. Graham.
Bull. American Astron. Soc., Vol. 1, 345 (1969). – Abstract AAS.

113.040 Intrinsic colors of the field horizontal-branch stars. A. G. D. Philip, L. E. Tifft.
Bull. American Astron. Soc., Vol. 1, 359 (1969). – Abstract AAS.

113.041 Photométrie des étoiles faibles. A. Lallemand.
Vistas in Astronomy, Vol. 11, 119 - 126 (1969).
En mettant en oeuvre l'électronographie, de grandes améliorations sont possibles dans les mesures photométriques.

113.042 A very red star in Perseus.
L. Balázs, I. Jankovits.
Observatory, Vol. 89, 237 - 239 (1969). – Letter.

113.043 The atmospheric extinction in photoelectric photometry. J. Stock.
Vistas in Astronomy, Vol. 11, 127 - 146 (1969).
Methods for determining the atmospheric extinction are considered. Results obtained at the Cerro Tololo Inter-American Observatory during more than two years are discussed.

113.044 Photoelectric photometry of the Wolf-Rayet type binary star HD 211853. A. A. Guseinzade.
Astrofizika, Vol. 5, 502 - 503 (1969). In Russian. – Engl. translation in Astrophysics, Vol. 5, No. 3.
The photoelectric observations of HD 211853 in a system near U, B, V made by the 20-cm telescope of the Crimean Astrophysical Observatory are presented.

113.045 Near infrared photometry of late-type stars.
R. G. Walker.
Phil. Trans., Ser. A, Vol. 264 (No. 1150), 209 - 225 (1969).

113.046 Infrared radiation associated with protostars.
A. G. W. Cameron.
Phil. Trans., Ser. A, Vol. 264 (No. 1150), 227 - 233 (1969).

113.047 A far infrared sky survey.
N. J. Woolf, W. F. Hoffmann, C. L. Frederick, F. J. Low.
Phil. Trans., Ser. A, Vol. 264 (No. 1150), 267 - 271 (1969).

113.048 Near infrared night sky background.
M. Harwit, K. Fuhrmann, M. Werner.
Phil. Trans., Ser. A, Vol. 264 (No. 1150), 273 - 278 (1969).

113.049 Some problems and instrumental features of infrared astronomy. A. E. Salomonovich.
Phil. Trans., Ser. A, Vol. 264 (No. 1150), 283 - 291 (1969).

113.050 On C.P.D. magnitudes and Melbourne diameters in $-71°$ to $-81°$.
A. N. Goyal, P. C. Gupta, R. S. Khandelwal.
Proc. National Inst. Sci. India, Ser. A, Vol. 35, 282 - 287

(1969).

The present investigation has been undertaken with a view to examine if any systematic error is shown by Melbourne diameters when they are compared with C.P.D. magnitudes. The stars were grouped according to their magnitudes. An empirical relation between magnitudes and diameters has been obtained.

113.051 UBV photometry of 137 stars in an area near the south galactic pole.   P. A. T. Wild.
Monthly Notes Astron. Soc. Southern Africa, Vol. 28, 123 - 130 (1969).

Presented are the results of new UBV photoelectric photometry of 20 bright comparison stars and 117 programme stars. The latter were selected from Luyten's Two-Tenths Catalogue (Luyten 1957).

113.052 Magnitudes, colours and coordinates of 175 ultraviolet excess objects in the field 13^h, $+ 36°$.
Lab. Nazionale Radioastronomia, Ist. Fis. "A. Righi" Univ. Bologna, Contr. No. 60, 33 pp. (1969).

We present here the magnitudes, colours and coordinates of 175 ultraviolet-excess objects in a 6.5×6.5 degree field centered at $13^h_1 + 36°$. These data were obtained as part of a program aimed at the discovery and study of radioquiet quasi-stellar objects by means of multicolour photometry.

113.053 Fotometria infrarossa. Limiti teorici al rapporto segnale-rumore nel caso generale.
G. Sedmak.
Atti XII Riunione Soc. Astron. Italiana, L'Aquila 1968, p. 29 - 31 (1969). – Abstract SAI.

113.054 Groupes physiques très jeunes. I. Les lois d'extinction par la matière interstellaire dans IC 1805.
G. Goy, A. Maeder.
C. R. des Séances, SPHN Genève, NS, Vol. 4, Fasc. 1, p. 26 - 35 = Publ. Obs. Genève, Sér. A, Fasc. 76/III (1969).

Dans le présent travail on restreint l'application de la méthode des «différences de couleur» à l'étude des étoiles O. La méthode prend alors une signification beaucoup plus précise. On montre que, compte tenu de tous les types de dispersions possibles ou prévisibles (précision des mesures, binarité, vitesse de rotation axiale, champ magnétique interstellaire et dispersion des couleurs intrinsèques), on peut affirmer que les six étoiles les plus chaudes de IC 1805 sont rougies selon des lois d'extinction individuelles.

113.055 Groupes physiques très jeunes. II. Etude photométrique de IC 1805.
G. Goy, A. Maeder.
C. R. des Séances, SPHN Genève, NS, Vol. 4, Fasc. 1, p. 35 - 37 = Publ. Obs. Genève, Sér. A, Fasc. 76/IV (1969).

Dans la publication I sur les groupes physiques très jeunes, on a montre l'existence de lois d'extinction interstellaire individuelles par l'intermédiaire d'un graphique utilisant les pseudocontinus. Dans le présent article on montre que par un choix judicieux des paramètres traditionnels de la photométrie en sept couleurs on peut arriver au même résultat.

113.056 BVRI photometry of 46 southern bright stars.
E. E. Mendoza V.
Bol. Obs. Tonantzintla y Tacubaya, Vol. 5, (No. 31), 57 - 58 (1969).

We have made four color photometric observations (BVRI system) of 46 southern bright stars, in order to enlarge, in right ascension, our previous catalogue (Mendoza, 1967).

113.057 Absorptionsbestimmungen in unmittelbarer Umgebung von Sternen in NGC 2264.   W. Götz.
MVS Sonneberg, Vol. 5, 86 - 87 (1969).

This investigation was made to complete previous determinations of absorption values for stars in NGC 2264. The published stars lie in the two-colour-diagram (U-B/B-V) to the right of or above the zero ago main sequence. They show a UV-excess but no H_a-emission. It is suggested that the extinction arises in the immediate surrounding of the stars and is not caused by a cloud of dust in the star cluster.

113.058 A search for faint violet stars in southern galactic latitudes.   S. Jaidee, G. Lyngå.
Ark.Astron. Vol. 5, 345 - 379 = Uppsala Astron. Obs. Medd. No. 165 (1969).

A search for faint violet stars on two-image Schmidt plates has yielded a catalogue of 296 stars over 540 square degrees. Some of these have been observed photometrically in *UBV* as well as in red.

113.059 Four southern photoelectric sequences.
M. P. FitzGerald, W. Wilson, J. E. Stegman.
Publ. Astron. Soc. Pacific, Vol. 81, 804 - 814 = Contr. Univ. Waterloo Obs. No. 3 = Contr. Cerro Tololo Inter-American Obs. No. 84 (1969).

UBV photoelectric magnitudes and color indices of 104 stars in four sequences in the galactic plane are presented. The sequence stars have been assigned MK classes from objective prism spectra.

113.060 Line-blanketing effects on the *uvbyβ* photometric system.   D. H. McNamara, D. J. Colton.
Publ. Astron. Soc. Pacific, Vol. 81, 826 - 833 (1969).

The equivalent-width data in the spectra of the sun, ξ Pegasi, 50 Andromedae, and HD 19445 are used to evaluate the effect that removal of the absorption lines has on the $(b - y)$, m_1, c_1, and β indices.

113.061 Infrared color indices of carbon stars.
J. Bahng.
Publ. Astron. Soc. Pacific, Vol. 81, 863 - 866 = Contr. Kitt Peak National Obs. No. 480 (1969).

Infrared magnitudes of carbon stars were measured at the wavelenghts of $1.2 \mu (j)$, $1.6 \mu (h)$, and $2.2 \mu (k)$.

113.062 Las magnitudes estelares.   A. L. de la Barra.
El Universo, Vol. 23, (No. 89), 109 - 113 (1969).
Popular article.

113.063 Polarisation südlicher OB-Sterne.
T. Neckel, G. Klare.
Mitt. Astron. Ges. No. 27, p. 149 - 151 (1969). – Conference paper.

113.064 Beobachtungen vom galaktischen Streulicht im südlichen Kohlensack.   K. Mattila.
Mitt. Astron. Ges. No. 27, p. 154 - 155 (1969). – Abstract AG.

113.065 Eine photographische und photoelektrische Durchmusterung nach Infrarotsternen in ausgewählten Feldern im Cygnus.   G. Ackermann.
Mitt. Astron. Ges. No. 27, p. 155 - 157 (1969). – Conference paper.

113.066 Note on the magnitude diameter relationship of 12013 stars in the Oxford Astrographic Zones $+28°$ and $+29°$.   A. N. Goyal, S. S. Mithal.
Proc. National Inst. Sci. India, Ser. A, Vol. 34, 360 - 363 (1968).

The Oxford diameters of 12013 stars in the Oxford Astrographic Catalogues $+28°$ and $+29°$ have been compared with Cambridge magnitudes.

Erste Ultraviolett-Aufnahmen aus dem Weltraum. Umschau, Vol. 69, 812, 814 (1969).

Stellar photometric data for various photocathode materials.   See Abstr. 034.025.

A theoretical investigation of focal stellar images in the photographic emulsion and application to photographic photometry.   See Abstr. 036.009.

Catalogue of nearby stars, edition 1969. See Abstr. 041.018.

Observational results concerning departures from LTE in late B stars.   See Abstr. 064.045.

Helium-weak stars.

See Abstr. 114.056.

Classification spectrale d'étoiles presque identiques dans la photométrie en 7 couleurs. See Abstr. 114.097.

Surface brightnesses in the U, B, V system with application on M_V and dimensions of stars. See Abstr. 115.001.

Infrared photometry of a helium star, HD 30353. See Abstr. 119.001.

Intermediate- and narrow-band photometry of Cepheids.   See Abstr. 122.094.

114 Stellar Spectra, Temperatures, Spectroscopy

114.001 Spectral analysis of a peculiar carbon star, WZ Cassiopeiae. M. Hirai.
Publ. Astron. Soc. Japan, Vol. 21, 91 - 110 (1969).

Absorption lines of atomic and molecular species were identified, and their equivalent widths were measured in the spectrum of WZ Cassiopeiae in the visual and infrared regions. Excitation temperature, micro-turbulent velocity and electron pressure were determined. These parameters indicate that the star is a supergiant. On the basis of coarse analysis, the abundances of twenty seven elements were obtained.

114.002 A new CH star, BD+42°2173.
Y. Yamashita.
Publ. Astron. Soc. Japan, Vol. 21, 119 - 120 (1969).

A carbon star, BD+42°2173, is classified as a CH star of the spectral type of C3, 1CH. The radial velocity also is found to be −76 ± 4 km/sec.

114.003 Line profiles in the spectrum of β Orionis.
A. B. M. Smit.
Bull. Astron. Inst. Netherlands, Vol. 20, 274 - 278 = Commun. Obs. Utrecht (1969).

Line profiles of various elements are presented, as well as the equivalent widths and the central depths. The agreement with the results of Svolopoulos (1966) is fairly good, but some strange discrepancies occur in the data for the hydrogen lines. An estimate has been made of the "microturbulence". A comparison is made between a computed profile and the observed profile of $H\gamma$. The central depth of the theoretical profile, predicted with a model of Mihalas (1965), is too large compared with the observed profile. Also the shape of the theoretical profile does not correspond to the empirical shape. This indicates that the atmosphere of a supergiant cannot be interpreted using a simple LTE theory of line formation.

114.004 MK spectral types for bright southern OB stars.
W. A. Hiltner, R. F. Garrison, R. E. Schild.
Astrophys. Journ. Vol. 157, 313 - 326 = Contr. Cerro Tololo Inter-American Obs. No. 54 = David Dunlap Obs., Univ. Toronto, No. 210 (1969).

MK spectral classifications are provided for all OB stars south of −20° and earlier than B8 which are listed in the *Catalogue of Bright Stars.*

114.005 Aus der Geschichte der astronomischen Spektroskopie. W. Seitter.
SuW, Vol. 8, 212 - 213 (1969).

114.006 Osawa's peculiar star HD 221568. K. Kodaira.
Astrophys. Journ. *(Letters),* Vol. 157, L59 - L62 (1969).

This letter briefly discusses the first scanner observations of the continuum of an extraordinary peculiar A star, HD 221568, and calls attention to the additional peculiarities of this object.

114.007 The infrared spectrum of the NML Cygnus object.
H. L. Johnson.
Commun. Lunar Planet. Lab. Vol. 8 (No. 139), 91 - 95 (1969). − Reprinted from Astrophys. Journ. *(Letters),* Vol. 154, L125 - L129 (1968).

114.008 The spectrum of two late-type stars.
R. Bajcár.
Contr. Astron. Obs. Skalnaté Pleso, Vol. 4, 63 - 92 (1969).

The spectra of two late-type stars (K0 III and G 8 IV)

are analysed and compared: HD 37160 is a high velocity star, HD 188512 a standard star. The lines in the region 4144– 3497 Å are fully identified. The analysis itself was made by the curve-of-growth method. This curve, obtained from lines in the UV region, served to determine a number of parameters of the atmospheres of both stars. $T_{exc} = 4315°$ (HD 37160) and $T_{exc} = 4225°$ (HD 188512), $\log P_e = 0.656$ (HD 37160) and $\log P_e = 0.425$ (HD 188512) were determined. The velocities of the atoms are higher in the atmosphere of HD 37160 than for normal stars. The relative representation of elements is summarized and the composition of the atmosphere is compared with the atmosphere of the sun.

114.009 Aus der Geschichte der astronomischen Spektroskopie (2. Teil). W. Seitter.
SuW, Vol. 8, 231 - 233 (1969).

114.010 Term analysis of Fe IV and identification of [Fe IV] in RR Telescopii. B. Edlén.
Monthly Notices, Roy. Astron. Soc. Vol. 144, 391 - 396 (1969).

From an experimental investigation of the fourth spectrum of iron we report the levels of $3d^4$ (^5D)$4s$ and $3d^4$ (^5D) $4p$ as well as the set of $3d^5$ levels. This set permits us to make a practically exhaustive prediction of the spectrum of forbidden Fe IV lines and to explain the origin of 17 emission lines in RR Telescopii.

114.011 MK classifications for F- and G-type stars. I.
E. A. Harlan.
Astron. Journ. Vol. 74, 916 - 919 = Lick Obs. Bull. No. 605 (1969).

New MK spectral classifications are given for 314 stars between magnitudes 5.0 and 7.5, mainly of types F and G. The assignments were made on slit spectrograms of dispersion 75 Å/mm at $H\gamma$.

114.012 Line intensities in the visual region of μ Cassiopeiae.
R. M. Catchpole, B. E. J. Pagel, A. L. T. Powell.
Roy. Obs. Bull. Greenwich–Cape, No. 154, 17 pp. (1969).

This paper lists equivalent widths and other relevant data for spectral lines of the mild subdwarf μ Cassiopeiae measured on four Palomar coudé exposures in the visual region taken by J. L. Greenstein in 1963. The analysis of the data has been published elsewhere.

114.013 Calcium auto-ionization lines in the spectra of M-supergiants. M. Ya. Orlov, M. H. Rodriguez.
Astron. Tsirk. No. 516, p. 1 - 2 (1969).   In Russian

114.014 The carbon, nitrogen, and oxygen abundances in four K giants. T. F. Greene.
Astrophys. Journ. Vol. 157, 737 - 756 (1969).

The abundances, relative to hydrogen, of carbon, nitrogen, and oxygen are determined in the K giants a Boo, a Ser, β Gem, and ϵ Peg. Use is made of the lines of C_2, CN, C I, and [O I] from spectrograms with dispersions of 2 and 6 Å mm^{-1}. Well-iterated LTE model atmospheres are employed in the reduction.

114.015 A search for neon in the spectra of peculiar A and B stars. A. I. Sargent, J. L. Greenstein, W. L. W. Sargent.
Astrophys. Journ. Vol. 157, 757 - 768 (1969).

We have obtained spectra in the red region, dispersion 6.8 Å mm^{-1}, for fourteen peculiar and five normal A and B stars, primarily in order to investigate the behavior of neon.

As expected on theoretical grounds, the Ne I lines behave like He I, C II, and Si III in the normal stars and increase in strength rapidly with increasing temperature for stars hotter than about B8. Neon was found in the spectra of only four of the peculiar stars, whereas nine of them are blue enough for Ne I to be expected.

114.016 **The infrared spectrum of Arcturus.**
E. F. Montgomery, P. Connes, J. Connes, F. N. Edmonds, Jr.
Astrophys. Journ., Suppl. Series, Vol. 19, 1 - 29 (1969). – Abstr. in Astrophys. Journ., Vol. 157, 1471 (1969).

Equivalent widths of 1036 stellar lines have been measured from high-resolution interferometric spectra of Arcturus covering the spectral range 8766 – 3978 cm^{-1} (11407 – 25129 Å). Approximately 60 percent of these lines have been identified.

114.017 **a^2 Canum Venaticorum and the oblique-rotator theory.** D. M. Pyper.
Astrophys. Journ. Suppl. Series, Vol. 18, 347 - 378 = Contr. Lick Obs. No. 290 (1969).

In order to test the oblique-rotator theory, a study of the spectrum, radial velocity, and magnetic variations of a^2 CVn has been made on spectrograms obtained with the 120-inch coudé spectrograph and the Zeeman analyzer at Lick Observatory. *UBV* photometric observations of this star were also obtained. The light and color variations do not agree with those predicted from the effective temperature changes during the cycle. The variations in line strength of the rare earths are also incompatible with a change in temperature. On the basis of the variations in line intensity and radial velocity, the elements are found to fall in three groups: the rare earths; the iron-peak elements Ti, V, Cr, Mn, and Fe; and Mg, Si, and Ca. It is shown that the observed changes in the effective magnetic fields, equivalent widths, and radial velocities for the first two groups can be successfully predicted by an oblique-rotator model. The light and color variations cannot be explained by the present model. Possibilities for further work on this subject are discussed.

114.018 **Infrared spectra of stars.**
H. Spinrad, R. F. Wing.
Annual Rev. Astron. Astrophys. Vol. 7, 249 - 302 (1969).

114.019 **Ultraviolet astronomy.**
R. Wilson, A. Boksenberg.
Annual Rev. Astron. Astrophys. Vol. 7, 421 - 472 (1969).

114.020 **The identification of G- and K-type giant stars at low dispersion.** A. R. Upgren.
Publ. Astron. Soc. Pacific, Vol. 81, 438 - 440 (1969).

114.021 **Far-ultraviolet spectrophotometry of bright stars in Orion.** F. E. Stuart.
Astrophys. Journ. Vol. 157, 1255 - 1264 = Contr. Kitt Peak National Obs. No. 443 (1969).

A 12-inch telescope and low-resolution grating spectrophotometer were flown on an Aerobee-rocket to obtain absolute vacuum-ultraviolet spectra of bright stars in Orion. Nineteen stars, including the Trapezium, were observed in the spectral range $\lambda\lambda$ 1050-2180. Agreement of observed spectra with a B1.5 V model was generally good. A rapid divergence for wavelengths below 1800 Å between observed data for an A3 star (β Eri) and a model was noted. Variation of the general slope of the spectra of the earlier-type stars with luminosity class can be observed. A large ultraviolet excess was found for 42 Ori and probably for BS 1923. Evidence is presented to indicate that in 42 Ori the excess is probably due to a diffuse glow around the star.

114.022 **An analysis of the peculiar A star HD 204411.**
W. L. W. Sargent, K. M. Strom, S. E. Strom.
Astrophys. Journ. Vol. 157, 1265 - 1278 (1969).

Well-widened, high-dispersion spectra were used to obtain the identifications and equivalent widths for over 1500 lines between 3700 and 4900 Å for the relatively cool Ap star, HD 204411. A model-atmosphere analysis of these data was performed. The large number of lines used in this study allowed us to discuss and rule out any significant departures from LTE in the level populations for the lines, or any large variation of the turbulent-velocity parameter with height. The results suggest a broadening of the Fe peak relative to H, an overall enhancement of the Fe peak relative to H, and a decrease in the ratio of light-element abundances to iron.

114.023 **Of and Be stars.** A. Slettebak.
Non-Periodic Phenomena in Variable Stars, IAU Colloquium, Budapest, 1968, p. 179 - 189 (1969).

114.024 **Photoelectric observation of line profiles with high time resolution in B and Be stars.**
J. B. Hutchings.
Non-Periodic Phenomena in Variable Stars, IAU Colloquium, Budapest, 1968, p. 191 - 196 (1969).

114.025 **Spectral evolution of the peculiar star MHα 328 – 116 (V 1016 Cyg) from 1965 to 1967.**
A. Mammano, L. Rosino.
Non-Periodic Phenomena in Variable Stars, IAU Colloquium, Budapest, 1968, p. 411 - 413 (1969).

Some informations are given on the spectral evolution of the peculiar star MHα 328 – 116 from 1965 to the end of 1967. It is shown that the degree of excitation is increasing. Some considerations on the nature of the object follow.

114.026 **Investigations of a region in Monoceros. III. MK classification of stars.** B. Karlsson.
Ark. Astron. Vol. 5, 241 - 248 (1969).

Spectral types on the MK system have been determined from slit spectrograms (dispersion 77 Å/mm at $H\gamma$) for 39 B, A and F type stars in a field around NGC 2264 (l^{II} = 203°, b^{II} = +2°).

114.027 **Spectrophotometry of F-stars. IV. Line profiles in the spectra of δ Boo, ν Her and 41 Cyg.**
T. A. Kipper.
Astrofizika, Vol. 5, 123 - 135 (1969). In Russian. English translation in Astrophysics, Vol. 5, No. 1 (1969).

A program for computing the profiles of the Balmer line H_γ and of the Ca II λ 3933 line is described. Using this program the profiles for 9 model atmospheres of F-stars are computed. The profiles are compared with the observed ones for δ Boo, ν Her and 41 Cyg. It is found that the profiles of Balmer lines are insensitive to changes in surface gravity. Using the profiles of Ca II λ 3933 line, the calcium abundances for δ Boo, ν Her and 41 Cyg are found.

114.028 **The composition and age of δ Pavonis.**
A. W. Rodgers.
Monthly Notices, Roy. Astron. Soc., Vol. 145, 151 - 162 (1969).

The temperature of δ Pav, G8 V, an old disc population star has been determined using hydrogen line profiles and near infra-red photometry. It is found that δ Pav suffers excess line blanketing for its temperature which is only 100°K cooler than the sun. Use of the derived effective temperature and well determined trigonometric parallax of δ Pav leads to an estimate of its age of $(8 \pm 3) \times 10^9$ yr, i.e. it was formed at an epoch between the formation of NGC 188 and M67.

114.029 Abundance determinations in late-type stars from Schmidt camera objective prism spectra.
W. B. Samson.
Monthly Notices, Roy. Astron. Soc., Vol. 145, 373 (1969). – Abstract. – The full text of this paper, see Publ. Roy. Obs. Edinburgh, Vol. 6, 225 - 239 (1969).

114.030 The spectra of stars in comet-like nebulae.
E. A. Dibay.
Astrofizika, Vol. 5, 249 - 268 (1969). In Russian.
English translation in Astrophysics, Vol. 5, No. 2 (1969).

Spectra of stars in comet-like nebulae obtained by means of a 125 cm reflector in 1964 - 1968 are described. Some cases of variability of the lines are found. The profiles of the emission lines in the spectra of eight stars have been studied. The motions in the envelopes have a complicated character and it seems that they are connected with the dynamics of surrounding nebulae. By means of photoelectric observations the position of stars connected with comet-like nebulae on the Hertzsprung-Russell diagram is analysed. It has been shown that some of these objects do not lie on the main sequence.

114.031 Total line and band absorption in the spectra of F−M stars. M. H. Rodriguez.
Astrofizika, Vol. 5, 269 - 281 (1969). In Russian.
English translation in Astrophysics, Vol. 5, No. 2 (1969).

Blanketing coefficients are given for 20 representative stars of the spectral classes F−M of different luminosities in the region $\lambda\lambda 6700 - 3575$ Å.

114.032 Line statistics for solar type stars.
E. Böhm-Vitense.
Journ. Quant. Spectrosc. Radiat. Transfer, Vol. 9, 1167 - 1190 (1969).

The wavelength interval from 3100 Å $< \lambda < 8200$ Å has been divided into 16 spectral regions. In each region, all of the lines have been combined to form one line with the same distribution of κ_ν as all the lines together. Values of $\Delta\lambda_m$ have been calculated for which $\kappa_\nu = \kappa_0 \exp(-m)$. Tables for $\Delta\lambda_3$, $\Delta\lambda_{10}$ and $\Delta\lambda_{15}$ as well as for κ_0 are given as a function of the gas pressure P_g and the temperature T.

114.033 Infrared and microwave astronomy.
P. A. Feldman, M. J. Rees, M. W. Werner.
Nature, Vol. 224, 752 - 758 (1969).

Recent observations of infrared and microwave emission from various astronomical objects, together with their theoretical implications, were discussed at a conference held in Cambridge from July 8 to 11, 1969. This rapidly developing field of astronomy promises to contribute significantly to our understanding of the formation and evolution of not only stars and galaxies but also the universe itself.

114.034 The variability of 21 Persei.
G. W. Preston.
Astrophys. Journ. Vol. 158, 251 - 260 (1969).

Radial velocities and line intensities in the spectrum of 21 Per vary periodically in the $2\overset{d}{.}88$ photometric cycle derived by Stępień. The velocity curve for the singly ionized rare earths consists of two branches that overlap near primary light maximum and overlap again one half-cycle later, when a weak secondary light maximum may occur. The range of the velocity variation is about 30 km sec^{-1}. The velocity variation for Ti II and Mn II resembles those for the rare earths, while lines of Si II, Sr II, and Fe II yield velocity curves of small amplitude with double waves. The magnetic field of 21 Per is weak and cannot be measured with precision because of the complicated kinematic effects. The interpretation of these results in terms of a rigid-rotator model is discussed briefly.

114.035 The Ba II line in late-type stars. G. A. Bakos.
Bull. American Astron. Soc. Vol. 1, 233 (1969). – Abstr. AAS.

114.036 Microturbulence in main-sequence stars.
F. H. Chaffee, Jr.
Bull. American Astron. Soc. Vol. 1, 236 (1969). – Abstr. AAS.

114.037 Ultraviolet spectrophotometric observations with the Wisconsin experiment package on OAO-2.
A. D. Code.
Bull. American Astron. Soc. Vol. 1, 237 (1969). – Abstr. AAS.

114.038 Analysis of the peculiar A star α^2 CVn.
J. G. Cohen.
Bull. American Astron. Soc. Vol. 1, 237 (1969). – Abstr. AAS.

114.039 Ultraviolet spectrum of Canopus taken from Gemini XI. Y. Kondo, K. G. Henize, C. L. Kotila.
Bull. American Astron. Soc. Vol. 1, 250 (1969). – Abstr. AAS.

114.040 Optical observations of the infrared source VY Canis Majoris. G. Wallerstein.
Bull. American Astron. Soc. Vol. 1, 265 (1969). – Abstr. AAS.

114.041 The Wolf-Rayet stars and mass loss.
A. B. Underhill.
Mass Loss from Stars, Trieste 1968, p. 17 - 25 (1969).

It is incontrovertible that Wolf-Rayet stars are losing mass, a typical rate of mass loss being near 10^{-5} $M_\odot$ per year. The outward directed velocity of the expanding shell has been estimated for 10 stars. The largest value found is 2500 km/sec; most values lie between 1000 and 1500 km/sec.

114.042 Mass loss from P Cygni. M. de Groot.
Mass Loss from Stars, Trieste 1968, p. 26 - 35 (1969).

It is found that many of the absorption lines are double, the hydrogen absorption lines even triple. This is attributed to line formation in different shells. In the outer shell variations with a period of 114 days lead to observed radial-velocity variations between −180 and −240 km/sec. A preliminary conclusion about the velocity field in the atmosphere of P Cygni is drawn and the mass loss is estimated.

114.043 Rocket observations of mass loss from hot stars.
D. C. Morton.
Mass Loss from Stars, Trieste 1968, p. 36 - 41 (1969).

Rocket observations have shown that the far-ultraviolet resonance lines have P-Cygni profiles in the spectra of many hot stars, including Of and Wolf-Rayet stars and OB supergiants. Velocity shifts as high as −3000 km sec^{-1} have been measured for the short-wavelength edges of some of the lines. Estimates of the rates of mass loss range from 10^{-8} to 10^{-6} $M_\odot$ year^{-1}.

114.044 Intrinsic polarization for objects with extended atmospheres. A. Kruszewski, G. V. Coyne, T. Gehrels.
Mass Loss from Stars, Trieste 1968, p. 42 - 48 (1969).

Intrinsic polarization in B emission-line stars and in red variables is discussed. The peculiar wavelength-dependence of the polarization in Be stars, particularly the decrease in polarization in the ultraviolet, is explained by scattering from electrons in an asymmetrical envelope together with selfabsorption in a hydrogen plasma. Red variable stars show large polarizations in the ultraviolet. Unlikely configurations of the scattering envelope, with the opaque cloud in front of the star, are required to explain the observed polarizations by Rayleigh scattering on molecules in an asymmetric envelope. Difficulties with a model of elongated graphite grains oriented by a magnetic field are also discussed.

114.045 Mass loss from OB supergiants.
J. B. Hutchings.
Mass Loss from Stars, Trieste 1968, p. 49 - 56 (1969).

Three main types of evidence in the spectra of OB supergiants are discussed, which lead to conclusions on the extent and mass motions of the outer envelopes of the stars. Illustrations are given from two southern stars. A method of computing strong line profiles in extended moving atmospheres is outlined, and it is shown how these profiles support the proposed atmospheric structure of the stars.

114.046 Preliminary general catalogue of early type emission stars.   F. C. Bertiau, M. F. McCarthy.
Ric. Astron. Specola Vaticana, Vol. 7, (No. 19), 523 - 632 (1969).

The present catalogue presents a list of 3216 stars mostly earlier than type F which have shown emission features and specifically one or more of the lines of the Balmer series in emission. It is a compilation and rearrangement of previous lists.

114.047 Evidence for mass loss from the F-type supergiant, 89 Herculis.   W. L. W. Sargent, P. S. Osmer.
Mass Loss from Stars, Trieste 1968, p. 57 - 63 (1969).

We describe changes in the spectrum of 89 Herculis (F2 Ia) in the interval 1961-67, which are ascribed to variable loss of mass from the surface. The star is remarkable in that some of the blue-displaced circumstellar absorption features have velocities of 150 km/sec.

114.048 A new outburst of the shell star 48 Librae.
A. B. Underhill, H. G. Geuverink.
Mass Loss from Stars, Trieste 1968, p. 64 - 71 (1969).

Spectrograms taken in 1967 and in 1968 of 48 Librae show that the shell spectrum is at present very strong with strongly asymmetrical lines. The radial velocity shown by different lines varies from about –130 km/sec to –3 km/sec. The observed velocities do not fit in with the cyclical pattern of changes observed between 1938 and 1962. A new outburst is indicated. The observations suggest that there is a strong radial-velocity gradient in the shell. The sharp components due to Na I and Ca II are probably formed in a circumstellar envelope which is escaping from the star.

114.049 Die Koeffizienten der kontinuierlichen Absorption bei hohen Temperaturen.   N. A. Sakhibullin.
Trudy Kazan. Gorod. Astron. Obs. No. 35, p. 3 - 20 (1968). In Russian.

Die Berechnung erfolgte für H, He I und He II bei Temperaturen von 90000° bis 700000°.

114.050 Spectrophotometry of B stars.
L. H. Aller, J. Jugaku.
Publ. Obs. Univ. Michigan, *Ann Arbor*, Vol. 9, No. 9, p. 203 - 273 (1969).

Although we shall present the analysis for several of these stars by model atmosphere methods, similar to those employed for γ Pegasi, we lay no particular emphasis on the numerical results, which will have to be revised as the models are improved. Our emphasis will be on the observational data, such as measurements of equivalent widths for lines of moderate to weak intensity and of profiles for hydrogen and helium lines.

114.051 The analysis of the low gravity halo star HD 214539.   A. Przybylski.
Monthly Notices, Roy. Astron. Soc., Vol. 146, 71 - 90 (1969).

In a coarse analysis the high-velocity metal-poor A0 Ib star HD 214539 has been compared with two normal giants η Leo and 13 Mon. It has been found that: 1. with a high degree of probability HD 214539 has a normal helium content; 2. silicon is under-abundant by a surprisingly low factor of 2 only; magnesium, which preceeds silicon in the periodic system of elements and, like it, is formed in the helium burning process, is under-abundant by a higher factor of 10; 3. the iron group is deficient by a factor varying between 13 and 30; and 4. sodium (formed through the *s*- and *p*-processes) is under-abundant by a factor of 30.

114.052 Infrared observations of possible protostars.
E. E. Becklin, G. Neugebauer.
Interstellar Ionized Hydrogen, Charlottesville 1967, p. 1 - 12 (1968).

114.053 The infrared spectrum of the cool dwarf Wolf 359.
R. F. Wing, W. K. Ford, Jr.
Publ. Astron. Soc. Pacific, Vol. 81, 527 - 529 (1969).

Wolf 359 is the nearest star to the sun after α Centauri and Barnard's star and is the coolest dwarf star for which fairly detailed spectroscopic observations can be obtained. The spectra of Wolf 359, Barnard's star, and several M giants have been observed from 6900 to 11 000 Å at a dispersion of 250 Å/mm.

114.054 Variations of silicon and helium lines in the spectrum of 56 Arietis.   W. K. Bonsack.
Publ. Astron. Soc. Pacific, Vol. 81, 531 (1969). – Abstract ASP.

114.055 The application of the Wilson-Bappu technique at moderate dispersion.   H. M. Dyck, M. C. Jennings.
Publ. Astron. Soc. Pacific, Vol. 81, 536 - 537 (1969). Abstract ASP.

114.056 Helium-weak stars.
M. Jaschek, C. Jaschek, M. Arnal.
Publ. Astron. Soc. Pacific, Vol. 81, 650 - 656 (1969).

A systematic survey was made for helium-weak stars and eight new objects were found.

114.057 An investigation of the NH bands in stellar spectra.
J. L. Schmitt.
Publ. Astron. Soc. Pacific, Vol. 81, 657 - 664 = Contr. Kitt Peak National Obs. No. 471 (1969).

The behavior of the NH bands at λ 3360 and λ 3370 in late-type stars has been investigated. A plot of NH strength vs. spectral type for both normal and strong-CN stars indicates that the strong-CN stars have little or no enhancement of NH over the normal stars. Thus the abundance of nitrogen seems to be very nearly the same in the two classes of stars.

114.058 Continuum energy distribution of O5–G0 stars in terms of spectral gradients.   A. Ardeberg.
Astron. Astrophys. Vol. 3, 257 - 269 (1969).

The purpose of the present work is to study the absolute energy distribution of stellar continua for wide ranges in spectral type and luminosity. This is accomplished through the use of spectral gradients covering a wave-number interval as wide as possible. Beside the choice of observational material, two main problems present themselves. Firstly, it is a question how to choose the wave-number base of the gradients. Secondly the influence of interstellar absorption must be taken into account. An additional difficulty is the reduction of the results to an absolute energy basis.

114.059   A unique emission line object.
  A. P. Cowley, W. A. Hiltner.
Astron. Astrophys. Vol. 3, 372 - 375 = Contr. Cerro Tololo Inter-American Obs. No. 87 (1969).

Line identifications have been made for the peculiar emission object CPD –56°8032. The ions primarily represented are C II, C III, O II, He I, Si III, and Si IV. The strongest features are due to C II and are seen only in emission. In contrast O II and He I show strong emissions with violet-displaced absorptions. Weak and sharp emissions of H and [O II] are observed.

114.060   The spectra of some Be stars.
  M. Jaschek, C. Jaschek, S. Malaroda.
Astron. Astrophys. Vol. 3, 485 - 487 = Veröff. Sternw. München, Band 7, No. 12 (1969).

Short descriptions of the spectra of 31 Be stars are given.

114.061   Submillimeter astronomy.   A. E. Salomonovich.
Priroda No. 9, p. 18 - 29 (1969).   In Russian.

114.062   The Ba II star ζ Cygni.
  F. R. Chromey, S. M. Faber, A. Wood, I. J. Danziger.
Astrophys. Journ. Vol. 158, 599 - 606 (1969).

A spectral analysis at high dispersion shows that the G8 II star ζ Cygni is a Ba II star showing a relatively small enhancement in the abundance of carbon, strontium, yttrium, barium, lanthanum, and cerium. The possible evolution of such stars is briefly discussed.

114.063   Atomic lines in the CH star HD 209621.
  G. Wallerstein.
Astrophys. Journ. Vol. 158, 607 - 612 (1969).

The spectrum of HD 209621 has been studied by using the equivalent widths of thirty-six lines measured on 13.5 Å mm^{-1} spectrograms. The atomic lines have been compared with the G9 III star ε Vir, two other CH stars, and RU Cam by curves of growth. After reduction of each comparison to a comparison with the standard star, ε Vir, we find that the metals are deficient in HD 209621 by a factor of 20 and that the electron pressure is lower by a factor of 10. The ratio of rare earths to metals is enhanced by a factor of 8 as compared with ε Vir.

114.064   Spectra of "infrared stars" from 2.8 to 5.1 microns.
  J. E. Gaustad, F. C. Gillett, R. F. Knacke, W. A. Stein.
Astrophys. Journ. Vol. 158, 613 - 618 (1969).

Low-resolution spectra from 2.8 to 5.1 μ have been obtained of the infrared stars NML Tau, CIT 3, CIT 6, and CIT 13. All of the spectra contain complex absorption features, but the resolution is insufficient for positive identification of the molecular bands.

114.065   Observations of the infrared object, VY Canis Majoris.
A. R. Hyland, E. E. Becklin, G. Neugebauer, G. Wallerstein.
Astrophys. Journ. Vol. 158, 619 - 628 (1969).

Infrared colors to 20 μ and optical and infrared spectra have been obtained for the irregular M variable star and bright OH source, VY CMa. Spectra at all wavelengths indicate that it is a supergiant with an effective temperature of 2500°– 3000°K. Compared with supergiants of the same temperature, VY CMa exhibits a large infrared excess and is one of the brightest objects known at 20 μ. Its OH emission, near infrared spectra, and far-infrared energy distribution are similar to those of the NML Cyg source; however, it is visually 10 mag brighter.

114.066   1 – 4 -micron spectra of four M stars and Alpha Tauri.
  R. I. Thompson, H. W. Schnopper, R. I. Mitchell, H. L. Johnson.
Astrophys. Journ. (Letters), Vol. 158, L117 - L122 (1969).

Infrared spectra of four M stars and α Tau, obtained with a Michelson interferometer, are presented at a resolution of 9 cm^{-1}. Absorption bands of $^{12}C^{16}O$ are observed in all of these stars, but no positive identification of $^{13}C^{16}O$ bands can be made. The intensities of the $^{12}C^{16}O$ bands in the M stars and in α Tau are compared, and the possibility of nonlinear curves of growth for the $^{12}C^{16}O$ bands is discussed. The high-temperature stellar H_2O absorptions in o Cet (Mira) are again observed.

114.067   1 – 4-micron spectra of four carbon stars and Sirius.
  R. I. Thompson, H. W. Schnopper, R. I. Mitchell, H. L. Johnson.
Astrophys. Journ. (Letters), Vol. 158, L55 - L60 (1969).

Infrared spectra of four carbon stars and Sirius, obtained with a Michelson interferometer, are presented at a resolution of 9 cm^{-1}. Bands due to C_2, $^{12}C^{16}O$, and $^{13}C^{16}O$ are found in the carbon stars, and the hydrogen Brackett series is found in Sirius. The carbon monoxide bands appear to be saturated in the carbon stars.

114.068   OH emission and the infrared star in VY Canis Majoris.   R. S. Booth.
Nature, Vol. 224, 783 - 784 (1969).

Strong OH-line emission was detected at 1665, 1667 and 1612 MHz from the IR star VY Canis Majoris. The emission at 1665 and 1667 MHz was found to be partially circularly polarized and some features were observed to vary in intensity over a period of 3 months. The emissions from the two main lines covered approximately the same velocity interval but there was no correspondence (above a 1°K limit) with the 1612 MHz emission.

114.069   Spectrophotometry of κ Cassiopeiae. III.
  B. Kovachev.
Izv. Sekts. Astron. Blg. AN, Vol. 3, 49 - 62 (1969).   In Bulgarian. – Abstr. in Referativ. Zhurn. 51. Astron., 12.51.208 (1969).

114.070   Spectrophotometry of 102 Herculis.
  K. Kovachev, B. Kovachev.
Izv. Sekts. Astron. Blg. AN, Vol. 3, 63 - 76 (1969). In Bulgarian. – Abstr. in Referativ. Zhurn. 51. Astron., 12.51.209 (1969).

114.071   Abundance determinations in late-type stars from Schmidt camera objective prism spectra.
W. B. Samson.
Publ. Roy. Obs. Edinburgh, Vol. 6, 225 - 239 (1969).

Sixty two stars of types F and G have been observed using a 40/60 cm Schmidt camera with an objective prism giving a dispersion of 400 Å/mm at the wavelength of Hγ. Following van den Bergh (1963), parameters corresponding to Fe abundance, CH-band strength and CN-band strength have been measured, along with a parameter corresponding to the effective temperature, for each star, with r.m.s. errors in the parameters of about 8% for three observations. The CH and CN parameters are found to be sensitive to luminosity, the CN parameter increasing with luminosity at about six times the rate of the CH paran.eter.

114.072   The spectrum of the Bp star HD 36916.
  M. Hack.
Astrophys. Space Sci. Vol. 5, 403 - 419 (1969).

The spectrum of the peculiar Bp star HD 36916 has been studied on plates with dispersion 9.7 Å/mm and compared with those of η Aur, B3V and ι And, B8V. HD 36916 presents the characteristics of the Si-λ 4200 stars: strong deficiency in helium and probably also in oxygen and nitrogen, strong excess of silicon and strontium; 3984 Hg II is present. Moreover this star also has characteristics which are not common to Si-λ 4200 stars but rather to Mn stars. The star is a member of the Sword subgroup of the Orion association.

114.073   **The Paschen discontinuity in B stars.**
A. E. Ringuelet.
Astrophys. Space Sci. Vol. 5, 459 - 468 (1969).

Measurements of the Paschen discontinuity in stars with $T_{eff} \geqslant 10^4$ K leads to the conclusion that the D_P/D_B ratio increases with temperature faster than expected. The increase of D_P/D_B with $(\log g)^{-1}$ is also steep.

114.074   **Variations photométriques et spectrophotométriques de la raie Hβ de l'étoile HD 109387 (κ Dra).**
G. Adam, J.-H. Bigay, A.-M. Delplace, M. Duval, R. Garnier, R. Herman, A. Peton.
Comptes Rendus, Acad. Sci. Paris, Sér. B, Vol. 269, 1332 - 1334 (1969).

On a mis en évidence, au photomètre photoélectrique, des variations certaines de l'intensité de la raie Hβ de l'étoile κ Dra; celles-ci semblent en corrélation étroite avec les variations spectrales du profil de la raie Hβ observée simultanément.

114.075   **Infrared astronomy.**
R. F. Webbink, W. Q. Jeffers.
Space Sci. Rev. Vol. 10, 191 - 216 (1969).

This review paper is a survey of infrared astronomy up to early 1969. The techniques and photometric standards are mentioned briefly, and results cover solar, lunar, and planetary observations. Point sources and extended sources both within and beyond the galaxy are included, ending with the problem of cosmic background radiation. It is concluded that great progress will be possible when large infrared telescopes are placed above the atmosphere in orbit for extended periods of time.

114.076   **Erratum: Carbon stars in a south galactic pole region. [Astron. Journ., Vol. 74, 373 - 374 (1969)].**
A. Slettebak, P. C. Keenan, R. K. Brundage.
Astron. Journ., Vol. 74, 1262 (1969). – See Astron. Astrophys. Abstr., Vol. 1, 114.070.

114.077   **A search for metal-deficient stars.**
H. E. Bond.
Bull. American Astron. Soc., Vol. 1, 336 (1969). – Abstract AAS.

114.078   **The ultraviolet flux of Capella.**
L. R. Doherty.
Bull. American Astron. Soc. Vol. 1, 339 - 340 (1969). – Abstract AAS.

114.079   **Far-UV model fittings for Sirius.**
D. Fischel, D. A. Klinglesmith, T. P. Stecher.
Bull. American Astron. Soc., Vol. 1, 342 - 343 (1969). – Abstract AAS.

114.080   **Interpretation of the CO bands in Arcturus.**
T. F. Greene.
Bull. American Astron. Soc., Vol. 1, 345 (1969). – Abstract AAS.

114.081   **Photoelectric measurements of the λ6495 line of Fe I in G and K stars.**   G. D. Gutsche.
Bull. American Astron. Soc., Vol. 1, 345 (1969). – Abstract AAS.

114.082   **Merak taken siriusly.**   D. W. Latham.

Bull. American Astron. Soc., Vol. 1, 352 (1969). Abstr. AAS.

114.083   **Infrared emission by mineral grains from red supergiants.**   N. Scoville, P. M. Solomon.
Bull. American Astron. Soc., Vol. 1, 361 (1969). – Abstr. AAS.

114.084   **A statistical study of the helium abundance of population I B stars.**   H. L. Shipman, S. E. Strom.
Bull. American Astron. Soc., Vol. 1, 362 (1969). – Abstr. AAS.

114.085   **A high-dispersion spectral analysis of the Ba II star HD 204075 (ζ Capricorni).**
J. L. Tech.
Bull. American Astron. Soc., Vol. 1, 364 - 365 (1969). Abstr. AAS.

114.086   **Early-type stars whose spectra have shown emission lines.**   L. R. Wackerling.
Bull. American Astron. Soc., Vol. 1, 367 (1969). – Abstr. AAS.

114.087   **Osmium lines in the spectrum of the peculiar A star 73 Draconis.**   B. N. G. Guthrie.
Observatory, Vol. 89, 224 - 226 (1969).

In using high dispersion Mount Palomar spectra unblended and slightly blended Os lines were found in 73 Dra. Lines of Th II and U II are suspected to be present, but need confirmation. Stars showing Os and Pt lines comprise only a small percentage of peculiar A stars.

114.088   **Absorption bands of titanium monoxide of α-system in stellar spectra of M class.**
M. E. Boyarchuk.
Izv. Krymskoj Astrofiz. Obs. Vol. 39, 114 - 123 (1969).
In Russian.

Absorption bands of titanium monoxide are investigated on 110 spectrograms of 28 stars of M0 - M8 spectral classes. The identification of titanium bands of α-system in stellar spectra is carried out. Intensities of band heads are estimated quantitatively. The intensity change of band heads of titanium monoxide with the spectral class is considered.

114.089   **The new Stockholm spectral survey of the southern Milky Way.**   L. O. Lodén.
Vistas in Astronomy, Vol. 11, 161 - 171 (1969).

This article gives a brief description of a new survey of the southern Milky Way which is being carried out at the Stockholm Observatory. Photographic and photoelectric material, obtained at the Boyden Observatory in South Africa, is used for the selection of interesting objects, and for a preliminary classification and photometry of these objects to facilitate subsequent detailed investigations.

114.090   **The abundance of lithium in cool stars, II.**
B. Warner.
Journ. Quant. Spectrosc. Radiat. Transfer, Vol. 9, 1637 - 1638 (1969).

Revised lithium abundances in late-type stars are given, based on the latest effective temperature scale.

114.091   **La composizione chimica delle stelle povere di metalli.**   M. Hack.
Atti XII Riunione Soc. Astron. Italiana, L'Aquila 1968, p. 101 - 102 (1969). – Abstract SAI.

114.092   **Preliminary results of the quantitative analysis of the Ap star HD 151199.**
N. G. Gokkaya.

Atti XII Riunione Soc. Astron. Italiana, L'Aquila 1968, p. 103 - 104 (1969). – Abstract SAI.

114.093 On the spectral duplicity of 68 Cygni.
E. A. Vitrichenko, P. N. Kholopov.
Astron. Tsirk. No. 535, p. 6 - 7 (1969).   In Russian.

114.094 Measurements of the absolute distribution of energy in stellar spectra.
B. N. Wainman, T. I. Tsaregradskaya.
Astron. Tsirk. No. 538, p. 6 - 7 (1969).   In Russian.

114.095 Some spectral features of late-type stars.
G. N. Dzhimshelejshvili.
Soobshch. AN Gruz. SSR, Vol. 55, 57 - 60 (1969).
In Russian. – Abstr. in Referativ. Zhurn. 51. Astron., 1.51.665 (1970).

114.096 Etoiles froides et structure galactique.
L. Martinet.
C. R. des Séances, SPHN Genève, NS, Vol. 4, Fasc. 1, p. 22 - 26 = Publ. Obs. Genève, Sér. A, Fasc. 76/II (1969).
Nous présentons ici quelques problèmes de structure galactique abordables à partir de l'étude des étoiles froides. Sous cette appellation nous grouperons les naines $G - M$, les géantes de classe spectrale plus avancée que $G5$, les variables à longue période et les étoiles carbonées.

114.097 Classification spectrale d'étoiles presque identiques dans la photométrie en 7 couleurs.
M. Golay, E. Peytremann, A. Maeder.
C. R. des Séances, SPHN Genève, NS, Vol. 4, Fasc. 1, p. 44 - 54 = Publ. Obs. Genève, Sér. A, Fasc. 76/V (1969).
L'article donne une liste de 106 groupes d'étoiles. Dans chaque groupe les étoiles représentent presque les mêmes valeurs des sept magnitudes monochromatiques tirées de la photométrie en sept couleurs de l'Observatoire de Genève. Les diverses classifications spectrales attribuées aux étoiles d'un même groupe sont données.

114.098 Emission object MHα 328 – 116 = V 1016 Cyg.
M. Kurpinska.
Inform. Bull. Variable Stars (I.A.U. Commission 27), Konkoly Obs., Budapest, No. 372 (1969).

114.099 Emission object V 1016 Cyg (MHα 328 – 116).
M. P. Fitzgerald, N. Houk.
Inform. Bull. Variable Stars (I.A.U. Commission 27), Konkoly Obs., Budapest, No. 400 (1969).

114.100 The Wolf Rayet stars: Informal perspective.
C. S. Beals.
Wolf Rayet Stars, Boulder 1968, p. 7 - 20 (1968).

114.101 The features of the system of Wolf Rayet stars.
L. F. Smith.
Wolf Rayet Stars, Boulder 1968, p. 21 - 100 (1968).
Classification, luminosities, distribution, association with OB associations and with H II regions, masses, interpretation of the classification system, binary stars, evolutionary status, source of instability, do the WR stars form a class? ; Discussion.

114.102 A survey of spectroscopic features of Wolf Rayet stars. L. V. Kuhi.
Wolf Rayet Stars, Boulder 1968, p. 101 - 179 (1968).
Detailed spectroscopic features of Wolf Rayet stars, identification of spectral lines, the continuous energy distribution, line profiles, line intensities, variations in line intensities and profiles, the effect of binary nature on spectra, intermediate objects, Wolf Rayet stars showing lines of both carbon and nitrogen, stars showing Wolf Rayet features and nebular lines,

central stars of planetary nebulae references; Discussion.

114.103 Spectroscopic diagnostics, interpretation, and atmospheric models. A. B. Underhill.
Wolf Rayet Stars, Boulder 1968, p. 181 - 236 (1968).
The types of spectroscopic information, the continuous spectrum, the line spectrum, spectroscopic diagnostics and interpretation, evidence concerning the temperatures of Wolf Rayet stars, evidence concerning the electron density in Wolf Rayet atmospheres, the types of physical processes to be considered, a model of a typical Wolf Rayet atmosphere, clues regarding the evolutionary stage of Wolf Rayet stars; Discussion.

114.104 A summary of problems, ideas, and conclusions on the physical structure of the Wolf Rayet stars.
R. N. Thomas.
Wolf Rayet Stars, Boulder 1968, p. 237 - 277 (1968).
Taxonomy of classical WR stars, specification of quasi–WR objects, spectral characteristics of the WR atmosphere, the structure of WR objects, and its effect on atmospheric features, the structure of quasi–WR objects and its effect on the atmospheric features; Discussion.

114.105 Spectrophotometric study of α² CVn. I.
Z. N. Chumak.
Trudy Astrofiz. Inst. Alma-Ata, Vol. 14, 50 - 53 (1969).
In Russian.
The analysis of hydrogen lines of the spectrum variable star α^2 CVn is made. Parameters $\lg n_e$ (n_m), $\lg n_e$ (H), $\lg N_{02}$ H, $\lg P_e$ (n_m) are found. $\lg g_{eff}$ is estimated by a comparison between contours of the lines of Hα and Hδ and of theoretical ones. $\lg g_{eff}$ is also found by a dynamical method.

114.106 The manganese mercury star π_1 Bootis.
J. W. Montgomery, L. H. Aller.
Proc. National Acad. Sci. USA, Vol. 63, 1039 - 1044 (1969).
Spectral-energy distribution measurements are combined with line intensity data for iron and manganese in two stages of ionization to obtain a fit with model atmospheres for $T_{eff} = 13.000°$K and $\log g = 4$. The influence of adopted T and g on the derived abundances is discussed. Although C, O, Mg, Si, Ti, Cr, and Fe appear to have nearly normal (i.e., solar) abundances, strontium appears to be enhanced in abundance by an order of magnitude, and scandium is about 50 times overabundant, while manganese and yttrium appear to be two order of magnitude overabundant.

114.107 Emission-line objects projected upon the galactic bulge. G. H. Herbig.
Proc. National Acad. Sci. USA, Vol. 63, 1045 - 1050 (1969).
Low-dispersion slit spectrograms have been obtained of 34 faint objects that lie in the direction of the galactic bulge and have the Hα line in emission upon a detectable continuum. Eleven of these are certain or probable symbiotic stars. A rough comparison with R CrB stars in the same area suggests that these brightest symbiotics in the bulge have in the mean $M_v \approx -3$ to -4, which suggest population II red giants rather than conventional population I M-type objects. The sample also contains a number of hot stars having H and [O II] or [O III] in emission, as well as four conventional Be stars, and six certain or possible planetary nebulae.

114.108 Aus der Geschichte der astronomischen Spektroskopie. W. Seitter.
SuW, Vol. 8, 264 - 266 (1969).

114.109 Investigations of a region in Monoceros.
B. Karlsson.
Medd. Lund Obs., Ser. I, No. 246, 6 pp. ((1969).

This paper is a summary of earlier papers. Their aim is to determine the space densities of stars of spectral type earlier than F8 and to study the interstellar extinction in a galactic field in Monoceros. The investigations involve the determination of luminosities and intrinsic colours on the basis of various derived spectral quantities from objective prism plates.

114.110 Photoelectric spectrophotometry of Hγ in early-type stars. A. D. Code, A. A. Hoag.
Publ. Astron. Soc. Pacific, Vol. 81, 848 - 853 = Contr. Kitt Peak National Obs. No. 495 (1969).

In the course of a spectrophotometric program (A. D. Code's) involving observations of extragalactic sources, the writers obtained spectrum scans of several early-type stars including three O stars in NGC 1893. The purpose of this communication is to report on the measurements of the equivalent widths of Hγ in these stars and to describe the technique employed.

114.111 Spectrophotometric investigation of the star κ Cassiopeiae. II. B. Kovachev, V. Dobrichev.
Izv. Fiz. in-t s ANEB, Vol. 18, 151 - 163 (1969).
In Bulgarian. — Abstr. in Referativ. Zhurn. 51. Astron., 2.51.530 (1970).

114.112 Quantitative Spektralklassifikation und ihre Anwendung auf Probleme der Entwicklung der Sterne und der Milchstraße. B. Strömgren.
Mitt. Astron. Ges. No. 27, p. 15 - 27 (1969). — Karl-Schwarzschild-Vorlesung 1969. Review article.

114.113 Dispersion of spectrograms in connection with abundance ratio determination of C^{12} to C^{13} in carbon stars. Y. Fujita.
Proc. Japan Acad., Vol. 44, 495 - 500 = Contr. Dep. Astron. Univ. Tokyo No. 102 (1968).

114.114 A note on the spectral classification of carbon stars. Y. Fujita.
Proc. Japan Acad., Vol. 45, 272 - 277 = Contr. Dep. Astron. Univ. Tokyo No. 109 (1969). — Note.

114.115 Scanner abundance studies. I. An investigation of supermetallicity in late-type evolved stars.

H. Spinrad, B. J. Taylor.
Astrophys. Journ. Vol. 157, 1279 - 1340 (1969).

In this paper we describe a new technique for obtaining differential abundances. We also describe results we have obtained to date from an application of this technique to K giants and subgiants. Evolved K stars with metal abundances greater than those of the Hyades exist in substantial numbers. This conclusion is derived from the individual strong-feature data, and is confirmed by blanketing measures and by spot checks with slit spectrograms. The abundances found range as high as 4 times the solar values for Ca, Mg, and Na. The cool evolved stars in M67 and NGC 188 have generally 3 times the solar abundances of the above elements; independent data on the hotter stars in M67 indicate that this is probably true of them, also.

114.116 Emission object HBV 475.
D. Crampton, J. Grygar.
IAU. Circ. No. 2174 (1969).

114.117 HBV 475. D. Crampton, J. Grygar.
IAU Circ. No. 2176 (1969).

114.118 HBV 475. Y. Andrillat.
IAU Circ. No. 2182 (1969).

Excitation temperatures of Fe I in the sun and A stars and systematic errors in the *gf* scale.
See Abstr. 071.078.

The period of the light variation of HD 173650.
See Abstr. 116.025.

Optical observations of very young stars.
See Abstr. 122.091.

Physical properties of cepheids and other supergiants from six-color fitting. See Abstr. 122.110.

OB stars near the supernova remnant RCW 86.
See Abstr. 125.003.

Two stars of the Large Magellanic Cloud showing emission lines of Fe II and [Fe II].
See Abstr. 159.009.

115 Stellar Luminosities, Masses, Diameters, HR-Diagrams and Others

115.001 Surface brightnesses in the U, B, V system with applications on M_V and dimensions of stars.
A. J. Wesselink.
Monthly Notices, Roy. Astron. Soc., Vol. 144, 297 - 311 (1969).

The surface brightnesses are expressed in the V system in terms of V_0 and d'' (the angular diameter) with the formula $s_V = V_0 + 5 \log d$. s_V is derived for 18 stars from observation and the relation between s_V and $(B - V_B)_0$ is obtained. The central problem is the determination of M_V from R (the linear radius) or vice versa. Lines of equal radius have been drawn in a HR diagram having $(B - V)_0$ and M_V coordinates and the radii of stars along the zero age main sequence are given. The mean absolute magnitude of δ Cephei has been derived in good accord with other modern determinations ($M_V = -3.56$). The absolute magnitudes M_V for the components of some eclipsing variables are derived and a mass-luminosity relation is drawn with small scatter. The dimensions of some supersupergiants belonging to the Magellanic Clouds are calculated; HDE 268757 is one of the largest known stars with a radius of 10.3 a.u.

115.002 Ca II K emission in southern late-type stars.
B. Warner.
Monthly Notices, Roy. Astron. Soc., Vol. 144, 333 - 350 (1969).

K-emission widths, intensities and derived absolute visual magnitudes are given for 200 southern G, K and M stars, measured on 15 Å mm^{-1} coudé spectra obtained at the Radcliffe Observatory.

115.003 Possible horizontal-branch stars at high galactic latitudes. IV. A. G. D. Philip.
Astron. Journ. Vol. 74, 812 - 813 = Contr. Cerro Tololo Inter-American Obs. No. 68 (1969).

Positions, B magnitudes, and spectra are given for ten new possible field horizontal-branch stars in high galactic latitude regions.

115.004 On the calibration of $M_V(K)$ for giants by means of trigonometric parallaxes. R. M. West.
Astron. Astrophys. Vol. 3, 1 - 4 (1969).

A reinvestigation of the problem of calibrating absolute magnitudes derived from H- and K-line emission reversals for giant stars by means of trigonometric parallaxes has been carried out. Systematic errors in the absolute magnitudes that arise when stars with small parallaxes are used in the calibration, have been computed. Three parameters of the problem, the true spatial distribution of the observed stars, the mean error of an observed parallax and the lower limit of the true parallaxes, have been varied within reasonable limits. Since the presently available calibration material consists of 33 stars only, the true values of these parameters are only known approximately. It is concluded that the use, in the calibration materials of stars with parallaxes down to 0".030, may lead to systematic errors in the absolute magnitude equal to or larger than 0^m.3. This is in contradiction to the fact that three calibrations by Wilson (1968) based on MK standards, Sun-Hyades, and trigonometric parallaxes agree rather well.

115.005 Distribution of luminous stars in the region of one high velocity hydrogen cloud.
S. N. Svolopoulos.
Astron. Nachr. Vol. 291, 129 - 130 (1969).

An objective prism survey of the area of the high velocity hydrogen cloud located at $l^{II} = 72°.6$, $b^{II} = +15°.9$ gave a low density of luminous stars in agreement to the expected value in the region's latitude.

115.006 Theoretische Zweifarbendiagramme für hohe galaktische Breiten. W. Pfau.
Astron. Nachr. Vol. 291, 155 - 176 (1969).

It is the aim of the present paper to give two-colour diagrams, which are valid for different distances z from the galactic plane. The diagrams are computed on the basis of theoretical considerations and take into account the differing part of members of the halo population. For several reasons the considerations are limited to the range 0.30 mag $\leq (B - V) \leq 0.73$ mag. The method proceeds from the relative frequency function of UV-excesses in the solar vicinity (different for stars with colour indices $(B - V)$ smaller and larger than 0.45 mag) and the relative frequency function of velocity components perpendicular to the galactic plane.

115.007 Remarques sur l'échelle de magnitudes absolues des étoiles O. L. Divan, M.-L. Burnichon.
Comptes Rendus Acad. Sci. Paris, Sér. B, Vol. 269, 235 - 236 (1969).

Le problème soulevé par la profondeur de 1000 pc trouvée pour le petit amas d'étoiles O, supposé cylindrique (moins de 10 pc de diamètre), situé dans IC 1805, est examiné. On attribue cette profondeur peu vraisemblable à l'incertitude qui règne sur les magnitudes absolues des étoiles O et l'on montre que la difficulté peut disparaître si l'on introduit un deuxième paramètre dans le calcul de ces magnitudes absolues (le paramètre λ_1).

115.008 On the possible distribution of mass among celestial bodies. M. G. Fracastoro.
Mem. Soc. Astron. Italiana, Nuova Serie, Vol. 40, 309 - 316 (1969).

According to the relationships which exist between mass and number of stars, asteroids and meteorites we make the hypothesis that the ratio $\Delta \log N / \Delta \log M$ be constant for all celestial bodies. One obtains a value of 6.8×10^{-24} g/cm^3 for the density in the solar neighbourhood. This value is considerably higher than the current one and in better agreement with that deduced dynamically at the distance of 8.2 kpc from the galaxy center.

115.009 Comparison between spectral-luminosity classes on the Mount Wilson and Morgan-Keenan systems of classification. M. P. FitzGerald.
Journ. Roy. Astron. Soc. Canada, Vol. 63, 251 - 259 = Contr. Univ. Waterloo Obs. No. 2 (1969).

Regression tables are presented for stars classified on both the Mount Wilson and Morgan-Keenan systems. These tables serve to show that a Morgan-Keenan spectral-luminosity class can be estimated from a Mount Wilson type to within two spectral sub-classes and to within one half a luminosity class.

115.010 The absolute magnitude–spectral type relation and scales of effective temperatures and bolometric corrections for O–B stars of the main sequence.
U. K. Dzervītis.
Latv. PSR Zinatņu Akad. věstis, Izv. AN Latv. SSR, No. 4, p. 39 - 49 (1969). In Russian. – Abstr. in Referativ. Zhurn. 51. Astron., 12.51.330 (1969).

115.011 **Accuracy of K-line luminosities and the masses of red giants.** B. E. J. Pagel, J. Tomkin.
Quarterly Journ. Roy. Astron. Soc. Vol. 10, 194 - 205 (1969).

Evidence is presented to show that the use of K-line luminosities gives an underestimate by about 1^m for G and K class III giants having a lower metal abundance than the Hyades. Recent estimates of masses and ages for red giants with known chemical composition are revised in the light of this result and shown to be quite consistent with other evidence.

115.012 **Some properties of K-giant field stars.** H. L. Helfer.
Astron. Journ., Vol. 74, 1155 - 1167 (1969).

Photometrically determined temperatures and metal abundances are given for approximately 175 early K giants belonging to the disk population. Approximately 1/3 of the stars have greater than solar metal abundance, while less than 5% have metal deficiencies of a factor of three or more compared to the sun. The high-velocity stars exhibit no preferred metal abundance in the range $+0.2 \geqslant [Fe/H] \geqslant -0.8$. Using K-line absolute magnitudes, space velocities and an effective mass index are also calculated for each star. The mass index is shown to be an observationally useful quantity. A class of massive metal-deficient stars definitely exists; it is also probable that some low-mass K giants exist.

115.013 **The absolute period-luminosity for cepheids.** S. Gaposhkin.
Bull. American Astron. Soc., Vol. 1, 344 (1969). — Abstract AAS.

115.014 **Surface brightness in the V system and angular diameters of stars.** A. J. Wesselink.
Bull. American Astron. Soc., Vol. 1, 368 - 369 (1969). Abstr. AAS.

115.015 **Criteri di luminosità e temperatura per stelle di tipo O.** M. Hack, R. Stalio.
Atti XII Riunione Soc. Astron. Italiana, L'Aquila 1968, p. 34 - 35 (1969). — Abstract SAI.

115.016 **Masse e posizioni nei diagramma evolutivo H–R di nove stelle variabili.**
A. Masani, A. Martini, E. Albino, G. Silvestro.
Atti XII Riunione Soc. Astron. Italiana, L'Aquila 1968, p. 91 - 96 (1969). — Abstract SAI.

115.017 **Photométrie des étoiles de type spectral A. I. Les étoiles A VI.** B. Hauck.
Bull. Soc. Vaud. Sci. Nat., Vol. 70, (No. 329), 179 - 180 =

Publ. Obs. Genève, Sér. A, Fasc. 76/VI (1969).

Three stars of spectral type A VI have been observed in the photometric system of the Geneva Observatory: HD 109995, HD 161817 and BD39°4926. The inspection of the different diagrams utilised in this system shows the possibility to make out these stars from those of population I.

115.018 **The period-luminosity relation: A historical review.** J. D. Fernie.
Publ. Astron. Soc. Pacific, Vol. 81, 707 - 731 (1969).

115.019 **Eine Nullpunktbestimmung der Perioden-Leuchtkraft-Beziehung von Cepheiden der Population I aus Eigenbewegungen und Radialgeschwindigkeiten.**
U. Geyer.
Mitt. Astron. Ges. No. 27, p. 127 - 128 (1969). — Abstract AG.

115.020 **Beobachtungen zur Leuchtkraftklassifikation von OB-Sternen durch Hβ-Indizes.**
U. Haug.
Mitt. Astron. Ges. No. 27, p. 148 - 149 (1969). — Abstract AG.

Note on the magnitude diameter relationship of 12013 stars in the Oxford Astrographic Zones +28° and +29°. See Abstr. 113.066.

On C.P.D. magnitudes and Melbourne diameters in −71° to −81°. See Abstr. 113.050.

The application of the Wilson-Bappu technique at moderate dispersion. See Abstr. 114.055.

A note on binary systems with undersize subgiant secondaries. See Abstr. 121.003.

Absolute magnitudes of Mira variables from statistical parallaxes. See Abstr. 122.101.

Short-period variability of B, A, and F stars. IV. Variability in the lower Hertzsprung gap. See Abstr. 122.103.

Incidence of short-period variability in the lower Hertzsprung gap. See Abstr. 122.107.

Subluminous stars. III. Luminosity calibration for subluminous stars and the space density of the blue subluminous stars south of declination −45°. See Abstr. 126.001.

116 Stellar Magnetic Field, Figure, Rotation

116.001 Carroll's method applied to small stellar rotations. A. Wilson.
Monthly Notices, Roy. Astron. Soc., Vol. 144, 325 - 332 (1969).

Carroll's Fourier transform method has been found to be accurate to two kilometres per second when applied to line profiles of solar quality. An upper limit of 3.5 km s^{-1} for the rotation velocity of Arcturus can be set using this method.

116.002 Partial resolution of Zeeman patterns in the spectrum of 53 Camelopardalis. G. W. Preston.
Astrophys. Journ. Vol. 157, 247 - 251 (1969).

Spectroscopic evidence is presented for a magnetic field of −15 kilogauss in the atmosphere of 53 Cam at phase 0.12 cycles prior to positive crossover.

116.003 The magnetic field variations of HD 188041. S. C. Wolff.
Astrophys. Journ. Vol. 157, 253 - 259 (1969).

Zeeman measurements of HD 188041 show that the amplitudes of the magnetic field variations are not identical for all elements. During the portion of the cycle spanned by the present observations, the fields derived from lines of Gd II, Mn I, Mn II, and Ca I were essentially constant, the fields of Cr I and Cr II changed by a factor of about 1.2, and the fields of Fe I, Fe II, Ce II, and Ti II changed by a factor of about 1.6. The elements which showed the smallest magnetic field changes showed the largest spectrum variations. Whenever an element was observed in two stages of ionization, the amplitudes of the magnetic and spectrum variations were the same for both. The observations can be explained most easily by the assumption that not all of the various elements are distributed uniformly over the surface of the star.

116.004 Stellar rotation. P. A. Strittmatter.
Annual Rev. Astron. Astrophys. Vol. 7, 665 - 684 (1969).

116.005 The problem of irregular variations in magnetic stars. T. Jarzębowski.
Non-Periodic Phenomena in Variable Stars, IAU Colloquium, Budapest, 1968, p. 227 - 237 (1969). − Introductory report.

116.006 Photometric search for periodicity among magnetic stars. K. Stępień.
Non-Periodic Phenomena in Variable Stars, IAU Colloquium, Budapest, 1968, p. 239 - 243 (1969).

116.007 Photometric research on magnetic stars at the Catania Astrophysical Observatory.
C. Blanco, F. Catalano, G. Godoli.
Non-Periodic Phenomena in Variable Stars, IAU Colloquium, Budapest, 1968, p. 243 - 251 (1969).

The results recently obtained at Catania from the observations of the magnetic stars SX Ari, 41 Tau, CU Vir, HD 173650, HD 184905, HD 219749, 8k Psc, HD 224801 are summarized.

116.008 Effects of prescribed circulations on magnetic fields. M. Maheswaran.
Monthly Notices, Roy. Astron. Soc., Vol. 145, 197 - 216 (1969).

Results of numerical computations performed to study the effects of prescribed circulations, within conducting spheres, on initially prescribed magnetic fields are presented in this paper. Various examples with different initial fields and different types of circulations in systems with compressible material and varying electrical conductivity have been considered. The structure of magnetic fields in uniformly rotating stars is discussed in terms of the results of the numerical computations.

116.009 The periodic variability of 78 Virginis. G. W. Preston.
Astrophys. Journ. Vol. 158, 243 - 249 (1969).

The occurrence of the crossover effect, the strength of the magnetic field, Henry's k index, and the radial velocity of 78 Vir all vary periodically in a period of 3^{d}7220. It is argued that there is no convincing evidence for a random component in the magnetic field.

116.010 The magnetic and spectrum variations of HD 188041. S. C. Wolff.
Bull. American Astron. Soc. Vol. 1, 266 - 267 (1969). − Abstr. AAS.

116.011 On the theory of rotating magnetic stars. G. A. E. Wright.
Monthly Notices, Roy. Astron. Soc., Vol. 146, 197 - 212 (1969).

The problem of a uniformly rotating magnetic star in radiative equilibrium is considered. It is found that in order to maintain radiative balance throughout the star, a magnetic field of P_1-type has to be strongly centrally condensed, but that the total magnetic energy is very much less than the centrifugal energy. The evolution of the field in a magnetic star is supposed to proceed by Ohmic decay through states of radiative equilibrium, i.e. states without meridian circulation. There is found to be a minimum flux, corresponding to zero surface field, below which no P_1-type solution can be found.

116.012 HR 2142 ejects a shell. G. Peters.
Publ. Astron. Soc. Pacific, Vol. 81, 548 - 549 (1969). − Abstract ASP.

116.013 Rotational velocities in NGC 2516. H. A. Abt, A. E. Clements, L. R. Doose, D. H. Harris.
Astron. Journ., Vol. 74, 1153 - 1154 = Contr. Cerro Tololo Inter-American Obs. No. 91 (1969).

The mean rotational velocities in the open cluster NGC 2516 are found to be similar to the unique mean rotational velocities in the Pleiades after eliminating the many Ap stars in the former cluster. This result supports the suggestion by Eggen that the two clusters have a common origin.

116.014 The mean surface magnetic field of β Coronae Borealis. G. W. Preston.
Astrophys. Journ., Vol. 158, 1081 - 1084 (1969).

Mean surface fields of 5900 ± 120 gauss and 6100 ± 150 gauss for β CrB have been derived from resolved Zeeman patterns on high-dispersion spectrograms near the phases of positive and negative cross-over respectively.

116.015 The period of the magnetic variations of HD 188041. S. C. Wolff.
Astrophys. Journ., Vol. 158, 1231 - 1233 (1969).

An analysis of the available observations of HD 188041 shows that the only period which will satisfactorily represent both the magnetic and spectrum variations of HD 188041 is $P = 224^d5$. All shorter periods have been eliminated.

116.016 A detailed rotational velocity study of the B2 component in 68 Herculis. R. H. Koch, S. Sobieski.

Bull. American Astron. Soc., Vol. 1, 350 (1969). — Abstr. AAS.

116.017 Models of magnetic stars.
J. D. Trasco.
Bull. American Astron. Soc., Vol. 1, 365 - 366 (1969). Abstr. AAS.

116.018 Hydrodynamic theory of stellar rotation. II. Stationarity of the rotation and meridional circulation.
V. V. Porfirjev.
Astron. Zhurn. Akad. Nauk SSSR, Vol. 46, 817 - 823 (1969). In Russian. English translation in Soviet Astron. AJ, Vol. 13, No. 4.

Without a magnetic field and internal friction a stationary rotation is possible if there is no meridional circulation. Formulas are given describing this kind of rotation. For stars with masses less than $\simeq 10 - 15\ M_\odot$ the angular velocity decreases in direction to the surface; for stars of larger masses this law is reversed. It is shown that the main suppositions of the theories of Eddington, Sweet, and Mestel of meridional circulation are contradictory and lead to incorrect results.

116.019 Photoelectric photometry of the magnetic variable HD 71866. C. Bartolini, P. Battistini.
Mem. Soc. Astron. Italiana, Nuova Serie, Vol. 40, 574 - 583 (1969).

New three colour photoelectric observations obtained at the Bologna Observatory during the years 1966 - 69 are reported. Within the limit of error, a constant period of $6\overset{d}{.}80005$ agrees with all published observations.

116.020 On the polarization of the radiation of magnetic-variable stars. N. S. Polosukhina.
Izv. Krymskoj Astrofiz. Obs. Vol. 39, 34 - 41 (1969). In Russian.

Results of measurements of polarization of magnetic stars in three colours with λ_{eff} = 3550, 4350, 5300 Å are given. The dependence of the degree of polarization (p) on the wavelength λ for the magnetic stars HD 71866, β CrB, HD 215441 is obtained.

116.021 Magnetic stars. Part II. Theory.
K. Stępień.
Postępy Astron., Vol. 17, 217 - 240 (1969). In Polish.

The article describes all major aspects of the theory of the magnetic stars. The origin of stellar magnetic fields and related theories: Fossil magnetism, hydromagnetic dynamo, Babcock's theory, thermal generation amplification mechanism due to electromagnetic instability. None of these theories

can describe properly the observations and the required theoretical conditions. Recent results on the relation between the angular velocity distribution inside the star and the internal structure of a star with a magnetic field are described.

116.022 Osservazioni di stelle magnetiche all'Osservatorio di Bologna. C. Bartolini, P. Battistini.
Atti XII Riunione Soc. Astron. Italiana, L'Aquila 1968, p. 63 - 64 (1969). — Abstract SAI.

116.023 Perdita di massa da stelle rapidamente ruotanti.
L. Nobili, L. Secco.
Atti XII Riunione Soc. Astron. Italiana, L'Aquila 1968, p. 111 (1969). — Abstract SAI.

116.024 The magnetic star 53 Camelopardalis.
R. Faraggiana.
Inform. Bull. Variable Stars (I.A.U. Commission 27), Konkoly Obs., Budapest, No. 388 (1969).

116.025 The period of the light variation of HD 173650.
E. W. Burke, Jr., J. B. Rice, W. H. Wehlau.
Publ. Astron. Soc. Pacific, Vol. 81, 883 - 887 (1969).

The star HD 173650 is a magnetic and spectrum variable of spectral type A0p. The photoelectric observations taken at the University of Western Ontario from 1960 to 1965 and a set of observations taken at Kitt Peak National Observatory in 1967 have been combined in order to improve the determination of the light curve and to establish an accurate period for use in a spectroscopic study of the star.

116.026 A list of magnetic null lines of astrophysical interest. J. D. Landstreet.
Publ. Astron. Soc. Pacific, Vol. 81, 896 - 899 (1969).

A list is presented of all the lines of neutral atoms and first second ions listed in the "Revised Multiplet Table" which are not split by a magnetic field. It is expected that these lines will be useful in the detailed study of atmospheric conditions in magnetic stars.

The excitation of nonspherical waves in differentially rotating stellar convective envelopes.
See Abstr. 064.010.

Rotational perturbation of a radial oscillation in a gaseous star. See Abstr. 065.001.

Effects of rotation on hydrogen parameters of early-type stars. See Abstr. 152.005.

117 Binary and Multiple Stars, Theory

117.001 Comparison of main-sequence binaries with theoretical models. S. Kříž.
Bull. Astron. Inst. Czechoslovakia, Vol. 20, 202 - 214 (1969).

The paper summarizes spectroscopic and photometric data on 26 eclipsing binaries for which the spectra of both components were observed. The absolute dimensions and masses of these binaries are newly computed. A comparison is made with Iben's (1967) and Horn's et al. (1969) theoretical models of main-sequence stars. For stars of smaller mass than $5\,M_\odot$ both series of models agree well with the binaries observed. Present observations of binaries do not permit a decision as to which of the used chemical compositions ($X = 0.71$ and $X = 0.60$) should be preferred. In the case of stars with a mass larger than $5\,M_\odot$, the theoretical models are overluminous in comparison with the binaries observed.

117.002 Alternate dynamical analysis of Barnard's star. P. van de Kamp.
Astron. Journ. Vol. 74, 757 - 759 (1969).

An alternate dynamical analysis of Barnard's star's motion over the interval 1938 – 1968 yields two companions in co-revolving, approximately coplanar, circular orbits with periods of 26 and 12 years, and masses of 1.1 and 0.8 times Jupiter, respectively.

117.003 The orbit of Eta Cassiopeiae. K. A. Strand.
Astron. Journ. Vol. 74, 760 - 763 (1969).

The orbit of η Cas has been computed with special attention to multiple-exposure photographic observations from 1914 to 1968. Even though the period is 480 years, the orbit can be considered well determined because the times of passage of the companion through the latus rectum points are accurately known. With a semimajor axis of $11.''994$ and a parallax of $0.''174$, the total mass is $1.42_\odot$, and the individual masses are $0.86_\odot$ and $0.56_\odot$ for a mass ratio of 0.394. No evidence has been found for a perturbation due to the existence of a previously suggested third companion.

117.004 Mass exchange in a massive close binary system. R. Kippenhahn.
Astron. Astrophys. Vol. 3, 83 - 87 (1969).

A binary system with a primary of $25\,M_\odot$ and a separation of $56\,R_\odot$ is followed through mass exchange after exhaustion of central hydrogen burning. The original primary star ends up as a star of $8.54\,M_\odot$ which almost completely consists of helium.

117.005 Gravity darkening in the components of close binary systems. A. Peraiah.
Astron. Astrophys. Vol. 3, 163 - 168 (1969).

The variation of temperature and brightness on the surface of the components of close binary systems has been investigated taking account of non-uniform rotation and tidal effects of the other component considered as mass point. The equations for such surfaces had been developed in terms of various parameters. When these equations are used to compute the temperature and brightness distributions by using different values for the parameters, it has been observed that temperature and brightness fall very rapidly between 45° colatitude and the equator. Higher temperatures tend to reduce gravity darkening due to tidal effect. The fall in brightness is steeper than that of temperature.

117.006 Temperature distributions on the surfaces of close binary stars. K.-Y. Chen, W. J. Rhein.
Publ. Astron. Soc. Pacific, Vol. 81, 387 - 398 (1969).

Close binary stars are assumed to be opaque, blackbodies with limb darkening. The increase of temperature due to radiative transfer between the components is calculated employing realistic geometry of spheres.

117.007 Photometric effects for highly distorted white dwarf secondaries in close binary systems. S. M. Ruciński.
Non-Periodic Phenomena in Variable Stars, IAU Colloquium, Budapest, 1968, p. 361 - 369 (1969).

117.008 On the reflection effect in close binaries. I. B. Pustylnik.
Non-Periodic Phenomena in Variable Stars, IAU Colloquium, Budapest, 1968, p. 423 - 426 (1969).

117.009 On the motion of gas in close binary systems. Yu. P. Korovyakovsky.
Astrofizika, Vol. 5, 67 - 73 (1969). In Russian.
English translation in Astrophysics, Vol. 5, No. 1 (1969).

The trajectories of gaseous jets in close binary systems are calculated in a three-dimensional case, dynamical effects of the gas pressure taken into account. The coordinates of the point of encounter of the gas stream and of the envelope of the main star are obtained. The relative velocity of the gas stream and of the envelope at the encounter ranges between the values of 200 and 300 km/sec for model I and between 900 and 1000 km/sec for model II.

117.010 The investigation of the dynamics of triple systems by the method of statistic tests. III. Case of components of different masses. J. P. Anosova.
Astrofizika, Vol. 5, 161 - 167 (1969). In Russian.
English translation in Astrophysics, Vol. 5, No. 1 (1969).

Numerical integration of the equations of triple systems for 300 random initial configurations has been carried out on an electronic computer. The components of the systems are assumed to be of different masses and motionless at an initial time. Three cases for the masses of the components have been regarded. In all cases the motion of components ended with the decay of the system. Decay took place after the close triple approach of components. The mean time T of the decay of the triple system is equal to $(27.8 \pm 3.5)\,\tau$ (τ is the mean time of the motion of the components through the system).

117.011 Dynamical formation of binaries. V. Szebehely.
Bull. American Astron. Soc. Vol. 1, 263 (1969). – Abstr. AAS.

117.012 Close-binary systems in the contracting phases. F. B. Wood.
Bull. American Astron. Soc. Vol. 1, 267 (1969). – Abstr. AAS.

117.013 The stellar three-body problem. R. S. Harrington.
Celestial Mechanics, Vol. 1, 200 - 209 (1969).

Two applications of von Zeipel's method to the stellar three-body problem eliminate the short period terms and establish two new integrals of the motion beyond the classical integrals. The remaining time averaged problem with only the second order Hamiltonian has one additional integral and can be solved. The motion with the third order averaged Hamiltonian included is more complex, in that there may be additional resonances, and the additional integral does not exist in all cases.

117.014 About the interpretation of gaseous streams in close binary systems. C. J. van Houten.
Mass Loss from Stars, Trieste 1968, p. 210 (1969). – Abstract.

117.015 Mass exchange in close binary systems with primary components of 30 $M_\odot$.
G. Barbaro, P. Giannone, M. A. Giannuzzi, C. Summa.
Mass Loss from Stars, Trieste 1968, p. 217 - 230 (1969).

The evolutions of the originally more massive components (primaries) of two close binary systems with very large masses have been computed during the phases of mass transfer to the companions and in the further detached stages at constant mass. The two double stars are composed initially of a primary of 30 $M_\odot$ and a secondary of 10 $M_\odot$ with separations of about 119 $R_\odot$ and 307 $R_\odot$, respectively.

117.016 On the time-scale of the mass transfer in close binaries. B. Paczyński, J. Ziółkowski, A. Żytkow.
Mass Loss from Stars, Trieste 1968, p. 237 - 241 (1969).

We show that the mass outflow from the more massive component of a close binary that fills up its Roche lobe, may take place on a dynamical time-scale if this component has a deep convective envelope. We suggest that the outbursts of the U Geminorum-type stars might be related to this phenomenon.

117.017 Mass exchange in close binaries of moderate mass and short periods. M. Plavec, J. Horn.
Mass Loss from Stars, Trieste 1968, p. 242 - 252 (1969).

The process of mass exchange in binary stars of short period is studied on a family of model binaries with primaries of 5 solar masses. Results obtained by means of stationary models are compared with actual computations of non-stationary model sequences. Changes in mass ratio, luminosity, period, masses, etc., are studied.

117.018 Evolution through mass exchange in close binary systems of total mass 2.5 $M_\odot$.
S. Refsdal, A. Weigert.
Mass Loss from Stars, Trieste 1968, p. 253 - 256 (1969).

Calculations show that the observed semi-detached binary systems of mass $M_1 + M_2 \approx 2.5\ M_\odot$ can be explained in terms of a mass exchange which starts after the central hydrogen in the original primary has been exhausted.

117.019 Mass exchange in close binaries of moderate period and mass. S. Kříž.
Mass Loss from Stars, Trieste 1968, p. 257 - 261 (1969).

Results of numerical computations are presented for the evolution of a binary consisting of components 13.8 $R_\odot$ apart having the masses 5 $M_\odot$ + 4 $M_\odot$. The mass exchange starts after the exhaustion of the central hydrogen. A possible connection between shell stars and products of mass exchange is indicated.

117.020 White-dwarf production in binary systems of large separation. D. Lauterborn.
Mass Loss from Stars, Trieste 1968, p. 262 - 266 (1969).

Numerical calculations are carried out for the evolution of a binary system with a primary of 5 $M_\odot$ and a secondary of 2 $M_\odot$, revolving round another in a circular orbit of 300 $R_\odot$. After finishing central helium burning, the primary starts to transfer mass to its companion. After the mass loss, the star of originally 5 $M_\odot$ has become a star of 1 $M_\odot$. This star has a carbon-oxygen core which is a well-developed white dwarf, and a very extended hydrogen shell.

117.021 Possible effects of the resonance in close binaries

on their mass exchanges. J.-P. Zahn.
Mass Loss from Stars, Trieste 1968, p. 267 - 270 (1969).

Present calculations on the forced non-radial oscillations of main-sequence stars show that the resonances in close binaries are much more probable than was assumed up to now. Some predictions are made about the behaviour of a star entering such a resonance.

117.022 Some problems concerning gas flows in close binary systems of dwarf stars. V. G. Gorbatzky.
Mass Loss from Stars, Trieste 1968, p. 271 - 273 (1969).

The influence of the stream flowing from the secondary on the disk-like envelope of the main star is briefly considered. It is found that the lifetime of the envelope after the flow cases must be less than 10^6 sec.

117.023 Les étoiles doubles, indicatrices d'une perte de masse séculaire des étoiles. J. Dommanget.
Mass Loss from Stars, Trieste 1968, p. 274 - 277 (1969).

Option parmi diverses hypothèses sur la formation des étoiles doubles.

117.024 The relative motion of two spheroidal rigid bodies. P. Lanzano.
Astrophys. Space Sci. Vol. 5, 300 - 322 (1969).

We consider two spheroidal rigid bodies of comparable size constituting the components of an isolated binary system. We assume that (1) the bodies are homogeneous oblate ellipsoids of revolution, and (2) the meridional eccentricities of both components are small parameters. We obtain seven nonlinear differential equations governing simultaneously the relative motion of the two centroids and the rotational motion of each set of body axes. We seek solutions to these equations in the form of infinite series in the two meridional eccentricities. The first part of the paper deals with the representation of the total potential energy of the binary system as an infinite series of the meridional eccentricities. In the second part, we expound a recurrent procedure whereby the approximations of various orders can be determined in terms of lower-order approximations.

117.025 Dynamical instabilities in semidetached close binary systems with possible applications to novae and novalike variables. G. T. Bath.
Astrophys. Journ. Vol. 158, 571 - 587 (1969).

The stability to mass loss on a dynamical time scale of the contact component of a semidetached binary system is investigated. The possibility of such explosively rapid mass loss and its effects on initial models in various regions of the H–R diagram are studied by adiabatic perturbation of zeroth-mode disturbances. The expansion of the outer layers is found to cause instabilities which give rise to two classes of behavior. Models of class I, unstable down to the He I zone, have resulting magnitude ranges, mass-loss values, and total energy emission typical of U Gem stars. Models behaving as class II, unstable down to the He II zone, have values more typical of novae.

117.026 The expected fraction of evolved close binaries among main-sequence stars of spectral type earlier than A5. E. P. J. van den Heuvel.
Astron. Journ. Vol. 74, 1095 - 1101 = Lick Obs. Bull. No. 603 (1969).

The expected fraction of close-binary remnants (CBR) among main-sequence stars of spectral type earlier than A5 is computed under the assumption of a constant stellar formation rate. A CBR is defined as the remaining main-sequence component of a binary that evolved with mass exchange. Only systems in which the mass exchange occurs shortly after core-hydrogen exhaustion of the primary are considered (Kippenhahn and Weigert's (1967) case B).

117.027 **The photometric proximity effects in close binary systems. II. The bolometric reflection effect for stars with deep convective envelopes.** S. M. Ruciński.
Acta Astron. Vol. 19, 245 - 255 (1969).

The condition of constancy of the entropy is used to derive the albedo (directly comparable with the reflection coefficient A_1) for stars having deep adiabatic convective envelopes. The bolometric albedo of a main sequence star of about one solar mass and a typical Algol secondary (with mass 1.45, luminosity 8.4, and radius 3.8, in solar units) is estimated to be 0.4 to 0.5 instead of unity for the strict radiative equilibrium; it depends on the l/H ratio and on the solution for the irradiated radiative photosphere.

117.028 **Contributions to the interpretation of the light curves of the close binary systems. II. The projections of the components on the plane perpendicular to the line of sight.** V. Ureche.
Stud. Cerc. Astron. Vol. 14, 53 - 61 (1969).

117.029 **Contributions to the interpretation of the light curves of the close binary systems. III. The loss of the light during the eclipses.** V. Ureche.
Bull. Astron. Inst. Czechoslovakia, Vol. 20, 312 - 317 (1969).

In two previous papers, we have analysed the degree of accuracy up to which the close binary system components may be approximated by two nonsimilar ellipsoids (Ureche, 1969a), and we have determined the components projections on the plane perpendicular to the line of sight (Ureche, 1969 b). In the present paper we shall calculate the loss of light during the eclipses of the close binary system components, within the limits of the same accuracy.

117.030 **Theoretical continuous and line spectra of stars in a close binary system.** P. Buerger.
Astrophys. Journ., Vol. 158, 1151 - 1160 (1969).

Continuous radiation emitted by rotationally and tidally distorted stars as they appear in close binary systems has been computed. Gray atmospheres with radiation incident at the surface were used. The equivalent width of a weak absorption line superposed on the atmosphere has also been computed. Hydrostatic equilibrium is assumed throughout. For a system of similar early-type stars, large variations in the ultraviolet flux are predicted as a function of separation, angle of inclination, and orbital phase.

117.031 **The stability of a two-planet model for Barnard's star.** R. S. Harrington.
Bull. American Astron. Soc., Vol. 1, 347 (1969). – Abstr. AAS.

117.032 **A study of the structure of rapidly rotating close-binary systems.** P. G. Martin, S. P. S. Anand.
Bull. American Astron. Soc., Vol. 1, 354 (1969). – Abstr. AAS.

117.033 **The distribution of angular momentum in planetary systems.** P. Brosche.
Observatory, Vol. 89, 206 (1969). – Letter.

117.034 **Occurrence of planetary systems in the universe as a problem in stellar astronomy.**
S.-S. Huang.
Vistas in Astronomy, Vol. 11, 217 - 263 (1969).

In many problems involving binaries, rotating stars, and planetary systems, the angular momentum serves as an important parameter in addition to the mass, and merits some special considerations. In this paper the orientations in space of stellar angular-momentum vectors are first discussed in the light of empirical data, and their origin is then explained in terms of simple models. It is found that the formation of planetary systems is closely related to the braking of stellar rotation and must be genetically different from that of binaries. For this reason the frequency occurrence and perhaps even the nature of planetary systems around the main-sequence stars later than F5 may be estimated from the rotational behavior of the main-sequence stars of early spectral types. According to this estimate the size of our own planetary system lies within the estimated range.

117.035 **On the frequency of double stars with different photometric effects.** G. Kakaras.
Bull. Vilnius Astron. Obs. No. 25, p. 39 - 55 (1969). In Russian.

All unseparated double stars can be devided into three groups according to their photometric effects [G. Kakaras, V. Straizys, Bull. Vilnius Astron. Obs. No. 23, p. 3 - 35, 36 - 44 (1969)]. In the present paper an attempt is made to determine the frequency of double stars of these groups.

117.036 **The dynamical evolution of triple-star systems.** R. S. Harrington.
Thesis, Univ. of Texas. Univ. Microfilms, Ann Arbor, Mi., 82 pp. (1968). – See Phys. Abstr. Vol. 72, No. 21879 (1969).

117.037 **On the dynamics of the gaseous jet in close binary systems of dwarf stars. I.**
V. I. Taranov.
Trudy Astron. Obs. *Leningrad*, Vol. 26 (= Uchenye Zapiski Leningr. Un-ta No. 347 = Seriya Matem. Nauk No. 44), p. 25 - 32 (1969). In Russian.

The shock wave in a gaseous stream is formed when it flows into the envelope of the primary star of a close binary system. The temperature and density in the shock wave are determined, the energy losses with the radiation and the influence of the magnetic field being taken into account. The position of the shock front is also found.

117.038 **Study of the dynamics of rotating triple systems.**
J. P. Anosova.
Trudy Astron. Obs. *Leningrad*, Vol. 26 (= Uchenye Zapiski Leningr. Un-ta No. 347 = Seriya Matem. Nauk No. 44), p. 88 - 91 (1969). In Russian.

Numerical integrations of the equations of motion of rotating triple systems for 100 random initial configurations have been carried out on an electronic computer. The components of the systems are assumed to have equal masses. In 94 cases the motions eventually led to the decay of the systems. In all the cases the decay took place after the close triple approach of the components. In 6 cases the decay of the system has not taken place during the period the motion was traced.

117.039 **On the possibility of formation of double systems in the General Relativity.** M. Abramowicz.
Postępy Astron., Vol. 17, 387 - 395 (1969). In Polish.

In the paper a singular solution to the Kepler problem in the General Relativity is applied to the theory of double system formation.

117.040 **Catalogue des différences de magnitude des composantes de 2379 étoiles doubles et multiples.**
S. Wierzbiński.
Acta Univ. Wratislaviensis No. 94 (Mat. Fiz. Astron. IX), 220 pp. = Contr. Wrocław Astron. Obs. No. 16 (1969).

117.041 **A statistical study of binary stars.**
W. D. Heintz.
Journ. Roy. Astron. Soc. Canada, Vol. 63, 275 - 298 = Sproul Obs. Reprint No. 191 (1969).

The distribution of elements of visual orbits, selection effects, and the discovery incompleteness are studied. It is found that 85% of the stars are members of double or multiple systems, and that the distribution of semi-axes major has a frequency peak near 50 a.u. Stepwise pre-stellar fragmentations probably account for all of the angular momentum initially available. Infra-red colour excesses may help to detect faint red companions.

117.042 **Multiple star Castor.**
B. Hacar.
Říše hvězd, Vol. 50, 225 - 228 (1969).   In Czech.

117.043 **Close binaries.**   O. Obůrka.
Říše hvězd, Vol. 50, 145 - 149 (1969).
In Czech.

117.044 **Apsidal motion in close binaries.**
J. U. Cisneros-Parra.
Mitt. Astron. Ges. No. 27, p. 171 - 172 (1969). — Abstract AG.

The precession and nutation of deformable bodies, III.   See Abstr. 042.008.

The Roche coordinates and their use in hydrodynamics or celestial mechanics   See Abstr. 042.024.

Probleme enger Doppelsterne.
See Abstr. 065.089.

118 Visual Binaries

118.001 Photometric determinations of magnitude differences for visual binaries. C. E. Worley.
Astron. Journ. Vol. 74, 764 - 767 (1969).

Visual photometric Δm determinations for 94 double stars are presented, and the accidental and systematic errors discussed.

118.002 Parallaxes and masses of the visual binary stars ADS 3475, 8862, 9617, and 16326.
W. D. Heintz.
Astron. Journ. Vol. 74, 768 - 773 (1969).

The Sproul astrometric plates for four visual binaries have been measured and yield parallaxes, mass ratios, and masses. The results for ADS 16326 are not reliable; the other stars fall close to the main sequence. The orbits of ADS 3475 and 8862 have been revised.

118.003 Mesures micrométriques des étoiles doubles, (19), sur la réfracteur Zeiss 65/1055 cm au cours de la période de 1965 à 1968.
P. M. Djurković, G. M. Popović, D. J. Zulević.
Bull. Obs. Astron. Beograd, Vol. 27, 1 - 21 (1969).

Les mesures comprennent quatre années de travaux d'observation, les années: 1965, 1966, 1967 et 1968. Trois auteurs présentent par cet ouvrage 1891 mesures de 542 systèmes doubles.

118.004 Die Spektren der Komponenten der visuellen Doppelsterne mit bekannten Bahnen der Hauptreihe des Hertzsprung–Russell Diagramms in Abhängigkeit vom Massenverhältnis. G. M. Popović.
Bull. Obs. Astron. Beograd, Vol. 27, 22 - 32 (1969).

Durch Anwendung eines festgesetzten Kriteriums wurden Doppelsterne mit bekannten Bahnen ausgewählt,die zur Hauptreihe des H–R Diagramms gehören. Mittels ihrer bekannten Massen und Spektren der A-Komponenten wurde ein empirischer Zusammenhang Masse–Spektrum abgeleitet. Mit diesem wurden die Spektren der B-Komponenten abgeleitet.

118.005 Bahnbestimmung von dreizehn visuellen Doppelsternen. G. M. Popović.
Bull. Obs. Astron. Beograd, Vol. 27, 33 - 54 (1969).

For the following 13 systems: STF 208 AB = ADS 1631, STF 577 = ADS 3390, A 2817 = ADS 5159, A 218 = ADS 5332, A 2146 = ADS 7677, HU 736 = ADS 8485, HO 260 = ADS 8887, A 2181 = ADS 9989, HU 481 = ADS 10017, A 1866 = ADS 10227 BC, HU 951 = ADS 12577, HU 83 = ADS 14492 and STF 2744 AB = ADS 14573 orbital elements, parallaxes, measurements and their representing ephemerides up to 1975, masses of components and other astrophysical values are shown.

118.006 Orbite de quatre étoiles doubles visuelles.
D. J. Zulević.
Bull. Obs. Astron. Beograd, Vol. 27, 55 - 62 (1969).

On donne pour la première fois les éléments préliminaires des couples ADS 6989 = Hu 120, ADS 8799 = Hu 572, ADS 8943 = A 1095 et ADS 16914 = Hu 1325. Pour ces orbites nous avons utilisé la méthode de Thiele–Innes–Van den Bos (Union Obs. Circ. No. 86, 1932). Les éléments de Campbell sont déduits des éléments de Thiele–Innes, résultat direct de la détermination.

118.007 Orbite de l'étoile double visuelle.
D. J. Zulević.
Bull. Obs. Astron. Beograd, Vol. 27, 63 - 64 (1969).
Concerning ADS 2538 = Aitken 980.

118.008 Trajectoire rectiligne du système binaire ADS 14645. S. Mali.
Bull. Obs. Astron. Beograd, Vol. 27, 65 - 66 (1969).

118.009 The new double stars discovered in Belgrade with the Zeiss refractor 65/1055 cm.–Supplement I.
G. M. Popović.
Bull. Obs. Astron. Beograd, Vol. 27, 67 - 70 (1969).

Twenty-seven measurements of 15 newly discovered pairs with Zeiss refractor 65/1055 cm in Belgrade and also the differential position of these pairs related to the neighbouring BD stars are given. Also the coordinates of these new pairs related on the epochs 1900, 1950 and 2000 are communicated.

118.010 Parallax and orbital motion of the double star 9 Puppis from photographs taken with the Van Vleck refractor. R. Grossenbacher, W. S. Mesrobian.
Astron. Journ. Vol. 74, 951 - 953 (1969).

Measurement and reduction of photographic plates taken with the Van Vleck refractor over the interval 1925 - 1968 yield $+0\rlap{.}''060 \pm 0\rlap{.}''005$ (m.e.) for the relative parallax and $+0\rlap{.}''111 \pm 0\rlap{.}''011$ for the semimajor axis of the photocentric orbit of 9 Puppis AB. The fractional mass of the companion is $B = 0.555$. The adopted value for the absolute parallax, $+0\rlap{.}''063 \pm 0\rlap{.}''006$, leads to masses of $0.64\,M_\odot$ and $0.81\,M_\odot$ for the primary and secondary components, respectively.

118.011 Micrometer measures of double stars.
G. F. G. Knipe.
Republic Obs. Johannesburg, Circ. Vol. 7 (No. 128), 177 - 183 (1969).

This list contains 544 measures of 512 pairs made with the $26\frac{1}{2}$-inch refractor, occasionally stopped down to 18 inches.

118.012 Micrometer measures of double stars.
J. L. Newburg.
Republic Obs. Johannesburg, Circ. Vol. 7 (No. 128), 184 - 186 (1969).

This list contains 382 measures of 163 pairs, all made with the $26\frac{1}{2}$-inch refractor.

118.013 Interferometer measures of double stars.
W. S. Finsen.
Republic Obs. Johannesburg, Circ. Vol. 7 (No. 128), 187 - 189 (1969).

The present list contains 245 observations of 77 double stars made with the author's eyepiece interferometer attached to the $26\frac{1}{2}$-inch refractor.

118.014 The orbit of β 1000 AB, C. J. L. Newburg.
Republic Obs. Johannesburg, Circ. Vol. 7 (No. 128), 190 (1969).

118.015 The orbit of β 738. W. S. Finsen.
Republic Obs. Johannesburg, Circ. Vol. 7 (No. 128), 190 - 191 (1969).

118.016 The orbit of Rst 2338. J. L. Newburg.
Republic Obs. Johannesburg, Circ. Vol. 7 (No. 128), 192 (1969).

118.017 The orbit of δ 85, ADS 4153. J. L. Newburg.
Republic Obs. Johannesburg, Circ. Vol. 7 (No. 128), 193 (1969).

118.018 **The orbit of I 1567.** J. L. Newburg.
Republic Obs. Johannesburg, Circ. Vol. 7 (No. 128), 193 - 194 (1969).

118.019 **The orbit of φ 357.** W. S. Finsen.
Republic Obs. Johannesburg, Circ. Vol. 7 (No. 128), 194 (1969).

118.020 **The orbit of HdO 296.** W. S. Finsen.
Republic Obs. Johannesburg, Circ. Vol. 7 (No. 128), 195 (1969).

118.021 **The orbit of Sellors 14.** G. F. G. Knipe.
Republic Obs. Johannesburg, Circ. Vol. 7 (No. 128), 196 (1969).

118.022 **Photometric observations of binary stars.**
G. F. G. Knipe.
Republic Obs. Johannesburg, Circ. Vol. 7 (No. 128), 197 (1969).

This list is a continuation of that published previously (Knipe 1966). The measures were made with the 9-inch refractor and, after January 1968, with the 20-inch reflector of the observatory. The reductions were made with the IBM 360 electronic computer of the CSIR using a programme written by Bertiau (1963). The data refer to the combined magnitude and colour of the pair.

118.023 **Magnitude and parallax of ADS 10092, $16^h 26^m.4$ $-6°50'$ (1900).** G. F. G. Knipe.
Republic Obs. Johannesburg, Circ. Vol. 7 (No. 128), 199 (1969).

118.024 **Orbite de l'étoile double visuelle ADS 1394 = h 3461 AB.** · R. R. de Freitas Mourão.
Mem. Soc. Astron. Italiana, Nuova Serie, Vol. 40, 295 - 299 (1969).

The physical and orbital elements of the binary star ADS 1394 = h 3461 AB are determined.

118.025 **Troisième catalogue d'éphémérides d'étoiles doubles.** P. Muller, C. Meyer.
Publ. Obs. Paris. 91 pp. (1969).
633 ephemerides for 610 objects are listed.

118.026 **A spectroscopic investigation of visual binaries with B-type primaries.** R. E. Murphy.
Bull. American Astron. Soc. Vol. 1, 254 (1969). – Abstr. AAS.

118.027 **MK classification of 142 visual binaries.**
J. W. Christy, R. L. Walker, Jr.
Publ. Astron. Soc. Pacific, Vol. 81, 643 - 649 (1969).

MK classifications are presented for 142 visual binaries contained in *A Catalog of Visual Binary Orbits* (Worley 1963). The accuracy of the classification is discussed.

118.028 **A spectroscopic investigation of visual binaries with B-type primaries.** R. E. Murphy.
Astron. Journ. Vol. 74, 1082 - 1094 = Contr. Kitt Peak National Obs. No. 476 = Warner and Swasey Obs. No. 189 (1969).

The absolute magnitudes of B stars have been investigated using visual binaries with B-type primaries. A new calibration is presented for the main-sequence stars of types B0.5 – B9 and new data on the absolute magnitudes of several classes of giants and supergiants are presented. The new main sequence is systematically fainter than that of Johnson and Iriarte (1958) but brighter than those of Weaver and Ebert (1964) and FitzGerald (1969). Four of the secondaries are peculiar A stars. Their presence in systems with main-sequence B stars implies that they are young objects.

118.029 **Observations of variable stars in visual binaries. I. ADS 9701.** O. G. Franz, R. L. Millis.
Bull. American Astron. Soc., Vol. 1, 343 - 344 (1969). – Abstract AAS.

118.030 **Photovisual magnitude differences of double stars.** K. A. Strand.
Bull. American Astron. Soc., Vol. 1, 364 (1969). – Abstr. AAS.

118.031 **Osservazioni di stelle doppie.**
T. Tamburini Job.
Mem. Soc. Astron. Italiana, Nuova Serie, Vol. 40, 543 - 548 (1969).

90 measures of 72 double stars, made at the Turin Observatory are given here. The instrument used is a Merz refractor: $A = 30$ cm, $F = 4,50$ m, enlargement 207 times.

118.032 **Orbita preliminare della binaria visuale Hu 1597.**
F. Job, T. Tamburini, M. A. Zaccone.
Mem. Soc. Astron. Italiana, Nuova Serie, Vol. 40, 549 - 551 (1969).

The orbital elements of the visual binary star Hu 1597, graphically deduced, are given.

118.033 **The visual double HR 3817 and 3 Centauri A.**
A. D. Thackeray.
Observatory, Vol. 89, 235 - 236 (1969).

Jaschek and Aguilar's discovery of peculiar lines in HR 3817 (HD 82984) is confirmed, especially for λ3984 Å (? *Hg* II), but they are more diffuse and weaker than in 3 Cen A. The possibility of time-variations has not been excluded.

118.034 **Measurements of double stars with a polarizing micrometer. VII.**
N. E. Kurochkin, G. A. Starikova.
Soobshch. Gos. Astron. Inst. Shternberga, No. 158, p. 49 - 51 (1969). In Russian.

Results of measurements of 35 double stars with a polarizing micrometer are given.

118.035 **Micrometric measures of 463 double stars.**
R. L. Walker, Jr.
Publ. U. S. Naval Obs., *Washington,* Second Series, Vol. 22, (Part I), 1 - 55 (1969).

This paper contains 1965 measures of 463 double star systems obtained with four different telescopes at the U. S. Naval Observatory and at Flagstaff, Arizona. Orbit residuals are given with the measures, along with pertinent comments. The double stars were selected on the basis of known or suspected rapid motion.

118.036 **Stadi evolutivi delle componenti di sistemi doppi visuali.** P. Giannone, M. A. Giannuzzi.
Atti XII Riunione Soc. Astron. Italiana, L'Aquila 1968, p. 43 (1969). – Abstract SAI.

118.037 **Antares som dobbeltstjerne.**
A. V. Nielsen.
Astron. Tidsskr., Årg. 2, 151 - 153 (1969).

118.038 **Der Doppelstern ADS 6126.**
J. Hopmann.
Anzeiger Österreich. Akad. Wiss., Math.-Naturwiss. Kl., 105. Jahrgang, p. 293 - 296 = Mitt. Univ.-Sternw. Wien, Vol. 14, No. 10 (1969). – Orbit determination.

118.039 **Etude d'une perturbation observée dans le mouvement relatif rectiligne des composantes du couple optique ADS 818 = Σ 80.** J. Dommanget.

Bull. Astron. Obs. Roy. Belgique, Vol. 6, (No. 8), 343 - 352 (1969).

118.040 **The system of α Ursae Majoris.** L. S. T. Symms. Roy. Obs. Bull. Greenwich–Cape, No. 157, p. 51 - 65 (1969).

By treating the pair as an unresolved astrometric binary and combining data from Allegheny, Greenwich and Herstmonceux parallax plates with the visual observations of position angle, a reliable photocentric orbit is obtained. These data in conjunction with the visual observations of separation and a new determination of the trigonometrical parallax give reliable values of the combined mass and the mass ratio.

118.041 **Orbites nouvelles.** P. Muller. Circ. Inform. (U.A.I. Commission des Etoiles Doubles), Obs. Meudon, No. 49 (1969).

118.042 **Etoiles doubles découvertes à Nice, lunette de 50 cm.** P. Muller. Circ. Inform. (U.A.I. Commission des Etoiles Doubles), Obs. Meudon, No. 49 (1969).

An astrometric study of L726-8. See Abstr. 111.005.

Proper Motion Survey with the forty-eight inch Schmidt telescope. XXI. Double stars with common proper motion. See Abstr. 112.017.

UBV observations of selected double systems, II. See Abstr. 113.011.

On the Hyades binaries. See Abstr. 153.032.

119 Spectroscopic Binaries

119.001 **Infrared photometry of a helium star, HD 30353.**
   T. A. Lee, K. Nariai.
Publ. Astron. Soc. Japan, Vol. 21, 67 - 70 (1969).

   Infrared data for HD 30353, a helium-rich, single-lined spectroscopic binary ($T_e \sim$ 11.000 °K, log $g \sim$ 1 for primary), are presented. The infrared colors suggest that the invisible component may be a K0-type supergiant. The large K–L index implies additional infrared radiation—possibly arising in a circumstellar envelope like that found for v Sgr.

119.002 **The frequency of spectroscopic binaries among high-velocity dwarf stars.** H. A. Abt, S. G. Levy.
Astron. Journ. Vol. 74, 908 - 916 = Contr. Kitt Peak National Obs. No. 445 (1969).

   A sample of 68 F- and G-type high-velocity dwarf stars have been studied for velocity variations and are compared with available data regarding low-velocity dwarf stars. It is concluded that short-period binaries are rare among all high-velocity dwarfs and are especially infrequent for the weakest-lined stars, although the frequency of long-period or visual binaries may be similar among high- and low-velocity stars.

119.003 **On the β Lyrae elements.** M. Yu. Skulsky.
Astron. Tsirk. No. 505, p. 3 - 5 (1969). In Russian.

119.004 **12 Camelopardalis and calcium emission in giant binaries.** H. A. Abt, R. J. Dukes, W. B. Weaver.
Astrophys. Journ. Vol. 157, 717 - 720 = Contr. Kitt Peak National Obs. No. 394 (1969).

   Bidelman predicted that 12 Cam is a spectroscopic binary because it shows the strong calcium emission that is evidently characteristic of giant binaries of short period. This prediction is confirmed, and orbital elements are derived. The strong calcium emission (1) does not originate near the secondary star, (2) varies slowly with time but not in phase in the eccentric orbit, and (3) does not significantly interfere with the Wilson-Bappu correlation of chromopsheric emission width with luminosity.

119.005 **The spectroscopic binary HD 112486.**
   R. Margoni, M. Perinotto, E. Nasi.
Mem. Soc. Astron. Italiana, Nuova Serie, Vol. 40, 301 - 307 (1969).

   As a part of a large program of study of the duplicity of Am stars, the double-line spectroscopic binary HD 112486 has been investigated on 60 spectra with dispersion of 42 Å/mm at Hγ. Its spectroscopical orbit has been derived.

119.006 **Light variations in Spica.**
   R. R. Shobbrook, D. Herbison-Evans, I. D. Johnston, N. R. Lomb.
Monthly Notices, Roy. Astron. Soc., Vol. 145, 131 - 140 (1969).

   The visual magnitude of a Virginis (Spica) has been measured over a period of 3 months. The light shows a 4.17036 hr period of amplitude 1.6 per cent showing that the primary of the binary star system is a β Canis Majoris variable. There is, in addition, a 3 per cent variation in brightness over the 4.014 day orbital period, which can be accounted for by aspect changes of the tidally distorted primary. Eclipses, if they occur, are probably less than 0.5 per cent in depth.

119.007 **Evidence of tidal effects in some pulsating stars. II. 16 Lacertae and β Cephei.**
W. S. Fitch.
Astrophys. Journ. Vol. 158, 269 - 280 (1969).

   Published observations of two β CMa stars are analyzed.

The star 16 Lac is a single-line spectroscopic binary with a circular orbit of 12^{d}096 period; the primary star pulsates in a primary period of 0^{d}169166. The star β Cep is also found to be a single-line spectroscopic binary with an orbit of large eccentricity of 10^{d}893 period. A simplified theory of resonant and nonresonant tidal modulation is presented. A summary is given of presently available information relating to tidal modulations of five β CMa stars and three δ Sct stars.

119.008 **Period and velocity curve of AE Aquarii.**
   C. Payne-Gaposchkin.
Astrophys. Journ. Vol. 158, 429 (1969).

119.009 **The double-line A star η Vir.** P. S. Conti.
   Bull. American Astron. Soc. Vol. 1, 237 - 238 (1969). – Abstr. AAS.

119.010 **The triple system HD 100018 (ADS 8189).**
   R. M. Petrie, A. H. Batten.
Bull. American Astron. Soc. Vol. 1, 257 (1969). – Abstr. AAS.

119.011 **The spectrographic orbit of H.D. 161701.**
   D. P. Hube.
Journ. Roy. Astron. Soc. Canada, Vol. 63, 229 - 232 = Commun. David Dunlap Obs. Univ. Toronto, Richmond Hill, No. 223 (1969).

   Orbital elements are derived for the relatively bright spectrographic binary, H.D. 161701. Two procedures are followed in computing differential corrections to the preliminary values of the orbital elements.

119.012 **Light variations in Ψ Orionis.** J. R. Percy.
   Journ. Roy. Astron. Soc. Canada, Vol. 63, 233 - 237 = Commun. David Dunlap Obs. Univ. Toronto, Richmond Hill, No. 236 = Contr. Kitt Peak National Obs. No. 437 (1969).

   Photoelectric photometry of the spectroscopic binary ψ Orionis indicates that this system is an ellipsoidal variable. The orbital period of 2.53 days satisfactorily represents the photometric data. Minima with depths of 0.030 and 0.025 magnitude are observed at the phases of predicted spectroscopic conjunction. There is no evidence that ψ Orionis shows any other kind of variation.

119.013 **The spectroscopic binary HD 206874.**
   R. S. Fisk, H. A. Abt.
Publ. Astron. Soc. Pacific, Vol. 81, 692 - 695 = Contr. Kitt Peak National Obs. No. 459 (1969).

   Tanner's suspicion that the original orbital period and eccentricity of this system are incorrect is confirmed, and new elements are derived. This system consists of two identical F2 IV stars probably rotating synchronously, but the system is probably not an eclipsing one.

119.014 **The early A stars. IV. Analysis of the double-line spectroscopic binary Eta Virginis.** P. S. Conti.
Astrophys. Journ., Vol. 158, 1085 - 1089 = Contr. Lick Obs. No. 300 (1969).

   The measured equivalent widths, corrected for the true continuum in each component, have been compared with the normal A star θ Vir. The two component stars do not have identical surface compositions. The secondary, of spectral type about A4, has deficient calcium, scandium, and titanium as in an Am star. In the primary, of spectral type A2, these elements are normal. Both stars show marginal overabundances of those heavy elements enhanced in Am stars.

119.015 **Spectral classification of A-type spectroscopic binaries.** H. A. Abt, W. P. Bidelman.
Astrophys. Journ., Vol. 158, 1091 - 1098 = Contr. Kitt Peak National Obs. No. 444 (1969).

MK spectral types have been determined or are quoted for ninety-eight of 101 known spectroscopic binaries with primaries in the range A2–F3. Half of these stars are metallic-line (Am) stars; most of the remainder are outside the domain (approximately A4–F1 IV, V) of the Am stars. The remaining nine normal stars in the domain have periods of either less than 2.5 or more than about 100 days. It is concluded that all stars in the range A4–F1, IV, V that are primaries of binaries with periods of approximately 2.5 – 100 days have metallic-line spectra.

119.016 **Orbital elements of spectroscopic binary HD 222317 = BD + 27°4588.** M. Imbert.
Astron. Astrophys. Vol. 3, 272 - 276 (1969). In French.

The orbital elements have been determined. The comparison between several series of observations made at different times give us the period with a good accuracy P = 6.2018 ± 0.0004 days.

119.017 **The double-line spectroscopic binary HD 12881.** R. Margoni, M. Perinotto.
Mem. Soc. Astron. Italiana, Nuova Serie, Vol. 40, 553 - 558 (1969).

The spectroscopic binary HD 12881 has been studied on 45 spectra with dispersion of 42 Å/mm at H_γ. An orbital solution is derived and the properties of the star are discussed. The work is a part of a large program devoted to the study of Am stars.

119.018 **The spectroscopic binary HD 184552.** E. N. Walker, D. H. P. Jones.
Observatory, Vol. 89, 202 - 205 (1969).

Reported are radial velocity measurements and the computed velocity curve.

119.019 **The spectrophotometric investigation of the close binary system HD 190918.** T. S. Galkina.
Izv. Krymskoj Astrofiz. Obs. Vol. 39, 44 - 62 (1969). In Russian.

The analysis of the composite spectrum of HD 190918 is carried out from spectrograms obtained in 1963 - 1964 (dis-persion 37 Å/mm) in the region λλ 4900 - 3600 Å. The spectral class of the absorption component is variable between O 9.0 and O 8.0, that of the emission component is estimated to WN 5.5. From the detailed investigation of the emission band λ4686 follows an orbital motion of the component with a period of about 105 days. A change of the emission intensity of λ 4686 He II has been discovered. Measurements of the emission band λ4686 He II showed its systematic displacement to the long-wave part of the spectrum by 110 km/sec relative to the absorption component.

119.020 **On some peculiarities of the atmosphere of γ UMi.** V. V. Leushin.
Izv. Krymskoj Astrofiz. Obs. Vol. 39, 108 - 113 (1969). In Russian.

The equivalent widths of hydrogen lines and those of some metals are measured. The variability with P = $0^d108449$ of lines $H_{11} - H_{20}$ and some metallic lines is discovered. The analysis of curves of growth for Fe I and Fe II is carried out. It is obtained: v_t (Fe I) = 1.8 km/sec, v_t (Fe II) = 3.6 km/sec. The electron pressures lg P_e = 1.66 for H and lg P_e = 0.95 for Fe are determined.

119.021 **Revised elements for the spectroscopic binaries μ Eridani and 57 Orionis.** G. Hill.
Publ. Dominion Astrophys. Obs. Victoria, Vol. 13, (No. 12), 323 - 328 (1969).

Radial velocities determined at the Lick, Yerkes, Dominion Astrophysical and McDonald observatories were combined to yield revised periods and new orbital elements for the spectroscopic binaries μ Eridani and 57 Orionis. These orbital elements are: μ Eridani – P = 7.35886 days, T = J. D. 2416392.46, ω = 150°, e = 0.26, V_0 = +23.3 km/sec and K_1 = 19.4 km/sec; 57 Orionis – P = 7.99687 days, T = J.D. 2416805.89, ω = 151°, e = 0.01, V_0 = 21 km/sec, K_1 = 70 km/sec and K_2 = 176 km/sec.

119.022 **19 Tauri.** H. Povemire.
IAU Circ. No. 2168 (1969).

119.023 **19 Tauri.** C. de Vegt.
IAU Circ. No. 2186 (1969).

The short-period variability of 14 Aur (HR 1706). See Abstr. 122.066.

120 Variable Stars: Catalogues, Ephemerides, Miscellanea

120.001 **Photométrie et étoiles variables.** A. Terzan.
L'Astronomie, 83ᵉ année, 287 - 296 (1969).

120.002 **On the research program concerning eclipsing variables at the observatories Nürnberg and Izmir.**
E. Pohl.
Non-Periodic Phenomena in Variable Stars, IAU Colloquium, Budapest, 1968, p. 471 - 472 (1969).

120.003 **The third catalogue of variable stars in globular clusters.** H. S. Hogg.
Non-Periodic Phenomena in Variable Stars, IAU Colloquium, Budapest, 1968, p. 475 - 479 (1969).

120.004 **Hoe goed is een waarneming en hoe slecht?**
G. W. E. Beekman.
Hemel en Dampkring, Vol. 67, 342 - 343 (1969).

120.005 **Anwendung der mittleren Kurve zur Bestimmung von Maxima und Minima bei veränderlichen Sternen.** I. Todoran.
Stud. Cerc. Astron. Vol. 14, 35 - 43 (1969).

120.006 **The calculation of periods of light variations of variable stars by electronic computers.**
P. N. Kholopov.
Soobshch. Gos. Astron. Inst. Shternberga, No. 158, p. 23 - 42 (1969). In Russian.
A method for calculation of periods of light variations of variable stars by electronic computers is described. It is used at the Department of Variable Stars at the Sternberg Astronomical Institute. The block-scheme and description of the program, as well as estimations of the computation time and some recommendations are given.

120.007 **Roczni Astronomiczny Obserwatorium Krakowskiego 1970. International Supplement No. 41.**
Under the supervision of K. Kozieł. 114 pp. Price zł 72.00 (1969). − Contents: Ephemerides of 759 eclipsing binaries (K. Kordylewski); RR-Lyrae-type variables (W. Zessewitsch, A. Szczepanowska); Auxiliary tables (S. Andruszewski, L. Orkisz).

120.008 **Ein Nomogramm zur Bestimmung der heliozentrischen Korrektur.** E. Mundry.
BAV Rundbrief, 18. Jahrgang, 25 - 30 (1969).

120.009 **Erforschung der veränderlichen Sterne.**
W. Wenzel.
Jenaer Rundschau (Jena Review), 13. Jahrgang, p. 330 - 333 (1968).

120.010 **Report of the Committee on Variable Stars in Clusters.** H. B. Sawyer Hogg.
Trans. IAU, Vol. 13A, 555 - 565 = Commun. David Dunlap Obs., Richmond Hill, No. 218 (1967). −Appendix II of the report of IAU Commission 27 (Variable Stars).

121 Eclipsing Variables

121.001 The spectroscopic orbital elements of eclipsing binary IZ Per = BV 224. I. Yavuz.
Astron. Astrophys. Vol. 2, 388 - 389 (1969). In German.

The spectroscopic orbital elements of the eclipsing binary IZ Per= BV 224 = BD + 53° 323 = HD 9234 were calculated from 22 spectra (dispersion 72 Å/mm at Hγ). Only one component was spectroscopically perceptible. For the determination of the orbital elements the method of Wilsing and Russell was used. The results are as follows: $V_0 = -41.2$ km/s, a sin i = 2.69×10^6 km, $e = 0.064$, $\omega = 102°8$, $T = T_0 + 0^d.299$.

121.002 Photoelectric times of minima of five eclipsing variables. I. Semeniuk.
Bull. Obs. Astron. Beograd, Vol. 27, 71 - 73 (1969).

Photoelectric observations in yellow colour and times of minima of CQ, Cep, BR Cyg, V477 Cyg, CO Lac and DR Vul are given.

121.003 A note on binary systems with undersize subgiant secondaries. J. V. Field.
Monthly Notices, Roy. Astron. Soc. Vol. 144, 419 - 423 (1969).

The observational data for the eighteen binary systems listed by Kopal as having 'undersized subgiant secondaries' are analysed to determine whether it is possible that these systems are in pre-main sequence contraction. It is found that only four of them could be in this phase.

121.004 Visual observations of EX Hydrae at minimum. W. S. G. Walker, B. F. Marino.
Roy. Astron. Soc. New Zealand Variable Star Sect. Circ. No. 138, 9 pp. (1969).

To assess the accuracy of visual observations a detailed study was made of EX Hydrae. Results in graphical form are presented showing that visual observations are in good agreement with predictions of eclipses of this binary star and also reveal an irregular flare like activity.

121.005 Narrow-band electrophotometry of the eclipsing binary CV Ser of Wolf-Rayet-type.
A. M. Tscherepashuk.
Astron. Tsirk. No. 509, p. 3 - 6 (1969). In Russian.

121.006 The radii of the components of 31 Cyg. K. T. Johansen.
Astron. Astrophys. Vol. 3, 179 - 185 (1969).

The eclipsing binary 31 Cyg has components of spectral type K3.5Ib and B4V. The supergiant consists of a star of almost sharp edge and an extended atmosphere. For the light curve of the primary minimum the eclipsing effect of the extended atmosphere has been removed and the radii of the components determined. In case of central eclipse the radius of the B star is found equal to 7.2 R_O, which is a value too large for the spectral type B4V. In case of a noncentral eclipse we find smaller values of the B star radius. Due to the uncertainty of the orbit dimensions and assumptions made, a definite value of the orbital inclination is not presented but it is concluded that the eclipse may not necessarily be central.

121.007 Narrow-band photoelectric photometry of the Wolf-Rayet eclipsing variable V 444 Cyg.
A. M. Cherepaschuk.
Soobshch. Gos. Astron. Inst. Shternberga, No. 161, 32 pp. (1969). In Russian.

Individual narrow-band (Δλ ~ 90 Å) and U, B, V photoelectric observations of the eclipsing variable V 444 Cyg are presented.

121.008 Lichtkurve und relative Dimensionen des Bedeckungssystems V 338 Herculis. K. Walter.
Astron. Nachr., Vol. 291, 225 - 229 = Mitt. Astron. Inst. Univ. Tübingen No. 119 (1969).

170 photoelectric measurements of V 338 Her were used to derive the light-curve and the relative dimensions of this Algol-type system. In the primary minimum, the bright component of spectral type A9 is partially eclipsed by a somewhat smaller component of low luminosity. In an addendum, the differences in amplitude against the observations of Veteš-nik are emphasized.

121.009 Das Bedeckungssystem AD Herculis auf Grund photometrischer Beobachtungen.
D. Korsch, K. Walter.
Astron. Nachr., Vol. 291, 231 - 237 = Mitt. Astron. Inst. Univ. Tübingen No. 113 (1969).

Photoelectric observations in B and V are used to determine complete light-curves of the Algol variable AD Her and to derive the relative dimensions of the semidetached system. In the phases following the secondary minimum the light-curves show disturbances which are interpreted as absorption effects by a gas stream going out from the subgiant component. Statements are given about the mode of absorption, the number of particles per cm^3, and the flux of masses between the components.

121.010 Ergebnisse der Beobachtungen von Bedeckungsveränderlichen. K. Locher.
Orion, Band 14, 134 (1969).

121.011 Binary stars among cataclysmic variables. X. Photoelectric observations of EM Cygni.
G. S. Mumford, W. Krzeminski.
Astrophys. Journ. Suppl. Series, Vol. 18, 429 - 442 = Contr. Kitt Peak National Obs. No. 415 (1969).

Photoelectric observations of EM Cygni, suspected of being an old nova, are presented. The data, gathered since 1962, indicate that this star is an eclipsing binary with a period of $0^d.29090942$. Primary eclipse appears to be partial, roughly 0.2 mag in depth. There is no evidence for a secondary minimum. Many features in the light curves are reminiscent of those found for such old novae and novalike variables as nova T Aur 1891 and U Gem. The star EM Cyg varies in brightness by about 2 mag in some 20 days. At maximum, the system is bluest as well as brightest in ultraviolet light. It seems likely that the greatest diminution of light at primary minimum may be attributed to an eclipse of a gaseous disk or shell surrounding one component.

121.012 The light variation and orbital elements of AG Virginis. L. Binnendijk.
Astron. Journ. Vol. 74, 1024 - 1031 = Contr. Kitt Peak National Obs. No. 474 (1969).

New photoelectric observations of AG Virginis are presented. A total of 451 observations in yellow light, 451 in blue light, and 444 observations in ultraviolet light were made on seven nights in 1968. In all three wavelength regions there is a large difference in the heights of the maxima of the light curve, and primary minimum is permanently distorted. Secondary minimum is caused by a total eclipse. A new internally consistent set of orbital elements has been derived.

121.013 The light variation and orbital elements of AM Leonis. L. Binnendijk.
Astron. Journ. Vol. 74, 1031 - 1037 = Contr. Kitt Peak National Obs. No. 475 (1969).

A total of 344 photoelectric observations in yellow light and 347 observations in blue light of AM Leonis are presented. This variable is the brighter component of the visual double ADS 8024. In the present observations, the light of the fainter visual double-star component was excluded. At primary minimum the system undergoes a total eclipse. New internally consistent elements are derived. A subluminous region on the larger star can be located.

121.014 The VV Cephei stars.   A. P. Cowley.
Publ. Astron. Soc. Pacific, Vol. 81, 297 - 331 (1969) (1969).
This paper summarizes the photometric and spectroscopic investigations of the 13 known VV Cephei-type binaries. The cool primaries are found to be luminous supergiants with irregular or semiregular light variations of less than a magnitude. Descriptions of the spectra of both the hot and cool stars as well as the peculiar emission features are given. The periods are characteristically long–of the order of a decade or more. Orbital studies indicate that the components have masses in excess of 30 solar masses. The location of these objects in associations or very near the galactic plane implies that they are recently evolved.

121.015 A new composite spectrum of the VV Cephei type.
R. M. Humphreys.
Publ. Astron. Soc. Pacific, Vol. 81, 440 - 443 = Contr. Kitt Peak National Obs. No. 428 (1969).

121.016 Photometric elements of the eclipsing system AY Camelopardalis.   P. Tempesti.
Mem. Soc. Astron. Italiana, Nuova Serie, Vol. 40, 345 - 354 (1969).
Photoelectric observations in V and B light allow to ascertain that the true period of AY Cam is twice the length reported in the General Catalogue of Variable Stars and to derive some other elements.

121.017 On a possible cause of brightness fluctuations in close binary systems of dwarf stars.
V. G. Gorbatzky.
Non-Periodic Phenomena in Variable Stars, IAU Colloquium, Budapest, 1968, p. 391 - 393 (1969).

121.018 Nearly contact binary HD 17514.
V. I. Burnashov, E. A. Vitrichenko.
Non-Periodic Phenomena in Variable Stars, IAU Colloquium, Budapest, 1968, p. 427 (1969).

121.019 Non-periodic phenomena in binary systems. Conventional binaries.   F. B. Wood.
Non-Periodic Phenomena in Variable Stars, IAU Colloquium, Budapest, 1968, p. 429 - 433 (1969). – Introductory report.

121.020 Photometric research on RS CVn at the Catania Astrophysical Observatory.
S. Catalano, M. Rodonò.
Non-Periodic Phenomena in Variable Stars, IAU Colloquium, Budapest, 1968, p. 435 - 441 (1969).
On the ground of extensive photoelectric observations of RS CVn made at Catania since 1963 several photometric peculiarities of this system are analysed.

121.021 Changes in the light curve of Beta Lyrae 1958 - 1959.
G. Larsson-Leander.
Non-Periodic Phenomena in Variable Stars, IAU Colloquium, Budapest, 1968, p. 443 - 455 (1969).
Photometric results obtained during the 35 days of the international programme on β Lyrae are compared with observations made in 1958, mainly at the Lick Observatory. From the minimum epochs of the two seasons a period of

12.9355 days is obtained. The total B magnitude is found about 0.10 mag. fainter and the colour about 0.05 mag. redder in 1959 than in 1958. A slow decrease in brightness during the 1969 campaign is indicated.

121.022 The O'Connell effect in some eclipsing variables.
E. F. Milone.
Non-Periodic Phenomena in Variable Stars, IAU Colloquium, Budapest, 1968, p. 457 - 464 (1969).

121.023 Sudden changes in the period of Algol.
T. Herczeg.
Non-Periodic Phenomena in Variable Stars, IAU Colloquium, Budapest, 1968, p. 465 - 470 (1969).

121.024 Photometric results from the 1959 international campaign on Beta Lyrae.
G. Larsson-Leander.
Ark. Astron. Vol. 5, 253 - 296 (1969).
The photometric material on β Lyrae obtained at fourteen observatories during the 1959 international campaign is presented and discussed in some detail. Magnitudes and colours for the comparison stars are derived and reduced to the B, V system. The method used for the reduction of the various series of observations to the same system is outlined. The resulting B, V and $B - V$ curves for β Lyrae are given; they cover an interval of 35 days with three primary minima. Comparisons are made mainly with the light and colour curves obtained by Wood and Walker from observations at the Lick Observatory in 1958. From the minimum epochs in 1958 and 1959 a period of 12.9355 days is obtained. A great number of photometric data are communicated.

121.025 A method for determining limb-darkening coefficients from the partial phases of complete eclipses.
R. E. Wilson.
Monthly Notices, Roy. Astron. Soc., Vol. 145, 367 - 372 (1969).
A method is described for determining the limb-darkening coefficients (x_g and x_s) of both components of an eclipsing system. Its chief advantages lie in the relative simplicity of the (non-iterative) calculations required, in the intuitive feeling it provides for the determinacy of the results, and in its heuristic value. The method is based on a relation between x_g and x_s which must be satisfied for paired observations, which differ in phase angle by $180°$, within the partial phases of complete eclipses. The method has been tested on one observed and five synthetic light curves. For the synthetic light curves, the results are in good agreement with the known correct answers.

121.026 The binary system TX Herculis.   R. A. Bozula.
Izv. Astron. Obs. Ehngel'gardta, *Kazan'*. No. 36, p. 240 - 267 (1968).   In Russian.

121.027 A study of observations of V566 Ophiuchi.
B. B. Bookmyer.
Bull. American Astron. Soc. Vol. 1, 234 (1969). – Abstr. AAS.

121.028 A comparison of linear and nonlinear laws of limb darkening for TW Draconis and other eclipsing binary systems.   M. L. Cooper.
Bull. American Astron. Soc. Vol. 1, 238 (1969). – Abstr. AAS.

121.029 On the structure of VW Cephei.
K.-C. Leung, I. Jurkevich.
Bull. American Astron. Soc. Vol. 1, 251 - 252 (1969). – Abstr. AAS.

121.030 Twelve eclipsing binaries with double-lined spectra.
D. M. Popper.
Bull. American Astron. Soc. Vol. 1, 257 - 258 (1969). — Abstr. AAS.

121.031 A frontal attack on eclipsing binaries.
D. B. Wood.
Bull. American Astron. Soc. Vol. 1, 267 (1969). — Abstr. AAS.

121.032 Mass loss from close binaries, 1941 - 68.
F. B. Wood.
Mass Loss from Stars, Trieste 1968, p. 149 - 155 (1969).

The history of studies of two types of mass loss — particle ejection and gradual loss caused by evolutionary expansion — is discussed briefly. Difficulties are encountered when we try to compare the true shapes of close binaries with theoretical models. In particular, the evidence at present indicates the W UMa systems are not 'contact' binaries as has been generally assumed, although the results of narrow-band observations or theoretical developments in rectification may change this picture.

121.033 General review of observational spectroscopic evidence for mass loss in close binaries.
J. Sahade.
Mass Loss from Stars, Trieste 1968, p. 156 - 158 (1969).

The evidence for mass loss in close binaries provided by spectroscopic observations is of two types; there is evidence for the existence of gaseous streams from one of the components and also for the existence of envelopes that surround a number of systems.

121.034 Spectroscopic study of the eclipsing system R Canis Majoris. M. Kitamura.
Mass Loss from Stars, Trieste 1968, p. 159 - 170 (1969).

Variation of the residual intensities of metallic and hydrogen lines of R CMa with phase is presented from measurement of its spectrograms obtained with dispersions of 10.3 Å/mm and 4.1 Å/mm at the Okayama Astrophysical Observatory. It is found from variation of the residual intensities of these lines, with the exception of the Ca-K line, that the duration of the eclipse is longer than expected from the photometric elements. The fractional loss of light of the eclipsed component at mid-eclipse has been derived from the ratio between residual intensities of the lines at mid-minimum and outside eclipse.

121.035 A gross secular expansion of the primary in RW Persei. D. S. Hall.
Mass Loss from Stars, Trieste 1968, p. 171 - 183 (1969).

RW Persei ($P = 13^{d}2$) is a binary in which a G or K subgiant eclipses a smaller, brighter A star. In this paper will be discussed observations, many of them new, which suggest that the primary component appears to have been expanding since the turn of the century.

121.036 Boss 5481 during the shell episode of 1965-67.
A. Mammano, A. Martini.
Mass Loss from Stars, Trieste 1968, p. 184 - 197 (1969).

Boss 5481, with a spectrum similar to that of VV Cep, has developed an absorption shell spectrum, observed at Asiago from 1965 to the end of 1966. The radial velocities of M and B components have been found to change, while the [Fe II] emissions remain stationary. Equivalent width of selected shell and stellar lines have been measured.

121.037 Mass motions in the system of VV Cephei.
K. O. Wright, S. J. Larson.
Mass Loss from Stars, Trieste 1968, p. 198 - 203 (1969).

A series of 65 spectra of the red region of VV Cephei have been obtained at Victoria between 1956 and 1968. Intensity tracings of the region near Hα have been made to study the emission and absorption profiles of this line. Additional absorption lines can be detected over a large part of the cycle. Their positions can be interpreted as gas moving from the M-type star towards the secondary star.

121.038 Periods of eclipsing novalike variables.
G. S. Mumford.
Mass Loss from Stars, Trieste 1968, p. 204 - 209 = Contr. Kitt Peak National Obs. No. 405 (1969).

Recent observations of minima of the novalike variables T Aurigae, EM Cygni, U Geminorum, DQ Herculis, EX Hydrae, V Sagittae, and WZ Sagittae are presented. A new period for EM Cygni has been derived; possible period changes for U Geminorum and DQ Herculis are discussed.

121.039 Mass exchange in the binary system AD Herculis obtained from photometric observations.
K. Walter.
Mass Loss from Stars, Trieste 1968, p. 211 - 212 (1969). Abstract.

121.040 Period changes of W Ursae Maioris systems.
T. Herczeg.
Mass Loss from Stars, Trieste 1968, p. 213 - 214 (1969). Abstract.

121.041 Evolution of close binaries and origin of Algol-type systems. J. Ziółkowski.
Mass Loss from Stars, Trieste 1968, p. 231 - 236 (1969).

Observational data for Algol-type systems are discussed basing on available evolutionary tracks. It is shown that the observed properties of the massive binaries (with total mass greater than $5\,M_{\odot}$) can be satisfactorily explained assuming that their contact components are burning hydrogen in the core.

121.042 Elemente der photometrischen Bahn von TT Aurigae. M. I. Lavrov.
Trudy Kazan. Gorod. Astron. Obs. No. 35, p. 60 - 80 (1968). In Russian.

Aus 222 (blauen) und 221 (gelben) Beobachtungen werden Lichtkurven abgeleitet und Haupt- bzw. Nebenminimum bestimmt. Daraus werden die photometrischen Elemente ($P = 1^{d}33273365$) abgeleitet. Ein Unterschied in den Randverdunklungskoeffizienten der Komponenten wird festgestellt. Durch Heranziehung früherer spektroskopischer Elemente (Joy, Sitterly, 1931) konnten die absoluten Systemgrößen berechnet werden.

121.043 Photographische, visuelle und photoelektrische Beobachtungen von MN Cassiopeiae.
M. I. Lavrov, N. V. Lavrova.
Trudy Kazan. Gorod. Astron. Obs. No. 35, p. 81 - 109 (1968). In Russian.

Nach der ermittelten Lichtkurve ist der Stern ein Doppelsystem vom Typ Algol mit der Periode $P = 1^{d}916929$. Die früher ermittelte Periode von $P = 0^{d}958462$ (M. Lavrov, 1959) mußte verdoppelt werden, da Nebenminima entdeckt wurden. Die Bahnform wird diskutiert.

121.044 UBV light curves of the eclipsing binary VZ Hydrae.
R. L. Walker, Jr.
Publ. Astron. Soc. Pacific, Vol. 81, 550 - 551 (1969). Abstract ASP.

121.045 Radial velocities of V448 Cygni.
H. L. Cohen.
Publ. Astron. Soc. Pacific, Vol. 81, 665 - 671 = Contr. Rosemary Hill Obs. No. 4 (1969).

Five spectrograms (63 Å/mm) of V448 Cygni, secured with the No. 1 36-inch telescope of the Kitt Peak National Observatory, were measured for radial velocity. The measured velocities of the B component fit the computed radial velocity curve of Petrie (1956) after the addition of a zero-point shift. However, the vague asymmetric lines of the O star make it difficult to make any meaningful measures of its motion. Hence, the mass ratio of V448 Cygni must be regarded as uncertain.

121.046 Hydrogen-line emission in the spectrum of U Cephei. A. H. Batten, P. G. Laskarides.
Publ. Astron. Soc. Pacific, Vol. 81, 677 - 684 (1969).

Although all attempts to observe $H\alpha$ emission in the spectrum of the eclipsing binary U Cephei during eclipse have so far failed, doubling of the hydrogen lines is observed in spectra obtained at quadratures. It is suggested that this is a partial filling of the absorption line by emission originating in the gaseous stream that flows from the cooler star to the hotter.

121.047 RS Columbae, a new W Ursae Majoris system. H. E. Bond, A. U. Landolt.
Publ. Astron. Soc. Pacific, Vol. 81, 696 - 699 = Contr. Louisiana State Univ. Obs. No. 28 = Contr. Cerro-Tololo Inter-American Obs. No. 83 (1969).

Spectra and UBV photoelectric observations show that the variable star RS Columbae is a short-period eclipsing system, rather than a 14-day cepheid.

121.048 Photoelectric light elements for the eclipsing binary AB Andromedae. A. U. Landolt.
Astron. Journ. Vol. 74, 1078 - 1082 = Contr. Louisiana State Univ. Obs. No. 29 = Contr. Kitt Peak National Obs. No. 473 (1969).

New times of minima derived from UBV photoelectric observations of AB Andromedae around the time of minimum light are presented.

121.049 Photometric study of the eclipsing binary RW Monocerotis. I. Infrared photometry and orbital solution. R. Brukalska, S. M. Ruciński, J. Smak, K. Stępień.
Acta Astron. Vol. 19, 257 - 286 (1969).

Nearly 1700 photoelectric observations in two near-infrared bands were collected for the study of the photometric properties of the components. The ultimate goal will be to determine the surface brightness distribution over the subgiant secondary. In this paper a "conventional" analysis of the primary minimum is reported and the geometrical elements of the system are given. From a preliminary analysis of the secondary minimum it follows that the "effective" limb darkening coefficient of the secondary component (as seen during that eclipse) is negative, what means that the gravity darkening is comparable, or larger than the limb darkening. Physical parameters of the components are also derived.

121.050 On the period of SS Cygni. J. Smak.
Acta Astron. Vol. 19, 287 - 290 (1969).

Radial velocity data now available for SS Cyg are insufficient for a unique determination of non-linear elements. New elements – alternative to those given by Walker and Chincarini (1968) – are presented. With these elements, the radial velocity variations of the hydrogen absorption lines during rising light turn out to be in phase with the radial velocity curve of the G-type component. This would imply that the G-type star is the seat of the outbursts.

121.051 Binaires avec éclipse observés photoélectriquement à l'Observatoire Astronomique de Bucarest en 1968. A. Dumitrescu.

Stud. Cerc. Astron. Vol. 14, 69 - 74 (1969). In Rumanian.

Au cadre du programme d'observations photoélectriques effectuées à l'Observatoire Astronomique de Bucarest on a déterminé les moments des minimes de plusieurs étoiles à éclipse parmi lesquelles: RX Her, AI Dra, WW Aur, V477 Cyg.

121.052 A search for eclipses in HD 104631. G. F. G. Knipe.
Republic Obs. Johannesburg, Circ. Vol. 7 (No. 128), 199 (1969).

121.053 Photoelectric observations of DO Cas during 1967. J. K. Gleim, L. Winkler.
Astron. Journ., Vol. 74, 1191 - 1196 = Contr. Kitt Peak National Obs. No. 494 = Rosemary Hill Obs. Contr. No. 7 (1969).

The β Lyrae eclipsing binary, DO Cas, was reobserved photoelectrically in blue and yellow light. Improved light curves were obtained and a consistent orbital solution obtained from them. The fainter component does not exhibit a spectrum, but it produces an oblateness of 0.115 in the component with the spectrum. A ratio of the sizes of the bright to faint component appears to be exceptionally large with a value of 2.22. The ratio of the duration of secondary totality to the duration of primary totality is approximately 3.

121.054 A study of V566 Ophiuchi. B. B. Bookmyer.
Astron. Journ., Vol. 74, 1197 - 1205 = Contr. Kitt Peak National Obs. No.490 (1969).

Photoelectric observations of V566 Ophiuchi, a W UMa-type eclisping variable system, were obtained with B and V filters on five nights in 1966. The resultant light curves are in agreement with those observed by Binnendijk in 1957 and confirm the fact that the system undergoes complete eclipses with secondary minimum occurring during the total eclipse. Three analyses of the photometric observations are presented.

121.055 A study of the eclipsing binary Beta Aurigae. L. G. S. Toy.
Astrophys. Journ., Vol. 158, 1099 - 1107 (1969).

The present study has led to the determination of the average effective temperature and surface gravity of the two components of the eclipsing binary system β Aur. By use of photoelectric spectrum scans in conjunction with detailed abundance analyses, an effective temperature $T_{eff} = 8750°$K and surface gravity log $g = 3.7$ were derived. The abundance analyses indicate similar chemical compositions for both components. Both members of this system appear to show abundance anomalies reminiscent of mild Am stars.

121.056 The Huang disk model applied to other eclipsing binaries. D. S. Hall.
Bull. American Astron. Soc., Vol. 1, 345 - 346 (1969). – Abstract AAS.

121.057 BM Orionis, the eclipsing binary in the Trapezium. D. S. Hall, L. M. Garrison, Jr.
Bull. American Astron. Soc., Vol. 1, 346 (1969). – Abstract AAS.

121.058 Longperiodic variations in the β Lyrae system. V. Ja. Aluseva.
Astron. Zhurn. Akad. Nauk SSSR, Vol. 46, 832 - 836 (1969). In Russian. English translation in Soviet Astron. AJ, Vol. 13, No. 4.

Longperiodic variations of spectral characteristics, light curves and integral brightness of β Lyrae in blue light are derived from spectrophotometric observations during 1961 - 1963, as well as from photoelectric observations (according to the international program of the years 1958 - 1959). The period of these variations is obtained to be $P = 1141^d$. The source of the variations is suggested to be connected with forces perturbing the orbital motions in the system.

121.059 Photoelectric observations of 31 and 32 Cygni in 1968. B. Cester.
Mem. Soc. Astron. Italiana, Nuova Serie, Vol. 40, 517 - 523 (1969).

During the eclipse of 1968, 32 Cyg was observed photoelectrically at Trieste together with 31 Cyg in the *UBV* system. The latter showed a progressive decrease in all colours of about 0.05 mag in *V* and 0.07 in *B* and *U*. 32 Cyg suffered a variation in *V* of about 0.07 mag during the eclipse, *B–V* varied of 0.11 and *U–B* of 0.60. The duration of the eclipse in *V* seems to have been longer than expected.

121.060 The period of the eclipsing binary AS Cam. R. W. Hilditch.
Observatory, Vol. 89, 143 - 146 (1969).

121.061 Observations of faint fast variable stars by means of a high sensitive television apparatus.
Yu. S. Efimov, V. V. Prokofieva.
Izv. Krymskoj Astrofiz. Obs. Vol. 39, 163 - 169 (1969). In Russian.

Observations of the eclipsing variable RW Tri are carried out by means of a high sensitive television apparatus with an additional brightness amplifier. It is shown that this apparatus allows to investigate very fast variations of brightness of faint stars with telescopes of moderate size. The mean square error of one measurement of magnitude differences of two faint stars is $\pm 0^m 08 - \pm 0^m 10$.

121.062 On the intrinsic polarization of the RY Persei radiation. O. S. Shoulov, G. A. Goudcova.
Astrofizika, Vol. 5, 477 - 485 (1969). In Russian. – Engl. translation in Astrophysics, Vol. 5, No. 3.

The average dependence of the polarization on phase has been established for the eclipsing binary RY Persei. This dependence is based on 230 individual polarization observations by N. M. Shakhovsky and O. S. Shoulov. The interstellar component of polarization has been determined from 24 stars in the region around RY Persei and the intrinsic polarization of the RY Persei radiation has been derived as a function of the phase. The intrinsic polarization in the RY Persei radiation has been interpreted in more details in terms of free electron scattering of light from the stars of the binary in the gaseous disk nebula surrounding the primary star. The mass of this nebula has been evaluated as $2.6 \times 10^{-11} M_\odot$.

121.063 V448 Cygni and V453 Cygni: Two eclipsing binaries in the P Cygni region. H. L. Cohen.
Thesis, Univ. of Indiana. Univ. Microfilms, Ann Arbor, Mi., 247 pp. (1968). – See Phys. Abstr. Vol. 72, No. 21908 (1969).

121.064 Three colour observations of HR 6283 (HD 152667). A. W. J. Cousins, H. C. Lagerwey.
Monthly Notes Astron. Soc. Southern Africa, Vol. 28, 120 - 122 (1969).

121.065 On the distribution of the electron temperature in the envelopes of close binary stars.
L. N. Ivanov.
Trudy Astron. Obs. *Leningrad*, Vol. 26 (= Uchenye Zapiski Leningr. Un-ta No. 347 = Seriya Matem. Nauk No. 44), p. 33 - 36 (1969). In Russian.

The electron temperature of the disk-like envelope of RW Tri is estimated using the U, B, V light-curves during the eclipse. It is found that the temperature of the outer layers of the envelope is of the order of 3×10^4 °K. The electron temperatures are smaller in the internal layers of the envelope.

121.066 The variability of the polarization of Z Vul. O. S. Shulov, G. A. Gudkova.
Trudy Astron. Obs. *Leningrad*, Vol. 26 (= Uchenye Zapiski Leningr. Un-ta No. 347 = Seriya Matem. Nauk. No. 44), p. 37 - 47 (1969). In Russian.

About 200 polarimetric observations of the eclipsing binary Z Vul were made photoelectrically in 1964–1967, and small variations with phase were detected in the polarization parameters. The observed polarization is the vector sum of the interstellar component and the intrinsic one, the latter being variable. Photometric, polarimetric, and spectral observations of 23 stars in the neighbourhood of Z Vul were carried out to find the unknown interstellar component and to derive the intrinsic polarization.

121.067 The infra-red photometry of U Cep. G. V. Hozov, N. A. Minajev.
Trudy Astron. Obs. *Leningrad*, Vol. 26 (= Uchenye Zapiski Leningr. Un-ta No. 347 = Seriya Matem. Nauk No. 44), p. 55 - 62 (1969). In Russian.

Infra-red photoelectric observations of the close binary system U Cep made in 1967 at the Astronomical Observatory of Leningrad are interpreted on the hypothesis of a gaseous stream in the system. This assumption explains a number of the observed effects.

121.068 Determinazione dell'area non occultata di Algol per mezzo della discontinuità di Balmer.
M. Fracassini, L. E. Pasinetti.
Atti XII Riunione Soc. Astron. Italiana, L'Aquila 1968, p. 36 - 39 (1969). – Abstract SAI.

121.069 Photometric elements of the binary system S Equ. S. Catalano, M. Rodonò.
Atti XII Riunione Soc. Astron. Italiana, L'Aquila 1968, p. 87 - 90 (1969). – Abstract SAI.

121.070 Evaluation of the light curves of eclipsing binaries with the aid of the fractional light-loss tables for atmospheric eclipses. I. B. Pustylnik.
Tartu Astron. Obs. Teated No. 23, p. 35 - 121 (1969). In Russian.

A review of spectroscopic and photometric data concerning atmospheric eclipses among the Algol type stars is given.

121.071 The variability of HD 128661. A. J. Harris.
Inform. Bull. Variable Stars (I.A.U. Commission 27), Konkoly Obs., Budapest, No. 365 (1969).

121.072 New elements for three eclipsing binaries. M. Kurutac, C. Ibanoğlu.
Inform. Bull. Variable Stars (I.A.U. Commission 27), Konkoly Obs., Budapest, No. 369 (1969).

121.073 1969 UBV observations of CG Cygni. E. F. Milone.
Inform. Bull. Variable Stars (I.A.U. Commission 27), Konkoly Obs., Budapest, No. 373 (1969).

121.074 The light curve of 442 Cas (S 9484) during the eclipse obtained by T. V. observations.
V. V. Prokofjeva, V. P. Epishev.
Inform. Bull. Variable Stars (I.A.U. Commission 27), Konkoly Obs., Budapest, No. 376 (1969).

121.075 New bright eclipsing binary. R. Zissell.
Inform. Bull. Variable Stars (I.A.U. Commission 27), Konkoly Obs., Budapest, No. 378 (1969).

121.076 Times of minima and light elements of S Velorum.

R. F. Sistero.
Inform. Bull. Variable Stars (I.A.U. Commission 27), Konkoly Obs., Budapest, No. 381 (1969).

121.077 PV Cassiopeiae – an eclipsing binary with eccentric orbit. E. Pohl.
Inform. Bull. Variable Stars (I.A.U. Commission 27), Konkoly Obs., Budapest, No. 386 (1969).

121.078 Minima of R CMa. G. K. Charyulu.
Inform. Bull. Variable Stars (I.A.U. Commission 27), Konkoly Obs., Budapest, No. 390 (1969).

121.079 OO Aql – an eclipsing binary with rapidly shortening period. E. Pohl.
Inform. Bull. Variable Stars (I.A.U. Commission 27), Konkoly Obs., Budapest, No. 391 (1969).

121.080 A spectroscopic study of the eclipsing binary R Canis Majoris. P. Galeotti.
Inform. Bull. Variable Stars (I.A.U. Commission 27), Konkoly Obs., Budapest, No. 392 (1969).

121.081 A spectroscopic study of the triple system VV Orionis. G. Beltrami, P. Galeotti.
Inform. Bull. Variable Stars (I.A.U. Commission 27), Konkoly Obs., Budapest, No. 393 (1969).

121.082 Minima of eclipsing variables. L. P. Surkova, N. V. Skatova.
Inform. Bull. Variable Stars (I.A.U. Commission 27), Konkoly Obs., Budapest, No. 394 (1969).

121.083 Radial velocity observations of the eclipsing system HD 128661. W. Gorza, J. F. Heard.
Inform. Bull. Variable Stars (I.A.U. Commission 27), Konkoly Obs., Budapest, No. 396 (1969).

121.084 The variability of BV 789. C. R. Chambliss.
Inform. Bull. Variable Stars (I.A.U. Commission 27), Konkoly Obs., Budapest, No. 397 (1969).

121.085 HD 216711, probably an eclipsing variable. K. Särg, S. Wramdemark.
Inform. Bull. Variable Stars (I.A.U. Commission 27), Konkoly Obs., Budapest, No. 398 (1969).

121.086 V-observations and light elements of Omega Cen V78.
R. F. Sistero, C. R. Fourcade, J. R. Laborde.
Inform. Bull. Variable Stars (I.A.U. Commission 27), Konkoly Obs., Budapest, No. 402 (1969).

121.087 The variability of BV 516. C. R. Chambliss.
Inform. Bull. Variable Stars (I.A.U. Commission 27), Konkoly Obs., Budapest, No. 408 (1969).

121.088 AD Herculis, ein Algolsystem mit Gasstrom. K. Walter.
SuW, Vol. 8, 261 - 264 (1969).

121.089 Steaua variabilă cu eclipsă Z Vulpeculae. H. Minţi, R. Dinescu.
Stud. Cerc. Astron., Vol. 14, 177 - 188 (1969).
In Rumanian.
Parameters of the eclipsing binary are determined from observations of different origin.

121.090 Variability of the eclipsing variable NQ Herculis. H. Minţi, A. Dumitrescu, H. Alexandrescu.
Stud. Cerc. Astron., Vol. 14, 189 - 191 (1969)

In Rumanian.
Observations from the period April–September 1969 are reported. For these observations we used the photoelectric photometer of the Bucharest Astronomical Observatory, whose accuracy was analyzed in the previous paper. The 277 observations indicated no variability with P = $0\overset{d}{.}870218$.

121.091 Observations of the Ca II K line in the spectrum of 32 Cygni at the 1965 eclipse.
K. O. Wright, K. H. Hesse.
Publ. Dominion Astrophys. Obs. Victoria, Vol. 13, (No. 11), 301 - 322 (1969).
The 1965 eclipse of 32 Cygni was well observed at Victoria and 46 spectrograms with dispersion 6.0 A/mm were obtained. Radial velocities were determined from most of the plates. Numerous components of the ionized calcium lines were observed on spectra obtained within two months of totality. Weak components at up to 1.5 A from the principal line were detected on intensity tracings; their lifetimes seem to range from a few days to, possibly, several weeks. The light ratio of the continuum of the B star to that of the K star was redetermined and found to be 0.4 at 3900 A. Spectroscopic observations seem to indicate that the grazing total eclipse lasts about two weeks, possibly with some variation from one eclipse to another because of atmospheric effects.

121.092 A *UBV* photometric study of MR Cygni.
D. S. Hall, R. H. Hardie.
Publ. Astron. Soc. Pacific, Vol. 81, 754 - 770 = Repr. Arthur J. Dyer Obs., Vanderbilt Univ. No. 47 (1969).
Photoelectric *UBV* observations of MR Cygni were obtained and the resulting three light curves solved. The $(U - B), (B - V)$ indices imply a spectral type of B3 for the primary; the old classification A0 is apparently wrong. Several lines of reasoning lead to a spectral type of B8 ± 1 for the secondary. A mass ratio of 1.8, typical of B3 and B8 main-sequence stars, leads to 7.5 $M_\odot$, 5.1 $R_\odot$ and 4.2 $M_\odot$, 3.7 $R_\odot$ respectively. MR Cyg is probably a system of two main-sequence stars, quite close, but normal and uncomplicated.

121.093 BM Orionis, the eclipsing binary in the Trapezium.
D. S. Hall, L. M. Garrison, Jr.
Publ. Astron. Soc. Pacific, Vol. 81, 771 - 794 = Contr. Kitt Peak National Obs. No. 485 = Repr. Arthur J. Dyer Obs., Vanderbilt Univ. No. 48 (1969).
Photoelectric *UBV* observations of BM Orionis were obtained by offsetting in a consistent manner which corrected for both nebulosity and scattered light. A complete light curve was obtained in *V*, with *U* and *B* observations made at maximum and minimum. A plot on the color-color diagram indicated that the hotter star is B2 or B3, in good agreement with spectral classifications. The cooler star is anomalous and can be considered an A1 star. With the mass function from Struve and Titus and with the $M:L$ relation applied only to the B star, and the resulting dimensions are 5.4 $M_\odot$, 2.5 $R_\odot$, $M_v = -0\overset{m}{.}7$ for the B star and 2.8 $M_\odot$, 8.5 $R_\odot$, $M_v = -1\overset{m}{.}1$ for the A star.

121.094 Hydrogen-line emission in the spectrum of U Cephei: A sequel. A. H. Batten.
Publ. Astron. Soc. Pacific, Vol. 81, 904 - 906 (1969).

121.095 Langfristige Periodenkontrolle für Bedeckungsveränderliche der Bamberger Liste.
H. Bauernfeind.
Veröff. Remeis-Sternw. Bamberg, Astron. Inst. Univ. Erlangen–Nürnberg, Vol. 8, No. 81, 96 pp. (1969).
Presented are the results of a discussion of the observations of 46 eclipsing variables.

121.096 **Lichtkurven und vorläufige Elemente der Be-
deckungsveränderlichen ES Lib und FF CMa.**
R. Knigge, U. Köhler.
Veröff. Remeis-Sternw. Bamberg, Astron. Inst. Univ. Erlangen–Nürnberg, Vol. 8, No. 83, 10 pp. (1969).
    Light curves and preliminary elements for both eclipsing variables.

121.097 **Orbital elements and period variation of the
eclipsing binary AP Leonis (BV 366).**
C. Bartolini, P. Battistini.
Pubbl. Oss. Astron. Univ. Bologna, Vol. 10, No. 3, 16 pp. (1969).
    797 V and 240 B photoelectric observations of the eclipsing binary AP Leonis were made at the Bologna Observatory during the years 1964 – 1968. New orbital elements derived by the Russell and Merrill method and the period

study are reported.

121.098 **Let us observe eclipsing binaries.**
O. Obůrka.
Říše hvězd, Vol. 50, 130 - 131 (1969).   In Czech.

121.099 **Besonderheiten and Lichtkurven von Algolsystemen.**
K. Walter.
Mitt. Astron. Ges. No. 27, p. 176 - 177 (1969). – Abstract AG.

Contributions to the interpretation of the light curves of the close binary systems. III. The loss of the light during the eclipses.   See Abstr. 117.029.

Eclipses of U Geminorum.
See Abstr. 122.073.

122 Physical Variables, Flare Stars, Pulsation Theory

122.001 Light variation of the A-type peculiar star HD 221568. S. Nishimura, K. Ichimura, K. Osawa. Ann. Tokyo Astron. Obs., Second Series, Vol. 11, 123 - 130 (1969).

Results of the photometric observations on the UBV system and on Strömgren's $uvby$ system of the Ap-type variable star HD 221568 from 1964 to 1968 are presented. The light appears to vary quite regularly with a period of about 159.3 days. No pronounced secular change is found in the light curves. There is some indication of minor short-period fluctuations in the light curves, but their regularity and period has not yet been well established.

122.002 Interstellar absorption and the intrinsic colors of the peculiar star HD 221568. K. Osawa, K. Ichimura, M. Shimizu. Ann. Tokyo Astron. Obs., Second Series, Vol. 11, 131 - 134 (1969).

Three-color photometry and low-dispersion spectroscopy were done of eighteen field stars around the peculiar star HD 221568. $E(B-V)$ was plotted against spectroscopic distance, and the most probable values of $E(B-V)$ and $E(U-B)$ of HD 221568 were estimated. The intrinsic colors of this peculiar variable star were found to be $B-V = +0.16$ and $U-B = -0.18$ at its reddest phase, and $B-V = -0.06$ and $U-B = -0.18$ at its bluest phase.

122.003 The light variation of the hydrogen-deficient star HD 30353. S. Nishimura, K. Ichimura, K. Osawa, K. Nariai. Ann. Tokyo Astron. Obs., Second Series, Vol. 11, 135 - 141 (1969).

Light variation of HD 30353 was observed from 1962 to 1968. Semi-regular variation with a period of about 30 days was confirmed. The amplitude is usually of the order of 0.17, 0.20 and 0.24 magnitudes in V, B and U, respectively. Occasional rise in amplitude was observed in January of 1966. Minor fluctuations also appeared from time to time.

122.004 Observations of the variability of the metallic-line star 28 Andromedae. S. Nishimura, E. Watanabe. Ann. Tokyo Astron. Obs., Second Series, Vol. 11, 142 - 156 (1969).

Observational data on the variability of the metallic-line A-type star 28 Andromedae for the years 1967 and 1968 are presented in detail. A new observation of the radial velocity using a "single-trailer" confirms the velocity variation and the phase relation between the light and velocity curves reported previously.

122.005 On the spectrum and nature of P Cygni. M. de Groot. Bull. Astron. Inst. Netherlands, Vol. 20, 225 - 273 = Commun. Obs. Utrecht (1969).

From a study of 36 high-dispersion spectrograms of P Cygni covering the period from 1942 to 1964 it is found that many of the absorption lines are double and the hydrogen absorption lines are triple. All absorption components are displaced to shorter wavelength. Wavelengths and identifications for all observable lines are given together with the radial velocities and equivalent widths for the hydrogen, the helium and some other important spectral lines. In the outer shell the radial velocity varies from −180 to −240 km/sec with a period of 114 days. It is concluded that the velocity of expansion of the atmosphere of P Cygni increases from 50 km/sec at the stellar surface to more than 200 km/sec at a distance of 3

stellar radii from the surface. The electron density and the geometrical thickness of the outer shell are determined. The mass loss is rather high: $2.0 \times 10^{-4} M_\odot$/year. Pulsational instability is suggested as the reason for the observed light variations.

122.006 Identification, structure, and variations of new TiO bands in the one-micron spectra of Mira variables. G. W. Lockwood. Astrophys. Journ. Vol. 157, 275 - 280 (1969).

The identification of seven new TiO bands in the region $\lambda\lambda 9725-10250$ Å in 45 Å mm^{-1} image-tube spectra of Mira variables is discussed. The bands appear in stars later than M6, and they strengthen toward minimum light. Rotational structure is observed. A relationship between band strength and temperature is demonstrated.

122.007 Strong optical polarization observed in VY Canis Majoris. S. J. Shawl. Astrophys. Journ. *(Letters)*, Vol. 157, L57 - L58 (1969).

Strong polarization in ultraviolet light was found for the OH emission source VY Canis Majoris. The polarization drops steeply, and its position angle rotates over 60° from the ultraviolet to the infrared.

122.008 Quantitative analysis of the δ Scuti variable δ Delphini. D. Reimers. Astron. Astrophys. Vol. 3, 94 - 109 (1969). In German.

The spectrum of δ Del is analysed using model atmospheres in non-grey radiative equilibrium. The equivalent widths of about 230 spectral lines are given. By fitting computed continua, colors, hydrogen lines, two hundred spectral lines etc. to the observations we obtain an effective temperature $T_e = 7100° \pm 100°$, a surface gravity $\log g = 3.8 \pm 0.2$ and a depth-independent microturbulent velocity $\xi = (7 \pm 0.5)$ km/s. It is not possible to attribute different velocityparameters to neutral respectively singly ionized elements. The abundances of the heavy elements up to Fe, with the exception of Mg, are reduced by a factor of 2 relative to the sun. Mg is underabundant by a factor of 4. Ni is slightly overabundant, Y and Zr are overabundant by a factor of two. Quantitative computations of the dependence of Strömgren m_1, $b-y$ and c_1 indices on metal abundance and microturbulent velocity have been made by detailed line statistics and model atmospheres.

122.009 Infrared observations of a preplanetary system. F. J. Low, B. J. Smith. Commun. Lunar Planet. Lab. Vol. 8 (No. 138), 87 - 89 (1969). − Reprinted from Nature, Vol. 212, 675 - 676 (1966).

122.010 New flare stars in the region around NGC 7023. II. L. V. Mirzoyan, E. S. Parsamian, N. L. Kalloghlian. Soobshch. Byurakan. Obs. No. 40, p. 31 - 34 (1969). In Russian.

In a region around NGC 7023 six new flare stars have been discovered on plates taken with the 40″ Schmidt camera of the Byurakan Astrophysical Observatory during 1967. The effective observing time is ~ 29^h. Coordinates and magnitudes of the stars are communicated.

122.011 Some remarks on the RW Aur type variables in Taurus. H. S. Badalian, L. K. Erastova. Soobshch. Byurakan. Obs. No. 40, p. 35 - 45 (1969). In Russian.

Results of observations of RW Aur variables in the Taurus association obtained from 1960 to 1966 are presented. Changes in the behavior of the variables are described. The color-magnitude diagram and histograms of light from some variables are constructed.

122.012 Photoelectric observations of RW Aurigae.
R. A. Vardanian.
Soobshch. Byurakan. Obs. No. 40, p. 63 - 68 (1969).
In Russian.

The results of three color (λ_{eff} = 5400, 4500, 3700 Å) photoelectric observations of RW Aur (in 1962) are presented. Its changes of brightness occur within 20 - 30 minutes. The brightness increase of 1^m0 in the blue region is followed by a color index decrease of 0^m4.

122.013 De U Geminorum sterren en aanverwante typen.
H. Feijth.
Hemel en Dampkring, Vol. 67, 238 - 241 (1969).

122.014 The flare star UV Ceti.   F. M. Bateson.
Roy. Astron. Soc. New Zealand Variable Star Sect. Circ. No. 139, 3 pp. (1969).

UV Ceti was monitored for 27 1/2 hours during the international observing session 1968 October 14 to 28. All observations were visual except on October 27 10 49.4 to 13 14.1 (U. T.) when it was monitored photo-electrically. The international observing session coincided with a period of very bad weather over the whole country. As a result such suspected flares as were observed visually do not exceed the errors of observation under the sky conditions prevailing. Three suspected flares were recorded photoelectrically but are considered of low reliability because of observing conditions.

122.015 The flare star YZ Canis Minoris.
F. M. Bateson.
Roy. Astron. Soc. New Zealand Variable Star Sect. Circ. No. 136, 8 pp. (1969).

YZ CMi was monitored for a total of 65 hours during the periods 1969 January 11-25; February 10-24. All observations were visual except for 2 1/2 hours of photo-electric monitoring by Andrews and Rowe at the West Melton Observatory. Tables of coverage, observers and suspected flares observed are given. The results are discussed.

122.016 The flare star V645 (Proxima) Centauri.
F. M. Bateson.
Roy. Astron. Soc. New Zealand Variable Star Sect. Circ. No. 137, 3 pp. (1969).

V645 Cen was selected for monitoring by the Section in 1969 March. Coverage was only 15 hours because of general bad weather. Four suspected flares were observed. The results are discussed.

122.017 Period-luminosity and period-radius relations for
δ **Scuti stars.**   M. S. Frolov.
Astron. Tsirk. No. 505, p. 1 - 3 (1969).   In Russian.

122.018 The spectrum of CH Cygni in autumn 1968.
G. N. Jimsheleishvili.
Astron. Tsirk. No. 505, p. 8 (1969).   In Russian.

122.019 The spectrum of the star AD Cygni and some peculiarities of S-C-type stars.
M. V. Dolidze.
Astron. Tsirk. No. 507, p. 6 - 8 (1969).   In Russian.

122.020 Polarimetric observations of YZ CMi and UV Cet during flares.   R. A. Vardanjan.
Astron. Tsirk. No. 508, p. 1 - 3 (1969).   In Russian.

122.021 Observations of BL Lacertae.

N. E. Kurochkin.
Astron. Tsirk. No. 508, p. 3 - 4 (1969).   In Russian.

122.022 Spectrophotometric investigations of CH Cygni.
N. L. Ivanova.
Astron. Tsirk. No. 508, p. 4 - 6 (1969).   In Russian.

122.023 Monochromatic light curves of CH Cygni.
E. B. Gusev.
Astron. Tsirk. No. 508, p. 6 - 7 (1969).   In Russian.

122.024 On the lines of intrinsic colour of variable super-giants.   E. N. Makarenko.
Astron. Tsirk. No. 509, p. 1 - 3 (1969).   In Russian.

122.025 Properties of the optical radiation of variable stars and quasars.   F. I. Lukatzkaja.
Astron. Tsirk. No. 511, p. 3 - 6 (1969).   In Russian.

122.026 Infrared spectrum of CH Cyg.
E. B. Gusev.
Astron. Tsirk. No. 511, p. 6 - 8 (1969).   In Russian.

122.027 On the reddening lines for the RR Lyrae stars.
O. E. Mandel.
Astron. Tsirk. No. 515, p. 3 - 5 (1969).   In Russian.

122.028 On the correlation of brightness and colour amplitudes for RR Lyr-type stars.   R. K. Kanisheva.
Astron. Tsirk. No. 515, p. 6 - 8 (1969).   In Russian.

122.029 Electrophotometric observations of AE Aur.
O. P. Abuladze.
Astron. Tsirk. No. 515, p. 8 (1969).   In Russian.

122.030 Spectrophotometric observations of the semiregular variable ST Her.   R. I. Tchuprina.
Astron. Tsirk. No. 516, p. 3 - 4 (1969).   In Russian.

122.031 The content of water vapor in some semi-regular variable star atmospheres.   E. B. Gusev.
Astron. Tsirk. No. 518, p. 2 - 4 (1969).   In Russian.

122.032 Rapid variations of the spectrum of γ Cassiopeiae.
I. D. Kupo, O. D. Dokuchaeva, T. S. Fetisova.
Astron. Tsirk. No. 519, p. 3 - 5 (1969).   In Russian.

122.033 Features of the observed pulsational instability strip for classical cepheids in the Galaxy.
N. N. Yakimova.
Astron. Tsirk. No. 521, p. 3 - 7 (1969).   In Russian.

122.034 One error in Kraft's analysis of classical cepheids.
N. N. Yakimova.
Astron. Tsirk. No. 522, p. 4 - 6 (1969).   In Russian.

122.035 HDE 310376: A rapid variable star similar to Scorpius XR-1.   R. E. Schild.
Astrophys. Journ. Vol. 157, 709 - 715 = Contr. Cerro Tololo Inter-American Obs. No. 72 (1969).

Spectroscopic and photometric data for the rapid irregular variable HDE 310376 are presented. The star is found to be similar to Sco XR-1 in many ways, but it remains undetected at X-ray wavelengths.

122.036 New evidence for the oblique-rotator model for
a^2 **Canum Venaticorum.**
K. Kodaira, W. Unno.
Astrophys. Journ. Vol. 157, 769 - 783 (1969).

We have reexamined the oblique-rotator hypothesis for this spectrum variable. Profiles of Si II $\lambda\lambda$4128 and 4130 were measured on polarimetric, high-dispersion spectra (1.5 Å mm^{-1}, four polarization components) of a^2 CVn. We determine the position angle of the line of force of the effective magnetic field a as well as the strength of the effective longitudinal magnetic field H_e^*. The observed H_e^*-curve differs slightly in form from that of Babcock and Burd but is in general agreement. Two possible solutions were obtained for the a-curve. One of them was found to show the rotatory behavior consistent with the observed H_e^*-curve and also with the oblique-rotator model proposed by Böhm-Vitense. Strong variations of the profiles of Eu II λ4129, Cr II λ4555, and λ4559 were observed. They coincided well with the predictions given by Böhm-Vitense for her model, suggesting again the validity of the oblique-rotator hypothesis for a^2 CVn.

122.037 Untersuchung schneller Helligkeitsschwankungen kleiner Amplitude bei instationären Sternen.
I. V. Shpychka.
Tsirk. L'vov. Astron. Obs. No. 43, p. 21 - 30 (1969).
In Russian.

122.038 Photoelektrische Beobachtungen von X Persei (1964 - 1965). I. V. Shpychka.
Tsirk. L'vov. Astron. Obs. No. 43, p. 31 - 37 (1969).
In Russian.

122.039 On the long period of Nu Eridani.
N. K. Rao.
Publ. Astron. Soc. Pacific, Vol. 81, 359 - 364 (1969).
    Observations of ν Eridani on 17 nights spread over a period of 736 days confirm the long period (n_L = 0.0633 cycles/day) of this star. The effect of tidal distortion appears to be present in both the fundamental and the first overtone oscillations. The fact that the two distortions are opposite in phase may be of importance for the theoretical models of pulsating stars.

122.040 *UBV* observations of long-period variable stars, VII.
A. U. Landolt.
Publ. Astron. Soc. Pacific, Vol. 81, 381 - 386 = Contr. Louisiana State Univ. Obs. No. 21 = Contr. Cerro Tololo Inter-American Obs. No. 73 (1969).
    Photoelectric observations of 25 long-period variable stars are presented on the *UBV* photometric system.

122.041 Infrared excess of RY Sgr. T. A. Lee, M. W. Feast.
Astrophys. Journ. *(Letters)*, Vol. 157, L173 - L176 (1969).
    A large infrared excess has been found for RY Sgr, an R CrB-type variable star. During a recent 8-month interval the visual brightness increased, while the radiation in the infrared, in the 2 - 3.4-μ region decreased. These observations support the general model for R CrB proposed by Stein et al., in which the infrared flux is ascribed to blackbody radiation from a circumstellar cloud of particles ejected at the time of deep minimum.

122.042 Rapid radio variations in BL Lac.
B. H. Andrew, J. M. MacLeod, J. L. Locke, W. J. Medd, C. R. Purton.
Nature, Vol. 223, 598 - 599 (1969).
    Large and rapid changes in the flux density of VRO 42.22.01 (BL Lac) have been revealed by frequent measurements of the radio source, at wavelengths of 4.5 and 2.8 cm, taken during 1968/69. The flux density of the source varies by almost a factor of two with a time scale as short as one month. The qualitative features of the outbursts are satisfactorily explained by a model which assumes an adiabatic expansion of the source.

122.043 Spectral analysis of the light curves of T Tauri stars
and other objects.   S. Plagemann.
Non-Periodic Phenomena in Variable Stars, IAU Colloquium, Budapest, 1968, p. 21 - 39 (1969).

122.044 Extremely young stars.   W. Wenzel.
Non-Periodic Phenomena in Variable Stars, IAU Colloquium, Budapest, 1968, p. 61 - 73 (1969). – Introductory report.

122.045 Photoelectric observations of 6 southern RW Aurigae variables.   W. Seggewiss, E. H. Geyer.
Non-Periodic Phenomena in Variable Stars, IAU Colloquium, Budapest, 1968, p. 85 - 92 (1969).

122.046 Hydrogen emission phenomena in T Tauri stars.
L. Anderson, L. V. Kuhi.
Non-Periodic Phenomena in Variable Stars, IAU Colloquium, Budapest, 1968, p. 93 - 101 (1969).

122.047 The evidence for variable infall of material in the ultraviolet excess stars.   M. F. Walker.
Non-Periodic Phenomena in Variable Stars, IAU Colloquium, Budapest, 1968, p. 103 - 105 = Contr. Lick Obs. No. 292 (1969).

122.048 Emission Ha-line profiles in several T Tau stars.
E. A. Dibaj, V. F. Esipov.
Non-Periodic Phenomena in Variable Stars, IAU Colloquium, Budapest, 1968, p. 107 - 109 (1969).

122.049 Flares of UV Ceti type stars.   R. E. Gershberg.
Non-Periodic Phenomena in Variable Stars, IAU Colloquium, Budapest, 1968, p. 111 - 126 (1969).

122.050 Cooperative 24-hour observations of UV Ceti-type stars.   P. F. Chugainov.
Non-Periodic Phenomena in Variable Stars, IAU Colloquium, Budapest, 1968, p. 127 - 130 (1969).

122.051 The classification of photometric light-curves of flares of UV Ceti stars.   V. S. Oskanian.
Non-Periodic Phenomena in Variable Stars, IAU Colloquium, Budapest, 1968, p. 131 - 136 (1969).

122.052 Multi-colour photometry of Orion flare stars.
A. D. Andrews.
Non-Periodic Phenomena in Variable Stars, IAU Colloquium, Budapest, 1968, p. 137 - 148 (1969).

122.053 Photoelectric research on flare stars at the Catania Astrophysical Observatory.
S. Cristaldi, G. Godoli, M. Narbone, M. Rodonò.
Non-Periodic Phenomena in Variable Stars, IAU Colloquium, Budapest, 1968, p. 149 - 157 (1969).
    In this paper the results recently obtained at Catania from the observations of the flare stars PZ Mon, YZ CMi, AD Leo, BD +55°1823, BD +51°2402 and EV Lac are summarized.

122.054 HD 160202 – an early-type flare star.
G. A. Bakos.
Non-Periodic Phenomena in Variable Stars, IAU Colloquium, Budapest, 1968, p. 159 - 160 (1969).

122.055 On the group of young stars in the solar vicinity.
M. A. Arakelian.
Non-Periodic Phenomena in Variable Stars, IAU Colloquium, Budapest, 1968, p. 161 - 163 (1969).

122.056 Flare stars near NGC 7023.
L. V. Mirzoyan, E. S. Parsamian.
Non-Periodic Phenomena in Variable Stars, IAU Colloquium,

Budapest, 1968, p. 165 - 167 (1969).

122.057 On the polarimetric study of UV Ceti type star flares. Yu. S. Efimov.
Non-Periodic Phenomena in Variable Stars, IAU Colloquium, Budapest, 1968, p. 169 - 171 (1969).

122.058 Preliminary results of a survey of nebular variables and flare stars. L. Rosino.
Non-Periodic Phenomena in Variable Stars, IAU Colloquium, Budapest, 1968, p. 173 - 177 (1969).

122.059 Some remarks on the spectral and light variability of P Cygni. L. Luud.
Non-Periodic Phenomena in Variable Stars, IAU Colloquium, Budapest, 1968, p. 197 - 201 (1969).

122.060 Spectral variations of P Cygni.
M. de Groot.
Non-Periodic Phenomena in Variable Stars, IAU Colloquium, Budapest, 1968, p. 203 - 213 (1969).
From a careful study of 35 high-dispersion spectrograms of P Cygni it is concluded that the spectroscopic data do not confirm the conclusion of Magalashvili and Kharadze that P Cygni is a W UMa type system. It is found that many of the absorption lines are double, the hydrogen absorption lines even triple. This is attributed to line formation in different shells. In the outer shell variations with a period of 114 days lead to observed radial velocity variations between -180 and -240 km/sec. A preliminary conclusion about the velocity field in the atmosphere of P Cygni is drawn.

122.061 Variations in the continuous spectrum of γ Cas.
N. L. Ivanova, I. D. Kupo, A. Ch. Mamatkazina.
Non-Periodic Phenomena in Variable Stars, IAU Colloquium, Budapest, 1968, p. 215 - 221 (1969).

122.062 Ry Sgr during the 1967-8 minimum: A blue continuum and other spectroscopic and photometric observations. M. W. Feast.
Non-Periodic Phenomena in Variable Stars, IAU Colloquium, Budapest, 1968, p. 253 - 256 (1969).

122.063 Tidal resonances in some pulsation modes of the Beta Canis Majoris star 16 Lacertae.
W. S. Fitch.
Non-Periodic Phenomena in Variable Stars, IAU Colloquium, Budapest, 1968, p. 287 - 296 (1969).

122.064 Variations dans le spectre de γ Pegasi.
J.-M. le Contel.
Non-Periodic Phenomena in Variable Stars, IAU Colloquium, Budapest, 1968, p. 297 - 301 (1969).

122.065 Is γ Bootis a spectrum variable star?
A. Baglin, F. Praderie, M. N. Perrin.
Non-Periodic Phenomena in Variable Stars, IAU Colloquium, Budapest, 1968, p. 303 - 306 (1969).

122.066 The short-period variability of 14 Aur (HR 1706).
C. Chevalier.
Non-Periodic Phenomena in Variable Stars, IAU Colloquium, Budapest, 1968, p. 306 - 311 (1969).

122.067 Instability of light curves and periods of longperiod cepheids belonging to the spherical component of the galaxy. O. P. Vasiljanovskaja, G. E. Erleksova.
Non-Periodic Phenomena in Variable Stars, IAU Colloquium, Budapest, 1968, p. 321 - 324 (1969).

122.068 Correlations between the irregularities of the pe-

riod and radial velocity, length of period, amplitude and spectrum of Mira-variables. P. Ahnert.
Non-Periodic Phenomena in Variable Stars, IAU Colloquium, Budapest, 1968, p. 325 - 329 (1969).

122.069 The irregularities in the light-changes of Mira Ceti.
P. L. Fischer.
Non-Periodic Phenomena in Variable Stars, IAU Colloquium, Budapest, 1968, p. 331 - 338 (1969).

122.070 On the intrinsic light-polarization in some late type stars. R. A. Vardanian.
Non-Periodic Phenomena in Variable Stars, IAU Colloquium, Budapest, 1968, p. 339 - 341 (1969).

122.071 Hot, very short-period eruptive binaries.
J. Smak.
Non-Periodic Phenomena in Variable Stars, IAU Colloquium, Budapest, 1968, p. 345 - 354 (1969).
Some problems connected with the interpretation of the photometric and spectroscopic observations of the U Geminorum type stars are discussed.

122.072 Photometric observations of the peculiar blue variable TT Ari = BD + 14°341.
J. Smak, K. Stępien.
Non-Periodic Phenomena in Variable Stars, IAU Colloquium, Budapest, 1968, p. 355 - 359 (1969).
Light variations of BD + 14°341 can be resolved into three, apparently independent activities: (1) periodic variations with $P = 0^d2658$ and an amplitude of about 0.15 mag.; (2) quasi-periodic fluctuations with periods between 14 and 20 minutes and variable amplitude; and (3) small-scale, apparently irregular fluctuations ("flickering") with a time-scale of the order of 1 min. It is suggested that the star is a hot, subluminous close binary system.

122.073 Eclipses of U Geminorum.
M. W. Mayall.
Non-Periodic Phenomena in Variable Stars, IAU Colloquium, Budapest, 1968, p. 377 - 380 (1969).

122.074 The standstills of light of Z Cam stars. Photoelectric and spectrographic observations.
M.-C. Lortet.
Non-Periodic Phenomena in Variable Stars, IAU Colloquium, Budapest, 1968, p. 381 - 382 (1969).

122.075 Analyse des courbes de lumière des étoiles du type U Geminorum. M. Petit.
Non-Periodic Phenomena in Variable Stars, IAU Colloquium, Budapest, 1968, p. 383 - 390 (1969).

122.076 Symbiotic stars. A. A. Boyarchuk.
Non-Periodic Phenomena in Variable Stars, IAU Colloquium, Budapest, 1968, p. 395 - 410 (1969).

122.077 Spectroscopic variations in WY, Gem, W Cep and HD 4174. A. Mammano, A. Martini.
Non-Periodic Phenomena in Variable Stars, IAU Colloquium, Budapest, 1968, p. 415 - 418 (1969).
Some late-type stars with forbidden lines, suspected to be binaries or to have combination spectra, are under observation at Asiago. In this paper the preliminary results for WY Gem, W Cep and HD 4174 are presented.

122.078 Theoretical light changes of W UMa stars with low mass ratio. H. Mauder.
Non-Periodic Phenomena in Variable Stars, IAU Colloquium, Budapest, 1968, p. 419 - 422 (1969).

122.079 The cepheids on the colour-colour plot.
N. S. Nikolov, P. Z. Kunchev.
Non-Periodic Phenomena in Variable Stars, IAU Colloquium,
Budapest, 1968, p. 481 - 484 (1969).

122.080 The masses of cepheids.
I. N. Latyshev.
Astrofizika, Vol. 5, 331 - 335 (1969). In Russian.
English translation in Astrophysics, Vol. 5, No. 2 (1969).

The masses of classical cepheids with periods up to 9^d
are in the interval 4 - 10 $M_\odot$, those with $P > 9^d$ are probably
larger. If the relation $P\sqrt{\rho}$ = const is assumed, the cepheids
obey the mass-luminosity law found for the upper part of
the main sequence. If the masses of the cepheids are independent of the periods, the concentration of mass to the center
of the star probably increases with the increase of the period.

122.081 About the T. V. photometry of faint variable stars.
A. N. Abramenko, V. V. Prokofjeva.
Non-Periodic Phenomena in Variable Stars, IAU Colloquium,
Budapest, 1968, p. 57 - 58 (1969).

**122.082 The anomalous behavior of aluminum oxide bands
in Mira variables.**
P. C. Keenan, A. J. Deutsch, R. F. Garrison.
Astrophys. Journ. Vol. 158, 261 - 268 (1969).

In spectra of normal M giants, absorption bands of AlO
occur with an intensity that is well correlated with the spectral
type as defined by the TiO bands. In Mira variables, however,
the AlO band can vary greatly from cycle to cycle and can be
abnormally strong or weak for the spectral type. This behavior
is not shared by other molecules, such as ScO and VO, which
vary predictably with spectral type. In some Mira variables
the AlO bands have been observed to go over into emission.
This occurred in Mira in 1924 and 1964, R Ser in 1960, R Psc
in 1966, R Crv and R Cas in 1967, and RT Lib in 1968. From
a survey of all spectrograms of Mira itself available at Mount
Wilson and Palomar Observatories, we find that the AlO bands
tend to be strong in absorption during "strong line" cycles
and tend to be weak or reversed in "weak line" cycles.

122.083 An interpretation of Delta Scuti stars.
K.-C. Leung.
Bull. American Astron. Soc. Vol. 1, 251 (1969). — Abstr.
AAS.

122.084 Balmer-line observations of α^2 CVn. H. J. Wood.
Bull. American Astron. Soc. Vol. 1, 267 - 268
(1969). — Abstr. AAS.

**122.085 Photoelectric observations of CH Cygni during the
explosive phases of 1967-68.** B. Cester.
Mass Loss from Stars, Trieste 1968, p. 329 - 335 (1969).
Observations made at Trieste in 1967 and 1968 of CH
Cygni reveal a strong variability, particularly in U colour.
From the observed colours, a suggestion is made about the
existence of a hot companion which could be a subdwarf.

**122.086 Spectroscopic evidence for mass loss from CH
Cygni.** R. Faraggiana, M. Hack.
Mass Loss from Stars, Trieste 1968, p. 336 - 345 (1969).
Some high-dispersion spectrograms of CH Cygni taken
in epochs at which only the M6 spectrum is visible and after
the explosions of June 1967 and July 1968 have been studied.
Strong negative radial velocities of the Ca II chromospheric
absorptions and the turbulent motions broadening the emission lines of H I, He I, Fe II, [Fe II] prove that mass is lost
from the star at a rate of the order of 10^{-8} solar masses per
year. Arguments are given in favour of the hypothesis that
CH Cygni is a close binary composed of an M giant and a blue
unstable subdwarf.

122.087 Differential velocities in the atmosphere of l Carinae.
J. A. Dawe.
Monthly Notices, Roy. Astron. Soc., Vol. 145, 377 - 387
(1969).

Observations, during the year 1962, suggest no significant departure of the radial velocity curve of l Carinae from
that given by Jacobsen (1934). Using the photoelectric measurements of Lake (1962), a mean radius for this star of
$(132.6 \pm 1.3) \times 10^6$ km is found by means of the Wesselink
method. Weak, though significant, differential motions have
been found in the atmosphere of this star from a study of the
relative Doppler displacements of the Fe I lines in its spectrum.
The profile of the Ca II line is complex, consisting of two absorption components.

**122.088 On the continuum and the absolute magnitude of
Eta Carinae.** R. Viotti.
Astrophys. Space Sci. Vol. 5, 323 - 332 (1969).

The excitation temperature, colour excess and continuum of η Car in the visible and near ultraviolet have been
determined from the study of the excitation of the ionized
iron emission lines. The excitation temperatures of Fe II and
[Fe II] are about 8000 K, value which is very much lower
than the mean ionization temperature of the elements in η
Car. E_{B-V} is about 1^m1; the absolute visual magnitude is
presently $-10^m5 \pm 1^m$. From the equivalent widths of the
emission lines the true continuum between 1.5 and $3.0\mu^{-1}$
has been derived. The corresponding $B - V$ is -0^m14, while
the Balmer jump is less than 0^m5. The continuum appears
mainly nebular in origin with a strong contribution of the
two-photon emission from the 2s-level of hydrogen in the
blue.

122.089 On the maximum of SZ Lyncis.
M. Wisse, P. N. J. Wisse.
Bull. Astron. Inst. Netherlands, Vol. 20, 333 - 334 (1969).
Note.

**122.090 Energy budget for the infrared radiation from Eta
Carinae.** B. E. J. Pagel. '
Astrophys. Letters, Vol. 4, 221 - 224 (1969).
When corrections are applied for interstellar and intrinsic absorption, the inferred ultraviolet radiation of η Carinae
carries about the right amount of energy to account for the
infrared emission by radiative heating of dust grains.

122.091 Optical observations of very young stars.
L. V. Kuhi.
Interstellar Ionized Hydrogen, Charlottesville 1967, p. 13 -
31 (1968).
I would like to discuss briefly some of the remaining
"optical" problems concerning the T Tauri stars and related
objects, and say a very few words also with regard to H II
regions and star formation.

**122.092 An unusual doubling of molecular lines in the spec-
trum of the long-period variable star R Leporis.**
J. G. Phillips, R. S. Freedman.
Publ. Astron. Soc. Pacific, Vol. 81, 521 - 526 (1969).
During the course of a study of molecular bands in the
spectra of carbon stars, our attention was drawn to the distinctly anomalous appearance of the spectrum of the long-
period variable star R Leporis. The spectrogram in question
was taken on September 26, 1964 when R Lep was near maximum light. Microphotometer measurements are reported and
the anomalous appearance of the spectrum described.

**122.093 Photoelectric observations of RZ Scuti, an unusual
binary star.** H. K. Hansen.
Publ. Astron. Soc. Pacific, Vol. 81, 540 - 541 (1969).
Abstract ASP.

122.094   **Intermediate- and narrow-band photometry of Cepheids.**   D. H. McNamara, D. Potter.
Publ. Astron. Soc. Pacific, Vol. 81, 545 - 546 (1969). Abstract ASP.

122.095   **The light-elements of an anonymous cepheid and on its probable membership in the galactic cluster NGC 6649.**   G. A. Tammann.
Astron. Astrophys. Vol. 3, 308 - 315 = Mitt. Astron. Inst. Univ. Basel No. 64 (1969).   In German.

Old and new observations are combined to derive the period ($P = 5\overset{d}{.}1180$) and light-elements of the suspected cluster variable in NGC 6649. The star is found to be a cepheid with a distance modulus of $(m - M)^0 = 11\overset{m}{.}58$ and an age of 5.0×10^7 years. Distance, colour excess, age, and the position of the star with the cluster area, strongly suggest that it is a member of the cluster although the distance of the cluster is only approximately determined due to insufficient observations.

122.096   **The light curve of l Carinae.**   A. Feinstein, J. C. Muzzio.
Astron. Astrophys. Vol. 3, 388 - 401 (1969).

From 178 photoelectric observations distributed over a 17 years interval, a period of $35\overset{d}{.}5330 \pm 0\overset{d}{.}00084$ has been derived for the brightest cepheid l Carinae. The residuals of these observations calculated from the best fitting Fourier series show periodicities of 190.1 and 29.92 days, both with an amplitude of the order of $0\overset{m}{.}02$ in V. There is another doubtful periodicity of 16.92 days. The principal period shows no continuous change in time larger than 0.001 day/year. Except for the secondary periods the curve seems to repeat itself closely over the whole time interval considered.

122.097   **Optical and radio study of BL Lacertae.**   C. Bertaud, B. Dumortier, P. Véron, G. Wlérick, G. Adam, J. Bigay, R. Garnier, M. Duruy.
Astron. Astrophys. Vol. 3, 436 - 442 (1969).   In French.

Frequent measurements of the flux of the variable object BL Lac have been made. The optical observations consist of two series: one series of photographs taken with Schmidt telescopes and one series of visual measurements. During the second semester of 1968, the brightness varied in an interval of 2 magnitudes and sometimes quick changes were observed. The interstellar absorption, in the direction of BL Lac has been estimated, by using the reddening of the photoelectric sequence stars. With the adopted value, $A_v = 1.2$, the position of BL Lac in the $(U - B, B - V)$ diagram is at the limit of the domain occupied by the quasars, near the variable object 3C 196. Also the slope of the continuous spectrum, is, after correction of the absorption, not significantly different of that of a quasar.

122.098   **On the interpretation of S Doradus.**   A. Martini.
Astron. Astrophys. Vol. 3, 443 - 451 (1969).

The purpose of this work is to give a detailed, spectroscopic study of S Dor, which is one of the brightest stars of the Large Magellanic Cloud, at the minimum and maximum phases and, as possible, to discuss a possible explanation for the behaviour of the star.

122.099   **Binary stars among cataclysmic variables. XI. Photoelectric and spectroscopic observations of the dwarf nova Z Camelopardalis.**   R. P. Kraft, W. Krzemiński, G. S. Mumford.
Astrophys. Journ. Vol. 158, 589 - 597 = Contr. Lick Obs. No. 296 (1969).

Z Cam is representative of a small subset of dwarf novae (U Gem stars) in which quasi-periodic outbursts of light are interrupted by intervals of relative quiescence. Like the more numerous "normal" group members, it is a semidetached spectroscopic binary (G1 + sdBe) of period $6^h 57^m 5$. We argue that the underlying blue star is a white dwarf with accretion heating as the energy source; this model is applied to U Gem stars as a group. Their lifetimes are estimated to be near 10^9 years, a value in agreement with the hypothesis that they descend from stars of the W UMa type.

122.100   **Der Lichtwechsel von Mira Ceti.**   P. L. Fischer.
Ann. Univ.-Sternw. Wien, Vol. 28, (No. 4), 137 - 209 (1969).

In this monograph about the visual light-variation of Mira Ceti it was tried to collect the whole material presented in the literature and kindly supplied by astronomical institutions.

122.101   **Absolute magnitudes of Mira variables from statistical parallaxes.**   M. L. Clayton, M. W. Feast.
Monthly Notices, Roy. Astron. Soc., Vol. 146, 411 - 421 (1969).

The McCormick proper motions are combined with the available radial velocities to derive absolute magnitudes for Mira variables as a function of period. The uncertainties in the results are estimated. The analysis is carried out both at mean maximum (M_m) and at mean light intensity (M_l) of the variables. Within the errors the absolute magnitudes vary smoothly with period from ~ 180 days to ~ 500 days (M_m, from -3.0 to -1.0 and M_l from -1.5 to $+1.0$). The shorter period Miras (~ 130 days) deviate from this period-luminosity relation. They have $M_m = -1.6$ and $M_l = -0.1$.

122.102   **Short-period variability of B, A, and F stars. III. A survey of Delta Scuti variable stars.**   M. Breger.
Astrophys. Journ. Suppl. Series, Vol. 19, 79 - 97 = Contr. Kitt Peak National Obs. No. 468 (1969). − Abstr. in Astrophys. Journ. Vol. 158, 431 (1969).

Two hundred and thirteen bright field stars were tested photoelectrically for short-period variability. Nineteen variables and one suspected variable were found. Narrow-band photometric and spectroscopic data have been collected for most of the thirty-nine δ Sct variables published so far. Determinations of absolute magnitudes for these stars are discussed, and new absolute magnitudes are calculated. Data on all stars inside the instability strip considered to be either variable or nonvariable have been collected, and a further list of suspected variables is given.

122.103   **Short-period variability of B, A, and F stars. IV. Variability in the lower Hertzsprung gap.**   M. Breger.
Astrophys. Journ. Suppl. Series, Vol. 19, 99 - 113 = Contr. Kitt Peak National Obs. No. 469 (1969). − Abstr. in Astrophys. Journ. Vol. 158, 431 - 432 (1969).

The lower instability strip extends from the main sequence upward to the RR Lyrae region. Hot and cool borders were established which intersect the main sequence at A4 and F2. In the instability strip, more than 20 percent of the stars show regular variability larger than 0.010 mag. These variables have normal masses and show a wide range in metal abundances and rotational velocities. A period-luminosity-color relation is established, with an average deviation of less than 0.3 mag in M_v. The observed coefficients of $\log P$ and $\log (T_e / T_{e\,\odot})$ agree adequately with the predicted relation between absolute magnitude, period, effective temperature and mass. It is proposed that the period differences among classical Cepheids, δ Sct stars, RR Lyrae stars, and dwarf Cepheids are caused not only by luminosity differences but also by large mass differences.

122.104   **Nine long period variable stars in the Cygnus cloud,**

VV 233-241. W. J. Miller.
Ric. Astron. Specola Vaticana, Vol. 7, (No. 18), 499 - 519 (1969).

A report is made on eight new long period variable stars and one old LPV, whose amplitudes range between a maximum of the 13th, and a minimum below the 19th photographic magnitude. The study is based on a total of 8136 Vatican, Hamburg, Heidelberg, Harvard and Mount Wilson and Palomar Observatory plates.

122.105 A (uvby) study of RR Lyrae field stars.
I. Epstein.
Astron. Journ., Vol. 74, 1131 - 1151 = Contr. Kitt Peak National Obs. No. 493 (1969).

Observations at minimum light in the four-color system of Strömgren and Perry of 38 variables, mostly of type RRa, are used to find interstellar absorption corrections in the four-color system. Correlations of the intrinsic metal index $(m_1)_0$ with kinematic properties of the variables indicate a separation of the RRa population into a relatively high metal-content group having a disk galactic distribution and into a low metal-content group having a spherical distribution. The low metal-content group variables show a period-luminosity relation in the sense of increasing luminosity with increasing period.

122.106 High-frequency stellar oscillations. II. G44–32 a new short-period blue variable star.
B. M. Lasker, J. E. Hesser.
Astrophys. Journ. (Letters), Vol. 158, L171 - L173 = Contr. Cerro Tololo Inter-American Obs. No. 93 (1969).

The proper-motion star G44–32 is a variable with amplitudes of about 2 percent and periods in the range from 10 to 27 min.

122.107 Incidence of short-period variability in the lower Hertzsprung Gap. M. Breger.
Bull. American Astron. Soc., Vol. 1, 337 (1969). – Abstract AAS.

122.108 The color-magnitude relation for novalike variables.
G. S. Mumford.
Bull. American Astron. Soc., Vol. 1, 355 (1969). – Abstr. AAS.

122.109 A mechanism of outbursts of U Geminorum stars.
Y. Osaki.
Bull. American Astron. Soc., Vol. 1, 357 (1969). – Abstr. AAS.

122.110 Physical properties of cepheids and other supergiants from six-color fitting.
S. B. Parsons.
Bull. American Astron. Soc., Vol. 1, 358 (1969). – Abstr. AAS.

122.111 New flare stars in the Pleiades region and in the Orion complex. L. Rosino, L. Pigatto.
Mem. Soc. Astron. Italiana, Nuova Serie, Vol. 40, 447 - 474 (1969).

Continuing the search for flare stars in program at Asiago, eighteen new flares have been discovered in the Pleiades with the 67 cm Schmidt telescope in 1968–69. Positions and magnitudes are given. Sixty new flares (of which 21 discovered during 1968–69) have been found in the Orion complex, including the Horsehead nebula. Identification charts are given. Positions and magnitudes are reported.

122.112 New photoelectric minima of W UMa and a possible interpretation of the period variations.
B. Cester.

Mem. Soc. Astron. Italiana, Nuova Serie, Vol. 40, 489 - 498 (1969).

Photoelectric times of primary minimum have been obtained at Trieste: the flat minimum lasts about 19 minutes but shows variations both in depth and in form. 369 minima observed since 1903 are considered to evaluate the period variations and their history can be divided into five time intervals. The presence of a third body is hardly acceptable.

122.113 A second series of observations of RU Cam.
B. Cester.
Mem. Soc. Astron. Italiana, Nuova Serie, Vol. 40, 525 - 528 (1969).

New observations of RU Cam made at Trieste from the end of 1967 and the first months of 1969 are reported. The oscillation amplitude still keeps very small (about 0.1 mag). The hypothesis of a beat phenomenon with a period 10 times the normal one is advanced.

122.114 The light variation and the period of RU Camelopardalis from November 1967 to August 1968.
P. Broglia, G. Guerrero.
Mem. Soc. Astron. Italiana, Nuova Serie, Vol. 40, 533 - 542 (1969).

Two colours photoelectric light curves of RU Cam concerning the interval November 1967 – August 1968 are presented. The mean brightness of the variable, constant during the 1966 – 1967 season, becomes gradually fainter of about $0.^m05$ in ten months. The amplitude of the light variation decreases slowly until March and is then followed by some peculiarities.

122.115 The intrinsic colour of l Carinae.
J. D. Fernie.
Observatory, Vol. 89, 206 (1969). – Letter.

122.116 On observations of the light polarization of AD Leo.
Yu. S. Efimov.
Izv. Krymskoj Astrofiz. Obs. Vol. 39, 42 - 43 (1969). In Russian.

Results of two-colour observations of polarization of AD Leo are given. A considerable difference between position angles of polarization planes in blue and yellow parts of the spectrum is found.

122.117 On stars of RW Aur type and irregular variables (Ia) of early spectral classes.
T. M. Rachkovskaya.
Izv. Krymskoj Astrofiz. Obs. Vol. 39, 96 - 107 (1969). In Russian.

Spectra of BU Tau and V923 Aql, stars of Ia type, and spectra of normal stars are compared on spectrograms obtained with dispersion 15 Å/mm. The difference in the spectra of comparable stars is explained by the fact that the variable stars BU Tau and V923 Aql have shells. The star V923 Aql has a deficit of hydrogen. Values of the electronic density $\lg n_e = 12.86$ and velocity $v_D = 36$ km/sec, obtained for the shell of V923 Aql, indicate that the shell of V923 Aql is more rarefied than the atmosphere of the supergiant 6 Cas A. Some results are deduced from a comparative investigation of 10 variable stars of RW Aur and Ia types. All investigated variables with the exception of 3 Be-stars are located at the lower boundary of the main sequence and below their comparison stars. This indicates the young age of these variables.

122.118 Flares of BD + 13°2618. N. I. Shakhovskaya.
Inform. Bull. Variable Stars (I.A.U. Commission 27), Konkoly Obs., Budapest, No. 361 (1969).

122.119 Spectroscopic investigation of the symbiotic stars AX Per, CI Cyg and BF Cyg. A. A. Boyarchuk.

Izv. Krymskoj Astrofiz. Obs. Vol. 39, 124 - 139 (1969).
In Russian.

Relative intensities of emission lines and the energy distribution in continuous spectra of AX Per, CI Cyg, and BF Cyg are obtained from 19 spectrograms (dispersion 80 Å/mm). The radial velocities, published earlier, are considered. It is shown that the data of observations can be explained by the supposition, that the stars investigated are binaries, consisting of a cold giant M5 III and a hot star of low luminosity. The both components are surrounded by a nebula with the mass of $\sim 10^{30}$ g and $n_e \sim 10^7$ cm^{-3}.

122.120 **An estimation of an upper limit of the X-ray flux of the flares of UV Cet type stars from a sudden cosmic noice absorption effect.**
R. E. Gershberg, Y. I. Neshpor, P. F. Chugainov.
Izv. Krymskoj Astrofiz. Obs. Vol. 39, 140 - 146 (1969).
In Russian.

A search for a sudden cosmic noice absorption during the flares of UV Cet type stars has been carried out. No effect is found. Supposing that the effect is smaller than the fluctuations of cosmic noice registrograms, the upper limit of a possible X-ray flux of the UV Cet type star flares is estimated. During the strong flare of EV Lac (18. 8. 1960) $F_X < 10^{-4}$ erg/cm^2 sec, and the ratio of a possible X-ray flux of the flare to the flux at the optical band B is $<10^5$. These estimations do not allow to make a choice between the possible mechanisms of the supposed X-radiation of stellar flares.

122.121 **Cepheids of W Vir type.** V. P. Tsesevich.
Zemlya i Vselennaya, No. 5, p. 86 - 89 (1969).
In Russian.

122.122 **On the light curves of flare stars.**
G. A. Gurzadian.
Astrofizika, Vol. 5, 383 - 392 (1969). In Russian. – Engl. translation in Astrophysics, Vol. 5, No. 3.

The usual law $J \sim e^{-\beta t}$ for the decrease of brightness after a maximum of a flare disagrees with the observations. In this paper a new interpretation of the light curves of the flare stars, based on the hypothesis of "rapid electrons", has been suggested. It is shown that the decrease of brightness is caused only by expansion of the shell of rapid electrons around the star; the effect of the energy loss of the electrons is negligible. The theoretical light curve is presented. These formulas explain the observed curves from the maximum till the end of the curve fairly well. The "dynamic parameters" of the flare are introduced; their numerical values are presented in a table for some flares.

122.123 **Two remarkable flare stars in the Pleiades.**
E. S. Parsamian, H. S. Chavushian.
Astrofizika, Vol. 5, 499 - 502 (1969). In Russian. – Engl. translation in Astrophysics, Vol. 5, No. 3.

The flare stars N 101 and H II 357 in the Pleiades flare with higher frequency than Pleiades flare stars in the mean. It is shown that both of them concern the very small group of flare stars in the Pleiades with $\nu_2^{-1} \sim 120^h$. The members of this group can be of early and late spectral types.

122.124 **Radio emission from flare stars near the Orion nebula.** O. B. Slee, C. S. Higgins, C. Roslund, G. Lyngå.
Nature, Vol. 224, 1087 - 1089 (1969).

Simultaneous optical and radio observations of flare stars near the Orion nebula indicate that many of the flares are probably accompanied by detectable radio emission at metre wavelengths. Three, out of a total of nine flares detected during the experiment, were each observed at two radio frequencies (136 MHz and 408 MHz). For two of these, the peak flux density was at least five times stronger at the lower

frequency, while the reverse was true for the third flare. Consideration of the optical energy emitted during a 3^m flare on a dK5 star (typical Orion flare star) indicates that such a flare is probably about a thousand-fold more energetic than a typical flare on the classical UV Ceti type flare stars near the sun.

122.125 **An unusual flare of AD Leo.**
A. H. Jarrett, J. P. Eksteen.
Monthly Notes Astron. Soc. Southern Africa, Vol. 28, 70 - 71 (1969).

122.126 **S Muscae.** A. W. J. Cousins.
Monthly Notes Astron. Soc. Southern Africa, Vol. 28, 104 - 105 (1969). – Note.

122.127 **Photometric observations of V 1216 Sagittarii.**
A. H. Jarrett, J. P. Eksteen.
Monthly Notes Astron. Soc. Southern Africa, Vol. 28, 131 - 133 (1969).

122.128 **On the age of the δ Cephei stars.**
I. Semeniuk.
Postępy Astron., Vol. 17, 399 - 403 (1969). In Polish.

122.129 **Sugli spettri di WY Gem, W Cep e EG And nel 1966-68.** A. Mammano, A. Martini.
Atti XII Riunione Soc. Astron. Italiana, L'Aquila 1968, p. 25 - 26 (1969). – Abstract SAI.

122.130 **Le variazioni delle velocità radiali e spettrali di CU Vir.** C. Aydin.
Atti XII Riunione Soc. Astron. Italiana, L'Aquila 1968, p. 27 - 28 (1969). – Abstract SAI.

122.131 **Lectures on variable stars.**
R. F. Christy.
Journ. Roy. Astron. Soc. Canada, Vol. 63, 299 - 307 (1969).
Chapter I. The linear theory. Presented is a synopsis of some aspects of the problem of the lunar theory of pulsation. Chapter II. The non-linear theory.

122.132 **Two groups of dwarf cepheids.**
M. S. Frolov.
Astron. Tsirk. No. 528, p. 3 - 4 (1969). In Russian.

122.133 **Variable stars in the region around the globular cluster M 92.**
R. G. Mnatsakanian, K. A. Sahakian.
Astron. Tsirk. No. 528, p. 5 - 7 (1969). In Russian.

122.134 **Maxima of long-period variable stars.**
A. G. Nudzhenko.
Astron. Tsirk. No. 528, p. 7 (1969). In Russian.

122.135 **DH and GM Aquilae.**
V. P. Tsessevich.
Astron. Tsirk. No. 529, p. 7 (1969). In Russian.

122.136 **AO Aquilae is a Mira Ceti star with variable period.**
V. P. Tsessevich.
Astron. Tsirk. No. 529, p. 8 (1969). In Russian.

122.137 **The search for Hα emission in spectra of rapid irregular variables.** G. A. Ponomareva.
Astron. Tsirk. No. 532, p. 1 - 3 (1969). In Russian.

122.138 **On the magnetic field of exploding variables.**
E. R. Mustel.
Astron. Tsirk. No. 534, p. 1 - 3 (1969). In Russian.

122.139   On the star No. 98 in NGC 2360.
T. I. Barblisvili.
Astron. Tsirk. No. 535, p. 8 (1969).   In Russian.

122.140   New carbon variable stars.
A. Alksnis, Z. Alksne.
Astron. Tsirk. No. 538, p. 7 - 8 (1969).   In Russian.

122.141   Radii of five RR Lyr type stars.   M. S. Frolov.
Peremennye Zvezdy, Byull., Vol. 16, 615 - 618
(1969).   In Russian.
Using Wesselink's method modified by the author, mean radii, absolute and relative amplitudes of the radial variations during a pulsation cycle were determined for the five RR Lyrae stars V499 Cen, SV Hya, VY Lib, HH Pup, and AF Vel. For V499 Cen the luminosity $M = +0^m\!.64$ was obtained.

122.142   On four variables of RR Lyr type in Ophiuchus.
O. E. Mandel.
Peremennye Zvezdy, Byull., Vol. 16, 628 - 649 (1969).
In Russian.
Elements were derived and light curves compiled for V773, V788, V822, and V829 Oph from visual and photographic observations. For the last three stars the Blazhko effect is probable.

122.143   Flares of UV Ceti.
S. S. Vykhrestiuk, T. M. Ishchenko, N. S. Komarov, V. I. Konyshev, Yu. E. Migach.
Peremennye Zvezdy, Byull., Vol. 16, 670 - 681 (1969).
In Russian.

122.144   Flare stars in the Pleiades region.
G. Haro, E. Chavira.
Bol. Obs. Tonantzintla y Tacubaya, Vol. 5, (No. 31), 23 - 34 (1969).
Following the search for flare stars in the Pleiades region previously reported (Haro, 1968), at the Tonantzintla Observatory we have continued the recollection and analysis of multiple exposure photographic material on the aforementioned region. The present paper deals mainly with the results corresponding to the observational period October 1965 – November 1967. A total number of 112 flare stars from which 32 have known proper motions are listed. The membership of these stars to the Pleiades is studied. Spectral types and some physical characteristics are discussed.

122.145   New flare stars in the Pleiades.
E. Parsamian, E. Chavira.
Bol. Obs. Tonantzintla y Tacubaya, Vol. 5, (No. 31), 35 - 40 (1969).
Photographic observations of the Pleiades region from 1968 October to 1969 January are described. The results for 17 flare stars are presented and discussed in detail.

122.146   H II 2411, a Hyades flare star.
G. Haro, E. Parsamian.
Bol. Obs. Tonantzintla y Tacubaya, Vol. 5, (No. 31), 41 - 44 (1969).
The collection of photographic material from February 1963 to January 1969 is analysed and the results are communicated. Among all the Pleiades field flare stars, H II 2411 was one of the more active.

122.147   A new "slow" flare star in Orion.
G. Haro, E. Parsamian.
Bol. Obs. Tonantzintla y Tacubaya, Vol. 5, (No. 31), 45 - 53 (1969).
We are dealing mainly with one "new" flare star of the "slow" type, which shows a large amplitude range of variation in the ultraviolet, and 3 more flare stars of the "fast" type

in its immediate vicinity. The area is localized in an obscure section, 1.25 degrees southwest from the Trapezium. Everything seems to indicate that these 4 flare stars and the irregular variable SW Ori, localized in the same small area, belong to the complex Orion association.

**122.148   Photoelectric observations of the flare stars YZ
CMi and AD Leo.**   R. B. Herr.
Inform. Bull. Variable Stars (I.A.U. Commission 27), Konkoly Obs., Budapest, No. 363 (1969).

122.149   Flare photometry of AD Leo.
A. H. Jarrett, J. P. Eksteen.
Inform. Bull. Variable Stars (I.A.U. Commission 27), Konkoly Obs., Budapest, No. 364 (1969).

**122.150   Evidence for very strong light-curve variation of
the Cepheid IU Cyg.**   G. A. Tammann.
Inform. Bull. Variable Stars (I.A.U. Commission 27), Konkoly Obs., Budapest, No. 366 (1969).

122.151   Remarks on the light-elements of CS Mon.
G. A. Tammann.
Inform. Bull. Variable Stars (I.A.U. Commission 27), Konkoly Obs. Budapest, No. 366 (1969).

**122.152   Report on the observations of the flare star AD Leo
obtained during 1969.**
S. Cristaldi, M. Narbone, M. Rodono.
Inform. Bull. Variable Stars (I.A.U. Commission 27), Konkoly Obs., Budapest, No. 367 (1969).

122.153   A comparison sequence to the flare star EV Lacertae.
A. D. Andrews, P. F. Chugainov.
Inform. Bull. Variable Stars (I.A.U. Commission 27), Konkoly Obs., Budapest, No. 370 (1969).

122.154   Flare photometry of V1216 Sagittarii.
A. H. Jarrett, J. P. Eksteen.
Inform. Bull. Variable Stars (I.A.U. Commission 27), Konkoly Obs., Budapest, No. 379 (1969).

122.155   UBV photometry of Gamma Pegasi.
M. Jerzykiewicz.
Inform. Bull. Variable Stars (I.A.U. Commission 27), Konkoly Obs., Budapest, No. 382 (1969).

122.156   An RR Lyrae star with a changing period.
J. Akyuz.
Inform. Bull. Variable Stars (I.A.U. Commission 27), Konkoly Obs., Budapest, No. 395 (1969).

122.157   EV Lac.
K. Osawa, T. Noguchi, T. Okada, K. Ichimura, E. Watanabe, K. Okida.
Inform. Bull. Variable Stars (I.A.U. Commission 27), Konkoly Obs., Budapest, No. 399 (1969).

122.158   Photoelectric observations of EV Lac.
K. L. Maslennikov, N. I. Shakhovskaya.
Inform. Bull. Variable Stars (I.A.U. Commission 27), Konkoly Obs., Budapest, No. 401 (1969).
The results of photoelectric observations of the flare star EV Lac carried out at the Crimean Astrophysical Observatory in the period of 6 - 26 August 1969 are given here.

**122.159   Observations of EV Lac during the international
campaign September 4 - 19, 1969.**
S. Cristaldi, M. Rodono.
Inform. Bull. Variable Stars (I.A.U. Commission 27), Konkoly Obs., Budapest, No. 403 (1969).

122.160 Observations of UV Cet during the international campaign October 3 - 18, 1969. ·
S. Cristaldi, M. Rodono.
Inform. Bull. Variable Stars (I.A.U. Commission 27), Konkoly Obs., Budapest, No. 404 (1969).

122.161 UV Ceti.
K. Osawa, T. Noguchi, T. Okada, K. Ichimura, E. Watanabe, K. Okida.
Inform. Bull. Variable Stars (I.A.U. Commission 27), Konkoly Obs., Budapest, No. 405 (1969).

122.162 Flare activity of UV Ceti.
A. H. Jarrett, J. P. Eksteen.
Inform. Bull. Variable Stars (I.A.U. Commission 27), Konkoly Obs., Budapest, No. 406 (1969).

122.163 Visual observations of the flare star EV Lacertae.
A. D. Andrews.
Inform. Bull. Variable Stars (I.A.U. Commission 27), Konkoly Obs., Budapest, No. 407 (1969).

122.164 Photoelectric observations of EV Lac.
P. F. Chugainov, N. I. Shakhovskaya.
Inform. Bull. Variable Stars (I.A.U. Commission 27), Konkoly Obs., Budapest, No. 410 (1969).

122.165 Photoelectric observations of UV Ceti.
P. F. Chugainov, N. I. Shakhovskaya.
Inform. Bull. Variable Stars (I.A.U. Commission 27), Konkoly Obs., Budapest, No. 411 (1969).

122.166 Spectrophotometry of EW Lac. I.
I. D. Kupo.
Trudy Astrofiz. Inst. Alma-Ata, Vol. 14, 32 - 49 (1969). In Russian.

Variations of the spectrophotometric gradients of EW Lac during the years 1963–1964 are given. It is shown that this quantity varies in the same sense in the photographic and visual region. An identification of the shell lines and a definition of electron density by the Inglis-Teller method was executed. Variations of the electron density were noted. It was found that the dependence between the logarithm of the inverse residual central intensity and the quantum number of the lines is composed of two parts with different slopes, what testifies that the hydrogen lines arise in two layers of the shell, abruptly differing in electron density. Line profiles of $H\beta$ and $H\gamma$ are given.

122.167 R Coronae Borealis. W. Wichmann.
VdS Nachrichtenblatt, 18. Jahrgang, 144 (1969).

122.168 SX Cancri. I. Todoran.
Stud. Cerc. Astron., Vol. 14, 105 - 116 (1969).
In Rumanian. - Photometric observations are presented and light curves derived in two colors.

122.169 Observaţii fotografice ale stelei variabile XY Eridani.
I. Todoran.
Stud. Cerc. Astron., Vol. 14, 117 - 122 (1969).

Photometric observations are presented, and a light curve is derived.

122.170 Photoelektrische Beobachtungen an SV Cephei.
W. Wenzel.
MVS Sonneberg, Vol. 5, 75 - 82 (1969).

Photoelectric UBV-observations of SV Cep, which is possibly a member of the T-association Cep T2 and of the OB-association I Cep, show quasi-periodic minima superposed on a long-term wave-shaped "normal light". In the course of these minima the colours B–V and U–B are essentially cons-

tant. Extinction processes in a rotating, nonspherical or cloudy envelop may cause the observed effects.

122.171 Photoelectric observations of RR Lyrae.
B. Onderlička, M. Veteśnik.
Mem.Observations Czechoslovak Astron. Soc., Czechoslovak Acad. Sci., No. 13, 17 pp. = Astron. Inst. Univ. Brno (Czechoslovakia), Publ. No. 8 (1968).

In 1961, photoelectric observations of RR Lyrae were started in University Observatory at Brno in order to extend the extensive photometric material of Walraven (1949) for a detailed study of the periodic terms in the light of this variable star. This paper contains the first part of our results, the observing material and a short description of our observing method.

122.172 A search for flare stars in the Orion nebula region.
C. Roslund.
Ark. Astron. Vol. 5, 381 - 385 = Uppsala Astron. Obs. Medd. No. 169 (1969).

A systematic search for flare type variables in the region of the Orion nebula has been carried out with the 20/26-inch Schmidt telescope of the Uppsala Southern Station at Mount Stromlo. Fifteen stars showing rapid flare-ups have been detected on plates taken in blue light during a total of nearly forty hours of effective observation time. Characteristics of the flares are given together with the positions of the stars.

122.173 Atmospheric depth effects during rising light of the 45-day cepheid, SV Vulpeculae.
T. C. Grenfell, G. Wallerstein.
Publ. Astron. Soc. Pacific, Vol. 81, 732 - 740 (1969).

We have studied spectroscopic changes in the 45-day cepheid SV Vulpeculae in the red region of the spectrum. Our radial velocities of Si II can be combined with measurements of other elements to indicate that a progressive wave is seen in the atmosphere during rising light. The observations clearly show that lines formed deepest in the atmosphere are first accelerated, and that the acceleration successively affects lines formed at higher levels. The profile of $H\alpha$ is complex and double, both at minimum light and again starting near maximum and lasting beyond phase 0.11. The $H\alpha$ radial velocities do not fit a simple kinematic model.

122.174 Photoelectric observations of BX Delphini.
B. Basu.
Publ. Astron. Soc. Pacific, Vol. 81, 834 - 839 (1969).

BX Delphini, classified by Hoffmeister as a classical cepheid with the shortest known period, was observed in the *UBV* system. The color-color diagram of BX Del lies entirely below that of the main-sequence stars, which confirms Preston's observation that BX Del is a strong-lined star with no spectral peculiarity.

122.175 Two new southern δ Scuti variables.
S. Demers.
Publ. Astron. Soc. Pacific, Vol. 81, 861 - 863 = Contr. Cerro Tololo Inter-American Obs. No. 82 (1969).

Two new variables of the δ Scuti type were discovered during a search among bright southern stars. HD 9133, with a period of 66 minutes, has one of the shortest periods in its class.

122.176 A new flare star in Pavo.
G. S. Mumford.
Publ. Astron. Soc. Pacific, Vol. 81, 890 - 894 = Contr. Cerro Tololo Inter-American Obs. No. 86 (1969).

The star S 5114 = K3π 5235 Pavonis underwent a flare some two magnitudes in extent in blue light on the night of July 8, 1969. The event lasted about 15 minutes. Both spectra and color indices indicate the object to be a late-type dwarf.

Hence it is likely a UV Ceti star rather than a U Geminorum variable as previously classified.

122.177 Langfristige Perioden-Kontrolle für Cepheiden der Bamberger Liste. H. Bauernfeind.
Veröff. Remeis-Sternw. Bamberg, Astron. Inst. Univ. Erlangen-Nürnberg, Vol. 8, No. 85, 21 pp. (1969).

Presented are the results of a discussion of the observations of 9 cepheids.

122.178 The radial velocity curve of SX Phoenicis. J. Stock, S. Tapia.
Inform. Bull. Southern Hemisphere, No. 14, p. 39 - 41 (1969).

The short-period variable star SX Phoenicis was discovered by Eggen in 1952. From 1964 until 1967 extensive photoelectric three-colour observations were obtained by Tapia. These, in combination with the earlier observations by other authors, are discussed and a light curve is derived.

122.179 Emission lines of Mn I in long-period variables. G. H. Herbig.
Bull. Soc. Roy. Sci. Liège, Vol. 37, 433 - 436 = Contr. Lick Obs. No. 287 (1968).

122.180 Some relation between the photometric characteristics of light curves and colour indices of cepheids in the UBV system. N. Nikolov, A. Nikolov.
Izv. Sekts. astron.B"lg. AN, Vol. 3, 125 - 136 (1969).
In Russian. − Abstr. in Referativ. Zhurn. 51. Astron., 2.51.557 (1970).

122.181 Intrinsic reddening of Eta Carinae. D. L. Lambert.
Nature, Vol. 223, 726 - 727 (1969), with comments by B. E. J. Pagel, p. 727 - 728 (1969).

Pagel has proposed a new method for the determination of the intrinsic and interstellar reddening for peculiar objects which is based on the intensities of the forbidden lines of ionized iron, and has applied the method to Eta Cariane. It is demonstrated that Pagel's method may be revised.

Synchronous three colour stellar photometry at the Catania Astrophysical Observatory. See Abstr. 034.027.

The formation and growth of carbon particles in the atmospheres of Mira variables. See Abstr. 064.037.

An explanation for the blue sequence of variable stars. See Abstr. 065.021.

Nonlinear periodic pulsations of stars. See Abstr. 065.099.

Proper motions of four variable stars of U Geminorum-type and of stars in their neighbourhood. See Abstr. 112.021.

Polarimetric observations of late-type dwarfs. See Abstr. 113.010.

Term analysis of Fe IV and identification of [Fe IV] in RR Telescopii. See Abstr. 114.010.

Mass loss from P Cygni. See Abstr. 114.042.

Observations of the infrared object, VY Canis Majoris. See Abstr. 114.065.

The absolute period-luminosity for cepheids. See Abstr. 115.013.

The period-luminosity relation: A historical review. See Abstr. 115.018.

Eine Nullpunktbestimmung der Perioden-Leuchtkraft-Beziehung von Cepheiden der Population I aus Eigenbewegungen und Radialgeschwindigkeiten. See Abstr. 115.019.

The double cepheid CE Cassiopeiae in NGC 7790: Tests of the theory of the instability strip and the calibration of the period-luminosity-color relation. See Abstr. 153.009.

A study of the periods of some variable stars of the globular cluster M 92. See Abstr. 154.016.

Period changes of RR Lyrae variables in the globular cluster Messier 5. See Abstr. 154.017.

Two-colour photometry of Cepheids near the centre of the Small Magellanic Cloud, based on plates taken by Dr. H. C. Arp. See Abstr. 159.004.

A comparative photometric study of population I Cepheids in the Magellanic Clouds, the galaxy and the Andromeda nebula. See Abstr. 159.005.

Further observations of Magellanic Cloud cepheids. See Abstr. 159.008.

123 Variable Stars: Lists of Observations, Individual Observations

123.001 **Waarnemingen van veranderlijke sterren in 1968.**
G. Comello, H. Feyth.
Hemel en Dampkring, Vol. 67, 288 - 296 (1969).

123.002 **Résultats des observations d'étoiles variables à éclipse.** K. Locher, R. Diethelm.
Orion Schaffhausen, Vol. 14, 109 - 110 (1969).

123.003 **The new variable star SVS 1610 UMa.**
B. V. Kukarkin.
Astron. Tsirk. No. 508, p. 8 (1969). In Russian.

123.004 **Six new variable stars in Virgo, Serpens and Libra.**
R. M. Russev.
Astron. Tsirk. No. 517, p. 7 (1969). In Russian.

123.005 **The elements of 12 variables in Aquila.**
V. P. Tsesevitch.
Astron. Tsirk. No. 517, p. 8 (1969). In Russian.

123.006 **Three variable stars in Taurus.**
N. B. Perova.
Astron. Tsirk. No. 520, p. 6 - 7 (1969). In Russian.

123.007 **About a new variable in the region of NGC 7023.**
E. S. Parsamian.
Astron. Tsirk. No. 524, p. 8 (1969). In Russian.

123.008 **Nine new variable stars.**
V. P. Tsesevich.
Astron. Tsirk. No. 526, p. 7 (1969). In Russian.

123.009 **CR Cephei und CZ Lacertae.**
M. B. Girnyak.
Tsirk. L'vov. Astron. Obs. No. 43, p. 10 - 13 (1969). In Russian.

123.010 **Beobachtungen des Veränderlichen SZ Lyrae.**
V. V. Golovatyj.
Tsirk. L'vov. Astron. Obs. No. 43, p. 14 - 16 (1969). In Russian.

123.011 **AZ Herculis.** A. T. Dul'tsev.
Tsirk. L'vov. Astron. Obs. No. 43, p. 17 - 20 (1969). In Russian.

123.012 **Photometric observatons of RV Pictoris.**
G. F. G. Knipe.
Republic Obs. Johannesburg, Circ. Vol. 7 (No. 128), 198 (1969).

123.013 **A minimum of U Ophiuchi.** G. F. G. Knipe.
Republic Obs. Johannesburg, Circ. Vol. 7 (No. 128), 198 - 199 (1969).

123.014 **Osservazioni di stelle variabili nei campi attorno a Omicron Persei e k Andromedae.**
G. Romano, M. Perissinotto.
Mem. Soc. Astron. Italiana, Nuova Serie, Vol. 40, 261 - 269 (1969).
Results of the photographic observations of 27 variable stars in two fields around o Persei and k Andromedae are reported in this paper.

123.015 **Variable star notes.** M. W. Mayall.
Journ. Roy. Astron. Soc. Canada, Vol. 63, 271 - 274 (1969).

123.016 **Nederlandse Verenigung voor Weer- en Sterrenkunde. Observations of variable Stars. Report No. 16.**
L. Plaut, H. Feyth.
Kapteyn Astron. Lab., Groningen – Netherlands. 9 pp. (1969).
2622 observations of 69 variable stars, January – June 1969.

123.017 **Variable star notes.**
M. W. Mayall.
Journ. Roy. Astron. Soc. Canada, Vol. 63, 221 - 224 (1969).

123.018 **Researches with the Schmidt telescopes. III. – Variable stars in the field of Gamma Cygni.**
G. Romano.
Mem. Soc. Astron. Italiana, Nuova Serie, Vol. 40, 375 - 413 (1969).
The results of photographic observations of 111 variable stars in the field of Gamma Cygni are given in this paper. Forty-four new variable stars have been discovered in plates taken with the Schmidt telescopes of the Padua-Asiago Observatory.

123.019 **Risultati delle osservazioni di stelle variabili ad eclisse.** K. Locher.
Orion Schaffhausen, Vol. 14, 159 - 160 (1969).

123.020 **Results of the Bamberg southern hemisphere sky patrol.** W. Strohmeier, R. Knigge.
Monthly Notes Astron. Soc. Southern Africa, Vol. 28, 75 - 78 (1969).

123.021 **RY Sagittarii.** R. P. de Kock.
Monthly Notes Astron. Soc. Southern Africa, Vol. 28, 102 (1969).
Light curve of the irregular variable from June 1967 to August 1969.

123.022 **Observations of variable stars in 1962.**
Mem. Japan Astron. Study Ass., Vol. 3, (No. 12), 143 - 166 (1969). In Japanese.

123.023 **Osservazioni fotoelettriche du RU Cam nel 1968.**
B. Cester.
Atti XII Riunione Soc. Astron. Italiana, L'Aquila 1968, p. 100 (1969). – Abstract SAI.

123.024 **Two new variable stars in Taurus.**
N. B. Perova.
Astron. Tsirk. No. 530, p. 6 - 7 (1969). In Russian.

123.025 **Elements of 42 long-period variables.**
V. P. Tsessevich.
Astron. Tsirk. No. 531, p. 6 - 7 (1969). In Russian.

123.026 **CSV 5879 is a new bright ultra-short period variable.** G. A. Lange.
Astron. Tsirk. No. 534, p. 7 (1969). In Russian.

123.027 **New variable SVS 1635 Tauri.**
A. N. Kulapova.
Astron. Tsirk. No. 539, p. 7 - 8 (1969). In Russian.

123.028 **Variable star notes.**
M. W. Mayall.
Journ. Roy. Astron. Soc. Canada, Vol. 63, 321 - 324 (1969).

123.029 **Study of four variable stars.** M. Gelfedinov.
Peremennye Zvezdy, Byull., Vol. 16, 619 - 627

(1969). In Russian.

Brightness estimations of CN Cep, V344 Cas, and the variable SVS 1509, the latter discovered by the author, were obtained on 170 plates, estimations of SVS 1508 on 86 plates of Sternberg Astronomical Institute. Hoffmeister's elements for CN Cep were corrected, elements for SVS 1509 and V344 Cas were derived, the type of SVS 1508 was not recognized.

123.030 Study of four red variable stars.
F. I. Lukatskaya.
Peremennye Zvezdy, Byull., Vol. 16, 650 - 662 (1969).
In Russian.

123.031 On DS Aquarii and V738 Sagittarii.
G. E. Erleksova.
Peremennye Zvezdy, Byull., Vol. 16, 663 - 670 (1969).
In Russian.

123.032 SY Aurigae. I. V. Sokolov.
Peremennye Zvezdy, Byull., Vol. 16, 682 - 683
(1969). In Russian.

123.033 Light curves of R CrA, S CrA, T CrA.
I. G. Sdanchuk.
Peremennye Zvezdy, Byull., Vol. 16, 683 - 696 (1969).
In Russian.

123.034 SS Cygni. G. Darsenius.
Astron. Tidssk., Årg. 2, 136 - 137 (1969).

123.035 Oversigt over Astronomisk Selskabs observationer af variable stjerner 1966.0 - 1968.0.
O. Klinting.
Astron. Tidssk., Årg. 2, 138 - 139 (1969).

123.036 The magnitude, color and spectrum of RV Sgr.
T.-K. Tan, B. Hidajat.
Inform. Bull. Variable Stars (I.A.U. Commission 27), Konkoly Obs., Budapest, No. 362 (1969).

123.037 T Cassiopeiae: Its light curve and its neighbor.
J. Ashbrook.
Sky Telescope, Vol. 38, 436 (1969).

123.038 On the star N 98 in NGC 2360.
T. J. Barblishvili.
Inform. Bull. Variable Stars (I.A.U. Commission 27), Konkoly Obs., Budapest, No. 374 (1969).

123.039 Possible outburst of HD 108486, observed 1965 April 5. G. Jackisch.
Inform. Bull. Variable Stars (I.A.U. Commission 27), Konkoly Obs., Budapest, No. 375 (1969).

123.040 V 348 Sgr brought up to date. N. J. Gregg.
Inform. Bull. Variable Stars (I.A.U. Commission 27), Konkoly Obs., Budapest, No. 377 (1969).

123.041 BD −6°4932, a new variable star.
D. S. Hall, A. D. Mallama.
Inform. Bull. Variable Stars (I.A.U. Commission 27), Konkoly Obs., Budapest, No. 383 (1969).

123.042 HBV 475: A new peculiar emission object in Cygnus.
L. Kohoutek.
Inform. Bull. Variable Stars (I.A.U. Commission 27), Konkoly Obs., Budapest, No. 384 (1969).

123.043 Observations of AB Comae. D. Hoffleit.
Inform. Bull. Variable Stars (I.A.U. Commission 27),
Konkoly Obs., Budapest, No. 385 (1969).

123.044 BD +18°4586. P. Moore.
Inform. Bull. Variable Stars (I.A.U. Commission 27),
Konkoly Obs., Budapest, No. 385 (1969).

123.045 Note on RU Cam. B. Szeidl.
Inform. Bull. Variable Stars (I.A.U. Commission 27),
Konkoly Obs., Budapest, No. 385 (1969).

123.046 Periods for twelve new variable stars in Sagittarius.
D. Hoffleit.
Inform. Bull. Variable Stars (I.A.U. Commission 27), Konkoly
Obs., Budapest, No. 387 (1969).

123.047 Etoiles variables nouvelles au nord de Beta Tauri.
A. Brun.
Inform. Bull. Variable Stars (I.A.U. Commission 27), Konkoly
Obs., Budapest, No. 409 (1969).

123.048 Bearbeitung von 105 Veränderlichen am Südhimmel, (Feld α Pavonis). I. Meininger.
MVS Sonneberg, Vol. 5, 83 - 85 (1969). – Concerning
S6824 - 6928.

123.049 Beobachtungen an Veränderlichen Sternen.
H.-J. Blasberg.
MVS Sonneberg, Vol. 5, 85 (1969).

123.050 Neuentdeckte Veränderliche (S 10504 bis 10618).
G. A. Richter.
MVS Sonneberg, Vol. 5, 88 - 93 (1969).

123.051 Photographische Beobachtungen von Veränderlichen auf Platten der Sonneberger Himmelsüberwachung. E. Splittgerber.
MVS Sonneberg, Vol. 5, 93 - 95 (1969).

123.052 Mitteilungen über Veränderliche der Bamberger Liste. H. Bauernfeind.
Veröff. Remeis-Sternw. Bamberg, Astron. Inst. Univ. Erlangen–Nürnberg, Vol. 8, No. 84, 14 pp. (1969).
Observations of 35 fainter variables in the southern sky.

123.053 Maxima of χ Cygni. F. Vaclík.
Říše hvězd, Vol. 50, 139 (1969). In Czech.
Visual observations obtained during the years 1959 – 1960 and 1968 are presented. Data of maxima, brightness of maxima and periods are given.

123.054 Observation of RU Camelopardalis.
F. Vaclík.
Říše hvězd, Vol. 50, 219 (1969). In Czech.
Visual observations obtained during March–June, 1969 are presented.

123.055 Probable new variable stars.
G. E. D. Alcock, M. J. Gainsford, J. E. Isles, J. Muirden, R. W. Payne, A. L. Smith, E. J. Voss.
British Astron. Ass. Circ. No. 510 (1969).

123.056 Bearbeitung von 150 Veränderlichen am Südhimmel, (Feld ε Pavonis). H. Geßner.
MVS Sonneberg, Vol. 5, 96 - 98 (1969). – Concerning
S6929 - 7078.

Bulletin of the Astronomical Institutes of the Netherlands 1921 - 1969. Index of variable stars.
See Abstr. 002.014.

124 Novae

124.001 Radiative transfer in novae.
J. P. Babuel-Peyrissac, G. Rouvillois.
Astron. Astrophys. Vol. 3, 31 - 41 (1969). In French.

We study radiative transfer in the interior of two concentric spheres in motion, supposing the density of the residual gas to be low. This scheme corresponds to a physical model often used in the study of novae with an internal star radiating like a black body and an external mantle in rapid expansion. Formulae allow full treatment of the problem on an electronic computer.

124.002 Nova in the Andromeda nebula (M 31).
A. K. Alksnis, A. S. Sharov.
Astron. Tsirk. No. 507, p. 1 (1969). In Russian.

124.003 Two novae in the Andromeda nebula (M 31).
A. S. Sharov, A. K. Alksnis.
Astron. Tsirk. No. 514, p. 1 - 2 (1969). In Russian.

124.004 On the frequency of nova outbursts in the Andromeda nebula. A. S. Sharov, A. K. Alksnis.
Astron. Tsirk. No. 517, p. 4 - 5 (1969). In Russian.

124.005 The radiation of novae before the light maximum.
V. V. Leonov.
Astrofizika, Vol. 5, 55 - 66 (1969). In Russian.
English translation in Astrophysics, Vol. 5, No. 1 (1969).

The sources of the radiation of a nova during the period before light maximum are considered. These sources are the heat energy of the envelope, the radiation of the star, the energy arising from the collision of the envelope with the matter ejected from the star. The calculated light curves are compared with observational data for CP Lac, V476 Cyg, DQ Her.

124.006 Fresh evidence concerning the type of ejection of novae. M. Friedjung.
Mass Loss from Stars, Trieste 1968, p. 303 - 310 (1969).

Reasons are firstly summarized for believing that ejection of matter by novae continues long after maximum light. Preliminary conclusions of a study of Balmer absorption-line profiles of nova Herculis 1963 are also given, the methods used being capable of indicating the nature of absorption systems.

124.007 A theoretical study of galactic novae.
V. V. Sobolev.
Vistas in Astronomy, Vol. 11, 181 - 188 (1969).

The present article briefly sets forth the results of some theoretical papers on the study of novae. The following aspects are considered: Detachment of the envelope from the star; The envelope-luminosity in the first period of the outburst; Motion of the envelope; Intensities of emission lines.

124.008 A nova in Sagittarius. F. M. Bateson.
Inform. Bull. Variable Stars (I.A.U. Commission 27), Konkoly Obs., Budapest, No. 389 (1969).

124.009 Nova in Sagittarius. M. P. Candy.
Inform. Bull. Variable Stars (I.A.U. Commission 27), Konkoly Obs., Budapest No. 408 (1969).

124.010 A nova in Corona Australis. N. Sanduleak.
Inform. Bull. Variable Stars (I.A.U. Commission 27), Konkoly Obs., Budapest, No. 368 (1969).

124.011 Probable nova in M 33.
C. Kowal.

IAU Circ. No. 2195 (1969).

Dynamical instabilities in semidetached close binary systems with possible applications to novae and novalike variables. See Abstr. 117.025.

124.100 Nova Vulpeculae 1968 No. 1

Lichtelektrische und spektrographische Beobachtungen der Nova Vulpeculae 1968 Nr. 1.
J. Dorschner, C. Friedemann, W. Pfau.
Astron. Nachr. Vol. 291, 217 - 223 (1969).

The results of photoelectric and spectrographic observations of Nova Vul 1968 No. 1 are described. U, B, V magnitudes and colours ranging from 1968 April 18 to June 15 are given. The distance of the nova was estimated to 1450 ± 400 pc (m. e.). Its interstellar absorption amounts to A_V = 1.2 ± 0.2 mag (m. e.). Spectra were obtained between 1968 April 18 and May 8 and were used to measure radial velocities for two absorption systems.

Nova Vulpeculae 1968 No. 1. I. Photoelectric observation at Ondrejov in 1968.
J. Grygar, L. Kohoutek.
Bull. Astron. Inst. Czechoslovakia, Vol. 20, 226 - 227 (1969).

Photoelectric observations of nova Vulpeculae 1968 N. 1. C. Bartolini, P. Battistini, C. delli Ponti, A. Guarnieri.
Mem. Soc. Astron. Italiana, Nuova Serie, Vol. 40, 529 - 532 (1969).

UBV photoelectric observations of nova Vulpeculae 1968 N. 1 made at the Bologna Observatory during 1968 are reported in this note.

Preliminary report on the spectrum of nova Vul 1968. A. Mammano, R. Margoni, L. Rosino.
Non-Periodic Phenomena in Variable Stars, IAU Colloquium, Budapest, 1968, p. 271 - 276 (1969).

Spectroscopic observations of nova Vul 1968 made at Asiago from April to August are reported in this paper. The nova belongs to the fast type. Absorption systems with velocities of –680, –800, –1380 and –2500 km/s have been observed. Some peculiarities of the emission components of CaII λλ8498, 8542 and 8662 (mult. 2) are pointed out. The evolution of the nova from the premaximum to the nebular stage is shortly described.

UBV observations of Nova Vulpeculae 1968 No. 1.
J. D. Fernie.
Publ. Astron. Soc. Pacific, Vol. 81, 374 - 380 (1969).

Seventeen UBV observations, including a premaximum observation, obtained between April 16 and December 18, 1968 are presented. The light curves in U and V, but not in B, are found to be almost strictly exponential. (B − V) appears to have been changing steadily from a time before maximum until 110 days after maximum. There are marked differences from the light and color curves of supernovae. It is estimated that the absolute blue magnitude at maximum light was –7.8, that the color excess is $0.^m 6$, and that the distance of the object is 1.3 kpc.

Nova Vulpeculae, 1968 No. 1.
P. Tempesti.
Yamamoto Circ. No. 1701 (1969). In Japanese.

Photoelectric observations of nova Delphini 1967 and nova Vulpeculae 1968. See Abstr. 124.102.

UBV observations of nova Delphini 1967 (= HR Del) and nova Vulpeculae 1968 No. 1. See Abstr. 124.102.

124.101   Nova Vulpeculae 1968 No. 2

Nova Vulpeculae 1968 No. 2 (Kohoutek).
L. Rosino, G. Chincarini, A. Mammano.
Astrophys. Space Sci. Vol. 4, 392 - 400 (1969).

Nova Vul 1968 No. 2 (Kohoutek) has been studied on Asiago material obtained before and after the announcement of discovery. The nova, fainter than 20 magnitude at minimum, brightened on July 16. The maximum (9.25 B) was reached on July 19. From the light curve the star can be classified as a normal fast nova. Objective prism spectra taken near maximum display the presence of absorption systems with radial velocities from −550 to −2200 km/sec. On slit spectrograms obtained at the end of October, the nova was found in the nebular stage with wide emission bands of H, He I, He II, N II, N III and forbidden lines of O III, N II, O I. The degree of excitation is slowly increasing. The nova is strongly reddened by interstellar absorption.

Spectroscopic observations of nova Vulpecula 1968 No. 2.   F. M. Stienon.
Publ. Astron. Soc. Pacific, Vol. 81, 613 - 618 (1969).

Spectra of nova Vulpecula 1968 No. 2 appear on two objective-prism plates taken with the 24-inch Burrell Schmidt of the Warner and Swasey Observatory. Near maximum light the rate of decay of the light curve predicts a photographic absolute magnitude of −9.6. With reasonable assumptions about the interstellar absorption, the distance of the nova is 7.6 kpc, about 250 pc above the plane of the Milky Way.

124.102   Nova Delphini 1967

B-, V-photometry of nova HR Delphini, 1967.
B. Mollerus.
Astron. Astrophys. Vol. 3, 376 - 377 = Veröff. Sternw. München, Vol. 7, No. 12 (1969).

Photoelectric observations of the nova Del 1967 in the B- and V-regions have been made between 1967, July 11 and 1968, Oct 19.

Photoelectric observations of nova Delphini 1967 and nova Vulpeculae 1968.   P. Tempesti.
Atti XII Riunione Soc. Astron. Italiana, L'Aquila 1968, p. 75 - 77 (1969). − Abstract SAI.

Lo spettro della nova Delphini nel 1968.
F. Ciatti, A. Mammano, L. Rosino.
Atti XII Riunione Soc. Astron. Italiana, L'Aquila 1968, p. 113 - 114 (1969). − Abstract SAI.

Narrow-band photoelectric observations of nova Del 1967 at Catania.   S. Catalano, S. Cristaldi, M. Rodonò.
Atti XII Riunione Soc. Astron. Italiana, L'Aquila 1968, p. 115 - 119 (1969). − Abstract SAI.

Nova HR Delphini 1967.   J. Hübscher.
BAV Rundbrief, 18. Jahrgang, 35 - 41 (1969).

Raies d'émission dans le spectre de la Nova Delphini 1967 entre 8400 et 9600 Å.

Y. Andrillat, L. Houziaux.
Comptes Rendus, Acad. Sci. Paris, Sér. B, Vol. 269, 546 - 548 (1969).

UBV observations of nova Delphini 1967 (= HR Del) and nova Vulpeculae 1968 No. 1.   J. Grygar.
Inform. Bull. Variable Stars (I.A.U. Commission 27), Konkoly Obs., Budapest, No. 371 (1969).

Infrared spectra of Nova Delphini 1967.
Y. Andrillat, L. Houziaux.
Mass Loss from Stars, Trieste 1968, p. 281 - 289 (1969).

Infrared spectra at 39 Å/mm of Nova Delphini 1967 have been obtained in the wavelength range λ 6500 − λ 8800. Typical microphotometer tracings are displayed and an estimate of the received flux is given for the most prominent lines at nebular stage. Loss of matter is very likely to occur.

Nova Delphini 1967-8.   J. B. Hutchings.
Mass Loss from Stars, Trieste 1968, p. 290 - 299 (1969).

A discussion of the peculiar, slow Nova Delphini 1967 is given, based on a large amount of data obtained in Victoria. A summary of the important changes in the spectrum is given, and consideration of the line profiles and line velocity curves leads to the conclusion that the photosphere attained a diameter in excess of 10^9 km for an extended period of time. After June 1968 no further matter was ejected and collapse of the photosphere left an auroral type spectrum with complex structure in its emission lines.

Polarization of Nova Delphini 1967.
B. Zellner.
Mass Loss from Stars, Trieste 1968, p. 300 - 302 (1969).

The light of Nova Delphini 1967 has intrinsic linear polarization, with both a small-amplitude fluctuation with a time-scale of days and a secular variation with a time-scale of months. A few observations of Nova Vulpeculae 1968 suggest that it also has variable polarization.

On nova Delphini 1967 and some slow nova characteristics.   W. C. Seitter.
Non-Periodic Phenomena in Variable Stars, IAU Colloquium, Budapest, 1968, p. 277 - 284 (1969).

Nova HR Delphini in the year 1968.
J. Grygar.
Říše hvězd, Vol. 50, 176 - 177 (1969).   In Czech.

The light curve obtained from observation during April − December 1968 is presented.

Nova Delphini.   W. Wichmann.
VdS Nachrichtenblatt, 18. Jahrgang, 143 - 144 (1969).

124.103   Nova RS Ophiuchi

Spectroscopic observations of the recurrent nova RS Ophiuchi from 1959 to 1968.
R. Barbon, A. Mammano, L. Rosino.
Non-Periodic Phenomena in Variable Stars, IAU Colloquium, Budapest, 1968, p. 257 - 260 (1969).

Spectra of RS Oph have been taken at Asiago with the 122 cm telescope during the 1959-67 minimum and on the occasion of the 1967-68 outburst. At minimum the variable has a composite spectrum, one component being of spectral type around M2, while the other component gives a blue continuum with emission lines of medium or high excitation. During the outburst the star has shown the same spectral evolution as in the 1933 and 1958, with the development of bright nebular and coronal lines, indicating an extremely high degree of ionization. Further details will be given in a forth-coming paper.

The postmaximum spectrum of nova RS Ophiuchi
1967.   'G. Wallerstein.
Publ. Astron. Soc. Pacific, Vol. 81, 672 - 676 (1969).

Measurements of spectrograms of RS Oph taken 110 days after the 1967 outburst are reported. Wavelengths, identifications, and intensities are presented. A continuum is present with a spectrophotometric gradient equivalent to that of a K0 star. Weak absorption lines confirm the presence of a cool source or a cool shell. The radial velocity of the emission lines is –47 km/sec and the absorption lines is –54 km/sec, neither of which differs greatly from velocities observed between outbursts.

Ne I in the sun and in RS Ophiuchi.
See Abstr. 074.026.

124.104   Nova T Pyxidis

Spectroscopic observations of the recurrent nova T Pyxidis during the 1967 maximum.
G. Chincarini, L. Rosino.
Non-Periodic Phenomena in Variable Stars, IAU Colloquium, Budapest, 1968, p. 261 - 269 (1969).

Spectroscopic observations of the recurrent nova T Pyx have been made at Asiago in the first months of 1967, during the slow decline of the star from maximum. The early spectra show wide emission lines of H, HeI, CII, NII, NIII, OIII, FeII, etc. with violet-shifted absorption components. The mean expansion velocity derived from the dark lines is about 1800 km/sec. Spectra taken in March and April indicate an increasing degree of ionization, as shown by the strengthening of the emission bands of HeII, NII, NIII, OIII. The absorption lines weaken or disappear. Although the forbidden lines of [OIII] and [FeX] 6374 are already present, the star has not yet reached its highest degree of ionization, as observed by Joy in 1944, when the observations were interrupted.

124.105   Nova WZ Sagittae

Photometric observations of nova WZ Sagittae and their interpretations.   W. Krzemiński, J. Smak.
Non-Periodic Phenomena in Variable Stars, IAU Colloquium, Budapest, 1968, p. 371 - 376 (1969).

A new model of the binary system WZ Sge is proposed. in which the secondary component contributes about 20 per-cent to the total light. The W UMa-type light curve (except for the primary eclipse) is explained as a result of the aspherical shape of the secondary. Both components are degenerate stars. Their effective temperatures are approximately 20000°K and 8000°K.

124.106   Nova DQ Herculis

About the nature of short-period light variations of DQ Her.   V. I. Taranov.
Astrofizika, Vol. 5, 337 - 341 (1969).   In Russian.
English translation in Astrophysics, Vol. 5, No. 2 (1969).

It is concluded that auto-oscillations of the front of a shock wave may be the reason of short-period light variations of DQ Her. For DQ Her an estimate of the period and of the amplitude of oscillations of light is given.

DQ Herculis: Synchronous photometry.
R. E. Nather, B. Warner.
Science, Vol. 166, 876 - 877 (1969).

Synchronous signal averaging, applied to the photometry of the stellar system DQ Herculis in order to study the 71.1-second pulsations discovered by Walker in 1956, yields a light curve which is a pure sinusoid, within the accuracy of measurement. The binary period is increasing, probably as a result of mass loss from the system.

To the problem of the motion of the interstellar medium under the action of a shell of a nova or supernova.
See Abstr. 131.118.

124.107   Nova RR Telescopii

Forbidden lines of Ni IV in the spectrum of RR Telescopii.   D. L. Lambert, A. D. Thackeray.
Astrophys. Space Sci. Vol. 5, 283 - 288 (1969).

The rich spectrum of the slow nova RR Telescopii contains several strong unidentified lines. The possibility is re-examined that some of these lines are due to forbidden transitions of [Ni IV]. Success is partial but the new identifications include four strong lines ($\lambda\lambda$ 5363, 5288, 5060, 5042) from the multiplet $a^4 F - a^2 G$.

124.108   Nova Coronae Australis

Nova Coronae Australis.
Sky Telescope, Vol. 38, 379 (1969).

125 Supernovae, Supernova Remnants

125.001 Search for an optical remnant of the Cassiopeia A supernova. S. van den Bergh, W. W. Dodd.
Nature, Vol. 223, 814 - 815 (1969).

Plates obtained with the 5m Hale telescope between 1951 and 1968 have been used to study the motions of the luminous filaments associated with Cas A. The centre of expansion is determined, and it is concluded that the explosion took place in AD 1667 ± 8 (m.e.) if the rate of expansion was uniform. No star has been found at the centre. Explanations for this fact are suggested.

125.002 Radio observations of the supernova remnant HB 21. J. W. Erkes, J. R. Dickel.
Astron. Journ. Vol. 74, 840 - 845 (1969).

A contour map of the supernova remnant HB 21 has been constructed from observations at a frequency of 2695 MHz. The integrated flux density at 2695 MHz is 160 ± 30 flux units. Variations of the spectral index across the remnant were found by combining the map at 2695 MHz with previously published maps made at frequencies of 178, 610.5, and 1430 MHz. The flux-density spectrum appears to be relatively straight in the central and southern parts of the supernova remnant, and has a mean slope of about -0.40. In the northern part of the supernova remnant, the spectrum is curved, with a turnover at about 250 MHz. Several possible interpretations of this spectral feature are discussed.

125.003 OB stars near the supernova remnant RCW 86. B. E. Westerlund.
Astron. Journ. Vol. 74, 879 - 881, 977 - 981 (1969).

The filamentary nebula RCW 86, identical with the non-thermal radio source MSH 14 - 63, is part of a supernova remnant. A group of OB stars is found near the radio source. The distance of the group is 2500 pc; this agrees well with the radio distance of the remnant. It is suggested that the remnant was formed by the explosion of a member of the group; the explosion occurred probably in 185 A. D.

125.004 OB stars near the supernova remnant RCW 103 and the galactic structure in Norma. B. E. Westerlund.
Astron. Journ. Vol. 74, 882 - 890 (1969).

The small emission nebula, RCW 103, is identical with a nonthermal radio source, Parkes 1613 - 50. It is most likely a remnant of a supernova of type II and a member of an OB association at a distance of 3900 pc. Another group of OB stars is found in the field at a distance of 1800 pc. Interstellar dust causes heavy absorption well within 1 kpc from the sun, and another strongly absorbing cloud appears at the distance of RCW 103. The available optical and radio data show that at $l^{II}= 332°$ portions of the Sagittarius arm and the Norma arm are seen. The latter contains most of the thermal radio sources in this direction as well as RCW 103 and its OB association.

125.005 A new estimation of distances of several supernova remnants. T. A. Lozinskaya.
Astron. Tsirk. No. 511, p. 1 - 3 (1969). In Russian.

125.006 Early supernova luminosity. S. A. Colgate, C. McKee.
Astrophys. Journ. Vol. 157, 623 - 643 (1969).

The diffusion of radiant energy from spherical expanding matter has been analytically and numerically calculated for masses and velocities of model supernova outbursts. The agreement with observation is satisfactory. The production of a large mass fraction of the radioactive isotope ^{56}Ni, which has been predicted from calculations of supernova nucleosynthesis, appears to be critical for the formation of the observed light curves. The radioactive energy from $0.25\,M_\odot$ of ^{56}Ni by the decay process ^{56}Ni $\rightarrow$ ^{56}Co supplies the radiant energy, 10^{49} ergs, during the "diffusive release" phase (5–20 days) of expansion near maximum. The subsequent decay process, ^{56}Co $\rightarrow$ ^{56}Fe, in conjunction with progressive γ-ray transparency of the expanding matter, gives rise to the long-time exponential light decay of 35 – 65 days.

125.007 Preliminary catalogue of supernovae, discovered till the end of 1967. M. Karpowicz, K. Rudnicki.
Publ. Astron. Obs. Warsaw Univ. [Warsaw University Press], Vol. 15, 189 pp. Price zł 16.00 (1969).

125.008 The radio structure of Tycho's supernova remnant. J. R. Dickel.
Astrophys. Letters, Vol. 4, 109 - 112 (1969).

Observations have been made of Tycho's supernova remnant at a wavelength of 2 cm with a resolution of 2 arc min. They show either a ring, or very irregular shell structure, for the remnant. Several interesting polarized features are noted.

125.009 On the presumed presupernova stage for type II supernovae. G. Barbaro, N. Dallaporta, C. Summa.
Non-Periodic Phenomena in Variable Stars, IAU Colloquium, Budapest, 1968, p. 41 - 49 (1969).

The physical conditions of stars in presupernova type II stage when the outburst is expected to be due to the Fe – He transition occurring in its core are reviewed. The arguments showing that the star must preserve a large envelope in this stage and therefore appear as a red supergiant are stressed, and a lower mass limit of about $10 \sim 14\,M_\odot$ for stars undergoing the outburst is confirmed on the basis of the more recent evaluations. Finally, the possibility that the presupernova type II stage could be represented by the small amplitude irregular and semiregular red variables with large masses belonging to young population I is briefly indicated.

125.010 Relation between white dwarfs and type I supernovae. G. S. Bisnovaty-Kogan, Z. F. Seidov.
Astrofizika, Vol. 5, 243 - 247 (1969). In Russian.
English translation in Astrophysics, Vol. 5, No. 2 (1969).

It is shown that the mass range of hot white dwarfs and their cooling times up to the moment of stability loss allow to present such white dwarfs, which can be used as models of type I supernovae in old clusters.

125.011 Supernova in Ursa Major. I. A. Dubyago, S. S. Tokhtas'ev.
Izv. Astron. Obs. Ehngel'gardta, *Kazan'*, No. 36, p. 268 (1968). In Russian. — $a = 11^h52^m 10^s$, $\delta = +53°32'$ (1855).

125.012 Supernova of AD 1437. D. K. Milne.
Nature, Vol. 224, 891 (1969).

The association of the radio source CTB 35 with the Chinese supernova record of AD 1437 is questioned.

125.013 Alleged supernova of AD 1006. C. M. Botley.
Nature, Vol. 224, 891 (1969).

125.014 A possible model of supernovae: Detonation of ^{12}C. W. D. Arnett.
Astrophys. Space Sci. Vol. 5, 180 - 212 (1969).

Stars of intermediate mass (about 4 to 9 solar masses) may ignite the $^{12}C + ^{12}C$ reaction explosively because of the high degree of electron degeneracy in their central regions. After the exhaustion of helium burning in the core of such stars, a helium-burning shell develops which is thermally unstable. Approximating this shell by suitable boundary conditions, the subsequent evolution of the core is examined quantitatively by standard techniques. An explosive instability due to ignition and detonation of $^{12}C + ^{12}C$ develops at a central density $2 \times 10^9 \, g/cm^3$. Subsequent hydrodynamic expansion is computed; final velocities of expansion up to 20000 km/sec are found. The star is totally disrupted; no condensed remnant is left. Such an explosion may be a plausible model for a significant fraction of supernovae.

125.015 Supernovae search at Corralitos Observatory.
J. R. Gallivan.
Publ. Astron. Soc. Pacific, Vol. 81, 539 (1969). – Abstract ASP.

125.016 The light of the supernova outburst.
P. Morrison, L. Sartori.
Astrophys. Journ. Vol. 158, 541 - 570 (1969).

We interpret the observed light as being principally fluorescence excited in the material as far as 1 lt-yr away from the explosion site by a strong ultraviolet pulse emitted either from the central explosion or from a dense coronal layer immediately surrounding it. This single assumption enables us to explain the major features of supernova light curves and spectra. In the case of supernovae of type I (SN I), even the simplest version of the theory provides a rather complete account of the observations. Most of the paper is devoted to an exposition of this account. The same explanation may apply to supernovae of type II also. The model for SN II, however, is more tentative than that for SN I.

125.017 On light curves of supernovae.
E. K. Grasberg, D. K. Nadezhin.
Astron. Zhurn. Akad. Nauk SSSR, Vol. 46, 745 - 746 (1969). In Russian. English translation in Soviet Astron. AJ, Vol. 13, No. 4.

It is shown that light curves of type II supernovae can be explained by the emerging of a shock wave into the extended atmosphere of a supergiant with a radius of 5000 - 10000 $R_\odot$.

125.018 Supernovae outbursts and generation of relativistic particles. L. E. Gurevich, A. A. Rumjantsev.
Astron. Zhurn. Akad. Nauk SSSR, Vol. 46, 1158 - 1164 (1969). In Russian. English translation in Soviet Astron. AJ, Vol. 13, No. 6.

A mechanism of the formation of relativistic particles in supernovae outbursts is considered. The acceleration of particles by radiation, emitted after the burst during brightness maximum, is estimated. The efficiency of the Fermi acceleration mechanism for particles, injected by radiation into a shell, is estimated. It is shown that the amounts of relativistic pro-

tons and electrons in the Crab nebula are comparable with one another. The mechanism of generation of relativistic electrons in the Crab nebula is indicated.

125.019 Supernova remnants and the galactic magnetic field. P. A. Shaver.
Observatory, Vol. 89, 227 - 230 (1969).

There are 18 well-resolved supernova remnants known at galactic latitudes smaller than $5°$ and outside the local spiral arm that are used for a preliminary statistical study of the distribution of radio emission from the shell structures.

125.020 One-dimensional polarization distributions over four supernova remnants at 1418 MHz.
G. A. Seielstad, K. W. Weiler.
Report AD–671151, California Inst. Technology, Owens Valley Radio Obs., Pasadena. 28 pp. (1968). – See Phys. Abstr. Vol. 72, No. 24804 (1969).

125.021 Experimental tests of the supernovae origin of cosmic rays. C. E. Fichtel, H. B. Ogelman.
Report NASA–TN–D4732, Goddard Space Flight Center, Greenbelt. 21 pp. (1968). – See Phys. Abstr. Vol. 72, No. 24805 (1969).

125.022 Supernova in NGC 1058.
L. Rosino.
IAU Circ. No. 2194 (1969).

125.023 Supernova in NGC 1058.
J. A. Hynek, J. R. Dunlap, Gallivan.
IAU Circ. No. 2195 (1969).

125.024 Supernova in NGC 1058.
F. Bertola, F. Ciatti, L. Rosino.
IAU Circ. No. 2196 (1969).

Supernova in the galaxy NGC 4472.
Astron. Tsirk. No. 514, p. 1 (1969).   In Russian.

Polarization of the Cygnus Loop at 11-centimeter wavelength.   See Abstr. 132.027.

Rotating collapsed objects, quasars and supernova remnants.   See Abstr. 141.039.

Observations of the distribution of polarized and non-polarized emission across the radiosource W 44 at 3.95 and 6.6 cm at Pulkovo.   See Abstr. 141.072.

6 cm observations of nonthermal radio sources near the galactic plane.   See Abstr. 141.081.

Synthesis of brightness distribution in radio sources. See Abstr. 141.154.

126 Low-luminosity Stars, Subdwarfs, White Dwarfs

126.001 Subluminous stars. III. Luminosity calibration for subluminous stars and the space density of the blue subluminous stars south of declination –45°. O. J. Eggen.
Astrophys. Journ. Vol. 157, 287 - 311 (1969).

The relatively large number of trigonometric parallaxes of white dwarfs now available, together with wide binaries containing a subluminous component and cluster parallaxes, makes possible a new calibration of the luminosities of subluminous stars. The tendency of these objects to form two sequences in the $(M_V, U - V)$-plane is confirmed and seems unlikely to be the effect of differential line blanketing alone. There is some confusion between the red subluminous stars and the M-type dwarfs with M_V near +10, especially among members of the Hyades cluster. The possibility arises that some subluminous stars in Hyades, and in the field, represent stars with masses less than one-tenth the solar mass that have become degenerate without passing through the normal evolutionary process (Kumar's "black dwarfs"). Photometry of 158 stars south of declination –45° and called "probable white dwarfs" by Luyten shows these objects to be nearly equally divided between subdwarfs and subluminous stars. The resulting space density of blue subluminous stars is similar to that found from a larger sample of Lowell proper-motion stars discussed in Paper I and leads to a conservative estimate of 1.5×10^{-3} pc^{-3}. Identification charts are provided for the subluminous stars south of declination –45°.

126.002 Differentially rotating Hamada-Salpeter white dwarf models. J. R. Gribbin.
Astrophys. Letters, Vol. 4, 77 - 79 (1969).

Pulsation periods are determined for degenerate, differentially rotating white dwarfs obeying the Hamada-Salpeter equation of state. Periods below 0.2 sec are obtained with moderate amounts of rotation, and no unstable models are found.

126.003 Some integral characteristics of the rotating white dwarfs and neutron stars.
V. V. Papoyan, D. M. Sedrakian, E. V. Chubarian.
Soobshch. Byurakan. Obs. No. 40, p. 86 - 97 (1969).
In Russian.

Integral characteristics of rotating white dwarfs and superdense configurations are calculated. It is shown that the angular velocities of neutron stars may exceed those of white dwarfs. The rotation in a small region of the central condensation corresponding to the maximum value of mass density breaks the stable equilibrium of white dwarfs. In this region neutron stars are much more stable against rotation. In metastable neutron stars rotation causes perturbations of the equilibrium of matter.

126.004 Subluminous stars. IV. Red subluminous stars and stars with a high ultraviolet excess.
O. J. Eggen.
Astrophys. Journ. Suppl. Series, Vol. 19, 31 - 56 (1969).
Abstr. in Astrophys. Journ., Vol. 157, 1471 - 1472 (1969).

A sample of 1000 southern, late-type proper-motion stars yields 419 objects with $B - V$ between + 0.35 and + 1.1 mag. These 419 stars contain (a) 300 stars which from the colors alone could be subdwarfs, but the tangential velocity for 10 percent of which would then be in excess of the total space motion observed in known subdwarfs, (b) 48 stars with ultraviolet excesses up to 0.1 mag larger than expected from the abundance effect, of which 30 percent, if subdwarfs, have excessive tangential velocity, and (c) 71 stars with still larger ultraviolet excesses that may all be subluminous objects. From the transverse velocity alone some 15 percent of the 419 stars

are probably subluminous.

126.005 Subluminous stars. V. Photoelectric (UBV) photometry of southern proper-motion stars.
O. J. Eggen.
Astrophys. Journ. Suppl. Series, Vol. 19, 57 - 78 (1969). –
Abstr. in Astrophys. Journ. Vol. 157, 1472 - 1474 (1969).

Some 1000 observations of 560, mainly southern proper-motion stars are given. These include additional observations of a few stars previously published in this series, as well as corrected values for a half-dozen stars previously misidentified. The present results, together with those listed in papers I, II, and III, give some 1800 magnitude and color observations of 1000 southern proper-motion stars. Comparisons are made with the available photographic magnitudes and colors. Most of the data presented here are discussed in paper IV.

126.006 The Lowell suspect white dwarfs.
J. L. Greenstein.
Astrophys. Journ. Vol. 158, 281 - 293 (1969).

A survey of eighty-six suspected white dwarfs in the Lowell GD lists shows that fifty-four are in fact, white dwarfs. These stars have a considerably smaller mean proper motion and a somewhat bluer mean color than the 202 stars observed by Eggen and Greenstein. Otherwise, the GD stars are normal white dwarfs of lower space motion than the EG sample. Line profiles, equivalent widths, and central absorptions are normal. The properties of these degenerate stars of the low-velocity population are shown in graphic form. Ten hot subdwarfs were found; twenty-two horizontal-branch stars or yellow subdwarfs were also found. A few especially interesting stars are noted, including GD 108, which seems to be of intermediate luminosity, and three very hot, probably helium-rich, O-type subdwarfs, GD 298, 299, and 300. The GD white dwarfs are largely population I kinematically, with a tangential velocity of 30 km sec^{-1}; some, therefore, are descended from recently evolved massive stars. The blue high-velocity white dwarfs are all descended from stars near the present turnoff point in the old disk and halo populations, i.e., at most 1.5 $M_\odot$. New spectra are given of twenty-two white dwarfs of larger proper motion. Several new suspected red degenerates (EG 254, 256) have been found. A new type of white dwarf with molecular bands may exist in EG 248, which shows broadened CH and possibly C_2 features. The EG list is continued to EG 266.

126.007 New subdwarfs. II. Radial velocities, photometry, and preliminary space motions for 112 stars with large proper motion. A. Sandage.
Astrophys. Journ., Vol. 158, 1115 - 1136 (1969).

Radial velocities have been measured for 112 stars of large proper motion selected from an unpublished photometric catalog which contains 300 new subdwarf candidates. Two particularly interesting velocity variables probably have blue subluminous companions which are brighter than white dwarfs but considerably fainter than main-sequence stars. Space motions for all stars are shown in a Bottlinger diagram. The asymmetrical-drift velocity and the dispersions in U and W increase systematically with decreasing metal abundance for the entire range of ultraviolet excess values from δ = 0.00 to δ = 0.31 mag. Many globular-clusterlike subgiants are present in the material. The existence of subdwarfs possessing high angular momentum which lead the sun in its galactic rotational velocity seems to be established. There may be a systematic variation of metal abundance with increasingly positive V velocity.

126.008 Slow rotation of white dwarfs and barionic stars.

V. V. Papoyan, D. M. Sedrakian, E. V. Chubarian.
Astrofizika, Vol. 5, 415 - 424 (1969).   In Russian. — Engl.
translation in Astrophysics, Vol. 5, No. 3.

The slow rotation of white dwarfs and barionic stars is
considered in the first approximation, considering the angular
velocity of the rotation Ω as a small parameter. The distribution of the nondiagonal component of the metric and of the
relativistic moment of inertia along the radius of the star is
found. The rate of the energy of rotation is evaluated. The
results obtained remain correct also in the second approximation. The results of the numerical integration are represented
in figures and in a table. The transition between the states of
hyperonic stars with different values of energy is considered,
when the number of barions and the angular momentum are
conserved. It is shown that during such transitions the amount
of energy of $\sim 10^{52}$ *erg* is radiated at the expense of the mass
difference. An energy of the same order is radiated owing to
the reduction of the energy of rotation. The possibility is
mentioned that this energy provides on one hand the observed luminosity and on the other the pulsation of the object.

126.009   The white dwarf CoD –38°10980.

J. B. Alexander, J. v. B. Lourens.
Monthly Notes Astron. Soc. Southern Africa, Vol. 28, 95 -
96 (1969). — Note.

126.010   Weiße Zwerge, Neutronensterne und der Endzu-
stand der Materie.   W. Deinzer.

SuW, Vol. 8, 224 - 229 (1969).

126.011   A blue object in Cygnus.

G. Ishida, M. Kondo, S. Nishimura, K. Osawa,
K. Ichimura.
Tokyo Astron. Obs. Bull., Second Series, No. 196, p. 227 -
2281 (1969).

Two-color image plates of a portion of the constellation
Cygnus were obtained at the Newtonian focus of the 188-cm
reflector of the Okayama Station. The blue object in question
has a faint visual companion which is about 5 seconds of arc
apart in the southeast direction. From spectrographic observations it seems that the blue object may probably be a white
dwarf of the type DAs or a hot subdwarf. The possibility of
being an early B-type main sequence star is also not completely ruled out.

Modellatmosphären für heliumreiche Weiße Zwerge.
See Abstr. 064.062.

**Treatment of pulsating white dwarfs including ge-
neral relativistic effects.**   See Abstr. 065.012.

**The effect of rotation on the pulsation periods of
degenerate white dwarf models.**   See Abstr. 065.014.

Fast evolution towards the white dwarfs.
See Abstr. 065.017.

Interstellar Matter, Gaseous Nebulae, Planetary Nebulae

131 Interstellar Space, Interstellar Matter, Polarization of Starlight

131.001 Search for microwave emission from the $^2\pi_{1/2}, J = 3/2$ state of OH. B. E. Turner.
Astron. Astrophys. Vol. 2, 453 - 457 (1969).

A search for three of the four microwave lines in the $^2\pi_{1/2}, J = 3/2$ excited state of OH has established upper limits of ~0.3 f.u. for the lines in emission in the direction of the 18 cm OH sources in W 3, W 28, Ori A, NML Cyg, CIT 3, and CIT 7. None of the pumping mechanisms for anomalous OH excitation predict detectable emission or absorption from this state, within the limits of the search.

131.002 A search at 15 GHz for compact H II regions in regions of possible star formation.
E. Churchwell, M. Felli, P. G. Mezger.
Astrophys. Letters, Vol. 4, 33 - 41 (1969).

Regions surrounding IR stars, sources of anomalous OH emission, T Tauri stars, $H\alpha$ emission regions of high surface brightness and small diameters, and young O-star clusters and associations have been surveyed with the NRAO 140-ft telescope at 15.4 GHz. In addition, four IR stars associated with anomalous OH emission have been surveyed at 1.4 GHz with the NRAO 300-ft telescope. Positive results have been obtained for three sources of anomalous OH emission and some of the $H\alpha$ emission nebulae and O star associations, respectively. In all other cases upper limits for the flux densities at 15.4 and 1.4 GHz are given.

131.003 Cold clouds and hot stars − An evolutionary sequence? C. Heiles.
Astron. Soc. Pacific, Leaflet No. 482, 8 pp. (1969).

131.004 Interstellar grains.
F. Hoyle, N. C. Wickramasinghe.
Nature, Vol. 223, 459 - 462 (1969).

Interstellar grains may be a mixture of graphite particles formed in carbon stars and of silicates in oxygen-rich giants. The astrophysical consequences of such a grain mixture are examined.

131.005 Observations of satellite-anomalous OH emission sources. B. E. Turner.
Astrophys. Journ. Vol. 157, 103 - 122 (1969).

The class II OH sources W28, W41, W43, W44, and Cas A differ from the class I sources (W3, W49, NGC 6334, etc.) by showing anomalous emission and absorption only in the satellite lines, but normal absorption in the main lines. The spatial distribution and Stokes parameters of these class II OH sources as well as of W33 and a new source W28 (A_2) have been observed; the results show that emission typically arises from more than one region within a source (W28, W44, W33, W43), or that it may be extended (W43, Cas A) and that several features (in W28 (A_2), W43, and Cas A) may not be physically associated with the continuum source at all.

131.006 Temperatures and OH optical depths in dust clouds. C. Heiles.
Astrophys. Journ. Vol. 157, 123 - 134 (1969).

OH microwave-line optical depths and excitation temperatures have been measured in five dust clouds. Optical depths in the 1667-MHz line range downward from 2.2; excitation temperatures are 4.5° K in one cloud and 5.4° K in another; lower limits of 4.7° and 9.6° K have been established in two more. Expected values of kinetic and OH excitation temperature were calculated for various values of cosmic-ray intensity and free-atom density in dust clouds. For each cloud two values of density are consistent with observations −e.g., either (roughly) 10^2 or 10^{-3} atoms per cubic centimeter.

131.007 Interferometric studies of interstellar sodium lines.
L. M. Hobbs.
Astrophys. Journ. Vol. 157, 135 - 163 = Contr. Lick Obs. No. 289 (1969).

Interferometric, photoelectric scans of one or both of the interstellar sodium D-lines in the spectra of seventy-seven stars have been obtained with a Pepsios spectrometer. The qualitative results are the following. (1) The tenfold improvement in resolution over that used by Adams shows that, for the sixty-five stars common to both programs, over 70 per cent of the lines listed by Adams as single are multiple. The few single lines remaining are confined almost entirely to stars within 200 pc of the Sun and a few high-latitude stars. (2) Doubling of the D-lines due to the presence of sufficiently narrow hyperfine-structure components is rarely observed, if at all. (3) A large variation in the interstellar Ca II/Na I abundance ratio is indicated by a comparison of these data with those of Adams for the calcium lines.

131.008 The profiles of the interstellar sodium D-lines.
L. M. Hobbs.
Astrophys. Journ. Vol. 157, 165 - 173 (1969).

The profiles of one or both of the interstellar sodium D-lines in seventy-seven stars are analyzed. The number of single lines suitable for analysis of individual interstellar clouds is small, mostly owing to overlapping of the generally multiple lines. Theoretical line profiles based on a simple, conventional model do fit acceptably the great majority of the thirty-three measured profiles which are found to be single from a priori conditions. The main data consist of a more restricted set of fifteen lines. Eleven of the fifteen lines indicate a Gaussian velocity distribution for the absorbers; four indicate an exponential distribution. The rms velocities within the Gaussian clouds require (with one possible, but unlikely, exception) that the clouds be in H I regions. Thermal broadening of the lines is negligible, if a temperature $T \sim 100°$ K is assumed; the turbulent velocity distribution is thus measured directly.

131.009 Radial velocities of the gas in H II regions and their associated stars. P. Murdin.
Astrophys. Journ. Vol. 157, 175 - 181 (1969).

The radial velocities of stars and gas in H II regions are generally well correlated and suggest that H II regions are not the result of chance encounters. There is, however, a systematic difference in radial velocity between the nebulae and the stars. This difference is not the result of the expansion of the individual nebulae; if it is not due to a systematic error in the system of radial velocities of B stars, it indicates that the population of gas is expanding with respect to B stars imbedded in it.

131.010 Distribution of ammonia density, velocity, and rotational excitation in the region of Sagittarius B2.
A. C. Cheung, D. M. Rank, C. H. Townes, S. H. Knowles, W. T. Sullivan III.
Astrophys. Journ. *(Letters)*, Vol. 157, L13 - L20 (1969).

Examination of 1.25-cm NH_3 inversion radiation in the Sgr B2 region shows an irregular distribution of NH_3 about 10' in size, with considerable variation in density, velocity, and state of excitation. Relative intensities of the (1,1), (2,2), and (3,3) inversion transitions do not appear to correspond to equilibrium conditions. Relaxation processes, which allow non-equilibrium, are examined.

131.011 Radio observations of the nebulae K3-50 and NGC 6857. R. H. Rubin, B. E. Turner.
Astrophys. Journ. *(Letters)*, Vol. 157, L41 - L44 (1969).

Observations of the high-frequency continuum in the direction of the nebulae NGC 6857 and K3-50 show that the peak of emission coincides with K3-50, generally believed to be a planetary nebula. We have measured radio recombination lines and OH emission, either previously detected in a planetary nebula. It is necessary to conclude that the classification of the object is uncertain and that it is possibly a small "compact" H II region, in which case the object remains unique, since it would be the first to have an optical counterpart.

131.012 An intermediate velocity cloud showing a velocity bridge to local matter. G. L. Verschuur.
Astron. Astrophys. Vol. 3, 77 - 82 (1969).

An intermediate velocity cloud of neutral hydrogen at $l = 102°5$, $b = +36°$ has been mapped with a 12'5 arc beam. The concentration occurs at −50 km/s but shows a bridge of hydrogen, subtending an area of about 20% of the cloud, which appears to stretch in velocity from that of the cloud to that of local matter. The presence of this bridge suggests that the concentration may be a relatively local phenomenon.

131.013 Magnetohydrodynamical models of a helical magnetic field in spiral arms.
M. Fujimoto, M. Miyamoto.
Publ. Astron. Soc. Japan, Vol. 21, 194 - 202 (1969).

A circular arm with elliptical cross-section is used as a model of the spiral arm, and it is demonstrated that interstellar gas may flow in a helical path along the axis of the arm. Interstellar helical magnetic lines of force can be in a stationary state, if such noncircular motion of the gas is superimposed on galactic rotation. Some observations on the interstellar helical magnetic field are dynamically explained on the interstellar helical magnetic field are dynamically explained by our model.

131.014 Radio observations of four diffuse H II regions.
N. Kaifu, M. Morimoto.
Publ. Astron. Soc. Japan, Vol. 21, 203 - 210 (1969).

Contour maps of radio emission from four diffuse thermal sources, W1 (NGC 7822), W5 (IC 1848 + S 26), W 13 (NGC 2175), and W16 (NGC 2244, the Rosette Nebula) are given with a resolution of 11 min. of arc at 4170 MHz. For all four sources there is good agreement between the radio and optical appearances. W5 may form an association with W4 and W3, near-by H II regions. W13 and W16 are ionization-limited and have shell structures.

131.015 Some very cold H I clouds found in emission.
G. L. Verschuur.
Astrophys. Letters, Vol. 4, 85 - 87 (1969).

Several very narrow neutral hydrogen emission lines have been found in the spectra of low and intermediate velocity neutral hydrogen clouds. At least two clouds must have kinetic temperatures of $\leqslant 30°K$.

131.016 Occultation positions of the 1665 MHz OH emission from G 0.7 - 0.0.
R. N. Manchester, W. M. Goss, B. J. Robinson.
Astrophys. Letters, Vol. 4, 93 - 98 (1969).

Lunar occultation measurements have given positions for five separate features of the circularly polarized 1665 MHz OH emission from G 0.7 - 0.0 (Sgr B2). The most intense left-hand and right-hand features appear to be separated by about 2 arc sec, while weaker features are displaced by up to 20 arc sec. For the stronger features the probable error in position is ± 2 arc sec in declination and ± 0.05s in right ascension.

131.017 The interstellar H_2^+ molecule.
T. P. Stecher, D. A. Williams.
Astrophys. Letters, Vol. 4, 99 - 102 (1969).

Interstellar extinction observations and the photodissociation cross section for H_2^+ give an upper limit of 3×10^{-4} cm^{-3} to the mean H_2^+ number density in interstellar space. Consideration of formation mechanisms suggests that sufficient H_2^+ may be present to contribute to interstellar extinction especially around young hot stars. The wavelength dependence of the extinction produced by H_2^+ is similar to that produced by grains except in the far ultraviolet where the extinction due to H_2^+ is relatively larger.

131.018 Siliziumkarbid als möglicher Bestandteil des interstellaren Staubes. C. Friedemann.
Astron. Nachr. Vol. 291, 177 - 186 (1969).

Assuming isothermic atmospheres for the carbon stars, the growth and motion of silicon carbide particles were computed for a set of densities of the atmospheres. It is shown that the particles are able to escape from the atmosphere in case the surface acceleration is sufficiently low. An estimation of the whole mass of the expelled particles points out that only a small fraction of the interstellar dust can be silicon carbide. Nevertheless it is possible that these particles could act as condensation cores in the interstellar space.

131.019 Interstellare Absorptionslinien und galaktische Struktur. H. Scheffler.
SuW, Vol. 8, 180 - 184 (1969).

131.020 The fragmentation of cosmic-ray nuclei in interstellar hydrogen. C. J. Waddington.
Astrophys. Space Sci. Vol. 5, 3 - 17 (1969).

In order to calculate the effects of traversal of interstellar matter on the charge spectrum of the cosmic radiation it is necessary to have values for the fragmentation parameters of nuclei of each element into all lighter elements. Most of these values have not been experimentally determined. As a consequence, they have been calculated from a semi-empirical mass spallation relation designed to fit the available partial cross-sections obtained from radio chemical determinations. This calculation has attempted to take into account the conditions that are peculiar to the cosmic ray problem. Values of the parameters are given for three characteristic energies and a comparison is made with the sparce experimental data. The effects of using these parameters in a calculation of the extrapolation of the charge spectrum through interstellar space are shown for some representative cases.

131.021 Diffuse interstellar absorption bands as due to quadrupole transitions enforced by a magnetic field. M. Rudkjøbing.
Astrophys. Space Sci. Vol. 5, 68 - 70 (1969).

The diffuse interstellar absorption bands at λ4890 and λ6180 are believed to belong to electric quadrupole transitions enforced by the presence of the interstellar magnetic field. Their intensities relative to the bands at λ4430 and λ4760 and their state of polarisation might be used for an investigation of the field.

131.022 On the temperature of interstellar grains.
G. B. Field.
Monthly Notices, Roy. Astron. Soc. Vol. 144, 411 - 418 (1969).

Cooling of grains by emission due to impurity atoms is evaluated. It is shown that recent estimates of this effect are too high, and consequently, that calculated grain temperatures are too low. At the higher temperatures calculated here, grains cannot retain mantles of solid H_2 in normal interstellar clouds. Because the temperatures calculated are minimum values, which are unlikely to be approached in nature, certainly formation of solid H_2 and probably formation of gaseous H_2 and of ice mantles, are likely to be restricted to dark dust clouds where the radiation field is weak.

131.023 A pumping mechanism for anomalous microwave absorption in formaldehyde in interstellar space.
C. H. Townes, A. C. Cheung.
Astrophys. Journ. *(Letters)*, Vol. 157, L103 - L108 (1969).

A pumping mechanism which produces an excess population in the lower state 1_{11} of H_2CO by molecular collisions is proposed. Calculation on the basis of a simplified model agrees reasonably well with the observed absorption at 4830 MHz by H_2CO of the 2.8°K isotropic radiation reported by Palmer and co-workers. Implications of various transition rates and column densities related to the observed H_2CO absorption are discussed.

131.024 Observations of the $^2\Pi_{1/2}$, $J = \frac{1}{2}$ excited state of OH in W3.
P. R. Schwartz, A. H. Barrett.
Astrophys. Journ. *Letters)*, Vol. 157, L109 - L110 (1969).

The $F = 2 \rightarrow 2$ line of the $^2\Pi_{1/2}$, $J = \frac{1}{2}$ state of OH has been detected in W3. The other three lines of this state could not be detected.

131.025 On the far-ultraviolet interstellar extinction law in the Orion nebula region.
G. R. Carruthers.
Astrophys. Journ. *(Letters)*, Vol. 157, L113 - L117 (1969).

A rocket measurement of the far-ultraviolet spectral intensity of θ Orions in the range 1000 - 1350 Å has shown that the flux is far too great to be consistent with predictions based on previous ultraviolet extinction measurements and the observed $E(B - V)$. Comparison with model atmospheres and with other stars observed indicates ultraviolet color excesses $E(1115 - V) = 0.37$ mag and $E(1270 - V) = 0.53$ mag, as compared with the average $E(B - V)$ of 0.276 mag. Hence, instead of varying as $1/\lambda$, the far-ultraviolet portion of the extinction curve appears to be flat, or even decreasing toward shorter wavelengths.

131.026 Interstellar polarization of starlight and the turbulent structure of the galaxy.
J. R. Jokipii, I. Lerche, R. A. Schommer.
Astrophys. Journ. *(Letters)*, Vol. 157, L119 - L124 (1969).

Fluctuations in the polarization of starlight provide strong support for the concept of a turbulent interstellar medium. We have analyzed theoretically the mean polarization and the variance of polarization about the mean as functions of distance R from the solar system. The mean increases linearly with R. For R less than the correlation length L the variance increases as R^2, whereas for $R \gg L$ the variance increases linearly with R. The observed mean polarization and the observed variance of polarization are found to be in excellent agreement with these theoretical deductions if the interstellar medium fluctuates irregularly with a correlation length of 150 pc.

131.027 Interstellar extinction in the ultraviolet. II.
T. P. Stecher.

Astrophys. Journ. *(Letters)*, Vol. 157, L125 - L126 (1969).

Interstellar extinction in the ultraviolet has been determined from the spectra of ζ and ϵ Persei. The results are presented.

131.028 The accretion of interstellar matter by a neutron star with magnetic field.
P. R. Amnuel, O. H. Guseinov.
Astron. Tsirk. No. 524, p. 3 - 5 (1969).   In Russian.

131.029 The general properties of the structure of absorbing matter.
G. A. Starikova.
Astron. Tsirk. No. 527, p. 7 - 8 (1969).   In Russian.

131.030 Observations of an unusual cold cloud in the galaxy.
K. W. Riegel, M. C. Jennings.
Astrophys. Journ. Vol. 157, 563 - 572 (1969).

Measurements of 21-cm line spectra indicate the existence of a large cold cloud of neutral hydrogen in the galaxy. The cloud is seen as a very strong hydrogen self-absorption feature, whose half-width is ~ 3.5 km sec^{-1}. We derive an upper limit to the cloud temperature of 42° K, and a probable value of 20° K if data on continuum-source absorption are used. The cloud extends over at least 20° of longitude in the direction of the galactic center, and is probably within 1 kpc of the sun.

131.031 A theoretical analysis of methods of interpreting radio-line data from H II regions.
R. M. Hjellming, M. H. Andrews, T. J. Sejnowski.
Astrophys. Journ. Vol. 157, 573 - 582 (1969).

The best form of analysis of data on radio recombination lines from H II regions is shown to require inclusion of both non-LTE effects and highly concentrated clumping. The principal physical effect of clumping is to reduce considerably the line enhancements predicted by the non-LTE theory. Theoretically, all radio recombination lines emitted by the same mass of ionized gas should, within the error limits, be interpretable in terms of a single electron temperature.

131.032 On the evolution of high-density, dust-filled H II regions.
W. G. Mathews.
Astrophys. Journ. Vol. 157, 583 - 599 (1969).

Recent observations suggest that massive stars may form within dense, dust-laden neutral clouds which become ionized as the star approaches the main sequence. An evolutionary dynamic model is presented for such a nebula, and the thermal radio continuum is given as a function of time. Radio time variations may be observable. The theory for sputtering of charged grains is reviewed, and the possibility of dust-limited H II regions is investigated. Dust inside HII regions will be sputtered away on short time scales if the electron density exceeds about 10^3 cm^{-3}. Radiation pressure on the dust within the surrounding neutral gas will affect the evolution of initially dusty, high-density H II regions, but not greatly. Finally, observations of W49 A, a compact H II region, are shown to be in conflict with the properties of the most massive stars as they are currently understood.

131.033 Recombination lines in thermal and non-thermal galactic sources.
D. K. Milne, T. L. Wilson, F. F. Gardner, P. G. Mezger.
Astrophys. Letters, Vol. 4, 121 - 127 (1969).

The continuum spectra of 54 well-resolved galactic sources are compared with the intensity of the H109α recombination line at 5009 MHz. We conclude that thermal sources generally exhibit a H109α line stronger than 3 per cent of the continuum whilst non-thermal sources show little or no recombination line.

131.034 Can solid hydrogen condense on interstellar grains?
J. M. Greenberg, T. de Jong.

Nature, Vol. 224, 251 - 252 (1969).

From a discussion of the temperature of interstellar grains and the vapor pressure of solid hydrogen it seems highly unlikely that a significant amount of solid hydrogen can exist in interstellar space.

131.035   A survey for galactic OH emission sources.
        B. E. Turner.
Astron. Journ. Vol. 74, 985 - 993 (1969).

A search for anomalous OH emission has been made in 87 galactic regions which comprise mainly three categories: (a) highly reddened stellar associations and clusters considered by Reddish (1967) to be regions of star formation; (b) Wolf-Rayet stars associated with nebulosity; (c) galactic emission nebulae. OH emission or absorption was mapped in the previously known sources W41 and W42, and detected in the peculiar object K3 - 50. The absence of OH emission in all other cases is discussed in terms of the nature of the sources, as far as is known.

131.036   Detection of interstellar $H_2 C^{13} O^{16}$.
        B. Zuckerman, P. Palmer, L. E. Snyder, D. Buhl.
Astrophys. Journ. *(Letters)*, Vol. 157, L167 - L171 (1969).

The $1_{11} - 1_{10}$ transition of interstellar $H_2 C^{13} O^{16}$ has been detected in the direction of the continuum sources Sgr B2 and Sgr A and probably W51. Comparison with $H_2 C^{12} O^{16}$ spectra obtained in these directions indicates the possible presence of regions of high C^{13} abundance near the galactic center. The rest frequency of the $1_{11} - 1_{10}$ transition has been determined more accurately than in a previous laboratory measurement.

131.037   Millimetre wave emission by interstellar dust.
        M. M. Litvak.
Nature, Vol. 223, 1143 - 1144 (1969).

Semiconductor properties of interstellar grains would allow sufficient absorption and emission efficiencies for observable effects at millimetre wavelengths. About 10^{-24} watts $m^{-2} Hz^{-1}$ at the top of our atmosphere would come from opaque, arc min regions with about 10^{12} grains per cm^2, while cooling of the grains by such emission results in temperatures of about $10°$K.

131.038   Clustering of cold hydrogen gas on protons.
        R. Clampitt, L. Gowland.
Nature, Vol. 223, 815 - 816 (1969).

It has been suggested that the existence of ion-clusters of H_2 in gaseous nebulae could account for absorption and polarisation of stellar radiation. Ion-clusters of H_2 have been ejected from solid H_2 by low energy electron impact. The primary nucleating centre is H^+ formed by dissociative ionisation of H_2; the most abundant ion is H^+_{15} which may be $H^+_3 (H_2)_6$. Ion-clusters up to H^+_{99} were resolved. A proposal by Reddish and colleagues that H_2 condenses on carbon grains at $3°$K means that the H_2 density in interstellar clouds may be high enough for ion-clusters to form by interaction with radiation.

131.039   Interstellar hydrogen atoms on graphite grains.
        H. A. J. McIntyre, D. A. Williams.
Nature, Vol. 223, 487 - 488 (1969).

Hydrogen atoms physically adsorbed on interstellar graphit grains have their resonance lines shifted by about 50 Å from λ1216 Å to longer wavelengths. Therefore, such a system cannot be responsible for the stellar ultraviolet flux minimum at wavelengths shorter than Lyman a.

131.040   Observations of the 21-cm hydrogen line toward high-latitude stars.
S. J. Goldstein, Jr., D. D. MacDonald.
Astrophys. Journ. Vol. 157, 1101 - 1118 (1969).

We have obtained 21-cm spectra in 213 positions toward early-type stars with galactic latitude numerically greater than $20°$ and declination greater than $-40°$. A velocity range of $±50$ km sec^{-1} with respect to the local standard of rest was covered with a resolution of 1.3 km sec^{-1}, and in most cases a resolution of 0.53 km sec^{-1} was also used. We give the velocities of 268 well-resolved features from these spectra. About half of the stars have published observations of interstellar calcium or sodium lines. We find that most of Adams's velocity measurements for calcium lines agree to within 2.5 km sec^{-1} with our measurements in the same direction. The integrated brightness temperatures found from the 21-cm spectra are the basis for a study of the galactic gas layer near the sun. On the assumption that the sun is at the center of an optically thin layer, stratified parallel to the galactic equator, our observations give 2.5×10^{20} atoms per cm^2 for the projected density of hydrogen atoms. If the layer is allowed to have an arbitrary pole, our observations fit best when the galactic coordinates of the pole are $l = 180°$, $b = 75°$, a value that differs by $4°$ from the pole of the Gould's belt system obtained by Shapley.

131.041   Electron temperatures of H II regions.
        R. H. Rubin.
Astrophys. Journ. Vol. 157, 1461 - 1463 (1969).

The very low electron temperatures found for H II regions from the best isothermal fit of the radioflux-density spectrum are shown to be lower limits. In view of this, the temperature structure of the Orion nebula is reexamined.

131.042   Rosseland and Planck mean absorption coefficients for particles of ice, graphite, and silicon dioxide.
S. A. Kellman, J. E. Gaustad.
Astrophys. Journ. Vol. 157, 1465 - 1467 (1969).

Rosseland and Planck mean absorption coefficients are computed for ice, graphite, and vitreous SiO_2, over temperature ranges relevant to early phases of star formation.

131.043   Intensities of radio recombination lines (II).
        M. H. Andrews, R. M. Hjellming.
Astrophys. Letters, Vol. 4, 159 - 164 (1969).

Strengths of hydrogen a-, β-, γ-, δ-, and ϵ-radio recombination lines emitted by H II regions are calculated for $40 < n < 225$, $T_e = 7500, 10000$, and $12500°$K, $E = 10^4$ to 10^8 pc cm^{-6}, and $N_e = 10^{4.5}, 10^{5.0}, 10^{5.5}$ cm^{-3} for a- and β-lines and $N_e = 10^2, 10^3, 10^4, 10^5$ cm^{-3} for γ-, δ-, and ϵ-lines.

131.044   The composition of the interstellar dust.
        J. E. Gaustad.
Astron. Soc. Pacific, Leaflet No. 483, 8 pp. (1969).

131.045   Galactic water vapor emission: Further observations of variability.
S. H. Knowles, C. H. Mayer, W. T. Sullivan III, A. C. Cheung.
Science, Vol. 166, 221 - 224 (1969).

Recent observations of the 1.35-centimeter line emission of water vapor from galactic sources show short-term variability in the spectra of several sources. Two additional sources, Cygnus 1 and NGC 6334 N, have been observed, and the spectra of W49 and VY Canis Majoris were measured over a wider range of radial velocity.

131.046   Grain temperatures in interstellar dust clouds.
        M. W. Werner, E. E. Salpeter.
Monthly Notices, Roy. Astron. Soc., Vol. 145, 249 - 269 (1969).

We have solved in detail the problem of radiative transfer in a spherical dust cloud, including the effects of scattering and of reradiation by the grains. The results have been used to calculate the grain temperature as a function of position in such a cloud, and to estimate such quantities as the cloud albedo and the radiant energy density in a cloud. Several grain models have been considered. The results indicate that grains of the types considered are not able to become sufficiently cold ($T \lesssim 4°K$) in the dust clouds commonly observed in the galaxy to permit the formation of solid H_2 mantles.

131.047 Distribution and temperature of interstellar electron gas. A. H. Bridle, V. R. Venugopal.
Nature, Vol. 224, 545 - 547 (1969).

We have used data on the variation of $R = N_e/N_H$ with galactic latitude, observations of interstellar absorption at 10 MHz, and directly measured pulsar distances to determine a model of the large scale properties of interstellar electron gas. A thick disk distribution for electrons is suggested by several different kinds of observational evidence.

131.048 Interstellar polarization by graphite-silicate grain mixtures. N. C. Wickramasinghe.
Nature, Vol. 224, 656 - 658 (1969).

The observed wavelength dependence of interstellar polarization may be understood on the basis of graphite-silicate grain mixtures.

131.049 The high-velocity hydrogen clouds considered as satellites of the Galaxy.
F. J. Kerr, W. T. Sullivan, III.
Astrophys. Journ. Vol. 158, 115 - 122 (1969).

The high-velocity hydrogen clouds at high latitudes have been widely discussed as material falling into the galactic disk from outside. The observed velocities show strong effects arising from the galactic rotation at the sun's position. When this rotation is removed from the calculations, the residual velocities are compatible with the clouds being in highly eccentric orbits around the Galaxy at distances of the order of 50 kpc. We propose new observations which may help to decide between galactic and extragalactic interpretations. A solar-motion solution for a sample of sixty-two high-velocity clouds with $|b| > 20°$ yields a velocity of 222 km sec^{-1} with respect to the local standard of rest toward $l = 136°$, $b = -4°$.

131.050 Infrared spectra of highly reddened stars: A search for interstellar ice grains.
R. F. Knacke, D. D. Cudaback, J. E. Gaustad.
Astrophys. Journ. Vol. 158, 151 - 160 (1969).

Infrared spectra extending to 3.5 μ have been obtained of the highly reddened supergiants VI Cyg No. 12, CIT 11, and HD 183143. No absorption band at 3.07 μ stronger than 10 percent of the expected absorption of ice grains can be fitted to the spectra of the first two objects, and none greater than 50 percent to the third. Theoretical calculations have been made of the absorption cross-sections for pure-ice grains and for composite grains consisting of ice mantles surrounding a graphite core. New laboratory spectra of the position, strength, and shape of the ice band confirm that these calculations are valid for interstellar temperatures. It is concluded that the interstellar grains contain very little ice.

131.051 Heating of H I regions by energetic particles. II. Interaction between secondaries and thermal electrons.
L. Spitzer, Jr., E. H. Scott.
Astrophys. Journ. Vol. 158, 161 - 171 (1969).

Direct heating of the thermal electrons by cosmic-ray primaries and secondaries is taken into account in the computation of the electron densities and temperatures in H I regions. The mean kinetic energy available for heating the gas per free electron produced increases from 3.4 eV for low n_e/n (H I) to 31 eV for n_e about equal to n (H I). When n_e/n_H exceeds about 0.1, heating of the thermal electrons by the primary energetic particles becomes the dominant source of thermal energy for the gas. If the cosmic-ray flux at low energies is assumed to be no greater than that observed near the

earth, the computed temperature and density are about the same as were found in paper I, in which direct heating of the thermal electrons was ignored. If the heating by cosmic rays is given a maximum value computed from the observed energy released in type I supernova shells, the temperature at low hydrogen densities is much increased, with about 10000° found for n_H equal to 0.1 cm^{-3}, and somewhat higher values at lower n_H. The value of n_e in this case varies very slowly with n_H, increasing from 0.01 to 0.02 cm^{-3} as n_H increases from 3×10^{-2} to 3 cm^{-3}.

131.052 Thermal properties of interstellar gas heated by cosmic rays. D. W. Goldsmith, H. J. Habing, G. B. Field.
Astrophys. Journ. Vol. 158, 173 - 183 (1969).

We present detailed calculations of thermal properties of predominantly neutral interstellar gas heated by low-energy cosmic rays. The heating rate does not depend on the form of the energy spectrum of the cosmic rays, but only on an integral over the spectrum. This integral is proportional to the ionization rate ζ. It is shown that two heating processes are important. Calculations of thermal equilibrium show that the kinetic pressure in the gas has a maximum at an atomic density of about 0.2 cm^{-3}. This leads to a model in which the interstellar gas is represented by two thermally stable phases in pressure equilibrium — a rarefied one at about 10^4 °K and a dense one at about 100° K. Finally, we discuss the influence on equilibrium temperatures of lowering abundances by accretion of trace elements on grains. Evidence is presented for underabundance of all trace elements observed in the interstellar gas.

131.053 Heating of H I regions by soft X-rays.
J. Silk, M. W. Werner.
Astrophys. Journ. Vol. 158, 185 - 192 (1969).

Recent observations of diffuse soft X-rays at 1/4 keV indicate that the energy input to H I regions due to photoionizations by soft X-rays and collisional ionizations by photoelectrons may be comparable in importance to cosmic-ray heating. Possible extrapolations of the flux of soft X-rays below 1/4 keV are considered, and some differences between heating by soft X-rays and heating by cosmic rays are discussed.

131.054 Measurements of the "corner" of the interstellar extinction law. J. W. Harris.
Nature, Vol. 223, 1046 - 1048 (1969).

The extinction law, which is shown by low resolution observations to have a rapid change of slope at about 4300 Å, has been examined at a resolution of 6 Å. It is suggested that the "corner" is a real feature of the law distinct from the nearby 4430 Å interstellar band. Within the errors of measurement, two straight lines intersecting at 4355 ± 17 Å best represent the extinction in magnitudes against reciprocal wavelength between 2.0 and 2.7 μm^{-1}.

131.055 The discovery of interstellar formaldehyde.
D. Buhl, L. E. Snyder, B. Zuckerman, P. Palmer.
Bull. American Astron. Soc. Vol. 1, 235 (1969). – Abstr. AAS.

131.056 Observations of microwave emission from the 1.35-cm line of interstellar water vapor.
A. C. Cheung, D. D. Cudaback, D. M. Rank, D. D. Thornton, C. H. Townes, W. J. Welch.
Bull. American Astron. Soc. Vol. 1, 236 - 237 (1969). – Abstr. AAS.

131.057 A theoretical model for the interstellar medium.
G. B. Field, D. W. Goldsmith, H. J. Habing.
Bull. American Astron. Soc. Vol. 1, 240 - 241 (1969). –

Abstr. AAS.

131.058 Molecular hydrogen formation on grains in H I regions. D. Hollenbach, E. E. Salpeter.
Bull. American Astron. Soc. Vol. 1, 244 (1969). – Abstr. AAS.

131.059 Preliminary low-temperature absorption and scattering data of organic powders simulating interstellar dust. F. M. Johnson, G. W. Hodgson.
Bull. American Astron. Soc. Vol. 1, 246 (1969). – Abstr. AAS.

131.060 Observations of H_2O line emission at 1.35-cm wavelength in the interstellar medium.
S. H. Knowles, C. H. Mayer, A. C. Cheung, D. M. Rank, C. H. Townes.
Bull. American Astron. Soc. Vol. 1, 249 - 250 (1969). – Abstr. AAS.

131.061 Ultraviolet observations of the 18-cm OH emission sources in NGC 6334 and Orion.
W. Liller, B. Zuckerman.
Bull. American Astron. Soc. Vol. 1, 251 (1969). – Abstr. AAS.

131.062 A study of high- and intermediate-velocity H I clouds with 12.5 spatial resolution. J. J. Rickard.
Bull. American Astron. Soc. Vol. 1, 259 (1969). – Abstr. AAS.

131.063 Observations of galactic OH. B. E. Turner.
Bull. American Astron. Soc. Vol. 1, 263 - 264 (1969). – Abstr. AAS.

131.064 Interstellar extinction and intrinsic color indices of stars in the ultraviolet, based on observations prior to December 1968. E. K. L. Upton.
Bull. American Astron. Soc. Vol. 1, 264 (1969). – Abstr. AAS.

131.065 Internal radial velocities of selected H II regions.
R. A. Williamson, R. B. Tully.
Bull. American Astron. Soc. Vol. 1, 266 (1969). – Abstr. AAS.

131.066 Detection of the $^2\pi_{3/2}$, $J = {}^5/_2$ state of OH at 5-cm wavelength.
J. L. Yen, B. Zuckerman, P. Palmer, H. Penfield.
Bull. American Astron. Soc. Vol. 1, 269 (1969). – Abstr. AAS.

131.067 Interstellar gas and field.
E. N. Parker, I. Lerche.
Comments Astrophys. Space Phys. Vol. 1, 215 - 219 (1969).
The present writing is a survey of the work that has been done on the whole picture of interstellar gas and field, including thermal effects, but with particular emphasis on the dynamical effects of the magnetic fields and cosmic rays.

131.068 The collapse of interstellar gas clouds – III. Numerical methods. D. J. Crampin, M. J. Disney, D. McNally, A. E. Wright.
Monthly Notices, Roy. Astron. Soc., Vol. 145, 423 - 433 (1969).
The finite difference numerical methods used in a solution of the equations of hydrodynamics with gravitation are presented. The accuracy and stability of the methods are discussed and it is shown that iterative procedures remain stable and accurate for an integration time comparable with the free-fall time determined by the initial central density of the cloud. This free-fall time imposes a natural limit to the conti-

nuance of the solutions for gas clouds having regions which are almost freely falling.

131.069 Entdeckungen von Molekülen im interstellaren Gas. Ein Überblick. H. Lambrecht.
Sterne, 45. Jahrgang, 129 - 132 (1969).

131.070 Cross-section for the excitation of highly excited hydrogen atoms by electrons and protons.
I. C. Percival, D. Richards.
Astrophys. Letters, Vol. 4, 235 - 237 (1969).
Accurate cross-sections for some important processes involving highly excited hydrogen which occur in the H II region are obtained in analytic form. The cross-sections are needed for the theory of radio recombination lines.

131.071 The collapse of interstellar gas clouds. – IV. Models of collapse and a theory of star formation.
M. J. Disney, D. McNally, A. E. Wright.
Monthly Notices, Roy. Astron. Soc., Vol. 146, 123 - 160 (1969).
Previously developed numerical methods for the solution of the equations of hydrodynamics with gravitation are applied to a series of cloud models to investigate the roles of heating, cooling and density distribution on the collapse. The effects of opacity are also investigated and it is shown that the increase of opacity with increasing central density is of critical importance for star formation. A theory of stellar evolution – from gas cloud to proto-star – is proposed and justified both in terms of the numerical models presented in this paper and by analytical linear wave flow approximations where appropriate. It is found that the smallest proto-stars which can be formed by this method have a mass $\sim 10^{32}$ g. A theory of the formation of interstellar gas clouds is also discussed.

131.072 A new class of compact, high-density H II regions.
P. G. Mezger.
Interstellar Ionized Hydrogen, Charlottesville 1967, p. 33 - 59 (1968).
What are the physical characteristics of those H II regions which are associated with OH emission clouds and what discriminates them from "normal H II regions" which do not show nonthermal emission of OH-lines? In a series of observations which were specifically aimed at this problem, my colleagues and I detected a new class of compact high-density H II regions with average electron densities of about 10^4 cm^{-3}, diameters of less than 0.5 pc, and total masses of a few solar masses.

131.073 Young stars, circumstellar clouds and globules.
V. C. Reddish.
Interstellar Ionized Hydrogen, Charlottesville 1967, p. 87 - 97 (1968).
The distribution of globules among young clusters and associations, The sharing of energy and angular momentum between a star and circumstellar cloud, Search for H II globules, The effect on the growth on an H II region of a dense dust cloud surrounding the exciting star.

131.074 Evolution of H II regions: Early stages.
W. G. Mathews.
Interstellar Ionized Hydrogen, Charlottesville 1967, p. 189 - 225 (1968).

131.075 Expansion stages in the dynamics of H II regions.
B. M. Lasker.
Interstellar Ionized Hydrogen, Charlottesville 1967, p. 227 - 243 (1968).

131.076 Space distribution of H II regions.

P. Murdin, S. Sharpless.
Interstellar Ionized Hydrogen, Charlottesville 1967, p. 249 - 267 (1968).

131.077 Optical work on the kinematics of galactic H II regions. J. S. Miller.
Interstellar Ionized Hydrogen, Charlottesville 1967, p. 269 - 282 (1968).

A more detailed account of this work is given in Astrophys. Journ. Vol. 151, 473 - 489 (1968).

131.078 Radio continuum observations of H II regions. Y. Terzian.
Interstellar Ionized Hydrogen, Charlottesville 1967, p. 283 - 312 (1968).

Observations, Galactic background, Physical parameters, Radio spectra, Complex thermal regions.

131.079 Comparison of optical and radio flux measurements of H II regions. W. Gebel.
Interstellar Ionized Hydrogen, Charlottesville 1967, p. 313 - 327 (1968).

131.080 The electron temperatures in H II regions. L. H. Aller.
Interstellar Ionized Hydrogen, Charlottesville 1967, p. 423 - 434 (1968).

131.081 Theoretical determinations of temperatures in H II regions. R. M. Hjellming.
Interstellar Ionized Hydrogen, Charlottesville 1967, p. 435 - 457 (1968).

131.082 Optical measurements of electron temperatures in H II regions. J. B. Kaler.
Interstellar Ionized Hydrogen, Charlottesville 1967, p. 459 - 475 (1968).

I am going to discuss the methods for the optical measurement of electron temperatures. There are basically three categories of measurement to consider; those methods involving forbidden lines, those involving line width, and those which use the recombination lines and continuum.

131.083 Radio measurements of electron temperatures. P. G. Mezger.
Interstellar Ionized Hydrogen, Charlottesville 1967, p. 477 - 505 (1968).

131.084 OH in the galaxy. H. Weaver.
Interstellar Ionized Hydrogen, Charlottesville 1967, p. 645 - 680 (1968). – Review article.

Categories of appearance of the OH line, Absorption lines, Emission lines, The general OH absorption, Anomalous OH emission in the vicinity of H II regions.

131.085 Interferometry of OH sources. J. M. Moran, Jr.
Interstellar Ionized Hydrogen, Charlottesville 1967, p. 681 - 694 (1968).

The very long baseline interferometer, The Haystack-NRAO interferometer, Hat Creek - NRAO interferometer, Fringe rate measurements, Polarization properties.

131.086 Theory of non-thermal emission from interstellar OH. A. E. E. Rogers.
Interstellar Ionized Hydrogen, Charlottesville 1967, p. 695 - 711 (1968).

131.087 Ultraviolet and infrared pumping of OH molecules. M. M. Litvak.
Interstellar Ionized Hydrogen, Charlottesville 1967, p. 713 -

745 (1968).

First, we discuss the general ultraviolet properties of OH, the photo-dissociation of OH, and two body formation of OH via preassociation and UV fluorescence. Then we discuss the saturated maser geometry and emission properties, the microwave line intensities for cases of anomalous absorption and of anomalous main line emission, due only to UV pumping. Finally, we discuss the anomalous satellite line emission and absorption due to far IR pumping and resonance radiation trapping.

131.088 Maser action in space. T. Gold.
Interstellar Ionized Hydrogen, Charlottesville 1967, p. 747 - 761 (1968).

I want to give some general considerations that would apply if a maser of almost any kind was at work. In the case of OH emission I myself have very little doubt that a maser is the right explanation, and I am persuaded that all other theories can be ruled out quite readily.

131.089 Interstellar polarization. G. V. Coyne, N. C. Wickramasinghe.
Publ. Astron. Soc. Pacific, Vol. 81, 533 (1969). – Abstract ASP.

131.090 Internal kinematics of a compact H II region. M. G. Smith, D. W. Weedman.
Publ. Astron. Soc. Pacific, Vol. 81, 550 (1969). – Abstract ASP.

131.091 On the fluctuations of the electron temperature in H II regions. R. Louise, G. Monnet.
Astron. Astrophys. Vol. 3, 270 - 271 (1969). In French.

In this paper, the electron temperature fluctuations influence on the profile-line method (Courtès et al., 1968; Louise and Monnet, 1969) is studied. The Orion nebula case is discussed. It is shown that the discrepancies between various methods is in qualitative agreement with the effect of both temperature fluctuations and departure from L.T.E. for radio frequency lines.

131.092 Interstellar electron temperatures from pulse delay measurements and hydrogen line spectra.
K. Rohlfs, U. Mebold, M. Grewing.
Astron. Astrophys. Vol. 3, 347 - 353 (1969).

Measurements of pulse delay can now provide direct information on electron densities for selected directions. By comparing integrated electron densities with the observed total hydrogen density in the same area, lower limits for the ionization ratio of the hot gas component can be obtained. This gives an estimate for the temperature of the gas. For high latitude objects temperatures of more than 10^4 °K were found, while low latitude objects gave values of a few 10^3 °K. It is not clear, however, whether this difference is genuine or only an apparent one.

131.093 Thermal instabilities in the interstellar gas. J. A. de Freitas Pacheco.
Astron. Astrophys. Vol. 3, 368 - 371 (1969).

The equilibrium conditions of the interstellar gas are calculated supposing heating by cosmic rays and cooling by electron impact excitation of impurities. The results obtained are much more sensitive to the adopted chemical composition of the interstellar medium than those presented by Field et al. (1969).

131.094 On the absorption and emission of light by interstellar grains. E. M. Purcell.
Astrophys. Journ. Vol. 158, 433 - 440 (1969).

The Kramers-Kronig relations are applied to the interstellar medium, which for this purpose is considered to be a vacuum sparsely populated by spheroidal grains. An integral over the spectrum of the observed extinction is thereby related to the static dielectric constant of the medium. This yields at once a lower bound on the fraction of the volume occupied by the grains. The same method is used to discuss the far-infrared emissivity of grains. The low-frequency emissivity of a grain cannot be arbitrarily increased by the inclusion of low-frequency oscillators. An upper bound on the attainable emissivity is derived, and from it a lower bound is derived on the equilibrium temperature of grains of a given size and shape.

131.095 Regional variations in the wavelength dependence of interstellar polarization.
K. Serkowski, J. W. Robertson.
Astrophys. Journ. Vol. 158, 441 - 447 (1969).

Polarimetric observations were made in the UBV spectral regions for 38 southern stars with large interstellar polarization. The distribution on the sky of all stars with known ratio p^V/p^B of the amount of polarization in the yellow spectral region to that in the blue is discussed. The largest values of p^V/p^B occur along the Milky Way at l^{II}= 290°-345° and in the relatively nearby associations in Ophiuchus-Upper Scorpius and in Orion. Stars nearer than 500 pc seem to have values of p^V/p^B higher than those of more distant stars.

131.096 Molecular and solid hydrogen in dense interstellar clouds.
P. M. Solomon, N. C. Wickramasinghe.
Astrophys. Journ. Vol. 158, 449 - 460 (1969).

Recent observational data point to a significant underabundance of neutral atomic hydrogen in dark interstellar clouds – suggesting extensive formation of molecular or solid hydrogen. Conditions for the formation of molecular hydrogen (gaseous and solid) at grain surfaces are discussed. It is expected that almost all dense clouds will be molecular regions. The process of recombination lowers the density and temperature and may lead to contraction of the clouds and the beginnings of star formation.

131.097 Regional studies of interstellar sodium lines.
L. M. Hobbs.
Astrophys. Journ. Vol. 158, 461 - 472 (1969).

The profiles of the interstellar sodium D_2-lines are compared for stars in each of five regions of the sky with typical dimensions of 10° by 15°. The five regions lie in Perseus, the Pleiades, Orion, Scorpius, and Cygnus. In Perseus, Orion, and Scorpius, one or more interstellar clouds are observed to have dimensions of at least 35 pc. Although most of the stars observed are OB stars with surrounding H II regions, the observed line widths require in most cases that the absorption occur primarily in foreground H I regions with a much higher total density.

131.098 Interstellar Lα absorption in β¹, δ, and π Scorpii.
E. B. Jenkins, D. C. Morton, T. A. Matilsky.
Astrophys. Journ. Vol. 158, 473 - 478 (1969).

Equivalent-width measurements of the Lα absorption in the rocket ultraviolet spectra of these stars reveal average interstellar hydrogen densities of 3.0, 2.4, and 1.4 cm^{-3} in the line of sight to the respective stars. The wide Lα lines in the Scorpius region could result from a dense sheet of gas in front of both ζ Oph and the Scorpius stars as proposed by Herbig.

131.099 Radiofrequency recombination lines from heavy elements: Carbon.
A. K. Dupree.
Astrophys. Journ. Vol. 158, 491 - 503 (1969).

The population of high levels of the carbon atom is evaluated for conditions expected in H II regions and planetary nebulae. Dielectronic recombination leads to significant departures of the population from its value in thermodynamic equilibrium. As a result, the intensities of radiofrequency transitions are considerably strengthened over equilibrium va-

lues. Such behavior is typical of complex atoms and can lead to observable spectra. The enhancement of carbon recombination lines, however, appears insufficient to explain the intensities of the observed "anomalous" recombination lines in H II regions.

131.100 An investigation of the spectra and time variations of galactic water-vapor sources.
D. Buhl, L. E. Snyder, P. R. Schwartz, A. H. Barrett.
Astrophys. Journ. (*Letters*), Vol. 158, L97 - L102 (1969).

Seven galactic water-vapor sources were observed during April and May of 1969 using the NRAO 140-foot telescope. Orion and W49 exhibited time variations on the scale of a few days, while variations in W75 (DR 21) and W3 occurred over a longer period. W3 OH and VY Canis Majoris appear to be constant over a period of several months. Linear polarization was observed in four sources, and a possible correlation between time variation and polarization is suggested by some of the data.

131.101 Stimulated emission of recombination lines in H I regions.
A. K. Dupree, L. Goldberg.
Astrophys. Journ. (*Letters*), Vol. 158, L49 - L53 (1969).

Radio-frequency recombination lines arising from ionized atoms in H I regions can have intensities which are greatly strengthened over their equilibrium values as a result of stimulated emission due to a background continuum. It is suggested that the "anomalous" recombination line is formed in this way. In addition, hydrogen and helium recombination lines from H II regions are expected to have components formed in H I clouds. Recombination lines may also be observable from H I clouds that are in the line of sight to nonthermal continuum sources. Such lines can be used to give a direct measure of the degree of ionization and abundances in the interstellar medium.

131.102 Interstellar extinction anomalies and the diffuse interstellar bands.
G. A. H. Walker, J. B. Hutchings, P. F. Younger.
Astron. Journ. Vol. 74, 1061 - 1066 (1969).

Interstellar extinction curves of 20-Å resolution have been obtained from photoelectric scanner observations in the range λλ4000 – 5000 for the five stars HD 37022, 46711, 183143, 198478, and 211971, and of 50-Å resolution for the stars HD 154043, 154368, 160529, 167971, in the range λλ4000 – 6500 from photoelectric data published by Willstrop. The curves show significant irregularities which vary from star to star and seem to be analogous to the irregularities in the available rocket ultraviolet observations of the extinction curve, and bear a close resemblance to certain laboratory absorption spectra of solid hydrocarbon mixtures recently reported.

131.103 Radio observations of interstellar formaldehyde.
J. B. Whiteoak, F. F. Gardner.
Proc. Astron. Soc. Australia, Vol. 1, 282 - 283 (1969). – Contribution ASA meeting.

131.104 Notes on the distribution of ionized interstellar gas.
V. A. Razin, I. P. Khizhnyakova.
Izv. vyssh. uchebn. zavedenij. Radiofizika, Vol. 12, 479 - 486 (1969). In Russian. – Abstr. in Referativ. Zhurn. 51. Astron., 12.51.385 (1969).

131.105 Possible identification of the diffuse interstellar absorption band at 4430 Å.
W. W. Duley, W. R. M. Graham.
Nature, Vol. 224, 785 - 787 (1969).

A series of laboratory experiments have examined the possibility that the interstellar band at 4430 Å might be due to the shifted 4226 Å resonance line of neutral calcium. It is assumed that calcium atoms are trapped in the dielectric mantles of interstellar grains. The spectra of atomic calcium trapped at 55°K in various hydrocarbon matrices have absorption bands in the 5500 - 3800 Å region. The Ca-benzene system exhibits a single broad band with asymmetric profile which shifts from 4420 Å to over 4500 Å with increasing Ca concentration.

131.106 Stability of solid hydrogen coated graphite particles.   K. S. K. Swamy, B. Donn.
Nature, Vol. 224, 788 - 789 (1969).

Calculations of the temperatures of 0.05 μ graphite particles with solid hydrogen coatings were carried out. Without impurities the temperature was about 17°K and with efficient far infrared impurity absorption a minimum temperature of 3°K was obtained. Based on hydrogen vapor pressure measurements, the lifetime of the coatings at 3°K is about one year. Purcell showed that efficient coating by impurities is not possible and therefore hydrogen mantles on graphite would be very unstable.

131.107 On the question of interstellar diamonds.   B. Donn, K. S. K. Swamy.
Nature, Vol. 224, 570 (1969).

The optical properties of diamond grains present several difficulties in reproducing the interstellar extinction. Laboratory and theoretical evidence indicate that diamonds could not nucleate under astrophysical conditions. Experiments suggest graphite, not diamond more likely to grow on seed crystals.

131.108 Wavelength dependence of polarization. XVIII. Interstellar polarization and composite interstellar particles.   G. V. Coyne, N. C. Wickramasinghe.
Astron. Journ., Vol. 74, 1179 - 1190 (1969).

Further observations in a continuing survey of the polarization of early-type stars in the galactic plane in the wavelength range 0.3 - 0.9 μ are presented. The observations are discussed in terms of a composite interstellar-grain model with a graphite core and dielectric mantle. Approximate solutions for the wavelength dependence of the polarization produced by an idealized model, consisting of a thin graphite disk with a radius of 0.05 μ imbedded in a dielectric oblate spheroid with a semimajor axis, $a = 0.3$ μ, and a semiminor axis, af, with $0.7 \leq f \leq 0.9$, are given.

131.109 Interstellar matter.   H. C. van de Hulst.
Cosmic Ray Studies, Bombay 1968, p. 122 - 130 (1969).

131.110 The dynamical behavior of the gaseous disk of the galaxy.   E. N. Parker.
Cosmic Ray Studies, Bombay 1968, p. 202 - 216 (1969).

131.111 Interstellar extinction in the ultraviolet.   R. C. Bless, A. D. Code, T. E. Houck, C. F. Lillie, B. D. Savage.
Bull. American Astron. Soc., Vol. 1, 335 (1969). – Abstract AAS.

131.112 Anisotropy of the velocity distribution of interstellar clouds.   S. W. Bruenn.
Bull. American Astron. Soc., Vol. 1, 337 (1969). – Abstract AAS.

131.113 Absorption and emission of recombination radiation by H I regions.   A. K. Dupree, L. Goldberg.
Bull. American Astron. Soc., Vol. 1, 340 (1969). – Abstract AAS.

131.114 Observations of excited state OH.   P. Palmer, B. Zuckerman.
Bull. American Astron. Soc., Vol. 1, 358 (1969). – Abstr. AAS.

131.115 Microwave detection of $H_2 C^{13} O^{16}$.   L. E. Snyder, B. Zuckerman, D. Buhl, P. Palmer.
Bull. American Astron. Soc., Vol. 1, 363 - 364 (1969). Abstr. AAS.

131.116 Interstellar extinction in the ultraviolet.   T. P. Stecher.
Bull. American Astron. Soc., Vol. 1, 364 (1969). – Abstr. AAS.

131.117 Interferometry of galactic H II regions.   W. J. Webster, Jr.
Bull. American Astron. Soc., Vol. 1, 368 (1969). – Abstr. AAS.

131.118 The problem of the motion of the interstellar medium under the action of the shell of a nova or supernova.
V. F. Djachenko, V. S. Imshennik, V. V. Palejchik.
Astron. Zhurn. Akad. Nauk SSSR, Vol. 46, 739 - 744 (1969). In Russian. English translation in Soviet Astron. AJ, Vol. 13, No. 4.

Solving numerically the Navier-Stokes equations for two fluids of completely ionized hydrogen plasma enables to describe the initial state of the motion of the interstellar medium and the forming of a shock wave. It is shown that the calculated picture of the motion is apparently realized for slow novae, such as Nova Her 1934. For shells of supernovae as well as for fast novae a considerable deceleration of the shells must be occurred, before the approximation of gas dynamics becomes valid.

131.119 On the intrinsic polarization of red giants and supergiants.   N. M. Shakhovskoy.
Izv. Krymskoj Astrofiz. Obs. Vol. 39, 11 - 33 (1969). In Russian.

Observations of polarization of stars μ Cep, R Sct, V CVn and R Aqr in three colours (U, B, V) are carried out. New methods of the analysis of multi-colour polarization observations based on the existence of correlation between parameters of the intrinsic polarization in different colours and on the use of a known dependence of the interstellar polarization on wave-length are worked out. For stars μ Cep, R Sct, V CVn, R Boo, RS Cnc, S CrB, U Her, R Leo, Z UMa the polarization of their proper radiation is separated from the interstellar polarization. The discovered regularities in the changes of the intrinsic polarization with wave-length and time confirm the supposition that the intrinsic polarization of red variable giants and supergiants originates from Rayleigh light scattering on neutral atoms and molecules in anisotropic (heterogeneous) extended atmospheres. It is shown that such a model can explain principal peculiarities of the intrinsic polarization when the optical thickness of a scattering shell is large enough.

131.120 Interstellar hydroxyl clouds.   J. M. Moran, Jr.
Science Journ. Vol. 5, No. 2, p. 60 - 65 (1969).

131.121 New OH radio emission sources in Cygnus. PP 6.   J. Elldér, B. Rönnäng, A. Winnberg.
Separate print Onsala Space Res. Lab., 6 pp. (1968).

A survey of the region L2 68 to L2 92 deg and B2 + 1.5 deg at 1612 MHz and 1665 MHz and right-hand circular polarization has revealed four new OH sources. – *JDM*

131.122 OH and formaldehyde radiation properties of the W75 region.

O. E. H. Rydbeck, J. Elldér, E. Kollberg.
Separate print Onsala Space Res. Lab., 4 pp. (1969).

The authors start off to investigate temporal variations in W75. Next thing they have discovered a two dip absorption feature which may mean "an expanding, contracting and also rotating gas ring or cloud." Close agreement is noted between the OH and formaldehyde absorptions. Finally they return to the original problem: "A certain temporal variation seems to remain in W75A." − *RXM*

131.123 Radio observation of galactic H II regions at 4170 MHz (2).
T. Ojima, T. Nakajima, Y. Ishizawa, K. Akabane, K. Miyazawa, T. Takahashi.
Journ. Radio Res. Lab., Vol. 15, (No. 82), 271 - 278 (1968).

Radio maps made at 4.17 GHz with a beamwidth of 11 arc min shown of W1, W5, 3C 152, W13 and W16. Also shown for comparison are the optical plates. −*BMT*

131.124 Interstellar polarization in the directions of globular clusters M3 and M13.
H. M. Dyck, R. D. McClure.
Astron. Journ., Vol. 74, 1177 - 1178 = Contr. Kitt Peak National Obs. No. 486 (1969).

Interstellar polarization in the directions of M3 and M13 has been measured and found to be less than 0.002 mag. The minimum value of the reddening derived from these measurements is $E(B − V) = 0.01$ mag, in good agreement with the low values of reddening recently determined by several other observers.

131.125 The profile of the H_2O radioline from W 49.
V. I. Ariskin, B. G. Kutuza, R. L. Sorochenko.
Astron. Tsirk. No. 529, p. 1 - 2 (1969).   In Russian.

131.126 Wavelength dependence of polarization.
Z. F. Seidov.
Astron. Tsirk. No. 533, p. 4 - 6 (1969).   In Russian.

131.127 High polarization in the direction of the region of RY Cam.   R. A. Vardanian.
Astron. Tsirk. No. 538, p. 3 - 5 (1969).   In Russian.

131.128 On the possibility of investigating the inhomogeneities of the interstellar plasma from scintillations of discrete sources.   L. M. Erukhimov, V. V. Pisareva.
Izv. vyssh. uchebn. zavedenij. Radiofizika, Vol. 12, 900 - 905 (1969).   In Russian. − Abstr. in Referativ. Zhurn. 51. Astron., 1.51.632 (1970).

131.129 Interstellarer Staub.
K.-H. Schmidt.
Jenaer Rundschau (Jena Review), 14. Jahrgang, 333 - 339 (1969).

131.130 A possible inter-relation between interstellar and interplanetary cosmic dust.
J. M. Greenberg.
Space Research IX, Proc. Tokyo 1968, p. 111 - 115 (1969).

In order to account for an injection of interstellar dust which maintains the total zodiacal light component of the interplanetary material we need of the order of 100 tons per second and this order of magnitude is not unreasonable. The physical and dynamical processes of this problem are discussed.

131.131 The possible OH source BC Cygni.
G. F. Gahm, R. R. Zappala.
Publ. Astron. Soc. Pacific, Vol. 81, 887 - 890 = Contr. Lick Obs. No. 304 (1969).

We have obtained Coudé spectrograms of BC Cygni at dispersions of 16 and 32 Å/mm on 103a-F plates with the 20-inch camera of the 120-inch telescope of the Lick Observatory. In addition the star has been observed on two nights with the prime-focus polarimeter of the Crossley reflector. Our observations therefore indicate that BC Cygni is a highly reddened supergiant with a normal spectrum and polarization.

131.132 The absorption spectrum of interstellar material.
G. H. Herbig.
Contr. Lick Obs. No. 264 [Reprinted from "International Conference on Spectroscopy", Bombay 1967], 16 pp. (1968). − Conference paper.

131.133 Dark clouds.   J. Klepešta.
Říše hvězd, Vol. 50, 134 - 137 (1969).
In Czech.

131.134 Interstellarer Wasserstoff (H_2).
S. W. Drapatz, D. C. Cartwright.
Mitt. Astron. Ges. No. 27, p. 151 - 154 (1969). − Conference paper.

131.135 Beobachtungen von 5 H II-Regionen mit einem Interferometer.
W. J. Webster, W. Altenhoff.
Mitt. Astron. Ges. No. 27, p. 179 (1969). − Abstract AG.

131.136 H_α- und [N II]-Beobachtungen von H II-Regionen im Cygnus X-Gebiet.
H. J. Wendker, H. R. Dickel.
Mitt. Astron. Ges. No. 27, p. 179 - 180 (1969). − Abstract AG.

131.137 Revidierte Wasserstofftemperaturen im interstellaren Raum.   U. Mebold.
Mitt. Astron. Ges. No. 27, p. 180 - 183 (1969). − Conference paper.

131.138 Ionisationsgrad und Temperatur des interstellaren Mediums in mittleren galaktischen Breiten.
K. Rohlfs, M. Grewing, U. Mebold.
Mitt. Astron. Ges. No. 27, p. 183 - 186 (1969). − Conference paper.

131.139 Ionisationsgrad und Temperatur des interstellaren Mediums in niedrigen galaktischen Breiten.
M. Grewing, U. Mebold, K. Rohlfs.
Mitt. Astron. Ges. No. 27, p. 186 - 190 (1969). − Conference paper.

131.140 OH and formaldehyde radiation properties of the W75 region.
O. E. H. Rydbeck, J. Elldér, E. Kollberg.
Astrophys. Journ. (*Letters*), Vol. 156, L141 - L146 (1969).

We have been recording the OH ground-state spectra of W75 from time to time since 1967 in order to estimate the magnitude and nature of these variations, using the Onsala 84-foot telescope. The OH ground-state spectra of the two sources W75A and W75B are described and temporal variations of the sources discussed.

Algonquin antenna picks up new radio source.
IEEE Spectrum, Vol. 6, 16, 18 (1969). − Concerning interstellar OH molecule.

Interstellar graphite and silicates.
Nature, Vol. 223, 445 (1969). − News note.

Interstellar clouds: Isotopic formaldehyde.
Nature, Vol. 224, 640 - 641 (1969). − News note.

Galactic Nebulae and Interstellar Matter.
See Abstr. 003.055.

Proceedings of the international symposium on the investigation of physical properties of the interplanetary medium by means of cosmic rays.
See Abstr. 012.026.

On the methods of multi-colour polarization observations of the stellar radiation.   See Abstr. 031.024.

On gravitational instability of the interstellar gas.
See Abstr. 061.004.

Magnetothermal instability in a rotating gravitating fluid.   See Abstr. 061.023.

The effects of star formation and evolution on the evolution of H II regions. ´ See Abstr. 065.060.

Dynamic action of radiation from a massive star.
See Abstr. 065.083.

Solar wind tail and the anisotropic production of fast hydrogen atoms.   See Abstr. 074.018.

Influence of interstellar matter on the density of atmospheric hydrogen.   See Abstr. 082.086.

Local interstellar reddening.
See Abstr. 113.006.

On the reddening lines for the RR Lyrae stars.
See Abstr. 122.027.

The distances of the galactic H II regions NGC 3576 and 3603.   See Abstr. 132.020.

Theoretical intensities of recombination lines.
See Abstr. 132.022.

Recent calculations on the interpretation of the radio frequency recombination line spectrum of hydrogen.
See Abstr. 132.023.

Reflection nebulae and the nature of interstellar grains.   See Abstr. 132.034.

On the luminescence of reflection nebulae in connection with the shape of the scattering indicatrix of interstellar dust grains.   See Abstr. 132.043.

Scattering of pulsar X-ray radiation by interstellar dust particles.   See Abstr. 141.049.

Pulsar distances, spiral structure and the interstellar medium.   See Abstr. 141.130.

On H II regions and pulsar distances.
See Abstr. 141.145.

Interstellar scintillations of pulsar radiation.
See Abstr. 141.147.

On the Faraday rotation in pulsars and in the interstellar medium.   See Abstr. 141.174.

Radio observation of galactic H II regions at 4170 MHz (II).   See Abstr. 141.193.

Galactic line emission from 1 to 10 keV.
See Abstr. 142.013.

Cosmic rays and the interstellar medium.
See Abstr. 143.034.

Solar motion with respect to the high-velocity H I clouds and to the local group.
See Abstr. 151.067.

The ratio of total to selective absorption in the galaxy.   See Abstr. 155.016.

Diffuse component of the cosmic far UV radiation and interstellar dust grains.   See Abstr. 155.017.

132 Emission Nebulae, Reflection Nebulae

132.001 The interpretation of hydrogen radio recombination lines. J. E. Dyson.
Astrophys. Space Sci. Vol. 4, 401 - 416 (1969).

Computations of the high level populations of hydrogen in gaseous nebulae are used to compare observations of radio recombination lines with theoretical predictions based on possible line enhancement. Attempts to confirm the existence of maser action from electron temperatures derived on the assumption of thermodynamic equilibrium are inconclusive. There is evidence that most of the low ($\lesssim$ 5000 K) derived temperatures can be increased by at least a few percent by allowing for line enhancement. Measured ratios of the peak temperatures of lines of the same frequencies originating from different upper quantum levels, indicate maser action if Stark broadening is not taken into account. The inclusion of Stark broadening allows confirmation of maser action only in the central regions of the Orion nebula in the case of the 137 β/ 109 α ratio, and in Orion, and, possibly, IC 1795 and M17, in the case of the 197 β/156 α ratio.

132.002 Internal motions in the Orion nebula. P. Lee.
Astrophys. Journ. *(Letters)*, Vol. 157, L111 - L112 = Contr. Louisiana State Univ. Obs., Baton Rouge, No. 23 = Contr. Kitt Peak National Obs. No. 451 (1969).

Slit-spectroscopic observations of a bright region near the Trapezium in the Orion nebula show evidence of matter with a mean radial velocity of approach of about 60 km sec^{-1} with respect to the nebula.

132.003 Observations of weak emissions in the regions of Ceti and Tauri. Z. V. Karjagina, V. E. Mozjaeva.
Astron. Tsirk. No. 519, p. 5 - 6 (1969). In Russian.

132.004 A study of Ha/[N$_{II}$] ratio in diffuse nebulae. G. Courtès, R. Louise, G. Monnet.
Astron. Astrophys. Vol. 3, 222 - 227 (1969).

We describe the different optical systems used to obtain the Ha/[N$_{II}$] ratio in H II regions and give first results on eight of them. From the Ha/[N$_{II}$] ratio and the electron temperature obtained from the line profiles the N(N$^+$)/N_e abundance ratio is deduced. The Ha/[N$_{II}$] variations within a nebula and from one nebula to another are discussed.

132.005 A spectroscopic study of the remarkable nebula NGC 6302. J. P. Oliver, L. H. Aller.
Astrophys. Journ. Vol. 157, 601 - 605 (1969).

The remarkable nebula NGC 6302, recently discussed by Minkowski and Johnson, is found to show very high-excitation features combined with the low-excitation radiation of [O II] at 3727 Å. The electron density is of the order of 0.3 – 1.0 × 10^4 cm^{-3}, and the electron temperature is in the neighborhood of 18000° K. Both temperature and density seem to vary from filament to filament. NGC 6302 appears to be one of the most highly excited gaseous nebulae known. The He/H ratio appears to be somewhat greater than the ratio ordinarily associated with gaseous nebulae or with the primordial value. NGC 6302 is probably not a planetary nebula, but the absence of nonthermal radio emission hints that it cannot be a supernova remnant, either.

132.006 Über die Farbcharakteristiken einer emittierenden Wolke aus molekularem Wasserstoff.
I. A. Klimishin.
Tsirk. L'vov. Astron. Obs. No. 43, p. 8 - 9 (1969).
In Russian.

132.007 Evolution of diffuse nebulae.

W. G. Mathews, C. R. O'Dell.
Annual Rev. Astron. Astrophys. Vol. 7, 67 - 98 (1969).

132.008 Helium abundances and the sizes of He II and H II regions. R. H. Rubin.
Astron. Journ. Vol. 74, 994 - 998 (1969).

The relative sizes of the ionized helium and ionized hydrogen regions in gaseous nebulae are examined using current model atmospheres for early-type stars. The problem is handled in a general context, and it is found that within an extremely narrow range of stellar effective temperatures, the ratio of the He II to H II volumes changes from nearly zero to one. How this affects the determination of the relative abundances of helium and hydrogen in diffuse nebulae is considered. Comparisons with existing observations of adjacent helium and hydrogen radio recombination lines indicate good agreement with theory. Additional observational tests are suggested — including a test for whether the Orion nebula is ionization bounded.

132.009 Faint emission lines of gaseous nebulae.
J. B. Kaler, L. H. Aller.
Astrophys. Journ. Vol. 157, 1231 - 1244 (1969).

We have reexamined tracings of the long-exposure plates used in the studies of NGC 7009, NGC 6572, and IC 4997, and have identified and measured the intensities of many faint spectrum lines that lie just above the plate-noise limit. We have also measured the intensities of a number of lines which were previously identified visually but for which no intensity measurements were originally made, and we have corrected some errors in the earlier work.

132.010 The light variations of the nuclei in Herbig-Haro object No. 2, 1946 - 1968. G. H. Herbig.
Non-Periodic Phenomena in Variable Stars, IAU Colloquium, Budapest, 1968, p. 75 - 83 = Contr. Lick Obs. No. 282 (1969).

132.011 An analysis of radio recombination lines emitted by the Orion nebula.
R. M. Hjellming, E. Churchwell.
Astrophys. Letters, Vol. 4, 165 - 171 (1969).

Radio recombination line data for the Orion nebula are analyzed using a simple non-LTE theory to obtain $\langle T_e \rangle$ = 11000° K, $\langle N_e \rangle$ = 2.2 × 10^4 cm^{-3}, and $\langle E \rangle$ = 1.8 × 10^7 pc cm^{-6} as parameters which predict, with reasonable accuracy, nearly all of the observed hydrogen a-, β-, γ-, and δ-lines strengths.

132.012 Physical conditions in the Orion nebula (NGC 1976) derived from observations of the excited hydrogen radio line H56a. R. L. Sorochenko, J. J. Berulis.
Astrophys. Letters, Vol. 4, 173 - 178 (1969).

The excited hydrogen radio line H56a has been observed in five points over the Orion nebula with a resolution of 1.9 arc min. The population of the 57th level deviates from local thermodynamic equilibrium and b_{57} = 0.88 ± 0.07. To account for the stimulated emission in the non-LTE populations one must postulate either an increased electron density — to permit clouds of electron gas — or a greater efficiency of collisional processes. The electron temperature T_e decreases and the velocities of the internal turbulent motions increase out to the edge of the nebula.

132.013 On the structure of reflection thin-filamentary nebulae. B. P. Artamonov.
Astron. Zhurn. Akad. Nauk SSSR, Vol 46, 978 - 984 (1969).
In Russian. English translation in Soviet Astron. AJ, Vol. 13, No. 5.

Conditions of the formation of the thin-filamentary structure in reflection nebulae are considered. The application of a mechanism of gutter-instability allows to obtain parameters of gas-dust clouds and galactic arms by the appearance of a system of thin filaments.

132.014   Relative intensities of hydrogen lines in the spectra of nebulae.   V. P. Grinin.
Astrofizika, Vol. 5, 213 - 221 (1969).   In Russian.
English translation in Astrophysics, Vol. 5, No. 2 (1969).

The equations determining the populations of excited levels of hydrogen are derived taking into account partial opacity in the Lyman lines. The redistribution of radiation between the lines as well as the frequency redistribution within the lines themselves are taken into consideration. The equations are solved numerically for the electron temperature T_e = 10^4 °K and for a set of optical thicknesses in Lα.

132.015   Excitation of high Balmer lines: Case C of Aller, Baker and Menzel.   M. J. Seaton.
Monthly Notices, Roy. Astron. Soc., Vol. 145, 91 - 93 (1969).

In case C of Aller, Baker and Menzel (1939) it is supposed that absorption of stellar radiation in spectrum lines contributes to the excitation of Balmer lines observed in gaseous nebulae. Calculations are made of the ratio, for various levels n of the H atom, R_n = (number of excitations by line absorption)/(number of excitations by recombination). It is found that these ratios are small and it is concluded that the case C mechanism cannot explain the anomalous intensities which have been reported for the high Balmer lines.

132.016   Recombination lines 158α and 198β in nine southern nebulae.
R. X. McGee, R. A. Batchelor, J. W. Brooks, M. W. Sinclair.
Australian Journ. Phys. Vol. 22, 631 - 640 (1969).

Recombination lines H 158α, He 158α, and H 198β (near λ 18 cm) have been observed in the H II regions Orion nebula, RCW 38 and 49, η Carinae I, RCW 57, PKS 1617 − 50 I in Norma, NGC 6334 and 6357I, and Omega nebula. Physical parameters such as ratio of line to continuum temperatures T_L/T_c, line halfwidths, median radial velocities, and intensity ratios $I(198\beta)/I(158\alpha)$ and $I(He)/I(H)$ are given.

132.017   Photoelectric observations of the H-alpha and [N II] emission in the Cygnus X complex.
H. R. Dickel, H. Wendker, J. H. Bieritz.
Bull. American Astron. Soc. Vol. 1, 239 (1969). − Abstr. AAS.

132.018   Narrow-band H-alpha observations of the Orion nebula.   R. R. Fisher, R. A. Williamson.
Bull. American Astron. Soc. Vol. 1, 241 (1969). − Abstr. AAS.

132.019   The helium abundance in galactic H II regions.
M. Peimbert, R. Costero.
Bull. American Astron. Soc. Vol. 1, 256 (1969). − Abstr. AAS.

132.020   The distances of the galactic H II regions NGC 3576 and 3603.   W. M. Goss, V. Radhakrishnan.
Astrophys. Letters, Vol. 4, 199 - 203 (1969).

A comparison has been made of H 126α velocities, OH velocities, and H I absorption velocities in the direction of the two prominent optical nebulae in Carina, NGC 3576 and 3603. Both objects are beyond the tangential point in this direction (3.6 kpc). The evidence suggests that NGC 3603 is at ∼ 8.4 kpc while NGC 3576 is just beyond the tangential point. NGC 3603 is one of the most distant optically visible H II regions in the galaxy. Its intrinsic properties are quite

similar to those of W49A.

132.021   Observations of radio frequency recombination lines.   P. Palmer.
Interstellar Ionized Hydrogen, Charlottesville 1967, p. 335 - 372 (1968). − Review article.

132.022   Theoretical intensities of recombination lines.
L. Goldberg.
Interstellar Ionized Hydrogen, Charlottesville 1967, p. 373 - 412 (1968).

A review is given of the effects of departures from thermal equilibrium on the intensities of radio recombination lines under conditions found in H II regions.

132.023   Recent calculations on the interpretation of the radio frequency recombination line spectrum of hydrogen.   J. E. Dyson.
Interstellar Ionized Hydrogen, Charlottesville 1967, p. 413 - 416 (1968).

132.024   Small-scale thermal homogeneity of the Orion nebula.   G. Münch.
Interstellar Ionized Hydrogen, Charlottesville 1967, p. 507 - 516 (1968).

The present communication reports in a preliminary fashion the observation of profiles for the λ4363 [O III] line in two areas of the Orion nebula where the N1 and N2 lines were observed photographically as complex or double.

132.025   A radio study of the H II region Orion B.
M. A. Gordon.
Astrophys. Journ. Vol. 158, 479 - 490 (1969).

Observations of continuum emission of the nebula and of 94α hydrogen recombination lines (7.792871 GHz) at twenty-one points give the distribution of mean electron temperature, emission measure, radial velocity, and dispersion of radial velocity over the nebula. The electron temperature ranges from 6000° to 8500° K. The nebula appears to rotate with a period of 10^6 years, and to expand (or contract) radially at approximately 6 km sec^{-1}. Such an expansion implies an age of 10^5 years.

132.026   Space distribution of small dark nebulae.
B. T. Lynds.
Proc. Fifth Berkeley Symposium on Mathematical Statistics and Probability, Vol. 3, 51 - 60 (1967). − Review article.

132.027   Polarization of the Cygnus Loop at 11-centimeter wavelength.   M. R. Kundu.
Astrophys. Journ. (*Letters*), Vol. 158, L103 - L106 (1969).

Polarization of the Cygnus Loop has been measured at 11-cm wavelength. Strong polarization of about 15 - 25 percent has been detected in the optically faint southern part of the source. The electric vectors are more or less perpendicular to the southern boundary of the source. No significant amount of polarization has been detected in any other part of the source.

132.028   Polarization of reflection nebulae associated with VY Canis Majoris and R Coronae Austrinae.
K. Serkowski.
Astrophys. Journ. (*Letters*), Vol. 158, L107 - L110 (1969).

Polarization of more than 40 percent was found in the blue spectral region for the nebula associated with the OH and H_2O emission source VY Canis Majoris and 13 percent for the nebula associated with R Coronae Austrinae. Colors of the nebulae are similar to those of the illuminating stars. Polarization in the blue spectral region of the T Tauri star R CrA increased during 5 weeks from 9 to 18 percent.

132.029 **Illumination of high-latitude nebulae by the central region of the galactic system.** K. A. Innanen.
Journ. Roy. Astron. Soc. Canada, Vol. 63, 193 - 199 (1969).

A recent galactic mass model and interstellar dust particle scattering data have been used to examine the suggestion that the integrated galactic light is illuminating those high-latitude nebulae with no apparent illuminating star. It is concluded that, to within an order of magnitude, this source of illuminations is reasonable, only if the nebular dimensions exceed 100 pc together with a nebular grain density exceeding 10^{-11} grains/cm^3. Thus these dimensions and densities impose rather stringent requirements.

132.030 **The electron temperature distributions and internal kinematics of seven diffuse nebulae.** P. Foukal.
Astrophys. Space Sci. Vol. 5, 469 - 492 (1969).

A Fabry–Pérot spectrophotometer is used to derive values of the intensity ratio Hα/[N II] at 98 points in the seven bright diffuse nebulae M8, M20, M16, M17, NGC7000, M42, IC434. The fraction of nitrogen in the singly ionized state is estimated in the different objects, and is found to be sufficiently constant within any one nebula so that the above intensity ratio may be used to derive accurate electron temperature distributions. The position of the peak of the nebular line, its excess non-thermal width, its shape and relative intensity are used to derive kinematical models of these objects.

132.031 **Broadband spectroscopy of the Orion nebula.**
E. Churchwell, P. G. Mezger.
Bull. American Astron. Soc., Vol. 1, 337 - 338 (1969). – Abstract AAS.

132.032 **On the color and polarization of reflection nebulae.** M. S. Hanner, J. M. Greenberg.
Bull. American Astron. Soc., Vol. 1, 347 (1969). – Abstract AAS.

132.033 **Interferometric investigations of the filamentary nebula S-22.** T. A. Lozinskaya.
Astron. Zhurn. Akad. Nauk SSSR, Vol. 46, 730 - 738 (1969). In Russian. English translation in Soviet Astron. AJ, Vol. 13, No. 4.

A series of observations of the filamentary nebula S-22 has been carried out with a Fabry-Pérot étalon and an image converter. The profiles of Hα and N II lines are investigated. The radial velocity of the nebula is measured by two methods ((–26 ± 6)km/sec, (– 44 ± 15) km/sec). An expansion of the nebula with a velocity of (35 ± 2) km/sec has been discovered. The problem of the origin of S-22 is discussed.

132.034 **Reflection nebulae and the nature of interstellar grains.** V. Vanýsek.
Vistas in Astronomy, Vol. 11, 189 - 216 (1969).

In the following paper the results of recent studies on reflection nebulae are discussed. An accurate photoelectric method, applied to measurements of brightness, colour and polarization of galactic nebulae, can now be used for the exploration of the physical nature of interstellar grains. The fitting of several theoretical models of interstellar grain-clouds to the observed colour distributions in bright reflection nebulae shows that the presence of dielectric grains in such objects is more probable than the presence of small particles with other physical properties.

132.035 **Temperatures and chemical abundances in H II regions, planetary nebulae and nuclei of galaxies.**
M. Peimbert-Sierra.
Thesis, Univ. California. Univ. Microfilms, Ann Arbor, Mi., 112 pp. (1967). – See Phys. Abstr. Vol. 72, No. 21852 (1969).

132.036 **Level populations of hydrogen and helium in gaseous nebulae.** P. D. Lee.
Thesis, Univ. of Illinois. Univ. Microfilms, Ann Arbor, Mi., 99 pp (1968). – See Phys. Abstr. Vol. 72, No. 27002 (1969).

132.037 **Excitation of forbidden lines in gaseous nebulae. I. Formulation and calculations for $2p^q$ ions.**
H. E. Saraph, M. J. Seaton, J. Shemming.
Phil. Trans., Ser. A, Vol. 264 (No. 1149), 78 - 105 (1968). See Phys. Abstr. Vol. 72, No. 27004 (1969).

132.038 **Expected infrared spectra of gaseous nebulae.**
D. E. Osterbrock.
Phil. Trans., Ser. A, Vol. 264 (No. 1150), 241 - 247 (1969).

132.039 **Chemical abundances in galactic H II regions.**
M. Peimbert, R. Costero.
Bol. Obs. Tonantzintla y Tacubaya, Vol. 5, (No. 31), 3 - 22 (1969).

New photoelectric observations of Orion, M8, and M17 are presented; from these data the chemical abundances of He, C, N, O, Ne, and S with respect to hydrogen are derived, taking into account the effect of temperature fluctuations over the volumes considered. The helium abundances derived for the Orion nebula, M8, and M17 are higher than the helium abundance derived from recent models of the sun, which implies, either that some helium enrichment has taken place since the sun was formed, or that the helium-to-hydrogen ratio was not homogeneous over the galaxy when the sun was formed.

132.040 **The detection of quarks in diffuse nebulae.**
L. Carrasco, M. E. Méndez.
Bol. Obs. Tonantzintla y Tacubaya, Vol. 5, (No. 31), 54 - 56 (1969).

The problem of detecting quarks in diffuse nebulae is studied, considering the existence of hydrogenic quark atoms. It is shown that the predicted upper limit of abundance, relative to hydrogen, is 1.5×10^{-8} if the infrared lines produced by the quark atom are used.

132.041 **Simple models for the approximation of optical properties of reflection nebulae.**
D. A. Rozhkovsky.
Trudy Astrofiz. Inst. Alma-Ata, Vol. 14, 3 - 11 (1969). In Russian.

Several simple models of a scattering medium illuminated by a star or by isotropic external diffuse radiation are described. The evaluated mean optical thickness of a real nebula with given integral brightness is based on a comparison with a corresponding model of the same mean integral brightness. Some extremal features of the homogeneous spherical model are considered.

132.042 **On the radiation of some typical nebulae with a spectrum C + E.** L. A. Pawlowa.
Trudy Astrofiz. Inst. Alma-Ata, Vol. 14, 12 - 17 (1969). In Russian.

The properties of several diffuse nebulae with composed spectra C + E are discussed. It is noted that spectra of these objects have a comparatively weak hydrogen emission and a considerably continuous luminescence produced by radiation of a central star reflected by dust particles.

132.043 **On the luminescence of reflection nebulae in connection with the shape of the scattering indicatrix of interstellar dust grains.** D. A. Rozhkovsky.
Trudy Astrofiz. Inst. Alma-Ata, Vol. 14, 22 - 31 (1969). In Russian.

The volume of interstellar matter effectively illuminated by a star depends on the factor of asymmetry of the indicatrix of scattering. The evaluation of the limiting volume for different properties of the indicatrix is given. As illustration optical properties of the halos around stars observed through clouds of the earth's atmosphere are considered.

132.044 Results of modeling the diffusion of radiation using the electronic computer BESM-3M.
A. W. Kurchakov.
Trudy Astrofiz. Inst. Alma-Ata, Vol. 14, 65 - 75 (1969).
In Russian.
Data for radiation intensity and fluxes of a spherical nebula with a star in its centre are given. Calculations are made for different optical thicknesses, particle albedos, and scattering indicatrices.

132.045 Observations of the 5μ source in Orion.
T. A. Lee.
Publ. Astron. Soc. Pacific, Vol. 81, 878 - 882 (1969).
Observations of the 5μ source in the Orion nebula made over a two-year interval show that the source has not varied significantly in this period. The object is confirmed to be nonstellar in nature and the temperature, as suggested by the energy distribution from 2.2μ to 5μ, is near 650° K.

132.046 A study of monochromatic radio emission of excited hydrogen. R. L. Sorochenko.
Vestn. AN SSSR, No. 4, p. 39 - 45 (1969). In Russian.
Abstr. in Referativ. Zhurn. 51. Astron., 2.51.650 (1970).

Galactic Nebulae and Interstellar Matter.
See Abstr. 003.055.

The effective temperatures of the O stars.
See Abstr. 064.041.

The spectra of stars in comet-like nebulae.
See Abstr. 114.030.

Remarques sur l'échelle de magnitudes absolues des étoiles O. See Abstr. 115.007.

Multi-colour photometry of Orion flare stars.
See Abstr. 122.052.

Radio emission from flare stars near the Orion nebula. See Abstr. 122.124.

OB stars near the supernova remnant RCW 86.
See Abstr. 125.003.

OB stars near the supernova remnant RCW 103 and the galactic structure in Norma. See Abstr. 125.004.

Search for microwave emission from the $^2\pi_{1/2}$, $J = 3/2$ state of OH. See Abstr. 131.001.

On the far-ultraviolet interstellar extinction law in the Orion nebula region. See Abstr. 131.025.

Recombination lines in thermal and non-thermal galactic sources. See Abstr. 131.033.

Ultraviolet observations of the 18-cm OH emission sources in NGC 6334 and Orion. See Abstr. 131.061.

Internal radial velocities of selected H II regions.
See Abstr. 131.065.

Stimulated emission of recombination lines in H I regions. See Abstr. 131.101.

Interferometry of galactic H II regions.
See Abstr. 131.117.

Distribution of brightness in polarization of Taurus A and brightness distribution of NGC 1976 at 9.55-mm wavelength. See Abstr. 134.004.

133 Planetary Nebulae

133.001 **Infrared emission from planetary nebulae.**
N. J. Woolf.
Astrophys. Journ. *(Letters)*, Vol. 157, L37 - L40 (1969).
Broad-band 11.5-μ observations are reported for eight planetary nebulae, four W-R and Of stars, and three symbiotic stars. The largest infrared excess is found for BD + 30°3639.

133.002 **A study of Ha/[N II] ratio lines in the Helix nebula.**
R. Louise.
Astron. Astrophys. Vol. 3, 29 - 30 (1969).   In French.
Being the largest planetary nebula, "Helix" (NGC 7293) is easily observed with the "Focal Reducer" interferometer. Thus, the radial velocity field has been obtained by Carranza et al. (1967). With the same techniques used by Mein (1968) and Courtès et al. (1969), the Ha/[N II] ratio lines are measured over the Helix nebula; the electron temperature can be derived from these ratio line measurements. First results show an increasing Ha/[N II] ratio to the outer parts of the nebula and consequently the electron temperature is decreasing with the distance to the exciting star.

133.003 **The planetary nebulae – IV.** L. H. Aller.
Sky Telescope, Vol. 38, 82 - 85 (1969).

133.004 **The planetary nebulae–V.** L. H. Aller.
Sky Telescope, Vol. 38, 152 - 155 (1969).

133.005 **On the Balmer decrement of planetary nebulae.**
S. V. Rublev.
Astron. Tsirk. No. 522, p. 1 - 4 (1969).   In Russian.

133.006 **Observations of planetary nebulae at 11 cm and 6 cm wavelengths.** J. C. Ribes.
Astron. Astrophys. Vol. 3, 156 - 162 (1969).
The 200 meter transit radiotelescope at Nançay was used to observe planetary nebulae at 2 wavelengths: 22 were measured at 11 cm, and 11 of them at 6 cm. 6 of the nebulae were not previously detected. The minimum detectable signal was about 0.01 f.u. Comparison is made with other surveys, and the effects of confusion are discussed. The thermicity of the spectra is established in all cases. The interstellar H$_\beta$ extinction is derived from comparison with optical observations.

133.007 **Erratum: Atmospheres and extended envelopes of central stars of planetary nebulae.** [Astron. Astrophys. Vol. 1, 180 - 192 (1969)].
K. H. Böhm.
Astron. Astrophys. Vol. 3, 256 (1969). – See Astron. Astrophys. Abstr. Vol. 1, 133.001.

133.008 **Spectrophotometric studies of gaseous nebulae. XIII. The high-excitation planetary NGC 4361.**
S. Heap, L. H. Aller, S. J. Czyzak.
Astrophys. Journ. Vol. 157, 607 - 610 (1969).
The nebula NGC 4361 is a planetary of one of the highest excitations known, and the photoelectric and photographic spectrophotometry yields intensities of the stronger lines. The electron temperature appears to be unusually high (near 24000° K), and most of the atoms are highly ionized. Trebly and quadruply ionized neon is more abundant than doubly ionized oxygen.

133.009 **Observations of 23 planetary nebulae at 408 MHz.**
A. E. Le Marne.
Australian Journ. Phys. Vol. 22, 545 - 547 (1969). – Short communication.

133.010 **The Bowen fluorescence mechanism in planetary nebulae and the nuclei of Seyfert galaxies.**
R. J. Weymann, R. E. Williams.
Astrophys. Journ. Vol. 157, 1201 - 1213 (1969).
The efficiency of the Bowen fluorescence mechanism in models of planetary nebulae and the nuclei of Seyfert galaxies has been calculated by using the Feautrier method to solve the equation of transfer for He II Ly-a and the Bowen lines. The calculated efficiencies, which do not show significant differences between planetary nebulae and Seyfert galaxies, range from about 40 to 50 percent for realistic models.

133.011 **Abundance in a halo planetary nebula.**
J. S. Miller.
Astrophys. Journ. Vol. 157, 1215 - 1223 = Contr. Lick Obs. No. 271 (1969).
Photoelectric measurements of emission-line intensities have been obtained for a planetary nebula of relatively low density and moderate excitation located near the north galactic pole. Considerations of the Hβ flux, apparent diameter, and electron density suggest a distance of approximately 20 kpc. The value N(He)/N(H) = 0.13 for this object is similar to that found in the planetary in M15 and disk planetaries in general. Its oxygen-to-hydrogen ratio is near the lower limit found for disk planetaries. The neon abundance is low in comparison with that found in typical planetaries, though it is similar to that measured for the one in M15.

133.012 **Spectrophotometric studies of gaseous nebulae. XIV. The bright moderate-excitation planetary IC 5217.**
S. J. Czyzak, L. H. Aller, D. Leckrone.
Astrophys. Journ. Vol. 157, 1225 - 1229 (1969).
Relative spectral-line intensities obtained by photographic and photoelectric spectrophotometry are given for the bright northern planetary IC 5217. The Balmer decrement suggests that the interstellar absorption must be small. Furthermore, the nebula must be relatively homogeneous.

133.013 **On the classification of emission-line spectra of planetary nuclei.** L. F. Smith, L. H. Aller.
Astrophys. Journ. Vol. 157, 1245 - 1254 (1969).
Emission-line spectra of nuclei of planetary nebulae are ordered into five classes according to whether they closely resemble spectra of population I Of and Wolf-Rayet stars or show marked differences from the latter. Spectra that are closely similar to those of population I stars are in the minority. One of the classes – which we call the O VI sequence – includes spectra which correspond to extremely high excitation; the nucleus of NGC 5189, which has been tentatively identified as an X-ray source, has a spectrum that falls in this class.

133.014 **On the expansion of planetary nebulae.**
V. V. Vityasev.
Astrofizika, Vol. 5, 83 - 95 (1969).   In Russian.
English translation in Astrophysics, Vol. 5, No. 1 (1969).
The statistical dependence between expansion velocities and radii of planetary nebulae is verified with a reliable positive correlation coefficient. The nebula expansion due to direct L$_c$- and diffuse L$_a$-radiation is considered. A noticeable role of L$_a$-radiation in the expansion of planetary nebulae is shown. Assuming the expansion of the shell under action of corpuscular pressure, the rate of corpuscular emission from nuclei of planetary nebulae is evaluated to be 10^{-8} solar masses per year.

133.015 **Radiative excitation in planetary nebulae.**

D. Van Blerkom.
Monthly Notices, Roy. Astron. Soc., Vol. 145, 75 - 90 (1969).

The recombination spectra of hydrogen and helium from several planetary nebulae and the Orion nebula cannot be explained by recombination theory in its present form. In this paper, we investigate the possibility that the deviations of the relative line intensities from theory are caused by direct photoexcitation of ground state atoms through absorption of stellar resonance line photons. The transfer equations for Lyman line and continuum radiation are solved in nebulae of different thicknesses by means of the normalized on-the-spot (NOS) approximation discussed in the text. When the central star radiates as a black body at all frequencies, the Balmer decrements are found to differ only slightly from those predicted by recombination theory under optically thick (case B) conditions. By inverting the problem, we find the spectral energy distribution of the central star which produces the observed intensities of the nebula emission lines.

133.016   **Abundances in a halo planetary nebula.**
J. S. Miller.
Bull. American Astron. Soc. Vol. 1, 253 (1969). – Abstr. AAS.

133.017   **Radio observations of the nebula NGC 6857-K 3-50.**   R. H. Rubin, B. E. Turner.
Bull. American Astron. Soc. Vol. 1, 260 - 261 (1969). – Abstr. AAS.

133.018   **[O I] λ6300 emission in planetary nebulae.**
R. E. Williams.
Bull. American Astron. Soc. Vol. 1, 266 (1969). – Abstr. AAS.

133.019   **A mechanism for the production of planetary nebulae.**   G. S. Kutter, M. P. Savedoff, D. W. Schuerman.
Mass Loss from Stars, Trieste 1968, p. 311 - 328 (1969).

C^{12} stars in the range 1.04 - 1.55 $M_\odot$ are evolved to simulate the core evolution of the possible precursors of planetary nebulae. The nuclear shell burning in stars above 1.2 $M_\odot$ advances to within about 0.2 $M_\odot$ of the surface, where the intense radiation interacts with the surface matter and causes mass loss. Comparison between our theoretical results and observations suggests that this may be a mechanism by which planetary nebulae are formed.

133.020   **Der Ionisationsgrad in den Atmosphären der Zentralsterne planetarischer Nebel.**
N. A. Sakhibullin.
Trudy Kazan. Gorod. Astron. Obs. No. 35, p. 21 - 43 (1968). In Russian.

Verf. weist nach, daß die Formel von Saha bei hohen Temperaturen und Drucken nicht anwendbar ist. Eine allgemeine Formel für den Ionisationsgrad des Wasserstoffs wird abgeleitet und das Energiegleichgewicht der freien Elektronen wird untersucht. Die Elektronentemperatur unterscheidet sich nicht wesentlich von der Strahlungstemperatur.

133.021   **The planetary nebulae – VI.**   L. H. Aller.
Sky Telescope, Vol. 38, 227 - 229 (1969).

133.022   **Relative intensities of Bowen lines.**
H. Nussbaumer.
Astrophys. Letters, Vol. 4, 183 - 186 (1969).

Transition probabilities for allowed transitions in O III are given. An apparent discrepancy between observed and calculated line intensities in planetary nebulae can be explained.

133.023   **H109α line observations of six planetary nebulae, DR 21 and IC 410.**   Y. Terzian, B. Balick.
Astrophys. Letters, Vol. 4, 195 - 198 (1969).

Hydrogen 109α recombination line observations have been made on six planetary nebulae, and on the H II regions DR 21 and IC 410. The recombination lines from planetary nebulae have not been detected. In particular previous observations of the line detection from NGC 7027 have not been confirmed. DR 21 and IC 410 show well defined lines.

133.024   **The ionization structure of planetary nebulae. – VII. The heavy elements.**   D. R. Flower.
Monthly Notices, Roy. Astron. Soc., Vol. 146, 171 - 185 (1969).

Computer models of planetary nebulae have been developed incorporating, in addition to H and He, the elements C, N, O and Ne. The inclusion of the heavy elements has little effect on the radiative transfer problem, however, their inclusion is critical to the thermal balance problem because of the contribution of collisional excitation of low-lying terms of ions of the heavy elements to the cooling of the nebula; this is discussed in detail.

133.025   **A comparison of radio-frequency and optical studies of selected planetary nebulae.**   L. H. Aller.
Proc. Astron. Soc. Australia, Vol. 1, 283 - 285 (1969). – Contribution ASA meeting.

133.026   **The planetary nebulae – VII, VIII.**
L. H. Aller.
Sky Telescope, Vol. 38, 306 - 309, 377 - 379 (1969).

133.027   **Optical positions of 32 planetary nebulae.**
L. A. Higgs.
Journ. Roy. Astron. Soc. Canada, Vol. 63, 200 - 202 (1969).

The 1950.0 equatorial co-ordinates of 32 planetary nebulae, north of declination -30°, have been determined from the Palomar Sky Atlas prints. These improved positions are required for centimetre wavelength radio observations of planetary nebulae.

133.028   **The ionization structure of planetary nebulae – VIII. Models of the nebulae NGC 7662 and IC 418.**
D. R. Flower.
Monthly Notices, Roy. Astron. Soc., Vol. 146, 243 - 263 (1969).

Using computer programs which are based on the theory developed in the previous papers in this series a number of models of the planetary nebulae NGC 7662 and IC 418 have been computed. An essential feature of the models is the inclusion of stellar atmosphere fluxes calculated by Hummer & Mihalas which fit the observational data for the central stars of these two nebulae. A detailed comparison is made between observed and calculated forbidden line intensities and values derived for the abundances of nitrogen, oxygen and neon relative to hydrogen in the nebulae.

133.029   **Hamburg Schmidt-camera survey of faint planetary nebulae. Galactic anticenter region.**
L. Kohoutek.
Bull. Astron. Inst. Czechoslovakia, Vol. 20, 307 - 312, 382 a, b (1969).

The second part of the spectral Schmidt-camera survey of faint planetary nebulae contains a large area of the northern Milky Way in the direction to the galactic anticenter, l^{II} 146° – 214°, b^{II} ± 10°. Except the 25 planetary nebulae known in this region 13 new objects were found. The appearance of half of these new nebulae is stellar in both the spectral plates and the Palomar Atlas. The measurements of non-stellar planetaries in the Palomar Atlas and the description of the spectra in the blue region of the two brightest new objects are given. The larger concentration of the new planetary nebulae towards the galactic equator indicates the approximate double average distance of these objects with respect to the known ones. In the Appendix five compact and small new H II regions have been described.

133.030 Errata: Catalogue of Galactic Planetary Nebulae.
[Published by Academia, Prague 1967.]
L. Perek, L. Kohoutek.
Bull. Astron. Inst. Czechoslovakia, Vol. 20, 381, 382a, b (1969).

133.031 Spectrophotometric studies of gaseous nebulae.
XV. The high-excitation planetary NGC 6741.
L. H. Aller, E. Krupp, S. J. Czyzak.
Astrophys. Journ., Vol. 158, 953 - 957 (1969).
Spectrographic data obtained at the prime focus of the Lick 120-inch telescope and photoelectric data obtained with the Cassegrain scanner at the 60-inch telescope at Mount Wilson are compared with earlier data to provide a set of line intensities for the relatively faint planetary nebula NGC 6741, which combines features of both high and low excitation.

133.032 Planetary nebulae as possible X-ray sources.
G. S. Khromov.
Astron. Zhurn. Akad. Nauk SSSR, Vol. 46, 747 - 749 (1969).
In Russian. English translation in Soviet Astron. AJ, Vol. 13, No. 4.
From a discussion of known mechanisms of X-ray emission it is shown that planetary nebulae cannot be observable X-ray sources in the region of 4 – 8 keV. From statistic considerations it follows that with the high accuracy of X-ray observations, the probability of accidental coincidences of the

sources with planetary nebulae in the region of the galactic equator is large. At the same time it is possible that hot nuclei of planetary nebulae will be observable in the region of energies of about 0.25 keV.

133.033 A possible new planetary nebula in Cygnus.
W. A. Sherwood.
Observatory, Vol. 89, 207 (1969). – Letter.

133.034 Planetary nebulae. V. P. Arkhipova.
Zemlya i Vselennaya, No. 4, p. 39 - 45 (1969).
In Russian.

133.035 On the formation of planetary nebulae.
K. S. K. Swamy, T. P. Stecher.
Publ. Astron. Soc. Pacific, Vol. 81, 873 - 875 (1969).
In this note we would like to discuss briefly a new mechanism which is physically more plausible than the ones suggested thus far.

133.036 Expandierende Atmosphären von Zentralsternen
Planetarischer Nebel.
J. Schmid-Burgk.
Mitt. Astron. Ges. No. 27, p. 235 (1969). – Abstract AG.

Radio observations of the nebulae K3-50 and NGC
6857. See Abstr. 131.011.

Radiofrequency recombination lines from heavy
elements: Carbon. See Abstr. 131.099.

134 Crab Nebula

134.001 Acceleration of relativistic particles in the Crab
nebula. F. C. Michel.
Astrophys. Journ. Vol. 157, 1183 - 1199 (1969).
On the basis of an acceleration mechanism proposed earlier, a model is developed to describe the synchrotron emission expected from objects such as the Crab nebula. We propose to transfer energy from magnetic fields into relativistic particles with essentially 100 percent efficiency, the electron energy being converted into synchrotron radiation also with 100 percent efficiency. This acceleration is shown to take place at neutral sheets within the magnetic field. On the basis of the acceleration mechanism and the observed properties of the Crab nebula (luminosity and spectral shape) we can express the bulk parameters (magnetic-field strength, individual and total electron energy, and background plasma concentration) in terms of the total area of the surfaces of the neutral sheets separating magnetic-field discontinuities.

134.002 The Crab explained.
Nature, Vol. 223, 1030 - 1032 (1969).
New notes on the Crab nebula and its pulsars.

134.003 Observation de pulsations dans le flux X de la
nébuleuse du Crabe 2.5 et 30 keV.
G. Ducros, R. Ducros, R. Rocchia, A. Tarrius.
Comptes Rendus Acad. Sci. Paris, Sér. B, Vol. 269, 932 - 934 (1969).
La découverte de deux pulsars au voisinage de la nébuleuse du Crabe, ainsi que l'observation d'éclairs lumineux en provenance de cette source, nous conduits à rechercher des pulsations dans le domaine des rayons X. Une expérience

faite en fusée, en décembre 1968, depuis la base de Kourou en Guyane française, a permis de déterminer le profil de la pulsation dont la période est 33.0929 ms.

134.004 Distribution of brightness in polarization of Taurus
A and brightness distribution of NGC 1976 at 9.55-
mm wavelength. K. J. Johnston, R. W. Hobbs.
Astrophys. Journ. Vol. 158, 145 - 150 (1969).
The polarized brightness distributions of Tau A and NGC 1976 were determined at a wavelength of 9.55 mm using a 1.6 beam. Flux densities of 313 ± 50 and 253 ± 37 f.u. were found for Tau A and NGC 1976, respectively, relative to Jupiter.

134.005 Observation of pulsed hard X-ray radiation from
NP 0532 from 1967 data. G. J. Fishman,
F. R. Harnden, Jr., R. C. Haymes.
Publ. Astron. Soc. Pacific, Vol. 81, 538 - 539 (1969).
Abstract ASP.

134.006 X-ray pulsar in the Crab nebula.
G. Fritz, R. C. Henry, J. F. Meekins, T. A. Chubb,
H. Friedman.
Publ. Astron. Soc. Pacific, Vol. 81, 539 (1969). – Abstract ASP.

134.007 On a peculiarity of the spectrum of radio emission
of the Crab nebula at decimeter wave-lengths.
V. A. Alekseev, É. D. Gatélyuk, D. A. Dmitrenko, A. A. Romanychev, N. M. Tsejtlin.
Izv. vyssh. uchebn. zavedenij. Radiofizika, Vol. 12, 168 - 172

(1969). In Russian. – Abstr. in Referativ. Zhurn. 51. Astron., 10.51.491 (1969).

134.008 Motions in the Crab nebula.
I. Halliday.
Journ. Roy. Astron. Soc. Canada, Vol. 63, 215 - 216 (1969).

134.009 Interstellar absorption of Crab nebula X rays.
F. D. Seward, R. J. Grader, R. W. Hill.
Bull. American Astron. Soc., Vol. 1, 361 (1969). – Abstr. AAS.

134.010 Search for polarization in the X-ray emission of the Crab nebula. R. S. Wolff.
Bull. American Astron. Soc., Vol. 1, 370 (1969). – Abstr. AAS.

134.011 A search for gamma-ray line emissions from the Crab nebula. A. S. Jacobson.
Thesis, Univ. California. Univ. Microfilms, Ann Arbor, Mi., 129 pp. (1968). – See Phys. Abstr. Vol. 72, No. 21858 (1969).

134.012 Investigation of the Crab nebula at 2.16 and 8.2 mm wavelength. V. A. Efanov, A. G. Kislyakov, V. I. Kostenko, L. I. Matveenko, I. G. Moiseev, A. I. Naumov.
Izv. vyssh. uchebn. zavedenij. Radiofizika, Vol. 12, 803 - 806 (1969). In Russian. – Abstr. in Referativ. Zhurn. 51. Astron., 1.51.618 (1970).

Supernovae outbursts and generation of relativistic particles. See Abstr. 125.018.

Optical polarization measurements of pulsar NP 0532. See Abstr. 141.005.

Optical observations of the Crab nebula pulsar. See Abstr. 141.014.

Spectroscopic observations of the pulsar NP 0532. See Abstr. 141.015.

Hydromagnetic plasma acceleration by rapidly rotating astrophysical objects. See Abstr. 141.017.

Dispersion measures of pulsars. See Abstr. 141.023.

Observations of the optical polarization of the pulsar NP 0532. See Abstr. 141.027.

On the nature of emission of the pulsar NP 0532. See Abstr. 141.035.

Precision measurement of the frequency decay of the Crab nebula pulsar, NP 0532. See Abstr. 141.058.

Polarized brightness distribution of Taurus A and Orion A at 9.55-mm wavelength. See Abstr. 141.094.

Pulsation period of NP 0532. See Abstr. 141.106.

NP 0532 stellar wind torques. See Abstr. 141.108.

Electromagnetic spectrum of NP 0532. See Abstr. 141.111.

Observations of the Crab pulsar during an occultation by the solar corona. See Abstr. 141.115.

Hard X-rays from the Crab pulsar. See Abstr. 141.117.

Search for rapid fluctuations in light from the Crab nebula pulsar. See Abstr. 141.124.

The period and hard-X-ray spectrum of NP 0532 in 1967. See Abstr. 141.127.

Estimate of the gravitational radiation from NP 0532. See Abstr. 141.129.

Further search for nanosecond structure in the light flashes from pulsar NP 0532. See Abstr. 141.150.

Search for rapid fluctuations in light from the Crab nebula pulsar. See Abstr. 141.166.

Anomalous temporal behaviour of NP 0532. See Abstr. 141.188.

A broad-band radio pulsar in the Crab nebula. See Abstr. 141.218.

Absorption of high energy gamma-rays in the vicinity of pulsar NP 0532. See Abstr. 141.226.

Observations of X-rays from Taurus X-1 and Cygnus X-1. See Abstr. 142.005.

Interstellar absorption of 10-Å X-rays. See Abstr. 142.008.

Spectrum of the Crab X-ray source from 4 to 40 keV. See Abstr. 142.029.

Radio Sources, Quasars, Pulsars, X Ray-, Gamma Ray-Sources, Cosmic Radiation

141 Radio Sources, Quasars, Pulsars

141.001 The radiospectra of a homogeneous sample of 4 C radio sources.
C. Fanti, R. Fanti, P. Londrillo, L. Padrielli.
Astron. Astrophys. Vol. 2, 477 - 483 (1969).

Using observations at 178, 408, 1420 MHz the radiospectra of 136 sources in the declination range 29° 30' to 34° 30' of the 4 C catalogue have been determined. From observations at 1420 MHz it is clear that the spectra of many sources flatten at frequencies below 408 MHz. There is evidence that this effect is more marked for sources with small angular diameter.

141.002 Distribution of redshifts of quasars.
B. M. Tinsley, T. N. L. Patterson, B. A. Tinsley.
Astrophys. Letters, Vol. 4, 55 - 56 (1969).

The observed complex distribution of quasar redshifts has been interpreted by Barbieri, Bonometto and Saggion (1968) as due to the combination of simple distributions of cosmological and gravitational redshifts. An error in their analysis is noted which invalidates their conclusions.

141.003 The detectability of quasars beyond $z \simeq 2$.
M. J. Rees.
Astrophys. Letters, Vol. 4, 61 - 64 (1969).

The apparent deficiency of quasars with redshifts substantially larger than 2 could be due to absorption by intervening atomic hydrogen, provided that the intergalactic gas were less highly ionized at early epochs than when $z \lesssim 2$. Some implications of this suggestion are discussed.

141.004 Attempt to identify optically PSR 0833-45.
J. E. Hesser, B. M. Lasker, D. R. Bochonko, D. E. Mook.
Nature, Vol. 223, 485 - 486 (1969).

The attempt was made with the 60-inch telescope on Cerro Tololo. No significant result was reached.

141.005 Optical polarization measurements of pulsar NP 0532.
W. J. Cocke, M. J. Disney, T. Gehrels.
Nature, Vol. 223, 576 - 578 (1969).

The linear polarization is of the order of 15 per cent and is variable, and the circular polarization is less than 5 per cent.

141.006 Galactic nuclei as collapsed old quasars.
D. Lynden-Bell.
Nature, Vol. 223, 690 - 694 (1969).

Powerful emissions from the centres of nearby galaxies may represent dead quasars.

141.007 Periodic intensity fluctuations in pulsars.
J. H. Taylor, M. Jura, G. R. Huguenin.
Nature, Vol. 223, 797 - 799 (1969).

Eight pulsars have been searched for signs of periodic modulation of the pulse amplitudes, such as has already been reported for two pulsars. Five more examples of this phenomenon have been found.

141.008 Do pulsars turn off?
J. P. Ostriker, J. E. Gunn.
Nature, Vol. 223, 813 - 814 (1969).

There are persuasive arguments, nearly independent of the chosen model, which indicate that pulsars do, in fact, turn off in a time short compared with the age of the galaxy. After indicating these arguments, we will show that this behavior is to be expected on the basis of established physical theory and the particular pulsar model which we and others have been developing, that the absence of long period pulsars can be quantitatively understood, and that fairly accurate predictions can be made for the rate of change of period P from other observed quantities.

141.009 Observations of radio sources at 4.3-mm wavelength.
R. W. Hobbs, H. H. Corbett, N. J. Santini.
Astron. Journ. Vol. 74, 824 - 826 (1969).

Measurements have been made at the positions of 23 radio sources at 4.3-mm wavelength using the 36-ft parabolic antenna of the National Radio Astronomy Observatory. The flux densities of the seven sources that were detected are given relative to Jupiter.

141.010 Flux densities of radio sources at a wavelength of 2.8 cm . L. H. Doherty, J. M. MacLeod, C. R. Purton.
Astron. Journ. Vol. 74, 827 - 832 (1969).

The flux densities of 146 radio sources at a wavelength of 2.82 cm (10.63 GHz) have been measured with the 46-m telescope at the Algonquin Radio Observatory. This list of sources includes all extragalactic sources in the 3CR catalogue (Bennett 1962) which have angular diameters less than 2.8 and peak flux densities greater than 0.4×10^{-26} W m^{-2} Hz^{-1} at 2.82 cm.

141.011 Declination measurements of selected 4C sources using the Arecibo 1000-ft reflector.
C. Hazard, D. L. Jauncey, S. Gulkis, D. C. Backer, A. E. Niell, J. Sutton, D. E. Harris, E. J. Gundermann Hardebeck.
Astron. Journ. Vol. 74, 833 - 839 (1969).

The Arecibo 1000-ft radio telescope is being used to measure declinations of sources in the 4C catalogue to an accuracy of 30''. The methods of calibration are described and the results are given for 200 sources. These new declination measurements are compared with the original 4C measurements.

141.012 The spectra of radio sources in the revised 3C catalogue.
K. I. Kellermann, I. I. K. Pauliny-Toth, P. J. S. Williams.
Astrophys. Journ. Vol. 157, 1 - 34 (1969).

Accurate values of the flux density are given for nearly all sources in the *Revised Third Cambridge Catalogue* at frequencies of 38, 178, 750, 1400, 2695, and 5000 MHz. These have been used to determine the spectrum of each source over this frequency range. It is concluded that the form of the spectra in many sources is determined not only by the energy distribution of relativistic electrons but also, as a result of self-absorption, by their spatial distribution. Sources identified with quasi-stellar objects have a wider dispersion of spectral indices than those identified with radio galaxies, particularly at the higher frequencies. Among the radio galaxies, the intrinsically strongest sources appear to have the steepest spectra. Unidentified sources have steep spectra similar to those of the stronger radio galaxies, and they are probably distant galaxies.

141.013 Polarization measurements of extragalactic radio sources at 3.12-cm wavelength.
G. L. Berge, G. A. Seielstad.
Astrophys. Journ. Vol. 157, 35 - 43 (1969).

The two-element interferometer at the Owens Valley Radio Observatory was used in late 1967 to measure the integrated polarization, both linear and circular, of thirty-two extragalactic radio sources at a wavelength of 3.12 cm. In addition, flux densities of 103 radio sources were also measured. There is a tendency for sources with inverted spectra at wavelengths between 11 and 3 cm to have inverted depolarizations in this range as well. No detection of circular polarization was made.

141.014 Optical observations of the Crab nebula pulsar.
E. J. Wampler, J. D. Scargle, J. S. Miller.
Astrophys. Journ. *(Letters)*, Vol. 157, L1 - L10 = Contr. Lick Obs. No. 303 (1969).

Optical observations of the pulsar NP 0532 show that the shape of the light curve was constant to a high degree of accuracy on three nights distributed over a 10-day interval; the shape also is in excellent agreement with a detailed curve reported by Warner, Nather, and MacFarlane. Light curves obtained in the infrared (~7400 Å) and ultraviolet (~3700 Å) are closely similar to each other. Detailed measurements of linear polarization as a function of phase are made. A simple geometrical model consisting of a rotation axis inclined 77° to the line of sight and a "polarization axis" located in or near the rotational equatorial plane is shown to fit the polarization data well. An upper limit of approximately 10 percent is obtained for the amount of circular polarization of the main and secondary pulses.

141.015 Spectroscopic observations of the pulsar NP 0532.
R. Lynds.
Astrophys. Journ. *(Letters)*, Vol. 157, L11 - L12 = Contr. Kitt Peak National Obs. No. 442 (1969).

New spectroscopic observations are presented that reaffirm the absence of discrete spectroscopic features for the optical pulsar NP 0532. For some of the spectrograms a special technique was employed that permits phase resolution of the spectra of the main pulse and interpulse. This resolution in phase makes it possible to place an improved upper limit on the strengths of spectroscopic features.

141.016 Coherent mechanisms of radio emission and magnetic models of pulsars.
V. L. Ginzburg, V. V. Zheleznyakov, V. V. Zaitsev.
Astrophys. Space Sci. Vol. 4, 464 - 504 (1969).

This paper is primarily concerned with the questions of models and the mechanisms of radio emission for pulsars, the polarisation of this radiation and related topics. For convenience and to provide a more complete picture of the problems involved, a short summary of the data on pulsars is also given. The paper contains the following sections: some facts about pulsars; the astrophysical nature of pulsars; coherent mechanisms of radio emission from pulsars; models of pulsars: magnetic, pulsating white dwarfs and neutron stars; the polarization of the radio emission from pulsars; a synthesized model of pulsars — magnetic, pulsating and rotating neutron stars.

141.017 Hydromagnetic plasma acceleration by rapidly rotating astrophysical objects. F. C. Michel.
Phys. Rev. Letters, Vol. 23, 247 - 249 (1969).

We have solved analytically for the acceleration of plasma away from a rotating magnetized object. Application of that analysis to rotator models for the radio, optical, and X-ray pulsar NP 0532 suggests that it has a magnetic moment of order 3×10^{28} G cm^3 and is losing mass at a rate order 2×10^9 g/sec. These parameters would provide particles in excess of 10^{11} eV and a 6×10^{-4} G field in the Crab nebula at 1 lt yr from NP 0532.

141.018 Dynamic spectra of pulsars in the frequency range 110 - 420 MHz.
G. R. Huguenin, J. H. Taylor, M. Jura.
Astrophys. Letters, Vol. 4, 71 - 75 (1969).

Observations of the dynamic spectra of 10 pulsars have been made in the range 110 - 420 MHz. Narrow-band emission features have been observed for 8 of them. The widths of the emission bands scalle approximately as frequency to the power 2.8, and not to the power $\geqslant 4$ as predicted by existing scintillation theories.

141.019 The redshift of the N-type radio galaxy 3 C 371.
T. Virtanen.
Astrophys. Space Sci. Vol. 5, 76 - 77 (1969).

A new identification for spectral lines of the N-galaxy 3 C 371 is proposed. The resulting redshift is $z = 0.40$.

141.020 Pulsars. R. van den Nieuwenhof, H. Rosenberg.
Hemel en Dampkring, Vol. 67, 227 - 232 (1969).

141.021 Ergebnisse des ersten Jahres Pulsar-Forschung.
G. A. Tammann.
Orion Schaffhausen, Vol. 14, 93 - 101 (1969). — Popular article.

141.022 A remark about the m (z) relation for quasars.
M. A. Abramowicz.
Acta Astron. Vol. 19, 225 - 227 (1969).

The large scatter of quasars in the log cz versus m plane makes impossible any comparison of the observational data with various theoretical models of the universe. With this in mind an attempt is made here to show that the scatter in the log cz versus m plane is correlated with the $B - V$ colour or — if we accept the cosmological interpretation of quasars — that the absolute magnitudes are also correlated with $B - V$.

141.023 Dispersion measures of pulsars.
K. Davidson, Y. Terzian.
Astron. Journ. Vol. 74, 849 - 854 (1969).

The relation between dispersion measure and distance for pulsars is considered. The nonuniformity of the interstellar medium makes this relation somewhat uncertain. The distribution of pulsars in the Galaxy is examined. It is found that pulsars tend to concentrate in a band from –10 - 0° in galactic latitude. The absence of pulsars from 6 - 25° in galactic latitude is also noted. The dispersion measures of several pulsars are discussed in detail, in particular, that for the Crab nebula pulsar.

141.024 Are quasi-stellar radio sources giant pulsars?
P. Morrison.
Astrophys. Journ. *(Letters)*, Vol. 157, L73 - L76 (1969).

Quasi-stellar radio sources are analogues to pulsars in every respect save that of scale, even exhibiting regular optical pulses. It is proposed that their common feature is a central, magnetized, spinning, condensed mass, whose luminous lifetime is governed by the rotational work done on charged particles by the moving magnetic field. The properties of the model are tabulated.

141.025 Are some quasi-stellar objects associated with clusters of galaxies?
J. N. Bahcall, M. Schmidt, J. E. Gunn.

Astrophys. Journ. *(Letters)*, Vol. 157, L77 - L79 (1969).

Five small-redshift quasi-stellar objects are listed that are within the approximate geometrical boundaries of catalogued clusters of galaxies. Spectra of four galaxies in one of the clusters show the same redshift as the associated quasi-stellar object B264.

141.026 Three radio sources with unusual intensity variations. J. L. Locke, B. H. Andrew, W. J. Medd.
Astrophys. Journ. *(Letters)*, Vol. 157, L81 - L86 (1969).

Variations in flux density of three sources (PKS 1510–08, PKS 0736 + 01, and NRAO 512) have been observed at 2.8- and 4.6-cm wavelengths. The variations are unusually rapid. It is suggested that the observed variations are caused by the prolonged injection of relativistic particles into a source which remains optically thin. The sources can be placed at cosmological distances only if the magnetic-field strength in the variable component is less than 10^{-6} gauss or if relativistic effects play an important part in the evolution of the sources.

141.027 Observations of the optical polarization of the pulsar NP 0532. Yu. S. Efimov, V. I. Pronik, N. M. Shakhovskoy.
Astron. Tsirk. No. 512, p. 1 - 3 (1969).   In Russian.

141.028 On the presumable flux of high-energy gamma-rays from the pulsar CP 1133.
A. A. Stepanian, B. M. Vladimirsky, I. V. Pavlov, V. P. Fomin.
Astron. Tsirk. No. 512, p. 3 - 4 (1969).   In Russian.

141.029 Pulsars and the classical astronomy. I. II.
D. Ya. Martynov.
Astron. Tsirk. No. 512, p. 5 - 7; No. 513, p. 1 - 3 (1969).
In Russian.

141.030 On the slow variations of radio emission of pulsars.
L. M. Erukhimov.
Astron. Tsirk. No. 513, p. 3 - 5 (1969).   In Russian.

141.031 Quasars – preclusters of galaxies.
B. A. Vorontzov-Veljaminov.
Astron. Tsirk. No. 513, p. 5 - 7 (1969).   In Russian.

141.032 Quasars and nuclei of galaxies.
N. E. Kurochkin.
Astron. Tsirk. No. 515, p. 1 - 3 (1969).   In Russian.

141.033 On the age of the pulsar PSR 0833-45.
I. S. Shklovsky.
Astron. Tsirk. No. 519, p. 1 - 3 (1969).   In Russian.

141.034 On the diffusion of relativistic electrons from discrete radio sources. V. V. Vajsberg.
Astron. Tsirk. No. 526, p. 3 - 5 (1969).   In Russian.

141.035 On the nature of emission of the pulsar NP 0532.
I. S. Shklovsky.
Astron. Tsirk. No. 527, p. 1 - 3 (1969).   In Russian.

141.036 On the forbidden line spectra QSS.
G. Mathez.
Astron. Astrophys. Vol. 3, 127 - 134 (1969).

We first recall the results of Burbidge et al. (1966) on the forbidden line spectra of QSS, and the physical conditions of the emitting regions that they have been deduced. We then show that, in these conditions, the ionization degree of an element is a function of the electron temperature only. Secondly, we adopt the cosmological interpretation of the redshifts, and show that this implies very high energy densities in the QSS. The metastable levels population becomes thus a func-

tion of three variables: the electron density N_e, the electron temperature T_e, and the radius R of the emitting region. By comparing the intensities of the forbidden lines, we can then derive for each ion the values of N_e and R, T_e being a priori set.

141.037 Radio sources and elliptical galaxies.
D. H. Rogstad, R. D. Ekers.
Astrophys. Journ. Vol. 157, 481 - 494 (1969).

One hundred ninety-one E and S0 noncluster galaxies with known redshift have been surveyed for radio emission at a frequency of 2640 MHz. The observations support the hypothesis that an elliptical galaxy must have an absolute photographic magnitude brighter than –20 to be a strong radio source. A comparison of the results of this survey with observations of radio galaxies in rich clusters indicates that cluster membership does not enhance the probability that an elliptical galaxy is a radio emitter. In addition to the strong radio sources, several E and S0 galaxies were found to have weaker radio emission associated with them, but their absolute radio luminosity was still on the order of 100 times larger than that for detected spiral galaxies.

141.038 Pulsar electrodynamics.
P. Goldreich, W. H. Julian.
Astrophys. Journ. Vol. 157, 869 - 880 (1969).

Gold has suggested that pulsars are rotating magnetic neutron stars which formed in supernova explosions. We have investigated the simplest such model, one in which the magnetic dipole moment is aligned with the rotation axis. We compare our model with the observed properties of the Crab pulsar (NP 0532) and CP 1919.

141.039 Rotating collapsed objects, quasars and supernova remnants. A. Cavaliere, F. Pacini, G. Setti.
Astrophys. Letters, Vol. 4, 103 - 105 (1969).

By analogy with the case of the Crab nebula, we suggest that the high energy activity of quasars and similar objects may result from the presence of a spinning, magnetic supermassive star.

141.040 Observations of variable radio sources.
B.-H. Grahl, M. Grewing.
Astrophys. Letters, Vol. 4, 107 - 108 (1969).

Six radio sources expected to be variable have been repeatedly observed during the first half year of 1968 at a frequency of 2.695 GHz. Additional measurements were obtained in February and March 1969. Combined with measurements quoted in Kellermann and Pauliny-Toth (1968) the recent results indicate pronounced variations in the flux from 3C 345 and PKS 1510-08 and minor changes for most of the remaining sources.

141.041 The unusual radio source OQ 208.
M. Ryle, G. G. Pooley.
Astrophys. Letters, Vol. 4, 137 - 138 (1969).

The radio source OQ 208 appears to be associated with a blue stellar object lying 10 arc sec from a 13 mag galaxy.

141.042 Quasistellar objects. M. Schmidt.
Annual Rev. Astron. Astrophys. Vol. 7, 527 - 552 (1969).

141.043 High-resolution observations of radio sources.
M. H. Cohen.
Annual Rev. Astron. Astrophys. Vol. 7, 619 - 664 (1969).

141.044 Search for red-shifted neutral hydrogen absorption in 3C 191. W. L. H. Shuter, J. F. R. Gower.
Nature, Vol. 223, 1046 (1969).

141.045 Unified model for pulsars.
H.-Y. Chiu, F. Occhionero.
Nature, Vol. 223, 1113 - 1116 (1969).

The action of a strong electric field on electrons at the surface of pulsars causes a population inversion, and a laser-type transition produces a continuous emission guided by the magnetic field into sharply defined beams.

141.046 Relation between the linear polarization and spectral index of quasars.
J. A. Gilbert, R. G. Conway, P. P. Kronberg.
Nature, Vol. 223, 1252 (1969).

Observations have been made at Jodrell Bank of the linear polarization of ninety-one radio sources at a frequency of 610 MHz (λ 49.1 cm). We report a striking correlation between the degree of linear polarization at λ 49 cm and the spectral index of twenty-three quasars included in the observations.

141.047 Quasi-stellar objects – A progress report.
G. R. Burbidge, E. M. Burbidge.
Nature, Vol. 224, 21 - 24 (1969).

After nine years, quasars are still puzzling astronomers. This review brings the story up to date. In particular, the following aspects are considered: Variability, distribution in sky, absorption and emission red-shifts, physical models, red-shifts and cosmology.

141.048 Pulsar PP 0943.
V. V. Vitkevich, Yu. I. Alekseev, V. F. Zhuravlev, Yu. P. Shitov.
Nature, Vol. 224, 49 (1969).

The discovery of this new pulsar is reported. The period is P = 1.093 ± 0.003 s., and the intensity varies considerably.

141.049 Scattering of pulsar X-ray radiation by interstellar dust particles. V. I. Slysh.
Nature, Vol. 224, 159 - 160 (1969).

Only a few percent of the optical flux from the Crab nebula pulsar is unaffected by interstellar dust particles. The rest is partly absorbed but mostly scattered into a wide angle and contributes to the diffuse interstellar light. Dielectric particles of a size comparable to the wavelength, are thought to be involved. It is shown in this article that interstellar dust particles may strongly affect the pulsed X-ray radiation from NP 0532.

141.050 Magnetic field decay in a neutron star and the distribution of pulsar periods. F. Pacini.
Nature, Vol. 224, 160 (1969).

Some comments are given on the distribution of pulsar periods in relation to the expected lifetime of the magnetic field in a neutron star.

141.051 Clusters of quasi-stellar objects. M. B. Bell.
Nature, Vol. 224, 229 - 234 (1969).

An analysis of 150 QSO red-shifts suggests that there are groupings of QSOs extending over large areas of sky. The inference is that QSOs are not at cosmological distances.

141.052 Pulsar periods and rapid changes in the terrestrial rotation rate. S. P. Maran, H. Ögelman.
Nature, Vol. 224, 349 (1969).

We wish to point out that the random fluctuations in the rotation rate of the earth are occasionally large enough significantly to affect the measured values of $\Delta P/P$ for pulsars.

141.053 Improved right ascensions for 0.9 steradians of the 4C catalogue. T. W. Clarke, R. H. Frater, M. I. Large, R. E. B. Munro, H. S. Murdoch.

Australian Journ. Phys., Astrophys. Suppl. No. 10, 36 pp. (1969).

The east—west arm of the Molonglo radio telescope has been used to determine more precisely the right ascensions of 4C sources from $10^h 30^m$ to $18^h 30^m$ and south of declination +18°. A description is given of the procedure used to analyse the data, which were digitally recorded on magnetic tape. For a large proportion of the stronger sources the accuracy of the right ascensions is approximately 2 sec of arc.

141.054 Quasi-stellar objects in the direction of clusters of galaxies. J. N. Bahcall.
Astrophys. Journ. *(Letters)*, Vol. 157, L151 - L152 (1969).

Twenty-one quasi-stellar objects are listed that are in the direction of clusters of galaxies. These objects are interesting candidates for absorption experiments.

141.055 On the production of QSO absorption spectra in Friedmann universes. R. C. Roeder.
Astrophys. Journ. *(Letters)*, Vol. 157, L153 - L154 (1969).

It is shown that the hypothesis of Bahcall and Spitzer concerning the origin of absorption lines in QSOs is apparently inconsistent with present observations.

141.056 A search for strong radio spectral lines in the range 20 - 25 GHz.
W. E. Howard III, H. Hvatum.
Astrophys. Journ. *(Letters)*, Vol. 157, L161 - L162 (1969).

The radio sources W3A, VY Canis Majoris, and W49 were searched for strong radio spectral lines in the frequency range 20 - 25 GHz on April 27 - 29, 1969, with the NRAO 140-foot radio telescope. No lines stronger than 3000 flux units were present in the three sources at the time of observation.

141.057 The spectrum of the cosmic radio background between 0.4 and 6.5 MHz.
J. K. Alexander, L. W. Brown, T. A. Clark, R. G. Stone, R. R. Weber.
Astrophys. Journ. *(Letters)*, Vol. 157, L163 - L165 (1969).

Measurements of the average spectrum of the radio background for a broad region near the north galactic pole have been obtained over the range 0.4 - 6.5 MHz on the basis of experiments on the Radio Astronomy Explorer satellite. The spectrum turns over below 4 MHz in a manner that supports the concept of a uniform admixture of galactic emission regions having a spectral index of 0.43 ± 0.05 and absorbing ionized-hydrogen regions having a mean electron density of about 0.03 cm^{-3} in the galactic disk.

141.058 Precision measurement of the frequency decay of the Crab nebula pulsar, NP 0532.
P. E. Boynton, E. J. Groth III, R. B. Partridge, D. T. Wilkinson.
Astrophys. Journ. *(Letters)*, Vol. 157, L197 - L201 (1969).

We are measuring the phase of optical pulses from NP 0532 against an atomic time standard. The first 6 weeks of observation give the pulsar frequency ν and its time derivatives $\nu_{,t}$ ($= d\nu/dt$), and $\nu_{,tt}$ in agreement with earlier measurements, but with improved accuracy. We find $(\nu_{,tt} \nu / \nu_{,t}^2) =$ 4.76 ± 0.65. This preliminary result is consistent with angular-momentum loss by gravitational radiation.

141.059 Possible dates of birth of pulsar from ancient Chinese records. T. Kiang.
Nature, Vol. 223, 599 - 601 (1969).

Prompted by the discovery of the Crab nebula pulsar, I examined the other known pulsars for coincidence in position with ancient Chinese novae records. (There are about 120 such records). Of 22 pulsars examined, possible coincidences, hence possible dates of birth were found for 15. The method of search was illustrated with the pulsar MP 1727, because in this case, the search led incidentally to the identification of an event in 1437 with the radio source CTB 35; it is not yet known whether the latter is a pulsar.

141.060 Pulsar slowdown rates for CP 0328 and HP 1506.
T. W. Cole.
Nature, Vol. 223, 487 (1969).

The periods of the pulsars CP 0328 and HP 1506 are linearly increasing with time by 60 and 167 nanoseconds per year. The lack of correlation between period and rate of period slowdown implies pulsars cannot all follow the same evolutionary path.

141.061 Volcano theory of pulsars.
F. J. Dyson.
Nature, Vol. 223, 486 - 487 (1969).

If the solid mantle of a neutron star has a hole in it, hot liquid material from the core may pour continually out to the surface and produce a localized beam of electromagnetic radiation. Such a volcano situated on the surface of a neutron star provides a possible model of a pulsar.

141.062 A general discussion of the distribution of brightness of extragalactic radio sources.
E. B. Fomalont.
Astrophys. Journ. Vol. 157, 1027 - 1045 (1969).

The paper concentrates on displaying general properties of radio sources rather than a discussion of individual sources. The angular scale and the morphological types of radio structures are discussed. Approximately one-third of the sources are true doubles composed of two similar components. A method is outlined for analyzing the structure of a large sample of sources by using the visibility function. The section gives the distributions of angular size, percentage and size of fine structure, "symmetry", and "compactness" of radio sources determined by use of this method. A comparison of the radio structure with optical structure, radio luminosity, and radio diameter suggests that the gross radio structure depends only on the radio luminosity. The spatial properties of the radio structure and the galaxy are discussed. For double sources the galaxy lies close to the midpoint of the two components, perhaps a bit closer to the stronger component. For the doubles with unequal diameters, the galaxy is definitely situated closer to, sometimes coincident with, the smaller-diameter component. It is suggested that some of these sources are similar to 3C 83.1 (NGC 1265). The galaxy is coincident with the core component of core-halo sources.

141.063 Arecibo occultation studies: List 3.
S. Gulkis, J. Sutton, C. Hazard.
Astrophys. Journ. Vol. 157, 1047 - 1053 (1969).

The positions and structures of ten radio sources are derived from occultation observations. Optical identifications are suggested for six of the sources.

141.064 On the possibility of detecting redshifted 21-cm absorption lines in the spectra of quasi-stellar sources.
J. N. Bahcall, R. D. Ekers.
Astrophys. Journ. Vol. 157, 1055 - 1064 (1969).

The expected strengths of redshifted 21-cm absorption lines are estimated by using the properties of the observed optical absorption lines in quasi-stellar sources. It is shown that the absence of O I and N I absorption lines in the observed optical absorption spectra of quasi-stellar radio sources indicates that 21-cm absorption lines wider than 100 kHz are likely to be weak unless the heavy-element abundance in the absorbing material is low. Lines narrower than 100 kHz may be strong; their optical counterparts would have escaped detection. A general expression is given for the spin temperature of neutral hydrogen when (following Field) Ly a excitation and de-excitation, 21-cm absorption and emission, and particle collisions are all included. The results are expressed

simply in terms of the strength and distance of the radio source and the separation between absorber and emitter.

141.065 On the acceleration of ultrarelativistic electrons and nuclei in nonthermal radio sources.
J. R. Burke, D. Layzer.
Astrophys. Journ. Vol. 157, 1169 - 1181 (1969).

Previous papers outlined a theory for the genesis of ordered magnetic fields and the acceleration of charged particles in nonthermal radio sources. The present paper gives a detailed analytic and numerical discussion of the acceleration process, which verifies the principal assumption underlying the previously given derivation of a differential energy spectrum of the form $E^{-\gamma}$ with $\gamma > 2$.

141.066 On the nature of pulsars. I. Theory.
J. P. Ostriker, J. E. Gunn.
Astrophys. Journ. Vol. 157, 1395 - 1417 (1969).

We present in this paper the initial installment of a quantitative exploration of one particular pulsar model. We first make plausible and then assume that the seat of the pulsar phenomenon is a rotating neutron star having a dipolar magnetic field which is not parallel to the rotation axis. Such stars may be expected to emit large amounts of magnetic-dipole and gravitational-quadrupole radiation, that these energy losses are associated with losses of angular momentum and increases in the rotation periods, and that the emitted low-frequency magnetic-dipole radiation is extremely efficient at accelerating charged particles to relativistic energies. An explicit expression for the period as a function of time allows us to calculate the age of the Crab nebula. We also determine the luminosity of the nebula and the highest-energy electrons presently being injected into it.

141.067 Irregular variations of radio sources.
M. V. Penston, R. D. Cannon.
Non-Periodic Phenomena in Variable Stars, IAU Colloquium, Budapest, 1968, p. 485 - 490 (1969).

141.068 A limit on the variability of the 22 MHz radiation from 3C 84 (NGC 1275).
R. S. Roger.
Astrophys. Letters, Vol. 4, 139 - 140 (1969).

Braude et al. (1969) have reported variability in the radio emission from 3C 84 below 25 MHz. However, measurements at 22 MHz over eighteen months reveal no variability greater than 12 per cent r.m.s.

141.069 On the absorption spectrum of PKS 0237-23.
G. Grueff.
Astrophys. Letters, Vol. 4, 141 - 142 (1969).

Eleven absorption lines measured in various spectrograms of the quasi-stellar source PKS 0237–23 by various authors are identified with sulfur lines redshifted to $z = 2.4516$.

141.070 On the origin of absorption lines in spectra of quasi-stellar objects.
I. S. Shklovsky.
Astron. Zhurn. Akad. Nauk SSSR, Vol. 46, 935 - 939 (1969). In Russian. English translation in Soviet Astron. AJ, Vol. 13, No. 5.

A critical analysis of absorption lines in spectra of quasars shows that they are probably formed in the neighbourhood of a quasar nucleus, but not in metagalactic space. The presence of some absorption systems in the spectrum of a quasar with Z_{abs} considerably smaller than Z_{em} is explained by "pushing out" magnetized gas clouds from the environment of a quasar nucleus by the pressure of relativistic particles. The acceleration of pushed out clouds is estimated and it is shown that for several thousands of years their velocities can reach the values observed from $Z_{em} - Z_{abs}$.

141.071 Variable radiation in the magnetodynamical model of quasars. II. Prediction of rapid variations of brightness. L. M. Ozernoy, V. E. Chertoprud.
Astron. Zhurn. Akad. Nauk SSSR, Vol. 46, 940 - 950 (1969). In Russian. English translation in Soviet Astron. AJ, Vol. 13, No. 5.

In the present article the "rapid" optical variability is considered in detail. An attempt is made to classify the expected types of "rapid" changes, including "flash up" processes. The expected properties of the autocorrelation function of brightness variations are predicted. In particular it is shown that the statistical properties of rapid brightness variations have to reflect some of the properties of the slow ones. This makes it possible to verify a model by new observations. A method of determining the relative contributions of turbulence and flash up to the brightness fluctuations is proposed.

141.072 Observations of the distribution of polarized and non-polarized emission across the radiosource W 44 at 3.95 and 6.6 cm at Pulkovo.
N. S. Soboleva.
Astron. Zhurn. Akad. Nauk SSSR, Vol. 46, 955 - 959 (1969). In Russian. English translation in Soviet Astron. AJ, Vol. 13, No. 5.

Informations on observations of the distribution of polarized and non-polarized radioemission of the supernova remnant W 44 at 3.96 and 6.6 cm with a knife directional diagram are given. As a result of measurements it is found that only the eastern part of the source is polarized. The maximal percentage of polarization at 3.9 cm is $(12 \pm 1.5)\%$, at 6.6 cm it is $(4 \pm 1.5)\%$.

141.073 Spectral peculiarities in the radioemission of W 41. V. I. Ariskin, I. I. Berulis.
Astron. Zhurn. Akad. Nauk SSSR, Vol. 46, 1126 - 1128 (1969). In Russian. English translation in Soviet Astron. AJ, Vol. 13, No. 5.

Measurements of the radioemission from W 41 were carried out at λ = 21.1 cm, 2.6 m, 3.5 m and 5 m at the Radioastronomical Station of the Physical Institute of the USSR Academy of Sciences. The results obtained indicate that in the direction to W 41 there are both a nonthermal radiosource with the spectral index a = 1.43 and angular dimension of 18' and a thermal one with dimension of 34'.

141.074 Observations of the structure of radio sources in the 3C catalogue – II. C. D. Mackay.
Monthly Notices, Roy. Astron. Soc., Vol. 145, 31 - 65 (1969).

The one-mile radio telescope at Cambridge has been used to map the structure of a further 60 radio sources in the 3C catalogue at 408 MHz and 1407 MHz. This paper presents contour maps of the sources together with a list giving the position, angular size and flux density of each component; the optical fields of the same regions of sky are also briefly described.

141.075 A pencil-beam survey of radio sources at 178 MHz. J. L. Caswell, J. H. Crowther.
Monthly Notices, Roy. Astron. Soc., Vol. 145, 181 - 196 (1969).

A sample region of the sky surveyed by the 4C interferometer has been observed by a pencil-beam instrument at the same frequency. Flux densities are given for 726 sources, 640 of which are also contained in the 4C catalogue. In addition to listing some new sources with angular sizes greater than a few minutes of arc, the present results have been compared with the 4C interferometer data to derive an upper limit for the number of sources omitted from the 4C catalogue due to partial resolution; the comparison also shows that the proportion of sources in the 4C catalogue whose flux densities are seriously underestimated (low by more than 30 per cent) is about 7 per cent.

141.076 Observations of the structure of radio sources in the 3C catalogue – IV. Correlation diagrams and the evolution of radio sources.
M. S. Longair, G. H. Macdonald.
Monthly Notices, Roy. Astron. Soc., Vol. 145, 309 - 325 (1969).

The radio structures of a complete sample of 200 radio sources from the revised 3C catalogue are now available with a resolution of $23'' \times 23''$ cosec δ from recent observations at 1407 MHz with the one-mile telescope at Cambridge. These results are taken together with other data from long baseline interferometry and studies of interplanetary scintillation to derive correlation diagrams of radio luminosity against surface brightness and of radio luminosity against total linear size at 178 MHz and 1407 MHz. A general procedure has been developed for testing theories in which the components of radio sources are ejected from the nuclei of radio galaxies and QSO's. In this analysis it is essential to incorporate the effects of cosmological evolution, which can considerably change the appearance of the predicted diagrams. It is shown that it is possible to account for a large part of the dispersion in the correlation diagrams as being due to the fact that sources are observed at different stages in their evolution.

141.077 The dynamics and structure of inertially confined plasma clouds. W. Christiansen.
Monthly Notices, Roy. Astron. Soc., Vol. 145, 327 - 345 (1969).

Models for the evolution of inertially confined plasma clouds are developed and discussed. It is shown that a plasmon can travel a distance $\sim (\rho_0/\rho_{IG})r_0$ before dispersing, where ρ_0 is the initial density of the plasma cloud, r_0 is the initial radius of the plasmon, and ρ_{IG} is the density of the intergalactic medium. It is concluded that in a closed universe, the components of giant double radio galaxies are probably not ejected with relativistic velocities. It is also shown, that the upper limit on the observed separation between the components of double radio galaxies of $\sim 3 \times 10^5$ pc, implies an upper limit on the intergalactic matter density of $\sim 3.5 \times 10^{-29}$ g cm^{-3}.

141.078 Estimation of intensities of the fluctuating component of the radio source 3C 273 at frequencies of 86 and 60 MHz.
T. D. Antonova, V. V. Vitkevitch, V. G. Panajian.
Astrofizika, Vol. 5, 283 - 289 (1969). In Russian. English translation in Astrophysics, Vol. 5, No. 2 (1969).

The paper presents data on scintillations of the quasar 3C 273 by irregularities of the interplanetary plasma. The intensities of the fluctuating component are determined by comparing the scintillation characteristics of 3C 273 and 3C 48 at wave-lengths of 3.5 and 5 m. The spectral index of the scintillating component is –0.95. The scintillations are ascribed to the core of the component A of 3C 273.

141.079 An estimation of angular dimensions of the radio source 3C 48 at 60 MHz from interplanetary scintillations. V. G. Panajian.
Astrofizika, Vol. 5, 291 - 296 (1969). In Russian. English translation in Astrophysics, Vol. 5, No. 2 (1969).

The characteristics of the intensity fluctuations of 3C 48 at 60 MHz are analysed to estimate its angular dimensions. The intensity fluctuation curve of 3C 48 is compared with that of the point source 3C 119. In the case of a symmetrical source with a Gaussian distribution of radio luminosities the angular diameter of 3C 48 is 0".5.

141.080 On the dependence of the spectral index of extragalactic radio sources on the flux density.

R. D. Dagkesamanski.
Astrofizika, Vol. 5, 297 - 304 (1969).   In Russian.
English translation in Astrophysics, Vol. 5, No. 2 (1969).

From the analysis of data for the flux densities of sources from the 3CR catalogue an increase of the mean spectral index is found when flux density is decreasing. The slopes of $\alpha - \lg S_{178}$ curves are calculated for some frequency regions.

141.081   6 cm observations of nonthermal radio sources near the galactic plane.   D. K. Milne.
Australian Journ. Phys. Vol. 22, 613 - 630 (1969).

Brightness distributions and flux densities at 5000 MHz are presented for 17 nonthermal sources (possible supernova remnants) together with their spectra derived from these and other observations. For most sources a comparison has been made between the brightness distribution at 5000 MHz and that obtained with comparable resolution at 408 MHz with the Molonglo 1 mile Cross.

141.082   Intensity variations of *PSR* 0833-45 at 1.720 MHz.   D. J. Cooke.
Nature, Vol. 224, 569 (1969).

Observations made at 1.720 MHz on July 20, 1969, indicate that the intensity of *PSR* 0833-45 had become extremely variable. The measurements were made at the Parkes Observatory using the 210-foot telescope.

141.083   Flux density variation in VRO 42 22 01 at 1,420 MHz.   J. F. R. Gower.
Nature, Vol. 224, 569 - 570 (1969).

Recent continuum flux density measurements confirm a change in intensity at 1,420 MHz and show that in late December 1968 the source intensity had dropped to below 50 per cent of the average of the 1964 and 1965 measurements, and has now returned to about 80 per cent of this value in a period of about three months.

141.084   Periodic clustering of red-shifts in the spectra of quasi-stellar and other unusual objects.
C. L. Cowan.
Nature, Vol. 224, 655 - 656 (1969).

Analysis of 178 z-values yields a strongly periodic system containing, among others, the period 1/6 and 1/16.

141.085   Spin up in neutron stars: The future of the Vela pulsar.   G. Baym, C. Pethick, D. Pines,
M. Ruderman.
Nature, Vol. 224, 872 - 874 (1969).

Assuming that the speed up of the Vela pulsar was caused by a "starquake", a number of predictions can be made, and the event can be taken as evidence that the interior of the pulsar is a superfluid.

141.086   Power spectrum analysis of the emission-line redshift distribution of quasi-stellar and related objects.
S. H. Plagemann, P. A. Feldman, J. R. Gribbin.
Nature, Vol. 224, 875 - 876 (1969).

An analysis of a recent list of 186 redshifts indicates that, at best, only marginal importance can be attached to the periodicities in the redshift distribution that have been found by others.

141.087   Synchrotron model for pulsars.   T. Takakura.
Nature, Vol. 224, 252 - 253 (1969).

The clear variation of elliptic polarization during the course of each pulse of pulsar CP0328 infers the synchrotron emission from a well-collimated beam of electrons as suggested by Clark and Smith. If this rotation model is correct, the duration τ of each pulse of such a pulsar may depend on the frequency as $\tau \propto f^{-1/2}$. The test observation is suggested. It

is also shown that wave amplification (negative absorption) of the synchrotron emission is possible, if the pitch angle of electrons is well defined around 90°.

141.088   Highly dispersed pulsar and three others.
M. I. Large, A. E. Vaughan, R. Wielebinski.
Nature, Vol. 223, 1249 - 1250 (1969).

Four new pulsars have been discovered at the Molonglo Observatory, Australia. They are MP 0254, MP 1154, MP 1706 and MP 1911. MP 1154, with a dispersion measure of ~ 300 cm^{-3} pc, has the highest dispersion measure of all pulsars discovered so far. This pulsar was found using a two channel dispersion removing system. The large dispersion measures of MP 1154 and MP 1240 (discovered earlier), place them at distances of at least 2 kpc and 1.5 kpc respectively. There is no evidence available as to the origin of these two pulsars.

141.089   Rotating neutron stars, pulsars and cosmic X-ray sources.   W. H. Tucker.
Nature, Vol. 223, 1250 - 1252 (1969).

Models for the X-ray source in the Crab nebula and Sco X-1 are discussed. The energy source in both cases is assumed to be a rapidly rotating, highly magnetized neutron star. It is suggested that the Crab nebula represents the case where the energy in the "neutron-star wind" is greater than the magnetic energy. The streaming protons and electrons deposit their energy far out into the nebula in a shock transition region which is responsible for the production of relativistic electrons and therefore an extended X-ray source. A compact source such as Sco X-1 represents the other extreme where the magnetic energy dominates so that the rotational energy is transferred to particles that remain near the neutron star resulting in a source with a small angular diameter.

141.090   Polarization of pulsating radio sources.
R. D. Ekers, A. T. Moffet.
Astrophys. Journ. (*Letters*), Vol. 158, L1 - L8 (1969).

The complete polarization characteristics of a number of pulsating radio sources have been determined at 13 cm. We have found both a stable average and a time-varying component of the polarized radiation. A change in the position angle of the linear component during the course of the pulse is a common property; however, the high degree of linear polarization and constant rotation of angle seen for PSR 0833 – 45 is atypical.

141.091   Radio emission of quasi-stellar objects.
K. R. Lang, Y. Terzian.
Astrophys. Journ. (*Letters*), Vol. 158, L11 - L13 (1969).

Observations at 318 MHz of thirty-eight quasi-stellar objects are reported. Radio emission has been detected from twenty-two of them. Previously most of these sources were known to be blue stellar objects with no detectable radio emission. In particular we note that the quasi-stellar object B264 has been shown recently to be a member of a cluster of galaxies.

141.092   Distribution of quasistellar radio sources on the sky.   H. Arp.
Bull. American Astron. Soc. Vol. 1, 232 (1969). – Abstr. AAS.

141.093   Pulsar electrodynamics.
P. Goldreich, W. H. Julian.
Bull. American Astron. Soc. Vol. 1, 242 - 243 (1969). – Abstr. AAS.

141.094   Polarized brightness distribution of Taurus A and Orion A at 9.55-mm wavelength.
R. W. Hobbs, K. J. Johnston.

Bull. American Astron. Soc. Vol. 1, 244 (1969). – Abstr. AAS.

141.095 Supersynthesis observations with the NRAO interferometer. G. Macdonald.
Bull. American Astron. Soc. Vol. 1, 252 (1969). – Abstr. AAS.

141.096 Changes in the spectrum of 3C273.
E. J. Wampler.
Bull. American Astron. Soc. Vol. 1, 265 (1969). – Abstr. AAS.

141.097 Pulsars and the origin of cosmic rays.
V. L. Ginzburg.
Comments Astrophys. Space Phys. Vol. 1, 207 - 214 (1969).
The author discusses several possibilities and models for cosmic ray production. Various kinds of galactic sources (supernova remnants, etc.) are considered in more detail.

141.098 PP 0943 – a new pulsar and the main characteristics of its radiowave emission.
Iu. I. Alexeev, V. V. Vitkevich, V. F. Zhuravlev, Iu. P. Shitov.
Dokl. Akad. Nauk SSSR, Ser. Mat. Fiz., Vol. 187, 1019 - 1021 (1969). In Russian.

141.099 Pulsars and local cosmic ray prehistory.
R. E. Lingenfelter.
Nature, Vol. 224, 1182 - 1186 (1969).
Pulsars have ages and distances which, if they are supernova remnants, would make them significant sources of local cosmic rays during the last 10^6 years.

141.100 The structure of 3C 9. B. G. Clark, G. K. Miley.
Astrophys. Letters, Vol. 4, 207 - 210 (1969).
The structure of the quasar 3C 9 has been measured at 408 MHz and 2695 MHz. The source is double with components whose spectral indices differ by 0.4. Morphologically, it resembles 3C 47.

141.101 The spectrum and angular structure of 3C 273 at low frequencies. S. J. Bell, A. Hewish.
Astrophys. Letters, Vol. 4, 211 - 213 (1969).
Observations of interplanetary scintillation show that 3C 273 contains two components possessing angular diameters of ~0.2 arc sec and ~2 arc sec at metre wavelengths. The compact source has a flux density which rises sharply at low frequencies.

141.102 Pulsar H-line absorption and dispersion in the interstellar medium.
R. N. Manchester, J. D. Murray, V. Radhakrishnan.
Astrophys. Letters, Vol. 4, 229 - 232 (1969).
Absorption measurements at $\lambda 21$ cm have been made on two Molonglo pulsars, PSR 0736-40 and PSR 0833-45, but distance limits cannot be placed for either pulsar from these measurements. Distance estimates from the dispersion measures are equally unreliable because of the heavy contribution to the dispersion from the Gum nebula.

141.103 Sources of radiation with relativistic streaming.
P. D. Noerdlinger.
Astrophys. Letters, Vol. 4, 233 - 234 (1969).
Sources of radiation like Woltjer's QSO model can exhibit apparent diameters far smaller than their true dimensions, and remarkably rapid time variations.

141.104 Interferometric observations with a baseline of 127 kilometres – I.
W. Donaldson, G. K. Miley, H. P. Palmer, H. Smith.
Monthly Notices, Roy. Astron. Soc., Vol. 146, 213 - 219 (1969).
This is the first of two papers describing observations made on 46 radio sources smaller than one second of arc with a long baseline interferometer using a radio link. In this paper the normalization and calibration procedures are described and the results are given on those sources whose fringe visibilities showed no significant variation with hour angle. Upper limits to the angular diameters of these sources, or their unresolved components, have been calculated using a circular Gaussian model.

141.105 Pulsar test of a variation of the speed of light with frequency. G. Feinberg.
Science, Vol. 166, 879 - 881 (1969).
The sharply defined optical and radio pulses from pulsars make possible a test of the variation of the speed of light with frequency, and of the possible existence of a photon mass. The data indicate that the mass of a real photon is less than 10^{-44} gram. Detection of extragalactic pulsars could allow a substantial improvement of this limit.

141.106 Pulsation period of NP 0532.
H. C. Goldwire, Jr., F. C. Michel.
Publ. Astron. Soc. Pacific, Vol. 81, 540 (1969). – Abstract ASP.

141.107 Some preliminary results of a survey of discrete source polarization at 610 MHz.
P. P. Kronberg.
Publ. Astron. Soc. Pacific, Vol. 81, 543 (1969). – Abstract ASP.

141.108 NP 0532 stellar wind torques. F. C. Michel.
Publ. Astron. Soc. Pacific, Vol. 81, 546 - 547 (1969). – Abstract ASP.

141.109 Pulsar-emission mechanisms.
F. C. Michel, W. H. Tucker.
Publ. Astron. Soc. Pacific, Vol. 81, 547 (1969). – Abstract ASP.

141.110 UBV observations of 3C 273, II.
M. S. Burkhead.
Publ. Astron. Soc. Pacific, Vol. 81, 691 (1969). – Note.

141.111 Electromagnetic spectrum of NP 0532.
B. Bertotti, A. Cavaliere, F. Pacini.
Nature, Vol. 223, 1351 - 1352 (1969).
The possibility is considered that the optical and X-ray emission from NP 0532 is due to incoherent synchrotron radiation close to the speed of light circle: we find $B \sim 10^6$ gauss, $n_e \sim 10^9 - 10^{10}$ cm^{-3} and typically $\gamma \sim 200$. The coherent radioemission could then result from the corotation of bunches of particles: the corresponding critical frequency is $\Omega\gamma^3$ where Ω is the basic rotation frequency of the star. The typical size of a bunch should be L $\sim$ few centimeters and each bunch should contain about $n_e L^3 \sim 10^{13}$ electrons.

141.112 Bremsstrahlung radiation in intense magnetic fields.
M. Simon, D. L. P. Strange.
Nature, Vol. 224, 49 - 50 (1969).
Chiu and Canuto have suggested that pulsar radio emission is due to electron bremsstrahlung in the intense magnetic field associated with a neutron star. We prove that their calculated emissivity results in a positive absorption coefficient for any electron distribution, and hence the proposed emission mechanism cannot lead to amplification.

141.113 Enhancements of interplanetary scintillation, corotating streams and Forbush decreases.
J. Burnell.

Nature, Vol. 224, 356 - 357 (1969).

During routine radio astronomical observations of the interplanetary scintillation exhibited by extragalactic radio sources anomalous enhancements of the scintillation were detected. A number of these show twenty-seven day recurrence, and appear to be associated with co-rotating features in the interplanetary medium. Correlation with various geomagnetic and other interplanetary phenomena is discussed.

141.114 Relativistic beaming of radiation from pulsars.
F. G. Smith.
Nature, Vol. 223, 934 - 936 (1969).

The shape and polarization of the radio pulses from pulsars are similar to those expected from synchrotron radiation, except that the pulse width does not vary with frequency. A rotating beam could instead be formed by the relativistic compression of radiation from an isotropic source in rapid orbital motion.

141.115 Observations of the Crab pulsar during an occultation by the solar corona.
S. J. Goldstein, Jr., D. D. Meisel.
Nature, Vol. 224, 349 - 350 (1969).

Accurate dispersion measurements at frequencies between 112 – 172 MHz of single pulses from NP 0532 were obtained with the 300 ft. radio telescope of the National Radio Astronomy Observatory. The observations set an upper limit to the electron content along the line-of-sight at a curtate distance of 9 $R_\odot$ of 3.7×10^{16} electrons/cm^2. The mean dispersion away from the sun, $(2.377 \pm .0001) \times 10^{17}$ Hz and the combined radio and optical observations of Conklin et al [Nature, Vol. 222, 552 (1969)] imply that the radio pulse and optical pulse are coincident within 0.40 msec. A peak flux density greater than 10^{-23} W/m^2/Hz was observed for one pulse at 170 MHz.

141.116 Nature of pulsars. S. C. Vila.
Nature, Vol. 224, 157 - 159 (1969).

Pulsars are interpreted as neutron stars with motions of rotation and precession of the rotation axis on a circular cone, the orientation of which is fixed in space. The radiation is synchrotron radiation from a collimated and bunched electron beam in the equator plane of the star. The motion of precession causes the main pulses and the rotation causes the second periodic pulsation discovered by Drake and Craft in AP 2015 + 28 and CP 1919.

141.117 Hard X-rays from the Crab pulsar.
F. W. Floyd, I. S. Glass, H. W. Schnopper.
Nature, Vol. 224, 50 - 51 (1969).

A balloon flight to measure the high-energy X-ray component of the radiation from the Crab nebula pulsar NP 0532 was conducted on 10 May 1969. Data contain pulsed signals at the same repetition frequency as the optical radiation. The proportion of the X-ray flux in the range 25 keV to 100 keV in the pulsed mode was determined to be 15 ± 5%. The pulsed component of the Crab nebula flux has a harder spectrum than the steady part, and the "interpulse" possesses a harder spectrum than the main pulse.

141.118 The radio spectrum of Virgo A from 1411.7 to 1423.8 MHz. R. J. Allen.
Astron. Astrophys. Vol. 3, 316 - 322 (1969).

The radio spectrum of Virgo A near 21 cm wavelength has been examined for evidence of "fine structure" (departures from a simple power law) using the large mirror radio telescope at Nançay. The method of observations must allow the measurement of spectral features which are expected to have widths up to several MHz and amplitudes of only a few parts in 10^3 of the average received signal. Special attention has therefore been paid to an evaluation of the instrumental effects and to

the calibration of the receiving system. The results are discussed and compared with those of Koehler and Robinson (1966).

141.119 On the number-magnitude relation for the quasistellar objects in the field 13^h, +36°.
A. Braccesi, L. Formiggini.
Astron. Astrophys. Vol. 3, 364 - 367 (1969).

Magnitudes and colours of about 300 ultraviolet-excess objects have been measured on plates obtained with the Palomar 48-inch Schmidt telescope. The four-colour (u, b, v, i) system employed permits the QSO's to be identified on the basis of their combined ultraviolet and infrared excesses. The number versus magnitude relation for the QSO's was found to be much steeper than should be expected in the $\lambda = 0$, $q_0 = +1$ relativistic cosmological model, but to be consistent with the prediction made when the cosmological evolution suggested by Maarten Schmidt (1968) is taken into account.

141.120 The radio spectra of 3C sources.
H. van der Laan.
Astron. Astrophys. Vol. 3, 477 - 480 (1969).

The major features of a large sample of radio spectra are discussed. Particular reference is made to a recent joint NRAO-Cambridge publications of 3C source flux densities. The following conclusions are reached: The many source spectra that are straight and smooth (in the log S_ν – log ν plane) are neither remarkable nor very informative; Many convex spectra would be straight but for the partial absorption of radiation at frequencies less than ~ 1 GHz. The fact that for convex spectra the median value of the spectral index a (750, 5000) is greater than for straight spectra is due to a selection effect operative when self-absorption is the cause of convexity; Although concave spectra are very rare, spectral data are consistent with the independence of component spectra in multiple sources; The absence of very steep spectra among 3C sources is the data's most remarkable feature. It has important implications for radio galaxy evolution.

141.121 Secondary electrons in radio sources.
G. C. Perola.
Astron. Astrophys. Vol. 3, 481 - 484 (1969).

This note presents a detailed analysis of the physical conditions required to explain the observations through electrons of secondary origin. The theory is able to predict the near absence of exponentially steep spectra and the presence of many convex spectra in a complete sample. However, this theory requires very dense haloes (10^{-1} atoms per cc) surrounding the galaxies; these would produce observable effects, which are not detected.

141.122 Some models for the emission-line region of 3C 48.
J. N. Bahcall, B.-Z. Kozlovsky.
Astrophys. Journ. Vol. 158, 529 - 533 (1969).

Models for the emission-line region of 3C 48 are derived by using calculated ionization distributions. Estimates are given of the parameters that characterize the emission-line region and of the relative abundance of H, He, O, Ne, and Mg.

141.123 Theoretical implications of the second time derivative of the period of the pulsar NP 0532.
A. Ferrari, R. Ruffini.
Astrophys. Journ. (Letters), Vol. 158, L71 - L75 (1969).

The expected value of the second time derivative of the period of NP 0532 is given, and possible relations with existing magnetic-dipole models are outlined. The amount of gravitational radiation to be expected is estimated.

141.124 Search for rapid fluctuations in light from the Crab nebula pulsar.
D. Hegyi, R. Novick, P. Thaddeus.

Astrophys. Journ. (*Letters*), Vol. 158, L77 - L81 = Contr. Columbia Astrophys. Lab. No. 4 (1969).

Observations have failed to reveal fluctuations in the light from the pulsar NP 0532 on a time scale from about 3×10^{-5} to 1×10^{-8} sec.

141.125   Quasi-stellar objects in the direction of rich clusters of galaxies.   J. N. Bahcall.
Astrophys. Journ. (*Letters*), Vol. 158, L87 - L89 (1969).

Fourteen quasi-stellar objects are listed that are in the direction of rich clusters of galaxies in Abell's catalog. The importance of making redshift measurements of galaxies with small angular separations from quasi-stellar objects with redshifts $\lesssim 0.2$ is discussed.

141.126   On the masses of quasi-stellar sources.
J. N. Bahcall, E. E. Salpeter.
Astrophys. Journ. (*Letters*), Vol. 158, L15 - L17 (1969).

A general procedure for determining the mass of a quasi-stellar object in a cluster of galaxies is described. Data already available allow one to conclude that the mass of B264 is less than $5 \times 10^{13} M_\odot$.

141.127   The period and hard-X-ray spectrum of NP 0532 in 1967.
G. J. Fishman, F. R. Harnden, Jr., W. N. Johnson III, R. C. Haymes.
Astrophys. Journ. (*Letters*), Vol. 158, L61 - L64 (1969).

In a previous letter (Fishman, Harnden, and Haymes 1969) we reported the detection of pulsed hard X-radiation from NP 0532 in the analysis of data obtained during a balloon flight conducted on 1967 June 4. A further analysis of these data has since been performed; the present letter discusses this analysis, which has produced additional information on the period and the energy spectrum of the pulsed radiation.

141.128   Spectra of extended extragalactic radio sources.
A. H. Bridle.
Nature, Vol. 224, 889 - 890 (1969).

It is shown that present observations of the total radio emission from extended extragalactic radio sources should not be interpreted as indicating spectral similarity among their components. More direct observations show significant spectral variations among the components of such sources in about half of those investigated. These variations are often not apparent in the total spectra due to structural complexity of the sources; some specific examples are discussed.

141.129   Estimate of the gravitational radiation from NP 0532.   H. J. Melosh.
Nature, Vol. 224, 781 - 782 (1969).

The gravitational radiation to be expected from the Crab nebula pulsar NP 0532 has been calculated on the basis of the oblique rotator model for the pulsars. The result of this calculation is that between 8×10^{32} and 8×10^{33} erg/sec should be emitted as gravitational radiation at 60.44 Hz, with a consequent flux of gravitational radiation of between 2×10^{-12} and 2×10^{-11} erg/cm^2 - sec at the earth.

141.130   Pulsar distances, spiral structure and the interstellar medium.   B. Y. Mills.
Nature, Vol. 224, 504 - 505 (1969).

Reasons are advanced for the assumption that the pulsars are immersed in a fairly uniform hot dispersive medium of $< n_e > \simeq 0.06$ cm^{-3} which extends to much greater heights above the plane than the dense H I clouds. This leads to distances which appear individually accurate to a factor of two or three; there is also an association of ten pulsars with the accepted location of the Sagittarius spiral arm.

141.131   Fifteen months of pulsar astronomy.
V. Radhakrishnan.
Proc. Astron. Soc. Australia, Vol. 1, 254 - 263 (1969). – Review article.

141.132   Pulsars and cosmic rays.
A. G. Fenton.
Proc. Astron. Soc. Australia, Vol. 1, 285 - 286 (1969). – Contribution ASA meeting.

141.133   A search for pulsed gamma emission from two pulsars.   J. G. Ables, J. M. Durdin, A. G. Gregory, B. J. Stone.
Proc. Astron. Soc. Australia, Vol. 1, 286 - 288 (1969). – Contribution ASA meeting.

141.134   On angular variations of the spectral index of nonthermal cosmic radio emission.
Yu. V. Tokarev.
Izv. vyssh. uchebn. zavedenij. Radiofizika, Vol. 12, 161 - 167 (1969). In Russian. – Abstr. in Referativ. Zhurn. 51. Astron., 10.51.532 (1969).

141.135   Investigation of radio brightness distribution over the source by means of a radio interferometer with autonomous heterodynes.   V. A. Alekseev.
Izv. vyssh. uchebn. zavedenij. Radiofizika, Vol. 12, 491 - 494 (1969). In Russian. – Abstr. in Referativ. Zhurn. 51. Astron., 11.51.444 (1969).

141.136   Coherent mechanisms of radio emission and magnetic models of pulsars.
V. L. Ginzburg, V. V. Zheleznyakov, V. V. Zajtsev.
Uspekhi fiz. nauk, Vol. 98, 201 - 236 (1969). In Russian.

141.137   Further search for high energy gamma rays from CP 1133.   W. N. Charman, R. W. P. Drever.
Nature, Vol. 224, 567 - 568 (1969).

Two Cherenkov night-sky light receivers, operated in coincidence, were used to search for pulsed, high-energy gamma rays from CP 1133. The arrival times of air shower events, initiated either by gamma-rays or by the cosmic ray proton background, were analysed for periodic effects at the 1.19 s period of CP 1133. No such effects were observed, the resultant upper limit for pulsed emission being about 8×10^{-9} photons m^{-2} s^{-1} at an energy of about 3×10^{13} eV, assuming a pulse width of 20 ms.

141.138   Optical variations in 3C 345.   T. D. Kinman.
Nature, Vol. 224, 565 (1969).

The claim by Hunter and Lu that 3C 345 varied by ~ 0.4 m$_{pg}$ in a few hours is analyzed and their observations are compared with those made at Lick Observatory. It is concluded that the evidence for this variation should be treated with considerable reserve since it depends on a single observation made at a large hour angle which is not corroborated by earlier or later observations.

141.139   Local theory for quasars.   M. Rowan-Robinson.
Nature, Vol. 224, 1094 (1969).

The local theory for quasars suggests that only a very small part of their large redshifts is of cosmological origin, so that the redshift is basically independent of distance. Nevertheless for each quasar of a particular type (luminosity, spectral shape, redshift) although its distance is not known, the ratio of its distance to the maximum distance it would be visible at before passing out of sight can be found simply from the inverse-square law. Hence a set of quasars complete down to known radio and optical flux-levels can be tested for uniformity of distribution in depth. Applied to 3C quasars this test shows that their distribution is decidedly non-uniform. If we are in a local cluster of quasars, then we are in a special position, at the bottom of a density-well. The more natural inference is that the number-density or luminosity of these objects was greater in the past, ruling out the steady-state cosmology.

141.140 High resolution observations of Cygnus A at 2.7 GHz and 5 GHz.   S. Mitton, M. Ryle.
Monthly Notices, Roy. Astron. Soc., Vol. 146, 221 - 233 (1969).

The maps obtained from observations with the one-mile telescope show that each component consists of a compact region of high surface brightness with an extension of decreasing brightness running towards the optical galaxy. A comparison with earlier observations shows little variation of structure with frequency except at frequencies <200 MHz; at low frequencies there is a relatively greater contribution from the extensive features, and the change in the distribution may be associated with the change in spectral index of the source in this frequency range.

141.141 Extension of the radio source luminosity function – I. Observations.   M. D. Windram, S. Kenderdine.
Monthly Notices, Roy. Astron. Soc., Vol. 146, 265 - 312 (1969).

The Cambridge one-mile radio telescope has been used to make observations at 408 MHz of 478 weak sources in 110 areas centred on 3C and other sources. For most areas the limiting flux density is $\sim 0.2 \times 10^{-26}$ W m^{-2} Hz^{-1}; in one area it is as low as 0.066×10^{-26} W m^{-2} Hz^{-1}. The positional accuracy is about 10'' arc for most sources, enabling optical identifications to be made with objects as faint as 19.5 m. Statistical analyses show that the weak sources are not related to the neighbouring 3C sources and hence form a valid basis for determining the radio source luminosity function.

141.142 Long baseline interferometer observations at 408 and 448 MHz – I. The observations.
N. W. Broten, R. W. Clarke, T. H. Legg, J. L. Locke, J. A. Galt, J. L. Yen, R. M. Chisholm.
Monthly Notices, Roy. Astron. Soc., Vol. 146, 313 - 327 (1969).

Ten series of long baseline interferometer observations have been made at 408 and 448 MHz. The measurements were made between pairs of stations selected from four observatories in Canada and one in England. The longest baseline used was 6833 km (9.3×10^6 wavelengths). Measured fringe visibilities are given for 33 radio sources. Details are also given of observations of a further 28 sources for which no fringes were detected.

141.143 Observations of the structure of radio sources in the 3C catalogue – III. The absolute determination of positions of 78 compact sources.   B. Elsmore, C. D. Mackay.
Monthly Notices, Roy. Astron. Soc. Vol. 146, 361 - 379 (1969).

Observations have been made with the Cambridge One-Mile telescope of 78 unresolved or slightly resolved radio sources in the 3C catalogue. Absolute positions have been determined for these sources with typical errors of $\sim 1'' - 1.5''$ arc in R. A. and $\sim 2''$ cosec δ arc in declination. Details of possibly associated optical objects are given.

141.144 Long baseline interferometer observations at 408 and 448 MHz – II. The interpretation of the observations.   R. W. Clarke, N. W. Broten, T. H. Legg, J. L. Locke, J. L. Yen.
Monthly Notices, Roy. Astron. Soc.,Vol. 146, 381 - 397

(1969).

Detailed model brightness distributions, deduced from long baseline interferometer observations at 408 and 448 MHz, are presented for 19 radio sources. Notes on the structure of 14 other sources are also included. The smallest scale structure which can be recognized from the observations is about 0.005 second of arc and it appears that none of the sources is completely unresolved. Upper limits to the internal magnetic fields are given for 12 different components of the sources. Most values lie in the range $10^{-3 \pm 1}$ G. Linear dimensions and equivalent brightness temperatures are also computed.

141.145 On H II regions and pulsar distances.   A. J. R. Prentice, D. ter Haar.
Monthly Notices, Roy. Astron. Soc., Vol. 146, 423 - 444 (1969). – Strömgren spheres; Associations and clusters; Supergiants and hot O-stars which are not members of associations or clusters; The cool stars; The density-temperature relation and the probability to belong to an association or cluster; The galactic H I layer; Electron dispersion and pulsar distances; The statistical contribution from H II regions; The contribution from H I regions; The contribution from single stars or clusters; The pulsar distances.

141.146 NP 0532.   D. W. Richards, J. M. Rankin, C. C. Counselman III.
IAU Circ. No. 2164 (1969).

141.147 Interstellar scintillations of pulsar radiation.   K. R. Lang.
Science, Vol. 166, 1401 - 1403 (1969).

Time fluctuations in the intensity of pulsed radiation from CP 0834, CP 1133, AP 1237, and CP 1919 have been investigated. Power spectra, modulation indices, frequency distributions, and decorrelation frequencies are consistent with scintillation theory. If it is assumed that these scintillations are due to irregularities in the interstellar medium that travel at a velocity of 20 kilometers per second, the irregularities have a scale size on the order of 10^4 kilometers and a distance from the earth of approximately 70 parsecs.

141.148 Radio observations in the short microwave region.   P. R. Foster.
Quarterly Journ. Roy. Astron. Soc. Vol. 10, 206 - 222 (1969).

This paper reviews the radio astronomical observations which have been made in the wavelength range 30 to 1 mm. The radio spectrum is well covered down to 6 cm where systematic sky surveys and detailed source studies have been in progress for some years, but at wavelengths shorter than this, coverage is poor. No more than a dozen radio sources outside the solar system have been observed in the range 1 to 8 mm. Reviewed are observations from sun, moon, planets, radio sources and of the microwave background radiation.

141.149 Quasars as the pre-clusters of galaxies.   B. Vorontsov-Velyaminov.
Astron. Soc. Pacific, Leaflet No. 485, 8 pp. (1969).

141.150 Further search for nanosecond structure in the light flashes from pulsar NP 0532.
J. V. Jelley, R. V. Willstrop.
Nature, Vol. 224, 568 - 569 (1969).

Two photomultipliers mounted at the prime focus of the 92 cm. Cambridge reflector were operated in coincidence with a resolving time of 20 n.sec. No coincidences were found in excess of the expected random rate, at a significance level of 5σ. Calibrating the system from the star KW 275 in Praesepe, the time-averaged magnitude of the light from NP0532, if bunched to 20 ns, is $m_B \geq 22$. It is concluded that NP0532 cannot be detected by existing Cherenkov light receivers. It is also shown that interstellar dispersion was negligible in this experiment; either from the interstellar plasma or from the neutral atomic hydrogen.

141.151 Distribution of pulsar duty cycles.
G. R. Henry, H.-J. Paik.
Nature, Vol. 224, 1088 - 1089 (1969).

On the basis of an extremely simple model we consider the fraction of time ("duty cycle") that radiation is received from a pulsar. The model fits the observed duty cycle distribution fairly well, and suggests that pulsars may radiate into a cone of about 6° half apex angle. Agreement with observation may be improved by any of several simple modifications of the basic model.

141.152 Variations of small quasar components at 2.300 MHz.
J. Gubbay, A. J. Legg, D. S. Robertson, A. T. Moffet, R. D. Ekers, B. Seidel.
Nature, Vol. 224, 1094 - 1095 (1969).

Comparison of interferometer observations at 8×10^7 wavelengths baseline made in 1967 November, 1968 May and 1969 June shows changes in fringe intensity for several sources. For 3C 279 the changes are in agreement with an expanding synchrotron source $\leq 0\overset{''}{.}001$ in size. For 3C 273 a substantial decrease in the total flux density over this time interval is not seen in the fringe intensity, implying an angular size for the variable component of $\geq 0\overset{''}{.}002$ and an expansion at relativistic velocities.

141.153 Linear polarization of pulsar PSR 0833–45 at 4.8 GHz.
F. F. Gardner, J. B. Whiteoak.
Nature, Vol. 224, 891 - 893 (1969).

Observations of the pulsar PSR 0833–45 at 6 cm wavelength have shown a double structure for the pulse, a drop in the polarization from 100% to 50% near the pulse maximum, and non-linear variation of polarization angle across the pulse. These are interpreted in terms of an oblique rotator model for the pulsar.

141.154 Synthesis of brightness distribution in radio sources.
D. E. Hogg, G. H. MacDonald, R. G. Conway, C. M. Wade.
Astron. Journ., Vol. 74, 1206 - 1213 (1969).

The results of the first synthesis of the brightness distributions of several strong radio sources with the NRAO three-element interferometer are presented. At 2695 MHz, the supernova remnant Cas A shows shell structure, while the remnant Tau A is more nearly a smooth ellipsoid, in agreement with previous work at lower frequencies. Structure of small angular scale is found within the components of the double sources Cyg A and Her A. Most of the flux from Vir A comes from an extended region surrounding the nucleus, but there are two point sources as well, one of which coincides with the nucleus, the other with the optically brightest knot in the jet.

141.155 Measurements of flux density and polarization of variable sources at a wavelength of 1.55 centimeters.
T. P. McCullough, J. A. Waak.
Astrophys. Journ., Vol. 158, 849 - 857 (1969).

The flux density of the 1.55-cm radiation from the quasi-stellar sources 3C 273, 3C 279, and 3C 454.3 and the Seyfert galaxy 3C 84 has been observed at intervals of about 2 months from 1966 May to 1968 May; measurements of linear polarization were made during the last half of this interval. Time variations of about 30 percent were found in the fluxes of these sources. In the case of 3C 273, two distinct maxima have been observed, and there is some evidence that the cloud of relativistic electrons which produced the later peak originated at an earlier epoch than the cloud which caused the first observed peak.

141.156 Particle acceleration during the 1966–1967 radio burst of 3C 273.
M. Simon.
Astrophys. Journ., Vol. 158, 865 - 869 (1969).

The data at 3.4 mm for the 1966–1967 variable component of 3C 273 indicate that the model of an expanding source emitting synchrotron radiation is breaking down at early stages of the evolution of the source. A value of 0.4 pc is derived for the intrinsic size of the source over which particle acceleration must have occurred.

141.157 The 21-cm absorption profiles of DR 21 and the source near NGC 6857.
A. R. Thompson, R. S. Colvin, M. P. Hughes.
Astrophys. Journ., Vol. 158, 939 - 951 (1969).

In the present paper we describe observations of the 21-cm H I absorption profiles of DR 21 and of another thermal source with similar characteristics near NGC 6857. As an aid in the interpretation of the observations of DR 21, an absorption profile of a nearby source, DR 23, was also derived. Studies of the H I absorption characteristics of sources near the galactic plane may be expected to reveal information on their distances and possibly on the presence of associated H I regions. Galactic coordinates, flux densities, and other details of the three sources are given. In the following discussions all radial velocities are measured relative to the local standard of rest, and the parameters given by Oort (1964) are used in calculating radial velocities expected from galactic rotation.

141.158 Collective bremsstrahlung from relativistic electrons as a possible mechanism in radio sources.
K. Papadopoulos, I. Lerche.
Astrophys. Journ., Vol. 158, 981 - 986 (1969).

The possibility of interpreting some of the radio emission from astrophysical plasmas as collective bremsstrahlung due to ultrarelativistic electron tails is discussed. It seems feasible that such a mechanism plays an important role in the radio emission from the "compact" source in the Crab nebula, quasi-stellar objects, and the planet Jupiter.

141.159 Quasars-interpretative.
W. H. McCrea.
Cosmic Ray Studies, Bombay 1968, p. 218 - 230 (1969).

141.160 Observations of pulsars.
J. R. Shakeshaft.
Cosmic Ray Studies, Bombay 1968, p. 231 - 244 (1969).

141.161 Radio source counts.
J. R. Shakeshaft.
Cosmic Ray Studies, Bombay 1968, p. 259 - 263 (1969).

141.162 Periodic variations in pulsar radiation intensity.
K. R. Lang.
Astrophys. Journ. (*Letters*), Vol. 158, L175 - L177 (1969).

Power spectra of pulsar-intensity fluctuations at 318 MHz indicate that seven pulsars have line structure in their intensity spectra. Since two of these pulsars, AP 1237 + 25 and AP 2303 + 30 are newly discovered, their characteristic parameters are presented. The power spectra of CP 1919 + 21 and AP 2015 + 28 do not show features corresponding to the alias of the class 2 pulsation that is thought to cause sub-pulse structure.

141.163 Accurate dispersions for thirteen pulsars.
S. J. Goldstein, Jr., J. T. James.
Astrophys. Journ. (*Letters*), Vol. 158, L179 - L182 (1969).

We determined the dispersions of thirteen pulsars with fractional accuracies between 2×10^{-3} and 1×10^{-4} from observations of strong individual pulses between 112 and 170 MHz. No variation with time or with pulse strength was detected. For six pulsars the dispersion obtained between 170 and 142 MHz differed from that obtained between 142 and 112 MHz by less than 1.0×10^{14} Hz, a value that sets an upper limit of about $10^7 \, cm^{-6}$ pc to the emission measure of dense plasma along the paths to these pulsars.

141.164 Sagittarius A: Observations of the galactic center at 3.3 mm.
M. M. Dworetsky, E. E. Epstein, W. G. Fogarty, J. W. Montgomery.
Astrophys. Journ. (*Letters*), Vol. 158, L183 - L187 (1969).

Observations at 3.3 mm of the radio source Sagittarius A have been used to obtain a flux estimate of 36 (+10, -5) $\times 10^{-26}$ W m^{-2} Hz^{-1}. This value is approximately half that predicted by extrapolation of results at longer wavelengths. The mean half-intensity diameter of Sgr A appears to be smaller ($\lesssim 1'.6$) at 3.3 mm than at longer wavelengths ($\sim 2' \times 3'$ at 30 cm), supporting the suggestion of Maxwell and Taylor that the apparent size of Sgr A decreases with increasing radio frequency.

141.165 Possible origin of absorption lines in the spectrum of QSO's.
J. Barnothy, M. Barnothy.
Bull. American Astron. Soc., Vol. 1, 334 (1969). – Abstract AAS.

141.166 Search for rapid fluctuations in light from the Crab nebula pulsar.
D. Hegyi, R. Novick, P. Thaddeus.
Bull. American Astron. Soc., Vol. 1, 347 - 348 (1969). Abstr. AAS.

141.167 Observations of the angular structure of extragalactic radio sources at 2995 MHz.
J. A. Högbom.
Bull. American Astron. Soc., Vol. 1, 348 (1969). – Abstr. AAS.

141.168 Statistical studies of radio sources in the Ohio list.
J. D. Kraus.
Bull. American Astron. Soc., Vol. 1, 350 - 351 (1969). Abstr. AAS.

141.169 On the photographic photometry of quasistellar objects.
P. K. Lü.
Bull. American Astron. Soc., Vol. 1, 353 (1969). – Abstr. AAS.

141.170 Pulsar stellar wind torques: NP 0532.
F. C. Michel.
Bull. American Astron. Soc., Vol. 1, 354 (1969). – Abstr. AAS.

141.171 Observations of the radio structure of quasars.
G. K. Miley, G. H. MacDonald.
Bull. American Astron. Soc., Vol. 1, 354 - 355 (1969). Abstr. AAS.

141.172 Magnetic decay and the maximum period of pulsars.
J. P. Ostriker, J. E. Gunn.
Bull. American Astron. Soc., Vol. 1, 357 - 358 (1969). Abstr. AAS.

141.173 Remarks on possible reasons of the secular increase of pulsar periods.
I. S. Shklovsky.
Astron. Zhurn. Akad. Nauk SSSR, Vol. 46, 715 - 720 (1969). In Russian. English translation in Soviet Astron. AJ, Vol. 13, No. 4.

Gold's hypothesis on the secular increase of pulsar periods is criticized. It is suggested that the powerful gravitational emission may be responsible for this effect in NP 0532.

The observed secular increase of the periods of some pulsars can be explained by their velocity of 10^9 cm/sec, which could be obtained during a supernova outburst due to the asymmetry of an ejected envelope.

141.174 On the Faraday rotation in pulsars and in the interstellar medium.
N. A. Lotova.
Astron. Zhurn. Akad. Nauk SSSR, Vol. 46, 1165 - 1168 (1969). In Russian. English translation in Soviet Astron. AJ, Vol. 13, No. 6.

The author discusses some experimental results: the systematic drift of the position angle and the variation of the polarization of emission of pulsars for the pulse emission time; the statistic variation of the polarization in a small region of the pulse with $\tau \simeq 10^{-4}$ sec. It is shown that all these effects can take place near the source. For estimating the magnetic field of the interstellar medium it is important to take into account the Faraday rotation in the corona of the source.

141.175 More precise statistical parameters for the optical variability of the quasar 3C 273.
L. M. Ozernoy, V. E. Chertoprud, S. D. Chuvachin.
Astron. Zhurn. Akad. Nauk SSSR, Vol. 46, 1317 - 1319 (1969). In Russian. English translation in Soviet Astron. AJ, Vol. 13, No. 6.

Revised data from smoothed photographic observations of the quasar 3C 273 are analysed. The numerical values of the statistical parameters of the optical variability are hardly changed compared with those published in a previous paper. All qualitative conclusions, including the existence of the quasiperiodic component of variability obtained there, remain completely valid.

141.176 Radio sources having widely separated components.
R. A. Hinder, N. J. B. A. Branson.
Observatory, Vol. 89, 178 - 184 (1969).

The present paper describes a statistical study based on the 3C, 4C and other surveys to determine whether there is a significant number of associated pairs of sources with separations lying between $15'$ and $5°$. It is concluded that such associated pairs, if any, form a very small class of the whole.

141.177 Accurate optical positions of radio sources.
J. R. Shakeshaft.
Observatory, Vol. 89, 209 - 210 (1969). – Letter.

141.178 Galactic sources in the 4C catalogue.
J. L. Caswell.
Observatory, Vol. 89, 230 - 234 (1969).

A comparison of the spatial density of 4C radio sources at high and low galactic latitudes shows that of the 103 4C sources with $|b^{II}| < 1°$, about 30 are probably galactic. Investigation of the individual sources responsible for the increased spatial density shows that the excess can be accounted for mainly by high-brightness features in extended supernova remnants.

141.179 Interplanetary scintillations. V. A survey of the northern ecliptic.
D. E. Harris, E. Gundermann Hardebeck.
Astrophys. Journ., Suppl. Series, Vol. 19, 115 - 144 (1969). Abstr. in Astrophys. Journ., Vol. 158, 432 (1969).

Source structure on the scale $0''.1 - 0''.5$ was examined at the Arecibo Ionospheric Observatory by interplanetary-scintillation observations of 500 Parkes and 4C sources. Slightly over half of all the sources displayed scintillations. For most of these, we were able to find rough values for the fraction of the source scintillating and upper limits to the diameter of the small component. The results corroborate earlier deductions that most scintillating sources are complex, with at least one component being very small and containing a significant fraction of the total intensity. Because a large number of sources were examined, the occurrence of scintillations could be correlated with source intensity, source identification, and source spectra.

141.180 A large-scale metagalactic magnetic field and Faraday rotation for extragalactic radio sources.
K. Kawabata, M. Fujimoto, Y. Sofue, M. Fukui.
Publ. Astron. Soc. Japan, Vol. 21, 293 - 306 (1969).

The Faraday rotation of emissions from 60 linearly polarized radio sources is investigated. It is concluded that the metagalactic magnetic field also contributes to the Faraday rotation. Distribution diagrams of the rotation measures against galactic latitudes show that scatters in the rotation measures are, at $|b^{II}|>35°$, significantly larger for the sources with large redshifts than for those with small redshifts. Combining the present results with other cosmological problems we have been able to impose some constraints on quantities associated with the cosmology.

141.181 Quasars: Finding a handle for a complex problem.
R. W. Holcomb.
Science, Vol. 166, 1609 - 1610 (1969). — News notes.

141.182 Peculiarities in the spectra of quasars.
B. V. Komberg.
Zemlya i Vselennaya, No. 4, p. 27 - 30 (1969). In Russian.

141.183 Improved positions and some optical identifications for 451 4C radio sources between declinations 4°
and 20°. D. Wills, J. G. Bolton.
Australian Journ. Phys. Vol. 22, 775 - 812 (1969).

Accurate positions have been measured at Parkes for 451 4C radio sources between declinations 4° and 20°. For most of the sources the r.m.s. uncertainty is ±10″ arc in each coordinate. Optical identifications are suggested for 94 sources, of which 22 are galaxies and 72 are possible quasi-stellar objects.

141.184 The linear polarization of radio sources between 11 and 20 cm wavelength. III. Influence of the
Galaxy on source depolarization and Faraday rotation.
F. F. Gardner, D. Morris, J. B. Whiteoak.
Australian Journ. Phys. Vol. 22, 813 - 819 (1969).

The dependence on galactic latitude of the 11 and 20 cm polarization of 355 extragalactic sources reveals no definite evidence of depolarization within the Galaxy. The distribution of Faraday rotation can be explained in terms of a field along the local spiral arm, deformed by magnetic "loops" in the solar neighbourhood.

141.185 The linear polarization of radio sources at 6 cm wavelength. F. F. Gardner, J. B. Whiteoak,
D. Morris.
Australian Journ. Phys. Vol. 22, 821 - 838 (1969).

Linear polarization measurements at 6 cm wavelength with the Parkes 210 ft telescope are given for 706 sources. Comparison of polarization parameters and flux densities with other published data is made. For the variable sources it is found that changes in polarized flux and position angle can accompany changes in intensity.

141.186 On the nature of evolutionary effects associated with quasi-stellar radio sources. I.
M. A. Arakelian.
Astrofizika, Vol. 5, 461 - 476 (1969). In Russian. — Engl. translation in Astrophysics, Vol. 5, No. 3.

The method of investigation of evolutionary effects applied earlier by M. Schmidt (V/V_m –method) to the sample of quasi-stellar radio sources from the 3CR Catalogue is used for two samples. The volume V_m is calculated for one of them

by means of optical luminosities only, and the radio luminosities are used for the other one. No evolutionary effect is found in the first case, but this effect is quite evident in the second case. Similar results are obtained by the investigation of quasi-stellar radio sources from the 4C Catalogue.

141.187 Rapid variations in the radio polarization of BL Lac.
E. T. Olsen.
Nature, Vol. 224, 1008 - 1009 (1969).

The non-thermal radio source BL Lacertae has been monitored since April, 1968 at a frequency of 8 GHz (3.75 cm) with the University of Michigan 85-ft paraboloid as part of a program to monitor the linear polarization of variable extragalactic radio sources. During this period two extremely rapid outbursts in the total flux density were observed. The polarized flux density has varied from 0.1 to 0.5 flux units, and the position angle has changed by 80°. Variations of 30° in the position angle were detected in the course of one week. From August, 1968 to September, 1969 the "quiescent" total flux density exhibited a slow rise similar to that reported in the Seyfert galaxy NGC 1275 (3C84).

141.188 Anomalous temporal behaviour of NP 0532.
R. Ramaty, S. S. Holt.
Nature, Vol. 224, 1003 - 1004 (1969).

By comparing an X-ray measurement of the period of the Crab Nebula pulsar, NP 0532, made in June 1967 with high-precisions measurements of the period and its first and second derivatives made in 1969, we have deduced that discontinuous speedups in the rotation rate of pulsar had to occur in the above mentioned time interval. The average magnitude of these speedups is of the order of 1 ns per 100 days.

141.189 Magnetorelativistic model of a pulsar pulse.
S. A. Kaplan, V. Ya. Eidman.
JETP Letters, Vol. 10, 203 - 206 (1969). [Translated from ZhETF Pis. Red. 10, No. 7, 320 - 323 (1969). In Russian.]

The character of pulsar radiation is usually connected with the existence of a definite directivity of the radiation pattern. To obtain an estimate of the aperture of the directivity pattern a new model of a pulsar is proposed.

141.190 Quasi-stellar objects.
D. W. Sciama.
Phil. Trans., Ser. A, Vol. 264 (No. 1150), 263 - 266 (1969).

141.191 Extragalactic radio sources in the infrared.
R. D. Davies.
Phil. Trans., Ser. A, Vol. 264 (No. 1150), 251 - 261 (1969).

141.192 Pulsars. M. L. Good.
Comments Nuclear and Particle Phys., Vol. 3, No. 3, p. 75 - 77 (1969).

141.193 Radio observation of galactic H II regions at 4170 MHz (II).
T. Ojima, T. Nakajima, Y. Ishizawa, K. Akabane, K. Miyazawa, T. Takahashi.
Journ. Radio Sci. Lab. Japan, Vol. 15 (No. 82), 271 - 278 (1968). — See Phys. Abstr. Vol. 72, No. 40134 (1969).

141.194 Polarization measurements of extragalactic radio sources at 3.12 cm wavelength.
G. L. Berge, G. A. Seielstad.
Observations Owens Valley Radio Obs., No. 2, 21 pp. (1969).

Interferometer at Owens Valley used to measure linear and circular polarization of 32 extragalactic sources. In addition flux densities of 103 sources were measured. — GD.

141.195 A general discussion of the brightness distribution of extra-galactic radio sources.

E. B. Fomalont.
Observations Owens Valley Radio Obs., No. 3, 32 pp. (1969).

141.196 Flux densities and positions of southern galactic sources at 1410 MHz. B. A. Manchester.
Australian Journ. Phys., Astrophys. Suppl. No. 12, 9 pp. (1969).

Flux densities and positions have been determined for sources in the 1410 MHz survey of the Southern Milky Way (Hill 1968) carried out with the Parkes 210 ft radio telescope.

141.197 Pulsars. S. Grzędzielski.
Postępy Astron., Vol. 17, 179 - 199 (1969). In Polish.

The article reviews the observational data on pulsars and their theoretical interpretation as published during the first year after the announcement of the discovery.

141.198 The relation absolute luminosity–colour index for 29 quasars. M. Abramowicz.
Postępy Astron., Vol. 17, 281 - 285 (1969). In Polish.

141.199 Pulsars. W. P. Hirst.
Monthly Notes Astron. Soc. Southern Africa, Vol. 28, 116 - 119 (1969). – Review article.

141.200 Line identifications in quasi-stellar objects. C. R. Lynds.
Contr. Kitt Peak National Obs. No. 284, [Reprinted from *"Beam-Foil Spectroscopy"*, Gordon and Breach, 1968, p. 539 - 574], 36 pp. (1968).

141.201 The radiospectra of an homogeneous sample of 4C radiosources.
Lab. Nazionale Radioastronomia, Ist. Fis. "A. Righi" Univ. Bologna, Contr. No. 56, 15 pp. (1969).

In this paper we present a set of radio data concerning a sample of 136 radiosources of the 4C catalogue (Pilkington and Scott, 1965) in the region $07^h00\ 00 - 18^h30^m00$ and $29°30' - 34°30'$, which have been observed at 408 MHz with the "Northern Cross Radiotelescope" and at 1420 MHz with the "Owens Valley Radio-interferometer". The purpose of this work was to obtain the radio spectra of an homogeneous sample of relatively faint radiosources and information about their radiostructure.

141.202 On the number magnitude relation for the quasi-stellar objects in the field $13^h + 36°$.
Lab. Nazionale Radioastronomia, Ist. Fis. "A. Righi" Univ. Bologna, Contr. No. 57, 11 pp. (1969).

The number magnitude relation is investigated for 300 ultraviolet-excess objects whose magnitudes and colors were measured on plates obtained with the Palomar 48-inch Schmidt.

141.203 Ipotesi sulla natura delle pulsars. F. Pacini.
Atti XII Riunione Soc. Astron. Italiana, L'Aquila 1968, p. 18 - 19 (1969). – Abstract SAI.

141.204 Progressi nel lavoro di riconoscimento delle 'quasars radioquiete' con il metodo dell'eccesso infrarosso.
A. Braccesi, L. Formiggini.
Atti XII Riunione Soc. Astron. Italiana, L'Aquila 1968, p. 65 - 67 (1969). – Abstract SAI.

141.205 Recenti risultati della 'survey' profonda e lavoro di identificazione ottica delle radiosorgenti.
G. Grueff.
Atti XII Riunione Soc. Astron. Italiana, L'Aquila 1968, p. 68 - 74 (1969). – Abstract SAI.

141.206 Suggested identifications of some MSH objects. E. M. Lindsay.
Irish Astron. Journ., Vol. 9, 53 - 56 (1969).

141.207 Possible explanation of the decrease of the period of the pulsar PSR–0833–45.
G. S. Bisnovaty-Kogan.
Astron. Tsirk. No. 529, p. 3 - 5 (1969). In Russian.

141.208 On the radio source Sgr A. V. N. Kurilchik.
Astron. Tsirk. No. 536, p. 4 - 6 (1969). In Russian.

141.209 Absolute measurements of the intensity of radiation of the discrete sources Cassiopeia A and Cygnus A at 30 – 60 cm. L. N. Bondar', M. R. Zelinskaya, S. A. Kamenskaya, V. A. Porfir'ev, V. L. Rakhlin, V. M. Rodina, K. S. Stankevich, K. M. Strezhneva, V. S. Troitskij.
Izv. vyssh. uchebn. zavedenij. Radiofizika, Vol. 12, 807 - 812 (1969). In Russian. – Abstr. in Referativ. Zhurn. 51. Astron., 1.51.560 (1970).

141.210 Remarks on the nature of pulsars. I. S. Shklovskij.
Vestn. AN SSSR, No. 8, p. 55 - 61 (1969). In Russian. Abstr. in Referativ. Zhurn. 51. Astron., 1.51.572 (1970).

141.211 Pulsarerna. G. Larsson-Leander.
Astron. Tidssk., Årg. 2, 113 - 131, 154 - 177 (1969). – Review article.

141.212 On the period changes of pulsars. L. Detre.
Inform. Bull. Variable Stars (I.A.U. Commission 27), Konkoly Obs., Budapest, No. 380 (1969).

141.213 Radiowave emission by the CP–1919 pulsar in the metre-wave range.
Iu. I. Alexeev, V. V. Vitkevich, Iu. P. Shitov.
Dokl. Akad. Nauk SSSR, Ser. Mat. Fiz., Vol. 187, 291 - 293 (1969). In Russian.

Results of observations of CP 1919 with a wide-band cross-array radio telescope DKR 1000 in Pushchino during April - July 1968 are presented. The passage time was 2 minutes. Regular observations have been made on 93 MHz, additional ones on 110, 106, 92, 73, and 61 MHz. The measured delays of pulses at the different frequencies give a maximum number of 3.8×10^{19} electrons/cm^2 on the path to the source. Intensity variations of the pulses with periods of about one year, several minutes, and 3 - 10 seconds have been found. It is pointed out that these short variations cannot be caused by interplanetary scintillation. A spectrum index 1.5 with a dispersion 0.5 has been derived from separate pulses. The energy distribution in the spectrum is continuous, there are no indications of fine structure. The authors favour the model of a rotating neutron star. *B. Onderlička*

141.214 Linear increase in periodicity of thirteen pulsars. G. C. Hunt.
Nature, Vol. 224, 1005 - 1006 (1969).

The periods of 12 pulsars are observed to increase linearly with time, although one pulsar (CP 0808) has an apparently constant period. It is shown that the rates of increase in period are consistent with the theory that pulsars are rotating neutron stars being slowed by magnetic breaking. However, the intrinsic luminosity is so great that there must be a source of energy for radiation other than the loss of rotational energy.

141.215 Pulsars as possible sources of super-heavy nuclei in the primary cosmic radiation. G. Silvestro.
Nuovo Cimento Lettere, Prima Ser., Vol. 2, 771 - 772 (1969).

The author suggests pulsars may be the source of super-heavy nuclei observed in cosmic rays. The process of formation and ejection of super-heavy nuclei elements from such objects is discussed.

141.216 , Pulsare als rotierende Neutronensterne.
M. Grewing, W. Priester.
SuW, Vol. 8, 258 - 260 (1969).

141.217 Die Entdeckung des ersten optischen Pulsars am Steward Observatory. M. J. Disney.
SuW, Vol. 8, 280 - 281 (1969).

141.218 A broad-band radio pulsar in the Crab nebula.
Journ. Astron. Soc. Victoria, Vol. 22, 66 (1969).

141.219 Optical variations of the radio source 0906 + 01.
G. H. Folsom, A. G. Smith.
Publ. Astron. Soc. Pacific, Vol. 81, 871 - 873 = Contr.
Rosemary Hill Obs. No. 3 (1969).

141.220 Studio preliminare per un programma di occultazioni lunari di radiosorgenti a medicina.
F. S. Delli Santi.
Pubbl. Oss. Astron. Univ. Bologna, Vol. 10, No. 2, 21 pp. (1969).
The expressions for the number of radiosources occulted by the moon per year and for the sensibility of a radiotelescope operating with a programme of lunar occultations of radiosources are given.

141.221 Pulsars. J. Grygar.
Vesmír, Vol. 48, 201 - 203 (1969). In Czech.

141.222 Pulsar "NP–0532" – a spectroscopic binary?
H. P. Dart, III.
Spectrosc. Letters, Vol. 2, No. 2, p. 31 - 35 (1969).

141.223 Pulsars. R. Wielebinski.
Mitt. Astron. Ges. No. 27, p. 29 - 30 (1969).
Abstract AG.

141.224 Pulsar-Theorien.
P. A. G. Scheuer.
Mitt. Astron. Ges. No. 27, p. 39 - 46 (1969). – Review article.

141.225 **Fast folding algorithm for detection of periodic pulse trains.** D. H. Staelin.
Proc. IEEE, Vol. 57, 724 - 725 = National Radio Astron. Obs., Green Bank, Repr. Ser. A, No. 111 (1969).
A fast folding algorithm is described which greatly facilitates the correlation of digital data with impulse trains. This algorithm is useful for detecting weak, noisy pulse trains of unknown period and phase.

141.226 Absorption of high energy gamma-rays in the vicinity of pulsar NP 0532.
B. McBreen.
Nature, Vol. 224, 893 (1969).
The purpose of this article is to point out that photon-photon absorption may remove high energy γ-rays from the vicinity of pulsar NP 0532 in the Crab Nebula.

141.227 Optical variations in quasars.
J. H. Hunter, Jr., P. K. Lü.
Nature, Vol. 223, 1045 - 1046 (1969).
A photographic patrol of selected quasars has been in progress during the past 2 1/2 years with the 40-inch telescope at Yale Observatory. We have already published an account of the behaviour of one of the patrol objects, 3C 454.3. The purpose of this article is to summarize our results for other selected quasars in the patrol programme.

141.228 Decrease of flux density of the radio source Cassiopeia A at 81.5 MHz.
P. F. Scott, J. R. Shakeshaft, M. A. Smith.
Nature, Vol. 223, 1139 - 1140 (1969).
This article describes some new observations in November 1967, April 1968 and June 1969 which, together with the results of Högbom and J. R. S., provide information at a frequency of 81.5 MHz over a span of 20 years.

141.229 BL Lacertae = VRO 42.22.01.
N. Visvanathan.
IAU Circ. No. 2170 (1969).

141.230 Quasi-sinusoidal components in arrival time of pulsar NP 0532.
D. W. Richards, G. H. Pettengill, C. C. Counselman, J. Rankin.
IAU Circ. No. 2178 (1969).

141.231 Apparent change in frequency of NP 0532.
P. E. Boynton, E. J. Groth III, R. B. Partridge, D. T. Wilkinson.
IAU Circ. No. 2179 (1969).

141.232 NP 0532.
D. W. Richards, G. H. Pettengill, J. A. Roberts, C. C. Counselman, J. Rankin.
IAU Circ. No. 2181 (1969).

141.233 Strong millimeter wave outbursts from NGC 1275.
E. E. Epstein, M. M. Dworetsky, J. W. Montgomery.
IAU Circ. No. 2195 (1969).

141.234 A fourth periodicity in MP 0031.
D. H. Staelin, M. S. Ewing, R. M. Price, J. M. Sutton.
IAU Circ. No. 2196 (1969).

More about NP 0532.
Nature, Vol. 223, 446 - 447 (1969). – News notes.

Do quasars cluster?
Nature, Vol. 224, 213 (1969).

Quasars: Significance of clusters.
Nature, Vol. 224, 640 (1969). – News notes.

Quasar redshifts not intrinsic.
Nature, Vol. 224, 843 (1969). – News notes.

X-ray pulsar.
Sci. American, Vol. 221, No. 1, p. 52 (1969).

A pulsar's slowdown.
Sky Telescope, Vol. 38, 283 (1969).

A technique for recording phase-resolved spectra of regularly-varying faint light sources.
See Abstr. 031.019.

A method for detecting weak radio sources in the presence of stronger sources in observations made with the Cambridge One-mile Telescope.
See Abstr. 033.034.

The formation of stars with particular application to temporary stars and quasars. See Abstr. 065.066.

Neutron starquakes and pulsar periods.
See Abstr. 065.104.

Electric fields in rotating, magnetic, relativistic stars. See Abstr. 066.055.

Search for seismic signals at pulsar frequencies. See Abstr. 081.016.

Properties of the optical radiation of variable stars and quasars.   See Abstr. 122.025.

Rapid radio variations in BL Lac. See Abstr. 122.042.

Optical and radio study of BL Lacertae. See Abstr. 122.097.

Search for an optical remnant of the Cassiopeia A supernova.   See Abstr. 125.001.

Occultation positions of the 1665 MHz OH emission from G 0.7 – 0.0.   See Abstr. 131.016.

OH and formaldehyde radiation properties of the W75 region.   See Abstr. 131.140.

On the fast variations of the optical emission from the nuclei of Seyfert galaxies and quasars. See Abstr. 158.017.

Compact radio source in the nucleus of M87. See Abstr. 158.057.

Light curve of the N-type galaxy 3C371. See Abstr. 158.076.

Aspects of radio galaxy evolution. See Abstr. 158.090.

Der Einfluß nichtthermischer Strahlung auf die Emissionslinienspektren von Seyfert-Galaxien und Quasaren. See Abstr. 158.101.

Intergalactic hydrogen along the path to Virgo A. See Abstr. 161.009.

Search for ghost images: A statistical test. See Abstr. 162.030.

An angular diameter-redshift test of cosmological models using observations of weak radio sources. See Abstr. 162.039.

142   X Ray-, Gamma Ray-Sources

142.001   Cosmic X-rays.
S. I. Salem.
Astron. Soc. Pacific, Leaflet No. 481, 8 pp. (1969).

142.002   Intensity of the soft X-ray background flux.
C. S. Bowyer, G. B. Field.
Nature, Vol. 223, 573 - 575 (1969).
A discussion of the measurements of soft X-rays from outside the galaxy which have been published by groups at the University of California and at the Naval Research Laboratory.

142.003   Cosmic background X-rays produced by intergalactic innerbremsstrahlung.   S. Hayakawa.
Progress Theoret. Phys. Japan, Vol. 41, 1592 - 1594 (1969). — Letter.

142.004   X-ray spectra of several discrete cosmic sources.
J. F. Meekins, R. C. Henry, G. Fritz, H. Friedman, E. T. Byram.
Astrophys. Journ. Vol. 157, 197 - 213 (1969).
Spectra of Sco XR-1, Cyg XR-1, and Cyg XR-2 are derived for the photon energy range of $1.5 - 13$ keV from proportional-counter measurements made on September 8, 1967. The spectrum of Sco XR-1 may be described by bremsstrahlung with a temperature of $(65 \pm 5) \times 10^6$ °K. No simple power-law (synchrotron) spectrum is consistent with this observation. Iron line radiation, if present, must be less than 5 per cent of the total emission in the $(1.5-13)$-keV region. The spectrum of Cyg XR-1 may be fitted by an energy power law of index 1.4 or 1.5; that of Cyg XR-2, by bremsstrahlung with a temperature of $(60 \pm 10) \times 10^6$ °K.

142.005   Observations of X-rays from Taurus X-1 and Cygnus X-1.   I. S. Glass.
Astrophys. Journ. Vol. 157, 215 - 222 (1969).
Measurements of the intensity and spectral distribution of high-energy X-rays from the Crab nebula and Cyg X-1 were made on October 2, 1967, using a 5000 cm² balloon-borne proportional counter array. Both spectra can be represented by power laws of the form $dN/dE = k(E/E_0)^{-\alpha}$ over the observed energy range from 20 to 70 keV. For the Crab nebula, k and α have the approximate values 0.0068 cm⁻² sec⁻¹ keV⁻¹ and 2.16, respectively, with $E_0 = 30$ keV. The corresponding values for Cyg X-1 are 0.008 and 1.8, with $E_0 = 30$ keV.

142.006   Models of X-ray stars.
O. P. Manley, S. Olbert.
Astrophys. Journ. Vol. 157, 223 - 246 (1969).
A radiation instability, peculiar to bounded, optically thin magnetoactive plasmas, is assumed to engender fluctuations in the magnetic field. The fluctuations are shown to lead to a variant of the Fermi acceleration mechanism which produces hard spectra of electron energies. A theory of galactic X-ray sources, based on this mechanism, appears to account for all the known and inferred properties of the X-ray star Sco X-1.

142.007   X-ray spectral data from GX3+1.
P. Gorenstein, R. Giacconi, H. Gursky.
Astrophys. Journ. Vol. 157, 463 - 464 (1969).
X-ray spectral data in the region $1-8$ keV is presented for GX3+1, a source which has been identified with an emitter of harder X-rays in the Sagittarius-Scorpio region.

142.008   Interstellar absorption of 10-Å X-rays.
S. Rappaport, H. V. Bradt, W. Mayer.

Astrophys. Journ. *(Letters)*, Vol. 157, L21 - L25 (1969).
Measurements of the 10-Å flux of X-rays from sources in the Scorpio-Sagittarius region give clear evidence for interstellar absorption, with a column density of at least 1×10^{22} H atoms cm⁻². In contrast, the X-ray flux from the Crab nebula shows no evidence for a cutoff down to 12 Å (1.0 keV). This yields an upper limit for the average gas density between the earth and the Crab nebula of 0.5 H atom cm⁻³.

142.009   X-ray flux from Centaurus X-2 in the energy range 2-20 keV.
U. R. Rao, E. V. Chitnis, A. S. Prakasarao, U. B. Jayanthi.
Astrophys. Journ. *(Letters)*, Vol. 157, L127 - L132 (1969).
Two rocket flights carrying X-ray payloads were conducted from Thumba Equatorial Rocket Launching Station, Trivandrum, India, on November 3, 1968, and November 7, 1968, respectively. The first evidence for the existence of low-energy X-ray flux in the energy range 2-20 keV from the Cen X-2 source since its reported extinction in May 1967 is presented. The observed flux has the same energy spectrum as that observed in the high-energy range by Lewin et al. in October 1967.

142.010   Energy spectrum and time variation of Sco X-1.
U. R. Rao, U. B. Jayanthi, A. S. Prakasarao.
Astrophys. Journ. *(Letters)*, Vol. 157, L133 - L137 (1969).
Two rocket flights carrying X-ray payloads were conducted from Thumba Equatorial Rocket Launching Station, Trivandrum, India, on November 3, 1968, and November 7, 1968, respectively. The energy spectrum of the Sco X-1 X-ray source observed during both flights in the energy range 2-20 keV is presented. A comparison of our results with the previous observations indicates that the flux of Sco X-1 exponentially decreased over the period $1965 - 1968$ with a time constant of about 4.1 years.

142.011   Analysis of the optical emission of Sco XR-1.
F. I. Lukatskaja, A. I. Emetz, G. U. Kovalchuk, I. K. Lankis.
Astron. Tsirk. No. 512, p. 7 - 8 (1969).   In Russian.

142.012   The cosmic gamma-ray spectrum from secondary-particle production in the metagalaxy.
F. W. Stecker.
Astrophys. Journ. Vol. 157, 507 - 514 (1969).
The purpose of this paper is to discuss the form and intensity of the spectrum of cosmic γ-rays resulting from the production and decay of $\pi°$-mesons produced in metagalactic cosmic-ray $p - p$ collisions. It is assumed that intergalactic space contains ionized hydrogen gas at a density of 10^{-5} cm⁻³, as is consistent with recent X-ray observations at 0.25 keV. The Friedmann solution to the Einstein field equations of general relativity is used as a description of our expanding universe, and a discussion is presented of the effects of redshift and spatial curvature on the generation and distortion of the local γ-ray spectrum from the decay of $\pi°$-mesons. Numerical calculations are presented for the Einstein—de Sitter solution, which is found to be an adequate model for these calculations. Two models are presented to represent the possible flux of metagalactic cosmic rays. In calculating metagalactic γ-ray spectra, the effect of γ-ray absorption at large redshifts is taken into account.

142.013   Galactic line emission from 1 to 10 keV.
J. Silk, G. Steigman.
Phys. Rev. Letters, Vol. 23, 597 - 600 (1969).
We calculate the flux of X rays produced by low-energy

cosmic-ray nuclei in HI regions. We consider electron capture to excited states by cosmic-ray nuclei of heavy elements, followed by cascades down to the ground state. It is found that the electron-capture processes may yield appreciable line intensities in the range 1–10 keV in the galactic plane.

142.014 Soft X-ray background flux.
A. N. Bunner, P. C. Coleman, W. L. Kraushaar, D. McCammon, T. M. Palmieri, A. Shilepsky, M. Ulmer.
Nature, Vol. 223, 1222 - 1226 (1969).

Measurements of the soft X-ray background flux show good qualitative correlation with 21 cm measurements of columnar hydrogen density but too little apparent absorption. A portion of the flux may originate in unresolved population II objects or may be extragalactic.

142.015 Evidence for a point source of high energy cosmic gamma rays.
G. M. Frye, Jr., J. A. Staib, A. D. Zych, V. D. Hopper, W. R. Rawlinson, J. A. Thomas.
Nature, Vol. 223, 1320 - 1321 (1969).

Two high altitude balloon flights from Parkes, NSW, have produced evidence for a point source of high energy gamma rays (above 50 MeV) from a point source in Sagittarius. The flights did not confirm the band source of gamma rays reported by Clark, Garmire and Kraushaar last year.

142.016 Evidence for a galactic component of the diffuse X-ray background.
B. A. Cooke, R. E. Griffiths, K. A. Pounds.
Nature, Vol. 224, 134 - 137 (1969).

Rocket measurements have shown an excess of radiation in the galactic plane in the Vela-Carina-Centaurus region. This article presents the evidence and considers the possible explanations.

142.017 On the distance to Scorpio XR-1.
G. Wallerstein.
Astron. Journ. Vol. 74, 999 (1969).

The published proper-motion data and the spectroscopic data from interstellar K-line strengths do not support the conclusion that Sco XR-1 is physically associated with the Sco-Cen association.

142.018 The recent appearance of a new X-ray source in the southern sky.
J. P. Conner, W. D. Evans, R. D. Belian.
Astrophys. Journ. (Letters), Vol. 157, L157 - L159 (1969).

The recent appearance of a strong X-ray source in the southern sky has been observed with detectors aboard two Vela 5 satellites. The satellite data indicate that the source appeared between 2330 U.T. July 6 and 0430 U.T. July 9, 1969, and is located at approximately R.A. $14^h 56^m$, decl. $-32° 15'$. During the first few days this source increased in strength and then began to decline. At its peak it was more than twice as strong as Sco XR-1 in the energy range 3 - 12 keV. The spectrum of this source has grown softer continuously since its appearance.

142.019 Observation of hard radiation from the region of the galactic center.
R. C. Haymes, D. V. Ellis, G. J. Fishman, S. W. Glenn, J. D. Kurfess.
Astrophys. Journ. Vol. 157, 1455 - 1459 (1969).

A balloon-altitude observation of Sagittarius was conducted on April 23, 1968, at photon energies between 31 and 544 keV. A power-law spectrum of the form $2.8 E^{-2.20 \pm 0.35}$ photons $cm^{-2} sec^{-1} keV^{-1}$ was detected, corrected to the top of the earth's atmosphere. If the entire flux is assumed to be from the source known as GX + 1, the corresponding number spectrum becomes $4.4 E^{-2.23 \pm 0.35}$ photons $cm^{-2} sec^{-1} keV^{-1}$;

a power law similar in shape to that of the Crab nebula appears to provide a better fit than an exponential to the spectrum over a broad energy range.

142.020 Probleme der Röntgen- und Gamma-Astronomie.
K. Pinkau.
Sterne, 45. Jahrgang, 81 - 95 (1969). – Lecture during the meeting of the Astronomische Gesellschaft, Nürnberg, 1968 Sept.

142.021 Balloon observations of cosmic X rays in the energy range 20 - 200 keV.
K. K. Rangan, P. D. Bhavsar, N. W. Nerurkar.
Journ. Geophys. Res. Vol. 74, 5139 - 5144 (1969).

A balloon-borne scintillation telescope was flown on March 31, 1968, from Hyderabad, India, to measure the isotropic component of the cosmic X rays. The spectrum of the primary diffuse X rays in the 20 – 130 keV interval is found to be $140 E^{-2.4 \pm 0.4}$ photons $cm^{-2} sec^{-1} ster^{-1} keV^{-1}$.

142.022 2–20 keV X-ray sky background.
E. A. Boldt, U. D. Desai, S. S. Holt, P. J. Serlemitsos.
Nature, Vol. 224, 677 - 679 (1969).

We report here on the results of a rocket flight launched from White Sands, New Mexico, on March 3, 1969, in which we investigated the diffuse sky background within 2 - 20 keV with a wide angle instrument especially suited to this purpose.

142.023 Primordial cosmic ray sources.
F. W. Stecker.
Nature, Vol. 224, 870 - 872 (1969).

New gamma ray observations have provided what may be evidence for primordial cosmic ray sources existing out to redshifts of the order of 100. These sources may be galaxies and quasars in their initial stages of formation. They may have provided the basic energy source for ionizing the intergalactic gas.

142.024 Results of gamma-ray balloon astronomy.
C. E. Fichtel, D. A. Kniffen, H. B. Ögelman.
Astrophys. Journ. Vol. 158, 193 - 206 (1969).

The results obtained from recent high-altitude balloon flights of the Goddard γ-ray telescope with a digitized spark chamber are reported and combined with earlier work. The data on the region near the galactic center are reanalyzed in terms of a possible line source in the galactic plane. The analysis leads to a value of $(2.3 \pm 1.2) \times 10^{-4} \gamma$ $(cm^2 sec rad)^{-1}$ above 100 MeV for a galactic-latitude interval of $-3°$ to $+3°$ in the region near the galactic center; this result in itself does not justify the claim of a detected flux, but it is consistent with the line intensity of $(4.1 \pm 0.7) \times 10^{-4} \gamma (cm^2 sec rad)^{-1}$ quoted from recent OSO-III results.

142.025 High-energy X-rays from Cygnus XR-1.
R. G. Bingham, C. D. Clark.
Astrophys. Journ. Vol. 158, 207 - 218 (1969).

Celestial X-rays (20–130 keV) from Cyg XR-1 were detected during a high-altitude balloon flight on 1967 May 3 from Palestine, Texas. The measured photon flux in the energy range 20.0 – 30.9 keV corrected to the top of the atmosphere is 0.0200 ± 0.0031 photons $cm^{-2} sec^{-1} keV^{-1}$ which confirms a recently reported high X-ray intensity from Cyg XR-1 during May 1967. Our results are consistent with a continuum energy spectrum, and they are equally compatible with either a blackbody at 1.2×10^8 °K or a hot, thin plasma at 3.5×10^8 °K.

142.026 Polarization of thermal X-ray sources.
J. R. P. Angel.
Astrophys. Journ. Vol. 158, 219 - 224 (1969).

It is shown that radiation from thermal X-ray sources

which are not spherically symmetric may show polarization of the order of 1 – 5 percent due to Thomson scattering.

142.027 Sudden changes in the intensity of high energy X-rays from Sco X-1. P. C. Agrawal, S. Biswas, G. S. Gokhale, V. S. Iyengar, P. K. Kunte, R. K. Manchanda, B. V. Sreekantan.
Nature, Vol. 224, 51 - 53 (1969).

During the study of cosmic X-rays in the energy range 20 – 120 KeV in the balloon flight from Hyderabad, South India on December 22, 1968, Sco X-1 was under observation for a period of about one hour and a half from 04 : 27 U.T. to 05 : 53 U.T. It was found that in the energy interval of 30 to 52 KeV the intensity of Sco X-1 showed a rapid decrease by a factor of about three over a period of about ten minutes. The intensity remained nearly at the same level during the subsequent forty-five minutes. This was followed by an increase in the intensity by a factor of about three and a decrease to nearly normal value; the rise and the fall of intensity occurred over a period of about thirty minutes. The last event can be classified as a flare in Sco X-1 and it seems likely that the first event is also of this type.

142.028 X-ray line emission from Sco X-1. J. R. P. Angel.
Nature, Vol. 224, 160 - 161 (1969).

Plasma models for Sco X-1 of rather high electron density have been recently proposed. If these models are correct, electron scattering will greatly reduce the apparent strength of X-ray emission lines.

142.029 Spectrum of the Crab X-ray source from 4 to 40 keV. L. W. Acton, R. C. Catura, P. C. Fisher.
Bull. American Astron. Soc. Vol. 1, 231 (1969). – Abstr. AAS.

142.030 The observation of several cosmic X-ray sources in the 20 – 250 keV range with a Ge (Li) detector. A. S. Jacobson, J. G. Laros.
Bull. American Astron. Soc. Vol. 1, 245 (1969). – Abstr. AAS.

142.031 Simultaneous X-ray and optical observations of Sco XR-1.
R. M. Pelling, J. L. Matteson, L. E. Peterson, H. M. Johnson, J. C. Golson.
Bull. American Astron. Soc. Vol. 1, 256 - 257 (1969). – Abstr. AAS.

142.032 The gamma-ray emission from the galactic center and the origin of galactic cosmic ray electrons.
M. S. Longair, R. A. Sunyaev.
Astrophys. Letters, Vol. 4, 191 - 193 (1969).

It is suggested that the intense γ-ray source observed in the direction of the galactic center may be the result of inverse Compton scattering of the photons of the intense infrared source at the galactic center. The total energy released from the galactic center ($\sim 10^{40}$ erg/sec) would be sufficient to supply the total radio background from the galaxy and the break in the cosmic ray electron spectrum would be naturally explained. The model has important consequences for the diffusion of cosmic ray electrons in the galaxy.

142.033 Model of Cygnus X-2. R. E. Wilson, S. Sofia.
Nature, Vol. 223, 1350 - 1351 (1969).

Several kinds of observations of the X-ray source Cygnus X-2 are interpreted in terms of a model consisting of a semi-detached binary star system in which accretion of infalling matter onto a white dwarf star is the X-ray generating mechanism. Since the infalling matter does not strike the small target star directly, a cloud of swirling gases, of radius approxi-

mately 10^5 km, is produced around the white dwarf. Due to heating by the emergent X-ray flux, this cloud is a source of high temperature continuum radiation which, together with large mass motions, accounts for the observation of very broad Balmer absorption lines. The unusual Balmer decrement is explained by filling-in of the low number Balmer lines by radiation from the G-type secondary component.

142.034 Identification of S5003 Cen with the new intense X-ray source in Centaurus.
O. J. Eggen, A. W. Rodgers.
Astrophys. Journ. (*Letters*), Vol. 158, L111 - L112 (1969).

The light variable S5003 Cen ($\alpha = 14^h 56^m5$, $\delta = -33°13$ [1950]) lies very near the intense X-ray source reported by Conner, Evans, and Belian and has colors as well as spectral features similar to those of other objects identified with X-ray sources.

142.035 Erratum: "Further simultaneous observations of the optical and X-ray spectra of Sco X-1".
[Astrophys. Journ. (*Letters*), Vol. 156, L67].
H. Mark, R. Price, R. Rodrigues, F. Seward, C. Swift, W. Hiltner.
Astrophys. Journ. (*Letters*), Vol. 158, L131 (1969).

142.036 A new X-ray source in the constellation Ara.
A. N. Bunner, T. M. Palmieri.
Astrophys. Journ. (*Letters*), Vol. 158, L35 - L36 (1969).

Two positions of celestial X-ray sources are confirmed, and a new source is reported in the Ara region. The spectrum of the new source is very soft.

142.037 Origin of the background X-radiation.
M. S. Longair, R. A. Syunyaev.
JETP Letters, Vol. 10, 38 - 40 (1969). [Translated from ZhETF Pis. Red. 10, No. 1, p. 56 - 59 (1969). In Russian].

A model explaining the spectral properties of the background X-rays is proposed. It is assumed that the X-ray quanta are produced by Compton scattering of radio, infrared, and optical quanta by the relativistic electrons produced with a power law spectrum $N(E) \sim E^{-\gamma}$.

142.038 Observations of a new X-ray source.
T. Kitamura, M. Matsuoka, S. Miyamoto, M. Nakagawa, M. Oda, Y. Ogawara, K. Takagishi.
Nature, Vol. 224, 784 - 785 (1969).

A new cosmic X-ray source was observed near Sco-X-1 at 1215 UT on August 7. It is very likely that we observed the source reported to have appeared between July 6 and 9 by Conner et al. (Astrophys. Journ. Vol. 152, L45 (1969)). Its position was $\alpha = 14^h 55^m \pm 10^m$, $\delta = -32° \pm 3°$. The X-ray intensity of the source in the energy range 2 ~ 25 KeV was 44 photoms $cm^{-2} sec^{-1}$ and its energy spectrum was fitted to a thermal radiation of 8×10^7 °K from a thin hot plasma.

142.039 Have the diffuse cosmic X-rays an anisotropic component? R. F. O'Connell, S. D. Verma.
Nature, Vol. 224, 505 - 506 (1969).

We have recently shown that if one assumes that the far-infrared radiation is galactic, then inverse Compton scattering of these photons with cosmic ray electrons would explain the flux and the energy spectrum of diffuse X-rays. We also predicted a large anisotropy which seemed to be at variance with experimental data which indicated that diffuse X-rays are isotropic within about 10 per cent. Recent measurements by Cooke, Griffiths and Pounds, however, now present evidence in support of the existence of anisotropy. Examining in more detail the predictions of our model regarding anisotropy, we find very good agreement with the experimental results of Cooke, et. al.

142.040 Low-energy cosmic X-ray measurements.
A. J. Baxter, B. G. Wilson, D. W. Green.
Canadian Journ. Phys., Vol. 47, 2651 - 2666 (1969).

An experiment is described to investigate cosmic X rays in the energy range 0.25 - 12 keV. The data-recovery system and methods of spectral analysis are considered. Results are presented for the energy spectrum of the diffuse X-ray component and its distribution over the northern sky down to 1.6 keV with a limited extension at 0.27 keV. At the lowest energies, the flux appears to increase more rapidly and exhibits some anisotropy in arrival directions related to the gross galactic structure. Spectral characteristics of the Crab nebula and Cygnus X-2 have also been determined.

142.041 Survey of possible sources of cosmic gamma rays above 50 MeV in the northern hemisphere.
G. M. Frye, Jr., C. P. Wang.
Astrophys. Journ., Vol. 158, 925 - 937 (1969).

In this paper we report the results of four high-altitude balloon flights in the northern hemisphere with a spark-chamber system which has the largest area-solid angle factor of any detector yet used. Good exposures were obtained. On none of these flights was any positive γ-ray signal ascribable to a point source detected above the diffuse atmospheric background, which is generated in the 2.5 g cm^{-2} of air above the balloon by the interaction of the primary cosmic radiation with atmospheric nuclei. The implications of these upper limits on cosmic γ-radiation are discussed for various astrophysical models.

142.042 X-ray and gamma ray astronomy.
B. Rossi.
Cosmic Ray Studies, Bombay 1968, p. 84 - 104 (1969).

142.043 Interpretations of the background component of X-rays and gamma rays. S. Hayakawa.
Cosmic Ray Studies, Bombay 1968, p. 105 - 121 (1969).

142.044 Balloon observations of a new-born X-ray source.
R. M. Thomas, G. Buselli, M. C. Clancy, P. J. N. Davison.
Astrophys. Journ. (*Letters*), Vol. 158, L151 - L154 (1969).

A position for the X-ray source discovered by Conner, Evans, and Belian has been estimated from balloon observations to be $\alpha(1950) = 14^h 57^m \pm 6^m$, $\delta(1950) = -30°.7 \pm 3°$; $l^{II} = 333°$, $b^{II} = +24°$. Estimates of the intensity of the source in different energy ranges are also given.

142.045 Search for line structure in the X-ray spectrum of Sco X-1.
S. S. Holt, E. A. Boldt, P. J. Serlemitsos.
Astrophys. Journ. (*Letters*), Vol. 158, L155 - L158 (1969).

Recently obtained data from a rocket-borne exposure to X-rays from Sco X-1 are shown to yield a positive indication of iron line emission at 3.25 σ at a level approximately an order of magnitude below previously published upper limits. Such emission is consistent with a bremsstrahlung source of universal elemental abundance, but the overall continuum shape indicates significant departures from an isothermal source.

142.046 X-ray line emission from Sco X-1.
J. R. P. Angel.
Bull. American Astron. Soc., Vol. 1, 333 (1969). – Abstract AAS.

142.047 X-ray survey experiment. P. C. Fisher.
Bull. American Astron. Soc., Vol. 1, 343 (1969). Abstract AAS.

142.048 A simultaneous measurement of the optical and

X-ray emission from Sco X-1.
A. Toor, F. D. Seward, L. R. Cathy, W. E. Kunkel.
Bull. American Astron. Soc., Vol. 1, 365 (1969). – Abstract AAS.

142.049 Metagalactic inverse Compton effect and cosmic X-ray background. M. Fukui, S. Hayakawa.
Progr. Theor. Phys. Japan, Vol. 42, 1129 - 1138 (1969).

It has been pointed out that the diffuse component of cosmic X-rays could be accounted for in terms of the inverse Compton collisions of relativistic electrons with universal blackbody photons. This hypothesis is examined in the present paper, taking into account evolutionary effects that result in decreases of the energy density of the universal blackbody radiation and of the production rate of relativistic electrons with cosmic age. The results obtained are: (1) the spectrum of X-rays bends at about 0.3 KeV according to the energy loss of electrons due to the inverse Compton effect and (2) the magnetic field strength in radio galaxies should be as low as $10^{-7} - 10^{-8}$ gauss if the electrons had an intensity high enough to explain the X-ray intensity and the same electrons were responsible for radio emission.

142.050 Galactic X rays and Gamma rays.
R. Cowsik, Y. Pal.
Phys. Rev. Letters, Vol. 23, 1467 - 1468 (1969).

A consistent explanation of the observed galactic flux of high-energy γ rays and the reported galactic X-ray flux is provided in terms of Compton scattering of cosmic-ray electrons with the submillimeter radiation observed recently. It is predicted that the flux of galactic X rays, relative to that of cosmic X rays, will increase with increasing energy. Thus, if an increased enhancement in the flux of high-energy ($\sim$200–keV) X rays in going across the galactic equator is not observed, the existence, on a galactic scale, of the submillimeter radiation will be ruled out.

142.051 The cosmic gamma-ray spectrum from secondary particle production in the metagalaxy.
F. W. Stecker.
Report NASA–TM–X–63254, Goddard Space Flight Center, Greenbelt. 24 pp. (1968). – See Phys. Abstr. Vol. 73, No. 5246 (1970).

142.052 A search for celestial sources of gamma rays of energy greater than 100 MeV. D. R. Hearn.
Report NASA–CR–95736, Smithsonian Astrophys. Obs., Cambridge, Mass., 77 pp. (1968). – See Phys. Abstr. Vol. 73, No. 5284 (1970).

142.053 Interpretatione del fondo di raggi X.
G. Setti.
Atti XII Riunione Soc. Astron. Italiana, L'Aquila 1968, p. 50 - 54 (1969). – Abstract SAI.

142.054 Stellar X-ray sources. L. Biermann.
Proc. Roy. Soc. London, Ser. A, Vol. 313, 357 - 366 (1969).

This paper attempts to provide order of magnitude estimates of the fluxes of quanta to be expected, on the basis of the solar data for stars of similar spectral type and luminosity and of theoretical inferences concerning the coronae of stars of other spectral and luminosity classes. The limited transparency of interstellar space makes it necessary to discuss also the value of the interstellar density between clouds. Finally, the observability of stellar flares will be discussed briefly, because such events would contribute quanta of higher energy than that of the quanta ordinarily to be expected from stellar coronae.

142.055 Possible mechanism of formation of the spectra of

cosmic X-ray sources.
Yu. N. Gnedin, A. Z. Dolginov, A. I. Tsygan.
JETP Letters, Vol. 10, 283 - 285 (1969). [Translated from ZhETF Pis. Red. 10, No. 9, 441 - 444 (1969). In Russian].

Various models have been developed to explain experimental data of the spectra of X-ray sources. The hypothesis that the radiation is due to bremsstrahlung does not explain the available data. For this reason the authors propose as a new model for Sco–XR 1 a relativistic-electron emitter in a magnetic field surrounded by an optically thick shell of cold plasma.

142.056   **Cosmic X-ray observations.**   H. Friedman.
Proc. Roy. Soc. London, Series A, Vol. 313, 301 - 315 (1969).

142.057   **Ground-based observations of X-ray sources – a short review.**   L. Gratton.
Proc. Roy. Soc. London, Series A, Vol. 313, 317 - 330 (1969).

142.058   **Discrete X-ray sources – a theoretical appraisal.**   G. Burbidge.
Proc. Roy. Soc. London, Series A, Vol. 313, 331 - 348 (1969).

142.059   **Interpretation of the cosmic X-ray background.**   D. W. Sciama.
Proc. Roy. Soc. London, Series A, Vol. 313, 349 - 355 (1969).

142.060   **Current developments in cosmic X-ray astronomy.**   K. A. Pounds.
Proc. Roy. Soc. London, Series A, Vol. 313, 367 - 380 (1969).

142.061   **Future requirements for cosmic X-ray astronomy.**   R. L. F. Boyd.
Proc. Roy. Soc. London, Series A, Vol. 313, 381 - 393 (1969).

142.062   **A mechanism for X-ray production in Sco X-1.**   O. P. Manley.
Proc. Roy. Soc. London, Series A, Vol. 313, 395 - 402 (1969).

142.063   **A spectral measurement of the cosmic X-ray background down to 2 keV.**
D. W. Green, B. G. Wilson, A. J. Baxter.
Space Research IX, Proc. Tokyo 1968, p. 222 - 225 (1969).

142.064   **Observations on discrete X-ray sources.**
C. Buselli, M. C. Clancy, P. J. N. Davison, P. J. Edwards, K. G. McCracken, R. M. Thomas.
Space Research IX, Proc. Tokyo 1968, p. 226 - 227 (1969).

142.065   **Photoelectric and photographic observations of Sco X–1.**
Tokyo Astron. Obs. Bull., Second Series, No. 195, p. 2275 - 2276 (1969).

142.066   **Balloon observations of high energy X-ray sources in the region of the galactic centre.**
W. H. G. Lewin, G. W. Clark, M. Gerassimenko, W. B. Smith.
Nature, Vol. 223, 1142 - 1143 (1969).

We surveyed the X-ray emission above 20 keV from the region of the sky from declinations –70° to –5° and from right ascensions 150° to 280°. We noted that the data showed several sources in a region near the galactic centre. We can now account for the bulk of the hard X-rays we observed in terms of four sources whose positions and spectra we report here.

142.067   **Possible identification of a new X-ray source.**   K. G. Henize, L. R. Wackerling.
IAU Circ. No. 2172 (1969).

142.068   **X-ray source.**
J. P. Conner, W. D. Evans, R. D. Belian.
IAU Circ. No. 2174 (1969).

142.069   **Identification of X-ray source.**
D. J. MacConnell, P. S. Hoover.
IAU Circ. No. 2180 (1969).

First gamma ray star?
Nature, Vol. 223, 1308 - 1309 (1969).

Polarimeter for celestial X rays.
See Abstr. 034.011.

Coronae around helium stars and X-ray sources.
See Abstr. 064.033.

The origin of the X-ray background.
See Abstr. 066.006.

Cosmic X-ray bremsstrahlung associated with supra-thermal protons.   See Abstr. 076.010.

Effect on the lower ionosphere of X-rays from Scorpius XR-1.   See Abstr. 083.016.

Ionospheric effect of X-rays from Scorpius XR-1.
See Abstr. 083.040.

Planetary nebulae as possible X-ray sources.
See Abstr. 133.032.

Interstellar absorption of Crab nebula X rays.
See Abstr. 134.009.

Search for polarization in the X-ray emission of the Crab nebula.   See Abstr. 134.010.

Rotating neutron stars, pulsars and cosmic X-ray sources.   See Abstr. 141.089.

A search for pulsed gamma emission from two pulsars.   See Abstr. 141.133.

Further search for high energy gamma rays from CP 1133.   See Abstr. 141.137.

Cosmic electrons and diffuse galactic X- and γ-radiation.   See Abstr. 143.027.

Sco XR-1 as a member of the upper Scorpius complex.   See Abstr. 152.007.

Synchrotron emission from the galaxy and the diffuse X-ray background.   See Abstr. 157.020.

Scattering of background X-rays by metagalactic electrons.   See Abstr. 162.022.

143 Cosmic Radiation

143.001 Observation of trans-iron nuclei in the primary cosmic radiation. G. E. Blanford, Jr., M. W. Friedlander, J. Klarmann, R. M. Walker, J. P. Wefel, W. C. Wells, R. L. Fleischer, G. E. Nichols, P. B. Price.
Phys. Rev. Letters, Vol. 23, 338 - 342 (1969).

Interleaved layers of nuclear photographic emulsion and plastic detectors, covering a total area of 21 m², were exposed to the primary cosmic radiation on high-altitude balloon flights. Flux values, in particles/m² sr sec, have been estimated to be $J(Z \geqslant 33) \geqslant 2.6 \times 10^{-5}$, $J(33 \leqslant Z \leqslant 40) \geqslant 1.9 \times 10^{-5}$, $J(Z \geqslant 70) \gtrsim 1 \times 10^{-6}$. These values refer to the top of the atmosphere, after extrapolation through 1.5 g/cm² of detector and 3.5 g/cm² of atmosphere, for particles with magnetic rigidities above 5 GeV.

143.002 Cosmic rays latitude survey at solar minimum. O. C. Allkofer, R. D. Andresen, E. Bagge, W. D. Dau, H. Funk.
Tellus, Vol. 21, 443 - 446 (1969).

The nucleonic component of the cosmic rays has been studied during two voyages on the german research vessel Meteor.

143.003 Are cosmic electrons anisotropic? J. A. Earl, A. M. Lenchek.
Astrophys. Journ. Vol. 157, 87 - 101 (1969).

Because the rate of energy loss due to synchrotron radiation depends upon pitch angle, the flux of relativistic cosmic-ray electrons tends to become strongly anisotropic as the electrons propagate in the galactic magnetic field. Explicit expressions for anisotropy spectra to be expected in several likely configurations are calculated as equilibrium solutions of an equation of pitch-angle diffusion.

143.004 On the possibility of cosmic rays being transient. J. R. Wayland.
Planet. Space Sci. Vol. 17, 1619 - 1628 (1969).

We have considered the possible observable effects that may be produced if the flux of cosmic radiation is time dependent. Thus we have solved the time dependent problem to find possible time variation in both the intensity and the momentum spectrum. It is concluded that cosmic radiation may be locally transient with a quasi-steady state on the large scale of our galaxy (or perhaps metagalaxy).

143.005 The low energy cosmic ray spectrum. R. M. Hjellming.
Astrophys. Letters, Vol. 4, 81 - 84 (1969).

The cosmic ray spectrum for energies less than 30 MeV/nucleon is determined from a knowledge of the ionization rate per hydrogen atom produced by low energy cosmic rays in the interstellar medium. It is suggested that a peak in the cosmic ray intensity between roughly 0.2 and 3 MeV/nucleon be identified with the high energy portion of particle emission from stars.

143.006 Flux and spectrum of primary hydrogen nuclei near geomagnetic equator. R. K. Puri, P. K. Aditya.
Journ. Geophys. Res. Vol. 74, 4787 - 4790 (1969).

In an emulsion stack exposed at an atmospheric depth of 3.8 g cm⁻² on March 16, 1965, the flux of primary hydrogen nuclei has been estimated equal to (103 ± 11) particles/m² ster sec, corresponding to a threshold rigidity of 16.8 Gv. The regression curve for the integral flux of primary hydrogen nuclei above 16.8 Gv versus neutron monitor count rate has been drawn. The ratio of primary hydrogen to helium nuclei measured simultaneously at the same location yields a value of 7.2 ± 1.0. The integral rigidity spectrum of primary hydrogen nuclei can be represented by a spectral index of $-(1.65 \pm 0.12)$.

143.007 Evidence of quarks in air-shower cores. C. B. A. McCusker, I. Cairns.
Phys. Rev. Letters, Vol. 23, 658 - 659 (1969).

In a study of air-shower cores using a delayed-expansion cloud chamber, we have observed a track for which the only explanation we can see is that it is produced by a fractionally charged particle.

143.008 On the energy dependence of the ratio of Li, Be and B to C, N, O and F nuclei in the primary cosmic rays. S. Biswas.
Astrophys. Space Sci. Vol. 5, 59 - 67 (1969).

A critical study of all available data on the energy dependence of the ratio of the intensities of Li, Be, and B to C, N, O and F nuclei (L/M-ratio) in cosmic rays is made. It seems that in a recent experiment by Von Rosenvinge et al. (1969), the flux of M-nuclei has come out higher in the energy interval of 200 – 400 MeV/nucleon and that of L-nuclei lower in the energy interval 400 – 600 MeV/nucleon; as a result they obtained a value of the ratio of L/M which is about 50% lower than all other investigators. The results of all other studies yield the best estimate of the ratio of L/M as 0.26 ± 0.03, 0.41 ± 0.03 and 0.26 ± 0.02 in the energy interval 50 – 150, 200 – 600 and > 1500 MeV/nucleon respectively.

143.009 Periodic solar time variations in the cosmic-ray muon component near sea level. R. M. Briggs, R. B. Hicks, S. Standil.
Journ. Phys. A, General Phys. Ser. 2, Vol. 2, 584 - 590 (1969).

Experimental results using eight scintillation counters obtained between 13th November 1966 and 5th May 1968 are presented. It is shown that the phase of the average diurnal anisotropy is consistent with the Parker and Axford theories alone, and its amplitude is dependent on the time in the solar cycle. Further, a semi-diurnal anisotropy is found, extra-terrestrial in the origin, with intensity maxima at right angles to the interplanetary magnetic field.

143.010 The propagation and anisotropy of cosmic rays. II. Electrons. D. G. Wentzel.
Astrophys. Journ. Vol. 157, 545 - 555 (1969).

Synchrotron radiation makes cosmic-ray electrons anisotropic. We evaluate the electrons' distribution function in pitch angle, in case the electrons steadily create – and are scattered by – hydromagnetic waves in an ionized plasma. We find that the radiation from radio galaxies is generated by nearly isotropic electrons. In our Galaxy, the anisotropy increases with energy and is 1 percent at $\sim 10^2$ GeV. However, the scattering of cosmic-ray electrons in the galactic disk is dominated by waves from other poorly known sources, primarily the streaming of cosmic-ray protons.

143.011 Leakage electrons from normal galaxies: The diffuse cosmic X-ray source. K. Brecher, P. Morrison.
Phys. Rev. Letters, Vol. 23, 802 - 806 (1969).

Diffuse cosmic X-rays arise from Compton collisions between galactic leakage electrons and the 2.7°K thermal background photons in extragalactic space. Assuming that the break seen in the electron spectra within normal galaxies is intrinsic to the cosmic-ray sources, we obtain detailed agreement with the observed X-ray spectral shape and intensity using only parameters obtained from radio observations. This serves to verify most of the properties of the microwave blackbody radiation.

143.012 Primary cosmic-ray electron energy spectrum from 10 to 200 MeV observed in interplanetary space.
C. Y. Fan, J. L'Heureux, P. Meyer.
Phys. Rev. Letters, Vol. 23, 877 - 880 (1969).

We have measured the quiet-time flux and energy spectrum of primary cosmic-ray electrons in the energy range from 10 to 200 MeV in interplanetary space. This investigation was carried out with an instrument on board the satellite OGO-5 and covers the period from 25 March to 25 April 1968.

143.013 Underground and surface measurements of the second harmonic of the cosmic ray daily variation and the upper limit to modulation.
A. Hashim, D. S. Peacock, J. J. Quenby, T. Thambyahpillai.
Planet. Space Sci. Vol. 17, 1749 - 1758 (1969).

The second harmonic of the cosmic ray daily variation has been explained in terms of a rising cosmic ray density gradient symmetric about the solar equatorial plane. The amplitude and rigidity dependence of the second harmonic is directly related to that of the solar cycle modulation. Hence studying the second harmonic of 100 GV primary particles provides information on the upper limit to the solar cycle modulation.

143.014 The diurnal variation of cosmic ray electrons.
G. M. Simnett.
Planet. Space Sci. Vol. 17, 1781 - 1793 (1969).

The cosmic ray electron intensity between 30 MeV and 1500 MeV has been measured with a system composed of a Cerenkov telescope and a lead-scintillator sandwich detector. Results are presented from a balloon flight made at Kiruna, Northern Sweden, in 1967. The electron measurements were made between 3.7 and 4.5 g cm^{-2} residual atmosphere from local midnight until after local noon. The results show that at energies between 30 MeV and 900 MeV the diurnal variation is 0 ± 25 per cent. This is compared with other results from Fort Churchill, Canada and with satellite observations.

143.015 Cosmic rays in the galaxy. P. Meyer.
Annual Rev. Astron. Astrophys. Vol. 7, 1 - 38 (1969).

143.016 Steady-state flux of positrons and the age of the galactic halo. S. Lal, K. Brunstein.
Progr. Theor. Phys. Japan, Vol. 42, 213 - 218 (1969).

In solving the continuity equation which describes the intensity of cosmic-ray electrons and positrons in interstellar space, one makes the a priori assumption that steady-state condition exists. In this paper we have investigated the condition under which this assumption is justified. We further show that if the cosmic rays are assumed to be confined uniformly throughout the galactic disc and the halo, then the age of the halo should be $\gtrsim 10^8$ years to account for the agreement between the calculated and observed flux of positrons in the primary component of cosmic rays.

143.017 Solar diurnal anisotropy of cosmic rays.
R. M. Jacklyn, S. P. Duggal, M. A. Pomerantz.
Nature, Vol. 223, 601 - 602 (1969).

A comprehensive analysis was carried out of measurements obtained over almost an entire solar cycle with a variety of detectors responding to primaries extending over the range 1 to 200 GV. Year-to-year changes in the amplitude of the daily variation were greater for experimental arrangements having higher mean rigidities of response. Furthermore, in contrast with neutron monitors, other detectors (inclined and underground meson telescopes) revealed marked changes in the phase. The average value of the upper cutoff rigidity above which the diurnal variation ceases was approximately 90 GV during the nine-year period 1958 - 1966, and the yearly averages ranged from about 55 GV to 100 GV. The need for exercising special caution in comparing available experimental results with the predictions of theoretical models of the diurnal anisotropy is indicated.

143.018 Observations of cosmic-ray electrons between 2.7 and 21.5 MeV.
G. M. Simnett, F. B. McDonald.
Astrophys. Journ. Vol. 157, 1435 - 1447 (1969).

Results are presented from the IMP-IV satellite on the intensity of electrons with energies of 2.7 - 21.5 MeV in interplanetary space during the period July 3 through August 27, 1967. The measured electron intensity is believed to be uncontaminated by solar electrons. The analysis procedure for background subtraction and the subsequent derivation of the electron spectrum are described in detail.

143.019 About the distortion of the outer anisotropic stream of galactic cosmic rays in the solar system.
A. V. Belov, L. I. Dorman.
Geomagn. Aeronom. Vol. 9, 613 - 616 (1969). In Russian.

Von der Gleichung der isotropen Diffusion ausgehend finden Verf. die erwartete Distorsion des äußeren anisotropen Stromes der galaktischen kosmischen Strahlung im interplanetaren Raum in Abhängigkeit vom Transportweg und dem Grad der Anisotropie. *D. Krahn*

143.020 Störmerbahnen energiereicher Teilchen der kosmischen Strahlung in einer Modellgalaxis.
K. O. Thielheim.
Zeitschr. Naturforschung, Vol. 24a, 1664 - 1665 (1969).

The influence of the interstellar magnetic field on the propagation of high energy particles of cosmic radiation is studied.

143.021 Etude des variations temporelles de la composante pénétrante du rayonnement cosmique à 1280 m équivalents d'eau en 1967 - 1968.
R. Regimbart, M. Scherer.
Comptes Rendus Acad. Sci. Paris, Sér. B, Vol. 269, 777 - 780 (1969).

Une étude de la distribution temporelle de la composante pénétrante du rayonnement cosmique a révélé l'absence de variations journalières et semi-journalières solaire et sidérale, la présence d'une variation lunaire semi-journalière et mensuelle importante ainsi qu'un effet de 27 jours.

143.022 A measurement of the spectrum of cosmic-ray electrons between 20 MeV and 3 beV in 1968. — Further evidence for extensive time variations of this component.
J. Rockstroh, W. R. Webber.
Journ. Geophys. Res. Vol. 74, 5041 - 5053 (1969).

The intensity and spectrum of cosmic-ray electrons from 20 MeV to 3 beV has been measured at a depth of 2.5 g/cm^2 at Fort Churchill in the summer of 1968 at a time when solar modulation effects had reduced the sea-level neutron monitor rate by ~12% below its sunspot minimum value. The results confirm and extend earlier measurements of the solar modulation of electrons made in 1965 - 1966. The energy dependence of the electron modulation is essentially the same in the 1966 - 1968 period as was measured in 1965 - 1966.

143.023 The survival of heavy nuclei in Colgate's supernova-acceleration model for cosmic rays.
J. H. Kinsey.
Astrophys. Journ. Vol. 158, 295 - 302 (1969).

Colgate's supernova-acceleration model for cosmic rays is examined to see if heavy nuclei can indeed survive such catastrophic acceleration. It is is assumed that all species of nuclei are collectively accelerated by a mechanism such as a plasma-wave instability in a shock wave, then the only factor remaining to dissociate heavy nuclei in Colgate's model is the blueshifted photon flux in the shock wave which moves radially outward. This would produce a sharp cutoff in the heavy component of cosmic rays by photodisintegration of the heavy nuclei for final cosmic-ray energies greater than 10^3 GeV per nucleon.

143.024 **Sur un effet attribué au changement de la composition chimique primaire des particules cosmiques dans l'intervalle de rigidité magnétique $10^{15} - 10^{17}$ V.**
P. Catz, J. Gawin, R. Maze, J. Wdowczyk, A. Zawadzki.
Comptes Rendus Acad. Sci. Paris, Sér. B, Vol. 269, 1056 - 1059 (1969).

Une analyse des mesures du rapport Nμ/Ne des grandes gerbes reçues pour différentes inclinaisons de leurs axes a permis de mettre en évidence une variation anormale de l'exposant a dans la relation $N\mu \approx Ne^a$ en fonction de Ne. Une telle anomalie est probablement due au changement de la composition chimique du rayonnement cosmique primaire.

143.025 **Effetto delle radiazioni stellare e fossile sullo spettro dei nuclei primari.** C. Castagnoli, G. Navarra, P. Penengo.
Atti Accad. Nazionale Lincei, Rend. Cl. Sci. fis. mat. nat., Serie Ottava, Vol. 46, 40 - 44 (1969).

The energy spectrum of the primary nuclei of the cosmic radiation is calculated by examining their collisions against the photons of the stellar light and of the 3° K blackbody radiation. The effect is studied in the energy range both below and above the pion production threshold. The results obtained on the breakdown of the He, C, Fe spectra are compared with the latest results on the proton spectrum.

143.026 **Large amplitude wave trains in the cosmic ray intensity.** A. Hashim, T. Thambyahpillai.
Planet. Space Sci. Vol. 17, 1879 - 1889 (1969).

The occurrence of an unusual class of large amplitude wave trains in the cosmic ray neutron intensity, which is distinctly different from the average diurnal variation as well as from other recognised types of large amplitude diurnal variations, is noted and the directional distribution in interplanetary space determined by the analysis of data from a number of 'high latitude' neutron monitors.

143.027 **Cosmic electrons and diffuse galactic X- and γ-radiation.** M. J. Rees, J. Silk.
Astron. Astrophys. Vol. 3, 452 - 454 (1969).

It is suggested that the γ-rays detected by Clark, Garmire and Kraushaar at ~ 100 MeV may be due to the Bremsstrahlung of cosmic ray electrons in the interstellar medium. A diffuse galactic X-ray flux is predicted, produced by inverse Compton scattering of starlight photons with these electrons. Further consequences are discussed.

143.028 **Galactic effects of the cosmic-ray gas.** E. N. Parker.
Space Sci. Rev. Vol. 9, 651 - 712 (1969).

On an astronomical scale cosmic rays must be considered a tenuous and extremely hot (relativistic) gas. The pressure of the cosmic-ray gas is comparable to the other gas and field pressures in interstellar space, so that the cosmic-ray pressure must be taken into account in treating the dynamical properties of the gaseous disk of the galaxy. The equations for the equilibrium distribution of the gaseous disk of the galaxy in the direction perpendicular to the disk are worked out. Perturbation calculations then show that the equilibrium is unstable, on scales of a few hundred pc and in times of the order 2×10^7 years. The instability is driven about equally by thermal effects, the magnetic field, and the cosmic-ray gas and is the major effect in forming interstellar gas clouds.

143.029 **Charge composition and energy spectrum of primary cosmic-ray electrons.**
J. L. Fanselow, R. C. Hartman, R. H. Hildebrand, P. Meyer.
Astrophys. Journ. Vol. 158, 771 - 780 (1969).

The flux, energy spectrum, and charge composition of the electron component of primary cosmic rays was measured in 1965 and 1966 in the range from 170 MeV to 14.3 BeV, and a finite flux of positrons was observed up to 4 BeV. For the first time, it has been possible to determine the energy spectrum of primary positrons above 220 MeV. To approximate this observed spectrum above 860 MeV by a power law requires an exponent of 2.6 ± 0.5, consistent with negligible modification of the source spectrum below 10 BeV by energy-loss processes.

143.030 **The isotopic abundances and energy spectra of ^{2}H, ^{3}He, and ^{4}He of cosmic-ray origin in the energy region ~ 10 - 100 MeV nucleon^{-1}.**
K. C. Hsieh, J. A. Simpson.
Astrophys. Journ. (*Letters*), Vol. 158, L37 - L41 (1969).

The IMP-4 satellite carried an instrument capable of resolving the H and He isotopes. The differential energy spectrum of ^{2}H from 6 to 60 MeV nucleon^{-1}, along with simultaneous measurements of ^{3}He and ^{4}He, has been obtained during the 1967 solar quiet times in interplanetary space. These measurements show that (1) an ^{2}H flux exists whose magnitude is not accounted for by current ideas for cosmic-ray transport in the galaxy; (2) by using the ^{2}H as a tracer for a component of galactic origin in interplanetary space it is found that the solar-proton component (1967) must be less than 15 percent of the interplanetary proton flux at 20 MeV; and (3) present forms of solar-modulation theory must be modified at low energies to account for the observations.

143.031 **Investigation of the solar wind with help of solar and galactic cosmic rays.**
I. V. Dorman, L. I. Dorman, L. I. Miroshnichenko.
Problemy kosmich. fiz. No. 4, p. 3 - 12 (1969).   In Russian.

A review of recent investigations of cosmic ray propagation in interplanetary space is given. Solar cosmic ray bursts supply important information about the inner and outer regions of the solar system. Some interesting features of the utmost regions can be determined from the 11-year galactic cosmic ray variation in connection with changes of solar activity. It is shown that the available estimations of the size of the modulation region depend strongly on the solar activity level and the propagation model involved.

143.032 **'Sidereal' cosmic-ray diurnal variations.** D. B. Swinson.
Journ. Geophys. Res. Vol. 74, 5591 - 5598 (1969).

A model is proposed in which a 'sidereal' variation can be produced in the vicinity of the earth as a result of the radial cosmic-ray density gradient and the interplanetary magnetic field; the model is tested by dividing the cosmic-ray data according to the direction of the interplanetary field and performing the same sidereal time analysis on these data. The results confirm the principal features of the model and demonstrate that the 'sidereal' variation can be produced in the inner solar system, correctly predicting the observed times of maximum in the northern and southern hemispheres, without recourse to assumptions about anisotropies of the particle flux in galactic space.

143.033 **Cosmic-ray intensity variations on January 26 - 27, 1968.** J. A. Lockwood, W. R. Webber.
Journ. Geophys. Res. Vol. 74, 5599 - 5610 (1969).

In this paper we present some observations on a Forbush

decrease on January 26, 1968, which was about 4.5% at the Mt. Washington neutron monitor. This event is interesting for two reasons. First, there were large anisotropies in the cosmic-ray flux, both within and perpendicular to the ecliptic plane during the event. Second, satellite data on the interplanetary field and the primary cosmic-ray flux outside any influences of the geomagnetic field are available, in addition to the neutron monitor counting rates.

143.034 Cosmic rays and the interstellar medium.
K. B. Fenton.
Proc. Astron. Soc. Australia, Vol. 1, 269 - 272 (1969). − Review article.

143.035 Recent evidence concerning the sidereal anisotropy in the charged primary cosmic radiation.
R. M. Jacklyn, A. Vrana.
Proc. Astron. Soc. Australia, Vol. 1, 278 - 280 (1969). − Contribution ASA meeting.

143.036 'Hysteresis' effect in cosmic ray modulation and the cosmic ray gradient near solar minimum.
S. R. Kane, J. R. Winckler.
Journ. Geophys. Res. Vol. 74, 6247 - 6255 (1969).

In this paper the total cosmic-ray ionization measurements made near the earth with the OGO 1 and OGO 3 ion chambers during the period September 1964 to December 1967 are presented and their implications regarding the energy dependence of the long-term modulation and the heliocentric gradient are investigated.

143.037 Cosmic-ray electrons between 12 MeV and 1 GeV in 1967. M. H. Israel.
Journ. Geophys. Res. Vol. 74, 4701 - 4713 (1969).

Observations of cosmic ray electrons in the energy range of 12 MeV to 1 GeV were made in 1967 in a series of high altitude balloon flights with a detector consisting of a scintillation counter telescope, gas Čerenkov counter, and lead-plate spark chamber. Three flights were launched from Ft. Churchill, Manitoba, in summer 1967, to measure the vertically incident primary electron flux; a fourth flight gave a direct measurement of splash albedo electrons. In April, 1967, return albedo electrons were observed on a flight launched from Palestine, Texas.

143.038 Observations on the abundance of nitrogen in the primary cosmic radiation.
J. A. Lezniak, J. F. Ormes, T. T. von Rosenvinge, W. R. Webber.
Astrophys. Space Sci. Vol. 5, 103 - 112 (1969).

New measurements of the intensity and spectrum of cosmic ray nitrogen nuclei made by instruments flown on balloons and on the Pioneer–8 space probe are reported. The nitrogen spectrum is found to be identical with that of the other medium nuclei, over the range of measurement from 100 MeV/nuc to > 22 GeV/nuc. The ratio of N to all M nuclei is found to be =0.125, constant to within 10% over this energy range. This ratio is extrapolated to the cosmic-ray source. Taking an average material path length of 4 g/cm^2 of hydrogen constant with energy, as required to make the abundance of L nuclei →0 at the cosmic-ray source, the resulting N/M source ratio is ⩽0.03. Thus, to the same degree that the so-called L nuclei are absent in the cosmic-ray sources, N nuclei are also absent. This nitrogen abundance is different from the estimated solar atmospheric abundance of ~0.10 for the N/M ratio. However, under certain conditions in the CNO bi-cycle that operates for the production of nitrogen in stellar objects a negligible production of nitrogen might be expected. It is suggested that these conditions exist in the cosmic-ray sources.

143.039 Zur Theorie der kosmischen Strahlung extrem ho-
her Energie. G. Helmis.
Monatsber. Deutsch. Akad. Wiss. Berlin, Band 11, 165 - 172 (1969).

143.040 Solar modulation of galactic cosmic rays, I.
L. A. Fisk, W. I. Axford.
Journ. Geophys. Res. Vol. 74, 4973 - 4986 (1969).

The equations governing cosmic-ray molulation allowing for convection, diffusion, and energy changes are approximated with simpler, more manageable equations that describe the particle behavior in a limited energy range. One of these equations determines an excellent approximation to the particle number density at energies above a few hundred MeV/nucleon, while others should describe the behavior of the number density and radial streaming at energies below about 50 – 75 MeV/nucleon. Analytic solutions to the exact equations are used to demonstrate that the approximate equations accurately describe the particle behavior in their respective limits.

143.041 Analysis of IQSY cosmic-ray survey measurements.
H. Carmichael, M. Bercovitch.
Canadian Journ. Phys., Vol. 47, 2073 - 2093 (1969).

In this, the last of a set of five papers reporting latitude surveys carried out in 1965 and 1966 at the time of and soon after the IQSY cosmic-ray maximum, the observations are reduced to a common atmospheric depth and at the same time the attenuation coefficients in the atmosphere for both the neutron monitor and the muon monitor are determined as functions of altitude and latitude.

143.042 Plasma instabilities of streaming cosmic rays.
E. Tademaru.
Astrophys. Journ., Vol. 158, 959 - 979 (1969).

A general expression is derived for the growth rate, due to particle-wave resonance, of plasma waves propagating in a spatially homogeneous, but anisotropic, relativistic plasma embedded in a uniform ambient magnetic field. Applying this expression to an idealized model of cosmic rays streaming parallel to the magnetic field in a cold background plasma, we find the system to be unstable against waves propagating at an arbitrary angle θ to the field direction, provided that the phase velocity of the waves is less than the streaming velocity times the cosine of θ.

143.043 Composition of cosmic rays. S. Biswas.
Cosmic Ray Studies, Bombay 1968, p. 45 - 64 (1969).

143.044 Composition and energy spectrum of prehistoric cosmic radiation. D. Lal.
Cosmic Ray Studies, Bombay 1968, p. 65 - 72 (1969).

143.045 On the propagation of cosmic rays.
Y. Pal.
Cosmic Ray Studies, Bombay 1968, p. 73 - 83 (1969).

143.046 Cosmic rays at energies beyond 10^{12} eV.
B. V. Sreekantan.
Cosmic Ray Studies, Bombay 1968, p. 151 - 180 (1969).

143.047 Cosmic ray electrons.
R. R. Daniel.
Cosmic Ray Studies, Bombay 1968, p. 181 - 201 (1969).

143.048 The cosmic-ray spectrum below 30 MeV/nucleon.
R. M. Hjellming.
Bull. American Astron. Soc., Vol. 1, 348 (1969). − Abstr. AAS.

143.049 Interaction between cosmic-ray electrons and cos-

mic-ray protons.　　D. B. Melrose, D. G. Wentzel.
Bull. American Astron. Soc., Vol. 1, 354 (1969). – Abstr.
AAS.

143.050　Decomposition and energy degradation of very high-energy cosmic rays by Doppler-shifted photons.
J. B. Pollack, B. S. P. Shen.
Bull. American Astron. Soc., Vol. 1, 360 (1969). – Abstr.
AAS.

143.051　On the possible mechanism of meteoric variation of the intensity of cosmic rays.　　S. A. Belsky.
Astron. Zhurn. Akad. Nauk SSSR, Vol. 46, 1330 - 1332 (1969). In Russian. English translation in Soviet Astron. AJ, Vol. 13, No. 6.

A possible mechanism of the effect of increase of intensity of cosmic rays during the maximum action of meteor streams is discussed. The effect is connected with the presence of regular magnetic fields in meteor streams, which are, in their way, traps for cosmic rays. The estimation of the magnetic field of the meteor streams leads to the value $\sim 10^{-5}$ e, that, by the order of value, coincides with values of magnetic fields observed in the interplanetary space.

143.052　Adiabatische Invarianz des Wirkungsintegrals für die Bewegung in nicht-regulären Kraftfeldern.
D. Pfirsch, K. Schindler.
Sitzungsber. Bayer. Akad. Wiss. Math.-Nat. Kl., Jahrgang 1968, p. 8* (1969). – Abstr.

143.053　Disintegration and energy degradation of very high-energy cosmic rays in intense photon fields.
J. B. Pollack, B. S. P. Shen.
Phys. Rev. Letters, Vol. 23, 1358 - 1361 (1969).

Very high-energy cosmic rays, on emerging from their places of origin, are subject to photodisintegration and energy degradation by blue-shifted photons. It is shown that the intense photon fields of supernova explosions, of quasistellar objects, and of one of the current pulsar models are able to cause complete disintegration of complex nuclei and significant energy losses of protons at energies consistent with observed changes in the cosmic-ray charge and energy spectrum, respectively.

143.054　Čerenkov light in extensive air showers and the chemical composition of primary cosmic rays at 10^{16} **eV.**　　A. S. Krieger, H. V. Bradt.
Phys. Rev., Second Series, Vol. 185, 1629 - 1635 (1969).

143.055　Measurement of the primary cosmic-ray proton spectrum between 40 and 400 GeV.
W. K. H. Schmidt, K. Pinkau, U. Pollvogt, R. W. Huggett.
Phys. Rev., Second Series, Vol. 184, 1279 - 1282 (1969).

The purpose of this experiment was to study the flux, composition, possible time variations, and nuclear interaction properties of cosmic rays at energies between 40 and 400 GeV.

143.056　Cosmic ray electron spectrum above 200 GeV.
K. C. Anand, R. R. Daniel, S. A. Stephens.
Nature, Vol. 224, 1290 - 1291 (1969).

On the basis of observations made on cosmic ray electrons of energy $\gtrsim$ 200 GeV recorded in an emulsion-lead sandwich assembly exposed in a balloon flight over India, very suggestive evidence is presented for a steepening of the energy spectrum around 200 GeV. Some important consequences that are likely to follow from this observation are the residence time of cosmic rays is $(0.5 - 1) \times 10^6$ years; they are essentially confined to the galactic disk; a mean number of $\approx$ 3 hydrogen atoms cm^{-3} in the galactic disk are required.

143.057　Stability of ^{7}Be in galactic cosmic radiation.
F. Yiou, F. Guchan-Beck.
Journ. Physique, Vol. 30, 401 - 405 (1969).　　In French.

143.058　About the dependence of the effect of increase before Forbush-decreases on the energetic spectrum of cosmic rays.　　L. I. Dorman, N. S. Kaminer, T. V. Kebuladze.
Geomagn. Aeronom. Vol. 9, 809 - 812 (1969).　　In Russian.

143.059　Asymmetrical effects in secular variations of cosmic rays.　　N. P. Chirkov.
Geomagn. Aeronom. Vol. 9, 738 - 740 (1969).　　In Russian.
Brief information.

143.060　About the modulation of intensity of galactic cosmic rays by the solar wind in the presence of an azimuthal asymmetry.　　A. V. Belov, L. I. Dorman.
Geomagn. Aeronom. Vol. 9, 972 - 981 (1969).　　In Russian.

143.061　On the 27-day recurrence of daily variations of cosmic rays.　　A. A. Danilov, S. O. Morozova.
Geomagn. Aeronom. Vol. 9, 1074 (1969).　　In Russian.
Brief information.

143.062　Phase jumps of diurnal variations in cosmic-ray intensity.
A. K. Pankratov, B. M. Vladimirsky, A. A. Stepanyan.
Cosmic Rays No. 10, Moscow, p. 27 - 29 (1969). In Russian.

The phase jumps of diurnal variations are considered on the basis of data of the world-wide network of cosmic-ray stations for the IQSY.

143.063　Periodic variations and their changes dependent on solar activity.
E. V. Kolomeets, G. A. Sergeeva, R. A. Chumbalova.
Cosmic Rays No. 10, Moscow, p. 43 - 45 (1969).　　In Russian.

Dependence of periodic variations on solar activity is investigated.

143.064　About the nature of the changes in cosmic-ray diurnal variation during magnetic storms with sudden commencement.
B. M. Vladimirsky, A. K. Pankratov, A. A. Stepanyan.
Cosmic Rays No. 10, Moscow, p. 46 - 49 (1969).　　In Russian.

The nature of changes of cosmic ray diurnal variations during magnetic storms with sudden commencement is considered.

143.065　About the structure of the interplanetary space according to the data on annual cosmic-ray variations.　　L. I. Dorman, A. A. Luzov, V. P. Mamrukova.
Cosmic Rays No. 10, Moscow, p. 162 - 166 (1969). In Russian.

The cosmic radiation intensity gradient for the plane normal to the ecliptic plane is computed and a dynamic description of the electromagnetic structure of interplanetary space and of its changes in the course of the solar activity cycle is given.

143.066　New possibility of explaining the complex form of the energy spectrum of ultrahigh energy primary cosmic rays.　　G. V. Kulikov, Yu. A. Fomin, G. B. Khristiansen.
JETP Letters, Vol. 10, 222 - 225 (1969). [Translated from ZhETF Pis. Red. 10, No. 7, 347 - 353 (1969).　　In Russian].

An investigation is made on the origin of ultra-high-energy cosmic rays. It is assumed that cosmic rays are of purely galactic origin and that a non-stationary component exists besides the stationary energy spectrum.

143.067 **Matter traversal of high–energy primary cosmic rays from antiproton measurements.**
S. Rosen.
Nuovo Cimento Lettere, Prima Ser., Vol. 1, 336 - 339 (1969).

This communication deals with that small portion of primary proton component which has energies in excess of 5.6 GeV and consequently contains the predecessors of antiprotons. The method consists in measuring and analyzing the antiproton flux in the primary galactic beam. Spallation processes are not involved.

143.068 **Composition of cosmic rays measured in Gemini XI.**
F. W. O'Dell, M. M. Shapiro, R. Silberberg, B. Stiller, C. H. Tsao, N. Durgaprasad, C. E. Fichtel, D. E. Guss, D. V. Reames.
Space Research IX, Proc. Tokyo 1968, p. 215 - 221 (1969).

143.069 **East–West asymmetry and charge sign ratio of primary cosmic-ray electrons at 8.3 GV rigidity cut-off.** B. Agrinier, Y, Koechlin, B. Parlier, J. Paul, J. Vasseur, G. Boella, C. Dilworth, L. Scarsi, G. Sironi, A. Russo.
Nuovo Cimento Lettere, Prima Ser., Vol. 1, 53 - 56 (1969).

143.070 **Measurement of the primary cosmic ray proton and alpha-particle spectra above 10 GeV/Nucl.**
K. Pinkau, U. Pollvogt, W. K. H. Schmidt, R. W. Huggett.
Mitt. Astron. Ges. No. 27, p. 143 - 148 (1969). – Conference paper.

143.071 **Experimental tests of the mechanism of radio emission from showers.**
F. G. Smith, A. D. Bray, R. A. Porter, W. S. Torbitt, J. V. Jelley.
Canadian Journ. Phys., Vol. 46, S230 - S233 = Astron. Contr. Univ. Manchester, Ser. II, Jodrell Bank Repr. No. 389 (1968).

143.072 **A comparison of the energy spectra of cosmic ray helium and heavy nuclei.**
T. T. Von Rosenvinge, W. R. Webber, J. F. Ormes.
Astrophys. Space Sci. Vol. 5, 342 - 359 (1969).

Recent observations of the spectra of cosmic ray helium, M, LH and VH nuclei in the energy range from $\lesssim 200$ MeV/nuc to > 22 GeV/nuc are reported. In this paper we shall compare these new measurements with our recent measurements of the helium spectrum covering the range from ~ 100 MeV/nuc to > 24 GeV/nuc (Ormes and Webber, 1968).

The astrophysics of cosmic rays.
Southern Stars, Vol. 23, 60 - 61 (1969).

Zusammensetzung der primären kosmischen Strahlung.
Umschau, Vol. 69, 558 - 559 (1969).

Cosmic Rays. Results of Researches on International Geophysical Projects. Articles No. 10.
See Abstr. 003.111.

Adiabatische Invarianz des Wirkungsintegrals für die Bewegung in nicht-regulären Kraftfeldern.
See Abstr. 022.018.

Statistical analysis of intensity increases of cosmic rays before Forbush-effects. See Abstr. 078.034.

A study of solar and cosmic radiation from the Venus 4 space probe. See Abstr. 078.037.

Recent advances in the study of fossil tracks in meteorites due to heavy nuclei of the cosmic radiation.
See Abstr. 105.075.

Fossil tracks in meteorites and the chemical abundance and energy spectrum of extremely heavy cosmic rays.
See Abstr. 105.110.

On the flux of low-energy particles in the solar system during the last 10 million years.
See Abstr. 105.111.

Cosmic protons interaction with interplanetary plasma. See Abstr. 106.016.

The fragmentation of cosmic-ray nuclei in interstellar hydrogen. See Abstr. 131.020.

Pulsars and the origin of cosmic rays.
See Abstr. 141.097.

Pulsars and local cosmic ray prehistory.
See Abstr. 141.099.

Pulsars and cosmic rays.
See Abstr. 141.132.

Pulsars as possible sources of super-heavy nuclei in the primary cosmic radiation. See Abstr. 141.215.

Primordial cosmic ray sources.
See Abstr. 142.023.

Universal cosmic rays and Harrison's inhomogeneity postulate. See Abstr. 162.007.

Universal cosmic rays and the matter-antimatter universe. See Abstr. 162.008.

Stellar Systems

151 Kinematics and Dynamics of Stellar Systems

151.001 **Hypothesis on the origin of the spiral structure of the galaxies.** L. S. Marochnik, A. A. Suchkov. Astrophys. Space Sci. Vol. 4, 317 - 326 (1969).

The spiral waves in a model galaxy consisting of differentially rotating and non-rotating subsystems are considered on the basis of participating phenomena. The subsystems involved represent the Populations I and II of normal spirals, respectively. The spiral waves in such a system are unstable in the Landau sense. Due to this instability they grow up to attain finite amplitude. This growth is stopped by a non-linear effect (the quasilinear effect), the steady state with waves of finite amplitude being established. The hypothesis is proposed that these waves should be identified with the spiral structure of the galaxies.

151.002 **Short-wavelength oscillations of cold-disk galactic models.** C. Hunter. Astrophys. Journ. Vol. 157, 183 - 196 (1969).

A WKBJ analysis is given of the free oscillations of cold, thin-disk galactic models. Asymptotic approximations to the frequencies and forms of oscillations of short wavelength both in and perpendicular to the plane of the disk are obtained. The short-wavelength oscillations in the plane are never steady but either grow or decay in amplitude with time. Moreover, provided the angular velocity of disk material decreases from the center outward, the non-axisymmetric oscillations that grow on the linear theory have the form of leading spirals. These results on the planar modes in a cold disk extend considerably beyond, but are compatible with, those obtained in earlier investigations.

151.003 **Effect of differential rotation on the gravitational instability of a stellar system.** H. Niimi. Publ. Astron. Soc. Japan, Vol. 21, 185 - 193 (1969).

The effect of a slight non-uniformity in the angular velocity on the local gravitational instability of a stellar system is investigated in the case when the disturbance is a plane wave with the wave vector parallel to the axis of rotation. The system is supposed to be in such an equilibrium state that the gravitational force is balanced by the centrifugal force. The wavelength of the disturbance is assumed to be so short that a local treatment may provide a good approximation. The conditions for instability of the system are derived.

151.004 **The importance of dynamical mixing for the process of relaxation of stellar systems.** I. L. Genkin. Astron. Tsirk. No. 507, p. 4 - 6 (1969). In Russian.

151.005 **Physical foundation of the dynamics of stellar systems.** F. A. Tsitsin. Trudy Astrofiz. Inst. Alma-Ata, Vol. 12, 3 - 16 (1969). In Russian.

General basic statistical and physical assumptions of stellar dynamics are considered. The necessity for an unambiguous determination of "volume" in the statistical theory of a discrete system is emphasized. For sufficiently large systems an estimate of the entropy is given. No functional relation exists between the entropy and the probability of state.

151.006 **Structure and dynamics of typical stellar systems.**

G. M. Idlis. Trudy Astrofiz. Inst. Alma-Ata, Vol. 12, 17 - 33 (1969). In Russian.

A critical survey of investigations on self-gravitating stellar systems (regular galaxies) carried out during the past decade is given. The following aspects are discussed: rotation as characteristic property of real stellar systems; existence of an equatorial plane of symmetry; three axial ellipsoid of the dispersion of stellar peculiar velocities; rotation and structural properties of galaxies gravitationally interacting or after accidental collisions.

151.007 **Papers on stellar dynamics.** I. L. Genkin. Trudy Astrofiz. Inst. Alma-Ata, Vol. 12, 34 - 74 (1969). In Russian.

The following aspects are discussed: The density waves in a homogeneous gravitating medium; The relaxation and viscosity in stellar systems; On the approximation of stellar systems by "equilibrium" states; The principles of constructing galactic models; The evolution of forms of dissipating stellar systems.

151.008 **Hydromagnetic stability of thin self-gravitating disks and spiral structure.** R. J. Hosking. Australian Journ. Phys. Vol. 22, 505 - 519 (1969).

The hydromagnetic stability of infinitesimally thin, current-carrying, differentially rotating disks is considered. The perturbation theory described is first order (linear) throughout. It is shown that azimuthal electric current produces instability for all (radial) wavelengths. The case of radial equilibrium electric current reduces to the non-hydromagnetic problem treated by Lin and Shu (1964), although hydromagnetic effects are expected for a disk of finite thickness or to second order due to azimuthal perturbation current.

151.009 **Sur l'équation de la chaleur et la structure spirale des galaxies-disque.** F. Nahon. Comptes Rendus Acad. Sci. Paris, Sér. A, Vol. 269, 823 - 825 (1969).

Nous reprenons la théorie gravitationnelle de Lin et Shu et nous montrons que les variations de la vitesse radiale sont régies par une équation de Fourier où on a permuté les variables d'espace et de temps et où le coefficient de diffusion est imaginaire.

151.010 **Models of clusters of point masses with great central red shift.** G. S. Bisnovaty-Kogan, Y. B. Zeldovich. Astrofizika, Vol. 5, 223 - 234 (1969). In Russian.

The isotropic self-similar solution of a kinetic equation with a self-consistent gravitational field is obtained in the case of Newtonian gravitation for power distributions of the density $\rho = \beta r^{-s}$, $s < 3$. An analogeous self-consistent solution is obtained in the case of $\rho = \beta r^{-2}$ for an anisotropic distribution function either in Newtonian gravitation or general relativity.

151.011 **Construction of models of stellar systems by a numerical method.**

T. A. Agekian, A. S. Baranov.
Astrofizika, Vol. 5, 305 - 316 (1969). In Russian.
English translation in Astrophysics, Vol. 5, No. 2 (1969).

A method of constructing models of stellar systems by a numerical experiment is suggested. The method is applied for the determination of density and potential functions and for the construction of the velocity ellipsoid in spherical quasi-stationary systems for the case of 5 bodies of equal masses.

151.012 Relativistic, spherically symmetric star clusters. III. Stability of compact isotropic models.
J. R. Ipser.
Astrophys. Journ. Vol. 158, 17 - 43 (1969).

In papers I and II of this series, methods were developed for studying, within the theory of general relativity, the stability against radial perturbations of collisionless, spherical star clusters. In this paper those methods are employed to diagnose numerically the stability of specific models for compact star clusters with isotropic velocity distributions. The clusters studied are: (i) clusters of identical stars with heavily truncated Maxwell-Boltzmann velocity distributions; and (ii) clusters whose densities and isotropic pressures obey polytropic laws of index 2 or 3. The calculations show that a cluster of either type is unstable against gravitational collapse if the redshift of a photon emitted from its center and received at infinity is $z_c \gtrsim 0.5$. The cluster is stable if $z_c \lesssim 0.5$.

151.013 Large-scale shock formation in spiral galaxies and its implications on star formation.
W. W. Roberts.
Astrophys. Journ. Vol. 158, 123 - 143 (1969).

The dynamical problems of shock formation and star formation in normal spiral galaxies are investigated. The motion considered is that of the continuum of turbulent gas composing the gaseous disk moving in a gravitational field consisting of a two-armed spiral field superposed on the Schmidt model for the Milky Way system. The possible existence of a stationary two-armed spiral shock pattern is demonstrated. It is suggested that galactic shock waves may very well form the triggering mechanism for the gravitational collapse of gas clouds, leading to star formation. If an upper bound of 30 million years is assumed for the process of formation and evolution of relatively massive stars initiated at the shock, it is shown that the possible locations of the regions of luminous, newly born stars and the H II regions lie on the inner side of each observable gaseous spiral arm of H I, extending from the sharp H I peak at the shock on the inner edge to approximately the center of the arm, in general agreement with observations.

151.014 Results of numerical experiments with orbits of groups of population II stars in a galactic mass model.
K. A. Innanen, F. House, R. V. Hodder, C. E. Smith.
Bull. American Astron. Soc. Vol. 1, 245 (1969). – Abstr. AAS.

151.015 A model for the formation of a spherical galaxy.
R. B. Larson.
Monthly Notices, Roy. Astron. Soc., Vol. 145, 405 - 422 (1969).

Numerical calculations have been made for a model representing the collapse of an initially gaseous proto-galaxy and the concurrent transformation of gas into stars. The assumed turbulent motions of the gas are represented by a simple model consisting of discrete colliding clouds, and the star formation rate is assumed to be given as a simple function of the density and turbulent velocity of the gas. The gas clouds and the stars are then treated separately by means of fluid-dynamical equations derived from the Boltzmann equation. It is found that, by assuming reasonable values for the various parameters of the model, it is possible in this way to reproduce reasonably well the observed properties of spherical and nearly spherical galaxies.

151.016 A note on the dynamics of the bending of the galaxy. K. A. Innanen.
Journ. Roy. Astron. Soc. Canada, Vol. 63, 260 - 263 (1969).

Evidence is presented which indicates that strong coupling of $\tilde{\omega}$ and z-motions must occur in the outer parts of the galactic system, following an initial perturbation in the z-direction. It is suggested that some of the observed radial motions of low-velocity, low-latitude neutral hydrogen in the outer parts of the galaxy might be evidence of such coupling.

151.017 Dynamics of self-gravitating gaseous spheres – II. Collapses of gas spheres with cooling and the behaviour of polytropic gas spheres. M. V. Penston.
Monthly Notices, Roy. Astron. Soc., Vol. 145, 457 - 485 (1969).

Numerical integrations of the full non-linear hydrodynamic equations are given for a number of different self-gravitating flows with spherical symmetry. The collapse of a sphere of gas to form a galaxy is studied under the effects of cooling by bremsstrahlung and recombination radiation and heating by cosmic rays and the results show that a state of free fall soon develops. The collapse of a Bok globule to form a star is also computed under the effects of cosmic ray heating and ionic cooling. In this case it is found that the flow is essentially similar to that obtained assuming isothermality. Finally and of most general significance, self-gravitating flows are classified according to various dimensionless ratios governing their behaviour. The effects of these ratios are discussed and a 'dimensionless cooling time' and a 'Jeans number' are defined. The effect on thermal instability of self-gravitation is also discussed and a criterion for instability is given.

151.018 Numerical experimental check of Lynden-Bell statistics – II. The core-halo structure and the role of the violent relaxation.
S. Cuperman, S. Goldstein, M. Lecar.
Monthly Notices, Roy. Astron. Soc., Vol. 146, 161 - 169 (1969).

In a previous paper, Lynden-Bell's statistical mechanics of collisionless stellar systems was checked by numerical experiments on four initial configurations of a one-dimensional self-gravitating system. In the present experiment the core-halo structure appearing in all four configurations considered by us was investigated; in addition, the fourth configuration was continued to a final state, and it was the state predicted by the statistics. A detailed analysis of the particle trajectories showed, that the configuration that reached the predicted final state, and one of those that did not, relaxed equally as violently. The other two configurations, by comparison, relaxed weakly.

151.019 Hydromagnetic stability of a galactic slab with halo. M. Aggarwal, S. P. Talwar.
Monthly Notices, Roy. Astron. Soc., Vol. 146, 187 - 196 (1969).

The problem of stability of a slab of an ideal hydromagnetic, self-gravitating liquid of uniform density embedded in a uniform outside medium is investigated using the normal mode technique. Two models of magnetic field are studied, namely a uniform field normal to the galactic slab and an orthogonal field geometry where the outside magnetic field is assumed equal but normal to the uniform field within the slab confined in its plane. Stability criteria are derived and the critical size of fragmentation of the slab determined in each case.

151.020 Optical study of ionized hydrogen.

G. Courtès, Y. Georgelin, Y. Georgelin, G. Monnet, A. Pourcelot.
Interstellar Ionized Hydrogen, Charlottesville 1967, p. 571 - 615 (1968).

With the Fabry-Perot interferometer, we obtain simultaneously and with a great dispersion (15 Å/mm) all the velocities in H II regions in a wide field (15' × 15'). Since 1955, the radial velocity of ionized hydrogen clouds of the galaxy has been measured at any longitude, with the same optical outfit. Observations are reported, and their contribution to the kinematics of the galaxy is discussed.

151.021 Sur le calcul de l'attraction dans les couches planes.
F. Nahon.
Comptes Rendus Acad. Sci. Paris, Sér. A, Vol. 269, 932 - 933 (1969).

Soit U le potentiel newtonien d'une couche étalée sur le plan $x = r \cos \theta$, $y = r \sin \theta$ avec la densité $\mu (r)$ et soit $K (r) = dU/dr$ la force d'attraction. Nous cherchons des solutions approchées de l'équation $K = 2i\pi\mu$ et nous les déduisons de l'équation analogue pour le potentiel logarithmique.

151.022 Formation of galaxies by thermal instability.
Y. Sofue.
Publ. Astron. Soc. Japan, Vol. 21, 211 - 220 (1969).

It is found that the mass of a proto-galaxy is restricted to $10^6 \, M_\odot \lesssim M \lesssim 10^{12} M_\odot$, under the assumption that a density excess of gas in an expanding universe is initiated by the thermal instability due to radiative cooling and is followed by non-linear gravitational contraction, resulting in a gravitationally bound system such as a galaxy. Such contraction takes place at cosmic age $t \sim 10^7$ years. The temperature of matter when the initial thermal instability occurs is required to be $10^6 - 10^{9} \, {}^\circ K$. This high temperature would be attained if relatively small kinetic fluctuation in cosmic background radiation causes turbulence in cosmic gas which heats the gas through its dissipation.

151.023 On the thermal instability of a uniformly rotating homogeneous medium. M. Saitō.
Publ. Astron. Soc. Japan, Vol. 21, 230 - 239 = Tokyo Astron. Obs. Repr. No. 362 (1969).

The stability of a uniformly rotating, self-gravitating homogeneous medium has previously been studied in connection with the formation of galaxies and the origin of large scale inhomogeneities of galaxies. In these investigations, however, the growing pattern of disturbances has not been discussed. In the present paper we shall re-examine the thermal instability of a uniformly rotating homogeneous medium, including the effects of self-gravitation and the azimuthal dependence of perturbations, and find a growing pattern of disturbances around the axis of rotation.

151.024 Dynamical effects of envelope stars on contraction of a star cluster. M. Fujimoto.
Publ. Astron. Soc. Japan, Vol. 21, 288 - 291 (1969).

A composite stellar system in which a huge and diffuse envelope surrounds a compact nucleus is used as a model of existing massive stellar systems with no angular momentum. An estimation is made of energy-inflow from the envelope to the nucleus through operation of binary encounters between member stars of the envelope and those of the nucleus.

151.025 Dynamics of self-gravitating gaseous spheres – III. Analytical results in the free-fall of isothermal cases. M. V. Penston.
Monthly Notices, Roy. Astron. Soc. Vol. 144, 425 - 448 (1969).

For a cold gas, the analytical solutions for collapse in various symmetries give density and velocity profiles at the instant when a singularity develops in the initially densest

part. The spherical results imply a density law proportional to $r^{-12/7}$. The planar and the cylindrical results are also given. The solutions before the singularity arises are similarity solutions with the density and velocity profiles retaining their shapes while altering their scales with time. We find a spherical collapse to form a galaxy is modified by the rise in central optical depth at a density of $\sim 10^{-20} \mathrm{g \, cm}^{-3}$. Flattening instabilities are resisted but not overcome by the more rapid growth of the density gradient which occurs in a planar collapse. A similarity solution for the spherical collapse of an isothermal gas is presented with a density profile proportional to r^{-2}. The collapse of a proto-star is halted by the rise in optical depth when the density reaches $10^{-18} \mathrm{g \, cm}^{-3}$.

151.026 How galaxies are born. A. A. Ryzmaikin.
Priroda No. 11, p. 52 - 55 (1969). In Russian.

151.027 Models of partially relaxed stellar disks.
F. H. Shu.
Astrophys. Journ. Vol. 158, 505 - 518 (1969).

The proposal by Lynden-Bell that violently changing gravitational fields serve as the primary mechanism of relaxation in a galaxy of stars is reexamined for disk galaxies. The point of view adopted is that the only relaxation mechanism operative for stars in the early life of such galaxies is an axisymmetric form of the Jeans instability discussed by Toomre. The most probable form of the distribution function which results for a disk of infinitesimal thickness is obtained from statistical considerations. An asymptotic method of solution based on this distribution is developed for the construction of galactic models from observed rotation curves.

151.028 A numerical test of the relaxation time.
E. M. Standish, Jr., K. Aksnes.
Astrophys. Journ. Vol. 158, 519 - 527 (1969).

The relaxation time of a simplified model of a stellar cluster is derived analytically and compared with many numerical examples, using up to 2500 field stars. The derivation closely resembles that of Chandrasekhar; it assumes independent binary encounters and, for the purpose of comparison, introduces a cutoff limit. Close agreement is found between the numerical and the theoretical values.

151.029 Effects of velocity dispersion on the evolution of a disk of stars. R. W. Hockney, F. Hohl.
Astron. Journ. Vol. 74, 1102 - 1104, 1119 - 1124 (1969).

A computer model of a thin disk galaxy has been used to study the stabilizing effect of velocity dispersion on the evolution of a disk galaxy. The motion of 50000 model stars are computed stepwise in time as they move in their mutual gravitational fields. A cold balanced disk is found to be violently unstable. A velocity dispersion of 27% of the velocity at the edge of the disk stabilizes the system. These results agree with the theoretical predictions of Toomre.

151.030 On physical effects at close encounters of galaxies.
V. M. Tomozov.
Vestn. Mosk. un-ta. Fiz., Astron., No. 2, p. 47 - 51 (1969). In Russian. – Abstr. in Referativ. Zhurn. 51. Astron., 10.51.405 (1969).

151.031 Some solutions of the problem of the structure of gravitating systems of particles with ellipsoidal distribution of velocities. A. D. Burdyugov.
Vestn. Mosk. un-ta. Fiz., Astron., No. 2, p. 96 - 100 (1969). In Russian. – Abstr. in Referativ. Zhurn. 51. Astron., 10.51.790 (1969).

151.032 Numerical tables on the random force.
T. Shimizu, Y. Baba.
Mem. Fac. Sci. Kyoto Univ., Ser. Phys., Astrophys., Geophys.,

Chemistry, Vol. 33, 121 - 142 (1968).

Numerical tables are given for evaluating characteristics of the random force or/and its time derivative.

151.033 The stability problem of oscillations along the axis of symmetry in a galaxy. I. The first order perturbations in non-resonance cases. P. Andrle.
Bull. Astron. Inst. Czechoslovakia, Vol. 20, 317 - 322 (1969).

The stability problem of oscillations along the axis of symmetry of a galaxy is studied by Jacobi's and "Lagrange's" equations. If there are no commensurabilities the amplitudes of ω and z-oscillations possess no secular terms. Secular terms among perturbations of t_ω and t_z occur if the powers of ω and z are even. The influence of the perturbing term $\beta\omega^2 z^2$ is studied as a special example. The upper and lower limits of both coordinates are found in this case and we may state: If we confine ourselves to the first order perturbations, then, the oscillations are stable.

151.034 Application of the density-wave theory to the spiral structure of the Milky Way system. I. Systematic motion of neutral hydrogen. C. Yuan.
Astrophys. Journ., Vol. 158, 871 - 888 (1969).

When a density-wave pattern is present in a disk galaxy, the motion of the interstellar gas will systematically deviate from the mean circular motion. The tangential component of this systematic motion furnishes a completely new method for identifying the spiral structure of the Milky Way system; this method is applied to the determination of the locations of the Sagittarius arm and the Norma–Scutum arm. A fairly detailed discussion is given to compare observational data with a theoretical spiral pattern for the Milky Way system based on the density-wave theory and the 1965 Schmidt model. The spiral pattern is estimated to travel at 13.5 km sec^{-1} kpc^{-1} for the Milky Way system.

151.035 Application of the density-wave theory to the spiral structure of the Milky Way system. II. Migration of stars. C. Yuan.
Astrophys. Journ., Vol. 158, 889 - 898 (1969).

Orbits and places of origin are calculated for twenty-five stars by using the Schmidt model for our galaxy. The results show that these stars are born in the spiral arms (as expected) only if the density-wave theory is adopted and the effect of the spiral gravitational field is included. Calculations are actually carried out with logarithmic spiral patterns, which are found to approximate extremely well those obtained theoretically from the dispersion relation of Lin and Shu with assigned pattern speeds. With the present observational evidence, the final value adopted for the pattern speed is 13.5 km sec^{-1} kpc^{-1}, and for the ratio of field strength, 5 percent.

151.036 Group velocity of spiral waves in galactic disks. A. Toomre.
Astrophys. Journ., Vol. 158, 899 - 913 (1969).

Studied here are density waves of the kind proposed especially by Lin to explain the spiral structures of disk galaxies. It is shown that any packet of such waves propagates radially (and toward increasingly short wavelengths) with a group velocity that is sufficient to obliterate it within a few galactic revolutions. This does not necessarily mean that the density-wave hypothesis is wrong. But it does imply that any existing spiral waves in the disk of a galaxy must somehow be replenished if their pattern as a whole is to persist. Three conceivable sources of such replenishment are also discussed in this paper.

151.037 On the locally ellipsoidal solutions of Liouville's equation in stellar dynamics. M. Trümper.
Astrophys. Journ., Vol. 158, 915 - 923 (1969).

Any locally ellipsoidal distribution of residual velocities is shown to be of the Chandrasekhar type $\Psi(Q+N)$ if it satisfies Liouville's equation (with gravitational potential only). Some of Chandrasekhar's results on ellipsoidal distributions are rederived in a simplified manner. Some properties of Q and N are discussed.

151.038 Equipartition and the formation of compact nuclei in spherical stellar systems. L. Spitzer, Jr.
Astrophys. Journ. (*Letters*), Vol. 158, L139 - L143 (1969).

In a spherical system composed of stars of two masses, m_1 and m_2, with m_2 greater than m_1, the heavier stars will lose kinetic energy with a time constant about twice the equipartition time, t_{eq}, and will gravitate toward the center. It does not seem to have been generally realized that, if the fraction of massive stars exceeds a critical value, these heavier stars cannot reach a condition of approximate equipartition, and the subsystem of such stars will continue to contract rapidly at the center of the system. The present letter analyzes in a simplified case the conditions required for this effect to occur and the rate at which such a subsystem will contract.

151.039 Two-point statistics of the gravitation arising from a random distribution of stars.
A. H. Marcus.
Bull. American Astron. Soc., Vol. 1, 353 (1969). – Abstr. AAS.

151.040 Computer simulation of galactic evolution. R. H. Miller, K. H. Prendergast,.W. Quirk.
Bull. American Astron. Soc., Vol. 1, 355 (1969). – Abstr. AAS.

151.041 Two population models of disk galaxies. M. L. West.
Bull. American Astron. Soc. Vol. 1, 369 (1969). – Abstr. AAS.

151.042 On the possible cause of origin of ring structure in disk galaxies. L. S. Marochnik, N. G. Ptitsina.
Astron. Zhurn. Akad. Nauk SSSR, Vol. 46, 762 - 774 (1969). In Russian. English translation in Soviet Astron. AJ, Vol. 13, No. 4.

The possibility of the co-existence of trailing (T) and leading (L) arms in disk galaxies is considered. The standing waves as a system of continuous or grained rings arise from the superposition of T- and L-waves in the infinitesimally thin disk; the co-existence of spiral and ring structures is possible. These spiral and ring waves are also possible along the z-coordinate. In the center of the system the spiral wave solutions have bar-like singularities of the same order in both the limiting cases (disk and cylinder). The obtained results qualitatively agree with the observational data for the ring and the spiral galaxies.

151.043 A relativistic stage of the evolution of stellar systems with non-elastic collisions.
M. A. Poduretz.
Astron. Zhurn. Akad. Nauk SSSR, Vol. 46, 787 - 796 (1969). In Russian. English translation in Soviet Astron. AJ, Vol. 13, No. 4.

A relativistic stage of the evolution of stellar systems from non-elastic collisions of stars is considered. At late stages of the evolution the rate of thermal relaxation is slower, because the distribution function of the stellar gas particles (stars) in the phase space differs from that in equilibrium. For finding the distribution function the kinetic equation has to be integrated. A complete system of equations and additional conditions (within the limits of the general theory of relativity), describing the system of stars and the central gaseous nucleus in the process of the evolution, is formulated.

151.044   On one cause of the origin of irreversibility in rotating stellar systems without star-star encounters.
S. G. Pomagaev.
Astron. Zhurn. Akad. Nauk SSSR, Vol. 46, 810 - 816 (1969). In Russian. English translation in Soviet Astron. AJ, Vol. 13, No. 4.

The kinetic equation, which in rotating stellar systems describes the variation of the distribution function of residual stellar velocities under the influence of arbitrary unstable fluctuations of the gravitational field, accidentally distributed in the space, is obtained. It is shown that the diffusion of stars in the space of velocities is accompanied by the increase of dispersion of residual velocities and stellar entropy, as well as by the tendency of the velocity distribution function of stars for the statistical equilibrium value. For cases of large and small velocities of stars the solutions of the kinetic equation are found.

151.045   Purely discontinuous random processes in a field of irregular forces. I.   I. V. Petrovskaja.
Astron. Zhurn. Akad. Nauk SSSR, Vol. 46, 824 - 831 (1969). In Russian. English translation in Soviet Astron. AJ, Vol. 13, No. 4.

The variation of different parameters of stellar motion, for example orbit integrals or velocity modulus, under the action of irregular forces in the stellar system is considered as a purely discontinuous random process. Kolmogorov-Feller's equations, describing this process, are recorded, taking into account the absorbing screen (critical velocity). The case of the variation of the velocity modulus of a star in the irregular field with lack of regular field is investigated in detail.

151.046   On the stability of stellar systems of the Orion Trapezium type.   G. N. Duboshin, A. I. Rybakov.
Astron. Zhurn. Akad. Nauk SSSR, Vol. 46, 895 - 906 (1969). In Russian. English translation in Soviet Astron. AJ, Vol. 13, No. 4.

The problem of the stability in Lagrangian sense of a stellar system of Orion Trapezium type within a vast stellar cluster or cosmic cloud is considered. It is supposed that stars of the system can be treated as material points, reciprocally attracted according to Newton's law. The first integrals of the system are derived; a relation analogous to the Lagrangian-Jacobi equation is deduced, and conditions of stability (necessary) and instability (sufficient) are obtained.

151.047   The purely discontinuous random processes in a field of irregular forces. II. The variation of the velocity modulus of a star.   I. V. Petrovskaya.
Astron. Zhurn. Akad. Nauk SSSR, Vol. 46, 1220 - 1227 (1969).   In Russian. English translation in Soviet Astron. AJ, Vol. 13, No. 6.

The method of solution of the second Kolmogorov-Feller equation proposed in a previous paper is used to investigate the evolution of the velocity distribution for a group of stars with a given initial velocity distribution and with masses equal to the average mass of a cluster star. The regular cluster potential is neglected. The velocity distribution of the group, the escape rate of stars, the amount of energy that is taken away by the dissipated stars, and the limit of the velocity distribution of the cluster for $t \to \infty$ are found.

151.048   Vlasov's equation and irreversibility in plasma physics and stellar dynamics.   I. L. Genkin.
Astron. Zhurn. Akad. Nauk SSSR, Vol. 46, 1228 - 1230 (1969).   In Russian. English translation in Soviet Astron. AJ, Vol. 13, No. 6.

A new statistical determination of the entropy that corresponds more exactly to the main (thermodynamic) determination, is proposed. The notion of the effective relaxation time is introduced.

151.049   Tidal interaction of galaxies and eventual formation of bars and tails.   N. Tashpulatov.
Astron. Zhurn. Akad. Nauk SSSR, Vol. 46, 1236 - 1246 (1969).   In Russian. English translation in Soviet Astron. AJ, Vol. 13, No. 6.

The outflow of matter from the vertex of a prolate homogeneous galaxy as result of its close encounter with another galaxy is treated hydromagnetically. It is shown that by sufficiently small pericentric distance nearly strait and rather short "bridges" between galaxies may be formed.

151.050   Non-axisymmetric oscillations of a self-gravitating disk.   M. Miyamoto.
Publ. Astron. Soc. Japan, Vol. 21, 319 - 336 (1969).

Non-axisymmetric free oscillations of a self-gravitating disk are investigated. The analysis is based on the assumption that the matter in the disk is divided into a number of concentric rings rotating differentially. The problem is reduced to an eigen-value problem. Unstable eigen-disturbances with two arms are numerically investigated.

151.051   The dependence of the velocity body of stars on space location.   K. Rudnicki.
Vistas in Astronomy, Vol. 11, 173 - 180 (1969).

151.052   The velocity distribution function for stars of small mass in star-clusters.   V. S. Kaliberda.
Astrofizika, Vol. 5, 433 - 441 (1969).   In Russian. – Engl. translation in Astrophysics, Vol. 5, No. 3.

The function of distribution of star velocities in non-rotating systems is found. It is supposed that all the stars of the field are of equal mass but the mass of the star under consideration is zero. The effect of multiplicity of star encounters is taken into account. The amount of energy taken away by the dissipated stars, and the rate of dissipation are estimated. It is concluded that the dispersion in mass accelerates the evolution of stellar systems.

151.053   On the relaxation of stars of flat subsystems of the galaxy due to spiral structure.
L. S. Marochnik.
Astrofizika, Vol. 5, 487 - 498 (1969).   In Russian. – Engl. translation in Astrophysics, Vol. 5, No. 3.

The spiral density waves (which are identified with the spiral pattern of the galaxy) grow due to Landau instability in a self-gravitating infinitesimally thin disk (or cylinder) consisting of a differentially rotating stellar population I and a non-rotating stellar population II. The relaxation of stellar population I is caused by the instability of the density waves. The velocity dispersion in the galactic plane in the solar neighbourhood doubles approximately in time of the order of $10^8 - 10^9$ years. The Schwarzschild velocity distribution is settled in the region of small velocities. In the region of large velocities the distribution function is superposed on the Schwarzschild distribution. It is possible that the growing spiral waves give rise to turbulence of the interstellar gas, but apparently the relaxation of the stellar population I in z-direction cannot be caused by them only.

151.054   Upper limit to radiation of mass energy derived from expansion of galaxy.
D. W. Sciama, G. B. Field, M. J. Rees.
Phys. Rev. Letters, Vol. 23, 1514 - 1515 (1969).

Loss of mass energy from the galaxy, whether by gravitational radiation or otherwise, should cause the galaxy to expand. Observations of stellar motions near the sun imply that the rate of radiation averaged over the last $\sim 10^8$ yr must be less than 200 $M_\odot$ yr^{-1}. Studies based on the 21-cm line have already yielded some evidence for galactic expansion, which could be a consequence of mass loss.

151.055 **Theory and results on collective and collisional effects for a one-dimensional self-gravitating system.**
F. Hohl.
Report NASA-TR-R-289, Langley Research Center, Langley Station, Va., 80 pp. (1968). – See Phys. Abstr. Vol. 72, No. 27000 (1969).

151.056 **Dynamics of self-gravitating systems: Structure of galaxies.** C. C. Lin.
SIAM Rev., Vol. 11, No. 2, p. 127 - 151 (1969). – See Phys. Abstr. Vol. 72, No. 47622 (1969).

151.057 **Statistical mechanics of stellar systems.**
D. ter Haar.
Journ. Phys. Soc. Japan, Vol. 26 Suppl., p. 25 - 29 (1969). See Phys. Abstr. Vol. 73, No. 2654 (1970).

151.058 **The dynamics of stellar systems. Part I.**
K. Rudnicki.
Postępy Astron., Vol. 17, 347 - 374 (1969). In Polish.
An introduction to the contemporary achievements of the dynamics of stellar systems is presented.

151.059 **A form of the gravitational potential for solving the equations of the motion in a plane by elliptic integrals.** G. G. Kuzmin, G. A. Malasidze.
Soobshch. AN Gruz. SSR, Vol. 54, 565 - 568 (1969). In Russian. – Abstr. in Referativ. Zhurn. 51. Astron., 1.51.720 (1970).

151.060 **Ein Beitrag zur Dynamik von Sternsystemen.**
W. Tscharnuter.
Anzeiger Österreich. Akad. Wiss. Math.-Naturwiss. Kl., 105. Jahrgang, p. 144 - 146 (1969). – Brief communication.

151.061 **Der Zustand eines quasistationären sphärischen Systems von Sternen verschiedener Masse.**
I. Mihăilă.
Stud. Cerc. Astron., Vol. 14, 91 - 104 (1969). In Russian.

151.062 **The solar motion of Ap stars.**
R. W. Day.
Publ. Astron. Soc. Pacific, Vol. 81, 866 - 871 = Publ. Dep. Astron., Univ. Texas, Ser. I, Vol. 2, No. 24 (1969).
The solar motion of Ap stars is calculated from radial velocities and proper motions, with galactic rotation effects included. The results indicate that the Ap stars are kinematically similar to normal late-B stars.

151.063 **Erratum: Kinetic theory of a self-gravitating stellar system with uniform rotation. [Phys. Fluids, Vol.**
11, 316 - 325 (1968)]. C.-S. Wu.
Phys. Fluids, Vol. 12, 264 (1969).

151.064 **The spectra of small oscillations of thin disk galactic models.** C. Hunter.
Stud. Applied Math., Vol. 48, 55 - 76 (1969).

151.065 **On types of gravitational instability in galaxies.**
A. A. Suchkov.
Dokl. AN Tadzh. SSR, Vol. 12, No. 5, p. 9 - 12 (1969). In Russian. – Abstr. in Referativ. Zhurn. 51. Astron., 2.51.729 (1970).

151.066 **Systematische Sternbewegungen im Bereich des Orion-Nebels.**
K. Ferrari d'Occhieppo, E. Göbel.
Mitt. Astron. Ges. No. 27, p. 128 - 129 (1969). – Abstract AG.

151.067 **Solar motion with respect to the high-velocity H I clouds and to the local group.**
G. de Vaucouleurs, W. L. Peters.
Nature, Vol. 223, 938 (1969).
Solutions were made for the solar motion with respect to the local group of galaxies. All solutions point to $l \approx 95°$, $b \approx -15°$, $V \approx 325$ km/sec. Solutions for the solar motion with respect to high-velocity H I clouds support the idea that the clouds are close satellites of the galaxy or loosely bound wisps of turbulent gas in the galactic corona.

151.068 **Contribution à l'étude du Groupe Local.**
C. Froeschle.
Ann. Obs. Besançon, Vol. 8, 7 - 17 (1969).
D'après l'ensemble des résultats obtenus, il semble bien que, parmi toutes les étoiles étudiées, seules les étoiles du Groupe A (c'est-à-dire les étoiles de type spectral O à B5 dont la "distance" est inférieure à 200 pcs) présentent les caractéristiques du Groupe Local. D'autre part, ces résultats montrent que compte-tenu de l'erreur systématique faite sur l'estimation des distances, ce Groupe ne s'étend pas au-delà d'une distance supérieure à 250 parsecs.

The high-velocity hydrogen clouds considered as satellites of the Galaxy. See Abstr. 131.049.

Dynamische Deutung der Altersverteilung offener Sternhaufen. See Abstr. 153.041.

Stellar groups in the old disk population.
See Abstr. 155.013.

152 Stellar Associations

152.001 **Zufallszahlengeber und scheinbare Sternverteilung.** J. Meurers.
Naturwissenschaften, Vol. 56, 457 (1969).

Nach einem Zahlengebergesetz von MacLaren und Marsaglia können Punktfelder mit zufälliger Punktverteilung erzeugt werden, die photographierten Sternfeldern gleichen und insbesondere auch Sternketten aufweisen.

U. Güntzel-Lingner

152.002 **Internal motions in the associations II Per and I Lac.** J. R. Lesh.
Astron. Journ. Vol. 74, 891 - 898 (1969).

New proper motions and spectroscopic data for the bright members of the associations II Per and I Lac are analyzed to determine whether these two groups are in a state of expansion. For II Per, the expansion age is calculated as 1.3×10^6 yr $\pm$ 9% (p.e.), in agreement with earlier investigations. The concentrated and dispersed regions of the Lacerta association are found to have mean distance moduli differing by about 1 mag. This effect appears to be real, and not the result of a systematic error in the calibration or of an intrinsically under-luminous main sequence in the younger subgroup. The concentrated subgroup, Ib Lac, has an expansion gradient corresponding to a kinematic age of 2.5×10^6 yr $\pm$ 19% (p.e.). It is unlikely that systematic errors in the proper-motion system could account for this result.

152.003 **Photometric data on Isserstedt's stellar rings and their physical parameters.** T. A. Uranova.
Astron. Tsirk. No. 523, p. 1 - 5 (1969). In Russian.

152.004 **Photoelectric studies of a nearby stellar ring in Aquila.** J. Isserstedt.
Astron. Astrophys. Vol. 3, 210 - 213 (1969). In German.

Photoelectric *UBV* photometry has been used to determine the distance of a stellar ring in Aquila. A distance modulus of $6\overset{m}{.}95$ and reddening $E_{B-V} = 0\overset{m}{.}08$ were obtained, yielding a minor diameter of 7.4 pc. Stars in the range $0\overset{m}{.}1 < B - V < 0\overset{m}{.}6$ exhibit an UV-excess similar to that observed in the very young open cluster NGC 2264.

152.005 **Effects of rotation on hydrogen parameters of early-type stars.** A. Gutierrez-Moreno, H. Moreno.
Publ. Dep. Astron. Univ. Chile, Obs. Astron. Nacional, Cerro Calan, Santiago, Vol. 1 (No. 5), 72 - 86 (1968).

The effects of rotation on the colors and on hydrogen parameters of early-type stars are analyzed, using photometric data obtained for Scorpio-Centaurus stars. It is found that rotation affects all hydrogen parameters, as well as the colors and the visual magnitude of the stars. The results agree with Collins and Harrington's theoretical conclusions.

152.006 **Interstellar matter in the region of the Perseus II association.** B. T. Lynds.
Publ. Astron. Soc. Pacific, Vol. 81, 496 - 520 (1969).

Observational data are assembled and discussed in order to determine the distribution of the interstellar gas and dust in the direction of the Per II association. It is concluded that most of the obscuration occurs in dark foreground clouds of variable opacities. Only one or two of the 17 Per II cluster members appear to lie in dark nebulae. The emission nebula NGC 1499 is considered to be a rim of a foreground dark nebula which is illuminated by the O7 star ξ Per. On the basis of the absence of detectable emission in the immediate vicinity of ξ Per, it is argued that the value of the electron density between NGC 1499 and ξ Per must be very low, probably less than 5 cm^{-3}.

152.007 **Sco XR-1 as a member of the upper Scorpius complex.** O. J. Eggen.
Astrophys. Journ. (*Letters*), Vol. 158, L31 - L34 (1969).

Three stars that share the proper motion of Sco XR-1, together with Sco XR-1 itself, are probably members of the Scorpius–Centaurus association of young disk stars. The resulting luminosity of Sco XR-1 is near +6 mag.

152.008 **Hβ observations of two subclusterings in the Orion OB 1 association.** A. Heiser.
Bull. American Astron. Soc., Vol. 1, 348 (1969). — Abstr. AAS.

152.009 **Some questions concerning the activity of variables in the T-association Ori T2.** V. S. Shevchenko.
Peremennye Zvezdy, Byull., Vol. 16, 606 - 614 (1969). In Russian.

The possible number of "microvariables" in the T-association Ori T2 was computed from empirical relations. 11 among 100 stars of the classes B0 – A8 are known as variables. A probability close to unity exists for discovering 3 more variables with $A^m \approx 0\overset{m}{.}3 - 0\overset{m}{.}5$ and 8 variables with $A^m \approx 0\overset{m}{.}1 - 0\overset{m}{.}3$ in the case of regular photoelectric observations during 2 – 3 seasons. Apparently, data on a great number of "microvariables" in the association Ori T2 obtained by Parenago are erroneous. The percentage of "microvariables" among G – M stars is comparatively small (of the order of 5 – 15% instead of 75 – 85% according to Parenago), most anew discovered variables should have the amplitude $A^m = 0\overset{m}{.}5 - 3\overset{m}{.}0$. A^m depends on the length of the interval and on the frequency of observations.

The spectrum of the Bp star HD 36916. See Abstr. 114.072.

Flare stars near NGC 7023. See Abstr. 122.056.

A new "slow" flare star in Orion. See Abstr. 122.147.

Photoelektrische Beobachtungen an SV Cephei. See Abstr. 122.170.

A search for globules in OB clusters and associations. See Abstr. 153.013.

Studies of local star streams. II. The Ursa Major stream. See Abstr. 153.037.

153 Galactic Clusters

153.001 Young stellar clusters.
I. P. Williams, A. W. Cremin.
Monthly Notices, Roy. Astron. Soc., Vol. 144, 359 - 373 (1969).

Using theoretical evolutionary tracks, masses and ages of the stars in the young stellar cluster NGC 2264 have been determined. This information has been used to obtain various cluster properties, such as the age distribution, the mass distribution and the mass-age correlation. The three clusters NGC 6530, IC 2602 and IC 5146 have also been investigated, but in less detail. These produce similar distribution diagrams to NGC 2264. It is concluded that there exists a large age spread in all the clusters and that the rate of formation of stars increases with time.

153.002 The H-R diagram of NGC 2516.
H. A. Abt, W. W. Morgan.
Astron. Journ. Vol. 74, 813 - 815 = Contr. Cerro Tololo Inter-American Obs. No. 71 (1969).

The open cluster NGC 2516 is found to contain an Mn II star, three Si II stars, and a pronounced shell star. The cluster seems to be unique in the variety of peculiar stars present, which include several showing the line at λ 3984 (identified by Bidelman with Hg II) in their spectra.

153.003 Ros 4, a distant open cluster associated with nebulosities.   R. Racine.
Astron. Journ. Vol. 74, 816 - 817, 847 (1969).

Photoelectric UBV photometry is reported for thirteen apparent members of Ros 4, an open cluster associated with nebulosities. A photometric distance of 2900 ± 300 pc is derived. The position of the cluster in the galaxy suggests that the local (Orion) arm exists as a continuous feature over a length of at least 4 kpc.

153.004 Four-color and H-beta photometry of open clusters. III. Praesepe.
D. L. Crawford, J. V. Barnes.
Astron. Journ. Vol. 74, 818 - 823 = Contr. Kitt Peak National Obs. No. 432 (1969).

Photoelectric observations are presented for 97 stars of the Praesepe open cluster, and are compared with standard relations valid for the Hyades cluster. Three stars are identified as definite nonmembers. The average color excess relative to the Hyades and nearby field stars is $E(b-y) < 0.^m01$; four stars are perhaps reddened about $0.^m04$. The F-type stars in Praesepe show the same effect noted in the Hyades; they deviate from the "zero-age" line in a c_1 vs β relation. The calculated distance modulus for the cluster is $6.^m1$, agreeing well with previous determinations. Rotational velocity effects on the observed parameters do not appear to affect any of the above conclusions.

153.005 The main sequence gap and red giant clump of NGC 6939.   R. D. Cannon, C. Lloyd.
Monthly Notices, Roy. Astron. Soc. Vol. 144, 449 - 458 (1969).

New photographic photometry is given for the open cluster NGC 6939. Using earlier proper motion measurements, a colour-magnitude diagram is obtained for those stars believed to be cluster members. The cluster appears to have an age of about 10^9 years, and has a low density region on the main sequence in agreement with a theoretically predicted gap. The red giant branch shows one very marked concentration which is tentatively identified with the horizontal branch found in globular clusters. There are a few anomalous bright blue stars, similar to those in other old clusters, well above the main sequence turn-off.

153.006 Photometric studies of southern galactic clusters. I. IC 2391.   C. L. Perry, G. Hill.
Astron. Journ. Vol. 74, 899 - 907 = Contr. Louisiana State Univ. Obs. Baton Rouge No. 24 = Contr. Dominion Astrophys. Obs. Victoria No. 134 = Contr. Cerro Tololo Inter-American Obs. No. 78 (1969).

Multicolor photometry is presented for 42 stars in the vicinity of the southern galactic cluster IC 2391. Independent analyses of UBV and $uvby$-Hβ data confirm that IC 2391, with a membership of 25 stars, is an unreddened cluster at a distance of 150 pc. These analyses show that the absolute magnitude calibrations and intrinsic color relations derived from both UBV and $uvby$-Hβ photometry are in good agreement. A nuclear age of 30×10^6 yr and a lower limit to the contraction age of 30×10^6 yr were obtained for the cluster. The $uvby$-Hβ data fit the standard relations of the Hyades Cluster although there is only limited overlap for the fainter members.

153.007 The composite colour-magnitude diagram for young clusters with giants and supergiants.
A. N. Sedyakina.
Astron. Tsirk. No. 523, p. 5 - 8 (1969).   In Russian.

153.008 Theoretically predicted color-magnitude diagrams for clusters and the observations.
B. M. Schlesinger.
Astrophys. Journ. Vol. 157, 533 - 544 (1969).

Stellar-evolution calculations by Iben have been used to predict the appearance of the color-magnitude diagram for clusters of six different ages. The predicted C-M diagrams are compared with observations. The relative number of stars in different parts of the C-M diagram receives special attention. Comparisons at particular ages help in the interpretation of the available observations. An examination of the relation between the luminosity of the stars at the top of the main sequence and that of the stars on the giant branch of young clusters reveals that some of the differences between the C-M diagrams of Magellanic Cloud clusters and of Galaxy clusters that are usually attributed to composition may be age effects. The possibility of a composition difference between M41 and M11 is discussed. If the distance moduli derived at the time of the original observations were correct, the differences between the C-M diagrams of M41 and M11 indicate a difference in composition. It is suggested that most yellow giants in clusters aged 5×10^8 to 10^9 years are burning core helium.

153.009 The double cepheid CE Cassiopeiae in NGC 7790: Tests of the theory of the instability strip and the calibration of the period-luminosity-color relation.
A. Sandage, G. A. Tammann.
Astrophys. Journ. Vol. 157, 683 - 708 (1969).

Separate light curves in B and V have been obtained for the double cepheid CE Cas a and CE Cas b. The components differ in color and luminosity, with component a brighter and redder than b by 0.07 mag in V and $B - V$. The observed ratio of the periods is $P_a/P_b = 1.15$, whereas the value predicted from $P\rho^{1/2} = Q$ using the photometric observations is $P_a/P_b = 1.16 \pm 0.04$, where the variation of Q across the instability strip is taken into account. Galactic cepheids in clusters and associations give P-L-C relations, which reproduce the observed M_V values for the thirteen calibrating stars to within ± 0.064 mag average deviation. The equations represent a correction of -0.05 mag to the previous calibration. Expressions for the instability strip in the $(M_V, B - V)$-plane for galactic cepheids are given.

153.010 On the problem of dynamics of open clusters.
O. V. Chumak.
Trudy Astrofiz. Inst. Alma-Ata, Vol. 12, 75 - 81 (1969).
In Russian.

The possibility of formation of open clusters as result of gravitational instability in a stellar gas is discussed.

153.011 Photometric studies of southern galactic clusters. II. IC 2602. G. Hill, C. L. Perry.
Astron. Journ. Vol. 74, 1011 - 1021 = Contr. Dominion Astrophys. Obs. Victoria, No. 141 = Contr. Louisiana State Univ. Obs. Baton Rouge, No. 26 = Contr. Cerro Tololo Inter-American Obs. No. 81 (1969).

Multicolor photometry is presented for 33 stars in the vicinity of the southern galactic cluster IC 2602. Independent analyses of UBV and $uvby$-Hβ data confirm that IC 2602, with a minimum membership of 31 stars, is reddened by $0.^m035$ and is located at a distance of 150 pc. These analyses show that the absolute magnitude calibrations and intrinsic color relations derived from both UBV and $uvby$-Hβ photometry are in good agreement. An upper limit to the nuclear age of 4×10^6 yr and a lower limit to the contraction age of 20×10^6 yr were obtained for the cluster. Hence star formation was not coeval in IC 2602. The $uvby$-Hβ data fit the standard relations of the Hyades cluster, although there is only limited overlap for the fainter members.

153.012 The open cluster NGC 559.
U. Lindoff.
Ark. Astron. Vol. 5, 221 - 229 (1969).

Magnitudes on the UBV system have been determined for stars in the open cluster NGC 559 from a combination of photoelectric and photographic measurements. The photoelectric observations were made with the 61 cm reflector of the Lund Observatory, while the photographic plates were taken at the Kvistaberg Observatory. The colour excess is $0.^m45$ for NGC 559 and the distance 1300 pcs. The age is estimated to 10^9 years.

153.013 A search for globules in OB clusters and associations. M. E. Sim.
Monthly Notices, Roy. Astron. Soc., Vol. 145, 375 (1969). – Abstract. – The full text of this paper, see Publ. Roy. Obs. Edinburgh, Vol. 6, 181 - 207 (1968).

153.014 Intermediate-band photometry for the α Per open cluster. D. L. Crawford, J. V. Barnes.
Bull. American Astron. Soc. Vol. 1, 238 (1969). – Abstr. AAS.

153.015 The distance of the galactic cluster NGC 2244.
T. K. Menon.
Bull. American Astron. Soc. Vol. 1, 253 (1969). – Abstr. AAS.

153.016 Possibility of mass loss in the red-giant stage from H-R diagrams of galactic clusters.
G. Barbaro, N. Dallaporta, G. Fabris.
Mass Loss from Stars, Trieste 1968, p. 89 - 105 (1969).

A statistical research on evolved stars beyond hydrogen exhaustion is performed by comparing the H-R diagrams of about 60 open clusters with a set of isochronous curves without mass loss derived from Iben's evolutionary tracks and time scales for population I stars. Interpreting the difference in magnitude between the theoretical positions thus calculated and the observed ones as due to mass loss, when negative, the results indicate that this loss may be conspicuous only for very massive and red stars. However, a comparison with an analogous work of Lindoff reveals that the uncertainties connected with the bolometric and colour corrections may invalidate by a large amount the conclusions which might be drawn from

such research.

153.017 The colour-magnitude distribution of giants in open clusters. U. Lindoff.
Mass Loss from Stars, Trieste 1968, p. 106 - 109 (1969).

153.018 Neutrino emission, mass loss, and termination of the giant branch in young clusters.
R. Stothers.
Astrophys. Letters, Vol. 4, 187 - 189 (1969).

Neutrino emission is shown to be able to account for the sudden termination of the giant branch in young clusters like NGC 1866, where the giants have originated from main-sequence stars of intermediate mass. However, an alternative interpretation based on rapid mass loss cannot yet be ruled out.

153.019 The open cluster IC 2581. T. L. Evans.
Monthly Notices, Roy. Astron. Soc., Vol. 146, 101 - 121 (1969).

Photoelectric and photographic UBV measures of 397 stars within 8' of the A7 supergiant HD 90772 are presented. Spectral classifications have been made for 10 stars. There are approximately 120 members brighter than $V = 15.5$, $M_v = +2.3$. These include two supergiants, an eclipsing binary (HD 90707) and two Be stars, one of them variable. Early-type stars in IC 2581 and the very similar cluster NGC 457 are well represented by theoretical models for 15 $M_\odot$, with an age of 10^7 years. The A–F supergiants HD 90772 and ϕ Cas, if members of the two clusters, have $M_v \sim -8.6$ and may represent a high luminosity stage in the evolution of a star of 18 $M_\odot$.

153.020 Spectroscopic studies of southern galactic clusters. I. IC 2391. C. L. Perry, H. E. Bond.
Publ. Astron. Soc. Pacific, Vol. 81, 629 - 636 = Contr. Louisiana State Univ. Obs. No. 25 = Contr. Cerro Tololo Inter-American Obs. No. 79 (1969).

Radial velocities and MK spectral types are presented for twenty stars in the vicinity of the southern galactic cluster IC 2391. Two new single-line spectroscopic binaries (HD 75029 and HD 75105) and two suspected variable-velocity stars (HD 74340 and HD 75202) are noted. The question of cluster membership is reviewed on the basis of multicolor photometry, proper motions, radial velocities, and MK spectral types with the result that, of the forty-two stars in the photometric observing program, only nineteen stars were selected as certain cluster members.

153.021 The stellar group Ba 7. A. Kiral.
Astron. Astrophys. Vol. 3, 327 - 330 = Mitt. Astron. Inst. Univ. Basel No. 60 (1969). In German.

The stellar group Ba 7 can be regarded as a physical group with a distance of 1600 pc and a reddening of $E(G-R)$ = 0.39 mag. These data have been determined on 48"-Palomar Schmidt plates by use of a three color photometry in the RGU-system.

153.022 New photometric data for the old galactic cluster NGC 188: The presence of a gap, chemical composition, and distance modulus. O. J. Eggen, A. Sandage.
Astrophys. Journ. Vol. 158, 669 - 684 (1969).

New photoelectric UBV photometry has been combined with rereduced previous photometry for the old galactic cluster NGC 188. A gap on the rising branch of the evolving main sequence has been found in a place similar, relative to the luminosity at the turnoff from the main sequence, to the gap previously found in M67. The gap in NGC 188 occurs at $4.33 \geq M_V \geq 4.53$. The C-M diagram extends from $V = 9.8$ $(M_V{}^0 = -1.3)$ to fainter than $V = 19$ $(M_V{}^0 = 7.9)$. The main sequence termination point occurs near $M_V \simeq +3.8$, $(B -$

$V)_{0,c} \simeq 0.62$. These values are 0.8 mag fainter and 0.08 mag redder than the termination point in M67. The two-color diagram shows the presence of a small ultraviolet excess relative to the Hyades for stars near the main sequence. A photometric modulus of $(m - M)_0 = 10.85 \pm 0.15$ is obtained by fitting the C–M diagram to the zero-age main sequence by the use of a new curve of evolutionary deviation.

153.023 Isochrones, ages, curves of evolutionary deviation, and the composite C–M diagram for old galactic clusters. A. Sandage, O. J. Eggen.
Astrophys. Journ. Vol. 158, 685 - 698 (1969).

Isochrones in the HR diagram are computed in the range $4.5 \geq M_{bol} \geq 2$ by using models due to Aizenman, Demarque, and Miller, and to Iben. The curves apply for chemical compositions of $X = 0.67$, $Z = 0.03$ for the first set of models, and $X = 0.708$, $Z = 0.020$ for the second. The phase of hydrogen exhaustion on the evolving main sequence, and the shallow departure from the zero-age main sequence of tracks with $M/M_\odot \gtrsim 1.3$ cause adjacent isochrones to intersect one another near $M_{bol} = +3$. Observational data shown suggest that the predicted crossover effect may occur. New curves of evolutionary deviation are calculated from the isochrones. Distance moduli for M67 and NGC 188, derived with these curves, are used to assign absolute luminosities to the cluster stars. Ages of $(5.5 \pm 0.5) \times 10^9$ years for M67 and $(8 - 10) \times 10^9$ years for NGC 188 are found.

153.024 Photometry of NGC 4103. A. J. Wesselink.
Monthly Notices, Roy. Astron. Soc., Vol. 146, 329 - 338 (1969).

The southern galactic cluster NGC 4103 has been studied photometrically. For 512 stars brighter than $V = 17^m$ magnitudes and colours are given. Spectral types are derived for 17 of the brightest using U, B, V photometry. Probably no stars having types earlier than B2 exist in the cluster. The mean colour excess $E_{B-V} = 0.30$. The apparent distance modulus was found to be 12.0. With an absorption of 1.0 the true distance modulus and the distance are 11.0 and 1600 parsecs respectively. An age of 27×10^6 years from the turn off point on the main sequence was found.

153.025 Membership of the Coma star cluster. A. N. Argue, C. M. Kenworthy.
Monthly Notices, Roy. Astron. Soc., Vol. 146, 479 - 488 (1969).

A search has been made for faint possible members of the Coma star cluster. Proper motions and $UBVI$ colours and magnitudes have been measured using a Schmidt telescope. As a result four new members are suggested, and three suggested by Trumpler rejected. Hence we uphold the conclusion drawn by Artyukhina, that the number of members fainter than Trumpler's limit of $m_{pg} = 12^m$ cannot be large; the number probably does not exceed ten. The paper includes comments on the use of the Schmidt as an astrometric telescope.

153.026 *UBV* photometry and spectral types in NGC 6611. W. A. Hiltner, W. W. Morgan.
Astron. Journ., Vol. 74, 1152 (1969).

UBV photometry and MK types are given for 15 of the brightest members of NGC 6611. A comparison with results of Walker shows good agreement in magnitudes but systematic differences in $B - V$ and $U - B$.

153.027 A photoelectric Hβ distance modulus of the open cluster NGC 6871. H. L. Cohen.
Astron. Journ., Vol. 74, 1168 - 1170 = Contr. Rosemary Hill Obs., Univ. Florida, Gainesville, No. 5 (1969).

Photoelectric Hβ observations are reported for ten bright OB stars (including the eclipsing binaries V448 Cyg and V453 Cyg) in the field of NGC 6871. The derived distance modulus

of NGC 6871 (11.4 mag) confirms values in the literature derived from wideband photometry. Both eclipsing binaries are placed at the same distance as the cluster (1.9 kpc). The Hβ observations, transformed into values of $W_{H\gamma}$, agree with published spectroscopic values of $W_{H\gamma}$ and disagree with published photoelectric $W_{H\gamma}$ values. Hence, the published H_γ modulus (9.8 mag) appears to be too small.

153.028 On the color-magnitude diagrams of NGC 2360 and NGC 3680. P. Demarque, R. H. Miller.
Astrophys. Journ., Vol. 158, 1037 - 1038 (1969).

Eggen's recent observations of the color-magnitude diagrams of the two galactic clusters NGC 2360 and NGC 3680 are interpreted by the method previously applied to M67 and NGC 188. The width of the gap in NGC 2360 is the same as that in M67, implying similar chemical composition. The gap in NGC 3680 is significantly wider, suggesting a much higher helium abundance. NGC 3680 may be a "super-metal-rich" cluster.

153.029 The Hyades red dwarfs and the distance of the cluster. O. J. Eggen.
Astrophys. Journ., Vol. 158, 1109 - 1113 (1969).

Red and infrared observations on the (R, I) system have been obtained for twenty-four red dwarfs in the Hyades cluster and twenty-five field stars of the young disk population with large trigonometric parallaxes. Proper motions of the Hyades members are available from two or more catalogs and were used to compute individual luminosities by the convergent-point method. The rms error of fit of the individual Hyades stars to the main sequence defined by the field stars is 0.2 mag, compared with the rms error of about 0.15 mag arising from errors in the proper motions.

153.030 The nearby, loose cluster Collinder 399. D. S. Hall, F. G. VanLandingham.
Bull. American Astron. Soc., Vol. 1, 346 (1969). – Abstr. AAS.

153.031 Photometric observations of the star cluster IC 166. M. S. Burkhead.
Astron. Journ., Vol. 74, 1171 - 1176 = Publ. Goethe Link Obs. Indiana Univ., Bloomington, No. 99 (1969).

Photoelectric and photographic photometry are presented for the cluster IC 166. The cluster exhibits the upper end of an evolved main sequence at $(B - V) = 1.15$, $m_V = 17.0$; a Hertzsprung gap from $(B - V) = 1.2$ to 1.6, $m_V = 17.0$; and a well-defined giant region from $(B - V) = 1.6$ to 1.8, $m_V = 17.0$ to 16.5. The color excess is poorly determined, but is assumed to be $E_{(B-V)} = 0.80$; the distance modulus is $m - M = 12.6$. IC 166 appears to be an intermediate-age cluster similar in age to NGC 7789 and NGC 752.

153.032 On the Hyades binaries. W. D. Heintz.
Observatory, Vol. 89, 147 - 149 (1969). – Letter.

153.033 The Boks' galactic cluster in front of the larger Magellanic Cloud. D. H. P. Jones.
Observatory, Vol. 89, 237 (1969). – Letter.

153.034 Observations of early-type stars of the cluster NGC 6913. R. M. Raznik.
Astron. Zhurn. Akad. Nauk SSSR, Vol. 46, 837 - 839 (1969). In Russian. English translation in Soviet Astron. AJ, Vol. 13, No. 4.

Photoelectric and spectral observations of early-type stars of NGC 6913 are carried out.

153.035 Photometry of stars in the field of the star cluster NGC 7142. A. S. Sharov.

Soobshch. Gos. Astron. Inst. Shternberga, No. 158, p. 43 - 48 (1969).   In Russian.

A catalogue of magnitudes and colours of 266 stars brighter than V = 15.^{m}5 in the field of the cluster NGC 7142 is given.

153.036   A search for A$_p$ stars in very young clusters.
         P. L. Bernacca, F. Ciatti.
Atti XII Riunione Soc. Astron. Italiana, L'Aquila 1968, p. 61 - 62 (1969). − Abstract SAI.

153.037   Studies of local star streams. II. The Ursa Major
         stream.   K. F. Ogorodnikov, I. N. Latyshev.
Astron. Zhurn. Akad. Nauk SSSR, Vol. 46, 1190 - 1200 (1969). In Russian. English translation in Soviet Astron. AJ, Vol. 13, No. 6.

Stars in the solar neighbourhood with motions equal to those of Ursa Major cluster were selected. Their number appeared to be by several times larger than that which might have been anticipated theoretically on the basis of the ellipsoidal velocity distribution. Thus the existence of the UMa stream has been confirmed. The space density of the stream in solar neighbourhood is 0.0025 stars per pc^3. The total number of stars belonging to the stream is estimated as high as 2000 or even larger. The H-R diagrams of the Hyades and UMa streams show significant differences. The age of the UMa stream is smaller than that of the Hyades. Galactic orbits of the streams were determined.

153.038   The ages of the Hyades, Praesepe, and Coma star
         clusters.   E. P. J. van den Heuvel.
Publ. Astron. Soc. Pacific, Vol. 81, 815 - 825 = Contr. Lick Obs. No. 304 (1969).

If one uses the revised T_e, $(B - V)$ relations and bolometric corrections for main-sequence stars derived by Morton and Adams (1968) and Iben's evolutionary tracks for Population I stars $(X = 0.71, Z = 0.02)$, it is found that the Hyades and Praesepe clusters are about 9×10^8 years old, while the Coma cluster has an age of about 6.5×10^8 years.

153.039   Two-color photographic photometry of three
         possible open star clusters.
R. M. Nelson.
Publ. Astron. Soc. Pacific, Vol. 81, 900 - 904 (1969). − Note.

153.040   Investigation of five open clusters.

Kh. Z. Ishmukhamedov.
Astrometr. Issled. p. 3 - 23 (1969).   In Russian.

153.041   Dynamische Deutung der Altersverteilung offener
         Sternhaufen.   R. Wielen.
Mitt. Astron. Ges. No. 27, p. 132 - 133 (1969). − Abstract AG.

153.042   Lichtelektrische UBV-Photometrie des galaktischen
         Sternhaufens NGC 2516.
J. Dachs.
Mitt. Astron. Ges. No. 27, p. 148 (1969). − Abstract AG.

153.043   Radiostrahlung offener Sternhaufen bei
         2.695 GHz.   R. Schwartz.
Mitt. Astron. Ges. No. 27, p. 178 - 179 (1969). − Abstract AG.

Die Verwendbarkeit des Großen Wiener Refraktors für die Bestimmung von Positionen und Eigenbewegungen. (Die absolute Eigenbewegung von NGC 6838).
See Abstr. 112.008.

Rotational velocities in NGC 2516.
See Abstr. 116.013.

The light-elements of an anonymous cepheid and on its probable membership in the galactic cluster NGC 6649.
See Abstr. 122.095.

New flare stars in the Pleiades region and in the Orion complex.   See Abstr. 122.111.

Two remarkable flare stars in the Pleiades.
See Abstr. 122.123.

On the star No. 98 in NGC 2360.
See Abstr. 122.139.

Flare stars in the Pleiades region.
See Abstr. 122.144.

New flare stars in the Pleiades.
See Abstr. 122.145.

H II 2411, a Hyades flare star.
See Abstr. 122.146.

154 Globular Clusters

154.001 Photometry of population II cepheids in globular clusters. I. M2.   S. Demers.
Astron. Journ. Vol. 74, 925 - 928, 983 = Contr. Cerro Tololo Inter-American Obs. No. 76 (1969).

U, B, V photographic photometry of the four variables in M2 with periods longer than one day is presented. The plates are calibrated with a new photoelectric sequence. Mean magnitudes and colors are determined from the light and color curves. As has been noted before, population II cepheids show little change of color with periods.

154.002 Three new variable stars in the globular cluster NGC 5466.   T. I. Gryzunova.
Astron. Tsirk. No. 526, p. 8 (1969).   In Russian.

154.003 The reddening, age difference, and helium abundance of the globular clusters M3, M13, M15, and M92.
A. Sandage.
Astrophys. Journ. Vol. 157, 515 - 531 (1969).

Reddening values of $E (B - V) = 0.00, 0.03, 0.12,$ and 0.02 mag are obtained for M3, M13, M15, and M92, respectively, from new photometry of horizontal-branch stars via the two-color diagram. The $B - V$ main sequence-turnoff colors agree to within 0.01 mag in all four clusters when corrections are applied for differential line blanketing. The result requires that M3 and M13, which have similar Z-values, have an age spread Δt of less than $\Delta t/t \simeq 0.03$. The same result applies separately to M15 and M92. The blue-boundary colors of the RR Lyrae instability strip in M3, M15, and M92 agree to within $\Delta (B - V)_{BE^{o.c}} = 0.025$ mag when reddening and blanketing corrections are applied. From Christy's models, the dependence of the blue boundary of the strip on Y, M_V, $M/M_\odot$, and $B - V$ is approximated by a linear equation which reproduces Christy's Y-values to within 0.004 ± 0.018 (A. D.) over a stated parameter range. Using the assumptions in the text, a mean helium abundance of $Y = 0.32 \pm 0.09$ (total range) is derived. The helium abundances for M3, M15, and M92 are the same to within the observational error, despite the large difference in metal abundance.

154.004 The reddening of M3, M13, M31, and M33 from photometry of late-type field stars.
R. D. McClure, R. Racine.
Astron. Journ. Vol. 74, 1000 - 1007, 1055 = Contr. Kitt Peak National Obs. No. 455 (1969).

A new method for the determination of the interstellar reddening of late-type stars is developed and applied in the direction of the high-latitude globular clusters M3 and M13, and of the two spiral galaxies M31 and M33. The color excess for both globular clusters was found to be zero: $E(B - V) = -0.01 \pm 0.02$ for M3, and $E(B - V) = 0.00 \pm 0.02$ for M13; further, we found $E(B - V) = 0.11 \pm 0.02$ for M31, and $E(B - V) = 0.03 \pm 0.02$ for M33. In the direction of M31, the absorbing material appears to extend to a distance of 200 ± 50 pc below the galactic plane.

154.005 Interstellar reddening for globular clusters M3 and M13.   D. L. Crawford, J. V. Barnes.
Astron. Journ. Vol. 74, 1008 - 1010 = Contr. Kitt Peak National Obs. No. 452 (1969).

Results of four-color and Hβ photometry of A- and F-type stars in the areal vicinity of M3 and M13 have been used to conclude that the space reddening, $E(b - y)$, in front of these clusters is nearly zero: $0\overset{m}{.}004$ for M3 and $0\overset{m}{.}011$ for M13.

154.006 Narrow-band and intermediate-band photometry
of globular star clusters.
S. L. Johnson, D. H. McNamara.
Publ. Astron. Soc. Pacific, Vol. 81, 415 - 425 (1969).

Narrow-band and intermediate-band photometry of the integrated light of globular star clusters in the *uvby* and β photometric systems are described. The observational results are used to derive the reddening of the clusters. Reddened-free indices are calculated and are shown to be related to the metal abundances of the member stars of the clusters.

154.007 Primeval globular clusters. II.   P. J. E. Peebles.
Astrophys. Journ. Vol. 157, 1075 - 1083 (1969).

The time variation of initially isothermal perturbations is computed numerically through the epoch of recombination of the primeval plasma. The mean mass of the primeval gas clouds, tentatively identified as proto-globular clusters, is computed for a range of possible choices of the parameters in the theory.

154.008 On the nature of some of the O−C diagrams of the RR Lyrae variables in M5.   C. Coutts.
Non-Periodic Phenomena in Variable Stars, IAU Colloquium, Budapest, 1968, p. 313 - 320 (1969).

154.009 A study of the variable gap in M5.
C. Coutts, R. Margoni, R. Stagni.
Bull. American Astron. Soc. Vol. 1, 238 (1969). − Abstr. AAS.

154.010 Stellar photometry in NGC 6791, M 92, and M 13 with the electronic camera.
H. D. Ables, A. V. Hewitt, G. E. Kron.
Publ. Astron. Soc. Pacific, Vol. 81, 530 (1969). − Abstract ASP.

154.011 The blue horizontal-branch stars of ω Centauri.
E. B. Newell, A. W. Rodgers, L. Searle.
Astrophys. Journ. Vol. 158, 699 - 709 (1969).

We have obtained photoelectric UBV photometry for twenty-six blue horizontal-branch stars in the halo globular cluster ω Centauri (NGC 5139). For twenty of these stars Hβ profiles have been determined from low-dispersion image-tube spectra. Atmospheric parameters are derived on the basis of the observed colors and line widths. For nine stars hotter than $\theta_e = 0.55$, we derive a mean mass-luminosity ratio of -2.02 ± 0.16. A distance modulus is estimated for ω Cen by fitting its C−M diagram to that of M92. The result, $(m - M)_{true}$ -14.0 ± 0.3 mag, is based on a modulus of 14.5 mag for M92 and leads to a mean luminosity $\langle \log L/L_\odot \rangle = 1.71 \pm 0.12$.

154.012 An analysis of the bright O star in the globular cluster M3.   S. E. Strom, K. M. Strom.
Bull. American Astron. Soc., Vol. 1, 364 (1969). − Abstr. AAS.

154.013 The RR Lyrae stars in M3 and ω Cen.
T. S. van Albada.
Bull. American Astron. Soc., Vol. 1, 366 (1969). − Abstr. AAS.

154.014 A model of globular clusters with inhomogeneous composition of stars.   V. M. Bagin.
Astron. Zhurn. Akad. Nauk SSSR, Vol. 46, 1201 - 1206 (1969).   In Russian. English translation in Soviet Astron. AJ, Vol. 13, No. 6.

A model of globular clusters, in which there are several groups of stars with different masses, is considered. Supposing a finite number of such groups, a closed system of equations describing stellar motions is obtained. For a cluster containing only two groups of stars this system can be integrated. From its solution follows that the total density of mass is distributed according to' Schuster's law, and at the periphery of the cluster the stars with smaller masses displace completely those with larger masses; this confirms qualitatively the results of observations.

154.015 The evolution of the stationary inhomogeneous model of a globular cluster.
V. E. Yakimov.
Soobshch. Gos. Astron. Inst. Shternberga, No. 158, p. 14 - 22 (1969).   In Russian.

The dynamic evolution within 0.75×10^9 years of the spatial distributions of two stellar groups in the general gravitational field of the model was studied. Their masses are different and correspond to giants and supergiants of a real cluster. The change of the distributions is the result of collective effects of stars of the general field of the model. It was found that the distributions of two types of stars, different at initial moments, did not change considerably within the evolution of the quasistationary model.

154.016 A study of the periods of some variable stars of the globular cluster M 92.
C. Bartolini, P. Battistini, E. Nasi.
Pubbl. Oss. Astron. Univ. Bologna, Vol. 9, No. 15, 25 pp. (1968).

The periods and the light curves of 12 variable stars of the globular cluster M 92, have been studied analyzing 176 plates obtained with the 60 cm reflector of Bologna Observatory from 1964 to 1966. Combining the results of the recent observations with Hachenberg's (1925, 1933, 1934) and Nassau's (1936), the periods were improved and the O-C diagrams for 9 variables were traced.

154.017 Period changes of RR Lyrae variables in the globular cluster Messier 5.   C. M. Coutts, H. S. Hogg.
Publ. David Dunlap Obs. Univ. Toronto, Richmond Hill, Vol. 3, (No. 1), 1 - 58 (1969).

The purpose of this investigation is to study period changes in RR Lyrae variables in the globular cluster M5. The study is based mainly on a collection of 167 plates taken between 1936 and 1966 at the David Dunlap Observatory. Studies of this type which have been carried out for other globular clusters are briefly discussed. A total of 66 RR Lyrae variables has been studied in M5. Of these, 16 have irregular periods, 18 have been constant, 20 have shown increases and 12 decreases in period during an interval of about seventy years. It seems not possible at present to attach any evolutionary significance to these changes.

154.018 Ratio of horizontal branch stars to red giant stars in globular clusters.
I. Iben, Jr., R. T. Rood, K. M. Strom, S. E. Strom.
Nature, Vol. 224, 1006 - 1008 (1969).

154.019 Age and initial helium abundance of globular cluster stars.   I. Iben, Jr., R. T. Rood.
Nature, Vol. 223, 933 - 934 (1969).

It has been shown that the number of horizontal branch stars relative to the number of red giant stars could be used together with other observable cluster characteristics to estimate the age and initial helium abundance of stars in globular clusters. In this article we repeat the analysis on the basis of main sequence and horizontal branch models constructed with new radiative opacities calculated by Cox and Stewart and of models at the red giant tip constructed by Eggleton.

155 Structure and Evolution of the Galaxy

155.001 Distribution of Wolf-Rayet stars in the Galaxy, and their evolution. Z. Mikuláŝek.
Bull. Astron. Inst. Czechoslovakia, Vol. 20, 215 - 222 (1969).

Three hypothesesof the formation and evolution of Wolf-Rayet stars are discussed. The distribution of the number of Wolf-Rayet stars is evaluated in dependence on galactic latitude and longitude. The relation between Wolf-Rayet stars and open star clusters is investigated, and the percentage of stars in physical connection with clusters is determined (21%). From their position with regard to clusters, the age of WN-type stars has been estimated at 3.2×10^6 years, and that of WC-type stars at 10.6×10^6 years. The relation of Wolf-Rayet stars and O associations is discussed; the age of the stars in connection with them is estimated at 3.6×10^6 years.

155.002 Unravelling the Milky Way. K. Hindley.
New Scient. Vol. 41 (No. 634), 221 (1969).—Popular article.

155.003 Giant M stars in the galactic anticenter. S. W. McCuskey.
Astron. Journ. Vol. 74, 807 - 811 (1969).

The space density of giant stars of spectral groups M0–M1, M2–M4, and M5–M8 has been evaluated as a function of distance at the galactic anticenter, $l^{II} = 186°$, $b^{II} = +1°$. Objective-prism spectra in the red, infrared, and V magnitudes are the observed data for 493 stars. The M stars decrease in number with distance from the sun in the anticenter directions; numerical results are given.

155.004 A finding list of early-type stars in regions of intermediate galactic latitude.
A. R. Upgren, R. T. Staron.
Astrophys. Journ. Vol. 157, 327 - 334 (1969).

A list is given of all stars of spectral class F2 and earlier, brighter than magnitude 11.5, found in four regions of intermediate galactic latitude. This objective-prism survey is part of a study of stellar densities aiming at the determination of the inclination to the galactic plane of surfaces of equal stellar density. The distribution of early stars fainter than 11.5, as well as other stars of later type, will be covered elsewhere. The conclusion is made that the stars included here do not appear to refute the model found by Oort but that the area covered is insufficient for a detailed analysis, as is also the case for other similar surveys cited.

155.005 1.65-19.5-micron observations of the galactic center. E. E. Becklin, G. Neugebauer.
Astrophys. Journ. *(Letters)*, Vol. 157, L31 - L36 (1969).

New observations of the nucleus of the galaxy in the wavelength region between 1.65 and 19.5 μ are presented. At 10 and 20 μ radiation originates in a source approximately 1 pc in diameter. The energy distribution of the galactic-center source is similar to that measured in the nuclei of Seyfert galaxies (Low and Kleinmann). Two possible mechanisms that could produce the infrared radiation are briefly discussed.

155.006 The infrared spectrum, diameter, and polarization of the galactic nucleus.
F. J. Low, D. E. Kleinmann, F. F. Forbes, H. H. Aumann.
Astrophys. Journ. *(Letters)*, Vol. 157, L97 - L101 (1969).

Observations of the galactic center have been made over the wavelength range 5 - 1500 μ. They include polarization measurements at 10 μ, diameter measurements at 10 and 22 μ, an upper limit to the flux at 1200 μ, a preliminary flux at 100 μ, and accurate flux measurements with a 25″-diameter beam at 5, 10, 11.5, 13, 18.9, 22, and 24.5 μ. Most of the energy radiated between 5 and 25 μ is produced by a source ~15″ in diameter, with linear polarization less than 2 percent and a spectral energy distribution similar to those found for other galaxies measured at these wavelengths, however, the power output from this 1-pc-diameter source is found to lie between 1.6×10^6 and 8×10^6 $L_\odot$, or from 2 to 10 times the power radiated by the stars in the same volume. The preliminary flux measurement at 100 μ strongly favors a nonthermal source. A reasonable fit to a 270°K blackbody distribution with an emissivity of about 0.001 is nevertheless possible if the 100-μ observation is excluded.

155.007 A comparison of the star-densities for disc and halo in the direction to the galactic poles by means of the three-colour-photometry. R. P. Fenkart.
Astron. Astrophys. Vol. 3, 228 - 235 = Mitt. Astron. Inst. Univ. Basel No. 59 (1969). In German.

By photometry in the RGU-system the populations of the disc and the halo were separated in the direction of SA 141 ($b_{II} = -86°$); the density-gradients were determined in different luminosity intervals. The density-gradients have been compared with those for SA 57 ($b_{II} = +85°$) which were obtained by reducing the results of an earlier paper (Fenkart, 1967).

155.008 The shape of the galactic spiral arms and parameters of galactic rotation, determined from observations of H II regions. G. Courtès, Y. P. Georgelin, Y. M. Georgelin, G. Monnet.
Astrophys. Letters, Vol. 4, 129 - 136 (1969).

It is shown that radial velocities and spectrophotometric distances for H II regions can be used as a link between galactic velocities determined from the λ21-cm line and the geometrical structure of the galactic arms as traced out by OB stars. The kinematic data from the interstellar matter (ionized hydrogen) and from the young stars (OB and cepheids) are the same, which indicates the predominance of gravitational forces. In the longitude range $l^{II} = 305–333°$, two maxima of velocities are found, and two arms are observed as clearly separated as the well-known Perseus arm and the local arm. Neutral hydrogen, H II regions, young clusters and O stars in the vicinity of the sun are located in four spiral arms. Those arms are separated by distances of about 2 kpc and their pitch angle is about 20°. The morphological type of the galaxy is discussed from these new data and it is found that the galaxy seems to be similar to a multi-arm Sc galaxy (de Vaucouleurs' type SAB (rs) cd). From these observations new galactic constants are also derived.

155.009 The large-scale distribution of hydrogen in the galaxy. F. J. Kerr.
Annual Rev. Astron. Astrophys. Vol. 7, 39 - 66 (1969).

155.010 Spatial distribution of B8 – A0 and early B stars in several regions of the Milky Way between 15° and 136°. N. B. Grigorieva.
Astron. Zhurn. Akad. Nauk SSSR, Vol. 46, 1029 - 1034 (1969). In Russian. English translation in Soviet Astron. AJ, Vol. 13, No. 5.

The reduction of Crimean Observatory data gave the possibility of obtaining the spatial distribution of B8 – A0 stars in eight regions. Spatial densities are given. For the first three regions the spatial distribution of early B (B0 – B3) stars is found. The data for B8 – A0 stars are completed up to 700 - 800 pc, for early B stars up to 2000 pc. Maximum stellar densities for different longitudes and distances are given.

155.011 **H II regions and larger scale galactic structure.**
 B. F. Burke.
Interstellar Ionized Hydrogen, Charlottesville 1967, p. 541 - 564 (1968). – Review article.

155.012 **Mise en évidence du bras de Persée dans une direction proche de l'anti-centre galactique.**
N. Martin.
Comptes Rendus Acad. Sci. Paris, Sér. B, Vol. 269, 1007 - 1008 (1969).
L'étude sur des clichés pris au grand prisme objectif d'un champ situé près de l'anti-centre galactique, montre un grand nombre d'étoiles B situées entre 2 et 4 kpc, ce qui paraît correspondre au bras de Persée. Ce résultat est confirmé par Y. Georgelin qui situe les trois nébuleuses NGC 2174, S 254 et S 257, vers 2.7 kpc.

155.013 **Stellar groups in the old disk population.**
 O. J. Eggen.
Publ. Astron. Soc. Pacific, Vol. 81, 553 - 593 (1969). – Review article.

155.014 **Is the galaxy losing mass on a time scale of a billion years?** D. W. Sciama.
Nature, Vol. 224, 1263 - 1267 (1969).
If the galaxy has been losing mass in gravitational waves for longer than 10^9 yr at the rate recently indicated by Weber, it should contain no stars in bound orbits with periods exceeding 10^9 yr. Analysis of stellar motions indicates that the distribution of periods does have a cut-off close to 10^9 yr although this may be for other reasons.

155.015 **The spiral structure of our galaxy – I.**
 B. J. Bok.
Sky Telescope, Vol. 38, 392 - 395 (1969).

155.016 **The ratio of total to selective absorption in the galaxy.** J. G. Ireland, K. Nandy.
Astrophys. Space Sci. Vol. 5, 438 - 443 (1969).
Values for the ratio of total to selective absorption are estimated from optical measurements on Cepheids, their radial velocities, and an assumed rotational velocity model of the galaxy. It is shown that there is no systematic variation of R with galactic longitude on the basis of these measurements.

155.017 **Diffuse component of the cosmic far UV radiation and interstellar dust grains.**
S. Hayakawa, K. Yamashita, S. Yoshioka.
Astrophys. Space Sci. Vol. 5, 493 - 502 (1969).
The diffuse far UV radiation ($\lambda\lambda$ 1350 - 1480 Å) observed in the sky region of $l^{II} \approx 180°$, $0° \leqslant b^{II} \lesssim 40°$ is analyzed in connection with the distributions of stars and dust grains as well as with optical properties of grains. Its intensity (starlight + scattered light) is about 6×10^{-7} erg cm^{-2} sec^{-1} sr^{-1} Å^{-1} in the direction of $b^{II} \approx 0°$ and $l^{II} \approx 180°$. The latitude dependence of the intensity is in approximate agreement with the plane parallel slab model of the galaxy with a reasonable set of parameters.

155.018 **Der Bau unseres Milchstraßensystems.**
 H. Elsässer.
Bild der Wissenschaft, Vol. 6, 35 - 43 (1969). – Review article.

155.019 **Early-type stars in a south galactic pole region.**
 A. Slettebak, R. K. Brundage.
Bull. American Astron. Soc., Vol. 1, 362 - 363 (1969). Abstr. AAS.

155.020 **On the diffuse galactic light.**
 J. L. Weinberg.
Bull. American Astron. Soc., Vol. 1, 368 (1969). – Abstr. AAS.

155.021 **The problem of the infrared brightness of the Milky Way.** A. S. Sharov.
Astron. Zhurn. Akad. Nauk SSSR, Vol. 46, 1207 - 1214 (1969). In Russian. English translation in Soviet Astron. AJ, Vol. 13, No. 6.
Data on the Milky Way infrared brightness are revised. It is shown that in the direction of the galactic centre this quantity is about 10^{-3} erg/cm^2 sec steradian and in the opposite direction 10^{-4} (λ = 9750 Å, $\Delta\lambda$ = 1500 Å).

155.022 **Galactic nucleus as a site of nucleosynthesis.**
 T. Ohnishi.
Publ. Astron. Soc. Japan, Vol. 21, 307 - 318 (1969).
A new model of the nucleosynthesis is proposed. It is hypothetically assumed that nucleosynthesis took place in the galactic nucleus and the elements were produced and accumulated in the evolutionary process of the early universe, galaxies, and stars. The age of the Galaxy according to this model is determined by means of cosmochronology in which the production ratios of r-process nuclides are used. The calculated Galactic age and the age of population II stars are $(17.5 \pm 2.0) \times 10^9$ years and $(11.0 \pm 2.1) \times 10^9$ years, respectively.

155.023 **Vintergatans gåta.**
 T. Elvius.
Astron. Tidssk., Årg. 2, 101 - 112 (1969).

155.024 **The large scale structure of our galaxy as derived from radio recombination line surveys.**
T. L. Wilson.
Mitt. Astron. Ges. No. 27, p. 47 - 53 (1969). – Review article.

Radioisotopes and the history of nucleosynthesis in the galaxy. See Abstr. 061.017.

Etoiles froides et structure galactique.
See Abstr. 114.096.

Distribution of luminous stars in the region of one high velocity hydrogen cloud.
See Abstr. 115.005.

Theoretische Zweifarbendiagramme für hohe galaktische Breiten. See Abstr. 115.006.

OB stars near the supernova remnant RCW 103 and the galactic structure in Norma. See Abstr. 125.004.

Magnetohydrodynamical models of a helical magnetic field in spiral arms. See Abstr. 131.013.

Interstellare Absorptionslinien und galaktische Struktur. See Abstr. 131.019.

Interstellar polarization of starlight and the turbulent structure of the galaxy. See Abstr. 131.026.

Space distribution of H II regions.
See Abstr. 131.076.

Optical work on the kinematics of galactic H II regions. See Abstr. 131.077.

Pulsar distances, spiral structure and the interstellar medium. See Abstr. 141.130.

Galactic effects of the cosmic-ray gas.
See Abstr. 143.028.

Application of the density-wave theory to the spiral structure of the Milky Way system. I. Systematic motion of neutral hydrogen. See Abstr. 151.034.

Ros 4, a distant open cluster associated with nebulosities. See Abstr. 153.003.

Infall of gas from intergalactic space. See Abstr. 161.008.

156   Galactic Magnetic Field

156.001   Direction of the nearby galactic magnetic field inferred from a cosmic-ray diurnal anisotropy.
K. H. Schatten, J. M. Wilcox.
Journ. Geophys. Res. Vol. 74, 4157 - 4161 (1969).

A twenty-year wave in the diurnal anisotropy component of galactic cosmic rays arriving at the earth from the asymptotic direction 128°E of the sun has been found by Forbush. This wave is interpreted in terms of enhanced magnetic reconnection between the nearby galactic field and the field lines in the polar regions of the heliosphere during one-half of the twenty-year solar magnetic cycle. This interpretation leads to the result that the component parallel to the solar rotation axis of the nearby galactic field is directed northward.

156.002   Origin of the magnetic field of the galaxy.
E. N. Parker.
Astrophys. Journ. Vol. 157, 1129 - 1135 (1969).

A recent formal calculation demonstrates that the individual Fourier components of a magnetic field are stochastic variables when the field is carried in a random turbulent velocity $V_t(r,t)$. The individual Fourier components randomwalk, with the result that their mean squares increase with time. The calculation is applicable to the galaxy when the magnetic field is sufficiently weak ($B \lesssim 10^{-6}$ gauss). Application to the gaseous disk of the galaxy indicates that a field of 10^{-6} gauss would be created in less than 10^9 years. The present orientation and large scale of the field along the galactic arm are presumably the result of the nonuniform rotation of the galaxy.

156.003   Faraday rotation, dispersion in pulsar signals, and the turbulent structure of the galaxy.
J. R. Jokipii, I. Lerche.
Astrophys. Journ. Vol. 157, 1137 - 1145 (1969).

We compute the mean and the variance about the mean of the Faraday rotation and signal dispersion for electromagnetic waves propagating in a turbulent medium such as the galaxy. With regard to the electron number density and the magnetic field as homogeneous random functions (with nonzero mean), we present the results in terms of the mean values and two-point correlation functions of these quantities. Application to a simple, statistically homogeneous, disk model of the galaxy and its turbulence is then considered.

156.004   Galactic magnetic fields.
L. Křivský.
Vesmír, Vol. 48, 248 - 249 (1969).   In Czech.

Supernova remnants and the galactic magnetic field. See Abstr. 125.019.

157 Galactic Radio Radiation

157.001   **A search for galactic $H_2{}^+$.**
W. L. H. Shuter, D. S. Sloan.
Canadian Journ. Phys. Vol. 47, 1233 - 1234 (1969).

An attempt was made to detect a radio-frequency spectral line from orthohydrogen ions in two galactic regions. An upper limit to the brightness temperature of $T_b \leqslant 0.25$ °K was established.

157.002   **Parkes hydrogen-line survey of the milky way. II. The section $l^{II} = 296°$ to $63\overset{\cdot}{.}5$, $b^{II} = -2°$ to $+2°$.**
F. J. Kerr.
Australian Journ. Phys., Astrophys. Suppl. No. 9, 147 pp. (1969).

This paper presents results of a survey with the Parkes 210-ft telescope and a 48-channel receiver of 21-cm line radiation from a major part of the galactic disk. It includes: (1) 34 velocity-longitude contour maps covering the galactic equator from $l^{II} = 296°$ to $63\overset{\cdot}{.}5$; (2) 98 velocity-latitude maps for various constant-longitude lines between 300° and 60°.

157.003   **On the large-scale structure of the radio emission of the thermal background component of the galaxy.**
V. I. Ariskin.
Astron. Tsirk. No. 517, p. 1 - 4 (1969).   In Russian.

157.004   **Observations at 178 MHz of the north galactic spur.**
D. J. Holden.
Monthly Notices, Roy. Astron. Soc., Vol. 145, 67 - 73 (1969).

The structural features of the north galactic spur at 178 MHz are described and compared with those noted by other observers.

157.005   **A measurement of the sky brightness temperature at 408 MHz.**   R. M. Price.
Australian Journ. Phys. Vol. 22, 641 - 654 (1969).

The absolute value of the background brightness temperature has been measured at a radiofrequency of 408 MHz. Observations were made with a large pyramidal "standard gain" horn aerial and the aerial temperatures were compared direct with the temperature observed with the receiver input connected to a matched resistive load of known temperature. For the region within 24° of the south celestial pole an average brightness temperature of 23.9° K was obtained. Observations of the northern regions indicate that values from previous determinations at northern latitudes are 4 – 5 deg K too high.

157.006   **Spectral and spatial distribution of cosmic noise observed by RAE-I.**
R. R. Weber, J. K. Alexander, R. G. Stone, T. A. Clark.
Bull. American Astron. Soc. Vol. 1, 265 (1969). – Abstr. AAS.

157.007   **The National Radio Astronomy Observatory 11 centimeter continuum survey of the northern galactic plane.**   W. Altenhoff.
Interstellar Ionized Hydrogen, Charlottesville 1967, p. 519 - 533 (1968).

It was decided to carry out a complete 2.7 GHz survey within 2° of the galactic equator from galactic longitudes 345°, through 0°, to about 240°. A parametric amplifier with 200°K noise temperature and 40 MHz bandwidth was used with the NRAO 140-foot telescope.

157.008   **Southern galactic continuum surveys at high frequencies.**   F. J. Kerr.
Interstellar Ionized Hydrogen, Charlottesville 1967, p. 535 -

539 (1968).

This is a short progress report on large-scale continuum surveys that are being carried out at Parkes with the 210-ft telescope.

157.009   **High-resolution observations at 408 and 5000 MHz of a region near $l^{II} = 312°$.**
P. A. Shaver, W. M. Goss.
Proc. Astron. Soc. Australia, Vol. 1, 280 - 282 (1969). – Contribution ASA meeting.

The purpose of this investigation has been to provide high-resolution maps of galactic radio sources; the half-power beam widths are ~3' at 408 MHz and ~4' at 5000 MHz. With these resolutions and wide range of frequency it has been possible to derive accurate spectral indices for over 200 discrete sources detected in the two surveys. We discuss in some detail a region near $l^{II} = 312°$ which serves as an example of the interesting objects that have been observed.

157.010   **A low-latitude survey from $l^{II} = 288°$ to 307° at 2650 MHz.**   B. M. Thomas, G. A. Day.
Australian Journ. Phys. Astrophys. Suppl. No. 11, p. 3 - 10 (1969).

This paper presents contour maps of the region of the Milky Way between longitudes 288° and 307°, latitudes ±2° at 2650 MHz. A list of sources with values of peak temperature and flux density is given.

157.011   **A low-latitude survey from $l^{II} = 307°$ to 330° at 2700 MHz.**   G. A. Day, B. M. Thomas, W. M. Goss.
Australian Journ. Phys. Astrophys. Suppl. No. 11, p. 11 - 18 (1969).

The results of a survey of the galactic plane at 2700 MHz from longitudes 307° to 330°, latitudes ±2° are presented as a contour map and a source list giving the positions and estimated flux densities for 117 radio sources. A computer-drawn ruled-surface picture of the area between $l^{II} = 324°$ and 328° is shown to illustrate the complexity of the region.

157.012   **A low-latitude survey from $l^{II} = 334°$ to 345° at 2650 MHz.**   B. M. Thomas, G. A. Day.
Australian Journ. Phys. Astrophys. Suppl. No. 11, p. 19 - 25 (1969).

This paper presents contour maps of the region of the Milky Way at 2650 MHz between longitudes 334° and 345°, latitudes ±2°. A list of sources with values of peak temperature and flux density is given.

157.013   **A low-latitude survey from $l^{II} = 345°$ to 5° at 2650 MHz.**   M. Beard, B. M. Thomas, G. A. Day.
Australian Journ. Phys. Astrophys. Suppl. No. 11, p. 27 - 34 (1969).

This paper presents contour maps of the region of the Milky Way between longitudes 345° and 5°, latitudes ±2° at 2650 MHz. A list of sources with values of peak temperature and flux density is given.

157.014   **An 11 cm map of a region in Vela.**
B. A. Manchester, W. M. Goss.
Australian Journ. Phys. Astrophys. Suppl. No. 11, p. 35 - 41 (1969).

This paper presents 11 cm contour maps of a region in Vela between right ascensions $08^h 52^m$ and $09^h 06^m$ and declinations -41° and -50°. The two intense H II regions RCW 36 and 38 are included in this survey.

157.015 **Continuum radio emission and magnetic fields in the galaxy.** J. R. Shakeshaft.
Cosmic Ray Studies, Bombay 1968, p. 131 - 141 (1969).

157.016 **21-cm radiation from high galactic latitudes.** C. R. Tolbert.
Bull. American Astron. Soc., Vol. 1, 365 (1969). — Abstr. AAS.

157.017 **On the extensive ionized complex of the matter at the periphery of the Galaxy.** V. I. Ariskin.
Astron. Zhurn. Akad. Nauk SSSR, Vol. 46, 750 - 754 (1969). In Russian. English translation in Soviet Astron. AJ, Vol. 13, No. 4.

On the basis of observations of the neutral hydrogen line and continuous spectrum of the galactic region in the direction of W41 and W42 it is assumed that the extensive ionized complex, in which W42 is situated, is extant at the periphery of the Galaxy. The hypothesis, that the emission of recombination lines in the direction of W41 and W42 originates from compact parts of HII regions, analogous to those found in W3, W49 and Cyg X-DR21, is established.

157.018 **The region of the Galaxy at the interval of longitudes $l^{II} = 20.8° - 32.8°$ at 21.1 cm.**
V. I. Ariskin, I. I. Berulis, R. L. Sorochenko.
Astron. Zhurn. Akad. Nauk SSSR, Vol. 46, 1149 - 1157 (1969). In Russian. English translation in Soviet Astron. AJ, Vol. 13, No. 6.

At the Radioastronomical Station of the Physical Institute of the USSR Academy of Sciences with the 22 m radiotelescope observations of this region were carried out both in the line of neutral hydrogen and in the continuous spectrum. The catalogue of the radioline profiles, isophotes of the brightness distribution of the neutral hydrogen line and of the continuous spectrum of the above-mentioned region of the Galaxy are given.

157.019 **A survey of the southern sky at 153 MHz.**
P. A. Hamilton, R. F. Haynes.
Australian Journ. Phys. Vol. 22, 839 - 841 (1969).

Recent low frequency surveys of the sky have shown the galactic plane in absorption as a region of low brightness. The analysis of this absorption has provided information on the distribution of ionized hydrogen in the galactic plane (Ellis and Hamilton, 1966). The extension of this analysis required a number of surveys at higher frequencies, where absorption effects are less evident, and this paper presents the result of such a survey at 153 MHz.

157.020 **Synchrotron emission from the galaxy and the diffuse X-ray background.**
P. A. Hamilton, R. J. Francey.
Nature, Vol. 224, 1090 - 1093 (1969).

The results of a low resolution, low frequency radio spectrum survey in the direction of the galactic poles show a marked difference in spectral index from north to south. The halo model emerging from the survey predicts a diffuse X-ray flux by inverse Compton effect which can be an appreciable fraction of the observed flux. The agreement in spectral index between the radio and available X-ray data is remarkable, and the possibility that appreciable amounts of the observed radio and X-ray fluxes originate from a common electron spectrum in the halo cannot be excluded.

Observations of the 21-cm hydrogen line toward high-latitude stars. See Abstr. 131.040.

158 Single und Multiple Galaxies

158.001 A note on the systematic velocity of M 31.
V. C. Rubin, S. D'Odorico.
Astron. Astrophys. Vol. 2, 484 - 488 (1969).

Earlier determinations of the central velocity of M31 are presented and briefly discussed. These values come either from observations of the central region, or from the assumption of symmetry of the rotation curve. The weighted mean velocity is $V_c = -300 \pm 4$ km/s for the optical values. Recent interferogram observations of the northeast arm of M31 by Deharveng and Pellett are reanalyzed in order to determine the central velocity under the assumption that V_{rot} is constant rather than V_{rot}/R. The dependence of the central velocity on these two different hypotheses and on the adopted inclination of the galaxy is examined. For the generally adopted value of the inclination $\xi = 77°$, a velocity $V_c = -295 \pm 4$ km/s is obtained.

158.002 Calculated [Fe X] and [Fe XIV] line strengths in a Seyfert galaxy model.   D. E. Osterbrock.
Astrophys. Letters, Vol. 4, 57 - 59 (1969).

The photoionization model of Williams and Weymann is extended to calculate approximately the sizes of the zones that emit [Fe X] $\lambda 6374$ and [Fe XIV] $\lambda 5303$, their temperatures, and luminosities. Excitation by resonance fluorescence is important, particularly for $\lambda 5303$. The calculated strengths of $\lambda 6374$ and $\lambda 5303$ are close to but below the observed values.

158.003 Das Geschwindigkeitsellipsoid der H II-Regionen in M33.   P. Brosche.
Astrophys. Space Sci. Vol. 4, 327 - 329 = Astron. Rechen–Inst. Heidelberg, Mitt. Ser. A, No. 36 (1969).

The peculiar radial velocities of H II regions in M33 varies with position angle in a manner that can be explained by an ellipsoidal distribution with minor axis in the direction of rotation. The amplitude of the variations, however, is too great as compared with theory or experience in the galaxy.

158.004 Infrared radiation from dust in Seyfert galaxies.
M. J. Rees, J. I. Silk, M. W. Werner, N. C. Wickramasinghe.
Nature, Vol. 223, 788 - 791 (1969).

The observed infrared radiation from Seyfert galaxies in the waveband 2.2–22μm may be emitted by dust grains which absorb energy from an intense optical or ultraviolet source at the galactic nucleus.

158.005 *UBVRIHKL* photometry of the central region of M 31.
A. R. Sandage, E. E. Becklin, G. Neugebauer.
Astrophys. Journ. Vol. 157, 55 - 68 (1969).

The nuclear region of M 31 has similar brightness distributions in BVR and $K(2.2\ \mu)$ wavelengths. No non-thermal infrared excess exists in the nucleus of M 31 above our limit of detection. A pronounced variation of $U - B$ exists across the central $\pm 60''$ (± 200 pc) which is not present in $B - V$ or redder colors. $U - B$ changes from 0.79 at the center to 0.60 at $r = \pm 60''$. Variation of the stellar luminosity function or metal abundance with distance from the center is a possible explanation, but neither is proved by the present data. Similar small-scale color gradients exist in M 32, M 81, and NGC 7331. The absolute energy distribution $I(\lambda)$ shows a broad maximum which extends between 0.45 μ and 1 μ. The surface brightness of M 31 at 2.2 μ averaged over the central ± 13 pc (7".62 diameter) is 2.2×10^{-28} W m^{-2} Hz^{-1} per square second of arc. This is fainter by a factor of 2.4 than the 2.2 μ surface brightness of the Galactic center averaged over the equivalent

linear diameter. The $I(r)$ profile for M 31 and the Galaxy is similar over the central ± 400 pc – a fact which, combined with the higher surface brightness of the Galaxy, suggests that the Hubble type for the Galactic system is closer to an early Sb than to an Sc of the M 33 or NGC 2403 type.

158.006 Observations of galaxies with large amounts of dust. I. The galaxy NGC 7625.
M.-H. Demoulin.
Astrophys. Journ. Vol. 157, 69 - 73 (1969).

NGC 7625 is a peculiar S0 galaxy with a large amount of dust irregularly distributed. The spectrum is a very early-type one with the Balmer lines well visible in absorption from Hβ to H 12. Emission lines usually found in spiral galaxies are intense. The region 3.5 kpc in diameter rotates as a solid body and has a mass of $8 \times 10^9\ M_\odot$. The total mass of the galaxy, which is 7 kpc in diameter, is $\simeq 1.6 \times 10^{10}\ M_\odot$, and $M/L \simeq 2.5$.

158.007 Observations of galaxies with large amounts of dust. II. Rotation and mass of NGC 3593.
M.-H. Demoulin.
Astrophys. Journ. Vol. 157, 75 - 80 (1969).

Direct photographs have shown the presence of large amounts of interstellar dust in the small peculiar S0 galaxy NGC 3593. Although the spectral type reveals a young stellar population, there is no photographic evidence of early Population I. The emission lines are intense. Calculation of the mass from the velocity curve gives $M \simeq 3 \times 10^9\ M_\odot$, which is almost 5 times smaller than the mass of a late-type spiral. In contrast, the central density is 2×10^{-21} g cm^{-3}, which is as large as that of the more massive spiral galaxies.

158.008 Observations of galaxies with large amounts of dust. III. Velocity field in NGC 3077.
M.-H. Demoulin.
Astrophys. Journ. Vol. 157, 81 - 85 (1969).

NGC 3077 is a small Irr-II galaxy. Spectra taken in various directions do not show any velocity differences larger than the precision of the measurements, except in the direction P.A. = 150°, and the small velocity gradient in this direction cannot be interpreted as rotation.

158.009 Observations of M82 in the optical infrared.
F. Bertola, S. D'Odorico, W. K. Ford, Jr., V. C. Rubin.
Astrophys. Journ. *(Letters)*, Vol. 157, L27 - L28 (1969).

A sequence of direct photographs of M82 from the blue to 1μ has been obtained. There is a small nucleus which becomes more prominent at longer wavelength. Infrared spectra reveal unbroadened, inclined emission lines of [S III] and He I $\lambda 10830$.

158.010 New infrared observations of the galaxy M82.
M. I. Raff.
Astrophys. Journ. *(Letters)*, Vol. 157, L29 (1969).

New infrared plates of M82 show several sources of radiation around the optical center. Its recent classification as a late-type spiral is subject to doubt.

158.011 A very high-velocity gas cloud near the nucleus of NGC 4939.
E. M. Burbidge, M.-H. Demoulin.
Astrophys. Letters, Vol. 4, 89 - 91 (1969).

A gas cloud about 750 pc from the nucleus of the spiral galaxy NGC 4939 has a radial velocity 700 km sec^{-1} less than that of the nucleus of the galaxy. The gas is most probably being ejected from the nucleus.

158.012 Reduced counts of galaxies along galactic coordinates. T. Kiang.
Commun. Dublin Inst. Advanced Studies, Ser. C, Dunsink Obs. Publ. Vol. 1, (No. 5), 109 - 132 (1968).

The counts of galaxies per square degree, published by the Lick Observatory (Shane and Wirtanen 1967) have been corrected for all known errors, and then selected to fill preassigned, approximately square, cells along the galactic coordinates. These numbers should be multiplied by a factor of 1.076 to restore the observed level. Average per square degree numbers, averaged over $1°$–latitude strips and over approximately $6°$–squares are given.

158.013 The colorimetry of barred spirals. IV. A. T. Kalloghlian.
Soobshch. Byurakan. Obs. No. 40, p. 15 - 30 (1969). In Russian.

The results of detailed colorimetric investigation of the barred spirals NGC 3351, 3367 and 3384 are presented. The Sb0 galaxy NGC 3384 becomes bluer towards its center, while NGC 3351 is relatively blue only in the central parts. The observed anomalies in color distributions and central structures of these galaxies are due to their youth. The results of the colorimetry are used to investigate the dependencies of relative intensities and mean surface brightnesses on color. The proposed new method allows to reveal some properties of the galaxies.

158.014 On the youth of groups of galaxies. H. M. Tovmassian.
Soobshch. Byurakan. Obs. No. 40, p. 57 - 62 (1969). In Russian.

Only 11% of single normal galaxies have detectable radio emission at 1400 Mc/s, while radio emission is detected from 30% of galaxies which are members of double or multiple galaxies. The difference is higher for the brightest members of the groups; 37% of them have radio emission. It means that the components of double or multiple galaxies are more often in an active stage of their evolution than single galaxies. This may be considered as additional evidence in favour of the youth of the members of some groups of galaxies.

158.015 A note on the mass-radius relation for elliptical galaxies. L. M. Genkina.
Bull. Astron. Inst. Czechoslovakia, Vol. 20, 303 - 304, with a reply from J. L. Sérsic, p. 305 (1969). – Letter.

158.016 Two new Seyfert galaxies. E. E. Khachikian, D. W. Weedman.
Astron. Tsirk. No. 506, p. 1 - 2 (1969). In Russian.

158.017 On the fast variations of the optical emission from the nuclei of Seyfert galaxies and quasars. V. N. Kurilchik.
Astron. Tsirk. No. 520, p. 3 - 5 (1969). In Russian.

158.018 On the variability of the polarization of the radiation from the nuclei of Seyfert galaxies NGC 1275 and NGC 4151 and the N-type galaxy 3C 371. V. A. Hagen-Thorn, M. K. Babadzhanjanz.
Astron. Tsirk. No. 526, p. 1 - 3 (1969). In Russian.

158.019 Masses and densities of galaxies. I. L. Genkin, L. M. Genkina.
Trudy Astrofiz. Inst. Alma-Ata, Vol. 12, 82 - 89 (1969). In Russian.

The masses of galaxies determined by direct methods are given in a new catalogue. Mass-luminosity relations are derived. Holmberg's color-density relation is criticized.

158.020 Statistical investigation of masses and mean densities of galaxies. I. L. Genkin, L. M. Genkina.
Trudy Astrofiz. Inst. Alma-Ata, Vol. 12, 90 - 96 (1969). In Russian.

A statistical study on the masses of galaxies is carried out. The fundamental mean characteristics (mass, dimension, angular momentum, etc.) of various types are given.

158.021 Energetic characteristics of galaxies of Hubble's sequence. L. M. Genkina.
Trudy Astrofiz. Inst. Alma-Ata, Vol. 12, 97 - 105 (1969). In Russian.

Potential energy, kinetic energy of rotation, thermal kinetic energy, complete and specific moment of impulse are found for galaxies in the "middle" of Hubble's sequence.

158.022 Luminosity and mass functions for field galaxies of various morphological types. L. M. Genkina.
Trudy Astrofiz. Inst. Alma-Ata, Vol. 12, 106 - 112 (1969). In Russian.

Using data from the Catalogue of Radial Velocities of Galaxies by Humason, Mayall and Sandage the luminosity functions of field galaxies for different morphological types were constructed. With these luminosity functions and mass-luminosity relations mass functions were calculated. Average masses were determined for all types.

158.023 Possible methods of the definition of real eccentricity of elliptical galaxies. E. K. Denisyuk, O. A. Tumakova.
Trudy Astrofiz. Inst. Alma-Ata, Vol. 12, 125 - 149 (1969). In Russian.

The changes of eccentricity of isophotes as functions of the distance from the center are discussed in detail. These functions are derived for 68 elliptical galaxies from the Palomar Atlas Prints. Galaxies with considerable eccentricities have been found. The eccentricity function may be used sometimes to distinguish between giant and dwarf elliptical galaxies.

158.024 Photography of the faint outer regions of galaxies. S. van den Bergh.
Astrophys. Letters, Vol. 4, 117 - 119 (1969).

Experimental use of Kodak IIIaJ emulsion behind filters which reduce the sky brightness is described. New data are given on M 51 and on the Seyfert galaxy NGC 4151.

158.025 Catálogo de nebulosas extragalácticas de la zona $-5°/-25°$ de declinación, seleccionadas para la determinación de un sistema absoluto de movimientos propios estelares. I. – Identificación, descripción y valoración de las nebulosas. M. López Palacios.
Separate print Inst. y Obs. de Marina, San Fernando, 13 pp. (1969).

This paper contains a catalogue of 158 extragalactic nebulae identified on 48 $2° \times 2°$ plates obtained with the Carte-du-Ciel astrograph in the $-5°$ through $-25°$ zone of the Pulkovo Program for absolute proper motions. Measured coordinates relative to the plate centre, appearance and evaluation are given. Complete records consist of 144 plates, three for each centre supplied by Tashkent Observatory.

158.026 IC 3258, a small extragalactic object with a blueshift. E. M. Burbidge, M.-H. Demoulin.
Astrophys. Journ. *(Letters)*, Vol. 157, L155 - L156 (1969).

Observations of IC 3258 show it to have an emission-line spectrum with a blueshift of -490 km sec^{-1}. If it is in the Virgo cluster, it has a high velocity relative to the average for the cluster (it might possibly be a large gas cloud ejected from the radio galaxies M84 or M87). If it is a foreground galaxy, it is a low-luminosity dwarf with a large random velocity.

158.027 High-energy X-ray study of M-87.
J. E. McClintock, W. H. G. Lewin, R. J. Sullivan, G. W. Clark.
Nature, Vol. 223, 162 - 163 (1969).

This paper describes the balloon flight of a large area, orientable X-ray detector and an observation of M-87 in the energy range from 20 keV to 65 keV. An upper limit is reported on the flux from M-87 which is lower than a previously reported measurement. To a 1σ level of statistical confidence and including systematic errors, this upper limit at 43 keV is $F_{M87} < 0.44 \times 10^{-2}$ keV/keV-cm²-sec.

158.028 Colors, linear polarization, and preliminary mapping of the magnetic field for the outer filaments in the exploding galaxy M82. A. Sandage, N. Visvanathan.
Astrophys. Journ. Vol. 157, 1065 - 1074 (1969).

The surface brightness, colors, and linear polarization of selected regions in the outer filaments of M82 have been measured in optical wavelengths. The measured filaments are very faint, ranging in surface brightness from $V = 22.4$ to $V = 25.0$ mag per square second of arc. The $(U - B, B - V)$-colors are much bluer than the main body of M82. Large linear polarization is present for each region and reaches a maximum value of 32 percent for patch J. The direction of the electric vector is perpendicular to the two main filaments on the north side of M82. The data support, but do not prove, the hypothesis that the large-scale filaments are emitting optical synchrotron radiation caused by relativistic electrons.

158.029 Integral properties of spiral and irregular galaxies.
M. S. Roberts.
Astron. Journ. Vol. 74, 859 - 876 (1969).

General properties of and relations among a sample of 98 spiral- and irregular-type galaxies are derived and discussed. Masses derived from optical and 21-cm measurements are compared and found to be in good agreement. Hydrogen masses are found to be dependent on the inclination of the system, implying an optical depth effect. A statistical correction to these masses is applied. The average ratios of hydrogen mass to luminosity and hydrogen mass to total mass vary systematically with galaxy type. The ratio of total mass to luminosity does not appear to vary significantly with structural type over the range Sa through Ir. The overall average is 7.5. The hydrogen mass-luminosity, hydrogen mass-total mass, and total mass-luminosity relations for spirals and irregular systems are given. Surface densities of luminosity, hydrogen and total mass are tabulated, and their variation with structural type is shown.

158.030 Katalog von Galaxien und Galaxienhaufen.
F. Zwicky.
Sterne, 45. Jahrgang, 115 - 116 (1969).

158.031 On the problem of radioemission of normal galaxies.
I. I. Pronik.
Astron. Zhurn. Akad. Nauk SSSR, Vol. 46, 951 - 954 (1969). In Russian. English translation in Soviet Astron. AJ, Vol. 13, No. 5.

According to the data of 22 near galaxies the mass-luminosity relation in the radio region was obtained: $L_r \propto M^{2.1}$ for I and Sc galaxies; $L_r \propto M^{1.5}$ for Sb, S0 and E. The relation $L_r \propto L^2_{pg}$ exists for normal galaxies. In the radio region this relation smoothly turns into a more steep one.

158.032 On the radioemission of close galaxies at 3.5 cm.
V. N. Kurilchik, A. E. Andrievsky, V. N. Ivanov, E. E. Spangenberg.
Astron. Zhurn. Akad. Nauk SSSR, Vol. 46, 1124 - 1125

(1969).   In Russian. English translation in Soviet Astron. AJ, Vol. 13, No. 5.

Results of investigations of the radioemission of some nearby galaxies at 3.5 cm (8.5 GHz) are given.

158.033 A new mass – luminosity relation for elliptical galaxies. I. L. Genkin, L. M. Genkina.
Astron. Zhurn. Akad. Nauk SSSR, Vol. 46, 1128 - 1130 (1969).   In Russian. English translation in Soviet Astron. AJ, Vol. 13, No. 5.

A list of elliptical galaxies, the masses of which are determined by the method of Poveda and in a uniform system, is given. The homogeneity of the listed masses allows to conclude on the shape of the mass – luminosity relation.

158.034 Spectral observations of Markarian galaxies with ultraviolet continuum. II.
D. W. Weedman, E. Ye. Khachikian.
Astrofizika, Vol. 5, 113 - 122 (1969).   In Russian.
English translation in Astrophysics, Vol. 5, No. 1 (1969).

About 80% of 18 investigated galaxies of Markarian's list show emission spectra. The emission lines of hydrogen and forbidden lines of oxygen, neon, sulphur and others are present in the spectra of these galaxies. Four Seyfert galaxies were discovered. Three of these galaxies show very large redshifts and absolute magnitudes, which are larger than those of the other known Seyfert galaxies and are close to those of some quasars. Some arguments in favour of cosmogonic activity in the nuclei of some of the Markarian galaxies and in their nonthermal radiation are presented. There is some evidence in favour of a correlation of the ultraviolet continuum with the emission spectra.

158.035 The Andromeda galaxy M 31. Preliminary model.
J. I. Einasto.
Astrofizika, Vol. 5, 137 - 159 = Tartu Astron. Obs. Teated No. 22 (1969).   In Russian. – English translation in Astrophysics, Vol. 5, No. 1 (1969).

The proposed model consists of four components: the nucleus, bulg, disc, and flat component. The masses of the components were derived from the velocity data, collected from optical and radio sources. The velocity dispersion and the mass-to-light-ratio, spectroscopically obtained for the centre of M 31, were also used. It was found that the circular velocity curve has a maximum $V = 380$ km/sec at the distance of 4' from the centre. The rotational velocity of the spheroidal component equals to only 125 km/sec in this region. The dynamical mass-to-light-ratio 17.3 is in good agreement with the spectroscopical one, 16.7. For the mass of the galaxy M 31 a value of 200×10^9 solar masses is found.

158.036 A model of the distribution of mass in M 31. II.
V. S. Sizikov.
Astrofizika, Vol. 5, 317 - 329 (1969).   In Russian.
English translation in Astrophysics, Vol. 5, No. 2 (1969).

A model of M 31 in the form of a non-homogeneous spheroid is calculated. The law of rotation of a subsystem of neutral hydrogen by van de Hulst, Raimond and van Woerden and the photoelectric data by Vaucouleurs et al. are taken as the basis of calculation. Adopting the value of 630 kpc for the distance of M 31, ellipticity, density, mass-luminosity ratio, the mass as function of the major semiaxis of the spheroid and the spatial potential are calculated.

158.037 Possible variations of $\lambda = 10\,\mu$m radiation from NGC 4151.   W. A. Stein, F. C. Gillett.
Nature, Vol. 224, 675 - 676 (1969).
The nucleus of the Seyfert galaxy NGC 4151 was observed at $\lambda = 3.5\,\mu$m and $\lambda = 11.5\,\mu$m on June 21, 1969, with one of the 36-inch telescopes at Kitt Peak National Observatory. The purpose of the observations was to look for possible

variations in the infrared flux. The source has varied by about a factor of three over a period of time between one and two years.

158.038 Infrared observations of M82 and M31, between 5 and 25 microns. D. E. Kleinmann, F. J. Low.
Bull. American Astron. Soc. Vol. 1, 248 - 249 (1969). – Abstr. AAS.

158.039 NGC 3783, a Seyfert-type galaxy. T. Page.
Bull. American Astron. Soc. Vol. 1, 256 (1969). – Abstr. AAS.

158.040 Blue condensations associated with elliptical and S0 galaxies. A. Stockton.
Bull. American Astron. Soc. Vol. 1, 262 (1969). – Abstr. AAS.

158.041 The jet of M87 as a galactic flare. P. A. Sturrock.
Bull. American Astron. Soc. Vol. 1, 262 - 263 (1969). – Abstr. AAS.

158.042 Observations of the nucleus of M82. S. van den Bergh.
Bull. American Astron. Soc. Vol. 1, 264 (1969). – Abstr. AAS.

158.043 Large scale distribution of H II regions in galaxies. W. W. Morgan.
Interstellar Ionized Hydrogen, Charlottesville 1967, p. 565 - 569 (1968).

We propose to outline the behavior of the more luminous H II regions in the nearer giant galaxies. For this purpose we shall use the Yerkes form-classification of galaxies.

158.044 Spiral arm features and the large scale distribution of neutral hydrogen in galaxies.
M. S. Roberts.
Interstellar Ionized Hydrogen, Charlottesville 1967, p. 617 - 640 (1968).

The distribution of neutral hydrogen in extra-galactic systems may be derived for systems whose minor axis is comparable to or larger than the antenna half-power beamwidth. The large scale neutral hydrogen distribution may be described as a ring. The optically visible arms of a spiral galaxy are defined primarily by H II regions and early-type stars. For M 31 the arms are embedded within the H I ring. This is not the case for several Sc-type systems for which sufficient data are available to yield the parameters of the ring distribution. In these systems the spiral arm features are more centrally located than the H I ring.

158.045 The velocity dispersion and mass of M 87.
J. C. Brandt, R. G. Roosen.
Publ. Astron. Soc. Pacific, Vol. 81, 531 - 532 (1969). Abstract ASP.

158.046 The brightest globular clusters in galaxies as distance indicators. G. de Vaucouleurs.
Publ. Astron. Soc. Pacific, Vol. 81, 533 - 534 (1969). Abstract ASP.

158.047 Magnitudes and color indices of globular clusters in the Fornax system. G. de Vaucouleurs, H. D. Ables.
Publ. Astron. Soc. Pacific, Vol. 81, 534 - 535 (1969). Abstract ASP.

158.048 Spectroscopic observations of the Seyfert galaxy I Zw 1535 + 55. J. B. de Veny, C. R. Lynds.
Publ. Astron. Soc. Pacific, Vol. 81, 535 - 536 (1969).

Abstract ASP.

158.049 On the stability of the Local Group.
W. Herbst, Jr.
Publ. Astron. Soc. Pacific, Vol. 81, 619 - 628 (1969).

The dynamics of the Local Group is investigated using the virial theorem in detail, with various assumptions about the orbits and the binding of subgroups. It is concluded that the Local Group is stable against disruption without the assumption of "missing mass".

158.050 The peculiar galaxy NGC 6052.
D. L. du Puy, J. B. de Veny.
Publ. Astron. Soc. Pacific, Vol. 81, 637 - 642 = Contr. Kitt Peak National Obs. No. 470 = Contr. David Dunlap Obs. No. 235 (1969).

Observational results are presented for the galaxy NGC 6052. Spectrograms at 130 Å/mm were measured for radial velocities at six points on the galaxy. The results suggest that the differences in velocity ($\Delta V \simeq 135$ km/sec) are not due entirely to rotation; an outflow of material may be partially responsible.

158.051 The compact galaxy Zw I 0120 + 34°.
A. P. Fairall, R. J. Angione.
Publ. Astron. Soc. Pacific, Vol. 81, 685 - 687 = Contr. McDonald Obs. Univ. Texas, No. 448 (1969). – Note.

158.052 Internal motions in NGC 5128. J. L. Sérsic.
Nature, Vol. 224, 253 - 254 (1969).

Spectroscopic and interferometric observations of NGC 5128 are discussed. It is found that the velocity field of the gaseous component is rotational while the movement of the stellar component shows a discontinuity of the order of a thousand km/sec, coinciding with the northern rim of the dark belt.

158.053 Neutral hydrogen content of small galaxies.
L. Gouguenheim.
Astron. Astrophys. Vol. 3, 281 - 307 (1969).

The neutral hydrogen content of 49 galaxies has been measured with the Nançay radiotelescope. The systemic velocities agree with optical measurements. The velocity of the sun with respect to the local universe, established from the radial velocities of 37 galaxies less than 6 Mpc away from the sun but outside the local group, is $V_0 = 250 \pm 40$ km s^{-1} towards $l^{II} = 130° \pm 20°$, $b^{II} = -9° \pm 8°$; Hubble's constant is found to be $H = 77 \pm 11$ km s$^{-1} \times$ Mpc^{-1}. Correlations have been found between several physical parameters. Particularly, a relation between the distance, the angular diameter, and the width of the 21-cm line profile of a galaxy allows determination of distances within a factor 1.7; this method is used to determine distances of 6 galaxies.

158.054 A neutral hydrogen study of the interacting galaxies NGC 4631 and NGC 4656. L. Weliachew.
Astron. Astrophys. Vol. 3, 402 - 417 (1969).

The binary system NGC 4631/NGC 4656 was investigated in the 21-cm line of neutral hydrogen with the Nancay radiotelescope. The resolution was 4' in right ascension, 24' in declination, and 59 km/s in velocity. Data were fitted by hydrogen distributions and velocity models by means of a computer program. The two galaxies are connected by a hydrogen link already detected elsewhere. This study shows in both galaxies important departures from circular symmetry on the sides connected with the link. The interpretation of our data, together with recent optical results, leads to a model involving a hydromagnetic interaction between the two galaxies. This interaction may result from a close approach of both galaxies.

158.055   **Companion galaxies on the ends of spiral arms.**
H. Arp.
Astron. Astrophys. Vol. 3, 418 - 435 (1969).

Photographic and spectroscopic observations are presented which show that companion galaxies on the ends of spiral arms of normal galaxies tend to have (1) high-surface brightness, (2) emission lines characteristic of excited gaseous material, and (3) early-type stellar absorption lines in their nuclei. One companion is shown to be expanding. Another is shown to be probably receding from the center of the larger galaxy. The hypothesis advanced is that these companions have been recently ejected ($10^7 - 10^8$ years ago) from the parent galaxy. It is concluded that they are short-lived, and that many are now in the process of expanding and ejecting secondary material. It is further suggested that ejection of material through the disks of rotating galaxies is generally important in the formation of spiral arms.

158.056   **Long-term behavior of the Seyfert galaxy 3C 120.**
P. D. Usher, B. S. P. Shen, F. W. Wright, H. Shapley, C. M. Hanley.
Astrophys. Journ. Vol. 158, 535 - 539 (1969).

The blue photographic luminosity of the nucleus of the Seyfert galaxy 3C 120 decreased by nearly 1 mag over a period of 4 years. The long-term behavior of 3C 120 is similar to that of the quasi-stellar object 3C 273B and provides another link in the phenomenological description of the two types of object. The time scale for the variability is an order of magnitude larger than the longest previously observed at radio and optical frequencies.

158.057   **Compact radio source in the nucleus of M87.**
M. H. Cohen, A. T. Moffet, D. Shaffer, B. G. Clark, K. I. Kellermann, D. L. Jauncey, S. Gulkis.
Astrophys. Journ. (*Letters*), Vol. 158, L83 - L85 (1969).

The compact radio source in M87 has been observed with a long-base-line interferometer composed of the 210-foot telescope at Goldstone, California, and the 85-foot telescope at Tidbinbilla, near Canberra, Australia. The observations suggest that the radio source is remarkably concentrated to the galactic nucleus and has a linear diameter of about 3 light-months.

158.058   **Variability of the Seyfert galaxy NGC 1275.**
D. M. Selove.
Astrophys. Journ. (*Letters*), Vol. 158, L19 - L20 (1969).

A decrease of 0.5 mag in the light from the galaxy has been observed.

158.059   **On the nuclear region of M82.**   A. B. Solinger.
Astrophys. Journ. (*Letters*), Vol. 158, L21 - L24 (1969).

The position of the hypothetical Seyfert-like nucleus originally proposed to explain the polarization pattern in M82 is calculated anew in a more quantitative manner. To within the errors involved, the resulting coordinates coincide with those of the infrared nucleus recently observed in this galaxy. The observations are discussed and found to support the Thomson–scattering hypothesis. New observations are suggested that may specify the physical conditions in the nuclear region more completely.

158.060   **A blast-wave model for the explosion in the galaxy M82.**   A. B. Solinger.
Astrophys. Journ. (*Letters*), Vol. 158, L25 - L30 (1969).

Radio and optical data reviewed in a previous paper (Solinger, 1969) provide the basis for consideration of a new model for the explosion of the galaxy. A hydrodynamic flow associated with the release of a large amount of energy over a very short time and in a very small volume is investigated.

158.061   **Two double galaxy systems.**   B. M. Lewis.
Proc. Astron. Soc. Australia, Vol. 1, 288 - 289 (1969). – Contribution ASA meeting.

158.062   **Distribution of mass in galaxies from data of radial velocities. II. The model of NGC 7331.**
V. S. Sizikov.
Vestn. Leningr. un–ta, No. 1, p. 140 - 147 (1969). In Russian. Abstr. in Referativ. Zhurn. 51. Astron., 12.51.394 (1969).

158.063   **On the origin of arms in spiral galaxies.**
H. Arp.
Sky Telescope, Vol. 38, 385 - 387 (1969).

158.064   **History of our understanding of a spiral galaxy: Messier 33.**   K. J. Gordon.
Quarterly Journ. Roy. Astron. Soc. Vol. 10, 293 - 307 (1969). – Review article.

158.065   **Evidence for the ejection of matter from the nucleus of the Seyfert galaxy NGC 4151.**
K. S. Anderson, R. P. Kraft.
Astrophys. Journ., Vol. 158, 859 - 864 = Contr. Lick Obs. No. 301 (1969).

At least three shell lines of He I ($\lambda_0 3889$) arising from the metastable $2^3 S$ state are found in absorption in the spectrum of the nucleus of NGC 4151. Corresponding shell lines of hydrogen are found or suspected to exist in at least Hβ and Hγ. Velocities of ejection of up to –970 km sec^{-1} are measured. If the mass of the nucleus is $\lesssim 10^9 M_\odot$, much of the mass will be ejected to infinity at an estimated rate of between 10 and $10^3 M_\odot$ yr^{-1}.

158.066   **Formation of gas clouds in galactic nuclei.**
T. Arny.
Bull. American Astron. Soc., Vol. 1, 333 (1969). – Abstract AAS.

158.067   **Calculated [Fe X] and [Fe XIV] line strengths in a Seyfert-galaxy model.**
D. E. Osterbrock.
Bull. American Astron. Soc., Vol. 1, 357 (1969). – Abstr. AAS.

158.068   **The stellar velocity field in M51.**
S. M. Simkin.
Bull. American Astron. Soc., Vol. 1, 362 (1969). – Abstr. AAS.

158.069   **Absolute energy curves of elliptical galaxies.**
A. E. Whitford.
Bull. American Astron. Soc., Vol. 1, 369 (1969). – Abstr. AAS.

158.070   **Spectroscopic evidence for rotation of the ionized gas outside of the stellar region of M82.**
J. D. Wray, H. M. Heckathorn III.
Bull. American Astron. Soc., Vol. 1, 370 (1969). – Abstr. AAS.

158.071   **Erratum: "New infrared observations of the galaxy M 82"** [Astrophys. Journ. (*Letters*), Vol. 157, L29 (1969)].   M. I. Raff.
Astrophys. Journ. (*Letters*), Vol. 158, L64 (1969). – See Abstract 158.010.

158.072   **Certain properties of the nucleus of the radio galaxy NGC 1275 (Perseus A).**   E. A. Dibaj.
Astron. Zhurn. Akad. Nauk SSSR, Vol. 46, 725 - 729 (1969). In Russian. English translation in Soviet Astron. AJ, Vol. 13, No. 4.

From optical and radio data the physical conditions in the nucleus of the radio galaxy Perseus A are analysed. Conclusions are drawn concerning especially the motion and structure of gas.

158.073 H II zones in central regions of 9 normal galaxies.
I. I. Pronik.
Astron. Zhurn. Akad. Nauk SSSR, Vol. 46, 755 - 761 (1969). In Russian. English translation in Soviet Astron. AJ, Vol. 13, No. 4.

Spectra of the central regions of 9 normal galaxies are obtained with high-speed spectrographs of the 2.6 m telescope of the Crimean Astrophysical Observatory. Measured equivalent widths of Hα emission lines have shown a correlation with colour indices of central galactic regions.

158.074 The luminosity function of field galaxies.
M. A. Arakelian, A. T. Kalloglian.
Astron. Zhurn. Akad. Nauk SSSR, Vol. 46, 1215 - 1219 (1969).   In Russian. English translation in Soviet Astron. AJ, Vol. 13, No. 6.

The galaxies with known redshifts or photometric distances are used for construction of the luminosity function of the galaxies which are not members of rich clusters. The logarithmic luminosity function is presented by segments of three straight lines. The comparison of the integral luminosity function of field galaxies and that of rich clusters shows that the ratio of bright galaxies in rich clusters and in the field rises with the luminosity.

158.075 Spectroscopic and photographic observations of compact galaxies – II.   R. Barbon.
Mem. Soc. Astron. Italiana, Nuova Serie, Vol. 40, 559 - 574 = Contr. Kitt Peak National Obs. No. 510 (1969).

Symbolic velocities, distances, absolute luminosities and linear dimensions for eight compact galaxies from the lists issued by Zwicky are derived from image tube spectra and blue color photographs obtained at the Kitt Peak and Asiago observatories. Preliminary data on an unusual object, IIZw40, with a very strong emission spectrum are presented. Considerations on the Hubble diagram and the surface brightness of compact systems are also given.

158.076 Light curve of the N-type galaxy 3C371.
P. D. Usher, R. D. Cannon, M. V. Penston.
Observatory, Vol. 89, 198 - 201 (1969).

158.077 Photometric and spectroscopic observations of globular clusters in the Andromeda nebula.
S. van den Bergh.
Astrophys. Journ., Suppl. Series, Vol. 19, 145 - 174 (1969). Abstr. in Astrophys. Journ., Vol. 158, 1243 (1969).

The 200-inch telescope has been used to obtain classification spectra, radial velocities, and UBV photometry for the brightest clusters in M31. The principal results of this study are: 1. The average metallicity of globular clusters in the Andromeda nebula is significantly higher than it is in the Galaxy. 2. There is no clear-cut evidence for a dependence of cluster metallicity on position. Some quite strong-lined clusters occur far out in the halo of M31. 3. Most of the stars in the inner halo of M31 are not extremely metal-poor. Available evidence indicates that the globular clusters in M31 are systematically brighter than those in the Galaxy and in M87. This suggests that considerable caution should be exercised in using globular clusters to determine the scale of extragalactic distances. Six different methods of distance determination yield $\langle H \rangle = 104$ km sec^{-1}Mpc^{-1}.

158.078 Les noyaux galactiques.
V. A. Ambartzumian.
L'Astronomie, 83e année, 425 - 438 (1969).

158.079 The galaxies with ultraviolet continuum. II.
B. E. Markarian.
Astrofizika, Vol. 5, 443 - 459 (1969).   In Russian. – Engl. translation in Astrophysics, Vol. 5, No. 3.

The second list of galaxies with ultraviolet continuum is presented. It contains data for 130 objects, detected during the investigation of spectra of faint galaxies at the Byurakan observatory.

158.080 Seyfert galaxies.   T. Kwast.
Postępy Astron., Vol. 17, 267 - 279 (1969).
In Polish.

The main characteristics of Seyfert galaxies are: small bright nucleus in the centre; strong emission lines, the widths of which are up to 10000 km/sec. Additionally, these galaxies emit strong infrared radiation, some of them are radio sources or variable. Shape, colour indices and bolometric luminosities of the Seyfert galaxies are intermediate between these features for the normal galaxies and quasars. According to the cardinal model of the Seyfert galaxy high temperature and strong emission arise in it because of either fall of matter or explosion of a supernova. Radio and infrared radiation is due to synchrotron emission of relativistic electrons spiraling in a compressed magnetic field. However, no theory agrees with the observations.

158.081 Polarimetric investigation of the galaxy NGC 7814.
V. A. Hagen-Thorn.
Trudy Astron. Obs. *Leningrad,* Vol. 26 (= Uchenye Zapiski Leningr. Un-ta No. 347 = Seriya Matem. Nauk No. 44), p. 48 - 54 (1969).   In Russian.

The results of the photoelectric and photographic observations of the galaxy NGC 7814 are given. It is found that in the region of the dark band the polarization is similar to that in our Galaxy.

158.082 The luminosity functions of galaxies and of their nuclei.   E. M. Nezhinsky, L. P. Osipkov.
Trudy Astron. Obs. *Leningrad,* Vol. 26 (= Uchenye Zapiski Leningr. Un-ta No. 347 = Seriya Matem. Nauk No. 44), p. 92 - 96 (1969).   In Russian.

The luminosity functions for the nuclei of the galaxies from the catalogue of Deutsch are derived by a method of the authors published formerly. It is found that the luminosity functions for galaxies of different types are quite similar.

158.083 B and V photometry of the irregular galaxy NGC 1156.   S. D'Odorico.
Atti XII Riunione Soc. Astron. Italiana, L'Aquila 1968, p. 40 - 41 (1969). – Abstract SAI.

158.084 On the optical variability of nuclei of Seyfert galaxies.   V. B. Ljutiy.
Astron. Tsirk. No. 528, p. 1 - 3 (1969).   In Russian.

158.085 Masses and mass to light ratio for galaxies.
B. A. Vorontsov-Velyaminov.
Astron. Tsirk. No. 534, p. 1 (1969).   In Russian.

158.086 On the peculiarity of the radio emission spectra of the Seyfert galaxy NGC 1068.
V. N. Kurilchik.
Astron. Tsirk. No. 536, p. 6 - 8 (1969).   In Russian.

158.087 On the problem of evolution of radio galaxies.
Yu. K. Melik-Alaverdyan.
Dokl. AN Arm. SSR, Vol. 48, No. 3, p. 134 (1969). In Russian. – Abstr. in Referativ. Zhurn. 51. Astron., 1.51.777 (1970).

158.088 Kompakte Galaxien – eine bemerkenswerte neue

Kategorie von Sternsystemen.
N. Richter.
Jenaer Rundschau (Jena Review), 14. Jahrgang, 324 - 329 (1969).

158.089 Der Andromedanebel auf Tautenburger Schmidt-Aufnahmen. F. Börngen.
Jenaer Rundschau (Jena Review), 14. Jahrgang, 330 - 332 (1969).

158.090 Aspects of radio galaxy evolution.
H. van der Laan, G. C. Perola.
Astron. Astrophys. Vol. 3, 468 - 476 (1969).

The radio luminosity function and radio spectral data each provide estimates of radio galaxy lifetimes. The former because radio galaxies are identified with the class of super-luminous elliptical galaxies of known spatial density, the latter thanks to inevitable energy losses due to magnetobremsstrahlung and inverse–Compton radiation. These estimates, in combination, severely limit the range of conceivable ways in which radio galaxies may evolve. The following modes of evolution are considered in turn: (1) the single burst origin; (2) the continuous supply of relativistic electrons by inelastic collisions of cosmic ray protons and the ambient gas; (3) continuous or quasi-continuous injection with accumulation till radiative losses balance injection gains; (4) noncumulative injection, where diffusion is so rapid as to prevent radiative losses affecting the observed radio spectrum. Only (4) conforms to observational constraints. The analysis implies that conditions sufficient for the occurrence of violent energy releasing events prevail in the candidate galaxies for most of the Hubble time.

158.091 A study of physical groups of galaxies.
E. Holmberg.
Ark. Astron. Vol. 5, 305 - 343 = Uppsala Astron. Obs. Medd. No. 166 (1969).

This paper presents the results of an investigation of 174 physical groups of galaxies, most of them presumably comparable to the Milky Way and M31 groups (= Local Group), and the M81 group. The groups selected are centered on prominent spiral galaxies, for which the distance moduli can be estimated. The survey work has been based on the Palomar Sky Atlas, the prints being evaluated down to the practical limit, as regards galaxies; the limiting diameter of the group members is 0.6 kpc, and the limiting absolute pg magnitude about –10.6. On account of the large disturbances from the background-foreground fields it has been necessary to restrict the survey to circular areas with a radius of 50 kpc around the central spiral galaxies. Areas of this size probably include about 30% of all the satellites. A summary of the observational data is given.

158.092 The color indices of the globular clusters of the Fornax dwarf galaxy. P. W. Hodge.
Publ. Astron. Soc. Pacific, Vol. 81, 875 - 878 = Contr. Cerro Tololo Inter-American Obs. No. 98 (1969).

New measurements of the UBV color indices of five globular clusters in Fornax indicate that they have normal values and that the previous report of abnormal color indices is in error. A total absolute magnitude of $M_V \simeq -7.0$ is derived for a sixth cluster, which appears to be a normal globular cluster.

158.093 Contribution à l'étude des galaxies singulières.
M.-H. Demoulin.
Thesis Sci. Phys., Paris [Centre Document., Centre National de la Recherche Scientifique, No. 3438], 74 pp. (1969).

158.094 Masses et moments angulaires des galaxies.
N. Heidmann.
Thesis Sci. Phys., Paris. [Centre Document., Centre National

de la Recherche Scientifique, No. 2716], 148 pp. (1968). – See Bull. Signal., Vol. 30, Section 120, No. 14580 (1969).

158.095 Extragalaktische Forschungen.
W. Bronkalla, H. Oleak, H.-J. Treder, P. Notni.
Jenaer Rundschau (Jena Review), 13. Jahrgang, p. 322 - 325 (1968).

158.096 Kinematics of NGC 5128.
J. L. Sérsic, G. Carranza.
Inform. Bull. Southern Hemisphere, No. 14, p. 32 - 38 (1969).

The observations discussed were made with an interferometer Perot-Fabry atached to the 1.54-m reflector at Bosque Alegre. The radial velocities were computed by comparison with suitable comparison Hα rings.

158.097 The subdivision of galaxies into spherical and flat subsystems. L. E. Gurevich, A. D. Chernin.
Trudy 6. Vses. ezhegodn. zimn. shkoly po kosmofiz., 1969. Ch. (Part) 1. Apatity, p. 32 - 34 (1969).   In Russian.
Abstr. in Referativ. Zhurn. 51. Astron., 2.51.734 (1970).

158.098 On the emission lines in the optical spectra of nuclei of radio galaxies.
Yu. K. Melik-Alaverdyan.
Dokl. AN Arm. SSR, Vol. 48, 212 - 214 (1969).   In Russian.
Abstr. in Referativ. Zhurn. 51. Astron., 2.51.764 (1970).

158.099 On the origin of emission lines in the optical spectra of nuclei of radio galaxies.
Yu. K. Melik-Alaverdyan.
Dokl. AN Arm. SSR, Vol. 48, 277 - 279 (1969).   In Russian.
Abstr. in Referativ. Zhurn. 51. Astron., 2.51.765 (1970).

158.100 Neutraler Wasserstoff in den Galaxien NGC 253 und M33. W. Huchtmeier.
Mitt. Astron. Ges. No. 27, p. 133 - 134 (1969). – Abstract AG.

158.101 Der Einfluß nichtthermischer Strahlung auf die Emissionslinienspektren von Seyfert-Galaxien und Quasaren. M. Grewing.
Mitt. Astron. Ges. No. 27, p. 139 - 143 (1969). – Conference paper.

Star formation in clouds of solid hydrogen grains – II. Some properties of model galaxies.
See Abstr. 065.038.

A photoelectric sequence near NGC 5128.
See Abstr. 113.016.

Ultraviolet spectrophotometric observations with the Wisconsin experiment package on OAO-2.
See Abstr. 114.037.

The Bowen fluorescence mechanism in planetary nebulae and the nuclei of Seyfert galaxies.
See Abstr. 133.010.

Galactic nuclei as collapsed old quasars.
See Abstr. 141.006.

The redshift of the N-type radio galaxy 3 C 371.
See Abstr. 141.019.

Quasars and nuclei of galaxies.
See Abstr. 141.032.

Radio sources and elliptical galaxies.
See Abstr. 141.037.

A limit on the variability of the 22 MHz radiation from 3C 84 (NGC 1275).   See Abstr. 141.068.

Periodic clustering of red-shifts in the spectra of quasi-stellar and other unusual objects.
See Abstr. 141.084.

Possible origin of absorption lines in the spectrum of QSO's.   See Abstr. 141.165.

The reddening of M3, M13, M31, and M33 from photometry of late-type field stars.
See Abstr. 154.004.

A comparison of some characteristics of field galaxies and galaxies in clusters.   See Abstr. 160.005.

Photoelectric spectrophotometry of B264.
See Abstr. 160.009.

Infall of gas from intergalactic space.
See Abstr. 161.008.

159 Magellanic Clouds

159.001 **Proven and probable members in the wing of the Small Magellanic Cloud.** N. Sanduleak.
Astron. Journ. Vol. 74, 877 - 878, 973 - 975 = Contr. Cerro Tololo Inter-American Obs. No. 80 (1969).

An objective-prism survey of the wing of the Small Magellanic Cloud was made using the Curtis Schmidt telescope on Cerro Tololo in Chile. An ultraviolet-transparent prism was used to photograph spectra of stars as faint as $m_{pg} \sim 13.5$ with a dispersion of 580 Å/mm at Hγ. A total of 47 stars showing definite or suspected high-luminosity spectral characteristics were found in the region of the wing. Of these, nine are known SMC members. Identification charts are provided.

159.002 **Erratum: Two-color composite photographs of the Magellanic Clouds.** [Astron. Journ. Vol. 74, 44 (1969)]. M. F. Walker, V. M. Blanco, W. E. Kunkel.
Astron. Journ. Vol. 74, 964 - 972 (1969).

159.003 **The probable origin of the Magellanic Clouds.** N. J. Rumsey.
Southern Stars, Vol. 23, 76 - 77 (1969).

159.004 **Two-colour photometry of Cepheids near the centre of the Small Magellanic Cloud, based on plates taken by Dr. H. C. Arp.** A. M. van Genderen.
Bull. Astron. Inst. Netherlands, Suppl. Series, Vol. 3, 221 - 298 = Commun. Obs. Leiden (1969). − Abstr. in Bull. Astron. Inst. Netherlands, Vol. 20, 335 (1969).

In the central region of the Small Magellanic Cloud 62 new variables have been discovered of which coordinates and finding charts are presented. On the basis of photo-electric standards by Arp photographic B and V systems were established. In these systems magnitudes and light-curves were derived for 105 Cepheids (including five with $P < 1^d$), two long-period variables and one eclipsing binary. The mean blue absorption in the central region amounts to 0.7 mag as against 0.2 mag in the northern part. The mean slope of the lines of constant period in the HR diagram is found to be 0.36, in good agreement with the value found from the relation $P\sqrt{p} = Q$ and the slope of the evolutionary path of the young SMC clusters NGC 330 and NGC 458. After correction for absorption the Cepheids fit well into the instability gaps of these young clusters. The total number of Cepheids in the Small Magellanic Cloud is estimated to be about 2000.

159.005 **A comparative photometric study of population I Cepheids in the Magellanic Clouds, the galaxy and the Andromeda nebula.** A. M. van Genderen.
Bull. Astron. Inst. Netherlands, Suppl. Series, Vol. 3, 299 - 326 = Commun. Obs. Leiden (1969).

A discussion is presented of BV photometry of population I Cepheids in the Magellanic Clouds, the galaxy and M31, collected from literature. Differences and resemblances in observed, theoretical and semi-theoretical P–L relations, in intrinsic colours and in other characteristics are discussed. The foreground absorption to M31 and the distance moduli of SMC, LMC and M31 are determined. It appears that the Cloud Cepheids are more luminous and bluer than the galactic and M31 Cepheids. Further a discussion is presented of other differences in the physical properties of the Cepheids in these four galaxies.

159.006 **Two strange clusters in the Large Magellanic Cloud.** S. Gaposhkin.
Bull. American Astron. Soc. Vol. 1, 241 (1969).−Abstr. AAS.

159.007 **Five-colour photometry of 12 cepheids in the Small Magellanic Cloud.** A. M. van Genderen.
Bull. Astron. Inst. Netherlands, Vol. 20, 317 - 332 (1969).

A discussion is presented of photo-electric five-colour observations of 12 SMC cepheids made in 1966. The periods of these variables range from 8 to 88 days. Their colour-curves, loops and intrinsic colours have been compared with those of galactic cepheids. The characteristics of their colour-curves differ in many respects from those for the galactic cepheids, and their intrinsic colours are on the average bluer by 0.06 in $V - B$, 0.13 in $B - L$ and 0.03 in $B - U$ (all in log intensity). A metal deficiency of the SMC cepheids could be the explanation.

159.008 **Further observations of Magellanic Cloud cepheids.** S. C. B. Gascoigne.
Monthly Notices, Roy. Astron. Soc., Vol. 146, 1 - 36 (1969).

Two-colour light-curves have been obtained, mostly by photoelectric photometry, for 13 SMC and 7 LMC cepheids, with periods in the range 1.24 to 6.69 days. P–L and P–C relations were derived. Reddening has been estimated from the colours of field B-stars, mostly faint. We found 0.02 for the SMC and 0.05 for the LMC. With this reddening the SMC cepheids are in the mean about 0.1 mag bluer than those in the galaxy, as opposed to the 0.2 found previously. Some implications of the above results are briefly discussed.

159.009 **Two stars of the Large Magellanic Cloud showing emission lines of Fe II and [Fe II].** C. Fehrenbach, E. Maurice, L. Prévot, M. Petit.
Astron. Astrophys. Vol. 3, 323 - 326 (1969). In French.

Many multiplets of Fe II and [Fe II] are identified in the emission spectrum of two stars in the Large Magellanic Cloud (S 22 of Henize and HDE 269217). A systematic difference between the radial velocities of the iron and hydrogen emission lines can be explained by an expansion of the Fe II layer. HDE 269217 shows also the hydrogen lines in absorption from H_8 to H_{20}. The expansion of the Fe II layer is larger and the expansion of the H absorption layer even more.

159.010 **A method for determining Magellanic Cloud membership using an objective prism and an absorption filter.** C. J. Butler, M. V. Norris.
Monthly Notes Astron. Soc. Southern Africa, Vol. 28, 107 - 114 (1969).

95 stars are examined for membership of the Large Magellanic Cloud on the basis of their radial velocities. The method used is that of an objective prism in conjunction with a neodymium chloride filter. Radial velocity is not quantitatively determined, but Cloud members and foreground stars divide into two separate groups in a frequency distribution.

159.011 **A method of distinguishing Magellanic Cloud membership.** P. A. Wayman.
Dunsink Obs. Repr. No. 54, [Reprinted from I.A.U. Symposium No. 30, p. 89 - 90], 2 pp. (1967).

159.012 **Polarisationsmessungen an den Magellanschen Wolken.** T. Schmidt.
Mitt. Astron. Ges. No. 27, p. 158 - 160 (1969). − Abstract AG.

The influence of metal content on the evolution of stars of five solar masses. See Abstr. 065.071.

On the interpretation of S Doradus. See Abstr. 122.098.

160   Clusters of Galaxies

160.001   Groups and clusters of southern galaxies.
A. R. Klemola.
Astron. Journ. Vol. 74, 804 - 806 (1969).

Descriptions of 44 probable groups and clusters of southern galaxies are given. These results are based upon examination of direct photographs taken with the Yale–Columbia 20-inch double astrograph.

160.002   On the luminosity function of the cluster of galaxies A 262.   A. T. Kalloghlian.
Soobshch. Byurakan. Obs. No. 40, p. 3 - 14 (1969).
In Russian.

Photographic magnitudes of all galaxies up to $18.^m0$ are determined in Abell's cluster No. 262, and its luminosity function is derived. This function does not increase monotonously but has a maximum or a flattening near 16^m and is then rising again. This is also the case for the distribution of the diameters. On the Palomar Sky Survey prints the galaxies up to 0.2 mm in diameter are counted, and the distribution function is given.

160.003   On the radio emission of clusters of galaxies.
H. M. Tovmassian, R. G. Mnatsakanian.
Soobshch. Byurakan. Obs. No. 40, p. 46 - 56 (1969).
In Russian.

136 clusters of galaxies which belong to the 5th distance group in Abells list were searched for radio emission and studied on the Palomar Sky Survey. Radio emission was detected in 26 clusters. In some cases the radio emitter is a giant D type or a peculiar galaxy, and in other cases the radio emitter was found to be a N type galaxy.

160.004   Dynamical evolution of clusters of galaxies – III.
S. J. Aarseth.
Monthly Notices, Roy. Astron. Soc., Vol. 144, 537 - 548 (1969).

Previous numerical studies of simulated galaxy clusters are extended to include rotation. The initial velocity distribution contains 26 retrograde and 74 direct orbits. The inclusion of rotation shortens the mean relaxation time and gives rise to a significant flattening. A large proportion of positive angular momentum is transferred to the halo but the flattening remains frozen in. After a total integration time of 139 initial crossing times the cluster contains 19 bodies representing 37.5 per cent of the total mass and only 13 per cent of the angular momentum. An upper limit of 370 initial crossing times is estimated for the time to lose all non-nuclear members inside $r \simeq 10$ by replacing the compact nucleus with one central body. Six further cases with different types of rotation have also been studied in order to explore the initial condition dependence.

160.005   A comparison of some characteristics of field galaxies and galaxies in clusters.   A. V. Zasov.
Astron. Tsirk. No. 520, p. 1 - 3 (1969).   In Russian.

160.006   On the kinematics of the Coma cluster of galaxies.
R. H. Gainullina.
Trudy Astrofiz. Inst. Alma-Ata, Vol. 12, 113 - 124 (1969).
In Russian.

Redshifts of 26 galaxies which are probably members of the Coma cluster are compatible with the hypothesis that the cluster members within 100' of the center rotate about the center on planar orbits slowly moving away. Mean masses of cluster members, rotation periods at a distance of 72' from the center and a lower limit for the age of the Coma cluster are estimated.

160.007   Superclusters of galaxies?
J. T. Yu, P. J. E. Peebles.
Astrophys. Journ. Vol. 158, 103 - 113 (1969).

We have analyzed the distribution in the sky of the "rich," "compact" clusters of galaxies, by means of a new statistical approach, in an attempt to find an independent test of the existence of superclusters in these data. Our results suggest that any tendency toward superclustering is scarcely above that to be expected for a random distribution. On the other hand, our results cannot rule out the existence of some superclusters. If there exist superclusters of the kind proposed by Abell, we place an upper limit of 10 percent on the possible fraction of the great clusters that might be found in such superclusters.

160.008   Brightest members of clusters of galaxies.
J. V. Peach.
Nature, Vol. 223, 1140 - 1142 (1969).

The absolute V magnitudes of 38 brightest members of clusters of galaxies have been analysed in the context of the statistical theory of Peebles. It is shown that this theory cannot explain the small dispersion of absolute magnitudes if Abell's measured slope for the bright end of the Coma luminosity function applies to all clusters; and that the distribution of the magnitudes of the brightest cluster members is compatible with the assumption that the absolute magnitudes are normally distributed and independent of cluster richness.

160.009   Photoelectric spectrophotometry of B264.
J. B. Oke.
Astrophys. Journ. (*Letters*), Vol. 158, L9 - L10 (1969).

The Braccesi object B264, which has been shown to be a member of a cluster of galaxies, is virtually indistinguishable in luminosity and spectral characteristics from the Seyfert or N-type radio galaxy 3C 120. Apart from luminosity, it is also very similar to the quasi-stellar source 3C 323.1.

160.010   Clustering of clusters of galaxies.
T. Kiang, W. C. Saslaw.
Bull. American Astron. Soc. Vol. 1, 247 (1969). – Abstr. AAS.

160.011   Age of the Coma cluster of galaxies from tidal-limited galaxian sizes.   T. W. Noonan.
Bull. American Astron. Soc. Vol. 1, 255 (1969). – Abstr. AAS.

160.012   The case of the missing mass.
G. R. Burbidge, W. L. W. Sargent.
Comments Astrophys. Space Phys. Vol. 1, 220 - 225 (1969).

A major problem of modern astronomy is the question of the true mass-energy content of the universe. In this discussion we shall survey the problems associated with clusters of galaxies. Three basic methods have been used to determine the mass of individual galaxies. They are: (a) Rotation curves, (b) Internal velocity dispersions, (c) Pairs of galaxies.

160.013   Intrinsic dispersion in the dimensions of clusters of galaxies and the angular diameter-redshift relation.
J. V. Peach, J. M. C. Beard.
Astrophys. Letters, Vol. 4, 205 - 206 (1969).

An estimate of the intrinsic dispersion in the diameters of rich clusters of galaxies indicates that the use of the angular diameter-redshift relation applied to rich clusters is a practicable method of determining the deceleration parameter, comparable in precision with the redshift-magnitude relation applied to brightest cluster members.

160.014 Significance of the first brightest galaxies in rich clusters. B. A. Peterson.
Publ. Astron. Soc. Pacific, Vol. 81, 549 (1969). – Abstract ASP.

160.015 The structure of the Virgo cluster as determined from supernovae. C. T. Kowal.
Publ. Astron. Soc. Pacific, Vol. 81, 608 - 612 (1969).

The apparent magnitudes of supernovae are used to determine the relative distances of members of the Virgo cluster of galaxies. It is found that the distances so derived differ markedly from the distances required by a multiple-group interpretation of the structure of the cluster. It is suggested that the cluster is essentially a single dynamical unit, and that the differences in galaxy velocities and luminosities, which were observed by de Vaucouleurs, must be explained by means other than differences in the distances of the galaxies.

160.016 Collapsed objects in clusters of galaxies. S. van den Bergh.
Nature, Vol. 224, 891 (1969).

Observations of tidally distorted galaxies in the Virgo cluster are used to show that collapsed objects, with masses in the range $10^8 M_\odot$ to $10^{12} M_\odot$, cannot account for the large cluster mass derived from the virial theorem.

160.017 Luminosity function and color-magnitude diagram of galaxies in a central zone of the Coma cluster. H. J. Rood.
Astrophys. Journ. Vol. 158, 657 - 667 (1969).

New photovisual magnitudes for 315 galaxies in a central region of the Coma cluster are presented. These data are combined with magnitudes and colors from several sources to construct luminosity functions and color-magnitude diagrams. Data for the Coma and Virgo clusters and for elliptical S0, and SB0 galaxies are compared. Evidence is discussed which indicates that, in the Coma core, concentration toward the cluster center tends to increase with increasing luminosity. This may be a consequence of partial dynamical relaxation.

160.018 Zwicky's Corona Borealis cluster counts. T. W. Noonan.
Astron. Journ. Vol. 74, 1105 - 1107 (1969).

A method for making allowance for an inhomogeneous background in galaxy-cluster counts is suggested and applied to a particular cluster. Zwicky's counts of galaxies within rings centered on the Corona Borealis cluster indicate the presence of 225 cluster members within a radius of about 30′ to the limit of the 18-inch Schmidt telescope ($m_{pg} \approx 19.2$), and of 850 members within a radius of about 50′ to the 48-inch Schmidt limit ($m_{pg} \approx 21.4$). There is a pronounced tendency for the brighter members to concentrate toward the cluster center.

160.019 Bright end of the galaxy luminosity function in clusters. P. J. E. Peebles.
Nature, Vol. 224, 1093 (1969).

The brightest members of clusters of galaxies have remarkably uniform absolute magnitudes. Two of the simplest possible interpretations are (1) the brightest member of a cluster is a special object, and it's luminosity is independent of the rest of the cluster; (2) the luminosity of the brightest member is determined by the same process that fixed the luminosities of all the other members, so in particular the luminosity of the brightest member is determined in a statistical manner by a continuous luminosity function. It is shown that the data published so far are consistent with the second hypothesis. Some tests are mentioned that might help rule out one or both of these simple ideas.

160.020 Investigation of the distribution of clusters of galaxies. A. I. Gusak.
Astron. Zhurn. Akad. Nauk SSSR, Vol. 46, 1231 - 1235 (1969). In Russian. English translation in Soviet Astron. AJ, Vol. 13, No. 6.

The work consists of a statistical investigation of the distribution of clusters of galaxies on the basis of data from Zwicky's "Catalogue of Galaxies and Clusters of Galaxies", (Vol. I and II). The result shows the existence of second order clustering. Coordinates of several possible second order clusters are given.

160.021 Clusters of galaxies, background density and segregation. B. I. Gorbachev.
Astron. Zhurn. Akad. Nauk SSSR, Vol. 46, 1321 - 1324 (1969). In Russian. English translation in Soviet Astron. AJ, Vol. 13, No. 6.

From published data for clusters in the constellations Com, Cnc, Hya, Per, and in the cluster Zw 97-8 distributions of surface densities of a number of galaxies in clusters are constructed; space densities are counted. The influence of accuracy of considering the background on the characteristics of the cluster is tested. The variation of the ratio of space densities of bright galaxies to faint ones rises with increasing population of the cluster.

160.022 Dependence between the mean density of clusters of galaxies and their morphological type. L. M. Ozernoy.
Astron. Tsirk. No. 536, p. 1 - 4 (1969). In Russian.

Are some quasi-stellar objects associated with clusters of galaxies? See Abstr. 141.025.

Radio emission of quasi-stellar objects. See Abstr. 141.091.

Cosmological implications of the possible non-existence of a second-order clustering of the galaxies. See Abstr. 162.002.

161 Intergalactic Matter

161.001 Search for a neutral-atomic-hydrogen link between M31 and M33. K. J. Gordon.
Astrophys. Letters, Vol. 4, 47 - 49 (1969).

An upper limit of 5×10^{19} atoms cm^{-2} is established for the projected surface density of neutral atomic hydrogen between the galaxies M31 and M33.

161.002 Intergalactic hydrogen.
P. J. E. Peebles.
Astrophys. Journ. Vol. 157, 45 - 54 (1969).

If the Universe contains an appreciable amount of atomic hydrogen not bound up in ordinary galaxies, it may be reasonable to suppose that this material is not uniformly distributed, but is instead incorporated in gravitationally bound clouds that happen to have been unable to complete the course of evolution to star systems. In this case, the available observational limits on the mean mass density due to intergalactic hydrogen must be corrected for the effects of self-absorption within the clouds. Taking this effect into account, the Gunn-Peterson test limits the abundance of atomic hydrogen to about 3 per cent of that needed to close the Universe, unless the clouds have unusual and perhaps unlikely properties. The more direct tests based on 21-cm emission and absorption apparently would permit the contribution by atomic hydrogen to be as large as 30 per cent of the critical mass density.

161.003 The possibility of existence of a dense intergalactic plasma. J. Bergeron.
Astron. Astrophys. Vol. 3, 42 - 56 (1969). In French.

The implication of the existence of a dense, hot, intergalactic plasma is discussed for Friedmann universes with a zero cosmological constant. The points of interest are the temperature of the intergalactic gas and the degree of ionization of its constituents. The most important observations to test the validity of the model are the isotropic X-ray emission around 3 keV and around 0.27 keV, and the lack of Lyman a absorption in the emission spectra of the QSO's. We take the intergalactic gas to be 90% hydrogen and 10% helium. The model assumes that the intergalactic medium is heated, and that the heating occupies a time short compared to the lifetime of the universe.

161.004 The spectrum of metagalactic inhomogeneities and peculiar velocities. L. M. Ozernoy.
Astron. Tsirk. No. 527, p. 4 - 7 (1969). In Russian.

161.005 Low-frequency intergalactic radio absorption theory and an experimental upper limit.
P. D. Noerdlinger.
Astrophys. Journ. Vol. 157, 495 - 505 (1969).

The theoretical values for free-free intergalactic absorption at low-frequency radio wavelengths as a function of redshift z are given for a wide variety of uniform, zero-pressure cosmological models. The dependence of absorption on redshift falls well below a linear law by the time $z = 2$, except for models of large negative deceleration parameter q_0. By seeking any secular tendency of the more distant radio sources with straight spectra down to 26.3 MHz to show absorption at 10 MHz, an upper limit of 0.15 is set on the optical depth at 10 MHz and $z = 2$. This is sufficient to exclude a few low-temperature, high-density cosmological models in which the gas is kept radiatively ionized.

161.006 The persistence of galactic H I irradiated by a hot intergalactic medium.
J. E. Felten, J. Bergeron.

Astrophys. Letters, Vol. 4, 155 - 157 (1969).

A recent argument that the neutral hydrogen observed in external galaxies precludes the existence of an intergalactic medium at $T_0 \sim 10^6 \,°K$ and $\rho_0 \sim 3H_0^2/8\pi G$ is incorrect. The ionizing flux can produce H II regions of emission measure $\sim 10^{-1} - 10^{-2} \, cm^{-6}$ pc above and below the H I disk of a galaxy. These regions are discussed in the Strömgren approximation; sample dimensions and densities are given.

161.007 The interaction of the ionizing background radiation with galaxies. Limits to the density of the intergalactic gas. R. A. Sunyaev.
Astron. Zhurn. Akad. Nauk SSSR, Vol. 46, 929 - 934 (1969). In Russian. English translation in Soviet Astron. AJ, Vol. 13, No. 5.

The background X-ray emission influences physical conditions in the interstellar medium by heating and ionizing. Owing to this heating, the formation of clouds of neutral hydrogen in the space between spiral arms and at the periphery of galaxies is impossible. The existence of helium and heavy elements in the interstellar gas increases the absorption of X-ray quanta. The observed distribution of neutral hydrogen at the periphery of galaxies contradicts the existence of intergalactic gas with a density exceeding one third of the critical one.

161.008 Infall of gas from intergalactic space.
J. H. Oort.
Nature, Vol. 224, 1158 - 1163 (1969).

There are several indications that a considerable flux of gas is constantly flowing into the galactic system, and that intergalactic gas probably makes up a large fraction of the mass of the universe. This discussion of the phenomena involved is a modified version of the Nuffield Lecture given at the Institute of Theoretical Astronomy, Cambridge, on July 7, 1969.

161.009 Intergalactic hydrogen along the path to Virgo A.
R. J. Allen.
Astron. Astrophys. Vol. 3, 382 - 387 (1969).

The continuum radio spectrum of Virgo A near 21 cm wavelength appears to be a featureless sloping line from -700 to $+1800$ km s^{-1}. If neutral atomic hydrogen is present in intergalactic space in quantities which are dynamically significant both for the Virgo cluster and for the deceleration of the universe, then some evidence of its absorption effects should have been found. The emission and absorption measurements may be combined to give the following results: If matter is present in amounts sufficient to gravitationally stabilize the cluster, then less than 1% of it is in the form of hydrogen atoms in the ground electronic state. If the density of hydrogen gas in intergalactic space is greater than 10^{-6} atoms cm^{-3}, then it must be more than 70% ionized regardless of its kinetic temperature. These two conclusions are dependent on a simple model for the spatial distribution of neutral hydrogen in the Virgo cluster and in the space between our galaxy and Virgo A.

161.010 Interaction of cosmic gamma rays with intergalactic matter. J. Arons, R. McCray.
Astrophys. Journ. (*Letters*), Vol. 158, L91 - L95 (1969).

X-rays and gamma rays emanating from large redshifts interact with intergalactic matter by Compton scattering and pair production. The maximum redshifts to observe discrete sources and diffuse background are calculated for evolutionary cosmological models of present densities 10^{-5} and 10^{-7} atoms cm^{-3}. Heating by the observed background and minimum equilibrium temperatures of the medium are estimated.

161.011   **Search for the intergalactic extinction.**
                R. Brukalska.
Acta Astron. Vol. 19, 301 - 306 (1969).

The numbers of very distant clusters in fields with the largest near and medium distant clusters in each volume of Zwicky's Catalogue are interpreted in terms of intergalactic extinction.

161.012   **Etude des possibilités d'existence de matière inter-galactique locale.**   J. Bergeron.
Thesis, 3e Cycle, Spéc. Astrophys., Paris. 31 pp. (1968). — See Bull. Signal., Vol. 30, Section 120, No. 9833 (1969).

A neutral hydrogen study of the interacting galaxies NGC 4631 and NGC 4656.   See Abstr. 158.054.

162 Structure and Evolution of the Universe, Cosmology

162.001 **Axially symmetric hot big-bang cosmologies.**
R. F. Carswell.
Monthly Notices, Roy. Astron. Soc., Vol. 144, 279 - 296 (1969).
An analysis of homogeneous axially symmetric anisotropic cosmological models containing radiation, matter and a uniform magnetic field is given. The results obtained are used to discuss the effect of magnetic fields and initial anisotropies on the isotropy of the 3°K background radiation and on the primordial helium production. It is shown that some anisotropic models with a small present day magnetic field and small microwave background anisotropy could have produced as little as about 5 per cent He^4 by mass in the early stages of their evolution.

162.002 **Cosmological implications of the possible non-existence of a second-order clustering of the galaxies.**
A. Gerasim.
Astrophys. Letters, Vol. 4, 51 - 54 (1969).
It is argued that if the non-existence of a second-order clustering of the galaxies, asserted by Zwicky, were interpreted as being due to a gravitational cut-off, there could not be such intense gravitational fields as are usually claimed by the relativistic cosmologies. The universe would be a quasi-Galilean-Euclidean one. The available observational data seem to suggest that in such a universe the light would be redshifted by an exponential decay process of the energy of the photon. Some consequences and possible objections are discussed.

162.003 **On the formation of condensation in an expanding universe.** M. Kondo.
Publ. Astron. Soc. Japan, Vol. 21, 54 - 66 (1969).
General solutions with arbitrary initial conditions are obtained analytically for the case in which the disturbance is one-dimensional and the wavelength is much longer than the Jeans wavelength. There are two families of solutions, one of which indicates a maximum-density region of perturbation undergoing free fall while the other indicates the opposite character. The density in the free-fall part becomes infinite within a finite time which is given as a function of initial conditions. There is no essential difference in the characteristics of the density-velocity phase plane between the case of the expanding universe and that of the static Newtonian universe.

162.004 **The interaction of matter and radiation in a hot-model universe.**
Ya. B. Zeldovich, R. A. Sunyaev.
Astrophys. Space Sci. Vol. 4, 285 - 316 (1969).   In Russian and English.
In this paper we continue the investigation initiated by Weymann as to the reason why the spectrum of the residual radiation deviates from a Planck curve. We shall consider the distortions of the spectrum resulting from radiation during the recombination of a primeval plasma. Analytical expressions are obtained for the deviation from an equilibrium spectrum due to Compton scattering by hot electrons. On the basis of the observational data it is concluded that a period of neutral hydrogen in the evolution of the universe is unavoidable. It is shown that any injection of energy at $t > 10^{10}$ sec (red shift $z < 10^5$) leads to deviation from an equilibrium spectrum.

162.005 **Density perturbation and preferential coordinate systems in an expanding universe.** K. Sakai.
Progress Theoret. Phys. Japan, Vol. 41, 1461 - 1469 (1969).
Under general coordinate conditions, the equation for density perturbation in an expanding universe has fictitious solutions. In order to exclude the fictitious solutions automatically, we adopt coordinate systems moving with the average distribution of matter; and we obtain some coordinate conditions which provide for these systems. Under these conditions, the equation for spatially periodic density perturbation becomes a second-order differential equation with respect to time and reduces to Bessel's differential equation when the equation of state is of the form p/ϵ = const (p = pressure, ϵ = energy density).

162.006 **Effect of various cross-section energy dependences in calculations of cosmological viscosity.**
R. A. Matzner.
Astrophys. Space Sci. Vol. 4, 459 - 463 (1969).
The Einstein Field Equations for homogeneous cosmologies are considered within the viscosity approximation. It is shown that some power laws $\sigma \sim T^m$ (where T is the temperature) for the cross-section σ lead to collisionless behavior near the initial singularity in a 'big bang' model. Under some circumstances one may have two viscous phases, as the cosmic fluid passes from collisionless to collisional to collisionless again, while the universe expands.

162.007 **Universal cosmic rays and Harrison's inhomogeneity postulate.**   P. D. Noerdlinger.
Phys. Rev., Second Series, Vol. 181, 2143 (1969).
Cosmologies of the type suggested by Harrison, in which initial baryon inhomogeneity leads to the formation of galaxies are shown to preclude the possibility that the bulk of cosmic rays are universal.

162.008 **Universal cosmic rays and the matter-antimatter universe.**   E. R. Harrison.
Phys. Rev., Second Series, Vol. 181, 2144 (1969).
The conflict between the universal theory of cosmic rays and the matter-antimatter theory of the universe, discussed by Noerdlinger, can be resolved by looking for antiparticles in the high energy end of the cosmic-ray spectrum.

162.009 **Electrodynamics of direct interparticle action. I. The quantum mechanical response of the universe.**
F. Hoyle, J. V. Narlikar.
Ann. Physics, Vol. 54, 207 - 239 (1969).
The path integral method of first quantisation is used to demonstrate that provided the universe is a perfect absorber along the future light cone the usual formulae for level shifts and for spontaneous transitions can be obtained in a steady-state model of the universe, but not in open Friedmann models.

162.010 **The evolution of density fluctuations in the universe. II. The formation of galaxies.**
M. J. Rees, D. W. Sciama.
Comments Astrophys. Space Phys. Vol. 1, 153 - 158 (1969).

162.011 **An open universe?**
R. A. Sunyaev, Ya. B. Zeldovich.
Comments Astrophys. Space Phys. Vol. 1, 159 - 164 (1969).

162.012 **Formation of gravitationally bound primordial gas clouds.**   K. Tomita.
Progr. Theor. Phys. Japan, Vol. 42, 9 - 23 (1969).
The formation of gravitationally bound systems from primordial gas is studied by means of Tolman's solution for dust-like matter. The critical values of density contract and its growth rate at an initial epoch are derived, which are ne-

cessary for an inhomogeneity to condense into a bound system before the appearance of the oldest stars. The matter distributions in an isolated inhomogeneity and an inhomogeneity included in a larger one are followed with time by assuming simple models for inhomogeneities, and it is shown how the bound region spreads outwards. Moreover, the minimum mass of fragments into which the gas clouds may break up is examined.

162.013 New information on the age of the universe.
A cosmological hypothesis of Soviet scientists.
I. Shklovsky.
Journ. British Astron. Ass. Vol. 79, 381 - 383 (1969).

162.014 Discrete sources and the microwave background in steady-state cosmologies.
C. Hazard, E. E. Salpeter.
Astrophys. Journ. *(Letters)*, Vol. 157, L87 - L90 (1969).

Attempts at interpreting the observed microwave background in terms of a suitable population of extragalactic radio sources are discussed for the steady-state cosmological model. For randomly distributed point sources the requirement on the number density n of sources would be $n > 3 \times 10^4$ Mpc^{-3}. Even for extended sources and cosmologies with irregularities the lower limit on n exceeds the number density of ordinary galaxies.

162.015 Paths in universes having closed time-like lines.
U. K. De.
Journ. Phys. A, General Phys. Ser. 2, Vol. 2, 427 - 432 (1969).

The equations of motion of charged particles in a few cosmological solutions are investigated. The solutions concerned have electromagnetic fields and closed time-like lines and the cosmic matter is also electrically charged. It is found that, unlike the Gödel universe, in these solutions particles may under some circumstances describe closed time-like lines.

162.016 A possible relationship between strong, electromagnetic, weak and gravitational interactions.
L. M. Stephenson.
Journ. Phys. A, General Phys. Ser. 2, Vol. 2, 475 - 476 (1969).

A possible causal relationship between the four interactions is developed, based on Dicke's interpretation of Mach's principle. The dimensionless coupling constants which are predicted from this relationship are of the correct magnitude.

162.017 Coriolis effects in the Einstein universe.
A. Lausberg.
Astron. Astrophys. Vol. 3, 150 - 155 (1969).

In order to study Thirring's problem in a bounded universe, the static spherical Einstein model is perturbed by the rigid-body rotation of a shell with finite thickness. Only Coriolis effects are investigated. It is shown that the dragging effect on the inertial frame at the center of the shell uniformly decreases with the distance, and increases with the thickness of the shell, in such a way that the complete dragging is reached when the shell covers the whole universe.

162.018 Development of the metagalaxy.
B. E. Laurent, L. Söderholm.
Astron. Astrophys. Vol. 3, 197 - 205 (1969).

The inner motion of an originally thin cloud of finite mass consisting of matter and antimatter is investigated. The general-relativistic system of equations for matter (and antimatter), radiation and gravitation is treated numerically. In a certain range of our parameters a violent explosion follows the contraction. The outward velocities in the solutions investigated so far do not however exceed 0.4 of the velocity of light. This may be too low to account for the observed Hubble red shift.

162.019 Radio sources opposite quasi-stellar objects and Einstein-Friedman's cosmology.
J. Audretsch, H. Dehnen.
Astron. Astrophys. Vol. 3, 252 - 255 (1969).

Assuming Einstein-Friedman's cosmology with vanishing cosmological constant ($\Lambda = 0$) the relations between the two redshifts of antipodal objects and the characteristic parameters of the universe are deduced. Comparison with observations yields: 1) in the case of elliptical space the hypothesis of the existence of antipodal objects is not in contradiction to Einstein-Friedman's cosmology ($\Lambda = 0$); 2) at present the direct verification of this hypothesis by means of both redshifts seems to be impossible.

162.020 Finite-range gravitation.
P. G. O. Freund, A. Maheshwari, E. Schonberg.
Astrophys. Journ. Vol. 157, 857 - 867 (1969).

The possibility that gravitation has a finite range of the order of the Hubble radius is explored. A unique Lagrangian formulation of the theory is obtained from the dynamical postulate that the symmetrical energy-momentum tensor is the source of gravitation. While locally the theory leads to extremely small deviations from Einstein's theory, the effects of finite range become decisive on the cosmological scale. A pulsating cosmological solution is explicitly worked out, and the cosmological-redshift formula obtained.

162.021 The finite rotating universe.
I. Ozsváth, E. L. Schücking.
Ann. Physics, Vol. 55, 166 - 204 (1969).

We construct on the Lie group $R \times S^3$ a left invariant metric, which satisfies the Einstein field equations with incoherent matter. We call the Riemannian space M_4, obtained this way, the finite rotating universe, since the normal subgroup S^3 constitutes the (finite) space sections of M_4, and the matter rotates. We discuss the geometry of M_4 and its relation to one version of Mach's principle.

162.022 Scattering of background X-rays by metagalactic electrons.   M. J. Rees.
Astrophys. Letters, Vol. 4, 113 - 115 (1969).

If the celestial X-ray and γ-ray background originates at large cosmological redshifts, the photons may have undergone scattering by metagalactic electrons. There would then be a characteristic distortion of the emitted spectrum, especially in the hard X-ray region.

162.023 Contributions to cosmology.   S. Silverman.
Phys. Today, Vol. 22, No. 10, p. 15, 17 (1969).
Letter.

162.024 Continuing excitement in cosmology.
Nature, Vol. 223, 1032 - 1035 (1969). – New notes.

162.025 The X-ray background in isotropic world models.
A. D. Payne.
Australian Journ. Phys. Vol. 22, 521 - 535 (1969).

This paper is an attempt to describe the diffuse X-ray background in terms of Compton radiation from cosmic ray electrons in intergalactic space. Similarities between the X-ray and radio source spectra suggest that fast electrons escape more or less freely from radio galaxies. It is assumed that the time scale of electron injection is small compared with the characteristic time of evolution of the universe. A least squares fit of the derived spectra to the experimental X-ray spectrum provides useful information on the epoch at which electron injection commences. Normalization requires either non-equilibrium conditions to exist within radio sources or the number of sources in unit coordinate volume to be a strong function of epoch. The results are consistent with a universe in a state of rapid expansion.

162.026 **Cooling of pre-galactic gas clouds by hydrogen molecule.** T. Matsuda, H. Satō, H. Takeda.
Progr. Theor. Phys. Japan, Vol. 42, 219 - 233 (1969).

According to the expanding hot universe model, neutralization of the cosmic plasma ceases at the stage of radiation temperature $T_r \simeq 4000°K$. After then, a density contrast in the uniform medium grows into a contracting gas cloud with mass greater than $10^{5 \sim 6} M_\odot$. Hydrogen molecules are formed in the cloud after the stage of $T_r \simeq 300°K$, and the thermal evolution is largely affected by the cooling through H_2, whose processes are studied quantitatively in this paper. The present paper is a preliminary one for a more interesting problem of galaxy formation.

162.027 **Limits on the cosmological deceleration parameter.** W. Rindler.
Astrophys. Journ. *(Letters)*, Vol. 157, L147 - L150 (1969).

For zero-pressure, big-bang Friedmann models of the cosmos it is shown that the known upper limit of σ_0 and the known lower limits of H_0 and t_0 impose, respectively, lower and upper limits on q_0, namely, $-10.1 < q_0 < +5.0$. These, in turn, delimit the values of Λ and k/R_0^2.

162.028 **Antimatter, galactic nuclei and theories of the universe. Speculation on the nature of the nuclei of galaxies.** F. Hoyle.
Nature, Vol. 224, 477 (1969).

The status of various charge symmetric cosmologies is reviewed. A theory of particle-antiparticle creation in galactic nuclei is outlined.

162.029 **Antimatter, galactic nuclei and theories of the universe. Antimatter and cosmology.** G. Steigman.
Nature, Vol. 224, 477 - 481 (1969).

The status of various charge symmetric cosmologies is reviewed. The following aspects are discussed: Annihilation products; steady state cosmology; Alfvén-Klein cosmology; big-bang cosmology; unsymmetric universe.

162.030 **Search for ghost images: A statistical test.** V. Petrosian, R. D. Ekers.
Nature, Vol. 224, 484 - 488 (1969).

For Lemaitre type universes, the fraction of radio sources with two diametrically opposite images included in a survey to a given flux limit is calculated to be small, because of the recent origin of the sources and the inhomogeneous distribution of matter in the universe.

162.031 **How important is steady state cosmology to classical and quantum electrodynamics?**
D. J. Leiter.
Nature, Vol. 223, 1145, with comments by F. Hoyle, J. V. Narlikar, p. 1145 - 1146 (1969).

Within the context of a new formulation of classical electrodynamics based on the paradigm that the basic building blocks of physical events are the mutual measurement interactions between "observer charges" and "observed charges", it is shown that the predictions of conventional electrodynamics and the Lorentz-Dirac equation for point charges is derivable from a Lagrangian formalism. However no direct self-interactions or "complete absorber assumptions" are needed or used. Hence this new electrodynamic theory, which offers a superior alternative to Maxwell-Lorentz and Wheeler-Feynman theory, does not require a steady state cosmology for its internal consistency. This means that the argument of Hoyle and Narlikar, that consistent local electrodynamics requires the steady state cosmology, is not on firm ground.

162.032 **The evolution of anisotropy in nonrotating Bianchi type V cosmologies.** R. A. Matzner.
Astrophys. Journ. Vol. 157, 1085 - 1100 (1969).

We consider the effect of neutrino viscosity and of free neutrinos in universes of Bianchi types I and V. We discuss the collisionless Boltzmann equation (which gives the evolution of the neutrino-distribution function) for these types and compare the type V results with those for type I previously given by Misner. We assume that dissipative effects lower ρ_β/ρ (the ratio of the anisotropy energy density to the matter energy density, both measured by their gravitational effect on the expansion) to a number of order unity before the electron neutrinos become collisionless. We present an approximate model universe of type V which begins its free evolution with $\rho\beta = \rho$ at a temperature of $\sim 10^{10} °K$.

162.033 **Lepton nonconservation and the early universe.** L. Oster.
Phys. Rev. Letters, Vol. 23, 987 - 988 (1969).

It is pointed out that the existence of a neutrinoless β-decay process would provide a basis for the customary assumption that the number of neutrinos and antineutrinos is equal in the universe.

162.034 **Elliptische Räume und antipodische Radioquellen.** M. von Reinhardt.
Naturwissenschaften, 56. Jahrgang, 511 (1969).

162.035 **The hypothesis about the initial spectrum of metric perturbations in Friedmann's model.**
Ja. B. Zeldovich, I. D. Novikov.
Astron. Zhurn. Akad. Nauk SSSR, Vol. 46, 960 - 964 (1969). In Russian. English translation in Soviet Astron. AJ, Vol. 13, No. 5.

The hypothesis is considered that at the beginning of cosmological expansion near the singularity all kinds of metric perturbations of Friedmann solutions are of the same order of magnitude and have the same spectrum. It is shown that at the time of cosmological hydrogen recombination the energy of long wave-length gravitational waves has the order of energy of acoustic waves and is by many times greater than the energy of vortex movements. The energy of long wave-length gravitational waves in this spectral region is greater by many orders of magnitude than the energy of equilibrium gravitational radiation in this spectral region.

162.036 **Observational dependences in a cosmological model of the universe with both matter and neutrino background.** A. M. Finkelstein.
Astron. Zhurn. Akad. Nauk SSSR, Vol. 46, 965 - 969 (1969). In Russian. English translation in Soviet Astron. AJ, Vol. 13, No. 5.

Within the framework of the evolutionary model of the universe with both matter and radiation (hypothetically neutrino background) the principal observational extragalactic dependences have been obtained: "apparent bolometric magnitude – redshift", "number of nebulae – redshift", "number of nebulae – apparent bolometric magnitude".

162.037 **Circular and radial trajectories in the Schwarzschild generalized field.** B. V. Prepelitza.
Astron. Zhurn. Akad. Nauk SSSR, Vol. 46, 1130 - 1132 (1969). In Russian. English translation in Soviet Astron. AJ, Vol. 13, No. 5.

The known results on the motion of test particles in the Schwarzschild field along circular and radial trajectories are partially extended to the cases: 1. $\Lambda \neq 0$ (Λ is the cosmological constant) and 2. of an electrically charged central body.

162.038 **Der Einfluß isotroper irreversibler Prozesse auf das Friedmannsche kosmologische Modell.**
G. Neugebauer, H. Strobel.
Wiss. Zeitschr. Friedrich-Schiller-Univ. Jena, Jahrgang 18, 175 - 180 (1969).

The system of Einstein's field equations and of balances of mass is given for a homogeneous fluid-mixture with viscosity, taken into account linear phenomenological equations for viscous pressure and for densities of chemical production. The Friedmann-type-differential equation including viscosity is solved with the equation $p = ae$ with respect to pressure and energy-density. Discussion, especially for a radiation filled universe, shows that the cycle of expansion and contraction becomes asymmetrical and the maximum world radius extends to larger and larger values.

162.039 An angular diameter-redshift test of cosmological models using observations of weak radio sources.
M. S. Longair, G. G. Pooley.
Monthly Notices, Roy. Astron. Soc., Vol. 145, 121 - 129 (1969).

Recent observations with the Cambridge One-mile radio telescope provide angular diameter data for sources having a wide range of flux densities. These enable cosmological models to be tested by comparing the number of sources of large angular diameter at different flux densities. On the basis of different world models, and the known distribution of the physical sizes of 3C sources, the expected numbers of extended sources in the 5C surveys are calculated. Significantly more sources of large diameter are found to be present in the 5C samples than are predicted by simple models; this excess may be explained by incorporating cosmological evolution.

162.040 Non-equilibrium processes in the early universe.
J. M. Stewart.
Monthly Notices, Roy. Astron. Soc., Vol. 145, 347 - 356 (1969).

In a recent paper Misner (1968) has suggested that neutrino viscosity is a highly efficient process for removing shear anisotropy during the early stages of the universe, so that the remarkable degree of isotropy observed in the microwave background would be expected to occur whatever the initial conditions. In this paper we suggest that the early universe may not have been in near thermal equilibrium as Misner assumed, so that under a wide range of initial conditions, arbitrarily large anisotropy could occur at any epoch. Therefore, the observed anisotropy does place some restrictions on early conditions in the universe.

162.041 Lemaître universe, galaxy formation and observations. K. Brecher, J. Silk.
Astrophys. Journ. Vol. 158, 91 - 102 (1969).

Several properties of the Lemaître universe are discussed. The possible growth of inhomogeneities from statistical fluctuations and the effect of galaxy formation on the stability of the quasi-static epoch are considered. It is concluded that the duration of the quasi-static epoch must be less than about 10^{11} years; otherwise, galaxy formation will induce a catastrophic collapse back to the initial singularity. Limits are derived on the redshift z_s at which the quasi-static epoch can occur and also on average cosmic-ray and electromagnetic fluxes produced following galaxy formation.

162.042 Effect of a constant magnetic field on the neutron beta decay rate and its implications for the production of He in the 'big-bang' expansion of the universe.
R. F. O'Connell, J. J. Matese.
Bull. American Astron. Soc. Vol. 1, 255 (1969). – Abstr. AAS.

162.043 Formation of protogalaxies and molecular processes in hydrogen gas. T. Hirasawa.
Progr. Theoret. Phys. Vol. 42, 523 - 543 (1969).

As one stage of galaxy formation we study the gravitational contraction of a cold gas due to the radiative cooling by hydrogen molecules. Assuming that the gas consists mainly of hydrogen atoms, we consider all possible atomic and molecular processes in the gas and find that H_2 molecules are formed most efficiently by way of H^- formation. The contraction of a gas cloud is studied in the wide range of initial conditions and it is found that sufficient amounts of H_2 are formed in the gas cloud so that it can evolve into a protogalaxy provided that the radiation temperature is less than $300°K$. We also examine the effect of He and D on the cooling and find that their effect is negligible.

162.044 Gravitational instability in the Brans-Dicke cosmology. H. Nariai.
Progr. Theoret. Phys. Vol. 42, 544 - 554 (1969).

In view of a possible relevance of the Brans-Dicke cosmology to the resolution of the primeval helium problem and the problem of galaxy formation, the gravitational instability in their cosmology is studied by the use of our Lagrangian gauge. It is shown that the density contrast consists, in general, of four independent terms, i.e. two terms resembling their counterparts in the general relativistic cosmology and the other two due to the variation of the gravitation "constant". At the later stage of the universe, the density contrast can grow in time. It is also shown that Mach's principle in the sense of Dicke provides us with some close connection between the last two terms and the time-growing term.

162.045 The correlation function for the distribution of galaxies. H. Totsuji, T. Kihara.
Publ. Astron. Soc. Japan, Vol. 21, 221 - 229 (1969).

The correlation function for the spatial distribution of galaxies in the universe is determined. The determination is based on the distribution of galaxies brighter than the apparent magnitude 19 counted by Shane and Wirtanen (1967).

162.046 Primordial helium production in "magnetic" cosmologies. G. Greenstein.
Nature, Vol. 223, 938 - 939 (1969).

If a primordial magnetic field in the range $10^{13} G \lesssim B \lesssim 10^{17} G$ existed at the epoch of primordial nucleosynthesis in a big-bang cosmology, the element abundances produced would be too large. If $B > 10^{17} G$ they are zero: if $B < 10^{13} G$ they are unaffected by the field.

162.047 Schwarzschild-Feld im Friedmann-Kosmos. K. Kramer.
Wiss. Zeitschr. Friedrich-Schiller-Univ. Jena, Jahrgang 18, 155 - 157 (1969).

It is investigated to what extent the Schwarzschild solution can be connected with the Friedman metric. The condition to be fulfilled is given by an unequality. The combined solution is presented explicitly in isotropic coordinates with the aid of elliptical functions.

162.048 Formation of gravitationally bound primordial gas clouds. II. K. Tomita.
Progr. Theoret. Phys. Japan, Vol. 42, 978 - 979 (1969).

In a previous paper we have studied the growth of a gas cloud from a small, but non-statistical fluctuation to a gravitationally bound system. The cosmological constant Λ has been taken to be zero. We treat the case with non-zero cosmological constant and examine the growth, particularly in the Lemaitre model.

162.049 On the relation between the Schwarzschild and Tolman coordinate systems.
Yu. M. Ajvazyan, M. E. Gertsenshtein.
Zh. ehksperim. i teor. fiz. Vol. 56, 830 - 834 (1969).
In Russian. – Abstr. in Referativ. Zhurn. 51. Astron., 9.51.713 (1969).

162.050 Superfluidity and superconductivity in the universe.

V. L. Ginzburg.
Uspekhi fiz. nauk, Vol. 97, 601 - 619 (1969). In Russian.

162.051 A hot universe. Ya. B. Zel'dovich.
Vestn. AN SSSR, No. 2, p. 36 - 43 (1969).
In Russian.

162.052 The possibility of galaxy formation in Lemaitre's
model. A. V. Byalko.
Pis'ma v ZhETF, Vol. 9, 483 - 487 (1969). In Russian. —
Abstr. in Referativ. Zhurn. 51. Astron., 10.51.792 (1969).

162.053 Quantized fields and particle creation in expanding
universes. I. L. Parker.
Phys. Rev., Second Series, Vol. 183, 1057 - 1068 (1969).

The spin-0 field of arbitrary mass is quantized in the expanding universe by the canonical procedure. Consistency of
the time development of the commutators provides a new
proof of the connection between spin and statistics. Particles
are spontaneously created in pairs. In the cases of a dust filled
universe and a radiation filled universe with flat 3-space, equations governing the expansion follow from the natural assumption that the reaction of the created particles back on the
gravitational field reduces the creation rate. These equations
are identical with those obtained from Einstein's field equations. Finally, we show that massless particles of arbitrary
nonzero spin are not created by the expansion.

162.054 Symmetrie und Kosmologie. H.-J. Treder.
Monatsber. Deutsch. Akad. Wiss. Berlin, Band 11,
226 - 230 = Sternw. Babelsberg, Inst. Relativistische und
Extragalaktische Forschung, Mitt. Neue Folge, No. 20 (1969).

162.055 Generation of magnetic fields in the primordial
fireball. E. R. Harrison.
Nature, Vol. 224, 1089 - 1090 (1969).

It is shown that magnetic fields are generated during the
radiation era of the early universe in regions that have rotation. These fields are weak compared with the present intensity of the galactic magnetic field and therefore must be am-
plified as the Galaxy forms and evolves.

162.056 A stationary model of the metagalaxy.
B. E. Laurent, B. E. Malm.
Ark. Fys., Vol. 38, 325 - 340 (1968). — See Phys. Ber., Vol.
48, No. 5 – 3526 (1969).

162.057 Multifluid cosmologies. L. P. Hughston.
Astrophys. Journ., Vol. 158, 987 - 989 (1969).

A new coordinate system which synthesizes all previous
work on multifluid cosmologies is presented. Using these coordinates in the Robertson–Walker, Bianchi type I, and some
nonrotating line elements, we investigate noninteracting mixtures of multicomponent fluids with the γ-law and polytropic equations of state.

162.058 The microwave background radiation.
W. H. McCrea.
Cosmic Ray Studies, Bombay 1968, p. 142 - 150 (1969).

162.059 Generation of seed magnetic fields in the radiation
era. E. R. Harrison.
Bull. American Astron. Soc., Vol. 1, 347 (1969). — Abstr.
AAS.

162.060 A hypothesis of the magnetic cosmological inhomo-
geneity. Ja. B. Zeldovic.
Astron. Zhurn. Akad. Nauk SSSR, Vol. 46, 775 - 778 (1969).
In Russian. English translation in Soviet Astron. AJ, Vol. 13,
No. 4.

A hypothesis of primary magnetic disturbances is combined with predictions of an anisotropic cosmological model related to nuclear reactions. Magnetic fields up to 10^{-7} Gauß are
admissible at present. The inhomogeneity of magnetic fields
can cause an inhomogeneity of the density of matter, sufficient for the isolation of galaxies and clusters having time to
take place at present.

162.061 Electrons and photons in an expanding universe.
I. Rozental, I. Shukalov.
Astron. Zhurn. Akad. Nauk SSSR, Vol. 46, 779 - 786 (1969).
In Russian. English translation in Soviet Astron. AJ, Vol. 13,
No. 4.

Conditions of the origin and distribution of electrons
and hard photons (X-ray emission and γ-quanta) in an expanding universe are studied. The distribution of electrons is investigated on the basis of the solution of an invariant kinetic
equation. The comparison of results of experiment and calculation yields the determination of some parameters of the
universe: average matter density, evolution and time of the
origin of sources of cosmic rays, and intensity of cosmic rays
in the universe.

162.062 The collapse of the universe: An eschatological
study. M. J. Rees.
Observatory, Vol. 89, 193 - 198 (1969).

If the universe is perpetually oscillating, and this contraction is merely a prelude to a subsequent re-expansion,
then plainly stars, galaxies and clusters must form anew in
each cycle. The catastrophic processes whereby the contracting cosmos reverts to primaeval chaos—together with some
speculations on how "genetic" information may survive from
one cycle to the next — are the subject of this article.

162.063 To the thermodynamics of the Boltzmann-uni-
verse (cosmological, stellar-dynamical and statistical aspects). F. A. Tsitsin.
Soobshch. Gos. Astron. Inst. Shternberga, No. 158, p. 3 - 13
(1969). In Russian.

The influence of the fluctuations of the thermodynamical entropy S on the characteristics of the cosmological
(Boltzmann's universe), stellar-dynamical and statistical systems is considered.

162.064 Cosmology and quantum electrodynamics.
P. C. W. Davies.
Nature, Vol. 224, 1102 (1969).

A quantum theory of the direct interparticle action electrodynamics of Wheeler and Feynman has been developed. The
method starts with a modification of the conventional S-mat-
rix perturbation expansion of quantum electrodynamics, from
which the Feynman propagator, togetherwith the usual atomic
transition rates and level shifts may be recovered when the
response of the universe is included. The formal structure is
closely similar to the conventional theory. The results confirm the recent work of Hoyle and Narlikar.

162.065 Traces of "photon eddies". L. M. Ozernoi.
JETP Letters, Vol. 10, 251 - 254 (1969). [Translated
from ZhETF Pis. Red. 10, No. 8, 394 - 398 (1969). In
Russian].

According to a hypothesis in the early universe, during
the phase of almost complete homogeneity, there existed local
dynamic motions of the photon gas and the plasma dragged by
it ("photon eddies") superimposed on the general cosmological expansion. It will be shown that the consequences of the
hypothesis, which pertain to the velocity and density spectra
of metagalactic structures, are confirmed by astronomical observations.

162.066 A class of homogeneous cosmological models.
G. F. R. Ellis, M. A. H. MacCallum.

Commun. Math. Phys., Vol. 12, No. 2, p. 108 - 141 (1969).
See Phys. Abstr. Vol. 72, No. 21839 (1969).

162.067 Exact solutions of Einstein's cosmological equations.
J. R. Trollope, B. E. Smith.
Nuovo Cimento, Vol. 59B, 125 - 136 (1969).

162.068 Classical dynamics in an expanding universe.
J. Kulhanek, G. Szamosi.
Nuovo Cimento, Vol. 60B, 86 - 92 (1969).

162.069 Inertial systems in an expanding universe.
N. Rosen.
Nuovo Cimento Lettere, Vol. 1, 42 - 44 (1969).

162.070 Numerical limitations on the cosmological constant and on the deceleration parameter in a closed radiation-type universe.
A. Agnese, M. La Camera, A. Wataghin.
Nuovo Cimento Lettere, Vol. 1, 45 - 46 (1969).

162.071 Inevitability of a point-singularity in a rotating Newtonian universe.   J. Pachner.
Phys. Letters, Vol. 29A, No. 3, p. 147 - 148 (1969).

162.072 Infrared astronomy and cosmology.
P. J. E. Peebles.
Phil. Trans., Ser. A, Vol. 264 (No. 1150), 279 - 282 (1969).

162.073 On a model of the expanding universe.
A. Wataghin.
Nuovo Cimento Lettere, Prima Ser., Vol. 1, 375 - 378 (1969).
Some new properties of the Friedmann solutions of the Einstein equations for the expanding or oscillating universe are discussed.

162.074 Covariant electromagnetic potentials and fields in Friedmann universes.   P. C. Peters.
Journ. Math. Phys., New York, Vol. 10, 1216 - 1224 (1969).

162.075 Inertia, relativity and cosmology.
Z. Horak.
Czech. Journ. Phys., Ser. B, Vol. 19, 703 - 720 (1969).

162.076 A stationary model of the metagalaxy.
B. E. Laurent, B. E. Malm.
Ark. Fys.,Vol. 38, 325 - 340 (1968). – See Phys. Abstr., Vol. 73, No. 2626 (1970).

162.077 A cosmological model in which 'singularity' does not require a 'matter singularity'.
L. C. Shepley.
Phys. Letters, Vol. 28A, 695 - 696 (1969).

162.078 Neuere Probleme der Kosmologie.
M. von Reinhardt.
Naturwissenschaften, 56. Jahrgang, 581 - 590 = Mitt. Astron. Inst. Bonn No. 100 (1969). – Review article.

162.079 Evolution of galaxies.
E. M. Burbidge, G. R. Burbidge.
Proc. Fifth Berkeley Symposium on Mathematical Statistics and Probability, Vol. 3, 1 - 18 (1967). – Review article.

162.080 Age distribution of galaxies.
W. H. McCrea.
Proc. Fifth Berkeley Symposium on Mathematical Statistics and Probability, Vol. 3, 19 - 29 (1967). – Review article.

162.081 Quadratic corrections to the Lagrangian gravitational field density, and singularity.

T. V. Ruzmajkina, A. A. Ruzmajkin.
Zhurn. ehksperim. i teor. fiz. Vol. 57, 680 - 685 (1969).
In Russian. – Abstr. in Referativ. Zhurn. 51. Astron., 1.51.795 (1970).

162.082 Dynamische Stabilität im de-Sitter-Raum.
O. Nachtmann.
Sitzungsber. Österreich. Akad. Wiss., Math.-Naturwiss. Kl., Abt. II, Vol. 176, 363 - 379 (1968).
The dynamics of quantized fields in de-Sitter space of positive curvature are investigated. It is shown that interacting fields in this space will lead to spontaneous creation. The rate however is very low at the present stage of the universe.

162.083 On the foundations of general relativity theory and the cosmological problem.   O. Klein.
Ark. Fys., Vol. 39, 157 - 170 (1969). – See Phys. Ber., Vol. 48, No. 12–283 (1969).

162.084 Exact Robertson–Walker cosmological solutions containing relativistic fluids.   J. P. Vajk.
Journ. Math. Phys., Vol. 10, 1145 - 1151 (1969).

162.085 General form of the Einstein equations for a Bianchi type IX universe.   M. P. Ryan, Jr.
Journ. Math. Phys., Vol. 10, 1724 - 1728 (1969).

162.086 The recent renaissance of observational cosmology.
D. W. Sciama.
Phys. Bull. (*G. B.*), Vol. 19, 329 - 336 (1968). – See Bull. Signal., Vol. 30, Section 120, No. 9298 (1969).

162.087 L'universo di De Sitter e la relatività proiettiva.
G. Arcidiacono.
Collect. Math., *Barcelona*, Vol. 19, No. 1 - 2, p. 51 - 71 (1968). – See Bull. Signal., Vol. 30, Section 120, No. 13806 (1969).

162.088 Cosmology yesterday and today.   W. Zonn.
Urania Kraków, Vol. 40, 194 - 197 (1969).
In Polish.

162.089 The universe and its evolution.
V. L. Ginzburg.
Fiz.-matem. spisanie, Vol. 12, No. 2, p. 111 - 124 (1969).
In Bulgarian.

162.090 Die extragalaktische Entfernungsbestimmung und die Eichung der Hubble-Konstanten.
G. A. Tammann.
Mitt. Astron. Ges. No. 27, p. 55 - 72 (1969). – Review article.

162.091 Probleme und Ergebnisse der modernen Kosmologie.   J. Ehlers.
Mitt. Astron. Ges. No. 27, p. 73 - 86 (1969). – Review article.

162.092 Zur Anisotropie der kosmischen Mikrowellen-Hintergrundsstrahlung.
G. Dautcourt.
Monatsber. Deutsch. Akad. Wiss. Berlin, Band 11, 231 - 235 = Sternw. Babelsberg, Inst. Relativistische und Extragalaktische Forschung, Mitt. Neue Folge, No. 21 (1969).
It is shown that collisionless radiation propagating in a gravitational field preserves an initial Planckian spectrum; the temperature, however, becomes inhomogeneous in general and depends upon the propagation direction. Application to the cosmic blackbody radiation suggests, that contributions to the anisotropy arising from the cosmological model as well as from an improperly chosen observer frame show

the same frequency dependence. They attain their maximum in the millimeter region.

162.093 **Singuläre Riemannsche Räume als global-kosmologische Approximationen.** G. Dautcourt.
Math. Nachr., Band 40, 333 - 341 = Sternw. Babelsberg, Inst. Relativistische und Extragalaktische Forschung, Mitt. Neue Folge, No. 22 (1969).

A bird's-eye view towards to universe may reveal the unexpected fact, that the large-scale world geometry is represented in a good approximation by a degenerate Riemannian metric $g_{\mu\nu}$ of rank 3 with $g_{\mu\nu} a^{\nu} = 0$: Let us assume a system of metagalaxies. Light signals emitted from the interior of a metagalaxy may not necessarily leave the metagalaxy if its density is high. From a suitable large-scale point of view the light signal then follows the world line a^{μ} of the metagalaxy. General relativity remains applicable even for this highly degenerate situation. A consistent 'ultrarelativistic' approximation procedure for the Einstein field equations shows that no global steady-state model of this type may exist; there are, however, many time-dependent solutions.

162.094 **Magnetic fields and highly condensed objects.** F. Hoyle.
Nature, Vol. 223, 936 (1969).

It has become clear that highly condensed objects are important energy sources, and the question evidently arises as to whether they augment the galactic field by dynamo action. This could happen through the winding by rotation of a field that emerges from the object.

162.095 **On the origins of galaxies.** R. Omnes.
Nature, Vol. 223, 1349 - 1350 (1969).

The evolution of condensations of matter and antimatter have been analysed during the cooling down of the universe, and the results are described here together with the consequences for galaxy and quasar formation.

An evolving universe.
Nature, Vol. 223, 1003 - 1004 (1969).

Problems of Modern Cosmogony.
See Abstr. 003.024.

Relativität und Kosmos. Raum und Zeit in Physik, Astronomie und Kosmologie. See Abstr. 003.120.

Äquivalenzprinzip und Abschirmung der Schwerkraft. See Abstr. 066.043.

Fluctuations in the microwave background radiation. See Abstr. 066.082.

Distortions of the background radiation spectrum. See Abstr. 066.083.

Clusters of quasi-stellar objects. See Abstr. 141.051.

On the production of QSO absorption spectra in Friedmann universes. See Abstr. 141.055.

The dynamics and structure of inertially confined plasma clouds. See Abstr. 141.077.

Local theory for quasars. See Abstr. 141.139.

A large-scale metagalactic magnetic field and Faraday rotation for extragalactic radio sources. See Abstr. 141.180.

The cosmic gamma-ray spectrum from secondary-particle production in the metagalaxy. See Abstr. 142.012.

Metagalactic inverse Compton effect and cosmic X-ray background. See Abstr. 142.049.

Leakage electrons from normal galaxies: The diffuse cosmic X-ray source. See Abstr. 143.011.

Primeval globular clusters. II. See Abstr. 154.007.

Infall of gas from intergalactic space. See Abstr. 161.008.

Author Index

BOTELHEIRO, A. P.
041.033
BOTLEY, C. M.
125.013
BOTTEMA, M.
032.018
BOUCHIAT, C.
066.071
BOURQUIN, L. B.
021.013
BOUSKA, J.
011.037 .038
051.040
094.239
102.048
103.109 .112
BOWELL, E. L. G.
010.022
BOWEN, P. J.
076.025
BOWERS, B. C.
076.014
BOWLES, K. L.
106.005
BOWMAN, C. D.
022.008
BOWMAN, M. R.
082.061
BOWMAN, R. L.
105.173
BOWYER, C. S.
142.002
BOYARCHUK, A. A.
022.083
122.076 .119
BOYARCHUK, M. E.
114.088
BOYCE, P. B.
097.053
BOYCE, W. M.
094.048
BOYD, R. L. F.
012.019
051.029
142.061
BOYER, R.
071.001
BOYER, T. H.
022.033
BOYNTON, P. E.
141.058 .231
BOZULA, R. A.
121.026
BOZYAN, F. A.
099.033
BRACCESI, A.
033.036
141.119 .204
BRACHET, G.
052.027
BRADLEY, D. J.
071.064
BRADT, H. V.
142.008
143.054
BRAEUNINGER, H.
034.100
BRAGINSKII, V. B.
066.063
BRAHDE, R.
072.076

BRANCH, D.
072.056
BRANDENBERGER, H.
035.010
BRANDT, J. C.
074.025 .062
158.045
BRANDT, L.
099.061
BRANDT, V. EH.
041.034
BRANLEY, F. M.
009.007
BRANSON, N. J. B. A.
141.176
BRAULT, J. W.
071.026 .027
BRAY, A. D.
143.071
BRAY, R. J.
073.046
BRECHER, K.
143.011
162.041
BRECKINRIDGE, J. B.
071.027
BREDOV, M. M.
105.182
BREENE JR., R. G.
082.010
BREGER, M.
122.102 .103 .107
BREIDO, I. I.
036.010 .015
BREIG, E. L.
082.013
BRETT, R.
105.080
BREUS, T. K.
091.044
BRICE, N.
084.246
BRIDGES, J. M.
022.090
BRIDLE, A. H.
131.047
141.128
BRIGGS, B. H.
033.008
BRIGGS, R. M.
143.009
BRIHAYE, C.
022.122
BRINKMANN, R. T.
082.048
BRINTON, H.
003.032
BRINTON, H. C.
083.003
BROADFOOT, A. L.
022.069
BRODSKIJ, B. I.
034.090
BROGLIA, P.
034.106
122.114
BROGLIO, L.
082.154
BROMANDER, J.
022.058

BRONKALLA, W.
158.095
BRONNIKOVA, N. M.
103.102 .116
BRONSHTEHN, V. A.
005.014
091.038
BRONSHTEN, V. A.
003.033
105.176
BRONSHTEN, W. A.
093.009
BROOKS, J.
105.005
BROOKS, J. N.
082.044
BROOKS, J. W.
132.016
BROSCHE, P.
044.041
091.039
117.033
158.003
BROTEN, N. W.
141.142 .144
BROUCKE, R.
021.004
052.030
BROWN, B. C.
022.111
BROWN, D. R.
073.029
BROWN, D. W.
102.036
BROWN, L. W.
141.057
BROWN, R. H.
032.031
BROWN, R. R.
093.016
BROWN, R. T.
022.062
BROWNELL, D. H.
066.059
BROWNELL JR., D. H.
066.059
BROWNLEE, D. E.
105.186
BROWN III, W. E.
082.011
BRUECK, H. A.
008.037
BRUECKNER, G. E.
080.019
BRUENN, S. W.
131.112
BRUIN, F.
075.028
BRUKALSKA, R.
121.049
161.011
BRUN, A.
123.047
BRUNDAGE, R. K.
114.076
155.019
BRUNER JR., E. C.
076.005
082.074
BRUNK, W. E.
053.024

FRIEDLANDER, M. W.
143.001
FRIEDMAN, H.
076.035 .039
134.006
142.004 .056
FRIEDMAN, M.
062.010
073.002
FRIEDRICH, H.
022.120
FRIENDS, J.
072.010
FRITZ, G.
134.006
142.004
FRITZ, T. A.
084.405
FROESCHLE, C.
151.068
FROGEL, J. A.
113.038
FROLOV, M. S.
003.013
122.017 .132 .141
FROLOV, P. M.
094.184
FROOME, K. D.
003.068
FROST, K. J.
076.032
FROST, M. J.
105.177
FRYE JR., G. M.
142.015 .041
FRYER, R.
094.006
FRYER, R. J.
053.011
094.026 .028
FUCHS, L. H.
105.138
FUERST, E.
077.056
FUERSTENBERG, F.
075.025
077.008
FUHRMANN, K.
113.048
FUJIKAWA, S.
103.113
FUJIMOTO, M.
131.013
141.180
151.024
FUJITA, Y.
114.113 .114
FUKUI, M.
141.180
142.049
FULCHIGNONI, M.
105.183
FULLER, B. D.
106.006
FULMER, C. V.
094.110
FUNG, A. K.
097.051
FUNG, P. C. W.
062.029

FUNICIELLO, R.
105.183
FUNK, H.
105.154
143.002
FUTAULLY, R.
054.006
FYMAT, A. L.
063.015 .016 .025
GABRIEL, A. H.
076.013
GADSDEN, M.
082.091 .092
GAFFNEY, J. E.
046.018
GAGNEPAIN, M.
035.028
GAHM, G. F.
131.131
GAINSFORD, M. J.
123.055
GAINULLINA, R. H.
160.006
GAIZAUSKAS, V.
010.023
071.038
GALE, W. A.
093.008 .027
GALEOTTI, P.
121.080 .081
GALKIN, L. S.
099.044
101.008
GALKINA, T. S.
112.002
119.019
GALL, R.
084.213
GALLIVAN
125.023
GALLIVAN, J. R.
125.015
GALT, J. A.
141.142
GAMBURG, S. S.
091.030
GAMJANINA, A. I.
041.010
GANDHI, J. M.
022.073
GANS, D.
098.033
103.110
GAPCYNSKI, J. P.
094.065 .146 .205
GAPOSHKIN, S.
115.013
159.006
GARAZDO-LESNYKH, G. A.
103.106
GARAZHA, V. I.
091.050
GARCIA, C. J.
074.035
GARCIA AGUDO, E.
105.184
GARDINER, G. W.
083.020
GARDNER, F. F.
131.033 .103
141.153 .184 .185

GARDNER, M. E.
082.037
GARNIER, R.
114.074
122.097
GARRISON, R. F.
114.004
122.082
GARRISON, R. L.
022.046
GARRISON JR., L. M.
121.057 .093
GARSTANG, R. H.
022.060
GARTHWAITE, K.
021.004
GARTON, W. R. S.
022.063 .077
GARY, B.
099.039 .042
GARZ, T.
071.003 .023
GASCOIGNE, S. C. B.
159.008
GASKELL, P. F.
003.022
GATELYUK, E. D.
134.007
GATEWOOD, G.
041.014
GATLAND, K.
003.031
GAUJARD, P.
071.013
GAULT, D. E.
094.111 .170 .173
105.167
GAUSTAD, J. E.
114.064
131.042 .044 .050
GAVRILOV, I. V.
003.072
GAVRILOV, V. P.
022.027
GAVRILOV, V. V.
054.012
GAWIN, J.
143.024
GAY, J.
071.002
GEBBIE, H. A.
022.110
GEBBIE, K. B.
073.017
GEBEL, W.
131.079
GEDEON, G. S.
052.008
GEHLICH, U. K.
064.006
GEHLOT, G. L.
066.019
GEHRELS, T.
098.002
099.022
114.044
141.005
GEISS, J.
074.065
105.002 .150

KRUSZEWSKI, A.
  114.044
KRZEMINSKI, W.
  121.011
  122.099
  124.105
KUBOTA, J.
  079.100
KUCHOWICZ, B.
  003.086
  061.024 .046
  105.195
KUCKES, A. F.
  077.059
KUECHEMANN, D.
  003.089
KUENZEL, H.
  072.023
  075.025
  077.049
KUHI, L. V.
  114.102
  122.046 .091
KUHLTHAU, A. R.
  022.014
KUIPER, G. P.
  071.010 .011 .012 .084
    .085 .086
  093.020 .021 .022 .026
  094.173
KUKARKIN, B. V.
  003.013
  123.003
KUKARKINA, N. P.
  003.013
KULAGIN, E. S.
  032.038
KULAGIN, S. G.
  045.001
KULAPOVA, A. N.
  123.027
KULESHOVA, K. F.
  072.074
KULHANEK, J.
  162.068
KULIKOV, G. V.
  143.066
KULIKOV, K. A.
  043.007
KULIKOVSKIJ, P. G.
  002.006
  003.087
  005.009
KULKARNI, P. P.
  083.030
KULLERUD, G.
  105.135
KUMAR, C. K.
  072.068
KUMAR, R. C.
  033.035
KUMAR, S. S.
  042.033
  064.050
KUMAR, V.
  022.045
KUNCHEV, P. Z.
  122.079
KUNDE, V. G.
  064.046

KUNDU, M. R.
  132.027
KUNERT, A.
  011.035
KUNG, H.-C.
  063.011
KUNKEL, W. E.
  142.048
  159.002
KUNTE, P. K.
  142.027
KUNZE, H.-J.
  022.037 .097
KUPERUS, M.
  074.054
KUPO, I. D.
  122.032 .061 .166
KUPRIANOVA, E. B.
  022.043
KURAT, G.
  105.102
KURCHAKOV, A. W.
  132.044
KURDGELAIDZE, D. F.
  061.036
  066.068
KURFESS, J. D.
  142.019
KURILCHIK, V. N.
  141.208
  158.017 .032 .086
KURILOV, V. A.
  083.052
KUROCHKIN, N. E.
  003.013
  118.034
  122.021
  141.032
KUROKAWA, H.
  079.100
KURPINSKA, M.
  096.003
  114.098
KURUTAC, M.
  121.072
KURYANOVA, A. N.
  035.023
  082.051
KUTTER, G. S.
  133.019
KUTUZA, B. G.
  131.125
KUTUZOV, S. M.
  033.032
KUZ'MENKO, K. N.
  005.024
KUZ'MIN, A. D.
  033.030
  093.043
KUZ'MIN, A. F.
  094.195 .196
KUZMIN, G. G.
  151.059
KUZMIN, V. A.
  080.043 .044 .045 .046
KUZ'MIN, V. I.
  009.010
KUZNETSOV, S. N.
  078.037
KUZNETSOVA, R. I.
  105.106

KUZNETZOVA, G. M.
  083.048
KVASHA, L. G.
  105.042
KVASHNIN, A. N.
  082.127
KVIFTE, G. J.
  084.006
KVIZ, Z.
  104.047
KWAST, T.
  158.080
KYLE, T. G.
  071.081 .089
  082.042 .044 .050 .096
LAAN, H. VAN DER
  141.120
  158.090
LABORDE, J. R.
  121.086
LA CAMERA, M.
  162.070
LACEY, J. D.
  003.057
  033.009
LACLAVERIE, J. J.
  042.006
LADD, A. C.
  051.030
LAFFINEUR, M.
  033.037
  074.014
LAFRAMBOISE, J. G.
  051.024
LAGERWEY, H. C.
  121.064
LAING, P. A.
  094.112
LAL, D.
  074.069
  105.075 .109 .111
  143.044
LAL, S.
  143.016
LALA, P.
  052.011
LALLEMAND, A.
  113.041
LAMAR, D. L.
  094.120
LAMBERT, D. L.
  071.071
  074.034 .047
  122.181
  124.107
LAMBIOTTE, J. J.
  082.075
LAMBRECHT, H.
  008.055
  131.069
LANCASTER, J. E.
  052.032
LANDAU, R. W.
  106.009
LANDINI, M.
  072.084
LANDMARK, B.
  084.014
LANDOLT, A. U.
  113.011
  121.047 .048

MANDELL, D. A.
062.015
MANDEL'STAM, S. L.
073.011
MANDRYKINA, T. L.
072.015
MANGENEY, L.
042.001
MANKA, R. H.
094.122
MANLEY, O. P.
142.006 .062
MANN, F. I.
052.032
MANNINO, G.
051.025
082.121
MANSUROV, S. M.
074.075
106.026
MANSUROVA, L. G.
074.075
MANUEL, O. K.
105.023
MAR, J. W.
003.017
MARAL, G.
084.047
MARAN, S. P.
022.069
141.052
MARCONERO, R.
074.078
MARCOTTE, L. P.
082.010
MARCUS, A. H.
065.084
091.016 .022 .024
094.032
151.039
MARCUS, E.
041.020
MARENIN, I.
022.079
MARENIN, I. R.
022.080
MARGOLIS, J. S.
099.015 .038
MARGONI, R.
119.005 .017
124.100
154.009
MARGOSHES, M.
113.013
MARGRAVE JR., T. E.
061.035
071.055 .079
MARIANI, F.
084.230
MARIANO, J.
097.062
MARIN, E.
043.002
MARINO, B. F.
121.004
MARIS, G.
073.058
MARK, H.
003.096
142.035

MARKARIAN, B. E.
158.079
MARKINA, A. K.
104.032
MARKOV, M. N.
082.039
MARKOVICH, M. Z.
102.008 .013 .045
103.105
MARLBOROUGH, J. M.
073.054 .055 .062
MAROCHNIK, L. S.
151.001 .042 .053
MAROUF, A.
044.021
MAROV, M. YA.
082.151 .161
093.025 .046
MARRACO, H. G.
036.020
094.030
MARSDEN, B. G.
102.039
103.003 .110 .112 .113
.114 .115 .120 .124
.126 .129 .130
MARSH, D.
054.008
MARTELLI, G.
083.014
MARTI, K.
105.081 .107
MARTIN, C. F.
081.018
MARTIN, L. J.
097.045 .046
MARTIN, N.
155.012
MARTIN, P. G.
117.032
MARTIN, T. Z.
082.081
MARTINET, L.
114.096
MARTINEZ-GARCIA, M.
022.053
MARTINI, .A
121.036
MARTINI, A.
115.016
122.077 .098 .129
MARTINS, P. DE A. P.
022.059
MARTRES, M. J.
073.080
075.012
MARTYNOV, D. JA.
103.110
MARTYNOV, D. YA.
093.035
141.029
MARYCH, M. I.
081.035
MASANI, A.
115.016
MASEIDE, K.
084.014
MASLENNIKOV, K. L.
122.158
MASLEY, A. J.
078.010 .017 .026 .027

MASON, B.
105.179
MASON, C. C.
094.256
MASON, H. P.
097.008
MASSANGIOLI, A. P.
035.016
MASSE, P.
078.019
MASSEVITCH, A. G.
012.021 .022
055.014
MASSEY, H.
012.019
MASUDA, A.
105.064
MASURSKY, H.
094.173
MATAS, V.
042.028
MATESE, J. J.
162.042
MATHER, R. S.
081.022
MATHERS, S. W.
034.008
MATHEWS, W. G.
131.032 .074
132.007
MATHEZ, G.
141.036
MATHISEN, O.
045.006
MATIAGIN, V. S.
054.020
MATILSKY, T. A.
131.098
MATJAGIN, V. S.
021.016
MATSOUKAS, D.
079.102
MATSUDA, H.
105.113
MATSUDA, T.
162.026
MATSUKOV, K. P.
103.108
MATSUMOTO, M.
063.003 .030
MATSUMOTO, T.
094.152
MATSUOKA, M.
142.038
MATSUSHIMA, S.
113.036
MATSUURA, O. T.
077.011
MATTEI, M.
034.002
MATTESON, J. L.
142.031
MATTIG, W.
071.067
072.004
080.054
MATTILA, K.
113.064
MATULAJTITE, S. P.
004.019

MOTTINGER, N. A.
094.081 .096
MOTTONI, G. DE
032.071
MOURAO, R. R. DE FREITAS
118.024
MOURILHE, I.
103.102 .103
MOYER, H. G.
052.038
MOYLE, L. D. T.
054.016
MOZER, M.
034.021
MOZJAEVA, V. E.
082.022 .024 .123
132.003
MRAZ, L.
105.015
MRKOS, A.
103.109 .110 .111 .112
.113 .114 .120 .122
MUCKE, H.
047.025
MUEHLFELD, R.
094.053
MUELLER, G.
105.127
MUELLER, G. E.
003.010
053.003
MUELLER, H. W.
105.149
MUELLER, O.
105.120
MUELLER, R.
075.004
MUELLER, R. F.
093.010
MUELLER, W. F.
105.157
MUENCH, G.
097.031
132.024
MUEUERSEPP, P.
005.013
008.114
MUGGLETON, L. M.
083.023
MUHLEMAN, D. O.
102.038
MUIRDEN, J.
123.055
MUKHINA, M. M.
033.029
MULHOLLAND, J. D.
042.032
094.081 .096
MULLER, A.
008.048
MULLER, A. B.
082.118
MULLER, P.
055.006 .007 .008 .016
094.144
118.025 .041 .042
MULLER, P. M.
094.076 .092 .126 .127
.156 .164
MUMFORD, G. S.
011.001

MUMFORD, G. S.
121.011 .038
122.099 .108 .176
MUNDRY, E.
120.008
MUNGALL, A. G.
035.025 .026 ,033
MUNIZ BARRETO, L.
008.101
MUNRO, R. E. B.
141.053
MURAKAMI, G.
044.028 .029
045.021
MURAWSKI, H.
003.062
MURCRAY, D. G.
071.081 .089
082.044 .050 .096
MURCRAY, F. H.
071.081 .089
082.096
MURDIN, P.
131.009 .076
MURDOCH, H. S.
141.053
MURIN, A. N.
105.086
MURPHY, R. E.
097.055
118.026 .028
MURRAY, B. C.
094.011 .128
097.003 .007 .030
MURRAY, J. D.
094.020
141.102
MURTY, S. S. R.
062.020
MUSATOV, A. A.
094.060
MUSEN, P.
042.027
MUSMAN, S. A.
073.034
MUSORIN, M. I.
034.081
MUSSINO, F.
044.006
MUSTEL, E. R.
078.028
082.122
122.138
MUTSCHLECNER, J. P.
065.072
080.023
MUZALEVSKY, Y. S.
072.073
MUZZIO, J. C.
122.096
MYRUP, L. O.
082.002
MYSIY, V. I.
104.038
NAAN, G. I.
003.141
NABOKOV, I. N.
034.063
NACHTMANN, O.
162.082

NACOZY, P.
102.023
NADEEV, L. N.
032.064
044.037
NADEZHIN, D. K.
125.017
NADKARNI, R. A.
105.076
NADUBOVITCH, YU. A.
084.041 .042
NAEBAUER, M.
055.013
NAEF, R. A.
010.036
098.005
104.012
NAGASAWA, S.
073.082
NAGASE, F.
076.041
NAGIRNER, D. I.
022.047
NAGNIBEDA, V. G.
077.055
NAGORSKAYA, I. A.
022.027
NAGY, A. F.
082.015
NAHON, F.
151.009 .021
NAIR, K. N.
084.219
NAJITA, K.
073.015
NAKAGAWA, M.
142.038
NAKAGAWA, Y.
062.007 .032
065.054
073.010 .022 .030
080.022 .033
NAKAJIMA, S.
066.003
NAKAJIMA, T.
131.123
141.193
NAKAMURA, Y.
094.001
NAKANO, T.
065.065
NANCE, R. L.
094.070
NANDY, K.
155.016
NAPARTOVICH, A. P.
064.023
NARBONE, M.
122.053 .152
NARIAI, H.
066.029
162.044
NARIAI, K.
064.032 .033 .051
119.001
122.003
NARLIKAR, J. V.
162.009
NASI, E.
047.010
077.052

SALPETER, E. E.
162.014
SALPETER, E. W.
022.115
SAMOKHIN, M. V.
084.261
SAMSON, W. B.
114.029 .071
SANAKULOV, EH. A.
041.038
SANAMIAN, V. A.
033.004
SANDAGE, A.
126.007
153.009 .022 .023
154.003
158.028
SANDAGE, A. R.
158.005
SANDIG, H.-U.
008.034
011.013
SANDNER, W.
081.010
SANDO, K.
071.014
SANDREA, A.
105.136
SANDULEAK, N.
124.010
159.001
SANFORD, P. W.
076.025 .040
SANTINA, R. E.
078.017
SANTINI, N. J.
141.009
SANZ, H. G.
105.022 .071
SAPIENZA, G.
075.021
SARABHAI, V.
084.219
SARAPH, H. E.
022.059
132.037
SARGENT, A. I.
114.015
SARGENT, W. L. W.
114.015 .022 .047
160.012
SARKADY, A. A.
034.092
SARRIS, E.
042.041
SARTORI, L.
125.016
SASAKI, T.
073.079
SASLAW, W. C.
160.010
SASTRY, CH. V.
077.037
SATAEVA, L. A.
034.077 .078 .079
036.017
082.136
SATO, H.
162.026
SATO, K.
045.020

SATO, T.
077.015
SATO, Y.
103.120
SATTAROV, I.
072.017 .061
SAUVAL, A. J.
071.073
SAUZEAT, M.
035.029
SAVAGE, B. D.
131.111
SAVEDOFF, M. P.
064.052
133.019
SAVENKO, I. A.
082.124
084.412
SAVUN, O. I.
084.412
SAWYER, C.
073.035
SAWYER HOGG, H. B.
120.010
SAYERS, J.
083.047
SAZONOV, V. N.
022.040
SCANLON, J. H.
092.004
SCARGLE, J. D.
141.014
SCARSI, L.
143.069
SCHAEDLER, J.
032.015
SCHAEFER, D.
082.085
SCHAEFER, J.
022.119
SCHAIFERS, K.
003.117
010.010
SCHALEN, C.
032.017
SCHARDT, A. W.
084.409
SCHARN, H.
052.012
SCHATTEN, K. H.
074.031 .041
079.103
094.130
106.013
156.001
SCHATZMAN, E.
007.000
065.061
SCHEFFLER, H.
131.019
SCHEIBER, L. C.
105.166
SCHERAGO, E. J.
031.016
SCHERER, L. R.
094.064
SCHERER, M.
143.021
SCHERRER, P. H.
071.070

SCHEUER, P. A. G.
141.224
SCHIELD, M. A.
074.023
SCHIELICKE, R.
034.083
SCHILD, R. E.
114.004
122.035
SCHINDLER, K.
022.018
143.052
SCHLESINGER, B. M.
065.071
153.008
SCHLUETER, D.
022.019
SCHMAHL, G.
034.082 .101
SCHMID-BURGK, J.
133.036
SCHMIDT, G.
084.205
SCHMIDT, H. U.
072.089
073.075
SCHMIDT, K.-H.
131.129
SCHMIDT, M.
141.025 .042
SCHMIDT, T.
159.012
SCHMIDT, W. K. H.
143.055 .070
SCHMIDT-KALER, T.
032.056
SCHMIED, L.
071.090
SCHMIEDER, B.
071.025
SCHMITT, J. L.
114.057
SCHMUS, W. R. VAN
105.124
SCHMUTZER, E.
022.104
SCHNEIDER, K. P.
095.001
SCHNEIDER, M.
052.043
SCHNETZLER, C. C.
105.006 .104
SCHNOPPER, H. W.
034.011
114.066 .067
141.117
SCHOFIELD, D.
097.031
SCHOLZ, D.
075.025
SCHOLZ, G.
074.007 .008
SCHOMMER, R. A.
131.026
SCHONBERG, E.
162.020
SCHORN, R. A.
093.031
SCHOVE, D. J.
011.002

SMITH, J. F.
051.030
SMITH, J. R.
033.005
075.024
SMITH, J. W.
033.008
SMITH, L. F.
114.101
133.013
SMITH, L. L.
093.028
SMITH, M. A.
064.045
141.228
SMITH, M. G.
034.035
131.090
SMITH, P. A.
012.020
SMITH, R.
065.048
SMITH, R. C.
064.053
SMITH, R. L.
084.017
SMITH, S. F.
003.076
071.046
SMITH, W. B.
098.010
106.005
142.066
SMITH, W. D.
083.002
SMITH, W. H.
022.013
SMITH-ROSE, R. L.
012.020
SMOLUCHOWSKI, R.
099.043
SMORODINOV, M. I.
094.060
SMRIGLIO, F.
034.001
044.007
SNIJDERS, R.
076.043
SNYDER, L. E.
131.036 .055 .100 .115
SOBERMAN, R. K.
082.142
SOBIESKI, S.
116.016
SOBOLEV, N. N.
022.043
SOBOLEV, V. V.
003.007
006.000
063.012 .013 .020 .023
.028
124.007
SOBOLEVA, N. S.
099.053
141.072
SODERBLOM, L. A.
094.013 .172
SOEDERHOLM, L.
162.018
SOFFEN, G. A.
034.030

SOFIA, S.
041.014
142.033
SOFUE, Y.
141.180
151.022
SOKOLOV, I. V.
123.032
SOKOLOV, V. A.
093.045
SOKOLOV, V. B.
052.018
SOLBERG JR., H. G.
099.009 .023
SOLINGER, A. B.
158.059 .060
SOLLOWAY, C. B.
052.034
SOLOMON, P. M.
114.083
131.096
SOLOMON, W. A.
099.002 .052
SOLONSKY, Y. A.
080.007
SOLOV'EVA, L. A.
041.037
044.034
SOLOVJEV, V. E.
074.070
SOMERVILLE, W. B.
010.022
SOMLO, P. I.
033.045
SONETT, C. P.
084.245
094.086
106.003
SOROCHENKO, R. L.
033.017 .027 .028
131.125
132.012 .046
157.018
SOROKINA, L. P.
099.065
SOSNOVA, A. K.
104.041
SOUKS, S. F.
074.051
SOULIE, G.
098.034
SOURK, C. K.
077.023
SOUSA NUNES, R. S. DE
032.066
SOUTHWICK, R. G.
074.062
SOUZA, H. DE
096.018
SOWINSKI, K. P.
094.175
SPADIN, P.
094.067
SPANGENBERG, E. A.
033.031
SPANGENBERG, E. E.
158.032
SPANNAGEL, G.
105.116
SPENCER, R. L.
094.176

SPERLING, F. B.
094.176
SPERLING, H. J.
042.019 .020
SPIEGEL, E. A.
094.020
SPINRAD, H.
101.011
114.018 .115
SPIRIDONOV, YU. G.
093.040
SPITZER JR., L.
131.051
151.038
SPLITTGERBER, E.
123.051
SPREITER, J. R.
084.241
SPRENKEL-SEGEL, E. L.
105.095
SREEKANTAN, B. V.
142.027
143.046
SRINIVASAN, B.
105.023
SRIVASTAVA, B. N.
022.103
SRIVASTAVA, K. M.
061.022
STABELL, R.
066.078
STAELIN, D. H.
022.121
141.225 .234
STAFEEV, A. M.
032.046
034.063
041.029
STAFEYEV, A. M.
041.013
STAGNI, R.
154.009
STAIB, J. A.
142.015
STALIO, R.
115.015
STALLKAMP, J. A.
097.073
STANDIL, S.
143.009
STANDISH JR., E. M.
151.028
STANILA, G.
045.010
STANIUKOVICH, K. P.
066.066
STANKEVICH, K. S.
141.209
STANKIEWICZ, A.
072.011
STANYUKOVICH, K. P.
091.038
STARIKOVA, G. A.
118.034
131.029
STARKOV, G. V.
084.045
STARODUBTSEVA, O. M.
093.052 .053
STARON, R. T.
155.004

SUNYAEV, R. A.
066.006 .082 .083
142.032
161.007
162.004 .011
SURKOV, E. P.
072.040 .043
SURKOV, EH. P.
072.014
SURKOV, YU. A.
094.061
SURKOVA, L. P.
121.082
SUSZEK, H.
104.006
SUTCLIFFE, D. S.
035.039
SUTTON, J.
141.011 .063
SUTTON, J. M.
141.234
SUVOROV, N. P.
066.038
SVENSSON, E. L.
032.009
SVESTKA, Z.
012.015
073.003 .045 .072
084.265
SVOLOPOULOS, S. N.
008.053
115.005
SWAMY, K. S. K.
131.106 .107
133.035
SWANSON, P. N.
033.039
SWANT, J. S.
082.010
SWARTZ, M.
076.036
SWEENEY, B. W.
054.016
SWEET, P. A.
073.012
SWEIGART, A. V.
093.006
SWENSON, G. R.
082.012
SWENSON JR., G. W.
033.007
SWENSSON, J. W.
071.020
SWIDER JR., W.
082.037
083.006
SWIFT, C.
142.035
SWIHART, T. L.
071.055
SWINGS, J.-P.
064.057
SWINGS, P.
093.051
SWINSON, D. B.
143.032
SWITZER, P.
073.039
SYMMS, L. S. T.
118.040

SYNGE, J. L.
066.064
SYROVATSKII, S. I.
022.034
073.076
SYUNYAEV, R. A.
142.037
SZAMOSI, G.
065.016
162.068
SZCZEPANOWSKA, A.
120.007
SZCZODROWSKA, B.
042.037
SZEBEHELY, V.
052.031
117.011
SZEIDL, B.
123.045
TADEMARU, E.
143.042
TAGLIAFERRI, G.
010.027
TAGLIAFERRI, G. L.
072.084
076.029
TAGO, A.
103.120
TAKAGI, S.
045.021
081.027
TAKAGISHI, K.
142.038
TAKAHASHI, T.
131.123
141.193
TAKAHASI, K.
082.158
TAKAKURA, T.
076.042
077.044
141.087
TAKEDA, H.
162.026
TALLANT, P. E.
072.033
TALWAR, S. P.
061.001 .023 .030
151.019
TAM, K.-K.
043.001
062.039
TAMBOVSKI, G. A.
034.012
TAMBURINI, T.
118.032
TAMBURINI JOB, T.
118.031
TAMHANE, A. S.
105.109 .111
TAMMANN, G. A.
011.028
103.124
111.001
122.095 .150 .151
141.021
153.009
162.090
TAN, T.-K.
123.036

TANDBERG-HANSSEN, E.
034.061
073.037
TANENBAUM, A. S.
071.056
TANG, C. C. H.
052.040
TANK, W. G.
082.014
TAPIA, S.
113.008
122.178
TAPIJA, R.
032.037
TAPLEY, B. D.
052.031
TARADY, V. K.
045.002
TARAFDAR, S. P.
064.025
TARANOV, V. I.
117.037
124.106
TARASHCHUK, V. P.
102.012
TARRIUS, A.
134.003
TASHENOV, B. T.
003.005
TASHPULATOV, N.
151.049
TATEVIAN, S. K.
055.014
TAUBENHEIM, J.
076.044
TAVASTSHERNA, K. N.
032.024 .037
TAYLER, R. J.
061.014
TAYLOR, B. G.
054.004
TAYLOR, B. J.
114.115
TAYLOR, D. J.
034.093
TAYLOR, G. E.
010.012
101.006
TAYLOR, G. J.
105.078
TAYLOR, G. N.
083.056
TAYLOR, J. H.
141.007 .018
TAYLOR, S. R.
012.003
105.008 .013
TAYLOR JR., H. A.
083.003
TAYLOR JR., H. P.
105.171
TCHENAKAL, V. L.
004.008
TCHERNEGA, N. A.
005.004
TCHERTOPRUD, V. E.
082.160
TCHUPRINA, R. I.
122.030
TEBBE, P. L.
113.003 .022

WALTER, K.
121.008 .009 .039 .088
.099
WALTER, L. S.
105.103
WAMPLER, E. J.
094.067 .181
141.014 .096
WAMPLER, J. M.
105.009
WANG, C. P.
142.041
WARD, D. F.
010.008
WARD, S. H.
094.229
106.006
WARDROP, .
103.120
WARES, G. W.
022.023 .081
WARNER, B.
022.017
114.090
115.002
124.106
WARNER, L. A. C.
014.002
WARNER, M. R.
094.082 .097
WARNOCK, J. M.
083.033
WARREN, E. S.
083.036
WARWICK, J. W.
077.031
099.004
WASSENBERG, W.
077.057
WASSERBURG, G. J.
061.005
105.003 .022 .025 .071
.123
WASSON, J. T.
105.079 .178
WATAGHIN, A.
065.057
162.070 .073
WATAGHIN, G.
162.073
WATANABE, E.
122.004 .157 .161
WATERFIELD, R. L.
100.008
103.109 .112 .114 .120
WATERS, A. C.
094.041
WATERWORTH, M. D.
034.047
WATSON, W. D.
065.006 .043 .075
071.077
WATTENBERG, D.
009.002
WATTS JR., R. N.
032.030
051.013
054.002
094.179
WAYLAND, J. R.
003.056

WAYLAND, J. R.
143.004
WAYMAN, P. A.
021.002
159.011
WDOWCZYK, J.
143.024
WEART, S.
072.037
073.042
WEART, S. R.
071.016
073.017
WEAVER, H.
131.084
WEAVER, W. B.
119.004
WEBB, R. H.
003.113
WEBBER, W. R.
143.022 .033 .038 .072
WEBBINK, R.
105.110
WEBBINK, R. F.
114.075
WEBER, E. J.
080.009 .024
WEBER, H.
105.153
WEBER, J. N.
094.162
WEBER, R. R.
141.057
157.006
WEBER, S. E.
074.079
079.103
WEBSTER, W. J.
131.135
WEBSTER JR., W. J.
131.117
WEDEKIND, J. A.
105.167
WEDEL, B.
036.001
WEEDMAN, D. W.
131.090
158.016 .034
WEEKES, K.
083.024
WEEKES, T. C.
003.088
WEEKS, R. A.
094.105
WEFEL, J. P.
143.001
WEHLAU, W.
009.016
WEHLAU, W. H.
116.025
WEHNER, H.
032.062
WEIGAND, A. J.
105.173
WEIGERT, A.
065.047 .089 .100
117.018
WEILER, K. W.
125.020
WEIMER, T.
008.002

WEINBERG, J. L.
094.186
155.020
WEINBERG, J. M.
034.059
WEISS, N. O.
072.038
WEISSKOPF, M.
034.050
WELCH, W. J.
131.056
WELIACHEW, L.
158.054
WELLS, R. A.
097.004 .043
WELLS, W. C.
143.001
WEMPE, J.
008.094 .111
WENDE, C. D.
076.008
077.034
WENDKER, H.
132.017
WENDKER, H. J.
131.136
WENTZEL, D. G.
143.010 .049
WENTZELL, R. A.
061.025
WENZEL, W.
120.009
122.044 .170
WERENSKIOLD, C. H.
031.018
WERNER, M.
113.048
WERNER, M. W.
114.033
131.046 .053
158.004
WESSELING, K. H.
033.048
WESSELINK, A. J.
115.001 .014
153.024
WEST, M. L.
151.041
WEST, R. M.
014.013
115.004
WESTCOTT, R.
032.039
WESTERHOUT, G.
012.009
WESTERLUND, B. E.
041.019
125.003 .004
WESTFALL, J. E.
094.250
WESTFOLD, K. C.
099.034
WESTHAUS, P.
022.024 .028
WESTPHAL, J. A.
099.006
WETHERILL, G.
074.069
WETHERILL, G. W.
105.026 .028 .133

WEXLER, R.
082.040
WEYMANN, R. J.
133.010
WHALING, W.
022.053
WHANG, Y. C.
094.141
106.022
WHIPPLE, F. L.
008.024
WHIPPLE JR., E. C.
051.014
WHITAKER, E. A.
094.174
WHITE, G. W.
003.092
WHITE, M. L.
107.005
WHITE, W. A.
076.036
WHITEOAK, J. B.
131.103
141.153 .184 .185
WHITE III, K. P.
073.023 .040 .056
WHITFORD, A. E.
158.069
WHITMARSH, R. B.
084.232
WHITNEY, A. R.
033.044
WHITTEN, R. C.
083.040
093.017
WICHMANN, H.
032.061
WICHMANN, W.
122.167
124.102
WICKRAMASINGHE, N. C.
131.004 .048 .089 .096
.108
158.004
WIDING, K.
076.023
WIEBER, D.
094.067
WIEDEMANN, E.
031.023
032.014
035.021
WIEHR, E.
034.017
WIELEBINSKI, R.
141.088 .223
WIELEN, R.
153.041
WIERZBINSKI, S.
117.040
WIESE, W. L.
022.090
WIGGINS, R. A.
081.016
WIIK, H. B.
105.105
WILCOX, J. M.
071.056
072.039

WILCOX, J. M.
074.041 .071
080.042
106.012
156.001
WILD, J. P.
074.084
077.038
WILD, P.
098.031 .032
103.114
WILD, P. A. T.
113.051
WILDEY, R. L.
094.047 .167
113.029
WILFORD, J. N.
003.112
WILHELMS, D. E.
094.142
WILKERSON, T. D.
074.005
WILKINS, G. A.
047.019
097.033
WILKINSON, D. T.
141.058 .231
WILKINSON, J. P. D.
051.020
WILLIAMS, D. A.
131.017 .039
WILLIAMS, I. P.
065.052 .066
107.008
153.001
WILLIAMS, J. P.
105.175
WILLIAMS, P. J. S.
141.012
WILLIAMS, R. E.
133.010 .018
WILLIAMS, W. J.
071.081 .089
082.044 .050 .096
WILLIAMSON, R. A.
131.065
132.018
WILLIAMSON, R. G.
081.018
WILLIAMS III, A. J.
022.002
WILLIS, J. P.
105.100
WILLMORE, A. P.
076.025
WILLS, D.
141.183
WILLSTROP, R. V.
141.150
WILSON, A.
116.001
WILSON, A. M.
071.004 .005
WILSON, B. G.
084.038
142.040 .063
WILSON, C. R.
084.016
WILSON, D.
105.070

WILSON, J. T.
105.200
WILSON, J. W.
082.075
WILSON, J. W. G.
083.047
WILSON, P. R.
071.049 .054
072.052 .064 .066
WILSON, R.
012.019
071.064
114.019
WILSON, R. E.
121.025
142.033
WILSON, R. N.
031.015
WILSON, R. W.
066.005
WILSON, S. J.
066.002 .048
WILSON, T. E.
082.074
WILSON, T. L.
131.033
155.024
WILSON, W.
113.059
WINCKLER, J. R.
084.211 .401 .403
143.036
WINDRAM, M. D.
033.034
141.141
WINER, I.
094.067
WING, R. F.
114.018 .053
WINKLER, H.
065.005
WINKLER, L.
121.053
WINN, F. B.
094.081 .096
WINNBERG, A.
131.121
WINTER, D. F.
094.119
WIRTANEN, T. E.
034.022
WISE, D. U.
094.143 .171
WISSE, M.
122.089
WISSE, P. N. J.
122.089
WITHBROE, G. L.
071.006 .019
076.024
WITKOWSKI, J.
007.000
WITT, G.
082.139
WITTE, L. DE
081.009
WITTMANN, A.
072.063
WLERICK, G.
122.097

Subject Index